Gravity and Magnetic Methods for Geological Studies

Principles, Integrated Exploration and Plate Tectonics

Gravity and Magnetic Methods for Geological Studies

Principles, Integrated Exploration and Plate Tectonics

Dinesh Chandra Mishra

Emeritus Scientist

National Geophysical Research Institute

(Council of Scientific and Industrial Research)

Hyderabad, India.

CRC Press

Taylor & Francis Group

Boca Raton London New York

CRC Press is an imprint of the

Taylor & Francis Group, an **informa** business

A BALKEMA BOOK

First published in paperback 2024

First published in India by
BS Publications, 2011

Published by CRC Press/Balkema
4 Park Square, Milton Park, Abingdon, Oxon, OX14 4RN

and by CRC Press/Balkema
2385 NW Executive Center Drive, Suite 320, Boca Raton FL 33431

CRC Press/Balkema is an imprint of the Taylor & Francis Group, an informa business

Publisher's Note
The publisher has gone to great lengths to ensure the quality of this reprint but points out that some imperfections in the original copies may be apparent.

ISBN: 978-0-415-68420-0 (hbk)
ISBN: 978-1-03-291739-9 (pbk)
ISBN: 978-0-203-71988-6 (ebk)

DOI: 10.1201/9780203719886

Visit the Taylor & Francis Web site at
http://www.taylorandfrancis.com

and the CRC Press Web site at
http://www.crcpress.com

Distributed in India, Pakistan, Nepal, Myanmar (Burma), Bhutan, Bangladesh and Sri Lanka
by **BS Publications**

Distributed in the rest of the world by
CRC Press/Balkema

Dedicated to

(i) Late Great Grandfather (Acharya Pt. Raghunandan Mishra- A Great Sanskrit Scholar), Late Parents (Shri Girish Chandra Mishra and Smt. Rajeshwari Devi) and Wife (Smt. Neelima Mishra).

(ii) All the Field Workers who sacrifice daily comforts of life for the subject.

"I (God-Lord Krishna) am Omni-Potent and Omni-Present
can neither be created nor destroyed",
So is the Gravity Field, Krishna also means attraction
(Quote is a Brief from Bhagwat Gita)

Scope of Book: Part I and II

Being simple and the most cost effective, gravity and magnetic methods are widely practiced for various exploration programmes. It was therefore, long overdue to describe their basic principles with examples and their various applications so as to use them judiciously. This book is planned as two parts. **Part I** describes the basic principles and methodology of the gravity and the magnetic methods of geophysical exploration with global and Indian examples, and **Part II** deals with their applications for geodynamics and integrated exploration of Indian plate that includes India and adjoining oceans and countries like Tibet, Pakistan, Bangladesh and Sri Lanka. **Part I** is largely based on lectures delivered by the author to students and research scholars in different universities and geoscientific organizations, and thus can be used by students and academics as **course material for gravity and magnetic methods** and their applications. It can also be used by professionals to refresh their knowledge of these methods and get acquainted with modern developments and their applications. **Part II** deals with the integrated exploration for Geodynamics and Seismotectonics with special reference to the Indian Plate including Indian Ocean, Himalaya, Tibet and Archean–Proterozoic Cratons and Mobile Belts. It also deals with the integrated exploration for Hydrocarbons, Minerals, Groundwater and some Environmental problems including Engineering and Archaeological sites. These sections are primarily based on professional work carried out by the author and lectures delivered by him in professional seminars and workshops and related works by other scientists and can be used as **reference book by professional geophysicists and geologists** for specific applications and case histories. It can also be used by students as course material for specialized courses on **Plate Tectonics, Geodynamics and Exploration Geophysics.** While dealing with applications, special emphasis is laid on an integrated approach using different geophysical data sets along with the geology of the region. However, as the subject matter of this book is primarily related to geophysics, the description of the geological aspect is limited to essentials that are important for the interpretation of geophysical anomalies. The interpretation and models presented in this book are those published by the author and others in refereed journals, but still may be alternative models that can be worked out by readers based on the data presented there. Some of the data sets discussed or referred to by the author may not be immediately useful but they have been included as part of information decimation so that prospective users can use them for their studies. In case of any clarification, author may be contacted.

Besides introduction, the subject matter of **Part I** is divided into three chapters. **Chapters 2 and 3** deal with **Gravity and Magnetic Methods,** respectively, describing their basic principles, instruments used for surveys, survey procedures, and various corrections to reduce the observed gravity and magnetic data to a common datum, so as to prepare profiles and maps. Being an important concept, **'Isostasy'** is dealt in detail and demonstrated on field data from Himalayas, Tibet and Western Ghat, India. Similarly, remnant magnetization plays an important role in the interpretation of magnetic data and therefore, a section on **'Rock Magnetism and Paleomagnetism'** has been introduced in Chapter 3, including its role in mineral exploration which normally does not find a place in books on 'Magnetic Methods'. Reversals of the geomagnetic field and it's relation to major catastrophic events in past, including mass extinction and climate changes, have been briefly described. Gravity and Magnetic surveys conducted from special platforms like **ships, aeroplanes and satellite** including their applications and limitations, are discussed. Field examples of gravity and magnetic anomalies observed over important geological and tectonic provinces related to **extensional and compressional (convergence)** regimes from the Indian continent and globally, are discussed. Some simple depth rules are provided and demonstrated on field data, so that, interpretation of these data sets can also be attempted when computing facilities are not available. Survey Procedures and Field Examples are described in detail for the benefit of students and young professionals. The purpose of describing field examples has also been to highlight those data sets for future applications by prospective users. **Table 2.2** briefly summarizes survey parameters of different types of gravity and magnetic surveys, including airborne and satellite surveys and their resolution and accuracies. **Table 2.3** provides details of various gravimeters and their accuracies to measure temporal variations in the gravity field at fixed stations that are useful for the study of co-seismic changes, environmental effects hydrological changes and volcanic mass changes.

Chapter 4 deals with **Special Data Processing Techniques,** common to both gravity and magnetic methods with some typical field examples of application of these techniques. Processing of gravity and magnetic data in frequency domain including low and high pass filters to separate large and small wavelength anomalies related to deep seated and shallow sources, continuation of fields and computation of derivative maps are emphasized in this chapter. It also describes inversion of gravity and magnetic anomalies and parameter estimation both in space and frequency domains, which can provide details of causative sources directly from the observed fields. These methods have been demonstrated on synthetic and field data. The most important application relates to 3-Dimensional basement computations that are useful for hydrocarbon exploration, directly from the observed potential field data that is demonstrated over airborne magnetic data from North Germany and Gravity data of Vindhyan basin, India. Thickness of the magnetic crust is estimated from the MAGSAT data of India using this technique. **Rheological properties** of lithosphere are important to assess their dynamic behavior which can be ascertained from their gravity anomalies and topography as described in Section 4.3 and demonstrated over gravity and elevation data from India and the Himalayas. **Softwares** compatible to PC related to inversion of gravity and magnetic anomalies for interpretation of these anomalies are referred to that are in public domain for students to practice on them. **Table 4.1** provides **the effective elastic thickness** for different geological provinces and its implications.

Part II deals with the integration of the gravity and magnetic data with other geophysical data and geology of the area for various geological problems related to the Indian plate. In any case, exploration is always a multidisciplinary approach and integration is an essential element of it. In this regard, modeling of gravity data constrained from available geophysical and geological data has the inherent advantage of integrating them in a comprehensive model. In this part, geophysical data from the Indian continent and adjoining oceans and countries like Pakistan, Sri Lanka, Bangaladesh and Nepal are extensively treated. **Satellite gravity and magnetic fields (CHAMP satellite)** are extensively used in Part II. **Chapter 5** briefly describes the important elements of **Plate Tectonics and Breakup of Indian Continent from Gondwanaland and its Drift History.** The related seafloor spreading magnetic anomalies and other important features of Indian Ocean, Arabian Sea and Bay of Bengal with some important reconstructions are discussed in this chapter. This is followed by the most important aspect of Indian plate, viz. **Collision of Indian and Eurasian Plates,** its consequences in formation of Himalaya and Tibet and related **seismicity** in **Chapter 6**. Its affect on environment and vice-versa **(orography)** is also briefly discussed in this chapter. The effects of the lithospheric flexure of the Indian plate under Himalaya and Tibet on seismotectonics along the Himalayan front are examined. Specific tectonics related to plate Boundary earthquakes along the Himalayan front with special reference to **The Kashmir and the Sichuan earthquakes** are discussed in this chapter.

The Extension of **Plate Tectonic Processes during Archean-Proterozoic Period** is examined in **Chapter 7** by considering gravity and magnetic fields of cratons and mobile belts of the Indian shield integrated with other geophysical, geological information available from these regions. Likely connection of Indian cratons to adjoining countries like **Madagascar, Antarctica and Sri Lanka during Archean-Proterozoic periods** is also discussed in this chapter. A comprehensive model of convergence of different Indian cratons during Archean-Proterozoic period is presented and **seismotectonics of some major earthquakes** of Indian continent, including the **Shillong earthquake of 1897** are described. Present day lithospheric structures, their nature under the Indian continent, and their role on the fast movement of the Indian plate and seismicity are also discussed in this chapter. **Seismotectonics** pertains to studies related to earthquakes, which is broadly discussed in case of different types of seismic activity in **Chapter 8**. Seismotectonics related to near plate boundary large earthquakes such as **the Bhuj earthquake of January 26, 2001** that are influenced by plate boundary activities is specifically discussed in this chapter and compared with **the New Madrid earthquake of 1811-12 in USA and Shillong earthquake of 1897** that is another near plate boundary great earthquake in India. Shuttle Radar Topographic Mission **(SRTM) Images** has been found to be extremely useful to delineate the tectonics of the Bhuj earthquake that can be used in other cases also specially in the inaccessible regions. **Co-seismic changes in elevation and gravity field** due to the Bhuj earthquake are also reported and analyzed. Specific tectonics related to **the Sumatra earthquake of 2004 and associated Tsunami** is described in this chapter. Temporal variation of the gravity field as a precursor study in Koyna, where seismic activity after monsoon rains are quite common, have been reported that show good correlation between the two.

Chapter 9 describes the application of gravity and magnetic methods integrated with other geophysical methods to exploration of **Hydrocarbons, Minerals, and Ground Water**. The application of gravity and magnetic methods for hydrocarbon exploration is basically of a reconnaissance type to limit the areas of interest for detailed exploration using seismic method. In this regard, the application of multidisciplinary surveys such as gravity, seismic and magnetotelluric surveys to delineate Subtrappean Mesozoic Sediments under Deccan trap in Saurashtra, Kachchh, Cambay, and Narmada-Tapti basins are discussed. Satellite derived Bouguer anomaly offshore west and east coasts, and Andaman-Nicobar islands are described for both **geodynamics and oil exploration**. Application of gravity method to delineate basement structures in the Himalayan foredeep such as the Upper Assam Shelf is demonstrated. Along with the exploration for hydrocarbons, specific geodynamics of these regions are also delineated and discussed using the same data set, emphasizing multiple applications of geophysical data. In this regard, the Deccan trap, that represents a large volcanic province on the Indian continent and offshore towards the west and its role in geodynamics of this region is described and discussed in great detail. Applications of gravity and magnetic surveys for minerals are both direct and indirect depending on their density and susceptibility, respectively. Sometimes, they are used to delineate rocks and structures such as volcanic plugs, faults, fissures, and fractures which might be associated with particular mineral. The delineation faults, fractures, and lineaments are also important for groundwater investigations in hard rock areas. In this regard, the uses of **the airborne magnetic lineaments** are emphasized. Some important case histories in this regard from **India and Africa** are discussed in this chapter.

The final **Chapter 10** under **Environmental Studies** and **Engineering Sites** related to **Near Surface Geophysics** deals with some major environmental problems like differentiation of subsurface **fresh water zones from saline water zones and detection of cavities in mines**. Engineering geophysics requires selection of suitable sites for large scale construction such as **dams, power**

plants, nuclear plants etc. which are also briefly discussed in this chapter and demonstrated through some examples. Civil engineers and students may find it useful. Gravity and magnetic signatures of impact craters such as **the Lonar crater, India** located in Deccan trap province, which is similar to the rocks exposed on **Mars and the Moon** is included to understand their effects on these celestial bodies. Effects on gravity and magnetic fields due to some exotic events such as **solar eclipse and lightning** are also described. Effect of solar eclipse is especially important as it is related to an important hypothesis of **the Alias Effect**. Examples to delineate Archaeological sites through geophysical methods are also included in this chapter. **Appendix I-IV** provide details of the geological time scale, important events of recent times (Pleistocene and Holocene), heat flow map of India and seismic zonation maps of SAARC countries- Pakistan and Bagaladesh, respectively.

At the beginning and the end of the book, an exhaustive **content list and bibliography of references and subject index** are, respectively provided.

D.C.Mishra

Acknowledgement

I express my most sincere thanks to the National Geophysical Research Institute (NGRI), Hyderabad, India and former Directors of NGRI who supported my research programme. This book is primarily related to airborne geophysical and gravity methods and their integration with other geophysical and geological information for various studies. Opportunities to work in these fields were provided by (Late) Dr. Hari Narain and Dr. H. K. Gupta without which it would not have been possible to write this book. I express my most sincere thanks to them. (Late) Dr. Hari Narian remained our guide and philosopher throughout his life. Thanks are also due to Prof. V. K.Gaur and Prof. D. Guptasarma for their support. I am also thankful to the Council of Scientific and Industrial Research and Dr. V. P. Dimri for Emeritus Scientist Scheme and their support after superannuation. Dr. Y. J. Bhasker Rao, Acting Director, NGRI made this book a 'Golden Jubilee Volume of NGRI' and I thank him for the same. I am thankful to (Late) Prof. H. S. Rathore, (BHU), Prof. D. Lal, (Late) Dr. P. W. Sashrabudhe, and (Late) Dr. C. Radhakrishnamurty (TIFR), Bombay for their guidance and support during doctoral work on 'Rock Magnetism'.

During my long innings of nearly 50 years in this field, several senior scientists at NGRI and in the country have supported me in various ways. Foremost among them are Dr. J. G. Negi and Dr. D. N. Avasthi and my most sincere thanks to them. I also thank Dr. P. V. Sanker Narayan, (Late) Dr. K. L. Kaila, Dr. R. K. Verma and (Late) Dr. M. N. Qureshy for their support. During this period, I collaborated with several international and national scientists and benefited considerably from their association. They are: (Late) Prof. A. Hahn and Mr. E. G. Kind of NLFB and BGR, Hannover, Germany; Prof. L. B. Pedersen of Uppsala, Sweden; (Late) Dr. D. Boie of PRAKLA, GMBH; Prof. H. J. Goetze and Dr. Sabine Schmidt from University of Kiel, Germany; and Dr. P. S. Naidu of I.I.Sc., Bangalore. I express my most sincere thanks to them. The comments of some of the reviewers of my research publications have been extremely informative and positive. I sincerely acknowledge them. The foremost among them are Dr. M. D. Thomas, Canada; Dr. Satish C. Singh, IPGP, France; Prof. M. Santosh, Kochi University, Japan; Dr. S. Sinha-Roy, Dr. B. P. Radhakrishna, Dr. K. S. Valdiya, and Dr. B. M. Reddy, India.

My colleagues in the gravity and magnetic group have always been supportive of me and I whole-heartedly thank them for the same. They are: Dr. B. Singh, Dr. S. B. Gupta, Dr. K. S. R. Murty, Dr. M. R. K. Prabhakar Rao, Dr. A. P. Singh, Dr. D. V. Chandrasekhar, Dr. V. M. Tiwari, Dr. Ch. D. Venkata Raju, Dr. V. Vijai Kumar, Dr. Kusumita Arora, Dr. J. R. K. Sarma, Mr. M. B. S. V. Rao, Mr. Ch. Ramasamy, Mr. G. Laxman, Dr. N. Kumar and Mr. H.K. Hodlur. Some of my colleagues, Dr. B. Singh, Dr. V.M.Tiwari, Dr. Ch. D. Venkata Raju and Dr. Kusumita Arora went through parts of the manuscript and made specific suggestions and also provided some significant literature for this book that are thankfully acknowledged. My students, Mr. R.P.Rajasekhar, Mr. M. Ravi Kumar and Mr. M. Singh helped in preparation of this manuscript and I thank them for the same.

I have always considered Prof. Y. Sreedhar Murthy and Mrs Y. Suguna Murthy as friends in hours of need who supported in various ways and provided some significant literature on this topic. I was amply supported by my colleagues and friends through various discussions and collaborations. They are, to mention a few: (Late) Dr. S.M. Naqvi, (Late) Dr. V. Babu Rao, Dr. D. Achuta Rao, Dr. P. K. Agarwal, Dr. B. S. P. Sarma, Dr. H. C. Tewari, Dr. S. V. S. Sarma, Dr. P. R. Reddy, Dr. U. Raval, Dr. R. N. Singh, Dr. K. Roychowdhary, Dr. S. B. Singh, Dr. T. Seshunarayana, Dr. M. S. Joshi, Dr. K. Mallick, Dr. S. K. Verma, Dr. B. K. Rastogi, and (Late) Dr. N. Krishna Brahmam (NGRI, Hyderabad); Dr. B. R. Arora (WIHG, Dehradun); Dr. Ravi Shankar, Dr. O. P. Mishra, Dr. B. D. Banerji, Dr. J. R. Kayal, Dr. U. C. Das, and Dr. T. S. Ramakrishna (GSI, Calcutta); Dr. B. Nagrajan (Survey of India, Dehradun); Dr. A. N. Bhowmick, and Dr. P. C. Chandra (CGWB); Dr. K. Rajendran (I.I.Sc., Banglore); Prof I. V. Radhakrishnamurty (Andhra, University); Dr. D. Gopal Rao, Dr. A. K. Chaubey, Dr. K. S. Krishna (NIO, Goa); Dr. M. Rajaram (IIG, Mumbai); Dr. J. S. Ray (PRL, Ahmedabad) and several others who are sincerely acknowledged. Discussions with Dr. U. Raval are always informative who is quick to suggest relevant references. Often I discussed various aspects of geology and geophysics with several scientists of this Institute like Dr. T. R. K. Chetty, Dr. Anil Kumar, Dr. J. M. Rao, Dr. G. V. S. P. Rao, Dr. D. V. Subba Rao, Dr. A. Pandey, Dr. S. S. Rai, Dr. R. K. Tiwari, Dr. S. N. Rai, Dr. V. K. Rao, Dr. M. M. Dixit, Dr. O. P. Pandey, Dr. Ram Babu, Dr. Ch. Rama Rao, Dr. P. Koteshwar Rao, Dr. T. Harinarayana, Dr. R. K. Chaddha, Dr. D. Srinagesh, Dr. Kirti Srivastva, Dr. M. Ravi Kumar, Dr. P. C. Rao, Dr. A. Manglik, Dr. P. Mandal, Dr. V. K. Gahalaut, Dr. K. Sain, Dr. Sukanta Roy and several others who are thankfully acknowledged. To keep physically fit, I am thankful to the medical facility at NGRI especially to Dr. A. Raval and Dr. M. Janardhan Reddy. Mr. B. M. Khanna (Incharge, Library) and Mr. V. Rajasekhar (Computer Graphics) have been extremely cooperative and I thank them for their help. Integrated geophysical exploration along various geotransects and hydrocarbon exploration in Saurastra, Kutch and Narmada-Tapti sections and Upper Assam, India were sponsored and financed by Department of Science and Technology, New Delhi (Late Dr. K. R. Gupta); Oil and Natural Gas Corporation, Dehradun (Dr. Y. B. Sinha, Dr. K. Chandra, Shri N. K. Verma, and Shri D. Sar); Directroate General of Hydrocarbons, New Delhi (Dr. A. Chandra, Dr. S. K. Srivastva, and Dr. S. V. Raju) and Oil India Ltd., Assam, (Shri K. K. Nath), respectively. I thankfully acknowledge them. A part of the manuscript was prepared when the author was working in a project from Ministry of Earth Sciences (MoES/PO(Seismo)/23(646)/2007) and I am thankful to them for the same.

Finally, my most sincere and affectionate thanks are due to my wife, Mrs Neelima Mishra and children and grand children: Avinash, Richa, Arushi, and Arnav Tiwari and Ritwik, Nivedita, Manya, Shashwat, and Shambhavi Mishra for keeping my morale high during long periods of eight years when this manuscript was being prepared. Last, but not the least, my most respectful tribute to Maa Saraswati, Hindu Goddess of knowledge for providing me strength to complete this task.

Permissions @

The following organizations, associations and journals permitted me to re-use figures published by them. Some figures published by them are also included in the manuscript under free access for 'fair use'. I express my most sincere thanks to them and to the authors of those articles that are referred to in the captions of the respective figures.

   (i)   Elsevier (All geoscientific Jounals: EPSL; JAES, Tectonophysics etc.)

   (ii)   American Geophysical Union (All geoscientific journals: JGR, GRL, Tectonics etc.)

   (iii) Geological Society of America (Bulletins, Memoirs, GSA-Today etc.)

   (iv)   American Association of Petroleum Geologists (Memoirs and Bulletins)

   (v)   United States Geological Survey, USA

(vi)   NASA, USA

(vii)   Society of Exploration Geophysics, USA (Geophysics)

(viii)   Nature Publishing Company, UK

(ix)   Geophysical Journal International, UK

(x)   Cambridge University Press

(xi)   Geoexploration

(xii)   Geophysical Prospecting

(xiii)   Pure and Applied Geophysics

(xiv)   Springer Verlag

(xv)   Episodes

(xvi)   National Geophysical Research Institute, Hyderabad, India

(xvii)   Association of Exploration Geophysicist, Hyderabad, India

(xviii)   Directorate General of Hydrocarbons, Delhi, India

(xix)   Survey of India, Dehradun

(xx)   Geological Survey of India (Memoirs, Records, Special Publications etc.)

(xxi)   Oil and Natural Gas Corporation, India

(xxii)   Oil India Limited, India and Burma oil Ltd. (Earlier)

(xxiii)   Geological Society of India, Banglore, India

(xxiv)   Water and Power Research Institute, Poona, India

(xxv)   Central Ground Water Board, India

(xxvi)   Indian Geophysical Union, Hyderabad, India

(xxvii)   National Disaster Management, India

(xxviii) Wadia Institute of Himalayan Geiology, India

(xxix)   Geological Survey of Pakistan

(xxx)   Geological Survey of Bangladesh

(xxxi)   Geological Survey of Srilanka

(xxxii)   Department of Mines and Geology, Nepal

(xxxiii) Institute of Seismological Research, India

(xxxiv)   SAARC Disaster Management center, New Delhi, India

(xxxv)   India Meteorological Department, New Delhi, India

Contents

Part I

Methodology with Global Examples
(Continental, Marine and Airborne – Satellite Surveys)

Chapter 1   Introduction

Chapter 2   Gravity Method

Chapter 4   Common Data Processing Methods and Parameter Estimation – Digital Signal Processing

Part II

Integrated Exploration of Indian Plate and Resources
(Geodynamics, Seismotectonics, Hydrocarbons, Minerals, Groundwater, Environment and Engineering Sites)

Chapter 5   Continental Drift and Plate Tectonics : Reconstructions, Gondwanaland Break-Up, Plumes and Drifting of Indian Plate

Chapter 6   Collision of Indian and Eurasian Plates and Seismotectonics: Himalayan and Tibetan Terrains

Chapter 7   Geodynamics of the Indian Continent and Seismotectonics: Isostasy, Archean-Proterozoic Cratons, Collision Zones, Rift Basins, Plumes and Lithosphere, and it's Flexure

Chapter 9   Resource Exploration and Geodynamics: Hydrocarbons, Groundwater, and Minerals

Chapter 10   Some Typical Environmental and Engineering Studies: Near Surface Geophysics

Part - I

Methodology with Global Examples
(Continental, Marine and Airborne-Satellite Surveys)

1

Introduction

Geophysics is the study of the earth, based on the principles and laws of physics. The study of the earth, based on the laws of the gravity and the magnetic fields, is known as the gravity and the magnetic methods of geophysical exploration. They are known as the natural methods as they employ the natural fields, namely, the gravity and the magnetic fields of the earth. Contrary to them, there are methods which employ artificial fields created specially for those surveys in an area such as electrical and seismic methods. As the gravity and the magnetic methods employ natural fields of the earth, they are the oldest geophysical methods used for the study of the earth and are easy to operate and cost effective compared to the other geophysical methods. Therefore, they are ideally suited for reconnaissance survey of large areas to limit the areas for detailed investigations. The gravity and the magnetic methods being directly related to the physical properties of the rocks, namely, the density and the susceptibility, respectively they are found to be very useful by field geologists and geophysicists in mapping and identification of various rock types. They are also used for direct detection of minerals with large contrast in density and susceptibility compared to country rock.

The earth has its own gravity and the magnetic fields, which gets modified in the presence of rocks of different properties. The earth's natural field F1 gets modified to F2 near a structure (Fig. 1.1) or anomalous body depending on its shape, size, depth and the physical properties like density or susceptibility in case of gravity and magnetic methods, respectively. The differences between the two fields (F2 - F1) is known as geophysical anomaly, namely, the gravity anomaly or the magnetic anomaly in the two cases. It depends on the configuration of the body, depth and physical properties of the causative sources. These fields are measured with the help of sensitive instruments at the surface of the earth or using different platforms, for example ship, helicopter, aeroplane and satellite depending on the target, their size, desired accuracy of the survey and accessibility to the survey area. The data is processed to obtain the gravity and the magnetic anomalies with respect to the ground position, which, in turn, are related to the surface or subsurface rocks, structures and their physical properties. The two most important characteristics of the anomalies are their spatial size and magnitude, which are popularly referred to as wavelength and amplitude, respectively. Broadly, the geological studies for which the gravity and magnetic methods have shown promise, are as follows:

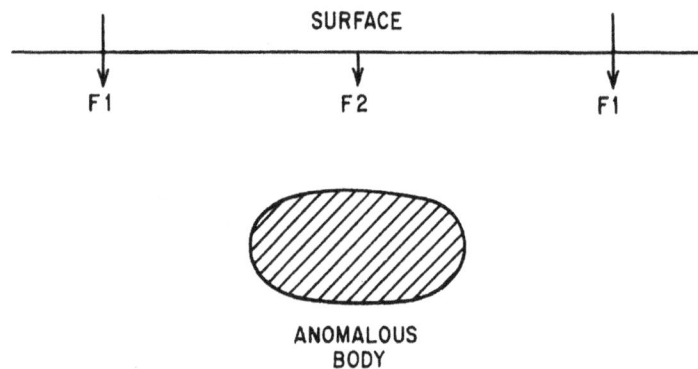

Fig. 1.1 F1 is the normal earth's gravity/magnetic fields which get modified to F2 in the vicinity of sub-surface anomalous bodies or heterogeneity. (F2-F1) is known as gravity/magnetic anomaly which depend on shape, size, depth and physical parameters, namely, density/ susceptibility of the anomalous body which can be derived from the observed/ measured anomaly.

1.1   Geological Studies and Gravity and Magnetic Methods

Gravity and magnetic methods are related to variations in density and susceptibility of rocks, respectively and produces complimentary images of structures which are integrated to provide their details. In fact, they are also integrated with all other available data sets to map subsurface structures. Their applications to various geoscientific studies are briefly described below while their detailed applications integrated with other geophysical - geological data sets are discussed and demonstrated in the forthcoming chapters.

1.1.1   Geodynamics and Plate Tectonics

Since 1912, when Alfred Wegener proposed the theory of continental drift that has explained several geological observations in a unified manner, it therefore, formed one of the most important aspects of geodynamics. However, it did not account for the forces responsible for drifting of the continents and was therefore, replaced by plate tectonics during 1960s, which accounted for these forces due to mantle convection (Uyeda, 1978).

Plate tectonics is presently one of the most important aspects of global geodynamics. Gravity and magnetic methods are used to study the different aspects of geodynamics and plate tectonics of a region. Some of their applications in this regard can be briefly described as follows. However, there are several other applications of geophysical methods in general and gravity and magnetic methods, in particular, that are outlined in the forthcoming chapters.

(i)   Plate Tectonics

Plate tectonic theory provides a unified model to explain most of the tectonic processes observed on the surface of the earth and subsurface. It is briefly outlined here to introduce this topic that is essential to discuss gravity and magnetic anomalies due to its certain aspects in Chapters 2 and 3, respectively. However, it is discussed in great detail in chapter 5 though same illustrations are referred here. Accordingly the earth's upper layer (lithosphere) is divided into different plates, which are separated by mid oceanic ridges and subduction/collision zones along plate boundaries as shown in Fig. 5.13. Plates may consist of both continental and oceanic parts representing both continental and oceanic lithospheres. It shows some major and some minor plates which are separated by ridges and subduction zones referred to as divergent and convergent plate boundaries where different plates diverge and converge, respectively. Some important mid-oceanic ridge systems are Mid-Atlantic Ridge, Indian Ocean Ridge system, East Pacific Rise etc., named after the oceans they occupy. Collision and subduction zones are found on other side of the plate accompanied by fold belts on continents and trenches in oceans, respectively such as Himalayan Fold Belt and Andaman-Sumatra-Java (Sunda) trench (HIM, ASJ; Fig 5.13). Besides these two features, the third important element is known as Transform faults, which are similar to strike slip faults along which the two plates slip past each other. San Andreas Transform fault system along west coast of USA is one such example (SAF, Fig. 5.13). Chamman fault in Pakistan (CH; Fig. 5.13) related to Pakistan Fold Belt between the Indian and the Eurasian plates is an example of transform fault related to the Indian plate.

Mid oceanic ridges are linear features where volcanic rocks wells up from inside the earth and spreads over the ocean bottom forming the ridges, which diverge the plates on either sides and are therefore known as divergent margins (Fig. 1.2). Mid oceanic ridges are, therefore characterized by mafic volcanic rocks and magnetic profiles across them show normal and reverse polarity of rocks located almost symmetric with respect to the ridge, which is discussed in detail in Section 5.2. These are known as sea floor spreading magnetic anomalies and their polarity indicate the polarity of earth's magnetic field at the time of their formation. On the other hand, along convergent margins plates on the surface of the earth converge and collide and in the process, one subducts under the other and is therefore known as subduction zones (Fig. 1.2). As shown in this figure, magma erupts and spreads at Mid Oceanic Ridges and pushes the oceanic lithosphere on either side as indicated by arrows. Once the oceanic lithosphere encounters a continental shelf as shown on either margins of this figure, it subducts below the continental crust as it is comparatively heavier (higher density) than the latter. The contact of the two is characterized by deepest parts of the ocean known as trenches where deep basins are formed. During its movement, it may encounter some localized sources of magma such as plume which may give raise to chains of volcanoes that are known as sea mounts in case of oceans. Once the zig saw puzzle of sea floor spreading magnetic anomalies were sorted out and continents were brought back in time, they appeared to join together. This gave rise to plate tectonic theory, which in most simple form suggests that the earth is made up of several plates, which move and collide with each other and on collision, form the mountain chains and depending on the density of rock types subduct one under the other. They are chanacterized by seismic actiurty due to intense tectonic activities at plate boundaries.

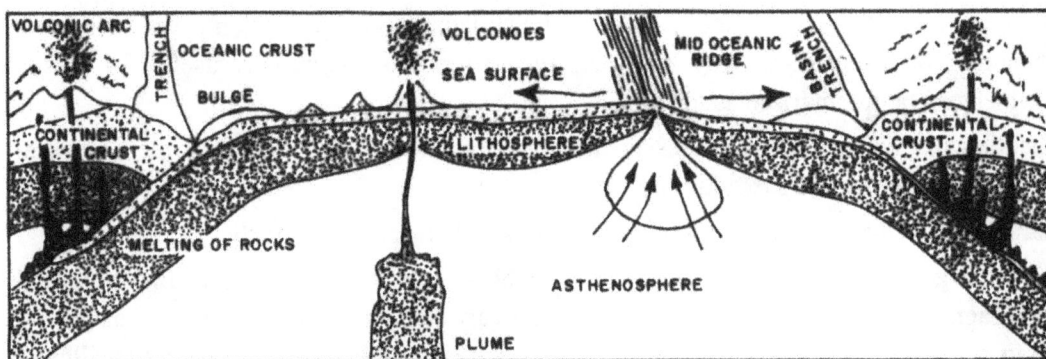

Fig. 1.2 A schematic section of Mid Oceanic Ridge where two plates diverge due to intrusion of magmatic material from asthenosphere forming oceanic crust. During plate motion, they encounter continental shelf where they would subside due to their higher density compared to continental crust. The subducted material melts at the depth giving rise to volcanic arcs. During plate motion, it may encounter plumes giving rise to volcanic chains.

Based on direction of forces in two cases, viz. mid oceanic ridges and subduction zones (Fig. 1.2), the tectonics related to them are termed as extensional and convergence tectonics. The subducted material at certain depth, melts due to frictional heat and high temperature to give rise to magma, which rises through fractures and faults giving rise to volcanic chains known as island arcs or magmatic arcs. However, in case of collision between two continental plates, such as Indian - Eurasian plates, the rocks are deformed as both are of almost same density. In such cases, the upper part of the crust forms the mountain chain through thrusting and folding, while its lower part slips one under the other causing thick crust and several related tectonic activities like earthquakes, volcanoes etc., Due to weight of the overriding plate, the subducting plate flexes and causes bulging of the subducting plate as it happens in case of cantilever beams in civil engineering (Fig. 1.2). This implies that while material is generated at mid oceanic ridges from within the earth, it is consumed at plate boundaries during subduction providing a mass balance in earth's system. Plate tectonics is important not only for tectonics and geodynamics but is also important for resource exploration. Most of the mineralized sections of base metals, precious metals (gold), chromites etc., occur along fold belts (mountains) that are formed due to collision of two plates as shown in Fig 1.2. In this regard, ancient fold belts of Archean- Proterozoic period (Appendix I) assume special significance as discussed in Chapters 7 and 9. It is also important for hydrocarbon exploration as most of the sedimentary basins are formed along fold belts (Fig 1.2) or along rifted margins that are essential elements of plate tectonics.

They are characterized by specific features which produce typical gravity and magnetic anomalies as discussed in sections 2.9 and 3.8 respectively. Gravity and magnetic methods used for various applications in plate tectonics are briefly as follows:

(a) Reconstruction of continents and their movement during different geological periods based on direction of magnetization and seafloor spreading magnetic anomalies.

(b) Crustal structures and physical properties of rocks (density and susceptibility) with depth.

(c) Continuation of large-scale structures from one continent to the other before their breakup based on their gravity and magnetic signatures.

(d) Mantle dynamics related to plate tectonics based on satellite gravity anomalies.

(ii) Crustal Structures

The top most layer of the earth is known as crust. Its structure and composition plays a vital role in geodynamics of a region. Gravity and magnetic methods are extremely useful for crustal studies, which can be summarized as follows:

(a) Delineation of deep seated structures in the upper mantle and the crust and their physical properties, viz density and susceptibility.

(b) Variation in the crustal thickness (depth to Moho) based on gravity anomalies.

(c) Curie point geotherm based on magnetic data, which is defined as the temperature beyond which magnetization in rocks cannot exist. It is equivalent to the Curie point of magnetite equal to $570°$ C. In some sections, it may coincide with Moho or may be deeper or shallower depending on heat flow in the region.

(d) Compensation of surface load and rheological properties of the crust and the lithosphere based on isostasy such as elastic thickness, flexural rigidity etc., based on gravity anomalies and topography which is described in Chapter 4.

(iii) Plume Tectonics

Plumes are large bodies of gaseous and fluids, which rise from inside the earth (Fig. 1.2) and give rise to large scale volcanic provinces in different parts of the world such as Deccan trap and Rajmahal trap in India, Karoo volcanics in Africa, Columbia flood basalt in USA and islands of Reunion, Kerguelen etc. Some of them are discussed in Chapter 5. Due to their high density and high susceptibility, gravity and magnetic methods are widely used for their studies, which are as follows:

(a) Delineation and demarcation of plume affected surface/subsurface regions

(b) Assessment of their physical properties like bulk density and bulk susceptibility and based on them identification of rock types.

1.1.2 Resource Exploration – Hydrocarbons, Minerals and Groundwater Exploration

The application of gravity and magnetic methods to resource exploration such as oil, mineral and groundwater exploration is generally indirect, which are basically used to limit the area of investigation and delineate structures that are important for this purpose. However, in case of exceptionally heavy and magnetic minerals, they can be used directly to explore them. Their specific applications for this purpose are as follows that are discussed with details in Chapter 9.

(i) Geological Mapping

Geological mapping is an important aspect of resource exploration. Regional gravity and magnetic surveys are found to be extremely useful for this purpose as they provide information about subsurface structures and depth extent of exposed structures. In this regard, airborne-surveys are found to be specially useful. Some of the usages of these methods for this purpose are as follows:

(a) Delineation and demarcation of different geological provinces and tectonic units such as basins, rifts, cratons, collision zones etc., and their three dimensional extensions.

(b) Delineate large scale structures such as lineaments, faults, fractures, joints, intrusives etc.

(c) Delineation of different rock types based on their physical properties, viz. density and susceptibility.

(ii) Hydrocarbon Exploration

Application of gravity and magnetic methods for hydrocarbon exploration is primarily reconnaissance in nature to limit the areas for detailed exploration. They are presently widely used along with other exploration strategies. Some of the usages of these methods in collaboration with other geophysical methods for integrated exploration programs in this regard are as follows:

(a) As a reconnaissance method to limit the area for detailed investigation by more involved and costly geophysical methods such as seismic surveys.

(b) Evaluate basement depth and three-dimensional basement configuration in sedimentary basins.

(c) Delineate structures such as faults, anticlines, synclines etc., in the basement and overlying sediments.

(d) Shallow structures in sediments for gas occurrences.

(iii) Mineral Exploration

Most of the minerals being characterized by specific density and susceptibility, these methods are widely used for this purpose, which can be briefly described as follows:

(a) As a reconnaissance method to limit the area for detailed investigation.

(b) Delineation of mineralized zones and the structures like lineaments, faults and fractures, which control the mineralization in an area.

(c) Direct detection of heavy/light and magnetic minerals such as iron ores etc., and their extensions depending on their physical properties, viz. density and susceptibility.

(iv) Groundwater Exploration

The application of gravity and magnetic methods for groundwater exploration are mostly indirect and serves as a complimentary method to electrical methods. Some of the usages of these methods for this purpose are as follows:

(a) Delineation of bed rock topography and structures.

(b) Delineation of faults and fractures in hard rock areas, which might be water bearing.

(c) Delineation of large scale lineaments specially airborne magnetic lineaments, which defines different hydrological regimes.

1.1.3 Environmental Studies – Seismotectonics and Near Surface Geophysics

Gravity and magnetic methods are widely used for environmental natural hazard assessment. There are two aspects to it, firstly the hazard assessment such as tectonics related to seismic activity known as seismotectonics and secondly for evaluation of engineering sites. Engineering geophysics is related to selection of suitable sites for large scale construction where gravity and magnetic methods are widely used to delineate shallow structures such as faults, fractures, lineaments etc., Environment is in fact a part of geosciences and gravity and magnetic methods are used for several studies related to environment.

(i) Seismotectonics

Seismic activity in a region depends on tectonics of region, which can be inferred to a great extent from gravity and magnetic surveys. Some of the usages of these methods for this purpose are as follows:

(a) To delineate the tectonics specially faults related to seismic activity.

(b) To determine the charactcristic parameters of faults related to seismic activity.

(c) Temporal variations in the gravity field due to dynamic changes at plate boundaries based on satellite gravity measurements.

(d) Coseismic changes based on gravity measurements before and after an earthquake.

(ii) Engineering Sites

These investigations are classified as Near Surface Geophysical Problems. Engineering geophysics involves the application of these methods to delineate local tectonics at the construction sites. Some of the usages of these methods for this purpose are as follows:

(a) Delineation of structure such as faults, fractures etc., at the site of large scale constructions such as power plants, dams, reservoirs etc.

(b) Bed rock investigation for its depth, nature and structures related to its stability.

(c) Nuclear waste disposal sites regarding their stability and absence of faults, fractures, lineaments etc., in these sections.

(iii) Climate Changes

Availability of high resolution satellite altimetry data for long periods (8-10) years have made it possible to study factors related to long range climatic changes. The following aspect of satellite altimetry is relevant for climatic changes:

(a) Mean Sea Level changes with time based on satellite altimetry.

(b) Changes in ice mass loss and its transfer in Greenland and Antarctica based on changes in the gravity field inferred from satellite altimetry which is connected to greenhouse effect and global warming.

(iv) Mining Geophysics

It involves the application of these methods for the problems related to mining. The application of these methods for this purpose are as follows:

(a) Delineations of faults, fractures, lineaments etc., similar to other applications discussed above.

(b) Delineation of voids, water fills and land fills etc., in the mining areas.

(iv) Volcanoes and Volcanic Activity

Volcanic rocks being usually characterized by specific density and susceptibility, gravity and magnetic methods can be effectively used for their studies. Some of the applications are as follows:

(a) Nature of volcanic activity in terms of their physical properties, viz density and susceptibility and probable composition.

(b) Onset of volcanic activity based on borehole gravity measurements in surrounding regions of known volcanoes.

(c) Surface/subsurface region affected by volcanic activity.

(v) Land Slides

Land slides in hilly terrains usually occur along fault planes and fractures, which can be delineated based on gravity and magnetic surveys as described above in case of other applications.

(vi) Impact Craters

Meteorites being igneous mafic or ultramafic rocks, they are characterized by high density and high susceptibility, which can be studied using these methods. Some of the applications are as follows:

(a) To infer their physical properties like density and susceptibility

(b) Their effects depth-wise on impact at the surface of the earth.

1.1.4 Different Modes of Surveys – Ground, Marine, Airborne and Satellite Surveys

In case of gravity method, the most popular mode of survey is ground gravity survey. In oceans, marine surveys or satellite altimetry is widely used for preparation of 2-dimensional gravity maps whose resolution, however, is limited due to height of satellite and data gaps between different passes.

In case of magnetic method, the most popular and useful mode of survey is airborne magnetic survey due to speed of survey and possibility of measuring field at closed interval in otherwise inaccessible regions. Airborne gravity surveys, however, are not very popular due to inaccuracies in the measurements of the gravity field from aeroplanes and cost involved in such surveys. However, due to improvements in technology and cost effectiveness, it is now being increasingly used to survey inaccessible regions as discussed in Section 2.6. Shipborne gravity and magnetic measurements are carried out for surveys in oceans. Satellite derived magnetic and gravity data are used for delineating deep seated structures in the lower crust and upper mantle specially in inaccessible regions such as hills (Himalayas), forests, oceans etc. (Sandwell and smith, 1997, Mishra et al., 2004, Reigberg et al., 2005). Satellite derived magnetic data are also used for estimation of depth to Curie point geotherm below which magnetization of rocks does not exist (Mishra and Venkatrayudu., 1985). In fact, gravity maps of oceans and delineation of large wavelength gravity anomalies (> 1000 km) are possible only by satellite based surveys, which enables to probe deep into the mantle.

Gravity Method

2.1 Introduction

Gravity method of the geophysical exploration depends on Newton's laws of gravitational force, which states that two masses attract each other depending on their mass and distance between them. The earth itself being a body of large mass, attracts all other bodies, which depends on their mass and the mass of the earth. The attraction due to the earth is called the gravity field of the earth and is one of the primary forces in nature. Depending on the latitude, every point on the surface of the earth is characterized by a normal value of the gravity field. The deviation of the earth's gravity field from its normal value is known as gravity anomaly at that place which is related to the rock mass surrounding it and in turn is related to the density of the rocks. In actual practice, instead of measuring absolute value of the earth's gravity field, which is a cumbersome process, the variation in the gravity field from one point to the other is measured, which is known as gravity anomaly. The variations in the gravity field of the earth are related to the variations in the density of the surface/subsurface rocks, which can be derived through application of this method.

2.2 Basic Principles

2.2.1 Forces of Gravity

Gravity is defined as the mutual attractive force between two masses. According to Newton's law of gravitation, the force of attraction between two masses m_1 and m_2 separated by distance r is given by (Fig. 2.1):

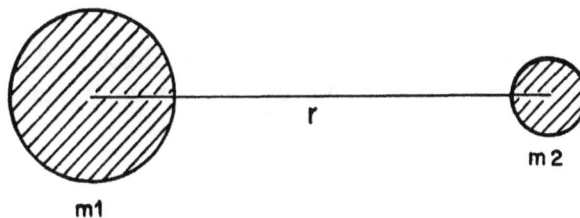

Fig. 2.1 m_1 and m_2 are two masses separated by a distance r.

$$F = - G\frac{m_1 m_2}{r^2} r_1 \qquad\qquad .....(2.1)$$

Where r_1 is a unit vector directed from m_1 to m_2 and − sign indicates that it is always attracted.

In C.G.S. units, F is in dynes, m in grams and r in centimeters. G is known as the gravitational constant and is given by $G = 6.67 \times 10^{-9}$ CGS units (dyne − cm^2/g^2) which is the force in dyne exerted between two masses of 1 gm each with their centers one cm apart.

In a simple form, equation (2.1) can be rewritten as:

$$F = G\frac{m_1 m_2}{r^2} r_1 \qquad \qquad .....(2.2)$$

2.2.2 Gravitational Potential and Laplace Equation

Potential is defined as the work done in moving a unit mass from infinity to its present position at distance r from the reference point. Therefore, the gravitational potential u is given by:

$$u = \int_r^\infty F dr \qquad \qquad .....(2.3)$$

Substituting for F from equation (2.2) for $m_1 = m$ and $m_2 = 1$. As per convention potential is always positive and therefore:

$$u = G\int_r^\infty \frac{m}{r^2} dr = \frac{Gm}{r} \qquad \qquad .....(2.4)$$

Similarly a force is defined as the derivative of potential with respect to distance. Therefore, as u is the gravitational potential, the gravitational force F is given by

$$F = -\frac{\partial u}{\partial r} \qquad \qquad .....(2.5)$$

Substituting for u in equation (2.5)

$$F = -\frac{Gm}{r^2} \qquad \qquad .....(2.6)$$

This is similar to equation (2.2) which suggest that potential and field are interchangeable. Gravity field of the earth being vertical in nature, it can be referred as:

$$g = -\frac{\partial u}{\partial z} \qquad \qquad .....(2.7)$$

where u is the gravitational potential and z is the vertical direction directed downwards. The components of the gravitational attraction in horizontal directions x and y is given by:

$$-\frac{\partial u}{\partial x} \quad \text{and} \quad -\frac{\partial u}{\partial y} \qquad \qquad .....(2.8)$$

These three components of gravitational field satisfy a unique relationship as:

$$\frac{\partial^2 u}{\partial x^2} + \frac{\partial^2 u}{\partial y^2} + \frac{\partial^2 u}{\partial z^2} = -4\pi G\rho \qquad \qquad .....(2.9)$$

where ρ is the density of the body. This is known as Poisson's equation which reduces to Laplace equation for $\rho = 0$, in a source free space. Therefore,

$$\frac{\partial^2 u}{\partial x^2} + \frac{\partial^2 u}{\partial y^2} + \frac{\partial^2 u}{\partial z^2} = 0 \qquad \qquad .....(2.10)$$

Laplace equation is therefore, a special case of Poisson's equation in source free space where $\rho = 0$.

Magnetic field and magnetic potential discussed in Chapter 3 also satisfy the Laplace equation and hence gravity and magnetic fields are known as potential fields. In simplified 2-dimensional form equation (2.10) can be written as:

$$\frac{\partial^2 u}{\partial x^2} + \frac{\partial^2 u}{\partial z^2} = 0 \qquad .....(2.11)$$

Solving equation (2.11) for a particular solution of type:

$$u(x, z) = X(x) \, Z(z) \qquad .....(2.12)$$

Substituting for u in equation (2.11)

$$z.X'' + xZ'' = 0 \qquad .....(2.13)$$

where

$$X'' = \frac{\partial^2 u}{\partial x^2} \quad \text{and} \quad Z'' = \frac{\partial^2 u}{\partial z^2}$$

Equation (2.13) can be written as:

$$(X''/x) + (Z''/z) = 0 \qquad .....(2.14)$$

This suggest a pair of solution as

$$X(x) = a_1 \exp(\pm i\, k\, x) \qquad .....(2.15)$$

$$Z(z) = a_2 \exp(\pm kz) \qquad .....(2.16)$$

where k is a constant which may be real, imaginary or complex.

The two solutions (2.15) and (2.16) can be combined as:

$$u(x, z) = C_k \exp \pm k\, (z + ix) \qquad .....(2.17)$$

A more general solution can be written as sum of all possible constants as:

$$u(x, z) = \sum_k C_k \exp k\, (z + ix) \qquad .....(2.18)$$

Equation (2.18) is the basis of representing potential field data as exponential functions which is the basis to apply modern signal processing techniques (Spectral method) as discussed in Chapter 4.

2.2.3 Gravitational Acceleration (g)

The force exerted on a mass at the earth's surface due to the attraction of the earth is known as the earth's gravitational force (F). Therefore, if the mass of a body is m and that of the earth is m_e and R is the radius of the earth, the force of attraction between them is given by equation (2.2) as:

$$F = \frac{Gm_e m}{R^2} \qquad .....(2.19)$$

This force will produce an acceleration, g and according to Newton's second law of motion, force is the product of mass and acceleration. Therefore,

$$F = mg \qquad .....(2.20)$$

where g is gravitational acceleration or acceleration due to the gravity field of the earth. From equations (2.19) and (2.20)

$$g = G\, m_e/R^2 \qquad .....(2.21)$$

g is force per unit mass which is equivalent to acceleration and is therefore, expressed as cm/s^2. In geophysical literature, it is called Gal after the famous physicist Galileo who first provided the value of g. The value of g at the surface of the earth has been measured and the worldwide average is found to be 980 Gal. Due to variations in the radius of the earth and its rotation, the gravity field changes at the surface of the earth with latitude, the maximum being at the pole and minimum at the Equator with a difference of approximately 5 Gal. The value at the Equator is 978.0318 Gal. Its variation measured from one place to other due to the variation in the density of rocks is very small and therefore, further smaller units as milliGal (mGal) = 10^{-3} Gal and micro Gal (μ Gal) = 10^{-6} Gal are used in geophysical prospecting. An intermediate unit as gravity unit (gu) equal to 10^{-4} gal is also used at some places.

Based on the numerical values of g and G, the mass of the earth (m_e) can be obtained for an average value of R = 6.37×10^8 cm = 6.37×10^3 km.

$$m_e = 6.14 \times 10^{27} \text{ gm} = 6.14 \times 10^{24} \text{ kg} \qquad \qquad .....(2.22)$$

2.2.4   Density of Rocks

Density (ρ) is an intrinsic property of materials, which is defined as mass per unit volume. Therefore, if m is mass and v is the volume:

$$\rho = m/v \qquad \qquad .....(2.23)$$

Its unit is g/cc or g/cm^3. A smaller unit of kg/m^3 equal to 10^{-3} g/cm^3 is used to describe fine variations in density. Different rocks have different densities, which is the basis of the gravity survey. The most commonly found rock on surface of the earth is granite and gnisses with a density of 2.65-2.70 g/cm^3 (2650–2700 kg/m^3) with an average density of 2.67 g/cm^3 (2670 kg/m^3). It is considered as the main constituent of the upper crust. The addition of mafic/ultramafic minerals increases the density. In brief, density of sedimentary rocks is less compared to igneous and metamorphic rocks. Table 2.1 presents the average density of important rock units generally found in the field.

A simple way of measuring density is based on Archimedes principle using a special balance. One such balance is known as Walker's Steel Yard balance. Rock samples are weighed in air (Wa) and water (Ww) which provides density (ρ) as:

$$\rho = Wa/(Wa-Ww) \qquad \qquad .....(2.24)$$

Due to heterogeneity present in rocks, their density varies considerably from sample to sample. Therefore, an average representative value for a particular rock type is obtained by measuring densities of several samples from that unit. This is referred to as bulk density of that rock type.

Table 2.1 The densities of some important rock types compiled from different sources:

1. Igneous Rocks	Range (g/cm^3)	Average
Acid Igneous (Average)	2.50 - 2.80	2.65
Andesite	2.40 - 2.70	2.60
Basalt	2.70 - 3.00	2.85
Basic Igneous (Average)	2.70 - 3.10	2.90
Diorite	2.70 - 2.90	2.80
Dolerite	2.80 - 3.00	2.90
Gabbro	2.80 - 3.00	2.90
Granite	2.50 - 2.80	2.65
Granodiorite	2.70 - 2.80	2.75
Rhyolite	2.40 - 2.70	2.55
Syerite	2.60 - 2.90	2.75
Trap (Deccan)	2.50 - 3.0	2.75

2. Matamorphic Rocks

Amphibolite	2.70 - 3.00	2.85
Eclogite	3.20 - 3.50	3.35
Gneisses	2.60 - 3.00	2.80
Granulite/Charnoclite	2.70 - 3.10	2.90
Lower Crust (Average) (Granulite)	2.80 - 3.00	2.90
Metamorphic Rock (Average)	2.50 - 3.10	2.80
Peridotite	2.80 - 3.40	3.10
Phyllite	2.60 - 2.90	2.75
Quartzites	2.50 - 2.70	2.60
Schists	2.40 - 2.80	2.60
Serpentine	2.40 - 3.10	2.78
Slate	2.70 - 2.90	2.80
Upper Crust (Average) (Granite and Gneisses)	2.60 - 2.80	2.70
Upper Mantle (Peridotite)	3.20 - 3.40	3.30

3. Sedimentary Rocks

Alluvium	1.90 - 2.00	2.00
Clays	1.60 - 2.40	2.00
Dolomite	2.30 - 2.90	2.60
Limestones	2.00 - 3.00	2.50
Rock Salt	1.80 - 2.00	1.90
Sandstones	2.00 - 2.40	2.20
Shales	2.00 - 2.80	2.40

4. Metallic Minerals

Copper		8.50
Gold	15.50 – 19.50	17.50
Silver		10.50

Oxides

Chromite	4.30 – 4.60	4.45
Hematite	4.90 – 5.30	5.10
Ilmenite	4.30 – 4.90	4.60
Magnetite	4.90 – 5.20	5.10
Uraninite	8.00 – 10.00	9.00

5. Non-Metallic Minerals

Bauxite	2.30 – 2.50	2.40
Coal (hard)	1.40 – 1.80	1.60
Coal (soft)	1.20 – 1.60	1.40
Diamond	3.40 – 3.60	3.50

Gas Hydrate (solid)	0.70 – 0.90	0.80
Gas	0.20 – 0.30	0.25
Graphite	1.90 – 2.30	2.10
Gypsum	2.20 – 2.60	2.40
Ice	0.80 – 0.90	0.85
Petroleum	0.60 – 0.90	0.75
Rock Salt (salt dome)	2.00 – 2.40	2.20
Water	1.00 – 1.00	1.00
6. Sulphides		
Arsenopyrite	5.90 – 6.10	6.00
Galena	7.40 – 7.60	7.50
Pyrite	4.90 – 5.20	5.10
Pyrrhotite	4.50 – 5.00	4.75

2.2.5   Variation of Density with Depth

The mass of earth as inferred in equation (2.22) is

$$m_e = 6.14 \times 10^{27} \text{ gm} \qquad \qquad .....(2.25)$$

If d is the average density of the earth.

$$m_e = v. \rho \qquad \qquad ......(2.26)$$

where v is the average volume equal to $4/3 \ \pi.R^3$. Substituting for m_e, v and ρ in equation (2.26)

$$\rho = 5.32 \text{ g/cm}^3 \qquad \qquad .....(2.27)$$

This is the average density of the earth, which is almost twice the average density (2.70 g/cm^3) of rocks exposed at the surface suggesting a general increase in the density of the rocks with depth inside the earth. The density inside the earth cannot be determined directly. It can only be inferred indirectly based on seismic velocities and composition of rocks derived from it at different depths.

Based on the velocities of waves inferred from earthquakes at different depths inside the earth, Bullen (1975) and Bott (1982) provided general composition of rocks and structures inside the earth and their densities dividing it into three major parts namely crust, mantle and core, which are further divided into two subdivisions each as lower and upper parts (Fig. 2.2). The differences in the thickness of continental (36-40 km) and oceanic crusts (6-8 km) and respective lithosphere may be noted. This figure shows the Lithosphere-Asthenosphere Boundary (LAB) at a depth of about 150-200 km under continents and about 10-100 km under Oceans, which represent the average depth world over. This is an important boundary for geodynamics defined by 1200-1300°C isotherm as it represents transition from rigid lithosphere to partially molten asthenosphere and therefore convection in the latter drives the lithospheric plates, which is primarily responsible for plate tectonics. Another important aspect of earth's interior is semi solid mantle, followed by fluid outer core and solid inner core.

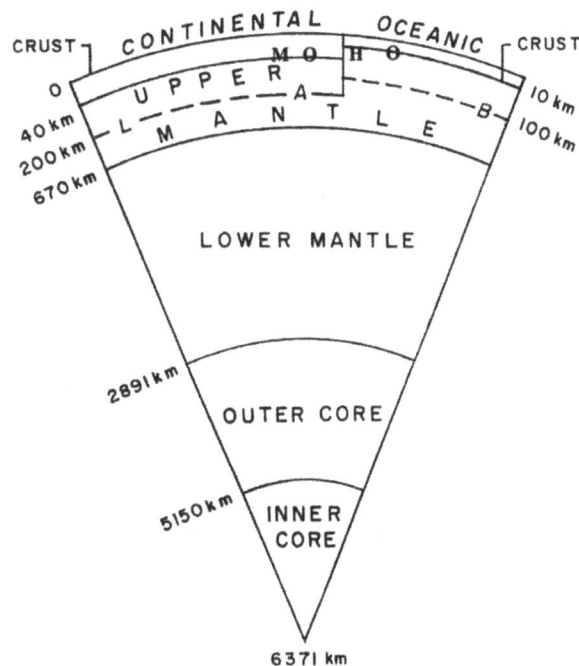

Fig. 2.2 Simplified layered structure of earth's interior under continents and oceans. Thick crust under continents (36 - 38 km) and thin crust under oceans (6- 8 km) are indicated by Mohorovicic discontinuity (MOHO). LAB- Lithosphere-Asthenosphere Boundary separate rigid lithosphere at the top of a partially molten asthenosphere below at a depth of 150-200 km under continents and 80-100 km under oceans.

Based on seismic velocities and inferred composition, density model with depth has been provided by several workers. One such model showing variations in density with depth based on Preliminary Reference Earth Model is reproduced in Fig. 2.3(a) (PREM; Dziewonski and Anderson, 1981). It shows a consistent increase in the density of the rocks from 2.7 g/cm³ at the surface to approximately 3.3 g/cm³ in the upper mantle below the Moho at a depth of about 40 km and small jumps at various discontinuities (200, 400, 670 km, Fig. 2.3(a)). It further increases to 5.9 g/cm³ at the base of the mantle at a depth of 2900 km where it shows a sudden jump of approximately 10.0 g/cm³ in the outer core increasing further downwards to approximately 12.0 g/cm³ at the base of the outer core, beyond which it remained almost constant up to the center of the earth (Fig. 2.3(b)). In fact these discontinuities inside the earth represent changes in seismic velocity, density and composition across them, which have been primarily used to define them. However while dealing with gravity anomalies recorded in gravity surveys the most important section of the earth is the crust and upper mantle, which represent the outer shell of the earth up to Mohoravicic discontinuity referred to as Moho and immediately below it. The Moho lies at an average depth of approximately 36-40 km under continents in relatively plane areas. The crust is generally divided in two equal parts namely upper and lower crust with bulk average densities of 2.7 g/cm³ and 2.9 g/cm³ respectively, which primarily represent felsic (granite and gneisses) and mafic (basalt) compositions, respectively. In fact, the variation of density in crust is continuous and therefore, crust can be divided even in three equal parts for modeling of gravity anomalies with bulk densities of 2.7, 2.8 and 2.9 g/cm³ as the upper, the middle and the lower crusts, respectively. The oceanic crust is primarily mafic in composition and can also be divided in two groups of basalt and sediments upper crust (2-3 km) and mafic/ultramafic (gabbros) lower crust (4-5 km) with bulk average density of 2.7-2.8 and 2.9-3.0 g/cm³, respectively. These are the average values on a worldwide basis, which may differ locally depending on local conditions.

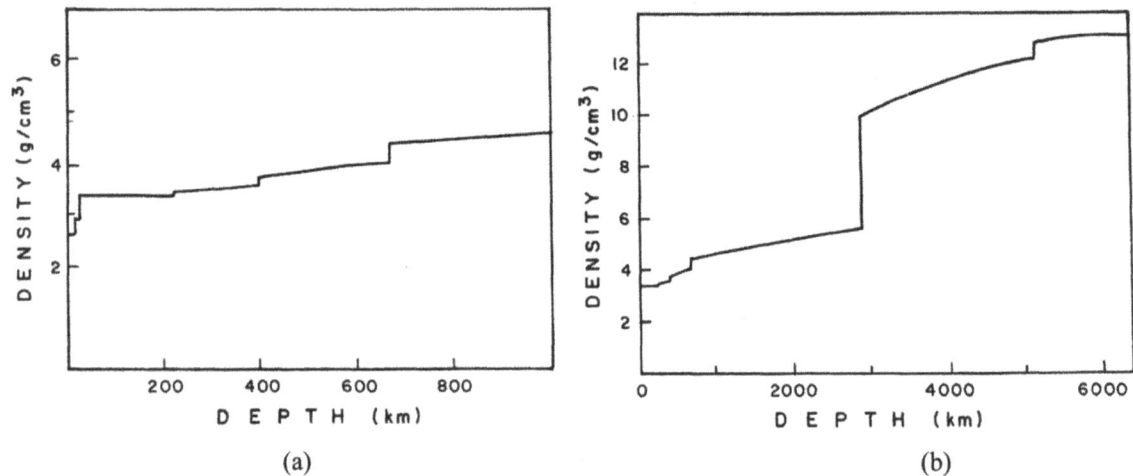

Fig. 2.3 Variation of density in side the earth with depth showing changes at various discontinuities based on Preliminary Reference Earth Model (PREM model, Dziewonski and Anderson, 1981). a) Density changes from surface up to 1000 km showing small changes at crustal and mantle discontinuities. b) Density variation from surface up to core of the earth showing continuous increase with depth with major jumps at mantle- core-inner core boundaries.

2.2.6   Thermal Structure of Lithosphere

Lithosphere and its thermal structure plays an important role in formation of surface tectonics. Further, it is also important to model the observed gravity and magnetic fields at the surface as both these physical properties, viz. density and susceptibility are controlled by temperature distribution. The average picture world over differ considerably from one part to the other depending on the thermal gradient due to heat sources such as plumes, volcanoes, radioactive minerals, etc., In general, the thermal gradient under shield is less (10 °-15 ° C/km) while it is more under oceanic crust and recent orogenic belts (20-30 ° C/km) (Section 5.2.7). It also decreases with depth due to presence of radioactive element in the upper part of the continental crust. Temperature distribution in crust and upper mantle is also important to understand the magnetization distribution depending on Curie point geotherm (Section 3.2.2) that is important to model magnetic anomalies (Chapter 3).

2.2.7   Density – Velocity Relationship

Propagation of seismic waves in rocks depends on the their physical properties including density and therefore, based on laboratory measurements of seismic velocities and densities of different rock types, several workers have provided empirical relationships between them, which can be used to infer density of rocks in case seismic velocities are known. Barton (1986) provided the relationship between the two parameters as given in Fig. 2.4, which shows a considerable scatter. However, the average values provided by solid line can be safely taken as representing the variation in bulk density with velocity. In general, the seismic velocity increases with depth inside the earth and thereby density also increases with depth. However, in certain sections in crust and upper mantle, the seismic velocities may decrease with depth and will be lower compared to the average value at those depths. Such zones are known as low velocity zones (LVZ) and correspondingly density also decreases in those sections, which have special significance for composition and geodynamics of those regions as described in Part II. Similarly, in certain parts of crust, the velocity and the density may be more compared to the worldwide average value, which are known as High Velocity Zones (HVZ) as is the case with volcanic provinces.

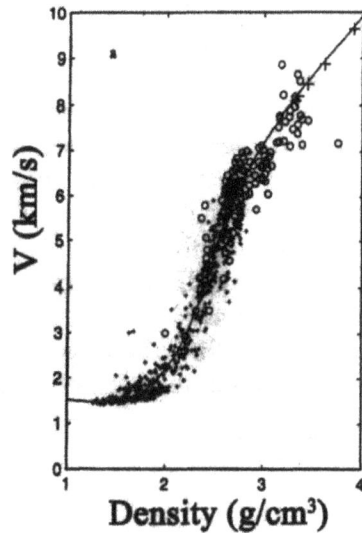

Fig. 2.4 Empirical relation between P-wave velocity (V) and density of rocks (Barton, 1986) which increases as velocity increases.

2.2.8 Variation of Earth's Gravity Field with Depth

The gravity field inside the earth is function of both density and pressure. The variation of gravity field and pressure inside the earth is given in Fig. 2.5. The gravity field remains almost constant up to the base of the lower mantle and then decreases fast to zero at the centre of the earth while pressure increases from surface downwards up to the centre of the earth where it is maximum (Dziewonski and Anderson, 1981). The decrease in the earth's gravity field in the earth's core is attributed to the high pressure.

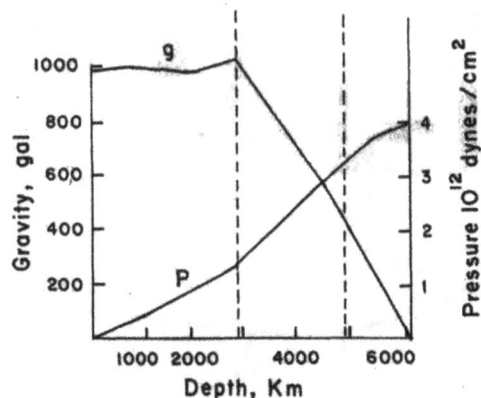

Fig. 2.5 Variation of gravity and pressure in side the earth with depth. Due to large pressure, gravity decreases with depth in core (PREM model; Dziewonski and Anderson, 1981).

2.2.9 Earth's Shape and Geoid

The earth's surface is approximated by an ellipsoid described by equation (2.35). It is a mathematical surface, which fits best with the earth's surface with out any undulations. It is therefore, an imaginary surface over oceans and under continents, which earth will assume if all oceans are filled and elevated lands are removed (Fig. 2.6(a)). It is a close approximation to an equipotential surface such that the earth's gravity field is normal to this surface and plumb line is vertical at every point. However, it is far from existing situation and the earth surface is undulating along with the surface/subsurface mass in homogeneity. Therefore, another equipotential surface, geoid is defined as an average sea surface over the oceans and under the continents, which is an equipotential surface and warped due to mass deficiency (water column) under the ocean and excess of mass (mountains etc.,) over the continent as shown in Fig. 2.6(a). It is referred to as reference geoid. It is further affected by local subsurface mass in homogeneity in a region (Fig. 2.6(b)). As shown in Fig. 2.6(b), due to subsurface positive mass, the geoid plane will pop up showing a positive anomaly while in case of negative mass, it will sink indicating a negative anomaly. The difference between the mathematical ellipsoid, which approximates the earth surface and the geoid as equipotential surface at a place is known as geoid undulation or geoid anomaly. It has acquired importance in recent years as geoid undulations can be obtained from satellite tracking data and used to infer deep seated density in homogeneity in lithosphere as described in section 2.6.

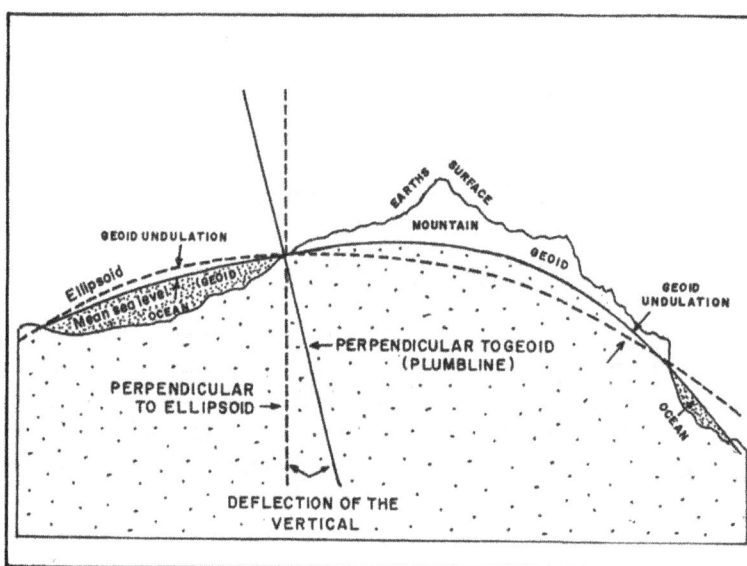

Fig. 2.6(a) Ellipsoid and geoid with latter being warped up due to surface mass caused by the presence of mountains and oceans. This changes the position of perpendicular line (plumb line) requiring corrections for any geodetic survey.

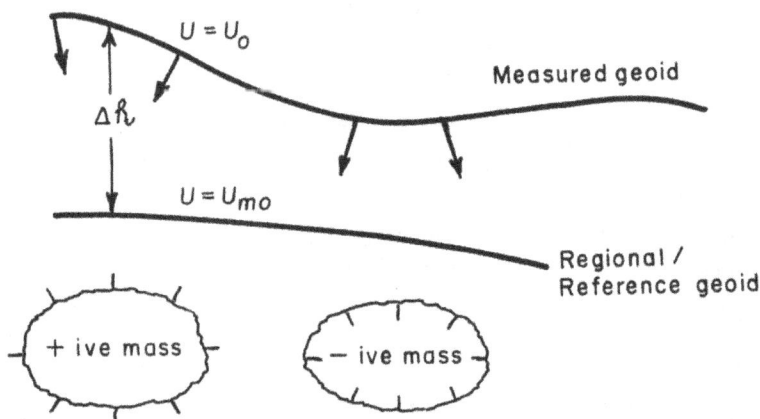

Fig. 2.6(b) Warping up of geoid due to local buried mass and its effect on local gravity field. Due to the effect of local mass, measured local geoid is different from regional/ reference geoid.

2.3 Instruments and Data Acquisition

2.3.1 Instruments – Gravimeters and Gradiometers

The earliest instrument used for gravity prospecting was the Torsion balance (Heiland, 1946) which was used to measure the horizontal and vertical gradients of the earth's gravity field. However, it was a cumbersome equipment and required almost 1-2 hours for one measurement compared to 5-10 minutes for a measurement using present day modern gravimeters. Therefore, with time, they became obsolete and presently only modern day gravimeters are used for gravity surveys. There are basically two kinds of gravimeters used for the measurement of the earth's gravity field, viz (i) Absolute gravimeters and (ii) Variometers. The former measures the absolute value of the earth's gravity field at a particular place while latter measures the variation in the gravity field from one place to the other, known as gravity anomaly. In geophysical exploration, we are generally interested in gravity anomalies and therefore, the variometers popularly known as gravimeters are primarily used for exploration purposes. The order of variation in-geophysical exploration is measured to an accuracy of 0.01 mGal to 0.001 mGal (one microGal), which require highly sensitive instruments.

(i) Absolute Gravimeters

These instruments are used to obtain the absolute value of the earth's gravity field at a particular place. There are basically two kinds of absolute gravimeters viz pendulums and free fall method. Pendulums based on a weight suspended from a thin fibre are the oldest instruments used for this purpose. It is based on the time taken by a pendulum for a full swing, which is described in any text book on physics for graduate students. The method of free fall is presently used to design modern gravimeters for measurement of absolute value of the gravity field. It has been lately modified to make several free falls of the weight wherein weight is automatically lifted to its original position and allowed to fall and the average time of several falls are used to obtain the gravity filed at that place. This modified version is known as rise and fall method.

(a) Rise and Fall Method

It measures the time taken by a mass to fall for a known distance in a sealed vacuum tube. In such a case the distance d traveled by mass in time t is given by Newton's law as:

$$d = ut + gt^2 / 2 \qquad \qquad .....(2.28)$$

where u is the initial velocity. In case of free fall u = 0

$$d = gt^2 / 2 \qquad \qquad .....(2.29)$$

Based on this principle, instruments were designed in which the falling object is again raised to initial position and allowed to fall. In this manner, time taken by object to fall several times are measured and averaged to obtain better accuracy. A Schematic diagram in Fig. 2.7(a) explains its principles (Lowrie, 1997). There are two levels of measurement using a laser source and a detector. The distance traveled at two levels for initial velocity u = 0 are given by:

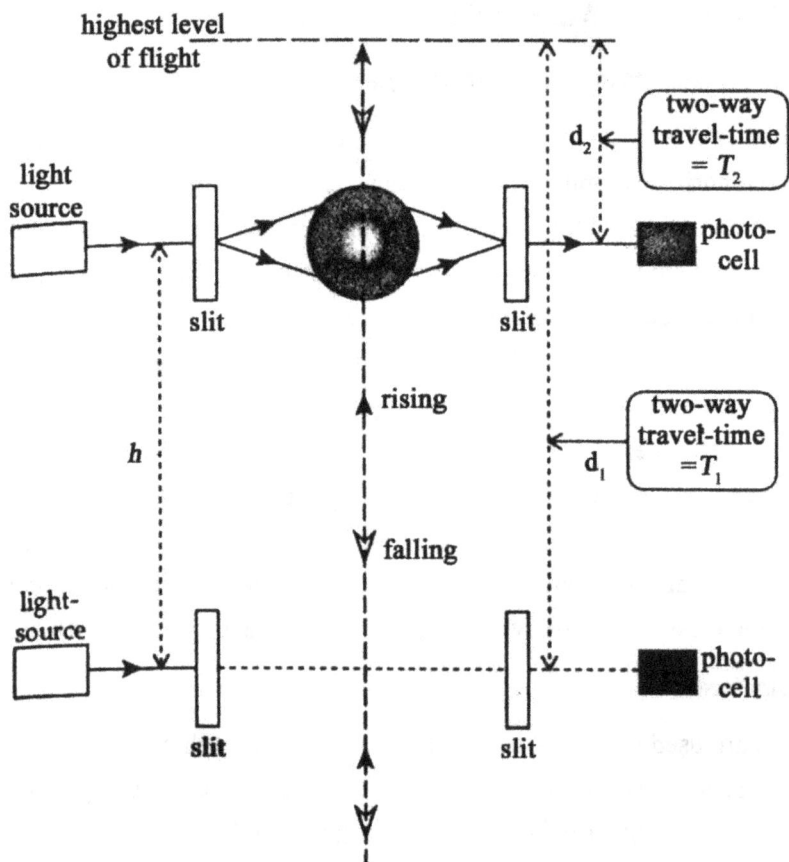

Fig. 2.7(a) Schematic diagram of Absolute Gravimeter (Lowrie, 1997) based on rise and fall method.

$$d_1 = g \, (T_1/2)^2 \, / \, 2 \qquad\qquad\qquad\qquad .....(2.30)$$

$$d_2 = g \, (T_2/2)^2 \, / \, 2 \qquad\qquad\qquad\qquad .....(2.31)$$

where T_1 and T_2 are two way travel time at two levels d_1 and d_2 (Fig. 2.7(a))

Distance between the two positions h = d_1 - d_2 = g $(T_1^2 - T_2^2)$ / 8 and

$$g = \frac{8h}{(T_1^2 - T_2^2)} \qquad\qquad\qquad\qquad .....(2.32)$$

As the desired accuracy in the measurement of absolute value of the earth's gravity field is 0.001 mgal (1 microGal), the measurement of time is presently carried out with the help of rubidium clock and distance is measured with the help of laser beams and the experiment is conducted in a vacuum tube. The accuracy of the instrument (Least count) is 1.0 microGal but the accuracy of instrument is affected considerably by atmospheric conditions such as pressure, temperature, humidity etc., for which specific corrections are made. The accuracy of the measurement, however, is improved by averaging the value over several falls and rise and for hours and days at the place of observation. An averaged value of 24-48 hours at a station can provide an accuracy of $\pm$ 2-3 microGal after corrections for atmospheric conditions. Due to high accuracy of measurements, the readings are corrected automatically for atmospheric conditions such as variations in pressure and temperature, which are sensed by sensors fitted with the gravimeter and requisite corrections are made by the computer attached with the gravimeter. Due to high accuracy, they are usefull to record temporal variation in the gravity field at a fixed station (Table 2.3, at the end of the chapter) Fig. 2.7(b) shows changes in the gravity value at the Gravity Observatory in NGRI Campus during different periods in a year, which shows a good correlation with water level changes in the region and therefore can be used to monitor ground level changes in a region. The average gravity value measured over one full day (24 hours) at this site is 978326943 ± 3 μGal. The instrument along with accessories for laser beam is quite bulky and the process of measurement is also cumbersome and time consuming and therefore, these instruments are basically meant to be used in laboratories and cannot be used in field for prospecting. They are used to provide the accurate value of the earth's gravity field at certain fixed places in a country, which are used as primary base stations to standardize variometers during gravity surveys. Other applications of these gravimeters are studies related to changes in elevation due to environmental hazards such as volcanoes' crustal deformation studies in seismically active regions or along coasts to monitor sea level changes (Groten and Becker, 1995; Tiwari et al., 2006). There are very few companies, which manufacture absolute gravimeter in the world and supply on specific orders. One such unit is FG5 (Niebauer et al., 1995) manufactured by Micro-g solutions, USA.

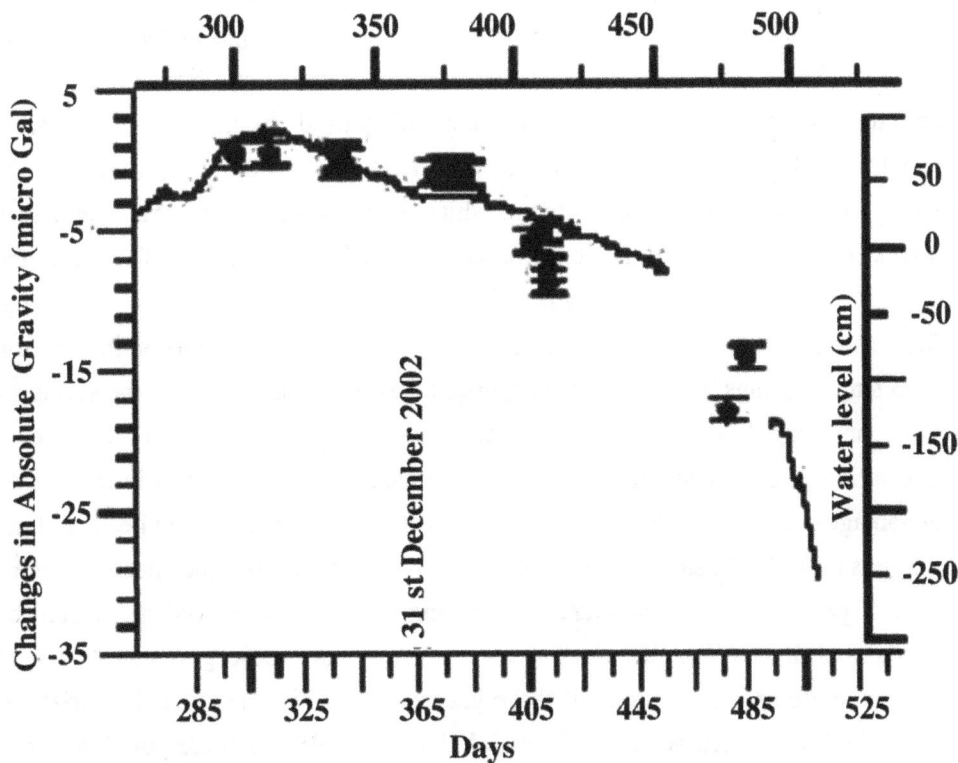

Fig. 2.7(b) Annual record of absolute gravity measurements at NGRI observatory showing seasonal variations in the gravity field (dots with vertical bars) and its correlation with changes in water level (continuous line).

(ii) Variometers: Relative Gravimeters

Gravimeters are variometers, which measures the variations in the earth's gravity field from one place to the other. It can be used to obtain the absolute value of the gravity field at a point station by measuring the variation in the gravity field with respect to a station where its absolute value is known. However, in geophysical exploration, we are basically interested in measuring the variations in the earth's gravity field, namely the gravity anomaly. A Gravimeter primarily consists of mass-spring system, which changes due to the variations in the earth's gravity field. During early periods of gravity surveys (1930s) changes in the spring was magnified by any mechanical or optical device and measured to a calibrated scale. They were known as stable gravimeters. However, these gravimeters could not provide the kind of accuracy desired in the gravity surveys and are therefore, replaced by unstable (astatic) gravimeters where the spring is brought to the original (null) position through a screw and the rotation of the screw is calibrated in terms of variations in the gravity field.

A schematic diagram of mass spring system of an (unstable) astatic gravimeter is shown in Fig. 2.8. The sensitive element in these gravimeters is a mass supported by a special spring called zero length spring, which is displaced due to changes in the gravity field. It is brought back to the null position by a measuring screw provided at the top of the gravimeter. The restoring force required to bring it back to the normal position is the measure of the changes in the earth's gravity field. The gravimeters based on this principle are most robust, portable and field worthy and are therefore, widely used for the gravity surveys. The most widely used gravimeters in this class are Worden Gravimeter and Lacoste-Romberg Gravimeter (LRG). The spring used in these systems is a special spring, known as zero length spring, which would collapse to zero length in the absence of external forces. In actual practice however, it is difficult to achieve. Therefore, there are special procedures to make them, and they remain trade secret of manufacturers. According to Hook's law, the tension in any spring due to a force is proportional to the extension in the spring. However, in zero length springs, the tension is proportional to the actual length of the spring as if it does not have any initial length so, is the name zero length spring. It is achieved by pre-stressing the spring of even sized coils such that they close on themselves as in case of door springs. The advantages of zero length spring lies in the fact that if it supports the beam and mass in the horizontal position, it will support them in any position after its displacement. In order to reduce the effect of pressure and temperature, the sensing elements are placed in vacuum tube with a suitable thermostat unit. The accuracy (least count) of both these gravimeters are 0.01 mGal. However, there are special versions of these gravimeters, which measure the variation in the earth's gravity field to an accuracy of 0.001 mGal (1 micro Gal) which are popularly known as microGal gravimeters. The LRG are manufactured and marketed by Scintrex Corporation Limited, Canada, as CG-5 gravimeters, which is an electronic version of earlier gravimeters with least count of 1 micro Gal. This gravimeter has a long range of 7000 mGal so as to cover the whole world without any scale adjustment.

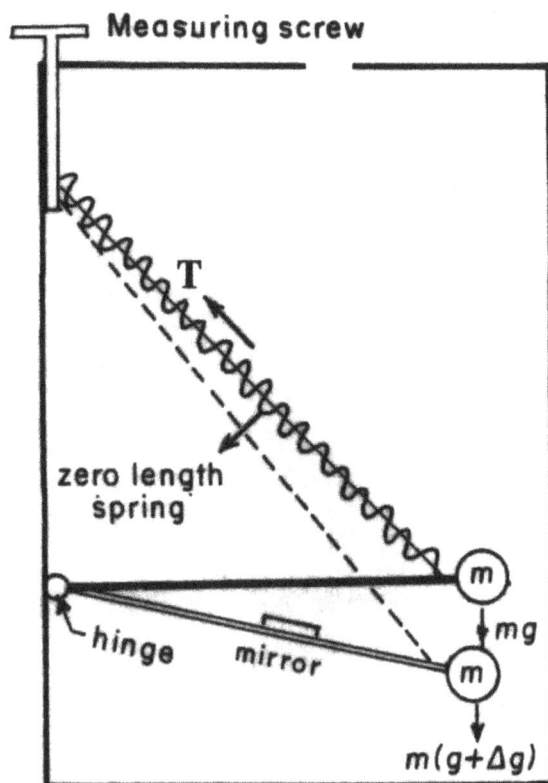

Fig. 2.8 Schematic diagram of Lacoste Romberg gravimeter showing mass (m) supported from zero length spring. A change in gravity field by Δg exerts tension T in the spring and changes its position which is brought back to initial position by measuring screw, which is calibrated in terms of changes in the gravity field.

There are special versions of these gravimeters for measurements on ships, aeroplanes and in boreholes, referred to as ship borne, air borne and borehole gravimeters which are briefly described in section 2.6.

(iii) Superconducting Gravimeter

Superconducting gravimeters are the most recent development in the series of instruments used to measure the earth's gravity field. These are also variometers based on the principles of superconductivity and can measure the earth's gravity field to an accuracy of 1 microGal. It is based on the idea to levitate a superconducting mass (ball) in a magnetic field (Fig. 2.9; GWR, 2004). As gravity changes, an electrostatic force is applied to the mass (ball) to keep it leveled. The basic elements are a superconducting sphere levitating above a pair of super conducting coils. The current in two coils are adjusted in a way that vertical gradient of the magnetic field supports the levitating mass in a defined position. If the gravity field changes, the sphere moves up and down and brought to the null position through the adjustment of magnetic field which is the measure of the variations in the gravity field. They are quite stable and provide the variations in the earth's gravity field to an accuracy of $\pm$ 2-3 micro Gal. However, they are as bulky and expensive as the absolute gravimeter mentioned above. Therefore, they are also used mainly for recording variations in the earth's gravity field at permanent stations and provide long term observations of time dependent gravity field related to crustal deformation studies.

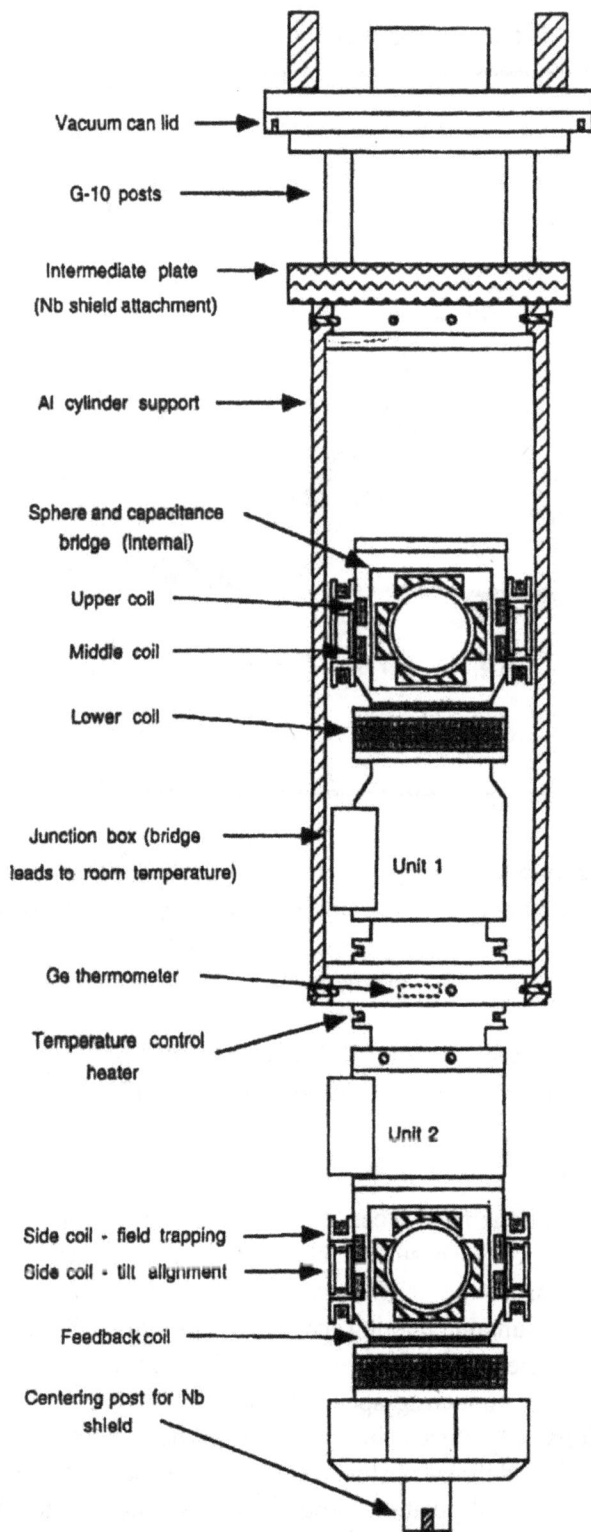

Fig. 2.9 Schematic diagram of Superconducting gravimeter showing two coupled units each containing niobium sphere that lavitate with change in the gravity field and three superconducting magnetic coils control its position. The other components are used to level and make measurements. (GWR Instruments Inc.).

The accuracy of the gravimeters discussed above vary from 0.01 mGal to 0.001 mGal (1 micro Gal) which represents their least count. But the accuracy of the gravity field measured by them in field cannot be better than 5-10 times of the least count of the instruments used due to various factors such as pressure, temperature and various corrections discussed in section 2.4. It also depends on whether the measurement is made in a controlled environment of a laboratory or in the field. They are useful to measure temporal variation in the gravity field (Tabel 2.3, at the end of the chapter)

(iv) Gradiometers

Torsion balance, which was primarily used in early stages of gravity survey, were in true sense gradiometers, which measured vertical and horizontal gradient of earth's gravity field at a station. They were quite useful in early days of oil exploration (1930's and 1940's; Bell, 1997; Jakosky, 1961). However, due to logistic problems in their operation in field, they are no more used in field. Gradient measurements do not require connected survey control (drift correction) and elevation of stations, which are two main primary concern in any gravity survey and thereby reduces the period and cost of any gravity survey. It only requires location of the station, which can be obtained from any normal GPS. It is therefore, specially suited for hilly and inaccessible terrains. Therefore, gravimeters have been some times used to obtain the gradients in field specially horizontal gradient by dividing the difference in field measured at three consecutive stations (Fig. 2.10(a)) through distance between them. In case, the two directions Y and Z (Fig. 2.10(a)) are approximately in a horizontal plane along the profile and perpendicular to it, the horizontal gradient can be obtained as $\sqrt{(a^2 + b^2)}$ where a and b are gradients in Y and Z directions. In order to obtain good signal d_1 and d_2 (Fig. 2.10(a)) should be more than 10-12 m. Though it is an approximate approach but in absence of ground gradiometers, this method of gradient survey can be followed in limited areas. Similarly vertical gradient of the gravity field was measured by designing two levels of platforms and recording gravity field over them and dividing the gravity field measured at those levels by difference in their heights. This however is a cumbersome process in field and there fore did not find favor with explorationists. The gradient is expressed in mGal/m or μGal/m and its standard unit is Eötvös unit (EU) which is equal to 0.1 μGal/m. A typical gradient anomaly due to shallow sources in oil exploration is in the range of ± 200 EU. Gradient measurements have been found to be more suitable for exploration of stratigraphic trap pinch outs for hydrocarbon exploration compared to normal gravity surveys (Fig. 2.10(b); Hammer and Anzoleaga, 1975). It is shown here that while normal gravity measurements provide similar decreasing field over pinch outs and a fault (Section 2.7), while the gravity gradient provide a high over the pinch outs which can be easily identified and separated from the fault anomaly. The disadvantage of gradient measurement, however, lies in its small order of anomalies and the theory related to their quantitative interpretation are not that well developed as in case of gravity anomalies. The interest in gradient measurement has revisited due to their application in airborne and future satellite surveys as described in section 2.6.

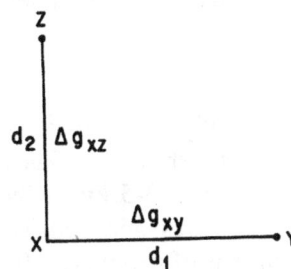

Fig. 2.10(a) Layout of stations for measurement of gradient using normal gravimeter. Three stations are laid at X,Y,Z and difference in gravity values divided by distance provide the gradient in two perpendicular directions along the profile and perpendicular to it. Z can be vertical axis which would provide vertical gradient.

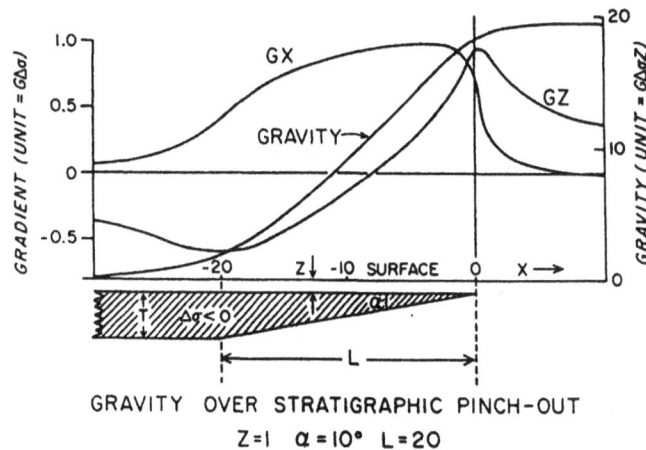

Fig. 2.10(b) Gravity and gradient anomalies of a stratigraphic pinch out. Gx and Gz are the horizontal and vertical gradients which show high over the pinch out, while the gravity field shows a gradient over it (Hammer and Anzoleaga, 1975) similar to that in case of a fault (Fig. 2.25(a)).

2.3.2   Field Operations and Survey Procedures – Geodynamic Studies and Resource Exploration

Field operations depend on the type of geological problems being investigated and desired accuracy of the gravity data. Based on equation (2.21), the factor which affects the earth's gravity field most is the distance from center of the earth (R). Therefore, the elevation of stations and its position should be precisely known. As discussed in the next section under the elevation correction, its effect is approximately 0.2 mGal/meter implying that an error of 1 meter in elevation of the stations will affect the gravity measurement by about 0.2 mGal which is quite significant, specially when one considers that accuracy of gravimeters are 0.01and 0.001mgal. Therefore, sufficient precautions are observed to record the elevation of the stations during a gravity survey depending on the desired accuracy of particular survey. Based on application of gravity surveys discussed in Chapter 1, different types of gravity surveys can be broadly classified into following three groups:

(i)   Regional gravity surveys for geodynamic studies and seismotectonics

(ii) Reconnaissance and detailed gravity surveys for hydrocarbon exploration.

(iii) Reconnaissance and detailed gravity surveys for mineral exploration and environmental studies including Engineering Geophysics, which are grouped under Near Surface Geophysics.

Survey details of different types of gravity surveys are described below and also while dealing with their applications in different chapters. They are briefly summarized in Table 2.2 at the end of this chapter.

(i)   Regional Surveys

As the name indicates, these surveys are basically regional in nature and are conducted in large regions for geodynamic studies including crustal structures. These surveys are carried out along roads and motorable tracks at a station spacing of 3-5 km, which delineates long wave-length anomalies related to large scale crustal structures and density in homogeneities. Anomalies recorded in these surveys are of the order of tens of mGal (10-100 mGal and more) spread over a few hundred kms (~100-1000 km). The accuracy of such surveys therefore, need not be very high as in case of exploration for oil and mineral described below. An accuracy of $\pm$ 1-2 mGal is sufficient for regional surveys. As these surveys are conducted at large station spacings, elevation of stations are controlled using Differential Geographic Positioning Systems

(Differential-GPS) with an accuracy better than 1 m in elevation. Position and location of stations in these surveys are controlled based on GPS and topo sheets on 1:50,000 (1 cm = 0.5 km) or preferably 1:25,000 (1 cm = 0.25 km) scale. Stations are plotted on these topo sheets and a base map is prepared over, which the gravity anomalies after making all corrections referred to in the next section are transcribed and contoured. The coordinates of stations from GPS measurements and the corresponding gravity values can also be entered to computer and contoured using any of the standard soft wares available for this purpose.

Seismotectonics: These are special kinds of studies related to tectonics, specially faults and intrusives associated with earthquakes. These studies encompasses from, largest wavelength (size) of gravity anomalies related to mega structures like mapping of subduction zones, collision zones to the smallest wavelength (size) of anomalies related to faults associated with specific earthquakes as discussed in Chapter 8. The former viz. mega structures are delineated and studied based on satellite or airborne gravity data as discussed in Section 2.6, while individual faults are delineated based on regional and detailed surveys discussed in this section for geodynamic studies and resource exploration depending on the size and extent of the causative fault.

(ii) Hydrocarbon Exploration

There are two aspects of gravity surveys for hydrocarbon exploration, viz (a) Reconnaissance survey and (b) Detailed/Integrated gravity surveys.

(a) Reconnaissance Gravity Survey

Gravity surveys for exploration of hydrocarbons are basically conducted as reconnaissance surveys to delineate basement structures and structures in sedimentary sections to delimit the areas for seismic surveys. It is generally carried out in a large area covering a specific basin along roads and tracks at a station spacing of approximately 2.0 km. It is planned in such a way that network of stations at about 2.0 km spacing are available in the entire region. The order of gravity anomalies expected from such structures are of a few mGal (1-10 mGal) in amplitude and therefore, the desired accuracy of such surveys are $\pm$ 0.1 mGal. The station elevation is therefore, obtained, by geodetic leveling or Differential-GPS providing the elevation to an accuracy of a few cm (<10 cm).

(b) Detailed/Integrated gravity survey

While conducting seismic surveys for exploration of hydrocarbons, it is a general practice to follow the same profiles by gravity surveys at closer stations spacing of 100-200 m. Such surveys are basically conducted for an integrated approach to supplement the information derived from seismic surveys and delineate structures within the sediment. The desired accuracy of such surveys are $\pm$ 0.01 mGal (10 μ Gal) and therefore, station location and their elevation to an accuracy of a few centimeter (< 5 cm) are usually obtained from geodetic surveys or dual frequency Differential GPS. These surveys are conducted using microGal gravimeters. In case of shallow gas occurrences, such surveys can be conducted over a grid with both profile and station spacing of 10-100 m.

(iii) Mineral, Ground Water and Engineering Geophysics

As discussed under introduction (Chapter 1), the application of gravity survey for mineral exploration primarily lies in delineating surface/shallow structures such as faults, fractures etc., which may be mineralized. Same is the case with engineering geophysics where shallow structures such as faults, fractures etc., are delineated to avoid those sections for large scale constructions such as dams, power plants etc., Even for ground water exploration, the application of gravity survey is indirect to delineate faults, fractures, weak zones etc., which might be water bearing. The survey procedure for application to these studies are therefore, similar in logistics. They fall under two categories, viz. reconnaissance and detailed/integrated surveys.

(a) Reconnaissance survey

This kind of survey is carried out to limit the area of interest from a larger area. It is conducted along profiles separated by a few hundred meters a part (100 - 500 m) at station spacings of a few tens of meters (10-100 m). If the area is large and inaccessible to lay down specific profiles, it is conducted along available roads and motorable tracks at a station-spacing of a few tens of meters (10-100 m). The order of anomalies recorded in such surveys are of a few milliGals (1 to 10 mGal) in amplitude and therefore, the desired accuracy of gravity survey is about ± 0.1 mGal. Due to desired accuracy of gravity data, elevation of stations in these surveys are obtained by geodetic leveling or Differential GPS, providing elevation to an accuracy of a few centimeters (< 10 cm). If the area is not large and accessible, a base line is laid out parallel to the general strike of the geological formation in the center of the block and profiles are laid out perpendicular to it at desired intervals (Fig. 2.11). Stations are marked along the profiles at regular intervals by geodetic survey or using differential GPS. Based on these reconnaissance surveys, promising zones and geologically significant structures such as faults etc., are delineated which are followed up by detailed gravity and other geophysical surveys such as electrical surveys for ground water investigations. In case of engineering site investigations, similar reconnaissance procedure to delimit the area as described above for mineral exploration is followed. However, in case of environmental studies related to seismotectonics long range profiles of few km in length across seismically active region can be recorded to delineate active faults. Alternatively gravity stations can also be recorded along road and tracks at 1- 2 km spacing as in case of regional gravity surveys for hydrocarbon exploration as described above and gravity maps can be prepared to assess the seismotectonics of the region.

(b) Detailed Gravity Survey

Detailed gravity surveys for mineral exploration and engineering geophysics are usually conducted in limited areas defined based on reconnaissance survey described above or known geology of the area. They are usually carried out in conjunction with other geophysical surveys like magnetic, electromagnetic and other electrical methods suited for the particular investigation and the information derived from them are integrated to provide the details of surface and subsurface structures or ore bodies. There are certain minerals/ores (Table 2.1), which have a higher density than the country rocks and therefore, can be directly delineated based on the gravity surveys. In other cases it is used to delineate details of significant structures such as faults, fractures etc., in association with other geophysical/geological data sets. This kind of survey is basically limited in areas of a few sq. meters and are carried out in grids with profiles and station spacing of a few meters (1-10 m) or may be even less depending on the size of the area and the target. A base line is laid out parallel to the general strike of the geological formations and profiles are laid perpendicular to it at almost same spacing of 1-10 m as per requirement of particular survey. Stations along the profiles are marked by the geodetic survey at the same interval as profiles forming a grid (Fig. 2.11). There are two kinds of these surveys namely normal gravity surveys (mGal range) as described above and microGal gravity surveys or high resolution gravity surveys, depending on the order of anomalies expected in a particular gravity survey.

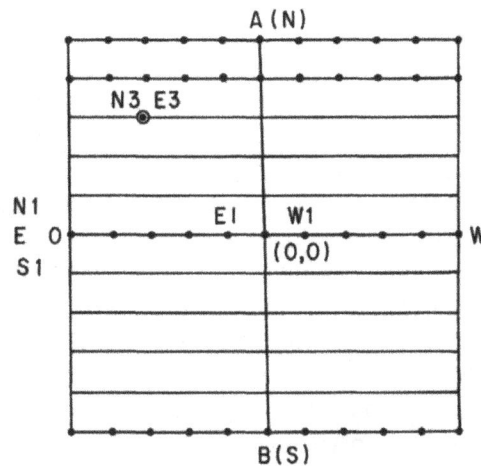

Fig. 2.11 Layout of profiles for detailed gravity survey in a grid form. A base line AB is laid out almost parallel to general strike in the area and profiles are laid perpendicular to it at desired spacing depending on size of target and nature of survey. Stations are recorded along the profiles at the same spacing as profiles providing grid pattern of stations. Stations are numbered from central point of base station E1,W1,................ and profiles are numbered as N1,S1,........ etc., such that the circled station is known as N3E3.

(iv) Micro Gal Surveys

In certain special cases, the observed gravity anomaly may be less than 0.1 milli Gal (0.1-0.01 mGal), which require survey accuracy of about 0.01 mGal (10 μGal). In such cases, special gravimeters with microGal range are used known as Micro Gal gravimeters with least count of 0.001 mGal (1.0 μGal) as described above. These surveys are usually conducted over a grid with profile and station spacing of 1-10 m as discussed above (Fig. 2.11). Similar grid surveys are also recommended for investigations related to shallow gas occurrences in sedimentary basins. Engineering site investigations also require similar kind of microGal surveys (Reynolds, 1997) along profiles or grids at profiles and station spacing of about 1-10 m. As the order of gravity anomalies in crustal deformation studies are also of same order (0.1-0.01 mGal) these observations are also made using microGal gravimeters. In this case, some permanent gravity stations are made in seismically active regions and repeat observation at these stations are made at constant interval of time (3-6 months). Even continuous observations can be made at permanent field stations using absolute gravimeters or super conducting gravimeters, which are best suited for crustal deformation studies. Seasonal changes in the gravity field due to environmental factors such as water level changes etc., are also recorded in the same manner.

2.3.3 Pre-Processing of Field Data

The gravity anomaly for prospecting are measured using gravimeters, which measure the variations in the gravity field between different stations and are highly sensitive. Since they are highly sensitive, they are affected by the changes in the instrument specially in the spring, which may be caused by mechanical and environmental changes such as jerks, pressure and temperature etc., Further as they measure variations in the gravity field, readings obtained from them are required to tie them with base stations where the absolute value of the gravity field is known. Therefore, certain corrections, for these factors are required in field to achieve the desired accuracy. In this regard, there are basically two corrections, viz. (i) Drift Correction and (ii) Base Loop Correction, which are applied during course of the field work in order to assess the effectiveness of the surveys.

(i) Drift Correction

When the repeat observations are taken by a gravimeter at the same point, it is found to be varying with time and this is known as drift of the gravimeter. The variation in the gravity field with time is primarily due to the tidal effect and changes in the gravimeter. The former being 0.02-0.03 mGal/hour, which is corrected by making Tidal Correction as per time and place of measurement as discussed in the next section. The latter which is also approximately of the same order is required to be corrected by taking repeat observations. The factors responsible for changes in the gravimeters are creep in the spring due to jerks, conditions of the battery used for temperature control etc., It is therefore, essential to observe the behavior of the gravimeter while being used in the field. For this purpose once in one or two weeks, the observations are recorded continuously through automatic recording or at intervals of 5-10 minutes from morning till evening at a fixed station in the field camp and plotted versus time as shown in Fig. 2.12(a). This figure shows observed gravity field at a particular station and tidal corrected field based on the location of the station and time of measurement. As one can see, fluctuations in the tidal corrected field are much less compared to the observed field. It provides the behavior of gravimeter during the period of observation. Based on this, one can choose the period when the drift of a particular gravimeter is minimum and linear so that it can be appropriately distributed over the readings recorded during that period. From Fig. 2.12(a), it is evident that the tidal corrected field is almost constant from 9.0 A.M. to 1.0 P.M. and therefore, this period is most suitable for use of this gravimeter for acquisition of gravity data. Depending on the type and accuracy of a particular gravity survey the time period to repeat gravity observations to correct for the drift of a gravimeter is decided. For example in case of oil and normal mineral exploration surveys, it is generally maintained at 2 hours while for high resolution (µGal) surveys and base stations it is about 1-2 hour. If the drift of the gravimeter is more than the permissible limit of 0.05-0.06 mGal/hour (including tidal variations), it should be thoroughly checked for scale adjustments, battery etc., However, if the drift of the gravimeter still persists beyond permissible limits, it is send to the manufacturers for thorough checking and renovation.

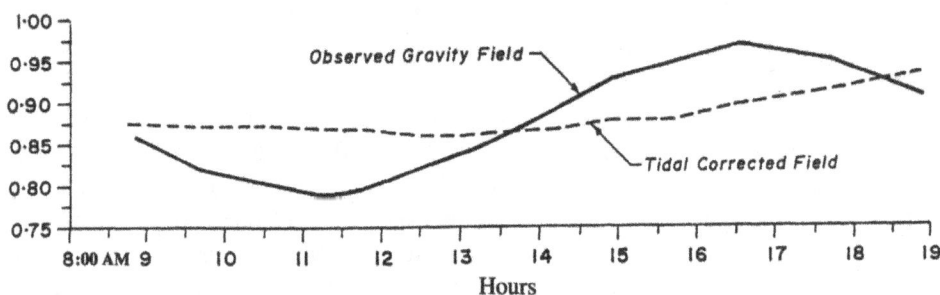

Fig. 2.12(a) Observed gravity field and tidal corrected filed at a field station from 8:00 AM to 7:00 PM showing drift of gravimeter. It also shows tidal corrected field with lesser fluctuation in the field during same period.

(ii) Base Loop Correction

Base stations are the places where the absolute value of the gravity field is known to an accuracy of 0.01 mGal (10 µGal) and are used to obtain the absolute value of the gravity field at the stations in the field by measuring its variations with respect to base stations. There are primary base stations in the country usually at fixed locations such as railway stations, airports etc., In any gravity survey, a network of secondary bases is established encompassing survey block, which is connected to any of the known primary base. To achieve the accuracy a particular procedure of measurements are followed. The bases are connected by looping three times between two bases A and B as A-B-A-B-A-B (A and B being two stations, Fig. 2.12(b)) and occupying them usually within an hour to ensure that the instrument drift is linear and minimum. This measurement procedure is referred as double tie of bases. This kind of looping provides three observations at every base, which are corrected for the drift of the gravimeter based on consecutive five readings. The consistent value from the three observations or the average of close by readings if they are not consistent is chosen for every base station. In case of large survey area, such as in case of regional survey or hydrocarbon exploration, bases are first established in the region which are subsequently used to tie the gravity stations for their absolute values and simultaneously used for drift correction so as to avoid reoccupying same base after 2 hours. In these cases, the survey is started from a base station and closed at some other base station after every 1-2 hours and the difference in the gravity field adjusted with respect to their absolute values is distributed to station occupied during that period.

Fig. 2.12(b) Establishment of base station by double tie such that at two stations A and B three independent values are obtained. Common value or their average value is taken as Base value.

The bases are established on permanent structures for easy identification during the survey and afterwards. Figure 2.13 shows a typical example of a network of bases which shows the differences in the gravity values between two bases along the arms with arrows showing towards higher values. These differences along the arms of a network (loop) are added and subtracted as the case may be depending on the direction of the arrows to obtain the difference at closing base of the loop. Ideally, it should be zero but in all practical surveys a difference is found, which is known as closed loop error (CE). This difference is divided by number of arms that gives the distribution error (DE). The arrow inside the loop points in the direction of the positive loop error (Mishra and Tiwari, 2010). In this case station number 1 is a primary base that is used to establish secondary bases in the survey block for ease of recording the gravity readings in the survey block. Initial checks for loop closure errors are done in the field itself in order to repeat the erroneous bases in case the loop error is found to be more than the permissible error. The permissible error for the closed loop error is 0.02 $\sqrt{n}$ mGal, where n is number of arms in a loop. Sazhina and Grushinsky (1971) have provided detailed description of these corrections under 'Adjustment of Gravity Networks' similar to geodetic networks. Having established the requisite number of base stations in an area, the gravity observations at various stations are tied to these bases within a maximum time limit of 1-2 hrs for drift correction during that period. In regional surveys where survey is conducted along roads at station spacing of 3-5 km, it may not be possible to make loops. However, as the desired accuracy of the gravity data in this survey is limited to a mGal, bases can be tied along the roads at about 20 to 30 km interval as A-B-A-B which provides two drift corrected readings at every base. The average of the two readings is taken as the base value. In case of mineral exploration, if the area is small and isolated far away from any known base, an arbitrary base is chosen near the camp with any arbitrary value, for example 0, 100, 1000 and all the readings are reduced to this arbitrary base value. This provides the gravity anomalies in the survey block with reference to the arbitrary base. In case desired, this arbitrary base can be tied to standard bases in the nearby region even after the survey and observed gravity values at stations are updated with reference to the absolute value of the gravity field at the base station accounting for the arbitrary value adopted previously for the survey.

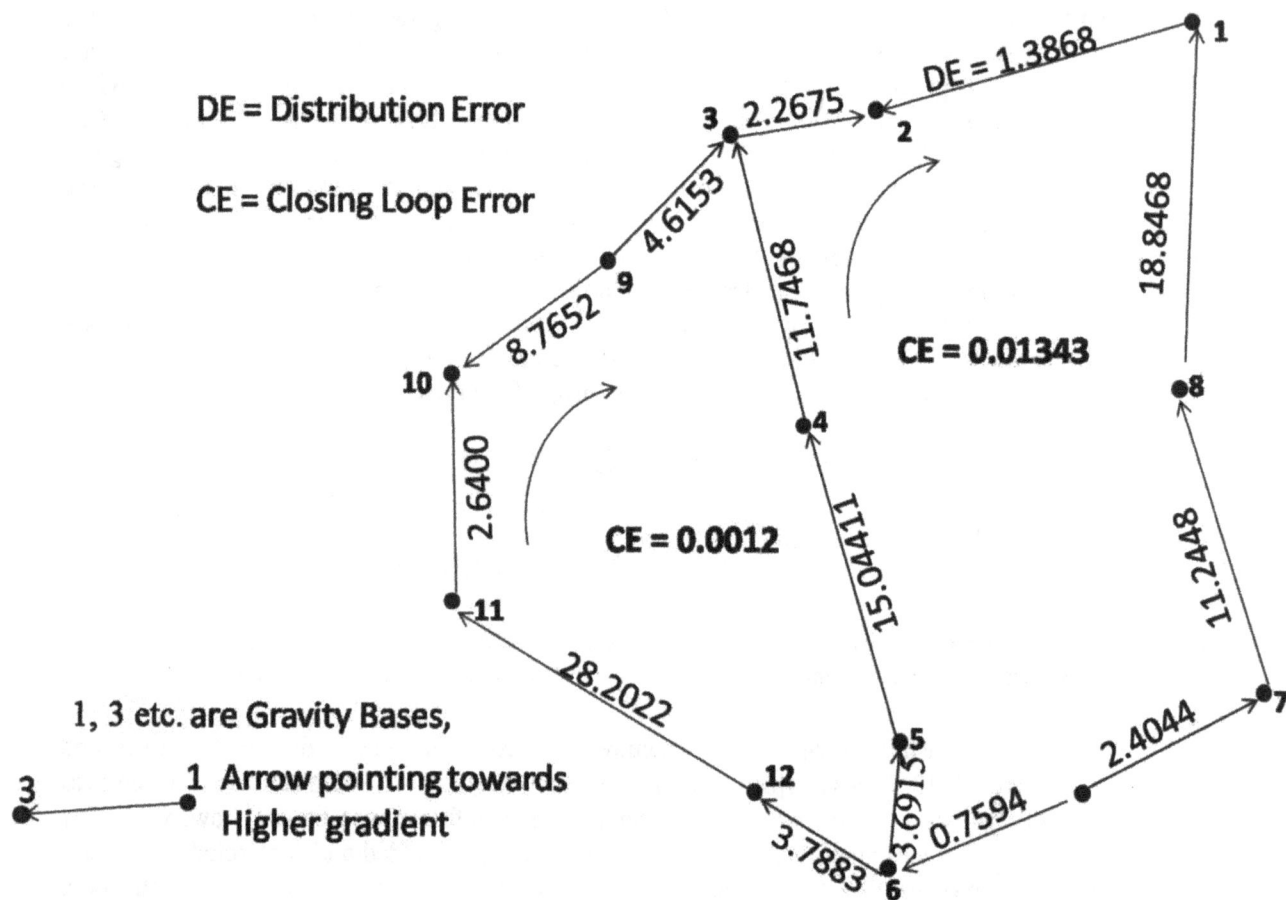

Fig. 2.13 A typical example of closed base loops. Closed loops of stations for loop correction with distribution error given along each arm that denotes differences in the observed field at adjoining bases. The closing loop error (CE) is the difference between the observed field at the first point of the loop and adjusted value after closing the loop divided by number of arms in the loop with arrows indicating toward the increasing values. Station numbers given at circle points are not in sequential order as they can be observed in different orders and different periods and loops can be formed subsequently and corrected.

2.3.4 High Resolution Surveys

As discussed above, the desired accuracy of gravity data for detailed surveys for exploration of mineral and hydrocarbons is ± 0.01-0.1 mGal. Data with such accuracies are referred to as high resolution data. In recording such data sets, some special precautions are followed which are discussed below:

(a) Gravimeters are regularly checked for their drift correction (Fig. 2.12(a)) at least once in two weeks which should not exceed 0.06 mGal/hr including tidal effect. However, tidal correction based on the latitude and the longitude of the station and time of measurement as discussed in the next section is applied to the observed gravity data on regular basis in order to reduce the uncertainties of diurnal correction based on repeat observations. Based on the drift of the gravimeter at a constant place, the period during which the variation is relatively least and linear is chosen for the gravity survey and repeat observations are made within 2 hours to correct for the drift of the gravimeters.

(b) Elevation of the station should be known to an accuracy of a few cm ($< \pm$ 10 cm) which is possible only by geodetic survey or using dual frequency differential GPS. However, in microGal surveys where accuracy of gravity data is 0.01 mGal (10 microGal), it should be known to an accuracy of $\pm$ 2-3 cm which is possible only by good quality geodetic leveling such as double tertiary geodetic surveys. The accuracy of geodetic leveling should be checked in the field based on whether it is single tertiary or double tertiary survey.

(c) If the area is large, base stations are first established in a loop at a distance of 10-15 km (Fig. 2.13) and loop error should not exceed 0.02 $\sqrt{n}$ mGal where n is the number of arms in a loop. In case loop error exceeds this value the individual bases in a loop should be checked and the particular base contributing more towards the loop error should be repeated.

(d) In spite of gravimeters being pressure compensated and maintained at constant temperature, while working in hot weather, precautions like use of umbrella and covering the gravimeter suitably by towels should be meticulously followed to avoid temperature effect on the measurement which is quite considerable. Hot period of mid day specially in tropical countries like India should be avoided for this kind of survey.

(e) To check the accuracy of recorded gravity value, 5-10% of gravity stations are repeated on different days and preferably by different observers which should be within twice the least count of the gravimeter (0.02 mGal) of the previously recorded value. Infact a particular base station near the camp is repeated every day before the start of survey to check the performance of the gravimeter, which is highly dependent on the condition of battery. In case the repeatability of this base station is not found within permissible limit of least count of the gravimeter (0.01 mGal), batteries and gravimeter are checked and only when the performance of gravimeter is satisfactory, the survey is resumed.

(f) Gravity stations are kept a few meters away from ditches, ponds, mound etc., which can affect the gravity readings due to terrain effect as discussed below.

2.4 Corrections and Reduction of Gravity Data

The measured gravity field at the surface of earth largely depends on the following factors:

(i) Earth tide

(ii) Latitude

(iii) Topography / Terrain

(iv) Elevation and

(v) Surface/subsurface density distribution.

Geophysicists and geologists are basically interested in the last factor, which is investigated in any exploration programme. However, contributions due to other factors (i-iv) are quite significant and are therefore, required to be first corrected before the effect due to the last factor is evaluated. These corrections are named after the factors for which the correction is made.

2.4.1 Earth Tide Correction

Due to the attraction of the sun and the moon (their gravitational forces) the earth's surface is affected in several ways, the most visible being the changes in the sea surface. It also deforms the earth's surface and therefore, its gravitational field. It is positive during high tides and negative during the low tides. It depends on relative position of the earth, the sun and the moon and therefore, depends on the place, date and the time. The differences in the gravitational field at a particular place between the high and the low tides may be as much as 0.3 mGal. This correction can be computed from the expressions describing this effect (Longman, 1959). There are standard computer packages for this purpose based on this expression which provides the tidal effect depending on the latitude and longitude of the place, the date and the time of observation.

2.4.2 Latitude Correction

It is known that the earth is not a perfect sphere, it is rather flattened at the poles and bulges at the equator suggesting a variation of earth's radii with latitude. Besides this, the rotation of the earth also causes an increase in the gravity field with latitude. These factors cause about 5 Gal difference between the earth's gravity field at the equator and the poles which is approximately 1 mGal/minute of the latitude. This is quite significant and the gravity data must be corrected for it. A reference ellipsoid is defined as the surface of the earth with excess land masses removed and oceans filled with similar land masses as discussed in section 2.2.8. On such a reference ellipsoid, the International Association of Geodesy in 1967 provided the expression for the theoretical value of the gravity field g_t as known as IGSN, 1972:

$$g_t\,(1971) = g_e\,(1 + 0.0053024 \sin^2 \phi - 0.0000058 \sin^2 2\phi)\ \text{Gal} \qquad .....(2.34)$$

where $g_t\,(1971)$ = theoretical gravity at latitude ϕ and

g_e is the earth's gravity field at equator = 978.0318 Gal

International Association of Geodesy modified this formula in 1980 (Vanicek and Krakiwsky, 1986) as follows (Mishra et al., 1992).

$$g_t\,(1980) = g_e\,(1 + a \sin^2 \phi + b \sin^4 \phi + c \sin^6 \phi)\ \text{Gal} \qquad .....(2.35)$$

where a, b and c are constants given by

    a = 0.0052790414
    b = 0.0000232718 and
    c = 0.0000001262

This is known as the 1980 Geodetic Reference System (GRS). Equations 2.34 and 2.35 provides the theoretical value of the gravity field at a particular latitude, to an accuracy of 4 and 0.7 microGal, respectively (Vanicek and Krakiwsky, 1986). Therefore, corresponding to a particular latitude at the place of observation, the theoretical value of the gravity field is computed based on these equations and subtracted from the observed gravity field, which provides the gravity anomaly at that place and automatically corrects for the variation in the gravity field with latitude and is therefore, referred to as the latitude correction.

2.4.3 Topography / Terrain Correction

This correction is applied for the undulating topography around the station such that after this correction, the observation point is presumed to lie in a plane area. In relatively plane areas, it is usually quite small which can be neglected but in areas with even moderate topography, this correction must be applied. As in Fig. 2.14, the observed gravity field at the point of observation B is affected by a component of attraction of mass both at A (higher topography) and C (lower topography) compared to the topography at the observation point at B. Therefore, based on this correction effects of hills and valleys are nullified at the point of observation as if the observation is taken in a plane area. In case of a hill (A) close to the point of the observation (B), there is a component of attraction towards A, which should be added to the observed field in order to nullify the effects of the hill that is opposite to the earth's gravity field. Similarly, in case of adjoining valley (C), its effect is missing from the observed field at B and therefore, the equivalent field should be added to the observed field in order to obtain the field for a plane area. Therefore, the topographic correction is always positive whether the point of observation is surrounded by a hill or a valley. A volume element (dv) situated at a distance r from the point of observation (Fig. 2.25(c)) produces a field (δg) equal to:

Fig. 2.14 The effects of surrounding terrains on a gravity station at B. The hills (A) have an attractive component at B which reduces the gravity field at B and the attractive component of B towards C is missing from the gravity observation at B suggesting positive corrections in both cases of stations at high and low altitudes.

$$\delta g = G\rho dv/r^2$$

where G = Gravitational constant and ρ = density

It's component in vertical direction, which would affect the observed gravity field is given by:

$$\delta g = (G\rho dv/r^2) \sin \alpha$$

where α = Angle between vertical and r (Fig. 2.25(c))

The effect of the total volume v is given by:

$$\Delta g = G\rho \int_v \frac{dv}{r^2} \sin\alpha \qquad\qquad ......(2.36)$$

The volume integral is difficult to evaluate analytically and therefore several schemes have been proposed to evaluate it numerically. Hammer(1939) was the first to suggest the use of a circular template of concentric circle divided into several compartments through radial lines to evaluate the terrain effect. The template on a transparent sheet is laid over the topographic map of the region with its center coinciding with the station where terrain effect is evaluated. The average elevation in each compartment is obtained from topographic map and their effect is computed by approximating the compartments to a vertical cylinder. The gravity field due to such a cylindrical body is given by equation (2.65) as follows.

$$\Delta g = 2\pi G\rho[z_2 - z_1 + (z_1^2 + r_1^2)^{1/2} - (z_2^2 + r_2^2)^{1/2}] \qquad\qquad .....(2.37)$$

where r_1 and r_2 are the distances of the inner and outer radii of the compartment.

ρ =density of the exposed rock usually taken as 2.67 g/cm^3 for upper crust

z_1 and z_2 are elevation to the top and bottom of the cylinder. In present case z_2= elevation of the station where terrain effect is evaluated

Hammer computed this factor (Equation 2.37) for a standard density of 2.67 g/cm^3 for different ranges of elevation and catalogued in a table form for convenience of users (Telford et al., 1990). The compartments in Hammer's template close to the center (point of observation) are smaller compared to those outside as elevation changes in them affect much more compared to the latter. This requires a better control on elevations in compartment close to the center. The total sum of the effect of all compartments of the template provide terrain effect. It is then shifted to the next station and process is repeated till the terrain correction for all the stations are obtained. On almost same principle automatic methods of evaluating terrain correction using appropriate softwares have been developed (Cogbill, 1990; Banerjee, 1998). The advantages of automatic methods of terrain correction lies in using Digital Elevation Model (DEM) available at $2' \times 2'$ grid worldwide. However, $2' \times 2'$ grids are quite large and can be used only for outer compartments for terrain correction and for inner compartments close to the stations, better controlled digital data is required through specially mapped topo sheets. Parker (1996) suggested a Fourier transform approach for terrain correction, which computes the gravity field due to given topography in digital format at the point where terrain effect is required. Some basic principles of Fourier transform and its application are discussed in Chapter 4. Terrain correction is required to be computed separately for every point of observation. It is a tedious process specially the computation of average elevation in each compartment and therefore, prospectors tend to ignore it. However, it is one of the most important factor affecting the accuracy of data and therefore, it must be applied even in case of moderately varying topography. The maximum distance up to which terrain correction should be applied has been a matter of discussion among geophysicists. Some suggest its application to apply it up to 167 km for which the earth's surface is presumed to be flat (Fig. 2.15). However, some others have suggested its aplication up to distance where its value reduces to less than the desired accuracy of the survey. In low relief, one must obtain a better controlled elevation data.

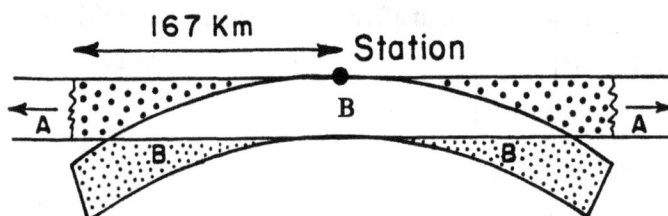

Fig. 2.15 The normal Bouguer correction is applied for the horizontal slab A while the earth's surface is spherical 'B'. The difference between the gravity effects of dotted parts of A and B is known as Bullard 'B' correction.

In order to reduce terrain effect in relatively plane areas, certain precautions can be observed. As terrain near the observation point affects most, this point can be selected in a plane area away from ditches and small hills. Surveys based on GPS have made it easy where the observer can choose stations as per best conditions and is not controlled by stations provided by geodetic surveys. The importance of proper site for gravity station can be gauged from the following effects (Leaman, 1998).

• A one meter high rock of 2×2 meter dimension will produce an effect of 0.05 mGal close to it while its effect will reduce to 0.02 mGal and 0.01 mGal at distance of 1.5 and 1.0 m away from it.

• Similarly a ditch or a culvert of same dimensions as high rocks above will also produce same effects.

It is therefore, advisable to record gravity stations 1-2 m away from exposed high rocks, ditches, culverts etc., in the field, in case it does not affect the survey specifications.

2.4.4 Elevation Correction

The earth's gravity field changes with the elevation due to factor R in the denominator of equation (2.21). Therefore, a suitable correction for variations in the elevation of different stations should be applied to reduce the observed field at a common datum. The common datum chosen is the mean sea level (msl), which is also the reference plane for measurement of elevation of stations. Once the topographic correction is applied, it is assumed that the effects of hills and valleys surrounding the station have been corrected and the point of observation lies on a plane with elevation equal to that of the station (A; Fig. 2.14). To reduce the gravity field at this station (A; Fig. 2.14) to mean sea level, two more corrections are required viz the first is due to changes in gravity field due to its elevation above msl and second is due to the effect of rocks between the plane of observation and msl (Bouguer slab; Fig. 2.15). The former is called free air correction and the latter is known as Bouguer correction.

(i) **Free Air Correction:** Based on equation (2.21), the ratio of the observed gravity field (g_o) at a station with elevation h and the reduced gravity field g, at mean sea level is given by

$$g/g_o = (R + h)^2/R^2 \qquad .....(2.38)$$

where R is the radius of earth

$$g = g_o (1 + 2h/R)$$

$$g - g_o = g_o (2h/R) \qquad .....(2.39)$$

Taking the average radius of earth R as 6,371 km and g = 981,000 mGal, free air correction, fc is given by

$$f_C = g - g_o = 0.3086 \text{ mGal / meter} \qquad .....(2.40)$$

This correction is always positive as field increases with decrease in the radius (R) of the earth. This is called free air correction as the point of observation is considered to be hanging in the air. This factor can be used to measure the height of a building or a hill by taking measurements at the bottom and top of the building or hill. However, earth is not a perfect sphere and the radius R changes with the latitude of the place, which must be accounted in equation (2.39). Several workers have suggested full formulas incorporating curvature of the earth for this purpose (Heiskanen and Moritz, 1967). However, the one given by Bhattacharji (1973) is widely used in this country (India):

$$f_C = 0.30855 \text{ h} + 0.00022 \cos 2\phi . \text{ h} - 0.000000072 \text{ h}^2 \qquad .....(2.41)$$

where f_C = free air correction in mGal, ϕ = Latitude and h = elevation in meters.

The first term in the expression is same as that given for a spherical earth (equation 2.40) and the additional second and third terms represent the effect of changes in radius (R) with latitude. Equation (2.41) can be referred to as complete free air correction. Contrary to it, the free air correction provided by equation 2.40 can be termed as simple free air correction.

(ii) **Bouguer Correction:**

Once the observed gravity field is reduced to msl, one must remove the effect of the rocks between the point of observation and the mean sea level. This correction is computed by assuming a slab between the point of observation and msl and its effect is subtracted from the observed field. This correction is referred to as Bouguer correction after the name of the geophysicist who first provided the formula for the gravity effect of a slab of thickness (t). The gravitational acceleration at the center of a vertical cylinder is given by equation 2.68.

$$\Delta g = 2.\pi.G.\rho\ (z_2\text{-}z_1)$$

where $(z_2\text{-}z_1)$ is the thickness (t) of the cylinder. Therefore,

$$\Delta g = 2.\pi.G.\rho.t \qquad\qquad .....(2.42)$$

where Δg = gravity effect due to slab of thickness t and density ρ.

G is the gravitational constant equal to 6.67×10^{-8} CGS units. Therefore,

$$B_C = 0.04188\ \rho\ \text{mGal/m}. \qquad\qquad .....(2.43)$$

The average density of basement rock is taken as 2.67 g/cm^3, therefore,

$$B_C = 0.1115\ \text{mGal/m} \qquad\qquad .....(2.44)$$

Bouguer correction is always negative as one must subtract the effect of rock material between the station and the mean sea level. Combining the free air and Bouguer corrections, the elevation correction is approximately equal to 0.2 mGal/m. This is equal to 2 μGal/cm. Equation (2.44) is known as simple Bouguer correction, which accounts for the horizontal slab (A, Fig. 2.15). However, the earth surface (B, Fig. 2.15) is curved and the gravity effects in two cases are different corresponding to dotted parts in the two cases. Therefore, the complete total Bouguer correction is given by:

$$B_C = 0.1115\ h + C \qquad\qquad .....(2.45)$$

where C is the correction factor due to curvature of the earth known as Bullard correction. Lafehr (1991) provided the value of this correction up to elevation 6300 m, which should be added to the simple Bouguer correction given by equation (2.44). Bhattacharji (1973) provided the following expression for this correction, which is used by different organizations in India (Krishna Brahmam and Subba Raju, 1972).

$$C = 0.000446\ h - 0.00000003115\ h^2 \qquad\qquad .....(2.46)$$

where C = curvature correction and h = elevation in meters.

2.4.5 Geoid (Indirect Effect) Correction

This is a special correction which is applied to the observed field in specific cases. It may not be important for surveys in small areas such as for oil and mineral exploration, as regional geoid is a smoothly varying field and caused by deep seated sources in upper/lower mantle. However, regional surveys for geodynamic studies and reconnaissance surveys for oil exploration, which involved large areas on almost continental scale, this correction assumes special significance. Geoidal and the ellipsoidal heights at the point of observation are different due to differences in these two reference surfaces (Fig. 2.6(b)). In gravity surveys, heights are measured with reference to mean sea level, which is represented by geoid and is warped due to mass heterogenity. It does not coincide with the theoretical reference ellipsoid (eq. 2.34) which approximates the earth surface and is used to correct for the theoretical gravity. This difference between the reference ellipsoid and geoid is known as geoid anomaly and needs to be corrected from the observed field. Götze and Li (1996) have given the following simple formula to correct it.

$$\text{Geoid (Indirect) correction} = 0.1119\ H. \qquad\qquad .....(2.47)$$

where H is difference between ellipsoid and geoid (geoid anomaly) in meters. It is almost equal to the effect of Bouguer slab corresponding to difference between geoid and the reference ellipsoid with −ve sign in case the geoid is above the reference ellipsoid and vice versa. H can be obtained from the Global geoid model obtained from satellite data as discussed in section 2.6. Another approach may be to obtain the field due to geoid variations in the survey block and subtract it from the observed field. Marsh (1979) has presented separately the gravity effect of Indian Ocean geoid low of degree 12 which affects the observed gravity field in the Indian Ocean and surrounding regions considerably and can be used to estimate this correction in this region

2.4.6   Free Air and Bouguer Anomaly Profiles and Maps (Images)

The computed gravity anomaly at the point of observation depends on the corrections applied to the observed gravity field. If all the corrections discussed in section 2.4 are applied, the corrected gravity field is known as Bouguer gravity anomaly or only gravity anomaly or Bouguer anomaly. Therefore, Bouguer gravity anomaly (Δg or in short g.) is given by

$$BA = \Delta g = g = g_o - g_t - t_{iC} + f_C - B_C + t_C \qquad \qquad .....(2.48)$$

where BA = Bouguer gravity anomaly or Bouguer/gravity anomaly denoted by Δg or simply g for convenience.

g_o = Observed gravity field.

g_t = Theoretical gravity field based on the latitude at the point of observation.

t_{iC} = Tidal correction

f_C = Free air correction and

B_C = Bouguer correction, which is usually computed for a crustal density of 2.67 g/cm^3 as discussed above in section 2.4.3.

t_C = Topographic correction

Bouguer anomaly as given in equation (2.48) is referred to as complete Bouguer anomaly. However, in case the terrain correction (t_C) is omitted in plain regions, it is referred to as simple Bouguer anomaly. However, terrain correction is a major source of error in the observed Bouguer anomaly and therefore even in case of moderately varying topography complete Bouguer anomaly is computed by incorporating terrain correction. In case the Bouguer correction is applied for density 2.67 g/cm^3, it is referred to as corresponding to density 2.67 g/cm^3. Bouguer anomaly can also be computed for any other density using the expression given in equation 2.43. In case, the exposed rocks are of different density as in case of sedimentary basins for oil exploration, the Bouguer correction is generally computed for lower density corresponding to bulk density of sediments ($\approx$ 2.0 g/cm^3) and in that case the computed Bouguer anomaly is referred to as corresponding to that density. Based on trial and error, Nettleton (1976) even suggested to compute the bulk density of near surface rocks, which would correspond to density for which short wavelength Bouguer anomaly showing positive correlation with topography disappears.

In case the Bouguer correction (BC) for the material between plane of observation and the msl is not applied, the corrected gravity field is called the free air anomaly. Therefore, the free air anomaly (FA) is given by:

$$FA = g_o - g_t - t_{iC} + f_c + t_C \qquad \qquad .....(2.49)$$

In most of the cases, both free air and Bouguer anomalies are routinely computed. In case of oceans, free air anomaly is used for interpretation as it refers to mean sea level and no material exists between msl and the plane of observation. There are some other uses of free air anomaly such as in case of isostasy, which is discussed in the forth coming sections.

Free air and Bouguer anomaly computed from the observed gravity field can be plotted as a profile with reference to distance from a fixed reference station (Fig. 2.16(a)) or plotted on a two dimensional map if data is recorded in a 2-dimensional plane (Fig. 2.16(b)). For preparing a map, data is transcribed on a two dimensional base map with stations plotted on it to an appropriate scale. In actual practice, a base map is first prepared with some towns, roads, rivers etc., plotted on it. Gravity stations along with the corrected gravity anomalies are transcribed on it and equal value contours are drawn

(Fig. 2.16(b)). Contour maps provide the pattern of the field in a plane which generally shows the positions of high and low values of the field and spacing of contours indicate rate of change in it. The contour interval is chosen depending on the accuracy of the gravity data, which should be approximately 2-3 times of the latter. For example, in regional surveys as the data accuracy is 1-2 mGal, the contour interval is generally 5 mGal. In reconnaissance surveys for oil exploration, it is generally 1 mGal while in detailed surveys for oil and mineral exploration it is 0.1 mGal for normal surveys and 0.01 mGal (10 microGal) for microGal surveys. Even if the data is available in map form, profiles perpendicular to strike direction indicated by elongated contours are drawn from the map for interpretation purposes. Both kinds of data sets show gravity "highs" and "lows" implying high density rocks below a gravity high and low density rocks below a gravity "low" showing excess and deficiency of mass, respectively. Gravity field is a smoothly varying field and therefore, irregular contours with sharp kinks indicate error in data or in processing, which can be used to ascertain the quality of data to a certain extent.

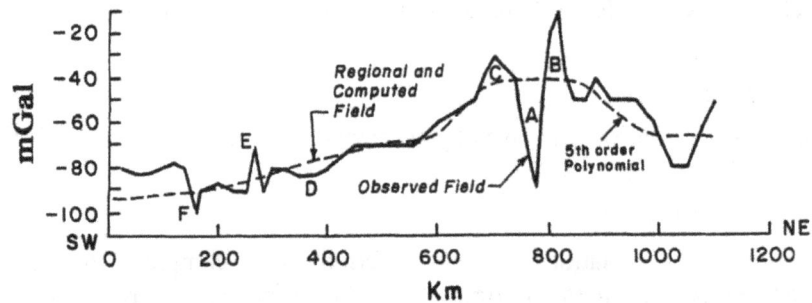

Fig. 2.16(a) A regional field drawn based on smoothing the observed field. The same regional field fits almost with the 5th order polynomial derived from the observed field using trend surface analysis (Chapter 4). The regional field represents a gravity high toward the NE decreasing towards the SW which is caused by deeper features. A, B, C, D, E and F represents the residual field caused mostly by exposed rocks.

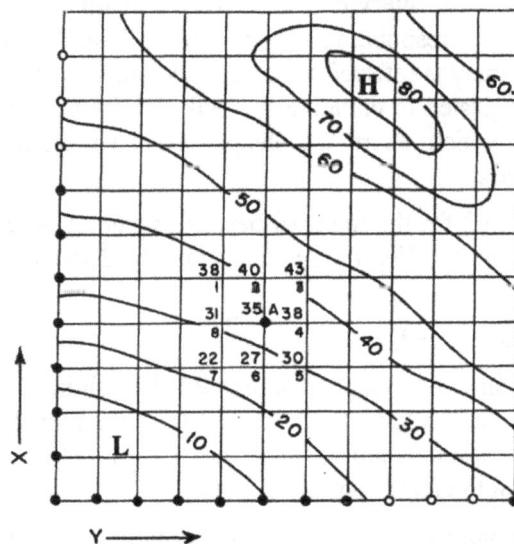

Fig. 2.16(b) Grid along X and Y axes superimposed over Bouguer anomaly map. The average of the field at grid points (1-8) surrounding a point A can be regarded as the regional field and its difference with respect to value at A is the residual field. In present case the average of field at stations 1-8 is 33.6 mGal and therefore residual field at A is 1.4 mGal. The gravity high and low are indicated as H and L.

Initially, contours were drawn manually by visual linear interpolation between two observations such that contours of specified values such as 1, 2, 5, 10 etc., pass in between the points. Some general principles of contouring are as follows:

(a) Contours passing through equal values should separate higher and lower values on either sides of it.

(b) They should never cross each other but they can form loops and end at the margins of the block.

(c) More the observations better are the contour maps and therefore, areas with sufficient data should be contoured.

Presently, there are several commercial software contouring packages which perform the same operation by computers. Most of the surveys record non-equispaced data except those recorded on ground in a grid format. Therefore, contouring by machine involves two operations, viz. **(i)** Interpolation of recorded data on a pre-specified grid and **(ii)** Preparation of contour maps from grided data. Interpolation implies obtaining field values at specified equispaced grid points using mathematical methods. There are several mathematical methods for this purpose but commonly used are minimum curvature and kriging methods. Minimum curvature method employs various forms of polynomial approximation (Crain, 1970; Briggs, 1974) which is briefly discussed in Section 4.1. It is fast and found to be suitable in regions with sufficient data points. Kriging is a relatively new approach and is found to be suitable even for regions with less data points but is relatively slow. It, however also provides the variance along with the interpolated values. It is a weighted linear function with weights depending on minimum variance (Gao et al., 1996). Commercial software packages for this purpose provide several other options for interpolation and one should choose them as per one's requirement.

Images: Bouguer anomaly maps can be presented as colour maps with different contour intervals filled with different colors automatically by computers. There are standard softwares for this purpose. Different sets of anomalies and their trends are easy to visualize in color maps. Gravity highs are indicated by warm colours (red) and lows by cool colours (blue). Smaller the contour intervals better are the representation of trends and anomalies. The colour Bouguer anomaly map of Godavari basin and Central India (Mishra and Venkatrayudu, 1991) and their images were prepared on colour graphic work station TEK-4125, which helped in delineating small structural trends, which are not visible in contour maps (Fig. 2.30(a)). Presently there are several commercial software packages available for this purpose. To prepare colored map or its image digital data is important. The maps available as contour maps can be digitized using AutoCAD, GIS or any other computer packages. One such scheme is described by Sreedhar Murthy et al., (1999) for digitization of analog maps. Some more details of preparation of images from contour maps are discussed in Section 3.4 while describing images of airborne magnetic data. A Bouguer anomaly map of India as image is given in Fig. 2.28.

2.5   Post-Processing Analysis and Isostasy

2.5.1   Regional and Residual Fields

The gravity field recorded at the surface of the earth is the cumulative effect of the sources at different levels. In gravity surveys, one largely records the gravity effect of sources from the surface down wards at least up to the Moho, which is a major density discontinuity. It is therefore, essential to separate the observed field in different groups originating from different levels. In general, they are separated in two groups namely the regional and the residual fields originating from deeper levels and shallow levels respectively. Horizontal layered models do not produce gravity anomalies, which are basically caused by perturbations in these layers or bodies of different densities embedded in them. Therefore, this approach of separating observed gravity field in two group viz. regional and residual fields representing deep seated and shallow sources has served satisfactorily over the years. As the name implies regional field is the component of the observed field, which is the characteristic of the region as a whole and is therefore, characterized by large wavelength (size) anomalies originating from deeper levels. On the other hand the residual field is the component of the observed field which are localized in nature and originate from shallow sources. Deeper levels primarily means crust-mantle boundary, viz Moho as it shows highest density contrast of 0.4 g/cm^3 compared to any other interface in the crust while shallow sources means individual bodies in upper crust referred to as sources of residual field. In geodynamics and crustal studies, therefore, the regional component of the observed field from deep seated sources is important, while in prospecting for oil, minerals and environmental studies, the residual field due to shallow sources assumes importance. There is a difference even in the mineral and the oil exploration as far as, depth of interest are concerned. While in mineral exploration, interesting zone is limited to a few hundred meters from the surface, in oil exploration, it extends up to the basement in sedimentary basins which may be on an average 3-5 km deep. In engineering problems too the zone of interest from surface is limited only to a few tens of meters. In seismotectonics it may extend from surface up to the upper mantle.

The easiest method to separate them is by visual inspection and drawing a regional field smoothly through the profile as shown in Fig. 2.16(a). This approach is mostly suitable along profiles and is based on assumption that deep seated sources of regional nature produce large wavelength smooth fields compared to shallow sources, which produces anomalies of limited extent. The smooth dashed line in Fig. 2.16(a) represents the regional field while its difference with the observed field provides residual field. The other method of regional-residual separation is the graphical method, which is suitable for the maps. A suitable grid (Fig. 2.16(b)) is laid over the map and the observed gravity field surrounding a particular grid point (A; Fig. 2.16(b)) is averaged and subtracted from the central value, which provides the residual field at that point. In this manner at all grid points a new grid is generated which is contoured to obtain the residual anomaly map representing shallow sources. This implies that surrounding values represents the regional field and therefore, its difference with the central point is the residual field at that station. In Fig. 2.16(b) the values surrounding the point A (1-8) provide an average value of 33.6 and therefore, its difference from the value at A (35 mGal) gives a residual field at A equal to 1.4 mGal. This method of regional-residual separation (grid method) is mainly used in case of detailed gravity surveys in small areas for mineral exploration and engineering site investigations related to near surface geophysics. These methods lack from any physical rational of representing the regional field by the observed gravity field in the immediate surrounding regions. The regional field in this case depends to a large extent on the length of profile or grid size being taken for regional-residual separation and individual interpreter.

Mathematical methods of Regional-Residual separation are discussed in chapter 4 in sections on "Trend surface analysis" and "Spectral Analysis". Another method of separating the component of the regional field originating from Moho is based on isostasy as discussed in the next section.

2.5.2 Isostasy

Bouguer anomalies observed over different parts of the world both over continents and oceans when plotted against topography / bathymetry, they provide in general an approximate linear relationship (Tsuboi, 1979) showing a negative correlation with height on continent and positive correlation with depth in oceans. This suggests that there exist mass deficiencies under continents related to height and mass excess under oceans related to depth indicating compensation of mass above and below msl at some depth. This compensation of excess and deficit of mass at subsurface levels related to topography is known as isostasy. There are primarily three different modes for compensation as described below:

(i) Airy's Root Model

The most common mode of compensation is Airy's root model in which the extra mass above msl is compensated just below it by mass deficiency at some level (Fig. 2.17(a)) known as the level of compensation which is usually regarded at the Moho. Moho is the most prominent large density discontinuity where compensation can easily take place. This implies that the crust is brittle and upper mantle is ductile and the crust is supported by hydrostatic pressure as in case of Archimedes principle of floating objects. This thickening of the crust below the mountains such as incase of Himalayas-Tibet, where the crust is thick ($\approx$ 70 km) and thinning under oceans where crustal thickness reduces to less than 10 km are related to isostatic compensation. Under continents in plane areas the average thickness of the crust is 33-35 km. This is most commonly considered mode of compensation as thickening of the crust under mountains and thinning of the crust under oceans is a well known fact established from seismic surveys in different parts of the world. As the compensation takes place just below the topography the shape and area of compensating mass is same as that of the topography. Therefore, if H is the height of the topography and a is its area, then the anomalous mass of the topography is ρ_C. H. a. where ρ_C is the density of the topographic load. Similarly, if T_C is the thickness of the compensating mass (crustal thickening) and ρ_c and ρ_m is the density of the lower crust and the upper mantle, respectively the compensating anomalous mass is $(\rho_m - \rho_C)$. T_C. a. Therefore, for Airy's model of isostatic compensation:

$$(\rho_m - \rho_C). \; T_C \;.a = \rho_C. \; H.a$$

or $\qquad\qquad\qquad (\rho_m - \rho_C). \; T_C = \rho_C. \; H \qquad\qquad\qquad\qquad\qquad$ .....(2.50)

Normally, the density of topography (crust) is 2.670 g/cm^3, which is used for Bouguer correction.

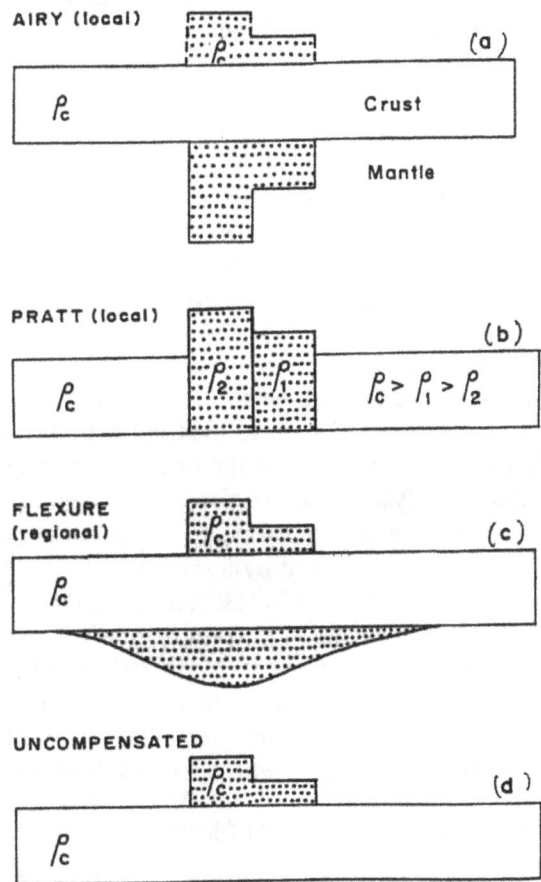

Fig. 2.17 Isostatic models (a) Airy's root model of varying crustal thicknesses; (b) Pratt's model of varying crustal density; (c) Regional model of flexure over a large area; (d) Uncompensated model.

Similarly ρ_c and ρ_m implying density of lower crust and upper mantle are equal to 2.9 and 3.3 g/cm^3, respectively). Therefore, using equation (2.50)

$$T_c = 6.7.H \qquad \qquad .....(2.51)$$

This implies that the thickness of the compensating mass at Moho (crustal thickening) for a full compensation will be 6.7 ($\approx$ 7) times of the topography. Therefore, if the topography above msl in an area is 1.0 km, the thickening of the crust for a full isostatic compensation in this model should be about 6.7 km. A crustal thickening of less or more than seven times (approximately) of topography is known as under or over compensation, respectively. W.A. Heiskanen subsequently provided sets of tables for calculating isostatic corrections based on this model and therefore, it is also known as Airy – Heiskanen model (Watts, 2001).

(ii) Pratt's Model

The second model of isostasy is, that due to Pratt (Fig. 2.17(b)) in which the compensation takes place due to lateral changes in the density at a certain depth, known as level of compensation. This implies that under mountains, the density of rocks are less and more under oceans compared to that under plane areas. However, there is no definite proof that the density of the subsurface rocks varies with elevation and therefore, this model of isostatic compensation is not very popular with geoscientists. However, in case of oceanic crust where lateral variation in density is considerable, this mode of compensation is considered viable. Both Airy's and Pratt models were inspired by geodetic measurements at foot hills of Himalayas and anomalous deflection of plumb line due to mountains and its root and were developed during middle of nineteenth century (Watts, 2001) for geodetic corrections much before the development of gravimeters and gravity surveys.

(iii) Regional Compensation (Flexural Model)

A third model is the regional compensation model, which is called flexural model (Fig. 2.17(c)). In this model, the compensation takes place over a large area instead of just exactly below the mountains. In this case the crust is deformed over a larger area compared to the topography and compensation is due to flexure of the crust. As in most of the cases, the topography is compensated over much larger area than the region occupied by it, it is presently most accepted model by geophysicists. In this case the upper layer (crust) behaves like an elastic plate and therefore, it supports the topographic load through bending over a larger area. The bending of the plate (crust) depends on its elastic properties such as flexural rigidity, which can be derived through coherence analyses between topography and the observed gravity field as described in Chapter 4, in section on "Spectral Analysis". This is also known Vening Meinesz model, who suggested it for the first time. However, as the crustal rocks in nature have some definite flexural rigidity, it appears that the compensation is caused by a combination of two factors, viz crustal root and flexural rigidity of crustal rocks. In such cases, Airy's crustal root model provides the maximum bound on the crustal thickness for flexural rigidity equal to zero. Flexural model has been found to be very useful in describing the rheological properties of the crust as discussed in Chapter 4. Fig. 2.17(d) shows an uncompensated model where the load of topography is supported by the strength of the crust. In this situation, flexural rigidity of crustal rocks tend to infinity, which is just opposite to Airy's root model where it tends to be zero.

2.5.3 Isostatic Compensation and Vertical Crustal Movement

Based on the nature of free air and Bouguer anomaly as positive, negative or close to zero, certain inferences can be drawn about state of isostatic compensation, which are useful for first hand interpretation. Isostasy basically occurs on large scale (Tsuboi, 1979) over a few hundreds of kilometers and therefore only regional aspects of elevation, free air and Bouguer anomalies should be considered to access isostasy. The most important aspect in this regard is the value of free air anomaly as zero which suggests that the effect of regional topography above msl is equal and opposite to that of the compensating mass indicating a fully compensated crust (Subba Rao, 1996). However, as isostasy relates to hilly terrain, the regional Bouguer anomaly where effects of regional topography as Bouguer slab has been removed, ought to be negative due to the effects of crustal thickening as compensating mass and shows a decreasing trend with increase in elevation (Fig 2.18; BA1 (Regional) and Regional Elevation in block 1 and 2). Depending on the nature of the regional Bouguer anomaly and free air anomaly vis-à-vis regional elevation (Fig 2.18), the following inferences can be drawn.

(a) In case along with the regional Bouguer anomaly being low (negative) showing inverse correlation with elevation as described above for hilly terrain, if regional free air anomaly is close to zero (FA1 (Regional); block 1), it indicate a fully compensated crust where the effects of regional topography as Bouguer slab is balanced by the effects of compensating mass. In such cases the crustal root would be $6.7 \times$ Regional Topography (block 3). Residual free air and Bouguer anomalies are related to shallow/local sources.

(b) In case along with the regional Bouguer anomaly being low (negative) showing inverse correlation with elevation as described above for hilly terrain, the regional free air anomaly is positive (FA2 (Regional); block 1) where the effects of the regional topography as Bouguer slab is more compared to that of the compensating mass, such cases are referred to as under compensated crust. In these cases, the crustal root is less compared to that of a fully compensated crust ($6.7 \times$ Regional Topography).

(c) In case along with the regional Bouguer anomaly being low (negative) showing inverse correlation with elevation as described above for hilly terrain, the regional free air anomaly is negative (FA3 (Regional); block 1) where the effects of the regional topography as Bouguer slab is less compared to that of the compensating mass, such cases are referred to as over compensated crust. In these cases, the crustal root is more compared to that of a fully compensated crust ($6.7 \times$ Regional Topography).

(d) Bouguer anomaly being high with positive correlation to regional topography in a hilly terrain, indicate total lack of compensation suggesting subsurface high density rocks or alternatively thin crust.

Regions belonging to group (a) are usually stable while those belonging to group (b) and (c) tend to subside or rise due to less or more buoyancy of the crustal root as is observed in case of sedimentary basins and fold belts (mountains), respectively. However, erosion from top and other tectonic activity keep on disturbing this balance that causes subsidence and uplift during different geological periods.

These are some approximate relationships to assess the isostasy in a region and can serve as thumb rules for preliminary assessment of gravity anomalies in a hilly region. However, the same can be quantified by computing isostatic anomaly which is difference between the observed anomaly and the gravity field due to full compensation based on Airy's root model. Based on Equation 2.51, the root of Airy's model for full compensation can be computed. The computed gravity field for this root is known as isostatic field and it's difference from the observed field is the isostatic anomaly. It's positive value suggests that crustal thickening is less (shallower Moho) compared to what it should be for a full compensation and negative value suggests vice versa. Positive isostatic anomaly, therefore, implies under compensation resulting into subsidence while negative values indicate over compensation resulting into uplift as has been discussed above based on isostasy. Positive and negative isostatic anomalies can also be caused by shallow low and high density bodies as is generally found in case of fold belts (mountains) and basins due to high density thrusted rocks and low density sediments, respectively (Chapter 7).

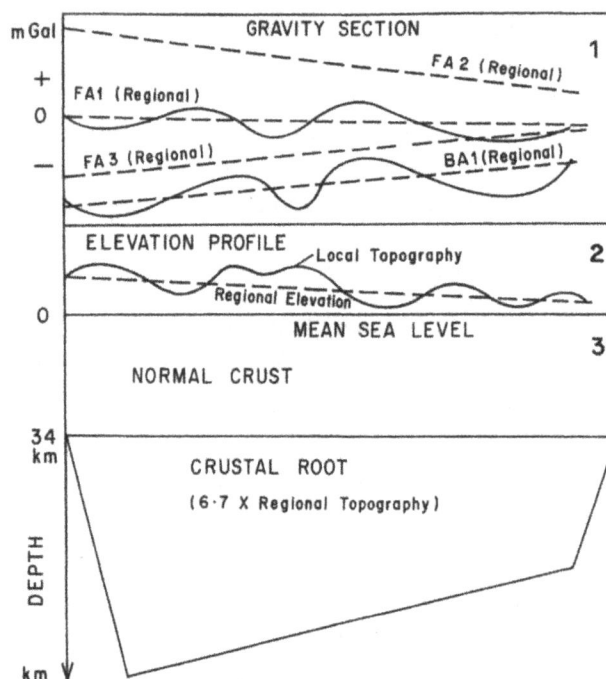

Fig. 2.18 : Fig. 2.18 A schematic diagram of isostatic compensation. Elevation is shown in block
(2) with dashed lines as regional elevation and superimposed over it is local topography. Schematic Bouguer
anomaly (BA) and free air anomaly for three cases (a-c) are shown in block (1). Case (a): The regional Bouguer
anomaly, BA1 (Regional) and Regional Elevation show an inverse correlation with negative values for Bouguer
anomaly indicating isostatic compensation. The nature of corresponding free air anomaly, FAI (Regional) is
close to zero indicating fully compensated crust. In such a case, the crustal root shown below the normal crust in
block (3) will be approximately 6.7 × Regional Topography. Case (b): The regional free air anomaly, FA2
(Regional) show positive regional field with inverse correlation to Bouguer anomaly, BA1 (Regional) which
indicates that the effects of topography is more compared to its root and is there fore under compensated crust.
Case (c): the regional free air anomaly, FA3 (Regional) is negative showing a positive correlation with Bouguer
anomaly indicating that the effect of root is more than topography and is therefore over compensated crust.

Fig. 2.36 shows Bouguer anomaly profile across Himalaya and Tibet (Rajesh and Mishra, 2003)
along 90° E longitude. It is based on some ground gravity data in the Indian Territory (NGRI, 1975)
and China (Sun, 1989) and satellite gravity data over Higher Himalayas and Tibet (EGM 96; Rapp and
Pavlis, 1990). It also shows elevation along this profile based on satellite data (Gtopo 30; Gesch et al.,
1999). This profile presents large negative Bouguer anomaly showing opposite correlation in shape
with the regional elevation and free air anomaly close to zero over Tibet. Regional elevation over
Tibet is about 5.0 km, which as per full compensated Airy's model of isostasy suggest a crustal
thickening (root) of 7 × 5 = 35 km. Normal crustal thickness of 35-36 km suggest a thick crust of
about 70-71 km under Himalayas and Tibet. It is interesting to note that as shown in Fig. 2.36, the
actual crustal thickness based on complete modelling of Bouguer anomaly constrained from seismic
INDEPTH profiles (Hauck et al., 1998) also suggest a crustal thickness of about 70-71 km indicating
full compensation for Tibet. However, Himalayan frontal thrusts (Chapter 6) show a positive isostatic
anomaly which are related to high density rocks along thrust belts as discussed above. It may be noted
that the crustal thickening as seven times of regional topography in a hilly region is an approximate
relationship and actual crustal thickness in a region depends on several other factors such as variation
of density in the crust etc. The correct estimate, therefore, requires complete modelling of Bouguer
anomaly as discussed in the fourth coming sections. However, an approximate estimate of crustal
thickness can be quickly obtained based on isostasy where topography is isostatically compensated as
per Airy's model.

The importance of isostatic compensation lies in the fact that in case of fully compensated crust, it is stable and does not record any vertical movements (uplift or subsidence). However, due to erosion from surface and vertical forces from down below such as magma rise etc., this balance is continuously disturbed and the region becomes unstable. Fast erosion in an otherwise compensated crust causes over compensation, which induces uplift in the region due to its buoyancy and whole process is further repeated. In case of under compensated crust, the region subsides and creates sedimentary basins. This situation may arise in an extensional regime, which may be induced due to volcanic activity or plate tectonic forces as discussed in Section 2.8.

2.5.4 Isostatic Regional and Isostatic Residual Anomaly- Deccan Trap, India

Whatever is the method of compensation, the effect of compensating masses at Moho are considerable and shadows the effect of shallow sources. It is therefore, essential to remove this effect, which as per the definition of the regional and the residual fields described above forms the main part of the regional field. The isostatic part of the regional field is known as isostatic regional and can be estimated from isostasy. Equation 2.50 states that in case of the Airy's model of full compensation, the effect of the Bouguer slab are equal to that of the compensating mass and in such cases free air anomaly will be zero. Alternatively, at places where free air anomaly is zero, the Bouguer anomaly represents the effect of compensating mass provided local effects have been smoothed out. Therefore, to remove the effect of local bodies regional elevation and corresponding regional free air and Bouguer anomalies are considered and Bouguer anomaly at places where free air anomaly is zero provides the effect of compensating mass termed as isostatic regional (Subba Rao, 1996). The isostatic regional is subtracted from the observed gravity field to provide isostatic residual anomaly which represent the effects of local bodies. Therefore,

$$\text{Isostatic Residual} = \text{Bouguer Anomaly} - \text{Isostatic Regional} \qquad \text{.....(2.52)}$$

Isostatic Residual anomaly therefore, mainly consists of the effects of sources up to the Moho and isostatic regional represents the effect of compensating mass at Moho, which is the major part of the regional field in an area. Bouguer anomaly in hilly regions is therefore, quite suitable for quantitative modelling based on isostasy as isostatic regional can be modeled due to thickening of the crust and the remaining isostatic residual anomaly can then be modeled due to shallow sources depending on the available geological and geophysical information from the region. The difference between the isostatic anomaly as described in the previous section and isostatic regional and residual as defined in this section may be noted. The former describes the state of isostatic compensation while the latter represents the present state of crustal thickness (depth to Moho) and anomalies due to other sources. However, having determined the depth to Moho, it can be used to ascertain the state of isostatic compensation.

An example of isostatic regional based on zero free air anomaly is demonstrated in Fig. 2.19, which shows topography, free air and Bouguer anomaly profiles across Deccan Trap in western India (Tiwari et al., 2000). The topography is smoothed to provide regional topography over which there are local undulations. Points on Bouguer anomaly profile corresponding to zero free air anomaly are joined, following the trend of the regional topography. This provides a smooth isostatic regional, over Bouguer anomaly profile which decreases consistently with minimum value under Western Ghats and indicates the effect of isostatic compensation under this region. This part of Bouguer anomaly (isostatic regional) can be modeled due to thickening of crust. The regional topography at the western Ghat is approximately 800 m above msl. Based on Airy's model of isostatic compensation, it is achieved by 5-6 km of crustal thickening, implying about 35-36 km thick crust at the eastern end of profile and 40-41 km under western Ghat, which is almost same as inferred from detailed modeling of this profile using seismic constraints (Tiwari et al., 2000).

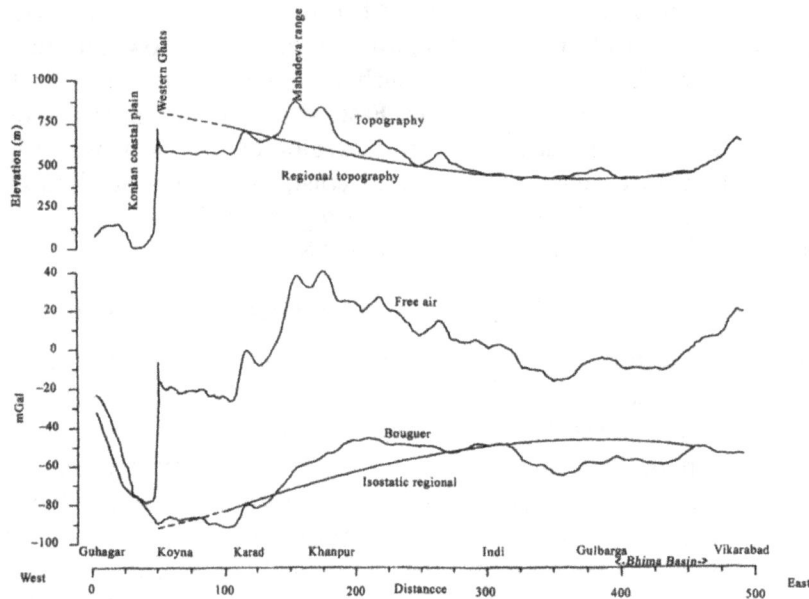

Fig. 2.19 Topography, free air and Bouguer anomaly across Deccan Trap, India showing regional topography and isostatic regional drawn based on points of Zero Free air Anomaly projected on Bouguer anomaly profile (Tiwari et al., 2000). It shows opposite correlation with regional topography and isostatic regional, which is consistently decreasing under Western Ghats indicating crustal thickening due to isostasy. R1 and R2 are isostatic residual anomalies caused by local sources.

2.6   Special Gravity Surveys – Marine, Airborne and Satellite Surveys

Initially, gravity surveys for exploration were mainly confined to the ground. But with developments of specialized gravimeters for applications in marine and airborne surveys and subsequent developments in geographical positioning system (GPS), new avenues were opened. Oceans are important both for geodynamic studies and resource exploration, which can be reached only through marine surveys. Importance of oceans for geodynamic studies can be understood from the fact that most of the elements of present day plate tectonics were discovered from observations in the oceans through marine geophysical surveys. In fact, most important features of plate tectonics, viz mid oceanic ridge systems, transform faults and subduction zones primarily occur in oceans (Section 5.2).

The gravity surveys in oceans are referred to as marine gravity surveys. The first marine gravity surveys were conducted as early as in 1930s by Vening Meinesz (1934) off Indonesian islands and by Matuyama (1936) off Japanese islands using a pendulum on a self designed stable platform in a submarine, which provided large amplitude linear gravity lows related to the trenches in these regions. Subsequently, after discovery of plate tectonics during 1960s, these gravity lows were identified as most important gravity anomalies in plate tectonics due to their association with subduction zones. There are certain places on the continent, which are inaccessible on ground such as hills, forests, marshy land etc., which can be surveyed only by airborne surveys. However, both these surveys viz marine and airborne surveys are more popular in case of magnetic method due to importance of marine magnetic surveys in the geodynamics due to sea floor spreading magnetic anomalies and ease of operation of airborne magnetic surveys as discussed in Chapter 3. Recently, satellite based surveys assumed importance due to fast data coverage and accessibility of otherwise inaccessible regions. Satellite based altimeter surveys over oceans have received wide application in preparation of gravity maps of oceans, which was otherwise not possible. It may be noted here that satellite gravity surveys do not directly measure the gravity field in satellite instead, it measures height of satellite over oceans

(Fig. 2.22), known as satellite altimetry and perturbations in its pre-specified tracks over continents known as satellite geodesy, are converted to the gravity field in the two cases as discussed in Section 2.6.6. Some important aspects of these specialized surveys are discussed in this section. When the gravity field is recorded from a moving platform, a special correction known as Eötvös correction is required to remove the effect of the speed of the platform from the observed gravity field.

2.6.1 Eötvös Correction

This correction arises due to speed of ship or aircraft, which is much less in former compared to latter. The attraction of the earth at a fixed point is reduced by the centrifugal force of the earth's rotation. Eastward movement of observer (ship/aircraft) will increase the centrifugal force and decrease the gravity field of the earth and vice versa in case of westward movement. This requires a correction for the eastward component of ship's velocity and is called Eötvös correction. It is given by

$$Ec = 7.503 \, v \cos \varphi \sin \alpha + 0.004154 \, v^2 \qquad \qquad .....(2.53)$$

where Ec = Eötvös correction in mGals, v = velocity of ship/aircraft in knots.

φ = Latitude and α = ship / aircraft course direction with respect to north.

This factor is primarily responsible for the reduction of the accuracy of ship and airborne gravity surveys when compared to ground gravity surveys.

2.6.2 Shipborne Marine Gravity Surveys

Traditionally marine surveys were started for mapping ocean floor during world war II and later this provided bathymetry of different oceans in great detail. These surveys delineated mid oceanic ridge system and trenches, which were subsequently attributed to plate tectonic processes. It was carried out using echo sounders, which transmitted regular signals from the ship and recoded back after reflection from ocean bottom. The half of two way travel time of reflected signal from ocean bottom recorded in ship multiplied by the velocity of signal provides the depth to the ocean bottom. In any marine survey, bathymetry is recorded along with other parameters in order to correlate the recorded data with the ocean bottom. Marine gravity surveys are conducted along profiles using specially designed gravimeters fitted in a ship. These surveys are basically conducted along with other geophysical surveys such as seismic, magnetic, bathymetry etc., for evaluating basement topography and crustal structures up to the Moho. Whenever seismic survey is conducted in ocean, gravity survey is conducted as a complementary survey. There are basically two kinds of these surveys viz regional surveys and detailed surveys. The regional surveys are conducted for geodynamic studies along ship tracks oriented approximately perpendicular to the general bathymetry in the region. These are basically individual long profiles or a few profiles at large intervals of 100-1000 km. The detailed marine gravity surveys are basically conducted for hydrocarbon exploration along continental shelf or in exclusive economic zones with closely spaced profiles of 1-10 km spacing depending on the size of the target. Data is recorded continuously on a digital tape as the ship cruises.

There are special problems related to marine gravity surveys out of which most important is exact location of ship and stabilization of gravimeters due to movement of ship and wave motion. The former viz; location of ship is nowadays controlled based on satellite navigation using GPS, which provides accuracy of a few meters in location. Initial modes of navigation were basically radio positioning devices using radio waves such as radar Shoran, Loran etc., These were based on radio waves from ships reflected from known targets on the shore and distance between them is product of half of two way travel time and velocity of radio waves. These systems provided an accuracy of ± 50–100 m. Subsequently, they are all replaced by high precision DGPS in present day surveys, which provides an accuracy of ± 1–5 m in position location. The stabilization of gravimeter, however, is still the major problem which is achieved through special design of the gravimeters. The gravimeters used in ship borne surveys are same as on the ground but special measures are taken for damping the sensing elements (mass) so as to reduce the effects of movements and jerks in the ship. There are several ship borne gravimeters but the Lacoste-Romberg gravimeters (LRG) fitted with a liquid filled device for damping is widely used in modern day marine gravity surveys and are referred to as LRG model 'S' Air-Sea gravimeter with an accuracy of ± 0.1 mGal.

Initially, before 1965, gravimeters on board the ship were suspended using rings as gimbal suspension, which could maintain near horizontal position of gravimeters in spite of ships movement and tilt in the stand. However, subsequently the gravimeters in ships were mounted on gyroscopically controlled platforms, which helped better in maintaining horizontal leveled platforms for meters (Jones, 1999). To reduce the effects of vertical accelerations, the readings are averaged over specified periods. The largest source of error in ship borne gravimetery is the vertical acceleration due to waves on the ship movements. However, noise due to waves is largely represented by high frequency noise in form of upward and downward movements. Therefore, by averaging data over certain period its effect can be minimized. Averaging over 4 minutes is sufficient in normal course, which can be increased to 12 minutes in rough weather. The same can also be achieved, using suitable filters. The speed of the ships varies in the range of 5-10 knots and therefore covers about 5-6 m in one second. In case the gravity anomalies of interest are larger than 500 m, the data may be filtered up to 100 sec to remove the high frequency noise. This provides an accuracy of about ± 1 mGal for wavelength of anomalies of 500 m or more, which is sufficient both for regional studies and hydrocarbon exploration. Because of use of filters to remove the noise caused by accelerations in case of ship and airborne surveys, the survey accuracy in these cases are dependent upon the wavelength of anomalies. With present day gravimeters and GPS used for location and elevation, specialized marine surveys can provide an accuracy of ± 0.2 mGal at wavelength 250 m and more.

2.6.3 Airborne Gravity Surveys

In certain cases, specially in inaccessible areas, gravity surveys are conducted by aeroplanes or helicopters. However, airborne gravity surveys are still in experimental stage due to problems in recording the gravity data from a moving platform and rapid changes in the earth's gravity field due to speed of the aircraft and changes in elevation.

Gravimeter reading in an aircraft = Effect due to geological in homogeneity

$$+ \text{Eötvös effect} + \text{Aircraft acceleration} \qquad .....(2.54)$$

In order to recover first factor on right hand side, which interests the geoscientists, the observed gravimeter reading should be corrected for the last two factors, namely aircraft acceleration and Eötvös effect. Eötvös correction caused by E-W component of velocity, which is described above has much larger influence in case of airborne survey due to its high speed. At a speed of 200-300 kmph it can be as high as 1000 mGal. Similarly, a change in elevation of 100 m will cause a change of 30 mGal in the gravity field. Therefore, these factors are required to be recorded precisely while conducting airborne gravity surveys. The development of modern navigational systems using high precession Differential GPS Systems (DGPS) has helped considerably to improve accuracy in airborne gravity surveys (Lane, 2004). Since early 1980s, the two most frequently used airborne gravimeter systems are by Naval Research Laboratory, USA (Brozena, 1991) for geodesy and Carson services Inc. for oil exploration (Hammer, 1983). Gummert (1997) has described the results of airborne gravity survey in a test area using the modified versions of Lacoste Romberg model 'S' Air-Sea gravimeter with special measures for its stability and has reported good comparison with ground gravity data continued upwards within an accuracy of ± 1 mGal. Even in high terrain of Swiss Alps, Klingele et al., (1997) and Verdun et al., (2003) have reported good comparison between airborne gravity data recorded at barometric altitude of 5100 m above sea level and ground gravity data continued to flight altitude. With present day technology, airborne gravity surveys can provide data to an accuracy of

± 1.0 mGal for wavelength of 10-20 km, which requires elaborate processing of data for elevation and Eötvös corrections.

Airborne gravity surveys are conducted along profiles perpendicular to general strike of region as in case of ground surveys. Subsequently, the recorded profiles in a region are tied together by recording tie lines, which are flown oblique to the recorded profiles and are used to correct for changes in the recorded field with time (Fig. 2.20). The gravimeters can be fitted in a helicopter or in a plane, which fly at speed of 100-200 km/hour at a height of 100-500 m above the highest topography in the region depending on the nature and amplitude of anomalies in the survey area. In hilly terrain as in case of Alps, Himalayas etc., it is flown at constant barometric height, higher than the highest peak in the region. Typical sampling rate in an airborne gravity survey is maintained at 10 Hz, which for an aircraft speed of 180 km/hour (50 m/sec) provides the gravity field at every 5 m interval.

The most important check in any airborne survey lies in comparing the value of tie lines with that along normal profiles. Tie lines are flown across set of profiles after normal profiles are recorded (Fig. 2.20) to check the accuracy of recorded data and also to make correction for variations in the instruments reading with time, which is known as drift correction in gravity surveys as described above under corrections and the same is referred to as diurnal variation in the magnetic surveys. Differences in two readings, 'Misties', are calculated at the intersections of tie lines with normal profiles and the difference is linearly distributed with time over the data recorded in that section. Sections of profiles with large misties are discarded and recorded again. This can be done based on quality weighting assignments using a variance criteria suggested by Mittal (1984). The method uses the normalized sum of squares of misties with other lines as the criterion for assigning a quality weighting to any line. This method also provides the DC shifts that have to be applied to each cross over value to reduce the misties to zero. Once the crossover values have been corrected, the other gravity data along the profiles between the intersections can be adjusted by distributing the DC component in a linear manner. Before adjustment and tie line correction, the standard deviations of crossover errors over Alps was high (15.34 mGal), which reduced considerably after this adjustment to a standard deviation of 0.012 mGal (Verdun et al., 2003).

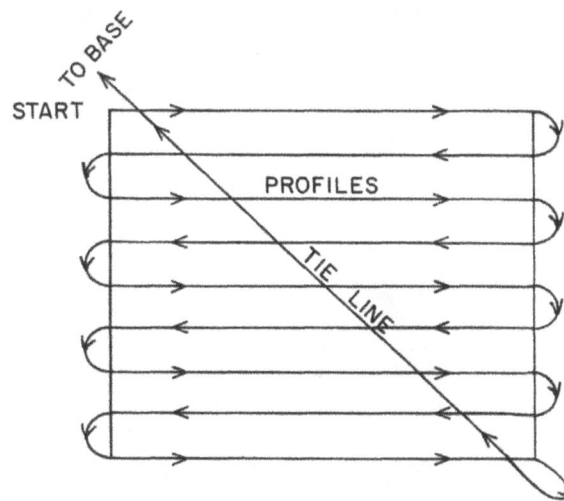

Fig. 2.20 Layout of profiles for Airborne survey. Profiles are flown almost perpendicular to general strike of the geological formations in the region at a height and profile spacing depending on the size of the area, target and the nature of the survey . After completion of set of profiles or block, tie lines are flown cutting across the recorded profiles for drift/ diurnal correction in airborne gravity/ magnetic surveys and check on the quality of the data.

An alternative method of conducting gravity survey has been found by towing gravimeters in the inaccessible regions and recording gravity data after lowering them at the surface (Seigal and McConnel, 1998). They have named this process as "Heligrav", which is found to be very useful for regional gravity surveys. In this case, a self-leveling gravimeter is towed by a helicopter and lowered at stations where gravity reading is desired. Preprogrammed GPS system guides the helicopter in the right direction and points the station location where the helicopter lowers the gravimeter and hovers till the gravity reading is automatically recorded. The elevation of helicopter is obtained from GPS system, which in turn provides elevation of stations for reduction of gravity data.

Further advancement in airborne surveys have come from remote controlled unmanned airborne vehicles (UAV), which has helped in reducing the flight heights and thereby, improving signal to noise ratio (Millegan, 2005). These are found to be extremely useful in locating small targets like Kimberlite pipes etc.

2.6.4 Airborne Gravity Gradiometry

Gravity gradiometry was revived in 1970s to overcome certain problems of airborne gravimetry. As it records the gradient of the field, the readings at different observation points are not coupled and therefore, repeat readings are not required to correct for the drift in the instruments and are less sensitive to elevation changes. The gradiometer basically consists of pairs of accelerometers and the difference of readings between them eliminates inertial accelerations and effects a separation of the gravitational signatures from them. It can thus be readily used on moving platforms. As it represents the gradient of the field, it is more sensitive to short wavelength anomalies (< 10 km) compared to the gravity field and is therefore, more suitable for exploration purposes. Bell Geospace Inc. USA has developed one such unit for use in the Oil and Gas Industry. The greatly increased bandwidth allows retention of high frequency, and short wavelength signals are generated by shallow features. This is because gradient falls as cube of distance from the target in contrast to conventional vertical gravity signal, which decays with square of distance. The increased sensitivity allows for greater resolution such that it can be integrated with seismic interpretation. When gradient of different components of gravity field in x, y and z directions are measured, it is known as Tensor Gradiometer, which measures nine components with x, y and z components of gradients in these directions (Fig. 2.21). The instrument contains three gravity gradient instruments each consisting of two opposing pairs of accelerometers arranged on a disc such that their axes are mutually perpendicular (Murphy, 2004). However, out of nine tensor components, only five are independent which can be used to determine the center of mass precisely (Fig. 2.21).

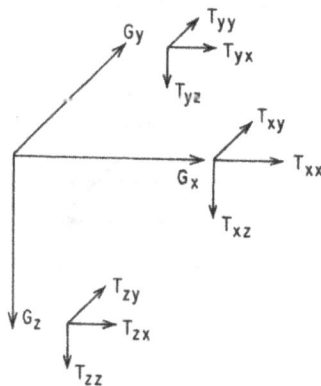

Fig. 2.21 Tensor gradiometer measuring in total nine components with five independent components used to infer mass anomalies. Gx, Gy, Gz are the gradients of the gravity fields in X,Y,Z directions and Txx, Tyy, etc., are their components in these directions. The five independent components are : any two of (Txx, Tyy, Tzz) , (Txy or Tyx), (Txz or Tzx) and (Tyz or Tzy), which are used to evaluate gradients.

Airborne gradiometer surveys are flown both for detailed and regional (reconnaissance) surveys with line spacing of 100 m-2.0 km depending on the average extent and depth of target in a region and type of survey. For example, closely spaced flight lines are required for exploration of small targets such as Kimberlite pipes while regional or reconnaissance survey for oil exploration can be conducted at 2 km spaced flight lines. Flight height varies from 100-500 m depending upon the expected anomaly and depth to the target. A low pass filter can be used to remove high frequency environmental noise of wavelength of about less than 500 m. Gradient is generally expressed in Eötvös unit such that 1 EU = 0.1 mGal/km. Due to high accuracy desired in gradient measurements, super conducting gravimeters with an accuracy of 10^{-1} EU are generally used in these surveys.

2.6.5   Advantages and Limitations of Airborne Surveys

There are certain advantages and disadvantages of airborne gravity surveys, which are mostly common to any airborne survey compared to ground surveys.

(a) Advantages

(i) Accessibility: Inaccessible regions such as hills, forests, marshy land and water/snow-covered areas can easily be accessed by airborne surveys.

(ii) Fast Coverage: Coverage is comparatively very fast at the rate of 100-200 km per hour in case of airborne surveys.

(iii) Better Data Coverage: The data along flight paths are recorded continuously in digital format, which are generally at 5 m interval as discussed above. They are processed automatically using specially designed softwares and Bouguer anomaly of large areas can be prepared fast.

(iv) Integrated Approach: Whenever airborne surveys are conducted, the aeroplane is generally fitted with some other equipment such as magnetometer, electromagnetic unit, spectrometer etc., and information derived from them can be integrated to provide a more viable model of surface/subsurface rocks and structures.

(v) Reduction in Terrain Correction and Surface Noise: The most tedious process and source of error in any gravity survey is terrain correction whose effect is considerably reduced in airborne surveys due to height of the aircraft and therefore, it is less sensitive to terrain correction. It has been suggested that at a height of 100-150 m, the effect of nearby topography represented by high frequency component is attenuated and therefore, Digital Elevation Model available at $2' \times 2'$ interval can be directly used for terrain correction in airborne surveys. The effect of these factors are further reduced in case of airborne gradiometer surveys, which is most recent development in this regard. Due to height of the aircraft, noise due to surface in homogeneities, are also attenuated at that height.

(b) Limitations

(i) Initial Investment and High Cost of Survey: Initial investment for equipment and aeroplane is very large. The cost of survey of moderate size area works out to be large. The cost of gravimeter (LRG Models) and required Softwares and training alone is quoted as US $ 1.0 Million. In case of contracted surveys, the cost is about US $ 100-150 per line km. However, in hilly terrain such as Alps and Himalayas, the cost of ground survey may be as high as airborne surveys and still ground survey may not be effective. Further, due to multidisciplinary nature of data recording in case of airborne surveys, it may work out cheaper for larger areas on a continental scale compared to recording different data sets on ground.

(ii) High Technology: The technology used in airborne gravity surveys are quite involved and specialized and is available only with few multinationals.

(iii) Loss of Information: Due to height of the aircraft, sometimes signals from shallow sources are suppressed, specially if they are feeble.

(iv) Location of Anomalies: In spite of good quality navigational equipment being presently available, such as differential GPS systems, exact location of airborne anomalies of small extents on ground is difficult. Therefore, airborne anomalies are often required to be followed up on ground to locate them precisely.

2.6.6   Satellite Gravity and Altimeter Surveys

The launch of Sputnic Satellite in 1957 and ability to measure perturbations in prescribed orbits of satellites provided a new opportunity to prepare worldwide geoid and gravity models. Gravity field from satellite are measured in two ways viz satellite altimetry and by tracking satellite orbits known as satellite geodesy. Satellite altimetry is presently the most important method to obtain the distribution of gravity field over the oceans. In an undisturbed region, the sea surface is a plane surface as shown in Fig. 2.6(a) and (b). However, due to surface/subsurface mass anomalies, it gets disturbed as shown in these figures. The difference between two surfaces is known as geoid anomaly, which is converted to gravity effect of the subsurface mass anomalies. Due to positive subsurface mass anomaly, the sea surface will prop up, while, due to negative mass anomaly, it sinks giving rise to positive and negative geoid anomalies (Fig. 2.6(b)). Orbiting the earth at a height of about 400-800 km, sensitive altimeters installed in a satellite (Fig. 2.22) measures the height of satellite above sea surface (h). Differences of the heights (h) with orbit height (H) provide geoid height (G$_H$) and its undulation due to mass anomalies, which is known as geoid anomalies.

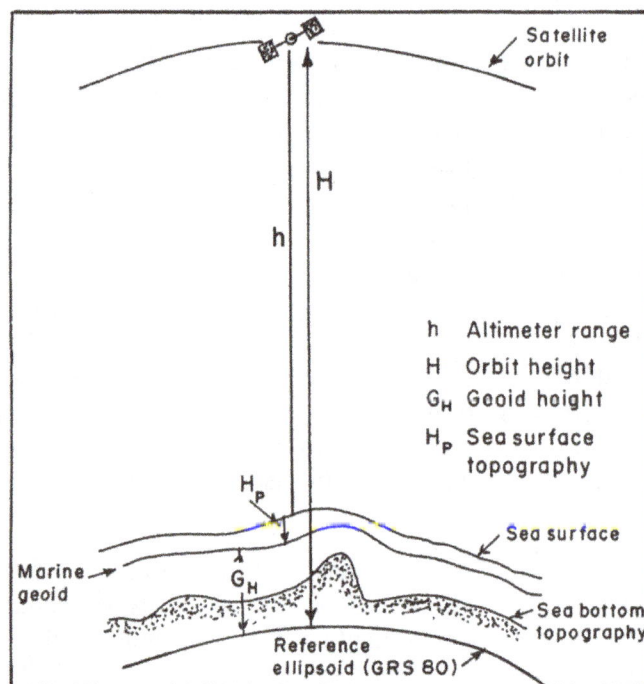

Fig. 2.22 Schematic diagram of Satellite measuring heights above the sea surface.

There have been several satellites in the past starting from GEOS-3 in 1975 followed by SEASAT, GEOSAT, ERS-1, TOPEX / POSEIDON, CHAMP etc., which were primarily launched by National Aeronautic Space Administration (NASA), USA and European Space Agency (ESA). In fact, radar pulses are transmitted from the satellite and received back after reflection from the nearest sea surface. The delay between transmitted and received signals provides the height of the satellite above sea surface. Each observation represents an area of sea surface providing an average height of the satellite in this zone thereby averaging the effect of the waves on the sea surface caused by wind, local disturbances etc., The differences in the transmitted and received signals also provide the variations in the sea surface due to local effects including wind, which is used to correct for these factors. The area of sea surface for one of the recent commonly referred to as Earth Research Satellite (ERS-1) launched in 1991 at a height of 800 km is 50 sq. km, which is the resolution of the data set obtained from this satellite. However, by combining data sets from different satellites such as GEOSAT and ERS-1, the resolution was improved to 20 sq. km (Sandwell and Smith, 1997). It was further improved by Andersen and Knudsen (2001) by combining all available marine gravity and satellite altimetry data to provide a model of gravity field over oceans with 1 arc min resolution (2 × 2 km) known as KMS, 2001. Due to high resolution, this data set can be used even for medium - short wavelength anomalies. This data set is specially suited for near coast coverage.

In reality the variations in sea surface at a particular time has two components one due to winds, tide etc., known as time dependent component and the other due to subsurface mass anomalies, which are independent of time. Therefore, recorded data is first corrected for effects due to the former to obtain the component due to the latter. Once the corrected geoid undulations are obtained, its Fourier transform (Chapter 4) provides the transform of gravity field whose inverse transform gives the observed gravity field (Chapman, 1979). As geoid and gravity anomalies vary inverse of distance and square of distance from the anomalous source, respectively the former is much more smoother compared to latter. Therefore, geoid anomalies record large wavelength features from greater depths such as lithosphere compared to gravity anomalies, which are normally effective to model sources up to Moho. Therefore, combination of two can be effectively used to model both shallow and deep sources up to lithosphere-asthenosphere boundary.

If Δh represents the geoid anomaly representing the geoid undulations, the corresponding gravity anomaly g_a is given by

$$F(g_a) = g. |k| .F(\Delta h) \qquad .....(2.55)$$

where $F(g_a)$ and $F(\Delta h)$ represents Fourier transform of free air gravity anomaly and geoid undulations respectively, g is normal value of the gravity corresponding to the latitude of the place as given by equation (2.35). $|k|$ is the one dimensional wave number given by k = 1, 2, 3, ... , n/2 where n is number of observations. The inverse transform of it provides the gravity anomaly (g_a).

To obtain the free air gravity anomaly corresponding to ellipsoid to which all gravity values are referred to, one must apply the elevation correction for geoidal undulations.

Therefore, $\qquad g_f = g_a + 0.2656. \Delta h \qquad .....(2.56)$

where 0.2656/m is the elevation correction for water of density 1.0 g/c.c. and Δh is the geoidal undulations in meters. The purpose of conversion of geoid anomalies to gravity anomalies relates basically to study the mass anomalies in the lithosphere, which in case of oceanic lithosphere is mainly confined to wavelength up to 60 km. Therefore, while using geoid anomalies to derive gravity field all the anomalies of wavelength larger than 60 km are suppressed to avoid large wavelength anomalies from the sources below the lithosphere. At this wavelength (60 km) one centimeter of geoid undulation approximately corresponds 1 mGal of gravity anomaly. As disturbances in sea surfaces are confined to short wavelength (a few meters) these wavelengths are also filtered out from the satellite altimetry data.

(i) Accuracy and Resolution of Satellite Altimetry Data

A comparison with marine gravity surveys suggests, that in general, the accuracy of satellite altimetry data is 4-7 mGal for random ship track (Sandwell and Smith, 1997). However, in general, it also depends on the wavelength of anomalies which increases with increase in the wavelength.

Yale et al., (1998) analyzed the gravity field derived from satellite altimeter in the Gulf of Mexico and ship borne gravity measurements, which is extensively covered by ship borne surveys for hydrocarbon exploration. Analysis of two data sets suggests the following significant points:

- They conclude that mean differences are generally less than 1 mGal

- There are good agreements for large wavelength features, however, small features are usually filtered out in satellite derived gravity maps.

- There is more noise in satellite derived gravity maps close to the coast say up to 20 km.

- Satellite gravity may define boundaries of basins and large scale structures but it is still difficult to define smaller structures within sedimentary basins such as salt domes, faults, anticlines etc., based on satellite altimetry.

- Based on number of such comparisons, they concluded that the general accuracy of gravity data derived from satellite altimetry is 3-7 mGal with resolution of 20-30 km depending on state of sea, number of tracks available in a region and proximity to coastline etc.

The best that satellite gravity can presently provide is the regional gravity picture, which along with local surveys can be used to model the regional and local tectonics in a region. However, before using any satellite derived gravity picture, one should exactly know, which filters have been used during processing of the data. However, with increasing accuracy of satellite altimetry, several new applications of this data have emerged (Fu and Cazenave, 2001).

Majumdar et al. (1998) prepared the free air anomaly map of Arabian Sea and Bay of Bengal from ERS-1 data and compared a profile across 90 East Ridge in Bay of Bengal with ship borne gravity profile. It shows good correlation between two data sets with maximum difference of 5-10 mGal in 60 mGal of free air gravity anomaly.

2.6.7 Satellite Geodesy and Gradiometry

Ocean surfaces being good reflectors, the above method of satellite altimetry provide a good estimate of geoid over them. However, the same is not true about continents, which do not provide good reflections to the transmitted signals. Therefore, over the continents, the second approach of tracking satellite orbits through laser beams or radio signals from stations on the ground or from satellite to satellite are used to track their orbits and perturbations in prespecified orbits to actual orbits are converted to geoid variations. The accurate tracking of satellites can be done by Satellite Laser Ranging (SLR) where a pulse of laser light is sent from the tracking stations on the surface of the earth and reflected signals from satellites provide its distance from the tracking stations. Distance of satellite measured from 3 stations or more provide its coordinates in the space. The accuracy of a single range measurement by SLR is about 1 cm.

The first experiment in this regard is known as Laser Geodynamic Satellite (LAGEOS) which was launched by National Aeronautics Space Administration (NASA), USA, in 1976 at a height of about 6000 km and has provided data for several years. It has been tracked by several laser tracking stations on different tectonic plates providing the rates of plate motion which has been confirmed from plate tectonic measurements such as sea floor spreading magnetic anomalies. The perturbations in the orbit of satellite besides other factors depend on geoid, which in turn provides the distribution of the Earth's gravity field. In this regard, the lower the satellite orbit the better the signal to noise ratio but such satellites do not last long due to atmospheric drag. Therefore, there is a trade off between accuracy of recorded data and the life of satellite implying the quantum of data recorded by it. In case of smaller life span of one year or less, temporal variation in gravity field cannot be studied, which is an important aspect of satellite geodesy.

Based on these considerations, satellites in the range of 250-350 km may last for about 6-12 months in orbit, while those in the range of 400-600 km may last up to 2-5 years. At these altitudes, the satellite makes about 15-16 revolutions of the earth per day at about 7.5 km/sec. Rapp (1989) showed that $1° \times 1°$ mean gravity anomalies could be determined to an accuracy of 5 mGal with altitudes of 240 km while $30' \times 30'$ mean gravity anomaly for same altitude will have an accuracy of 11 mGal. This implies that as resolution increases the accuracy of data drops to almost half.

There are three different kinds of satellite missions presently in operation (NRC, 1997), namely

(i) **Global Positioning Systems (GPS)** on board a low altitude orbiting satellite along with an accelerometer to record atmospheric effects for corrections (Reigber et al., 1996). CHAMP (Challenging Mini Satellite Pay Load) satellite launched in 2000 by European Space Agency at a height of 450 km belonged to this category (Reigber et al., 2005). It was a combined mission to measure both gravity and magnetic fields. Its main goal has been to improve the gravity field from GPS tracking and accelorometer data on board (Schrama. 2003).

(ii) **Satellite-Satellite tracking (SST)** wherein two satellites chase each other in the same orbit. Mutual tracking between them provides differential accelerations acting on them. Gravity Recovery and Climate Experiment (GRACE) Satellite launched by USA and German Space Agencies in 2002 belonged to this category (Tapley and Bettadpur, 2004). GRACE in this series is a dedicated Satellite mission to map the global gravity field every 30 days for about 8-10 years with a spatial resolution of about 400 km. Two satellites were launched in a near circular orbit at 500 km separated by about 220 km to provide accurate tracking data so as to measure geoid height accuracy of 1-2 cm uniformly over continents and oceans. In certain cases even a better accuracy of 2-3 mm at a spatial resolution of 400 km has been claimed (Tapley et al., 2004). This accuracy is good enough to record seasonal variations due to hydrological cycle opening a new range of applications of satellite geodesy. As GRACE satellite has a long span of life of 8-10 years, it is ideally suited for the study of transient variations in geoid and gravity field such as sea level changes, changes along subduction zones and ice mass loss in a region (Velicogni and Wahr, 2006) etc., Study of changes along subduction zones may lead to better understanding of processes related to them and associated seismic activity in the region. An example of mean sea level changes inferred from satellite data in the Indian ocean is given in Section 2.8.3 (Fig. 2.29).

(iii) **Space-Borne Gravity Gradiometer (SGG)** measurements, which is basically a futuristic proposal such as GOCE satellite to be launched by European Space Agency (Rummel et al., 2006). In this case, gravimeters are installed along three mutually orthogonal base lines (few cm long) which form the basis for tensor gravity gradient measurements (Fig. 2.21) as incase of airborne surveys described above. To obtain high precision of differential acceleration measurements cryogenic super conducting gravimeters described in Section (2.3) are used for this purpose.

The common features of these missions have been to measure static gravity field and its variation with time. The static field is mean over a period of 1 year and temporal variation is computed for a period of one month. The first kind of mission, namely GPS measurements provide accurate measurements of large wavelength features (400-500 km) while the second one provides both large and medium wavelength anomalies. The third one, namely gradient measurements being less sensitive to altitude of the satellite and terrain effects as discussed above may provide high resolution data for medium and short wavelength anomalies. However, by combining satellite gravity data with surface gravity data over continents and satellite altimetry data over oceans, the resolution of the world gravity models are considerably improved as described below.

Satellite Gradiometry

The Gravity steady state Ocean Circulation Explorer (GOCE) Satellite, which is launched by European Space Agency is a dedicated gravity field mission which combines 3-D full tensor gravity gradiometry and high-low satellite to satellite traking between GOCE and GPS satellites of US and GLONASS satellites of Russia. The mission objectives are to determine accurate earth gravity field globally with better spatial resolution. Some of the achievable objectives are (Rummel et al., 2006; Han et al, 2006):

- Determine gravity anomalies of about 1-2 mGal

- Determine geoid to an accuracy of less than 1 cm with a resolution of 65 km.

- Record global gravity anomalies of wavelengths below 100 km half wavelength

- Operation period 17 months with at least 12 months of data recording at a flight altitude of about 260-250 km.

2.6.8   Global Models of Earth's Geoid and Gravitational Field

Based on the different data sets of gravity field obtained from satellite and ground surveys, several attempts have made to provide worldwide gravity models by joining them through spherical harmonic analysis (Bowin, 1985). The larger the number of spherical harmonic coefficients, the better is the resolution of the models. However, with increase in number of coefficients (referred to as degree and order) the computations increases enormously. Spherical harmonic analysis of spherical surfaces is similar to Fourier spectral analysis of plane surface as described in Chapter 4. The gravitational potential u at geoid can be expressed as spherical harmonic series expansion as:

$$u = -G\frac{M}{r} \sum_{n=0}^{\infty} (R/r)^n \sum_{m=0}^{n} (C_{nm}\cos m\phi + S_{nm}\sin m\phi) P_{nm}\cos\theta \qquad .....(2.57)$$

where G and r are the universal gravitational constant and mean equatorial radius of the earth, M is the mass of the earth, θ, φ and R are geocentric latitude, longitude and radius, respectively. C and S are normalized spherical harmonic coefficients for degree n and order m and P_{nm} is the Legendre polynomial. A detailed treatment of this equation is beyond the scope of this book and those interested may refer to Heiskanen and Moritz (1967) for further details. The purpose to give this expression here is to familiarize the reader with degree and order of spherical harmonic analysis as it is often used for the world geoid and gravity models based on satellite data.

Initially, some gravitational models were provided with degree and order 10-30 based on only satellite data, which however, were limited in resolution (Bowin, 1985). These models provided only large wavelength anomalies related to plate tectonics. Subsequently with availability of high speed computers, order and degree were increased to 180 and 360, which provided data with 1° × 1° and 30′ × 30′ resolution, respectively (Rapp and Pavlis, 1990). However, data set beyond order and degree 30 were found to be contaminated by large errors. Simultaneously with development of large computational facilities, geoscientists started combining ground data with satellite data to provide better models of earth's gravitational field. This gave rise to the following three types of Global Geopotential Models (GGM).

(i) **Satellite Only GGM**, were developed based only on satellite data. Earlier GGMs prepared during 1970s and 80s belonged to this class of models with limited order and degree due to computational limitations. Marsh (1979) and Bowin (1985) provided models up to order and degree 12 and 30 respectively, which delineated only large wavelength features. The former was based on altimetry data and provided model for the oceans, which showed the largest gravity low in Indian Ocean. The gravity low of Indian ocean is a long wavelength feature caused by deep seated sources presumably in lower mantle, which affects the observed gravity field in this entire region including south India. Therefore, the observed gravity field may be required to correct for this long wavelength gravity field observed in this region. Bowin (1985) models were based on both satellite tracking and altimetry data over continents and Oceans, respectively providing a worldwide model which was referred to as Goddard Earth Model (GEM-9) on the name of Goddard Space Center, USA. He computed geopotential model for order and degree 30 and provided geoid undulations and gravity field, which delineated mainly large wavelength features related to plate tectonics such as mid oceanic ridges, subduction zones, orogenic belts etc., Reigber et al. (2002, 2003) based on only CHAMP satellite data provided a model EIGEN-1 and 2 with resolution $5° \times 5°$ which is considered as presently best available only satellite model of large wavelength features with minimum error. EIGEN-2 (European Improved Gravity Field Model of the Earth by New Techniques, Reigber et al., 2003) claims to provide an accuracy of 10 cm and 0.5 mGal in geoid and gravity field respectively, in half wavelength resolution of 550 km.

(ii) **Combined GGMs** are derived from combination of satellite tracking data, satellite altimetry over oceans, land data and ship track data and even airborne gravity data (Rapp, 1997). Under this class EGM-96 (Lemoine et al., 1998) with (50×50 km) resolution, which combined satellite tracking and altimeter data from several satellites with land data is the most popular and widely used model and is freely available on INTERNET. (http://cddisa.gsfc.nasa.gov/926/egm 96/egm96.html). Subsequently, Sandwell and Smith (1997) combined satellite altimetry data from GEOSAT and ERS-1 and provided a high resolution geopotential model over the oceans. The geoid and the free air anomaly map of Indian and adjoining countries and oceans are presented in Chapters 5 and 6, which shows all the major tectonic features of this region. Anderson and Knudsen (2001) combined all available altimetry data from different Satellites [GEOSAT, ERS, Satellite Repeat Mission (SRM)] with ship track data available along coasts and provided a geopotential model, KMS 2001 over oceans with a resolution of $1' \times 1'$ (2×2 sq. km) which has resolved even medium wavelength and some short wavelength anomalies.

(iii) **Tailored GGM** accounts for new acquired data sets with combined GGM previously computed GGMs. For example, combining EIGEN2 from CHAMP only data (Reigber et al., 2002) up to order and degree 2-32 and beyond this order and degree 33-360 from EGM96, a marginally improved geopotential model for New Zealand and Australia (Amos and Featherstone, 2006) has been developed. This suggests that as data from future satellites GRACE, GOCE etc., are incorporated in the geopotential model; it is likely to improve the existing models considerably.

(iv) **Residual Geoid and Gravity Field from Satellite Data:** As discussed above, the large wavelength features represent the regional field in a region, while medium-short wavelength features represent the residual field. As satellite data contains signals of different wavelengths, depending on order and degree of the harmonic coefficients as discussed above (Equation 2.57), fields related to lower order and degree can be used to represent regional field. Similarly, difference of fields between data of higher order and degree and lower ones can be referred to as residual field. This is one of the simple methods to obtain residual geoid and residual gravity field in a region for detailed modeling. Applications of satellite gravity and geoid anomalies for Indian Ocean, Himalaya and Indian continent are described in Chapters 5, 6 and 7, respectively.

2.6.9 Advantages and Limitations of Satellite Surveys

(i) **Accessibility:** It provides gravity and magnetic data over regions, which are inaccessible by any other means.

(ii) **Multiple Data Set and Lithospheric Structures:** Satellite gravity provide two data sets firstly, the geoid undulations and secondly, the gravity field and both fields can be simultaneously modeled to obtain density structures up to lithosphere-asthenosphere boundary. The former is sensitive to deeper lithospheric structures while the latter is best suited for crustal structures and therefore, density structures from surface up to lithosphere-asthenosphere boundary can be suitably modeled, which otherwise is usually limited only up to the Moho when gravity data alone is modeled. Besides, it also provides topography/ bathymetry in a region, which can be used in conjunction with above data sets.

(iii) **Universal Appeal:** The greatest advantage of satellite based surveys lies in their universal appeal, which can be used for large areas without any limitations due to geographical boundaries. Therefore, Satellite gravity can provide global picture of gravity field, which may reflect large scale lithospheric density structures related to plate tectonics. Potts et al. (2005) and Kabban et al. (2005) have provided density distribution in lithospheric mantle under USA and worldwide, respectively based on gravity models computed from CHAMP Satellite gravity data. They separated the terrain correlated and terrain de-correlated fields from the observed gravity model. The former represents the isostatically balanced crust while the latter represents mantle dynamics. This is a novel use of satellite based gravity data and principles of isostasy wherein gravity anomalies of wavelength larger than those due to isostasy can be assigned to deeper levels in mantle. In other types of surveys, anomalies of the largest wavelength are assigned to isostasy and small ones are attributed to shallow sources. This is because satellite based surveys offer an opportunity to record anomalies of the largest wavelength of continental/tectonic plates scale which is not possible in any other type of survey.

(iv) **Time Varying Field:** Variation of gravity field in a region with time can be obtained from Satellite data known as time varying gravity field which can be used for crustal deformation studies such as uplift, subsidence etc., (Table 2.3 at the end of this chapter). It is important in case of seismically active zones and plate boundaries such as Himalayas, Alps, Andes etc., The variation in altimetry data over oceans provide sea level changes with time (Section 2.8.3 Cabanes et al., 2001; Tiwari et al., 2004), which can be attributed to thermal effects providing heat budget of an ocean column. It can also be related to inter annual climate variability and compared to the effects of major events like El Nino-Southern oscillations etc., Velicogna and Wahr (2006) and Luthcke et al. (2006) have used GRACE data to estimate ice mass loss during different periods in Greenland, which is important for the study of climatic changes.

(v) **Cost:** In most of the cases, these data sets are available on INTERNET free of cost.

The major Limitations of Satellite data, however, lies mainly in its initial cost and limited accuracy and resolution, which makes it unsuitable for small amplitude, short wavelength anomalies due to shallow targets. This, however, may be fulfilled to some extent by gradient measurements from GOCE and other future satellites.

2.6.10 Comparative Study of Different Gravity Surveys

We have discussed the following different types of gravity surveys for measuring the gravity field in regards to different geoscientific problems.

(i) **Ground Gravity Surveys**

(a) Geodynamic studies: Regional surveys

(b) Oil exploration: Reconnaissance survey and detailed ground survey including microGal gravity surveys.

(c) Mineral and Ground exploration and Engineering Geophysics and Seismotectonics: Reconnaissance survey and detailed ground survey including microGal gravity surveys.

(d) Variation of gravity field with time (Table 2.3 at the end of this chapter)

(ii) Ship borne marine survey

(a) Regional surveys for geodynamic studies

(b) Reconnaissance surveys for oil and minearl exploration

(iii) Airborne survey

(a) Regional surveys for geodynamic studies

(b) Reconnaissance surveys for oil and mineral exploration

(c) Detailed surveys for oil and mineral exploration

(iv) Satellite Altimetry and Geodesy Surveys

(a) Regional surveys for geodynamic studies

(b) Temporal variation of gravity field for sea level changes

Each method is best suited to the investigation of different density distributions depending on the target and wavelength and amplitude of gravity anomalies produced by them. For example, the smallest wavelength and amplitude of anomalies are best recorded by microGal gravity surveys on the ground. The gravity anomalies next in wavelength and in amplitude are best recorded in the detailed gravity surveys on the ground while the large wavelength gravity anomalies are recorded by regional gravity surveys (> 10 km) on the ground, airborne (> 10 km) and satellite surveys (> 100 km). The marine, airborne and satellite surveys are specialized surveys which are used for specific purposes and in specific regions. However, largest wavelength gravity anomalies on continental scale of wavelength > 1000 km are only recorded in satellite surveys and are therefore best suited for deep seated density structures in upper and lower mantle specially the geoid anomalies. Table 2.2 at the end of this chapter summarizes all the methods discussed above in brief, based on nature of targets and expected gravity anomalies from them along with their accuracies and resolutions. Magnetic method discussed in Chapter 3 is similar to gravity method and therefore, this table also summarizes different kinds of magnetic surveys along with the gravity surveys, which readers can refer after familiarizing with them in Chapter 3. Table 2.3 at the end of this chapter provides the details of time varying gravity field, instruments used and their accuracies and their possible applications.

2.7   Interpretation

The Bouguer and free air anomaly profile and maps computed from the observed gravity field reflect subsurface density structures. Some information about their sources can be derived from comparison with geology / tectonics of the region and their magnitude and nature which is referred to as qualitative interpretation. Besides, they can be modeled for their causative sources which are referred to as quantitative interpretation.

2.7.1   Qualitative Gravity Signatures of Extensional and Compressional Tectonics – Their Applications and Wilson Cycle

Based on the critical examination of Bouguer anomaly maps and their comparison with geomorphology, geology and tectonics of the region, certain inferences can be drawn regarding the nature and the details of the causative sources of the gravity anomalies in a region as has been demonstrated in the next section from the Bouguer anomaly map of India. Gravity field being a vertical field, the causative sources are located just below the gravity anomalies. Based on their amplitudes and width, some preliminary inferences about nature of their sources can be readily drawn as given below for different applications.

(i) Geodynamic Studies

The most important aspect of geodynamic studies are related to crustal structures and nature of tectonics in a region. Gravity "highs" and "lows" can immediately indicate whether the source is a high density or a low density body and elongated direction of contour indicates its strike direction. In regional gravity survey, thickening and thinning of the crust is reflected as large wavelength (width greater than 100 km) gravity "lows" and "highs", respectively. As discussed in Section 1.1 on plate tectonics, the most important tectonic regimes can be classified as a) extensional and b) compressional regimes. Mid oceanic ridges and subduction zones are the best examples of extensional and compressional tectonics, respectively, which produce typical gravity signatures. However, these processes are related to plate boundaries, but similar tectonic forces may also operate within the plates as demonstrated in Fig. 2.23 (a) to (c). We shall now examine qualitatively nature of gravity anomalies in these two cases.

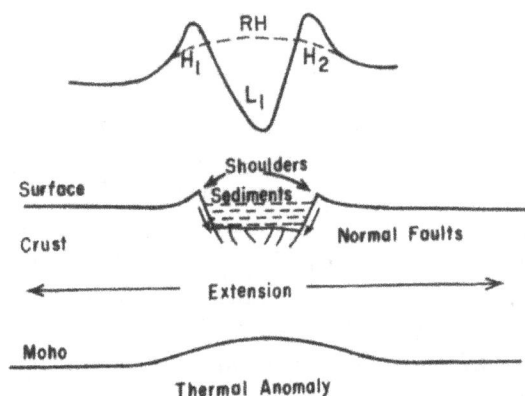

Fig. 2.23(a) Schematic representation of extensional tectonics and their gravity signatures. Extensional tectonics caused by a thermal anomaly in the lithosphere causes upwarping of Moho over a broader area causing a regional gravity high (RH). The presence of sediments in the rift basin flanked by normal faults produces gravity low L1 and gravity highs H1 and H2 represent high density rocks along shoulders.

(a) Extensional Tectonics: Rift Basins

Extensional regions are zones in the earth's crust, which have experienced extension during geological past and tectonics related to it, are referred to as extensional tectonics. Such zones are characterized by specific features, which would produce specific gravity signatures. A block diagram of rift basins related to extension is given in Section 8.1 which are also known as grabens. Extension in crust usually occurs due to thermal anomaly in the lithosphere below the crust, which might be related to volcanic activity giving rise to mafic intrusive rocks in extensional regimes (Fig. 2.23(b)). These intrusives originating from asthenosphere are characterized by both felsic and mafic intrusives known as bimodal volcanics which is a typical characteristic of such regimes. Such rifts are known as active rifts and represent early stages of continental breakup. However, in cases where there may not be intrusives but still due to subcrustal thermal anomaly or any other causes (Plate tectonic forces), the asthenosphere may be upwarped that causes thermal anomaly in a limited section which would induce extension in the region and such rifts are known as passive rifts (Fig. 2.23(a)). The most important examples of active rifting are mid oceanic rift system and rifted margins (Fig. 1.2) along the continental margins due to extension and intrusives. However, there may be cases that continents do not break apart but still there may be extension in a localized space giving rise to rifting in a limited space due to thermal anomaly in the lithosphere below the crust. Such zones over the continents are characterized by linear sunken parts of the crust bounded by normal faults on two sides or at least on one side, which is subsequently covered by sedimentary rocks and are known as continental rifts or grabens (Fig. 2.23(a) and (b)). Due to extension and faulting, there are uplifts along the shoulders of the rift basin, which are attributed to compression along the sides of the rift basin. Therefore, extensional regimes are characterized by following features, which produces specific gravity signatures. Fig. 2.23(a) and (b) are schematic diagram of continental rift basins and associated gravity signatures for passive and active rifts, respectively.

(i)   Due to subcrustal mafic intrusives, it may show a broad regional gravity high. Even if there are no intrusives, the subcrustal thermal anomaly would also produce crustal upwarp causing pop up of Moho in the central part, which would produce a broad regional gravity high (RH, Fig. 2.23(a)) in case of passive rifts. An example of Godavari rift basin, India is given in Section 2.9 (Fig. 2.30(a)), which does not show significant intrusives and therefore, can be regarded as passive rifts. In case of old rifts such as this, the Moho upwarp may subside with time and may acquire normal crustal thickness.

(ii) Due to presence of sediments in the central part of the rift basin, it would produce a gravity low in the residual anomaly of the observed gravity field. Therefore, the broad gravity high due to asthenospheric upwelling and crustal upwarping is primarily reflected as gravity highs along the shoulders of the rift basin while in the central part, it is characterized by the gravity low. This is schematically shown in Fig. 2.23(a) with gravity highs H1 and H2 along the shoulders of the rift basin and a gravity low L1 over top of it. Similar gravity signatures are also expected over mid oceanic rift and ridge system as the central part of this system represent a rift structure and sediments are deposited over them which give rise to gravity lows, while ridges along the shoulders produce gravity highs.

(iii) In case of active rifts, if the effects of mafic intrusives are large compared to the deposited sediments, it would produce a central gravity high, H1 with flanking lows, L1 and L2, along the margins of the basin and gravity highs, H2 and H3 along the shoulders as shows in Fig. 2.23(b). An example of East African rift basin as an active rift is discussed in Section 2.9.

(iv) In most of the cases world over, the continental rift basins are linear features of 80-120 km wide.

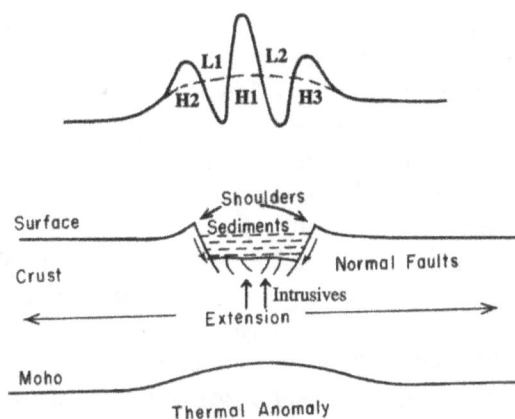

Fig. 2.23(b) Schematic representation of extensional tectonics with intrusives and their gravity signatures. In case of active rift basins, intrusives in the central part provide a gravity high H1, flanked by gravity lows, L1 and L2 due to sediments. Gravity highs H2 and H3 represent shoulder highs.

As rift basins subsequent to their formation are occupied by thick sediments, they are important for hydrocarbon occurrences. Due to tectonic activities associated with rift basins, they are important for seismic activity. Invariably, away from plate boundaries, they are most active regions in otherwise stable continental regions (SCR). It is discussed in further details in Chapter 8.

(b) Compressional (Convergence) Tectonics: Horst

Fig. 2.23(c) is a schematic diagram of compression within a plate, which would produce folding and faulting. In case, central part of the crust between two reverse faults pops up, it is known as a horst which is just opposite of a rift structure as discussed above. A block diagram of horst is given in Section 8.1. They are characterized by reverse faults and fold belts while rift basins are characterized by normal faults and depressions. Due to compression and erosion, the lower crustal rocks are exposed, which produces a gravity high H1 over the horst structures. However, in case sediments are deposited on sides of the horst, gravity lows L1 and L2 will be produced. It may be noted that the gravity signatures of a horst is just opposite to that of a rift (Fig. 2.23 (a)).

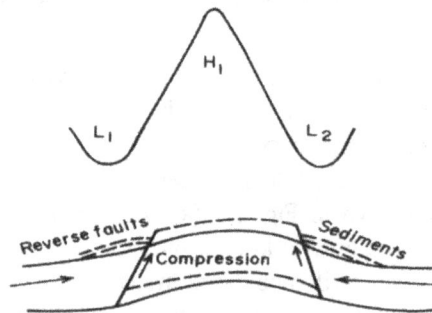

Fig. 2.23(c) Schematic representation of compressional tectonics and gravity signature. Compression causes folding and reverse faults giving rise to horst which produces a gravity high H1. In case sediments are deposited on sides, this would produce gravity lows L1 and L2.

(c) Collision and Subduction Zones

As discussed above, the subduction zones where two plates converge and collide also represent compressional tectonics (Fig. 2.23(d)). In such cases, one plate subducts under the other and parts of subducting and the overriding plates are thrusted upwards resulting into mountain building. Due to convergence and subduction, the crust in the collision zones shortens and thickens giving rise to gravity low L1, while thrusting and high density intrusives of volcanic arcs (Fig. 1.2) give rise to a gravity high H1 (Fig. 2.23(d)). Erosion with time exposes lower crustal rocks which enhances the observed gravity high H1 and reduces the gravity low L1 due to reduced crustal thickness. This implies that in case of present day collision zones, the observed gravity low, L1 due to crustal thickening dominates as in case of Himalaya (Fig. 2.36) while in case of old collision zones of Archean-Proterozoic Period, the gravity high, H1 predominates as described in case of Grenville province of Canada and Appalachian Mountains, USA in Section 2.11 and Indian cratons and fold belts in Chapter 7. It may be noted that in ancient collision zones, the gravity high occurs over the younger (Proterozoic) collision zones while a broad low is observed over older (Archean) cratons as described in detail in Chapter 7. As thrust zones bring rocks upwards from depth, they act as conveyor belts for transport of important minerals to the surface and mineralized zones are invariably associated with them as described in Chapter 9.

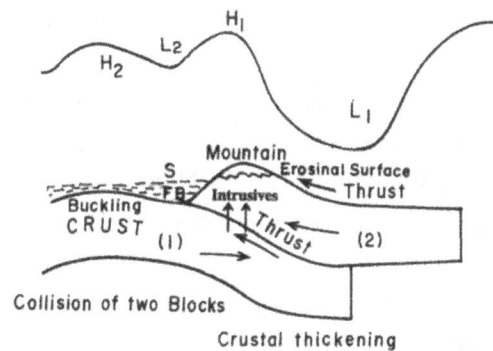

Fig. 2.23(d) Schematic representation of compressional (collision) tectonics. In case of collision between two plates and subduction of one under the other due to compression , the crust thickens which produces a broad gravity low, L1 and thrusting of high density rocks and intrusives related to arc magmatism produce gravity high (H1). Erosion with time enhances H1 and reduces the gravity low L1. Due to sediments on the subducting plate in the Foreland Basin (FB), a gravity low L2 is observed. Another small gravity high H2 is produced due to crustal buckling caused by the flexure of the subducting plate. H1 and L1 are referred to as paired gravity anomaly of collision zones, especially in case of Archean – Proterozoic collision zones.

The combination of gravity high (H1) and gravity low (L1) are referred to as paired gravity anomalies related to ancient collision zones. Such pairs of gravity anomalies are typical of collision zones in the Archean–Proterozoic terrains world over as discussed in Chapter 7. A gravity low, L2 (Fig. 2.23(d)) is observed over the sediments in the frontal part of the mountain, known as foreland basins (Fig. 1.2) as Ganga basin is related to Himalaya. A small gravity high (H2) will be observed farther from mountain chain due to buckling of the subducting plate caused by its flexure under the overriding plate. It is a small amplitude gravity high which will be easily masked under the effect of sediments if present along the frontal part of the mountain chain over the subducting plate. In such cases separation of anomalies becomes important which is discussed in Chapter 4. Compressional zones such as plate boundaries are centers of intense seismic activity as they are regions of high stress. Further details of these aspects are discussed in Chapter 6 and 7 in regard to Himalayan orogeny and Archean-Proterozoic collision zones in the Indian continent.

(d) Wilson Cycle

There may be extension and compression one after the other in the same region which usually happens quite often. This is known as Wilson cycle after its originator Prof. J. T. Wilson who proposed a simplified model of extension and compression in plate tectonic framework. Accordingly, there are five stages as follows:

(i) Uplift stage that breaks the continents and produces uplift on a regional scale. It is probably indicated by shoulder uplifts during extension as discussed above.

(ii) Rifting stage due to extension and uplift. Rifting continues and at some stage, sea floor spreading starts and develops into an ocean such as formation of Indian and other oceans. However, if seafloor spreading does not take place, sediments are deposited in the rifted part and continental rift valleys are developed. Such cases are referred to as failed or aborted rifts as they do not develop into oceans.

(iii) Subsequently, sea floor spreading starts declining and finally stops and continents on either side starts converging towards each other. This is the start of compressional phase.

(iv) Due to convergence, the oceanic crust subducts under the continents and is destroyed in due course of time.

(v) Finally, continents collide leading to mountain building through thrusting as in case of Himalayas. This is known as obduction, opposite of subduction. At this stage, in some cases, oceanic crustal rocks are obducted over continents.

The continents are sutured to form one landmass as India and Asia along Himalayan collision zone.

This is a simplified model of plate tectonics which has gone through various stages of changes as we know it today as discussed in Chapters 5 and 6.

(ii) Geological Mapping

Sediments being of lesser density (2.0-2.4 g/cm^3) compared to gneisses (2.67 g/cm^3) forming upper part of the crust, sedimentary basins are characterized by gravity "lows" spread over the entire basin. Granite intrusions/batholiths being of small dimensions of lesser density (2.6-2.65 g/cm^3) compared to gneisses (upper crust), show circular gravity "lows" limited in dimensions. Mafic dykes being of higher density (2.7–3.0 g/cm^3) are seen as linear gravity "highs" extending along its strike direction. Sharp linear gravity gradients specially running for large distances indicate contacts separating rocks of different densities, which may represent faults. These features and their gravity signatures are demonstrated in the next section, while describing the Bouguer anomaly map of India and other field examples.

(iii) Hydrocarbon Exploration

Gravity surveys for Hydrocarbon exploration over sedimentary basins provide gravity lows of different amplitude (1-10 mGal) and width (1-100 km) based on basement structures and presence of low density sediments. Basement highs produce relative gravity highs and basement depressions produce gravity lows. However, the amplitude and width of gravity anomalies due to basement are less compared to those due to crustal thinning and thickening and more than those encountered in mineral exploration. Therefore, they are grouped under medium amplitude and wavelength anomalies. It is discussed in more details in Chapter 9 while describing integrated exploration of sedimentary basins.

(iv) Mineral exploration and Engineering Geophysics

Gravity anomalies in these cases are usually of small amplitude (0.1-1.0 mGal) and small width (less than 1 km) except in case of specific minerals with high density. Even in these cases, the amplitude of observed gravity anomalies are less due to their small concentration in associated rocks. Alignment of small amplitude and small wavelength anomalies in a particular direction implies fracture zones and faults which are important for mineral and ground water exploration and engineering surveys. These cases are discussed in more detail in Chapters 9 and 10.

(v) Some Approximate Estimates

The gravity anomalies contain much more information about the causative sources such as their dimensions, depth, densities etc., which, can be inferred only by quantitative interpretation discussed in the next sections. However, some limits on depth or thickness of a body can be obtained from simple calculations using the formula for a slab given in Equation (2.42) for computing Bouguer correction. This relationship has been found to be extremely useful for quick estimate of thickness of exposed sediments in a sedimentary basin or approximate thickness of the crust in a region as demonstrated below. The gravity field due to a slab of thickness t is given by:

$$\Delta g = 2\pi.G.\rho. t \qquad\qquad .....(2.58)$$

where ρ is the density contrast and t is the thickness in meters

Substituting the value of π and G

$$\Delta g = 0.04185 . \rho.t. = 0.042 . \rho . t \qquad .....(2.59)$$

Fig. 2.36 shows a Bouguer anomaly profile across Himalayas and Tibet adopted from Satellite gravity data (Rajesh and Mishra, 2003). Bouguer gravity anomaly over Himalayas and Tibet along this profile is approximately - 520 mGal. Therefore, crustal thickening for a bulk density contrast of 0.4 g/cm^3 between lower crust and upper mantle in this region is given by Equation (2.59) as t $\sim$ 31 km.

Taking 38 km as the normal crustal thickness under the Indian continent, suggest a crustal thickness of 69 km under Tibet, which is close to that provided by complete modelling of Bouguer anomaly (70 km; Fig. 2.36). As this example relates to crustal thickening under Himalayas and Tibet, which covers a large area, it is safe to assume a horizontal slab for the crustal thickening in this region for which the expression given in Equation 2.58 is valid. The same Equation (2.59) can also be used to determine the bulk density contrast if the thickness is known. The same approach can be used to roughly estimate the thickness of sediments in a basin by using the residual anomaly instead of the observed field. However, this thumb rule provides only rough estimates and should be corroborated from complete quantitative modelling of Bouguer anomaly as discussed below.

2.7.2 Quantitative Interpretation and Modelling: Simple Ideal Models

The simplest approach for quantitative interpretation of gravity anomalies is trial and error method. In this regard, the first approach is to approximate the geological bodies by simple physical models like sphere, cylinder, contact etc., for which the expression for gravity field is known and compute its field for appropriate density contrasts. The computed field is compared to the observed field and body parameters and density contrast is varied to geological feasible parameters to obtain a good match between them. The expressions for the gravity field due to some simple models (Figs. 2.24 and 2.25) along profiles across them are given below, which can be used for computation of the gravity field due to them. The gravity profiles and maps are known as one and two dimensional data sets, respectively. However, when they are modeled and causative sources of respective anomalies are plotted with respect to x- axis (Profile) or x and y axes (Map) and vertical (z-axis) they are referred to as two and three dimensional bodies, respectively.

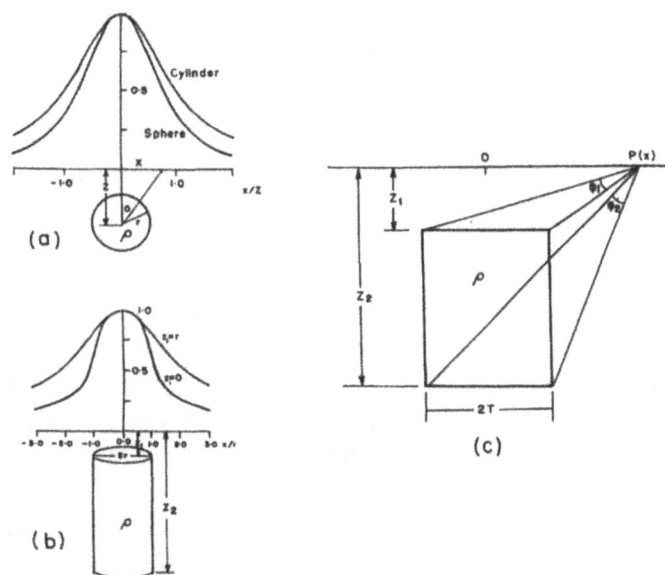

Fig. 2.24 Gravity field due to (a) sphere and horizontal cylinder showing the decrease of maximum field to half at a distance of 0.75z for sphere and z (depth) for cylinder (b) vertical cylinder and (c) prism model

However, geological bodies cannot be correctly approximated by such simple shapes and therefore, there are expressions for 2 dimensional arbitrary shaped bodies (Talwani et al., 1959). These bodies are referred to as two dimensional as their extent along strike direction is considered to be infinite. However, infinite extent is a misnomer and therefore, they should be large enough to be treated as infinite large, viz. larger than perpendicular to the strike direction. Bodies limited in all directions are referred to as 3-D bodies (three dimensional). Radhakrishnamurthy and Mishra (1989) and Radhakrishnamurthy (1998) have extensively dealt with gravity field due to simple and 2 and 3 dimensional arbitrary shaped bodies and have provided requisite softwares for this purpose. In these cases the subsurface geological models are defined by the polygonal cross-section or layers and the gravity field due to them are computed along a profile almost perpendicular to them and compared with the observed field. The configuration of the bodies and their densities are varied within the geologically plausible limits till a good match between the observed and the computed fields are obtained. There are expressions even for 2.5 dimensional bodies which are limited in strike direction (Webring, 1986; Rasmussen and Pedersen, 1979). Based on the extent of body along strike direction with reference to the profile, these expressions compute the gravity effects of arbitrary shaped limited bodies. The whole scheme can also be made as an inversion scheme, which is discussed in Chapter 4. While modeling the causative sources for the observed gravity field, they are constrained from all available geological and geophysical information from a region and therefore, it inherently integrates all this information and provides an integrated model of the subsurface structures. Further, as Moho is the most prominent density discontinuity with the largest density contrast of about -0.4 g/cm^3 between lower crust and upper mantle across it, the observed large wavelength gravity anomalies are primarily modeled due to variations in the Moho depth.

(i) **Sphere:** This is the simplest model whose gravity field can be easily computed. The gravity field along a profile (x) due to a sphere (Fig. 2.24(a), Radhakrishnamurthy, 1998; Parasnis,1986) is given by

$$g(x) = G.\rho. \ v.z/ \ (x^2 + z^2)^{3/2} \quad\quad\quad .....(2.60)$$

where G is the gravitational constant, ρ is the density contrast between the spherical mass and the surrounding, z is the depth to the center of the sphere and v is the volume of the sphere given by $4/3 \ \pi r^3$, r being its radius. It is a symmetrical anomaly along the anomaly axis with its maximum located at the top of the center of the sphere where x = 0. Therefore

$$g_{max} (0) = G\rho \ v/z^2 \quad\quad\quad .....(2.61)$$

The maximum field reduces to half its value at a distance x_1 given by

$$g_{max}/2 \ (x_1) = G.\rho \ v.z./(x_1^2 + z^2)^{3/2} \quad\quad\quad .....(2.62)$$

Combining (2.61) and (2.62):

$$z = 1.3 \ x_1 \quad\quad\quad .....(2.63)$$

Therefore, the depth to the sphere can be obtained from the distance between the maximum anomaly and where it reduces to half of it. Such simple relations, which expresses the characteristics of the observed fields, in terms of parameters of the causative sources, are known as the thumb rules and are extremely useful in finding quickly some of the significant parameters of the sources. In actual practice, it is difficult to find a geological body as a sphere. However, bodies, which are limited in extent with almost same dimensions along all three axes, can be approximated to a sphere and its depth can be approximately determined from the above relationship. Such bodies provide near circular anomalies which help in identifying the nature of their sources.

(ii) Horizontal cylinder

If a geological body is elongated in the horizontal direction (strike), and limited on other sides, it can be approximated to a horizontal cylinder which would provide gravity anomaly similar to a sphere (Fig. 2.24(a)) and its gravity field along x-axis perpendicular to strike direction is given by (Radhakrishnamurthy, 1998):

$$g(x) = 2\ G\ \rho\ vz/(x^2 + z^2) \qquad .....(2.64)$$

where v is its volume given by πr^2, r being its radius

The maximum anomaly is located at the top of the axis of the cylinder. In this case, the depth, $z = x_1$, x_1 being the point where maximum anomaly reduces to half its value.

(iii) Vertical cylinder

The bodies which extend in vertical direction and limited in horizontal direction such as volcanic pipes, salt domes etc., can be approximated to a vertical cylinder (Fig. 2.24(b)). The maximum anomaly is observed at the top of the central axis of the cylinder which decreases symmetrically sharply on either sides. The gravity anomaly due to a vertical cylinder over its axis is given by (Radhakrishnamurthy, 1998):

$$g(x) = 2\ \pi.G.\rho\ [(z_2 - z_1) + (z_1^2 + r^2)^{1/2} - (z_2^2 + r^2)^{1/2}] \qquad .....(2.65)$$

The expression is further simplified if $z_1 = 0$ implying the body is exposed. In such a case, the gravity field reduces to

$$g(x) = 2.\pi\ G.\rho\ [(z_2 + r - (z_2^2 + r^2)^{1/2}] \qquad .....(2.66)$$

Further, if $r \to \infty$ implying a horizontal slab

$$g(x) = 2.\pi.G.\rho.\ z_2 \qquad .....(2.67)$$

This is same as expression used for Bouguer slab of thickness, $z_2 = t$ as given

in Equation (2.42) and used for Bouguer correction.

Further, if $r \to \infty$, equation (2.65) reduces to

$$g(x) = 2.\pi.\ G.\rho\ (z_2 - z_1) \qquad .....(2.68)$$

This expression is same as that for the Bouguer slab at a depth of z_1. This implies that maximum field due to a horizontal slab is same whether it is exposed or buried at some depth z_1. This is the basis for computation of crustal thickness based on Bouguer slab formula as discussed in section 2.7.1.

(iv) Prism:

The gravity field due to a prism (Fig. 2.24(c), Radhakrishnamurthy and Mishra,1989) is given by:

$$g(x) = 2G\rho\left\{(z_2\varphi_2 - z_1\varphi_1) + 0.5\left[(x+T)\ ln\left(\frac{(x+T)^2 + z_2^2}{(x+T)^2 + z_1^2}\right) - (x-T)\ ln\left(\frac{(x-T)^2 + z_2^2}{(x-T)^2 + z_1^2}\right)\right]\right\}$$

$$.....(2.69)$$

where

$$\phi_2 = arctan[(x+T)/z_2] - arctan\ [(x-T)/z_2]$$

and

$$\phi_1 = arctan[(x+T)/z_1] - arctan\ [(x-T)/z_1]$$

(v) Contact (Fault)

This is one of the most important models as it is used to model faults which are widely encountered in geophysical surveys. The gravity field along x-axis due to a contact (Fig. 2.25(a); Radhakrishnamurthy and Mishra, 1989; Radhakrishnamurthy, 1998; Raju et al., 1998) is given by

$$g(x) = 2G\rho \left[\{(x\text{-}d) \sin \delta - z_1 \cos \delta\} + \{\sin \delta \ln r_2/r_1 + \cos \delta \, (\phi_2 - \phi_1) \} + z_2\phi_2 - z_1\phi_1 \right] \quad \dots\dots(2.70)$$

where

$$r_1 = \sqrt{[x\text{-}d)^2 + z^2_1]}$$

$$r_2 = \sqrt{[\{(x\text{--}d) + (z_2\text{--}z_1)\cot \theta\}^2 + z^2_2]}$$

$$\phi_1 = \pi/2 + \tan^{-1} (x\text{--}d)/z_1, \text{ and}$$

$$\phi_2 = \pi/2 + \tan^{-1} \{(x\text{--}d) + (z_2\text{--}z_1)\cot \theta\}/z_2.$$

The various symbols along with a typical anomaly due to a fault are given in Fig. 2.25. The maximum anomaly due to a contact is $2 \pi G\rho \, (z_2\text{-}z_1)$ corresponding to a slab as discussed above and minimum anomaly is zero and therefore, contact (faults) produces basically a gradient with high values located towards the block with high density. The origin of a fault is located at the inflexion between the maximum and the minimum anomaly. This is the most widely applicable model in several geological problems.

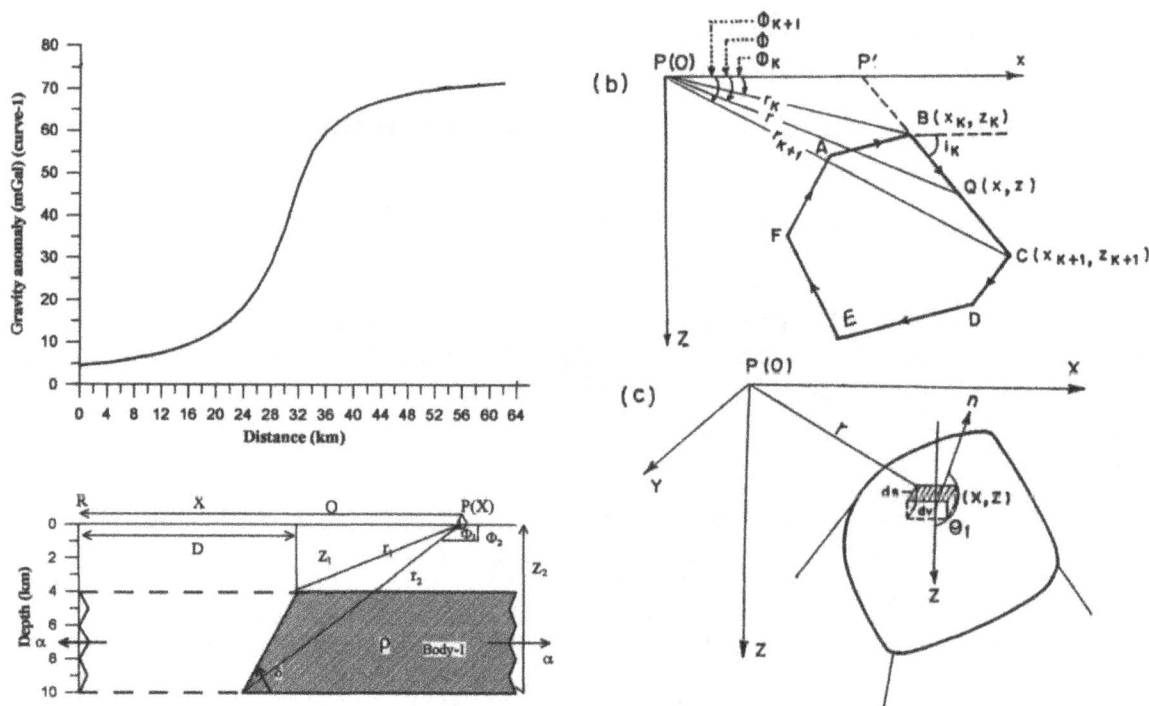

Fig. 2.25 Gravity field due to (a) A fault model and its gravity field as gradient. b) Model showing two dimensional arbitrary shaped bodies with a polygonal cross section. The vertices of polygon are defined by x and z coordinates. c) Volume and surface elements dv and ds and normal with respect to z axis for a three dimensional model.

2.7.3 Gravity Field due to Arbitrary Shaped Two Dimensional Models

Quite often, it is difficult to approximate geological bodies in nature to any well defined physical models discussed above. In such cases, the geological bodies are approximated to two dimensional bodies of arbitrary shape with polygonal cross-sections, which are defined by the apexes of the polygons in x, z planes (Fig. 2.25(b)). Talwani et al. (1959) first provided a suitable scheme to compute the gravity field due to such a model. The gravity field due to each polygonal cross section is computed and summed to obtain the field due to the whole body. The gravity field due to a cross-section of a polygon (Fig. 2.25(b)) is given by Radhakrishnamurthy and Mishra (1989).

$$g(x) = 2G\rho \left(x_k \sin i_k - z_k \cos i_k\right) . \left[0.5 \sin i_k \, ln \, (r_{k+1}/r_k) - \cos i_k \, (\phi_{k+1} - \phi_k)\right] \qquad .....(2.71)$$

where r_k, r_{k+1}, ϕ_k, ϕ_{k+1} and other symbols are explained in Fig. 2.24(b) and are given by:

$$r_k = \sqrt{(x^2_k + z^2_{k)}},$$
$$r_{k+1} = \sqrt{(x^2_{k+1} + z^2_{k+1)}},$$
$$\sin i_k = (z_{k+1} - z_k)/r,$$
$$\cos i_k = (x_{k+1} - x_k)/r,$$

$$\phi_k = \pi /2 - \arctan (x_k/z_k), \qquad \text{if } z_k \neq 0,$$
$$= \pi /2 . (x_k/ |x_k|), \qquad \text{if } z_k = 0,$$
$$\phi_{k+1} = \pi /2 - \arctan (x_{k+1}/z_{k+1}), \qquad \text{if } z_{k+1} \neq 0$$
$$= \pi /2 . (x_{k+1}/ |x_{k+1}|), \qquad \text{if } z_{k+1} = 0,$$

and

$$r = \sqrt{[(x_{k+1} - x_k)^2 + (z_{k+1} - z_k)^2]}$$
$$x_{n+1} = x_1, \text{ and } z_{n+1} = z_1.$$

n is the number of vertices in a polygon.

In these cases, the subsurface geological models are defined by the polygonal cross-section and the gravity field due to them are computed along a profile almost perpendicular to them and compared with the observed field. The configuration of the bodies and their densities are varied within the geologically plausible limits till a good match between the observed and the computed fields are obtained.

2.7.4 Gravity Field due to Three Dimensional Models

Sometimes geological bodies, which are limited in extent, cannot be approximated by two dimensional models and therefore, it is essential to compute the gravity field due to a 3-dimensional body. However, a good geological control is essential in such cases to develop an initial model which is subsequently modified based on the differences between the computed and the observed fields in a plane (x, y). This implies that we use a contour map or digital data in a plane to compare the computed and the observed fields and their differences are adjusted by changing the parameters of the body and its density contrast. There are several schemes for this purpose (Talwani and Ewing, 1960; Götze and Lahmeyer, 1988) etc.,) which have been reviewed by Li and Chouteau (1998) in a comparative study. Most of the earlier methods used aggregates of prisms to define the bodies and compute the gravity field due to them by summing up the effects of individual prisms. Codes for the computation of gravity field due to 3-dimensional bodies by dividing the body into vertical prismatic bodies and summing the effect due to all such bodies have been described by Radhakrishnamurthy and Mishra, (1989) and Radhakrishnamurthy (1998). However, the gravity field due to polyhedra can also be computed directly by evaluating the volume integral. Most of the schemes depend on transformation of volume integral to surface and line integrals. The gravity field due to a small volume element dv at a distance r (Fig. 2.25(c)) is given by (Barnett, 1976, Singh and Gupta Sarma, 2001):

$$g = G\rho \iint_{v} \int \frac{\delta}{\delta Z}\left(\frac{1}{r}\right) dv \qquad \qquad ....(2.72)$$

This is transformed to surface integral as

$$g = G\rho \int_{S} \int \cos(\text{n, z})\frac{1}{r}d_s \qquad \qquad .....(2.73)$$

where cos(n, z) is the angle between outward normal of the surface element with respect to z axis (Fig. 2.25(c)). The surface integral is expressed as sum of contributions from the individual facets of polyhedra and therefore

$$g = G\rho \sum_{i=1}^{m} [\cos(ni,z)\iint_{si} \frac{1}{r}ds] \qquad \qquad .....(2.74)$$

where i denotes the number of facets

The surface integral in the above expression is transformed to line integral using Stokes theorem by transforming the coordinate system such that the z direction coincide with the direction of outward normal on the surface s_i and the limits of integration are the coordinates of vertices defining the polyhedra. Singh and Guptasarma (2001) transformed the surface integral to line integral using Stokes theorem and evaluated gravity field combining line integral and solid angle subtended by each facet at the observation point. They have also provided a code for the computation of gravity field due to three dimensional polyhedra, which can be used by students as an exercise. Schemes for two and three dimensional modeling of gravity field and their interface with graphic softwares are quite involved and there are several commercial softwares available for this purpose such as GM-SYS, 2000 IGMAS (Götze and Lahmeyer, 1988) etc., which can be readily used for this purpose. However, students may try to prepare their own codes or use those referred to above to model the observed gravity field due to two and three dimensional arbitrary shaped bodies.

2.7.5 Ambiguity in Gravity Interpretation

The expression for the gravity fields due to different types of bodies given above basically consists of two groups of unknowns, viz the first group is related to configuration and depth of the bodies and the second is the density contrast with respect to surroundings. Therefore, the computed gravity field is the product of values assigned to these two groups of unknowns. An increase, decrease in the value of one group of parameters can be balanced by changes in the other set so as to obtain the same computed field. For example, an increase in depth can be balanced by increasing the density contrast or vice - versa. Therefore, the same gravity field can be produced by more than one set of values for different parameters. This is known as inherent ambiguity in gravity interpretation and can be resolved only based on geological and other geophysical information from the region. Skeels (1947) and Roy (1962) have shown that the same gravity field can be produced by different subsurface bodies and density distribution at different depths, respectively. Solution of Laplace equation for gravitational potential given in equation 2.18 suggests that while this equation consists of unknowns along two axis x and z, the data set is available only along x-axis, which makes it indeterminate providing several solutions to it. In fact, such inherent ambiguity exists, in all geophysical methods (Roy, 1962) as unknowns are more compared to the known measurements. Such under determined systems are dealt in Chapter 4 under 'Generalized Inversion and Parameter Estimation'. Therefore, an integrated approach using data from more than one method is emphasized in geophysical exploration as demonstrated in Part II.

Inherent ambiguity of gravity field is demonstrated in Fig. 2.25(d) (Radhakrishna Murthy, 1998), which shows gravity field due to a contact (fault) with density contrast ρ. Curve (1) is related to body (1) when density contrast of + ρ is towards the right and curve 2 is related to body 2 when it is located towards the left. Same curves will be obtained by snapping their positions with negative density contrasts. It will, therefore, be difficult to resolve the sign of the density contrast purely on the basis of the observed gravity field. It can however, be resolved in case it is known that on which side is the higher density body based on the geology of the region. The same is true about magnetic field discussed in Chapter 3 as both are potential fields governed by the Laplace equation given in equation (2.11). The only difference lies in the physical parameters, which in case of magnetic field is susceptibility instead of density in case of gravity field.

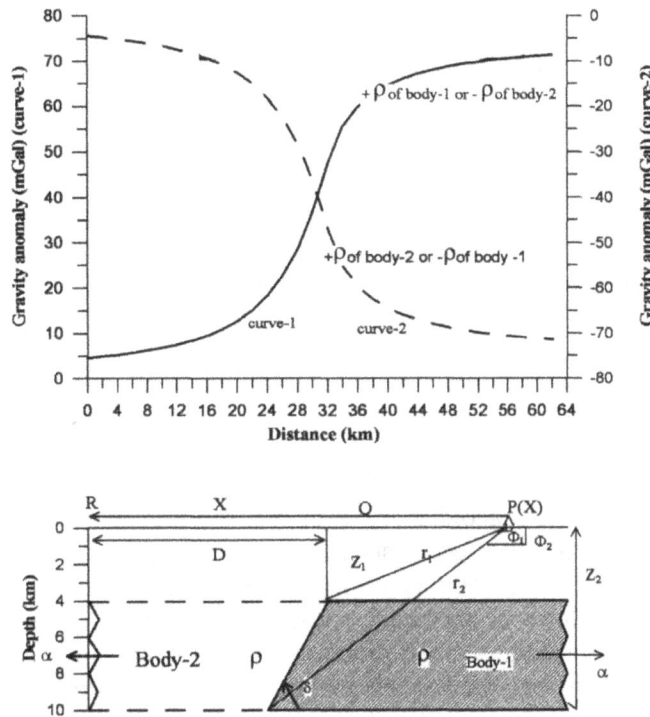

Fig. 2.25(d) Gravity anomalies over a contact (fault) demonstrating ambiguity in gravity interpretation. Similar gradient anomaly is produced due to +ρ density contrast towards the right and –ρ density contrast towards the left and vice versa.

2.7.6 Geoid Modeling and Terrain Effect

The difference in elevation between measured geoid and the regional/reference geoid Δh is referred to as the geoid anomaly (Fig. 2.6(b)). The reference geoid itself includes an averaging over density anomalies with in the earth. The difference between potential at measured and reference geoids (the potential anomaly ΔU) can be related directly to the geoid anomaly Δh (Fig. 2.6b).

$$\Delta U = U_{mo} \text{ (measured potential)} - U_o \text{ (reference potential)} \qquad .....(2.75)$$

According to Turcotte and Schubert (2002), it can be expressed in terms of geoid anomaly as:

$$U_0 = U_{m0} + \left(\frac{\partial U}{\partial r}\right)_{r=r_0} \Delta h$$

Because $\Delta h/a \ll 1$.

where r is the radial distance from the center of the earth and r_0 is the distance to the reference geoid.

The radial derivative of the Potential is the acceleration of gravity on the reference geoid, hence: $\left(\dfrac{\partial U}{\partial r}\right)_{=r_0} = g_0$

$$\Delta U = -g_0 \times \Delta h \qquad \qquad .....(2.77)$$

This is known as Brun's Equation.

where g_0 is the acceleration on reference geoid. This can be obtained from equation 2.35 providing theoretical value of gravity field (g_t) with reference to 1980 Geodetic Reference System. The gravity anomaly Δg is related to ΔU as (Moritz, 1980)

$$\Delta g = -\left(\dfrac{\partial \Delta U}{\partial r}\right) - \dfrac{2\Delta U}{r} \qquad \qquad .....(2.78)$$

where r is the radius of the earth, g_0 is the value of gravity field on reference geoid which can be obtained from equation 2.35 providing theoretical value of gravity field with reference to 1980 Geodetic Reference System. However, before modelling geoid field, it is corrected for the topography effect similar to terrain correction in case of the observed gravity field as described below.

(i) Terrain Effect on Geoid

The global geoid undulations obtained from satellite data represents the effect of total mass distribution at the data point on the reference geoid. This includes both the surface mass in terms of topography-bathymetry and subsurface mass distribution. The surface mass being close to the reference plane affect the reference geoid more than the subsurface mass distribution. It is, therefore, imperative to correct for the effect of surface mass on the geoid in order to evaluate the effect of subsurface mass distribution. This is similar to terrain correction described in section 2.4.3 in regard to gravity field. However, it is applied differently in this case. Here the topography-bathymetry data are used to generate a three dimensional surface which would represent the distribution of surface mass in that region. For this purpose the standard Digital Elevation Models (DEM) available on INTERNET can be used or better controlled elevation model can be locally obtained from topo sheets etc., The geoidal undulation due to this mass is then computed and removed from the observed geoid (Ebbing et al., 2001). There are schemes even to correct for the isostatic effects of topography and bathymetry in the region from geoid anomaly. Dixon and Parke (1983) and Mckenzie and Bowin (1976) have suggested Fourier transform methods to correct the geoid data for these effects.

2.8   Applications: Gravity Anomaly and Related Maps of India

Gravity anomaly and some related maps of India are described below to familiarize the reader with them and their application for geodynamic studies and resource exploration.

2.8.1   Important Maps for Interpretation of Gravity Anomalies

As gravity anomalies are controlled by elevation, rock types and structures, following maps are important to understand significance of the observed gravity anomalies.

(i) Tectonic and Elevation Maps of India

Elevation and geomorphology are controlled to a great extent by recent tectonic activities and sometimes even by older tectonic events such as faults, fractures etc., and therefore, good correlations exist between them. A simplified tectonic map of India (modified after, ONGC, 1968) is given in Fig. 2.26. There are some more detailed tectonic maps of India, published by Oil and Natural Gas Corporation, India and Geological Survey of India but they are quite complex and are difficult to reproduce in a book format. However, for a comparison with geophysical anomalies, author has found this simplified map quite adequate. An image of the elevation map of India is presented in Fig. 2.27(a). The two maps viz tectonic and elevation maps together show following significant features.

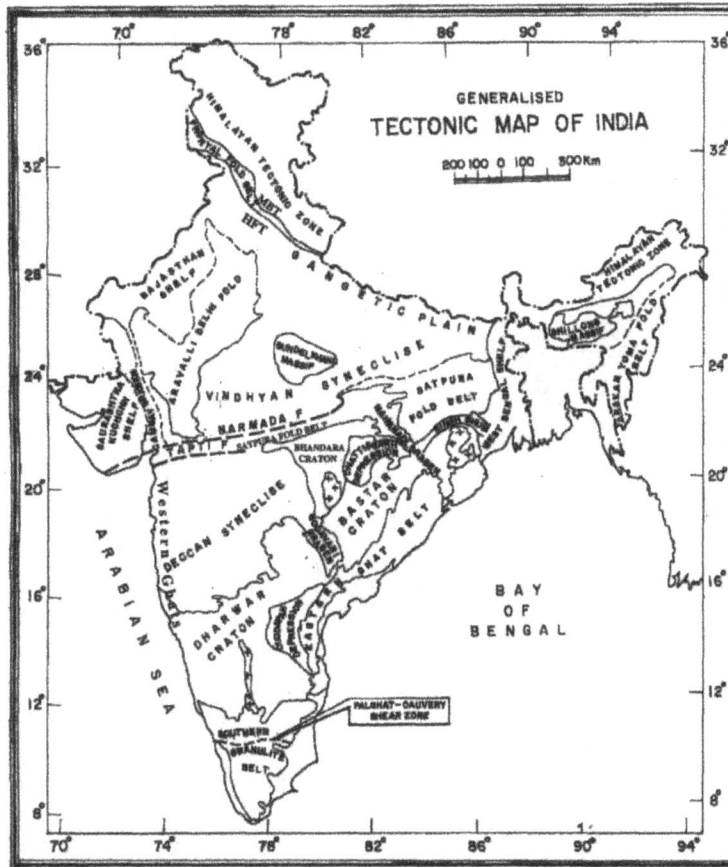

Fig. 2.26 A simplified tectonic map of India (Modified after ONGC, 1968).

(a) Towards the north, the tectonic map is characterized by the Himalayan Tectonic Zone, which consists of high mountains of Western and Eastern Himalayas bounded by frontal thrusts towards the south referred to as the Main Boundary Thrust (MBT) and the Himalayan Frontal Thrust (HFT) (Fig. 2.26) all along the frontal part.

(b) Towards further north, it is characterized by the Main Central Thrust (MCT) and the Indus Tsangpo Suture Zone (ITSZ) between the Indian and the Eurasian plates where Tibet starts as shown briefly in Fig. 2.35(b) and described with details in Chapter 6.

(c) South of the HFT is the Ganga basin (Indo-Gangetic Plains; Fig. 2.26), which is occupied by several major rivers such as Sindh, Sutlej, Yamuna, Ganga and Brahmaputra etc., originating from the Himalaya.

(d) Further south, the Central India is occupied by the Satpura Fold Belt (Mountains, Fig. 2.26) bounded by the Narmada fault towards the north, which almost follows the course of the river Narmada. North of the Narmada fault are the Vindhyan syneclise (basin) and the Bundelkhand massif (craton). It is interesting to note that rivers related to Satpura Fold Belt such as the rivers Narmada and Tapi represent deep seated faults and flow towards the west, while all other rivers towards the north and the south of them flow eastwards.

(e) The Western India is occupied by The Aravalli-Delhi Fold Belt and the Rajasthan Shelf, while the eastern India is occupied by the Shillong massif and the Bengal shelf.

(f) The north eastern part is known as the Assam shelf largely occupied by the Brahmaputra river bounded by the Eastern Syntaxial bend of the Himalayan Tectonic Zone and the Arakan Yoma Fold Belt.

(g) South of Satpura Fold Belt are the Deccan Syneclise and the Dharwar, the Bhandara and the Singhbhum cratons separated by Gondwana rift basins (Godavari and Mahanadi Grabens).

(h) The Dharwar craton south of the Deccan Syneclise and the Godavari rift basin (graben) extends southwards up to high grade rocks of the Southern Granulite Terrain.

(i) The eastern and western parts of India are occupied by the Eastern Ghat Fold Belt and the Western Ghats (Fig. 2.27(a)).

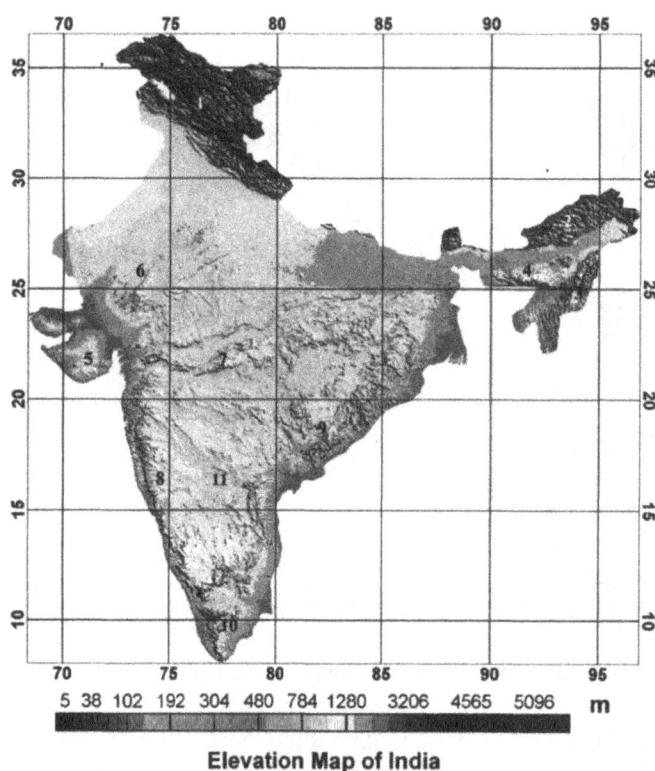

Elevation Map of India

Fig. 2.27(a) An elevation map of India. Numbers (1-11) represent important units of the country as follows. (1) Western Himalayas (2) Eastern Himalayas (3) Arkan Yoma mountains (4) Shillong plateau (5) Saurashtra – Kutch shelf (6) Aravalli – Delhi mountains (7) Satpura mountains (8) Western Ghats (9) Eastern Ghats (10) Southern Granulite Terrain (11) Dharwar Craton (Peninsular Shield). [Colour Fig. on Page 761]

These various tectonic elements of India and their evolution are discussed in more details in Chapter 7. It may be noted that mountains are mostly occupied by the fold belts suggesting that they are synonymous and are related to convergence and collision tectonics (Fig. 2.23(c) and (d)). In general, elevation of the peninsular shield varies from 700-1000 m, while that of fold belts (mountains) described above varies from 800-1200 m except some specific mountain peaks in South India and Western Ghats. Elevation of any region can be obtained from Digital Elevation Model (Web Site).

(ii) Geological map of India

As gravity anomalies are controlled by density of rocks, geology plays an important role in the interpretation of gravity anomalies. A simplified geological map of India (GSI, 1993) is given in Fig. 2.27(b). The south Indian shield is largely occupied by Archean gneisses and high grade granulite and Charnockite rocks with lower Proterozoic supracrustals and granitoids. Some supracrustal rocks of middle-late Archean are also present in the western part. This region also consists of some basins occupied by metasediments of Proterozoic period in which most prominent is the Cuddapah basin south of Hyderabad. Western India is composed of Mesozoic volcanics known as Deccan trap which extends up to central India and North West India. The most prominent feature of central India is the Satpura Fold Belt (SFB, Fig. 2.26), which consists of Proterozoic metasediments and Archean gneisses covered by Deccan trap in certain sections. North of the SFB is Proterozoic sediments of the Vindhyan basin bounded by Bundelkhand massif at its northern margin. Further north is alluvium of Ganga basin leading to Himalayas which is largely composed of post Cenozoic sediments and Proterozoic-Mesozoic metasediments along Himalayan frontal thrusts. North West India is covered by Archean Proterozoic gneisses and metasediments of Aravalli-Delhi Fold Belt (ADFB, Fig. 2.26) and Mesozoic sediments and alluvium west of it. East India is composed of Proterozoic metasediments of Eastern Ghat Fold Belt (EGFB, Fig. 2.26) and Mesozoic sediments of Gondwana rift basins, viz. Godavari, Mahanadi and Damodar basins. North East India consists of Proterozoic metasediments and granitoids of Shillong massif (Fig. 2.26) and Proterozoic metasediments along the Himalayan frontal thrusts.

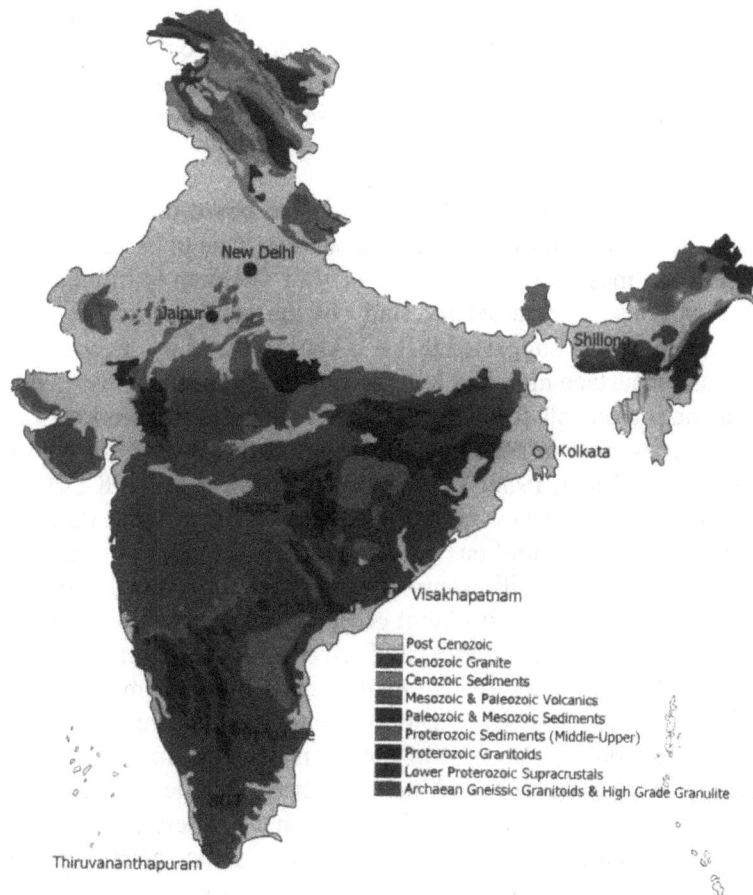

Fig. 2.27(b) A simplified geological map of India (GSI,1993). [Colour Fig. on Page 762]

2.8.2 Gravity Studies in India – Geodynamics and Exploration

Gravity surveys in India were started as early as in 1865 by British scientists using pendulums. But in a planned manner it was started by the Survey of India and the Geological Survey of India for geodesy and resource exploration programmes during 1950s after independence. National Geophysical Research Institute, (NGRI), Hyderabad, initiated a national programme of preparation of a country-wide regional gravity maps of India in 1964 (Hari Narain et al., 1964) which gave rise to series of gravity maps of India, published in 1975 (NGRI, 1975) on 1:5 Million scale (1 cm = 50 km) with 10 mGal contour interval. Subsequently, these maps were updated in 2006 by incorporating new gravity data acquired by different organizations in India which provided gravity maps on 1: 2 Million (1 cm = 20 km) scale with better resolution of 5 mGal contour interval (GSI-NGRI, 2006). Region-wise gravity studies were conducted by different institutions and workers and detailed gravity maps were prepared for those regions such as Godavari basin (Mishra et al.1987), Cuddapah basin (Krishna Brahmam et al.1986), Aravalli-Delhi Fold Belt (Reddi and Ramakrishnan 1988) etc., Sedimentary basins were largely covered by Oil and Natural Gas Corporation. The gravity measurements in India are reduced to the National Base value of 979,049.09 mGal based on IGSN71 at Survey of India, Dehradun, which was tied to the International reference station at Potsdam, Germany. Initially, the base stations for gravity surveys in the country were established by Survey of India whose details can be obtained from Gulatee (1956). Some base stations along a N-S line starting from Delhi to Trivandrum at major airports were established by Manghnani and Woolard (1963). Subsequently, NGRI established base stations throughout the country whose details can be obtained from publications related to them available in NGRI (Qureshy and Krishna Brahmam, 1969). Now with availability of absolute gravimeter (FG5), NGRI has established a base station at the Gravity Observatory, NGRI (Tiwari et al., 2006) and planned to establish primary base stations in the country with an accuracy of about 5-6 microGal. Bouguer anomaly maps of India vis-à-vis tectonics are described below to show their applications to geodynamics and resource exploration.

(i) Bouguer Anomaly Map of India

The Bouguer anomaly map of India (NGRI, 1975) was prepared based on gravity data at 5-7 km interval along major roads as described above for regional surveys. Due to large station spacing and altimeters used for the elevation measurements the accuracy of this map is limited to ±2-3 mGal. This has been one of the most useful data set for study of the geodynamics of this country. A critical examination of this map clearly suggested that it is characterized by several gravity "highs" and "lows" implying surface/subsurface high and low density rocks, respectively as their causative sources. There are several sharp gradients implying contacts/faults. Several workers have used this map for interpretation of gravity data in terms of geology and tectonics on a continental scale (Naqvi et al., 1974; Qureshy and Warsi, 1980; Verma, 1985; Mishra et al., 1987, 2000; Shreedhar Murthy, 1999; Mishra, 2002, 2006). The new Bouguer anomaly map of India (GSI-NGRI, 2006) were prepared based on uniform distribution of data at 5.0 km and are terrain corrected representing complete Bouguer anomaly with accuracy better than ±1.5 mGal. This map is based on about 51,000 gravity stations acquired by different geoscientific organizations of the country as referred to in brochure of these maps. To get better perspective of linear features vis-à-vis tectonics and geology (Figs 2.26 and 2.27b), the Bouguer anomaly map of India as image illuminated from the north is presented in Fig. 2.28 with some gravity contours superimposed on it for ready reference about amplitude of anomalies. As most of the anomalies are linear oriented NW-SE to NE-SW, their image illuminated from north would highlight them. Detailed Bouguer anomaly map of India as image with contours is also presented in Figs 7.8a and b where detailed interpretations of the most of the observed anomalies are attempted. However, some interesting anomalies are discussed below qualitatively to demonstrate that some information about causative sources can be derived directly from the Bouguer anomaly maps without recourse to any major computations. For further details, it is advised to refer Part II (Chapter 6, 7, 8 and 9).

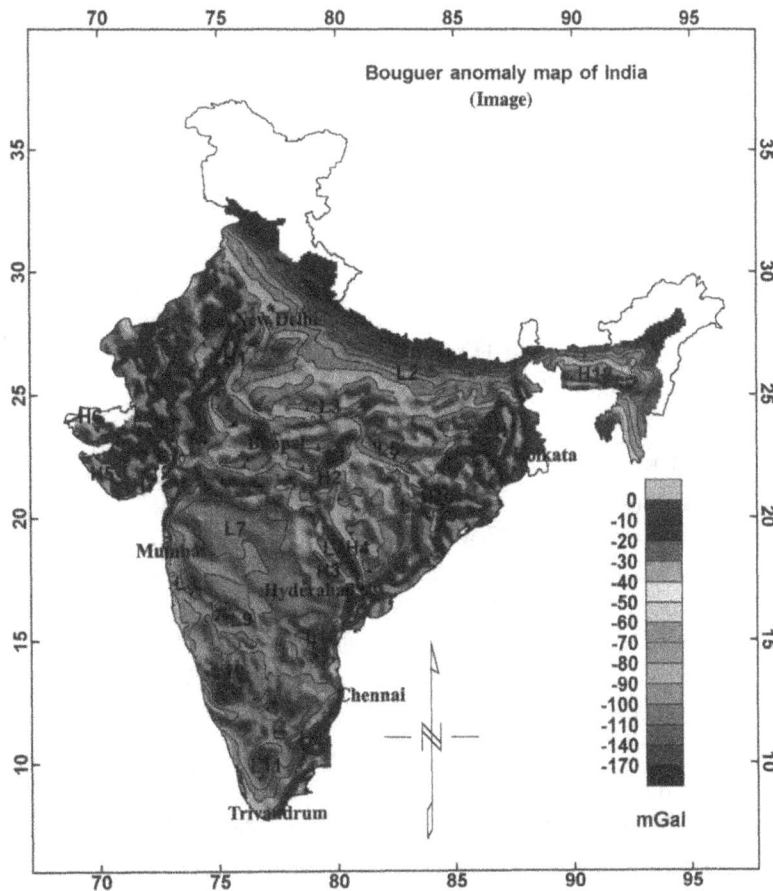

Fig. 2.28 Image of the Bouguer anomaly map of India illuminated from the north. H and L denote gravity highs and lows (Mishra et. al., 2008). [Colour Fig. on Page 763]

(a) Starting from the north, Himalayas are characterized by large wavelength and large amplitude gravity lows (L1) with sharp gradient towards the south indicating crustal thickening northward caused by collision of Indian and Eurasian plates due to compressional tectonics (Fig. 2.23(d)). Sharp gravity gradients at southern margin of Himalayas indicate faults, which may represent here the system of faults/thrusts related to the evolution of Himalayas such as Main Central Thrust, Main Boundary Thrust and Himalayan Frontal Thrust (Fig. 2.26).

(b) The Ganga Basin, south of Himalaya, is represented by gravity lows (L2), which can be attributed to low density sediments in this section and can be used to delineate basement structures which are important for Hydrocarbon exploration that is given in Section 7.4.2. The lowest of these anomalies occur all along Himalayan front that is known as foredeep where sediment thickness is maximum.

(c) Towards the west, the Aravalli-Delhi Fold Belt (Fig. 2.26) is represented by linear gravity highs (H1). The Aravalli mountains being topography high should normally be characterized by gravity lows due to crustal thickening caused by isostasy as in case of Himalayas. However, it is characterized by gravity highs which indicate high density rocks at surface and subsurface levels as explained in section 2.7.1 for old collision zones between two plates (Fig. 2.23(d)). Such thrust zones due to compressional tectonics are important for mineral exploration and Delhi Fold Belt is characterized by several occurrences of base metals and other minerals as discussed in Section 7.5.

(d) The gravity highs, H5 and H6 over Saurastra and Kutch extending to the Western Rajasthan suggest the presence of surface/subsurface high density rocks. These may be in the forms of mafic intrusive that are discussed in Chapter 7.

(e) The Bundelkhand massif and the Vindhyan basin (Fig. 2.26) are represented by gravity lows, L3 and south of it which are caused by low density granite/ gneisses and sediments of the basin, respectively.

(f) The Satpura Fold Belt (Fig. 2.26) is also characterized by linear gravity highs (H2) that extend from the west coast of India up to Bengal basin towards the east. These gravity highs can be interpreted due to high density mafic intrusive of Satpura orogeny representing an old collision zone. Gravity highs, HI and H2 appear to join together in Western India to form a composite orogeny as described in Section 7.6.3.

(g) The Southern Peninsular shield (Fig. 2.27(a), Dharwar Craton and Southern granulite belt) is primarily characterized by a major large wavelength gravity low, L7 that can be primarily attributed to the thickening of the crust under the Western Ghats and Southern Granulite Terrain (L11) due to isostasy (Fig. 2.27(a)) or low bulk density of crustal rocks due to several granite intrusions (Fig. 2.27(b)) in this region that is discussed with details in Sections 7.8 and 7.9. Within this large wavelength gravity low, L7 there are several high amplitude linear gravity lows, L8, L9 and L10 extending from the west coast of India up to the east coast that may represent transcontinental lineaments with granite intrusive bodies as several such bodies are exposed in this region (Fig. 2.27(b)).

(h) The Eastern Ghat Fold Belt is largely characterized by gravity highs (H9 and H10) that extend along the east coast of India. They may represent high density rocks of the Eastern Ghat Fold Belts (EGFB) and thin crust along the coast. The gravity gradient west of these gravity highs represent a deep seated fault/thrust on a continental scale related to the evolution of the EGFB that is discussed with details in Section 7.10.

(i) The set of gravity lows and highs, L4 and H3 and H4 represent the Godavari rift basin representing the Gondwana sediments and the shoulder highs as defined in Section 2.7 in regard to extensional tectonics (Rift Basins). Further details are discussed in Section 7.7. Similarly, another Gondwana rift basin, the Mahanadi basin also provides gravity lows, L5 with small high along the shoulders.

(j) The Shillong plateau (Fig. 2.27(a)) being high land should have shown a gravity low due to isostasy but it shows gravity highs, II12 indicating subsurface high density rocks as described with modeling in Section 7.6.4.

2.8.3 Satellite Gravity and Altimetry Maps – Sea Level Changes in Indian Ocean

Satellite data have provided new avenues for the study of gravity field over inaccessible regions such as Himalayas, oceans etc., which are extremely important for the study of plate tectonics. As discussed in section 2.6, satellite gravity data provides a regional picture and therefore, it should be accordingly used for regional studies. As satellite altimetry provides changes in sea surface, it is directly used to monitor sea level changes that is discussed below as an example. The same data is used to derive geoid undulation, which in turn provides the free air anomaly map as described in Section 2.6. These data sets for India and contiguous regions of Indian plate are discussed in Chapters 5, 6 and 7.

Satellite Altimetry: Sea Level Changes in Indian Ocean

Besides conventional applications of satellite gravity for geodynamics and tectonics, it offers unique opportunities to study the variations in the gravity field with time which has given rise to some special applications of this data set. Monitoring of sea level changes and ice mass budgeting are important both for socioeconomics and climate studies. Socioeconomics as it affects the population living along coast lines and close to them. They are important for climate studies as they are related to greenhouse effect and global warming. Conventionally, long term sea level changes were monitored using tide gauge measurements which have limited spatial distribution near coasts and measured changes in sea level with some reference point on coast. Satellite altimetry for the first time provided global coverage of changes in sea surface over a long period of 8-10 years. TOPEX/POSEIDON (T/P) launched jointly by American (NASA) and French Space Agencies provided high resolution altimetry data for precisely measuring long term sea level variations (Nerem and Mitchum, 2001). However, to derive sea level changes from satellite altimetry requires several corrections for waves height, wind speed etc., which are specialized topics (Fu and Cazenova, 2001) and can not be dealt here.

Tiwari et al. (2004) analyzed T/P data for Indian ocean for period 1993-98 and provided sea level trends for this period (Fig. 2.29) which shows a maximum change of about +27 mm/yr in NW part and −7 to −8 mm/yr in the Arabian Sea and the Bay of Bengal. This pattern shows a good correlation with changes in sea level due to changes in temperature below sea surface in those regions. This implies that changes in sea level inferred from altimetry data can be related to changes in temperature which in turn related to heat budget under water column and El Nino–Southern Oscillations (ENSO) events controlling the climatic conditions including the monsoon. A similar approach is also used to estimate ice mass loss and its transport in Greenland and Antarctica which are related to greenhouse effect and global warming controlling the climatic condition world over (Zwally and Brenner, 2001; Zwally et al., 2002). As GRACE satellite has a long span of life of 8-10 years, it is ideally suited for the study of transient variations in geoid and gravity field which are useful for sea level changes, ice mass loss and changes along subduction zones in a region (Velcogni and Wahr, 2006). Changes in mass budget along subduction zones may lead to better understanding of processes related to them and associated seismic activity in the region (Section 8.4).

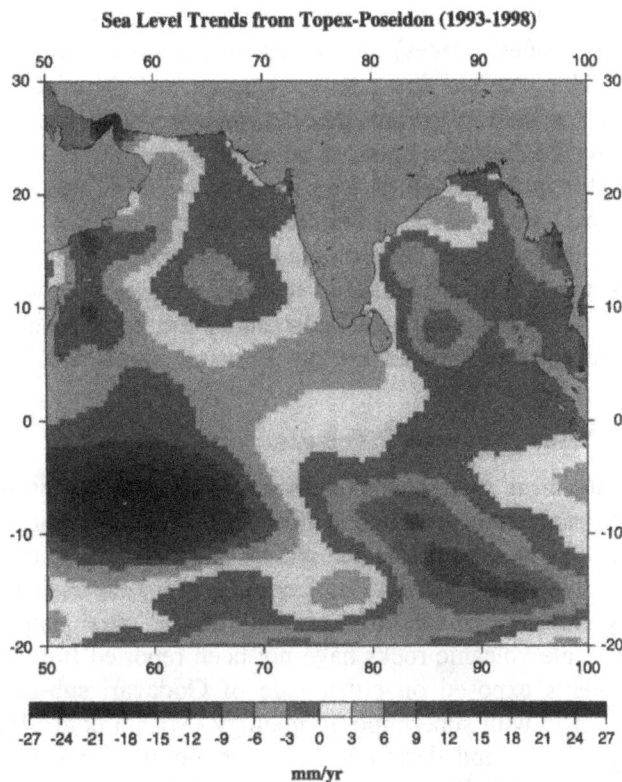

Sea Level Trends from Topex-Poseidon (1993-1998)

Fig. 2.29 Sea Level Trends for period 1993-98, derived from Topex / Poseidon satellite data (Tiwari et al., 2004). [Colour Fig. on Page 764]

2.9   Field Examples and Gravity Signatures – Extensional Tectonics

In this section, important case histories of extensional tectonics (Fig. 2.23 (a) and (b)) are discussed to familiarize with nature of gravity anomalies related to them and their interpretation. Most important examples of extensional tectonics are continental rift basins and continental margins as described below.

2.9.1   Passive and Active Rift Basins – Godavari and Mahanadi Basin, India and Continental Margin off West Coast of India

There are two kinds of rift basins and continental margins, viz. passive and active. As described in Section 2.7.1, the active rift basins are associated with large scale volcanic rocks and are related to them for their evolution while passive rifts are formed due to extension related to other causes such as lithospheric upwarping due to plate tectonic forces etc., There may be some small amount of volcanic rocks even in case of passive rifts as intrusive. Extension induces lithospheric upwarping, which in turn induces further extension and normal faulting at the surface leading to formation of rift basins. Rift basins normally lead to continental break up and formation of oceans as in case of Red Sea rift described in Section 3.8.1. In case they do not lead to break up of continents, they are known as failed rifts as in case of continental rift basins. In this regard, continental margins are best examples of rifting.

There are two kinds of continental margins, viz. active and passive but they differ in definition compared to rifting by same name as described above. Most of the continental margins are being associated with intrusives, therefore, both active and passive margins are characterized by volcanic rocks but they differ in regard to plate tectonic processes. Active margins are characterized by plate tectonics and related activities such as subduction and volcanic arc etc., such as Andaman-Sumatra Islands, West Coast of USA (Fig. 5.13) etc., while passive margins do not show these activities on any large scale and occur along edges (opposite sides) of opening oceans such as east and west coasts of India, east coast of USA etc., This implies that passive margins are associated with extensional tectonics while active margins are related to present day compression/convergence. As most of the passive margins are associated with different amount of volcanic rocks, some with large amount of volcanic rocks due to plumes or hot spots and others with small amounts of intrusives due to melting of upper mantle rocks locally, these two types are differentiated as volcanic and non-volcanic passive margins, respectively. The two, however, can be distinguished based on extent and amplitudes of gravity and magnetic anomalies which is larger in case of volcanic margins compared to non-volcanic ones. For example, western and the Eastern margins of India can be termed as volcanic passive margin as the break up along these margins were caused by the Reunion plume and Kerguelen hot spot respectively (Chapter 5).

(i)   Godavari Gondwana Rift Basin- A Passive Rift Basin

Godavari basin is one of the most prominent Gondwana (Permian) basins in India, which is regarded as a rift basin. As described in section 2.7, rift basins are sunken part of the crust, which are linear features bounded by normal faults on one or both sides running for hundreds of kilometers. Sunken port is subsequently filled by sediments forming the basins. The Godavari rift basin is characterized by thick sediments of Permian-Triassic time comprising sandstone, limestone, chert etc. Large scale volcanic rocks have not been reported from this section. There are Proterozoic, metasediments exposed on either side of Godavari sub-basin. The Bouguer anomaly map of Godavari basin with superimposed geology (Mishra et al., 1987, Mishra et al., 1989) is presented in Fig. 2.30(a) and described below to familiarize with nature of gravity anomalies observed over continental rift basins in general and Godavari rift basin in particular, which is an important tectonic element both for geodynamics studies of this region and hydrocarbon exploration.

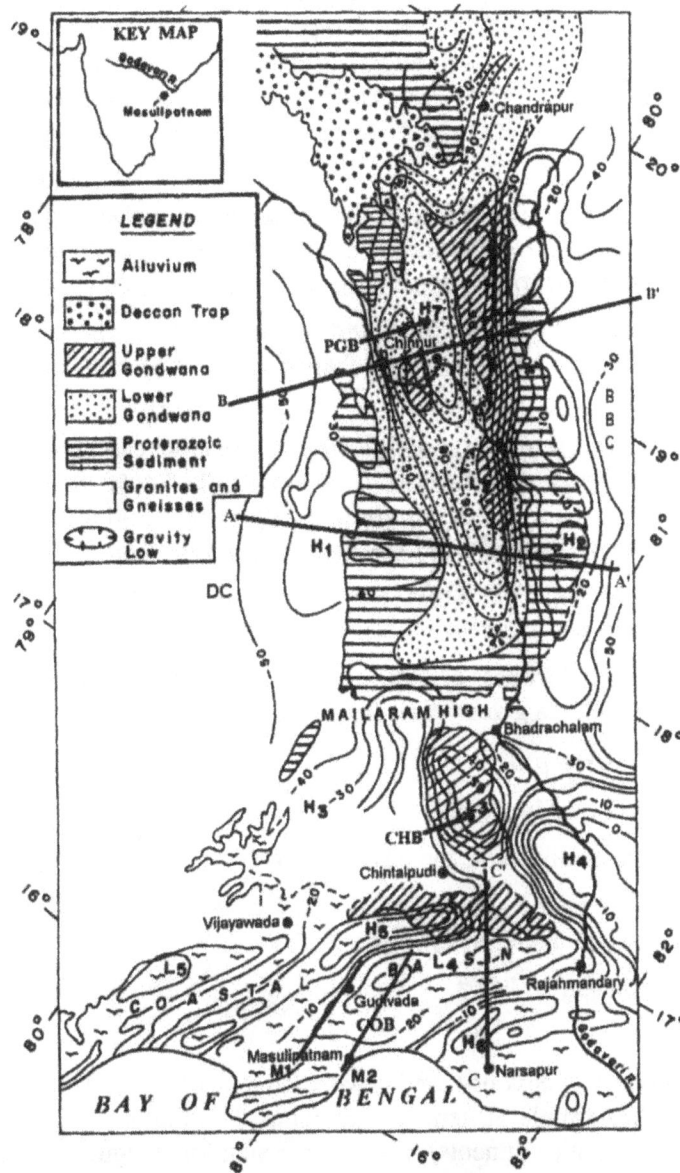

Fig. 2.30(a) Bouguer anomaly map of Godavari basin, India with superimposed geology. L1 and L2 represents main Godavari sub basin while L3 and L4 represents Chintalpudi and Coastal sub basins. H1, H2, H3 and H4 represent shoulder highs typical of continental rift valleys. The gravity gradient between L1, L2, L3, towards the south and H2, H4 towards the north is sharp and linear indicating master fault of the basin. The coastal sub basin, L4 is almost parallel to east coast of India. H7 around Chinnur represents a median high, another signature of a rift valley. It shows various profiles modeled in different chapters. Star indicates the Bhadrachalam earthquake of 1969 of magnitude 5.3 described in Section 7.7.1.

DC- Dharwar Craton; BBC- Bhandara-Bastar craton and CHB- Chintalpudi subbasin.

(a) Fig. 2.30(a) is characterized by linear central gravity lows (L1 and L2) over the Godavari basin and adjoining sub parallel "highs" (H1 and H2) along shoulders which are typical signatures of continental rift basins as shown in Fig. 2.23(a).

(b) Central gravity low striking NW-SE corresponds to the Gondwana Sediments of almost similar strike.

(c) Based on nature and amplitude of gravity lows, Godavari basin can be broadly divided into two sub-basins namely the Pranhita-Godavari sub-basin (L1 and L2) and Chintalpudi sub-basin (L3) separated by an exposed basement high known as Mailaram "high". Coastal part of this basin along east coast (L4) strike NE-SW same as the coastline and Eastern Ghat, which might be attributed to subsurface structures such as Eastern Ghats forming the basement in this section.

(d) Adjoining highs, H1 and H2 coincides with the Precambrian metasediments and basement gneisses and lower crustal granulite rocks whose densities are higher (2.7-2.8 g/cm^3) compared to the normal density of the upper crust and may be caused by them.

(e) Gradient between the central gravity "low" and the adjoining "high" towards the north is sharp and straight for 400-500 km indicating a faulted margin which is the master fault of the rift basin.

(f) The Bouguer anomaly 'low' is approximately –50 mGal in Godavari sub–basin (L1 and L2) and –30 mGal in Chintalpudi sub–basin (L3), therefore, thickness of sediment should be almost double in Godavari sub-basin compared to those in Chintalpudi sub–basin if the density of sediments in two sub-basins are assumed to be almost same.

(g) There is a small "high" around Chinnur (H7) representing exposed basement "high" which is a median "high" of the rift basin and form a typical signature of rift basins world over.

(h) The approximate minimum thickness of sediment in Godavari basin can be obtained by using the formula for Bouguer slab as given below (Equation 2.43).

$$\Delta g = 0.04188 \, \rho.t \qquad\qquad .....(2.79)$$

where ρ is the density contrast in g/cm^3.

The maximum gravity anomaly inside the basin is $- 100 - (- 50) = -50$ mGal.

The density of Gondwana sediment is approximately 2.35 g/cm^3. Therefore, the density contrast from surrounding (gneisses) and upper crustal rocks = $2.35 - 2.70 = - 0.35$ g/cm^3.

Therefore, $-50 = 0.042 (- 0.35). t$

or $t = 3333$ m $= 3.33$ km   .....(2.80)

This provides the minimum thickness of Gondwana sediments in Godavari sub–basin based on thumb rule of Bouguer slab formula. It generally provides the minimum thickness of sediment due to limited extent of the basin (sediments), while expression for Bouguer slab assumes infinite extent of the plate. The maximum bound on thickness is double the minimum thickness, which implies about 6.6 km in case of Godavari basin. This implies that actual thickness of sediment may lie in between 3.3 and 6.6 km. However, to obtain the correct estimate of thickness of sediments, complete modeling of the gravity anomalies are required which provides a thickness of about 5.0 km (Sections 4.5.4 and 7.7.1). The actual computed depth in the present case is almost average of estimates of minimum and maximum thicknesses of sediments in this basin based on Bouguer slab formula. Therefore, the Bouguer slab formula provides only a rough estimate which should be collaborated from other evidences. The absence of volcanic rocks to any significant amount in this basin suggests it to be a passive rift basin.

(ii) Active Continental Rift Basins and Crustal Model

Fig. 2.30(b) shows a schematic section of crustal model under an active rift basin based on studies on such rift valleys world over. As stated above (Section 2.7), rift basins are formed due to extensional tectonics and are therefore, characterized by lithospheric upwelling which in turn causes crustal thinning and in most cases, low density layer in the middle crust. This process would produce following typical density structures (Fig. 2.30(b)):

(i)   Low density layer in upper mantle (-0.20 g/cm^3)

(ii) High density layer in lower crust ($+0.35$ g/cm^3)

(iii) Low density layer in the middle crust (-0.28 g/cm^3) and

(iv) Low density sediments at the surface (-0.4g/cm^3).

Using these density contrasts a typical gravity profile at the surface is computed and given in Fig. 2.30(b). Presence of under plated crust and crustal upwarping suggests it to be a typical crustal model of an active rift system. It is compared with an observed gravity profile across Mahanandi basin, India (L5, Fig. 2.28) which shows good correspondence with the computed profile. However, they may not truly represent the crustal/upper mantle structures under Mahanadi basin as they have not been constrained from any independent method such as seismics and therefore represents a schematic section of density structures under an active rift. This model is used here to demonstrate the expected crustal and upper mantle structures under an active rift basin. The quantitative values of density contrast may vary for different rift valleys. These contrasts are typically for an active rift basin. In case, there are independent evidences of mafic intrusive in the crust, they can be added in the model with configuration and bulk density inferred for them as in case of Red Sea Rift discussed in Section 3.8. Mahanadi basin (L5, Fig. 2.28) across east coast of India is another Gondwana basin of Permo-Triassic period. It appears that this basin represents an active continental rift basin while the contemporary Godavari basin as described above represents a passive continental rift basin. This difference in the structures and mechanism of formation of two contemporary rift basins separated only 300-400 km apart (Fig. 2.26) can be attributed to the proximity of Kerguelen hot spot to Mahanadi basin as discussed in Section 5.6.

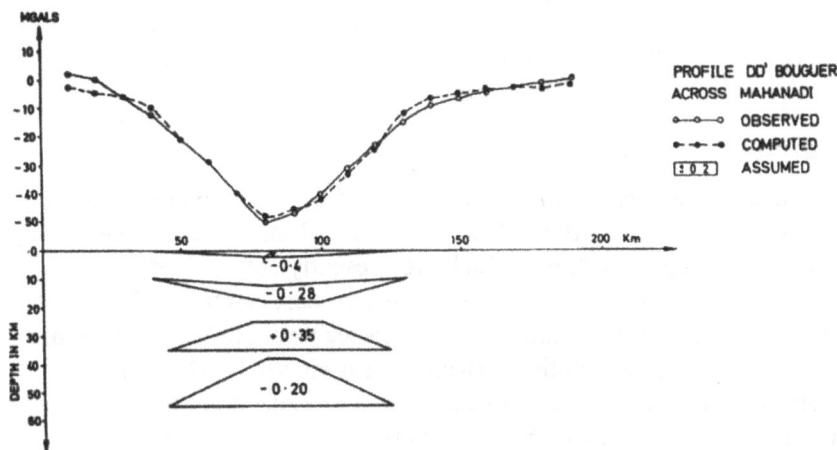

Fig. 2.30(b) A typical Bouguer anomaly profile across an active rift basin compared with a gravity profile across the Mahanadi Gondwana rift basin.

(iii) Continental Margin Offshore West Coast of India – Gravity and Seismic Profiles

Continental margins represent transition from continental to oceanic crusts and therefore, assume special significance. As seaward sides are characterized by basins, they are important for hydrocarbon exploration. Passive margins are best example of extensional tectonics (Section 2.7, Fig. 2.23(b)), as continents break apart along these sections through extension, which are related to intrusives. Therefore, margins are usually associated with large scale intrusions and igneous activity. A gravity profile across offshore west coast of India is shown in Fig. 2.31 along with the seismic section. It shows seawards gravity high (H1) which coincides with the uplift of a marker reflector (white dashed line) in the seismic section, which indicates intrusive. This kind of gravity high offshore along the shelf is observed throughout the offshore west and east coasts of India (Section 5.5) and along margins of most of the continents world over. These gravity highs are related partly to thinning of the crust in transitional section from continental to the oceanic crust and partly due to intrusive related to extension and rifting phase. Further, importance of these anomalies about geodynamics and hydrocarbon exploration are given in Sections 5.5 and 9.6, respectively.

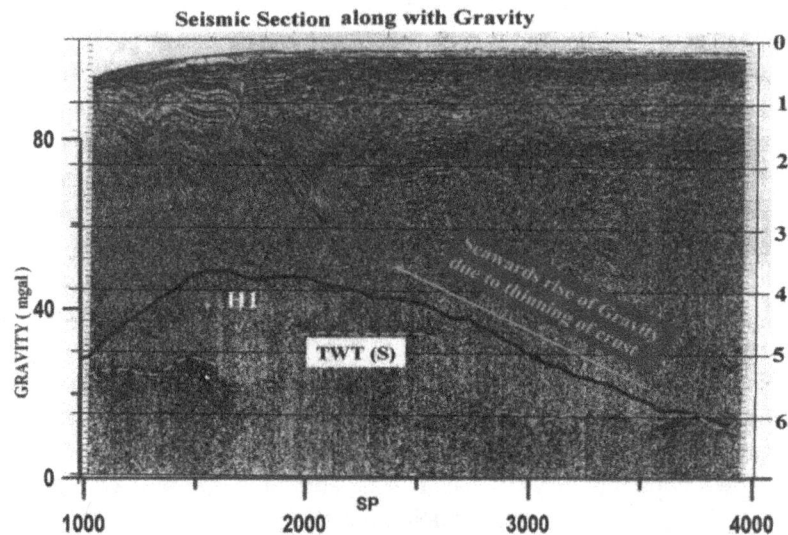

Fig. 2.31 A gravity profile across the western margin of India superimposed over a seismic section (Sar 2008). It shows sea ward rise of the gravity field (H1) over the continental margin due to thinning of the crust and intrusive related to the extension and rifting of the Indian continent along the west coast of India that is referred to as Deccan trap activity.

2.9.2 Active Rift Basins – East African Rift System

Active rift basins are characterized by volcanic rocks and thick sedimentary section, which are important for hydrocarbon occurrences such as Cambay basin in India (Section 9.4). They are also zones of intense seismic activity specially those, which are presently active. East African Rift system is a long rift system of about 3000 km long in two parts, western part extending from lake Malawi to lake Victoria and eastern part from south of Nairobi to the Red sea through Afar triangle in Ethiopia. It is presently an active rift system along which the African plate is slowly breaking into two parts. As it is presently active, several investigations have been carried out to understand the processes related to active rifting and breakup of continents. Several gravity models constrained from seismic investigations have been proposed (Mechie et al., 1997; Ravat et al., 1999). Ravat et al. (1999) based on detailed seismic investigations and an elaborate method of converting seismic velocities to density provided a density model along the KRISP seismic profile (Fig. 2.32). This figure shows the main geology of the central part of the rift (Kenya rift; Prodehl et al., 1994) and profiles of seismic investigations. Either sides of the rift valley are characterized by cratons and orogenic belts, which in itself suggests that it has been a weak zone from Archean-Proterozoic period. The rift valley itself is occupied by large amount of rift volcanics and sediments suggesting its active nature related to some subsurface hot spot or plume, which has given rise to these volcanics and eventually resulted into the breakup. It has been suggested that Afar plume (Ebinger and Sleep, 1998; Ebinger and Casey, 2001) has been responsible for volcanics along East African rift and its evolution. The density model along

KRISP profile D as given by Ravat et al. (1999) is reproduced in Fig. 2.33, which suggests that the density structure of the whole lithosphere under the rift system is modified due to sublithospheric thermal anomaly and intrusives. The central part of the upper mantle under the rift system is modified showing low density due to high temperature (200-375°C higher). This thermal anomaly causes upwarp in Moho and other horizons of the crust resulting into a central high (H1) and flanking lows (L1 and L2) and highs (H2 and H3) along shoulders which are typical of active rift basins as shown schematically in Fig. 2.23(b). The central high, H1 is known as median high and gravity highs H2 and H3 are referred to as shoulder highs, which are characteristic of active rift valleys world over.

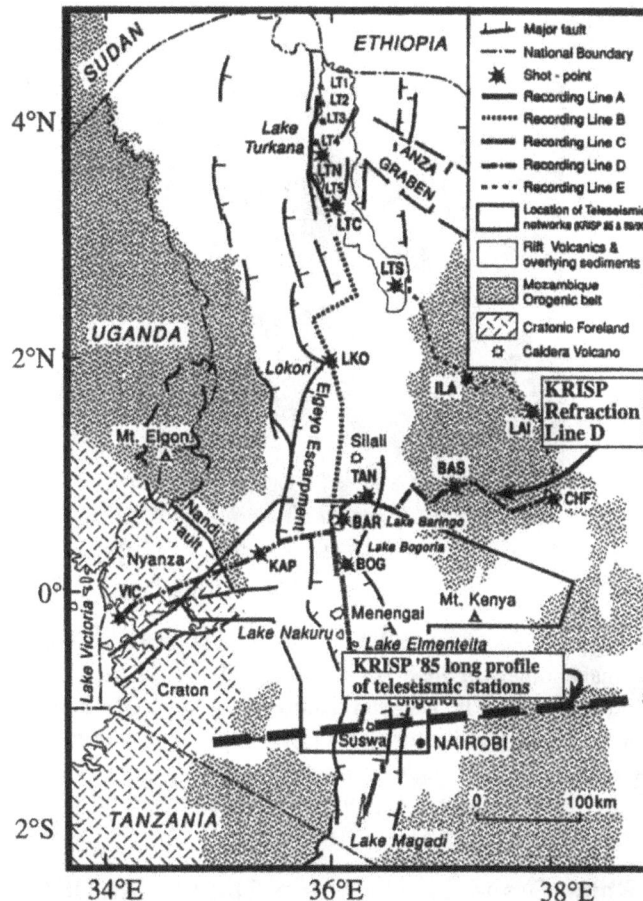

Fig. 2.32 General geology of Kenya rift, a part of East African Rift System and KRISP Refraction line D along which the crustal section and density models are shown in the next figure (Fig. 2.33) (Prodehl et al., 1994).

High density layer (3.0 g/cm^3) along the Moho with its density and velocity (7.5 km/sec) being in between that of lower crust (2.9 g/cm^3 and 6.9 km/sec) and upper mantle 3.3 g/cm^3 and 8.1 km/sec) is known as underplated crust and are found in volcanic provinces due to spread of these rocks along the Moho during intrusion. This is typical of active rift basins as discussed above. The top layer with a density of 2.5 mg/m^3 represents rift volcanics and sediments. In geological time it may develop into a sea as Red sea rift breaking African continent in two parts.

Cambay rift basin in Western India also represents an active rift basin. However, due to its importance for hydrocarbon occurrences, it is described in Section 9.4.

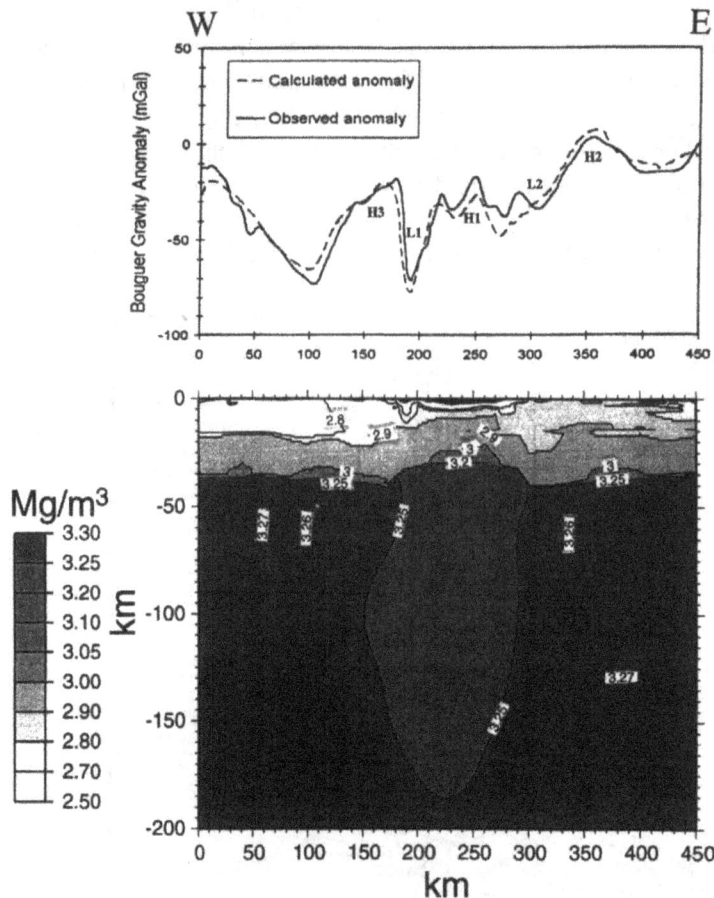

Fig. 2.33 Crustal density model along profile AA' across Kenya rift based on gravity modeling constrained from seismic section and velocities. Thermal anomaly in the lithosphere has reduced the density and caused upwarp of density interfaces in the crust under the rift basin. Magmatic intrusion has caused a high density 3.0 mg/m³ (g/cm³) layer along the Moho, known as underplated crust. This layer is typical of volcanic provinces, indicative of active rift. The surface layer of density 2.5 mg/m³ represent volcanic rocks mixed with sediments (Ravat et al., 1999).

2.9.3 Global Examples – Rhine Graben, Europe; Rio Grande Rift, USA, and Baikal Rift, Russia

Convergent margins though represent compressional tectonics but they are also associated with rift basins due to volcanics associated with them. Generally, they are associated with island arcs related to convergent margins and are known as back arc basins (Fig. 1.3) as they form on the back side of island arcs. However, if they continue to long distances, they are related to plate tectonic forces causing lithospheric unwrap in the entire regions while its initiation might have been done by island arc magmatism. Rhine graben in Europe and Baikal rift in Russia are examples of this kind of rift systems. Besides these, several small scale rift systems randomly distributed but mostly parallel and perpendicular to convergent margins have been reported over Tibet which shows an eastward movement (Chapter 6) due to convergence of the Indian and the Eurasian plates (Yin, 2006).

(i) Rhine Graben, Europe

Graben implies rift basin and therefore, Rhine rift basin is known as Rhine graben. It is located in the northern foreland part of the Alps (Section 2.10) and is related to it for its evolution. Though

Alps itself is due to convergence between European and African plates, in case of such convergence, localized extensional regimes are created perpendicular to the convergence zone, which produces normal faults and rift basins (Fig. 2.34(a)). Such rift zones are also found in Tibet almost perpendicular to Himalayan collision zone, between Indian and the Eurasian plates. Baikal rift system in Siberia is also attributed to this convergence. These rift basins and related structures are caused by plate tectonic forces. Its association with Cainozoic volcanic rocks reported from a particular section of Rhine graben suggests that it may be an active rift. However, significant amount of volcanic rocks as in case of East African rift system as described above have not been reported from this region. This suggests that it might have been initiated by convergence related volcanism but its extension northwards must be related to plate tectonic forces causing lithospheric upwarp. Small amount of volcanic rocks can be generated even due to asthenosphere-lithosphere upwarp and melting of mantle rocks due to decompression. The Rhine graben is still developing and therefore, since its inception at about 40-45 Ma (Prodehl et al., 1992), it has developed into a long narrow rift basin starting almost from north of Alps near Mediterranean sea extending up to North sea near Rotterdam (Fig. 2.34a). Based on detailed gravity modeling constrained from large scale seismic investigations, Prodehl et al. (1992) provided crustal structure (Fig. 2.34(b)), which shows all characteristics of continental rifts due to extensional tectonics (2.23a). It shows a central gravity low (L1) and shoulder highs, (H1 and H2) and a median high, H3 while modeled crustal structure shows thinning of the crust (25 km) related to asthenosphere-lithosphere upwarp and high density intrusive rocks (2.68 and 2.82 g/cm^3) along the shoulders. High resolution teleseismic travel time tomography in the Egar rift in Central Europe (Plomerova et al., 2007), part of European Cainozoic Rift System east of Rhine graben suggested broad low velocity anomaly similar to asthenosphere-lithosphere upwarping instead of columnar type of velocity anomaly indicative of plumes. However, small plumes under French massif and South Germany were postulated where volcanic rocks are exposed. This indicates that this kind of rift basins might be initiated by volcanism but their propagation to large distances depend primarily on asthenosphere-lithosphere upwarping which might be related to plate tectonic forces in these regions.

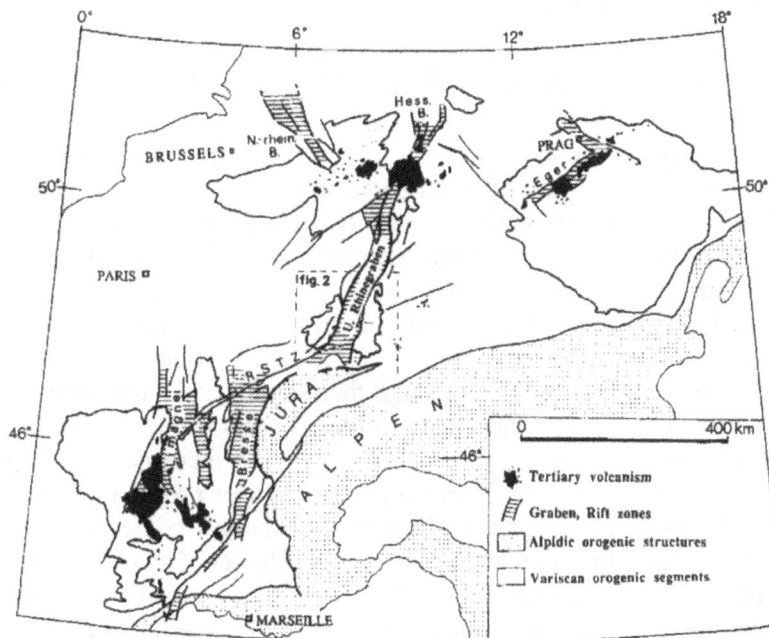

Fig. 2.34(a) The European Cenozoic rift system (Rhine graben) between the North Sea and the Mediterranean Sea (after Echtler et al., 1994).

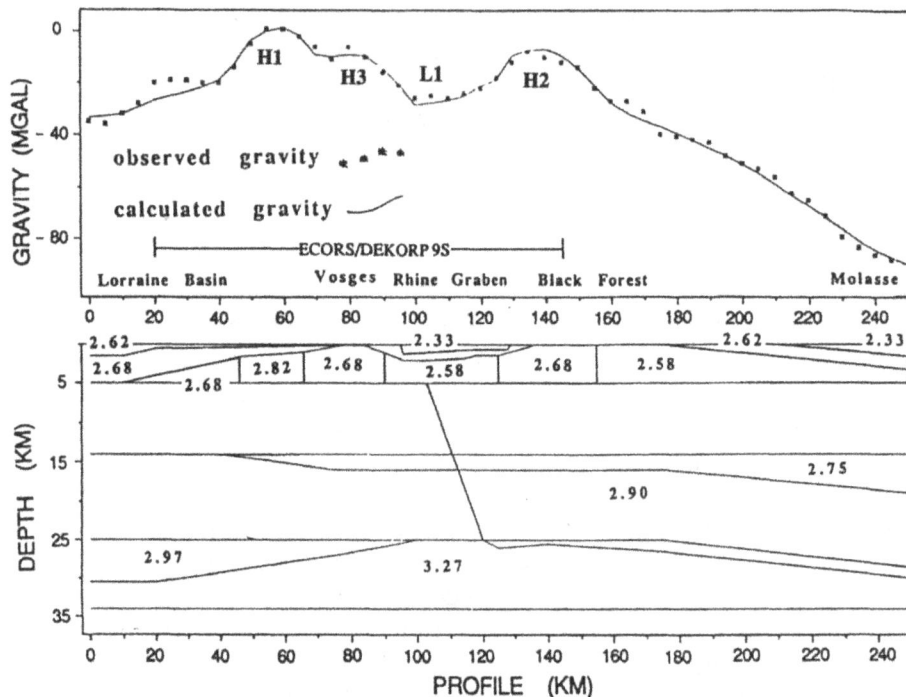

Fig. 2.34(b) Gravity profile across Rhine graben showing a central gravity low L1 and shoulder highs H1 and H2, H3 is a medium high, typical of rift basins. Density crustal model along the ECORS/DEKORP- 9S seismic line shows low density upper mantle and Moho upwarping up to 25 km(Prodehl et al., 1992).

(ii) Rio Grande Rift, USA and Baikal Rift, Russia

Similar crustal structures with lithospheric upwarp have also been reported in case of Rio Grande rift, USA (Fig. 2.35(a)) and Baikal rift, Siberia (Fig. 2.35(b)). The former has shown a crustal thickness of 28-30 km and low density and low velocity in the upper mantle (Cordell, 1978; Sinno et al., 1986). However, the Baikal rift (Fig. 2.35(b)) is more complex showing a gravity high in the southern part and low in the northern part (Inset, Fig. 2.35(b)) related to thin (~30 km) and thick (~50 km) crust in the respective sections (Petit and Deverchere , 2006). This rift is located at the eastern margin of Siberian craton indicating a weak plane, along which it has developed. This rift basin has developed in two stages, viz. firstly, strike slip faulting during Oligocene and secondly, formation of rift basin during Miocene. Both these periods are important for the convergence of Indian and Eurasian plates; the first one marks the collision of two plates and second one is related to the collision of their continental parts. Therefore, the two stages of formation of Biakal rift might be related to these events of Himalayan orogeny. It is interesting to note that crustal upwarp is noticed only in the southern part of the Baikal rift, thereby suggesting that once the rift basins are initiated, they can propel themselves for long distances. This might be mode of formation of convergent margin related rift basins in two stages, viz. a strike slip faulting followed by rift formation. The Rio Grande rift in the western USA is located east of Basin and Range province which are also related to plate tectonic process along the western margin of North American plate. It occurs along the margin of Colorado plateau, which is also a weak plane. Similar gravity signatures have also been reported for Proterozoic rift systems such as Mid-continent gravity high of USA and adjoining gravity lows as discussed in Section 3.8 indicating almost similar gravity signatures for rift systems throughout the geological period.

Fig. 2.35(a) Location of Rio Grande rift along eastern margin of Colorado plateau (Cordell, 1978).

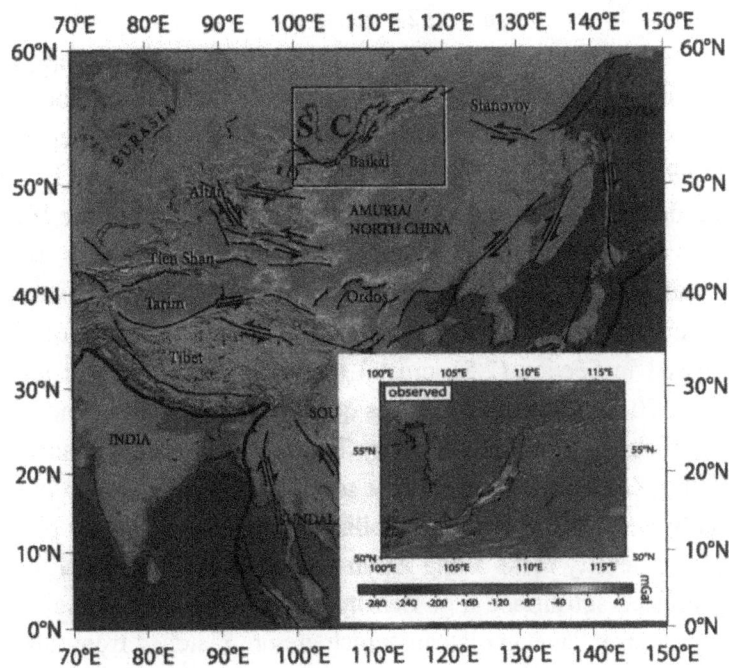

Fig. 2.35(b) Topographic shaded relief map of Himalayan convergence zone and Baikal rift along eastern margin of Siberian Craton (SC). Inset shows grey shaded map of Bouguer anomaly of Baikal rift, showing high over the southern part and low over the northern part (Petit and De'verche're, 2006).

2.10 Compressional (Convergence) Tectonics: Recent Convergent Margins and Subduction Zones

Convergent margins and subduction zones are the best examples of compressional tectonics (Fig. 2.23(d)). They are associated with thrusting and shortening leading to mountain building and crustal thickening. Due to thrusting of subsurface rocks, these regions are important for mineral occurrences and are seismically active regions. There are basically three types of convergent margins depending on types of plates involved in it as described below.

2.10.1 Continental-Continental Convergence – Himalaya and Tibet and Alps Mountains

In these cases, the colliding plates on both sides are continental of almost same density and therefore, subduction of one under the other becomes difficult resulting in deformation of crustal rocks in the form of folds, thrusts and nappes. Two cases of such collisions are described below.

(i) Indian and Eurasian Plates: Himalaya and Tibet

Himalaya represents a collision zone between two major plates, namely Indian plate towards the south and Eurasian plate towards the north (Figs. 2.35b) whose effects are seen over a large area up to Baikal rift in the north. This collision gave rise to Himalayan orogeny that has been discussed with details in Chapter 6. However, crustal model along a profile is given below to introduce the effects of continent-continent collision. Satellite free air anomaly over Himalaya (Fig. 6.6) along a profile 90° E longitude is converted to Bouguer anomaly (Rajesh and Mishra, 2003) using the elevation data along this profile from GTOPO-30 (Gesch et al., 1999) and controlled from available ground gravity data over Tibet and India (Sun, 1989; NGRI, 1975) for medium-short wavelength anomalies. The Bouguer anomaly along this profile and elevation data are shown in Fig. 2.36, which shows almost a constant elevation of 5000 m and almost constant regional Bouguer anomaly low of about – 525 mGal over Tibet, respectively indicating that Bouguer anomaly low may be caused by crustal thickening due to isostasy. The Bouguer anomaly along this profile is modeled using a three layered crustal model constrained from the results of INDEPTH seismic and magnetotelluric surveys as discussed in chapter 6 (Zhao et al., 1993; Hauck et al., 1998). Modelling is carried out using software (SAKI) due to Webring (1986). The computed field due to crustal section given at the bottom of the Fig. 2.36 matches quite well with the observed field. As this crustal model is constrained in parts from the results of seismic investigation and isostatic considerations, it can be safely assumed as truly representing the crustal section in this region. This model provides a crustal thickness of about 70-71 km under Tibet. As discussed in Section (2.5), crustal thickness required for a full compensation under Tibet is also about 70-71 km, which matches quite well with the thickness of the modeled crust. The high density rocks along Main Boundary Thrust (MBT), Main Central Thrust (MCT) and Indus-Tsangpo Suture Zone (ITSZ) towards the south suggest thrusting of high density rocks along them, which is a general characteristic of collision zones between two plates as shown in Fig. 2.23(d). Similar high density rocks along Altyn Tagh and Kunlun faults towards the north (Fig. 6.5(a)) suggest thrusted subsurface high density rocks along them. Small wavelength and small amplitude anomalies over Tibetan plateau, which are not matched from the computed field, may represent surface or shallow density structures, which require detailed investigation based on local geology.

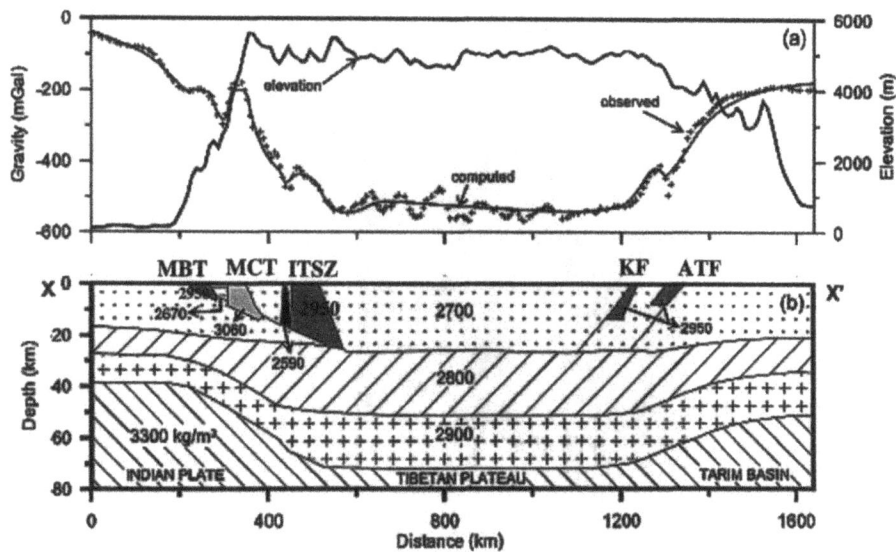

Fig. 2.36 A Bouguer anomaly and elevation profile along 90 °E longitude over Himalaya and Tibet. An inverse correlation between the Bouguer anomaly and elevation, and free air anomaly being close to zero qualitatively suggest isostatic compensation. The crustal model and the corresponding computed gravity field are shown for comparison with the Bouguer anomaly. Densities of various layers are given in the model in kg/m³. High density shallow bodies are associated with thrusts as follows: MBT- Main Boundary Thrust; MCT-Main Central Thrust; ITSZ-Indus Tsangpo Suture Zone; KF- Kunlun Fault and ATF- Altyn Tagh Fault.

(ii) European and African Plates: Alps Mountains

Alps Mountains also represent a continent-continent collision between European plate, a part of the Eurasian plate and African plate. However, collision and subduction is highly complex in this case due to presence of several small micro plates such as Antolian, Adriatic, Meditteranean blocks in between the two large plates. Several studies of the gravity field of different parts of the Alps have been reported (Holliger and Kissling, 1992; Braitenberg et al., 2002; Lillie et al., 1994). Ebbing et al. (2006) have provided crustal and lithospheric density structures across Eastern Alps in Italy (Fig. 2.37(a); Tauern Window) based on gravity and geoid data (Fig. 2.37(b)) constrained from the results of seismic investigations. It shows a gravity low primarily related to crustal thickening as in case of recent orogenic belts (Fig. 2.37(b)). Crust in the central part thickens to 52-53 km, which is much less compared to Himalayan-Tibet belt and Andes Mountains as described above and below, respectively. The lithospheric keel at about 220 km is anomalous and probably represents the effect of convergence from both sides. It may represent the subducted part of lithosphere as shown by (Holliger and Kissliing, 1992) in central Swiss Alps where lower crust and Moho of Northern Alps have been shown to subduct under upper and lower lithosphere of Southern Alps showing subduction from the north to the south. In present case also the faults related to Tauern window suggest subduction from the north to the south. A Moho map of Europe (Cloetingh et al., 2006) does not show any large scale crustal thickening as in case of Himalayas and Tibet as described above. Maximum crustal thickness of about 50 km is reported from eastern Alps north of Italy and Adriatic Sea. Absence of large scale crustal thickening and absence of plateau development in this section as in case of Himalaya and Tibet and Andes suggest more of internal deformation due to occurrences of small micro plates in between European and African plates. It can also be attributed to slow movements of the two plates involved in convergence in this case. Modeling of geoid data has provided density structure of lithosphere, which is important for geodynamic studies.

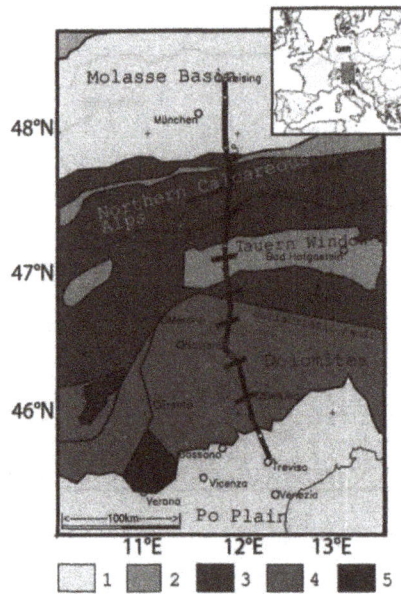

Fig. 2.37(a) Geology of the Eastern Alps with TRANSALP seismic profiles used to model density structure. 1= Molasse basins, 2= Penninic units, 3 = Austroalpine and Helvesic units, 4= Southern Alps, 5 = Plutonic intrusions. (Ebbing et al., 2006).

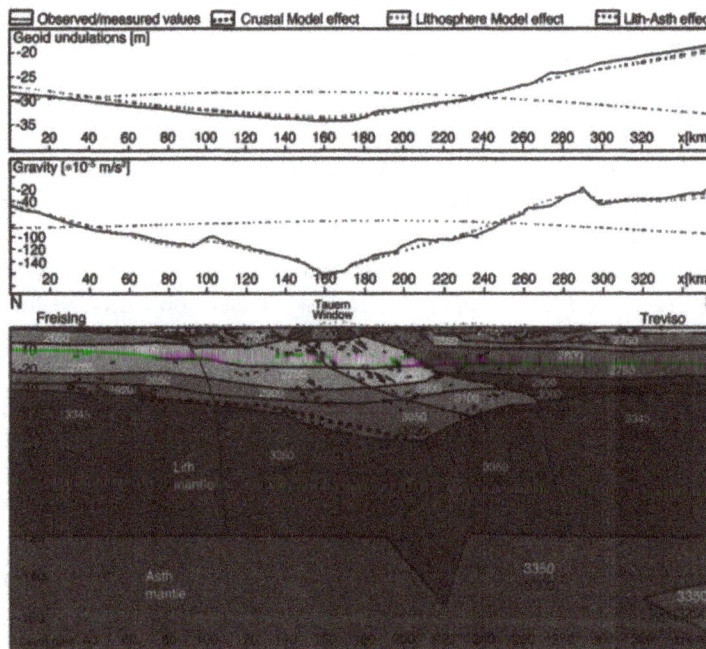

Fig. 2.37(b) Cross section through 3-D density model of geoid (topography reduced) and gravity field showing asthenosphere-lithosphere boundary with a keel at 220 km due to convergence. Crustal thickness up to 50 km is shown under Southern Alps due to convergence and collision. White numbers indicate density values in kg/m³. Alternative densities (black) for the mantle are used to demonstrate the combined effect of crustal geometry and lithosphere-asthenosphere boundary. Black lines are line drawings of vibroseis experiments for Moho (TRASALPS; Working, 2002; Ebbing et al., 2006) and dashed lines are from receiver function analysis (Kummerow, 2002).

2.10.2 Oceanic-Continental Convergence:
Nazca and South American Plates and Andes Mountains

In this case, the convergence occurs between an oceanic part of a plate and continental part of the other plate unlike previous case, where both parts are continental. In this case, due to higher density of the oceanic part, it subducts under the continental crust as shown in Fig. 1.2. In these cases, the margins of the continents such as western margin of South America as described below qualifies for being active margin.

Andes Mountain, along west coast of South America between Nazca and South American plates, is same as Himalayas between Indian and Eurasian plates. In this case, oceanic crust of the Nazca plate is subducting under the South American plate, therefore, it represents a typical oceanic-continental convergence. This region has been extensively studied by scientist world over (Oncken et al., 2006) as it forms a typical ocean-continent interaction. Gravity anomaly of Andes was studied in great detail (Grow and Bowin, 1975). Recently large scale magnetotelluric measurements and seismic investigations have led to delineation of crustal structures in great detail, which suggested a crustal thickness of about 60-70 km under Altiplano-Puna regions (Yuan et al, 2002) which is similar to Tibet plateau in case of Indian and Eurasian plate convergence. A gravity profile along 22° 17'S starting from oceanic part to continental central Chilean margin is given in Fig. 2.38 (Prezzi et al., 2009), which shows a gravity low (L1) over the oceanic part related to fore arc basin and trench. The gravity high H1 is modeled due to thickening of the high density oceanic crust along continental margin and crustal thinning along the margin due to erosion of crust related to subduction. The gravity low L2 is related to crustal thickening of about 75 km. The Altiplano-Puna region shows high elevation and large gravity low (–400 mGal, Prezzi et al., 2009) indicating the effect of isostasy. An elevation of about 4-5 km in this section suggest a crustal thickness of 65-70 km based on Airy's root model as described above in Section 2.5, which in fact has been suggested based on magnetotelluric and seismic studies. It may be noted that in case of Himalaya and Tibet also, the elevation is approximately isostatically compensated, which is a characteristic of recent orogenic belts. The lithosphere-asthenosphere boundary based on geoid data suggests lithosphere thickness of about 80 km on oceanic side and ~160 km towards the continent. They have introduced subducted slab lithosphere of 3.38 Mg/m^3 (g/cm^3) on the oceanic side and asthenospheric wedge of 3.25 Mg/m^3 (g/cm^3) on the continental side that are related to the subduction process. The difference in gravity anomalies of continental-continental convergence as a central gravity low (Fig. 2.36) with that of oceanic-continental convergence as gravity high with flanking lows may be noted. Such cases where margin of the continent is involved with plate tectonic processes are termed as active margins. Andaman-Sumatra-Java Subduction Zone (Chapter 6) is another case of oceanic-continental convergence that is described in great detail in Section 8.4 with regard to Sumatra earthquake of 2004 and related tsunami that is an extension of Himalayan collision with zone between the Indian and the Eurasian plates.

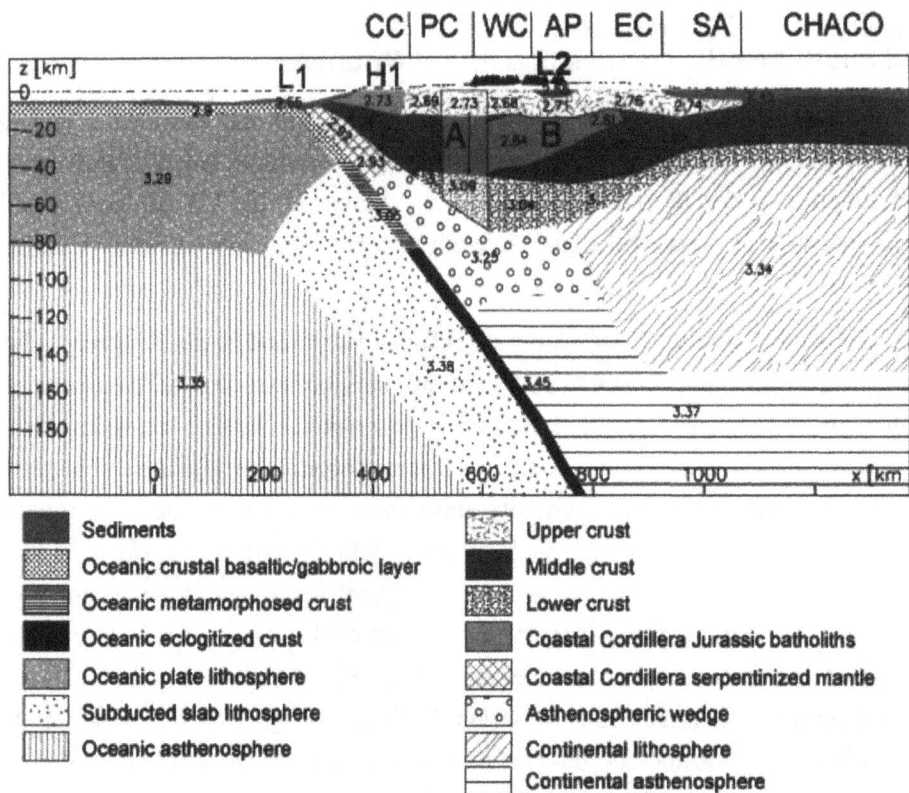

Fig. 2.38 Cross section through 3D density model across Chile trench and Andes along 22°17' S with layer densities as given in the figure in Mg/m³ (g/cm³). It shows subduction of oceanic Nazca plate (Prezzi et al., 2009) and provides a gravity high, H1 in the central part due to thickening of oceanic crust related to subduction flanked by gravity lows, L1 and L2 due to sediments in the trench (fore arc basin) and crustal thickening up to ~75 km, respectively. Subduction zone coincides with the hypocenter of earthquakes almost from 10 km up to 130 km (Hackney et al., 2006).
WC: Western cordillera, AP: Altiplano-Puna, EC: Eastern Cordillera, SA: Subandean Ranges, A: Atacama block, B: Altiplano-Puna Partial melting zone and Triangles: Active volcanoes on surface over partial melting zone.

2.10.3 Oceanic-Oceanic Plateau/Ridge Convergence: Kermadec Ridge and Trench (Mariana Type Subduction)

There are situations when oceanic plates encounter linear oceanic plateau whose crustal density structure is similar to that of the oceanic crust differing only in thickness such as in case of large subduction zones (Fig. 5.13) in Pacific Ocean. In these cases, both colliding plates have almost similar crustal structures. One such situation exists along New Zealand-Kermadec ridge between Australian and Pacific plates. A gravity profile across Kermadec trench and ridge is given in Fig. 2.39 along with the computed crustal section and densities of the various layers (Davy, 1990). It has been constrained from seismic sections and shows the model of subduction under the ridge. This type of collision shows a sharp gravity high (H1) over the ridge (Plateau) and low (L1) over the trench where the subduction starts. Both the gravity high and low are almost of same amplitude and form truly a paired gravity anomaly characteristic of compressional tectonics as given in Fig. 2.23(d). The gravity high is due to high density and thicker ridge side and the gravity low is due to the trench and sediments in it. It may be noted that in the present case the subduction is at a very high angle (70-80°) and therefore, it is known as Mariana type of subduction in comparison to Mariana arc along Philippines islands (Fig. 5.13). Such subduction zones generate low stress and are therefore, less prone to seismic activity, compared to previous ones such as Himalaya and Andes which are seismically very active. However, both kinds of Himalaya and subduction can exist across same subduction zone making some sectors of a subduction zone more prone to seismic activity compared to others. In fact, angle of subduction may change considerably from one part to the other in a subduction zone.

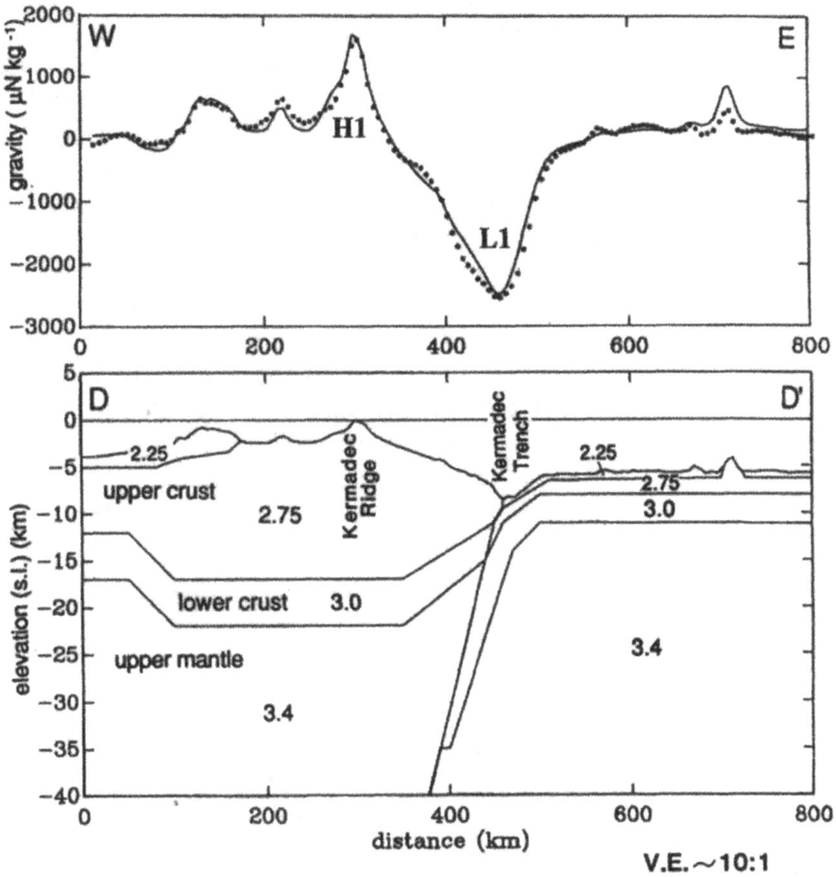

Fig. 2.39 A two dimensional density model across Kermadec ridge and trench east of New Zealand between Pacific and Indo-Australian plates (Davy, 1990). Both sides show almost similar density structures differing only in thickness and therefore, provide sharp gravity high and low due to high angle subduction known as Mariana type subduction causing low stress and therefore, less prone to seismic activity.

2.11 Ancient Collision Zones: Proterozoic and Paleozoic Convergence

Based on integrated geophysical and geological study of Archean-Proterozoic terrains world over, it has been established that plate tectonics was operative at least since early Proterozoic period with some modifications in its operational details such as plates might have been smaller in dimension and due to more heat flux, there might be more vigorous movement of plates with low angle subduction or almost horizontal movement providing transpressive zones in Archean Greenstone Belts. The fold belts or mobile belts of these periods are identified as collision zones between cratons on either side of them. However, their geophysical signature differs considerably from present day collision zones. Some of them, in case of Indian shield, are briefly discussed in Section 2.8 while describing Bouguer anomaly map of India and further details with crustal models based on integrated data sets are given in Chapter 7. Gravity signatures over some important ancient collision zones are described below to understand differences with those of recent times as described above.

2.11.1 Grenville Province, Canada

Grenville Province in the Canadian shield represents one of Meso Proterozoic collision zone (1.1 Ga), which is characterized by a linear gravity low at its contact with superior province towards the north (Thomas, 1985). It is followed by a prominent high towards the south (Fig. 2.40). However, as explained in Section 2.7 (Fig. 2.23(d)), the gravity lows along ancient collision zones are less prominent and associated gravity highs are more prominent due to uplift and erosion compared to recent collision zones (Fig. 2.36). As explained, ancient collision zones are characterized by paired gravity anomaly consisting of a high and a low over the suture caused by high density intrusive Section 7.1 and crustal thickening, respectively. It has also been suggested that high would be of large amplitude and small wavelength (width) compared to the gravity low. These are known as paired gravity anomalies consisting of a high and a low associated with ancient collision zones. It may be noted that the gravity highs are of larger amplitude and small wavelength compared to the associated gravity low. The subsurface density structure (Fig. 2.40) suggests a higher bulk density of collision zone and produces the gravity high (H1) while crustal thickening produces the gravity low (L1). The large wavelength nature of gravity low over adjoining older cratons can be explained due to low bulk density of cratons. Similar, gravity highs due to high density rocks and associated gravity lows over adjoining cratons have also been observed in case of Proterozoic collision zones in India as described in Chapter 7. The bulk high density of ancient collision zones can be attributed to thrusting of high density lower crustal rocks and collision related high density intrusives like anorthosites etc. .

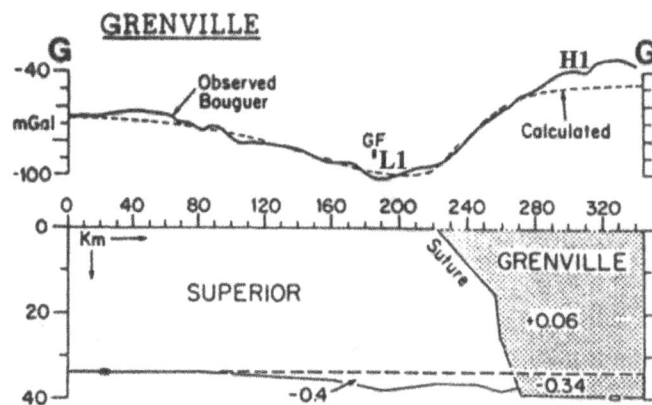

Fig. 2.40 A gravity profile across superior and Greenville provinces in Canada showing paired gravity anomaly L1 and H1 due to crustal thickening and high bulk density of the crust across the suture between the two provinces of Meso Proterozoic period (Thomas, 1985).

2.11.2 Appalachian, USA and Caledonian, Europe Orogenies: Wilson Cycle

The most important and well established ancient collision zone relates to collision of North America and Europe, which gave rise to Appalachian and Caledonian orogeny, on respective sides during Paleozoic period. A Cambrian-Ordovician suture (Whittington, 1973) was predicted based on similarity of fauna on either side of it, with those of North America and Europe and lack of it after that as shown in Fig. 7.1. Subsequently, Appalachian section along the east coast of the USA was an ocean, which is suggested by large sedimentary sections and large caves like Luray caves in dolomite rocks of early Ordovician period. Later convergence gave rise to Appalachian and Caledonian orogeny on two sides. A further opening of Atlantic Ocean between two plates at about 180-200 Ma is well documented based on sea floor spreading anomalies which is still continuing. This kind of opening and closing of oceans are known as Wilson Cycle, a simplified hypothesis proposed by

Prof. T.J. Wilson to explain formation of sedimentary basins and orogenic belts. The cycles of opening and closing in case of Appalachian-Caledonian suggest a cycle of about 150-200 Ma, which may or may not be applicable in different cases. However, subsequent studies of basins and fold belts suggest this process is more complex than simple opening and closing of oceans. This collision zone being intermediate to present day and Proterozoic collision zones represents a connection between the two. The Bouguer anomaly of a part of Appalachian Mountains and adjoining regions is given in Fig. 2.41, which shows a pair of linear gravity highs (H1, H2 and H3) and lows (L1 and L2), which are similar to those reported above for Greenville province and can be considered to be caused by similar sources. In this case also, it may be noted that amplitude of gravity highs is larger compared to the adjoining lows which is much less compared to those of recent collision zones as described above. Similar linear gravity highs and lows extend through out the east coast of North America, which coincide with the Appalachian Mountains and adjoining Appalachian basin, respectively (Jachens et al., 1989) and can be attributed to high bulk density of crustal rocks and intrusives and crustal thickening, respectively. Crustal thickening up to 50 km in this part has been reported by Taylor (1989). In case of Caledonian Orogeny, the primary gravity field is set of linear gravity lows (L1; Fig. 2.42; Ebbing and Oleson, 2005) related to Precambrian rocks metamorphosed and thrust due to Caledonian Orogeny and crustal thickening of 10-15 km has been reported in the eastern part of Caledonian Orogeny compared to the western part (Kinck et al., 1993). The gravity highs related to such orogenies as discussed above and observed in case of Appalachian orogeny are not observed in this case, which might be due to absence of high density intrusives in this section that was confined mainly in the Appalachian section. However, small gravity highs H1 (Fig. 2.42) along continental edge and eastern margin of Caledonian Orogeny almost sub-parallel to the gravity low, L1 may represent some high density intrusives related to Appalachian-Caledonian Orogeny and gravity low is related to Caledonian Orogeny might be synonymous with that observed over Appalachian basin (L1 and L2; Fig. 2.41). The continental shelf off Norway, however is subsequently affected by rifting of Greenland from Norway at about 55 Ma in response to Icelandic plume and therefore signatures of older events offshore is difficult to decipher.

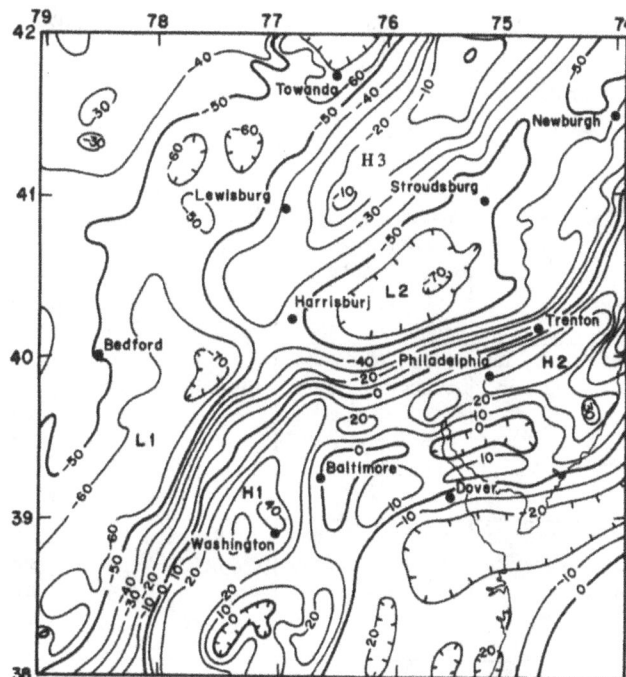

Fig. 2.41 Bouguer anomaly map of a part of east coast of USA showing gravity highs and lows, H1, H2 and H3 and L1,and L2, respectively over Appalachian Mountains and Appalachian basin, representing high bulk crustal density and crustal thickening, similar to Greenville Province (Fig. 2.40). Such gravity highs and lows occur all along the east coast of USA. (Bouguer Anomaly map of USA, USGS, 1964).

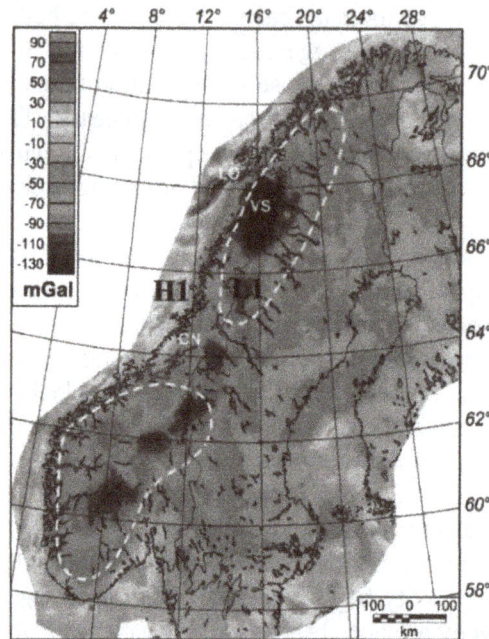

Fig. 2.42 Bouguer anomaly map of Fennoscandia providing linear gravity lows (L1) over Caledonian orogeny and small high west of it along the coast (Ebbing and Olesen, 2005). [Colour Fig. on Page 764]

2.12 Comparitive Study of Different Gravity and Magetic Surveys and Expeced Anomalies for Different Targets and Temporal Variations in Gravity Field

Table 2.2 Details of Gravity and Magnetic Surveys and Expected Anomalies for Different Targets

Sl. No.	SURVEYS	PURPOSE/ TARGETS	LAYOUT AND SPACING	ACCURACY	REMARKS: RESOLUTION
I.	GROUND SURVEYS				
1.	**Regional Surveys**	Geodynamics: Large scale structures and their density/ susceptibility distribution	Along roads, station spacing 3-5 km. In hills can be reduced to 1-2 km. Magnetic: 1 km	± 1.0 mGal Elevation accuracy: < ±1.0 m. Magnetic: ± 5.0 nT	DGPS, for elevation and position location. Width(W)of Anomaly=100-1000km Amplitude (A) =10-100 mGal
2.	**Hydrocarbon Exploration: Reconnaissance Survey**	To limit areas for detailed investigation.	Along roads and tracks at about 2.0 km spacing (uniformly). Magnetic: 1 km	± 0.1 mGal Elevation: < 10 cm. ± 5 nT	DGPS with dual frequency system W=1-100 km A=1-10 mGal
	Detailed Survey	Along seismic profiles or grid surveys in limited areas	Profiles and station spacing: 100-500 m.	± 0.02 mGal (20 µGal) Elevation: ± 1-5 cm. < ± 2 nT	High Precession Geodetic survey or DGPS dual frequency survey. W=1-10 km A=0.1-10 mGal
	Special MocroGal Surveys	Small Targets (Gas occurrences)	Profile and station spacing: 10-100 m. (Grid Survey)	± 10 µGal Elevation:± 1-2 cm ± 1-2 nT	High Precession geodetic survey. W=100 m-1 km A=100-1000 µGal

Table 2.2 *Contd...*

3.	**Mineral Exploration, Environmental and Engineering Geophysics: Reconnaissance Survey**	To limit area for detailed surveys	Profile spacing: 100-1000 m. Station spacing: 10-100 m.	± 0.1 mGal Elevation: ± 10 cm. ± 5 nT	DGPS with dual frequency system. W=10-100 m A=1-10 mGal
	Detailed Investigation (MicroGal Survey)	To locate target and its extent	Grid survey with spacing 1-10 m.	± 10 µGal. Elevation: ± 1-2 cm. < ± 1 nT	High precession geodetic survey or DGPS dual frequency survey. W=1-10 m A=10-1000 µGal
4	**Crustal Deformation and Subsurface Mass Transfer**	Temporal elevation changes (Table 2.3)	Repeat fixed stations indicate uplift/ subsidence	< ± 10 µGal (Table 2.3)	Subsurface mass transfer such as hydrological changes or changes in oil fields etc.,
II	**MARINE SURVEYS:**				
5.	**Regional Surveys**	Geodynamics: mid oceanic ridges, subduction zones etc.,	Individual long Profiles with spacing: 10-100 km.	± 1 mGal. (wavelengths 500 m and more). ± 1 nT	High Precession DGPS Dual frequency system. Position Location: ± 5 m.
6.	**Hydrocarbon Exploration Reconnainance/ detailed Surveys**	Along Seismic profiles. To locate target and its extent	Profile spacing 1-10 km Resolution: <1 km	< ± 1.0 mGal (Wavelength 250 m and more) ± 1 nT	Same as above with position location: ± 1 m.
III	**AIRBORNE SURVEYS:**				
7.	**Regional Surveys**	Geodynamic studies, large scale structures	Profile spacing: 5-10 km. Height: 1-5 km above msl.	< ± 1-2 mGal. (wavelengths 10-20 km) < ± 1 nT	Same as above with Position location: ± 5 m.
8.	**Hydrocarbon / Mineral Surveys Reconnaissance surveys**	Delineate structures for detailed surveys	Profile spacing: 1-2 km. Height: 10-100 m above highest topography or a.m.s.l.	± 1 mGal. (wavelengths 10-20 km.) Resolution about 1 km ± 1 nT	Same as above with Position location: ± 5 m.
9.	**Detailed Survey**	Locate the target and its extension	Profile spacing 10-100 m. Height 10-100 m. Resolution:<1 km	< ± 1 mGal < ± 1 nT	Same as above with Position location: ± 1 m.
IV	**SATELLITE SURVEYS (REGIONAL)**				
10.	**Magnetic Satellite (MAGSAT)**	Long Wavelength Magnetic Anomalies and External Magnetic Field Model	550-300 km ≈ 400 km Launched by NASA in 1979.	Resolution: $2° × 2°$ ± 5 nT	W = 100-1000 km A = 1-10 nT Amplitude
11.	**Satellite Altimetry**	Free air / Anomaly Map of Oceans (ERS-1)	800 km	± 5 mGal Resolution: 20 X 20 km	W =100-1000 km A = 10-100 mGal
12.	KMS-2001	Improved gravity model over oceans	Combining ERS-1 Geosat and ship track data along coasts	< ± 5 mGal (Widely used gravity data over oceans)	1 arc min 2 X 2 km

Table 2.2 *Contd...*

13.	Satellite Tracking (GPS)	Worldwide gravity model EGM-96	400 km (Combining all available data sets over oceans and continents)	± 5 mGal (Widely used worldwide gravity model) Resolution: 30' × 30' (50 × 50 km)	**W = 100-1000 km** **A = 1-100 mGal**
14.	**CHAMP / GRACE Satellite Gravity and Magnetic Measurements (Satellite-Satellite Tracking)**	Long Wavelength Magnetic and Gravity Fields and their variation with time	400-450 km + Airborne Mag EMAG 3 (4km above geoid)	± 1 mGal and 10 cm in geoid (EIGEN-2 Model-Resolution: 5° × 5° Block Magnetic: ± 2-3 nT 3.7 × 3.7 km (EMAG 3)	(iii) Climatic changes (ii) Sea Level Changes (iii) Ice mass loss balances (iv) Mass changes at Subduction zones for seismic activity
15.	**Satellite Gradiometry (Future Programme)**	High Precession 3-D Full Tensor Gradient Measurements	250 km	± 1-2 mGal. Geoid= ± 1 cm (Resolution ≈ 65 km)	GOCE Satellite to be launched in 2010 by ESA.

mGal, μGal and Gamma are units of gravity and magnetic fields, respectively.

1 mGal = 10^3 μGal; 1 mGal = 10^{-5} m/s^2; 1 Gamma = 1 nT (nano-Tesla) = 10^{-5} Orersted or Gauss.

Note: Magnetic method is described in Chapter 3, but the survey procedure for gravity and magnetic method being similar, both methods are briefed in the same table for comparison.

Table 2.3 Types of surface gravity measurements and their potential applications in studying time varing gravity field (Tiwari V.M., Pessonal comunication)

	Absolute gravity meter (AG: FG 5)	Superconducting gravity meter (SG: GWR type)	Relative gravity meters (CG: Scintrex, CG3M, CG5M)
Precision	≤1 μGal	0.00 1 μGal	I μGal
Accuracy	1–3 μGal	0.1 μGal	10 μGal
Drift	0	0.01—0.05 μGal/day (variable)	3–10 μGal/day (variable)
Setup time and Operation	AG is a transportable instrument that operates in a controlled temperature environment. It can be Set up in 1–3 h. 1–3 days measurements are desirable to get best results	Setup ofSG requires a few days and some weeks to stabilize. It is housed at one place and needs calibration and drift monitoring by AG for better accuracy	CG is a field insturment that can be Set up in a few minutes. A network of stations is used for measurement loops. Requires repeated measurements at a reference Station every few hours
Dominant noise	Microseismic and man-made noise	Environmental effects	Environmental effects, microseismic noise, vehicle shaking, drift
Possible applications	Tidal, hydrological changes, pre-co-post seismic changes, glacial isostatic adjustment, tectonics, oil, gas, and geothermal reservoir monitoring, volcanic mass changes	Tidal, polar motion, hydrological changes, atmospheric and ocean loading's core processes, pre-co-post seismic changes, free oscillation of the glacial isostatic adjustments, volcanic mass changes	Tidal, hydrological changes, co-seismic changes for large quakes, volcanic mass changes

Magnetic Methods

3.1   Introduction and Basic Principles

The tendency of a freely suspended magnet to orient itself in a particular direction suggested beyond doubt that the earth has its own magnetic field, which is basically bipolar in nature. The idea that the earth behaves like a magnet started first with the work of Sir William Gilbert as early as in the early seventeenth century, who considered the earth as a spherical magnet. Subsequent works of Gauss in the nineteenth century brought out the important features of the earth's magnetic field. As the first approximation, the earth's magnetic field can be considered to arise from a short powerful bipolar magnet at the center of the earth inclined 10.5° from the axis of rotation (Fig. 3.1(a)). However, careful studies suggest that it consists of a relatively small non dipole component due to processes above the earth's surface which is comparatively irregular and rapidly changing. The former is known as the earth's main field and latter as external field of the earth.

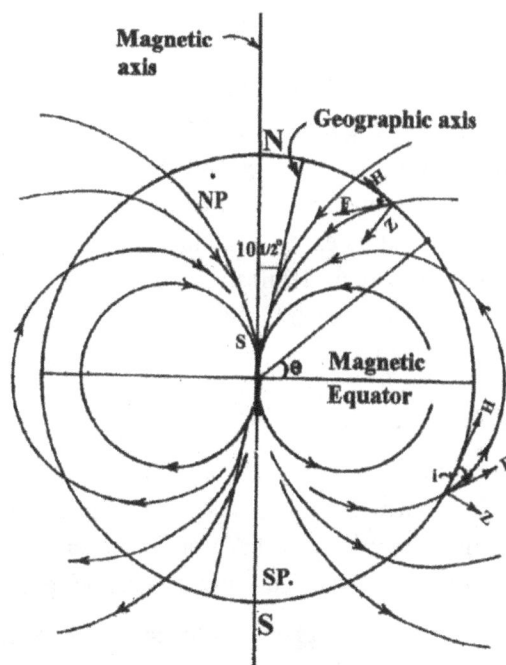

Fig. 3.1(a) Magnetic lines of force of the best fitting dipole at the center of the earth inclined 10.5° from geographic axis. Geomagnetic poles, N and S are imaginary points where this dipole intersects the surface of the earth. North and south magnetic poles, NP and SP are points where inclination (I) of earth's magnetic field is +90° and -90° respectively while at magnetic equator it is zero. θ is the magnetic latitude or inclination of earth's magnetic field at that point. It also shows Total intensity of earth's magnetic field (F) and its Horizontal (H) and vertical (Z) components on the surface of the earth.

Magnetic method has been widely used for geodynamics and resource exploration. It is based on the differences in the magnetic characteristics of various rock types, which can be measured and used to differentiate them. In geodynamics, the most important applications relate to determination of paleopositions of various continents based on direction of magnetization in rock formations of different periods. It thus provides the drift history of different continents, which provided the best proof of continental drift/plate motion. Another application of magnetic method in geodynamics lies in delineating sea floor spreading magnetic anomalies across mid oceanic ridges based on marine magnetic surveys, which provided the basis for modern plate tectonics. In resource exploration, it has been used over large basins for hydrocarbon exploration to delineate basement structures and delimit areas for more detailed and involved exploration using seismic methods. In case of mineral exploration, it is used as direct method of detection of magnetic minerals such as iron ores etc., while in case of base metals or other deposits, it is used indirectly to delineate structures like faults etc., with which these minerals might be associated. Similar is the case for ground water exploration in hard rock areas and near surface geophysics for selection of engineering sites where it is used to demarcate faults, fractures etc., which might be water bearing and can be followed up by electrical surveys for further confirmation. In selection of sites for any large scale construction like dams, nuclear plants etc., or for nuclear waste disposal, sites with faults and fractures are avoided.

In case of environment, the paleo positions of continents obtained from direction of magnetization in rocks of different periods indicate paleoclimatic conditions during those periods and can be used to correlate with flora and fauna (fossils) of that period (Blackett, 1961). This information also helps in mineral exploration, which are confined in specific climate and latitude based on their paleolatitude positions. In this regard, direction of magnetization and paleo position of lake sediments can even provide information about recent climatic changes like glaciation, desertification etc. The study of magnetic behavior of soil can guide about its fertility and its suitability for specific crops. Some bacteria produce magnetic minerals, which can be detected based on magnetic measurements. Pollution of environment by steel plants and its extent can also be detected based on magnetic surveys (Evans and Heller, 2003). Magnetic surveys can delineate archeological sites associated with graves, klinks etc., Some of these applications are demonstrated in section 3.8 under 'Field Examples' and in Part II.

3.1.1   Vectorial Representation

The earth's main magnetic field is referred by total magnetic intensity or the earth's total field and is denoted as T or F. It is a vector field represented with respect to angle D and I between geographic north and magnetic north and magnetic north (meridian) and the earth's magnetic field known as declination and inclination of the earth's magnetic field, respectively (Fig. 3.1(b)).

At any point on the earth's surface, the magnetic field can be conveniently represented by a Vector (T). This vector starts at the point of observation and extends in the direction of the line of force of the earth's field. Its vector length is proportional to the strength of the total intensity of the field. The total intensity can be resolved into two components in the plane of the magnetic meridian (magnetic north) and vertical to it known as horizontal component (H) and the vertical component (Z), respectively. The angle between the field (T) and the vertical (Z) is known as magnetic inclination (I) or dip of the earth's field at that place. The horizontal component can be further projected on geographical north (X) and east (Y) with reference to the angle between the magnetic north and the geographical north known as declination D at that place.

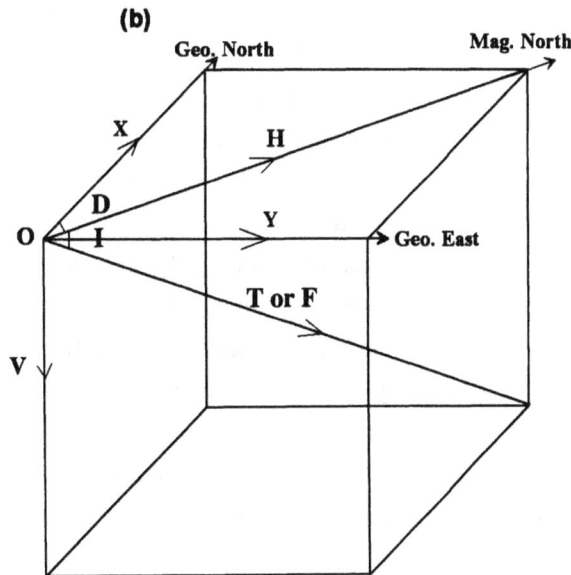

Fig. 3.1(b) Vectorial representation of magnetic filed in the northern hemisphere. F or T represents the total intensity of earth's magnetic field while H and Z are its component towards magnetic north and vertical to it. X and Y are components of H towards geographic north and east. I and D represent the inclination and declination of earth's magnetic field with respect to magnetic north and geographic north respectively.

The quantities T, H,Z, D and I are called the geomagnetic elements which are related as follows (Fig. 3.1(b)):

$$H = T \cos I$$
$$Z = T \sin I$$
$$T = (H^2 + Z^2)^{1/2}$$
$$T = Z \sin I + H \cos I$$
$$X = H \cos D = T \cos I \cos D$$
$$Y = H \sin D = T \cos I \sin D \qquad \qquad .....(3.1)$$
$$H = (X^2 + Y^2)^{1/2}$$
$$\tan I = Z/H$$
$$\tan D = Y/X$$
$$T = (X^2 + Y^2 + Z^2)^{1/2}$$

When the earth's field or its components are referred with respect to angles I and D, they are known as vector field while their numerical values are referred to as scalar field. These are the relations between different elements of the earth's magnetic field. The same relations hold between their anomalies that are usually denoted as ΔZ, ΔH and ΔT. However, as we deal only the magnetic anomalies in geophysics, they can be simply denoted as Z, H and T

3.1.2 Geomagnetic Pole and Magnetic Dip Pole

The earth's magnetic field is a dipole in nature with its north and south poles. These two points on the globe are known as geomagnetic poles, which are referred to as N and S geomagnetic poles similar to geographical poles. Geomagnetic poles are the imaginary points on the surface of the earth where the inclined dipole passing through the center of the earth intersects the surface of the earth (Fig. 3.1(a)).

As the dipole is inclined 10.5° from the earth's axis, the geomagnetic poles in 1995 was located at 79.3° N, 71.4° W (Northwest Greenland) and 79.3° S, 108.6° E (Antarctica). However, a freely suspended magnet like a dip needle does not remain vertical at these geomagnetic poles implying that the earth's magnetic field is not truly vertical at these poles instead, has a horizontal component to it. It is caused by the non-dipole component of the earth's magnetic field. Shifted from these points, there are points on the earth's surface, where the earth's magnetic field is truely vertical; these points are called magnetic dip poles or simply magnetic poles. Therefore, magnetic poles are the points on the surface of the earth where the horizontal components of dipole and non-dipole part of the earth's magnetic field cancels each other making the earth's magnetic field vertical. This implies that at magnetic poles, the inclination of the earth's magnetic field (I) is equal to 90°. They are shifted from geomagnetic poles and for 1995 epoch are located approximately at 78.9° N, 254.9° W, and 64.7° S, 138.7° E (McElhinny and McFadden, 2000). Contrary to this, the points where the earth's magnetic field is fully horizontal implying vertical component as zero and I = 0°, is called magnetic equator (Fig. 3.1(a)).

Equation 3.1 states that at the magnetic poles where I = 90°; T = Z and at the magnetic equator where I = 0°; T = H.

3.1.3 Variation of Geomagnetic Elements

The magnetic field at any point on the Earth's surface may be specified by three parameters eg. the total intensity T, declination D and inclination I or the two horizontal components X and Y and the vertical component Z. The variation of the magnetic field over the earth's surface is best illustrated by isomagnetic's charts i.e., maps showing equal values of the specific magnetic elements, which are periodically issued by geomagnetic observatories. Contours of equal intensity total field or in any of its components X, Y, Z and H are called isodynamics and that of equal declination and equal inclination are called isogonics and isoclinics.

Referring back to the vectorial representation in Fig. 3.1(b), the total field T becomes horizontal when I = 0° and becomes vertical when I = 90°. Therefore, by implication inclination varies from I = 0° at magnetic equator to I = 90° at magnetic poles. However, as we have two halves of the globe divided by magnetic equator, the inclination in the northern hemisphere is referred to as + ve and in the southern hemisphere, it is referred to as -ve. It is also referred to as downwards and upwards, respectively depending on how a magnetic + ve pole dips in northern and southern hemisphere, respectively. Therefore, inclination varies from 0 at the equator to +90° at the north magnetic pole and -90° at the south magnetic pole. The maximum value of the earth's magnetic field is about 0.6 (Oersted) at the north magnetic pole and 0.7 oersted at the south magnetic pole. The minimum value is about 0.25 oersted off South America. Variations of magnetic elements for a particular period and region can be obtained from IGRF code given at that site of International Association of Geomagnetism and Aeronomy (IAGA, 2006) based on latitude and longitude of the place. The total intensity and inclination and declination of the earth's magnetic field over Indian continent are given in Fig. 3.2 (a), (b) and (c) respectively, (IAGA, 2006), which can be used for interpretation of magnetic anomalies as discussed in Section 3.7.

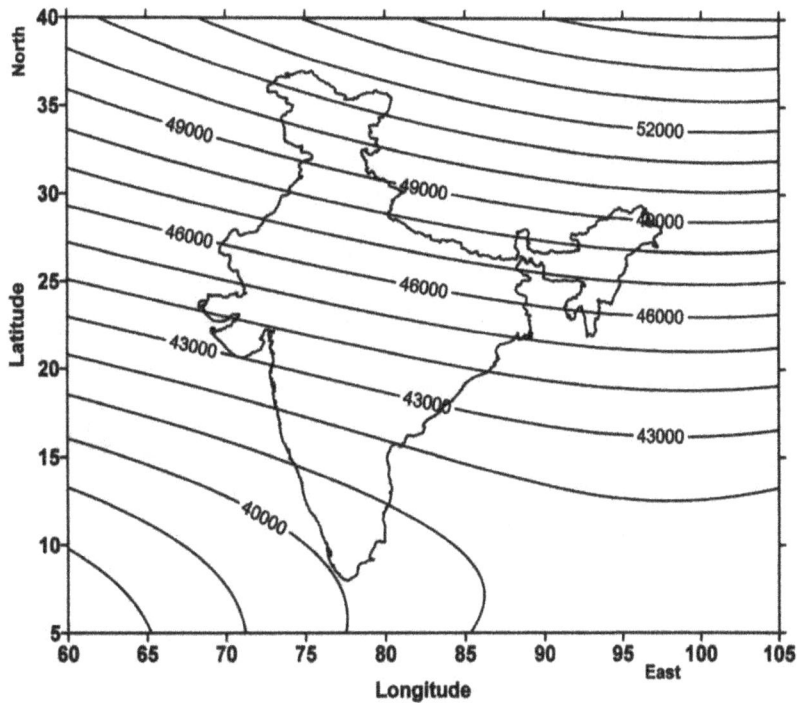

Fig. 3.2(a) Total intensity of earth's magnetic field over India which varies from about 40,000 nT to 51,000 nT increasing towards north (IGRF-10, IAGA , 2006).

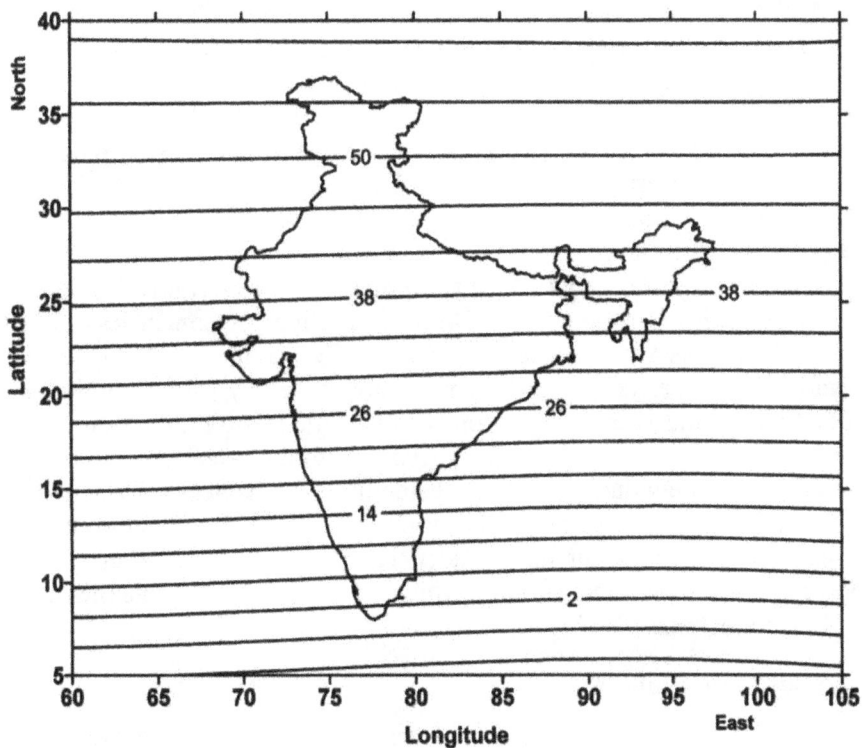

Fig. 3.2(b) Inclination of earth's magnetic field over India which varies from 0^0 N - 54^0 N increasing towards the north. Magnetic equator corresponding to I = 0^0 passes through south of India (IAGA, 2006).

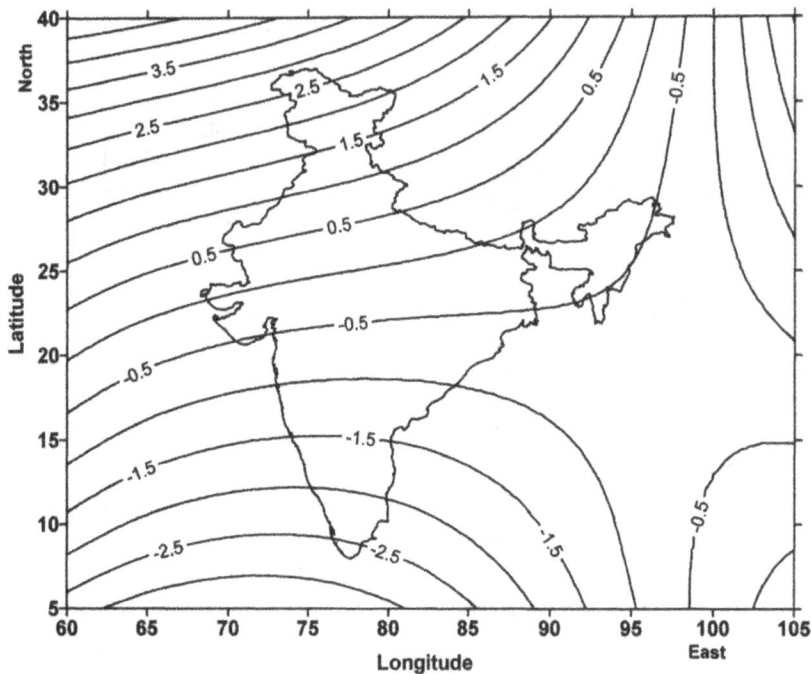

Fig. 3.2(c) Declination over India changing from almost -2.5^0 to 2.5^0 with magnetic north and geographic north being same in central India where $D = 0^0$. (IAGA, 2006).

3.1.4   Secular and Transient Variations of Geomagnetic Elements

The variations in the magnetic elements over longer periods are recorded continuously at geomagnetic observatories throughout the world. These records of the earth's geomagnetic elements show slow and continuous changes over long periods which are known as secular variation of the earth's magnetic elements. These variations are cyclic in nature, which show several large cycles over hundred of years. Besides cyclic variations over long periods, the magnetic elements also change from day to day and hour to hour of the day known as diurnal variations of the geomagnetic element. The daily records of the magnetic element show that the variations are slow and smooth on certain days while on some other days, they show large and irregular variations. The former is known as quiet days and the latter as disturbed days associated with magnetic storms. Diurnal variations are much less during nights compared to the day time known as the Lunar and the Solar daily variations. The magnetic storms, which causes violent variations in the geomagnetic elements are caused by solar activities, which last from a few days to a few weeks and years. The regions of magnetic storm on the Sun are known as Sun spots, which show a cycle of 11 years. During magnetic storms, there are large and sudden variations in the earth's magnetic field due to sources outside the earth namely solar activities during this period. Such periods are not suitable for magnetic surveys. A period of high sunspot activity and associated magnetic storm is likely to occur during 2010-12. The diurnal variations in the earth's magnetic field is a source of error in magnetic surveys, which is corrected by suitable measurements as discussed in Section 3.4 on corrections.

3.1.5   Origin of Geomagnetic Field

The major component (99%) of the earth's magnetic field originates from inside the earth while a very small fraction of approximately 1% originates from outside. They are referred to as the internal field or the main field and the external field respectively. The transient variations described above are associated with the external field, which originates from ionospheric part of the earth's atmosphere.

The internal field was initially considered to originate from permanent magnetization inside the earth, which is ruled out as it requires a very high intensity of magnetization of rocks to account for the known earth's magnetic field and none of the rocks are known to possess this kind of high magnetization. Moreover, only a small fraction of rocks (basic and ultrabasic rocks) occurring in the crust are supposed to possess magnetization which rules out the earth's magnetic field being caused by magnetization of rocks inside the earth. Therefore, several other alternatives were considered to account for the main field (internal field). Presently, the most acceptable view point is the dynamo theory, which proposes that the earth behaves like a self exciting dynamo. The earth's outer core is composed mainly of iron and nickel in a fluid state where convection in the presence of electric currents generates magnetic field. Elsasser (1946) and Bullard (1949) suggested that the electrically conductive iron core produces electric currents necessary to maintain the earth's magnetic field. It is now confirmed that magnetic field is generated by convection in the fluid iron core by numerical simulations which can generate self sustaining magnetic field with characteristics similar to the earth's magnetic field including reversal of the field (Buffett, 2000). The nature and amplitude of the earth's internal magnetic field is poorly understood. Recently, Buffett (2010) based on tidal dissipation, estimated the earth's magnetic field in outer fluid core as 25 oersted that is almost 50 times the average filed on the surface of the earth. This requires a strong convection in the outer fluid core indicating strong heat source to maintain it that is attributed partly to the residual heat from the earth's initial heat source when it formed from the molten state and partly from radioactive decay of long lived elements like potassium, uranium and thorium. Changes in the geomagnetic field can be attributed to changes in the convection pattern. One of the causes for changes in the convection pattern can be extra spin of inner solid core at a rate of about 0.3-0.5 degree per year with respect to other parts of the earth (Zhang et al., 2005). However, function of self exciting dynamo in earth's core to maintain the earth's main magnetic field is a complex subject and those interested may refer the books and literature as cited above.

3.1.6 Magnetic Force and Pseudogravity

As we know that if a magnetic material is placed in a magnetic field, it acquires its own magnetization due to induction. The rocks and minerals under the influence of the earth's magnetic field get magnetized depending upon their magnetic characteristics, which if properly measured can be used for their detection and exploration. However, to describe the magnetic properties of rocks and minerals we require certain measures, which are described below in the form of definitions.

Coulomb's law describes the magnetic force between two poles m_1 and m_2 separated by r cm as:

$$F = (m_1 m_2 / \mu r^2)\, \mathbf{r_1} \qquad\qquad .....(3.2)$$

It is similar to Newton's law of gravitation described in the previous Chapter. The pole strength m_1 and m_2 is expressed in emu and F is the force in dynes, $\mathbf{r_1}$ is a unit vector directed from m_1 towards m_2 and is used to show the direction of the force. It is important as magnetic force is both attractive and repulsive unlike the gravity field. It is attractive if the poles are of opposite sign and repulsive if they are of the same sign. It has no numerical value and therefore can be droped from the expressions as given below. μ is the permeability of the medium surrounding the poles and describes the dimensionless quantity with its value as 1 in vacuum. For practical purposes its value in air is also taken as 1. Therefore

$$F = (m_1 m_2)/\mu r^2 \qquad\qquad .....(3.3)$$

If two poles m_1 and m_2 each of strength 1 emu, are placed 1 cm apart in a vacuum (or in air), the force between them, is 1 dyne. The sign convention adopted for $\mathbf{r_1}$ is that a positive pole is attracted towards the earth's north magnetic pole and a negative pole is attracted towards the earth's south magnetic pole. The terms north-seeking and south-seeking are also used in place of positive and negative poles.

Equation (3.3) is similar to the gravitational force defined in equation (2.1) with a difference of mass instead of pole strength in case of magnetic. Poisson's relation expresses the actual relationship between the gravitational and magnetic potential as follows:

$$V = \frac{M}{G\rho}\frac{\partial u}{\partial \alpha} = -\frac{M}{G\rho}g_\alpha$$

Where G is the gravitational constant and ρ is the density, M is the magnetization in direction α and $g\alpha$ is the component of gravity in the same direction. Differentiate this expression in vertical direction:

$$\frac{\partial V}{\partial z} = \frac{M}{G\rho}\frac{\partial g_z}{\partial z}$$

.....(3.3)

$\partial V/\partial z$ is the vertical component of the magnetic field and $\partial g_z/\partial z$ is the derivative of the earth's gravity field, which can be computed as described in Section 4.2. Using this expression one field can be transformed to the other and vice-versa. In case, the vertical gradients of two fields are computed as given in Section 4.2.4, M/ρ can be computed from Equation 3.3 or M or ρ in case the other is known.

3.1.7 Magnetic Field Strength

The magnetic field (M) is measured in terms of force (F) on a unit pole. Therefore,

$$M = F/m_1 = m_2 / \mu r^2$$

.....(3.4)

M is measured in gauss or Oersteds that is dynes/unit pole. However, magnetic field recorded in magnetic surveys are quite small and therefore, a smaller unit of gamma or nanotesla (nT) equal to 10^{-5} Oersted is used.

3.1.8 Magnetic Moment

Magnetic poles always exist in pairs and the fundamental magnetic property is the pole strengths, +m and -m separated by a unit distance. Magnetic moment is defined as the magnetic field due to unit poles separated by 1 cm apart. Therefore, magnetic moment due to poles of strength +m and –m separated by distance l cm is given by:

$$M = m$$

.....(3.5)

As magnetic moment of a material is related to spinning electrons, which also produces electric field, it is expressed as Am^2.

3.1.9 Intensity of Magnetization

Intensity of magnetization (I) is defined as the magnetic moment per unit volume is given by:

$$I = M/v$$

.....(3.6)

It is expressed as magnetic moment / unit volume, which as equal to A / m.

3.1.10 Magnetic Susceptibility

A magnetic body placed in an external magnetic field is magnetized by induction in the direction as that of the external field. The intensity of magnetization (I) is proportional to the strength of the field and its direction is same as that of the external field. Therefore if T is the external field,

$$I = kT$$

.....(3.7)

Where k is known as susceptibility and is the property of rocks or materials which controls its magnetization. In case of rocks or minerals acquiring magnetization due to the earth's magnetic field, T is the total intensity of the earth's magnetic field.

The susceptibility k is given by

$$k = I / T \qquad \qquad .....(3.8)$$

Therefore, in case, rocks have acquired magnetization in the earth's magnetic field and its intensity (I) is measured in laboratory, its susceptibility k can be determined, which is an intrinsic property of rocks as density in case of gravity field. It is, therefore, most important characteristic of rocks for magnetic methods and the whole principle of magnetic method is based on it. Its value for various rocks and minerals generally varies from 10^{-1} to 10^{-6} emu. Susceptibility is the fundamental parameter in magnetic prospecting. The magnetic response of rocks is determined by the amount of magnetic minerals in them, specially magnetite since the latter have susceptibility values much larger than any other mineral. The susceptibility of some important rock types and minerals are given in Table 3.1. In general, the igneous basic rocks and metamorphic rocks have a higher susceptibility value than felsic igneous and sedimentary rocks.

Table 3.1 Bulk Magnetic Susceptibilities of Common Minerals and Rock Types

ROCK TYPE	SUSCEPTIBILITY RANGE X 10^3 emu	AVERAGE X 10^3 emu	Curie Temperature ^{0} C
Minerals:			
Magnetite	100 – 1000	500	570
Titanomagnetite	100 – 500	200	150
Ilmenite	1 – 10	5	233
Pyrhotite	1 – 200	100	320
Maghemite	100 – 500	300	> 600 and < 675
Hematite	1 – 100	50	675
Goethite	1 – 10	5	120
Rocks			
Sed. Rocks	0.001 – 0.1	0.01	
Acidic Igneous Rocks	0.001 – 0.1	0.01	
Basic Igneous Rocks	0.1 – 10	1.0	
Metamorphic Rocks	0.01 – 1.0	0.1	

3.1.11 Magnetic Induction (B)

The magnetic bodies acquire magnetization through induction when placed in an external field (T). Magnetization is acquired by spinning electrons and how that is oriented with respect to adjoining electrons. Magnetization by induction amounts to lining up the dipoles of the magnetic material in direction of the external field which is often referred to as the magnetic polarization. Magnetic induction is the total fields in the body namely both the induced field ($4\pi I$) and the inducing fields (T). Therefore,

$$B = T + 4\pi I \qquad \qquad .....(3.9)$$
$$B = T + 4\pi kT$$
$$B = T (1 + 4\pi k)$$

If
$$B = \mu T$$

Therefore,
$$\mu = 1 + 4\pi k \qquad \qquad .....(3.10)$$

μ is permeability defining the characteristics of the body for a magnetization due to induction. This equation expresses the relation between the permeability and the susceptibility of a substance. B is expressed in gauss and therefore, μ is gauss / oersted

3.1.12 Magnetic Units

As stated above, the unit of magnetic field is oersted or gauss which is the force between two unit poles separated one cm. apart. The earth's total intensity magnetic field is of the order of 0.5 oersted and therefore, in magnetic prospecting, the field measured is usually much smaller than the earth's magnetic field and therefore, a smaller unit of magnetic intensity or field strength, known as gamma or nano-Tesla (nT), is introduced. 1 gamma or nT = 10^{-5} Oersteds

3.1.13 Magnetic Materials

Since magnetic anomalies are caused by the magnetic minerals contained in the rocks, it is necessary to discuss these minerals and their characteristics. As described above, the magnetic properties of materials depend on the orientation of various spinning electrons in them. Depending on these orientations, materials can be classified in different groups as follows:

(i) Diamagnetism

Diamagnetic substances show negative susceptibility and permeability implying k and $\mu < 0$. Because negatively charged electrons in these substances align in a direction opposite to an externally applied inducing field. Some examples of diamagnetic substances are graphite, gypsum, marble, quartz and salt.

(ii) Paramagnetism

Paramagnetic substances show small value of positive susceptibility (k) and permeability (μ) and therefore, they get aligned in the direction of the external inducing field giving rise to small net magnetization to the substance/rock in the direction of inducing field.

(iii) Ferromagnetism

Some of the paramagnetic substances such as iron, cobalt, nickel etc., display strong magnetic interaction between atoms with spinning electrons aligning all in one direction and are therefore, characterized by large value of susceptibility and permeability. They show significant amount of magnetization in the presence of an inducing field. Such substances are called ferromagnetic substances (Fig. 3.3(a)). The susceptibility of diamagnetic and paramagnetic materials is mostly less than 10^{-3} emu and that of ferromagnetic substances varies from 10^{-1}-10^{-3}. Magnetization in these substances decreases with temperature and vanishes at a temperature known as Curie point when heated. Conversely, these substances acquire magnetization when cooled below Curie point in the presence of a magnetic field. Some ferromagnetic substances are divided in magnetic sub domains with their magnetic moments in opposite directions but still show a net magnetization in an inducing field, called ferrimagnetic substances (Fig. 3.3(b)). However, if the magnetic moments in different domains cancel each other such that net resultant magnetic moment is zero, these substances are called antiferromagnetic substances (Fig. 3.3(c)).

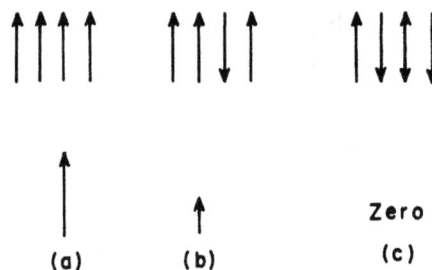

Fig. 3.3 Schematic representation of magnetic moments in (a) Ferromagnetic (b) Ferrimagnetic (c) Antiferromagnetic materials.

(iv) Magnetization due to Induction: Hysteresis Loop

When a magnetic material is placed in an inducing field, it shows a magnetization, which increases with the increase in the inducing field (Fig. 3.4). However, it reaches a saturation level, beyond which the magnetization does not respond to increase in the inducing field. If the inducing field is reduced, the magnetization reduces but does not come to initial zero level even if the inducing field is made zero. The magnetization of material when inducing field becomes zero is known residual magnetism. In case the inducing field is made negative implying that it is applied in opposite direction the residual magnetism becomes zero at a specific negative field known as coercive force. In case the inducing field is further increased in opposite direction (negative), the whole cycle is repeated in asymetric manner as shown in Fig. 3.4 with opposite residual magnetism and coercive forces. This whole cycle is known as hysteresis loop and is important to know the relationship between magnetization and the inducing field.

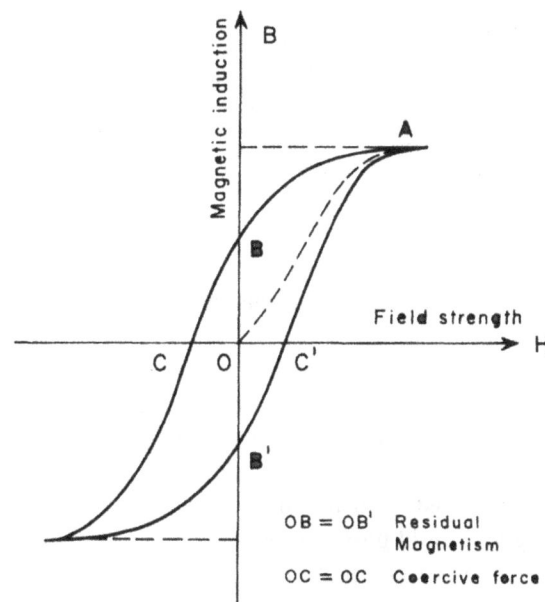

Fig. 3.4 Hystersis loop showing variation of magnetization due to inducing field, which shows an asymmetric pattern with changes in the field in opposite direction.

3.2   Rock Magnetism and Paleomagnetism

As stated above, when a magnetic material is placed in a magnetic field, it acquires a magnetization in the direction of the induced field which is known as induced magnetization. Rocks containing magnetic minerals acquire magnetization in the presence of the earth's magnetic field in a direction parallel to it, which is known as natural magnetization or natural remnant magnetization (NRM). Direction of magnetization in rocks can, therefore be defined with respect to declination and inclination as in case of the earth's magnetic field. In case it is acquired in present day earth's magnetic field, it is known as induced magnetization. The earth's magnetic field, however, is known to have been changing its direction in the geological past and therefore, quite often rocks display magnetization acquired during the time of their formation, which is known as remnant magnetization acquired in the direction of the earth's magnetic field prevalent at the time of their formation. Rock magnetism being important both for geodynamic studies and resource exploration based on magnetic method, its basic principles are described below. Magnetization in rocks depends on the type and quantity of mafic minerals present in them. Mafic minerals in rocks are defined with respect to ternary diagram of Feo-Fe$_2$o$_3$-Tio$_2$ (Fig. 3.5). Susceptibilities of some important mafic minerals and rocks are given in Table 3.1, which suggest maximum susceptibility for magnetite and therefore, its quantity in a particular rock controls its magnetization.

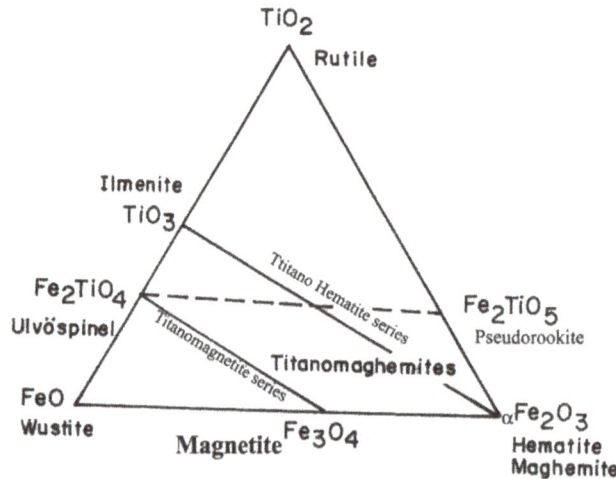

Fig. 3.5 Ternary diagram of important mafic minerals associated with igneous rocks.

3.2.1   Remnant Magnetizations

In most cases, rocks display magnetization in the direction of the present day earth's geomagnetic field, known as induced magnetization. However, as discussed above, rocks may also possess remnant magnetization which was acquired previously during the time of their formation or subsequently, under the influence of the earth's magnetic field prevalent at that time. Such magnetizations are known as remnant magnetization and the subject dealing with this kind of magnetization is known as paleomagnetism.

3.2.2   Thermo Remnant Magnetization (TRM) and Curie Point Temperature

This is the magnetization which is prevalent in the igneous basic rocks and are acquired in the presence of the earth's magnetic field at the time of their formation. When a magnetic material cools down in the presence of a magnetic field below a particular temperature, it acquires a magnetization parallel to the ambient fields as described above. The igneous rocks therefore, when cool down in the presence of the earth's magnetic field during their formation, acquire a magnetization parallel to the earth's magnetic field at that time. Similarly, if a material is heated, it loses its magnetization at the same specific temperature, which is called Curie point temperature. Curie point temperature is different for different minerals such as, for magnetite, it is 570 °C and for hematite, it is 675 °C (Table 3.1), above which these minerals and rocks containing them do not possess any magnetization and become non-magnetic in nature. The remnant magnetization acquired in this manner is known as thermoremnant magnetization and is most common in the igneous rocks and is particularly consistent for a particular rock and period. By implication, as temperature increases with depth inside the earth, rocks below the level corresponding to the Curie point of magnetite, are non magnetic in nature. This level is known as Curie point geotherm. As temperature gradient in the earth under normal continental shield is about 10 ° - 15 °C/km, this geotherm lies at a depth of about 40-60 km, quite often coinciding with Moho. However, it can be shallower or deeper in a region depending on the temperature gradient, which can be inferred from the heat flow in a region. Regions with high heat flow shall show shallower Curie point geotherm compared to regions with low heat flow. An approximate estimate of its depth can be obtained from average thermal gradient.

3.2.3 Detrital Remnant Magnetization (DRM)

This magnetization is acquired by sedimentary rocks during deposition due to alignment of their magnetic fraction under the influence of the then prevalent earth's magnetic field. The direction of DRM may be slightly different from the aligning field due to effects of dip of the beds, currents at the time of deposition, consolidation of sediments etc. However, effect of dip is most prominent among these factors, which is corrected from the measured direction of magnetization. It is a considerably weak kind of magnetization and therefore, there may be large scatter among the measured field from the same rock formation and does not provide consistent direction of magnetization compared to TRM in case of igneous rocks.

3.2.4 Chemical Remnant Magnetization (ChRM)

Magnetization acquired in rocks during chemical changes under the influence of the then prevalent earth's magnetic field is known as ChRM. This kind of magnetization is generally encountered in metamorphic rocks or sometimes even in sedimentary rocks due to chemical reactions.

3.2.5 Paleomagnetic Studies and Stability Tests

As stated above this study relates to the magnetization of rocks acquired by them at the time of their formation. It has several applications as described below:

(i) **Applications**

Paleomagnetic studies are important for several geological problems as described below:

(a) **Geodynamics:** The measurement of magnetic directions from rocks can provide information about the earth's magnetic field prevalent at the time of formation of these rocks and provide strong tool for the study of geodynamics of the earth namely continental drift, polar wandering and platetectonics etc. Due to bipolar nature of the earth's magnetic field, the pole position during formation of rocks can be computed based on the direction of magnetization measured in rocks, which in turn can provide the paleopositions of the rocks in case poles are considered to be almost fixed as discussed below. This is the main basis of obtaining the paleopositions of continents, which for most of the continents shows a different position during different geological periods implying continental drifting. Further details of continental drifting, specially in context of Indian continent are discussed in Chapter 5.

(b) **Exploration:** Paleopositions of continents are also important for mineral exploration as some minerals form in specific climatic conditions confined to specific latitudes. For example, Sheldon (1964; 1982) used the information that equatorial conditions are important for formation of the phosphate rocks, to explore these deposits in different countries in suitable rock formations when these continents were close to equator in geological past inferred from paleomagnetic studies. Based on paleolatitude of Indian continent described in Chapter 5 and suitable rock formations, phosphate rock was discovered in Rajasthan near Jaisalmer (Sheldon, 1966; Chauhan and Sisodia, 1989). Similarly, evaporates consisting of gypsum, anhydrite, halite and potassium salts are formed by intense evaporation of enclosed saline lakes and such conditions occur between 10°-30° in tropical conditions. Evans (2006) has shown that most of the evaporate deposits of the world formed when those regions were located close to the equator. This in fact confirms the geocentric axial bipolar nature of the earth's magnetic field and application of paleomagnetic studies for exploration of minerals occurring in specific climatic conditions.

(c) Magnetostratigraphy: Based on magnetic directions of well dated rocks, the relative ages of other rocks showing similar direction can be assigned which is known as magnetostratigraphy.

(d) Interpretation of Magnetic Survey Data: Paleomagnetism is not a direct concern of the exploration geophysicist. However, the information obtained through this discipline concerning direction of remnant magnetization in rocks can be used for the interpretation of magnetic anomalies observed in the magnetic surveys. It is particularly so because during magnetic surveys in field the total field viz. resultant of the remnant magnetization and the induced magnetization are measured. Therefore, knowledge of remnant magnetization in rocks, if any present can be used for proper interpretation of the observed magnetic anomalies in field by considering the resultant magnetization of the causative sources as discussed in section 3.7 and demonstrated in Fig. 3.21.

(ii) **Measurement:** For paleomagnetic studies, oriented samples of rocks are obtained in the field and measurements are made in the laboratory. The best rock formations for paleomagnetic studies are igneous mafic/ ultramafic rocks which show significant magnetization and can be measured accurately in the laboratory. The second in preference are undisturbed sedimentary rocks with good quantity of mafic components such as ferruginous sandstones etc., which however, are weakly magnetized (susceptibility$<10^{-3}$ emu) compared to igneous rocks (susceptibility$>10^{-3}$ emu). In case of metamorphic rocks, due to different stages of metamorphism, it becomes difficult to assign period for the measured direction of magnetization.

Having selected the right rock formations for paleomagnetic study, oriented samples are obtained from them in the field. This implies marking magnetic north on the sample and at least two horizontal lines on sides of it so as to orient it in the laboratory in the same manner as in the field. Oriented core samples obtained from the field samples are used to measure the direction of magnetization in laboratory using suitable magnetometers. Initially, self designed astatic magnetometers were used to measure the direction of magnetization in cored samples. However, presently there are several commercial magnetometers like spinner, fluxgate, optically pumped magnetometers etc., (Campbell, 1997) for this purpose which are more sensitive compared to the former. Paleomagnetic studies relates to the measurement of extremely weak fields in presence of strong earth's magnetic field and therefore, it is measured in a field free space by creating field opposite to the earth's magnetic field at the center of a set of Helmholtz coils or by any other means. These instruments are highly sensitive to measure small magnetic fields due to susceptibility of 10^{-5} - 10^{-6} emu. In paleomagnetic studies, generally three components of magnetization vector viz X, Y and Z are measured, which provides declination, inclination and intensity of magnetization, as given in equations 3.1. This kind of studies carried out world over have provided large data sets about paleopositions of different continents and plates (Nagata, 1953, Uyeda, 1958, McElhinny and McFadden, 2000; Runcorn, 1956). Paleomagnetic results from Indian rocks have been provided by several workers such as Blackett et al., (1960), Athavale et al., (1963), Mishra (1965), Sahasrabudhe and Mishra (1966), Radhakrishnamurthy et al., (1967), Klootwick (1979), Bhimasankaram (1975), Bhalla et al., (1979), Verma and Mittal (1974) and Vandamme et al., (1991). These results with regard to Indian plate are discussed in Chapter 5.

(iii) Stability of Remnant Magnetization

The most important aspect of paleomagnetic study relates to the stability of remnant magnetization implying that it has not changed since it has been acquired during the formation of the particular rock. As the magnetization of rocks acquired at the time of formation can change under the influence of different tectonic activities, which might have operated subsequently in that region, it is important to test whether the measured direction of magnetization is stable. In case of unstable directions of magnetization, it cannot be relied upon and should be discarded for any paleomagnetic studies. It can happen that some samples from same rock formation show stable direction of magnetization while some others show unstable directions and therefore, only the directions from former samples should be used for paleomagnetic studies. There are basically two methods to test the stability of direction of magnetization in the laboratory viz Alternating Current (A.C.) demagnetization and Thermal demagnetization. However, some simple rules like ratio of remanence and induced magnetization known as Königsberger ratio and consistency of directions from different samples and sites can also be used to infer stability of measured direction of magnetization in rocks.

(a) **Königsberger Ratio :** Königsberger ratio (Q_n) is defined as the ratio of remnant magnetization to that of induced magnetization.

$$Q_n = I_r / I_i$$

where I_i and I_r are intensity of induced and remnant magnetization, respectively measured in the laboratory. I_i is measured by creating a field equal to present day earth's magnetic field of about 0.5 Oersted using Helmholfz coils and I_r is measured in a field free space to avoid induced magnetization. For a stable remnant magnetization Q_n should be greater than one. However, larger its value better are the samples for paleomagnetic study and in general its value should be about 10 for stable remnant magnetization.

(b) **Consistency Test: Stereogram Plot :** The other and most important test is the scatter in direction of magnetization obtained from different samples at a site and among different sites from a particular rock formation. The measured direction of magnetization of all samples (4-5) from a particular site showing consistent direction of magnetization are averaged and these averaged directions of magnetization, viz. declination and inclination (D and I) from all sites in a particular rock formation are checked for their consistency. Declination is plotted on an equal area projection stereogram. Closely packed concentration of points indicates consistent directions of magnetization and suggests stable magnetization as shown in Fig. 3.6. In case the stereogram shows two directions of magnetization normal and reverse in same rock formation as given in Fig. 3.6 (Mishra, 1965; Sahasrabudhe and Mishra, 1966), it further confirms the directions of magnetization to be stable. The two averaged direction of magnetization in this case are N0°E, +30° (downwards) and S6°W and -30° (upwards) representing normal and reverse magnetization, respectively, which further confirm stable direction of magnetization in these rocks in spite of being weekly magnetized sediments (Vindhyan sandstones of Neoproterozoic period, Radhakrishnamurthy and Mishra, 1966). In case of basic intrusions in sediments, consistent direction of magnetization of intrusive rocks and baked sediments surrounding it also suggest stable direction of magnetization as they must have been acquired at the same time.

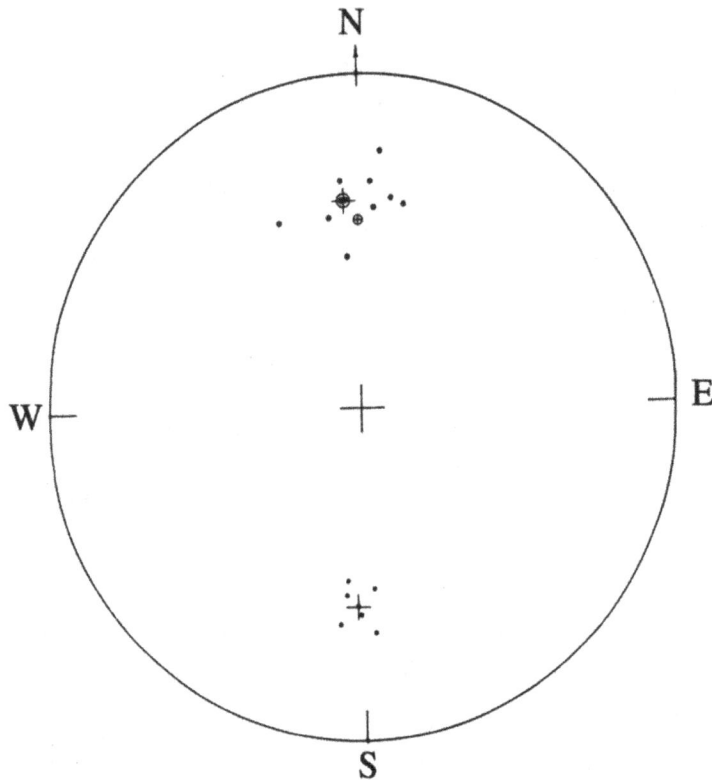

Fig. 3.6 A typical stereogram plot of declination of magnetic directions showing clusters of directions of remanant magnetization from different sites. This is an example of Vindhyan (Kaimur) sediments in India, showing two concentrations of directions indicating stable remnant magnetization in these rock formations (Mishra, 1965; Sahasrabudhe and Mishra, 1966). The two averaged directions of magnetization are N 0^0 E , $+30^0$ and S6^0 W, -30^0, representing normal and reverse magnetization, which further confirm the stability of remanant magnetization.

(c) **A.C. Demagnetization** : In alternating current demagnetization technique, the sample is subjected to an alternating field while rotating the sample randomly and the direction of magnetization is measured. The alternating field is increased slowly in steps up to 400-500 oersted and direction of magnetization is measured at every step till it shows consistency in the intensity and direction of magnetization. The unstable component of magnetization known as secondary magnetization vanishes when the alternating field is increased while stable component known as primary magnetization persists till the end. Two cases are shown in Fig. 3.6(a) and (b) (Mishra, 1965), which show both stable and unstable cases, respectively as the A.C. field is increased. Fig. 3.6(a) shows almost consistent directions of magnetization (declination and inclination) with increase in the field as shown at the bottom of this figure while it changes as field is increased from 264 to 396 oersted in case of Fig. 3.7(b).

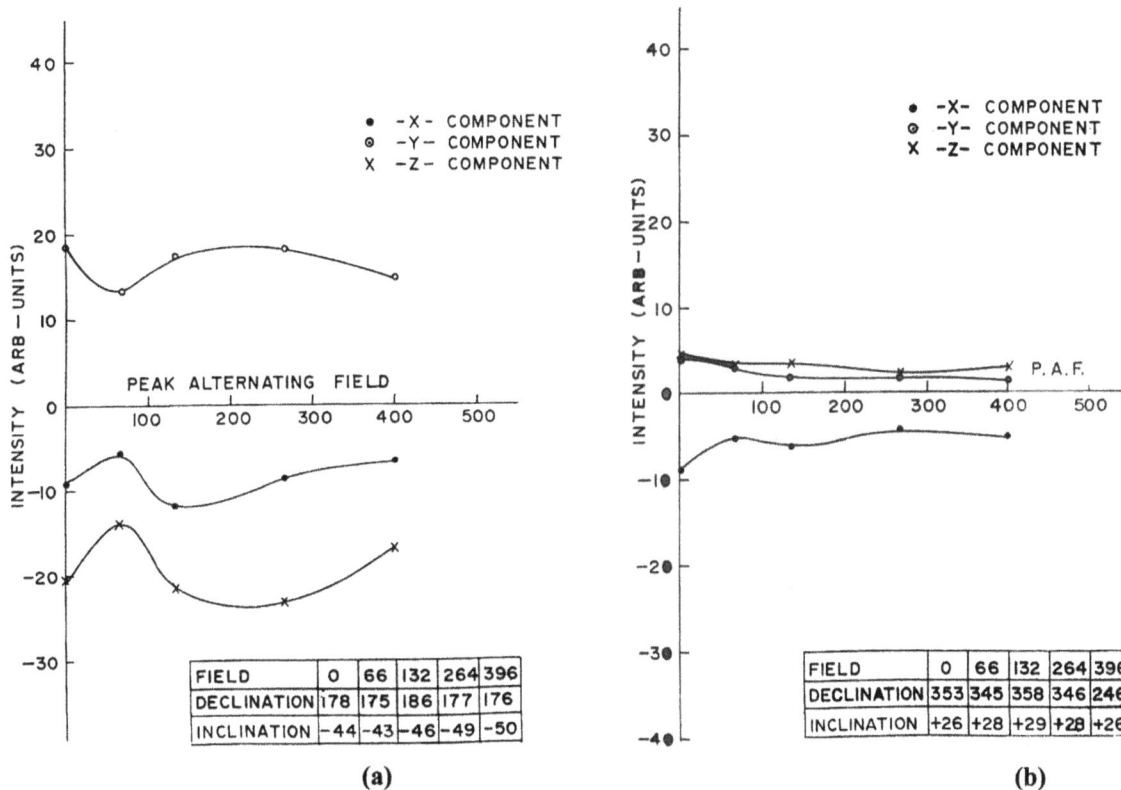

Fig. 3.7 (a) A typical plots of alternating current demagnetization of representative samples (Mishra, 1965) with inset showing direction of magnetization after sample is subjected to different A.C fields in Oersteds. It shows almost consistent directions of magnetization as field increases and therefore represents stable remanant magnetization. **(b)** AC demagnetization of a sample showing direction of magnetization (declination) changes as field increases from 264 to 396 Oersted, indicating unstable natural remnant magnetization (NRM) and therefore such samples are not suitable for paleomagnetic studies.

(d) **Thermal Demagnetization and Curie Point Temperature:** Thermal demagnetization uses the heating process to demagnetize the samples and obtain the primary magnetization in them. In this case the samples are heated to a particular temperature and cooled in a non-magnetic space and direction of magnetization is measured. The temperature is increased in steps up to Curie point till the samples becomes non-magnetic in nature. The plots of thermal history of magnetization (Fig. 3.7, Mishra, 1965) also provide the nature of secondary and primary magnetization and their Curie point temperature, which helps in identifying the magnetic minerals of the rock formation. Fig. 3.8(a) shows a kink, which on interpolation provide Curie point of about 570 °C and primary magnetization provides a curie point of about 670 °C, which suggest the presence of both magnetite and hematite in this sample (Table 3.1). However, impurities or mixing of different mafic minerals such as presence of titanium in solid solution with hematite can also lower the Curie point temperature, which however can be ascertained based on the rock types. In case of another sample, Fig. 3.8(b) shows both the heating and cooling curves. The heating curve shows a kink with Curie point as 150 °C and 550 °C, which may represent titanomagnetite and magnetite, respectively. However, the cooling curve basically represents only one Curie point of 550 °C related to magnetite indicating that the magnetization due to titanomagnetite with Curie point of 150 °C was unstable. Curie point temperatures of some common mafic minerals are given in Table 3.1 which can be used for their identification.

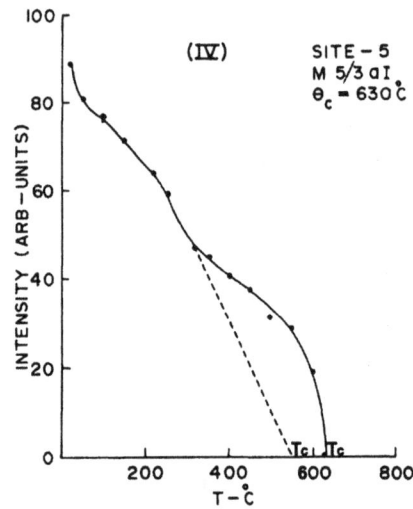

Fig. 3.8(a) A typical Thermal demagnetization curve showing the decay of magnetization with temperature. It shows a curie point of about 650^0 C and the intermediate kink is related to Curie point of about 550^0 C which indicate the presence of hematite and magnetite, respectively in this sample (Mishra, 1965).

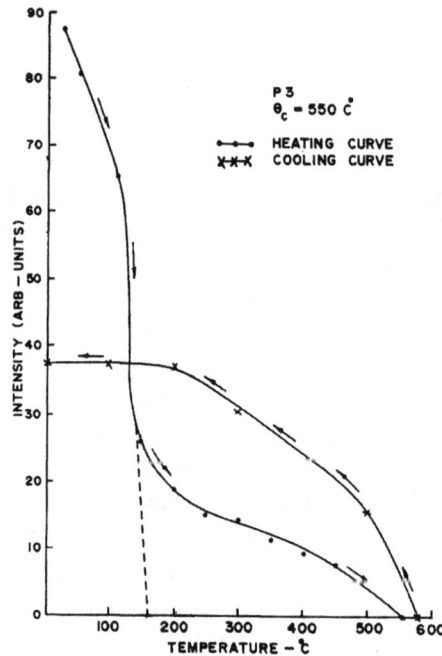

Fig3.8(b) A Typical thermal demagnetization curve showing the effects of heating and cooling. Heating curve shows two Curie point temperatures of 150 °C and 530 °C which are related to maghemite and magnetite, respectively. The cooling curve shows only one curie point temperature of 575^0C related to magnetite indicating that natural remanant magnetization associated with maghemite might be unstable one.

3.2.6   Apparent Polar Wander Path (APWP), Continental Drift, Reversal of Earth's Magnetic Field, Magnetostratigraphy, and Catastrophic Events

Considering the earth's magnetic field as geocentric dipole during geological past and direction of magnetization obtained from rocks of different periods being caused by the magnetic dipoles at the time of their formation, the latitudes and longitudes of the magnetic poles can be obtained from them, which provide APWP when plotted with respect to the geological period for that continent. Alternatively, the poles have remained fixed throughout the geological periods and continents have drifted to provide different direction of magnetization during different geological periods, which is attributed to continental drifting. The computed poles for different periods when plotted show a gentle curved path, which is called APWP. Sharp changes in APWP are interpreted as changes in plate motion. Similar APWP for different blocks in a plate suggest their relative position remained the same throughout the geological periods. However, to identify APWP from continental drift, it cannot be differentiated based on data from the same block as it implies that reference plane is in motion and can be done by comparing data from different blocks and different plates. Though it is difficult to ascertain between polar wandering and continental drift, based on paleomagnetic results, as they are related to each other, results from other geoscientific investigations are overwhelming in favour of latter, viz. continental drift/plate motion. But direct observations related to magnetic pole suggest its shifting since seventeenth century and therefore, movement of poles during geological periods can not be ruled out.

The magnetic pole position can be determined from the measured direction of magnetization as follows (Creer et al., 1954; Blackett et al., 1960):

$$\sin \lambda_p = \sin \lambda \sin \lambda_o + \cos \lambda \cos \lambda_o \cos \psi \qquad \qquad .....(3.11)$$

$$\sin (L_p - L_o) = (\sin \psi . \cos \lambda) / \cos \lambda_p \qquad \qquad .....(3.12)$$

where λ_o and L_o are present day latitude and longitude of the rock formation and λ_p and L_p are latitude and longitude of the magnetic pole and λ is the ancient magnetic latitude at the time, the rocks acquired their magnetization and is given by

$$\tan \lambda = 0.5 \tan I \qquad \qquad .....(3.13)$$

where I is the measured inclination.

ψ is called rotation, which is obtained from measured declination (D) of the direction of magnetization as:

$$\psi = D \text{ or } D - 180° \text{ if } 0° < D < 180° \qquad \qquad .....(3.14)$$

$$\psi = D - 360° \text{ or } 180° \text{ if } 180° < D < 360° \qquad \qquad .....(3.15)$$

The information about earth's magnetic field in geological past is extended to recent times based on measurements on lake sediments and archeological objects such as pottery, bricks etc., known as archeo-magnetism which show magnetization during the time of their baking. Paleomagnetic studies including that of lake sediments are important for paleoclimatic studies and are used to confirm the paleoclimatic indicators like flora and fauna of that period occurring in their natural habitat. Archeomagnetic studies also help in understanding the climatic conditions depending on paleolatitudes of particular civilizations and throws light on their period and growth. Irving (1959), Runcorn (1961) etc., have shown that there are appreciable differences between the APWP for different continents and discrepancies increases for period more than 50 Ma. This indicates that polar wandering alone can not explain the paleomagnetic data otained from different continents and along with it the process of continental drift is required to explain these data satisfactorily (Blackett, 1960).

Typical APWP for Indian continent is given in Fig. 3.9(a) and (b). Fig. 3.9(a) and (b) shows APWP for Neoproterozoic period (750 Ma) to recent times (Rao et al., 2003, Athavale et al., 1980) and Paleo-Neoproterozoic periods (Rao and Mishra, 1997), respectively, which suggest consistent movement of N-magnetic pole for Indian continent throughout the geological period. It suggests a pole position close to present day N-magnetic pole during 750 Ma corresponding to Malani rhyolite which drifted southwards close to the south pole during 600-450 Ma and again moving northwards to occupy present day pole position (Fig. 3.9(a)). This figure also shows a sudden change in the direction of APWP before occupying the present position (R; Fig. 3.9(a)), which may indicate reversal of the earth's magnetic field at that time (Redfern, 2001). In case we consider fixed pole position during the geological period, the APWP can be interpreted due to drifting of continents, which also suggest a northern hemisphere position for Indian continent during Meso- Neoproterozoic period drifting to southern hemisphere and back to northern hemisphere (Mishra, 1965; Radhakrishnamurthy et al., 1967). This indicates a good correspondence between APWP and continental drift history. During Paleo-Neoproterozoic period, the APWP for Indian continent is mainly confined to southern hemisphere indicating the position of Indian cratons during that period. Paleomagnetism of Indian rocks related to continental drift is described in more details in Chapter 5.

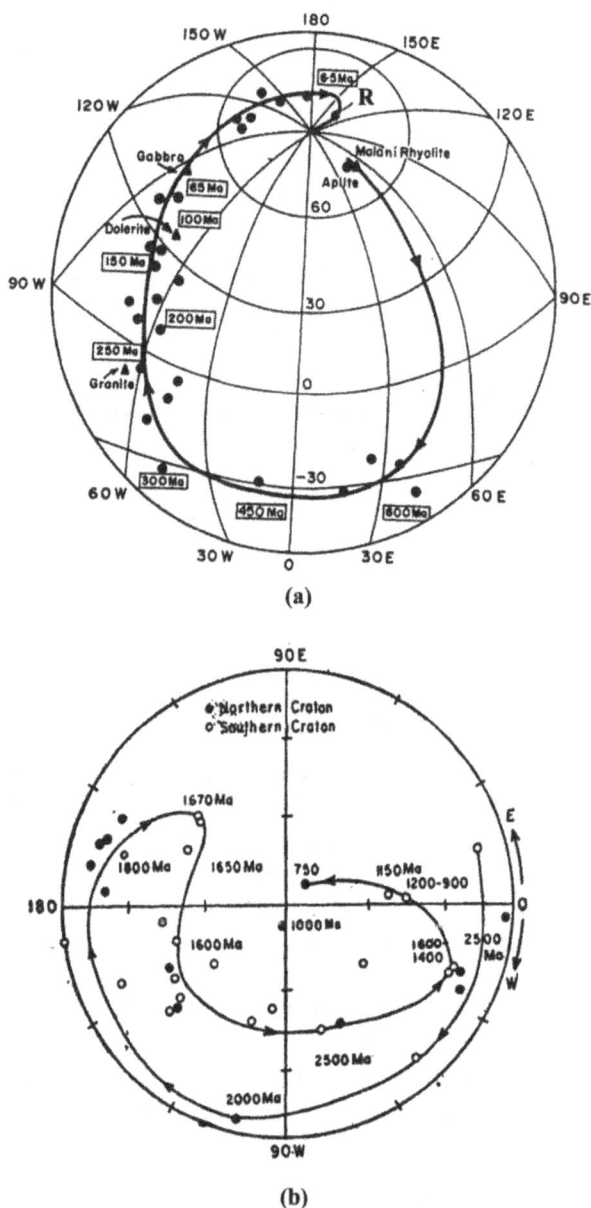

(a)

(b)

Fig. 3.9 Apparent Polar Wandering Path (APWP) curve for Indian rocks. **(a)** Paleozoic period (Athavale et al., 1980), R shows reversal of earth's magnetic field before present day field **(b)** Proterozoic period (Rao and Mishra, 2003). For most of the Proterozoic period, it is located in southern hemisphere.

(i) **Reversal of Geomagnetic Field:** During paleomagnetic measurements, it has been found that both normal and reverse magnetization considering present day earth's magnetic field as normal direction exist in same rock type of almost same period. For example, Fig. 3.5 shows two concentrations of direction of magnetization when plotted in a stereogram. The average direction for two groups are D = N 0° E and I = +30° (downwards) and D = S 6° E and I = -30° (upwards). These directions suggest that former with reference to present day earth's magnetic field is normal magnetization while the latter represents a reverse magnetization. These directions of magnetization relate to the Kaimur series of upper Vindhyan rock formations (Mishra, 1965) of Neoproterozoic period. These directions of magnetization suggest that these are stable directions of magnetization as they exist in pair of normal and reverse magnetization in same rock type. Further, implications of these directions of magnetization to geodynamics are discussed in Chapter 5. The reverse direction of magnetization in rocks can be explained in two ways viz, reversal of earth's magnetic field during the period these rocks were formed or self reversal of magnetic field in these rock formations. However, former is most common and has been observed world wide in rocks of different periods.

Self reversal as suggested by Neel (1955) and Uyeda (1958) can be caused due to specific chemical reactions which however are rare. The best proof of reversal of the earth's magnetic field in the past has come from marine magnetic surveys in oceans providing sea floor spreading magnetic anomalies across mid oceanic ridges, which apparently gave the idea of plate tectonics as discussed in Chapter 5. It might be caused due to changes in the dynamo current in the molten earth's core, which produces the earth's main field. The most recent reversal of the earth's magnetic field (S-N) occurred about 700,000 years ago (Redfern, 2001), which was inferred from marine magnetic anomalies across Mid Atlantic Ridge in South Atlantic Ocean. Reversal pattern can be seen even in APWP as in case of Indian continent (R, Fig. 3.9(a)), which shows a sudden change in the direction of APWP before occupying the present day position. Courtillot et al., (2007) has also suggested connection between the earth's magnetic field and climate.

(ii) **Magnetostratigraphy:** The change in polarity of the earth's magnetic field during different geological period is universal which implies its occurrence over different continents at the same time. This makes possible to use it for stratigraphic correlation depending on polarity of magnetization in rocks, which is known as magnetostratigraphy. Based on dating of rocks with normal and reverse polarity in different continents, the polarity of rocks during different geological periods was dated which are known as magnetic chron. Based on magnetic polarity of rocks and approximate stratigraphy, their period of formation can be assigned based on magnetic chron as given below.

Marine magnetic anomalies recorded along profiles across mid oceanic ridges provided alternating positive and negative magnetic anomalies (Vine and Matthews, 1963), which were attributed to normal and reverse magnetization of earth's magnetic field during formation of these rocks. As is known that mafic rocks well up from mantle at mid oceanic ridges (Fig. 1.2) and spread on either sides to form new crust, these alternating polarity of earth's magnetic field in most of the cases are placed symmetrically on either side of it and are therefore, named as sea floor spreading magnetic anomalies. This in turn implied that earth's magnetic field has flip flopped in geological past and its record is maintained in the form of sea floor spreading magnetic anomalies. When these oceanic magnetic anomalies across a particular mid oceanic ridge is recorded along several profiles almost parallel to each other and presented as black and white strips for normal and reversed polarity respectively, they provide a striped pattern in the survey block related to changes in the polarity of earth's magnetic field and are therefore also known as Zebra type magnetic anomalies. These magnetic anomalies were observed in almost all oceans of the world and were subsequently dated based on cored samples from the ocean bottom. Once they were dated from different oceans, they themselves provided a tool for dating based on polarity and width of the observed magnetic anomalies, which are known as magnetic chron. As oldest magnetic anomalies observed in oceans belong to 160-180 Ma, magnetic chron extend up to that period from the present, which have been named as C1N, C1R... C34N up to middle Cretaceous and M0, M1, M28 afterwards. Fig. 3.10(a) provides the details of these magnetic chrons (Hoffman, 1988). Similar scales with some changes have also been provided by Kent and Gradstein (1986), Cande and Kent (1995) and McElhinny and McFadden (2000) etc. Based on known normal and reverse polarity of dated continental rocks, the magnetic chron have been extended back wards in time up to Meso-Proterozoic period 1100 Ma (Champion et al., 1988), though they may not be reliable due to absence of continuous record of normal and reversely magnetized rocks over the continent.

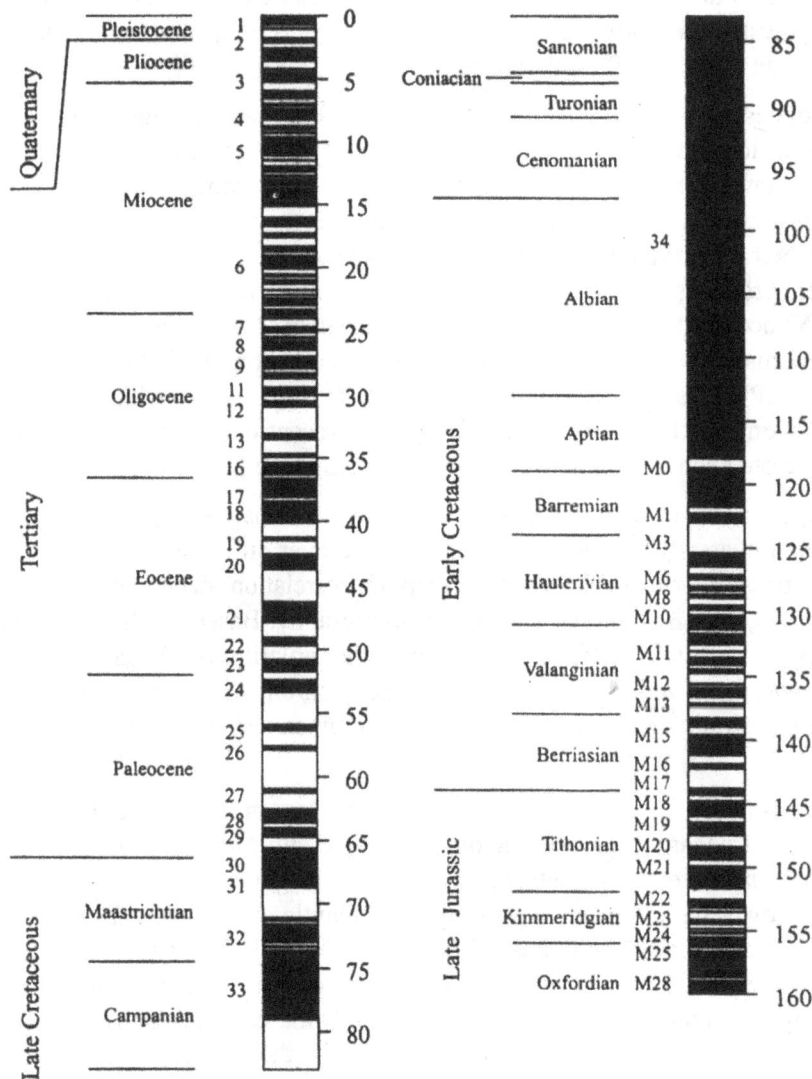

Fig. 3.10(a) Magnetic polarity with geological period up to 160 Ma (Magnetic Chrons), with black and open blocks indicating normal and reverse polarity respectively. They are referred to as C1N, C1R,........C34N up to middle Cretaceous and after that M 0,M1, M28 (Kent and Gradstein, 1986; Cande and Kent, 1995).

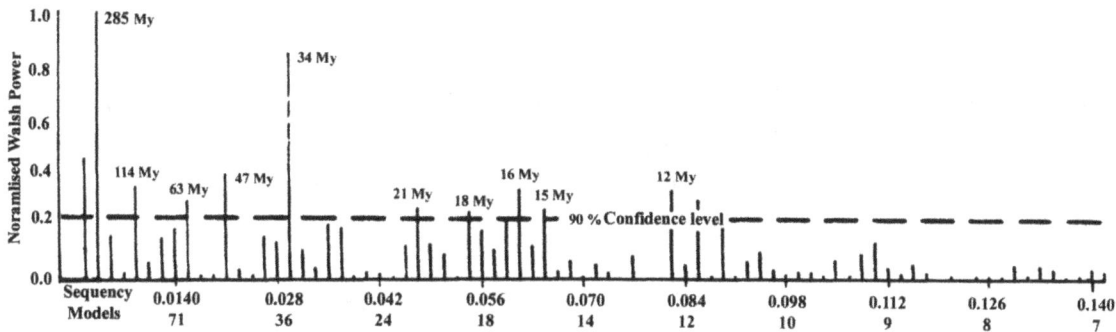

Fig. 3.10(b) Walsh power spectra of geomagnetic reversals during Phanerozoic period (Negi and Tiwari, 1983).

(iii) Reversals, Periodicity and Catastrophic Events: Reversals of geomagnetic field during geological past provide a unique data set which connects to the processes deep inside the earth. Erickson, 2001 have shown coincidence of magnetic reversals with several earth processes specially catastrophic events. These events, therefore have been analyzed using various methods including time series analysis (Chapter 4) along with other data sets to look for any periodicity between them. Crain et al., (1969) based on time series analysis of geomagnetic reversals suggested two periodicities of 75 My and 300 My and associated these cycles to movement of the solar system. Negi and Tiwari (1983) analyzed reported reversals of geomagnetic field during Phanerozoic period (0-560 My) using Walsh transform, suitable for rectangular functions as is the case with geomagnetic polarity series and suggested prominent peaks at 285 and 34 Ma. They suggested that the peak at 285 Ma coincides with a similar peak in galactic motion of the solar system. Pal and Creer (1986) also reported similar peak at 32.1 My in geomagnetic reversals record up to 80 My. However, some argued that no such periodicity exist in geomagnetic reversals (McFadden, 1987). Subsequently, prominent peaks at 30 ± 5 My in different data sets related to several geological activities (Table 3.2, Tiwari and Rao, 2003) were reported which suggested that they might be correlated with each other and related to some common cause. As one can see, it relates to several geological events including catastrophic events like climate changes, mass extinction including that of Dinosaurs at 65 My which forms two cycles of above periodicity. As geomagnetic field originates from convection in core and some correlation also exists between reversals, mantle convection and global tectonic activities (Courtillot and Besse, 1987), it is likely that common cause for them lies in changes in core and mantle convection pattern and their interaction and plate tectonic processes such as changes in plate motion etc. Common periodicity in so many varied geological events suggests a coupled earth system of geosphere, atmosphere, biosphere and hydrosphere. Short term periodicity of changes in geomagnetic field in range of about thousand years and less (Barton, 1983) might be related to the extra spin of the earth's inner solid core which it makes in 700-800 years compared to other parts of the earth (Zhang et al., 2005). Similar periodicity of 634 years has also been reported from sea level changes. Several other causes for this periodicity in different data sets have been postulated (Tiwari, 2005).

Table 3.2 Review of 30 ± Myr periodicity in various Earth processes (Tiwari and Rao, 2003)

Period of record inMyr	Geological events	Dominant periodicity inMyr
0-570 and	Geomagnetic Reversals	32 ± 3
0-165	Geomagnetic Reversals	30 ±3
0-250	Flood basalts	33 ± 3
0-570	Orogenic events	30.6
0-180	Sea floor spreading	26.7, 18.4
0-200	Sea level changes	28.2
0-245	Ocean Anoxic events	26.1,39.7
0-258	Evaporite deposits	28.2
0-160	Phosphorus burial rate	33 ± 3
0-140	Abundance of dolomite Fluctuations	25 ± 2
0-250	CO2 variations	33 ± 3
0-260	Mass extinctions	26, 32
0-260	Impact crate ring	33 ± 3
0-120	CTBF	35 ± 5

(a)   Mass Extinctions: Dinosaurs

Similar periodicity in various natural processes as discussed above suggests that they might be related in some aspects. For example, extinctions in geological past are considered to be related to volcanic activity, meteoritic (bolides) impact, climatic changes etc. In fact, climatic changes are related to both volcanic activity and bolides impact as large amount of ashes and dust generated in these processes affect climatic conditions such as changes in oxygen content, ozone layer, sun's radiation etc. Climatic changes affect sea level changes, global warming etc., which are important factors in any catastrophe. In fact, all these factors are affected by plate tectonic processes which in turn, appear to be related to mantle convection while geomagnetic reversals are presumably related to convection in the core and two might be coupled. There are six major extinctions reported in geological past but every time some species survived indicating that there must have been some thing specific to species which perished besides the general cause of extinction during that time. It has also been found that the species which survived were entirely different form those perished during that period. For example, dinosaurs became extinct at K/T (Cretaceous-Tertiary) boundary at 65 Ma though fish, mammals and some plants survived this extinction. This could be due to large size of dinosaurs, sometimes as big as 21 m in length and 33 ton in weight which made them difficult to move during any catastrophic events. It is also noted that before extinctions, decline of those species had already started making catastrophic events as only one cause for it.

Ice ages and its aftermath invariably related to mass extinctions such as those occurred at 670 Ma and 440 Ma (Ordovician). As recent as 11,000 years back, during early Holocene after Pleistocene glaciation and 6000 years back after neoglaciation extinctions has been reported. The greatest mass extinction occurred at the end of Permian (250 Ma) and Cretaceous (65 Ma; Dinosaurs) that coincided with massive volcanic eruption in Siberia and India (Deccan Basalt and its counter part in the Arabian sea). However, the one at 65 Ma (Dinosaurs) also coincided with a large meteorite impact at Chicxulub in USA and both might have contributed to climatic changes during that time and extinction of

Dinosaurs. Volcanic eruptions may lead to poisonous environment due to global warming (Kump et al., 2005) and drought (Flynn and Krause, 2000) that led to extinctions of Dinosaurs. It, therefore appears that two major causes for extinctions in the geological past have been large scale volcanic eruptions and meteorite impacts leading to climatic changes such as global warming poisonous gases in atmosphere etc., Further, the meteorite impact and volcanic eruption might have been much more in oceans which has not been investigated in detail. In such cases marine life will be severely affected and that may cause floods and climatic changes to affect even continental flora and fauna as has been portrayed in Hollywood film, 'Deep Impact', for meteorite impact in oceans. Oceans being 80% of earth's surface and during early period of earth's history marine flora and fauna being much more compared to those on the continent which became extinct with time, this effect appears to be more pronounced.

3.3 Instruments and Data Acquisition

The instruments used for measuring magnetic fields in surveys for exploration are known as magnetometers. Two kinds of instruments are primarily used to measure the magnetic field in magnetic surveys, one which measures the variation in the earth's magnetic field and the other which measures its absolute value as in case of gravity method. The former is known as variometers and the later as magnetometers. The variometers measure the variations in the vertical or horizontal components of the earth's magnetic field. The most common instrument in this class are Schmidt vertical and horizontal balance variometers. These instruments measure variations in the vertical or the horizontal component of the earth's magnetic field. Their accuracy is of the order of 5 gamma (nT). However, the Schmidt type variometers are of historical importance and have now been replaced by more sensitive magnetometers, which measure the earth's total intensity magnetic field to an accuracy of 0.01 to 0.1 gamma. The most common instrument in this class is nuclear precision magnetometers and optical pumping magnetometers, which are briefly described below. Readers may refer the text books and publications given below for further details of these instruments.

3.3.1 Nuclear Precision Magnetometers

This magnetometer measures the total intensity of the earth's magnetic field (T) to an accuracy of 0.1 gamma and are most widely used magnetometers for magnetic surveys. These instruments are based on the principle of nuclear precision in the presence of the earth's magnetic field. Protons being most common nuclei, it was used in the first magnetometer of this kind and therefore they are also known as Proton precision magnetometers. Protons (hydrogen nuclei) being easily available in water, a sample of pure water is used as sensing element. When such a sample is placed in a magnetic field generated by electric currents in a coil around the sample, all the protons get aligned (polarized) in the direction of the magnetic field generated by the coil and the water sample possesses a net magnetic moment. For greater effect, this field is generated approximately perpendicular to the earth's magnetic field. Once this field is switched off, all the protons polarized due to this magnetic field reorients themselves in the direction of the earth's magnetic field. While reorienting they presses around the earth's magnetic field and the frequency of precision is proportional to the earth's magnetic field. The frequency of precision is measured by recording the electric current generated by it in a coil wound around the sensing element. The proton presses at an angular velocity (ω) which is known as Larmor frequency and is proportional to the earth's magnetic field and is, therefore given by:

$$\omega = \gamma T \qquad\qquad .....(3.16)$$

Where γ is the constant of proportionality, known as gyromagnetic ratio of the Proton. It is equal to the ratio of magnetic moment to spin angular momentum and is known to an accuracy of 0.25×10^{-4}.

$$T = \omega / \gamma = 2\pi f / \gamma \qquad\qquad .....(3.17)$$

The factor $2\pi / \gamma = 23.4874 \pm 0.0018$ gamma/HZ and frequency f is measured from the induced voltage in a coil around the sample (protons). A block diagram of this magnetometer has been given in all text books on Exploration Geophysics such as Telford et al., (1990). The advantages of proton precision magnetometers lies in their ease of operation in the field and its sensitivity to an accuracy ± 0.1 nT.

3.3.2   Optical Pumping Magnetometers

The optical pumping magnetometers are based on the principle that when electron falls from a higher energy level to a lower level in the earth's magnetic field, they emit radiations, which is proportional to the field itself. Therefore if all the electrons of a medium are brought to a higher level by irradiating them by light or radio frequency and is withdrawn subsequently, they fall back to the lower level (ground level). During this process, they release some of the previously absorbed energy and presses around the earth's magnetic field at Larmor frequency as in case of Proton precision magnetometers described above which is a measure of earth's magnetic field. This process of raising electron to higher energy level is known as optical pumping and therefore its name. In case of simple potential energy levels there are three levels A1, A2 and B (Fig. 3.11(a)). Initially all the electrons are located in the ground level A1. However, if it is irradiated with a light beam from which spectral line A2 and B has been removed, the atoms in level A1 will absorb the energy and rise to A2 or B levels but those in level A2 will not be excited. When the excited atoms fall back to ground state, they may return to any lower level A2 or A1. However in A1 they will again be pumped to higher level and will return back to any lower level. Eventually all the atoms will be pumped from the ground level A1 to higher level A2 (Fig. 3.11(b)). At this stage the medium is transparent to the irradiating beam. Now if a radio frequency signal with energy corresponding to the difference between A1 and A2 is applied, the pumping effect is nullified and the atoms return to the ground level A1. This frequency of the signal to nullify this effect is given by $f = E / h$.

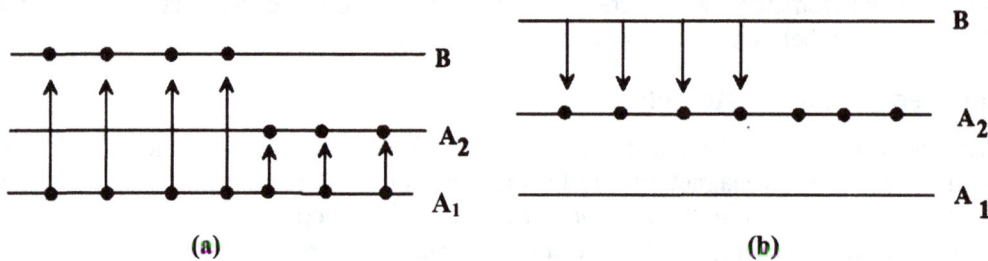

Fig.3.11 Energy levels of hypothetical atom used in optical pumping magnetometers (a) during pumping (b) after pumping.

where E is the difference in the energy level and h is Plank's constant given by 6.62×10^{-34} Joule-sec.

The frequency f is related to earth's magnetic field as given above.

$$T = 2\pi f / \gamma \qquad \qquad .....(3.18)$$

where γ is the gyromagnetic ratio. Elements, which are used for optical pumping, are cesium, rubidium, sodium and helium and accordingly the instruments are named after these elements. Using this principle magnetometers have been designed, which can measure the earth's magnetic field to an accuracy of $\pm .01$ nT.

3.3.3   Data Acquisition: Ground Surveys

Depending upon the geological problems, as discussed in Chapter 2 on 'Gravity Methods', magnetic data is recorded along profiles or in two dimensional plane. In fact, most of the ground gravity and magnetic surveys are conducted together at same stations and therefore, details of field operations are same in the two cases as discussed in Section 2.3 for gravity surveys. However, some specifics of magnetic surveys are discussed below. Like gravity surveys magnetic surveys are also categorized broadly in three categories, namely

(a) Regional surveys for large scale geological mapping and crustal magnetization.

(b) Reconnaissance and detailed integrated surveys for Hydrocarbon exploration

(c) Reconnaissance and detailed integrated surveys for Mineral exploration and Engineering and environmental geophysics – popularly known as near surface geophysics.

Ground magnetic surveys are by far the oldest geophysical method practiced for the exploration of minerals and hydrocarbons. Surveys are conducted both along the profiles and the perpendicular grids as described in section 2.3 for gravity surveys. Station intervals in both the cases depends on the problems which can be a few meters in case of mineral exploration to a few kilometers (1-2 km) for oil exploration and regional geological investigations as discussed in case of gravity surveys. In these surveys the magnitude of the total intensity or any one of its components, the horizontal or the vertical are measured. However, in most of the present day surveys total magnetic intensity using proton precision or optical pumping magnetometers are measured. Some specific details of these magnetic surveys are summarized in Table 2.2 at the end of Chapter 2 along with details of similar gravity surveys.

Magnetic field being bipolar in nature and anisotropy in susceptibility of rocks being more compared to density, it varies more sharply from station to station. It is therefore, advisable to keep station spacing less in case of magnetic surveys for regional studies and reconnaissance surveys for hydrocarbons to a maximum of 0.5-1.0 km. This implies that for better representation of magnetic field, it is advisable to introduce one more magnetic station between adjacent gravity stations spaced 1-2 km apart. However, in most cases, to avoid the laying of additional station and to save survey time, geophysicist usually record magnetic and gravity data at the same stations. For other surveys related to near surface geophysics as described in (c) above and discussed in section 2.3.2, station spacing is any way less of the order of a few meters to few hundred meters and therefore, same spacing can be used both for gravity and magnetic surveys. For this reason, airborne magnetic surveys, where data is recorded at a much closer spacing of 1-3 m, is preferred. As ground magnetic data is affected to a large extent by surface in homogeneity of rocks, it may be advisable to average data from adjoining stations using 3 or 5 points moving average to reduce noise specially in case of surveys over Archean-Proterozoic terrains for mineral exploration. There are some filters available for this purpose such as Naudy filter (Naudy and Dreyer, 1968) or any other digital filter described in Chapter 4. Due to these reasons, it is advisable to deal ground magnetic data, which show large fluctuations at adjacent stations as profiles instead of maps. In profiles, if there are fluctuating or spurious readings, they can be smoothed and adjusted based on adjoining readings, while in case of maps, they may disturb the whole map.

3.4   Corrections and Presentation of Magnetic Data

3.4.1   Diurnal Correction

The most important correction in a magnetic survey is diurnal correction, which arises due to the transient variations in the earth's magnetic field. It is corrected in two ways (i) by repeating the first reading of the day after the day's survey at the same station and considering a linear change in the earth's magnetic field during this period or (ii) in case of more sensitive and accurate surveys a continuous record of the earth's magnetic field or readings at interval of 5-10 minutes is made at a fixed station near the base camp using a second magnetometer. This provides a better correction for the transient variations in the earth's magnetic field. However, as typical variations in the earth's magnetic field on quiet days are about 60-70 gamma over a period of 5-6 hours during the day even the former method yields good results unless it is a high resolution survey. However, in case of magnetically disturbed days, when transient variation in the earth's magnetic field show fluctuations of hundreds of gamma, the survey is called off. Due to sharp fluctuations in the earth's magnetic field on such days, it may not be possible to observe consistent readings. A typical transient variation of the earth's magnetic field is shown in Fig. 3.12(a), which was recorded at a ground station at every 5 minutes interval. Based on this record, diurnal correction can be made to magnetic readings recorded on that day in that area.

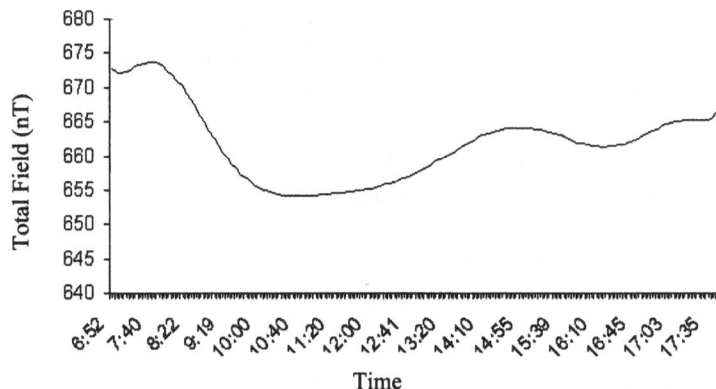

Fig. 3.12(a) A typical diurnal variation of magnetic field recorded at a field station using Proton precision magnetometer showing the changes in field during the day with maximum change of 20 nT. Diurnal correction reduces the particular day's data to the first reading of day.

3.4.2 Datum and Base Level Corrections

Besides hourly variation during a day, the earth's magnetic field also varies from day to day, which requires a correction for this change in case the survey in a region is spread over more than one day. For this purpose, a datum level is chosen, for example the first reading on the first day of the survey or any other value. Diurnal correction implies that the reading on that particular day is reduced to the first reading at base station on that day. Therefore, the difference between selected datum level and the first reading of a particular day to which all other readings of that day have been reduced through diurnal correction is added or subtracted to the reading on that day to bring all the recorded data in a particular survey to a common datum.

Due to base level (datum) differences, the surveys carried out during different periods (epochs) are required to be corrected for this factor while considering / joining them together. It can be done by adding / subtracting the difference of two datum values to one of the data sets. It is, therefore advisable to specify the datum level of a particular magnetic survey on the map itself. While joining two blocks of data, some specific trends such as N-S, E-W etc., may appear at the junction of two blocks, which may not fit with general geology of the area.

3.4.3 International Geomagnetic Reference Field (IGRF) Correction

After Diurnal correction, another correction, which needs to be applied to the magnetic data, is the IGRF correction. IGRF implies International Geomagnetic Reference Field, which pertains to the normal earth's magnetic field. In order to obtain the anomalous part of the observed magnetic field, the normal earth's magnetic field at the point of observation is subtracted from the observed magnetic field. As discussed above, the total intensity of earth's magnetic field is presented as isomagnetic's charts (Fig. 3.2(a)) and its value with reference to latitude and longitude of the area can be obtained and subtracted from the observed field to obtain the anomalous component. There are computer packages for this purpose, which provide the normal earth's magnetic field based on the latitude and longitude of the place, which can be subtracted from the observed field (Section 3.03). Isomagnetic's charts for total intensity magnetic field over India (IGRF10, IAGA, 2006) is given in Fig. 3.2(a), which can be used for correction of IGRF based on latitude and longitude of survey block. As IGRF describes the regional earth's magnetic field, this correction automatically removes the first order regional field from the observed field.

3.4.4 Data Presentation

After proper corrections have been made to the field data, they are presented in a suitable form for visual inspection, which can be correlated easily with local features such as geology, topography etc., As in case of gravity data, magnetic data are also presented as:

(i)   Profiles
(ii) Equal value contour maps and
(iii) Images

There are also smoothing processes and helps in identification of bad quality data, which do not fit in specific pattern with respect to surrounding values.

(i) **Profiles:** Profiles are one-dimensional presentation of magnetic data with respect to their position on the ground. With reference to a fixed position, the corrected magnetic readings are plotted at a scaled distance for the observation point, which makes it possible to know the value of the magnetic field and its variation in the direction of measurement. Some typical magnetic profiles are presented in Section 3.8, which can be followed for presentation of magnetic profiles. Similarly, one can record several profiles parallel to each other and plot them one over the other with reference to the central point, which is known as stacked profiles that demarcates the horizontal extent of the anomalous zone (Fig. 3.12(b)). In case, the profiles are zig zag along a road/track, the observation points are projected on a central straight line and plotted as profiles.

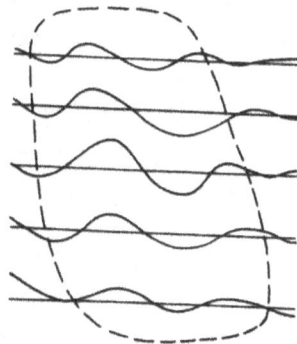

Fig. 3.12(b) Typical magnetic profiles stacked together to delineate potential zones marked by dashed lines around central zone. Magnetic anomalies with maximum amplitude are recorded along the central profile decreasing on either sides.

(ii) **Equal Value Contour Maps:** Contour maps represent the distribution of some physical quantity on a scaled two-dimensional map of the area. In other terms, it is a projected or a scaled replica of a physical process. It is prepared by plotting the field values at the observation points in a plane, and by joining the equal values as described in the previous chapter for gravity data. Presently, there are several standard computer packages for preparation of contour maps.

(iii) **Coloured Maps and Images:** Digital data obtained in gravity and magnetic surveys after processing and correction can also be presented as coloured maps and images as discussed in section 2.5.1 for gravity maps. This implies presentation of specific intervals of data in different colours or as continuous change in the shade of a colour from one range of contour values to other higher or lower ranges. Higher values are represented by shade of warmer colours (Red) and lower values by shades of blue. These are basically special methods to present the maps such that particular set of anomalies are highlighted. There are standard commercial softwares for this purpose. Kowalik and Glenn (1987) have suggested four types of images.

(a) Small scale grey level maps

(b) Artificial illumination

(c) Local contrast stretch and

(d) Directional filtering

However, in actual practice, the first two approaches are most commonly used. In grey level maps, specific contour intervals are represented as changes in grey level and displayed on computer screen in small scale such that major regional features can be visualized as continuous features with differences in shades of grey. Contour maps show sudden changes in values at contour intervals such as 5, 10 nT etc., while grey levels change continuously. Therefore, any pattern in grey scale is easy to visualize by human eye. In artificial illumination, contour map is illuminated from a particular angle and its image is produced either in different shades of grey scale or colours. It usually enhances linear features and can therefore, be easily identified on such a map. In directional filtering, the features with specific directions are optically enhanced such that they are highlighted on the map. As most of earlier data sets are in analog form, they are first required to be digitized before their colour maps or images could be prepared. There are several commercial softwares for this purpose such as AutoCAD etc., which can digitize large maps in relatively short times. One such scheme is described by Sreedhar Murty et al., (1998) and they provided grey scale image of aeromagnetic map of Kimberlite zone west of Cuddapah basin (Fig. 3.13(b)), which shows linear features associated with known volcanic pipes around Wajrakarur, which may represent a fracture or fault zone occupied by mafic rocks. There are several such fracture zones visible on this map, which were difficult to visualize on total intensity contour map. Similar processes can be used to prepare image of any data set more so geophysical data such as gravity, magnetic, electrical etc. However, closely spaced data sets provide better images and therefore, it is widely used for airborne magnetic and gravity data.

3.5   Data Processing

3.5.1   Regional – Residual Fields

The geophysical parameter such as magnetic field recorded at the surface of the earth is the cumulative effect of the causative sources at different levels as in case of gravity method discussed in the previous chapter. In geophysical literature, we broadly classify them into two groups referred to as the regional and the residual fields. Regional, as the terminology indicates, is the characteristic of the region as a whole suggesting deep seated sources which are of regional nature. Contrary to this, the residual field is the component of the observed field, which remains after the regional effect is removed from the observed field thereby suggesting the effects due to shallow sources. Both these components of the observed field are important depending on the problems being investigated in particular cases. In mineral or hydrocarbon exploration, the residual anomaly after the removal of the regional field, is important while in geodynamics and regional investigations for crustal structures, the regional component of the observed field assumes importance and is interpreted in terms of the deep seated sources. However, the term residual field for mineral and hydrocarbon exploration assumes different connotations. In case of mineral exploration, the depth of investigation is limited to a few hundred meters, while incase of hydrocarbon exploration, the magnetic anomalies from the basement under sedimentary basins assumes important, which is usually a few km (3-5 km) deep. Therefore, in case of former viz mineral investigation, the magnetic anomalies from the basement are treated as the regional field. However, in case of latter, the total magnetic field observed over sedimentary basins after IGRF correction can be treated as basement anomalies.

As referred to above, in case of mining and oil exploration, the residual component of the observed field assumes importance, which is separated from the observed field and interpreted in terms of causative bodies. However, the problem of regional and residual separation is not very strictly followed in case of magnetic surveys. In case of gravity surveys, all the rocks due to their density distribution, produce gravity anomalies. However, in case of magnetic surveys, most of the rock formations are non magnetic and therefore, the magnetic anomalies observed are primarily due to individual sources, which are magnetic in nature. However, in case of survey of large areas such as in case of aerial surveys, the problem of regional field may arise as the area of investigation extends over large area in which the earth's total magnetic field is likely to vary. This aspect is taken care by IGRF correction described above. However, in case, a part of regional field still remains in the data after IGRF correction, it can be removed by any of automatic methods of separating regional and residual fields described in Chapter 4. Graphical and Grid methods as described in the previous chapter in section 2.5.2 for gravity method can also be used for this purpose, specially in case of mineral investigations as small areas are involved and close grid spacing would provide local residual anomalies, important in such surveys.

3.6   Special Magnetic Surveys

Besides ground magnetic surveys, there are special magnetic surveys carried out from different platforms. Depending on the platforms used for magnetic surveys, they are known as (i) Ship borne or Marine magnetic survey (ii) Airborne magnetic survey and (iii) Satellite magnetic survey. Some specific details of these methods are summarized in Table 2.2 in Chapter 2 along with details of similar gravity surveys.

3.6.1 Marine Magnetic Surveys

Marine magnetic surveys are most common geophysical methods in ship borne surveys in ocean, which have provided significant results in geodynamics and plate tectonics in the form of sea floor spreading magnetic anomalies across mid oceanic ridges (Vine and Matthews, 1963) as discussed in Chapter 5. The marine magnetic surveys are conducted in the oceans with the help of ships using proton precession or optical pumping magnetometers briefly described in the previous section. The magnetic sensor (called 'fish') is towed behind the ship usually 500-1000 ft behind using cables to minimize the magnetic effect of the ship. The movement of ship is usually slow of the order of 10 knots/hour. The navigation was previously controlled using electronic systems, which were known from their trade name as SONAR, LONAR etc. However, in present day's surveys, mostly GPS is used for position location. In most of the marine surveys data are usually recorded along irregular profiles. Therefore profile analysis is more common in marine surveys than regular grids which are logistically difficult to record. Some marine magnetic data from Bay of Bengal and Arabian Sea related to Indian plate are presented in Chapter 5.

(i) Diurnal Correction

Transient variation of geomagnetic field in sea is more compared to continents and therefore, diurnal correction plays an important role in marine surveys. In case of marine surveys, besides transient variation of the geomagnetic field, the electromagnetic induction also causes the transient variation in the magnetic field. Electrically conducting sea water produces its own magnetic field, which varies depending on the amplitude and direction of the waves. Diurnal correction, in case of marine surveys, were attempted based on records from the nearest geomagnetic observatories on the continents. However, they do not truly represent the variation of geomagnetic field at sea. Therefore, in high resolution surveys, sometimes effort was made to record the local geomagnetic field in moored buoys near the survey area and use these records for diurnal correction. However, due to problems in laying and recovering of moored buoys, this procedure is not common. Sometimes, diurnal correction can be estimated from crossover points of tie lines as discussed in case of airborne surveys by recording a tie line across recorded ship tracks. However, due to movement of ship, the exact location of crossover points and recorded fields at those points are difficult to compare and may introduce error. To avoid diurnal correction, sometimes, it is advisable to record horizontal gradient of the total field by towing two sensors at known distance apart, which is not affected by the transient variation of the earth's magnetic field.

3.6.2   Airborne Multi-Parametric Surveys – High Resolution Surveys and Advantages

Airborne geophysical surveys are usually multi-parametric surveys in which magnetic, spectrometer and electromagnetic units are generally fixed in the same aircraft. However, airborne multi-parametric surveys are desired mainly for mineral investigations, while for geodynamic studies and hydrocarbon exploration, mostly the airborne magnetic surveys are carried out. Now a days airborne gravity surveys can also be conducted along with the airborne magnetic surveys but that makes the survey costly due to cost involved in the former.

(i) Airborne Magnetic, Spectrometer and Electromagnetic Surveys

A typical photograph of a plane with sensors for magnetic, radiometric and electromagnetic surveys is given in Fig. 3.13(a). Due to importance of airborne geophysical surveys for various geoscientific problems, some of its details are provided below. Airborne surveys are most widely used geophysical surveys for quick appraisal of economic potentiality of a region. As the name implies, the survey is conducted from above the ground using planes or helicopters as per requirement of the survey. The survey details are provided in the next section. Due to large areas of investigation and closely spaced data points, considerably huge amount of data are generated, which require automatic methods of processing and interpretation, using computers. Airborne magnetic survey is basically a reconnaissance type of survey to delineate potential zones to be followed up by more specialized methods. It is primarily used for (a) mineral investigations, which have been associated in part with magnetic minerals, (b) in hydrocarbon exploration, it is used to delineate basement structures in large basins and (c) it is also used for 3-dimensional geological mapping and geodynamics studies through 'Regional Surveys'.

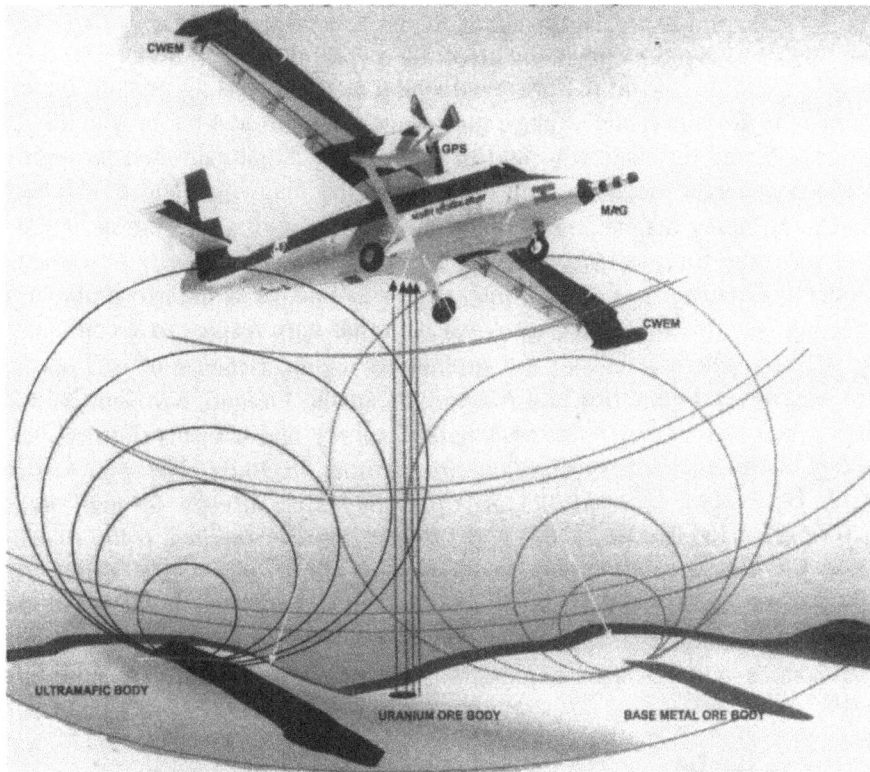

Fig. 3.13(a) Photograph of a typical multi parametric airborne geophysical survey (AMSE, 2002)

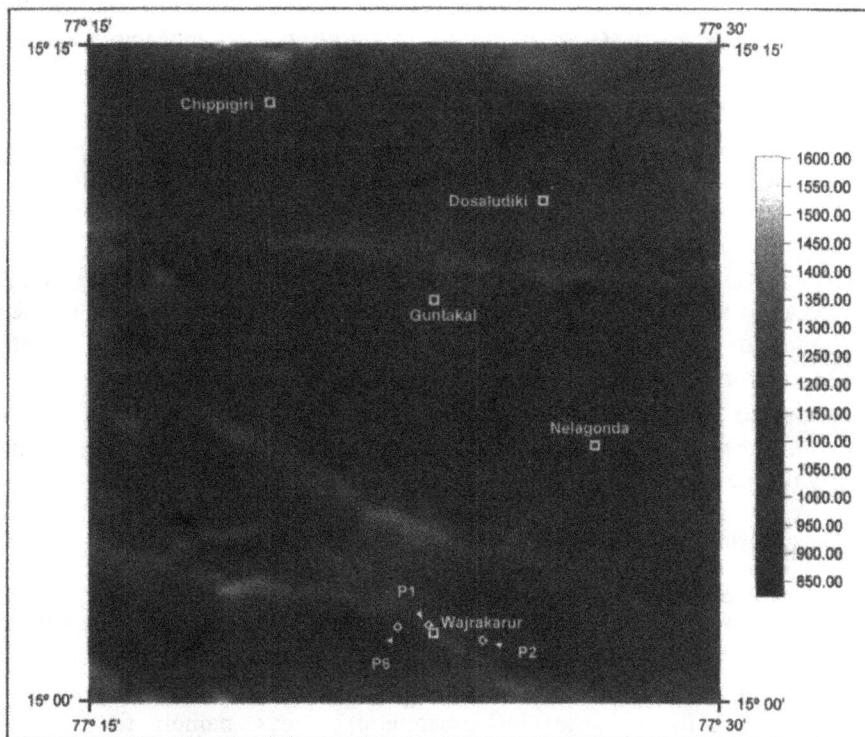

Fig. 3.13(b) A typical gray shaded relief of airborne total intensity magnetic field in south India west of Cuddapah basin. Volcanic pipes around Wajrakarur are associated with linear features delineated in this map (Sreedhar Murthy et al., 1998).

In earlier airborne surveys, proton precision magnetometer was used and the sensing element was towed behind the aircraft to avoid its magnetic effect by a special cable of about 100 ft (~ 30 m) long, which was also used to transmit the observed signal to the recording units in the plane. In the present days, there is also provision to place the sensing element in a boom which is attached to the tail of the aircraft and the magnetic field of the aircraft is compensated through various electrical coils, which generate magnetic field opposite to the magnetic field of the aeroplane. Presently, the optical pumping magnetometers (Cesium vapor), with an accuracy of ± 0.01 nT, are generally used for these surveys. After making all requisite corrections as described below, the data is presented as contour maps of total intensity or as images as described above. In case of mineral prospecting, profiles are stacked one over the other with respect to specific anomalies as shown in Fig. 3.12(b), which delineates the anomalous region. Because of its importance to geoscientific problems several countries like Australia, Canada, Finland, Sweden, Norway etc., have covered their whole country by airborne magnetic survey and it forms the first basic data along with geology to initiate any exploration programme. In India, National Geophysical Research Institute, Hyderabad initiated the airborne magnetic surveys through indigenous development in 1966-67 (Hari Narain, 1969) and covered specific regions, some of which are described in section 3.8. Subsequently regional country wide survey was undertaken by Airborne Mineral Survey and Exploration (AMSE) Wing of Geological Survey of India in collaboration with National Remote Sensing Agency, Hyderabad. Some applications of airborne magnetic surveys for geodynamics and resource exploration are given in section 3.8 and further in Chapters 7, 8 and 9.

(ii) Airborne Spectrometer Survey

It is a specialized survey to explore radioactive elements. These surveys are conducted using a sensitive gamma ray spectrometer, which measures Total Count, U, Th and K and provides the concentration of radio active elements in the region in counts per second above the reference background. However, gamma rays have very limited depth of penetration and therefore, these systems are primarily useful for surface deposits. Therefore to explore subsurface deposits some indirect methods are used such as airborne electromagnetic surveys as described below. Electromagnetic surveys are extremely useful to delineate paleochannels, which may be associated with radioactive elements. Airborne spectrometer surveys have been widely used for exploration of radioactive minerals by Atomic Minerals Division of Department of Atomic Energy, India (Kak et al., 1997). It can also been used to monitor environmental radiation. If radioactive elements are associated with faults and fracture zones, it can be used to delineate and trace them. An example of airborne spectrometer survey over parts of Chitradurga schist belt in Dharwar craton is described in section 3.8 along with the airborne magnetic survey, which shows higher radiation related to younger granite intrusives compared to schist belts and gneisses and their association with lineaments and fractures (Figs. 3.34 and 3.35). Some more specific case histories related to airborne spectrometer surveys are discussed in Section 9.10

(iii) Airborne Electromagnetic Survey

It is a specialized survey used for exploration of base metals. The electromagnetic method is based on principles of electromagnetic induction. An alternating magnetic field in a coil or cable induces electric currents in a conductive body, which is the measure of the conductivity. The former is known as the primary field and the latter as the secondary field. There are basically two kinds of units in operation for airborne electromagnetic surveys, namely time domain and frequency domain systems. Transient time domain electromagnetic (INPUT) system is based on the principles of electromagnetic induction, when a steady current in a wire loop is switched off, a time varying magnetic field is generated, which in turn induces eddy currents in conductive rocks below. The rate of decay of eddy currents (secondary field) is directly related to conductivity of the rocks. A primary wave is generated through direct current pulses in a coil located over the wings of the aircraft or towed below it and secondary currents generated by surface/subsurface conductive bodies due to induction of primary wave are measured when primary current is switched off. Secondary currents are measured after switching off the primary current in order to avoid their interference during measurements, which is done after necessary gaps of time, that provides decay or discharge rates at certain time. One such unit was developed in NGRI (Guptasarma et al., 1976) and used for multi parametric airborne geophysical surveys. The sensor elements to measure secondary field is towed behind the aircraft by a 70-80 m long cable as a bird as in case of aeromagnetic surveys. Its basic principle is demonstrated in Fig. 3.14(a), which shows that after the primary current is put off; secondary induced voltage is measured at six time intervals 1-6. The first channel immediately after the current is switched off reflects surface noise, which is subtracted from all other channels in order to remove the effect surface in homogeneity. The rate of decay and its persistence up to the last measuring channel is measure of conductivity of rocks underneath. Qualitatively, the slower decay rate and persistence of signal in more channels indicate conductive ore bodies. A typical example for decay rates of three different kinds of conductors, good, medium and bad, at different time intervals, is given in Fig. 3.14(b). A good conductor shows slow decay rate compared to medium and poor conductors. An example of multi-parameter airborne geophysical survey and its result in Sonrai Copper Belt, U.P., India (Mishra et al., 1978) is described in Chapter 9. In the present day systems, by increasing the primary field and number of channels, the depth of penetration can be increased to almost 600-800 m.

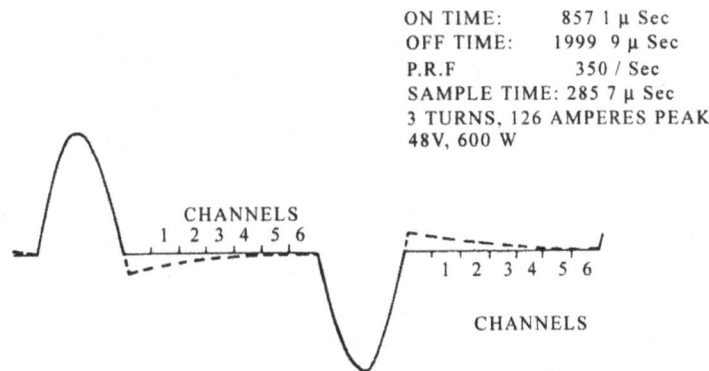

Fig.3.14(a) A schematic diagram of waveform and its decay in airborne time domain INPUT system. Secondary current induced due to conductive bodies are measured at six time intervals marked as 1-6. In set shows the details of the waveform.

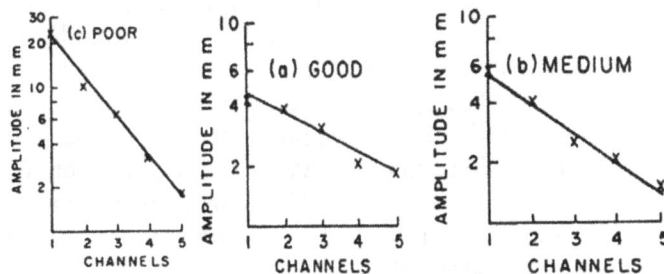

Fig. 3.14(b) Decay pattern of time domain E.M system for (a) good (b) medium (c) poor conductors (Mishra et al., 1978). Slow decay rate indicates conductive bodies.

In case of frequency domain system, there are as many transmitter and receiver coils as number of frequencies used to measure the secondary field housed in the bird towed behind the aircraft by a 30-50 m long cable. For example, in one such system known as Humming bird system, there are five sets of transmitter, receiver and bucking coils corresponding to five frequencies from 880 Hz to 34133 Hz housed in the bird (Vijaya Gopal et al., 2004). Bucking coils are used to attenuate the amplitude of primary field at the receiver coil while measuring the secondary field. In this case alternating current is used which due to short cycle time polarizes the ground for a short time and therefore its effect on measurement of secondary current is minimized. Both in phase and quadrature components are measured and interpreted in terms of surface/subsurface conductive bodies. Airborne electromagnetic system besides delineating zones of conductive minerals such as base metals has also been used to delineate aquifers and paleochannels due to their conductive nature. A similar survey along Pakistan-India western border in Cholistan delineated subsurface paleochannels with pools of water related to the Vedic Saraswati of 2-3 B.C. Delineation of paleochannels is important for investigations related to radioactive elements like uranium, which gets concentrated in these channels. It can also be used to demarcate fresh water zones from saline water along coasts as latter shows higher conductivity. Application of airborne electromagnetic survey for mineral exploration is demonstrated in Section 9.10 through specific case histories.

(iv) Survey Parameters

Airborne magnetic survey for reconnaissance mineral exploration are generally conducted at 1 km profile spacing at a height of approximately 500' (~150 m) from ground which in case of oil exploration are increased to 2 km and 1000' (~300 m) respectively. The regional surveys for geological mapping are conducted with profile spacing as large as 2-5 km and terrain clearance of 500-1000 m depending on the geological problems and the area. Long range aeromagnetic profiles for geodynamic studies are flown at an altitude of 9000-10000' (~3000 m) above m.s.l. In case of undulating topography efforts are made to maintain constant terrain clearance by changing the flying height of the aircraft. In case of flat areas as in case of sedimentary basins for oil exploration, a constant height above m.s.l. can also be maintained. In case of hilly terrain to reduce the flying height contour flying can be undertaken to maintain a constant terrain clearance in which the flight path follows a particular topographic contour. In case the survey demands, closely spaced profiles and less terrain clearance, helicopter-borne surveys are recommended, which can record close profiles at tens of meters (10-100 m) interval with a terrain clearance of 10-20 m. Such detail airborne surveys are presently quite common for mineral investigations such as Kimberlite pipes, base metal exploration etc.

The profiles perpendicular to the general strike of the rocks are laid out on aerial photograph or topographic map of the region on an enlarged scale preferably 1:250,000 scale (1 cm = 2.5 km). The flight lines are generally flown in alternating opposite directions as given in Fig. 2.20(a). These profiles termed as flight paths are followed by the pilot with the aid of navigational instruments in the plane and the requisite data is recorded automatically. Sampling rates in case of airborne magnetic survey is 10-100 samples / second but it is generally maintained at 10 samples/ sec unless closely spaced data is required for some specific mineral exploration. With average speed of helicopter of 120 km/hour (≈ 30 m/sec), it provides data at about 3 m interval. Incase of aeroplane with speed of about 240 km/hr, it provides data at about every 6 m interval. However, it can be reduced by increasing sampling rate as per requirement. The sampling rate for airborne electromagnetic survey can vary from 10-40 samples/sec but it is also usually maintained at 10 samples/sec. Spectrometer reading is recorded at a rate of one sample/sec. Besides this, the modern surveys incorporate electronic navigational device such as GPS, which helps pilot in following the profiles (flight path) precisely and records the distance traversed along the particular flight path, which are subsequently used to fix the recorded magnetic data to the ground position.

(v) Data Reduction

In any airborne survey, there are basically two kinds of data namely, geophysical data such as magnetic / radioactive etc., and navigational data, which are tied together to provide the ground coordinates to the recorded geophysical data. In earlier surveys, the data were recorded continuously on an analog chart, which consisted of earth's magnetic field recorded in units of hundreds and tens of gamma and altimeter recorded the height above the surface along with the time mark and the fiducial reference points. The same fiducial marks were introduced in the 35 mm film exposed during the flying. As the data were recorded in analog format, the data reduction was basically manual. Based on the 35 mm film, the flight paths with fiducial marks are recovered on a photo mosaic of the area or enlarged topo sheets. This is known as the base map on which all the recorded data were first transcribed. In present day surveys GPS provides the latitude, longitude and altitude of the plane in digital format, which are used to process the airborne data using computers. However, presently the whole process is automated.

(a) **Diurnal Correction:** The variation of earth's magnetic field from time to time and day to day is called diurnal variation and the correction to account for this variation is known as diurnal correction as discussed above. In airborne survey, the simplest approach to correct for this variation is to record the earth's magnetic field continuously at the base station during the period of survey. The observed field in the airborne survey is reduced to a common datum by subtracting or adding the difference between the recorded field at that time and the chosen datum. This reduces all the recorded data of a particular survey to a common datum implying that the variation in the earth's magnetic field has been corrected. This kind of diurnal correction has inherent drawback that the survey is usually conducted hundreds of kilometer away from the base station on the ground where the diurnal recordings are made. In this respect the use of tie lines as described in section 2.6.3 in connection with airborne gravity surveys is found to be more accurate. Briefly, in this method tie lines are flown perpendicular to the flight lines connecting all the previously flown flight paths and the magnetic field along these tie lines are recorded. The difference in the field values recorded at the intersection of flight lines and the tie lines are distributed in a loop which accounts for the diurnal correction (Fig. 2.20(a)). Tie lines can be flown after 2-3 days of survey making loops to cover the area flown during that period.

(b) **Heading Error and Correction:** As flight lines are alternatively oriented in opposite directions (Fig. 2.20(a)), there is a difference in magnetic readings due to eddy currents and effect due to the aircraft etc., which requires to be corrected. For this purpose, in survey area same flight line is flown in both directions such as N-S and S-N and average change in the level of magnetic field between two data sets is subtracted or added to data from one direction so as to bring them to the same level.

(c) **IGRF Correction:** IGRF correction is important in airborne magnetic surveys as large areas are covered in such surveys. It is applied in same manner as in case of ground magnetic surveys described in Section 3.4.

(vi) High Resolution Surveys

High resolution surveys imply the detection of anomalies of small magnitude and extent. Such surveys are usually carried out for direct detection of minerals or structures with which they are associated such as faults, fractures, volcanic pipes etc. This demands a high accuracy in data recording and its processing. The accuracy of a conventional airborne survey is usually $\pm$ 2-3 nT. However, with advent of sensitive magnetometers such as optical pumping magnetometers, better accuracy of ±1 nT can be achieved. The high resolution surveys are usually carried out for specific exploration programmes in small areas, which are already demarcated based on local geology and reconnaissance geophysical surveys such as normal airborne magnetic surveys. The profile spacing and terrain clearance in such surveys are reduced to ten of meters depending on the exploration problem and flight conditions in the region. Drift correction, including diurnal correction is made through tie lines flown across the recorded profiles and the intersection points are picked up as accurately as possible which is vital for the overall accuracy of the data. Due to high cost of flying, most of the airborne surveys presently conducted belong to this category so as to acquire high quality data.

Due to high resolution airborne magnetic surveys, several new applications of this method have been suggested (Nabighian et al., 2005). Grant (1984/85a,b) suggested its use for deciphering geological and ore environment such as presence of calc-alkaline series of magmatic rocks or metamorphic grades depending on amplitude of magnetic anomalies and general geology of the region. Greenstone belts are source of base metals world over and consist of iron mineral but are deficient in magnetite. Therefore, they produce weak magnetic anomalies over large distances and can be delineated based on high resolution surveys. Clark (1999) suggested to decipher magnetic petrology of igneous intrusions based on aeromagnetic anomalies. Donovan et al., (1979) suggested using aeromagnetic surveys for detection of digenetic magnetite over oil fields, which help in direct exploration for hydrocarbons. They suggested that leakage from a reservoir can react with ferric oxides or hematite in surrounding rocks and produce magnetite, which can be detected as small anomalies recorded in aeromagnetic surveys. This kind of anomalies can also detect leakage from oil pipe lines. Now a days, it is also be used for ground water resource assessment (Blakely et al., 2000a; Mishra et al., 1998) environmental pollution (Smith and Pratt, 2003) and seismic hazards (Blakely et al., 2000b) etc.

High resolution surveys require constant checking of the quality of data. The following two checks are generally followed in the industry (Reeves, 2010; Terraquest, 2010) so as to re-fly in case of bad data quality.

(a) Noise Level

Noise levels are decided based on the fourth difference (FD) that is defined as

$$FD = X_{t-2} - 4 X_{t-1} + 6 X_t - 4 X_{t+1} + X_{t+2}$$

Where X is measured amplitude of five consecutive readings from t-2 to t+2.

The tail and wing tip magnetometers will have a down line fourth difference noise envelope of 0.5 nT. If the noise of any of the magnetometers exceeds this limit for a distance of more than 3 km along the flight line, it requires to be reflown. The noise levels are measured before any filtering of data. Data losses in the magnetic data that exceeds 10% of the readings over 120 seconds interval requires to be re flown.

(b) Magnetic Figure of Merit

This test is required to check the compensation mechanism of the aircraft and magnetic field produced by it during maneuvering while flying. The aircraft flies a square pattern along the four survey flight lines and control lines directions at high altitude over a magnetically quite area and period and perform pitch ($\pm5^0$), roll ($\pm10^0$) and yaw ($\pm5^0$) maneuvers for a period of 6 seconds along the four lines. The sum of the RMS peak to peak residual noise amplitudes in the total compensated signal resulting from the twelve maneuvers is referred to as the Figure of Merit (FOM) Index that should be less than 1.2 nT for tail stinger sensor. This test is repeated every time when any modification in the aircraft or measuring instruments are done.

(vii) Advantages of Airborne Survey : Most of the advantages and limitations of airborne magnetic surveys are same as described in section 2.6.2 for airborne gravity survey. However, airborne magnetic survey is much more popular compared to gravity survey due to ease in operation of magnetometers from a moving platform.

The following advantages of this method make it ideally suited for any investigation:

(a) Due to anisotropy in susceptibility of exposed rocks, magnetic surveys are affected to a great extent by heterogeneous magnetization of exposed rocks, which is specially more in case of volcanic provinces and Archean-Proterozoic metamorphic terranes that are important for mineral investigation. Magnetic fields due to such heterogeneous magnetization are characterized by high frequency components, which get attenuated at the height of the aircraft.

(b) Speed of the Coverage can be approximately 100-150 km/hour in a helicopter-borne survey and 200-250 km/hour in regular airborne surveys using aero planes.

(c) Accessibility to areas which are otherwise inaccessible such as mountains, forest, marshy land etc.

(d) Cost effectiveness for airborne survey over large areas are worked out on the basis of effective cost per line-kilometer. The cost of airborne magnetic survey in India by an Indian agency like NGRI is about Rs. 1500 / line km while combined airborne magnetic, spectrometer and electromagnetic surveys cost about Rs. 4000 / line km. However, on an average the survey area should be larger than approximately 10,000 line km for a cost effective airborne survey.

(e) Multi-parametric survey makes it more efficient and cost effective.

(f) Recording at close interval provide data along the profiles at close interval of 1-3 m.

(g) Digital recording of data sets are suitable for automatic processing, interpretation and map preparation, which has reduced the time gap between data acquisition and its utilization.

(h) Due to closely spaced data points, linear features like lineaments, faults, fractures etc., are better delineated in airborne magnetic surveys that are important for resource exploration like groundwater, minerals etc., as discussed in Chapter 9.

Main limitation of airborne surveys lies in its initial cost and cost of survey. Several applications of airborne multi-parameter surveys for mineral investigations have been demonstrated in Section 9.10.

3.6.3   Satellite-Borne Magnetic Surveys (MAGSAT) – India and Himalaya

Satellite magnetic survey was conceived to record the earth's magnetic field from space without any limitations to national boundaries. The worldwide high altitude airborne magnetic profiles (Alldredge et al., 1963), which recorded small wavelength magnetic anomalies related to crustal blocks and oceans superimposed over large wavelength dipole anomalies of main field originating from the core of the earth can be considered as precursor to satellite magnetic. The first satellite magnetic data was recorded from POGO Satellite during 1970-71, which provided worldwide magnetic anomaly maps at an average altitude of 540 km (Regan et al., 1975) which did show correlation with major features of the earth such as plate boundaries, collision and subduction zones etc., It also showed correlation with trans continental features like Himalayas, Narmada-Son lineament in central India etc., (Mishra, 1977). These results encouraged NASA to launch a truly dedicated Satellite to record magnetic field of the earth in 1979, which was called MAGSAT. It recorded magnetic field of the earth during October, 1979 to June, 1980 at altitudes of 525 km and 325 km with an average altitude of about 400 km. It provided scalar and vector magnetic measurements over the entire globe in near polar orbits using flux gate magnetometers. The measurement accuracy of magnetic field was expected to be 6 nT. It was used to improve the model of the earth's main magnetic field what we refer to as International Geomagnetic Reference Field (IGRF) and also for the study of crustal magnetic anomalies for solid earth geophysics (Langel et al., 1982). The most important aspect of MAGSAT has been its truly international nature as data obtained from it were distributed to the scientists world over. This resulted in voluminous literature related to the application of MAGSAT data in the form of special issues of

Geophysical Research Letters (1982) and Journal of Geophysical Research (1985) and several subsequent publications, which can be referred to by those interested in details of satellite magnetic anomalies over different parts of the world. MAGSAT data provided for first time a global model of magnetic crust (Hahn et al., 1984).

With success of MAGSAT, some more magnetic satellites were launched recently such as Oersted Satellite in February, 1999 at altitude of 600-800 km by Danish Space Agency, which was basically meant for accurate models of main field as resolution of crustal component was affected by its high altitude. Subsequently, CHAMP Satellite (Maus et al., 2002) dedicated to measurements of both earth's gravity and magnetic field as described in section 2.6 was launched in July, 2000 by German Space Agency in Collaboration with NASA, USA. It was a low altitude Satellite with initial altitude of 455 km reducing to 300 km. It has a longer lie span and therefore, provided much larger data set compared to any other satellite. It could, there be used for several new studies like transient variations in fields with time or due to ocean tidal flow (Tyler et al., 2003) and changes in fields due to on going major tectonic activities such as those at subduction zones etc., which might be relevant to seismic activity in those regions (Purucker and Ishihara, 2005). It provided the distribution of magnetic field world over with an accuracy of 5 nT at an average altitude of 438 km (Maus et al., 2002). However, still the separation of continent-ocean boundary based purely on CHAMP data might be difficult (Hemant and Maus, 2005) though in some cases where bulk susceptibility contrast is good specially in case of volcanic margins, it might be possible to separate them. Curie point geotherm under Antarctica using CHAMP satellite magnetic data has provided the estimate of heat flux and its effect on ice sheet and glaciers (Maule et al., 2005). One can obtain this data set from internet at site http:op.gfz-potsdam.de/champ/main CHAMP.shtml, which provides several other variants of the field such as at 400 km of average altitude of satellite and continued field to ground surface and its components like ΔZ, ΔH etc., CHAMP satellite magnetic map over Indian continent, Himalaya and Tibet and adjoining oceans are given in Section 6.2.5.

(i) MAGSAT Total Intensity Anomaly Map of India, Himalayas and Tibet

This data set (MAGSAT) for Indian continent were obtained by Indian Scientists of different organizations and analyzed for different purposes (Mishra and Venkatrayudu, 1985; Negi et al., 1986; Arur et al., 1983; Singh and Rajaram, 1990). Langel et al., (1982) published world wide scalar magnetic anomaly maps using 2° x 2° grid. Mishra and Venkatrayudu (1985) reprocessed the same data for 1° – 1° grid and provided a scalar total intensity magnetic anomaly map for Indian continent (Fig. 3.15), which was used to infer Curie point geotherm over the entire continent and its correlation with major tectonic elements of the country (Mishra, 1986a). This map (Fig. 3.15) shows a magnetic low of < – 8 nT over Himalayas and Tibet increasing south wards > 5 nT over Indian Peninsular Shield. It has resolved several new magnetic anomalies compared to 2° × 2° data set, which are of great significance for geodynamic studies as described below. The correlation of these anomalies with tectonics is discussed in the next section. It may be interesting to note that the magnetic anomalies observed over Himalayas and Indian Peninsular Shield in POGO Satellite data were of same nature and amplitude of – 8 and + 4 nT (Mishra, 1977), respectively as compared to MAGSAT but differed in their details, which confirm the accurate representation of magnetic field in this map. The total intensity magnetic anomaly provided by CHAMP satellite at about 400 km over India and adjoining regions (Maus et al., 2002) shows almost the same type of magnetic anomalies as obtained from the MAGSAT (Fig. 3.15) but with difference in amplitude. The MAGSAT data is used here to delineate large scale deep seated tectonics such as Curie point geotherm.

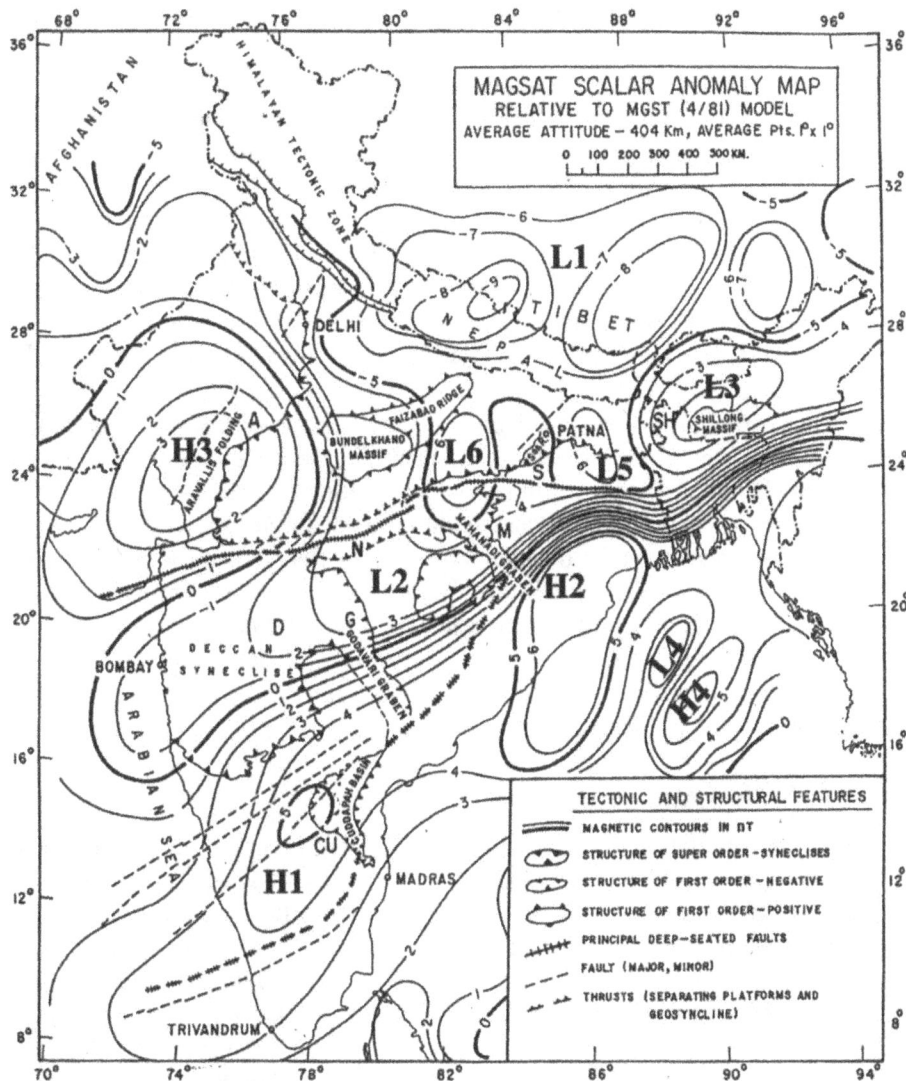

Fig. 3.15 MAGSAT total intensity map of India, Himalayas and Tibet at an average height of 400 km and $1° \times 1°$ averaged data super imposed over a regional tectonic map of India. L1-L6 and H1-H4 are magnetic lows and highs (Mishra, 1986a).

(ii) MAGSAT Total Intensity Magnetic Anomaly over India and Surrounding Regions and Large Scale Tectonics

The total intensity magnetic anomaly map of Indian continents superimposed over major tectonic elements of the country (Fig. 3.15) suggests the following important correlations (Mishra, 1986a):

(a) A large wavelength magnetic low over Himalayas and Tibet (L1), suggest the presence of large scale mafic rocks under it, for induced magnetization, which might be due to mafic/ultramafic intrusives in this section related to Himalayan orogeny (Chapter 5).

(b) Large-medium wavelength magnetic highs over South Indian Peninsular Shield and Eastern Ghat Fold Belt (H1 and H2) and Aravalli Fold Belt (H3) also indicate mafic nature of the crust in these sections as they represent Proterozoic collision zones (Mishra, 2006;

Chapter 6). However, magnetic highs related to them suggest predominance of remnant magnetization.

(c) The magnetic low L3 over Shillong plateau and surrounding regions suggest large scale mafic intrusives in the form of lower crustal rocks at shallow depth in this section and may even represent the extension of Satpura Mobile Belt (L2) east wards (Rajsekhar and Mishra, 2006).

(d) The two relative magnetic lows (L5 and L6) across Narmada-Son Lineament extending from Ganga basin south wards indicate the interaction of South Indian Peninsular Shield with north Indian shield and extension of ridges of Peninsular shield under Ganga basin such as Faizabad ridge and Mungher-Saharsa ridge east of Patna as shown in Fig. 3.15. These are relatively short wavelength anomalies, which are reflected only in $1° \times 1°$ MAGSAT map (Fig. 3.15).

(e) The sharp magnetic gradient extending from NE India (South of Shillong Massif) to west coast of Bombay indicates contact of sharp susceptibility contrast in the crust. There is no known feature in Indian geology/tectonics, which coincides exactly with this gradient. Therefore, it appears to represent different features in different sections. For example in east India, it may represent mafic oceanic crust under Bangladesh and mafic rocks of Singhbhum and Satpura Fold Belt, which is a major crustal boundary between the northern and the southern cratons representing Proterozoic collision zone as described above.

(f) These are some of the inferences described as an example. Different workers can use this basic data from their own perspective. However, while using this map, it should be borne in mind that it is based on $1° \times 1°$ degree averaged data.

3.7   Interpretation of Magnetic Anomalies

There are two approaches to interpretation as in case of gravity method namely (i) Qualitative and (ii) Quantitative. The former implies deriving information about geology and tectonics of the area and causative sources based on visual inspection of their nature and amplitude while the latter deals with the modelling of magnetic anomalies to derive unknown parameters of the causative sources. Magnetic sources are bipolar in nature and therefore, produce both positive and negative anomalies simultaneously. In this respect, it differs considerably from gravity field, which is unipolar and therefore, produces either a positive (high) or negative (low) gravity anomaly exactly over the source depending on the density contrast between the causative body and the surrounding rocks. This implies that even a negative (low) magnetic anomaly suggests a magnetic body with positive susceptibility while a gravity low indicates negative density contrast with surrounding rocks. Besides the intensity of magnetization of causative sources, the relative amplitude and disposition of magnetic highs and lows depend on (i) Direction of magnetization viz. inclination of earth's magnetic field for induced magnetization (ii) Dip of the body and (iii) Profile azimuth. These considerations make the interpretation of magnetic anomalies quite involved. It is therefore, essential to understand the effect of these factors on observed magnetic anomalies, which is discussed in the next section.

3.7.1   Changes in Total Intensity Magnetic Anomaly

The observed total intensity magnetic anomalies depend to a large extent on the inclination of magnetization vector, which for induced magnetization is the inclination of the present day earth's magnetic field, dip of the body and profile azimuth with respect to strike of the body. Therefore, total intensity magnetic fields for different values of these parameters are given below to understand their effects on the observed field. The computed fields with respect to these parameters presented here relate to induced magnetization in northern hemisphere, which by convention shows north towards the right. The same curves are also valid for southern hemisphere with convention of south towards right. To simplify, remnant magnetization is not considered. In such cases, the inclination of resultant magnetization influences the observed magnetic anomaly instead of induced magnetization alone.

(a) Inclination of Earth's Magnetic Field (Induced Magnetization) : The effect of inclination on total intensity magnetic anomaly is qualitatively demonstrated in Fig. 3.16 (a), (b) and (c) at equator, pole and an intermediate inclination of about 45° along a N-S profile. It is demonstrated here to understand the nature of observed magnetic anomalies in a simple manner. Fig. 3.16(a) demonstrates the nature of magnetic field at equator. Due to horizontal nature of earth's magnetic field, it will induce - ve poles on the southern side of the body and + ve poles on opposite side, which would produce its own field opposite to the earth magnetic field at the surface causing a magnetic low over it. At magnetic pole, the earth's magnetic field is vertical (Fig. 3.16(b)) which would induce -ve pole at the top of the body and + ve pole at the bottom of the body which in turn would produce its own field in the same direction as that field at the surface, producing a magnetic high at the top of the body. At an intermediate magnetic latitude of about 45° inclination of the earth's magnetic field (Fig. 3.16(c)) it would induce -ve and +ve poles on its sides such that its field would oppose earth's magnetic field towards the north causing a magnetic low in this direction and a magnetic high towards the south. Fig. 3.16(b) and c suggests that – ve pole produces magnetic high at the earth's surface as it would attract a + ve pole through which the magnetic field is measured and vice versa.

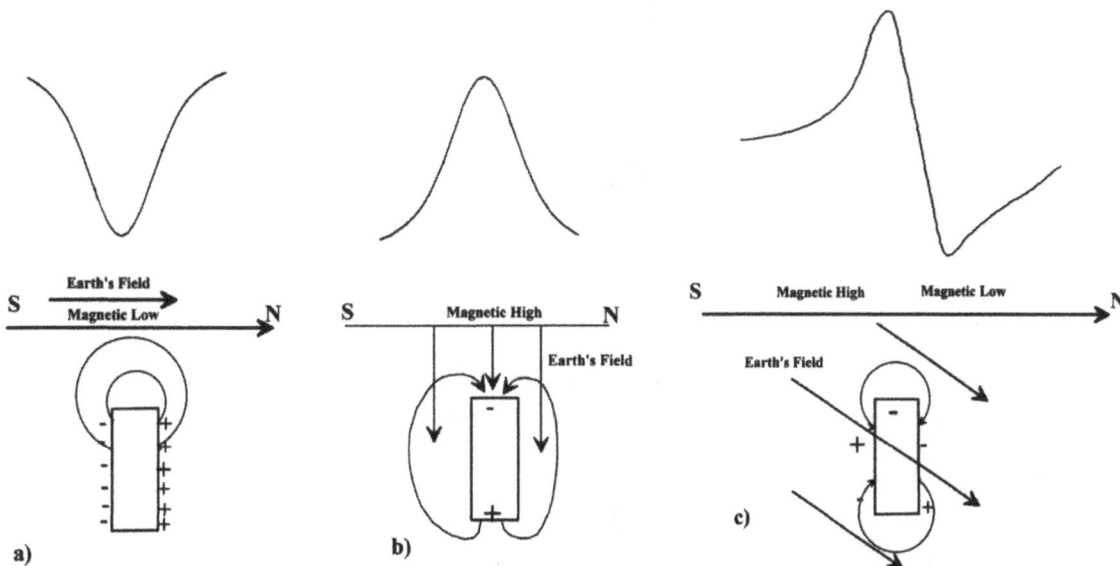

Fig. 3.16 Magnetic anomalies based on inclination of the Earth's magnetic field for induced magnetization **(a)** at magnetic equator corresponding to $I = 0^0$, magnetic low is observed **(b)** at magnetic pole corresponding to $I = 90^0$, a magnetic high is observed **(c)** at intermediate Inclination of about $I = 45^0$ provides a pair of magnetic high and low with low towards the north.

Now, we shall demonstrate the same by computing magnetic field for different inclination of the earth's magnetic field. Fig. 3.17(a) shows the total intensity magnetic field with variation in the inclination of the earth's magnetic field for I = 0°, 30°, 60° and 90° for a vertical dyke of some specified magnetization, depth and width. As is apparent, the total magnetic field at equator (I = 0°) is a symmetric low over the body approaching almost to zero on either sides similar to the one given in Fig. 3.16(a). This is because the total field is horizontal at the equator. With increase in inclination; I = 30°, 45° and 60°, it produces an asymmetric anomaly with equal amplitude for high and low for I = 45° similar to Fig. 3.16(b). For I = 90°, it is again a symmetric field with a high over the body as the total field is vertical at the North pole similar to Fig. 3.16(b). It may be noted that as inclination increases, the magnetic low is reduced and is located towards the north while magnetic high increases as inclination increases. However, this configuration of magnetic low and high may change due to remnant magnetization and dip of causative sources. In case of remnant magnetization, the resultant of induced and remnant magnetization vector should be considered. Therefore, in case the magnetic low of the observed magnetic anomaly in the Indian continent is located towards the north for an almost vertical body, it can be assumed to be caused by induced magnetization.

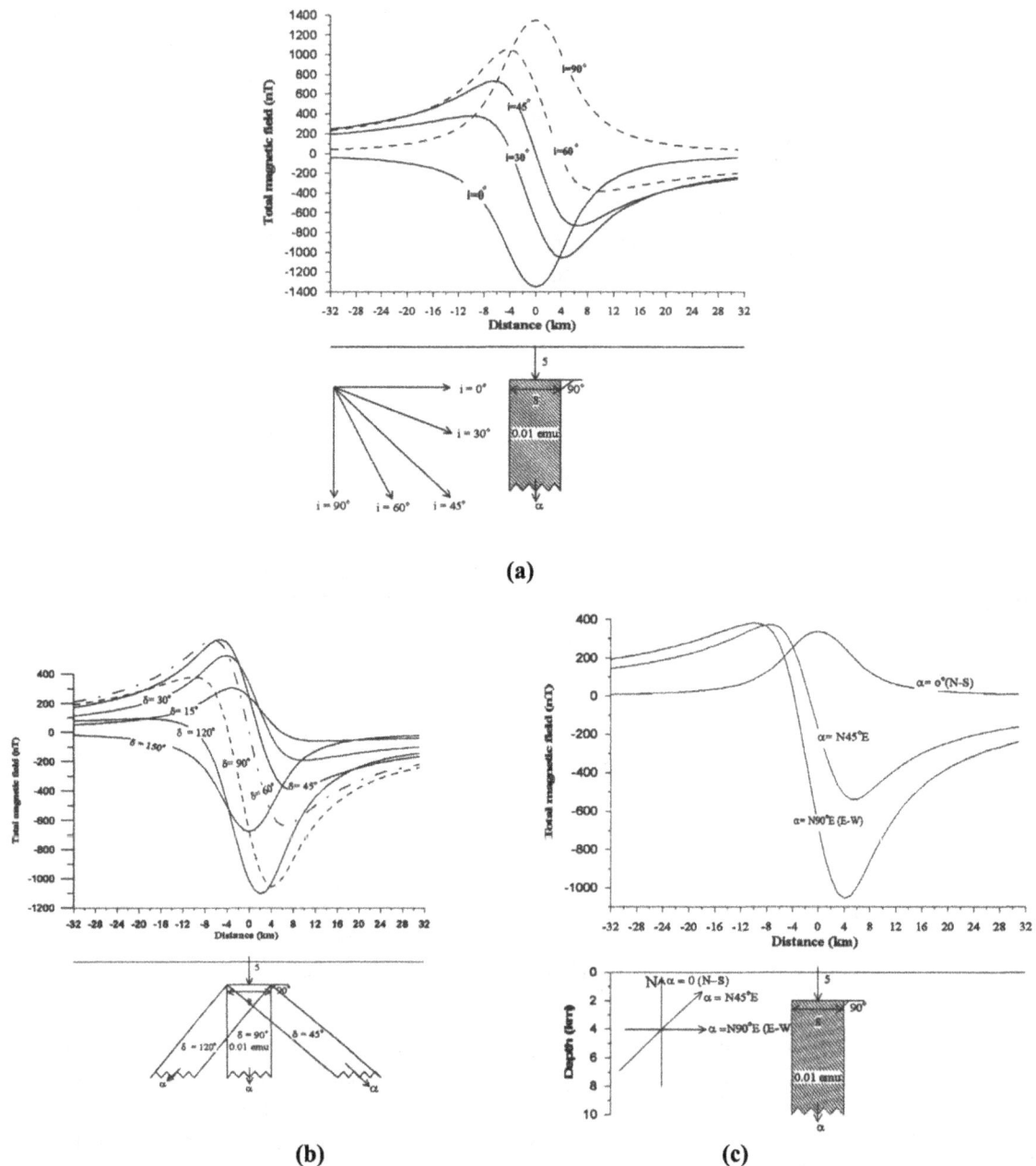

(a)

(b)

(c)

Fig. 3.17 Total intensity magnetic field along a N-S profile with respect to **(a)** inclination of earth's magnetic field **(b)** dip of body for $I = +30^0$ inclination and **(c)** profile azimuth for $I = +30^0$ inclination and vertical body. North is taken towards right hand side in northern hemisphere. The same is valid in southern hemisphere with south towards right hand side (Venkata Raju, 2004).

(b) Dip of the Body: Fig. 3.17(b) shows the variation in the computed total magnetic field for different dips (δ) of the body. The field is computed for an N-S profile and the earth's magnetic field inclination as 30°. An inclination of 30° is used as it is almost an average inclination for Indian continent. When dip of the body is 90° for a vertical body, the amplitude of magnetic low is more than that of corresponding magnetic high and is located towards north of the magnetic high. As dip decreases to 60°, 30° and 15° the relative amplitude of magnetic low decreases and magnetic high increases to the extent that for dip equal to 15° implying an almost horizontal sheet, mainly a magnetic high of minimum amplitude is observed. Contrary to it, when dip increases to 120° and 150° the relative amplitude of magnetic high decreases and the magnetic low increases to the extent that for dip equal to 150° mainly a magnetic low is observed with minimum amplitude. This example illustrate that dip of the body besides inclination of the inducing field plays an important role in producing relative amplitude of magnetic lows and highs.

(c) Profile Azimuth : Fig. 3.17(c) shows the effect of orientation of profile for a vertical body at 30° inclination of the earth's magnetic field. The amplitude is maximum for an E-W profile with a large magnetic low and is minimum for a N-S profile. For profile azimuth equal to 45°, the computed field is asymmetric with relative large amplitude for magnetic low. This is the reason that a N-S profile is not preferred in magnetic surveys. Even if the strike of the body is E-W, the survey profiles are laid 45° from N instead of N-S profiles.

The nature of these model anomalies are of great help in qualitative interpretation of observed magnetic anomalies and can be used as master curves to compare with the field curves for qualitative interpretation.

3.7.2 Qualitative Interpretation

After necessary corrections including the separation of IGRF regional field, the residual anomalies are presented either in a profile from or in the form of contour maps known as "Residual anomaly profile/map". The first thing now is to derive information, which are immediately apparent from visual inspection of the map. In this regard, the two most important factors are described below.

(i) Nature of Magnetic Anomalies : Information about the geology of the area suggests the type of problem involved in the exploration i.e., whether the survey is meant for mineral exploration, oil exploration or regional studies. Depending upon this, one should undertake a detailed examination of magnetic data.

In case of mineral exploration, two problems are generally encountered, viz. to work out the extension of exposed features and to obtain the details of concealed features. In case of former one can examine the nature of anomalies first on the exposed part and look for similar signatures in the adjacent region. The characteristics, which are found to be helpful in this regard, are strike, shape, amplitude etc., of the recorded anomalies. In second case when the magnetic sources are buried, it is generally used to delineate the target and obtain the parameters such as depth, width, dip etc., of the causative sources. Sources can be delineated easily by observing the density of the magnetic contours and amplitude of magnetic anomalies and unknown parameters can be obtained through model studies discussed in the next section. However, a general rule of thumb is that the sharper (large gradient) the anomaly, the shallower it is and the larger is the amplitude of magnetic anomaly, the more is the mafic component.

In oil exploration, magnetic method is used to delineate the basement configuration which in turn controls the structure in the over lying sediments. Depending upon the favorable structures and thickness of the overlying sediments further detailed investigations are recommended. Therefore, the magnetic method in oil exploration is generally of reconnaissance type. Knowledge about the configuration of magnetic highs and lows for basement high and low in the area of investigation with respect to present day inclination of the earth's magnetic field as given in Fig. 3.17(a) is very much helpful in this regard and demonstrated in Section 3.8. Basement highs generally behave like prismatic bodies and their magnetic anomalies are approximately similar to that of dykes of limited extent along profiles.

Regional surveys are generally conducted to delineate large scale structures such as faults, fractures etc., These large scale features on magnetic map can be delineated through alignment of similar trending anomalies as discussed in 3.8 under 'Field Examples'. In regard to regional surveys, it differs considerably from regional gravity surveys, which are conducted for crustal structure and depth to Moho. Magnetic surveys do provide the magnetic structure of the crust in terms of mafic intrusives etc., but depth of investigation in this case is limited by curie point geotherm beyond, which magnetization of rocks does not exist. However, even to get information from curie point geotherm is difficult as it represents the lower pole of the magnetic sources while the upper pole is formed at the top, which masks the effect of the lower pole. Therefore, the deepest source, which can be safely detected in magnetic surveys are the basement in sedimentary basins at a depth of about 5-6 km. It might be possible to record signal from deeper sources, viz. Curie point geotherm in satellite magnetic surveys as it records mainly large wavelength anomalies originating from deeper levels. The depth to the base of magnetic crust (Curie point geotherm) for Indian continent based on MAGSAT (Fig. 3.15) is given in Chapter 4.

(ii) Amplitude of Magnetic Anomalies and Rock Types

Depending on the nature and amplitude of the magnetic anomalies recorded in a particular area, a fair guess can be made about the nature of the causative sources. For example, igneous rocks are more magnetic than metamorphic and sedimentary rocks. A qualitative guide about rock types, depending on the order of magnetic anomalies, are given below.

Case a: Largest magnetic anomalies with a range > 10,000 nT are generally caused by extensive magnetite deposits often of economic importance (iron ores).

Case b: Anomalies of the order of 1,000 - 10,000 nT are generally caused by volcanic mafic rocks rich in magnetite such as basalts, dolerite dykes etc.

Case c: Magnetic anomalies in the range of 300-1000 nT can be caused by bodies of Class b, containing less of magnetite while structures such as fault fractures etc., associated with mafic rocks can also produce magnetic anomalies in the range of 300-500 nT. These anomalies are generally characterized by specific pattern suggesting intrusions and specific structures. However, this order of anomalies can also be caused by metamorphic rocks in fold belts or convergence zones but they do not show any specific pattern due to heterogeneous distribution of magnetic minerals in them or variation in magnetization (anisotropy).

Case d: Magnetic anomalies of 100-300 nT with large-medium wavelength are generally associated with sedimentary basins and arise due to basement undulations depending on their depth and mafic component in them. These anomalies do not generally show any pattern as they are caused by isolated undulations in basement and are generally broad in nature.

Case e: Magnetic anomalies ≤ 100 nT are caused by structures like fold, faults etc., with in the sediments or basement under them with less of mafic components. It can also be caused by felsic intrusive bodies with less mafic components.

Besides amplitude of the magnetic anomalies, their disposition or alignment in a particular fashion is usually used to infer structures like fissures, faults and fracture planes. The nature of the magnetic contours are even suggestive of the depth of the causative sources such as smooth contours and low gradient along the flanks of anomalies suggest a deeper source compared to high gradient regions. This criteria is very much helpful in delineating deep basins compared to surrounding platform, which are usually characterized by sharp gradients and high values.

3.7.3 Quantitative Interpretation – Characteristic Points and Curve Matching

Quantitative interpretation implies estimation of the unknown parameters of the causative sources from the observed magnetic field such as depth, bulk susceptibility etc. The purpose of depth estimation of magnetic sources primarily relates to the depth to top of the body as the effect of second pole at the bottom of it is masked due to effects of poles at the top of it. However, while modeling the full extent of the causative sources using two or three dimensional sources as discussed in the next Section (3.7.4), the total extent of the sources can be considered. There are basically three approaches to it:

(i) Characteristics Points

It is based on some important characteristics of the observed field such as maximum anomaly, its gradient etc. The most important characteristic of the observed field is the maximum anomaly, its gradient and points where it decays to half or zero. Peters (1949) was first to suggest some of the depth rules from the observed magnetic anomalies. He suggested that in case of a tabular body such as dykes the horizontal distance between maximum gradient (slope) and half of it, is equal to 1.6 h where h is depth to the body.

Vacquier et al., (1963) provided important characteristic points related to depth indices for a prismatic body which implies vertical thick dykes limited in depth extent and have been used widely for basement magnetic anomalies. One of the suggestions made by them related to the application of curvature map (horizontal derivative map) where zero contour indicates edges of the causative source and distance between curvature minima/maxima and zero contour can be used to estimate depth.

(a) Line of Poles

In case the magnetic charges are distributed along a line (Fig. 3.18(a)), the magnetic field due to a small element dx at a point, P in direction r is given by (Jakosky, 1961).

$$dF = mdx / r^2 \qquad \qquad .....(3.19)$$

where m is the pole strength per unit length. Its component vertical to it in direction d is given by:

$$dF_d = (md / r^3)\, dx \qquad \qquad .....(3.20)$$

As profiles are usually planned perpendicular to the strike (line of poles), the total field can be obtained by integrating it, given by:

$$F_{max} = \int_{-\infty}^{\infty} \frac{md\,dx}{r^3} = \frac{2m}{d} \qquad \qquad .....(3.21)$$

This expression provides the maximum anomaly, which would be observed if the body can be represented by line of magnetic poles. This implies thin bodies such as a sheet or thin dyke of thickness less than its depth. This expression can be used to determine approximately the depth of the body (d) if magnetization (m) is known or vice-versa. In case of observed magnetic anomaly is the intensity of magnetization (I), which is given by:

$$I = k\,T \qquad \qquad .....(3.22)$$

where T is the earth's magnetic field equal to approximately 0.5 Oersted and k is susceptibility. In case of igneous rocks, if the susceptibility is equal to 1×10^{-3} emu, then

$$m = 10^{-3} \times 50{,}000 = 50 \text{ nT} \qquad \qquad .....(3.23)$$

If the observed maximum anomaly is 100 nT, then

$$d = 1 \text{ k.m.} \qquad \qquad .....(3.24)$$

This is a very approximate relationship as direction of magnetization etc., has not been considered but can be used as a thumb rule for quick estimation of depth.

(b) Line of Dipoles

Fig. 3.18(b) shows a typical distribution of dipoles, which may be applicable to geological bodies like thick dyke etc. The magnetic field due to such a dipole can be simply obtained by substituting for the magnetic moment, mb in equation (3.21), where b is the width between line of poles (Jakosky, 1961).

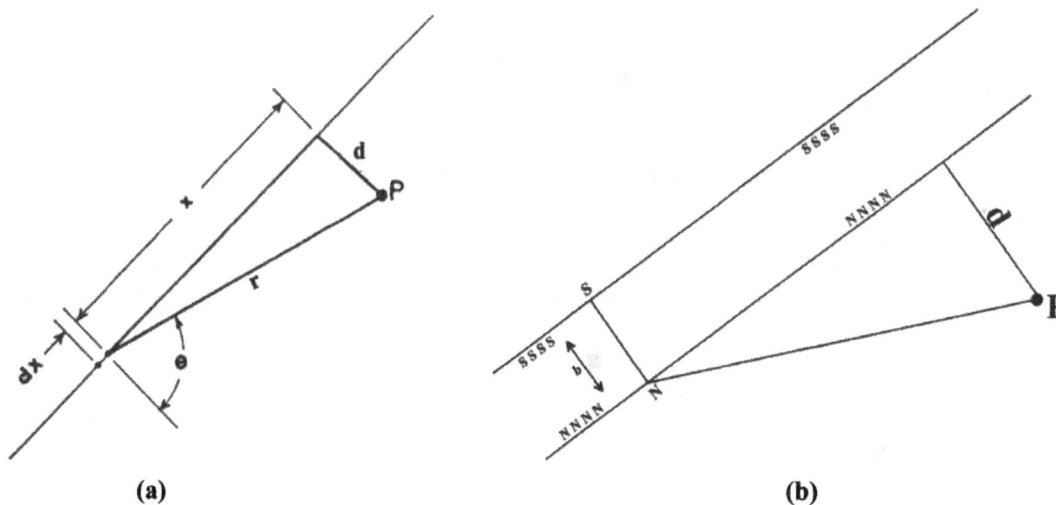

Fig. 3.18 Magnetic field due to (a) line of poles (b) line of dipoles.

$$F_{max} = 2mb\,/\,d \qquad \qquad .....(3.25)$$

where b = distance between the poles and incase of magnetic survey it will represent width of the body. Using same approach as given above for line of poles, one can obtain the approximate depth of dipoles (thick dyke) by assuming distance between ½ max anomaly points as the width of the body.

Some simple depth rules have been given in almost all text books on geophysical exploration, which can be easily followed by field workers in the field. Smellie (1956) has considered asymmetry of observed magnetic anomalies at different magnetic latitudes and given some simple depth rules based on distance between maximum and half of maximum anomaly on either side of the maximum point, which is known as half width. A set of standard curves given by him due to line of poles in northern hemisphere is reproduced in Fig. 3.19 for its demonstration in the next section.

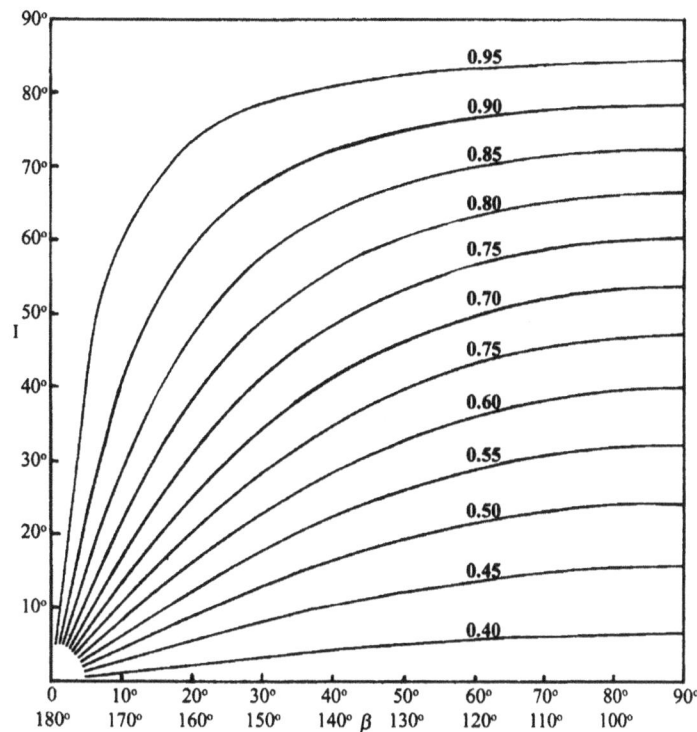

Fig, 3.19 Standard curves for constant K_2' related to half width of anomaly towards south due to line of poles in northern hemisphere (Smellie, 1956).

(ii) Curve Matching : It is based on comparison between already computed field due to specific bodies with known parameters called standard curves and the observed magnetic anomaly. The most important curve matching technique for the magnetic field due to tabular bodies has been described by Gay Jr. (1963). He provided standard curves for these bodies, which have been extensively used for interpretation of magnetic anomalies due to thin and thick dykes. As full anomaly curve is used in this case, it provides better estimates of unknown parameters compared to depth rules based on a few characteristic points described above. The expression used by him to provide the standard curves for tabular bodies is given below under computation of magnetic fields (Section 3.7.4). However, due to limitations in availability of standard curves covering the whole gamut of geological bodies with different unknown parameters, it is advisable to compute the fields due to specific bodies and compare it with the observed field for the best fit as described below in section 3.7.4.

(a) Application: Airborne Total Intensity Profile Panna Diamond Belt

An airborne total intensity magnetic profile across Panna Diamond belt is shown in Fig. 3.20 (NGRI, 1969; Mishra, 1987). The airborne total intensity map of this region is discussed in section 7.6.2 in regard to its application to geodynamics and mineral exploration. This profile AA' is extracted from the total intensity map prepared from data flown at 500' (~152 m) height and interpreted through characteristic points of Smellie (1956) and curve matching using the standard curve for tabular bodies (Gay, Jr., 1963) as described below. It may be noted that magnetic low along this profile is located towards the north suggesting induced magnetization for almost a vertical body as causative source and amplitude and width of anomaly indicate thick intrusive body as a dyke. The parameters obtained from matching this magnetic anomaly with standard curves (Gay, Jr, 1963) are given below (Achuta Rao, 1981).

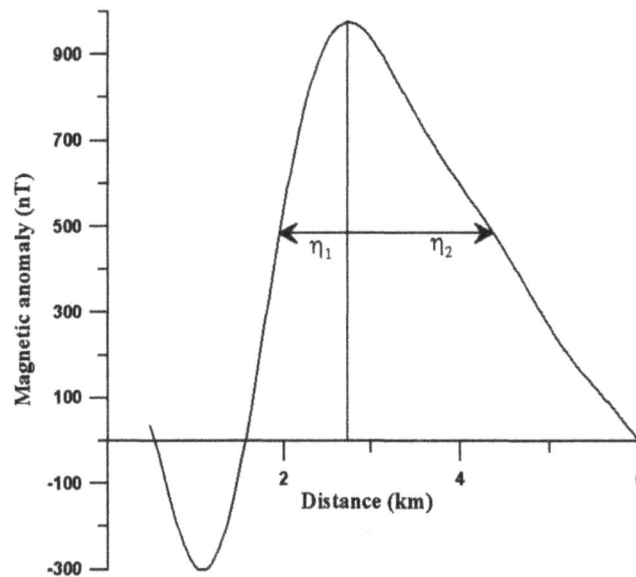

Fig.3.20 Airborne total intensity profile across Panna diamond belt in central India flown at height of 500'
(~152 m) and its interpretation using standard curves for tabular bodies and characteristic
curves given in Fig. 3.19.

Depth below plane of observation:		503 m
Depth below surface:	$503 - 152$ =	351 m
Susceptibility contrast:	6.5×10^{-3} emu	
Dip of the body:		78°

Interpretation of same profile using characteristic points was carried out for line of pole
(Fig. 3.20). The magnetic profile provides half widths, $\eta_1 = 1.3$ cm, $\eta_2 = 4.0$ cm with reference to
46,000 nT as the value of undisturbed field (base) in the region.

Therefore $\eta_1 / \eta_2 = 0.32$

The standard curves for line of poles (Fig. 3.19) provides $K_2 = 0.45$ for inclination $I = 34°$ in this
region and $\beta = 50°$, (angle between north and strike of the body, elongated contours) which
provides depth below plane of observation as 0.45×4.0 cm $= 1.8 \times 0.4$ km (as per scale of
Fig. 3.20) = 720 m. Depth below surface is therefore 720-152 = 568 m.

Difference between the depth estimated from curve matching (351 m) and characteristic points
(568 m) can be attributed to the assumptions involved in the two methods. However, both
suggest a shallow source for this anomaly. It may be noted that the total curve matching has
provided all the unknowns of the causative sources while characteristic points provide only its
depth indicating preference of curve matching for the interpretation of magnetic anomalies.
These methods are now mainly of historical importance. Still they can be used when
computational facilities are not available to provide rough estimates. Presently, computation of
fields due to specific bodies is mainly practiced for interpretation of magnetic anomalies as
discussed below.

3.7.4 Computation of Magnetic Field Due to Simple Models

There are two approaches in this class. Firstly, a two dimensional approach in which the magnetic
field can be modeled along a profile or a three-dimensional approach where magnetic field due to a

three dimensional body is computed in a plane and compared to the observed field. The latter, namely three-dimensional modelling is a very involved process due to several unknowns about source configuration and induced and remnant magnetization and is therefore, mainly used for specific geological problems where it is extremely essential. Some pertinent references related to modelling of magnetic field due to three-dimensional arbitrary shaped bodies are as follows, which can be referred by those interested in it: Talwani and Ewing (1960), Coles (1976), Barnett (1976), Coggan (1976), Pedersen (1978), Guptasarma and Singh (1999), Furness (1994) etc.

Two-dimensional modeling implies modelling along profiles and determination of unknown parameters of the causative sources along that profile. Selection of profile is most important aspect of two-dimensional modelling. Firstly, it should be almost perpendicular to the general strike of the body, which is indicated by general geology of the region and also by the elongated contour pattern of the anomaly. Secondly, it should pass through maximum observed anomaly and cut across essential elements of the composite anomaly. For example, in case a particular magnetic anomaly is defined by its positive (high) and negative (low) centers, in that case the selected profile should pass through both of them and preferably through their maximum values. As stated above, the magnetic anomalies are influenced by direction of present day earth's magnetic field and direction of remnant magnetization. Therefore, modelling of magnetic anomalies is a quite involved process. However, the most common approach to model magnetic anomalies is to approximate the causative sources by some simple physical model like a dyke as a tabular body and compute its field and compare with the observed field.

The expression for magnetic field due to simple bodies and related computer codes are given in several text books on this method, which can be used for this purpose (Radhakrishnamurthy and Mishra, 1989, Radhakrishnamurthy, 1998). The magnetic field due to a tabular body has been dealt exhaustively by Gay Jr. (1963) who has also provided the standard curves for this purpose, which can be readily used for the interpretation of anomalies due to such bodies. Expressions to compute magnetic field due to arbitrary shaped two dimensional bodies have been given by Bhattacharya (1964, 1980), Talwani (1965) and several others which can be used for this purpose. Expression for 2.5 dimensional arbitrary shaped bodies which are limited in extent have also been provided by Rasmussen and Pedersen (1979), Menichetti and Guillen (1983) and Webring (1986) etc. The code provided by Webring (1986) has been found to be very useful as it computes both the gravity and the magnetic fields for 2.5 dimensional bodies and is freely available on Website of USGS, USA. In case of arbitrary shaped bodies, the causative sources are approximated by polygonal cross sections, which are defined by their vertices and the magnetic field is computed with respect to specific susceptibility contrast with surrounding rocks. Some of these schemes can be simultaneously used on magnetic and gravity field along a specific profile.

Expressions to compute magnetic fields due to some simple bodies are given below and related computer codes are referred which can be used to compute the fields due to such bodies. Oceanic magnetic anomalies are modeled differently due to sea floor spreading magnetic anomalies by tabular bodies of different widths (Bott and Hutton, 1970).

(i) Expression for Magnetic field due to some Simple Models

With availability of fast computing facilities and commercial softwares, the magnetic field due to arbitrary shaped bodies, are computed and compared with the observed field. Using least squares approach, the observed field can be directly inverted in form of causative sources as described in Chapter 4. However, as number of unknown increases, the uncertainties of estimation also increase. In this respect, simple models with fewer unknowns are attractive and can be used for first hand interpretation of magnetic anomalies to be followed up by more advanced interpretation, if so required. Expressions to compute magnetic field due to some simple bodies are described below. Some of them have common characteristics for magnetic and gravity fields which can be used to compute the two fields simultaneously. While modeling magnetic field due to simple models like tabular bodies, it is assumed that the first pole is located at the top of the body and second pole is far away assuming at infinity as its effect will be less which can be ignored. In such cases the depth to the body implies depth to the top. One can model for limited body in depth extent by taking second pole at the bottom but this exercise complicates the process. In such cases, one can use the expression for 2 dimensional bodies along a profile referred to above. These expressions are also useful to develop inversion schemes for the two fields as given in Chapter 4. The model and the symbols used in expressions are given in Fig. 3.21(a). Y-axis represents the strike of the body and X-axis is the profile direction with α as the angle with respect to magnetic north. Due to interaction of induced and remnant magnetization and strike of the body, Hood (1964) defined an effective inclination and declination, which are important in case of magnetic anomalies. If I_o and α are the inclination of the earth's magnetic field (T) and profile azimuth with respect to north, respectively and J_o and A represents the inclination and declination of the resultant magnetization including both induced and remnant magnetization. Then the effective inclination of the induced and the resultant fields I_1 and J_1, respectively are given by (Hood, 1964).

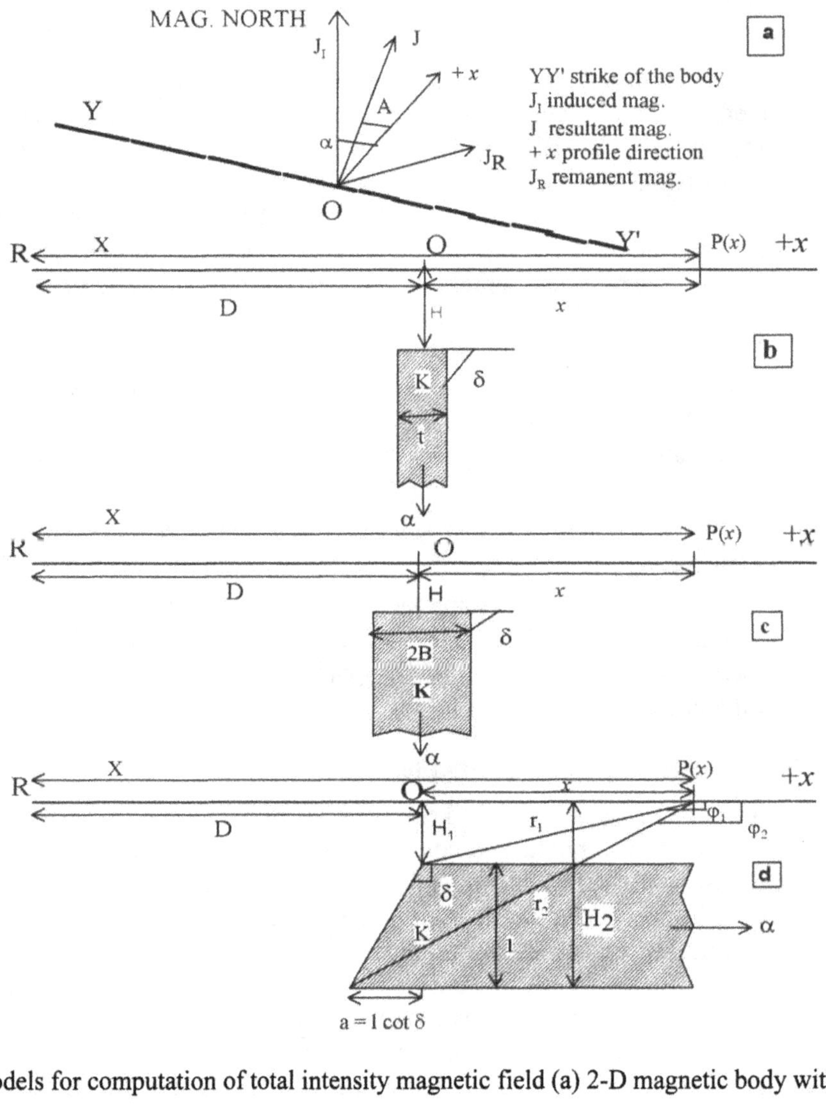

Fig. 3.21 Models for computation of total intensity magnetic field (a) 2-D magnetic body with magnetization

vectors (b) Thin sheet (c) Tabular body (dyke) (d) contact (Venkata Raju, 2004).

$$\tan I_1 = \tan I' / \cos \alpha \qquad \qquad \dots\dots(3.27)$$
and
$$\tan J_1 = \tan J' / \cos A \qquad \qquad \dots\dots(3.28)$$

These factors are used in the forth coming section to define magnetization in the models.

(a) Thin Sheet Model

It represents a linear vertical or inclined source whose thickness is much less compared to depth. It is used to represent mineralized shear zones, fractures, faults, thin intrusive bodies etc. Thick dykes with widths less than depth can also be represented by this model expressed as ensemble of thin sheets. The general expression for total intensity magnetic fields due to such a model (Fig. 3.21(b)) along a profile perpendicular to it is given by (Venkata Raju, 2003).

$$\Delta T(x) = P_1 \frac{x\sin Q + Z_1 \cos Q}{x^2 + Z_1^2} \qquad \qquad \dots\dots(3.29)$$

where x and Z1 are distances along profile from central point over the body and depth, respectively and P_1 and Q are known as amplitude factor and index parameter, respectively, which are given by:

$$P_1 = 2kT\beta\, (1-\cos^2 I_o \sin^2 \alpha)^{\frac{1}{2}} (1-\cos^2 I_o \sin^2 A)^{\frac{1}{2}} \text{ and} \qquad \dots\dots(3.30)$$
$$Q = I' + J' - \delta - 90° \text{ and } \beta = t \text{ where } t = \text{thickness of sheet} \qquad \dots\dots(3.31)$$

Similarly, the horizontal derivative of gravity anomaly over a thin sheet along a profile perpendicular to strike is given by (Stanley, 1977).

$$g(x) = P_2 \frac{x\sin Q + Z_1 \cos Q}{x^2 + Z_1^2} \qquad \qquad \dots\dots(3.32)$$

Here again P_2 and Q are known as amplitude factor and index parameter, respectively, which are given by:

$$P_2 = 2G\rho\beta \text{ and } Q = 180 - \delta \qquad \qquad \dots\dots(3.33)$$

where ρ is the density contrast and $\beta = t$ where $t = $ thickness of sheet

One can find the similarity between the two expressions viz (3.29) and (3.32), which can be coded into a common computer package. However, one may notice that expression for gravity field in equation (3.32) is related to horizontal gradient of gravity field, which can be obtained by dividing the differences in the gravity field of consecutive stations by horizontal distances.

(b) Long Tabular Bodies (Dyke) Model

This is a very popular model for magnetic interpretation as most of the intrusives and even basement anomalies can be represented by it. The total intensity magnetic field (ΔT) due to such a model (Fig. 3.21(c)) is given by (Gay, Jr., 1963).

$$\Delta T(x) = P_1 [(0.5) \sin Q \ell n \frac{(x+B)^2 + Z_1^2}{(x-B)^2 + Z_1^2} + \cos Q \{ \tan^{-1}\frac{(x+B)}{Z_1} - \tan^{-1}\frac{(x-B)}{Z_1} \}] \qquad \dots\dots(3.34)$$

Where P_1 and Q are same for thin sheet and tabular bodies (dyke) model with half width of the dyke, B = sin δ. Venkata Raju (2003) has provided a code for this expression along with the inversion scheme, which can be obtained from his publication. An example of modelling a total intensity magnetic profile across Panna Diamond belt described in the previous Section 3.7.2 using the expression (3.34) and inversion scheme discussed in Chapter 4 is given in Chapter 4 under 'Field Examples'.

(c) Sloping Contact (fault) Model

The sloping contact (Fig. 3.21(d)) is used to model the contact of rocks of different susceptibility density, which generally specify a fault. It is one of the most important geological models encountered in the field and provide specific gravity and magnetic anomalies like step with sharp gradient over the contact as given in Fig. 2.25(a).

The general expression for total intensity magnetic anomaly (ΔT) due to such a body is given by (Venkata Raju, 2003):

$$\Delta T(x) = P_3 \left[(0.5)\cos Q \, \ell n \frac{(x-a)^2 + Z_2^2}{x^2 + Z_1^2} + \sin Q \left\{ \tan^{-1}\frac{x-a}{Z_2} - \tan^{-1}\frac{x}{Z_1} \right\} \right] \quad \ldots..(3.35)$$

where

$Z_2 = $ Bottom Depth
$P_3 = 2kT\beta \,(1-\cos^2 I_o \sin^2 \alpha)^{\frac{1}{2}} \,(1-\cos^2 I_o \sin^2 A)^{\frac{1}{2}}$ $\ldots..(3.36)$
$Q = I_1 + J_1 - \delta - 90°$ $\ldots..(3.37)$

and

$a = l\cot\delta$
$l = $ thickness of contact

3.7.5 Automatic Methods of Depth Estimation

(i) Werner Deconvolution

With development of airborne magnetic surveys, several computer based automatic methods were developed to analyze large volume data provided by these surveys. The most important in this class was provided by Werner (1953) who expressed the magnetic field due to thin two dimensional dyke as linear equation in unknown parameters of the dyke whose solutions provided the unknown parameters of the dyke. This method is known as Werner Deconvolution. Hartman et al., (1971) extended this approach to derive more unknowns using derivative of the total intensity magnetic field. Briefly the magnetic field (T) observed along x axis for a dyke can be expressed as:

$$F(x) = \frac{A(x - x_0) + Bz}{(x - x_0)^2 + z^2} \quad \ldots..(3.38)$$

A and B are functions of the field strength, susceptibility and the geometry of the body (dip, strike, magnetic inclination etc.,), x_0 is the horizontal distance along the traverse over top of the dyke and z is the depth to its top.

This expression can be expressed as:

$$x^2 F = a_0 + a_1 x + b_0 F + b_1 x F \quad \ldots..(3.39)$$

where

$a_0 = -Ax_0 + Bz$
$a_1 = A$
$b_0 = -x_0^2 - z^2$ and $b_1 = 2x_0$

This implies that

$x_0 = \frac{1}{2} b_1$ and $z = \pm \frac{1}{2}\sqrt{(-4b_0 - b_1^2)}$

There are four unknowns and using the field at four observation points, the simultaneous equations in these four unknowns can be solved and x_0 and z can be obtained. In this manner the window of four points can be shifted along profile and almost a continuous set of x_0 and z can be obtained and plotted along the profile. The geological meaning of these parameters, however, should be provided by interpreter based on the observed magnetic anomalies for that magnetic inclination and geology / tectonics of the region. Hartman et al., (1971) introduced the effect of interference due to surrounding bodies by incorporating an interference polynomial in the above equation (3.38) as follows:

$$F(x) = \frac{A(x - x_0) + Bz}{(x - x_0)^2 + z^2} + C_0 + C_1 \; x + \ldots\ldots\ldots + C_n x^n \qquad \ldots\ldots(3.40)$$

Now, there are $(n + 5)$ unknowns and $(n + 5)$ points are required to evaluate these unknowns. In practice a first or second order polynomial is sufficient for interference factor implying that 6 or 7 points are used to evaluate the source parameters.

(ii) Euler Deconvolution

Thompson (1982) suggested a method based on Euler's homogeneity relationship for automatic interpretation of airborne magnetic profile. Its advantage lies in its independence from any specific geology or type of sources like dyke etc.

For any general type of magnetic source such as point source, dipole etc., the total magnetic intensity in (x,y) plane caan be expressed as function of its distance with respect to the source (x_0, y_0, z_0).

$$T(x,y) = f[(x - x_0), (y - y_0), z_0] \qquad \ldots\ldots(3.41)$$

Euler's equation for this function is given by:

$$(x - x_0)\frac{\partial T}{\partial x} + (y - y_0)\frac{\partial T}{\partial y} - z_0\frac{\partial T}{\partial z} = -N\,T(x,y) \qquad \ldots\ldots(3.42)$$

The gradients in x,y,z directions can be calculated in frequency domain as given in Section 4.2 or measured and used in this equation. In a two dimensional case of a profile $\partial T/\partial y = 0$. Therefore,

$$(x - x_0)\frac{\partial T}{\partial x} - z_0\frac{\partial T}{\partial z} = -NT(x) \qquad \ldots\ldots(3.43)$$

or
$$x_0\frac{\partial T}{\partial x} + z_0\frac{\partial T}{\partial z} = x\frac{\partial T}{\partial x} + NT(x) \qquad \ldots\ldots(3.44)$$

Having calculated the gradient $\partial T/\partial x$ and $\partial T/\partial z$, the only unknowns are x_0, z_0 and N. The former two represent location and depth of the equivalent source and the latter (N) represent the type of source which best represents anomaly and varies from 1.0 for line of poles to 3.0 for point dipole and 2.0 for point pole and line of dipoles, that can be fixed based on the nature of anomalies as described in Section 3.7.3. The vertical gradient is determined in frequency domain as discussed in Section 4.2. In case of two dimensional maps there are four unknowns x_0, y_0, z_0 and N. A 3 X 3 window in this case can be used to estimate these parameters, which provides over determined system of nine equations and are solved using principles of least squares as described in Chapter 4. The least squares solution of the over determined set of equations also yields estimate of the standard deviation in parameter z, which can be used as 'error bar' to check its reliability.

There are some more automatic methods like analytic signal (Nabighian, 1972, Roest et al., 1992) but all these methods suffer from converting the information derived from them to geologic/tectonic models. A simple method of depth estimation using first order magnetic derivatives was proposed by Salem et al., (2007). There are commercial softwares, which perform these operations on given data sets but as enough precautions are not taken in identifying the full anomaly due to causative sources, the results obtained from them become difficult to interpret in terms of regional local geology. However, a smart interpreter can always use them in conjunction with other methods.

3.8   Field Examples: Geodynamic Studies and Mineral Exploration

Some field examples to demonstrate the nature of magnetic anomalies for different geological and tectonic settings are presented below. Readers may choose suitable ones for them depending on their requirement and interest. Some depth estimates using simple depth rules as described above are also presented. More involved quantitative estimates of magnetic anomalies are described in Chapter 4 after inversion schemes have been introduced in that chapter. Specific applications of magnetic surveys for geodynamic studies and resource exploration are discussed in Part II, where they are integrated with other geophysical and geological information for comprehensive interpretation.

3.8.1   Extensional Regimes – Red Sea Rift, Continental Margins of India, Norway and USA, and Proterozoic Rift Basin, USA

Extensional and Compressional Tectonics as discussed in Section 2.7.1 and demonstrated in Figs. 2.23(a-d) are two most important aspects of geodynamics, which are related to extension and convergence, respectively. As stated in that section, the extensional tectonics are caused by sub-lithospheric thermal anomaly and give rise to rift basins. They are usually associated with intrusives, which in case of new rifts as in case of Red sea rift given below are represented by several vertical to sub vertical intrusives in the central part that may be exposed or subsurface. In such cases, the expected magnetic anomalies will be similar to those demonstrated for vertical bodies in different geomagnetic latitudes (inclination) in Figs. 3.16 and 3.17. Corresponding to geomagnetic latitudes in India, one should get a pair of low and high with low located towards the north (Fig. 3.16) for a vertical body with induced magnetization. Further as there are several intrusives due to extensional tectonics, one may expect more than one pair of magnetic highs and lows representing sub-vertical mafic intrusives.

(i)   Red Sea Rift

Red Sea rift is a classical example of active rifting during recent times where process is still in progress. A typical magnetic and gravity profile across central axis Red Sea (Fig. 3.22(a); Drake and Girdler, 1964) shows Bouguer gravity high in the central part as discussed in Section 2.7 (Fig. 2.23(b)). However, free air anomaly is close to zero indicating isostatic compensation and high Bouguer anomaly suggests a crustal thinning as is the case with active rifts (Fig. 2.23(b)). Isostatic compensation in this case suggests that isostasy operates on a few million time scale and therefore, recently developed tectonic units are also likely to be isostatically compensated. It also shows pairs of magnetic lows and highs of almost 1500 nT (gamma), which indicate large scale mafic intrusives related to rifting as discussed above. This figure also shows the importance of Bouguer anomaly in oceans, which reflects the central gravity high instead of free air anomaly specially in regions of large bathymetry changes. Central gravity high is reflected as a linear high in satellite altimetry confirming its importance for geodynamic studies even for medium sized tectonic features.

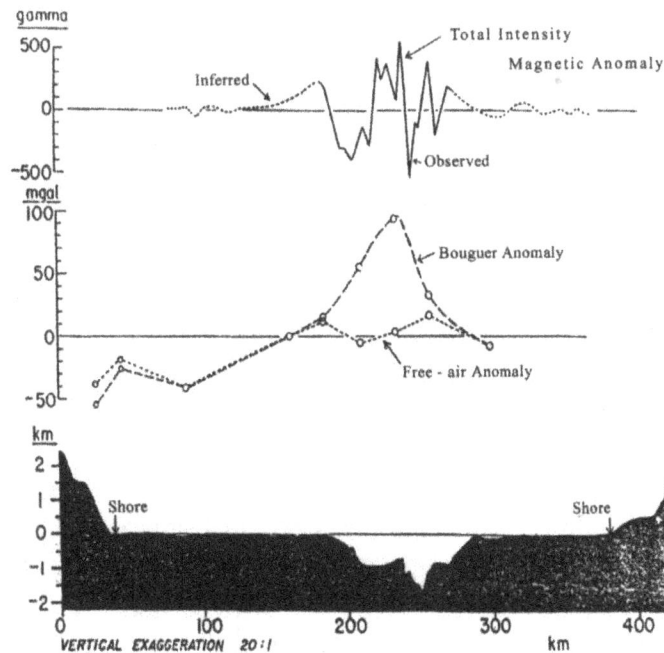

Fig. 3.22(a) Marine gravity and magnetic profile across Red sea central trough near 16^0 N showing Bouguer anomaly high and high amplitude magnetic anomalies related to intrusives in the central part associated with extension and rifting (Drake and Girdler, 1964). Free air anomaly close to zero indicate isostatic compensation.

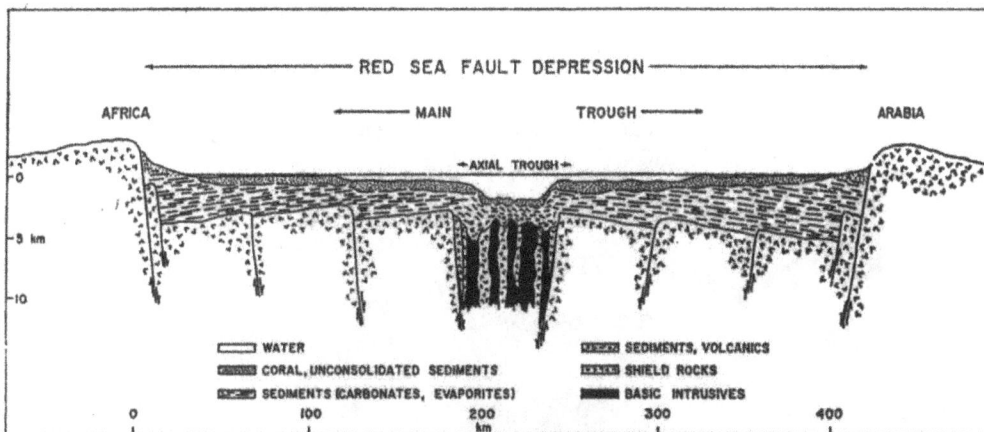

Fig. 3.22(b) Structural section of the northern part of the Red Sea based on geophysical surveys. It illustrates the rifting process between Africa and Arabia related to extensional tectonics and associated basic intrusives during recent times (Drake and Girdler, 1964).

Fig. 3.22(b) is an illustration of rifting of Africa and Arabia along the Red sea rift due to extensional tectonics and its structure based on large scale geophysical surveys in the entire section (Drake and Girdler, 1964). The basic intrusives in the axial trough were delineated primarily based on marine magnetic surveys along several profiles across the Red Sea, which mapped details of each intrusive in this section. Their magnetic signatures were also supported from marine gravity as given above and seismic surveys, which suggested high density (3.0 g/cm^3) and high velocity (7.1 km/sec) rocks indicating basaltic rocks that form the oceanic crust. It also shows the formation of axial trough through vertical intrusives as demonstrated in Fig. 2.23(b) and platform on either sides. As discussed in section (2.7.1) such rifts with basic intrusives are referred to as active rifts. However, there are also passive rifts, which are formed due to sub lithospheric upwelling caused due to some thermal anomaly or due to any other cause such as plate tectonic forces etc. In such cases, however, the magnetic anomalies due to intrusives are absent and they may originate from basement, which would be considerably subdued (200-300 nT) and wide.

(ii) Volcanic Passive Margin Offshore West Coast of India and Norway

As discussed in Section 2.9.1, passive margins are the best examples of extensional tectonics as they invariably represent extension and breakup of continents along them. They are therefore, associated with formation of basins and are important for hydrocarbon exploration that is discussed with details in Chapter 9. They are associated with linear gravity highs as in case of offshore west and east coast of India, representing changes in the crustal thickness and intrusives representing extensional rifting phase. Fig. 3.23 shows a gravity and a magnetic profile along with seismic section across continental shelf offshore west coast of India. It shows a gravity high (H1) coinciding with a bulge in a marker seismic horizon, which has been interpreted as caused by intrusives related to the extensional phase. A small residual gravity high between shot point 2000 to 2500 coincide with the bulge of the marker horizon and appears to be caused by intrusive related to the regional gravity high. Interestingly in this section, there is a pair of magnetic high and low (MH1 and ML1), which coincide with the gravity high and appears to be caused by the mafic intrusive in this section. Mafic intrusives in low geomagnetic latitudes would produce a pair of magnetic low and high (Section 3.7). Coincidence of magnetic anomaly with broad gravity high confirms the occurrence of intrusives related to extensional rifting phase in this section.

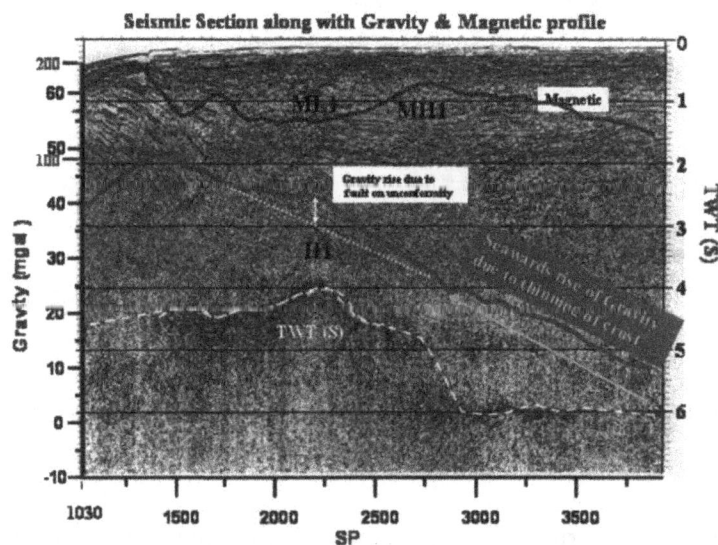

Fig. 3.23 A magnetic and gravity profile along with seismic section showing a gravity high and a pair of magnetic high and low related to intrusives associated with extension and rifting offshore west east of India (Sar 2008).

(a) Laxmi Ridge and Laxmi Basin

Fig. 3.24(a) describes the general tectonics of Laxmi basin and adjacent region offshore west coast of India (Krishna et al., 2006) superimposed over satellite free air gravity anomalies (Sandwell and Smith, 1997). It shows ridges and basins are associated with gravity highs and low, respectively related to high and low density of rocks in respective sections and bathymetry. However, the free air anomaly being close to zero in certain sections suggests isostatically compensated crust and most of the observed anomalies in this map are due to crustal sources. The gravity low associated with Bombay high, a major oil producing field in India is marked as BH, which appears to be caused by sediments in this section and crustal thickening. Further details of Bombay high is discussed in Section 9.6.1. The integrated crustal model along profile RE-11 based on gravity modeling (Fig. 3.24(b), Krishna et al., 2006) is constrained from seismic studies in this region. This model shows the crustal structure from continental shelf to oceanic crust towards the west across continental shelf, Laxmi basin, Laxmi ridge and part of the oceanic crust. Laxmi basin shows a broad gravity high with its central part showing a sharp gravity high (H1) related to Moho upwarping and Panicker ridge, respectively as shown in Fig. 2.23(b) for extensional regimes caused by rifting. Seismic velocity of 7.4 km/sec above the Moho in the western part indicate underplated crust which is typical of volcanic provinces.

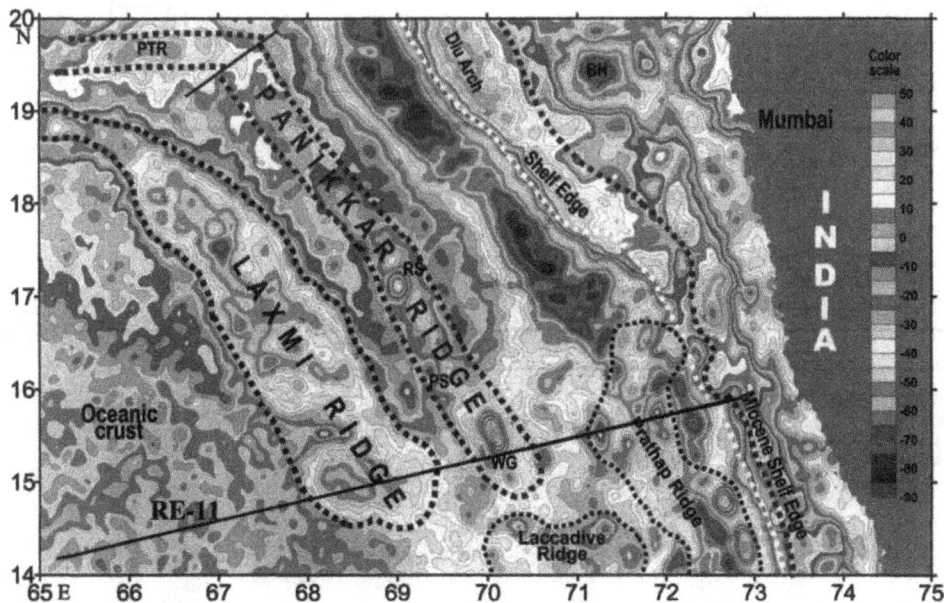

Fig. 3.24(a) Map of satellite free-air gravity anomalies [Sandwell and Smith, 1997] and interpreted geomorphic features. PTR, Palitana Ridge and BH, Bombay High structure (Krishna et al ., 2006). [Colour Fig. on Page 765]

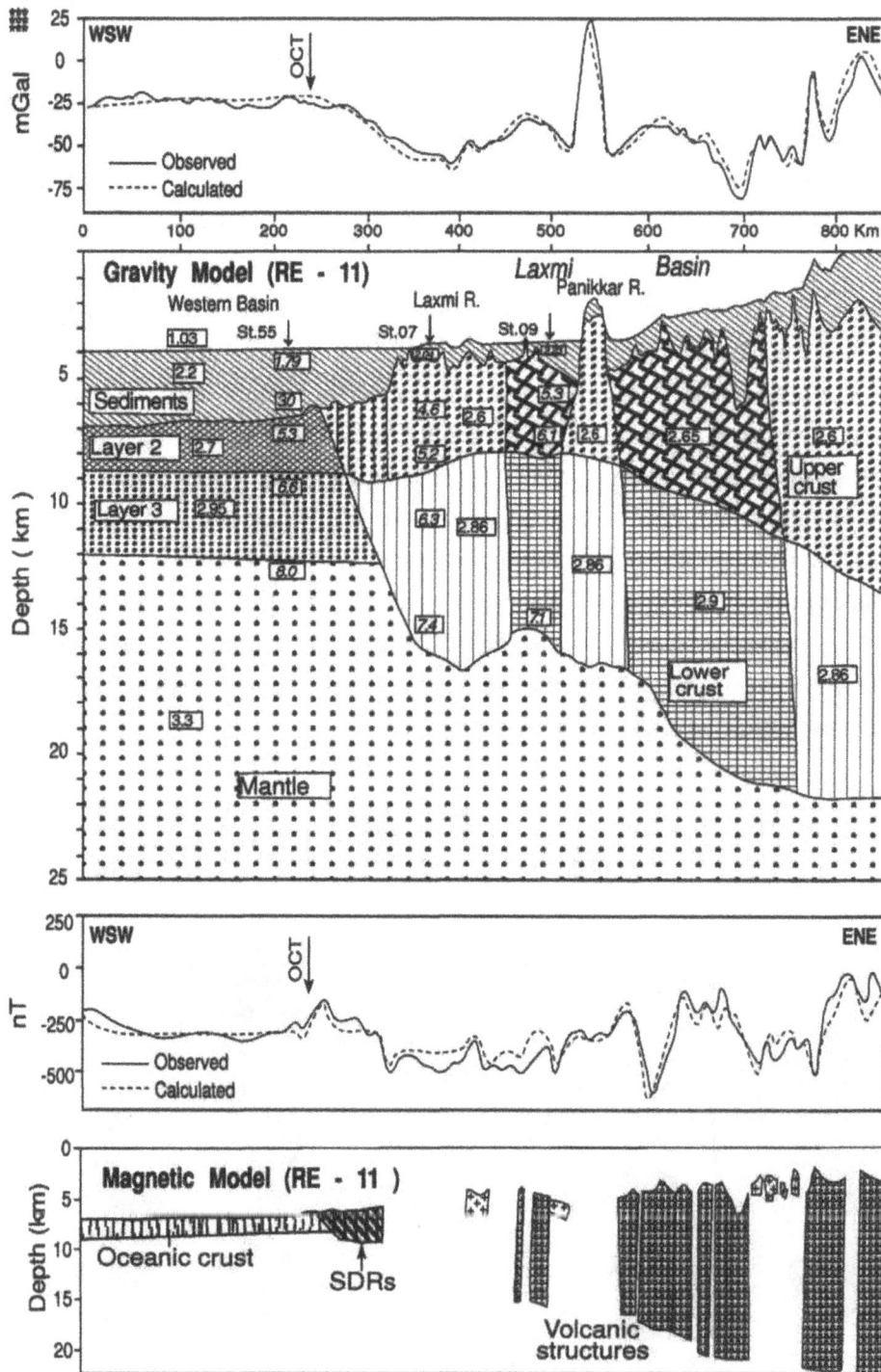

Fig. 3.24(b) Modeled crustal structure across the continental shelf, offshore west coast of India across Laxmi Basin and Laxmi Ridge from free-air gravity and magnetic anomalies. The Laxmi Basin basically consists of stretched continental crust, as inferred from seaward dipping reflectors and magmatic intrusions from magnetic anomalies, which are modeled for remanant magnetization corresponding to Deccan trap. Long-range sonobuoy (stations 55, 07 and 09, close to profile RE-11) velocity results are projected onto the crustal model. The values below the stations represent the velocities in km/s (Krishna et al., 2006). SDRS in bottom figure indicate the position of sea ward dipping reflectors just prior to oceanic crust.

It also shows sharp magnetic anomalies at the bottom of the figure indicating intrusives. The crustal section under Laxmi ridge shows a transitional crust with thickness varying from 16-22 km and several intrusives. As suggested above, that combination of gravity high and magnetic anomalies related to intrusives along continental passive margins suggest extensional rifting phase, which has apparently given rise to Laxmi basin. Comparing the crustal structures given in this model (Fig. 3.24(b)) with that of extensional rift basins (Fig. 2.23(b)), it shows a crustal upwarp in the center and a median ridge (Panicker ridge) in the center. It also shows shoulder highs in the form of Laxmi ridge and Pratap ridge along the shelf edge. Laxmi ridge in the present case forms Continent-Ocean-Boundary section 5.5, (Mishra et al., 2004). The magnetic intrusive along the coast at the bottom of the figure represent the first phase of intrusions that initiated the extension and the rifting while the intrusions shown as SDRS indicate the position of the seaward dipping reflectors (SDRS, Hinz, 1981) along the oceanic crust represent its final phase as has been generally the case in case of volcanic rifted margins as described below in case of Voring plateau offshore Norway. Seaward dipping reflectors in a seismic section across the southern part of the Arabian sea are shown in Sections 5.4 and 9.1 (Figs. 5.33b and 9.1c) that are the best signatures of the volcanic rifted margins. They are formed due to volcanic extrusion in submarine conditions due to chilling effects of cold water on hot erupting magma. The direction of remnant magnetization used by authors to model magnetic anomalies relates to normal and reverse direction of magnetization reported from Deccan trap. These signatures suggest Laxmi basin formed in response to extension and rifting. These processes, appear to be related to Reunion plume, which was responsible for Deccan trap eruption in ocean and over Indian continent.

The satellite based Bouguer anomaly map offshore west coast of India (Sections 5.5 and 9.6,) also shows two sets of linear gravity highs almost parallel to the coast, one set being close to the coast while the other being towards the Arabian Sea along Chagos-Laccadive Ridge suggesting two episodes of volcanic eruption related to inner SDRS and outer SDRS as described above.

(b) Offshore West Coast of Norway – Voring Plateau and Basin

A similar crustal structure has been reported in case of Voring plateau and Voring basin offshore west coast of Norway in NE Atlantic (Fig. 3.25(a); Mjelde et al., 2007). This figure shows one of the most detailed crustal structures based on seismic investigations under a continental margin which provides an insight into their evolution. It shows a gravity high related to extensional and rifting phases as described above which also coincides with a magnetic high related to volcanic mafic rocks (Tsikalas et al., 2002).This figure provides crustal structure of Voring escarpment and west of it while towards the east under Fenris graben similar listric faults are reported as eastern most fault (F) in this figure as shown in a schematic diagram (Fig. 3.25(b)). An underplated crust of density 3.1-3.2 g/cc and velocity 7.4 km/sec and basalt flows at the surface as in case of Laxmi Ridge and Laxmi Basin suggest it to be a volcanic passive margin. They have identified two groups of seaward dipping reflectors (SDRS), inner and outer related to initial extrusion of dykes etc., and final phase of rifting and break up at about 55 Ma. This rifting might be in response to Icelandic plume as break up along west coast of India was in response to Re union plume. This crustal section and its schematic model (Fig. 3.25(b)) show lithospheric and crustal upwarping, typical of extensional and rifting phases. It is similar to the typical crustal section model for volcanic rifted margins given by Menzies et al., (2003) who have also shown inner and outer SDRS coinciding with transitional continental and oceanic crusts, respectively and are now typically identified with volcanic margins.

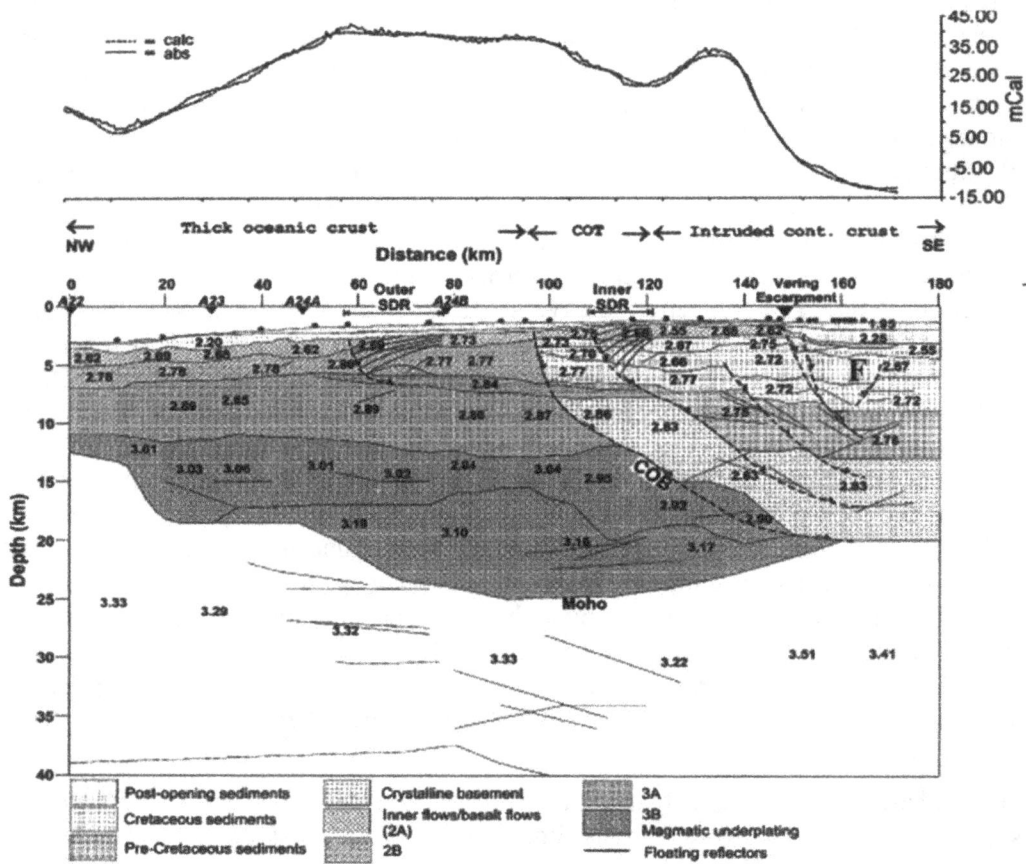

Fig. 3.25(a) Crustal structure of continental margin off west coast of Norway; Voring Plateau and associated rifting based on seismic studies and gravity modeling. Layer of density 3.1-3.2 g/cm^3 with corresponding velocity of 7.4 km/sec represent a underplated layer above the Moho.

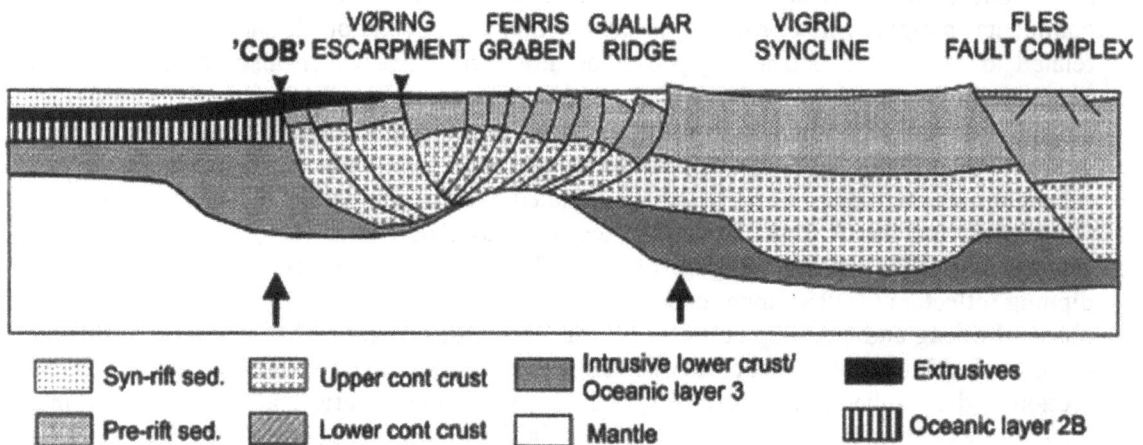

Fig. 3.25(b) A schematic section of crustal structure of Voring escarpement and Fenris graben showing upwarping of Moho and listric faults, typical section of volcanic passive margin.

(iii) Volcanic-Non-volcanic Passive Margin Offshore East Coast of India

As non-volcanic passive margins are also characterized by intrusive rocks, they show gravity highs and related magnetic anomalies but are generally of lesser extent and amplitude compared to volcanic passive margins.

(a) Gravity and Magnetic Anomalies along a Seismic Profile

A gravity and magnetic profile along with the seismic section offshore Krishna-Godavari basin (east coast of India, Fig. 2.26) is given in Fig. 3.26 (Maheshwari and Sar, 2004; Sar, 2008). It shows a gravity high, HI and set of magnetic anomalies, Ml and M2 related to the continental breck as shown in the seismic section. These anomalies, specially the magnetic anomalies suggest intrusives along the continental break related to extension and the rifting. In this magnetic latitude, magnetic lows are prominent magnetic anomalies for magnetic bodies. It, therefore appears that magnetic anomaly, M1 may correspond to initial phase of rifting while M2 may correspond to its final phase as described above in case of the offshore west coast of India and offshore Norway. A similar magnetic anomaly (Fig. 5.47) close by offshore Godavari-Krishna basin was modeled by Venkata Raju et al., (2002) using a trapezium type of body similar to the continental shelf, which suggests a magnetic susceptibility of 1.2×10^{-2} emu units indicating a mafic body. It is modeled using remnant magnetization as $D = 310°$ and $I = 67°$, which is similar to reported direction of magnetization for Rajmahal trap of Meso Cretaceous age (Rao and Rao, 1996) exposed in the Bengal basin (Table 5.1). Volcanic of same age has been considered to be responsible for the breakup of India from Antarctica as discussed in Chapter 5 (Mishra, 1984), which separated almost at same time (Norton and Slater, 1979). This indicates that the northern part of the eastern margin of India offshore Bengal and Mahanadi Basins where large extent and amplitude of gravity highs and magnetic anomalies related to intrusive (Mishra, 1984, Fig. 5.50) have been reported, can qualify for volcanic passive margin while the southern part of this margin south of offshore Godavari-Krishna basins (Fig. 2.26(a)) where volcanic rocks are limited may represent a non-volcanic passive margin (Section 5.6). Chand et al., (2001) have drawn the same inference based on effective elastic thickness as discussed in Chapter 4. This suggests that part of a continental margin affected by large scale volcanism can represent volcanic margin while remaining part affected only by some intrusions locally can be termed as non-volcanic margin. Ninety East Ridge, being far from the east coast of India, it makes it difficult to judge the nature of rifted margin offshore east coast of India. There are gravity highs offshore east coast of India (Sections 5.4 and 9.1) but due to gravity highs of Eastern Ghat Fold Belt (Section 7.10) in the same section, it becomes difficult to separate them and assign their sources.

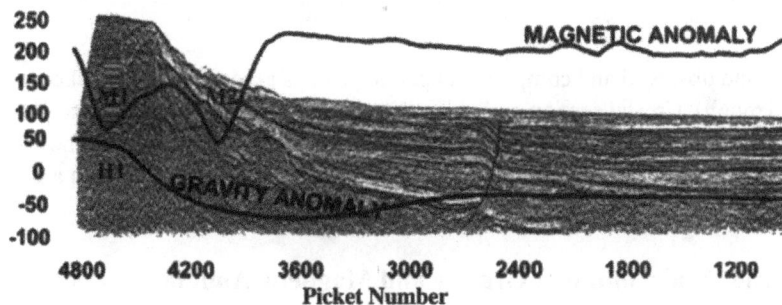

Fig. 3.26 A total intensity magnetic and gravity profile across offshore east coast of India. Gravity high, H1 and set of magnetic anomaly M1 and M2 are related to intrusives associated with extension and continental breakup (Maheshwari and Sar, 2004).

(iv) Passive Margin: East Coast Magnetic Anomaly, USA

To examine the well known East Coast Magnetic Anomaly associated with continental margin offshore East Coast of USA, a magnetic profile along with the crustal section based on detailed seismic studies across east coast of USA in Baltimore Canyon Trough are reproduced in Fig. 3.27 (Talwani and Abreu, 2000). This crustal section shows a continental crust towards the left and Oceanic crust towards the right with a transitional crust in between with high velocity and seaward dipping reflectors suggesting intrusives related to extension and rifting in this part. The magnetic field computed with intensity of remnant magnetization (susceptibility) of 0.005 emu matches quite well with the observed field. High intensity of magnetization suggests mafic intrusive rocks. This represents a typical crustal section and associated magnetic anomaly over a passive margin. A gravity high is also observed over the extensional rifting part in this case (USGS, 1982) coinciding with the magnetic anomaly, which is another signature of extension and rifting as discussed above.

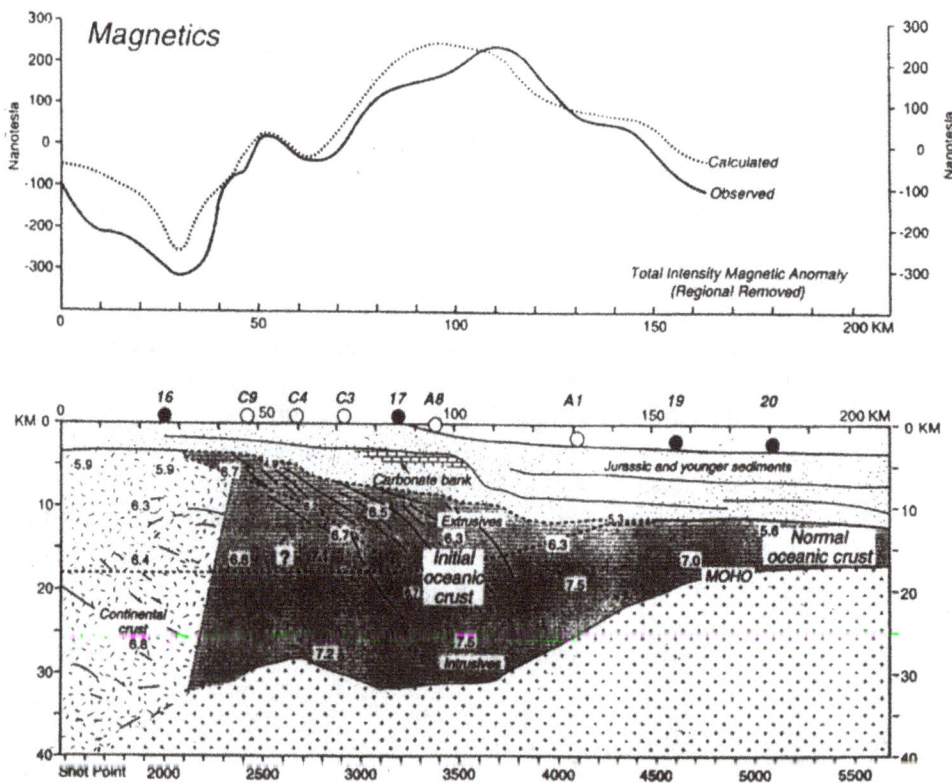

Fig. 3.27 Crustal section and observed and computed magnetic profiles across offshore east coast of USA (Baltimore Canyon Trough). Crustal section and velocities are controlled from seismic studies. Volcanic extrusives with seaward dipping reflectors indicate extension and rifting. This is well known east coast magnetic anomaly occurring through out along continental margin coinciding also with a gravity high. (Talwani and Abreu, 2000).

(v) Proterozoic Rift Basin: Mid Continent Gravity and Magnetic Anomalies, USA

Mid continent gravity high, which is a linear gravity high of large amplitude related to Keweenawan mafic belt in Western Iowa and adjoining part of Nebraska, USA is a well known rift basin of Meso Proterozoic period. The gravity and airborne magnetic anomalies of this section are given in Fig. 3.28(a) and (b) (Zeitz et al., 1966), which shows a linear gravity high,

H1 flanked by gravity lows L1 and L2 characteristic of active rift basins as given in Section 2.7 (Fig. 2.23(b)). It also shows a similar linear magnetic low indicating remnant magnetization for the Keweenawan mafic unit. Otherwise, due to present high geomagnetic latitudes (inclination), the magnetic anomaly for induced magnetization would be a simple high. The gravity high, H1 is related to the mafic unit while adjoining lows are related to the Precambrian clastic sediments. The shoulder highs of the rift basin are not prominent in this case as it is an old rift system. However, H2 and H3 can be considered as shoulder highs. This shows the similarity of structures between recent (Section 2.9) and ancient rift basins with some modifications.

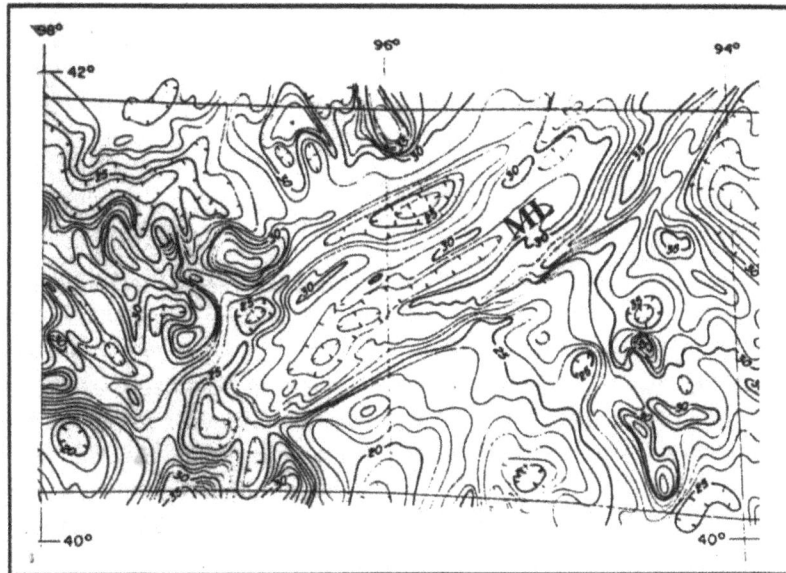

(a)

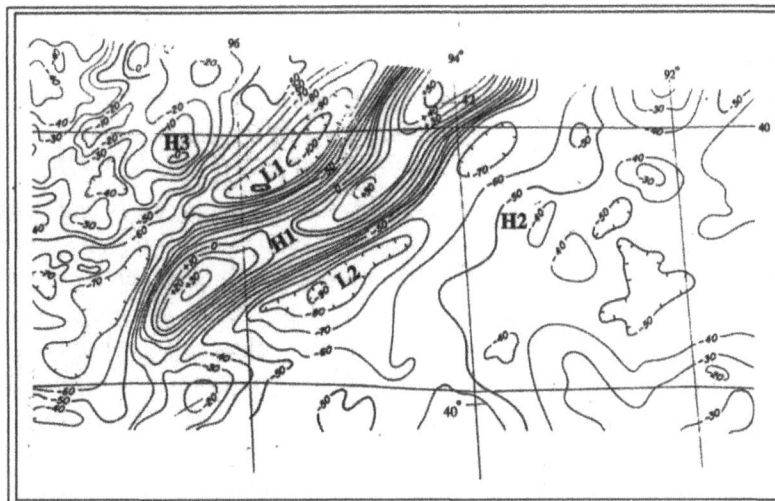

(b)

Fig. 3.28 Airborne total intensity (a) and Bouguer anomaly map (b) of mid continent rift system of USA (Zeitz et al., 1966) showing gravity high, H1 associated with Keweenawan mafic unit and adjoining lows, L1 and L2 associated with Precambrian clastic rocks are typical of active rift basins. Gravity highs, H2 and H3 may represent shoulder highs. Magnetic map shows a magnetic low (ML) corresponding to gravity (H1) due to mafic unit indicating remnant magnetization in reverse direction with respect to present day earth's magnetic field.

3.8.2   Convergence Tectonics – Grenville and Appalachian Orogenies, Canada and USA

Compressional regimes are characterized by thrusting (Fig. 2.23(c) and (d)), which mostly shows high dip of 60°-70° and therefore these causative bodies are often inclined and show heterogeneous magnetization due to thrusted metamorphic lower crustal rocks. They are, therefore characterized by haphazard nature of anomalies but they are usually narrow linear anomalies as they occur in narrow bands. In regard to amplitude of the observed magnetic anomalies, it depends on the mafic components as discussed in section 3.7.2. The order of magnetic anomalies due to mafic intrusives in case of the extensional regimes is generally more compared to the compressional regimes as former is likely to have more mafic components compared to the latter where magnetic anomalies are primarily due to lower crustal rocks. Further, compressional regimes usually show linear compact or sometimes even erratic magnetic anomalies due to heterogeneous magnetization compared to extensional regimes.

(i)   Grenville and Appalachian Orogenies, Canada and USA.

(a)   Airborne Magnetic and Gravity Anomalies

Grenville and Appalachian orogenies which represent typical convergent tectonics during Meso Proterozoic and Paleozoic periods, respectively. Here we investigate their magnetic fields along with their gravity fields in same sections. Thomas (1992) and Hinz and Zietz (1985) have analysed their magnetic signatures. Thomas (1992) has given an aeromagnetic map showing a large linear magnetic high and a small low north of it over the Grenville province, Canada, east of Superior craton. He termed them as bipolar nature of magnetic anomalies and attributed them to Proterozoic magmatic arcs and oceanic mafic rocks along collision suture in comparison to present day suture zone. The effect of crustal thickening does not reflect in magnetic data as in case of gravity data due to limitation on depth of magnetic properties of rocks based on Curie point geotherm.

Hinz and Zeitz (1985) have shown linear magnetic highs and lows, which extend all along the east coast of USA and attributed them to igneous and metamorphic rocks of Appalachian Orogeny of Paleozoic period. They suggested that magnetic anomalies depend to a large extent on the grade of metamorphism. Hatchar and Zeitz (1980) attributed these magnetic anomalies to lower crustal rocks of Blue Ridge and Inner Piedmont, parts of the Appalachian Orogeny, which are responsible for gravity high in convergent tectonics (Section 2.7; Fig. 2.23(d)).

Fig. 3.29(a) and (b) (Zeitz et al., 1966) presents airborne magnetic map and corresponding gravity map of a part of Appalachian Orogeny. The airborne magnetic map is based on twenty profile flown at a height of elevation between 3000-16000', mostly between 11000-16000' with a relative accuracy of $\pm$ 2 nT. These maps are reproduced here as to get proper perspective of these anomalies. The Bouguer anomaly map shows a major gravity high, H1 and a gravity low, L1 related to Pre Mesozoic (Paleozoic) rocks of Appalachian orogeny and Appalachian basin, respectively. Corresponding to these magnetic highs and lows there are similar type of magnetic anomalies M1 and M2, the former a linear bands of magnetic anomalies, usually associated with compressional tectonics (Section 3.8.1) and the latter circular and semicircular anomalies, associated with the basement in sedimentary basins (Section 3.7.2). Magnetic anomalies however are combination of magnetic highs and lows due to magnetization vector and susceptibility contrast of the rocks as discussed in Section 3.7. Magnetic anomalies, M1 are linear bands of compact magnetic anomalies of moderate amplitude, typical of compressional tectonics (Section 3.8.1). Depth estimates from the magnetic anomaly M2 suggested a depth of 8.3 km below flight level and approximately 3.5-4.0 km below sea level (Mishra, 1970) which confirms to the depth of basement in this section. This illustrates the nature of the observed magnetic anomalies in conjunction with the gravity anomaly over ancient collision zones. More such cases of gravity and magnetic signatures of Proterozoic collision zones, related to Indian continent, are discussed in Chapter 7.

West of Appalachian basin is the subsurface extension of Grenville boundary, which is characterized by gravity high, H2 and set of magnetic anomalies M2. As referred to above, it represents a collision zone of Meso Proterozoic period and is characterized by pair of gravity anomaly H2 and L2, almost sub parallel to each other and appearing to form pair with each other. Basement rocks in this region (central Ohio) appears to be affected by Grenville orogeny, which is exposed northwards in Canada.

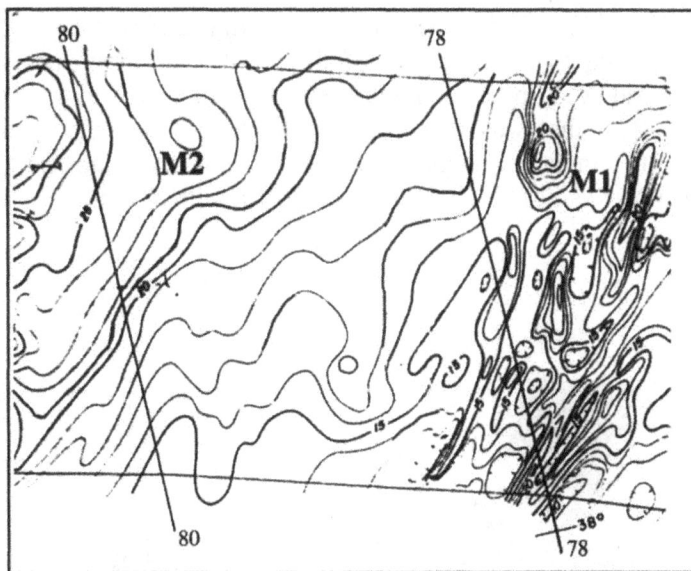

(a)

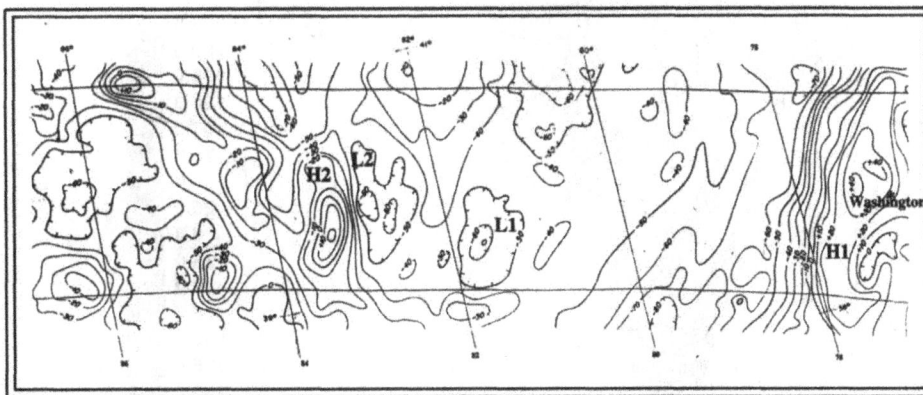

(b)

Fig. 3.29 Airborne magnetic **(a)** and Bouguer anomaly map **(b)** of Appalachian orogeny (Zeitz et al., 1966) showing gravity high, H1 related to high density intrusives of the orogeny and gravity low, L1 associated with Appalachian basin caused by crustal thickening. Gravity high, H2 and low, L2 are related to subsurface Grenville boundary, which may be southward extension of Grenville orogeny in Canada. Magnetic anomalies, M1 and M2 are related to Appalachian exposed igneous and metamorphic rocks and basement under Appalachian basin. Linear and compact nature of magnetic anomalies are indicative of compressional (collision) tectonics.

3.8.3 Satellite-Airborne Magnetic Map of India (EMAG2)

A magnetic map forms the basic data like geology, tectonic and Bouguer anomaly maps of a region for the present day geoscientific investigations related to geodynamics, hydrocarbon and mineral investigations. Under countrywide regional mapping and mineral investigations programme, parts of the Indian shield were covered by airborne magnetic surveys. Total Intensity map of these regions and their analysis were published by Reddi et al., (1988), Mathew et al., (2001), Mishra and Vijaya Kumar (2005) etc. Subsequently, Rajaram et al., (2006) compiled all the data set available from different sources including vertical intensity data for certain regions and some magnetic data from continental shelves off shores. However, several gaps still existed in this map that makes it difficult to check on the continuity of structures. Moreover, due to merging of several data sets acquired during different periods and different agencies, noise level appeared to be quite high. Subsequently, airborne magnetic data from different parts of the world were processed together to make a World Magnetic Anomaly Map (Korhonen et al., 2007) that was merged with the long wavelength component from CHAMP satellite data to prepare a world Magnetic Anomaly Map (http://geomag.org/ models /EMAG.html1) known as EMAG-2 (Maus et al., 2010) with a resolution of 2 arc minute

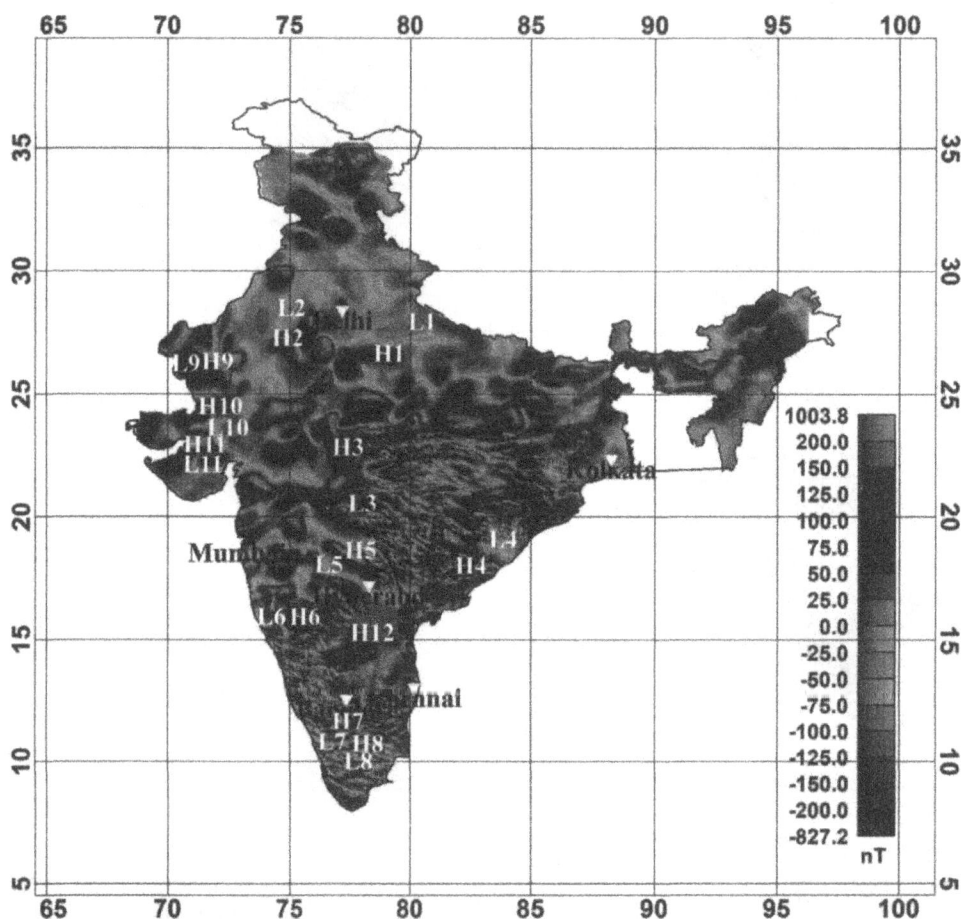

Fig.3.30 EMAG 2: A 2 arc-minute (3.7 km) resolution Satellite-Airborne magnetic map of India compiled from airborne magnetic data from different sources acquired for World Digital Magnetic Anomaly Map (Korhonen et al, 2007) of Commission of the World Geological Map (CWGM, http://ccgm.free.fr/) and long wavelength CHAMP lithospheric field model MF6 EMAG2 (Maus et al., 2008). All data sets are merged at 4 km altitude. H and L denote magnetic highs and lows (Maus et al., 2010). [Colour Fig. on Page 765]

(3.7 km). The whole data set was brought to 4 km altitude above geoid. Same authors have recently brought out a new version of this map as EMAG-3 by incorporating marine magnetic data in oceans but over the continent EMAG-2 and 3 are almost same. The part of this map for Indian continent is given in Fig. 3.30 that reflects most of the geology and tectonics of this country. Most of these features also find reflection in the Bouguer anomaly map (Fig. 2.28) and their combined analysis may provide additional information. As this part is located in low geomagnetic latitudes, any magnetic feature will produce a combination of magnetic high and low. Some of its significant features are discussed below. In comparison to the tectonic map of India (Fig. 2.26), the following features are easily discernible.

(a)   Magnetic high and low, H1 and L1 are related to basement ridges and depressions of the Ganga basin.

(b)   Proteozoic fold (mobile) belts are reflected as bipolar magnetic anomalies such as Aravalli-Delhi fold belt (H2 and L2), Satpura fold belt (H3 and L3 extending from the west coast to eastern margin of the Indian plate) and Eastern Ghat fold belt (H4 and L4 along the east coast of India). These fold belts are also reflected as paired gravity anomalies in Bouguer anomaly map (Fig. 2.28) suggesting high density and high susceptibility rocks associated with them. These fold (mobile) belts and their geophysical anomalies are discussed in more details in Chapter 7 where geodynamics of Indian continent are discussed.

(c)   There are several linear magnetic highs and lows in Dharwar craton (Section 7.8) of the Indian Peninsular Shield such as H5, L5 and H6, L6 that are related to large lineaments, some of which also find reflection in the Bouguer anomaly map (Fig. 2.28). In fact such linear features are better reflected in the airborne magnetic maps due to closely spaced data sets compared to Bouguer anomaly maps.

(d)   There are several curvilinear magnetic anomalies in the Southern Granulite Terrain like H7, L7 and H8 and L8 etc., that are related to various shear zones (Section 7.9).

(e)   Magnetic highs and lows, H9 and L9, H10 and L10 and H11 and L11 appears to related to volcanic plugs of Deccan trap origin and prior to it during break up of Africa and India (Section 5.3) as several such plugs are found in this part that have provided gravity highs in Bouguer anomaly map (Sections 7.11, 9.2.6 and 8.2.5). The latter ones (H11 and L11) from Saurastra belong to Deccan trap eeruption. However, those from Western Rajasthan and Kutch (H9 and L9 and H10 and L10) may belong to earlier event of the break up of Africa and India that has given rise to several Mesozoic basins in this part.

(f)   Cuddapah basin is characterized by a large magnetic high, H12 that extends westwards beyond the limits of Cuddapah basin. This magnetic high is attributed to mafic sills and mafic basement of the western part of Cuddapah basin as discussed in Section 7.10.1 based on detailed airborne magnetic map of this basin. This anomaly extends westwards as there are several mafic dykes outside the basin whose magnetic effects have merged with those from the basin.

3.8.4   Long Range Airborne Magnetic Profiles – Manglore-Madras, India

Examples of some long range airborne magnetic profiles are given in this section in order to show their utility for different geological problems. Long range high altitude airborne magnetic profiles have been widely used for geodynamic studies and crustal magnetization (Agocs, 1958; Alldredge, 1963; Zeitz et al., 1966; Mishra, 1984). In India, we recorded such profiles across some important geological provinces for geodynamic studies, one of which is discussed below as an example while some others are dealt in Chapter 7 while describing different geological provinces.

(i)   Manglore- Madras Profile across Indian Shield: Continental Margins and Cratons

Airborne total intensity magnetic profile from Manglore-Madras was recorded at an altitude of 9000 ft ($\approx$ 2730 m) above mean sea level. The magnetic profile along with the Bouguer anomaly and topography is plotted in Fig. 3.31 (Hari Narain et al., 1969; Mishra, 1970). It also shows the exposed geology and flight path at the bottom of the figure. It shows some major anomalies marked as A to I, which are discussed below.

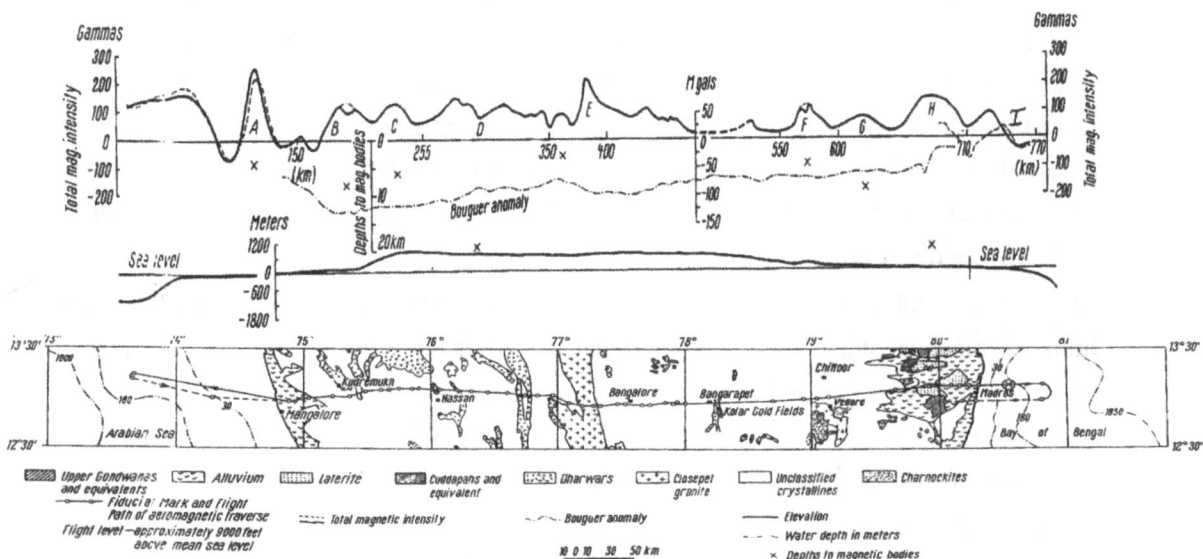

Fig.3.31 Airborne total intensity magnetic profile across Dharwar Craton, India along 13⁰ N parallel (Mangalore – Madras). A-I are magnetic anomalies with A and I being related to intrusive mafic rocks along continental margins during breakup of Indian continent related to extensional tectonics. Magnetic anomalies D and E are typical contact type of anomaly separating western and eastern Dharwar cratons and Closepet granite respectively, other magnetic anomalies may represent local bodies.

(a) The two anomalies, which are geodynamical of great significance are A and I, which appear to represent mafic intrusive bodies along the west and the east coasts of India, respectively. They may represent intrusives related to the breakup of Indian continent along these margins. A and I are a typical contact (fault) kind of anomaly which may represent continental margin offshore east and west coasts of India.

(b) The magnetic anomaly H, is accompanied by a gravity anomaly (Bouguer anomaly), which typically represents a fault kind of gradient anomaly that suggest a mafic intrusive associated with a fault or a thrust in this region and might be related to evolution of the Eastern Ghat Fold Belt.

(c) The other important magnetic anomalies, D and E, are related to mafic rocks of Hassan and Chitradurga schist belts. Magnetic anomaly E is a fault kind of magnetic anomaly, which is accompanied by a gravity high and therefore, appears to represent a set of anomalies related to changes of geological Terranes from the west to the east known as the Western Dharwar craton and the Eastern Dharwar craton, respectively. Magnetic anomaly B coincides with Kudremukh iron ore deposite and appears to be caused by it. The corresponding broad gravity low, however, is related to crustal thickening under the Western Ghats due to isostasy as explained in section 2.8 under Field Examples of Gravity Surveys and demonstrated in Fig. 2.19.

(d) Other magnetic anomalies may represent mafic rocks and their extent up to different depths. Depth estimates of these magnetic anomalies (Mishra, 1970) suggest that most of these anomalies are located in the upper crust.

(e) One may notice well defined smooth magnetic anomalies over Peninsular shield (Dharwar craton) along this profile (F and G) inspite of exposed rocks being gneisses, which usually provides irregular erratic anomalies due to heterogeneous magnetization. This is because the present airborne magnetic profile was recorded at high altitude where surface noise due to heterogeneous magnetization of rocks gets attenuated.

3.8.5 Mineral Exploration: Airborne Magnetic and Spectrometer Surveys of Chitradurga Schist Belt, India

This schist belt occurs at the eastern margin of the western Dharwar craton (Fig. 2.26) in Karnataka, South India, and are largely composed of Late Archean (2.7-2.5 Ga; Anil Kumar et al., 1996) Chloritic and Hornblendic schist. They mainly include a succession of basic volcanic flows, agglomerates, tuffs, traps, pillow lavas and laminated cherts and ferruginous quartzites. There are several dolerite dykes oriented in different directions (Naqvi, 1973). Schist belts world over are important for economic minerals and so is the Chitradurga schist belt which is considered to host several economic minerals such as copper, lead, antimony, manganese, pyrite, iron ore etc.

Due to economic importance of this region, the first airborne multi parameter geophysical survey in this country with indigenous developments was carried in about 2200 sq km. in this region (NGRI, 1968; Hari Narain, 1969). The survey was conducted at flight line spacing of 1 km with a terrain clearance of about 500' (~ 152 m). The data was processed manually and the total intensity map of a part of this region is presented in Fig. 3.32 along with superimposed geology (Mishra, 1978). This was one of the most significant magnetic anomalies recorded in this region besides several other magnetic anomalies. It is a large amplitude magnetic anomaly of about 2000 nT coinciding with volcanic (trap) rocks with center of magnetic high and low coinciding with bands of ferruginous quartzites specially magnetite quartzites and mineralized zones. The general undisturbed value of the earth's total magnetic field in this region is about 41,000 and the magnetic low of this anomaly is located towards the north indicating an induced magnetization. Weathered traps, in general do not produce such well defined large magnetic anomalies and therefore, this high amplitude magnetic anomaly and coincidence of its central part with hills of magnetite quartzites suggest it to be primary source of this anomaly. Airborne magnetic data along profile AA' (Fig. 3.33) is interpreted using standard curves for tabular bodies (Gay, Jr., 1963), which provided following parameters:

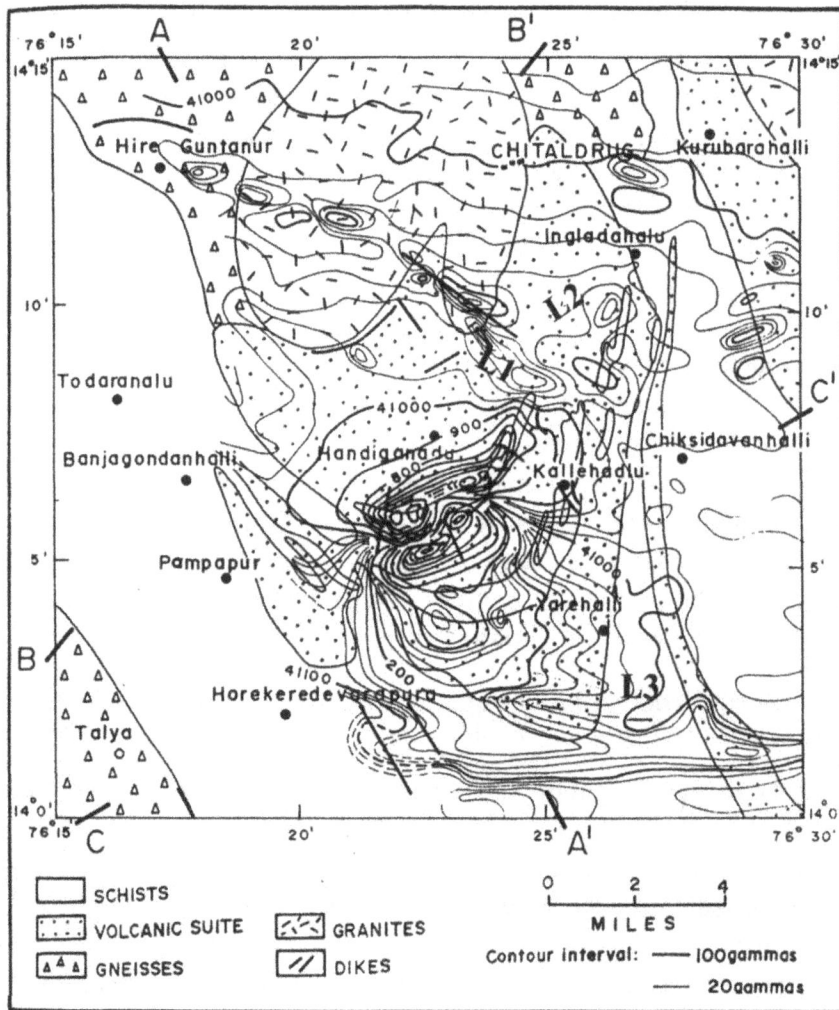

Fig. 3.32 Airborne total intensity map of part of Chitradurga Schist belt, Karnataka, India, flown at terrain clearance of 500' (~152 m) showing a pair of magnetic low and high associated with magnetite quartzite which also shows maximum susceptibility in this region. This is a typical magnetic anomaly observed in low geomagnetic latitudes for induced magnetization with magnetic low towards the north. L1-L3 are magnetic lineaments important for mineralization.

Depth below plane of observation	= 740'
Flight level	= 500'
Depth below surface	= 440'
Bulk susceptibility	= 1.2×10^{-3} e.m.u.

This susceptibility value indicates concentration of magnetite, which is associated with ferruginous quartzites. Besides this large magnetic anomaly, the magnetic lineaments L1, L2 and L3 show alignment of small amplitude magnetic anomalies indicating lineaments which are quite significant for the purposes of mineralization. The results of airborne spectrometer survey is given in Fig. 3.34 which shows concentration of radioactive elements in several blocks and block R1 south of Chitradurga coincides with the lineament L1 indicating this lineament to be important for concentration of radioactive elements. In this manner, airborne multi-parametric geophysical survey is an important tool for mineral investigations.

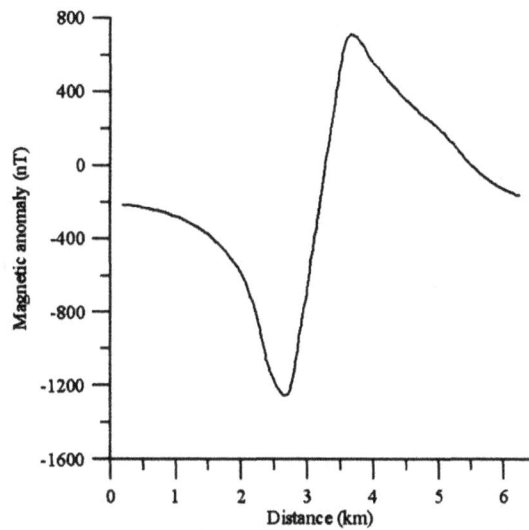

Fig. 3.33 Total intensity magnetic profile AA' from the map given in Fig. 3.32 and its interpretation based on standard curves for tabular bodies.

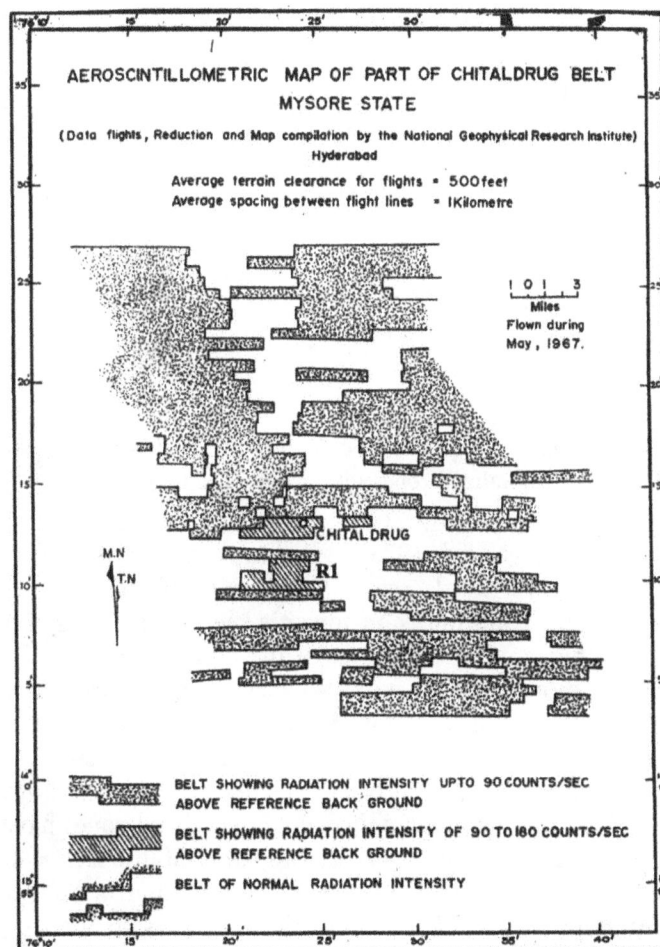

Fig. 3.34 Airborne Spectrometer map of part of Chitradurga Schist belt, Karnataka, India showing concentration of radioactive elements R1 along magnetic lineament L1. (Fig.3.32).

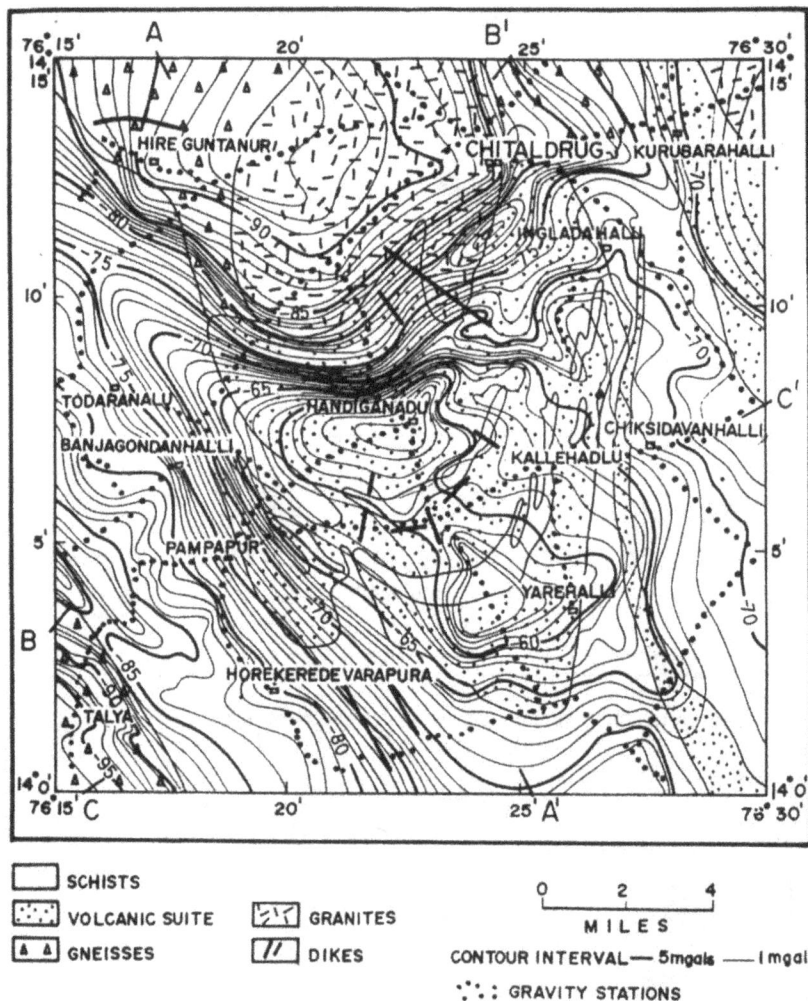

Fig. 3.35 Bouguer anomaly map of part of Chitradurga Schist belt, Karnataka, India showing gravity highs over volcanic suit (trap) rocks and low over intrusive granite(Naqvi, 1973) while airborne magnetic anomaly is located over the mineralized zone, with in the volcanic suite.

Fig. 3.35 is the Bouguer anomaly map (Naqvi, 1973) of the same section of Chitradurga schist belt which shows gravity high of about 10-15 mGal over volcanic suite of rocks and a gravity low of almost same amplitude over exposed granite intrusion towards the north. It may be noted that the gravity high is shifted northwards compared to the center of the magnetic high and low (Fig. 3.32), which form a composite magnetic anomaly. This indicates that their sources are shifted in space, though they may be related in their genesis. The gravity high is spread over the entire volcanic suite and is therefore, caused by it and the center of the two gravity highs may represent the centers of the thickest section of the volcanic suit of rocks. The airborne magnetic anomaly is however, limited in parts of volcanic suite where bands of magnetite quartzites and mineralized zones have been reported. It is therefore, evident that airborne magnetic anomaly is more successful in delineating mineralized zones compared to gravity anomaly, which is spread over a large section in the present case. The airborne magnetic map has delineated several such magnetic anomalies in this region, which are important for their mineral potential.

4

Common Data Processing Methods and Parameter Estimation – Digital Signal Processing

Some methods to process and model gravity and magnetic data are common to both, which are described in this chapter. They are basically mathematical methods of data processing and are briefly described here to understand their applications to gravity and magnetic fields. Their applications to different kinds of potential field data are demonstrated through field examples in course of description and at the end of the chapter.

4.1 Regional and Residual Separation

As described in section 2.5.2 and 3.5.1, the separation of regional and residual fields from the observed field play an important role, in processing of both gravity and magnetic data. As described in those sections, regional and residual fields represent the characteristics of these fields in the whole region and in localized sections of that region, respectively. The former is generally caused by deeper sources compared to the latter, which is considered to be caused by shallow sources. However, both the components, viz. the regional and the residual components of the observed field are important for different purposes. The regional field is primarily used for geodynamic studies as it is related to the whole region and residual field is used for exploration of mineral, hydrocarbon etc., and near surface geophysics related to environmental studies in a localized area. By definition the regional field is the component of the observed field, which is not desired for exploration of minerals and hydrocarbons and should therefore be removed before interpretation. The two components, however supplement each other and used in conjunction for specific purposes. In early days of developments of these methods, the graphical and grid methods, described in previous chapters were used for this purpose. Presently with availability of digital data and automatic methods of processing and interpretation, mathematical methods of polynomial approximation or wavelength filtering as described below are used.

Regional-Residual separation is more serious in gravity field compared to magnetic field as discussed in section 3.5.1. In case of gravity field, all the surface and subsurface bodies/layers have a specific density and produces their gravity fields. Therefore, it is essential to separate the observed fields at least in two groups, viz. the regional and the residual fields caused by deep seated and shallow sources, respectively. In case of magnetic surveys, however the magnetic field is caused primarily by specific magnetic sources as all surface/subsurface bodies are not magnetic in nature. Therefore, after IGRF correction for earth's main magnetic field, usually magnetic anomalies due to individual sources are observed, which can be used directly for modeling and evaluation. In fact, IGRF itself is considered as regional field in magnetic surveys, which is corrected by standard method as discussed in section 3.4.3. However, in case there are more than one magnetic sources in a region, such as intrusives in a sedimentary basin where there will be combined magnetic anomalies due to basement and shallow intrusives, in such cases, any of the following standard methods for regional-residual separation can be used. In this case, magnetic anomalies due to basement will be broader and smaller in amplitude compared to those due to intrusives. However, in applying any of these mathematical methods, they should be applied judiciously guided by known regional geology and tectonics as these methods were initially developed by workers of different disciplines for different purposes and therefore, interaction of an interpreter at every stage is absolutely important.

4.1.1 Polynomial Approximation

In this method, the regional and the residual fields are represented by low and high order surfaces, respectively. The observed gravity anomaly is approximated by a power series. The observed gravity and magnetic fields F(x) along x-axis can be represented by

$$F(x) = a_0 + a_1 x + a_2 x^2 + ... a_n x^n \qquad .....(4.1)$$

where n is the order of the polynomial being used to approximate the regional field and a_0, a_1, a_2, ... are the coefficients and in present case are regional gravity/magnetic fields at observation points.

The coefficients of different orders (n) are evaluated using the principles of least squares as described in section 4.5.1. One of the low order trends (upto order of 2 - 5) is selected based on visual inspection as demonstrated in the forthcoming sections as the regional field and its difference from the observed field is the residual field. However, the regional field obtained in this way is a mathematical surface and may not be correctly related to geology of the region. The selection of the order of polynomial (n) to represent the regional field is quite arbitrary and depends considerably on the experience of the interpreter. If the depth to the shallow sources such as basement is known in certain sections from seismic profiles or borehole information, some constraints on the order of polynomial can be imposed which will provide the right magnitude of the residual field at these points. In case of two dimensional data surfaces of different order are approximated in x,y direction over the grided data. Grant (1957) was first to introduce this approach for regional-residual separation, which has now become a standard method for this purpose. In actual practise polynomials of different orders can be computed from the observed gravity and magnetic fields and the one, which may explain some known features of regional geology can be chosen from them as representing the regional field. It is relatively easier to visualize the regional field vis-à-vis the observed field along profile as demonstrated in next section under applications. It is therefore, advisable to separate regional and residual fields along profiles or in case of maps, check the separated regional field with respect to the observed field along some profiles taken from the two maps.

4.1.2 Applications – A Gravity Profile across Himalayas

Fig. 4.1 shows a gravity profile from New Delhi to Garhwal Himalaya (Mishra et al., 2006) picked from the Bouguer Anomaly Map of India (Fig. 2.28). It shows a gravity gradient leading from a gravity high (H1) to gravity low (L1) under Himalayas, which has been attributed to crustal thickening under Himalayas and Tibet (Section 2.8.1). Superimposed over this large wavelength regional gravity high and low are small wavelength gravity high and low (H2 and L2) of 10-15 mGal at about 200 km mark over Main Himalayan Thrust (MBT) and Himalayan Frontal Thrust (HFT), respectively. This is clearly a case of gravity anomalies (H2 and L2) due to shallow sources superimposed over the regional gravity low due to deeper sources (crustal thickening). A regional field shown in Fig. 4.1 is drawn based on smoothing of the field (graphical method). The observed field along this profile is also used for polynomial approximation and the same curve drawn as per smoothing of the field is also obtained for 5th order polynomial, which is regarded as the regional field in the present case and small amplitude and wavelength gravity high and low (H2 and L2) are the residual fields caused by shallow sources related to a high density intrusive along MBT and thick sediment along Himalayan front south of it. Their geodynamic implications are discussed in Chapter 6.

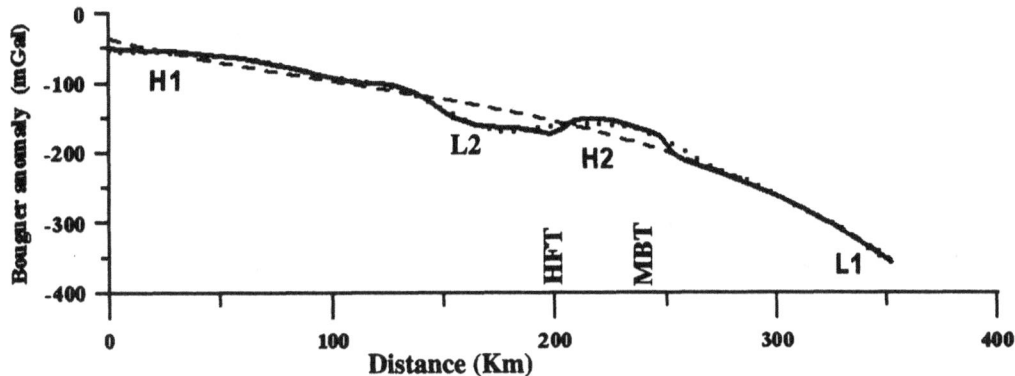

Fig. 4.1 A gravity profile from New Delhi to Garhwal Himalayas showing regional field (H1 and L1) approximated by polynomial of order 5 by dashed curve, which almost coincide with regional field as per visual inspection. H2 and L2 are residual gravity high and low due to high density intrusive along Main Boundary Thrust (MBT) and thick sediments along Himalayan Frontal thrust (HFT).

4.1.3   Finite Element Approach – Gravity Map of Epicentral Zone of Latur Earthquake of 1993

Some workers (Mallick and Sharma, 1999) have used finite element method to define regional field over a gravity map. In this method they define eight nodes outside the anomalous zone and the regional field in the region is determined using weighted shape function at these nodes as discussed below.

Fig. 4.2 is the Bouguer anomaly map of epicentral zone of Latur earthquake on September 30, 1993 (Mishra et al., 1994), which is a part of Bouguer anomaly due to Kurduwadi lineament in Maharashtra (Mishra et al., 1998). This map shows small order of gravity highs and lows with epicenter of this earthquake located at the junction gravity high H1 and Low L1 south of Yekundi. A gravity profile CC' adopted from this map is shown in Fig. 4.3, which also shows the isostatic regional field based on zero free air anomaly as discussed in section 2.5.5. This regional field, has a geological and geophysical significance as it is based on well known theory of isostasy. The residual field after subtracting the isostatic regional from the observed field shows all the essential anomalies of the original gravity map, viz. L1, H1, L2, H2 and H3. The epicenter lies at the gradient between L1 and H1, which implies a contact (fault) separating low density rocks towards the south and relatively high density rocks towards the north. This gradient partially coincides with high conductivity body at depth of 7.0 km (Fig. 4.2) as obtained from magnetotelluric studies (Sarma et al., 1994; Gupta et al., 1996) and has therefore, been attributed to fluid field fractured zone along Yekundi fault, which triggered this earthquake.

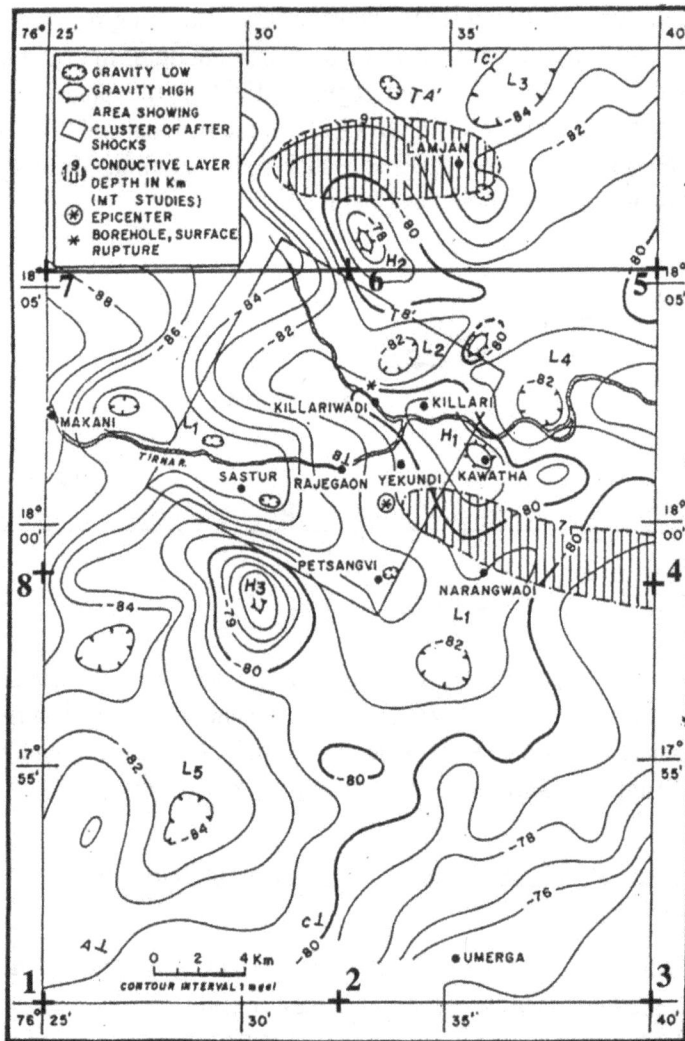

Fig. 4.2 Bouguer anomaly map of epicentral zone of Latur earthquake (September 30, 1993) showing small amplitude gravity highs H1-H3 and gravity lows L1-L5 with the epicentre of this earthquake located at the gradient between H1 and L1 (Mishra et al., 1994). It also shows two conductive bodies delineated from magnetotelluric studies (Gupta et al., 1996). The symbol + at the margins are the nodes of the finite element to estimate the regional field.

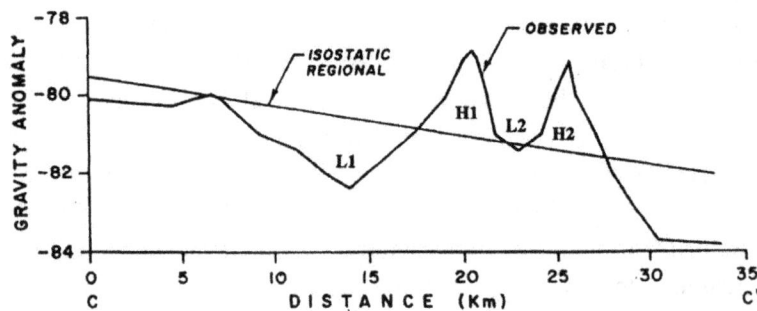

Fig. 4.3 Gravity profile CC', from Fig. (4.2(a)) showing gravity lows and highs L1, H1, L2, H2 and isostatic regional drawn based on zero free air anomaly.

The digital data of Fig. 4.2 is also used for separation of regional and residual fields using finite element method (Mallick and Sharma, 1997). They defined 8 nodes elements (1 through 8, Fig. 4.2) at the margin of the rectangle out side the anomalous zone. The gravity values at these nodes are defined by weighted shape functions, which represent the regional gravity field. The resulting regional and residual fields based on this method are given in Fig. 4.4(a) and (b), respectively. The regional field looks like a second or third order polynomial surface decreasing in amplitude from about -87 mGal in NW corner to –78.5 mGal in the SE corner. The residual field shows gravity highs and lows as in the original map with reduced amplitude. However, some small wavelength anomalies such as around Kawatha usually related to shallow features have disappeared in this residual map (Fig. 4.4(b)) compared to the original map (Fig. 4.2) and the gradient of the gravity field across Yekundi fault associated with the epicenter of the Latur earthquake is also considerably reduced. This is not the case when regional field was estimated based on zero free air anomaly as given above (Fig. 4.3). These features of regional-residual fields separation are brought out here to caution the users that any method for this purpose should be used cautiously and the separated fields should be examined vis-à-vis observed field and the general geology and tectonics of the area. Regional-residual separation still remain an arbitrary process and therefore utmost precaution by interpreter is essential and their general characteristics should conform with the observed field.

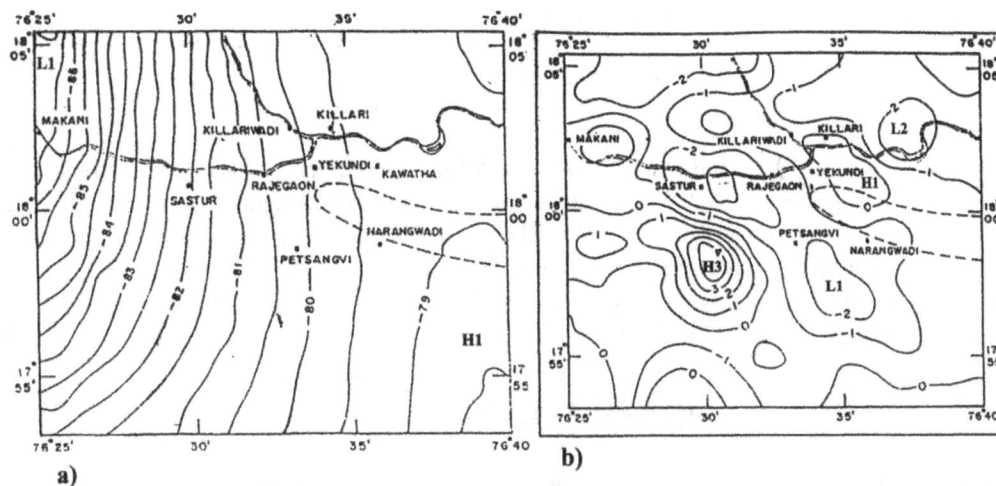

Fig. 4.4 (a) Regional and **(b)** Residual gravity fields separated from the observed field based on finite element method (Mallick and Sharma, 1999). Regional field strike N-S decreasing towards the west, which is the effect of crustal thickening under the Western Ghats due to isostasy. The residual field is oriented NW-SE and shows almost same order of anomalies (H1 and L1) as obtained from zero free air anomaly confirming the two independent methods of regional residual separation.

4.2   Spectral Analysis

4.2.1   Fourier Series Representation and Linear System

Some basic principles of this topic related mainly to potential field data are described below. Those interested in details can refer to the books related to it as given below.

Fourier series implies presentation of data in waveform as function of sine and cosine waves. The solution of Laplace equation governing potential field in the form of exponential function as given in equation 2.18 is the basis for representation of gravity and magnetic fields as Fourier series representation. This in turn implies sinusoidal time/space varying periodic data. The simplest data of this type is a sine wave (Fig. 4.5(a)), which can be expressed as:

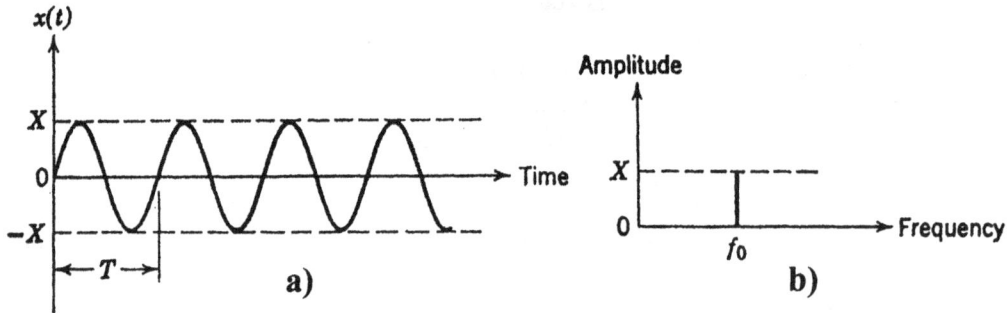

Fig. 4.5 A sine wave (a) and its spectrum (b), which is constant at a specific frequency f_0

$$x(t) = X.\text{Sin} (2\pi ft + \theta) \qquad .....(4.2)$$

where $x(t)$ = instantaneous sinusoidal data, X = amplitude, f = cycles per unit time, known as frequency and θ = phase angle with respect to origin of time in radians.

In case of a continuous data set:

$$x(t) = X.\sin 2\pi ft \qquad .....(4.3)$$

Therefore, in Fourier series representation, the two important characteristics are amplitude X and frequency f as shown in Fig. 4.5(b) for a sine wave, which shows a constant amplitude X at a frequency f_0. The time interval required for one full cycle of data is called time period T and number of cycles per unit time is called the frequency f. However, in nature it is difficult to conceive a purely sinusoidal data. Most of the data sets are complex periodic data, which can be expressed as combination of sine and cosine waves as follows:

$$x(t) = \frac{a_o}{2} + \sum_{n=1}^{\infty} (a_n \cos 2\pi f_n t + b_n \sin 2\pi f_n t) \qquad .....(4.4)$$

where a_o is constant equal to its mean value and is known as D.C. component.

f = fundamental frequency given by 1/T where T is time period and

a_n and b_n are Fourier coefficients known as amplitude of sine and cosine waves given by:

$$a_n = \frac{2}{T} \int_{-T/2}^{T/2} x(t) \cos 2\pi f_n t \, dt \qquad n = 0, 1, 2,..... \qquad .....(4.5)$$

$$b_n = \frac{2}{T} \int_{-T/2}^{T/2} x(t) \sin 2\pi f_n t \, dt \qquad n = 1, 2, 3..... \qquad .....(4.6)$$

The combined amplitude X_n is given by:

$X_n = \sqrt{(a_n^2 + b_n^2)}$ and is known as amplitude spectrum and phase $\theta_n = \tan^{-1} (b_n / a_n)$ where n = 1, 2, 3...

This wave form provide an spectrum of constant amplitudes $X_1, X_2, ..., X_n$ at frequencies $f_1, f_2, f_3, ... f_n$ as given in Fig. 4.6(a). The X_n, and θ_n can be represented as shown in Fig. 4.6(b). A function, therefore can be fully defined once the X_n and θ_n is known for all the frequencies present in it. An alternative way to express equations (4.4) and (4.5) is in form of complex function as:

(a)

(b)

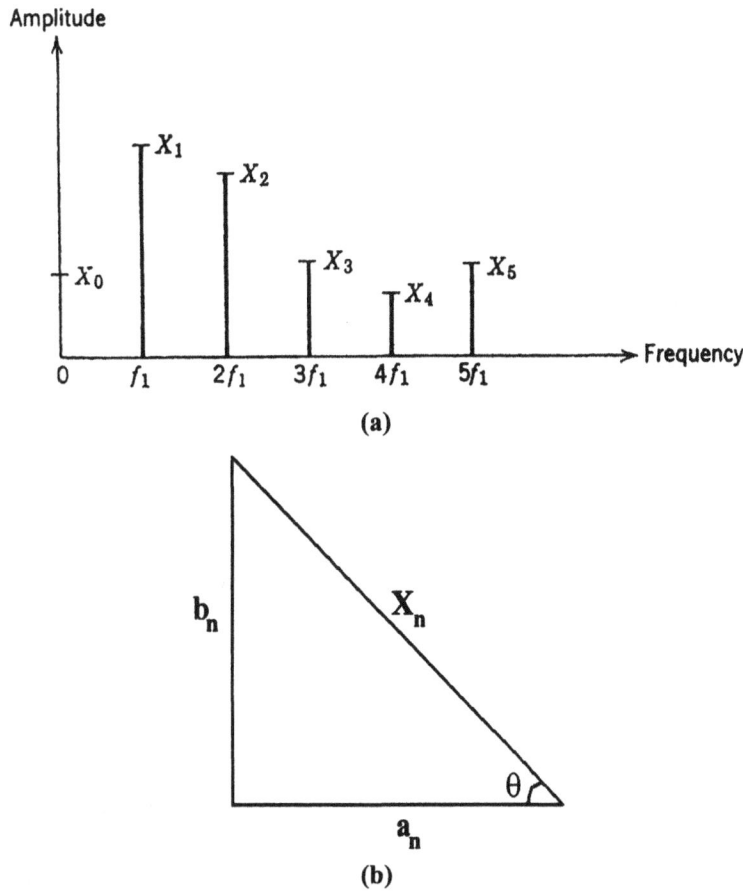

Fig. 4.6 **(a)** Spectrum of a complex periodic data **(b)** Amplitude spectrum (x_n) and phase (θ) of complex periodic data such that $x_n^2 = = a_n^2 + b_n^2$ and $\theta = \tan^{-1} b_n/a_n$

$$x(t) = \sum_{n=-\infty}^{\infty} X_n(f) \exp i(2\pi f_n)t \qquad \qquad .....(4.6)$$

and

$$X_n(f) = \frac{1}{T} \int_{-T/2}^{T/2} x(t) \exp - i(2\pi f_n t)dt \qquad \qquad .....(4.7)$$

$X_n(f)$ is the Fourier coefficients of function x(t) in frequency domain and is complex in nature. Equations 4.6 and 4.7 are known as Fourier pairs, the latter is called the direct transform and the former as the inverse transform.

(i) Linear Systems

Most of the applications of signal processing technique to geophysical data is based on the assumption of the earth behaving as a linear system and time variant field can be considered as variation in space in case of geophysical data.

Linear system is defined as a system such that if y_1 is the output of a linear system for an input x_1 and y_2 is output for input x_2 then in case of a linear system, linear combination of inputs provides the linear combination of the two outputs (Fig. 4.7). In case of such a system, if x(t) is the input at a particular time and y(t) is the output and h(t) expresses the response of system due to an impulse at t = 0, known as impulse response function of the system or transfer function (Kanasewich, 1975), then

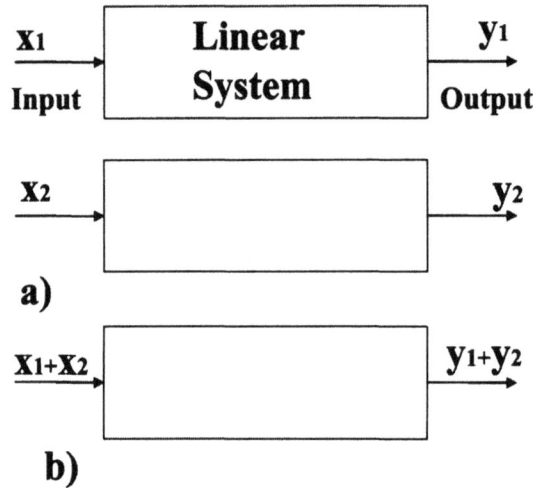

Fig. 4.7 **(a)** Linear system showing outputs y_1 and y_2 for input x_1 and x_2. **(b)** Linear combination of two inputs provide similar linear combination of two outputs.

$$y(t) = \int_{-\infty}^{\infty} h(\tau)x(t-\tau)d\tau \qquad \qquad .....(4.8)$$

where τ is time lag in the response of the system. This expression states that output at a particular time is the function of combined inputs before that time with a time lag depending on the response function of the system. As input in negative time axis has no meaning it can be expressed as:

$$y(t) = \int_{0}^{\infty} h(\tau)x(t-\tau)d\tau \qquad \qquad .....(4.9)$$

It is briefly written as convolution of input function and impulse response function as:

$$y(t) = h(t) * x(t) \qquad \qquad .....(4.10)$$

where * is referred to as convolution in the real (time) domain, which is equivalent to multiplication in frequency domain (Kansewich, 1975). Therefore

$$Y(f) = H(f).X(f) \qquad \qquad .....(4.11)$$

where $Y(f)$, $X(f)$ and $H(f)$ are Fourier Transform of output, input and impulse response function of the system known as frequency response function. This implies that in case of linear system, the transform of output and input can be used to obtain system characteristics $H(f)$ and vice versa. Another important relationship in this regard is obtained by multiplying both sides of the above equation by transform of the input function as:

$$Y(f).X(f) = H(f).X(f).X(f)$$

or $\qquad \qquad C_{xy}(f) = H(f). X^2(f) \qquad \qquad .....(4.12)$

where $C_{xy}(f)$ is the cross spectrum between input and output signal and $X^2(f)$ is the power spectrum of the input signal.

This provides the frequency response function of a system whose inverse transform can provide the impulse response function in real domain. The relationship has been widely used in describing the isostasy and rheology of a region as discussed below in Section 4.4.

4.2.2   Discrete Fourier Transform – Fast Fourier Transform and Geophysical Data

Fourier Series primarily represent continous data with respect to time. Equations 4.6 and 4.7 are Fourier transform pairs, valid for a continuous function. In case the data is discrete as in case of geophysical data, it is sampled at a specific constant interval (Fig. 4.8(a)), which amounts passing the continuous signal through an infinite Dirac comb (Kanasewich, 1975), which is given by Dirac delta function:

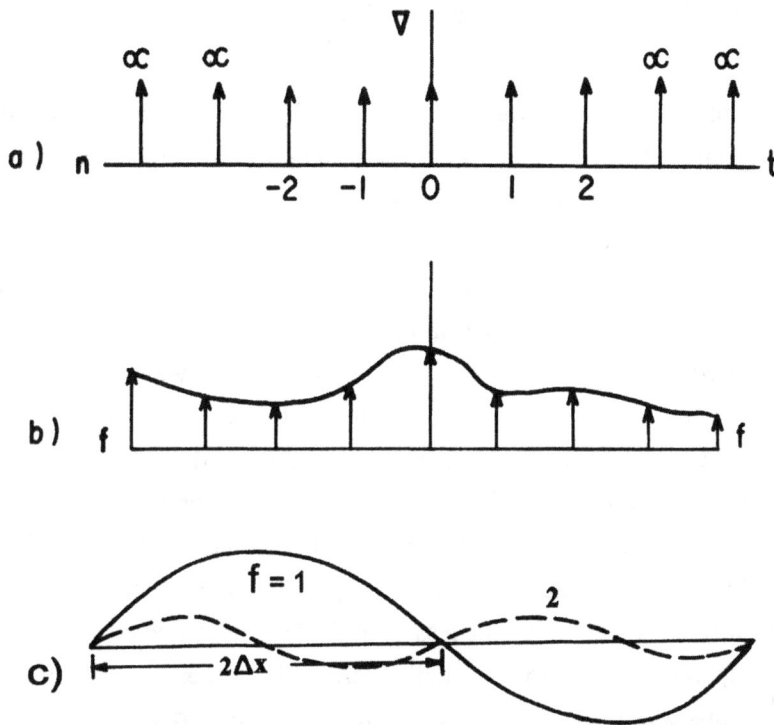

Fig. 4.8 (a) A continuous function sampled using Dirac delta functions at discrete points at constant interval or spacing (b) The transform of continuous function sampled at specific interval is also a continuous function defined at discrete frequencies sampled by Dirac Comb equivalent to amplitude modulation. (c) If Δx is sampling interval the largest wavelength in present case is $4.\Delta x$ of frequency 1. Similarly frequency 2 has a wavelength of $2.\Delta x$ suggesting wavelength and frequency are inversely related. In thic case smallest (Nyquist) wavelength $(\Delta\lambda) = 2\Delta x$ and Nyquist frequency $(\Delta f) = 1 / 2. (\Delta x)$, therefore sampling interval $(\Delta x) = 1/ (2. \Delta f) = \Delta\lambda/2$.

$$\nabla(t, \Delta t) = \sum_{n=-\infty}^{\infty} \delta(t - n\Delta t) \qquad \qquad .....(4.13)$$

The Fourier transform of this function is also a delta function in frequency domain as (Blackman and Tukey, 1958):

$$\nabla\left(f, \frac{1}{\Delta t}\right) = \frac{1}{\Delta t} \sum_{n=-\infty}^{\infty} \delta\left(f - \frac{n}{\Delta t}\right) \qquad \qquad .....(4.14)$$

This amounts to sampling the Fourier transform of the function with a similar Dirac Comb in frequency domain (Fig. 4.8(b)). This is known as Discrete Fourier Transform. Some geophysical data such as gravity and magnetic data are expressed in space with respect to distance and same principles can be applied by changing time period with wavelength λ and frequency f in such cases are given by $f = 1/\lambda$ instead of $1/T$. This is known as Discrete Fourier transform in space (Naidu and Mathew, 1997).

(i) Nyquist Frequency and Gibb's Phenomena: Discrete Fourier Transform (DFT) is band limited upto a minimum and maximum of certain period or length in space depending on station interval or data spacing, which imposes following conditions on the data set to be truly represented by Fourier transform.

(a) The first condition in this regard relates to the representation of minimum wavelength or the highest frequency present in the data based on data sampling which is known as Nyquist wavelength / frequency. In practice, equispaced digital data is generated from the recorded data by interpolation or alternatively the profiles and maps are digitised at equal intervals depending on stations spacing and depth of investigation. However, the sampling interval should be such that it represents the highest frequency present in the data set known as Nyquist frequency (Δf), which is equal to $1/\Delta\lambda$; where $\Delta\lambda$ is the Nyquist wavelength, shortest wave length present in the data. The shortest wavelength included in the digital data with Δx as sampling interval is $2.\Delta x$, and its frequency is $(1/2\Delta x)$. Therefore, for proper representation of data, $\Delta\lambda = 2\Delta x$ and Nyquist frequency $(\Delta f = 1/\Delta\lambda) = (1 / 2\Delta x)$ or sampling interval $(\Delta x) = \Delta\lambda / 2$ or $1 / 2\Delta f$. This implies that the sampling interval for proper representation of a given data set should be half the Nyquist wavelength ($\Delta\lambda/2$). In case sampling interval is larger than the Nyquist wavelength, the power/energy of the waves with wavelength less than $2. \Delta x$ gets reflected in the waves with higher wavelength compared to Nyquist wavelength, which is known as spectral folding.

(b) The second condition relates to the maximum wavelength present in the data set. As in case of geophysical data, we consider limited length of profile or block of data, the maximum wavelength of the data set is fixed by their size and the data set inherently repeats itself in space beyond available data set. In case there are big differences in the first and last value of data set, it creates discontinuity at the borders of the data set whose transform is different from a smoothly varying field. This phenomena is known as Gibb's phenomena and is reduced by applying suitable windows such as cosine taper window etc., (Bendat and Piersol, 1993). Another approach, which is followed in practise, is to add a few data at the end of the data set to be analysed through cubic interpolation between last and the first value so that sudden discontinuity at the border of the data is avoided, which is demonstrated in Fig. 4.9. The first values x_o and y_o repeat themselves after n observation points on both axes after three station intervals and three points on either axis are added through cubic interpolation between x_n, y_n and x_o and y_o such that data matrix for transformation becomes $(n + 3) \times (n + 3)$.

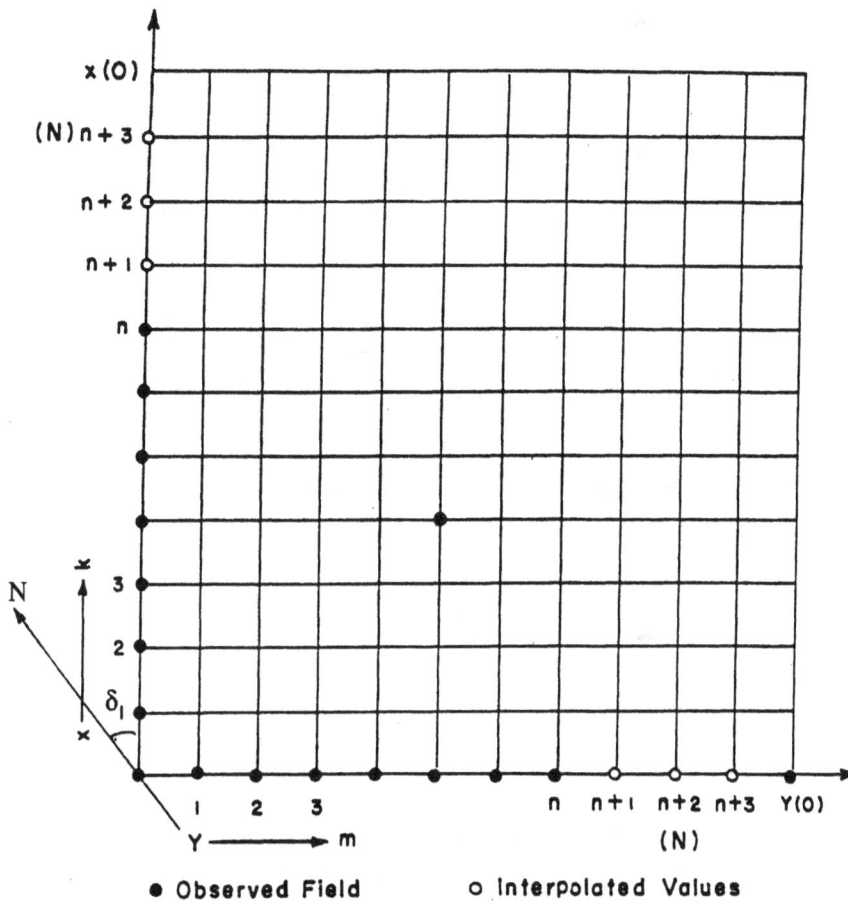

Fig. 4.9 Two dimensional data along x, y grid for Fourier transformation. k and m are respective frequencies. The actual data set is numbered 0, 1, 2, ..., n and the values n + 1, n + 2 and n + 3 are interpoled values between last value of respective rows and columns and its first value placed beyond them with same station interval through cubic interpolation to avoid Gibbs phenomena. Finally (n + 3) × (n + 3) values are used in transformation. δ_1 is angle of x-axis with magnetic north.

4.2.3   Fourier Transform of Gravity and Magnetic Fields and Filters

The solution of Laplace equation governing potential field in the form exponential functions (Equation 2.18) suggest that the observed magnetic and gravity field F(x,z) at the surface along a profile (x-axis) at a level z with positive z axis downwards representing depth can be represented by discrete Fourier transform in frequency domain as follows:

$$F(x,z) = \sum_{f=-n/2}^{n/2} F(f).\exp[2\pi f(ix + z)/\lambda] \qquad \qquad .....(4.15)$$

Where f is the frequency and n is the number of observations at equidistant points at Δx sampling interval, z is the plane of observation and λ is the wavelength given by n.Δx. The transform of this field [F (f)] is given by:

$$F(f) = \frac{1}{n}\sum_{x=1}^{n} F(x)\exp(-2\pi.f.ix/n) \qquad \qquad .....(4.16)$$

Where F(f) is the Fourier transform of the observed field F(x) and is defined by its amplitude spectrum as given above corresponding to frequencies f and the corresponding power or energy spectrum is F^2

$$F^2(f) = F(f).F^*(f) \text{ where } * \text{ denotes complex conjugate.}$$

In this manner, the digital data is transformed into frequency domain providing the amplitudes at various frequencies (wavelengths) present in the data set. In case of two dimensional data equations, (4.15) and (4.16) take the following form in x and y directions perpendicular to each other (Hahn, 1965), which represents the direction of gridded data from the map (Fig. 4.9).

$$F(x,y,z) = \sum_{m=-n/2}^{n} \sum_{k=n/2}^{n} F(k,m)\exp(2\pi[i(kx+my)+z\sqrt{k^2+m^2}]/\lambda \qquad ......(4.17)$$

where F(x,y,z) is the observed field at n x n grid points and z is the plane of observation, k and m are frequencies along x and y axis (Fig 4.9) given by:

k, m = - n/2 . . . . + n/2 when n is even

= (n- 1) / 2 . . . . + (n-1) / 2 when n is odd.

$$F(k,m) = \frac{1}{n^2} \sum_{x=1}^{n} \sum_{y=1}^{n} F(x,y)\exp[-2\pi i(kx+my)/n] \qquad .....(418)$$

In case of maps, the two-dimensional version of discrete Fourier transform is used and the computed amplitude spectrum can be averaged in concentric circles for similar frequencies to provide the variation of F(f) with f known as radial spectrum (Naidu, 1970; Hahn et al., 1976). As two dimensional data involves large data matrix, it can be transformed using Fast Fourier Transform as briefly described below (Colley and Tuckey, 1965; Naidu, 1970; Kanasewich, 1975). In these cases $f = (k^2 + m^2)^{1/2}$ where k and m are frequencies along x, y axis (Fig. 4.9). Fig. 4.9 also shows that the observed field is nxn which however is enlarged to N × N by adding 3 equidistant points beyond the data set to be analyzed by cubic interpolation between last and first point of each row and column. This is done to reduce Gibbs phenomena as described above, which is caused due to large differences in the field values at the last and the first points as the data set is repeated like a chess board in Fourier transformation. The same can also be achieved by multiplying with cosine function as windows at the border of the data set (Bendat and Piersol, 1993). The averaged radial spectrum is given by (Mishra and Naidu, 1974).

$$R(f) = \frac{1}{2\pi} \int_{0}^{2\pi} F(f\cos\theta, f\sin\theta)d\theta \qquad .....(4.19)$$

Where f cos θ = k and f sin θ = m

The averaging of spectrum in concentric circles is demonstrated in Fig. 4.10. The plot of log of amplitudes spectrum versus frequencies (Equations 4.16) provides straight line segments (Fig. 4.35 and 4.36) corresponding to sources distributed at different levels as described in Section 4.3 for layered models. The first segment corresponding to low frequencies represent relatively deeper sources compared to other segments corresponding to higher frequencies. In practice, 2-3 layers can be effectively separated in a spectral plot.

Top grid (each cell shows A(f) value over f value):

PQ \ M	0	1	2	3	4	5	6	7	8
K=8	.72/8.00	.80/8.06	1.25/8.25	.54/8.54	.78/8.94	.38/9.43	.64/10.00	.30/10.63	.21/11.31
7	1.19/7.00	1.51/7.07	1.28/7.28	2.72/7.62	1.22/8.06	.39/8.60	.88/9.22	.49/9.90	.21/10.63
6	.29/6.00	2.43/6.08	3.32/6.32	2.90/6.71	1.43/7.21	.93/7.81	1.92/8.49	.82/9.22	.18/10.00
5	.89/5.00	3.13/5.10	.70/5.39	1.35/5.83	1.46/6.40	2.76/7.07	2.34/7.81	1.13/8.60	.52/9.43
4	1.87/4.00	8.79/4.12	9.17/4.47	8.58/5.00	3.57/5.66	1.19/6.40	2.22/7.21	.66/8.06	1.41/8.94
3	5.12/3.00	9.57/3.16	12.67/3.61	10.51/4.24	5.86/5.00	3.49/5.83	1.16/6.71	.60/7.62	1.30/8.54
2	6.33/2.00	8.97/2.24	11.31/2.83	17.66/3.61	5.37/4.47	5.76/5.39	2.41/6.32	2.76/7.28	1.27/8.25
1	9.53/1.00	13.66/1.41	4.94/2.24	.91/3.16	2.41/4.12	3.18/5.10	3.11/6.08	2.22/7.07	2.20/8.06
0		8.69/1.00	10.48/2.00	1.69/3.00	2.26/4.00	1.62/5.00	1.01/6.00	.69/7.00	.85/8.00

Bottom grid:

PQ \ M	0	1	2	3	4	5	6	7	8
K=8	.72/8.00	.65/8.06	.26/8.25	1.90/8.54	.29/8.94	.61/9.43	.54/10.00	.19/10.63	.27/11.31
7	1.19/7.00	.65/7.07	1.03/7.28	1.82/7.62	1.66/8.06	.88/8.60	.40/9.22	.46/9.90	.33/10.63
6	.29/6.00	.72/6.08	.50/6.32	1.91/6.71	2.79/7.21	.43/7.81	.52/8.49	.53/9.22	.45/10.00
5	.89/5.00	3.30/5.10	7.46/5.39	2.95/5.83	.81/6.40	2.30/7.07	.83/7.81	.77/8.60	.45/9.43
4	1.87/4.00	4.65/4.12	8.46/4.47	3.36/5.00	2.71/5.66	.49/6.40	1.03/7.21	.55/8.06	.70/8.94
3	5.12/3.00	2.27/3.16	8.35/3.61	9.80/4.24	4.67/5.00	3.05/5.83	.13/6.71	1.51/7.62	.49/8.54
2	6.33/2.00	8.57/2.24	11.79/2.83	11.22/3.61	6.55/4.47	6.12/5.39	.98/6.32	1.96/7.28	1.45/8.25
1	9.53/1.00	20.52/1.41	11.17/2.24	25.70/3.16	6.70/4.12	8.75/5.10	1.45/6.08	1.03/7.07	1.53/8.06
0		8.69/1.00	10.48/2.00	1.69/3.00	2.26/4.00	1.62/5.00	1.01/6.00	.69/7.00	.85/8.00

A(f)	19.22	12.68	9.40	8.42	5.28	2.17	1.61	1.26
f	1.00	1.97	3.04	3.95	4.99	6.02	6.95	8.01

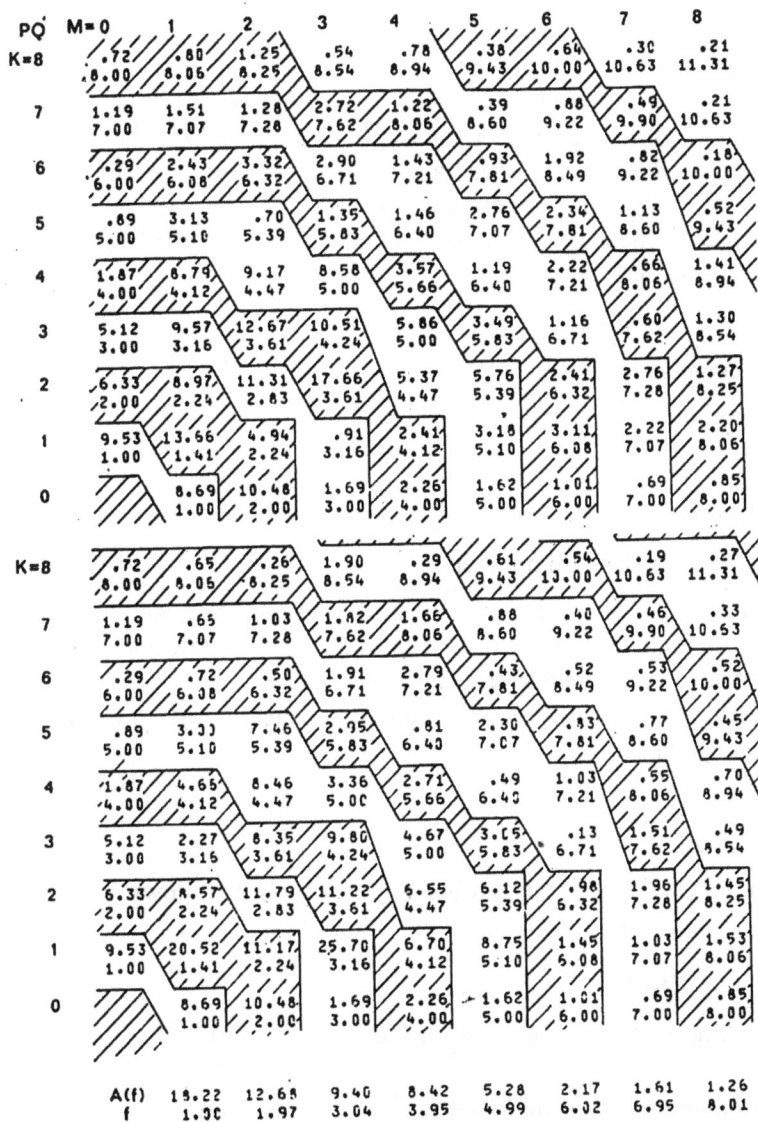

Fig. 4.10 Averaging of amplitude spectrum of two dimensional data to obtain radial spectrum in one direction for combined frequency $f = (k^2 + m^2)^{1/2}$. Averaged amplitude spectrum A(f) and frequencies (f) in various strips are shown at the bottom.

Following precautions are observed while computing the Fourier Transform of the observed field in actual practise.

(i) The data set to be transformed should contain more anomalies of different magnitude, shapes and sizes to represent the anomalies statistically. This condition is better satisfied in case of airborne magnetic maps, which usually contain anomalies of different wavelength compared to Bouguer anomaly gravity maps.

(ii) The essential elements of anomalies should not be intersected at the borders and there should not be large differences between the anomalies on the opposite side of borders to avoid Gibbs phenomena.

(iii) The data set should preferably represent one geological unit.

(i)   Fast Fourier Transform and Computation of Spectrum of Two Dimensional Data

Due to large amount of computation involved in computation of spectrum of two dimensional data, several ingenious approaches were suggested for this purpose. For example N x N data matrix requires N^4 operations for its Fourier transformation in conventional manner, which is quite time consuming. Therefore, Fast Fourier transform scheme of Cooley and Tuckey (1965) is used, which requires only $4N^2 \log_N^2$ operation for this purpose. However, the number of operations can be further reduced by utilizing the following properties of Fourier transform (Naidu, 1970). In one dimensional data set.

$$F(k) = F^*(k) \text{ and} \qquad\qquad\qquad .....(4.20)$$

$$F(k = 0 \text{ or } \pi) = \text{a real number}$$

Similarly, in case of two dimensional data

$$F(k,m) = F^*(-k,-m)$$

$$F(-k,m) = F^*(k,-m) \qquad\qquad\qquad .....(4.21)$$

$$F(k = 0 \text{ or } \pi, m = 0 \text{ or } \pi) = \text{a real number.}$$

These relations implies that Fourier transform is same in opposite quadrant and therefore, in practice, one needs to compute transform for one fourth of total frequencies present in the data and the rest can be obtained based on above characteristics. Further, reduction in operations can be done by making a complex sequence of two columns of digital data and transforming them together using FFT algorithm, such that:

$$f(x) = f_1(x) + if_2(x) \qquad\qquad\qquad .....(4.22)$$

and $$F(k) = F_1(k) + iF_2(k)$$

Using equation (4.20), $F_1(k)$ and $F_2(k)$ can be obtained separately as follows:

$$F_1(k) = 0.5[F(k) + F^*(-k)]$$

$$F_2(k) = 0.5i[F(k) - F^*(-k)] \qquad\qquad\qquad .....(4.23)$$

In this manner following steps can be followed to transform a two dimensional data set conveniently.

(a) Two consecutive columns such that 1 and 2, 3 and 4 etc., can be used to form complex sequences and transform using Cooley and Tuckey (1965) algorithm. Using equations (4.23) transforms of each column separately can be obtained.

(b) Next the transform of each row is obtained separately as they are complex sequences and obtain their real and imaginary parts.

The total number of arithmetic operations required to transform a data set of N × N in this manner are $4N^2 \log_2 N$ which is much less compared to N^4 operations required to transform by conventional method. However, Fast Fourier Transform (FFT) can be applied only to the data sets, which are 2^n, viz. 4, 8, 16, 32, 64, 128 etc., which is a limitation of FFT. However, it can be circumvented by putting zeros at the end of a given data set to comply with this condition. The other alternative to comply with it to fill the remaining data gap by cubic interpolation between last and first data repeated after requisite number of data to comply with 2^n condition. This takes care of Gibbs phenomena automatically. However, presently there are several softwares, which can transform data set with any number of points along sides without any limitations of being 2^n or even if data is rectangular with different numbers of rows and columns. There are several commercial soft wares to transform grided datasets. However, students may attempt to transform two dimensional data as an exercise using FFT given in MATLAB and above properties of the transform, which would help them to understand the various operations in this regard. There have been several discussions (Regan and Hinze, 1976; Nielsen and Pedersen, 1979) about the use of a suitable windows to avoid discontinuity at the borders while computing the spectra of magnetic and gravity fields. Regan and Hinze (1976) have advocated the use of a rectangular window, while Nielsen and Pedersen (1979) have suggested the use of a cosine taper or an exponential window (Bendat and Piersol, 1993). The type of window, however, used in these cases should largely depend on the effect they will produce on the computed spectrum. If the field on either end of the profile is the same such that there is no discontinuity at the borders, the application of a rectangular or a cosine window might not cause any difference on the computed spectrum. However, if the field on either ends of the profiles is different as is generally the case with statistical models such as basement studies, the application of a rectangular window will create discontinuities at the borders, thereby resulting in some distortions in the computed spectrum. The amount of distortion, however, depends upon the difference in the field on either ends of the profile. Further, details can be obtained from references referred to above.

(ii) Digital Filtering: Regional – Residual Separation

In case of causative sources, located at two subsurface levels and $F_1(f)$ and $F_2(f)$ being the amplitude spectrum of gravity and magnetic fields due to them, the power spectrum of the composite field $F(f)$ is given by (Hahn et al., 1976).

$$F^2(f) = F_1^2(f) + F_2^2(f) + 2F_1(f) \, F_2(f) \cos [A_2(f)-A_1(f)] \qquad .....(4.24)$$

Where $A_1(f)$ and $A_2(f)$ are phases of the gravity and magnetic fields due to individual layers. In case, the expectancy value $E [\cos A_2(f)-A_1(f)]$ is zero,

$$F^2(f) = F_1^2 (f) + F_2^2(f) \qquad .....(4.25)$$

$$F_1^2(f) = F^2(f) - F_2^2(f) \qquad .....(4.26)$$

This suggests that $F_1(f)$ or $F_2 (f)$ can be obtained from $F(f)$, the computed spectrum of the observed field, if the other is known or $F_1(f)$ and $F_2(f)$ have different frequency characteristics. However, in actual practice, it is a difficult proposition to know the spectrum $F_1(f)$ or $F_2(f)$ of individual sources before hand. Therefore, the usual characteristics of regional and residual fields originating from deeper and shallow levels represented by low and high frequency bands respectively are used to separate them from the observed field. Using this characteristic low and high pass filters (Fig. 4.11) with reference to a cut off frequency f_c are designed to separate the regional and the residual fields, respectively. f_c can be decided based on the frequency characteristics of the linear segments of the spectral plot.

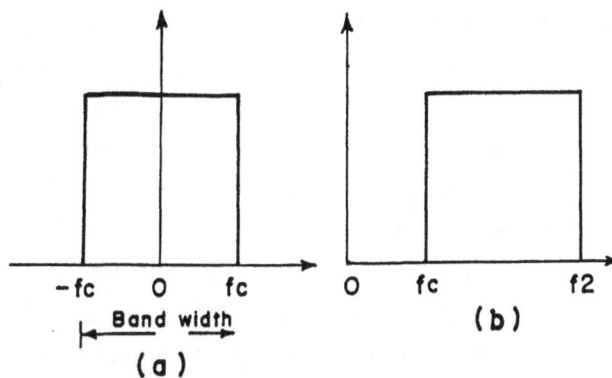

Fig. 4.11 A schematic representation of band pass filters **(a)** A low pass filter with frequency range $0 - f_c$; f_c is known as cutoff frequency. **(b)** A high pass filter with frequency range $f_c - f_2$.

The low pass filter retains lower frequencies (0-f_c, Fig. 4.11(a)) while rejecting the higher frequencies, therefore the filter characteristics are

$$F(f) = 1 \text{ for } -f_c \leq f \leq f_c \text{ and } F(f) = 0 \text{ for } -f_c > f > fc \qquad \text{.....(4.27)}$$

Similarly, a high pass filter between frequencies fc and maximum available frequency f_2 (Fig. 4.6(b)) can be designed as

$$F(f) = 1 \text{ for } f_c \leq f \leq f_2 \text{ and } F(f) = 0 \text{ for } f_c > f > f_2 \qquad \text{......(4.28)}$$

However, in practice, it is difficult to assign frequency bands for sources originating from several levels and therefore usually low pass filter is designed for the regional field with low frequency band and the residual field is obtained by subtracting it from the observed field. These filters are rectangular functions (Fig. 4.11(a) and (b)) whose transform can be multiplied with the transform of the field given by equation (4.18) and its inverse transform using equation (4.17) provides the desired field. The transform of rectangular function being oscillatory in character, it is advisable to apply a cosine window function to taper down the amplitudes at the border slowly (Bendat and Piersol, 1971).

The selection of cut off frequency is arbitrary as order of trend surface in polynomial approximation and depends largely on the experience of the interpreter. It is, therefore advised to design 3-4 alternatives of these filters and through trial and error based on other available information and experience decide the appropriate cut off frequency/wavelength. As stated above, the plot of log F(f) versus f provides straight line segments (Fig. 4.35 and 4.36) in specific frequency bands which has been considered to originate from different levels.

(iii) Application of Filters for Regional-Residual Separation: Bouguer Anomaly Map of Godavari Basin

The Bouguer anomaly map given in Fig. 2.30(a) is chosen to demonstrate the application of wave band filtering. This map was digitized at about 5.5 km interval and several filters were computed with harmonics 1, 2 etc., corresponding to first harmonic of wavelength 222 km is given in Fig. (4.12), which may represent the regional gravity field. This regional field indicates a gravity low in the central part of the basin, which was chosen as the regional based on the results of a seismic profile across Godavari basin (Kaila et al., 1990), which suggested crustal thickening under the basin. The regional field in case of gravity data mainly originates from Moho as it is the most prominent density discontinuity with maximum density contrast. The corresponding residual field after subtracting regional field from the observed field is given in Fig. 4.13, which shows almost all essential elements of the observed field with a gravity low in the central part due to low density sediments and shoulder highs due to high density intrusives in the upper crust (Mishra et al., 1993). This study suggests that one has to be very cautious in separating regional and residual fields using digital filters and requires constant interaction of interpreter in deciding appropriate waveband for this purpose based on known information from the region. It is therefore, apparent that in case of gravity data, a sharp cutoff low frequency filter of first or at the most second harmonics (frequency 1 or 2) are required to obtain the regional field.

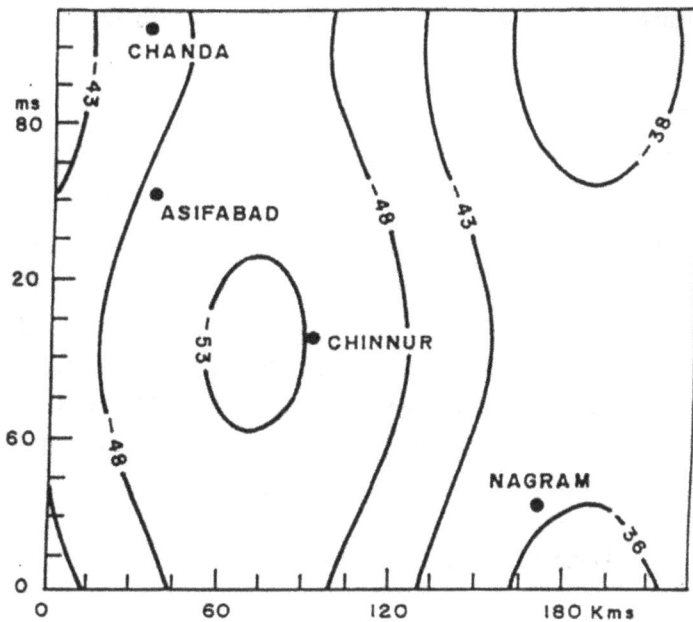

Fig. 4.12 A low pass filtered Bouguer anomaly of Godavari basin (Fig. 2.30(a)) for first harmonic (frequency-1) of wavelength 360 km, which provides the regional field showing a gravity low indicating crustal thickening (Mishra et al., 1993). Regional field in gravity data primarily originates from Moho as it is most prominent density discontinuity at relatively deeper level.

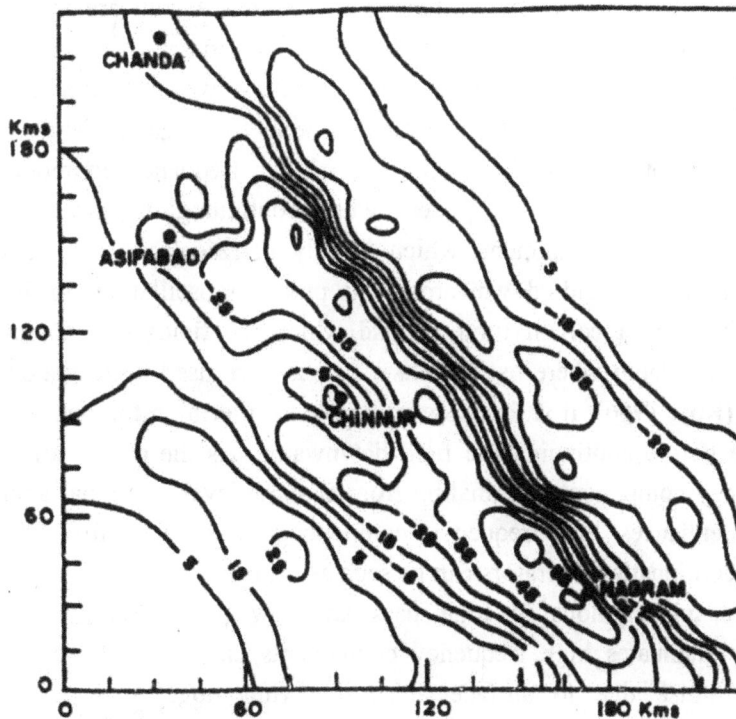

Fig. 4.13 Residual Bouguer anomaly of Godavari basin showing a gravity low in the central part due to sediments and highs along the shoulders due to high density intrusives. It consists of all the essential elements of the observed field, viz. central low and shoulder highs, caused by shallow bodies.

4.2.4   Analytical Operations in Frequency Domain and Applications

The observed gravity field can be modified through some mathematical operations, which enhances certain components of the observed field while suppressing the other components. These operations are called analytical operations. The two most important operations are Continuation and Derivatives of the observed field.

(i) Continuation of the field: The field observed at one plane can be continued mathematically upwards or downwards to any other plane. Several schemes have been provided for this purpose (Roy, 1967; Trejo, 1954; Agarwal and Lal, 1972), Grant and West, 1965). However, the one carried out in frequency domain discussed below is most simple to apply (Hahn, 1965; Karasewich and Agarwal, 1970). As it is evident from equation (4.15 and 4.17), the observed field at a level z_0 can be continued to another level (z_1) by simply multiplying the Fourier coefficients of the observed field at level z_0 by:

$$\sum_{f=-n/2}^{+n/2} \exp[2\pi . f . (z_0 - z_1) / \lambda$$

z being positive downwards. Therefore, the continuation operators for $\Delta z = z_0 - z_1$ are as follows:

Upward Continuation: $\sum \exp(2\pi f(-\Delta z) / \lambda$ ....(4.29)

Downward continuation: $\sum \exp(2\pi f . \Delta z . / \lambda)$ .....(4.30)

Equations (4.29 and 4.30) are known as continuation operators, which are multiplied to the transform of the observed field and their inverse transform provides the continued field along same profiles at that height (z_1). These equations shows that the downward continuation operator increases exponentially with depth and is therefore unstable for large Δz while continuation upward is an stable operation. Further, the amplitudes of high frequency component increases much more than the low frequency component while continuing downwards (Fig. 4.14). Therefore, the high frequency component, which usually represents noise gets much more enhanced while continuing the fields downwards. This produces oscillations in the downward continued field and imposes a certain limit beyond which the fields can not be continued downwards. This limit on depth where oscillations start has been used to estimate the depth to the causative sources (Roy, 1966). It is therefore advisable to filter high frequency components, which represent noise before continuing the field downwards. As the continuation downward enhances high frequency components originating from shallow levels (residual anomalies) and continuation upward enhances low frequency components originating from deeper levels (regional anomalies), continuation operators can be seen as means of enhancing the anomalies of interest and therefore, synonymous to high pass and low pass filters, respectively. As continuation upwards attenuates high frequency components, airborne magnetic and gravity surveys reduces the noise due to surface anisotropy (inhomogeneity) in magnetization (susceptibility) and density, respectively. This is specially true for airborne magnetic surveys as anisotropy in susceptibility of rocks is quite common to give rise to erratic ground magnetic data, which gets attenuated in airborne magnetic surveys.

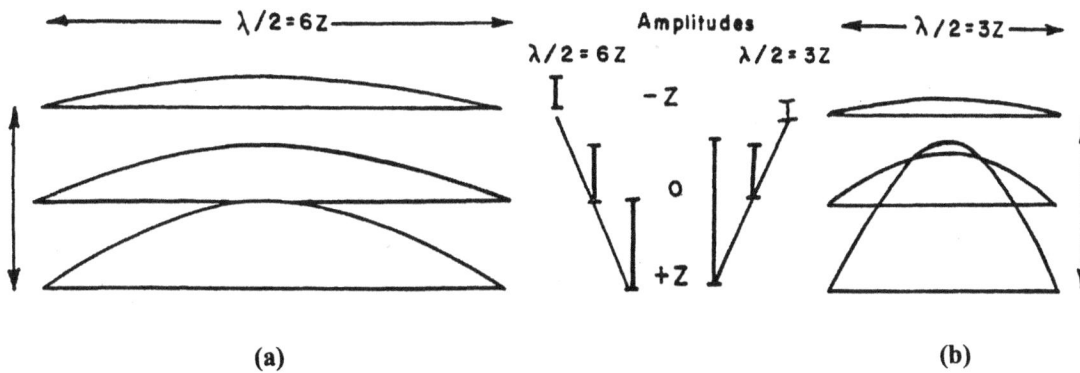

Fig. 4.14 Continuation upwards and downwards for two cases of **(a)** large wavelengths ($\lambda / 2 = 6\ Z$) and **(b)** short wavelength $\lambda / 2 = 3\ Z$, which shows continuation to $+ z$ and $-z$ levels from observational level (0). Vertical bars in center represent amplitudes of the field at different levels. In continuation downward, the amplitude increases, which is more for small wavelength (high frequency component) compared to large wavelength (low frequency component). This increment in amplitude of high frequency components causes oscillations in the downward continued field beyond certain depths causing instability. Continuation upwards reduces the amplitude of small wavelength components attenuating them compared to low frequency components due to deeper sources and therefore high frequency surface noise is reduced in upward continued field.

(ii) Upward Continuation and Low Pass Filter: Bouguer Anomaly Map of Deccan Trap, India

The Bouguer anomaly map of Deccan trap is given in Fig. 4.20(a) (Tiwari and Mishra, 1999), basically shows two linear gravity lows L1 and L2 known as Koyna and Kurduwadi gravity lows, respectively and some short wavelength anomalies. The digital data from this map is continued upwards at different heights to check the deep seated and shallow anomalies. One such continued field at 38 km is given in Fig. 4.15(a) (Tiwari and Mishra, 2006), which shows a regional gravity low related to crustal thickening in this region due to isostasy. The same digital data of this map (Fig. 4.20(a)) is also filtered using a low pass filter of wavelength >250 km, which corresponds to harmonics 1 and 2 (Fig. 4.15(b)). This map also shows a major gravity low similar to the one given in Fig. 4.15. It is interesting to note that shape of the filtered gravity low coincides with the topography high (Fig. 2.27(a)) indicating that it is caused by crustal thickening in response to isostasy. The two maps showing an overall similarity provide credence to each other and also suggest that continuation upwards operation is synonymous to a low pass filter operation. The continued field shows reduced amplitude as it represents field at that height while low pass filtered map is the field observed at the surface. This example also reiterates that sharp cutoff low pass filter of 1 or 2 harmonics are required to separate the signal from Moho as discussed in previous Section (4.2.3) in case of Bouguer anomaly map of Godavari basin.

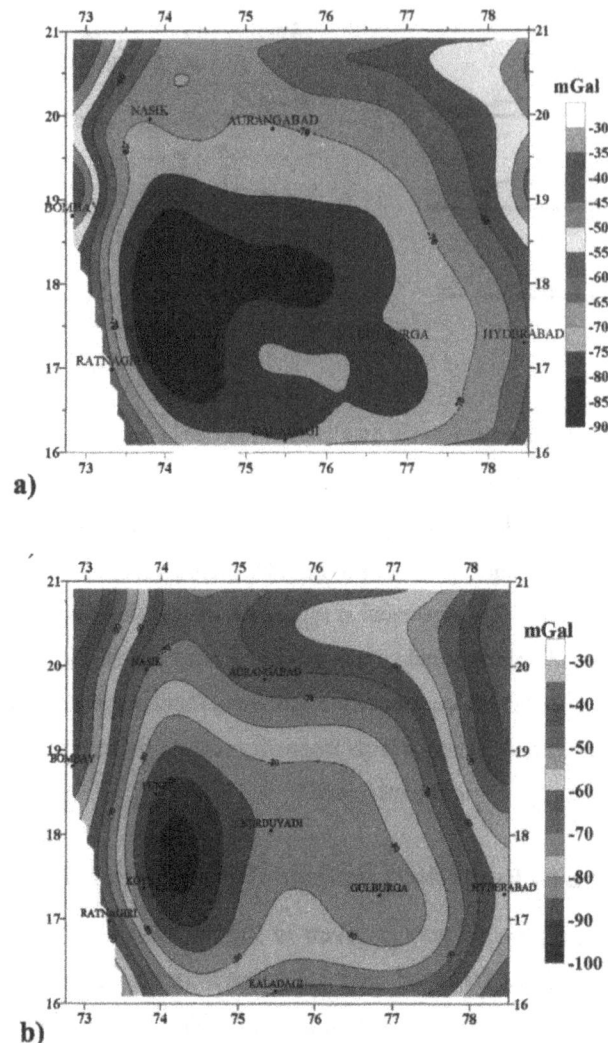

Fig. 4.15 (a) Upward continued Bouguer anomaly of Deccan trap (Fig. 4.20(a)) at 38 km height showing a regional gravity low related to crustal thickening under the Western Ghats in response to isostasy.
(b) A low pass filtered Bouguer anomaly map of Deccan trap for wavelength > 250 km showing a regional gravity low similar to Fig. 4.15(a).(Tiwari and Mishra, 2006).

Because of their application in enhancing regional or residual components of the observed field, the continuation upwards and downwards are regarded as low and high pass filters and are therefore, widely used in practice. There are several schemes to continue field even from uneven surfaces (Parker and Huestis, 1974) which are useful in case of surveys in hilly uneven terrain.

(iii) Derivative Maps: Horizontal and Vertical

The derivative of the magnetic and gravity fields is the rate of change of field in a particular direction and has been popular with the geophysicist as it suppresses low frequency components of regional field and enhances high frequency components of residual field delineating shallow bodies. In this respect, it is similar to continuation of fields downwards. As in case of continuation of fields, several sets of coefficients have been provided by different workers to compute the first and second derivatives of the gravity field (Nettleton, 1976, Grant and West, 1965). However, the operations in frequency domain for this purpose as in case of continuation operator are simple and easy to perform (Bharttacharya, 1964). Differentiating equation (4.15) in the same plane (z):

$$\frac{\partial F(x,z)}{\partial z} = \sum_{f=-n/2}^{n/2} F(f).(f)\exp(2\pi f)(ix+z)/\lambda$$

$$= f.F(x,z) \qquad\qquad .....(4.31)$$

$$\overline{F}\left[\frac{\partial F(x,z)}{\partial z}\right] = f.F(f)$$

or $\quad \dfrac{\partial F(x,z)}{\partial z} = F^{-1}[f.F(f)]$

where F and F^{-1} represent Fourier transform of gradient and inverse transform.

In equation (4.31), (f) can be regarded as a filter function for the first derivative. Similarly, for second derivative up to n^{th} derivative, the filter functions are $f^2 \ldots f^n$. The first vertical derivative expresses the slope or gradient of the field in vertical direction, while second derivative expresses the slope of first derivative or curvature of the original field. Therefore, the second derivative enhances high frequency components more than the first vertical derivative.

Due to multiplication of f and f^2, the high frequency components get much more accentrated compared to low frequency part and therefore high frequency part representing noise should be filtered out before computing the derivative maps.

This process therefore involves following three steps:

  (a) Fourier transform of the observed field.
  (b) Multiplication of Fourier coefficients by filter function $(f, f^2$ etc.)
  (c) Inverse transformation to obtain derivative maps in space.

Same steps are also required in case of continuation map using appropriate filter function as given above.

Horizontal derivative of the field can also be computed as follows:

$$\frac{\partial F(x,z)}{\partial x} = \sum_{f=-n/2}^{n/2} F(f).(if).\exp(2\pi f)(ix+z)/\lambda \qquad\qquad .....(4.31a)$$

Similarly Fourier transform of second horizontal derivative is given by

$$F\left[\frac{\partial^2(F(x,z))}{\partial x^2}\right] = -f^2.F(f)$$

$$[\partial^2.F(x,z)/\partial.x^2] = F^{-1}[-f^2.F(f)] \qquad\qquad .....(4.31b)$$

where F and F^{-1} are Fourier transform of horizontal gradient and inverse transform of it.

This implies that the filter function for second horizontal derivative is $-f^2$ while for second vertical derivative, it is f^2 and they differ only in sign.

As maximum gradient along the observed gravity profiles indicate boundaries of the causative sources, the first horizontal derivative provide maximum value at the borders and second horizontal derivative is zero at the borders of the sources, which are easy to mark compared to based on gradient of the observed field. In this regard, magnetic field can also be used by transforming it first to pseudo gravity as defined in section 3.1 and a scheme for the same is provided by Blakely (1996).

The amplitude of horizontal gradient (HG) of the gravity or Pseudo gravity in plane can be determined as follows:

$$HG(x,y) = \left[\left\{ \frac{\partial g\,(x,y)}{\partial x} \right\}^2 + \left\{ \frac{\partial g\,(x,y)}{\partial y} \right\}^2 \right]^{1/2}$$

It can be evaluated using simple finite difference by taking two adjacent reading and dividing their differences by the distance and expressing them in mGal/km. An example of the application of pseudo gravity and its horizontal derivative is shown in Fig. 4.16(a) (Blakely, 1996). It shows a pair of magnetic anomaly, which gives simple high when transformed to poseudo gravity and its horizontal gradient (derivative) provide highs at the margins of the body, which can be easily identified. Jachens and Zoback (2000) have used this approach to map the extension of San Andreas fault system along west coat of USA in the Ocean based on airborne magnetic map of this region.

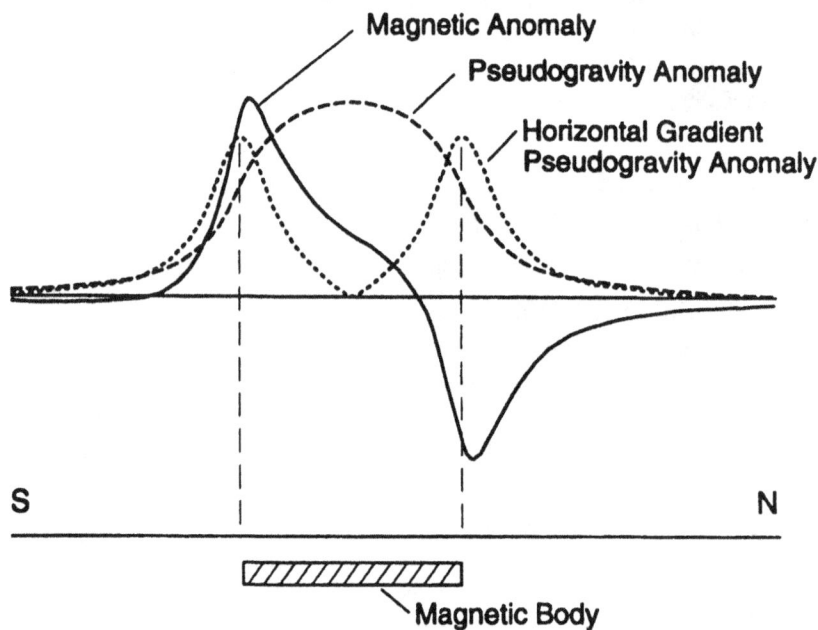

Fig. 4.16 (a) Edges of a magnetic body based on horizontal gradient of psedogravity computed from the magnetic anomaly (Blakely, 1996).

(iv) Reduction to Pole

Magnetic fields observed at latitudes different from those at equator and pole are complex in the form consisting of both highs and lows as discussed in the previous chapter. Therefore, it is difficult to separate the field observed due to individual sources and interference of anomalies from adjoining regions make them further complex. Therefore by this process, the observed magnetic anomalies are reduced to pole such that the earth's magnetization vector is vertical as if the observed field is recorded at the pole. In frequency domain suitable filters have been provided for this purpose. Kanasewich and Agarwal (1970) and Agarwal and Kanasewich (1971) have described frequency domain operations in detail to obtain the magnetic field observed at any latitude to that corresponding to magnetic pole and trend analysis, respectively. Those interested may refer the reference given above or Blakely (1996) for these studies. Some details of reduction to pole is given in next section while describing computation of basement relief from the observed magnetic fields. The low latitude problem where the pole reduced field is considerably disturbed and shows unproportionately high amplitude for north-south waves is also discussed in that section. Some workers use reduction to equator, which also simplify the observed magnetic anomalies and makes their interpretation simple.

(v) Analytic Signal

In search for automatic methods for the interpretation of magnetic anomalies Nabighian (1972, 1974) defined analytic signal, which has been used widely to define the magnetic sources. Automatic methods of potential field interpretation became very popular with airborne magnetic data due to large amount of data acquired in such surveys and problems in separating individual magnetic anomalies due to bipolar nature of these anomalies and height of the aircrafts where anomalies from neighbouring sources trend to interfere with each other. He defined following two characteristics of the the Fourier transform of the horizontal derivative of the observed magnetic field (total intensity, vertical or horizontal component) (a) It is casual function (defined only for positive frequencies) whose real and imaginary parts are the Hilbert transform of each other and (b) The inverse transform of the real part of it is the original function itself (T(x)) while imaginary part is the vertical derivative of the field and corresponds to Hilbert transform, H(x). Therefore, if T(x) is the observed field and $\Delta T(x)$ is the horizontal derivative

$$\Delta T(x) = \frac{\partial T(x)}{\partial x}$$

It is transformed to $\Delta T(f)$, which is a causal function given by:

$$\Delta T(f) = 2\Delta T(f) \text{ for } f > 0$$
$$= \Delta T(f) \text{ for } f = 0$$
$$= 0 \text{ for } f < 0$$

As given above in (a) its inverse transform is given by

$$\Delta T(f)^{-1} = \Delta T(x) - iH(x) \qquad\qquad .....(4.31c)$$

where $\qquad$ H(x) is the Hilbert transform of $\Delta T(x) = \partial \Delta T/\partial z$ as given above in (b)

Analytical signal is the amplitude of the above transformation (Equation 4.31c)

Analytic signal therefore, can be computed in two ways (a) by computing Hilbert transform of the observed field and adding it to the filled itself or (b) by Fourier transforming the field (T(x)) and doubling all values for k > 0 and setting zero for k < 0 and inverse transforming it.

$$(\text{Amplitude})^2 = a(x) = T^2 + H^2 = \alpha^2 / (h^2 + x^2) \qquad\qquad .....(4.31d)$$

Where α denotes the location of the sources and h is its depth and Phase $\theta(x) = - \tan^{-1} H/T$

For a single sheet type of body, the amplitude of this analytic signal or any of its derivative is bell shaped symmetric function with maximum exactly over the edges of the magnetic sheet.

Roest et al., (1992) extended this work for three dimensional cases, which can be referred in case of analysis of two dimensional data (map). In case of several sheets, the amplitude of the analytic signal (or its derivatives) is made up of several bell shaped symmetric functions, each showing maximum over corners of the polygon. Anand and Rajaram (2004) have used analytic signal to define the magnetic sources of Satpura Fold (Mobile) Belt based on airborne total intensity magnetic map (SFB; Fig. 3.30(b)) as given in Fig. 4.16(b) (i-iii). Fig. 16(b)(i) shows the airborne total intensity map of the central part of Satpura Fold Belt (SFB, Section 7.6) whose analytic signal (Fig. 4.16(b); ii) sows maxima (Red colour) south of south Narmada fault and specifically in the south eastern part, which is geodynamically quite significant as it indicates the concentration of igneous and metamorphic rocks of SFB in this section. It is also confirmed from second vertical derivative map of same data (Fig. 4.16(b); iii), which shows minima over the same region where analytic signal has provided maxima.

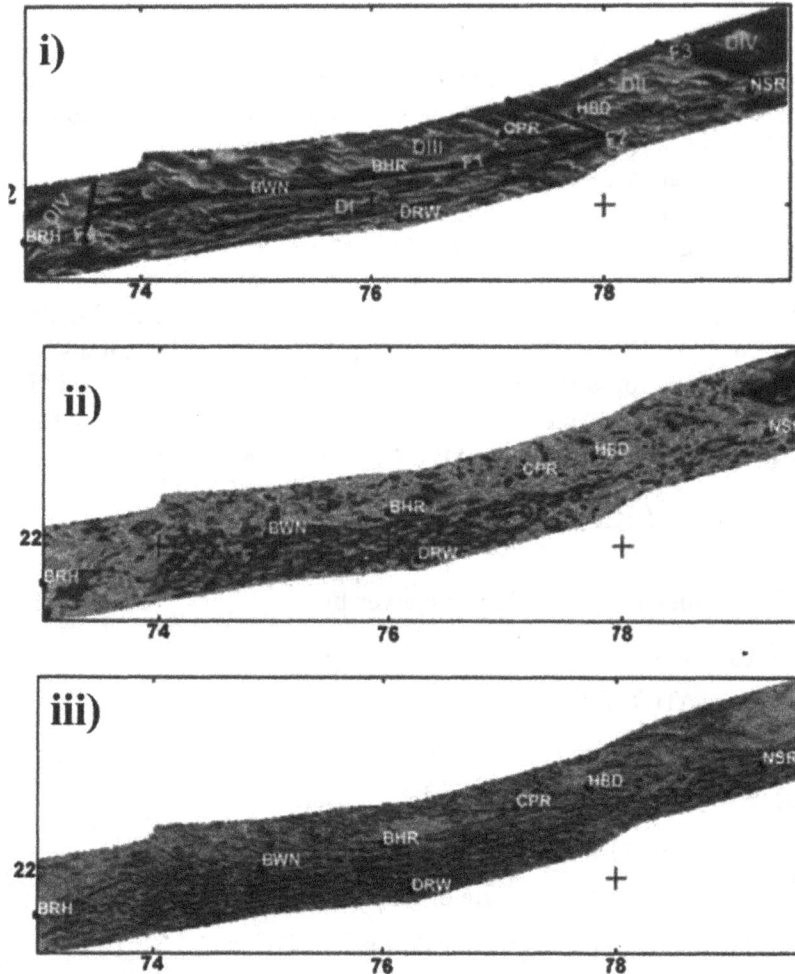

Fig. 4.16 (b) Total field anomaly map and its transformations. Red depicts highs and blue lows. **(i)** Total field aeromagnetic anomalies with superposed interpreted major faults. F1 indicates Barauni Suktha Fault (BSF).
(ii) Analytic signal map, maxima (highs) are located over the magnetic sources south of BSF.
(iii) Second vertical derivative of total field, showing minima over magnetic sources south of BSF. HBD—Hoshangabad, NSR-Narsingpur, BRH- Baruch, BWN-Barwani, BHR-Bharwaha, CPR-Chhipaner, DRW-Dorwa. (Anand and Rajaram, 2004). [Colour Fig. on Page 766]

4.3   Fourier Transform of Gravity and Magnetic Fields: Layered Model and Basement Relief

A horizontal layer may produce gravity/magnetic anomalies in two ways viz. (i) due to variations in its density/susceptibility (Fig. 4.17(a)) or (ii) due to changes in its relief (Fig. 4.17(b)). The transform of gravity field due to a horizontal interface at a depth, z_0 (Fig. 4.17(a)) for a density variation $\rho(x)$ along it can be expressed as (Mishra and Pedersen, 1982).

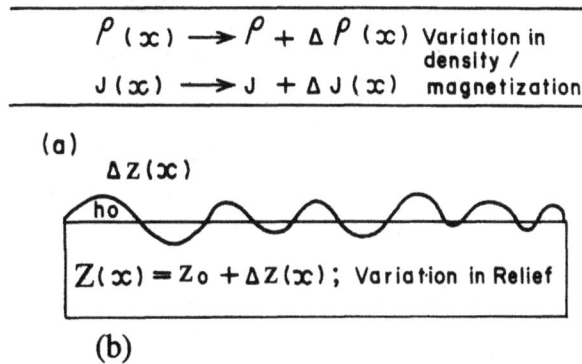

Fig. 4.17 The gravity and magnetic fields due to a thin layered model can be produced in two ways. (a) Due to variations in density [ρ(x)] or magnetization [J(x)]. (b) Due to variations in the relief of the layer [Δz(x)]for a constant density or magnetization.

$$g(k) = 2\pi G\rho \int_{-\infty}^{\infty} \rho(x)\exp-(ikx)dx \int_{z0}^{\infty}\exp-|k|zdz \qquad .....(4.32)$$

where G is the Gravitational constant and $|k| = 2\pi/\lambda.k$ in discrete Fourier transform for grided data as given in Equation (4.15).

The integral in z may be readily evaluated and the integral in x gives Fourier Transform of density distribution [ρ(k)]. Therefore

$$g(k) = 2\pi G\rho(1/|k|)\exp-|k|z_0\,\rho(k) \qquad .....(4.33)$$

Next let us consider a model with variable interface (relief, Fig. 4.17(b)), z(x) with a constant density contrast (ρ₀). The transform of gravity field due to such a model is given by:

$$g(k) = \frac{2\pi G\rho}{|k|} \int_{-\infty}^{\infty}\exp-(ikx).\exp-|k|.z(x)dx \qquad .....(4.34)$$

The exponential in z(x) can be expanded as Taylor's Series as::

$$e^{-|k|z(x)} = e^{-|k|[z+\Delta z(x)]} = e^{-|k|z}(1-|k|\Delta z(x) + (1/2)|k|^2[\Delta z(x)]^2...) \qquad .....(4.35)$$

Where Δz(x) is the fluctuation in relief over the average depth z₀. Taking into account only the first order term as approximation when $\Delta z(x) \ll z_0$

$$g(k) = 2\pi G\rho\exp-|k|z_0.\Delta z(k) \qquad .....(4.36)$$

The corresponding expressions for magnetic field in vertical direction V can be obtained as derivative of equations (4.33) and (4.36) and substituting magnetization for gravitational constant and density.

$$V(k) = 2\pi J\exp-|k|z_0\,\Delta J(k) \qquad .....(4.37)$$

$$V(k) = 2\pi J|k|\exp-|k|z_0\,\Delta z(k) \qquad .....(4.38)$$

where ΔJ(k) and Δz(k) is the transform of changes in the magnetization and relief of the layer, respectively and J is the average magnetization per unit length. The same is also true for total intensity T(k), which would incorporate a term for direction of magnetization as described in next Section 4.3.2 on 'Basement Relief'. These are simplified expressions just to show the relationship between transform of the gravity and magnetic fields and transform of variation in density and susceptibility and relief of the causative layer.

(i) Apparent Density and Magnetization (Susceptibility) Maps

Equations (4.33), (4.36), (4.37) and (4.38) look like linear system (Section 4.2.1) with transform of gravity and magnetic field as output and transform of variations in density and susceptibility or relief as input with remaining terms on right hand side as response function of the system, which are constants and can be computed. Equations (4.33) and (4.37) can be used to map the variations in density or magnetization distribution of the causative layer. In these cases $\Delta\rho(k)$ and $\Delta J(k)$ viz transform of variations in density and magnetization are obtained from the transform of the field using remaining terms on right hand side of these equation as filter functions. This represent variation in density or magnetization over the average bulk density ρ or magnetization J. As these are computed values of density or magnetization, they are referred to as apparent values. Muniruzzaman and Banks (1989) and Nicolosi et al., (2006) have given schemes to map density and magnetization and magnetization vector of basement and crustal structures from gravity and magnetic anomalies. This scheme is specially important to map the variation in magnetization of Archean-Proterozoic terrenes, which show considerable variation in magnetization. Based on these maps structures like faults, fractures etc., and compositional changes such as mafic, felsic etc., can be easily delineated, which are significant for mineral exploration and near surface geophysics.

(ii) Average Depth of Layers

Equations (4.36) and (4.38) suggest that in case $\Delta z(k) = 1$, ln g(k) or ln V(k) with respect to k provide a straight line whose slope is equal to the average depth to the sources, viz. relief . The same is also true for total intensity T(k) which would incorporate a term for direction of magnetization on right hand side as response function, which is discussed in the next section for computation of basement relief or can be obtained from Spector and Grant (1970). Therefore, taking ln of both sides of Equation (4.36) for $\Delta z(k) = 1$,

$$\ln g(k) = \ln (2\pi G\rho) - (2\pi/\lambda).k \cdot z_0$$

$$= A - (2\pi/\lambda).k \cdot z_0 \qquad\qquad .....(4.38a)$$

Where A is a constant. In case of total magnetic intensity (T) , it is expressed as:

$$\ln (T(k)/k) = B - (2\pi/\lambda).k \cdot z_0 \qquad\qquad .....(4.38b)$$

Where B is a constant in case of magnetic field which depends on the direction of magnetization. This implies that in case of the magnetic field, a modified spectrum can be used by dividing the computed amplitudes of the spectrum for specific frequencies by that frequency. Mishra and Pedersen (1982) used this modified spectrum to compare the depth estimates from spectral decay of gravity and magnetic fields due to same sources. Okubo et al., (1985) have provided full expression for constant B and used this modified spectrum to obtain depth to the center of the magnetic sources (z_0) considering the sources to be a thick body instead of a thin layer while spectral decay of the observed magnetic field T(k) as such provided depth to the top (z_t). They combined the two estimates as $(2z_0 - z_t)$ to obtain the depth to the bottom of the magnetic sources which represented Curie point depth where magnetization of rocks vanishes.

$\Delta z(k) = 1$ implies random distribution of the sources. However, in actual practice, the sources are not randomly distributed but in general are correlated which introduces some slope to the spectral decay implying that the slope of spectral plots of ln g(k) or ln T(k) versus k provide higher estimate of the average depth to the sources. In case of magnetic, it is however, balanced to some extent by $|k|$ in the numerator (equation 4.38), which has a tendency to decrease the slope of the spectrum (Hahn et al., 1976). However, it is advisable to correct the computed spectrum for this factor by dividing the amplitude at respective frequencies by that frequency. Due to absence of this factor $|k|$ in the numerator of the expression for the transform of the gravity field, the presence of $\Delta z(k)$ introduces more slope in the transform of gravity field providing a higher depth estimate compared to the magnetic field. Estimate of $\Delta z(k)$ for specific bodies have been provided by different workers, which can be used to correct the computed spectrum for this factor. In case of an ensemble for two dimensional prisms as sources, it has been termed as shape factor and is a product of two sinc functions, viz [sin (af . cos θ) / (af . cos θ)].[sin (bf . cos θ) / (bf . cos θ)] where a and b are dimensions of prism and $\theta = \tan^{-1} k/m$ (Spector and Grant, 1970). An ensemble of long tabular bodies (dykes) provide a singular sinc function [sin af / af] where a is the width of the dyke (Mishra et al., 1980). Due to presence of sinc functions, transform of potential field due to these bodies are oscillatory in character and thereby averaging over certain frequency bands are essential to obtain smoother spectrum. The simplest spectrum is obtained for magnetic field due to horizontal cylinder where the shape factor is one and computed spectrum can be divided by k to obtain the plot of ln T(k) versus k whose slope is equal to depth (Murthy and Mishra, 1980). Several workers have provided expressions for transform of potential field due to different bodies (Bhattacharya, 1966; Odegard and Berg, 1965; Parker, 1973, 1975; Pedersen, 1978 etc.,) and have provided expressions to compute the field due to specific bodies in frequency domain.

In case the sources are distributed at different depths (layers), they are separated as linear segments in different wave number bands as discussed in section on digital filtering and their slopes are equivalent to their respective depths (Equations 4.36 and 4.38). However, there is always a mixing of energy from one wave band to other which makes it difficult to separate them by linear segments specially at the junctions of two segments but still efforts can be made to separate sources through linear segments in their spectral plots. Here again there is a difference in the gravity and magnetic fields. In gravity field, there is a continuous change in the density of subsurface rocks and therefore one normally gets exponentially decaying curve and interpreter tends to approximate it by several straight line segments (3-4) with their slopes related to depth to the causative sources which in fact may not represent different layers. However in case of magnetic field, there are magnetic bodies at specific depths, which will reflect as different segments in the spectral plot with their slopes related to depths to their sources. This makes spectral depths of magnetic field more reliable compared to gravity fields. Depth from the computed spectrum (slope) can be obtained as follows:

If A_{k1} and A_{k2} are amplitudes at frequencies k1 and k2, then,

$$A_{k1} = \exp (2\pi/\lambda). (-k1). z_0 \text{, and}$$
$$A_{k2} = \exp (2\pi/\lambda). (-k2). z_0 \text{, therefore}$$
$$A_{k1} / A_{k2} = \exp (2\pi/\lambda) . (k2\text{-}k1).z_0$$
$$z_0 = \lambda/(2\pi . (k2\text{-}k1)) . \ln (A_{k2}/A_{k1}) \qquad\qquad .....(4.38c)$$

In case of power spectrum computed using FFT, depth to the sources are half of the spectral slope or half of the value obtained from Equation (4.38c) as it is square of the amplitude spectrum.

There were several attempts to improve the depth estimates based on spectral slope (decay) such as application of scaling factors (Fedi et al., 1997; Maus and Dimri, 1995) and use of wavelets (Moreau et al., 1997; Martelet et al., 2001). However, estimation of average depth of causative sources by spectral decay or application of scaling factors should be in accordance with the general geological information from the region and should be confirmed from direct modelling along a profile by placing the appropriate sources at those depths and varying their physical parameters (density / susceptibility) with in geologically feasible limits. In this manner, the estimation of average depths based on spectral decay or scaling factor is mainly an approximate estimate to be used as initial value which should be followed up by more detailed modeling methods.

(iii) Stability of Computed Spectrum

In order to check the stability and reliability of computed spectrum and average depth determined from it, one may compute several spectrum drawn from same data set such as original data, averaged in different groups of 2 × 2, 3 × 3 etc., in case the data matrix is large enough or portion wise of the original data and compare the computed spectrums for same wavelength. It is advisable to select the common linear segments from these spectrums, which may truly represent the subsurface sources. It may kindly be kept in mind that data size should be 8-10 times larger than the depth to be investigated.

The linear segments on the spectral plot of energy versus frequencies are generally drawn by visual inspection due to the following considerations:

(a) g(k) or V(k)/T(k) in Equation (4.36 and 4.38) most often being the product of sinc functions, some of the points on the plot fall much below the general level corresponding to the zeroes of the sinc functions, and are, therefore, discarded while approximating it by linear segments as demonstrated below in section on 'Applications of Depth Estimates'.

(b) Due to the presence of the noise towards the end of the spectrum, the significant band width used varies from one plot to the other and is decided by a careful examination of individual plots.

(c) The number of the linear segments fitted to the spectral plot varies from one plot to the other. With these precautions it is generally possible to draw reliable linear segments. However, small changes in slope do not change the depth estimate to any large extent. For example a change of 5° in the slope of power spectrum will affect the estimate as follows:

$$\Delta d = (\lambda / 4\pi) \tan 5°$$

where λ is the wavelength.

If $\lambda = 100$ km then $\Delta d = 0.7$ km.

4.3.1   Applications of Depth Estimates

(i) Theoretical Models

(a) Deterministic Model: Dyke

The model used for this study is a prismatic body of 16 km in horizontal extent lying at a depth between 5 and 6 km. Other constants used for the field computations are density contrast = 0.3 gm/cm3; susceptibility = 5 x 10^{-2} C.G.S. units; Earth's magnetic field = 40,000 nT; inclination = 90°; declination = 2°; profile orientation 90° from N.

The magnetic and gravity fields were computed for 64 km long profiles with a field value at every 0.5 km interval, resulting in 128 values for transformation. The amplitude spectrum of the magnetic and the gravity fields from such a model is plotted in Fig. 4.18(a) and (b). These plots clearly show the exponential decay governed by the depth superimposed over the sinc functions, corresponding to the factor $\Delta z(k)$ *as* discussed above. The asymptotic straight line of the sinc functions provides the slope given in the insert of the diagram, which is almost equal to the depth in case of magnetic, and slightly higher estimate in case of gravity as discussed above. However, the depth estimate in case of gravity field can be improved by dividing the computed spectrum by 1/k to make it comparable with the spectrum of the magnetic field (Equation 4.36 and 4.38) as given in the insert of Fig. (4.18(b)). The width of the sinc functions can be used to estimate the width of the body. If Δka is the width of the sinc functions and 2a that of the body:

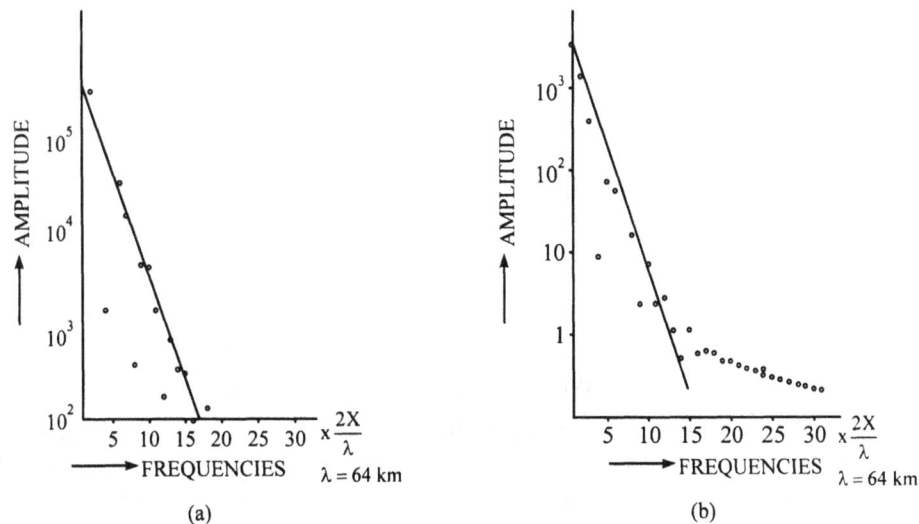

(a)   (b)

Fig. 4.18 Amplitude spectrum versus frequencies of magnetic (a) and gravity (b) fields due to a deterministic model-tabular body (dyke). Spectral slope providing the depth to the top of the body are 5.2 km and 6 km respectively, which in case of spectrum of the gravity field is more due to shape factor $\Delta z(k)$ that in case of magnetic field, which is balanced to some extent by factor $|k|$ in the expression. Zeros of Sinc functions due to shape factor are clearly visible (Mishra and Pedersen, 1982).

$$\Delta k.a = \pi$$

From Fig. (4.16), $\Delta k = 4.2\pi / 64 = \pi / 8$

Therefore a = 8 km and width of body = 16 km, which is same as that of the theoretical model

Hahn, Kind and Mishra (1976) and Mishra and Pedersen (1982) have demonstrated the application of spectrum to compute the average depth to the causative sources on theoretical statistical models and found that spectral depths usually represent the maximum bound on the depth, which should be used accordingly. It may be noted that in case the power spectrum of gravity and magnetic fields ($g^2(k)$ and $V^2(k)$) are computed from Fast Discrete Fourier Transform, the spectral decay provide twice the depths while in case of amplitude spectrum, the spectral decay is equivalent to depth.

(b) Statistical Relief Model

Statistical distributions are generally used to model the subsurface relief representing the basement or large-scale discontinuities such as the Moho etc. Statistical distribution of relief to test the average depth based on spectral decay was generated using random numbers. It differs from the deterministic case in mainly two respects, (1) the signal from different sources (parts of the relief) are mixed, and (2) the profiles are mainly situated entirely over the causative sources instead of extending beyond them as in case of deterministic models.

Three different kinds of relief were used for this purpose. The first two were generated from random numbers using a low pass cosine filter between (1) 10-20% and (2) 5-10% of the total frequency band. A third representing very smooth and large features, as sometimes encountered in the field, was drawn by hand. To obtain the subsurface relief from random numbers we first generated 2048 numbers and transformed it resulting in 1024 frequencies. The amplitudes so obtained were multiplied by the following function between frequency ranges k_1 and k_2 with $k_1 = 100$ and $k_2 = 200$ in the first instance, and $k_1 = 50$ and $k_2 = 100$ in the second example.

$$F(k) = .1 \text{ when } k < k_1$$

$$F(k) = 1/2 \{\cos[(k - k_1/\pi(k_1 - k_2)] + 1\}$$

when $k_1 < k < k_2$ …..(4.39)

$$F(k) = 0$$

when $k > k_2$

The parts of the two reliefs (1) and (2) incorporating 128 points were placed at a depth between 5-6 km, and the corresponding magnetic and gravity fields were computed using the same value for the constants as given above for the deterministic model.

Prior to the transformation, the fields were multiplied by a 10 point cosine window at the ends in order to remove the discontinuities at the borders caused by Gibbs phenomena. The corresponding amplitude spectra of magnetic and gravity fields for one of the reliefs (2) are plotted versus frequency in Fig. 4.19(a) and (b). Two linear segments separated by a hump are easily discernible. The depths derived from the slope of the two segments given in the insert of diagrams are almost same. In fact, both of them correspond to the same sources but displaced in frequency band due to sinc function caused by the shape factor $\Delta z(k)$ and non-linearity, due to the second and higher order terms in equation (4.36) and 4.38). In case the slopes of the two segments are entirely different, they have to, be interpreted in a different manner such as due to different layers depending upon the prior geological information from the area.

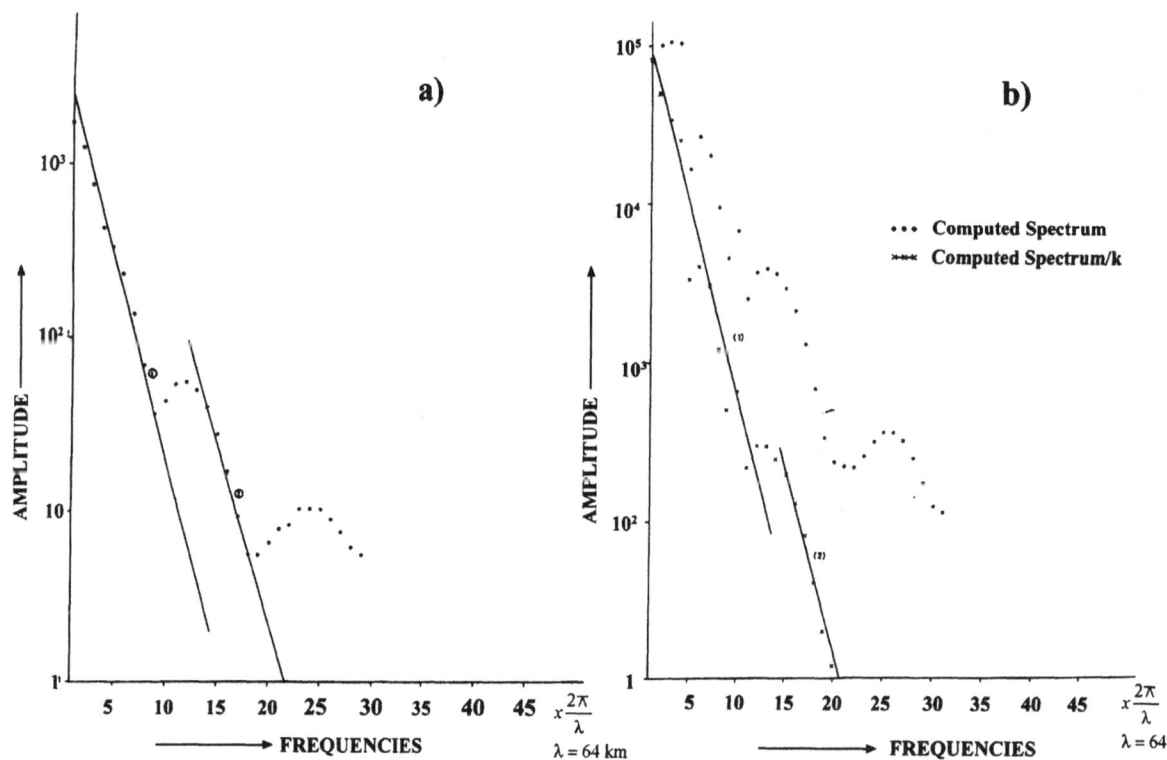

Fig. 4.19 Amplitude spectrum of magnetic **(a)** and gravity **(b)** fields due to a statistical relief model. Spectral slope equal to depth are 5 and 5.6 km, respectively (Mishra and Pedersen, 1982).

4.3.2   3-Dimensional Basement Relief Model using Harmonic Inversion

As discussed above, Equation 4.36 and 4.38 can be treated as linear system where the transform of the observed gravity $(g(k))$ and magnetic fields $[V(f)$ or $T(f)]$ on left hand side are related to the transform of the relief $(\Delta z(k))$ through filter function on right hand side. Therefore, using this filter function, the relief of the layer producing the gravity and magnetic anomalies can be computed at an average depth of z_o. In case of sedimentary basins, these anomalies are primarily caused by the relief of the basement and therefore, this method is found to be very useful in estimating basement relief under sedimentary basins for hydrocarbon exploration (Hahn, 1965; Hahn, Kind and Mishra, 1976; Gunn, 1976; Goldflam et al., 1977). In case of two dimensional data set, it provides directly the basement depth at every point of observation.

For two dimensional cases frequency $f = (k^2+m^2)^{1/2}$ as given in equations 4.17 and 4.18

where k, m = 0, ±1, ±2,........., ±n/2 when n is even and k, m = 0, ±1, ±2,........., ±(n-1)/2 when n is odd.

In case of computation of basement relief from total intensity magnetic field, Hahn (1965) has provided the following expression for variations in the relief (Δz). He represented variation in relief by a relief waves and converted the observed field waves into relief wave through magnetization of the relief.

$$\Delta z\,(k,m) = [\,|\,\Delta T(k,m,z_o)\,|\,]\,/\,4\pi^2 J(\sin^2 I + \cos^2 I \cos^2(\theta+\delta))(k^2+m^2)^{1/2}/\lambda \qquad \text{.....(4.40)}$$

where $\Delta z\,(k,m)$ is the amplitude of relief wave and $\Delta T(k,m,z_o)$ is the amplitude of observed anomalous field continued downwards to the average depth of the relief, z_o, $\theta + \delta$ = angle between normal plane of the wave and magnetic meridian (Fig. 4.9) with θ as is the phase of various waves = $\tan^{-1} m\,/\,k$, J is the bulk magnetization in nT of the causative source (relief- basement) given by kT where k being susceptibility and T is earth's magnetic field, I is the inclination of inducing earth's magnetic field. In case of remenent magnetization, the effective inclination as defined in equation 3.27 and 3.28 is used.

Having computed the basement relief directly from the observed gravity and magnetic fields, same expression as equation (4.40) can be used to compute the field ΔT backwards at the plane of observation. The difference field between the observed and the computed fields can be further used to compute the relief in the same manner as above and added to the previously computed relief to provide a better model. Through further iteration the basement model can be improved till the left out field is minimum, which can be judged based on root mean square value. Parker (1973) and Oldenburg (1974) have given an iterative scheme to compute the density interface based on the density contrast across the interface and Fourier Transform of the observed gravity field. Based on same scheme Gometz-Ortiz and Agarwal (2005) provided a code using MATLAB functions, which is available on Web sites given in their publication.

The factor in the denominator of equation (4.40) represents the filter function with direction of magnetization, which reduces the observed field to pole. As it contains sin I in the denominator, at magnetic equator $(I = 0)$ or close to it the filter function becomes indeterminate and causes oscillations in the reduced to pole field. It largely affects N-S oriented waves as for these waves $\theta \to 0°$ and therefore second term in denominator also tends to zero. In such cases the amplitude of N-S or near N-S waves can be reduced to half or one third so that they do not blowup and problems related to low latitude regions can be circumvented in this manner. However, Gunn (1995) have provided a scheme to reduction to pole which works for all latitudes. It therefore, implies that computation of relief of the causative layer from the magnetic field, inherently involves following operations:

(a) Transformation of the IGRF corrected observed magnetic field

(b) Removal of high frequency component using a low pass filter of specific wave band as it involves continuation of the observed field downwards to the average depth of the relief.

(c) Continuation of the observed field downwards to the average depth of the relief

(d) Reduction of the observed field to the magnetic north pole

(e) Conversion of this pole reduced field to a relief model based on magnetization (J) in nT given by J = kT, where k is susceptibility and T is earth's magnetic field.

Applications of this method in geodynamics and exploration are demonstrated in Sections 4.6 and 4.7 under 'Field Examples'.

4.4   Admittance Analysis and Coherence

The principles of isostasy discussed in section 2.5.3 is one of the most important aspects of geodynamics based on gravity data and topography of the region as the state of stability of the region can be inferred based on these two data sets. However, as discussed in that section, it can be only inferred qualitatively whether a region is isostatically balanced (compensated) or it is under or over compensated. It can be quantified based on correlation between topography (elevation data) and its gravity field in frequency domain which is known as coherence/ admittance analysis. Most of the tectonic activities observed at the surface of the earth have deep origin in lithosphere and asthenosphere and their respective strength and interaction. Depending on their strength and elastic properties, they behave to the loads of mountain over them, which causes flexure. These parameters related to their behavior can be estimated based on this method, which in turn are related to their stability. In isostatic compensation, there are apparently three scenarios. Firstly, as in case of Airy's model of isostatic compensation (Fig. 2.17(a)), any excess or deficient mass at the surface is compensated at the base of the crust (Moho) by an equal and opposite mass. In this case, the crust as such does not play any role in compensation and is said to have zero strength. On the other hand, there may be cases where excess or deficient mass at the surface is not reflected at the base of the crust and is mainly supported by its strength, in such cases, the strength of the crust is said to tending to infinity (Fig. 2.17(d)). The intermediate cases where compensation may be regional over a large area due to limited strength of crust/lithosphere and is known as regional compensation (Fig. 2.17(c)). In such cases the lithosphere flexes under surface load due to its elastic behavior and its flexure is the measure of the compensation, which can be estimated based on correlation between the observed gravity field and the topography (elevation).

The strength of crust depends on several factors such as thickness, rock types and previous tectonic history indicated by flexural rigidity and thermal history of the region. In general, it provides the strength of the lithosphere in a region and inturn, the effective elastic thickness, provides the depth to the first transition of brittle-ductile deformation in the crust. The brittle part is usually the zone in which rupture takes place even by small stress accumulation as the presence of small penny type of fractures reduces the strength of the rocks in this part of the crust and is therefore important for seismic activities.

The admittance function Z(k) is the transfer function between the observed gravity field and the topography (Section 4.2.1) and is therefore, given by the ratio of Fourier transform of the gravity (g(k)) and the topography (t(k)) in a region (Equation 4.11).

$$Z(k) = g(k) / t(k) \qquad\qquad .....(4.41)$$

This is related to rigidity of the crust which is discussed below.

4.4.1 Isostasy in Frequency Domain and Effective Elastic Thickness (EET)

The transform of gravity effect due to a surface layer with relief is given by Equation (4.36) as:

$$g(k) = 2\pi G\rho \exp(-|k|z).z(k) \qquad .....(4.42)$$

where k is the wave number related to wavelength λ as $k = 2\pi / \lambda$ and z is the average depth to the relief and z(k) is the transform of the relief itself. In case topography is expressed as a relief, equation (4.42) can be used to express the transform of the gravity field due to topography h(x) as:

$$g_1(k) = 2\pi G\rho_1 \exp[-|k|h_o].h(k) \qquad .....(4.42)$$

where ρ_1 is the bulk density of topography; h_o is average height of topography and h(k) is the transform of the relief of the topography h(x). Similarly the transform of the gravity effect of isostasic root z(x) at Moho can also be expressed as:

$$g_2(k) = -2\pi G\rho_2 \exp[-|k|z].z(k) \qquad .....(4.43)$$

Negative sign indicates the negative nature of the gravity field due to isostatic effect.

where ρ_2 is the density contrast of the crustal root with surrounding (upper mantle) expressed as $\rho_2 = \rho_m - \rho_c$ and z(k) is the Fourier Transform of the root z(x) and z is the average depth = h + t, t being crustal thickness. The combined effect of topography and its root can be given by:

$$g_3(k) = 2\pi G\rho_1 h(k).\exp(-k h_o)[1-(\rho_2 / \rho_1)(z(k) / h(k))\exp(-kt)] \qquad ....(4.44)$$

where $g_3(k)$ represents the transform of combined effect of topography and its root.

In rheological model, the earth's crust is considered as an elastic plate and topography is regarded as a load, which causes deformation of the plate as isostatic root of the earth's crust. The deformation of such an elastic plate (isostatic root) z(x) is given by (Hetenyi, 1946)

$$D\frac{\partial^4}{\partial x^4}r(x) + \rho_3\, gr(x) = \rho_1 gh(x) \qquad .....(4.45)$$

where $\rho_1 = \rho_t - \rho_{air}$ (continental loading), $\rho_1 = \rho_t - \rho_{water}$ (for oceanic loading), $\rho_2 = \rho_m - \rho_c$ where as ρ_t = density of topography, ρ_m = density of upper mantle and ρ_c = density of the crust and $\rho_3 = \rho_m - \rho_{in\,fill}$, $\rho_{in\,fill}$ is the density material occupying the flexed crust and D is flexural rigidity given by:

$$D = ET_e^3 / 12(1-\sigma^2) \qquad .....(4.46)$$

where E is the Young's modulus, T_e is the elastic plate thickness and in case of earth's crust is known as effective elastic thickness (EET) and σ is the poissan's ratio. The solution of above equation (4.45) in frequency domain is given by (Banks et al., 1977):

$$z(k) = [\rho_1 / \rho_3][1 + (k^4 D / \rho_3 g)]^{-1} h(k) \qquad .....(4.47)$$

This equation is like a linear system between transform of isostatic root and topography with quantity in the bracket on right hand side as filter function, which also incorporates flexural rigidity (D).

(i) In case $D \to 0$, it represents Airy's model and is known as Airy loading.

$$Z(k) \to (\rho_1 / \rho_3).h(k) \qquad ....(4.48)$$

The transform of combined effect of load and root is given by combining equations (4.47) and (4.44) as follows:

$$g_3(k) = 2\pi G\rho_1 h(k).\exp(-kh_o)[1-\{\rho_2 / \rho_3\}.\{\exp(-kt)\}] \qquad .....(4.49)$$

Admittance in this case is therefore,

$$Z(k) = g_3(k) / h(k) = 2\pi G\rho_1.\exp(-k.h_o) [1-\{\rho_2 / \rho_3\}.\{\exp(-kt)\}] \quad .....(4.50)$$

In case crustal rocks filling the flexed part of the beam (crust) is same as normal crustal rocks and therefore $\rho_2 \to \rho_3$ and admittance for Airy's model ($Z_a(k)$) is given by

$$Z_a(k) = 2\pi G\rho_1.\exp(-kh_o).[1-\exp(-kt)] \quad\quad .....(4.51)$$

(ii) In case of regional compensation when $0 < D < \infty$, the admittance for regional compensation ($z_r(k)$) is given by (Karner, 1982).

$$Z_r(k) = 2\pi G\rho_1.\exp(-kh_o)[1-\{\rho_2 / \rho_3\}.\{1 + (k^4D / \rho_3g)\}].\exp(-kt) \quad .....(4.52)$$

(iii) For uncompensated topography $D \to \infty$, and the admittance is given by

$$(k) = 2\pi G\rho \exp(-kh_o) \quad\quad .....(4.53)$$

(i) Admittance

As defined above admittance in frequency domain can be defined as function of the transform of gravity and topography as follows:

$$Z(k) = g(k) / t(k) \quad\quad ....(4.54)$$

However, as both gravity and topography is not continuous function and its discrete version is used they both contain significant noise, which distorts their spectrum. Therefore, several methods were proposed to compute these functions with least noise. One such estimate is based on cross spectral techniques (McKenzie and Bowin, 1976) where in numerator and denominator is multiplied by transform of the topography.

$$Z(k) = g(k).h(k) / h(k).h(k) \quad\quad .....(4.55)$$

$g(k).h(k)$ is the cross spectrum of the observed gravity field and the topography and $h(k).h(k)$ is the power spectrum of the topography as shown in equation 4.14. Therefore,

$$Z(k) = C_{gt}(k) / P_t(k) \quad\quad .....(4.56)$$

$C_{gt}(k)$ and $P_t(k)$ are cross spectrum of the gravity field and the topograph and power spectrum of topography, respectively can be conveniently computed as follows:

$$C_{gt}(k) = \frac{1}{N}\sum_{m=1}^{N} g_m(f)h_m^*(f)$$

$$P_{gt}(k) = \frac{1}{N}\sum_{m=1}^{N} h_m(f)h_m^*(f)$$

Where m denotes the number of profiles or segments of a profile used for averying. The larger the number of m, the better the estimates of cross correlation and power spectrum as described above under computation of spectrum. This may be the reason to prefer two dimensional data.

$g(k)$ and $h(k)$ represents discrete Fourier Transform of the observed gravity anomaly and topography and asterisk denote their complex conjugate.

McKenzie and Bowin (1976) also suggested to compute the coherence (γ) of observed gravity field and topography and the phase of admittance which are given by:

$$\gamma^2 = [C_{gt}(k).C_{gt}^*(k)] / P_g(k).P_t(k) \quad\quad .....(4.58)$$

The uncertainty of coherence can be estimated from

$$\delta = [(\gamma^{-2}-1) / 2N]^{1/2} \quad\quad .....(4.59)$$

And phase of admittance is given by

$$Q(k) = Z(k) / Z^*(k) \quad\quad .....(4.60)$$

(ii) Theoretical Admittance: Surface and Sub-surface Loading

As $\rho_2 \rightarrow \rho_3$ and $\exp(-kh_o) = 1$ for free air anomaly as it is observed over the topography, equation (4.52) for regional compensation can be stated as (Mckenzic and Fairhead, 1997):

$$Z_r(k) = 2\pi G \rho_1 [1 - \exp(-kt_c) / A] \qquad \dots\dots(4.61)$$

Where

$$A = \varepsilon \; ; \varepsilon = 1 + Dk^4 / g(\rho_m - \rho_c) \text{ and}$$

$$D = E T_e^3 / 12(1 - \sigma^2)$$

k is 2π / wavelength (λ) is wave number

G is gravitational constant

E is Youngs modulus (= 10^{11} NM)

σ is Poisson's ratio (= 0.25)

g is acceleration due to gravity = 9.8 m/s^2

ρ_c and ρ_m are average crustal density and density of material below flexed elastic plate.

t_c is effective depth of compensation.

T_e is effective elastic thickness (EET)

In case of sub-surface loading, expression for A is given by (Forsyth, 1985 and Bechtel et al., 1987).

$$A = [f^2\rho_c^2 + \varepsilon^2(\rho_m - \rho_c)^2] / [\varphi t^2 \rho_c^2 + \varepsilon(\rho_m - \rho_c)^2] \qquad \dots\dots(4.62)$$

Where $\varphi = 1 + (Dk^4 / g\rho_c)$; and f is the ratio of subsurface to surface loading.

Misfit function H_f is defined as:

$$H_f = [\sum (Z_o - Z_c / \Delta Z_o^2)^2]^{1/2} \qquad \dots\dots(4.63)$$

Where Z_o and ΔZ_o are the observed admittance and its standard deviation in frequency domain. 2 H_{min}^f is the measure of uncertainty in the estimate of T_e and $G_f = H_{max}^f / H_{min}^f$, is goodness of fit.

(iii) Coherence: Surface and Sub-surface Loading

However, there are even sub-surface mass inhomogeneity, which would cause sub-surface loading and affect the isostatic balance, which has been considered by Forsyth (1985), Bechtel et al., (1987). The theoretical coherence assuming uncorrelated top and bottom loads is given by Forsyth (1985). If H_t and H_b are relief of the surface (topography) and sub-surface loads, and W_t and W_b are their corresponding effects at the compensation level, respectively, then their total effects are

$$H = H_t + H_b$$

$$W = W_t + H_b \qquad \dots\dots(4.64)$$

The initial amplitude of the topography and its load involved in surface and sub-surface loading are

$$H_i = H_t - W_t \text{ and } W_i = W_b - H_b \qquad \dots\dots(4.65)$$

The amplitude of density interface relief is found from downward continuation of Bouguer anomaly.

$$W(k) = g(k).\exp(kz_m) / 2\pi (\rho_m - \rho_c).G \qquad \dots\dots(4.66)$$

Where g(k) is the transform of the observed gravity field.

The theoretical coherence in such cases are given by (Bechtel et al., 1987).

$$\gamma^2 = <H_t W_{t+} W_b H_b>2 / <H_t{}^2 + H_b{}^2> <W_t{}^2 + W_b{}^2> \qquad .....(4.67)$$

Where H_t and W_t are contribution to topography and flexure due to surface loading and H_b and W_b are corresponding contributions due to subsurface loading.

A typical plot of coherence versus wavelength is shown in Fig. 4.21(b), which shows the variation of coherence with crustal thickness for regional compensation. For large wavelength, the coherence between the observed free air gravity field and topography is one which apparently corresponds to Airy's compensation while for small wavelength the coherence is minimum, which corresponds to uncompensated topography. In the intermediate wavelength, it shows a sharp change in coherence and is dependent on the thickness of the part of the crust which shows elastic behavior and is known as Effective Elastic Thickness (EET), which can be obtained by comparing computed admittance or coherence with theoretical curve for different EET as discussed above. In fact, EET in a particular case suggest the strength of the crust. Larger its value, stronger is the crust, which defines the rheology of the crust and is an important parameter to understand the geodynamics of a region. The wavelength where it changes from compensated to uncompensated topography is also important to understand the geodynamics of a region. In stable cratonic part, the effective elastic thickness is large while in case of orogenic belts, it is quite small due to weaker crust compared to former. It is also related to thermal behavior of a region. Regions with high heat flow are characterized by a weak crust and lower EET compared to regions with low heat flow. Rift valleys, volcanic margins, island arc etc., are therefore regions of low EET. Some typical values of EET for different geological provinces, continents and oceans are given in Table 4.1. Though there are large variations in Te for similar tectonic provinces in different parts of the world, in general, it is more for cratons and stable parts of the continental crust and low for orogenic belts, rifts and continental margins. In case of oceanic lithosphere, its value is low for mid oceanic ridges and more for trenches suggesting a correlation with the age and thickness of oceanic crust. Some of the discrepancies in values for same tectonic units obtained by different workers can be attributed to different schemes followed by them for this purpose.

Table 4.1 Summary of results of Te from Spectral (Admittance and Coherence) Studies
(Modified after Rajesh, 2003)

N	FEATURE	Te (km)	Te Error (km)	Reference
(a)	**Continents & Continental Margins**			
1	North America	7.5	2.5	Banks et al., (1977)
2	Australia	1.0	1.0	Mcnutt and Parker (1978)
3	North America	7.0	2.0	Cochran. (1980)
4	East Africa	27..5	2.5	Forsyth. (1985)
5	Siberia	0.0	x	Kogan and Mcnutt (1987)
6	Urals	90	0.0	Mcnutt and Kogan (1987)
7	Tien Shan	45	5.0	Mcnutt and Kogan (1987)
8	Cucasus	60.0	20.0	Mcnutt and Kogan (1987)
9	East Africa	25.0	x	Bechtel et al., (1987)
10	Urals	85.0	x	Mcnutt et al., (1988)
11	Pamirs	20.0	x	Mcnutt et al., (1988)
12	Tien Shan	40.0	x	Mcnutt et al., (1988)
13	East Africa Cratons	77.0	13.0	Ebinger et al., (1989)
14	Appalachians	39.0	x	Bechtel et al., (1990)
15	Alps west	34.5	4.5	Macario et al., (1995)

Table 4.1 Contd...

16	Brazil	85.0	65.0	Ussami et al., (1993)
17	Siberian Shield	15.5	3.5	*Mckenzie and Fairhead (1997)*
18	Canadian Shield	80.0	20.0	Wang and Mareschal, (1999)
19	Western United States	9.1	0.3	Mckenzie and Fairhead (1997)
20	West US Colarado plateau	22.0	10.00	Lowry and Smith (1995)
21	Indian Subcontinent (Northern Part)	23 – 26		Rajesh & Mishra (2004)
22	Indian Subcontinent (Southern Part)	12 – 16		Rajesh & Mishra (2004)
23	Deccan Trap, India Offshore-Chagos-Laccadive Ridge	08 – 10 2-6 km		Tiwari & Mishra (1999) Tiwari et al., (2007)
24	Himalaya, Tibet & Tarim basin	35, 20, 30 km, respectively		Rajesh et al., (2003)
25	Eastern Margin of India Northern Part Southern Part	10 – 25 km 05 – 0 km		Chand et al., (2001)
26	Western Margin of India	08 – 15 km		Chand and Subrahmanyam (2003)
27	Ganga Basin-Himalaya-Tibet	50 ± 10 km		Tiwari et al., (2006) BA+Topo.
28	Western Syntaxis, Himalaya	53 km		Tiwari et al., (2008) (BA+ Topo.)
29	Ganga Basin and Western Himalaya	30 km		Tiwari and Mishra (2008)
30	Aravalli- Delhi Fold Belt, India	50 km		Tiwari and Mishra (2008) FA + Topo.
31	Satpura Fold Belt, India	40 km		Tiwari and Mishra (2008)
32	Eastern Ghat Fold Belt	30 km		Tiwari and Mishra (2008)
33	South America Archean Proterozoic Continent Continental Margin, South America	100 ± 15 km 30 km		Tassara et al., (2007)
34	Eastern Alps	2-18 km increasing northwards		Braitenberg et al., (2002)
35	Fennoscandia: Southern Part Northern Part	20 km 0 km (Airy's)		Ebbing & Olesen (2005)
36	Canadian Shield	60-125 km Increases with age		Pilkington (1990)
(b)	**Oceans: Mid Oceanic Ridges & Trenches**			
37	Mid Atlantic Ridge	10	03	Cochran (1979)
38	East Pacific Rise	4.0	2.0	Cochran (1979)
39	Juan de Fuca Ridge	4.0	1.0	McNutt (1979)
40	Chile Trench	27.0	2.5	McNutt (1984)
41	Kermadec Trench	40.7	0.7	McNutt (1984)
42	Mariana Trench	51.3	5.0	McAdoo and Martin (1984)
43	Central Indian Ocean	12.5	2.5	Karner and Weisal (1990)
44	90 East Ridge	0-22 km		Tiwari and Mishra (2008)
45	Laccadivie Ridge	2-6 km		Tiwari et al., (2007)

Several schemes have been used for spectral estimation such as Peridogram method using Fast Discrete Fourier Transform as has been discussed above. However, subsequent to it, several other approaches were developed such as multitaper method wherein the input signal is multiplied by series of orthogonal eigentapers of different shapes (Simons et al., 2000; Simons et al., 2003). The application of all these techniques is to obtain most stable spectrum. The use of multitapers on the computation of spectrum and admittance has been demonstrated by Simon et al., (2000, 2003). However, the effective elastic thickness estimates by two methods, viz. admittance from the transfer function between the free air gravity and the topography and coherence between the Bouguer gravity anomalies and topography differ significantly and makes it difficult to correlate with the dynamic condition of the region. Mckenzie (2010) has attributed it to dynamically maintained gravity and topography arising from mantle convection and post glacial recovery that generate long wavelength (>500 km) gravity and topography. In case these anomalies are modeled as supported from elastic forces, it may lead to over estimation of the effective elastic thickness.

Effective elastic thickness implies the thickness of that part of the crust, which is elastic in nature involving in the regional compensation. Due to elastic nature of this part of the crust, it is involved in accumulation of stress and strain and breaking due to them when their values reaches to a critical point. Therefore, one of the interpretation of EET relates to changes from elastic to ductile part of the lower crust which assumes importance to explain seismicity in a region. As rocks can bend and break under accumulated stress in the upper brittle part of the crust, the EET is important to define the part of the crust important for seismic activity. This is supported by occurrences of most of the earthquakes in upper part of the crust (Maggi et al., 2000). In case of Himalayas, Rajesh and Mishra (2003) and Mishra and Rajsekhar (2006) have shown the relation between EET and seismogenic layer of the earth's crust.

4.4.2   Applications – EET, Sedimentary Basins and Backstripping

Applications of Effective elastic thickness (T_e) has been wide spread during recent times as it provided a tool for rheological studies under different geological settings (Watts, 2001). However invariably two methods of computations based on Bouguer coherence and free air admittance provide different values in the same area. The values in same areas may be as widely different as 100 km and 10 km as discussed below for Deccan trap region. This problem to some extent may be related to spectrum computations as discussed in Sections 4.2 and 4.3 which is required in both the cases. Invariably the Bouguer coherence method yields higher values compared to free air admittance (Perez-Gussinye and Watts, 2005). However, certain inferences can be qualitatively drawn based on the relative values of EET. A higher EET value indicates a stronger crust in a region compared to regions with lower value of EET computed using same approach. This implies that older cratons provide higher values compared to orogenic belts specially of recent times. High heat flow regions such as rift valleys and rifted margins provide smaller values compared to stable regions without intrusives. It is, therefore also related to seismically active regions which are characterized by low values. This brings to the other definition of EET that it represents the brittle part of the crust where rocks fail with increase in stress and strain.

(i)   Effective Elastic Thickness under Deccan Trap India

Watt and Cox (1989) estimated the EET for Deccan trap, which was very high (110 km) and attributed it to the flexural rebound caused by erosion after the eruption of Deccan trap for the development of plateau (uplift). Tiwari and Mishra (1999) analysed the free air and elevation data along three profiles (I, II and III, Fig. 4.20(a)) recorded at a close interval of 0.5 km and provided the estimate of EET under Deccan trap. These profiles are shown in Fig. 4.20(a). The Bouguer anomaly map (Fig. 4.20(a)) is characterized by two major gravity lows, L1 and L2 related to crustal thickening under the Western Ghat in response to isostasy and some felsic intrusive as discussed in Section 4.2.4.

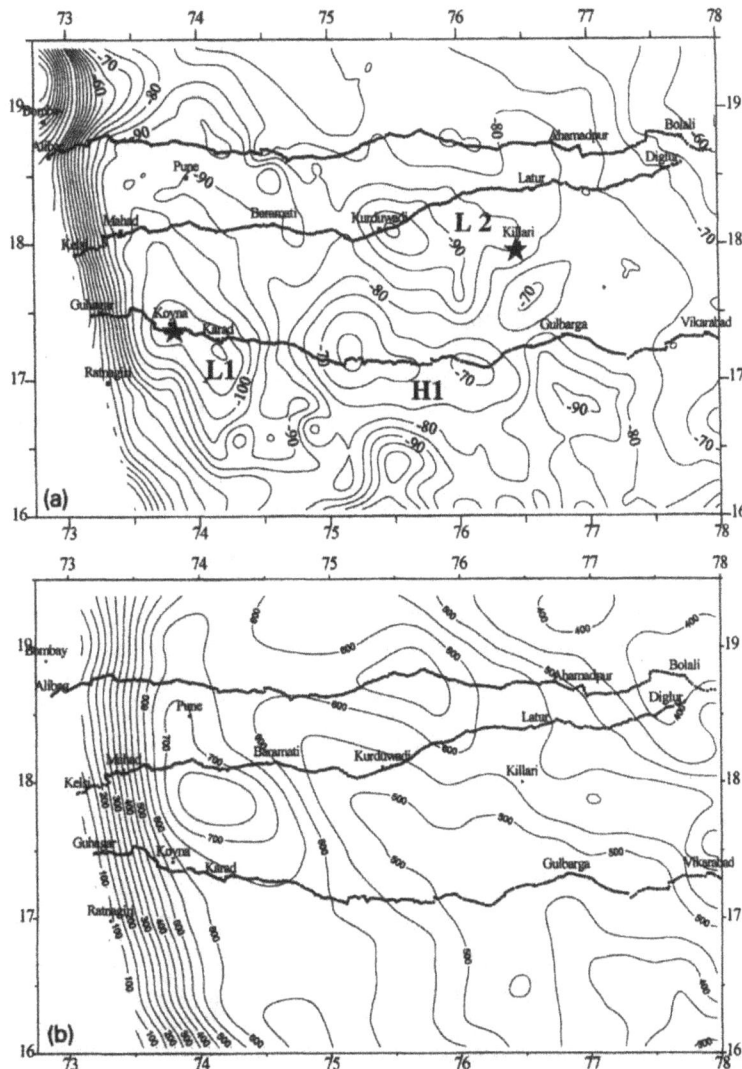

Fig. 4.20 (a) Bouguer anomaly map of Deccan trap, India showing two major gravity lows, L1, L2 referred to as Koyna and Kurduwadi gravity lows, related to crustal thickening under the Western Ghats and low density felsic intrusives along the Kurduwadi lineament, respectively. Gravity high, H1 represents intrusives. **(b)** Elevation map of Deccan trap India showing high elevation of about 750 m along the Western Ghats east of the Koyna (Tiwari and Mishra, 1999).

First the power in the observed free air gravity anomaly is compared with power in the topography as given in Fig. 4.21(a) in order to estimate the wavelength for which they are comparable and use that wave band for admittance analysis and estimation of EET. It shows comparable power in the two data sets for large wavelength up to wave number 0.1, which are involved in isostatic compensation. The computed coherence for Bouguer anomaly and topography using equation (4.67) for equal amount of surface and subsurface loading with a standard deviation of one is plotted in Fig. 4.21(b) along with theoretical coherence assuming uncorrelated surface and subsurface loading at Moho for EET = 5, 10 and 15 km, which suggest an EET of 8-10 km. Standard deviation of coherence in present case is quite large as it depends on the number of profile used for averaging the spectrum. The computed coherence based for an equal amount of surface and subsurface loading lies between theoretical coherence for Te = 5 and 12 km. The scattered value of coherence may be indicative of lateral variation of EET as suggested by Forsyth (1985).

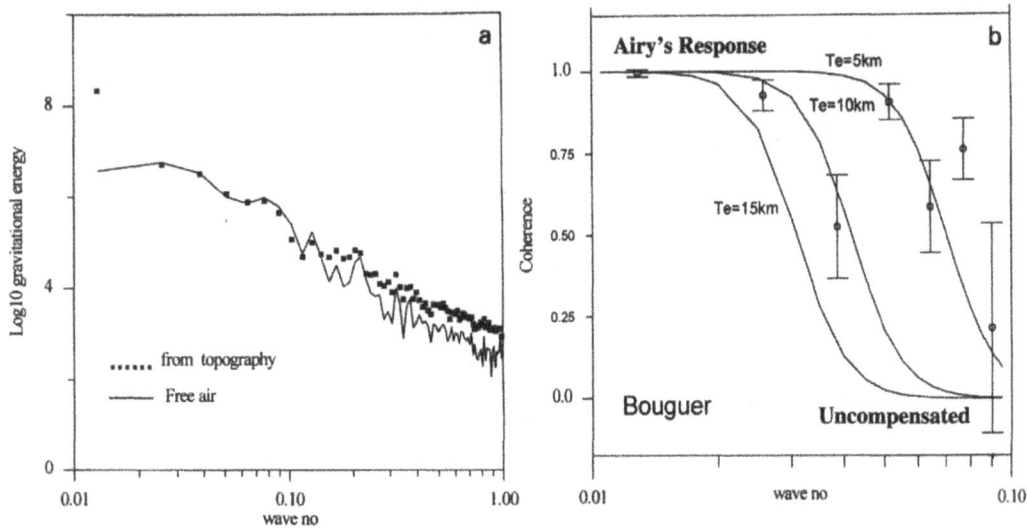

Fig. 4.21 **(a)** Power spectra of free air gravity anomaly and topography showing comparable power in large wavelength up to 0.1 wave number. **(b)** Coherence between Bouguer gravity anomaly and topography with standard deviation of 1.0 , showing a coherence of 1 for large wavelength related to Airy's model and minimum coherence for small wavelength related to uncompensated part of crustal density inhomogeneity. Theoretical coherence for effective elastic thickness (Te) = 5, 10 and 15 km are given, assuming uncorrelated surface and subsurface loading at Moho (Tiwari and Mishra, 1999).

The observed admittance based on equation (4.61) and bestfit elastic model for T_e = 8 km, T_c = 35 km and T_e = 11 km and T_c = 16 km are given in Fig. 4.22(a). Both the models are comparable to the observed admittance at short wavelength but at long wavelength only the former shows a better fit. The misfit function given in Fig. 4.22(b) also suggest a T_e of 8 km for which it shows the minimum value. Therefore, the model which best fits with observed admittance is for EET as 8-10 km and confirm the result obtained previously from (Fig. 4.21(b)) for free air gravity anomaly. This suggests a weak crust in this section, which might be due to different phases of rifting along west coast of India and eruption of Deccan trap (Section 3.8.1 and 5.4.2). A low value of EET for Deccan trap region is subsequently confirmed by Jordan and Watts (2001).

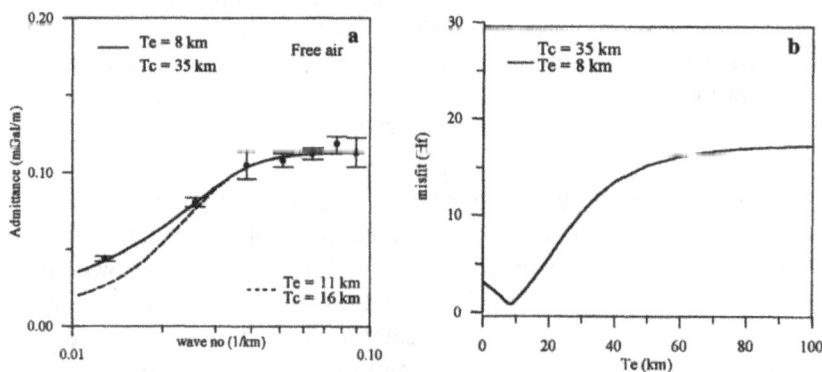

Fig. 4.22 **(a)** Admittance from free air anomaly and topography and best fitting elastic plate model for Te = 8 and Tc = 35 km and Te = 11 and Tc = 16 km. The former shows a better fit at all wave number. **(b)** The misfit function showing minimum for Te = 8 km (Tiwari and Mishra, 1999).

(ii) Post Glacial Isostatic Rebound: Ice Age and Global Warming

Vertical crustal movements in ice covered glaciated regions after deglaciation is referred to as isostatic rebound. Due to load of ice, first there is a flexure of the lithosphere due to isostasy and when ice starts melting on large scale, the region experiences an uplift related to isostatic rebound. It is similar to erosion of mountains but erosion is a slow process (Fig. 4.23(a)). This phenomenon has been felt on a large scale in Fennoscandia, Greenland and northern North America and Canada, which were covered extensively by ice cap during last ice age about 18,000-12,000 years ago. Since the melting of this ice cap, there is an uplift experienced in this entire region which is known as isostatic rebound. Melting of ice sheets has given rise to large rivers and several lakes and uplifts of North America and Fennoscandia such as Niagara River and Niagara falls in North America and Canada. During deglaciation, such large quantities of water was released that most of the water in this entire region is fossil water and only a small percentage of less than 1% is renewable on annual basis. Milankovitch cycle suggests that ice age of glaciations period is followed up by general warming which lasts for about 10-20 thousand years (Berger, 1988) and we are presently in this phase of global warming which provides isostatic rebound on a large scale where ever the glaciers are melting such as Arctic, Antarctica, Himalayas etc., Maximum present day uplift due to isostatic rebound is about 0.9 cm/yr in Gulf of Bothnia between Sweden and Finland, where the ice sheet was thickest during last ice age. The reported uplift (Ekman and Mäkiner, 1996) is almost elliptic in shape following the shape of Fennoscandia with its center located in the Gulf of Bothnia. This uplift can be used to compute the viscosity of the asthenosphere and mantle below it, which can give rise to the measured uplift. It provides a viscosity of 4×10^{19} Pas (Pascal second) for upper mantle (asthenosphere) and 10^{21} Pas for lower mantle (Cathles, 1975) while uplift of about 1.0 cm/yr in Canada provides a slightly higher estimate of about 10^{21} Pas for upper mantle and $10^{22} - 10^{23}$ Pa for lower mantle, which is based on the spectral study of the gravity field of Hudson Bay (Simons and Hager, 1997). Mathematical formulation to compute viscosity is given in papers referred to above.

(iii) Mountain Building and Sedimentary Rift Basins

Sedimentary basins and mountains on the surface of the earth are two most visible and important tectonic units. As explained in Section 2.7.1 and demonstrated through field examples in Section 2.9 and 2.10, basins are formed due to extensional tectonics while mountains are formed in compressional regimes due to convergence. However, both extensional and compressional regimes are also accompanied by compression and extension in certain specific sections giving rise to both types of situations in the two cases. For example, in case of extensional regimes along the sides of extensional regimes, shoulder highs are created due to compression along the margins (Fig. 2.23(a) and (b)) and basins are created on sides of mountains due to extension (Figs. 2.23c and d). It has also been stated in Section 2.5 on 'Isostasy' that over compensation causes uplift while under compensation causes subsidence.

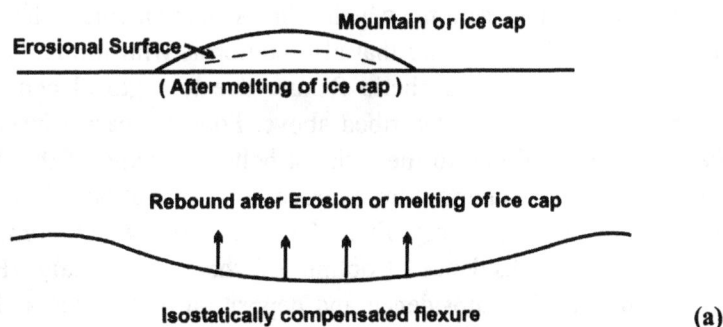

Mountain or Ice cap

Erosional Surface

(After melting of ice cap)

Rebound after Erosion or melting of ice cap

Isostatically compensated flexure   **(a)**

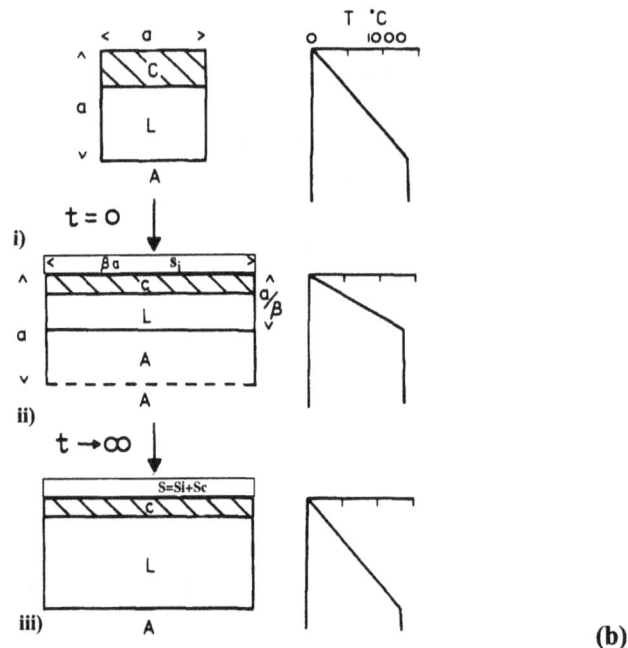

Fig. 4.23 **(a)** Isostatic rebound after melting of ice cap or erosion of mountain. **(b)** Basin formation due to stretching: **(i)** At time t = 0, a = initial length and c = crustal thickness, and L = Lithosphere. Right hand side shows the normal temperature distribution with depth. **(ii)** Stretching by a factor β causes expansion and subsidence, S_i and reduction in crustal thickness and lithospheric thickness by a factor β and A=Asthenosphere that rises due to stretching. The temperature gradient is disturbed as given on right hand side. **(iii)** Due to cooling of lithosphere, it acquires equilibrium as before and final subsidence S = S_i + S_c, where S_c is subsidence due to cooling (Mckenzie, 1978).

Lithospheric flexure due to regional isostatic compensation is spread over much larger region than surface load contrary to Ary's model (Fig. 2.18) where compensating mass is just below the surface load. Most of the thrust and fold belts show over compensation, such as in case of Andean type margin (Section 2.10), it can occur due to volcanic intrusions (island arcs) from upper mantle, which has given rise to mountain ranges. However, in case of Himalaya, it happens due to thrusting of high density rocks (Fig. 2.35) forming mountain ranges which are even presently rising. Erosion may further disturb the isostatic balance causing further uplift as incase of isostatic rebound.

However, positive Airy's isostatic anomaly invariably over recent thrust belts (Himalaya) and Archean Proterozoic fold belts in India (NGRI, 1975) and world over suggest that they may be isostatically under compensated as per Airy's model but their uplift and high density rocks in upper crust indicate over-compensation, which shows that it must be due to regional compensation based on lithospheric flexure that also conforms with formation of deep foreland basins along the thrust belts. The lithospheric flexure due to regional compensations can be computed from admittance analysis as described above. Foreland basins in convergence zones are formed on the flexure formed due to these thrust belts and some of the deepest basins are formed in this manner, which are important for hydrocarbon exploration. The extensional regions are accompanied by lithospheric upwelling (Fig. 2.23(a) and (b)), which normally causes under compensation giving rise to subsidence. Cooling of thermal anomaly (Plumes, hotspots, intrusions etc.,) results into further subsidence and deposition of sediments. The same process operates also in case of rift basins and rifted continental margins. Mathematical formulation to compute lithospheric flexure based on the load or subsidence under different geological conditions is given by Watts (2001).

A mathematical framework of basin formation in response to stretching of continental lithosphere (extension) and subsequent cooling is provided by Mckenzie (1978). It involves brittle failure of upper crust and ductile (pure shear) stretching of lower crust and upper mantle (Fig. 4.23(c)). Fig. 4.23c shows the subsidence (S_i) caused by the stretching (i) of the lithosphere, which disturbs the thermal equilibrium (ii). Stretching by a factor β causes corresponding reduction in crustal and lithospheric thickness. This initial subsidence can be obtained by assuming isostatic balance before and after the subsidence due to lithospheric flexure in the entire region. The stretched lithosphere further subsides (S_c) due to cooling of the lithosphere (iii) to attain the thermal equilibrium. The total subsidence $S = S_i + S_c$ can be obtained by considering the thermal subsidence as exponential with time constant equal to the time constant of the cooling of oceanic lithosphere. This is a case of limited extension giving rise to subsidence with or with out normal fault on one side of the basin or on both sides of the basin. However, the most important objection to this model arose from the absence of processes, which can give rise to this kind stretching for large basins required for their formation without any thermal anomaly in the lithosphere. It may however be valid in case of passive rift basins.

(iv)   Basins with Large seale Extension

In certain cases, large scale extension over 600-800 km have been observed as in case of Basin and range province, USA, and the Aegean Sea. Wernicke (1985) proposed a simple shear model for basin and range province in USA. The model described above relates to pure shear where there is a uniform changes in lithosphere and crust implying same amount of extension and warping of lithosphrere and crust along a plane (Fig. 4.23(c)). The same is also shown in Figs. 2.23a and b. Simple shear, however relates to non uniform changes in lithosphere and crust along a plane and extension is accommodated along several normal faults that join with the master fault in depth (Fig. 4.24(a)). In this case simple shear results into a large lithospheric scale detachment plane and extension along it created Basin and Range province type of tectonics in a large region (Wernicke, 1985). In case of Basin and Range province extension along detachment plane creates a crustal break away followed by extensional allochthons (Fig. 4.24(a)). Due to lithsopheric upwarping along this plane and extension, crust thins and thinnest crust is found under Core Complex Range. Wernicke and Axen (1988) and Buck (1988, 1991) suggested that either side of the main normal fault responds to the loading and unloading that rotates at shallow depth causing new normal faults while at large depths, it remains straight. Fig. 4.24(a) shows that with increase in extension more normal faults are generated that increase the region of extension. This figure also shows that on the western side of the Basin and Range Province, high gravity and low topography are observed while towards the east, its opposite of high topography and low gravity are observed. Opposite correlation between topography and observed gravity suggest an over all isostatic compensation. However, the absence of lowest gravity in the central part where thinnest crust is observed suggest under compensation in this section that should show subsidence. However, absence of subsidence in this section suggest that there is an upward force that balances the isostatic downward force which may be related to thermal upward force as advocated by Wernicke (1985). Low effective elastic thickness (Te) in such cases also indicates the presence of thermal anomaly (Buck, 1988). The uplift however is seen in the eastern part that may be synonymous with the shoulder uplift in case of rift basins (Fig. 2.23(a) and (b)).

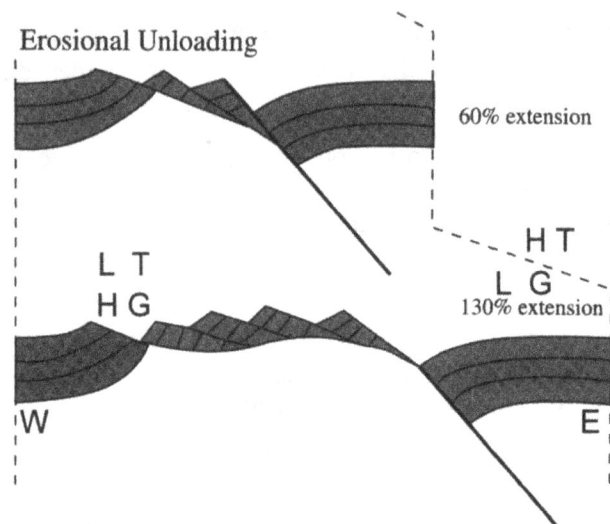

Fig. 4.24 (a) Model of lithospheric scale detachment fault following normal faults causing extension in large area as seen in case of Basin and Range Province. HT- High Topography and LG- Low Gravity towards the east and LT- Low Topography and HG- High Gravity towards the west suggest over all isostatic compensation but is not applicable in the central part where thin crust is reported, USA (modified after Wernicke and Axen, 1988; Buck, 1988; Watts, 2001).

Kusznir et al., (1991) combined both pure and simple shear models with flexural cantilever beam to explain formation of sedimentary and rift basins such as Grand Banks, Viking graben in North Sea etc., However, lithospheric and crustal up warping due to extension as shown in case of rift basins and passive margins (Section 2.7.1; 2.9, 3.8.1) are essential elements to all the models used to explain the evolution of sedimentary basins. Beaumont (1978; 1981) considered sedimentary basins formed on an elastic lithosphere with different flexural rigidity due to load at one end, which is similar to those encountered in foreland basins along thrust belts such as Ganga basin along Himalayan frontal thrust. He suggested that larger the flexural rigidity lesser is the subsidence.

One of the most important observations in case of rift basins are the shoulder uplift (Section 2.7; Fig. 2.23(a) and (b)), which are related to crustal thinning caused by normal faulting. The magnitude of this uplift and basin subsidence and underlying crustal structure are related to flexural rigidity of the lithosphere, the amount of extension and dip of the major basement faults along, which the extension has taken place. Based on model calculation using standard values for these unknowns, shoulder uplifts of a few meters to 1.5 km is predicted, which appear to correlate with uplift along several known rift basins (Egan, 1992).

(v) Backstripping

One of the consequences of isostatic loading and subsidence in a sedimentary basin is to study its effect on basement, which is known as Backstripping. This implies to determine the position of basement in absence of water and sediment loading. As it requires unloading of sediment and so is the name Backstripping. Depending on types of isostatic compensation, there are two types of Backstripping viz. Airy's and flexural. Backstripping process based on Airy's model is shown in Fig. 4.24(b) (Watts, 2001), which shows both loaded and unloaded cases implying existing present day situation and what would have existed before the sedimentation of layer S_i* had started. Equating the load of two columns at depth of compensation provides the depth to the basement (Y_i) in unloaded case as (Watts, 2001)

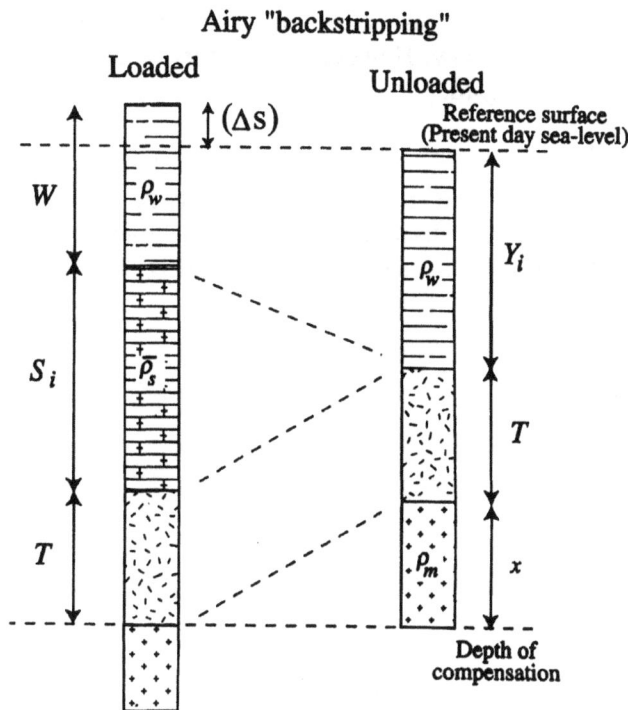

Fig. 4.24 (b) Backstripping based on Airy's isostatic compensation (Watts, 2001).

$$Y_i = W + S_i \left[\frac{\rho_m - \rho_s}{\rho_m - \rho_w} \right] - \Delta s \, \frac{\rho_m}{\rho_m - \rho_w}$$

Y_i = Depth to the basement in Sea before deposition of sediment

W = Present day water column

ρ_m = Density of the upper mantle

ρ_s = Density of the sediment

ρ_w = Density of the water column

Δs = Change in global Sea level at the time of deposition of sediment, which can be obtained from seismic sections.

This is known as backstripping equation, which can be evaluated based on stratigaphic well data. Mathematical formulation for backstripping based on flexural model is complex and those interested may refer Watts (2001).

4.5   Linear Inversion and Parameter Estimation

Inversion methods relates to the estimation of unknown parameters of the causative sources directly from the observed field using the principles of least squares and is therefore, known as least squares inversion. The first attempt in this direction was due to Bott (1975) who configured the sedimentary basins based on the observed gravity field along a profile using this principle. In this case at each point of observation thickness of sediment is determined using simple Bouguer slab formula, as described in section 2.7, which becomes the initial model for computation. The difference between the computed field from initial and successive models and the observed field is reduced in least squares sense through iterations and the final model is obtained. In this case, the sediment thickness at each observation point is the unknown and therefore the number of unknowns are equal to that of the observation points and can be solved by least squares methods. Some basic principles involved in the inversion of potential field are discussed below. However, for further details, readers are referred to books and references cited below in this regard.

4.5.1   Least Squares Inversion

In case there are n number of observations referred to as y_n (n = 1, 2, ... , n) and m are the unknowns Z_m (m = 1, 2, ... , m) of the causative sources such as thickness of sediment at each observation point in case of a sedimentary basin as described above (Fig. 4.25(a)). The two in case of linear relationship can be expressed as:

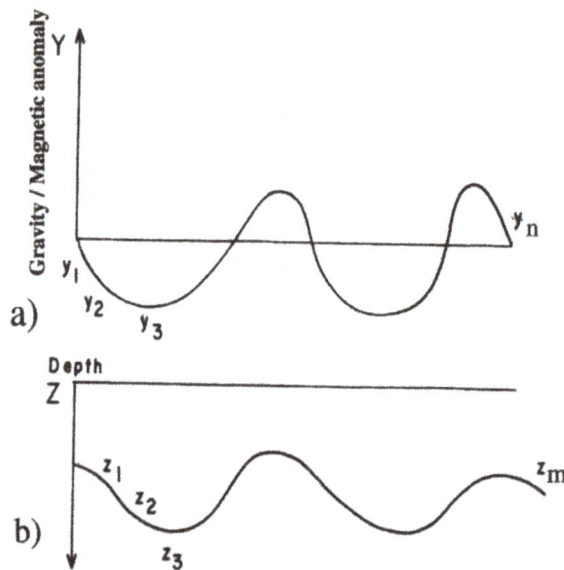

Fig. 4.25 (a) A schematic section of a basin as a causative source and y_n (n = 1, 2, ... n) representing observations and Z_m (m = 1, 2, ... n) being the unknown depths at m points to be determined. In least square case n = m while over and under determined systems have n > m and n < m, respectively.

$$Y_1 = A_1 (Z_1 . . . . Z_m)$$

$$Y_n = A_n (Z_1 . . . . Z_m) \qquad .....(4.68)$$

These sets of equation can be expressed in matrix form as:

$$Y_n = A_{(n,m)} (Z_m) \qquad .....(4.69)$$

However, the system of equations in most of the cases are non linear which in the simplest form can be linearized using Taylor's Series expansion at some initial values $(Z_m^°)$ and ignoring higher order terms, then the equation 4.69 reduces to:

$$Y_n = A_n (Z_m^°) + [(\partial A_n / \partial Z_m) | Z_m^°].\Delta Z_m \qquad .....(4.70)$$

or
$$Y_n - A_n (Z_m^°) = [(\partial A_n / \partial Z_m) | Z_m^°].\Delta Z_m \qquad .....(4.71)$$

where ΔZ_m is the increment to unknowns Z_m over its value $Z_m^°$.

This expression at all points of observation can be expressed in matrix form as:

$$Y_n = A_{(n,m)} (Z_m) \qquad \qquad \dots (4.72)$$

The matrix A of dimension n x m is obtained through linearization of a non linear system and the final solutions of Z_m are obtained after iterations with reference to the minimum objective function in a least squares sense which is given by:

$$E = \sqrt{\sum_{i=1}^{n} [(Y_o(i) - Y_c(i)]^2 / n} \qquad \qquad \dots (4.73)$$

Where $Y_o(i)$ is the observed gravity / magnetic fields

$Y_c(i)$ is the computed gravity / magnetic fields

For minimum objective function, the least squares solution is given by

$$Z = [A^T A]^{-1} A^T Y \qquad \qquad \dots (4.74)$$

The approach is applicable only for cases when n > m known as over determined systems and the rank of A is m, otherwise $[A^T A]^{-1}$ can not be evaluated. However, in cases for n < m, known as generalized inversion using singular value decomposition (Lanczos, 1961; Parker, 1977; Weidelt, 1974) should be used.

4.5.2 Generalized Inversion: Singular Value Decomposition and Resolution Matrix

As mentioned above, in cases, when the unknowns are more than the observations, the least squares solution is not applicable and those for generalized inversion using singular value decomposition is used. In these cases, the matrix A is factored as:

$$A = U W V^T \qquad \qquad \dots (4.75)$$

Where $UU^T = I_n$; $U^T U = V^T V = I_m$ and W = diag $(\sigma_1, \sigma_2, \dots, \sigma_m)$

U is a n x m matrix whose columns consist of m of total of n ortho normal data eigen vectors, which are not associated with null eigen values of AA^T. The matrix V is a m x m matrix in model space whose columns contain the m orthonormal parameters eigen vectors of $A^T A$. W is a m x m diagonal matrix containing m non singular values such as $W_i (i = 1, \dots, m)$ or non nagative square roots of the eigen values $W_i^2 (i = 1, \dots m)$ of $A^T A$. They are known as singular values arranged in decreasing order. The inverse of A is given by:

$$A^{-1} = [U W V^T]^{-1} = V W^{-1} U^T \qquad \qquad \dots (4.76)$$

and $\qquad \qquad Z = V W^{-1} U^T Y \qquad \qquad \dots (4.77)$

To avoid singularities in $A^T A$ of least squares solution and W^{-1} of singular value decomposition, a constant known as Murquardt parameter (Marquardt, 1963) can be added along the main diagonal of N to increase the DC level of eigen values such that none of its eigen values are zero. This is also known as ridge regression method.

Resolution Matrix

As discussed above, the eigen values in generalized inversion are arranged in decreasing orders and small eigen values are discarded in order to reduce the error in least squares formulation and fast convergence. This reduces the resolution in model space, which can be judged by resolution matrix (Pedersen, 1977) given by:

$$R = V V^T \qquad \qquad \dots (4.78)$$

R is an identity matrix of order m and is a measure of the uniqueness of the solution Z. The least squares solution implies m x m unit matrix as resolution matrix. However, in generalized inversion, as small eigen values are rejected, it can be defined by the resolution matrix, which would show resolution in model space depending on its closeness to identity matrix for specific parameters. The resolution of the solution can however be improved by constraining the variations in parameters based on the geology / tectonics of the region. This can be achieved using suitable weighting matrices as suggested by Jackson (1972), Pedersen (1977) etc., As some of the low eigen values are rejected in case of generalized inversion, resolution is reduced and error in model parameters is improved. This is same as trade off between resolution and error in model space as given by Backus and Gilbert (1970) in case of linear inverse theory.

4.5.3   Inversion of Gravity Field Due to Simple Bodies (Fault Model)

Modelling of gravity anomaly due to simple body like a fault (Fig. 4.25(b), Venkata Raju, 2004) involves unknowns for body parameters, like density contrast (ρ), its location from a reference point (d), depth to top and bottom (z_1 and z_2) and inclination (θ). The initial parameter can be provided by known geology and other geophysical information from the area. Based on the initial values, the gravity field at specific intervals along the profile are calculated using the expression for gravity due to such bodies and compared with the observed field. To compare the observed and the computed fields, an objective function E defined as root mean square error between the two fields is defined, as given in equation (4.73):

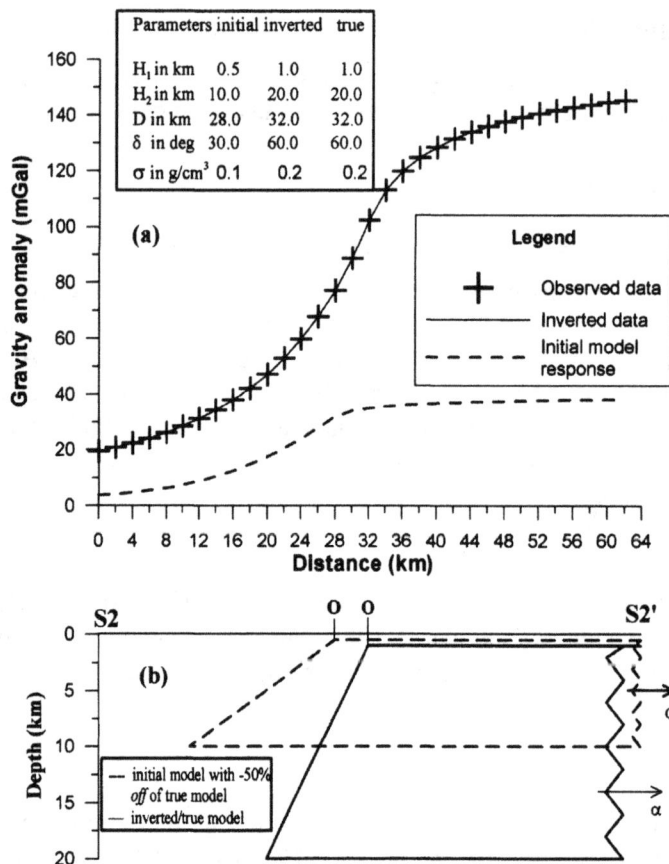

Fig. 4.25 (b) Inversion of theoretical gravity field due to a fault for unknown parameters as given in the inset in first column. The computed field is inverted with initial values for these unknowns with -50% error in their theoretical values and computed parameters are given in second column, which is almost same as their actual values given in first column (Venkata Raju et.al., 1998).

$$E = \sqrt{\sum_{x=1}^{n}[g_o(x) - g_c(x)^2]/n} \qquad\qquad .....(4.79)$$

Where $g_o(x)$ is the observed gravity field, $g_c(x)$ is the computed gravity field and x = 1, 2, ... n are the number of observations.

The error, E, should be minimum for a good match between the observed and the computed fields. If $(\delta\rho, \delta z_1, \delta z_2, \delta\theta,$ and δd are the increments or decrements to the initial values for the better match, the difference δg between the observed and calculated anomaly can be expressed as:

$$\partial g = \frac{\partial g(x)}{\partial \rho}\delta\rho + \frac{\delta g(x)}{\partial z_1} + \delta z_1 + \frac{\partial g(x)}{\partial z_2}\delta z_2 + \frac{\partial g(x)}{\partial \theta}\delta\theta + \frac{\partial g(x)}{\partial d}\delta d \qquad .....(4.80)$$

Equation (4.80) can be expressed in matrix form for all the observations (n) as:

$$\mathbf{E = AP} \qquad\qquad .....(4.81)$$

Where **E** is the error matrix of order $n \times 1$ obtained by subtracting the calculated anomaly from the observed anomaly, **A** is the matrix of order n × 5 (Jacobian matrix of partial derivatives) obtained by taking the partial derivatives of the expression for the gravity and magnetic anomalies due to bodies being considered with reference to unknown parameters and **P** is the matrix of order 5 x 1, which contains increments or decrements of the parameters. The parameter matrix **P** can be obtained by the following equation:

$$\mathbf{P = A^{-1}E} \qquad\qquad .....(4.82)$$

A may not be a square matrix and therefore its transpose, $\mathbf{A^T}$, is used to evaluate the parameter matrix, P and is therefore given by

$$\mathbf{P = [A^T A]^{-1} A^T E} \qquad\qquad .....(4.83)$$

Most of our geophysical problems are over determined, i.e. (the number of observations are more than the number of unknown parameters) and therefore equation (4.83) provides least squares solution to the parameter matrix. However, if initial solution is not close to the actual solution, convergence becomes difficult. To overcome this problem, Marquardt factor λ is introduced and equation (4.83) is modified as:

$$\mathbf{P = [A^T A + \lambda I]^{-1} A^T E} \qquad\qquad .....(4.84)$$

I is an identity matrix, and λ is the Marquardt's damping factor (Marquaedt, 1963). When λ=o, the algorithm is equivalent to Newton's Gauss method. This method is best suitable for good initial solution. However, if the initial solution is far from the true solution then it leads to a divergent solution. If λ is very large then this algorithm is close to steepest descent method, and takes a long time to converge even for poor initial solution. In this method, one can assign λ = 1 or 0.1 as the initial value. By dividing this value by three each time λ is successively reduced, if the objective function is less than the existing value. Otherwise it is multiplied by a factor 2. The modified parameters are given by $\rho = \rho +\delta\rho$,$z_1 = z_1 + \delta z_1$, $z_2 = z_2 + \delta z_2$, $\theta = \theta + \delta\theta$, and d = d+ δd.

By using modified parameters, theoretical anomaly is again calculated. The parameters may be further modified in an iterative way till the objective function E, is minimum. The program is terminated either after the specified number of iterations are completed or when the Marquardt parameter assumes a very large value or the r.m.s. error is reduced to a minimum value. In practice, gravity or magnetic anomaly along a profile perpendicular to strike direction can be digitised at suitable station interval from a reference point (from one end). The distance versus field (Bouguer anomaly) is given to the computer as input data. The initial parameters of the model and Marquardt parameter (λ=0.1 or 1.0) are also given as input to the computer. Computer modifies the solution through iteration till one of the conditions specified above is satisfied. The final parameters and the corresponding computed field are obtained as the output. Further details on inversion scheme including singular value decomposition, resolution matrix and information density matrix can be obtained from Lanczos (1961) Jackson (1972), Parker (1977), Pedersen (1977) etc., Several workers have given schemes and computer codes for inversion of gravity / magnetic anomalies due to different type of bodies like faults, dykes etc., and also for arbitrary shaped bodies. Some of them are Radhkrishnamurthy (1998), Fedi and Rapolla (1999), Venkata Raju (2003) etc.

4.5.4   Applications – Godavari Basin, India and Flexural Model

The best approach towards quantitative interpretation of gravity and magnetic anomalies is to model profiles using simple models like tabular body (dyke), faults, sedimentary basin etc., using inversion schemes for them as discussed above provided the anomalies due to individual sources can be separated. This is quite often true in case of magnetic anomalies as observed magnetic anomalies are mainly due to individual sources as all bodies do not produce magnetic anomalies. Having obtained the first hand information about their sources from inversion due to individual bodies, modelling using complex bodies can be carried out, in case required based on available information from the region and schemes for 2 or 2.5 dimensional arbitrary shaped multiple bodies. This is specially so in case of gravity anomalies as all bodies produces gravity anomalies.

(i)   Theoretical Example of Gravity Field due to a Fault

The scheme given by Venkata Raju (2003) and described above was tested for a fault model on theoretical gravity field due to a fault as given in Equation 2.70. The gravity field for the parameters given in the inset of Fig. 4.25(b) was computed using equation (2.70) and plotted in the same figure. The distance versus computed field along with initial solution with - 50% error in the actual value of the parameters are given as input to the computer for inversion. Final parameters obtained through inversion and the computed field with these parameters are given in the same figure, which are almost same as the actual value of the parameters used for the computation of the field, which is quite satisfactory. Same result was also found when initial values of parameters were set with ±50% error. Using this scheme gravity profiles AA' across Godavari basin (Fig. 2.30(a)) India was inverted and depth to the basement was estimated as described below (Venkata Raju et al., 1998)

(ii)   Bouguer Anomaly Profile: Godavari Basin

The Bouguer anomaly profile AA' across Godavari basin (Fig. 2.30(a)) is modelled in Section 7.7.1 using the usual trial and error method for arbitrary shaped 2.5 dimensional multiple bodies. In that case total subsurface crustal structure was modelled by using multiple sources. In such cases where there are more unknowns due to multiple bodies and their fields interfere with each other, inversion of the field may not be the best choice. In such case, it is always advisable to separate the observed field as the residual and the regional fields and in case they are caused by individual sources, inversion schemes in such cases can be applied to obtain the parameters of the causative sources. In the present case of gravity profile AA' across Godavari basin (Fig. 4.26(a)), a regional field as shown in this figure is separated from the observed field to obtain the residual field as a central low (Fig. 4.26(b)). As it is known that central gravity low coincides with the Gondwana sediments bounded by faults, it is largely caused by them. This residual field is, therefore modelled (i) using a basin mode (Fig. 4.25(a)) with depth to the basement as unknowns at every point of observation (Radhakrishnamurthy and Mishra, 1989; Radhakrishnamurthy, 1998) and (ii) using fault model on either flanks of it as described above (Venkata Raju et al., 1998). Fig. 4.26(b) provides the depth to the basement at every point of observation based on basin model for density contrast of 0.35 g/cm^3. The maximum depth to the basement is about 5.0 km. In second case, the gravity anomalies on either flanks CC' and BB' (Fig. 4.26(a)) are approximated to a fault and inverted for same density contrast (Fig. 4.26(c), Venkata Raju et al., 2002), which provided almost similar configuration of the basin with maximum depth to the basement of about 4.9 km. It may be noted that modelling of the same profile using multiple bodies (Section 7.7) has also provided almost same depth to the basement, which indicates that under simple conditions where sediments are not much disturbed depth to the basement can be estimated reasonably correct by following any of the standard methods of modelling gravity field as described above.

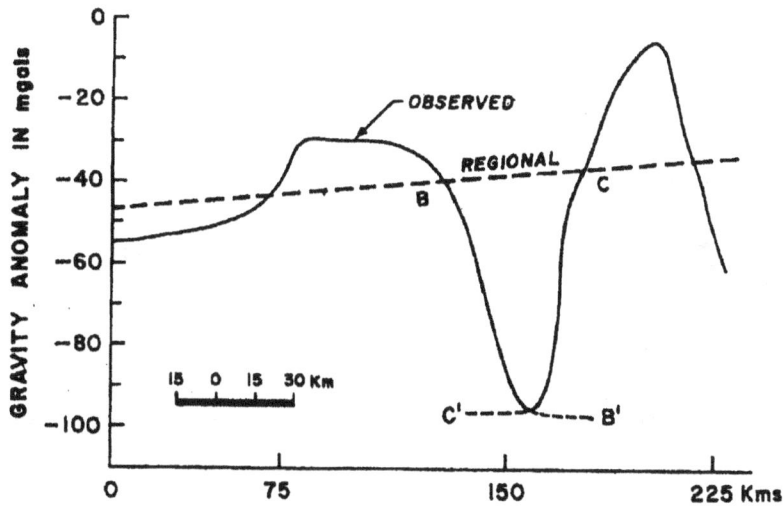

(a)

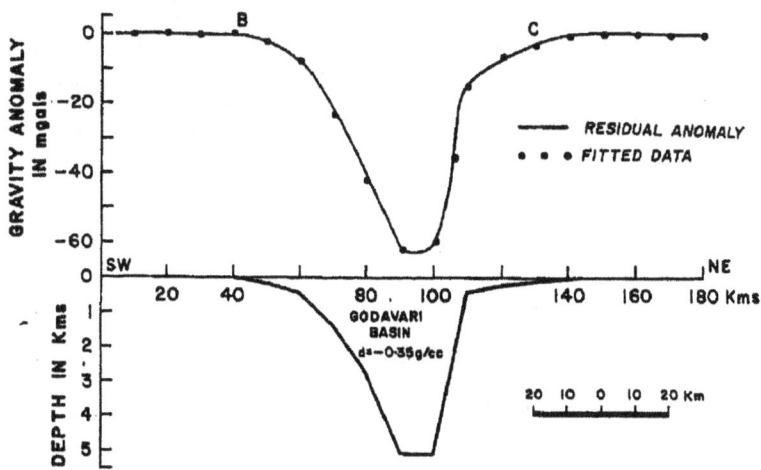

(b)

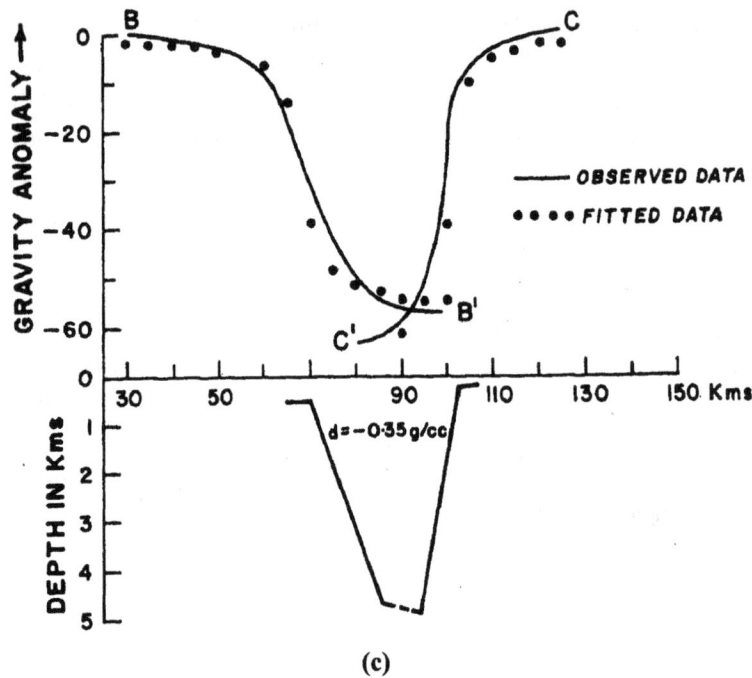

(c)

Fig. 4.26 **(a)** A Bouguer anomaly profile AA' across Godavari basin (Fig. 2.30(a)) showing a gravity low in the centre due to Gondwana sediments and flanking highs along the shoulders due to intrusives. A regional field is drawn based on the field values at the margins of the basins, B and C and joining them together. **(b)** The residual gravity low in the center can be independently modelled using a basin model as given in Fig. 4.25(a), providing depth at each observation point or equally spaced digitised data. It provides a maximum depth of 5.0 km for the basement. **(c)** The Bouguer anomaly profile AA' (Fig. 4.26(a)) can be considered as composed of two parts BB' and CC' representing the contacts of Gondwana sediments with the adjoining gneisses on either sides of the basin. These components BB' and CC' appear as the gravity field due to a contact as given in Fig. 4.24(b) and can be inverted using linear inversion scheme as discussed in Section 4.5 with appropriate density contrast. It provides maximum thickness of sediment in the basin as 4.9 km almost same as obtained from basin model in Fig. 4.25(b) (Venkata Raju et.al., 1998).

(iii) Basin Modelling Using Flexural Model: Godavari Basin

The gravity profile across Godavari basin discussed and modelled in Section 2.9 using conventional method of modelling gravity field, are modelled here using flexural model due to simple stretching as discussed by Egan (1992). Accordingly, the flexural isostatic response of lithosphere as thin elastic plate due to stretching is given by (Turcotte and Schubert, 2002)

$$\frac{\partial^2}{\partial x^2}D + \frac{\partial^2 \omega_x}{\partial x^2} + (\rho_m - \rho_s)g\omega_x = L_x$$

where L_x is the load due to sediment.

ρ_m and ρ_s are the densities of mantle rocks = 3.3 and sediments, respectively.

D is the flexural rigidity as defined above and given by

$$D = (ET_e^3)/[12(1-v^2)]$$

Where E is young modulus, T_e is the effective elastic thickness as described above and

ω_x is the flexural uplift/subsidence and

υ is the Poisson's ratio = 0.25

Using this expression Egan (1992) has given a scheme to model crustal bulge as negative load at Moho due to normal faults that characterizes rift systems. Accordingly,

$$\frac{\partial^2}{\partial x^2}D + \frac{\partial^2 \omega_{sx}}{\partial x^2} + (\rho_m - \rho_{air})g\omega_{sx} = -\rho_c g S_x$$

where ρ_{air} and ρ_c are densities of air and crust and S_x is simple shear crustal thinning. It does not assume any thermal anomaly and isostatic adjustment takes place due to mantle upwelling.

Using this scheme, gravity profile (Fig. 4.26(a)) across Godavari basin is modelled as shown in Fig. 4.27(a) in order to confirm the amount of subsidence and faults configuration of basin as inferred from conventional modelling given above. Fig. 4.27(a) is a model along profile AA' (Fig. 2.30(a)) with northern fault as the master normal fault with extension (E) = 7 km and T_e = 10 – 20 km. It provides the width and depth of the basin almost same as that observed at the surface and obtained previously. The second model (Fig. 4.27(b)) is related to a profile (Fig. 2.30(a)) that represents the observed field across median high (Chinnur high) showing two lows on either side of a central high besides shoulder highs. It is modelled introducing two more faults within the basin, on either side of the central high for T_e = 20 km that is obtained from published literature (Rajesh and Mishra, 2004). The computed model shows thinning of the crust and shoulder highs in response to simple isostatic adjustment which are typical characteristics of a continental rift systems. The width and depth of the basement in the basin also matches quite well with that exist along this profile. E is the amount of extension which in this case is 7 km along boundary fault and 1 and 2 km along faults related to median highs. It may be noted that the listuric faults can be modelled with initial lower dips. Hodgetts et al., (1998) extended this scheme to 3-D flexural modelling.

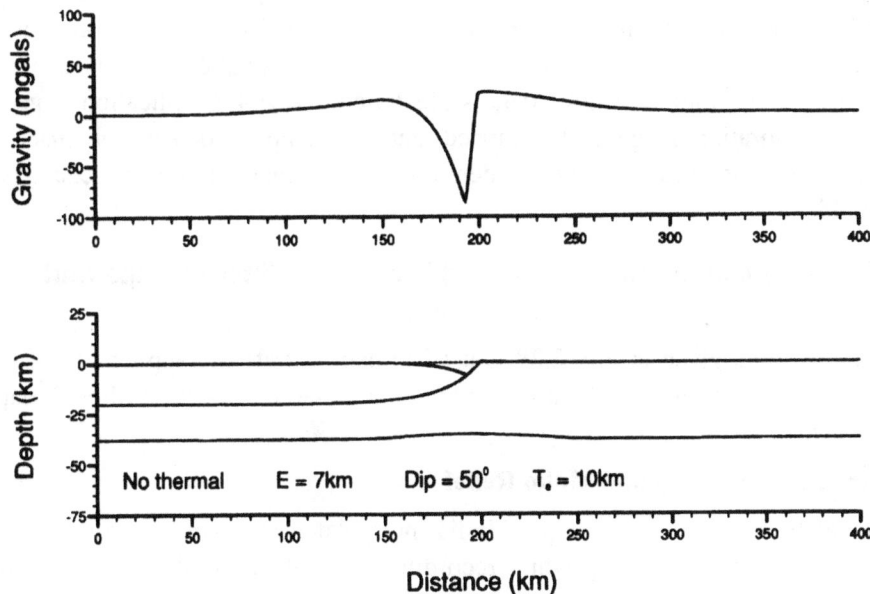

(a)

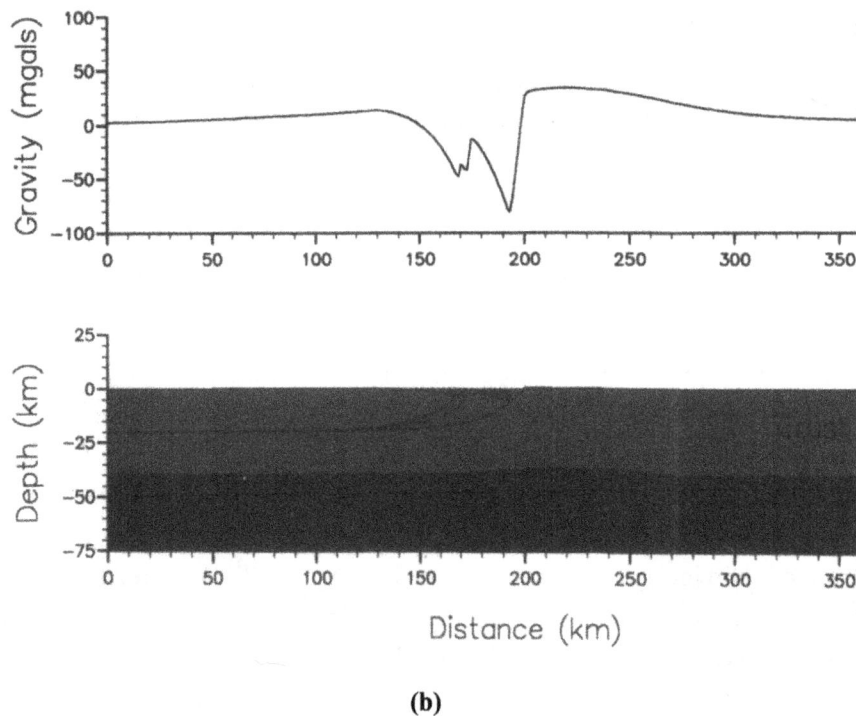

(b)

Fig. 4.27 (a) Flexural model of Godavari basin along profile AA' (Fig. 2.30(a)) for Dip = 50°, Extension (E) = 7 km across the northern master fault and Te = 10 km Showing a crustal upwarping. **(b)** Flexural model along a profile across median high (Fig. 2.30(a)) for E= 7, 1 and 2 km across northern master fault and faults associated with median high, Dip = 50° and Te = 20 km. It also shows a crustal bulge.

4.6   Field Examples: Harmonic Inversion and Geodynamic Studies

Harmonic inversion as discussed in Section 4.3 has been found to be very useful for modelling gravity and magnetic anomalies due to basement as illustrated in Section 4.7. However, it is also useful for geodynamic studies and its application to computation of Moho from Bouguer anomaly map of India and thickness of magnetic crust from MAGSAT data is illustrated below. Its application to geological mapping and mineral exploration is generally carried out by mapping density or susceptibility variation in Archean-Proterozoic terranes, which is demonstrated by mapping density changes in the Indian Peninsular Shield.

4.6.1   Bouguer Anomaly Map of India – Low and High Pass Filtered Maps and Moho Relief

Bouguer anomaly map of India given in Fig. 2.28 is used to demonstrate the application of low and high pass filters and obtain the depth of Moho and its variations based on low pass filtered map using harmonic inversion scheme.

(i)   Low Pass Filtered Gravity Map and Moho Relief

The digital data of Bouguer anomaly map of India interpolated at 5 km interval, which on an average represents the station spacing while recording this data is used for regional/residual separation using a low pass filter as described above (Mishra et al., 2004). The radial spectrum of this data set is given in Fig. 4.28(a), which shows two linear segments related to deeper regional and shallow residual sources. As discussed in section 2.8, the regional field in case of Bouguer anomaly maps originates from Moho as it represents the largest density discontinuity while residual field represents exposed and shallow sources within the crust. The two segments intersect at the wave number 0.02 corresponding to frequency of 50. Using this wave number, low and high pass filters are designed and low pass filtered Bouguer anomaly map of India is presented in Fig. 4.28(b). One can note the difference between the Bouguer anomaly map of India (Fig. 2.28) and this low pass filtered map where small wavelength anomalies of original map have disappeared indicating its regional nature. Most of these anomalies are discussed with details in Chapter 7. However, some important features are described here to show the utility of low pass filtered regional maps:

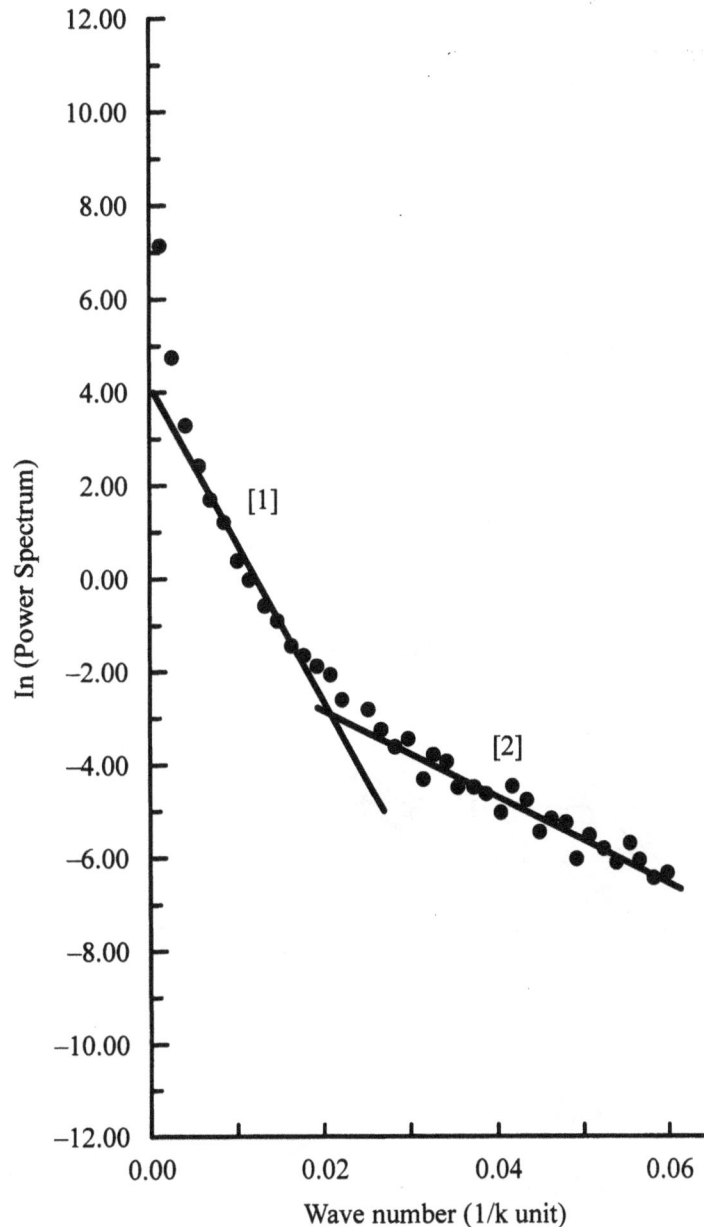

Fig. 4.28(a) Radial spectrum of Bouguer anomaly map of India (Fig. 2.28) showing two linear segments 1 and 2 related to deeper sources (Moho) and surface/ shallow sources, respectively.

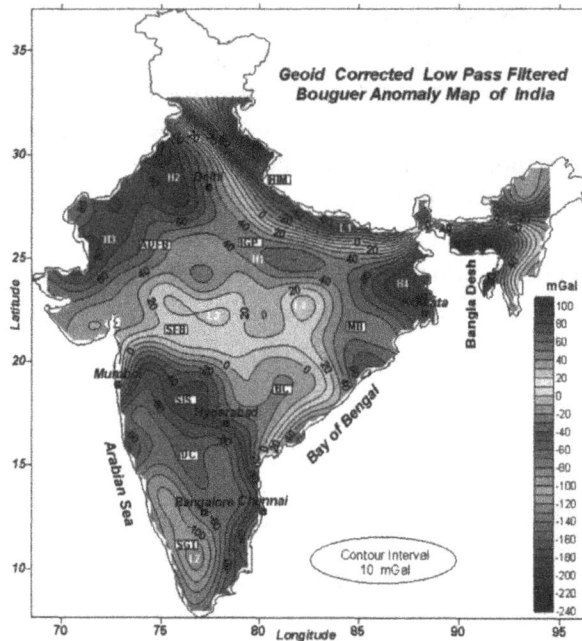

Fig. 4.28(b) Low pass filtered regional Bouguer anomaly map of India with wave number < 0.02 (first segment of spectrum, Fig. 4.28(a)). L1-L2 are two major gravity lows suggesting crustal thickening in these sections and H1-H4 are gravity highs related to crustal bulge associated with lithospheric flexure due to Himalayas. ADFB=Aravalli Delhi Fold Belt, BC = Bastar Craton, DC=Dharwar Craton, HIM-Himalayas, IGP = Indo Gangetic Plains, MB = Mahanadi Basin, SFB = Satpura Fold Belt, SGT = Southern Granulite Terrain, SIS = South Indian Shield (Mishra et al., 2004). [Colour Fig. on Page 767]

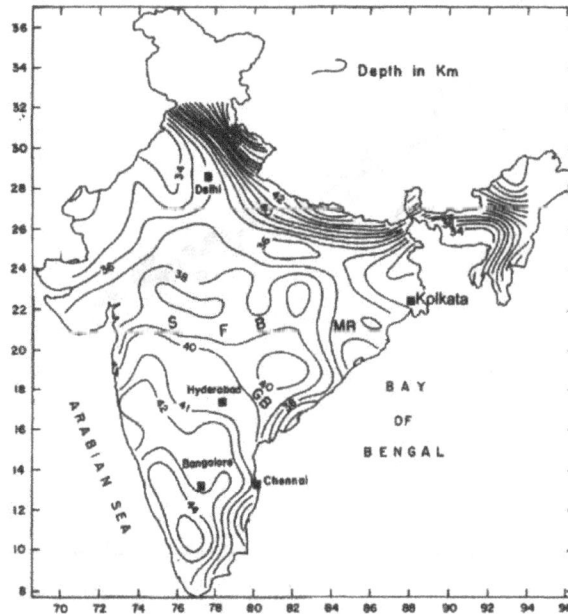

Fig. 4.28 (c) Moho relief computed from low pass filtered Bouguer anomaly map of India for a bulk density contrast of –0.4 g/cm^3 (– 400 kg/m^3) between lower crust and upper mantle. It largely conforms with seismic Moho except in certain sections where bulk density contrast used for computation may be different (Mishra et al., 2004).

(a) It shows a large wavelength gravity low (L1) over Himalaya and Ganga fore deep related to crustal thickening caused by Himalayan orogeny as discussed in section 2.10 and thick sediments of the Ganga basin.

(b) Gravity highs (H1-H4) is partially related to the crustal bulge in the foreland Ganga basin caused by flexure of the Indian lithosphere under Himalaya and high density rocks in these sections (Mishra et al., 2004)

(c) Towards south, the gravity field gradually decreases up to Satpura Fold Belt (L3, SFB), which further decreases to the gravity low L2 over Southern Granulite Terrain related to crustal thickening (Mishra and Rao, 1993). Further details of these anomalies are given in Section 7.8 and 7.9.

(d) Towards the west, the Aravalli-Delhi Fold Belt (ADFB) is reflected as a wide gravity gradient that suggests deep seated density discontinuity that extends from surface in the lithosphere. Further details of density structure of this region are provided in Section 7.5.

(e) The gravity high in the eastern part (H4) is related to high density rocks of Singhbhum craton and Shillong Plateau (Fig. 2.26).

As the low pass filtered gravity anomalies originate mainly from Moho, the variations in Moho as relief was computed for a bulk density contrast of $- 400$ kg/m^3 between the lower crust and upper mantle using harmonic inversion scheme given in Section 4.3. The computed Moho map is presented in Fig. 4.28(c), which shows a thick crust of about 50-55 km under Himalayas and 45 km under Southern Granulite Terrain and shallow crust of 36-38 km in the central part under Ganga basin and Satpura Fold Belt. There is an over all agreement between crustal thickness given in Fig. 4.28(c) and those given based on deep seismic studies in different parts of the country based on seismic studies discussed in Chapter 6 and 7. However, some noted discrepancies under Satpura Fold Belt and Eastern Dharwar Craton can be attributed to the computations based on entire data set for a constant bulk density contrast at Moho, which may not be uniform for the entire continent. It therefore, provides only a first hand picture of Moho that should be confirmed from better controlled data set of individual geological provinces. In case, the crustal thickness is estimated part-wise with appropriate density contrast between lower crust and upper mantle depending upon the tectonic regimes, a better match between Moho computed from the low pass filtered gravity anomaly and seismic Moho can be obtained as has been shown below in case of Deccan trap.

(ii) High Pass Filtered Gravity Map and Shallow Sources

The same digital data of Bouguer anomaly map of India, which was previously used for low pass filtered anomaly map was filtered using a high pass filter for wave number > 0.02. This filtered map which apparently represents the residual field is presented in Fig. 4.29(a). This map compared to previous regional map (Fig. 4.28(a)) shows several small wavelength anomalies, which are primarily due to shallow sources in the upper crust.

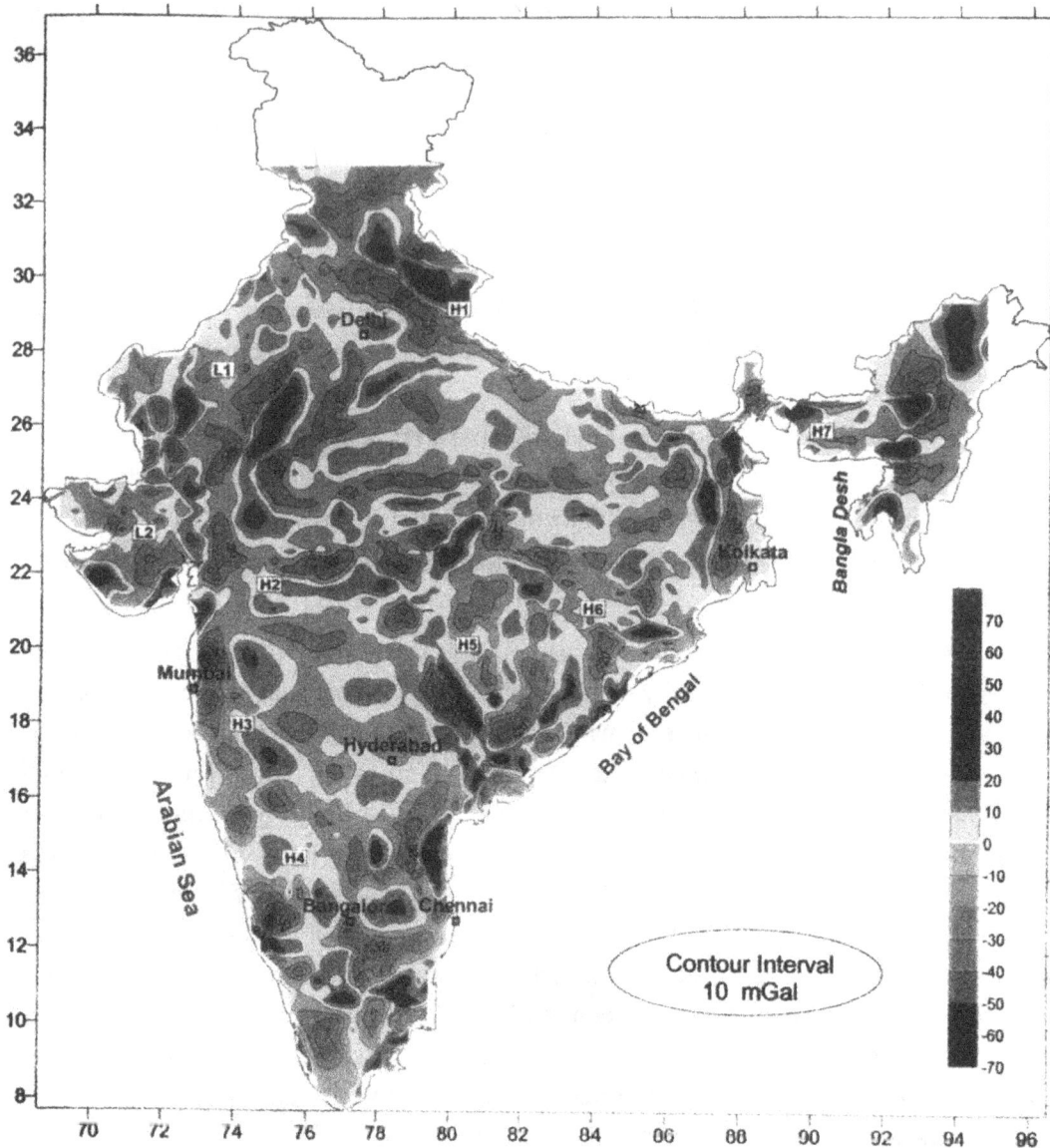

Fig. 4.29(a) High pass filtered residual Bouguer anomaly map of India (Fig. 2.28). It shows several gravity highs and lows due to shallow and exposed sources, which can be used to delineate these bodies (Mishra et al., 2004). [Colour Fig. on Page 768]

A discussion of individual anomalies is beyond the scope of the present description. However, some gravity anomalies, which are conspicuous in this residual map (Fig. 4.29(a)) are discussed below:

(a) The linear gravity highs (H1) north of Delhi were completely masked under the influence of the observed gravity low due to the Himalayas (Fig. 2.28), which is clearly delineated in the present map. It represents the high-density Proterozoic rocks and intrusive thrusted along the various thrusts in the Himalaya, which are extremely important for geodynamics point of view as discussed in Chapter 6.

(b) The gravity highs over Satpura Fold Belt (H2) can be explained by the reported high-density intrusives in this region as it represents a Proterozoic collision zone (Section 7.6). Similar gravity highs are also observed in the western part over the Aravalli-Delhi Fold Belt (ADFB, Fig. 4.28(b)) due to high density rocks related to Proterozoic collision zone (Section 7.5).

(c) The gravity highs (H3 and H4) lying west of Hyderabad and north of Bangalore could be delineated only after the large-wavelength deep-seated gravity low (L2; Fig. 4.28(b)) is filtered out. These gravity highs coincides with the shear zone between the Western and the Eastern Dharwar Craton and represents high-density intrusive along this shear zone (Section 7.8).

(d) The gravity highs, H5 and H6 are associated with Godavari and Mahanadi rift basins, which may be attributed to high density intrusives along their shoulders (Section 2.9).

(e) The gravity highs H7 in NE India is related to high-density surface and subsurface rocks of Shillong plateau (Section 7.6)

(g) The gravity lows (L1 and L2) west of Delhi and over parts of Rajasthan, Kutch and Saurashtra represent sedimentary basins, which are delineated after gravity highs in the original map (Fig. 2.28) due to intrusive in this region are filtered out.

(h) There are several other anomalies which readers can attribute to sources based on exposed rocks and tectonics of those regions.

(iii) Moho Relief under Deccan Trap, India

In case, the gravity field from Moho has been separated from the observed field using low pass filter, it can be used to obtain the relief of Moho using same scheme as described above in Section 4.3. As discussed in Section (4.2.4), the low pass filtered Bouguer anomaly map of Deccan trap truly represents the sources at Moho, it can be used to demonstrate the application of this scheme to obtain Moho configuration. Therefore, the low pass filtered Bouguer anomaly over Deccan trap (Fig. 4.15(b)) is used to compute relief of Moho at an average depth of 38 km for a density contrast of -350 kg/m^3 between lower crust and upper mantle (Fig. 4.29(b)). The standard density contrast between lower crust and upper mantle varies from -400 to -350 kg/m^3 and in this case a lower contrast is chosen as lower crust may show a slightly higher density compared to a normal lower crust due to infusion of high density rocks caused by eruption of Deccan trap in this region. The computed Moho configuration shows a deeper Moho up to 41 km under the Western Ghat due to its isostatic effect decreasing to 37-38 km in the eastern part, which is confirmed from deep seismic sounding profiles in this region (Kaila et al., 1981). This shows the application of harmonic inversion scheme to map Moho variations in a region from Bouguer anomaly if some constraints are available on its average depth. However, in such cases first the separation of field due to Moho should be confirmed through low pass filter and only large wavelength anomalies should be used to obtain the Moho configuration. It may be noted that in the present case wavelength larger than 250 km related to harmonic 1 and 2 are used to obtain the gravity field due to Moho.

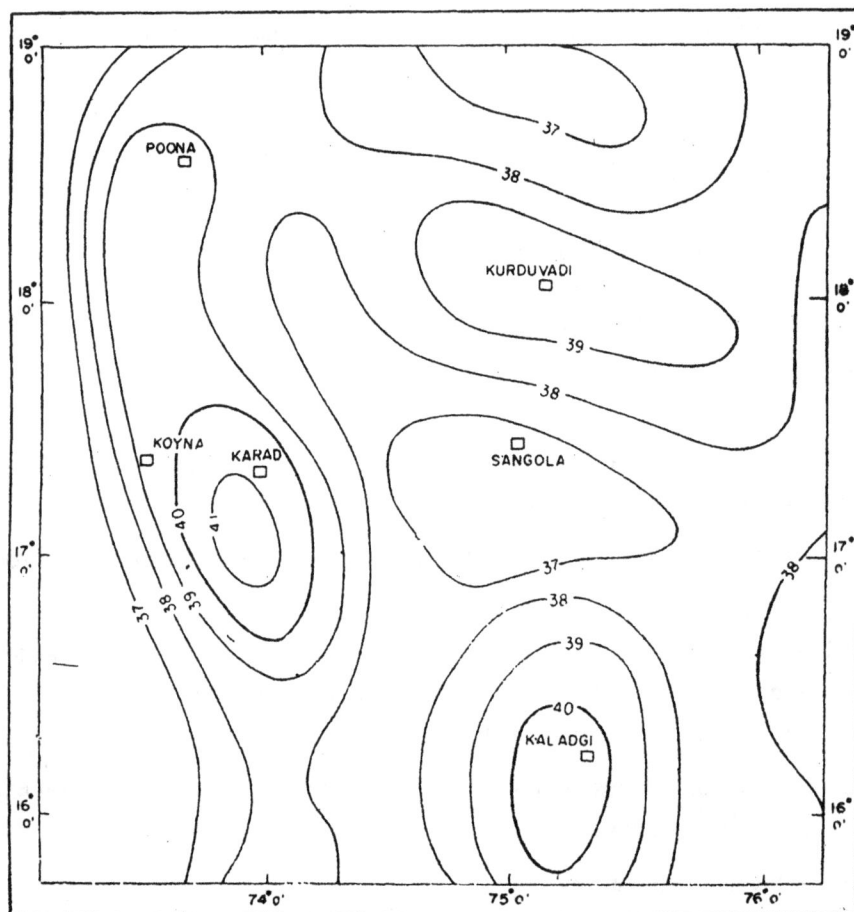

Fig. 4.29 (b) Moho relief computed from low pass filtered Bouguer anomaly of Deccan trap, India (Fig. 4.15(b)) for a density contrast of -0.35 g/cm^3 between lower crust and upper mantle. It shows crustal thickening up to 41 km under Western Ghat in response to isostasy (Mishra, 1989).

4.6.2 Magnetic Crust Based on MAGSAT data

Magnetic anomalies observed in satellite magnetic are primarily from deep seated sources and therefore, these anomalies can be modelled for the extent of magnetic crust in a region. The bottom of magnetic crust approximately represents the Curie point geotherm. In this regard the harmonic inversion scheme discussed in Section 4.3 can be used to map the variation in magnetic crust. However, for this purpose a constant magnetization is required to assume for the whole region, which may not be valid. However, it provides an approximate estimate of thickness of magnetic crust and depth to the Curie point geotherm. Using this scheme and digital data of total intensity map of India (Fig. 3.15), magnetic crust for this region is computed.

(i) Magnetic Crust under Indian Continent, Himalayas and Tibet

As the MAGSAT data is recorded at an altitude of 400 km, it is basically related to deep seated magnetic anomalies, which in most cases on a continental scale represent the base of the magnetic crust, namely Curie point geotherm. The observed magnetic anomalies also suggest large wavelength anomalies related to deep seated sources. Using the harmonic inversion scheme described above, the observed MAGSAT anomalies (Fig. 3.15) are transformed to the subsurface relief of magnetic crust at an average depth of 35 km for a magnetization of 200 nT. Hahn et al., (1984) have considered 1.5-2.0 A/m (200 nT) for the average magnetization of upper and lower crust based on reported susceptibility of different kind of rocks, which form the bulk of the crust. Therefore, a bulk susceptibility of 200 nT was considered for bulk magnetization of crustal rocks. The computed map of thickness of the magnetic crust is given in Fig. 4.30 (Mishra and Venkatrayudu, 1985), which shows the following important features. Thickness of magnetic crust implies the depth below, which magnetization does not exist (Section 3.2) or can not be recorded. This depth however cannot be considered to truely represent the Curie point geotherm as some magnetization must exist at this level to be recorded at satellite height. Therefore, it can be referred to as level where magnetization decreases considerably that cannot be recorded. In fact, Curie point geotherm may lie below this level.

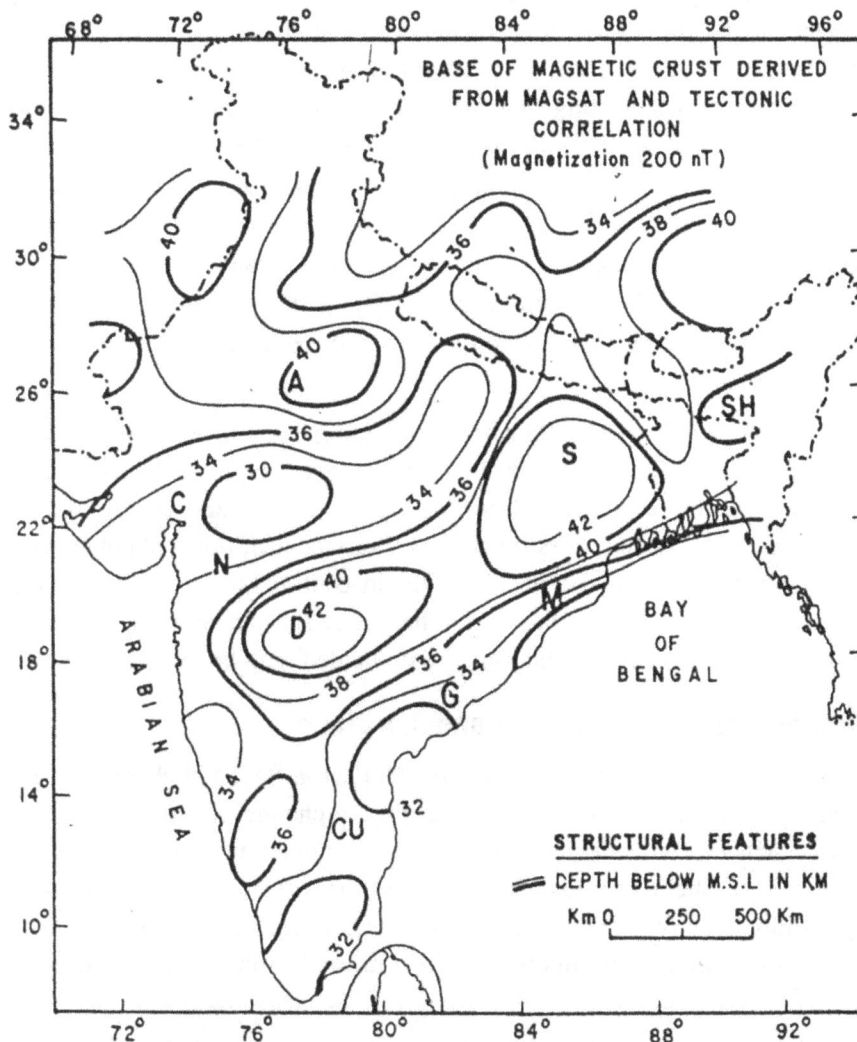

Fig. 4.30 Magnetic crust computed from MAGSAT map of India (Fig. 3.15) using harmonic inversion scheme. It signifies Curie point geotherm below, which magnetization does not exist. Thin magnetic crust is found under Himalayas as it is an orogenic belt where heat flow and subsurface temperature gradient is likely to be more. Thin magnetic crust along east and west coast of India may be attributed to rifting along these coasts (Mishra and Venktarayudu, 1989).

(a)   It shows a shallow magnetic crust under Himalalya and South Tibet of 34-36 km, which can be attributed to the large heat flow in this region as it represents an orogenic belt of recent origin.

(b)   Another section of thin magnetic crust is reported from South India, which consist of Eastern and Western Dharwar Craton and Southern Granulite Terrain (Fig. 2.26). Depth to the magnetic crust in these sections approximately coincides with the Moho indicating Moho being a major magnetic discontinuity similar to density discontinuity. These regions generally show low-moderate heat flow but there are certain sections in this region, which show high upper mantle high heat flow (Roy and Rao, 2000), which might have raised the Curie point geotherm in some parts. However, some mismatch due to assumption of uniform induced magnetization can not be ruled out.

(c)   Thick magnetic crust under Aravalli (A), Satpura Mobile Belt (S), Deccan Volcanic Province (D) and Shillong Plateau (SH) can be attributed to cold crust as these sections represent Archean-Proterozoic basement rocks with several large fractures and faults, which might have helped in dissipating internal heat through them. It is interesting to observe thick magnetic crust under the Deccan Volcanic Province (D) indicating that the initial heat has been dissipated that is also confirmed from moderate heat flow in this section (Roy and Rao, 1999).

(d)   Shallow Curie point geotherm along east and west coast of India (32-34 km) indicate higher temperature gradient in these sections, which can be attributed to rifting of continents along them that are usually related to thermal sources like plumes, volcanic intrusions etc.

(e)   The Cambay basin and Narmada-Tapti section (N and C) in Western India show a relatively shallow Curie point geotherm, which is in accordance with high heat flow reported from these regions (Roy and Rao, 2000).

Curie point geotherm estimated from steady state heat conduction equation and velocity of P waves in central India across Satpura Fold Belt and Godavari basin (Ramana et al., 2003; Rai et al., 2003, 2006) is in good agreement with those provided from MAGSAT data (S and G, Fig. 4.30). It also coincides quite well with Curie point geotherm estimated from surface heat flow data obtained from P_n seismic velocity, which also provides a depth of 39-40 km in central India (Sharma et al., 2005). Maule et al., (2005) have used the MAGSAT data to estimate heat flux underneath the Antarctic ice sheet and areas of high heat flux obtained from this analysis coincided with known volcanic centers.

4.6.3   Crustal Structure: Faeroe- Shetland Channel, Iceland

A few marine magnetic profiles in the northwestern part of the Faeroe-Shetland channel extending up to the Faeroe outer shelf, and a regional gravity profile across the channel are analysed to verify the principles discussed above (Mishra and Pedersen, 1982). Due to its position at the junction of several major tectonic units, the Faeroe-Shetland trough has attracted the attention of several investigators (Bott, 1975; Talwani and Eldholm, 1977) from time to time. The determination of average depth based on spectral decay is demonstrated in this case in order to show its application to marine data for crustal studies.

The Faeroe shelf is characterized by Tertiary basalt lavas, which disappear below the sediments on the outer shelf. The channel is about 1-1.5 km deep, underlain by 1.5 km thick sediments, in general and at places characterized by several large sedimentary basins as deep as 5 km (Korsakov, 1974; Bott, 1975). Bott (1975), on the basis of gravity studies, suggested a thinner crust of the order of 19.3 km for this region. Magnetic profiles described by Nielsen et al., (1979) are characterized by sharp fluctuations towards Faroe outer shelf superimposed over a broad anomaly towards Faeroe Shetland trough. A typical recorded profile along with its computed amplitude spectrum is presented in Fig. 4.31(a) and (b). The recorded digital data at 50 m interval (2048 data points) was transformed using Fast Fourier Transform. As the computed spectrum was oscillatory in character, running averages of five frequencies, two on either side, are obtained, which provided a stable spectrum. The other 3 profiles analysed in the present study also provided similar spectra. The amplitude spectrum in Fig. 4.31(b) can be approximated by two linear segments, the low-frequency part providing a depth of 5 km, and the higher-frequency part 0.5 km. As is seen from the magnetic profile presented in Fig. 4.31(a), this profile is characterized by a broad anomaly towards the channel side and short-wave length fluctuations as the Faeroe shelf is approached. Therefore, the low-frequency segment provides the depth of the magnetic basement in the channel (trough), whereas the high-frequency segment provides the depth of the basaltic layer towards the Faeroe outer shelf. This, being the average depth is in conformity with depths in the two regions as provided by seismic studies as described above.

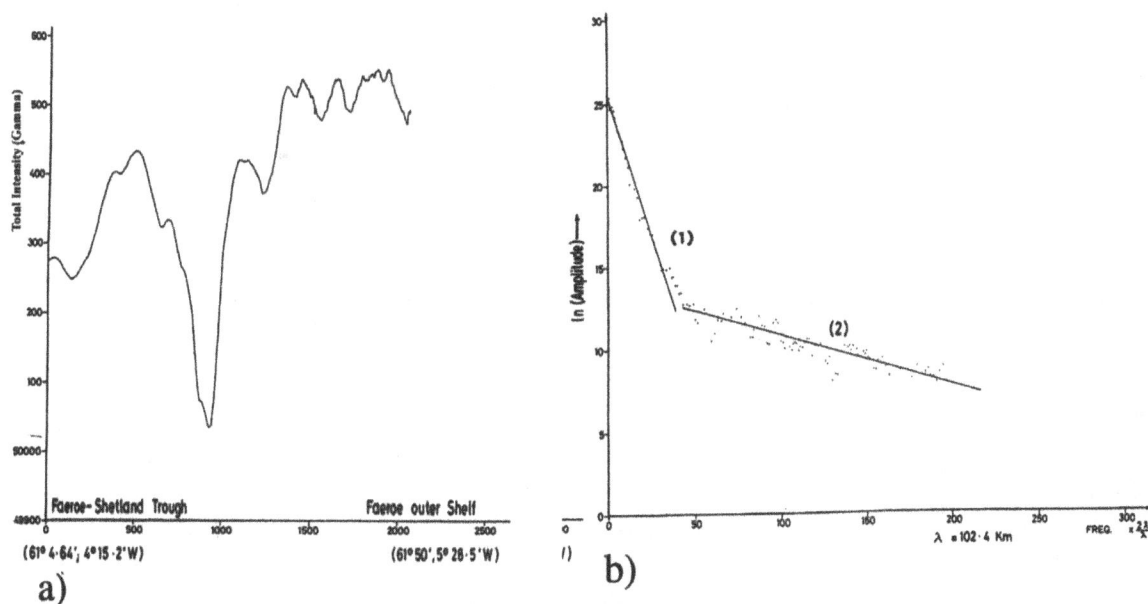

Fig. 4.31 (a) A marine magnetic profile from the Faeroe outer shelf to the Faeroe-Shetland trough sharp magnetic anomalies at Faeroe shelf are due to mafic intrusives, while large wavelength features in trough are due to mafic basement. (b) Amplitude spectrum versus frequencies for the magnetic profile showing two linear segments corresponding to slope (depth to sources) equal to 5 and 0.5 km related to the basement in the trough and basaltic rocks along the shelf, respectively (Mishra and Pederson, 1982).

A gravity profile across the northwestern part of the channel along with its amplitude spectrum is given in Fig. 4.32(a) and (b). The amplitude spectrum in general provides two linear segments and the corresponding slopes are 20 km and 6 km. These figures coincide well with the depth of the Moho and the basement in this region referred to above. The shallow part of magnetic spectrum due to basalt on the shelf does not find reflection in gravity spectrum as it may not produce enough signals in gravity survey while in magnetic survey it will be prominently reflected due to large susceptibility variations. Similarly, the signals from Moho do not find reflection in the spectrum of the magnetic profile as due to large depth and Curie point geotherm being shallower; it may not produce any significant magnetic field.

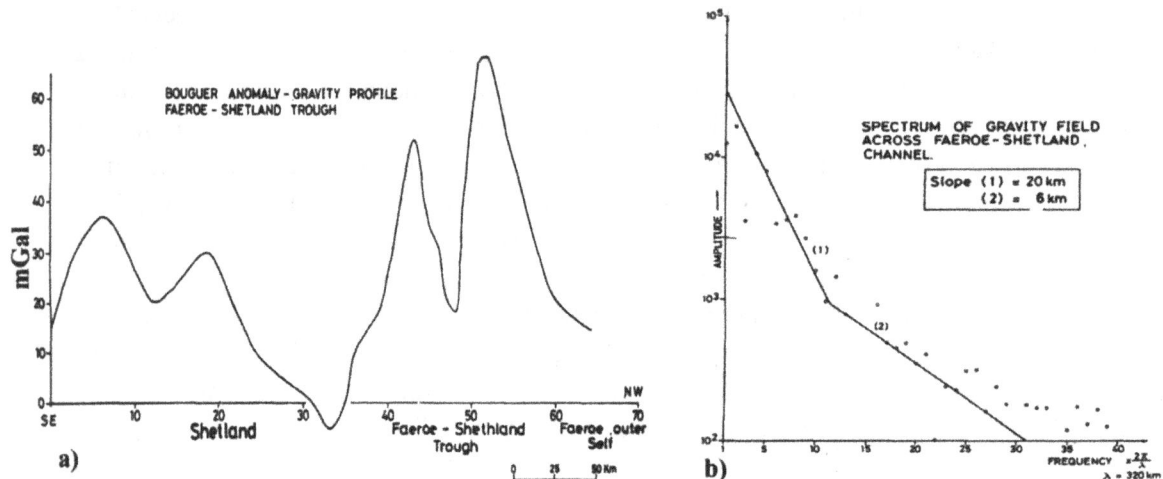

Fig. 4.32 **(a)** Bouguer anomaly profile across Faeroe Shetland channel **(b)** Amplitude spectrum versus frequencies for the gravity profile showing two linear segments with slope (depth to the sources) of 20 and 6 km related to Moho and the basement in this region(Mishra and Pederson, 1982).

4.6.4   Effective Elastic Thickness – Himalaya and Tibet

As discussed in Section 4.4, effective elastic thickness (Te) is an important parameter for geodynamics of a region, which primarily suggest the thickness of the elastic part of the crust that decides about its strength. Larger Te indicates stronger crust.

(i) Himalaya and Tibet

There have been several efforts towards computation of EET for Himalaya and Tibet as given in Table 4.1. Jin et al., (1994) based on 1-D coherence provided a value of Te = 40-50 for this region. Subsequently Jin et al., (1996) provided a value of 90 km for foreland Ganga basin and 30-35 km for Tibton plateau increasing northwards 40 to 45 km for Tarim basin (Fig. 2.26(b)). Cattin et al., (2001) however, provided a lower value of Te = 40-50 km for Ganga basin. Tiwari et al., (2006) provided a value of 50±10 km for combined Ganga basin and Himalayas and Tibet, which may represent an average value for the entire region. Rajesh, Stephen and Mishra (2003) analysed the Bouguer anomaly and the Topography of Himalaya and Tibet (Figs. 4.33a) in three distinct windows, viz. Ganga basin and Himalayas, Himalaya and South Tibet and North Tibet and Tarim basin with some overlapped portions. The topography data is Gtopo 30 while gravity data were obtained from different sources (NGRI, 1975; Jin et al., 1974; Cettin et al., 2001 and Sun, 1989). The computed coherence using multitaper scheme by Simon et al., (2000) and predicted coherence for different Te are given in Fig. 4.33 (b) for three windows (Fig. 4.33(a)), W_1, W_2 and W_3. It shows a lowest value of 20 km for Himalaya and Tibet, which increases to 35 km for Ganga basin southwards and 30 km for northern block including Tarim Basin. The lowest value over Himalayas and Tibet can be attributed to collision tectonics and related thrusts and intrusive, which have rendered a weak crust in this section. The Te in recent orogenic belts is always less due to orogenic activities and the same is true about Himalaya and Tibet. The effective elastic thickness of Himalaya and Asian plate along western syntaxial and eastern syntaxial bends towards west and east of the central region (Fig. 6.5(a)), respectively suggest about 23 km (Rajesh and Mishra, 2003), which is almost same as that for the central part indicating almost similar strength for the entire collision zone between the Indian and the Eurasian plates.

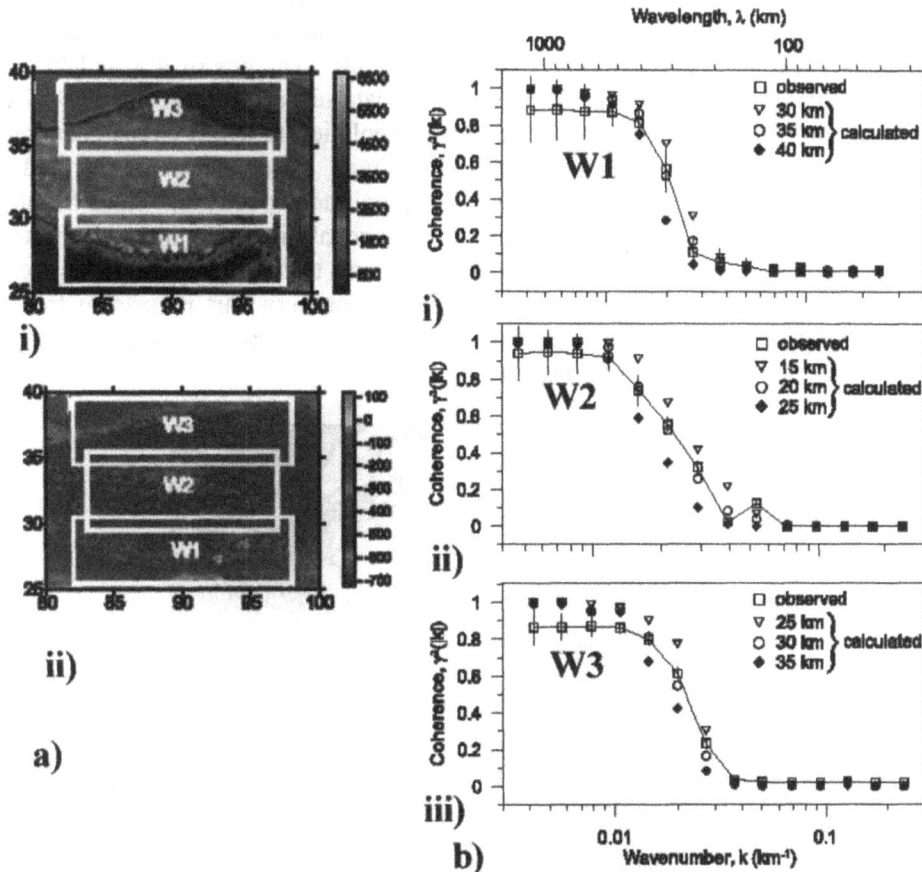

Fig. 4.33 (a) Topography (i) and Bouguer anomaly (ii) images of Himalayas and Tibet. W1, W2 and W3 are the data windows primarily representing Indian plate (Ganga basin and Himalayas), Tibet and Tarim basin, respectively. (b) Effective elastic thickness based on multi taper coherence estimates for W1, W2 and W3 data sets given in (i), (ii) and (iii), respectively (Rajesh et al., 2003).

4.7   Field examples: Basement Model and Geological Mapping

4.7.1   3-D Basement Relief: Airborne Magnetic Map of North Germany

The airborne total intensity map of North Germany is given in Fig. 4.34 (Hahn et al., 1976). The area was flown at a height of about 700 m a m.s.l. with profile spacing of 2.2 km. This map largely shows magnetic lows in the north related to basement structures, which indicate thick column of sediments in this section. The digital data used for analysis was also grided at 2.2 km, which provided a data matrix of 147 × 109. A data set of 125 × 109 is chosen from this data and a square matrix of 125 × 125 is made by adding 19 columns by cubic interpolation between last and first value of each row. It is further enlarged into 128 × 128 by cubic interpolation in order to transform it using Fast Descrete Fourier Transform scheme (Naidu, 1970). The computed averaged radial spectrum is given in Fig. 4.35, which basically shows two linear segments with slopes (depth) equal to 15.1 and 4.2 km. The depths to magnetic sosurces below m.s.l. are, therefore 15.1-0.7 = 14.6 km and 4.2-0.7 = 3.5 km. The deeper segment at about 14.6 km apparently represents the basement while the second segment at a depth of about 3.5 km represents intrusive with in the sediment, which are observed in total intensity map in southern part of this map west of Hannover. In order to check the reliability of the computed spectrum 96 x 96 data set from the northern and the southern parts with some overlap in the central part are chosen and transformed in a conventional manner by taking averages of 4 x 4 data points making matrix of 24 × 24 data points at interval of 8.8 km. These data sets are enlarged to 27 × 27 data points by adding 3 points to each row and column by cubic interpolation to avoid Gibbs phenomena. These data sets for northern and southern blocks are transformed by conventional method and computed spectrums are given in Fig. 4.36. These plots of ln (amplitude spectrum) versus frequency also provide slopes equal to about 15 km, which confirms the reliability of computed spectrums (Fig. 4.35 and 4.36) and depths estimated from them. It is interesting to note that second segment at depth of about 4.2 km shown in Fig. 4.35 due to intrusive in the southern part is missing from the spectrum of these blocks (Fig. 4.36) when computed by averaging of 4 x 4 points. This might be due to increase in station spacing of 8.8 km due to averaging.

Fig. 4.34 Airborne total intensity magnetic anomalies over North Germany recorded at 700 m above surface with profile spacing of 2.2 km. It shows a linear N-S magnetic high in the centre of the map with magnetic lows located on either side of it (Hahn et al., 1976).

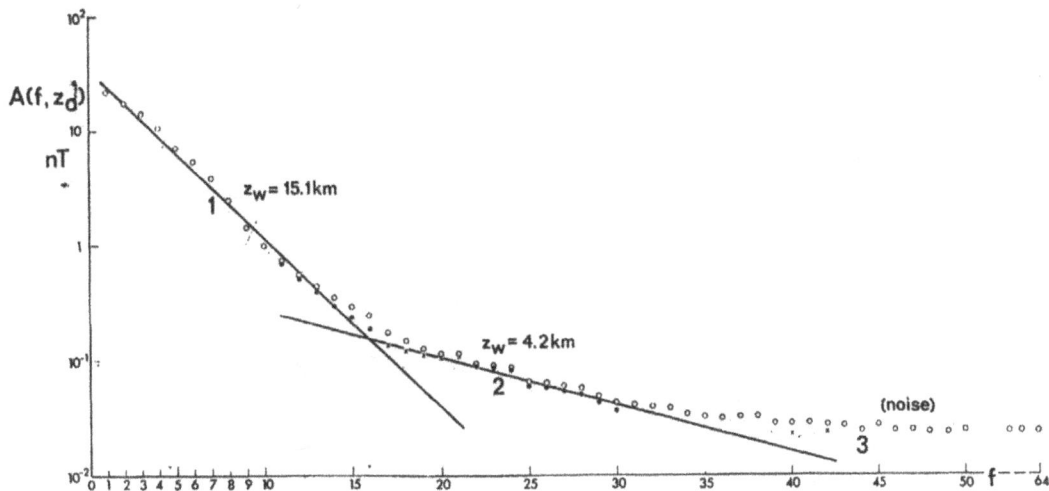

Fig. 4.35 Amplitude spectrum versus frequency for the airborne magnetic data of North Germany. It shows two data linear segments with slopes of 15.1 and 4.2 km providing depth below msl of 14.4 and 3.5 km. They represent average depth to the basement and intrusives with in the sediments, respectively.

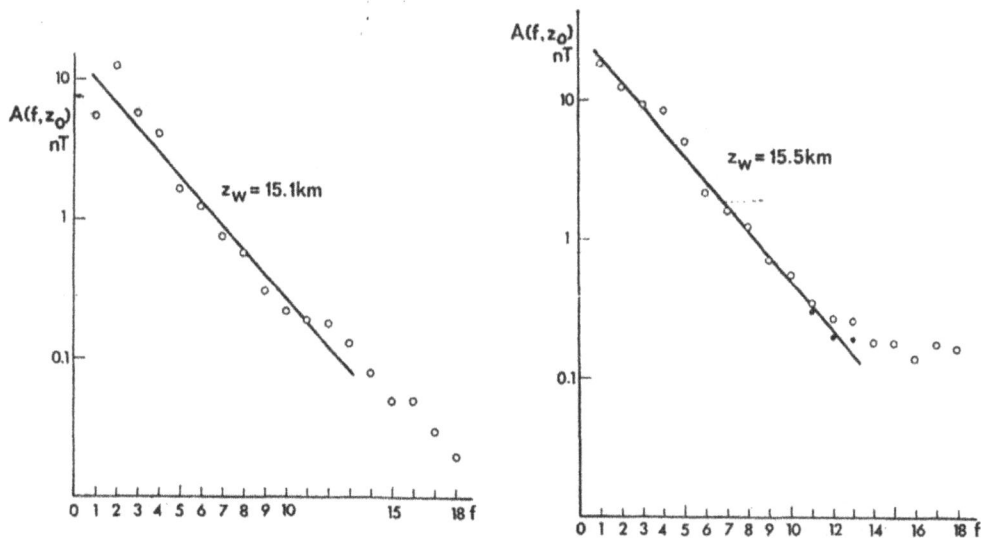

Fig. 4.36 Amplitude spectrum versus frequency of two blocks of data towards the north and the south drawn from the larger data set and averaged into 4 × 4 data points. They provide linear segments with almost same slope as first segment in Fig. 4.35 suggesting a stable spectrum. The second segment of Fig. 4.35 is missing from these plots due to large data spacing caused by averaging.

The digital data of this map is transformed to basement relief (Fig. 4.37) using Equation (4.40) for an average depth of relief Z_o as 13 km and magnetization J equal to 2 Am^{-1} (200 nT). Average depth of relief is taken as 13 km as spectral decay provide a higher estimate due to shape factor and represent maximum bound on the depth as described above. The computed relief shows the deeper parts of basin on either side of a N-S ridge west of Hannover and passing through Hamburg and Kiel. The maximum thickness of sediment is about 15-16 km, around Bremerhaven and east of Kiel. Further, higher values of basement depth east of Kiel appears to be caused by edge effect as it lies on the eastern margin of the data sets and cannot be relied upon. It may be noted here that the computed ridge in the map extending from the west of Hannover to Kiel corresponds to the observed magnetic high in this section (Fig. 4.34) and magnetic lows in the total intensity map corresponds to basement depressions in the map. This is because this region lies in the high geomagnetic latitude where the observed field for induced magnetization due to a basement high (ridge) will approximately correspond to a magnetic high. The computed relief (Fig. 4.37) was subsequently confirmed from large scale magnetotelluric survey in this region (Losecke et al., 1979) which also provided an average depth of basement as 13 km and maximum depth as 15-16 km.

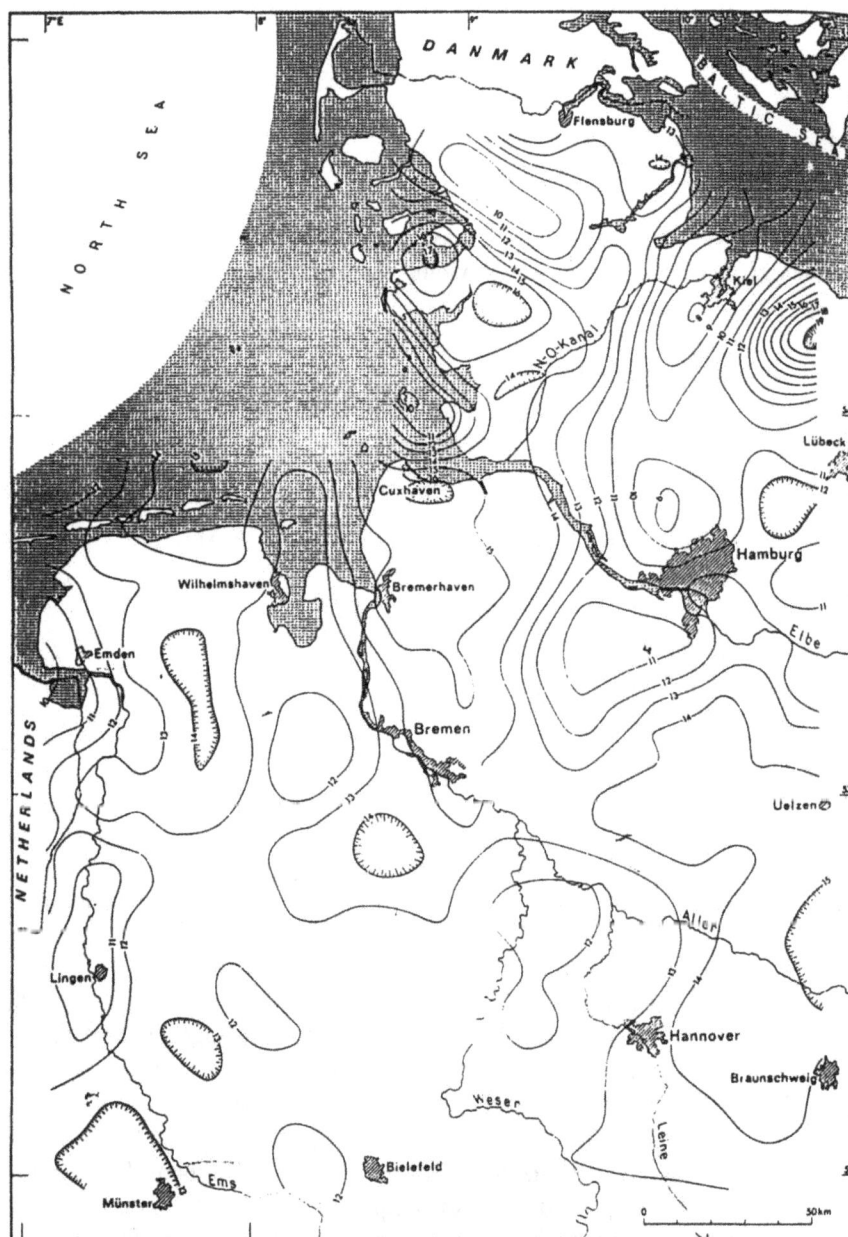

Fig. 4.37 Basement relief computed from total intensity magnetic anomalies of North Germany showing variation in depth from about 10 km to 16 km. The N-S magnetic high in the centre of the map is reflected as a basement ridge with depressions on either sides.

4.7.2   Geological Mapping: Apparent Density Map of Peninsular Shield, India

The digital data of Bouguer anomaly of Peninsular shield of India (Fig. 2.28) is used to separate regional and residual fields based on zero free air anomaly values as discussed in Section 4.1. As density or magnetization variations are determined for surface rocks, the residual Bouguer anomaly is transformed to density variation for an average bulk density of 2.75 g/cm^3 (2750 kg/m^3) (Singh et al., 2003) using the formulation given in Equation (4.33). The computed map (Fig. 4.38) shows a low density of 1.9-2.0 g/cm^3 (G) in the NE corner related to low density sediments of Godavari basin to a highest value of 2.78-2.80 g/cm^3 west of Bangalore and along east coast of India in the southern part, which reflect high density of schist belts and basement rocks in these sections, respectively. As 2.75 g/cm^3 is used to compute the density variation, densities higher than this are referred to as high density sections while those lower than this density are referred to as low density sections. As expected, it is just a replica of residual Bouguer anomaly map showing low density for low Bouguer anomalies and high density for high Bouguer anomalies. This map, however can be used more efficiently to delineate different rocks types depending on their density compared to Bouguer anomaly map. For example high density rocks in the basement along east coast of India (E) may represent the high density intrusives in the basement similar to those found in the Eastern Ghat north wards such as anorthosite, carbonatite etc., However, knowledge of exposed rock types in the region helps considerably in delineating their extensions and rock types. Further the high density section coinciding with Bhavani and Mettur shear zones (BMSZ) joins with the Eastern Ghat near Chennai (Madras), whose northern margin appear to represent major crustal boundary, which does not find prominent reflection in the original Bouguer anomaly map (Fig. 2.28). The high density body surrounding Bangalore appears to be some special feature and may represent shallow lower crustal section related to the collision of Western and Eastern Dhawar Craton and Shouthern Granulite Terrane towards the south that is discussed with details in Sections 7.8 and 7.9. This map is further discussed in Section 7.9.4 for Archean-Proterozoic collision tectonics in this section.

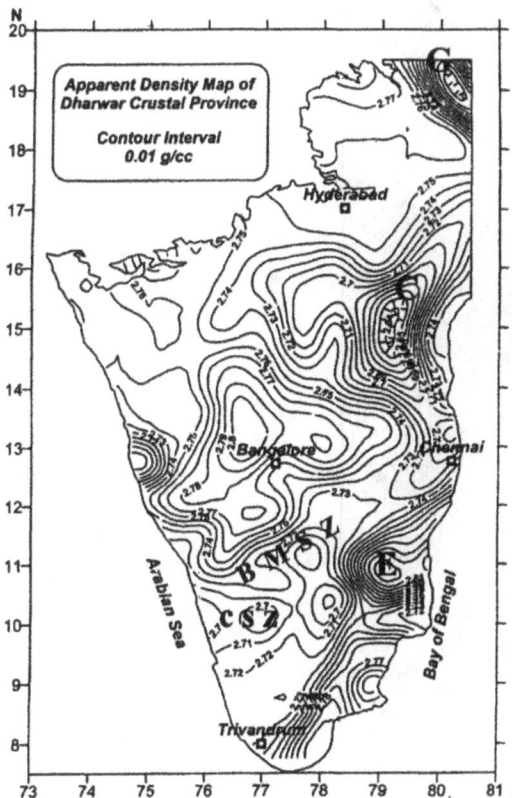

Fig. 4.38 Apparent density map based on the harmonic inversion of residual Bouguer anomaly of Peninsular shield of India with reference to bulk density of 2.75 g/cm^3 showing zones of high and low densities. G = Godavari Basin, C = Cuddapah Basin, E = Southward Extension of Eastern Ghat, CSZ = Cauvery Shear Zone and BMS = Bhavani-Mettur Shear Zone (Singh et al., 2003).

4.7.3 Gravity and Magnetic Investigations in German Continental Deep Borehole (KTB) - Variscan Orogeny

Deep boreholes provide an opportunity to test the results obtained from the measurements of gravity and magnetic fields at the surface and compare them with those obtained from borehole measurements. German Deep Continental Drilling Programme provided one such opportunity where a pilot bore hole of 4000 m deep and main bore hole of ~10000 m was made in southern part of Germany (Oberpfalz, Eastern Bavaria). It was carried out after detailed geological and geophysical investigations at the surface and was logged for various geophysical parameters. This section represented a paleosuture of Variscan fold belt of Devonian/Carboniferous period.

The Bouguer anomaly and helicopter borne magnetic map observed at the surface is given in Fig. 4.39(a) and (b) (Bosun et al., 1993a) which shows an elongated (NW-SE) gravity high of about 20 mGal and magnetic anomaly 200-300 nT of similar trend that are usually observed over mafic intrusives. As indicated in these figures, KTB is located at the gradient of those anomalies indicating a contact that separates rocks of different densities and magnetization with high density and susceptibility rocks towards the south-west of it. The adjoining gravity high and low caused by mafic and felsic intrusive, respectively in old terrane represent typical paired anomalies due to paleosutures as discussed in Chapter 7. In the bore hole, the rock metabasites (metamorphosed basalts) of high density 2.8 g/cm³ extends almost from shallow levels up to its bottom intersected by faults. However, due to presence of several faults, its density is altered considerably as shown by density log given below. Metabasite is located towards the SW of the bore hole while the NE section is occupied by granite and gneisses as seen in the bore hole section below (Fig. 4.41(a)).

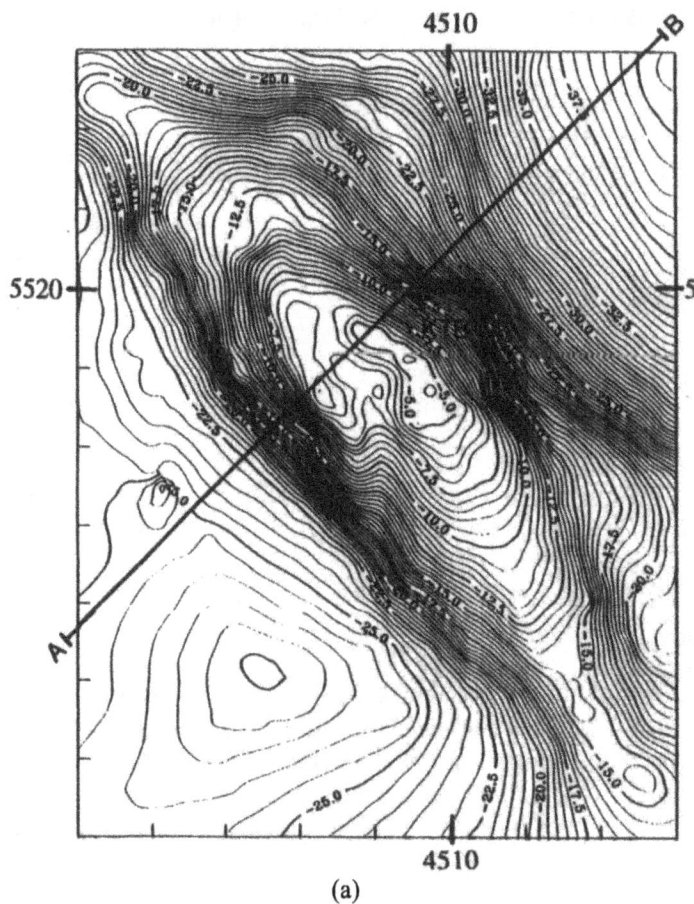

(a)

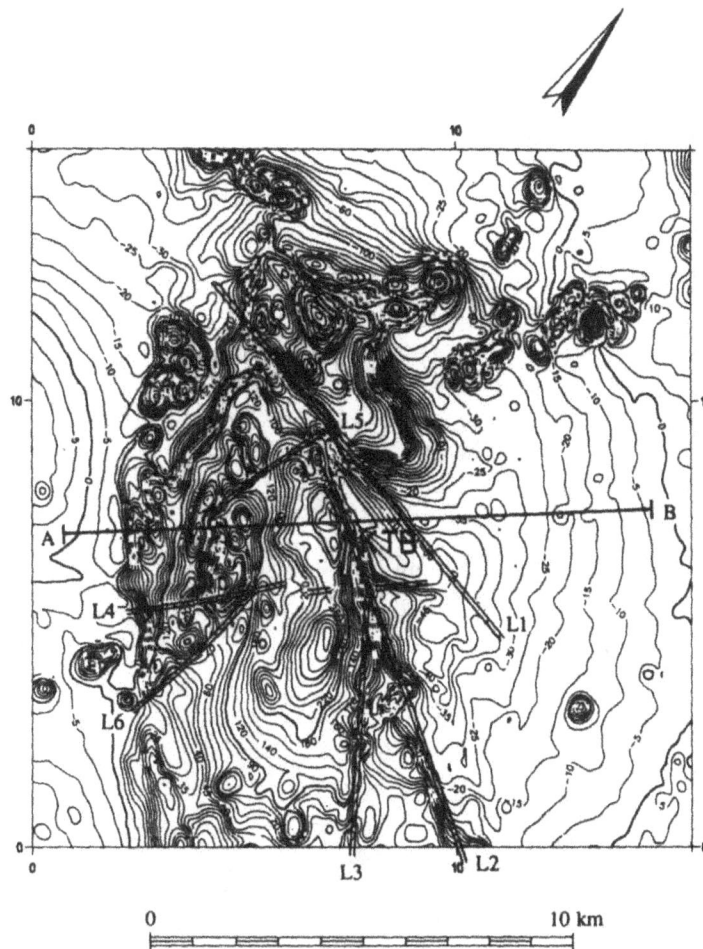

(b)

Fig. 4.39 (a) Bouguer anomaly map of KTB showing a linear high (NW-SE) of about 20 mGal. KTB is located at the gradient of this anomaly indicating a contact of high density body towards the SW. (b) Helicopter borne total intensity magnetic map recorded at a height of 70 m above ground and flight line spacing of 200 m with KTB located in between high amplitude magnetic anomalies towards the south-west and low amplitude towards the north-east (Bosum et al., 1993a).

The amplitude spectrum of gravity and helicopter borne magnetic fields recorded at the surface are given in Fig. 4.40(a) and b (Bosum et al., 1993b) which shows average depths of causative sources as 3.6 and 0.9 and 1.8, 0.7 and 0.25 km, respectively. The first segment of spectrum is affected by the size of the data set being used for transformation (largest wave length) and therefore, difference in the depth estimates for the first segment from two methods is quite logical. However, the spectrum of the regional airborne magnetic data of Germany (Fig. 4.35) does provide a segment at a depth of 3.5 km that may be synonymous to that obtained from the spectrum of the gravity field in the present case. Fig. 4.35 also shows a deeper segment corresponding to 14.4 km that was interpreted as basement in North German sedimentary basin. Other segments provide depths to shallow sources. The first segment (Fig 4.40(a)) of spectrum (3.5-3.6 km) can be attributed to a metabasalt unit drilled at a depths of 3500 and 3650 m with high magnetization of 15-16 x 10^{-2} A/m. As this section is traversed by a fault, magnetic anomalies are attributed to alteration of magnetic minerals by fluids to pyrrhotite due to chemical processes.

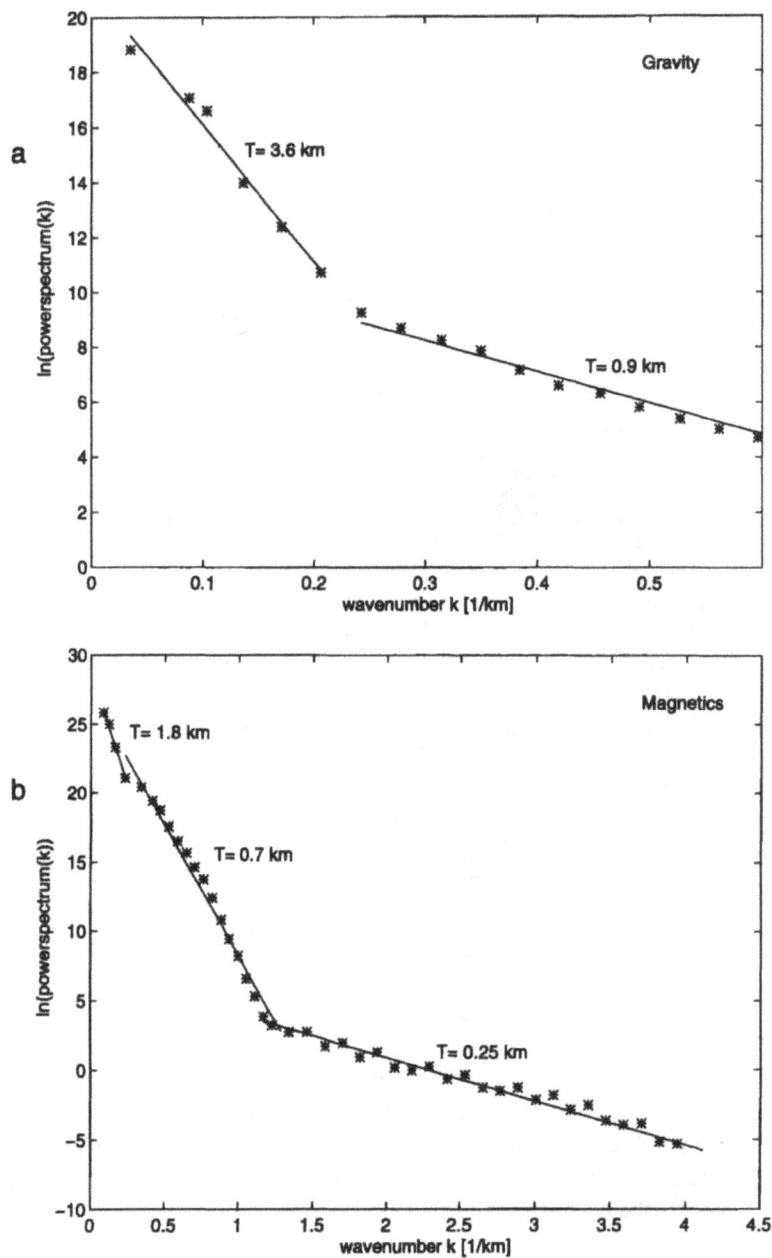

Fig. 4.40 (a) Spectrum of gravity field showing two linear segments with slopes equal to 3.6 and 0.9 km for respective sources. Block size being 20 x 20 km, deeper sources do not find reflection in computed spectrum. (b) Spectrum of magnetic field with slopes showing depths of 1.8, 0.7 and 0.25 km. (Bosum et al., 1993b).

Bosum et al., (1993a, b, 1997) have modeled the gravity and the magnetic field observed at the surface using multiple bodies inferred from the borehole logs. Fig. 4.41(a) and (b) gives the borehole section and measured gravity field with depth that shows anomalies of ±1.5 mGal. It also shows calculated gravity and measured and calculated densities from the observed field, respectively that matches quite well with those observed. It shows the gravity low (L) and high (H) primarily associated with the upper and the lower parts of metabasalt separated by a fault, the former might be more weathered compared to latter and so is the change in density. There will also be influence of low density granite on the gravity low observed in the top layer. High density zone is at depths of about

2.5-4.5 km that may correspond to the first segment of the spectrum of the gravity field (Fig. 4.40(a)) with average depth of 3.6 in the spectrum while the second segment of spectrum provide the same depth (~0.9 km) as that of the gravity low, L. There are considerable variations in density as it is metamorphosed due to fluids percolating through several faults. They have encountered metabasalt up to bottom of the main bore hole (~ 9000 m) and extended it in their computed model almost up to 13-14 km that is reflected as first segment in the spectrum of airborne magnetic data of Germany (Fig. 4.35) as described above. These deep seated sources did not reflect in the present spectrum (Fig. 4.40(a)) as it is related to a small area of about 20 x 20 km that normally cannot reflect sources below 3-4 km. As this bore hole is located over a fault separating metabasalt on one side and gneisses and granite on the other side and is close to a paleo-suture of the Variscan fold belt related to Variscan orogeny, the metabasalt in the upper crust may represent the then oceanic crust that is thrusted along the suture and metamorphosed subsequently and granite may represent subduction related magmatism of that time. It provides good constraints to model gravity and magnetic fields from paleo-sutures and collision zones from other parts of the world.

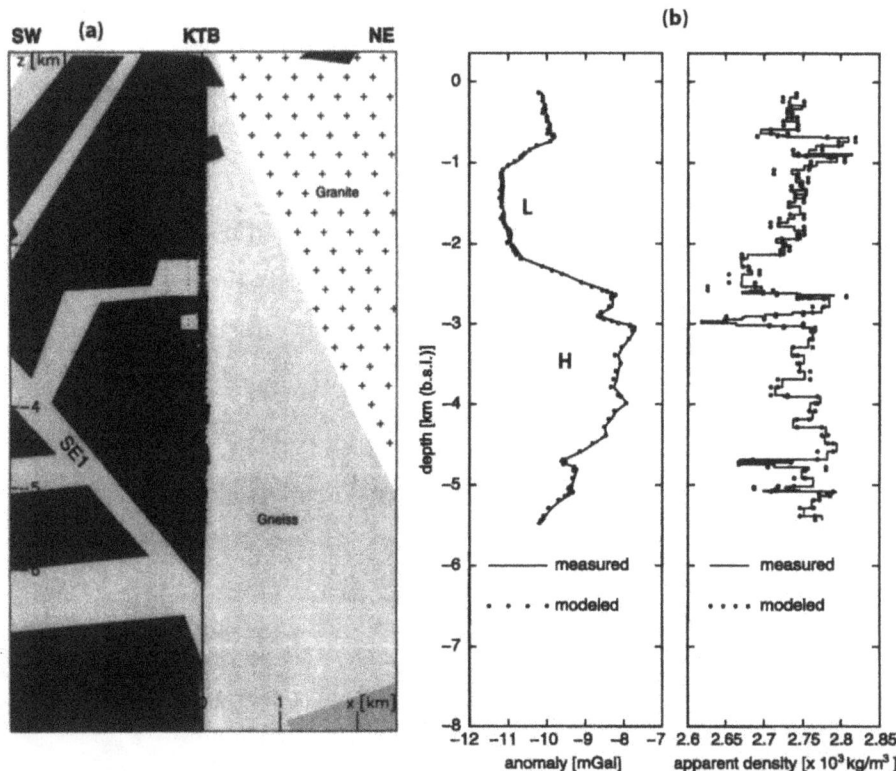

Fig. 4.41 (a) Borehole section up to 8.0 km showing metabasites (metamorphosed basalt) towards the SW and granite and gneisses towards the NE with borehole located on the contact that may represent a paleosuture of Variscan orogeny. (b) Observed and computed gravity anomalies in the borehole for densities: granite- 2600 kg/m3; metabasites- 2900 kg/m3 and gneisses- 2730 kg/m3. Apparent density computed from the observed field based on Harmonic inversion as described above and measured densities in the bore hole are also shown that matches quite well.

Part - II

Integrated Exploration of Indian Plate and Resources

(Geodynamics, Seismotectonics, Hydrocarbons, Minerals, Groundwater, Environment and Engineering Sites)

Continental Drift and Plate Tectonics : Reconstructions, Gondwanaland Break-Up, Plumes and Drifting of Indian Plate

5.1   Continental Drift

Since Wegener proposed his ideas in 1912 that continents are not fixed and in fact have been continuously on move, scientists world over have been trying to explain various observables in its framework. In its simplest form, he stated that all the continents were joined together to form a super continent at about 250 Ma ago named as 'Pangea'. This super continent moved first from the north to the south before disintegrating into different continents. Motivation for such a hypotheses came from (a) the visual fit between coastline of some of the continents such as S. America and Africa (b) similar fossil records in continents presently widely separated and (c) fossil and fauna of tropics being found in rocks presently occurring in high latitudes and vice versa. Based on these considerations, Wegener (1929) provided reconstruction of continents for different geological periods. This hypothesis was successful to some extent as it provided a common explanation for the formation of oceans and orogenic belts due to movement and collision of continents during drifting. However, it did not account for forces responsible for movement of the continents. In this regard, the first idea about the forces responsible for movement of continents came from the works of Holmes (1944) who suggested mantle convection as a possible mechanism. Accordingly, continents were pulled apart by ascending currents giving rise to growing rifts, which eventually formed oceans while in areas of descending currents mountain ranges, were formed due to collision of continents. Subsequently Maurice Ewing, Harry Hess and others, through their exploration in oceans throughout the world, presented the maps of ocean floor (bathymetry) which suggested that most of the deepest parts of the oceans known as trenches do not occur in the middle of the oceans as perceived earlier due to sagging instead they occur along margins of the continents, which were characterized by mountains along the coast and there are large ridges in the middle of the ocean like mountains on the surface of the earth.

Based on these findings, Hess (1962) first postulated sea floor spreading hypothesis which suggested that molten rocks (magma) wells up along mid-oceanic ridges and spreads to form ocean floor and finally sinks along trenches. One may add here that the deepest ocean is found in Mariana trench (11,034 m) along coast of Philippines, which is more than the highest peak on the earth (Mt. Everest, 8,848 m above m.s.l). In the Indian Ocean, the deepest section is along Sumatra-Java trench. Vening Meinesz had discovered the linear gravity lows almost parallel to islands of Sumatra-Java during expeditions in 1930's along this trench indicating deepest section of the ocean which were the first marine surveys in the world. These findings were in true sense precursor to plate tectonics which was subsequently formulated during 1960's, that can be termed as modified version of continental drift. The most important evidence for continental drift came from paleomagnetic studies of rocks of different periods and different parts of the world. As described in section 3.2, paleo position of continents can be computed from remanent direction of magnetization in rocks of different periods, if it represents the magnetization acquired during their formation known as remanent magnetization.

This, however, is based on assumption that the magnetic poles were fixed during geological past. In case, the magnetic poles have also moved during this period, one can compute polar wandering path describing the positions of magnetic poles during geological period as described in Section 3.2. Both, viz. polar wandering and continental drift are related to each other and can be derived one from the other and vice-versa. Paleomagnetic studies are therefore extremely useful for geodynamic studies.

5.1.1 Drift History and Apparent Polar Wander Path (APWP) of India and Adjoining Continents

In India, the paleomagnetic studies were initiated in 1950s by Tata Institute of Fundamental Research, Bombay, in collaboration with Imperial College of Science, London. Initially, igneous rocks such as Deccan Traps, Rajamahal trap etc., were studied (Athavale et al., 1963) followed by sedimentary rocks (Mishra, 1965; Verma and Bhalla, 1968; Bhalla et al., 1979). Subsequently, based on these directions of magnetization, drift history of Indian continent during geological past was provided by several workers, such as Blackett et al., (1960), Mishra (1965), Radhakrishnamurthy et al., (1967). Blackett (1961) compared the paleopositions of some continents belongings to Gondwanaland like India, Australia, Africa and South America derived from paleoclimatic studies and paleomagnetic studies and found a good correlation between them supporting each other. Subsequently, several igneous sedimentary and metamorphic rocks from different parts of the continent were used for paleomagnetic studies, which provided polar wandering path and drift history of the continent. Klootwijk (1979) has provided a detailed review of paleomagnetic studies of rocks of greater India including that of Himalayas and Pakistan. Most of the workers used paleomagnetic measurements to derive the paleo position of poles instead of continents due to uncertainty in the rotation of the continents as it provides only the paleolatitudes. However some workers have used these data sets to obtain drift history of the continent (Athavale et al., 1963; Mishra, 1965; Radhakrishnamurthy et al., 1967; Bhalla et al., 1979). Mishra (1965) and Radhakrishnamurthy et al., (1967) based on stable directions of magnetization for rocks of different period given in Table 5.1, provided drift history of Indian continent since Proterozoic period (Fig. 5.1) which suggest that Indian continent has been moving from northern to southern hemisphere and back to northern hemisphere in the geological past indicating the mobility of the continents.

Table 5.1 Paleomagnetism of Some selected rocks formations and Paleolatitudes of Indian Continent

Formation	Age X10⁶yrs	Mean Site Location		No of Samples	Mean Direction		Polarity	Precision*	Ancient Lat of Nagpur (degrees)
		Lat ⁰N	Long ⁰E		A⁰Eof N	D +iv e Downwards (degrees)			
1	2	3	4	5	6	7	8	9	10 Ref.
1. Deccan Traps and Dykes	64-68	19	73	150	335	-50	N	10	29S
				300	150	55	R	5	34S*
2.Rajamahal and Sylhet Traps	118-120	25	88	170	323	-64	N	7	50 S 1*
		25	91	36	332	-59	N		44S
3. Kamti Sand Stone (Gondwana)	250-300			52	306	-50	N	5	28S 2*
4. Upper Vindhyan Quartzites	600	24	83	42	0	30	N	6	13N 3*
				18	186	-32	R	7	14N
5. Bijawar Traps	800-1200	26	78	7	70	3	N	18	3 S 4*
B.H.Q		24	85	5	270	-12	R	-	9 S
B.H.J		24	81	5	289	-9	R	-	8 S

* 1. Athavale et al., (1963) 2. Verma and Bhalla (1968) 3. Mishra (1965) 4. Mishra (1965)

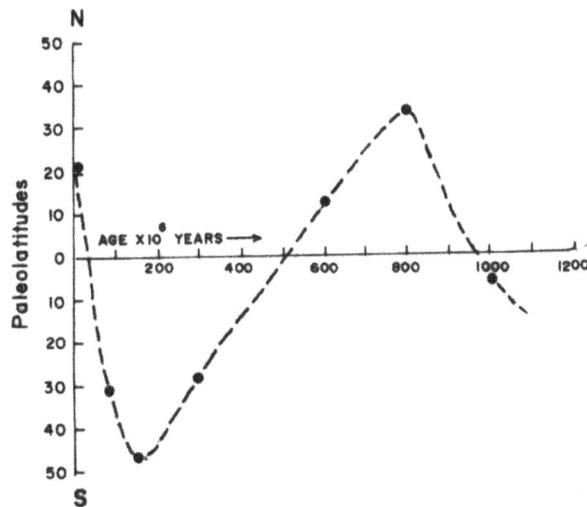

Fig. 5.1 Paleolatitudes of Indian continent drawn based on Table 5.1 suggesting its position near equator during Paleo-Meso Proterozoic period moving to northern hemisphere during Neo Proterozoic period. During Meso Cretaceous (118-120 Ma) it was close to South Pole (50 ^{0}S) and from there it drifted north wards to occupy present day position (Referred to Nagpur 21^{0}N). (Mishra, 1965; Radhakrishnamurthy et al., 1967).

Similarly, the Apparent Polar Wandering Path (APWP) referred to South Pole as described in Section 3.2 computed from paleomagnetic measurements for Indian continent since 155 Ma is given in Fig. 5.2 (Klootwijk et al., 1986). It suggests an equatorial position for South Pole during 155 Ma, which moved from there to occupy the present day pole position in Antarctica. In case, it is presumed that position of S-Pole was fixed in Antarctica, as present day position during geological past, in that case, Indian continent would have required to move from near S-Pole at 155 Ma (50^{0}S) as given above to present day position in due course of time as shown in Fig. 5.1. Both are, therefore, synonymous to each other. In this manner, when different continents were placed back in time, they appeared to join together and from there broke and drifted apart to occupy the present day positions.

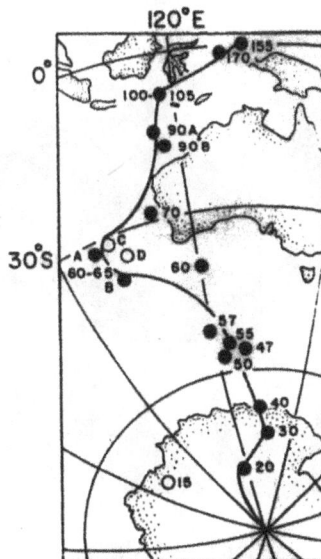

Fig. 5.2 Apparent Polar Wandering path (APWP) for Indian continent Since 155 Ma, with reference to S. pole located presently at Antarctica moving southwards from equator, (Klootwick et al., 1986). It is synonymous with the drift curve for the same period (Fig. 5.1) in case poles are fixed.

5.1.2   Reconstructions – Paleozoic-Mesozoic, Rodinia and Precambrian Reconstructions

In this manner, when continents are placed back in time, it is known as reconstruction of continents during that time. There have been several reconstructions for different periods including Archean-Proterozoic periods but for older period's uncertainties increases. Some important reconstructions are described below.

(i) Paleozoic-Mesozoic Reconstruction

The most important reconstruction relates to Pangea at about 250 Ma which was initially conceived by Wegener and later provided by several authors (Lawver et al., 1992, Lawver, 2003; Scotese et al., 1988, Scotese, 1997; McElhinny and McFadden, 2000). Reconstruction of Gondwanaland during 250-255 Ma (Permian, Fig. 5.3) shows the super continent of Pangea consisting of northern part as Laurasia and southern part as Gondwanaland with Paleo Tethys Ocean in between and large Panthalassic Ocean which subsequently became Pacific Ocean surrounding it. Gondwanaland consisted of Australia-New Zealand, India-Africa and South America joined around Antarctica and Laurasia consisted of Laurentia (North America and accreted land mass), Europe, Kazkistan and Siberia. There are some small continents known as micro continents (micro plates in plate tectonics terminology as described in the next section) of North China, South China, Indo China, Tibet etc., which subsequently drifted and joined together to form the present day configuration. The Tethys Ocean in between Gondwanaland and micro plates of Tibet, Iran and Turkey is also referred to as Neo Tethys Ocean. Since then, the super continents of Laurasia and Gondwanaland have broken a part and drifted to occupy present day position. Reconstruction related to Meso-Proterozoic period (Rodinia) is given in the forth coming section as an example of continents during earlier periods. It is interesting to observe that in most of the cases, the margins of the continent are characterized by mountains (Fig. 5.3, 1-8) that represent ancient fold belts, collision zones of those periods (Proterozoic-Paleozoic). This suggests that quite often, the continents have been rifting and colliding almost along the same lines that is demonstrated with details in Chapter 7.

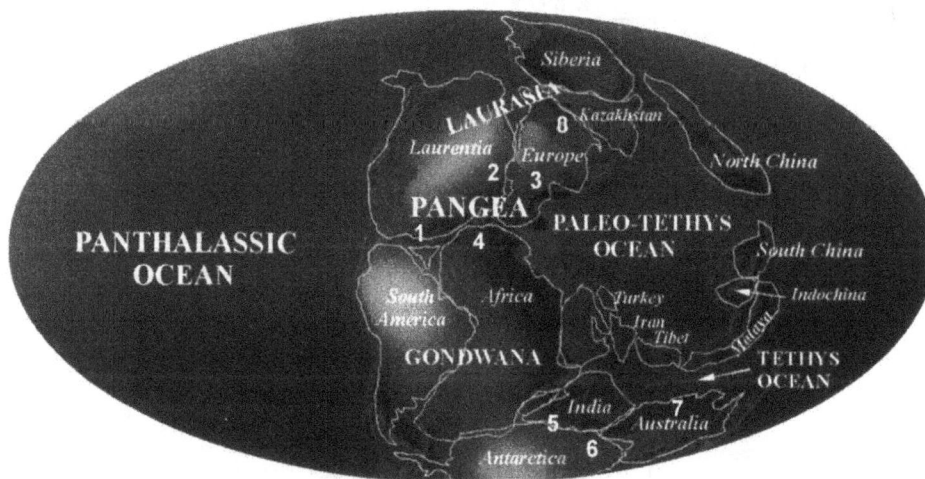

Fig. 5.3 Reconstruction of Pangea during Permian period (~255 Ma) showing the super continents of Laurasia and Gondwanaland (Modified after Scotese et al., 1988, ; http:// www.scotese.com). Numbers 1-8 indicate fold belts (mountains) of ancient times (Proteozoic-Paleozoic period) as follows: 1- Quachita Ranges, 2- Appalachians, 3- Caledonides, 4- Mauritanides, 5-Eastern Ghats, 6- Napier-Prince Charles Mountains, 7- Hamersley Range, 8- Urals. [Colour Fig. on Page 769]

(ii) Rodinia and Precambrian Reconstructions

Rodinia reconstruction of Grenville age (1.3-1.1 Ga) is important as signatures of this orogeny have been reported from several continents and cratons world over. In the Indian continent, it is an important tectonic event in formation of Eastern Ghat orogeny along east coast of India and similar orogeny along northern coast of East Antarctica (Enderby Land). It is discussed in more detail in Section 7.10. This reconstruction has been provided by several authors such as Hoffman 1991, Piper, 2000 etc., (Fig. 5.4). It primarily pertains to agglomeration of cratons in various continents, which show Grenville orogeny. It shows Laurentia in the center and other continents/ cratons are spread along its margin, which, on collision gave rise to orogenic or mobile belts of that period along their margins as shown in the figure. Due to uncertainties of paleomagnetic and age data and various tectonic activities since then, it is difficult to suggest a date for its breakup but it might have remained like that up to end of the Proterozoic period.

Fig. 5.4 Reconstruction of Neo-Proterozoic super continent of Rodinia showing Grenville orogeny (~1.3-1.1Ga) along margins of the cratons due to their collision (Piper, 2000). East Coast of India in conjunction with the coasts of Enderby Land, East Antarctica shows Greenville orogeny of Eastern Ghats (Section 7.10).

5.1.3 Drift History of Eurasia and Gondwanaland

The continent of Europe and Northern Asia (China, Russia etc.,) are together referred to as Eurasia while India-Australia-Antarctica-Africa and South America have been clubbed together as Gondwanaland during geological history as discussed above. Drift history of Eurasia and Gondwanaland is important for plate tectonics of this region and is therefore, described separately here. The drift history of Eurasia and Gondwanaland during last 400 Ma based on paleomagnetic results from these regions are depicted in Fig. 5.5 (McElhinny and McFadden, 2000). This figure shows that the two might have joined together since about 400 Ma when they separated and moved apart. Gondwanaland first moved south wards up to about 38° S at about 350 Ma before moving northwards while Eurasia moved northwards and collided with Gondwanaland at about 50 Ma. This collision gave rise to Alps-Himalayan mountain chains as discussed in Chapter 6.

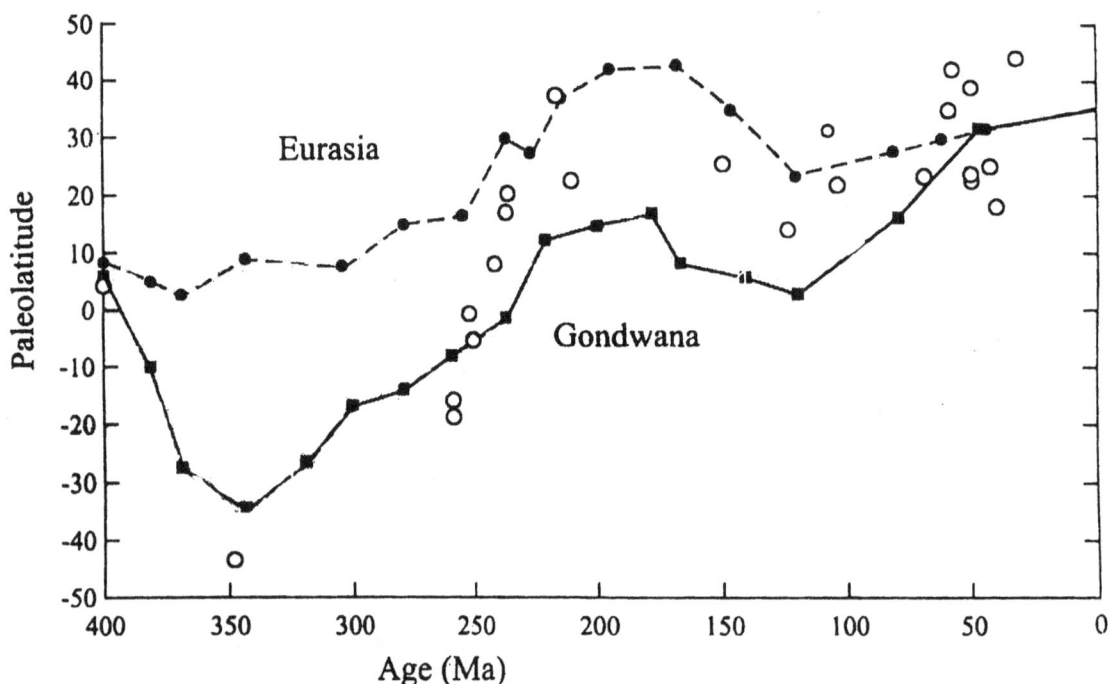

Fig. 5.5 Drift history of Eurasia and Gondwanaland drifting apart at about 400 Ma and colliding at about 50 Ma (McElhinny and McFadden, 2000).

5.2   Plate Tectonics

Continental drift visualized the movement of continents ploughing through oceans, which implied that continents and oceans were separate. This was difficult to visualize due to intimate connections between the two and gave rise to plate tectonics, which considered the two together and defined plates consisting both of continents and oceans. The first idea in this regard came from sea floor spreading magnetic anomalies across mid oceanic ridges (Vine and Matthews, 1963), which in simplest form provided magnetic anomalies of normal and reverse polarities on either sides of mid oceanic ridges placed almost equidistant from the ridge as described below.

5.2.1   Sea Floor Spreading Magnetic Anomalies

Bathymetry and magnetic surveys in various oceans during 1950s suggested that most often the central parts of the oceans are characterized by ridges and deepest parts are along the continents known as trenches, which in fact should have been otherwise in case oceans were formed due to sacking. Further, magnetic anomalies across mid oceanic ridges presented a striped pattern of magnetic highs and lows, which were placed almost equidistant on either sides from the mid oceanic ridges and were observed along several profiles across the ridges. They were typically 20-30 km wide and hundreds of km long with amplitudes of several hundred gammas indicating mafic rocks. Based on experience of interpretation of magnetic anomalies observed over continents as discussed in section 3.7 and 3.8, they could be explained only due to strips of mafic rocks polarized alternately normal and reverse giving rise to magnetic highs and lows or vice versa. Such organized change in polarity can occur only if the earth's magnetic field has also changed its polarity alternately, normal and reverse when these rocks were formed as discussed in Section 3.2 for reverse magnetization. Based on measurement of direction of magnetization of continental volcanic rocks of recent origin, it was found that the nature of magnetic anomalies observed across these ridges matched quite well with those observed on continents. This has a far reaching consequence about origin of ocean floor, which implied that they are related to mid-oceanic ridges and formed due to eruption along these ridges and spreading along the ocean floor. This came to be known as Vine-Mathews-Morley hypothesis of sea floor spreading magnetic anomalies, who first observed them across mid oceanic ridges. Subsequently, these magnetic anomalies were dated based on the magneto- stratigraphy as described in Section 3.2, which provided the age of the mid oceanic ridges and the ocean floor and suggested that the age of the ocean floor increases as one goes farther from it. Further, heat flow measurements indicated that maximum heat flow was found over the mid oceanic ridges, which decreased as one goes farther from the ridges providing minimum over the trenches where oldest oceanic floor was encountered as discussed in the next section.

Fig. 5.6 presents a schematic model of mid oceanic ridge (MOR). Positive and negative signs indicate polarity of magnetization as normal and reverse, respectively and number indicates the strips of these anomalies. The strip (1) is oldest and strip (5) is youngest indicating that strip (1) erupted first through mid oceanic ridges and displaced as subsequent eruptions for strips 2-5 took place and magnetized in the direction of the earth's magnetic field prevalent at the time of their eruption. This provided almost equi-spaced strips of almost same width on either side of the mid oceanic ridges and the oldest magnetic anomalies lie farther from the ridge. Fig. 5.7 (Heirtzler et al., 1966) is a typical example of stripped pattern of Sea floor spreading magnetic anomalies observed across Mid-Atlantic Ridge. Due to their pattern, they are also referred to as Zebra- type magnetic anomalies. Based on the width of these strips and period, one may compute the rate of spreading. The most important results are related to the oldest ocean floor in the world, which are less than 200 Ma and are found along trenches close to several coastlines. This made to think that probably, along trenches, the ocean floor is subducting underneath the continents. This implies that ocean floors are created along mid oceanic ridges and destroyed along the trenches.

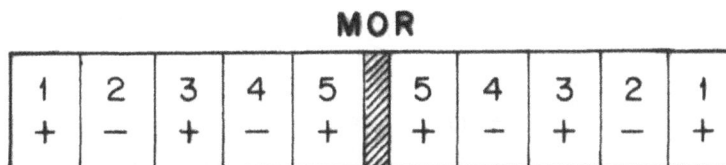

MOR

1	2	3	4	5		5	4	3	2	1
+	−	+	−	+		+	−	+	−	+

Fig. 5.6 Schematic section of a mid oceanic ridge with sea floor spreading magnetic anomalies 1-5 on either sides with normal [+] and reverse polarities [-].

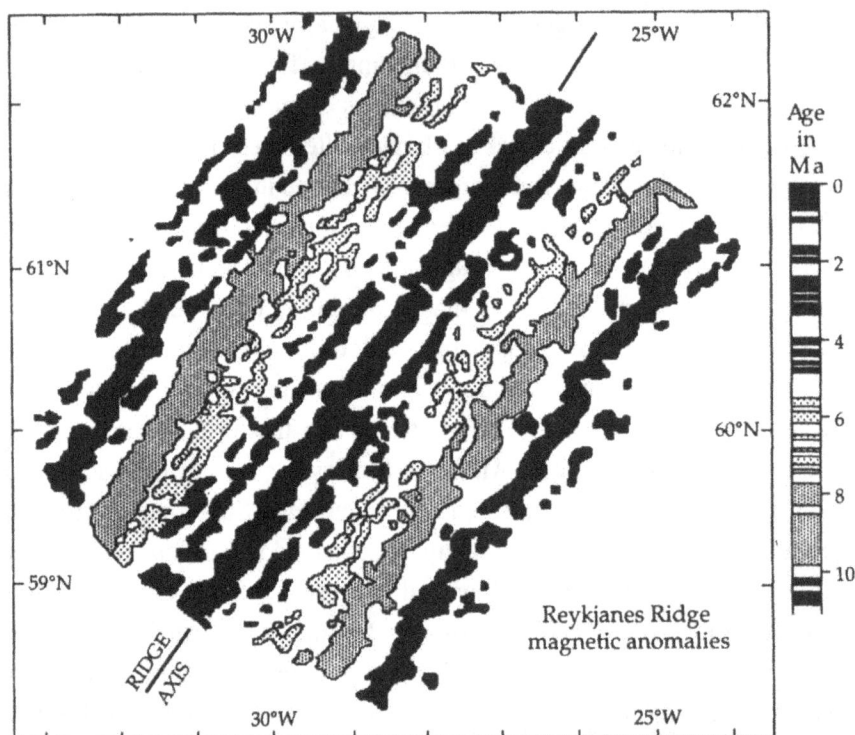

Fig. 5.7 Sea floor spreading magnetic anomalies across Mid-Atlantic Ridge showing stripped pattern related to normal (filled) and reverse (open) magnetization as per their ages. Youngest ocean floor occurring along the ridge axis and oldest farthest from the ridge (Heirtzler et al., 1966).

Modeling of sea floor spreading magnetic anomalies is an involved process. One of the earliest efforts to model these anomalies were due to Pitman and Heirtzler (1966) who modeled the magnetic anomalies across Pacific-Antarctica ridge and also provided a model for the evolution of these ridges. Fig. 5.8 (McKenzie and Slater, 1971) shows seafloor spreading anomalies in the Indian ocean south of Sri Lanka. This figure shows the oldest magnetic anomaly of 33, 32 on magnetography scale that are related to Late Cretaceous and represent a time period just prior to Deccan trap eruption. After this stage, the spreading axis shifted from the Indian Ocean to the Arabian Sea that might be related to this eruption. This figure also shows the effect of the strike of the ridge on the observed magnetic anomalies. However, it is same for N45^{0}E and N45^{0}W. The modeling of seafloor spreading magnetic anomalies involve following stages:

(a) Block models depending on the width of the anomalies are used for the computation of the magnetic field.

(b) Remanent magnetization in x and z direction based on the reconstruction where these rocks formed is obtained.

(c) Using this remanent magnetization and induced magnetization in the present day earth's magnetic field at that place, the magnetic field due to blocks prepared as per given in (a) is computed and compared with the observed field and difference field is obtained.

(d) The difference field is adjusted by varying the magnetization and width of the blocks.

Observed at 40° N

Observed at 20° N

Observed at 20° N, magnetic equator

Observed at equator

Observed at 20° S

Observed at 40° S

1000 nT

27 28 29 30 31   32   33

100 km

Ridge striking due north

Ridge striking N45°E

Ridge striking N45°E/W

1000 nT

27   28   29   30   31   32   33

100 km

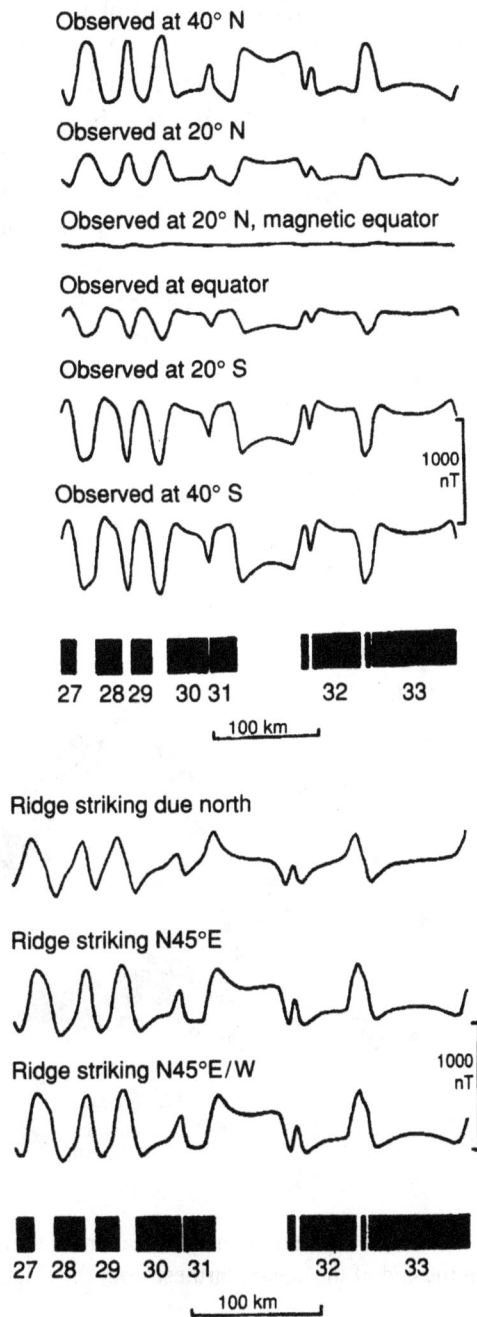

Fig. 5.8 Observed magnetic profiles in the Indian Ocean across various latitudes and modeled sea floor spreading magnetic anomalies with alternating magnetization. It also provides their ages in magnetography scale with oldest magnetic anomaly corresponding to 33, 32 related to Late cretaceous just prior to Deccan trap eruption in the Indian lithosphere (McKenzie and Slater, 1971). It is posulated to coincide with jumping of spreading ridge from Indian Ocean to the Arabian Sea that might have been caused by this eruption. It also shows the effects of strike direction on the recorded magnetic anomalies.

Further details of modeling of these anomalies can be obtained from Mckenzie and Sclater (1971). The half rate of spreading across ridges varied from 10 mm/yr in North Atlantic Ocean to 40-60 mm/yr in the Pacific Ocean (Vine, 1966). This provides the rate of plate motion on either side of these oceans, which is an average over several million years. This is half rate of plate motion as it is measured based on the width of the sea floor spreading magnetic anomalies on one side of the mid oceanic ridges. Subsequently, geodetic measurements of separation between continents on either side of Atlantic provided a rate of 1.7 cm/yr, which is quite close to 20 mm/yr as given above based on magnetic anomalies, that is an average over several million years. Fig.5.9 (Scotese et al., 1988) provides the age of ocean floor inferred from sea floor spreading magnetic anomalies. This also presents a striped pattern similar to seafloor spreading magnetic anomalies. The oldest magnetic anomalies of mid Jurassic (M25-155 Ma) are found along the coast lines, which also coincide with trenches as described above. Contrary to it, the youngest ocean floor coincides with mid oceanic ridges, which however may not occur always in the middle of the ocean. Based on rate of spreading, the nature of mid oceanic ridges changes as discussed in the next section.

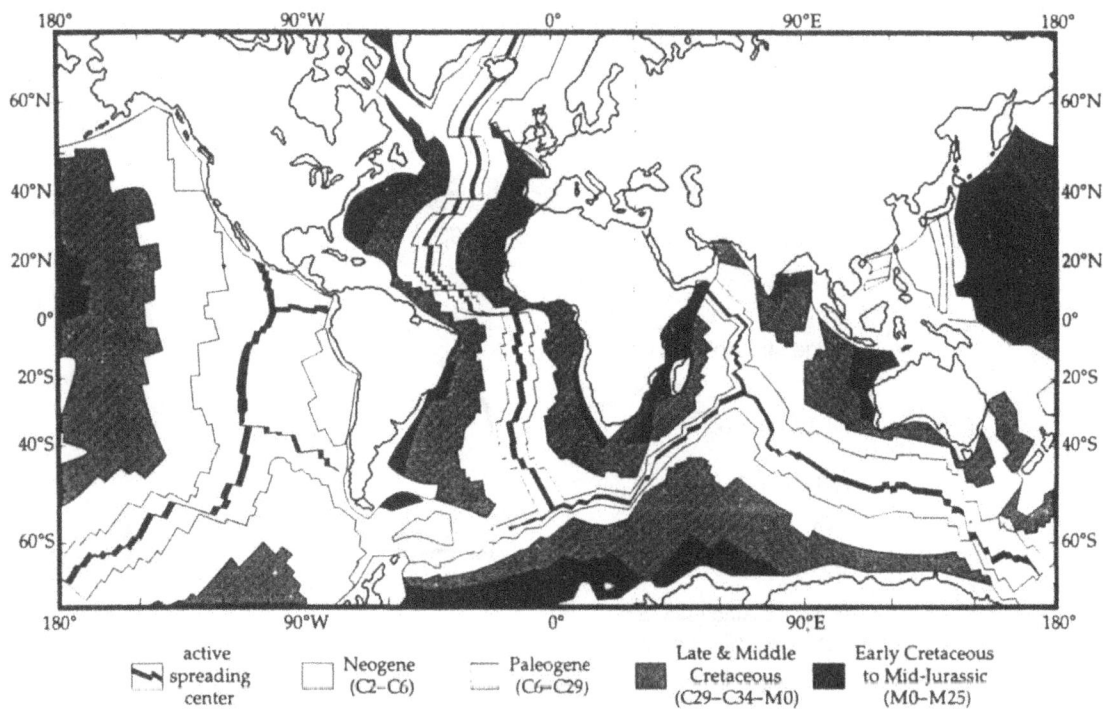

Fig. 5.9 Age of the ocean floor based on sea floor spreading magnetic anomalies. Youngest rocks are found along mid oceanic ridges and oldest, at the end of the ocean, farthest from the ridges. (Scotese et al., 1988).

5.2.2   Heat Flow and Depth of Ocean Floor with Age

Due to volcanic eruptions, maximum heat flow is observed over the mid oceanic ridges where ocean floor is created and is also shallow in this section. Similarly, oldest ocean floor are found along trenches which depict minimum heat flow and is deepest in the entire section. This shows that some relationship exists between heat flow, age and depth of the ocean floor. Heat flows (loss) at mid oceanic ridges depend on several factors such as rate of spread, nature of magma etc., and therefore, it is difficult to predict heat flow at ridges and in adjoining regions. However, it stabilizes at about 55 Ma and after that shows an asymptotic relationship flattening with age (Fig. 5.10). A Global Depth and Heat Flow Model (GDHI) expresses as the following relationship between them (Lowrie, 1997). If q is the heat flow in mWm^{-2} and t is the age in Ma, then (Stein and Stein, 1992)

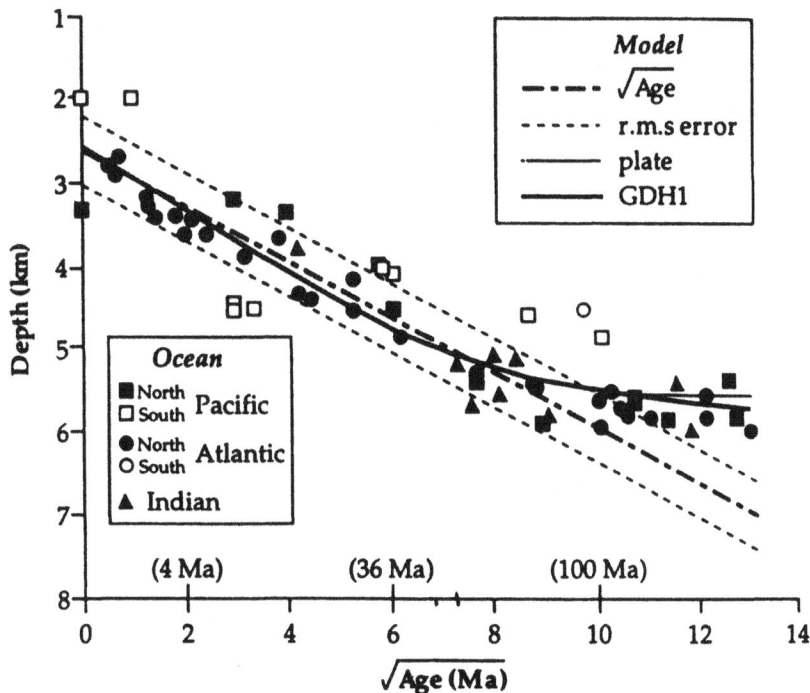

Fig. 5.10 Depth versus √age (Ma) of the ocean floor showing a linear relationship upto about 80 Ma. Global depth and heat flow model (GDH1) is also shown for different oceans which decreases with age and becomes asymptotic at about 55 Ma (Johnson and Carlson, 1992) .

$$q = 510 / \sqrt{t} \ (t \le 55 \ Ma) \qquad \qquad .....(5.1)$$

$$q = q_s [\ 1+2\exp\ (-k\pi^2 t\ /\ a^2)]\ (t > 55 \ Ma) \qquad \qquad .....(5.2)$$

where q_s is the asymptotic heat flow over old oceanic crust = 48 mWm^{-2}

a is the asymptotic thickness old oceanic lithosphere = 95 km

and k is its thermal diffusivity = 0.8 x 10^{-6} m^2 s^{-1}

Equation (5.2) reduces to

$$q = 48 + 96 \exp\ (-0.0278t)$$

In case, it is assumed that the depth of the ocean floor is caused due to subsidence from cooling, it should be equivalent to √t based on cooling in a half space where t is the age of the lithosphere. It is normally valid up to 80 Ma (Fig. 5.10, Johnson and Carlson, 1992) when it shows a linear relationship with age.

5.2.3   Mid Oceanic Ridges and Subduction Zones

From above consideration, it appears that ocean floor is created at mid oceanic ridges which move horizontally and subducts at the trenches and section of earth surface between these two processes is known as a plate. Earth is composed of several plates (7 large ones and several smaller ones), which are in continuous motion changing their configuration during the geological past. However, motion of plates takes place on a spherical surface of the earth around pole of rotation, therefore, the plates breaking apart and drifting in time do not show parallel boundaries. In fact, they may show conjugate configurations such as east coast of South America and west coast of Africa but they will not depict parallel coastlines.

Behavior of plates in a plane is schematically represented in Fig. 5.11, which describes the various activities related to plate motion horizontally and vertically in depth for following three cases.

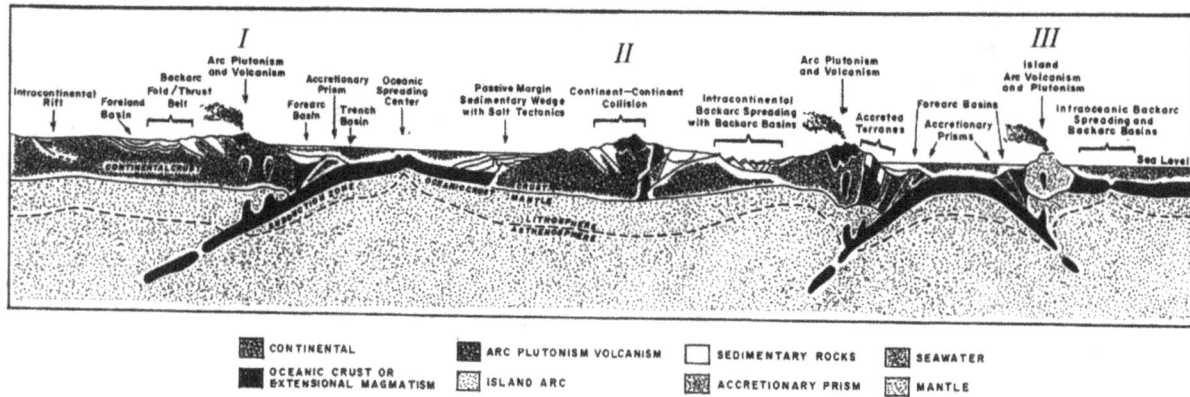

Fig. 5.11 A schematic section of plate tectonic processes at mid oceanic ridges and subduction zones. The left side **(I)** demonstrate the formation of oceanic crust and its subduction along a continental margin and associated processes. Such continental margins are known as active continental margins. Central part **(II)** shows a case of continent-continent collision while right side **(III)** is a case of oceanic plate subducting under a continent on one side and island arc or oceanic plateau on the other side and related activities. (Modified after Greggory, 1986).

(I) This figure shows mid oceanic ridges as oceanic spreading centers at the left side (I) where ocean floor is created through extrusion of magma. It is pushed on either side due to new material being extruded at mid-oceanic ridges. When the ocean floor encounters thick continental crust, it subducts, under it due to high density of oceanic crust as shown on left margin of Fig. 5.11. Before subduction, a trench is formed where parts of oceanic crust and sediments are deposited known as trench basin and accretionary prism. On subduction, the oceanic crust, consisting of mafic rocks and sediments with fluid, melts at subsurface asthenospheric level to produce magmatic volcanic rocks which erupt over the continents and are known as arc volcanism (magmatism) due to their shape as presently seen in case of modern subduction zones such as Sumatra-Java islands. The section just before the island arc over the continental shelf is known as forearc basin and accretionary prism which are differentiated based on the nature of sediments found in them. Accretionary prisms are largely characterized by oceanic crustal rocks and sediments, for example Andaman-Nicobar islands while fore arc basin consists of eroded parts of arc volcanism and oceanic sediments. There are different stages in the development of fore arc basins such as incipient stage, starved stage and fully developed stage depending on sediments in them (Dickinson, 1977). The continental margin in front of island arcs in this case is known as active margin due to its association with plate tectonics processes such as subduction, island arcs etc. At the back of the arc volcanism, there are fold and thrust belts formed due to compressional forces. These are low angle thrusts in direction of compression along which the older rocks are thrusted upwards to form fold and thrust belts in the form of mountain chains. At the back of fold and thrust belts, foreland large basins and intra continental rift basins are formed such as Rhine Graben in case of Alps and Baikal rift in case of Himalayas (Section 2.9) The crust under island arcs and fold and thrust belts thickens due to addition of material during these processes. It is a case of low angle subduction.

(II) On the other side of mid oceanic spreading center is the case of a plate consisting of both oceanic and continental parts, (II, Fig. 5.11) colliding with continental part of another plate, it is known as continent-continent collision, which gives rise to crustal shortening due to convergence and under thrusting. Thrust and fold (orogenic) belts are formed on both the plates due to thrusting of rocks from both sides and the two continents merge together to form a suture. In this case, the materials (rocks) between two plates are squeezed up due to convergence forming a central core complex. As prior to this collision there would have been ocean between the two plates, therefore the squeezed rocks of core complex consist of oceanic crust and associated sediments. Oceanic crustal rocks found in the present day orogenic belts are known as ophiolites, which is one of the best signatures of orogenic belts. This whole process can be presently seen in case of Himalayan collision zone, which represents a typical continent-continent collision as described in the next Chapter 6. The other side of the continent at its contact with oceanic crust is known as passive margin where sediments eroded from the continent are deposited to form a sedimentary wedge. In such collisions, crust thickens due to under thrusting of rocks from both sides and crustal shortening.

(III) Towards the right side of Fig. 5.11 (III) is shown oceanic crust subducting under continental crust on the one side and oceanic plate on the other side with a preexisting island arc or oceanic plateau giving rise to arc volcanism on either sides. These are high angle subduction whose consequences are discussed in the next section. Due to subduction and arc volcanics on both the sides, there are extensional regimes on either side beyond arc volcanism forming into continental and intra oceanic back are basins. These are referred to as active margins as defined previously (Section 3.8). With continuing subduction the total ocean subducts and continents collides with the arc volcanic rocks increasing the volume of continental rocks. Kohistan arc in western Himalaya is an example of this process which is discussed in the next chapter on Himalaya. The arc volcanic rocks and associated sediments with it which are accreted to the continental crust on the other side after subduction of oceanic plate is known as allocthonas or suspect terrane as it has not formed where it exit after collision. In this case, oceanic crust subducts on either sides as it has a higher density compared to continental crust and arc volcanic rocks. The arc volcanism in this case are narrow islands and form chain of islands on the oceanic plates known as islands arcs as in case of Japanese islands. Intra oceanic back arc spreading centers in this case form rift basins within the oceans and are important for hydrocarbon occurrences.

In this manner, plates move, collide, subduct and accreted to form new and new configurations of oceans and continents during geological past. Plate tectonics has been able to explain most of the observables on the surface of the earth such as formation of basins, mountains etc., and regarded as an unified theory to explain most of the geodynamics phenomena. It is also important for resource exploration as most of the mineralized zones like that of base metals, precious metals etc., are associated with the fold belt, specially ancient ones that represent collision zones of those periods as discussed in Chapters 7 and 9. Therefore, identification of ancient collision zones and sutures are key to their exploration. As sedimentary basins are also formed due to plate tectonics processes, its understanding is crucial to hydrocarbon exploration.

5.2.4 Delineation of Plates on Earth's Surface

As plate boundaries, viz. mid oceanic ridges and subduction zones as defined in previous section are associated with tectonic movements, they are associated with earthquakes. Mid oceanic ridges being shallow in nature are associated with shallow earthquakes while subduction zones being deep in nature are associated with both shallow and deep focus earthquakes. When the epicenters of all earthquakes world over, viz. shallow and deep focus earthquakes were plotted, they defined plate boundaries. One such plot for the period 1978-1989 is given in Fig.5.12 (Fowler, 2005) which shows them aligning along narrow bands defining the plate boundaries. In this manner, various plates on the surface of the earth were defined, some of which are large and some small known as micro plates (Fig. 5.13; Fowler, 2005). These plates on surface of the earth are mobile as continents were considered to be mobile in continental drift hypothesis with the difference that plates consist of both continental and oceanic parts while previously they were considered separate from each other. As movements of plates can be visualized only through convection as suggested previously by Holmes for continental drift hypothesis, the whole lithosphere of a plate comprising of continental and oceanic lithosphere is involved in the motion through convection in the asthenosphere. The boundary between the lithosphere and the asthenosphere is defined as the thermal boundary where rocks become ductile facilitating the large scale convection. It is defined by 1200-1400^{0}C isotherm where partial melting of rock facilitate formation of convection cells and approximately lies at a depth of 200 km under the continents and 100 km under ocean. However, it varies considerably from region to region depending on thermal gradient in a region. Further details of it are provided in forthcoming section on thermal structures.

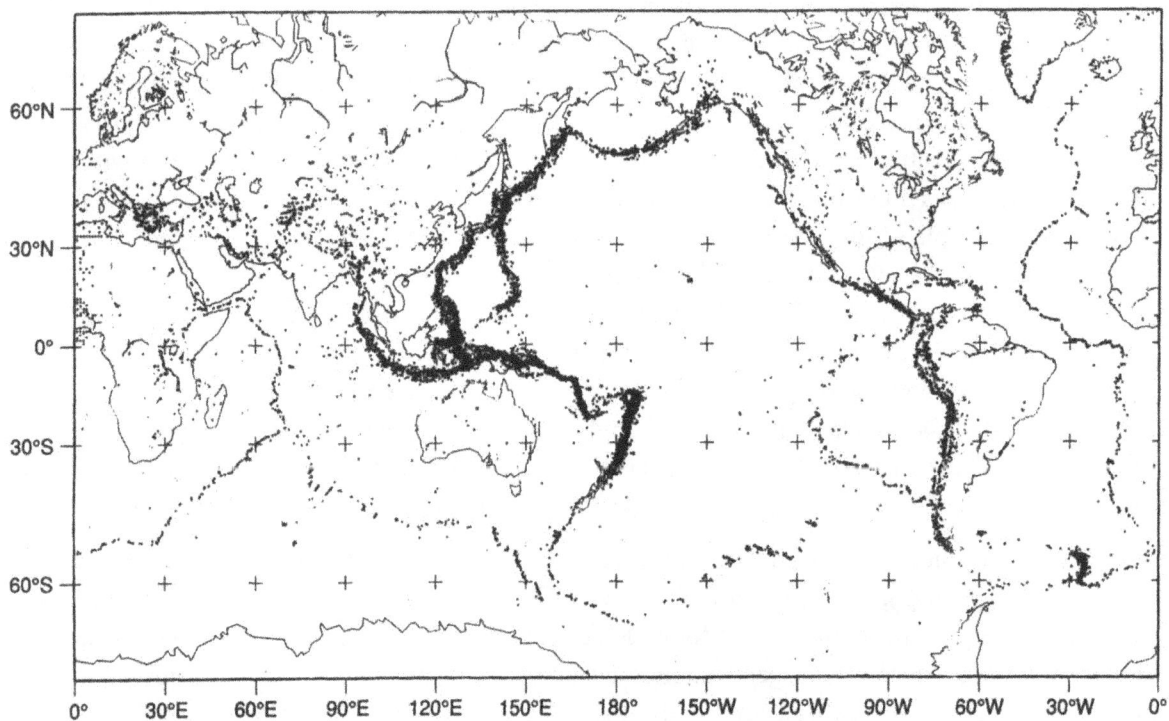

Fig. 5.12 Distribution of shallow and deep focus earthquakes during 1978-89 (from ISC catalogue defining the boundaries of the plate as they are more seismogenic compared to the interior of the plates (Fowler, 2005; ISC catalogue).

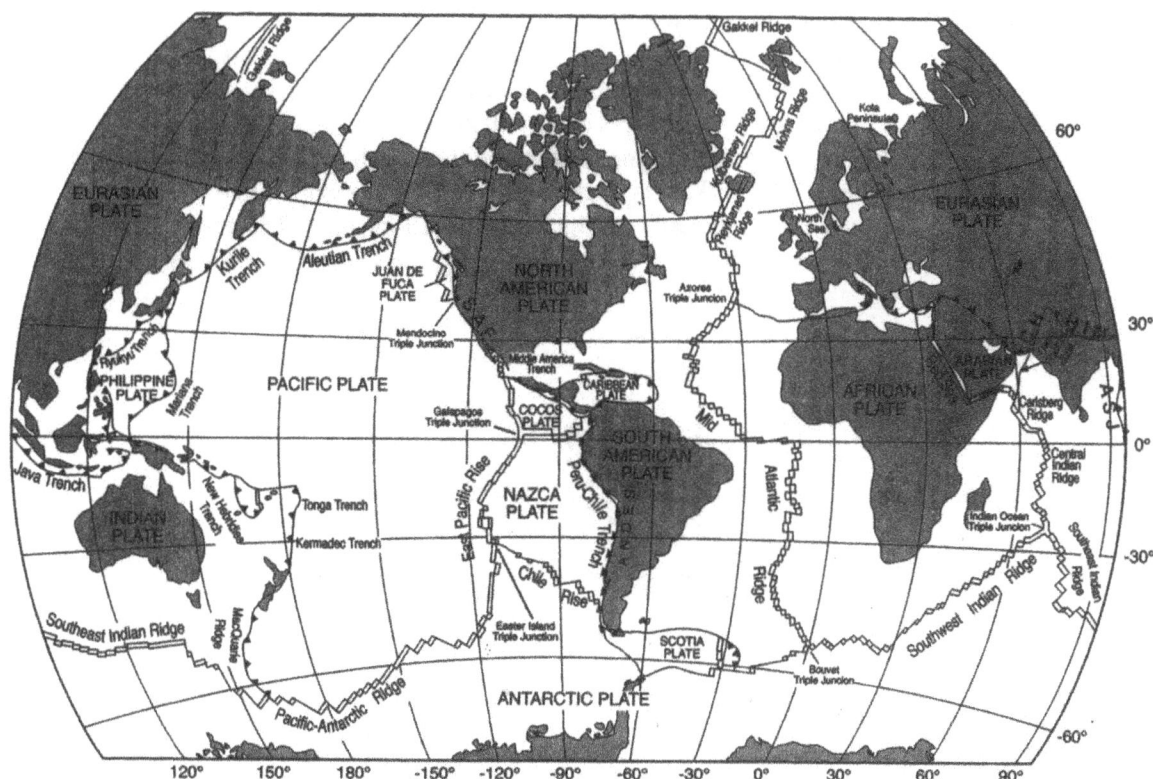

Fig. 5.13 Plates defined by mid oceanic ridges and trenches-subduction zones over surface of the earth defined by the earthquakes in Fig. 5.11. (Fowler, 2005). CH- Chamman Fault, HIM- Himalayan Fold Belt; ASJ-Andaman-Sumatra-Java (Sunda) Trench and SAF- San Andreas Fault System.

5.2.5 Mapping of Subduction Zones: Wadati-Benioff Seismic Zone

During 1930s Wadati and Benioff have reported that the foci of deep focus earthquakes related to upper mantle lie on a plane that is inclined downward from ocean to the continents which in most cases coincided with trenches as shown in a schematic diagram in Fig. 5.14. As one of the necessary conditions for occurrences of earthquakes is cold brittle rocks, which normally does not exist in upper mantle, so crustal rocks must be reaching at these depths in some manner to give rise to deep focus earthquakes. Added to it, mid oceanic ridges are characterized by shallow earthquakes up to a few kilometers as rocks below this level are hot and ductile in these sections. This information when combined with other information from sea floor spreading magnetic anomalies that the oldest ocean floor is less than 200 Ma and occur along trenches gave rise to modern day plate tectonics theory which briefly states that ocean floor is created at mid oceanic ridges and consumed as subducted plates along trenches. Hasegawa (1989) in fact mapped two distinct levels of hypocenters of deep focus earthquakes associated with top and central parts of the subducting slab across Japan Trench as schematically shown in Fig. 5.14. However, this kind of differentiation could not be noticed in case of all trenches till Brudzinski et al., (2007) demonstrated in case of several trenches based on better resolved data from local seismic networks. They also suggested that the separation of two Benioff zones increases with age of the subducting plate and is consistent with dehydration of associated rocks at those depths. Kawakatsu and Watada (2007) suggested that along with the subducting slab, water is also transported into deep mantle through hydrated oceanic crust as a serpentinite layer at the top which may act as a lubricant and aid into deep focus earthquakes. It may be noted that earthquakes are concentrated in subducting plate along two planes, top and in the middle. This was explained due to high temperature of surrounding rocks, which would produce extension in the upper part of cold subducting plate causing strain and normal earthquakes while middle part being cold will be in state of compression and produce thrust earthquakes. Maximum seismic activity however, occurs where subducting plates bends (Fig. 5.14), as this section experiences maximum extension. Mckenzie (1969) has suggested that the earthquakes are restricted to those regions of the mantle which are colder than a definite temperature and shown that a 70-100 km thick plate subducting into mantle can remain cool down to 600-700 km, which is the reported depth of the deepest earthquakes. The input of large volumes of cold material can also control the convection and cause downward movements in the mantle near island arcs.

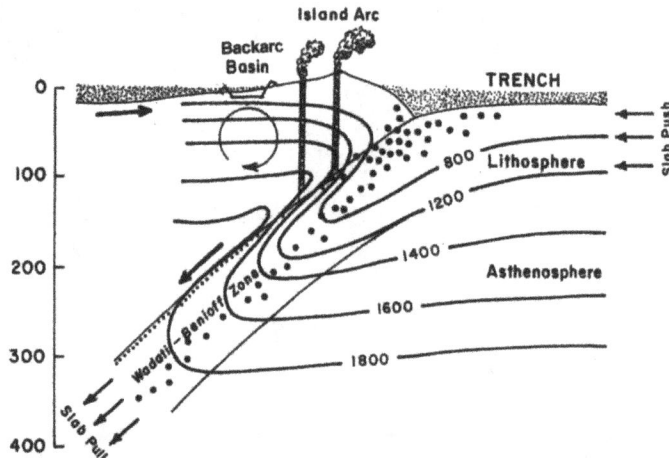

Fig. 5.14 A schematic section of a subducting slab across a trench showing two distinct levels of hypocenters of deep focus earthquakes known as Wadati-Benioff Zone. Cold subducting slab modifies the isotherm with lowest temperature isotherm in the central part of the slab its depth depending on the speed of subduction. Volcanic arcs are formed due to melting of the subducting slab at depth of 120-140 km in the mantle which comes up as volcanoes. Convection cells at the back of the islands arcs causes extension and formation of back arc basins (Fig. 5.11). It also shows slab push and pull forces on a subducting plate caused by creation of new oceanic crust at mid-oceanic ridge and mass of the subducting slab, respectively.

5.2.6   Thermal Structure of Subducting Slab and Back Arc Basins: Delamination and Foundering of Lower Crust

Fig. 5.14 also shows the isotherms which gets modified due to subducting cold slab. Central part of the slab is coldest which coincides with hypocenters of deep focus earthquakes originating from the central part of the slab. Due to higher temperature in the asthenosphere, the slab at the upper part starts melting at about 120-140 km and causes partial melts which moves upwards due to its buoyancy and causes volcanoes to erupt as arc magmatism at the surface as discussed in the previous section. As these volcanoes originate in the upper mantle through melting of oceanic crust comprising mafic and ultramafic rocks and passes through continental crust of felsic nature, their composition is generally intermediate between mafic and acidic rocks, known as andesite. This implies that basement rocks of island arcs are andesites formed due to volcanoes from melting of the subducting slabs. As it is shown in Fig. 5.13, Pacific plate is primarily an oceanic plate surrounded by trenches and subduction zones such as Mariana trench, Japan trench etc., which has given rise to several island arcs such as islands of Japan, Philippines etc. Due to occurrences of so many trenches and islands arcs around Pacific plate, there are always some active volcanoes in this section and therefore, it is known as ring of fire.

Changes in isotherm also produce convection cells; under the over riding plate in the mantle (Fig. 5.14) which causes extension and produce back arc basins. It is caused by mantle flow due to viscous coupling with subducting slab and thermal buoyancy due to temperature gradient between subduction zone including islands arc and adjoining continents (Curie and Hyndman, 2006). These convection cells are also responsible for erosion and delamination of lithosphere causing thinning and extension in back arc regions. Foundering of lower crust under arcs also causes trench parallel anisotropy of seismic waves showing fast shear wave seismic velocity parallel to the strike of the trench (Behn et al., 2007). Due to subducting plate being cold compared to surrounding lithosphere, it display lower density which should be accordingly used while modeling the gravity anomalies of subduction zones as shown in some of the models given in section 2.10 under 'Field Example'. Transportation of water along the subducting slab transforms peridotite of mantle into serpentinite (~10% water) causing low density which should be accounted while modeling gravity anomalies due to them.

5.2.7 Thermal and Density Structure of Continental and Oceanic Lithospheres

A general distribution of temperature with depth under Canadian Cordillera (mobile belts) and craton are given in Fig. 5.15. (Hyndman et al: 2005, Saltus and Hudson. 2007), showing a larger thermal gradient of about 26-27 ^{0}C/km under mobile belts (orogenic belts) compared to about 9-10 ^{0}C/km under cratons. This figure also shows the rock xenoliths corresponding to Curie point isotherm at depths where they intersect 580 ^{0}C isotherm, which is about 60 km under continents and 22 km under mobile belts. This diagram is presented here as it provides an important constraint in modeling of large wave length magnetic anomalies (Section 3.7) viz. depth to the Curie point isotherm and understand the formation of convection cells between subduction zones and adjoining continents (cratons). It may however, vary from one region to the other based on the thermal gradient but this represents an average figure. Based on the large wave length magnetic anomalies from the lower crust, they inferred mafic intrusive rocks in the lower crust under orogenic belts making it weak and depleted strong upper mantle. Extrapolation of temperature gradient in the upper mantle under cratons (Fig. 5. 15) suggest that the temperature of 1200-1300 ^{0}C would reach at depth of about 200-220 km which defines the lithosphere-asthenosphere boundary. It may be noted that lithosphere asthenosphere boundary in case of continents (cratons) also represent a low density and low velocity zone due to change from solid to molten state at this boundary. In global earth model, the density of asthenosphere is 3260 kg/m^3 while that for lithospheric mantle is 3300 kg/m^3 suggesting a density contrast of – 40 kg/m^3. This is useful to model geoid anomalies. Temperature under mobile belts (orogenic) suggests that the temperature related to lithosphere-asthenosphere boundary reach at a depth of 60-80 km which however varies depending on temperature gradient in a region. In case of oceanic lithosphere as temperature gradient varies considerably from mid oceanic ridges (new crust) to trenches (old crust), the lithosphere thickness also varies considerable. Afonso et al., (2007) have provided thermal and density structure of oceanic lithosphere including that of mid-oceanic ridges along with the rock types (Fig. 5.16). It suggests an average lithosphere thickness of about 100 km for old oceanic crust which decreases to about 40 km under mid-oceanic ridges. They also suggested a density contrast of, +40 kg/m^3 between the plate and adiabatic mantle, in case of oceanic lithosphere which is almost half of the value considered previously (70 kg/m^3).

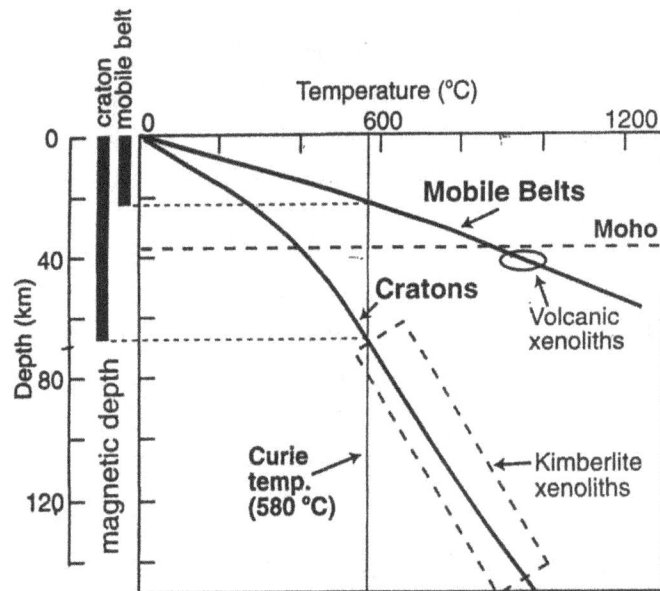

Fig. 5.15 Thermal structure of craton (N. American craton) and mobile belts (Canadian Cordillera) (Hyndman et al., 2005) and depth of Curie point isotherm (580 °C) along with Xenoliths associated with them (Saltus and Hudson, 2007). It suggests average thickness of lithosphere (1200-1300 °C) under mobile belts as 60 km and under cratons as 220-240 km.

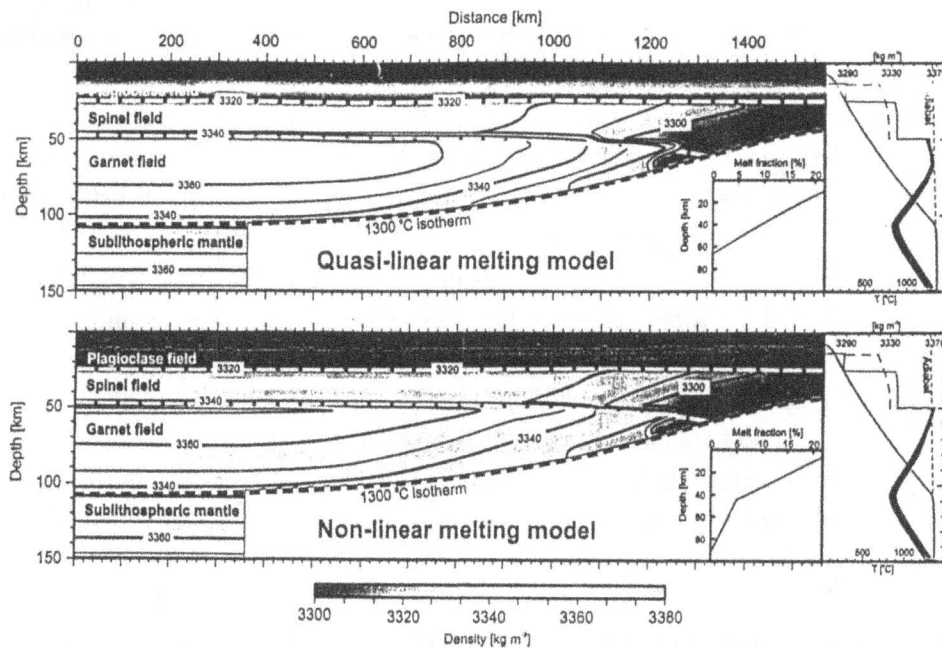

Fig. 5.16 Thermal and density structure of oceanic lithosphere along with important rock types related to main discontinuities. (Afonso et al., 2007). It shows a lithospheric thickness of 40 km near mid oceanic ridges (right side) and about 100-110 km farthest from at (left side) where oceanic crust is old with minimum thermal gradient.

5.2.8   Plate Margins – Divergent (Submersibles), Convergent, Transform Faults and Triple Junctions

It is now clear that plate margins are defined by mid oceanic ridges and subduction zones, which are also known as divergent and convergent plate margins. Some times two plates my slip past each other along a common margin and such margins are known as transform faults. The junction of three plates is known as triple junction as discussed below

(i)   Divergent Plate Margins: Submersibles, Mineral Deposits and Missing Crust.

These are the margins of plate where they move away from each other due to eruptions from mid oceanic ridges. They are characterized by flows, pillow lavas and tholeiitic basalt known as Mid-Oceanic Ridge Basalt (MORB). Oceanic core complex of mid oceanic ridges primarily consist of gabbroic sequences surrounded by serpentinized periodite due to hydro thermal activity. Extension at mid oceanic ridges are accompanied by intrusion of dykes and huge amount of hydrothermal solution which are usually associated with mineral deposits like iron, copper sulphides etc. They are associated with shallow earthquakes (less than 10 km) as discussed above due to ductile nature of rocks below a certain depth caused by heat from the extruding magma. Depending on the rate of spreading, they are referred to as (a) fast spreading ridge (9-18 cm/yr) such as East Pacific Rise (b) moderate spreading ridge (5-9 cm/yr) such as South East Indian Ridge and (c) slow spreading ridge (1-5 cm/yr) such as Mid Atlantic Ridge. They are associated with crustal magma chamber and axial bounding faults that reaches to the magma chamber. It could be as shallow as 3 km beneath the sea floor (Singh et al., 2006). Rate of spreading controls the formation of that ocean such as its size and also the nature of the mid oceanic ridges. For example, slow spreading ridges are characterized by a strong and thick lithosphere and crust and give rise to a rift like structure in the central part with uplifted shoulders, which are usually absent from other two types of ridges. Other most important characteristics of these ridges are large fracture zones perpendicular to them, which are attributed to eruption over a spherical earth surface which is rotating.

Most important information about ocean floor and overlying sediment were obtained from Deep Sea Drilling Project (DSDP) which was carried out in oceans throughout the world. Samples from these sites provided the nature of sediments and composition of ocean floor along with their ages which also helped in standardizing magnetostratigraphy scale as given in section 3.2. Another direct observation about mid-oceanic ridges related to submersible vehicles which made possible scientists to observe these ridges from close by. Direct observation of ridges started first with diving of scientist as early as 1975 which was subsequently replaced by manned submersible vehicles. There are now unmanned submersibles presently in operation which has been found to be extremely useful in sampling from the ridges and recording other observations such as temperature etc., through remote operated robots. One such vehicle used by Woods Hole Oceanographic Institute, USA was named as Alvin which was used in exploration of East Pacific Rise, Gorda Ridge and Juan De Fuca Ridge during 1980's (Lonsdale, 1977; Macdonald et al., 1980; Hekinian et al., 1983; Malahoff, 1983; CASE, 1985). Using these submersibles, they observed for the first time the volcanic vents, eruption of magma and black smokers with boiling water and temperature as high as 300-400 °C in specific sections while temperature of surrounding water column will be about –1 to –2 °C. There are axial volcanoes with calderas as big as 4 × 6 sq km which is bigger than most of calderas found on the surface of the earth. The reported missing crust along a section of the Mid-Atlantic Ridge might be an old Caldera where upper mantle rocks are exposed. Such sites offer unique window for future investigations. It may also represent a section where sea floor spreading did not take place and new crest did not form. There may be several such sections along mid-oceanic ridges world over which form the sites for future investigations (http://www.noc.soton.ac./ur/gg/classroom@sea/jc0071).

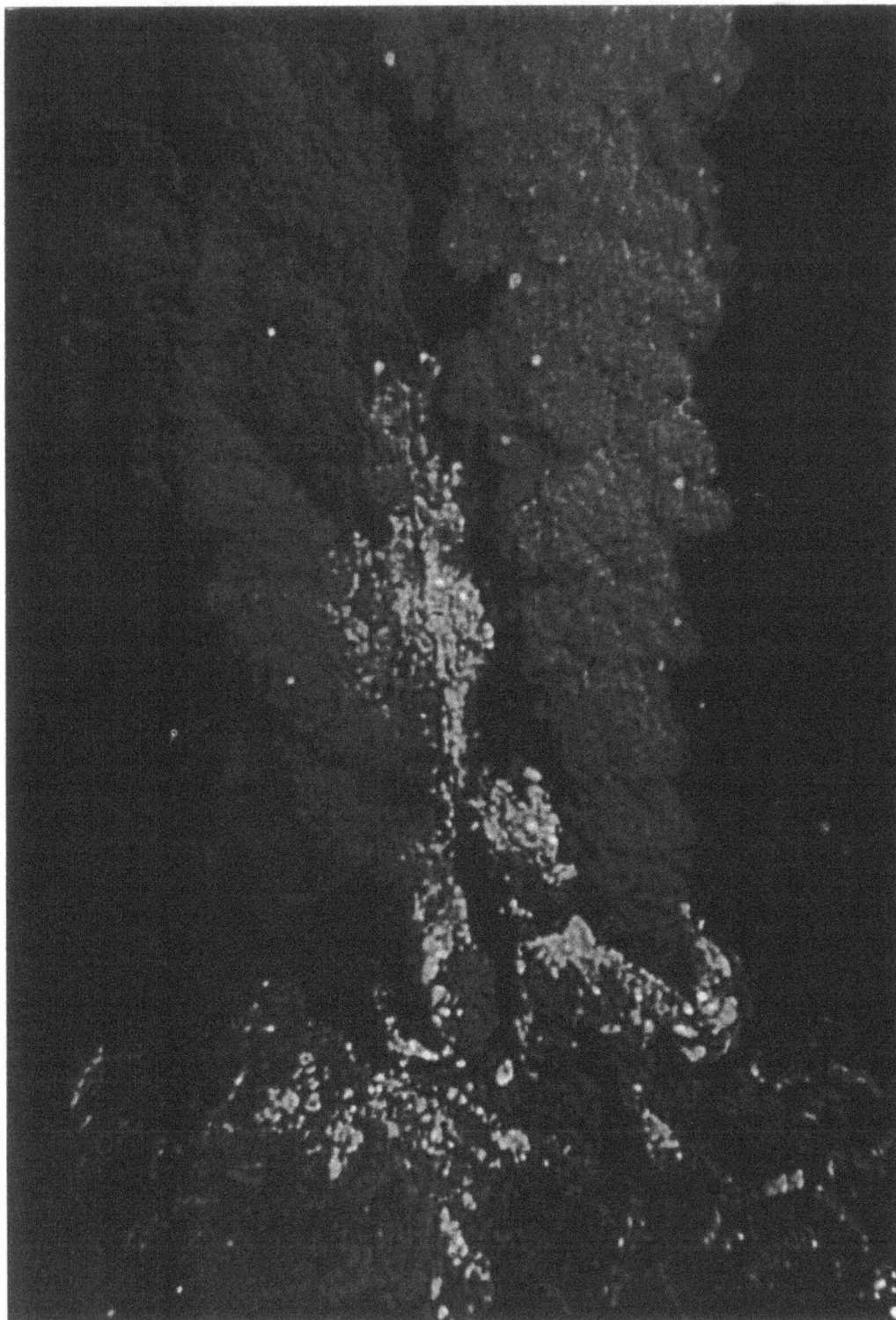

Fig. 5.17 Black smokers at East Pacific Rise photographed using a submersible vehicle. Such black smokers are treasure trove of minerals (sulphides and gold deposits) and biological species in spite of absence of sunlight, high temperature and hazardous gases in these sections. (Sleep and Woolery, 1978; Malahof, 1983, Macdonald et al., 1980; Kious and Tilling, 1999 that was photographed by Dudley Foster from RISE experiment and courtesy of William R. Normack, USGS as quoted by Kious and Telling). [Colour Fig. on Page 769]

Nautile

Video Snapshot (dive GNT05)

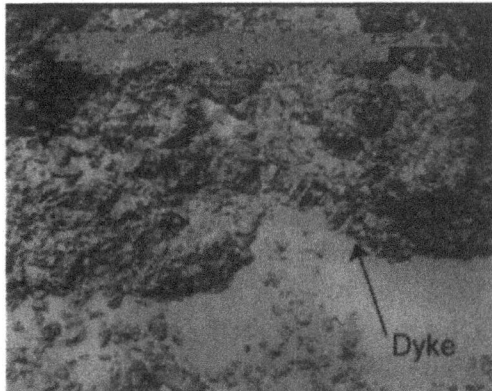

1. Dyke in breccia showing near-vertical colling joints (05h473)

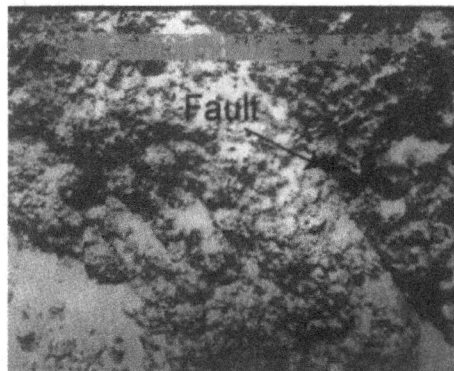

2. Fault separating pillows from breccia (05h49 m)

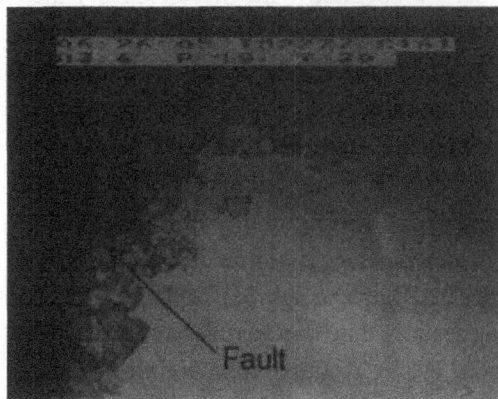

3. Vertical fault scarp trending NNW-SSE (O6h26m)

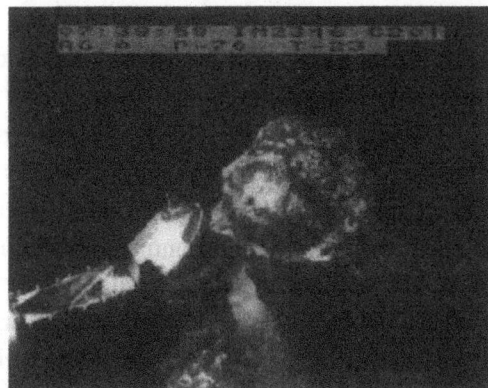

4. Sampling GNT05-3 (07h38m)

Fig. 5.18 *Contd...*

5. Lave tubes over breccia (09h14m)

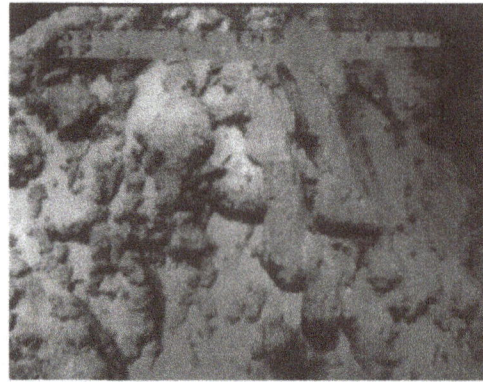

6. Lave tubes over pillow and breccia (09h16m)

Fig. 5.18 Nautile submersible vehicle of French Cruise in Indian ocean and (1-6) photographs of tectonic features of Central Indian Ocean Ridge as explained there in. Photograph 4 shows a robot hand towards the left sampling from the ridge. (Courtesy Dr. A. K. Choubey, NIO, Goa).

Black smokers are formed due to percolation of sea water into fractures of mid-oceanic ridges which are transformed into steam due to high temperature inside and ejected out along with minerals present there. Initially, it was thought that due to plate tectonic forces, seawater is forced downwards through large faults along the planks of the ridge where it is heated. When it comes close to the volcanic rocks/vents, it re-emerges at middle of the ridge. However, recent detailed investigations by Prof. Tolstroy of Columbia University and his team at East Pacific Ridge using seismometers to record micro seismicity recorded these activities at 2500 m depth. They suggested that water percolate down even through minute fissures instead of large faults to a depth of 1000-1500 m and come into contact with magma chamber where it is heated and disgorged along the ridge through several vents. These vents may be as far as 2-3 km from the place where it goes down below. They also suggested that micro seismic activity is caused by the physical stress of cold water passing through hot rocks. It may also contain gases coming out of the magma of the mid-oceanic ridges. Its colour usually depends on the minerals, present in them. They discovered massive sulphide and gold deposits associated with these ridges which made them economically important. Fig. 5.17 shows a black smoker associated with East Pacific Rise. Similar vents with hot hydrothermal circulation were also observed along different mid oceanic ridges (Lonsdale, 1977; Sleep and Wolery, 1978; Hekinian et al., 1983; Macdonald et al., 1980, Kious and Tilling, 1999). Black smokers are treasure trove both for minerals and biological species. Most surprising, however was the discovery of large number of organisms of different species in these vents who thrive in spite of absence of sun light and high temperature and chemically hostile environment. Some of them were new species, unknown so far (Corliss, 1989). The number of species found along these ridges and hot springs out numbered found anywhere on the surface on the earth at one place suggesting that life might have started at similar sites in early history of the earth (Miller and Bada, 1988). Another submersible by name 'Nautile' (Fig. 5.18) has been used by French scientists to explore Central Indian Ridge where Indian scientist from NIO, Goa, India, also participated along with them. Fig. 5.18 shows some typical photographs of tectonics and rock types associated with this ridge (Chaubey, 2007). Most interesting are lava tubes and volcano domes (calderas) suggesting the association of volcanism with mid oceanic ridges.

(a) Typical Gravity Anomalies

Regional free air anomaly is high over the ridges while Bouguer anomaly is a large low indicating mass deficiency under them which may be related to hot thin crust under them (Fig. 5.19). These are typical gravity anomalies observed over the mid oceanic ridges and corresponding model of relatively low density rocks under them caused by high temperature. Gravity anomalies in case of Indian Ocean and Arabian Sea are demonstrated in the forth coming sections.

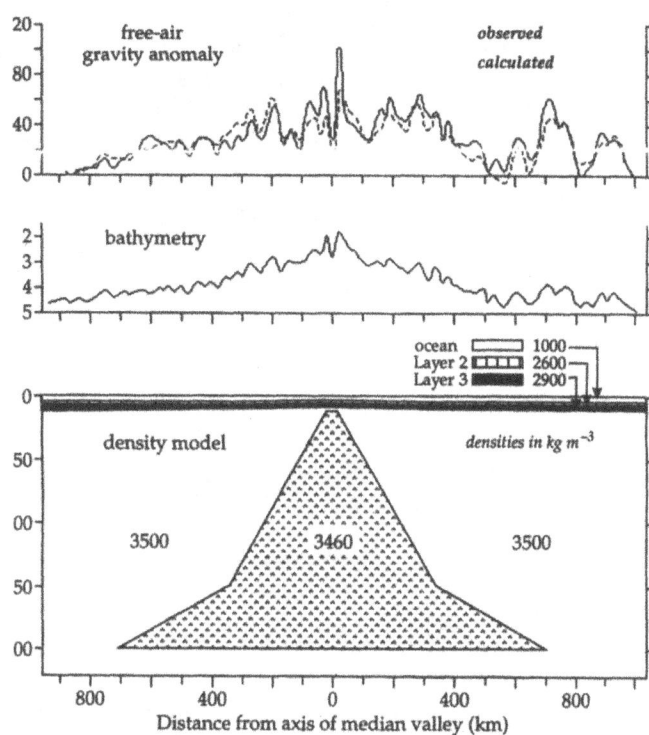

Fig. 5.19 Free air gravity anomaly and bathymetry over mid Atlantic Ridge and modeled section showing low density rocks due to high temperature. (Keen and Tramontini, 1970).

(ii) Convergent Plate Margins and Subduction Model along West Coast of South America: Under plating, Serpentinite, Eclogite and Flexure

Convergent plate margins are basically characterized by subduction of oceanic plate under another continental plate or island arcs. Angle of subduction may vary depending on the thickness of the two plates and speed of their motion. In this regard, there are two types of subduction, popularly known as (a) low angle subduction (< 45°), which are known as Andean type subduction based on the angle of present day subduction along west coast of South America (b) high angle subduction (> 60°), which is also referred to as Mariana type subduction based on the present day subduction along Mariana trench, along east coast Philippines. Gravity anomalies related to them are discussed under 'Field Examples' in Section 2.10 and subducting plates are modeled based on gravity anomalies. The former is seismically more active compared to the latter, which produces less stress and strain due to high angle subduction. A model of low angle subducting oceanic plate along West Coast of South America where Nazca oceanic plate is subducting is given in Fig. 5.20 that most of the essential elements of convergent plate margins and subduction. (Cloeting et al., 2007) by integrating all the available geophysical and geological information that displays most of the essentials of convergent plate margins and subduction. Fig. 5.20(a) shows depth migrated seismic section which is converted to geodynamic model (Fig. 5.20(b)) along with isotherms. As it represents a typical low angle subduction, it is seismically active and hypocenters of earthquakes are shown in the figure. One can check the importance of Wadati-Benioff Zone in the inset which even suggests a break in the subducting plate coinciding with hypocenters of deep focus earthquakes and the lowest temperature isotherm of the subducting slab. Arc volcanism is shown as volcanic front which has also given rise to underplated crust as a high density layer (7.1 km/sec). Hydrated mantle represents transport of fluids along with the subducting plate in form of serpentinites (~10 % water in peridotite) which is associated with low velocity layer at the top of the subducting slab as suggested by Kawakatsu and Watada (2007). The fluid layer at the top of the subducting plate facilitates the occurrence of seismic activity as a lubricant for the adjoining rocks to slip and produce earthquake including deep focus earthquakes. Basalt of subducting oceanic plate at depth may show phase changes to form eclogite which is characterized by relatively high velocity and high density. The layer of seismic velocity 7.6-7.8 km/sec along subducting plate may represent this rock type (eclogite) in Fig. 5.20(b). Grow and Browin (1975) have modeled the same subduction zone based on a gravity profile and showed, a thin layer of high density at the top of subducting lithosphere representing eclogite.

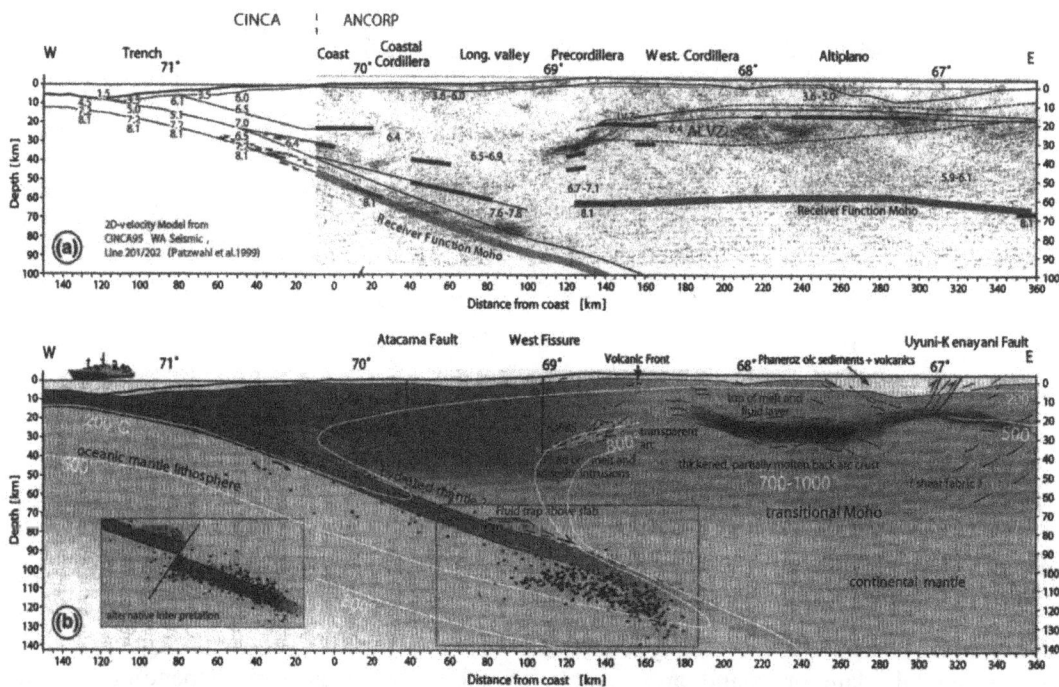

Fig. 5.20 (a) A seismic section across Central Andes where oceanic Nazca plate is subducting under South American plate with seismic velocities showing low and high velocity zones. **(b)** A typical geodynamic Model of a subduction zone based on seismic section in (a) and isotherm with hypocenter of earthquakes. Distribution of these hypocenters suggests a break in the subducting plate (inset) in accordance with the Wadati-Benioff Zone (Cloetingh et al., 2007). Hydrated mantle suggest the serpentinization of peridotite rocks and high velocity along the subducting plate (7.6-7.8 km/sec) may indicate eclogite transformation of basalt. It also shows under plated crust of intermediate velocity (7.1) km/sec caused by arc volcanism and mid crustal low velocity zone (LVZ) caused by fluids and partial melts.

The most important aspect of convergent margins is modifications of isotherms due to subducting cold plate (Fig. 5.14). The forces which may be acting on subducting plates are several but most important are ridge push (RP) and slab pull (SP) forces as shows in Fig. 5.14. However, there are other forces like mantle drag etc., opposing slab pull forces. The most important vertical force operating on the subducting plate is load of the over riding plate which causes flexure of the subducting lithosphere as shown in Fig. 5.21(a) as outer rise. It is a schematic representation of a subduction zone with load of the over riding plate shown as L and flexure of subducting plate which is reflected as outer rise just prior to the trench in bathymetry. It can be compared with cantilever beam (Fig. 5.21(a)). The same is known as lithospheric/ crustal bulge under Ganga basin as described in Section 6.1 in regard to Himalaya. Uplift or bulge can be modeled using flexural rigidity of the beam or subducting lithosphere. Turcotte et al., (1978) have modeled and provided the extent of the outer rise along several trenches which matched quite well with those observed in bathymetry.

(a) Typical Gravity and Magnetic Signatures

Fig. 5.21(b) gives a simplified schematic section of an ocean-continent collision and subduction that shows a gravity low and a magnetic high due to sepentinized mantle wedge (Christensen et al., 2009) and thickening and thinning of the upper crust and a gravity high (H) due to outer rise. It is a simplified version of a model of Cascadia convergent margin based on gravity and airborne magnetic anomalies given by Blakely et al., (2009).

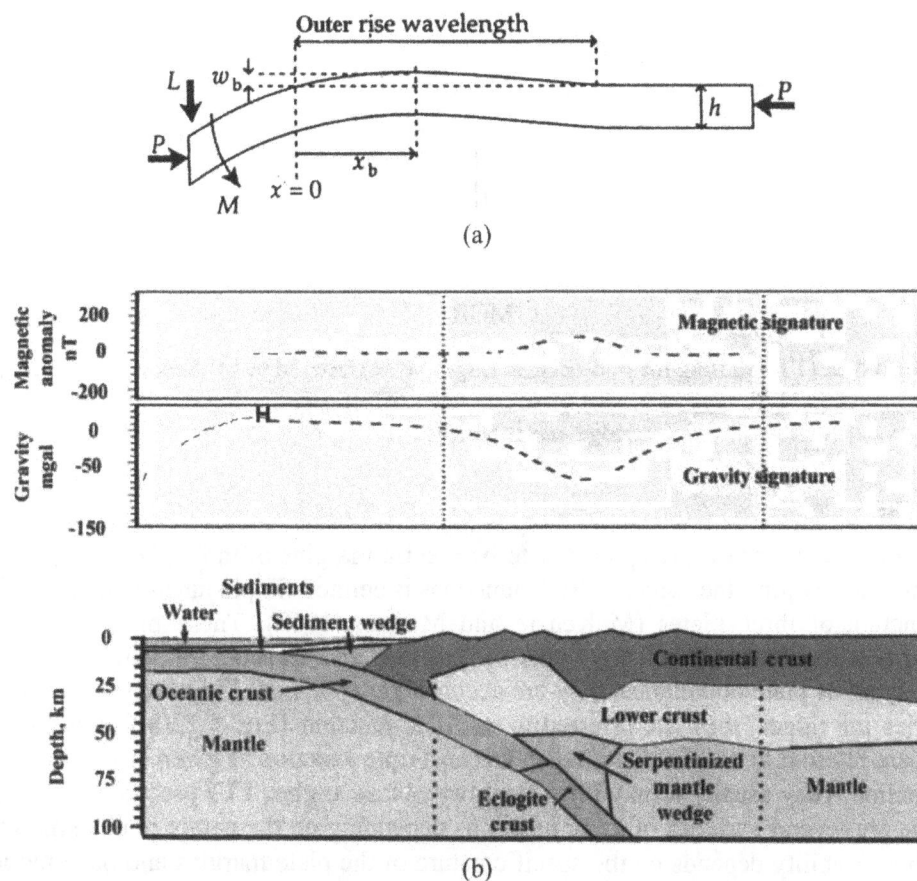

(a)

(b)

Fig. 5.21 (a) The outer rise or bulge in a subducting slab caused due to lithospheric flexure (Turcotte et al., 1978) which is similar to a cantilever beam under the load of over riding plate. **(b)** A schematic section of a Ocean-Continent collision and subduction and likely gravity and magnetic signatures. Gravity low and magnetic high are caused by serpentinized mantle wedge and gravity high, H is due to outer rise as shown in Fig. 5.21(a) (Modified after Christensen et al., 2009).

(iii) Transform Faults

The concept of Transform faults was introduced by Wilson (1965) to account for plate margins, which show horizontal shift of mid oceanic ridges along them. It was primarily postulated to explain San Andreas fault system along the west coast of USA, which connects East Pacific Rise of Cocoas plate to Juan De Fuca Plate in the north (Fig. 5.13) and is seismically active. Transform faults shifts the mid oceanic ridges without disturbing the fabric of the plate as shown in (Fig.5.22a). The Mid Oceanic Ridge (MOR) is shifted towards the right along TT' from a position M to M' and sea floor spreading magnetic anomalies are observed across the right part of the MOR in the same manner as left part of the MOR. It is interesting to note that the seismic activitiy are reported only from the section MM' indicating that this fault does not affect the other parts of the plate. Gregg et al., (2007) have suggested strong correlation between spreading rate along oceanic transform faults and gravity anomalies. Intermediate and fast slipping transform faults depict more negative gravity anomalies than adjacent ridge segment due to crustal thickening under the ridge part of the transform fault though a part of gravity low under the trough which show maximum deformation can also be attributed to porosity and serpentinization of peridotite as described above.

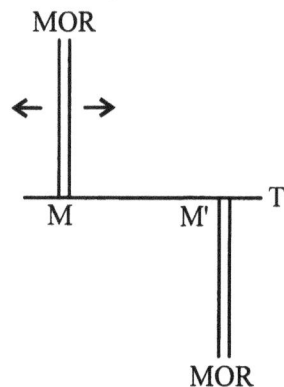

Fig. 5.22(a) Transform fault (TT') shifting the mid-oceanic ridge (MOR) from M to M' Seismic activities are observed only in section M M' suggesting that transform faults affect only the section where shift in the ridge axis is noticed.

(iv) Triple Junctions

Triple junctions are junctions of three plates or to be precise margins of three plates. Depending on the nature of plate margins, the nature of triple junctions is defined. Triple junction was initially defined as junction of three plates (McKenzie and Morgan, 1969). These plate boundaries meeting to form junction can be of any type, viz. ridges, and trenches or transform faults. Depending on types of plate boundaries, they are accordingly specified. For example, in case all three boundaries are ridges, they are referred to as RRR junction (Fig. 5.22(b)) or in case of trenches, they are referred to as TTT. The Indian Ocean Triple Junction as given in Fig.5.13 is an RRR triple junction. They usually form with at least two obtuse angles. TTT junctions are always unstable. There are several varieties of these junctions depending on the nature of plate margins, R, T and F whose stability depends on the specific nature of the plate margins and plate motion. Due to differences in plate velocities, triple junctions may shift with time (Fowler, 2005). It may be noted that in case of RRR junctions, there is compression all around the junction while it is extension in the individual ridges. This situation changes with types of plates forming the junction. There are several such combinations of plate boundaries, some of which are stable and some unstable depending on the interaction of different types of plate boundaries (Fowler, 2005). Subsequently, this term was being used to denote the junction of any three tectonic elements of fundamental nature such as rift basins, fold belts etc., specially in the context of plume based origins (Burk and Dewey, 1973). This concept originated to signify large lithospheric upwarp as it happens in case of active rift basins, it is likely to give rise to three sections of extensions placed almost at about 120^0 from each other, with maximum lithospheric up warp coinciding with the junction. These extensions may give rise to rift basins which would propagate independently for large distances till they encounter some fundamental older structures. Some of these rift basins may develop into oceans or may remain rift basins which are known as failed arms of triple junctions. This led to demarcation of several such triple junctions world over such as in Central India where Gondwana Rift basins (Godavari and Mahanadi basins) encounter Satpura Fold Belt (Fig. 2.26(a)) etc. As these triple junctions are invariably associated with plumes, they are characterized by large amount of mafic volcanic rocks of high density and susceptibility and therefore, cause large amplitude regional gravity highs and magnetic anomalies which can be distinguished from their amplitude and associated rock types. With in the regional gravity highs, there may be several local highs and lows related to intrusive and sediments, respectively as shown in Fig. 5.19.

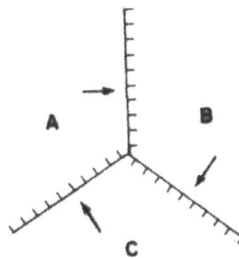

Fig. 5.22 (b) A triple junction of three ridges known as RRR junction.

5.3   Mantle Plumes and Break up of Gondwanaland

Break up of continents are attributed to igneous activity as suggested previously in Sections 2.9 and 3.8 in Part 1. There are two types of such activities, viz. plumes which give rise to large volcanic provinces and the second one related to igneous activity on limited scale due to decompression in mantle or local thermal sources. However, break ups due to plumes are more common (Morgan, 1981) which are known as volcanic passive margins while the second one is known as non-volcanic passive margin. Break up of Gondwanaland and causes leading to it are important for geodynamics of the Indian plate and are there fore, discussed here in some detail.

5.3.1   Mantle Plumes

Plumes are positions on surface of the earth where hot magma wells up from inside the earth and affects the plate moving over them. These are known as hot spots. According to plate tectonics, volcanoes occur both at the mid oceanic ridges (divergent margins) and convergent margins. However, there are intraplate volcanoes which are explained due to plumes (Morgan, 1971). Morgan (1972) also suggested mantle convection due to plumes causing plate motion. They also give rise to chain of volcanic islands in oceans such as Hawaii islands on Pacific Plates and, Chagos- Laccadive-Maldive islands on the Indian plates etc. They give rise to large volcanic provinces over the continents and oceans such as Deccan Volcanic Province in India, Ontang-Java plateau in Pacific plate and Kerguelen plateau north of Antarctica etc. Formation of rocks at mid oceanic ridges and their subduction along trenches are two most important paradigm of plate tectonics. To budget the mass subducting along subduction zones, it was suggested that subducting rocks may sink inside the earth and reach up to the crust-mantle boundary (Fig. 5.23(a)) and are circulated in side the earth in molten state through convection and rises again through plumes and give rise to volcanoes. Some of them facilitate convection cells in the asthenosphere which rises at mid oceanic ridges (Fig. 5.23(a), Pirajno, 2007; Davies and Richards, 1992). They have been considered responsible for changes in plate motion (Sharp and Clague, 2006) and mid oceanic ridges as discussed in section 5.4 in case of Central Indian Ridge due to Reunion plume. The same plume has also given rise to Deccan trap offshore west Coast of India and in western India and formed the western margin of the Indian continent. Another plume that has been important for the Indian continent is Kerguelen hotspot that was responsible for the break up of India from Antarctica during middle Cretaceous (~120 Ma) as discussed in the forth coming sections. Fig. 5.23(b) given by Raval and Veeraswamy (2007) demonstrates their respective positions. Beside small narrow plumes as suggested in Fig. 5.23(a), Campbell and Griffiths (1990) gave the idea of a super plume rising from core mantle boundary and affecting a large area of thousands of kilometer on the surface of the earth. Kellogg et al., (1999) based on numerical models of thermo mechanical convection suggested a compositionally different layer in the lower mantle below 1600 km which is about 4% denser compared to upper mantle and suggested plumes rising upwards from this layer giving rise to volcanoes at the surface and facilitate formation of convection cells in the asthenosphere. There have been several discussions regarding existence of plumes and super plumes (McNutt, 2006; Hoffman and Hart, 2007) but from all arguments, it appears as a viable mechanism to explain intraplate volcanism and large igneous provinces. The most important evidences for mantle plumes have come from tomography experiments from velocity images of upper and lower mantle (Romanowicz and Gyno, 2002). Montelli et al., (2004) have mapped velocity images of some of the known plumes such as Hawaii, Kerguelen etc., as low velocity zones up to depth of 2800 km almost up to core-mantle boundary. Nolet et al., (2007) have shown low velocity zones under Iceland and Hawaii from transition zone to the base of the lithosphere indicating the existence of plumes at least in the upper mantle while the evidence for their extension below the transition zone is weak.

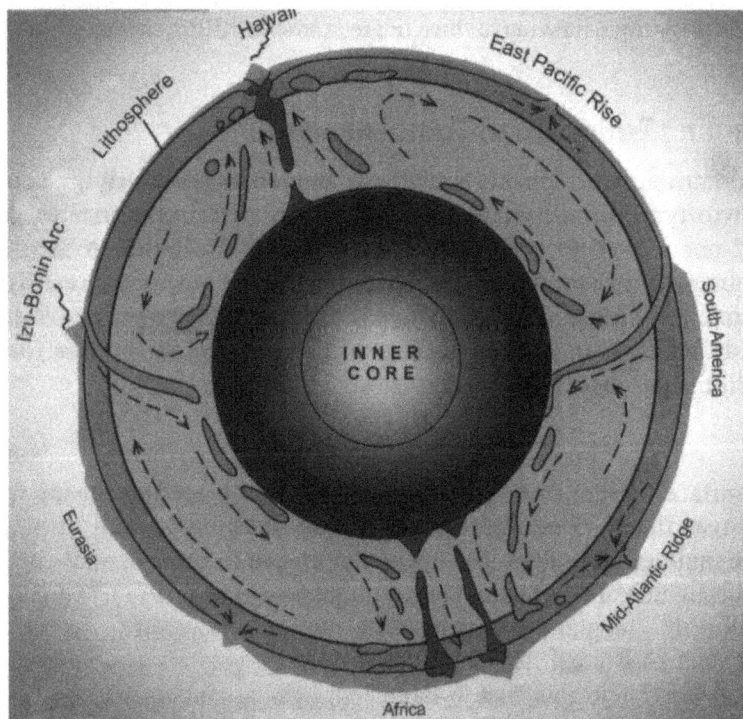

(a)

(a) A schematic model showing relationship between subducting slabs which reaches up to base of the mantle and rises as plumes which sets convection cells in asthenosphere that drives the plates and rises at mid-oceanic ridges. (Davies and Richards, 1992, Pirajno, 2007). [Colour Fig. on Page 770]

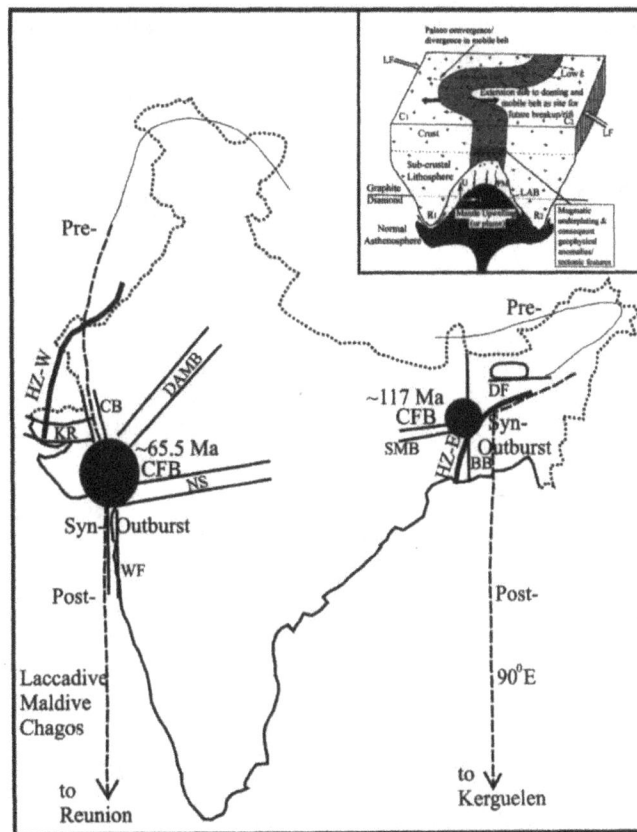

(b)

Fig. 5.23(b) Trace of Reunion and Kerguelen hot spot and their respective positions on Indian continent where maximum effects are visualized. DAMB- Delhi-Aravalli Mobile Belt, NS-Narmada-Son lineament, KR- Kutch Rift, HZ-W and HZ-E- Hinz Zone West and East, CB- Cambay Basin, SMB- Satpura Mobile Belt and CFB-Central Fold Belt, DF- Dauki Fault. Inset shows rise of a plume causing extension, rift and break up (Raval and Veeraswamy, 2007).

The concept of plumes has been extended to Archean–Proterozoic period based on occurrences of Komatites, (Herzburg, 2007). Such plumes and convection cells must have also helped in dissipating the earth's initial heat through conduction of heat through base of the lithosphere and thickening it since early history of the earth. However, there are some alternative views about origin of igneous provinces such as upwelling of asthenospheric melt (Foulger et al., 2005) which can be attributed to delamination of lithosphere (Elkins-Tanton, 2005) or decompression in the lithosphere. Even plate flexure has been considered to give rise to volcanoes (Hirano et al., 2006).

Some of the characteristics of regions affected by plumes are as follows which can be used for their identification:

(i) Progressive increase in the age of rocks of volcanic chain located away from the plume source/ hotspot (Fig. 5.24). This figure relates to the break up of Seychelles-Mascarene plateau from India at about 65 Ma at the time of eruption of Deccan trap which gave raise to volcanic islands of Chagos, Lacadive, Maldives etc. It shows progressive decrease in ages from about 68 Ma (Basu et al., 1993) in N-W India to about (65 Ma, Baksi, 1987) 40, 20 and 0 over Seychelles-Mascarene plateau and Reunion Island (Campbell and Davies, 2006) suggesting that these volcanic eruptions took place at Reunion islands, where this plume is located. Details of this eruption leading to the break up of Seychelles along the West Cost of India are given in the next section. A detailed account of geochemistry, paleomagnetism and geodynamics of Deccan trap is given in Section 9.2 while describing the Deccan Volcanic Province of Saurastra and subtrappean Mesozoic sediments

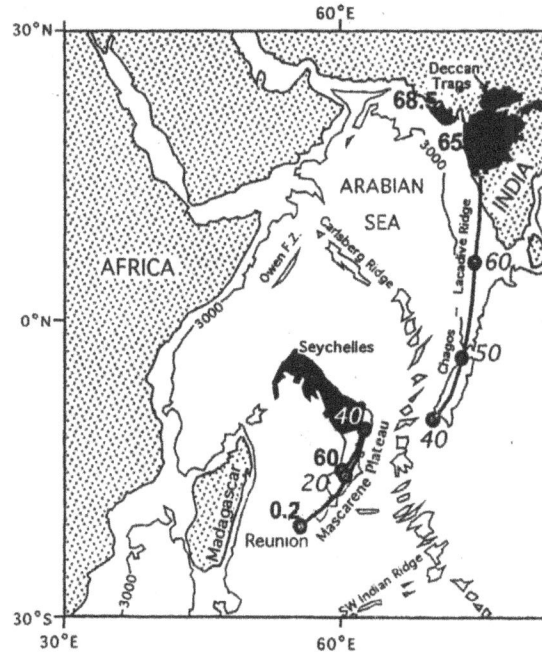

Fig. 5.24 Trace of Reunion plume with decreasing age towards the source of the plume, viz. Reunion Island. It has been responsible for the break up of Seychells-Mascarene plateau from the west coast of India and gave raise to Chagos-Lacadive-Maladive Ridge and Deccan Volcanic province over the Indian continent. (Modified after Campbell and Davies, 2006).

(ii) A regional uplift of moderate amplitude, larger than the existing volcanic province

(iii) A hot magma chamber in the mantle of limited extent based on the seismic tomography.

(iv) An under plated crust which implies rocks of intermediate velocity (>7.0 km/s) and density (>2.9 g/cm^3) between the lower crust (6.9 km/sec, 2.9 g/cm^3) and the upper mantle (8.1 km/sec, 3.3 g/cm^3) along the Moho.

(v) Presence of seaward dipping reflectors in seismic section along rifted continental margins as discussed in case of West Costs of India and Norway in Section 3.8 while discussing their gravity and magnetic signatures and demonstrated in a schematic section (Fig. 5.27) and in a seismic section in section 9.1

Some of these points along with rock types associated with plumes are discussed in detail by Condie (2001), Campbell and Davies (2006) and Pirajno (2007) etc.

5.3.2 Breakup of Gondwanaland

Storey (1995) has considered three stages in the break up of Gondwanaland (Fig. 5.25(a)).The first stage (a) at 200 Ma represents the scenario before the break up while the second stage (b) is related to Early-late Jurassic (180-160 Ma) when west and East Antarctica (India, Australia and New Zealand) separated from each others (Fig. 5.25(b)). The third stage is related to Early Cretaceous times (120-130 Ma) when South America, Africa and Indian plates separated from Antarctica (Fig. 5.25(c)). The fourth stage (Fig. 5.25(d)) relates to Late Cretaceous times (100-90 Ma) when Australia and New

Zealand separated from Antarctica. Gaina et al., (2007) based on Sea floor spreading magnetic anomalies in the Enderby basin off East Antarctica suggested same period of about 120 Ma for break up of India-Antarctica and India-Australia. According to Storey most of these events are related to large scale volcanic eruption from different plumes as given in these Figs. 5.25(a), (b), (c) and (d). Separation of Madagascar and Seychelles from the Indian continent was related to Marion and Reunion Plumes at about 88 Ma and 65 Ma, (Fig. 5.26 (a) and (b), Storey, 1995) respectively, the latter being associated with the eruption of Deccan volcanic province along west cost of India. Fig. 5.26(a) also shows the position of the Kerguelen hot spot that was responsible for the break up of India at about ~120 Ma from Antarctica and gave rise to Large Volcanic Province of Kerguelean Island and contemporary magmatic rocks of Ninety East ridge in the Bay of Bengal (Fig. 5.35) and Rajmahal and Sylhet traps over the Indian continent (Section 7.6). The periods for these events are inferred based on sea floor magnetic anomalies in the Indian Ocean and direct dating of samples.

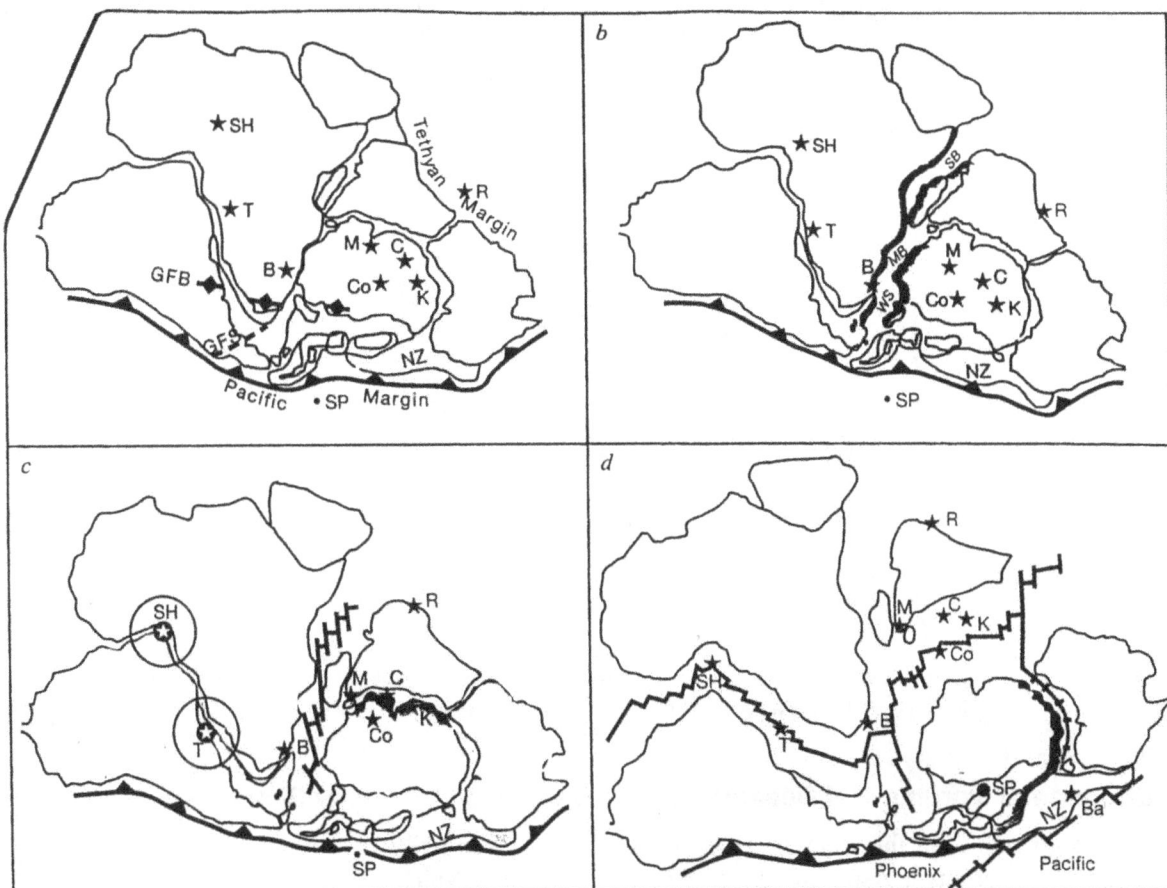

Fig. 5.25 Continents of Gondwanaland and plumes and hot spots responsible for their break up and initiation of sea floor spreading at (a) 200 Ma (b) 160 Ma (c) 130 Ma and (d) 100 Ma related to positions before the break up (a) and the break up of West and East Antarctica (b) while (c) and (d) relates to the break up of Africa and the India and India and Australia from Antarctica respectively. Plumes, hotspots are referred to as: Ba = Balleny; B = Bouvet; C = Crozet; Co = Conrad; K = Kerguelen; M = Marion; R = Reunion; SH = St Helena; T = Tristan da Cunha. Some tectonic/ geographical units are: GFS= Gastre Fault system; GFB= Gondwana Fold Belt, MB = Mozambique basin, NZ = New Zealand continental block, SB = Somali basin; SP= South Pole; WS = Proto Weddell Sea. ((Lawver, et al., 1992; Storey, 1995).

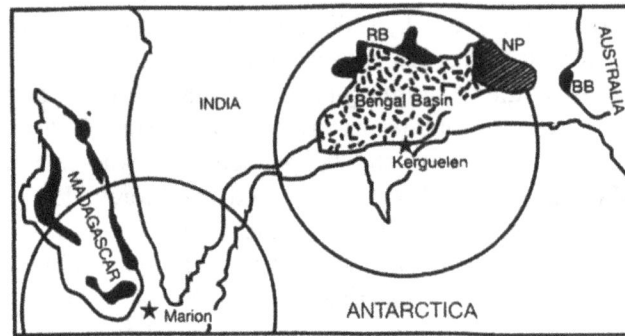

Fig. 5.26(a) Break up of Indian plate from Antarctica due to Kerguelen plume/ hotspot at about 118 Ma and Madagascar from the Indian plate at about 88 Ma. RB= Rajmahal Basalt, BB= Bunburg basalt, NP= Naturalistc plateau (Kent, 1991; Muller et al., 1993; Storey, 1995).

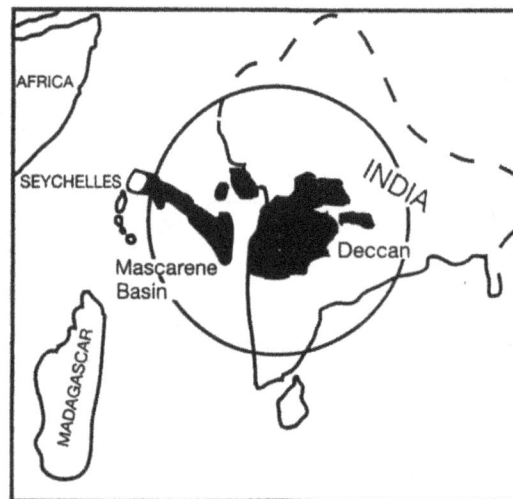

Fig. 5.26(b) Break up of Seychelles-Mascarane plateau from the Indian plate at about 65 Ma due to Reunion plume (Fig. 5.24) (White and McKenzie, 1989; Storey, 1995).

5.3.3   Continental Margins and Oceanic Crust – Density and Velocity Structures

Break up of continents form the continental margins which are primarily two types, viz. active and passive continental margins. Passive margin is further classified as volcanic or non-volcanic passive margins.

Continental Margins which are involved with plate tectonic processes (in Section 5. 2) such as subduction along West Coast of S. America (Fig. 5.20), West Coast of Andaman-Sumatra islands, West Coast of U. S. A (Transform fault) etc., are known as active margins. Contrary to it, the margins which are devoid of plate tectonic processes are known as passive margins such as offshore West and East Coasts of India. Passive Margins can form in two ways as discussed above, firstly, due to plume related break up forming a large volcanic province along continental margins. Such margins are called volcanic passive margins such as offshore West Coast of India. Secondly continents can also break up due to localized igneous activity from the upper mantle caused by decompression or local thermal sources. Such continental margins are known as non-volcanic passive margins. However, it may be noted that igneous activity is important for break up of continents in both the cases, but they differ in their source and magnitude in the two cases.

Break up of continents cause reduction in the crustal thickness and produces several specific structures between continent and the deep ocean formed due to break up and subsequent spreading which are given in Fig. 5. 27. The entire section between coast line and ocean is known as continental margin which is further divided into continental shelf, shelf break etc. Due to igneous activity, continental shelf is characterized by mafic rocks and shows extensional tectonics in form of normal faults, developing into graben like basins and there by represent stretched lithosphere. Several large scale oil bearing basins such as Bombay High offshore West coast of India have been discovered over the continental shelf of volcanic passive margins that is discussed in Section 9.6. There are normally two stages of intrusions similar to rift basins as discussed in Section 2.7 and 3.8 that are referred to as intrusion (1) and intrusion (2), as given in Fig. 5.27. The first stage of intrusions in the form dykes etc., are found close to the coastline while second stage form the main phase of igneous activity along the shelf break and is primarily responsible for the break up. These intrusions originating from plumes or upper mantle and passing through crust consist of both mafic and felsic rocks known as bimodal volcanics which is a characteristic of intrusives related to rifting or continental margins. These being bimodal volcanics give rise to oscillating characters of magnetic anomalies, positive and negative with in short intervals due to heterogeneity in their composition and maybe difficult to separate from the sea floor spreading magnetic anomalies close to continental breaks. Therefore, sufficient precautions should be taken to model sea floor spreading magnetic anomalies in these cases. These igneous intrusions show Seaward Dipping Reflectors, (SDR) in the seismic section as inner SDR and outer SDR (Menzies et al., 2003) as shown in practial examples in Figs 3.24b. 3.25a and 3.27. They are demonstrated in an actual seismic section offshore west coast of India in Section 9.1. However outer SDR are more common. They are formed due to volcanic flows under submarine conditions which gets solidified layer-wise in presence of cold water due to quenching and produce reflectors dipping towards the ocean due to rifting. It may be noted that the same volcanic flows at the surface such as Deccan Volcanic Province do not produce reflections due to their massive nature. Sedimentary deposits on the continental shelf and slope may show sectoral characteristics along the strike direction due to deposits formed from the rivers bringing material from different sources such as sandstones, shales, phyllites etc., occurring in sections strike-wise. In addition, the slope and deep sea fan deposits show sorted nature of sediments due to turbidities which cause heavier and larger grains to settle first.

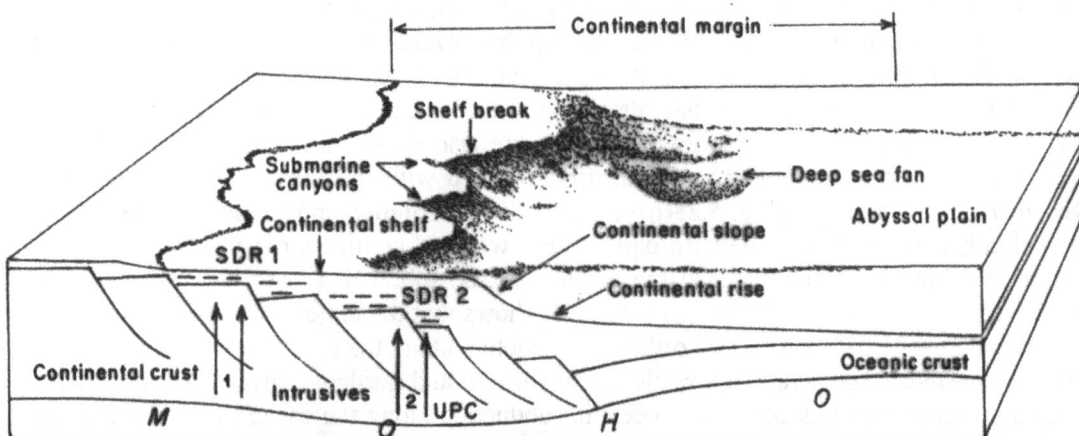

Fig. 5.27 A schematic diagram of passive continental margin with continental shelf, shelf break and associated canyons. It also shows the intrusive related to break up which are usefully found in two groups near the coast and the shelf break giving rise to two sets of Seaward Dipping Reflectors (SDR) over the basement and under plated crust (UPC). SDR1 and SDR2 are known as inner and outer SDRs, respectively.

Oceanic crust beyond the continental shelf is also mafic in nature as it is formed from sea floor spreading process. Based on seismic investigations and some deep bore holes in ocean, the oceanic crust is divided primarily in three layers known as Layer 1 (sediments); Layer 2 (basalt) and Layer 3 (gabbro) whose further division and physical characteristics are given in Table 5.2. A typical cross section of the oceanic crustal rocks and related gravity anomalies are given in Section 6.3.4 (Fig. 6.54(a)) across Muslim Bagh Ophiolite along the Weastrn Fold Belt (Pakistan).

Table 5.2 Oceanic Crustal Layers with their Velocities and Densities

Layers	Thickness (km)	Velocities (km/s)	Density gm/cc
Sea water	2-4 km (average)	1.5	1.03
Layer 1; sediments	0.5-1 km	2	1.8-2.0
Layer 2			
Layer 2A; Extrusive Basalt	0.5 km	3.0-4.5	2.7
Layer 2B; Intercalated Sediments	0.5 km	4.5-5.7	2.4-2.8
Layer 2C; Sheeted Dykes	1-2 km	5.5-6.5	2.9
Layer3			
Layer 3A Gabbroic Rocks	2-4 km	6.5-7.0	3.0
Layer 3B			
Mafic Cumulate	2-4 km	7-7.5	3.1
M	O	H	O
Upper Mantle Peridotite		8.1	3.3

5.3.4 Cenozoic Reconstructions of Gondwanaland

Reconstruction of Continent during this period being important for present day plate tectonics, it is discussed below along with the drift history of Eurasia and Gondwanaland with special reference to the Indian continent. Ricou (1996) has given the reconstruction of these continents for different periods, which is quite revealing as given below.

Fig. 5.28(a) shows the reconstruction for 67 Ma just prior to the eruption of Deccan trap (65 Ma), which shows the Indian and the African plates moving northwards towards Eurasian plates and occanic crust is shown subducting along thc northcrn margin of thc Eurasian platc. Thc micro continents of Tibet, North and South China had already accreted to Europe and Siberian plate to form Eurasian plate by 130-140 Ma. North and South America are also shown moving westwards. In this manner by 55 Ma the western part of Indian plate first collided with the Eurasia forming Western Syntaxis. Reconstruction for 45 Ma (Fig. 5.28(b)) after collision of Indian and Eurasian plates shows that the subduction had stopped in the western part while it was still continuing in the eastern part as ocean-continent collision which had initiated the bending of the subduction zone along the Eastern Syntaxial Bend. A reconstruction for 18 Ma (Fig. 5.28(c)) shows that the under thrusting of the Indian plate under Eurasian plate had started in the north and subduction along the eastern part of Eurasia was still continuing as ocean-continent collision while in the western and northern parts, it was continent-continent collision. Because of this continuing oceanic subduction along the eastern part, the Eastern Syntaxis of Himalaya is formed and the entire collision zone took a southward turn along the Eastern Syntaxial Bend and subduction along Sumatra-Java subduction zone continued. Due to continued subduction in the eastern part along Java trench, Indian continent experienced an anticlockwise rotation as can be seen from the position of Indian continent in Fig. 5.28(b) and (c). The anticlockwise rotation is still continuing and might be responsible for intense folding and deformation zone in the Indian Ocean between the Indian and the Australian plates (Fig. 5. 35).

Fig. 5.28(a) Reconstruction of continents at 67 Ma showing northward movement of Indian and African plates and subduction of Tethys ocean under the Eurasian plate. It also shows the separation of Indian and Australian plates towards the east and North and South American plates from Africa and European plates towards the West and development of subduction zones along West Coasts of North and South America. It is used to demonstrate the development of subduction zones with special reference to Indian plate (Ricou, 1996).

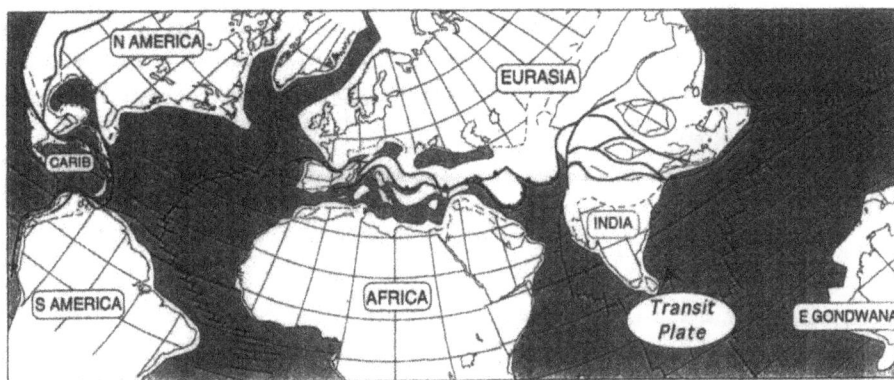

Fig. 5.28(b) Reconstruction at about 45 Ma showing junction of Indian and Eurasian plates in the Western part of the subduction zone while in the eastern part, the oceanic crust continues to subduct causing an anticlockwise rotation of the Indian continent (Ricou, 1996).

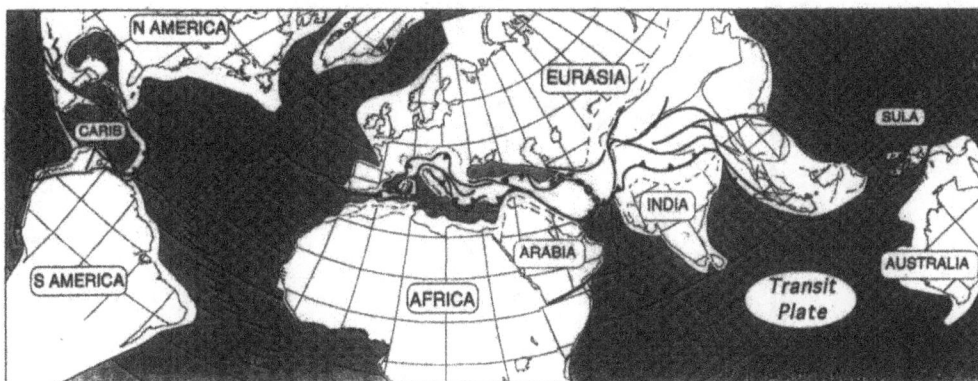

Fig. 5.28(c) Reconstruction of continents at 18 Ma showing under thrusting of Indian continent and subduction of Tethys oceanic crust under the eastern part was still continuing causing southward bending of the subduction zone which gave raise to Eastern syntaxial bend of Himalayas along upper Assam. This causes further (anticlockwise) east ward rotation of the Indian continent. (Ricou, 1996)

5.3.5 Cenozoic Movement of Indian Continent

Drift history of Indian continent during Cenozoic period is important for plate tectonics in this region and formation of Himalayas, which is geodynamically the most important event in this region and is therefore, described separately below. As given in Fig. 5.1, it has moved from about 50° S to the present day position during last 110-120 My. Prior to it, it was in northern hemisphere during Precambrian-Cambrian period, crossing equator during this period. Using paleolatitudes for Deccan trap at the start of Cenozoic period and measured direction of magnetization in rocks of Himalayas, which are of later period, Klootwijk et al., (1986) provided a drift history of the Indian plate for Cenozoic period (Fig. 5.29) based on India-Africa relative movement data fixed to hot spot frame in the Atlantic ocean. It was found to be in good agreement with paleolatitude observations from DSDP cores on the Indian plate presented by them. This figure shows how this continent moved from about 30° S to present day position. They also suggested a counterclockwise rotation of the northern part of the Indian continent with respect to Australia during Miocene time when several tectonic activities from Himalaya (Chapter 6) have been reported. Anticlockwise rotation of India with respect to Java trench as envisaged above, might be responsible for intense deformation zone in the Indian Ocean between India and Australia known as Central Indian Deformation Zone as discussed in Section 5.4.4 and shown in Fig. 5.35. Initially, Indian and Australian plates were together known as Indo-Australian plate but subsequent to the discovery of this deformation zone, the two are treated separately with diffused plate boundaries (Fig. 5.35). It is interesting to note that seismic activity has been reported from this section and its junction with Java Trench (Fig. 5.35) has been site for several great earthquakes including that of December 26, 2004, which produced one of the largest Tsunami waves as discussed in Chapter 8.

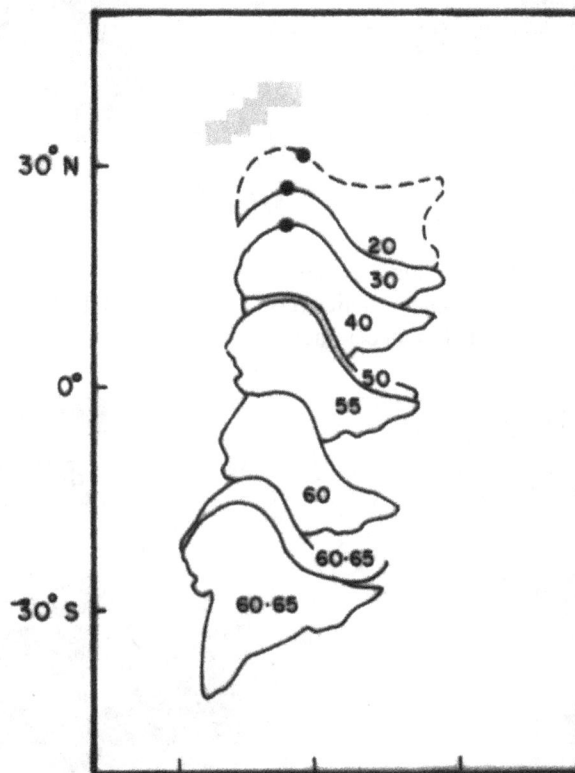

Fig. 5.29 Paleo positions of Indian continent based on paleomagnetic data (Klootwick and Peirce, 1975) showing its movement from 30°S during 60-65 Ma (Table 5.1) to 30°N (present day position), (Modified after Klootwijk et al., 1986).

Another reconstruction of Indian continent for this period has been provided based on sea floor spreading magnetic anomalies in the Indian Ocean and paleomagnetic measurements on samples from Deep Sea drilling projects site in the Indian Ocean (Fig. 5.30; Molnar and Tapponier, 1975; Kious and Tilling, 1999) which also shows an almost similar drift history. The speed of the Indian plate at about 55 Ma reduced from ~18 cm/yr to ~11 cm/yr that further reduced to 6.4 cm/yr at about 40 Ma (Molnar and Tapponier, 1975) indicating the collision of the NW part might have taken place during first reduction of the speed (Section 6.3) while the collision along the entire Himalayan arc might have taken place during second reduction in speed that slowed down its speed.

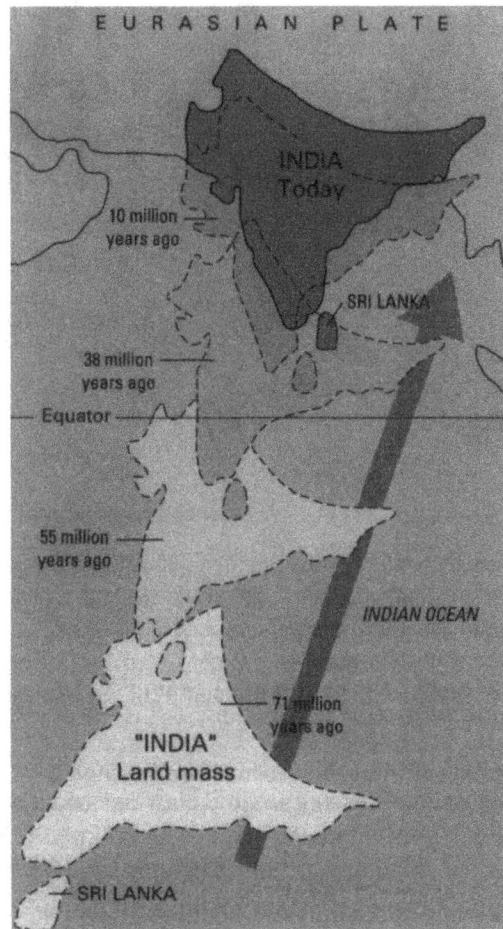

Fig. 5.30 Paleoposition of Indian continent based on sea floor magnetic anomalies in the Indian Ocean showing equatorial position at 55 Ma. It has drifted almost 6000 km from 40°S to the present position (Kious and Tilling, 1999; Modified after Molnar and Tapponier, 1975).

5.4   Indian Ocean

As indicated above, in reconstructions during different geological periods and plate tectonic elements in Fig. 5.13, the Indian Ocean is relatively quite complex with several spreading centers subduction zones, ridges and triple junctions. Fig. 5.31 (McKenzie and Sclater, 1971) is a simplified map showing the important plate tectonic elements with epicenters of earthquakes plotted on it for a specific period showing their inter relationship and some important sea floor spreading magnetic anomalies with their numbers as per the magneto stratigraphy scale (Fig. 3.10(a)). A more detailed distribution of these magnetic anomalies are given by Schlich (1982) Miles and Roest (1993), Miles et al., (1998), Owen

(1983) etc. There are primarily three main ridges, viz. Central Indian Ridge, South West Indian Ridge and South East Indian ridge forming a triple junction (RRR) known as Indian Ocean triple junction. The Carlsberg ridge towards the west and Andaman-Sumatra-Java (Sunda) trench towards the east and associated subduction zone and several aseismic ridges make this region tectonically quite complex and interesting.

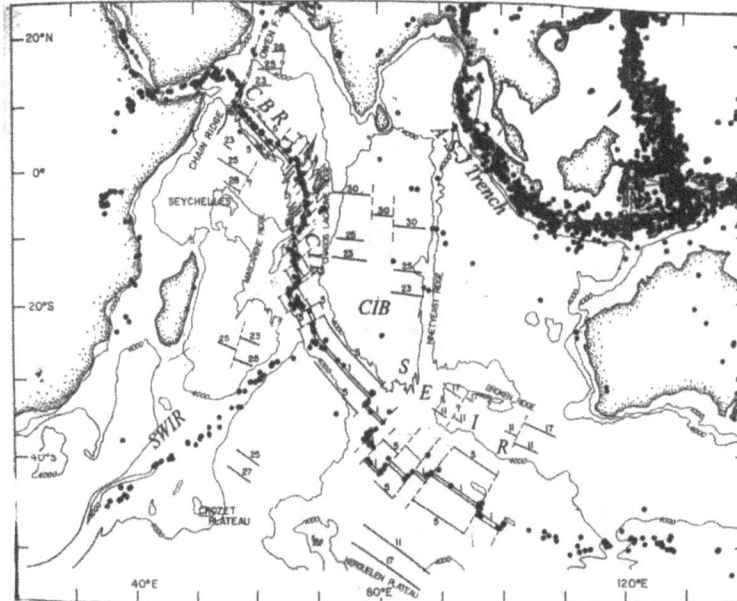

Fig. 5.31 Indian ocean and some important sea floor spreading magnetic anomalies showing oldest magnetic anomaly 30 south of India and next set of magnetic anomalies 28 in the Arabian Sea and Somali basin north of Seychelles across Carlsberg ridge. Plot of seismic activities suggest that the Andama-Sumatra-Java (A-S-J) trench is seismically most active section (Mckenzie and Slater, 1971). CBR = Carlsberg Ride, CIB = Central Indian Basin, CIR = Central Indian Ride, SWIR = South West Indian Ride, SEIR = South East Indian Ridge.

The oldest magnetic anomaly reported in the Indian ocean are M25-M22 (Owen, 1983) related to Jurassic-Cretaceous period north and west of Australia and between Madagascar and East Africa, which shows that the breakup of Africa from Gondwanaland was initiated at about 160-150Ma and remaining parts of Gondwanaland (India, Madagascar and Australia) moved south wards. Next was breakup of Madagascar and India at about M0 (~118 Ma) along a spreading ridge, which eventually developed into south West Indian ridge. Finally, Australia separated from Antarctica along South East Indian ridge at about anomaly 34 (84 Ma) completing the breakup of Gondwanaland. As shown in (Fig. 5. 31), the oldest magnetic anomaly in northern part of the Indian ocean across Central Indian Ocean Ridge is 30 corresponding to about 66-68 Ma. As per the paleo latitude position the Indian continent was at about 34° S at that time (Fig. 5.1). This is also about the time the large scale Deccan trap eruption from Reunion plume took place giving rise to basaltic flows which form the chain of Lacadive-Maladive islands (Fig. 5. 24). However, subsequently some older magnetic anomalies (34) have also been discovered in this region which is discussed separately in the next section while discussing western part of the Indian Ocean. After magnetic anomaly 30(~67-68 Ma) we find the next stage of magnetic anomalies from 28(~63-64 Ma) on wards in the Arabian Sea. This suggests that probably the Central Indian Ridge jumped in to the Arabian Sea during eruption of the Deccan trap (65 Ma) forming Carlsberg ridge where the oldest magnetic anomaly 28 (~63-64 Ma) has been reported. Further details of events of the Western and the Eastern Indian Ocean including that of the Arabian Sea and the Bay of Bengal are discussed in the next sections.

5.4.1   Western Indian Ocean and the Arabian Sea

A detailed picture of sea floor spreading magnetic anomalies in this region has been compiled by Chaubey et al., (1998) as given in Fig. 5.32(a). It shows the oldest magnetic anomalies across Central Indian Ridge in the Indian ocean as 34 off shore Madagascar and 32 in the north and off shore Seychelles-Mascarene plateau indicating that break up Madagascar from the Indian continent and this plateau which at that time was attached to the Indian continent took place at magnetic Chron 34 which is related to the Cretaceous quite zone from about 83 to 118 Ma. Storey (1995) provided a date of about 88 Ma for this break up (Fig. 5.26(a)) and that for India-Antarctica break up as 118 Ma caused by Marion and Kerguelen hot spots, respectively. However dating of volcanic rocks from St. Marry island offshore West coast of India provided dates of 85-91 Ma (Pande et al., 2001) which is similar to the Late Cretaceous volcanic province of Madagascar (~84-92 Ma) (Torsvik et al., 2000) . Some dykes of similar age (87-91 Ma) have also been reported from the West Coast of India (Kumar et al., 2001) confirming the approximate period of break up of Madagascar from India as 88-90 Ma. This break up is suggested to be caused by Marion plume. However absence of large scale volcanics of this period from West Coast India indicate that it may represent even a non volcanics margin due to local igneous activity from mantle without involvement of any plume. However several authors as suggested above prefer a plume model. It may, however, happen that the large scale volcanic rocks of this period might be buried under Deccan trap of latter origin offshore west coast of India.

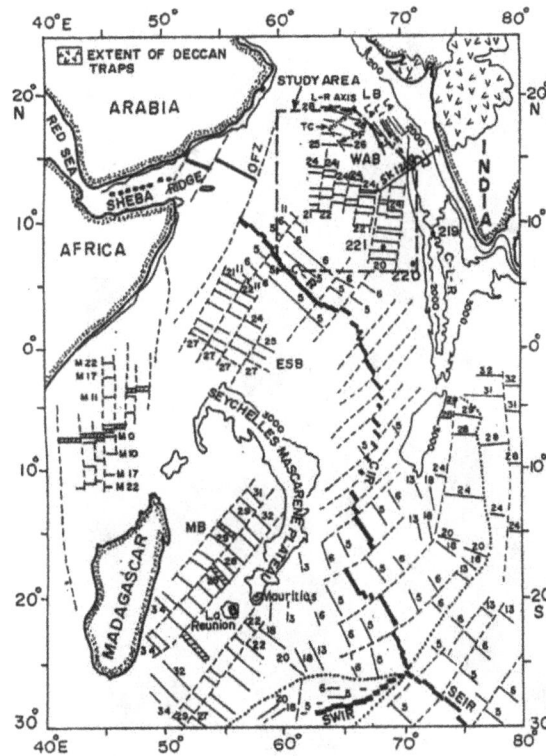

Fig. 5.32(a) Tectonic map of Western Indian Ocean and Arabian sea along with sea floor spreading magnetic anomalies (Choubey et al., 1998) which shows oldest magnetic chrone 32 South of India and 34 North of Madagascar and south of Seychelles-Mascarane plateau indicating the break up Madagascar from India at 88 Ma. The second set of magnetic chrone 28 (63-64 Ma) back wards are seen in Arabian Sea, and Somali basin across Carlsberg Ridge which are just after the eruption of Deccan trap suggesting that the spreading centre of Central Indian Ridge had shifted to Arabian sea as Carlsberg Ridge along a transform fault which might have been triggered due to volcanic eruption from Reunion plume. SK 1207 is a gravity-magnetic profile given in Fig. 5.32(b). LR = Laxmi Ridge and CLR = Chagos-Laccadive Ridge.

Another set of sea floor spreading magnetic anomalies in Fig. 5.32(a) are 28 close to Laxmi Ridge in North Arabian Sea and 27 off shore Seychelles-Mascarene plateaus and in the eastern Somali Basin offshore east coast of Africa. These appear to be related to the break up of latter which was part of the then West Coast of India. Therefore, the Laxmi Ridge offshore West Coast of India and offshore Seychelles-Mascarene plateaus and Somali basin form the conjugate section where seaward dipping reflectors have also been reported (Hinz, 1981), further confirming that they represent volcanic rifted margins as discussed above. This along with the trace of Reunion plume given in (Fig. 5.24) suggest that this break up was caused by Reunion plume at about 63-64 Ma which gave rise to Chagos-Laccadive –Maldive Ridges and Deccan Volcanic province over the Indian continent. Magnetic Chrons 27-21 in the eastern Somali basins suggest that spreading in this section took place during 61-46 Ma. The motion of Indian plate with respect to Somali basin plate is also estimated as 6.3 cm/yr (Royer et al., 2002) similar to that for the Indian plate after 55 Ma as given in Section 5.3. East of Laxmi Ridge, there are magnetic lineation L1-L4 in the Laxmi Basin which represent stretched lithosphere as discussed in Section 3.8.1. Choubey et al., (1998) have suggested that they may even represent sea floor spreading magnetic anomalies between magnetic Chron 28 to 24 related to Early Tertiary. This implies that while Reunion plume was interacting with lithosphere in this section, sea floor spreading was also taking place in this section.

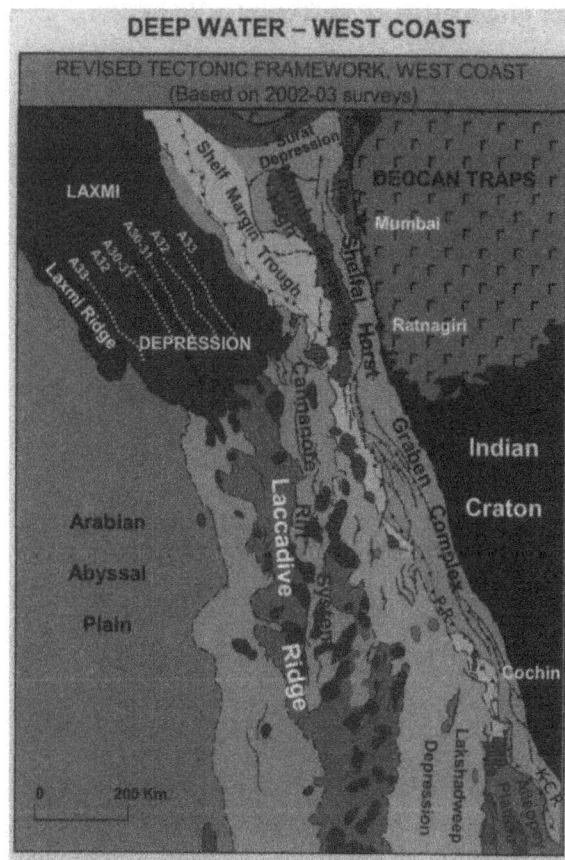

Fig. 5.32(b) Continental shelf offshore west coast of India showing horst and rift structure caused by Deccan trap eruption shown in green colour occupied by Lacadive ridge. It shows sea floor spreading magnetic anomalies in the northern part in the Laxmi basin (DGH, 2005).
KCR- Kori-Comorin Ridge and PR- Pratap Ridge. [Colour Fig. on Page 771]

5.4.2   Continental Shelf Offshore West Coast of India, Laccadive Ridge and Seaward Dipping Reflectors

Important tectonics offshore west coast of India is given in Fig. 5.32(b) (DGH, 2005) that shows horst and rift structures. These structures are formed due to Deccan trap eruption from Re Union plume and break up of Seychelles from the west coast of India during Late Cretaceous as described above. This map also shows sea floor spreading magnetic anomalies in the northern part, Laxmi Basin as postulated by Bhattacharya et al., (1994). A regional free air anomaly map of the continental shelf offshore west coast of India based on Satellite data is given in Fig. 5.40 that is discussed below. It shows two linear gravity highs almost parallel to the coast, one along the coast and the other along the continental margin corresponding to two linear sections of ridges shown in Fig. 5.33(a). This is typical of volcanic rifted margins world over as discussed in Chapter 3 (Section 3.8) due to large scale magmatic activity from a plume or hot spot. The features close to the coast represent the first phase of activity while the other away from the coast represents the main activity causing rifting. A detailed account of geodynamics and geochemistry Deecan trap is given in Section 9.2 while describing the Deecan Volcanic Province of Saurastra and subtrappean Mesozoic sediments. Continental shelf offshore west coast of India in the central part related to Laxmi ridge and Laxmi basin and in the southern part has been described in Sections 3.8 and 2.9, respectively which shows a wide continental shelf in the central part reducing southwards. As described in those sections, it represents a volcanic margin with seawards dipping reflectors and Laxmi Basin representing the stretched part of the continental shelf as found world over in case of volcanic margins. Specific examples are offshore west coast of Norway with Voring Plateau and Voring Basin similar to Laxmi Ridge and Laxmi Basin as described in section 3.8.

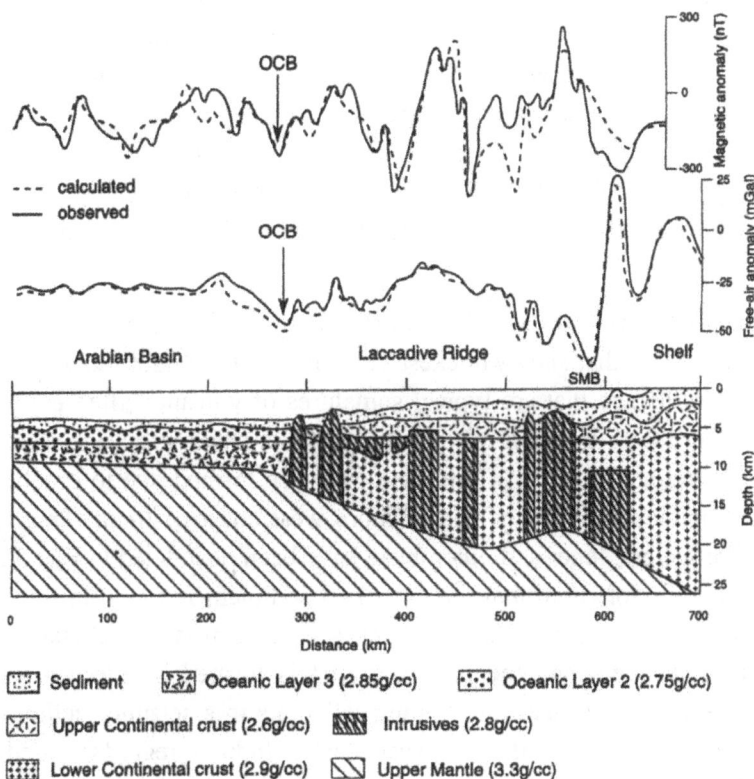

Fig. 5.33(a) Gravity and Magnetic profiles (SK 1207; 5.32a) across continental shelf off West Coast of India and computed crustal model showing mafic intrusives related to Deccan trap eruption resulting in to a volcanic passive rifted margin and stretched lithosphere between shelf and Arabian basin (Chaubey et al., 2002).

In the southern part the most important section offshore is the Chagos– Laccadive–Maladive Ridge which is considered to be part of Deccan trap eruption from the Reunion plume at about 65 Ma which resulted in the break up of Seychelles-Mascarene plateau from the Indian continent. The plume trace of Reunion plume is shown in Fig. 5.24 which shows the ages of rocks decreasing southwards towards the source of the plume and these Ridges form the trace of the plume. The effective elastic thickness (EET) along the plume trace suggest an EET of 2-6 km under Chagos-Lacadive-Maldive Ridge and those under Reunion –Mauritius Islands is higher (8-30) km (Tiwari et al., 2007). Almost similar Effective Elastic Thickness of 5-15 km over central Indian Ridge (Radhakrishna, 1996), 10-13 km along West Coast of India (Chand and Subrahmanyam, 2003) and over Deccan Volcanic province and Peninsular Shield, India (Section 4.4, 8-10 km) have been reported. These values suggest that this entire region was affected by volcanic activity and related effects of rifting and Chagos-Lacadives-Maldive Ridge with minimum EET formed the plume trace close to the spreading centre (Ridge).

Chaubey et al., (2002) have described the nature of continental shelf in the central part based on modeling gravity and magnetic data constrained from seismic studies along a profile extending from the shelf to the deep ocean basin across Laccadive ridge. This profile marked as SK 12-07 is shown in Fig. 5.32(a). Gravity and Magnetic data and the computed model are shown in Fig. 5.33(a) which shows a relatively thick (~20 km) transitional crust between shelf and the oceanic crust and several mafic intrusives related to Deccan Trap eruption along the shelf. The intrusives are in two groups, viz. along the shelf and along Laccadive ridge which forms the western margin of the shelf as suggested above in regard to volcanic margins (Fig. 5.27). It shows an upwarp of Moho in the central part that is typical of rifting and represents stretched lithosphere. Crustal structure is almost similar to that given across Laxmi basin north of this region as described in Section 3.8 with an uplift of Moho in the central part and thick crust under the Laccadive Ridge forming the shoulder as Laxmi Ridge in the previous case. Seismic velocities of 7.15-7.30 km/sec above the Moho reported by them represents under plated crust as is generally the case of volcanic margins shown in Fig. 5.27. A detailed discussion about continent-ocean boundary and crustal structure of this region has been given in Section 5.5.

A schematic geological section across Laccadive Ridge based on seismic profiles in this section is given in Fig. 5.33(b) (DGH, 2005) that shows thickest volcanic section under this ridge with sheeted dykes and seaward dipping reflectors that are typical signatures of volcanic rifted margins. Seaward dipping reflectors are formed due to eruption of volcanic rocks in shallow marine conditions due to chilling effect of cool water on hot magma. In this case, they are formed due to eruption of Deccan trap that was responsible for the formation of the offshore west coast of India. This figure clearly shows two sets of dipping reflectors, inner and outer those have been postulated in Fig. 5.27 indicating two phases of volcanic eruptions for rifted margins. Thickest sediment is found in the Lakshadweep depression that is prospective for oil exploration as discussed in Section 9.6. Cretaceous-Paleocene sediments are found fault controlled over the basement that is formed by Deccan trap indicating eruption of Deccan trap just prior to it. Seaward dipping reflectors in a seismic section are shown in Fig. 9.1(c), east of the Laccadive ridge and offshore west coast of India in regard to rifted margins and oil exploration. Fig. 9.1(c) clearly shows seaward dipping reflectors in the basement (Red) that is formed by Deccan trap in this section which are different in nature from those of the sediments overlying the basement.

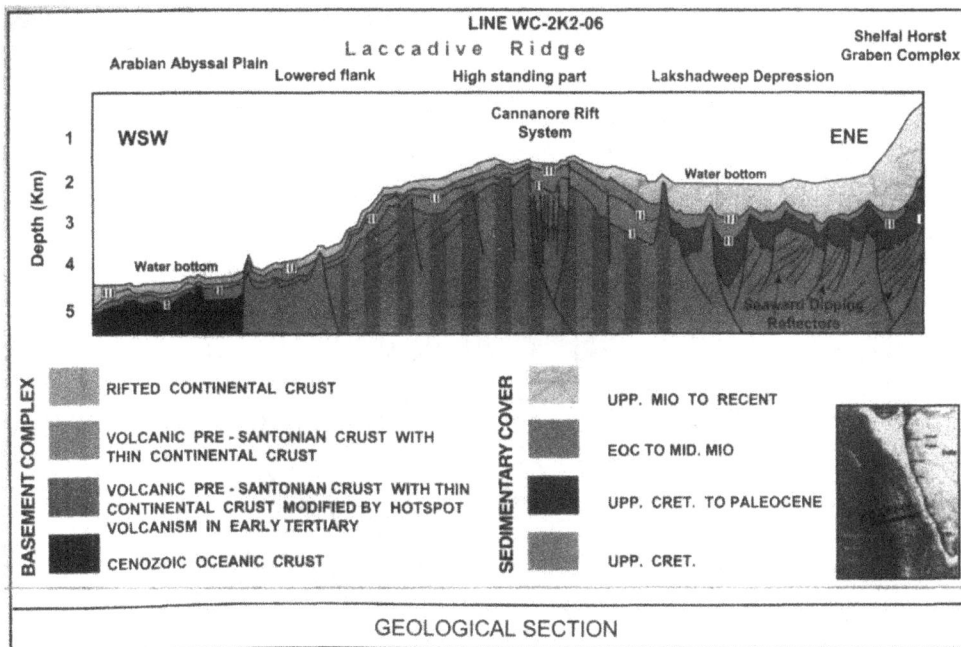

Fig. 5.33(b) A schematic geological section across Lacadive ridge offshore west coast of India showing rifted margin formed due to Deccan trap volcanic intrusive and seaward dipping reflectors caused by them indicating extension and rifted margin (DGH, 2005). [Colour Fig. on Page 772]

5.4.3 North Arabian Sea: Airborne Total Intensity Map

Southern part of the Arabian Sea and sea floor spreading anomalies of this section are described above along with the continental shelf off shore West Coast of India. The Northern section is covered by thick sediments of Indus fan and its magnetic characteristic are described below.

The total intensity airborne magnetic map of north Arabian Sea is presented in Fig. 5.34(a), which was covered by an airborne magnetic survey from Karachi as base and radial flight lines spreading in to ocean (Taylor, 1968). It shows the application of airborne magnetic surveys over ocean for quick coverage. It has delineated two trends of magnetic anomalies due to Murray ridge, SW of Karachi (0 Block) and Saurashtra ridge, SW of Saurashtra (M and N blocks) (Mishra, 1981). The magnetic anomalies in southern part AA' has been associated with the extension of Narmada-Son lineament in the Arabian Sea (Mishra, 1977). The zone between the Murray and the Saurashtra ridges are relatively free of magnetic anomalies and represents the section of thick sediments from river Indus north of it known as Indus fan. Murray ridge is an aseismic ridge, which represents the western plate boundary of the Indian plate and Saurashtra ridge may represent a fracture zone, which are occupied by mafic igneous rocks related to Deccan Trap eruption which is prevalent off shore West Coast of India as described above and covers most of Saurashtra over the continent (Section 9.2). The magnetic anomalies observed in this map are well defined lows and highs, which can be attributed to intrusives in this section. Spectral analysis of this data (Fig. 5.34(b)) provides a three layer model with spectral depths as 45.6, 21.6 and 8 km. which indicate approximate average depth to Curie point isotherm, Moho and the basement. Depth to the basement is confirmed from a seismic profile across Murray ridge (Closs et al., 1974). A depth of 21.6 km for Moho indicates a transitional crust as described below in Section 5.5 based on satellite gravity data. A higher depth estimate for Moho can also be attributed to thick sediments of Indus fan in the northern section. Even the depth to Curie point isotherm (580^0 C) as 45 km suggest a temperature gradient of 15 ^{0}C/km. which lies in between those corresponding to cratons and oceanic crust as discussed above (Section 5.2) indicating to a transitional crust.

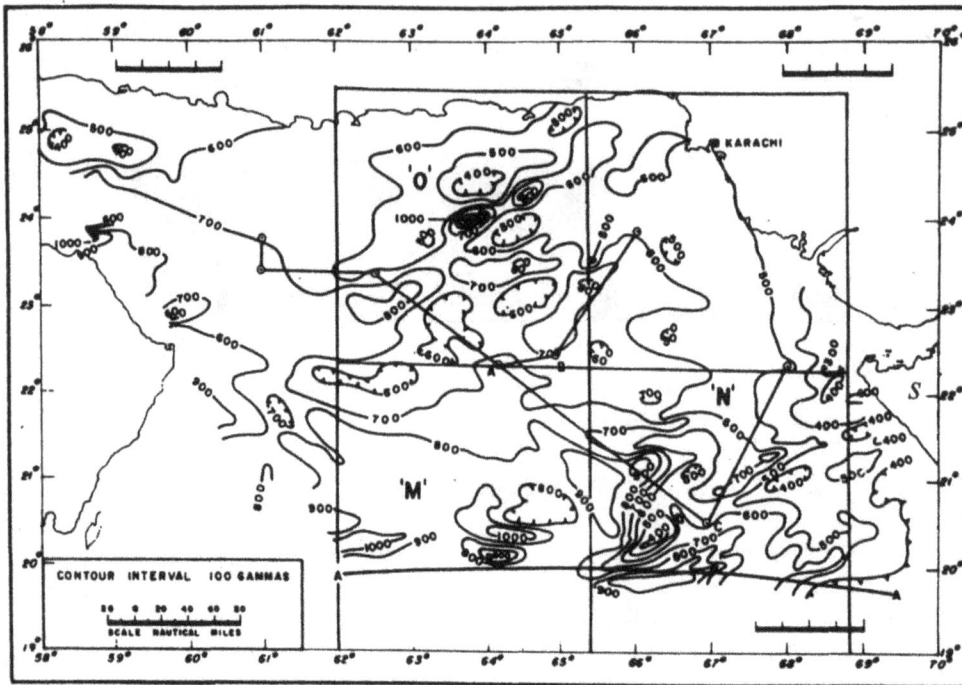

Fig. 5.34(a) Airborne total intensity map of North Arabian Sea (Modified after Taylor, 1968). It shows two sets of magnetic anomalies related to Murray Ridge off shore Karachi and Saurastra Ridge off shore Saurastra (S). These are typical magnetic anomalies due to mafic intrusives and might be related to Deccan trap eruption. (Mishra, 1977).

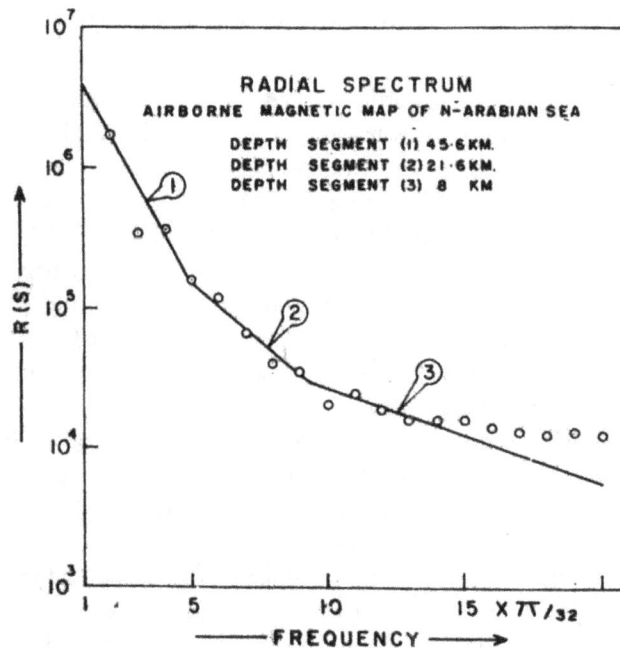

Fig. 5.34(b) Spectral analysis of Airborne magnetic map of North Arabian Sea showing three linear segments related to average depths at 45.6, 21.6, and 8 km which may represent Curie point isotherm, Moho under transitional crust and the basement respectively (Mishra, 1981).

5.4.4 Eastern Indian Ocean and Bay of Bengal

As shown in Fig. 5.31, eastern part of the Indian Ocean is dominated by south East Indian Ocean ridge which shows the oldest magnetic anomalies of 17 corresponding to about 38 Ma. This shows the period of separation of western part of Australia from Kerguelen plateau. The other important tectonics of this section is an almost N-S oriented Ninety East Ridge extending almost from 40^0 S up to Equator and even further north as a subsurface ridge up to Coasts of India and Bangladesh as discussed below in section 5.5. Fig. 5.35 (Fischer et al., 1982) provides a detail bathymetry and tectonics of this region showing 90 East Ridge with respect to Java trench and several other fractures almost parallel to it. It also shows Deep Sea Drilling Project and Ocean Drilling Project sites in this section. This map also shows the Central Indian Ocean Deformation Zone between the Indian and the Australian plates as diffused central plate boundaries. It is characterized by tectonic deformation, over thrusted basement and sediments, seismic activity and high heat flow (Verzhbitsky and Drolia, 1998). It might have developed due to plate tectonic movements of the Indian plate and anticlock wise rotation with respect to Java trench as discussed in the previous section. Another interesting feature in this section is the Eighty Five East Ridge. A regional free air anomaly map offshore east coast of India is given in Fig. 5.40 that shows one set of gravity highs along the coast and the other west of the Eighty five East Ridge as in case offshore west coast of India. These gravity highs may be related to magmatism related to the break up of India from Antarctica indicating volcanic rifted margin as discussed in Section 5.5.

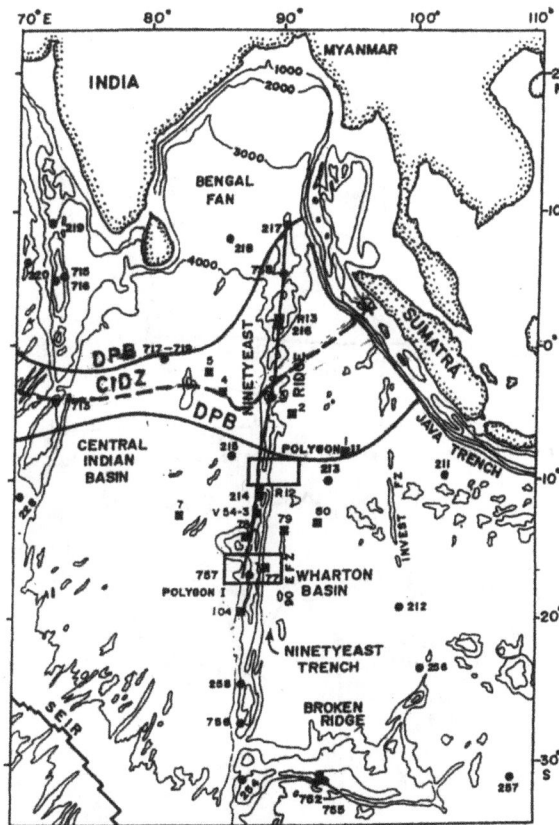

Fig. 5.35 Bathymetry of Eastern Indian ocean showing Ninety East Ridge and site of Deep Sea Drilling Project and Ocean Drilling Program. There are several fracture zones and ridges east of 90 East Ridge and parallel to it extending up to Java (Sunda) trench (Fisher et al., 1982, Krishna et al., 2001). It also shows the Central Indian Deformation Zone (CIDZ) indicated by Diffused Plate Boundaries (DPB) between the Indian and the Australian plates.

5.4.5   Bay of Bengal, Eighty Five East Ridge and Offshore Andaman Islands – Marine Magnetic and Seismic Profiles

Four magnetic profiles recorded in Bay of Bengal approximately along 10°, 14°, 17° and 20° N by Scripps Institute of Oceanography along with the bathymetry (Rao and Rao, 1986) are given in Fig. 5.36(a). Bathymetry shows sudden change in the depth of the ocean floor at about 100 km from the coast line suggesting a narrow continental shelf in this section compared to the offshore west coast of India. The Eighty Five East Ridge does not find reflection in any of the profiles except in the northern most profile (20^0 N) while Ninety East ridge is prominently reflected in the southern profiles. A seismic section across the Eighty Five East Ridge is given in Fig. 5.36(b) (DGH, 2005) that shows its depth as ~7-8 km and over lying Tertiary sediments. It also shows the structural highs and wedge out prospects for hydrocarbons that is discussed in Sections 9.1 and 9.6. Another seismic section offshore Godavari-Krishna basin along the East Coast of India (Fig. 5.36(c), DGH, 2005) shows sharply dipping continental margin with thick Tertiary sediments followed by Mesozoic sediments. It also shows several structural highs, wedge outs and bright events that are significant for oil exploration as described in Sections 9.1 and 9.6. It is interesting to note that this section does not show seaward dipping reflectors as offshore west coast of India (Fig. 5.33(b) and 9.1(c)) that is due to formation of the East Coast of India happened much earlier during break up of Gondwanaland due to Kerguelen hot spot (Section 5.3) during Middle Createceous and reflectors related to that event must be further below that has not been penetrated so far.

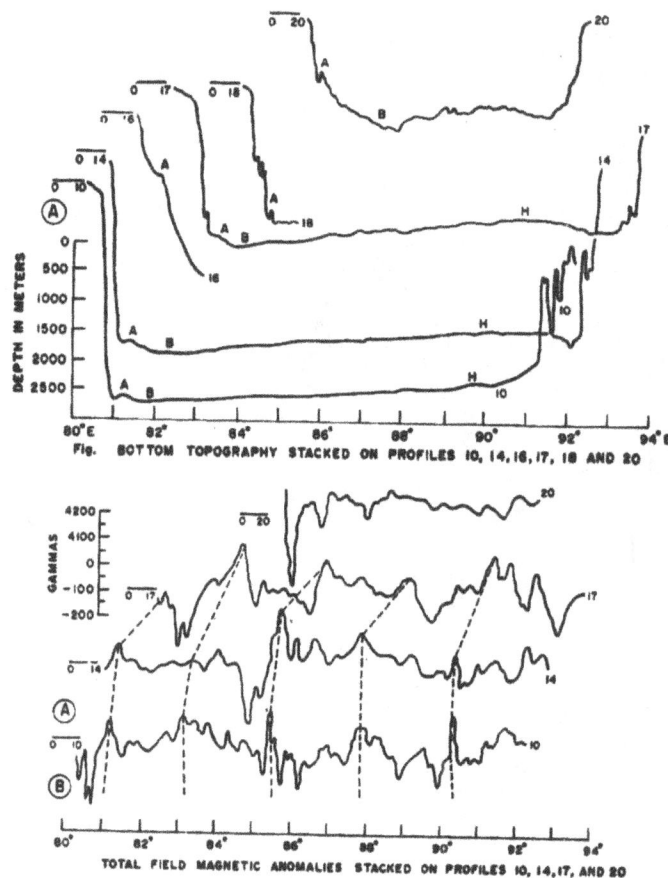

Fig. 5.36(a) Bathymetry and marine magnetic profile in Bay of Bengal (Rao and Rao, 1986) along 10, 14, 17 and 20^0 N profiles showing oscillatory character, typical of sea floor spreading type of magnetic anomalies (Mishra, 1991).

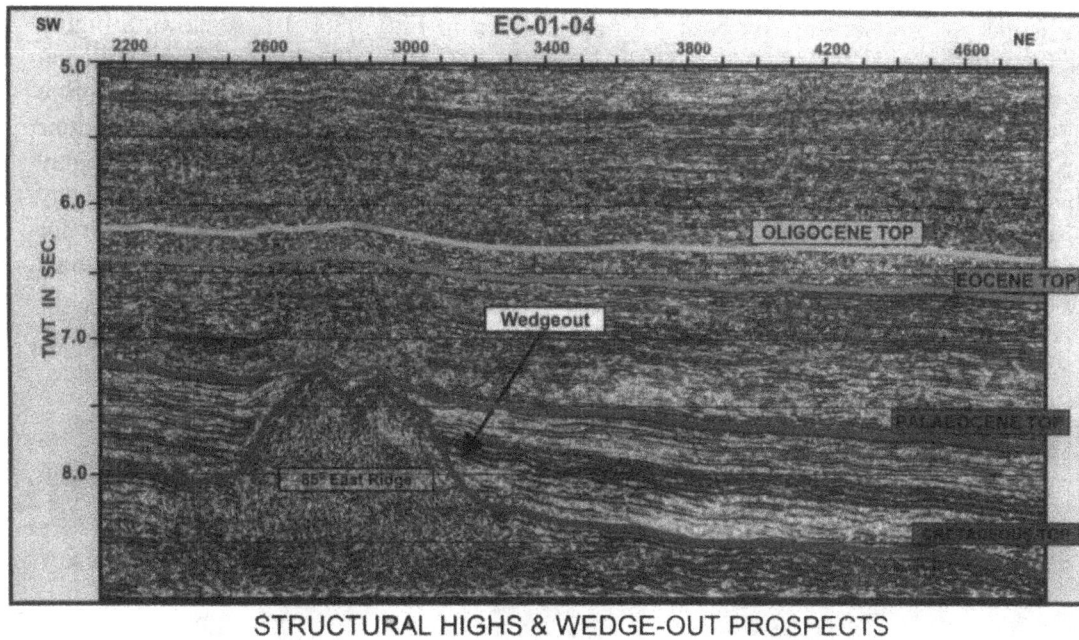

STRUCTURAL HIGHS & WEDGE-OUT PROSPECTS

Fig. 5.36(b) A seismic section across Eighty Five East Ridge showing a depth of ~7-8 km and the stratigraphy of Tertiary sediments in this section. It also shows the structural high and the stratigraphic wedge out as suitable sites for oil traps (DGH, 2005). [Colour Fig. on Page 772]

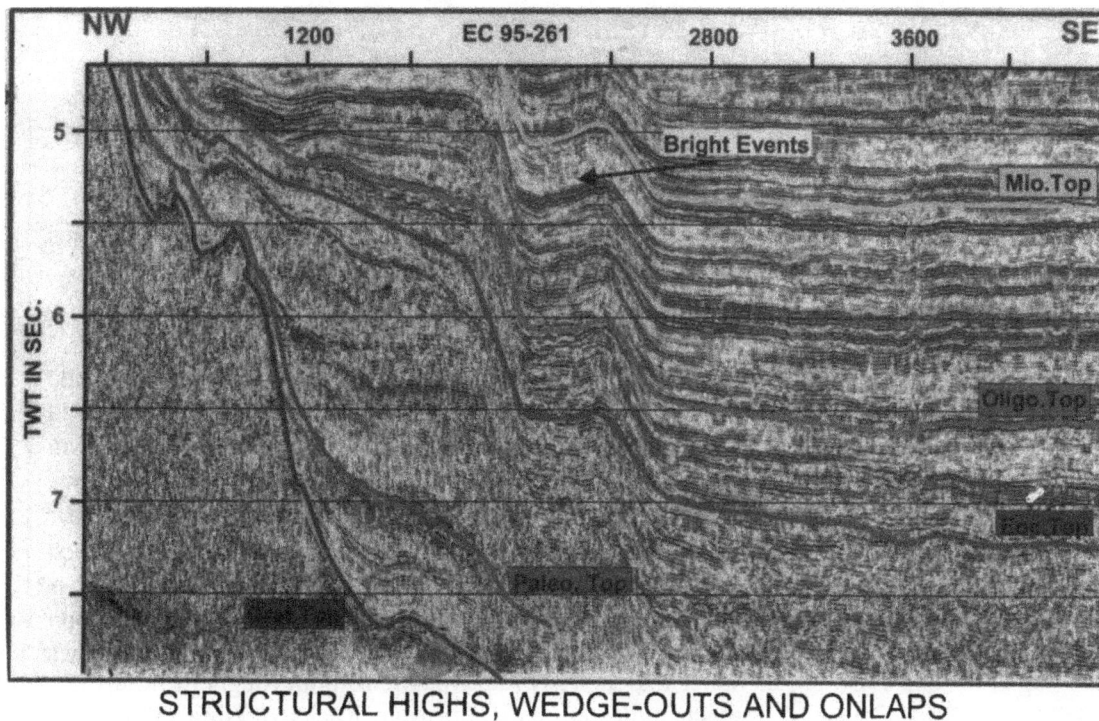

STRUCTURAL HIGHS, WEDGE-OUTS AND ONLAPS

Fig. 5.36(c) A seismic section offshore Godavari-Krishna basin along the east coast of India showing Tertiary stratigraphy and a sharply dipping continental margin and structures favorable for oil exploration (DGH, 2005). [Colour Fig. on Page 773]

Another seismic profile offshore Andaman islands is given in Fig. 5.36(d) that shows a highly disturbed section with several folds, faults and shale diapirism as a result of convergence and collision of the Indian plate and Burma micro plate along Andaman-Sumatra-Java subduction zone. It, therefore represent a forearc basin that is usually characterized by such tectonics. Differences in seismic sections Figs 9.1c, 5.33c and d across the offshore West and the East coasts of India and Andaman islands may be noted. The first two sections related to rifted margins due to extensional tectonics show a relatively undisturbed section and fracture zones are limited to the volcanic rocks forming the basement for subsequent sedimentation while the third one shows a highly disturbed section due to compressional tectonics and overlying sedimentary section is also faulted and folded.

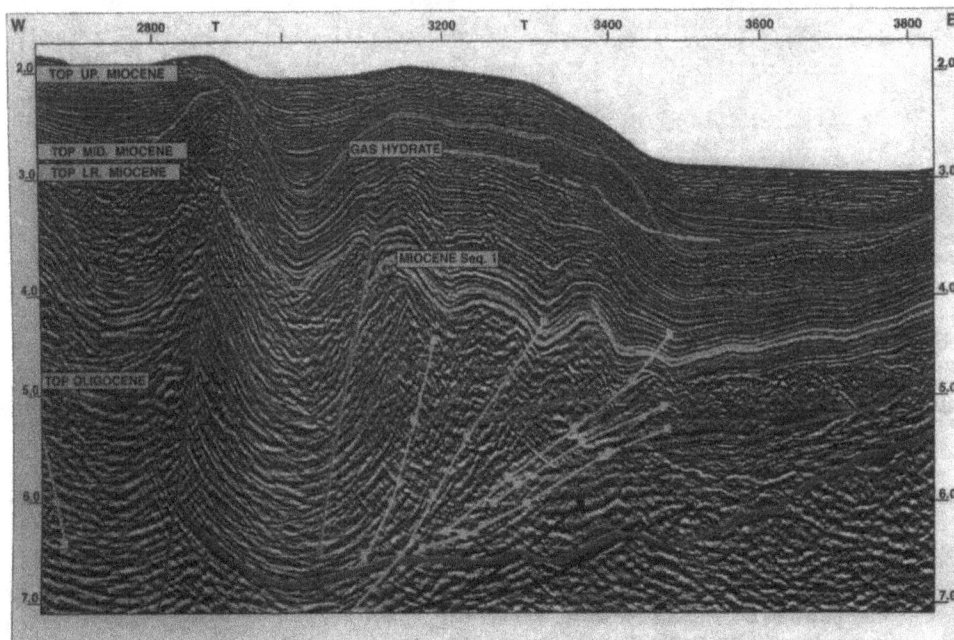

SHALE DIAPIRISM DUE TO COMPRESSIONAL TECTONICS, FORE ARC BASIN

Fig. 5.36(d) A seismic section offshore Andaman islands showing a highly disturbed section with several faults, folds and shale diapirism that are typical of compressional tectonics near subduction zones.
[Colour Fig. on Page 773]

The magnetic profiles (Fig. 5.36(a)) show oscillatory characters typical of oceanic magnetic anomalies, which might have been caused by sea floor spreading kind of magnetic anomalies due to volcanic rocks of different polarity. Data along these four profiles were digitized at about 2 km interval and individually transformed to frequency domain. The power spectral plot versus frequency for data along two profiles along 10° N and 14° N are given in Fig. 5.37(a), which provides linear segments with depth of magnetic basement as 8 km along 10°N and 10 km along 14°N, which approximately coincides with depth to the basement based on seismic profiles (Curray et al., 1982) as discussed in the next section. The digital data along these profiles are transformed to magnetization distribution by spectral analysis (Section 4.3) and given along a profile in Fig. 5.37(b), which shows positive and negative magnetization, the latter indicating reverse magnetization. Magnetization varies from about 2.5×10^{-3} to -2.5×10^{-3}, which indicates mafic Intrusive Rocks (Mishra, 1991). The magnetization distribution repeat itself on either side of 85° indicating that it might have been a spreading ridge sometimes in geological past. However, in absence of any direct dating of rocks, period for this spreading ridge is difficult to assign. However, the most important tectonic event in this section has been the breakup of India from Antarctica during Late Cretaceous (118-120 Ma) and these magnetic anomalies may correspond to it. Late Jurassic–Early Cretaceous magnetic anomalies have been reported from eastern Indian Ocean adjacent to Australia (Fullerton et al., 1989; Royer and Sandwell, 1989). Ramana et al., (1997) have also mapped Cretaceous magnetic anomalies across 85 East Ridge. As indicated below in section 5.5, the trace of gravity anomalies due to 85 East Ridge connects it to the Afanasy-Nikitin seamount in the Indian Ocean. Paul et al., (1990) have suggested low elastic plate thickness of 2-5 km indicating that this seamount might have formed close to a spreading axis as suggested above for 85 East ridge.

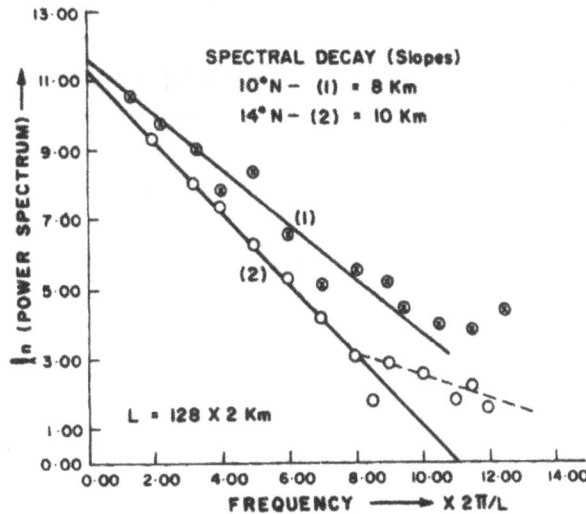

Fig. 5.37(a) Spectrum of magnetic profile along 10^0 and 14^0 N showing depths of 8 and 10 km, respectively which represent depth to the basement. Increasing depth northwards indicate thickening of sediments (Mishra, 1991).

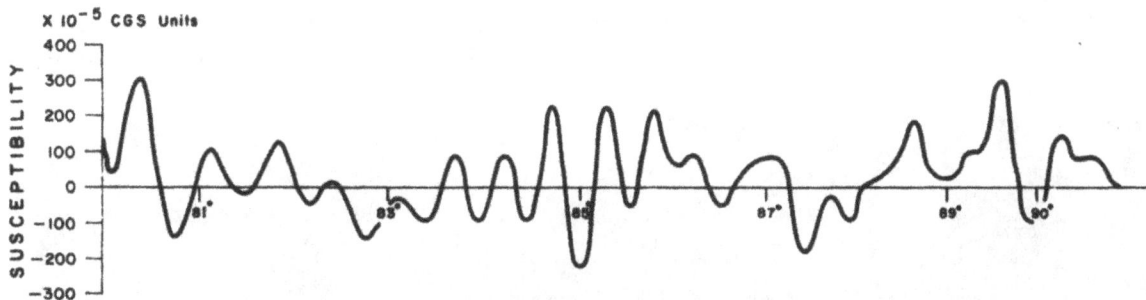

Fig. 5.37(b) Magnetization distribution along a profile computed from magnetic data (Fig. 5.36). It shows positive and negative magnetization varying alternately across 85 East Ridge which may represent normal and reverse magnetization. (Mishra, 1991).

5.4.6 Ninety East Ridge – Crustal Section and Effective Elastic Thickness

The most important element of Eastern Indian Ocean and Bay of Bengal are 90 East Ridge and Andaman-Sumatra-Java trench. Important tectonic elements of this region are given in Fig. 5.35 (Fischer et al., 1982) which shows a N-S linear bathymetric feature corresponding to this ridge and a curvilinear Java trench related to subduction zone. Density Model (Fig. 5.38(a)) along 17^0 S across this ridge in the Indian Ocean, constrained from a seismic profile suggest a thick crust of about 22 km

(Tiwari and Mishra, 2003) and a thick under plated lower crust (~ 6 km) indicating its volcanic origin, based on gravity and seismic investigations, Krishna et al., (2001) have also provided a similar thickness of about 22 km in the central part of this ridge. The low density body in the upper mantle west of the root of the ridge may indicate serpentinization of peridotite rocks of upper mantle. Serpentinization is an important process which may reduce the density of upper mantle rocks due to absorption of water in existing peridotite rocks upto 10% to form serpentinite. Crustal thickening in this case is about 12 km. The maximum topography of this ridge in this section is about 2.6 km which as per Airy's model of isostatic compensation (Section 2.5) would provide crustal thickness of of 2.6 (2.32-1.03)/(3.35-3.05) ~11.2 km that is almost approximately same as that given in the above figure. Based on admittance computations (Section 4.4), Tiwari and Mishra (2003) suggested on effective elastic thickness of close to zero km (Fig. 5.38(b)) which also suggests Airy's root model of compensation. Fig. 5.35 also show some major fracture zones almost equally spaced at 1° interval east of 90 East Ridge, which almost extend up to Java trench. Their almost equal spacing suggest that they might be related to some basic earth processes such as eruption over a rotating sphere as described above for fractures perpendicular to Mid Oceanic Ridges. These major fracture zones are also subducting along Java trench which might be the cause for this section to be seismically active as discussed in Chapter 8.

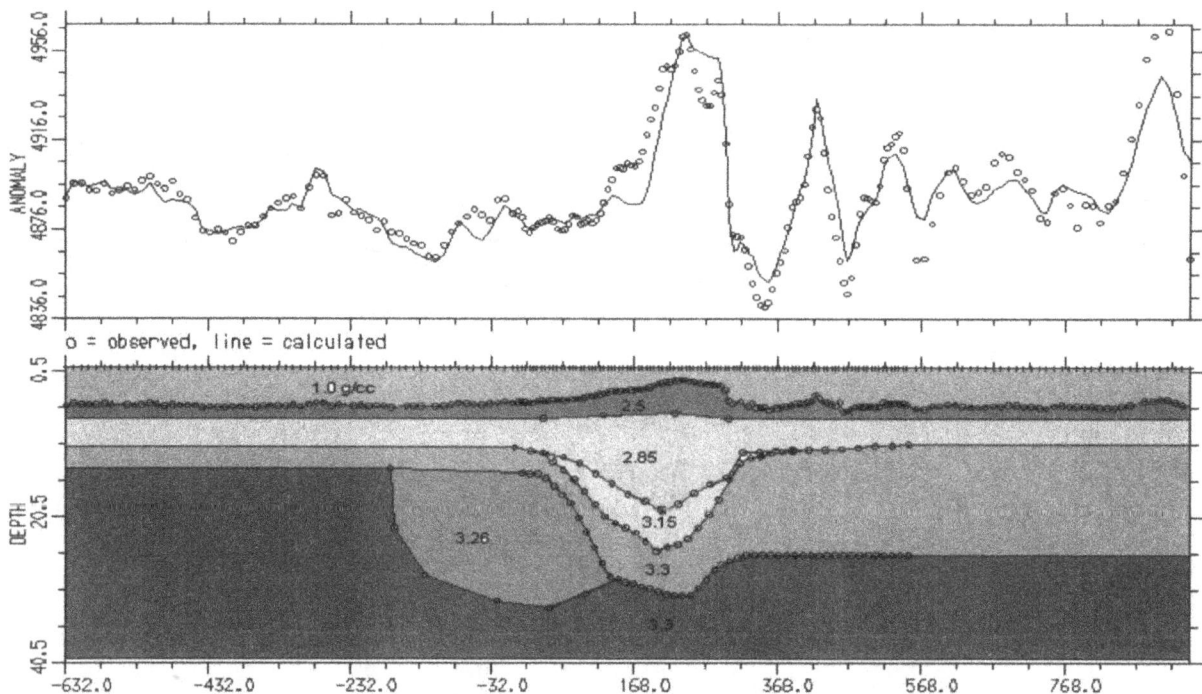

Fig. 5.38(a) A satellite free air gravity profile across Ninety Ridge along 17⁰S latitude and modeled crustal section showing a thick (~22km) crust with under plated rocks (~6 km).The low density body towards the east may represent hydrated mantle in form of serpentinization of mantle rocks. (Tiwari and Mishra, 2003).

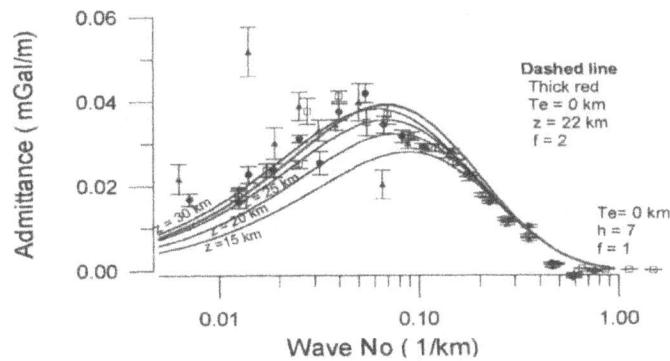

Fig. 5.38(b) Model of compensation of 90 East Ridge along 17⁰S profile showing effective elastic thickness of 0-2 km which fits best with the observed admittance implying Airy's model of root compensation (Tiwari and Mishra, 2003)

5.5   Arabian Sea and Bay of Bengal: Satellite Derived Gravity Maps

Arabian Sea and Bay of Bengal are extensions of Indian Ocean adjoining Indian continent. They are of great significance in formation and evolution of Indian continental margins and are, therefore discussed below in some details.

Satellite altimetry and tracking derived gravity data sets over India and adjoining countries and Oceans, viz. the Arabian Sea, Bay of Bengal and Indian Ocean north of equator was retrieved from Sandwell and Smith (1997) available on INTERNET. This data set included 2' × 2' grid of mean bathymetry/ elevation, the mean free air gravity anomaly and the standard deviation of the gravity anomaly. They are referred to as IGSN-71 system (Heiskanen and Moritz, 1967). The bathymetry over oceans and elevation over the continents from this data set is plotted in Fig. 5.39, which shows deep ocean of Arabian Sea and Bay of Bengal in the southern parts reducing in depth towards the coast line. The prominent ridges and basins of Arabian Sea and Bay of Bengal and important geomorphologic features of continents are clearly reflected which are marked as 1-24 in Fig.5.39 and explained in Table 5.3 (Mishra et al., 2004).

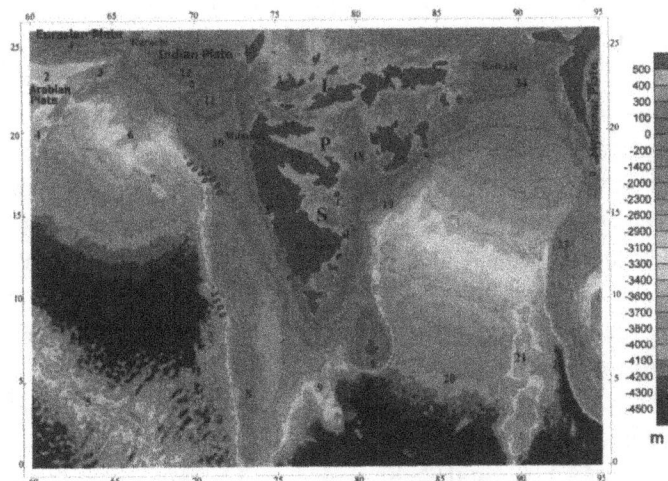

Fig. 5.39 Bathymetry of Arabian Sea and Bay of Bengal and topography of adjoining continent (Smith and Sandwell, 1994). Annotated features marked as 1-24 are explained in Table 5.3; IPS= Indian Peninsular Shield (Mishra et al., 2004). [Colour Fig. on Page 774]

Table 5.3 Major geologic and geomorphic features and corresponding gravity anomalies of Peninsular India and adjoining oceans

Features numbers (fig. 2.31)	Geologic/ geomorphic Features	Gravity signatures and corresponding numbers (fig. 2.32)
1.	Makran coast	High (H1)
2.	Gulf of Oman	Low (L2)
3.	Murray ridge	High (H3)
4.	Owen fracture zone	Low (L4)
5.	Carlsberg ridge	Low (L5)
6.	Saurashtra ridge	High (H6)
	North Arabian basin	Low (L6)
7.	Laxmi ridge	Low(L7)
8.	Chagos-Laccadive ridge	High (H8)
9.	Comorin ridge	High (H9)
10.	Western continental shelf	High (H10)
11.	Saurashtra	High (H11)
12.	Kutch	High (H12)
13.	Delhi-Aravalli Fold Belt	High (H13)
14.	Satpura ranges	High (H14)
15.	Indian Peninsular Shield	Low (L15)
16.	Western ghats	Low (L16)
17.	Eastern ghats	High (H17)
18.	Godavari basin	Low (L18)
19.	Basin along continental shelf off east coast of India	Low (L19)
20.	Eighty Five East ridge	High (H20) in the south, Low (L20) in the north
21.	Ninety East ridge	High (H21)
22.	Andaman trench	Low (L22)
23.	Andaman and Nicobar Islands	High (H23)
24.	Bangladesh coast	High (H24)

5.5.1   Bouguer Anomaly of India and Free Air Anomaly Map of Adjoining Oceans

The free air anomaly map of the oceanic part are merged with Bouguer anomaly of the continental part, India (NGRI, 1975), Bangladesh (Rahman et al., 1990), Southern Pakistan and Sri Lanka (ESCAP, 1976). The terrestrial and marine gravity data along the coast were merged by regridding the available data together at 1.0 km interval. The merging of free air anomaly of oceans with Bouguer anomaly of continent is justified on the ground that mean sea level is used to reduce the gravity data over continents and therefore, the two are analogous. The resultant map is given in Fig.5.40 (Mishra et al., 2004) showing gravity highs (H) and lows (L) which are identified by same numbers for different features as given in Fig. 5.39 preceded by nature of gravity anomaly H or L. The gravity anomalies corresponding to different units in Fig.5.40 are also listed in Table 5.3. Such a gravity map of oceanic part would not have been possible from any number of ship borne surveys but for Satellite altimetry. The comparison of Fig. 5.39 and 5.40 and gravity anomalies listed in Table 5.3 corresponding to various geomorphological features brings out the following significant points.

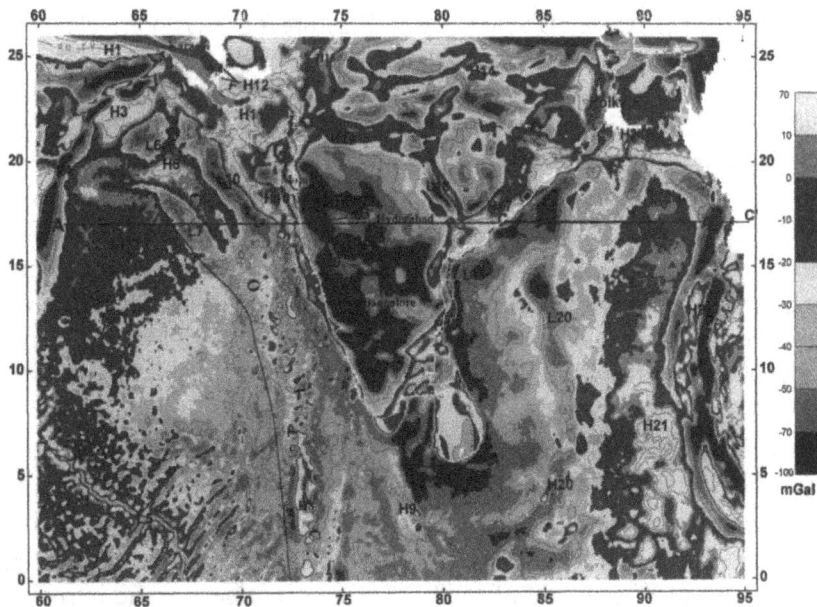

Fig. 5.40 Gravity map of Peninsular India and adjoining oceans prepared from satellite altimetry derived free air anomalies over oceans (Sandwell and smith, 1997) and surface Bouguer anomalies over India and adjoining countries (India, NGRI, 1975; Bangladesh; Rahman et al., 1990; Pakistan and Sri Lanka; ESCAP, 1976). Important gravity highs (H) and lows (L) are annotated for the features listed in Fig. 5.39 and explained in Table 5.3. Also shown is the profile AA', modeled in Fig.5.43. BB' and CC' represent the extent of the seismic profiles over the continent and the Bay of Bengal (Fig. 5.42(b) and (c)) used to constrain the gravity model. The extent of continent-ocean transition crust (COTC) in the eastern Arabian Sea is indicated by a line (Mishra et al., 2004).
[Colour Fig. on Page 774]

(i) The continental shelf off west coast of India shows two sets of linear gravity highs H10 and H8 towards the west with gravity lows in between, two sets of linear gravity highs indicate intrusive related to the break up of continents as suggested in Section 5.3 (Fig. 5.27). Those close to the continents are the first phase of intrusives in the form of dykes while the second one represents the main phase of volcanic activity causing the break up and lows in between represents the stretched lithosphere with basins at the top. The second set of gravity highs (H8) represent the Laccadive-Maladive islands which were caused by Deccan Trap eruption from Reunion plume as described above. The gravity high (H8) joins with the gravity high (H6) which in fact coincides with Laxmi basin due to crustal thinning and mafic intrusives (Section 3.8) and represent stretched lithosphere along with the gravity low L10 next to it.

(ii) Laxmi ridge (L7) separates the deep ocean towards the west from shallow parts towards the east. It is characterized by a gravity 'low' though as a ridge, it should have been characterized by a gravity 'high' which suggest a crustal root to it, due to isostatic compensation. Western limit of gravity anomalies due to ridges, H8 and L7 indicate the limit of continent-ocean transition crust (COTC). The gravity anomalies due to these ridges and intrusives (H10, H8 and L7) merge into the gravity high due to Murray ridge which represents Western boundary of the Indian plate. A part of gravity high, H10 extends over the continents in Saurastra (H11) and Kutch (H12) where Deccan Trap and contemporary volcanic plugs of alkaline complexes are exposed. A part of the gravity high (H10) also extends along Gulf of Cambay (East of Saurastra) and over the continent along Cambay rift basin which were evolved due to its interaction with continental lithosphere in this section as discussed in Section 2.9.2. This region with several alkaline complexes (Chapter 9) and reported occurrence of the oldest rocks related to Deccan trap eruption (68.5, Fig. 5.24) may represent the first burst of the Reunion plume.

(iii) Towards north-west the Murray Ridge (H3) and Owen Fracture Zone (L4) are characterized by prominent gravity highs and lows. They are oriented NE-SW and separate the Indian plate towards the east and the Arabian plate towards the west and the latter joins to the Carlsberg ridge in the Indian Ocean. The gravity high, H3 due to Murray Ridge, joins with the Chamman fault and Kirthar Ranges of, the Himalayan orogeny near Karachi.

(iv) Carlsberg ridge is characterized by a NW-SE oriented gravity 'low' (L5), which separate the Indian plate and the African plate towards the NE and the SW respectively. The ridges and fractures perpendicular to this ridge are nicely delineated as narrow linear gravity highs and lows which in larger map can be identified separately. The central rift over this ridge which is a characteristic of mid oceanic ridge systems is reflected as a linear gravity low in the central part. This infact highlights the importance of satellite altimetry which has delineated even small features associated with mid oceanic ridges.

(v) Gravity low, L2 and gravity high, H1 represents the Makran trench and coast with volcanic arcs which represents a subduction zone, between the Arabian and the Afghanistan-Iran micro plates.

(vi) In Bay of Bengal, the most prominent feature is the NE-SW oriented gravity high (H17) along East Coast of India and N-S oriented gravity high (H21) corresponding to the Ninety East Ridge. As discussed above, in case of West Coast of India, they may represent mafic intrusives related to the break up of Indian plate from Antarctica. In between, the large gravity 'low' corresponds to stretched crust with sediments in the Bay of Bengal.

(vii) Beyond 90 East Ridge toward the east is the N-S oriented linear gravity 'lows' of Andaman trench (L22) and adjoining gravity highs (H23) for islands which represent accretionary prism in plate tectonics paradigm (Fig. 5.11). The gravity highs at the eastern margin of the map represent Andaman Sea with mafic rocks and volcanic island arc such as Barren islands which is an active volcano in this region.

(viii) The gravity 'high', H24 over coasts of Bangladesh and West Bengal suggest large volume of mafic rocks underneath which may represent extension of 90 East ridge as under plated crust related to volcanic rocks during breakup of India from Gondwanaland during middle Cretaceous, that also formed 90 East Ridge (H21).

(ix) Gravity highs coinciding with the Sylhet and Rajamahal, traps in the East India north of Kolkata, (H 25) has been dated as 118 Ma similar to 90 East Ridge in composition (Mahoney at al., 1983; Kumar et al., 2003). Part of gravity high, H17 in the northern part both onshore and offshore Mahanandi basin is also related to subsurface mafic rocks and under plated crust due to intrusion from Kerguelen hot spot as discussed in the next section (5.6.1). It is, therefore likely that gravity highs, H 21 (northern part), H24, H25 and H17 forms the first burst of Kergulen hot spot.

(x) Part of the gravity 'highs' (H17) over the Eastern Ghats (Continental part) along east coast of India appears to be related to high density mafic rocks of this unit, which has been thrusted from lower crust and upper mantle as described in chapter 6.

(xi) The Gravity Low, H20 is related to 85 East Ridge which indicate crustal thickening similar to Laxmi Ridge in the Arabian Sea. This gravity low extends south wards and joins with the gravity high, H 20 which coincides with the broken hills and forms the trace of 85 East Ridge. Here it bends Westward and follows the axis of gravity low joining with the gravity high centered at 0^0 N and 82.5 °E forming another bathymetry ridge which merges into Afanasy-Nikitin Seamount (Krishna , 2003).

(xii) The Gravity high, H9 in the Indian Ocean is related to a ridge which might be part of mafic intrusives along west coast of India defining the eastern limit of these intrusives. It has been named as Camorin Ridge (Fig. 5.32(b)).

(xiii) The large gravity low south of Sri Lanka represents the effect of well known geoid low in this region whose sources lie in the mantle as described in section 7.11.

Gravity anomalies over Indian Peninsular Shield have been discussed in detail in the next chapter. However, some limited discussions of these anomalies are given here to show their inter-relationship with anomalies and sources in adjoining oceans.

5.5.2 Satellite Derived Bouguer Anomaly Map of India and Adjoining Oceans

Satellite tracking derived free air anomaly over India and adjoining continents and satellite altimetry derived free air anomaly of Arabian Sea, Bay of Bengal and Indian Ocean north of equator obtained from INTERNET (ttp://baltica.ucsd.edu) were transformed to Bouguer anomaly using satellite derived elevation/ bathymetry data from the same file as given in Fig. 5.39. The Bouguer correction over the continent is computed for a crustal density of 2670 kg/m^3 while in case of oceans, it is computed for a density of 2540 kg/m^3 corresponding to bulk density of crustal rocks minus that of sea water. The satellite derived free air anomaly over the continent is transformed to Bouguer anomaly to compare it with Bouguer anomaly over the Indian continent computed from surface data (Fig. 5.40) and check the validity of satellite derived gravity data over continents. The resultant computed Bouguer anomaly map of the whole region is given in Fig. 5.41, which shows a good correlation for large wavelength anomalies with Bouguer anomaly derived from surface data over the continent (Fig.5.40). For example H11-H14 and L15, L16, L18 (Fig. 5.40) find similar reflection in Fig. 5.41. However, short wavelength feature of Fig. 5.40 do not find reflection in Fig. 5.41 indicating the limitation of Satellite data for short wavelength features over the continents. In the oceans, regional Bouguer anomaly map show broad features. Detailed Bouguer anomaly maps offshore west and east coasts of India and Andaman-Nicobar islands are given in Fig. 9.38, 9.40(a) and 9.41(c), respectively that show several additional anomalies compared to Fig. 5.41 related to basement. These maps are also derived from satellite data but are specially processed for medium-short wavelength anomalies originating from the basement in regard to oil exploration. Some important points to be noticed in the Bouguer anomaly map of continent and ocean relates to following.

(i) Continuous increase of the field from continent to deep ocean basins (DOB) with intermediate values over the ridges (L5, L7, L8 and L21) suggest an over all isostatic compensation with thickest crust under continent decreasing to the minimum value under ocean basins and intermediate values under oceanic ridges.

(ii) Both the gravity maps (Fig. 5.40 and 5.41) show a change in strike of anomalies along a gradient marked as COTC (Continent-Ocean Transition Crust) in the Arabian Sea, which suggest a transitional crust east of it with strike of anomalies similar to those of Indian Peninsular shield. Western limits of L8 and L7 representing the ridges in this part coincide with the COTC. The strike of anomalies west of it, is similar to those of oceanic basins in the Arabian Sea. Bathymetry (Fig. 5.39) also shows a change in pattern on either sides of the COTC.

(iii) In Bay of Bengal, both bathymetry and gravity contours depict trends similar to Eastern Ghats and off shore East Coast of India extending almost up to 90 East Ridge. This indicates that crust under Bay of Bengal may also represent a transitional crust.

(iv) Central Indian Deformation Zone (CIDZ) in the Indian Ocean (Fig. 5.35) is reflected as relative gravity highs which does not find reflection in the free air anomaly (Fig. 5.40). This suggests the importance of Bouguer anomaly in case of oceans.

Gravity anomalies over oceans and continents discussed above suggest that while satellite altimetry data are capable of recording medium and relatively short wavelength features ($\approx$10 km), the satellite tracking data over continents record mainly the large wavelength features (>100 km)

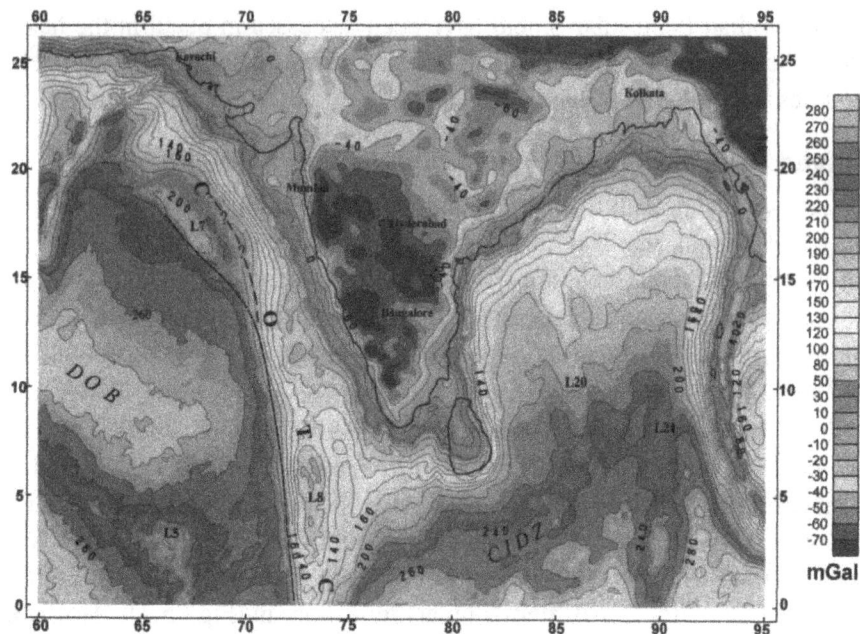

Fig. 5.41 Simple Bouguer anomaly maps of the Indian Peninsular Shield and adjoining oceans prepared from the free air anomaly from satellite tracking over continent and altimetry over oceans. The dashed line in the Arabian Sea emphasizes the change in the trends on either side, indicating the COTC. The Bouguer gravity field increases consistently from relative gravity low over the continent and oceanic ridges (L5, L7, L8, L20 and L21) to relative gravity high over the deep ocean basins (DOB) indicating the effect of isostatic compensation. Central Indian Deformation Zone (CIDF) is indicated by trend of relative Bouguer high which does not find reflection in the free air anomaly maps (Mishra et al., 2004) and can be delineated based on this map.

[Colour Fig. on Page 775]

5.5.3   Modelling of Gravity Anomalies along 17° N – Arabian Sea, Indian Continent and Bay of Bengal

In order to model the crustal structure across this region, viz Arabian Sea, Indian continent and Bay of Bengal, a gravity profile along 17°N is modelled constraining it from all available geophysical/ geological information from this region (Mishra et al., 2004). Density Modelling along this profile is described below in detail to demonstrate the use of constraints to model gravity data along a profile.

(i)   Constraints Used for Gravity Modeling

(a) The Arabian Sea

Naini and Talwani (1982) have given structural frame work of the Arabian Sea based on magnetic and seismic refraction investigations. Structurally, they divided the north Arabian Sea into three major units that include the Laxmi ridge (L7 Fig. 5.40) and the eastern and western basins on either side of it (Fig. 5.42(a)). In this model, the Moho depth is approximately 22 km under the Laxmi Ridge compared to 12 and 17 km in the western and the eastern basins, respectively. The seismic velocities for different crustal layers in the north Arabian Sea reported by Naini and Talwani (1982) are averages of layer velocities that are obtained from several seismic recording in the Arabian Sea.

a

b

c

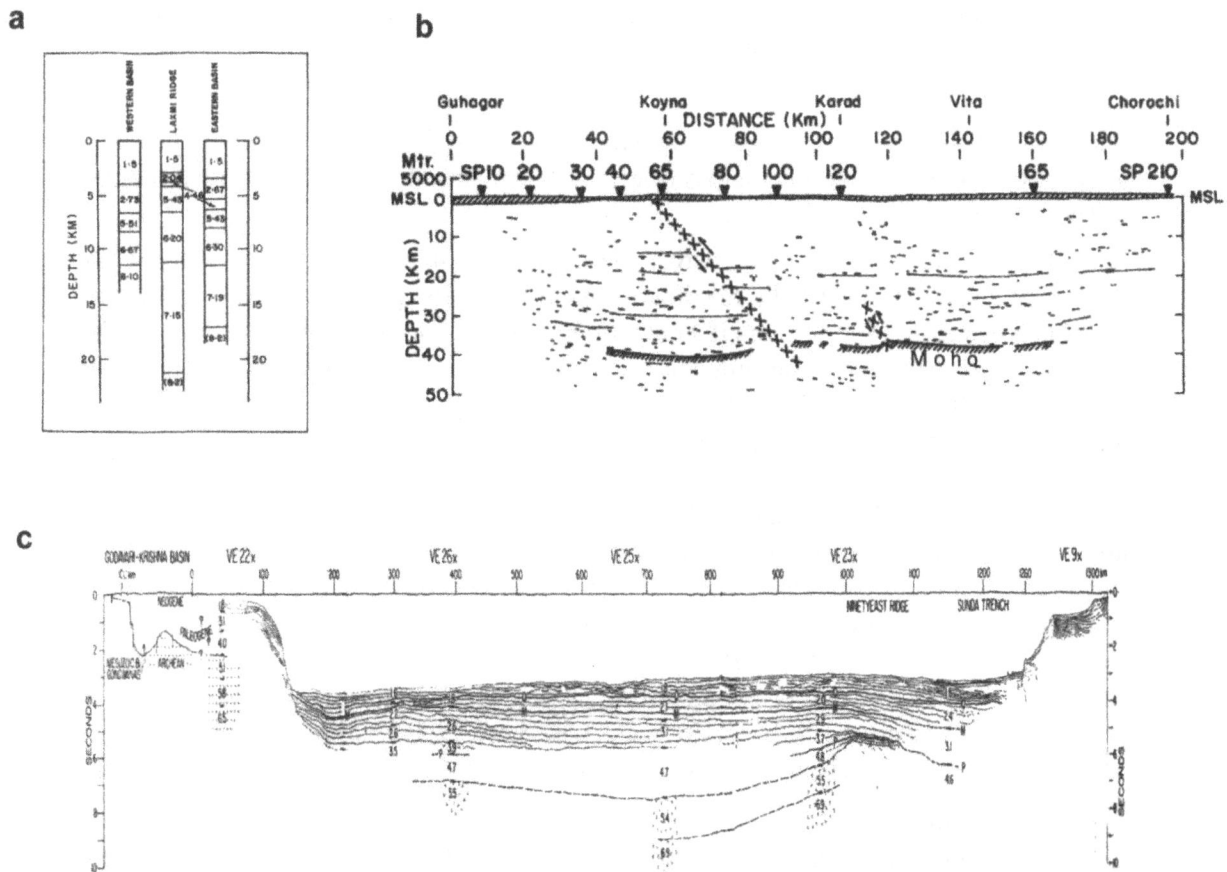

Fig. 5.42 Seismic constraints for gravity modeling: (a) Average crustal columns with different seismic velocities in the Arabian Sea towards the east and the west of the Laxmi ridge (Naini and Talwni, 1982). Layer with velocity 4.46 km/s under Laxmi ridge and eastern basin represents the Deccan trap that is exposed in NW Peninsular India and form the continental shelf off west coast of India. The other layers represent the sedimentary and crystalline components of the crust as discussed in text. (b) Crustal section near the west coast close to 17^0 N showing the thickening of crust under the Western Ghats (Kaila et al., 1981). (c) A seismic section along 17^0 N latitude showing the velocity structure in the Bay of Bengal (Curray et al., 1982).

The top layer of velocity 1.5 km/s represents the water column, which is underlain by Tertiary sediments of velocity 2.04-2.73 km/s. Beneath these sediments, there is a layer of velocity 4.46 km/s under the Laxmi ridge and the eastern basin, which appears to represent continuation of the Deccan Trap exposed in the Western India as discussed above. This layer is absent under the western basin. The layers with velocities 5.51 and 6.67 km/s in the western basin appear to represent layer 2 and 3, respectively, of the oceanic crust. The layer with velocities 6.2-6.3 and 7.15-7.19 km/s under the Laxmi ridge and the eastern basin may represent the crustal layers, although the latter may also reflect underplated lower crust.

The intermediate layer with velocity 5.43 km/s under the Laxmi Ridge and the eastern basin may represent Mesozoic sediments and volcanics that have been found in oil wells on the adjoining Indian continent in Saurashtra and Kutch (Sections 8.2 and 9.2; Singh et al., 1997; Srinivasan and Khar, 1995). Mesozoic sediments in Kutch consist of limestone that raises the seismic velocities and the corresponding densities of these layers. Mesozoic sediments are also reported in the north Arabian Sea off Karachi based on seismic data (Hinz., 1981). Mesozoic sediments and volcanics of almost similar velocities are also reported from the Somali basin along the east coast of Africa (Miles et al., 1998) that was conjugate to the west coast of India.

(b) The Indian Continent

The constrained used for modeling density structure over the continent come from the result of deep seismic sounding (DSS) across Deccan volcanic province (DVP), gravity modeling and perceived tectonics in the region.

A DSS profile (Kaila et al., 1981) across the DVP (Fig.5.42b) passes almost along 17^0N (BB', Fig. 5.42) and suggested a maximum crustal thickness of 40 km under the Western Ghats that decreases to 35 km on either side. It provides 2-km-thick Deccan trap under the Western Ghats, which thins towards the east and is exposed up to approximately 350 km from the cost. The DSS profile also suggests a low velocity zone in the upper mantle along the west coast of India compared to more normal P-wave mantle velocity at that depth (Krishna et al., 1991). A low velocity zone in the upper mantle along the west coast of India has also been reported from tomography (Kennett and Widiyantoro, 1999). An underplated lower crust and a low density zone in the upper mantle along the west coast have also been suggested from gravity modeling (Section 2.5, Tiwari et al., 2001). Based on teleseismic receiver function analysis, a crustal thickness of 35 km has been suggested near the east coast of India in this region (Prakasam and Rai, 1998). Gravity modeling suggest that anomaly L18 is caused due to 3.0-5.0 km thick low density Gondwana sediment that is underlain by lower crustal high density rocks which are exposed along adjoining Eastern Ghats and have produced continental part of anomaly H17 (Fig. 5.40) (Mishra et al., 1999).

(c) The Bay of Bengal

The gravity modeling in the Bay of Bengal (CC', Fig. 5.40) uses constraints from the seismic section along 17^0N given by Curray et al., (1982, Fig. 5.42c).Here seismic investigation suggested almost 3.0 km of water underlain by sediments up to depth of 7.0 km . The sediments are underlain by basement of velocity 5.4 km/s followed by lower crustal velocity 6.9 km/s to a depth of approximately 9.0 km. Crust in the region is thick to the extent of 23-24 km as suggested by Brune and Singh (1986) based on seismological studies. The seismic section reveals basic rocks corresponding to Ninety East ridge and a small basement uplift corresponding to Eighty five East ridge and a depression between these uplifts. A crustal density model across the Eighty five East ridge showing basic intrusion have been proposed based on gravity data (Subrahmanyam et al.,1999).

(ii) Crustal structure along 17^0N

A 3700 km long profile AA' shown in the Fig. 5.40 along 17^0 N was chosen for modeling (Fig. 5.43). We have avoided modeling of the Owen fracture zone towards the west and subduction zone towards the east due to involved complexity and unavailability of sufficient constraints. A smoothing filter was applied to the satellite-derived free air data over the oceans to remove very short wavelength anomalies and enhance the regional aspects of the model. The initial model and densities were fixed according to the seismic layers and their velocities discussed above and generalized density values for the various rock types. The density-velocity relationship (Barton, 1986) as given in section 2.2 was used to convert the given seismic velocities of the various layers to approximate density values. The initial model, however are modified based on comparison between the observed and the computed gravity fields.

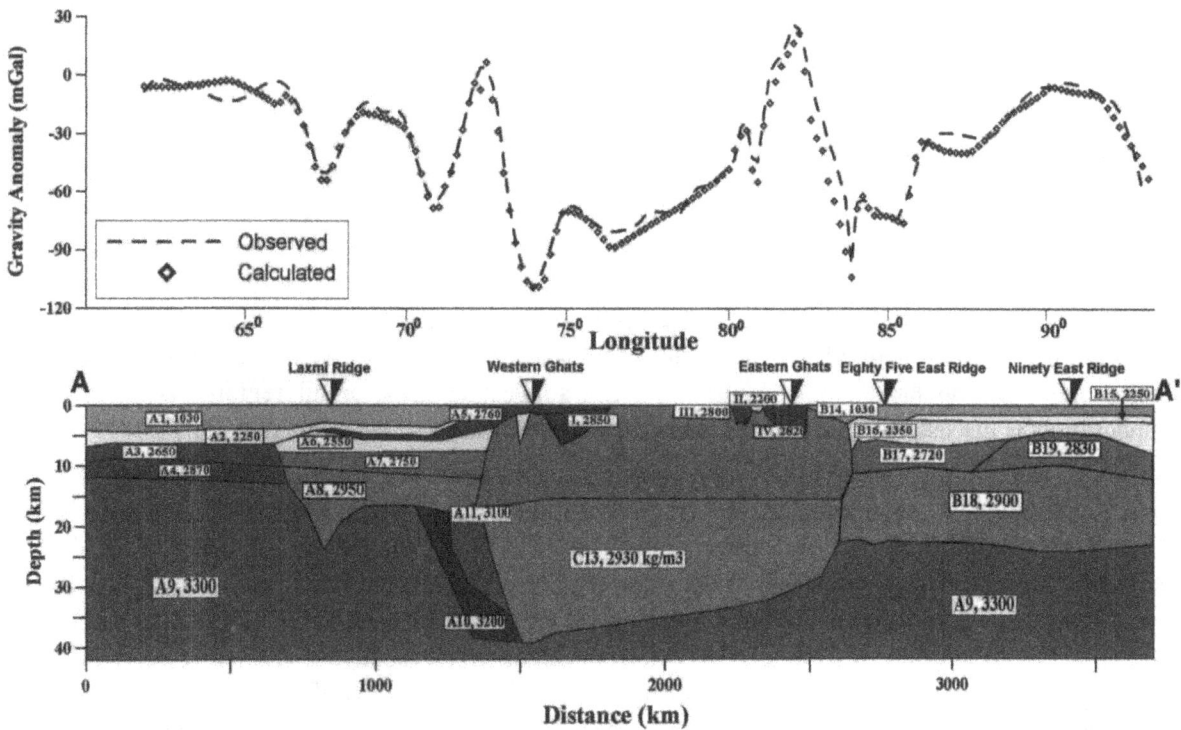

Fig. 5.43 Gravity field along profile AA' (Fig. 5.40) and the inferred crustal model. The computed gravity field due to this model is shown for comparison with observed field. Crustal layers of the Arabian Sea, Indian continent and Bay of Bengal are marked by identifiers A1-A11, C12-C13 and B14-B19, respectively, along with their densities in kg/m^3. Subsurface limited and exposed bodies over the continent are marked by identifiers I to IV. (Mishra et al., 2004)

The crustal model in Fig. 5.43 across the Arabian Sea consists of units A1-A11 based on the constraints discussed above. Units A1 and A2 are the water column and Tertiary sediments, respectively. Units A3 and A4 represent oceanic crustal layers 2 and 3 in the Western Basin with average bulk densities of 2650 and 2870 kg/m^3. Reduced density for layer 2 (A3) indicate mixing of sediment with basalt and layer 3 (A4) represents typical gabbroic layer of oceanic crust. Unit A5 representing the Deccan Trap extend over the Laxmi Ridge on the west and continues over the continent through the continental shelf off west coast of India in this section The average density for the Deccan Trap, which appears to be more than the predicted value from velocity-density relation, was taken as per laboratory measurements (Tiwari et al., 2001)

Units A7 and A8 under the Laxmi Ridge and the Eastern Basin typically represent the upper and the lower continental crust of bulk densities of 2750 and 2950 kg/m^3, respectively .The intermediate layer A6 with velocity 5.43 km/s marks the Mesozoic sediment that is found in oil wells in adjoining Saurashtra and Kutch as described above. Density for unit A6 is also controlled from surface and borehole measurements in Kutch (Chandrasekhar and Mishra, 2002; Srinivasan and Khar, 1995), which suggests a bulk density of about 2550 kg/m^3. Units A9 represents upper mantle rocks with density 3300 kg/m^3, which is common throughout the section. Units A10 and A11 represent the low velocity that extends under the continent as suggested above from seismic studies. These units are presumably related to the eruption of Deccan Trap and interaction of reunion hotspot with the Indian lithosphere.

The units C12 and C13 under the Indian Shield represent the upper and lower continental crusts respectively. Small-scale bodies (I, II, III and IV) of limited extent over the continent are constrained from surface geology and local tectonics, and were so introduced as to obtain a good match between the observed and the computed fields. The body I represent high density rocks in the basement, which may represent the extension of Archean schist belts with high density metasediments and iron formation exposed to the south of Deccan Trap (Mishra, 2002). Towards the east additional bodies represent the Gondwana sediments of the Godavari basin (II) with up thrusted lower crustal rocks along its shoulders (III) and Eastern Ghat (IV), respectively (Mishra et al., 1999). The most important aspect of crustal model over continent is the change in the crustal thickness from about 33 km along the east coast of India to 40-41 km under Western Ghats where regional elevation is about 1-1.2 km suggesting an over all regional isostatic compensation as discussed in Section 2.5 along the same profile recorded over continent.

The Bay of Bengal, the units B14-B19 are modeled according to the seismic section discussed above to include the water column (B14), Tertiary sediments (B15), Mesozoic Gondwana sediments (B16) and the upper and lower crusts (B17 and B18 ,respectively). Units B16 is a basic body corresponding to the Ninety East ridge. The crustal thickness in the Bay of Bengal is relatively more as also suggested from seismological study (Brune and Singh, 1986). A thick lower crust (12-13 km) in this section (B18) is introduced based on the higher seismic velocity of 6.9 km/s reported by Curray et al., (1982) shown in Fig. 5.42(c) at a depth of 8-10 km. However, in an alternative model Fig. 5.43, B18 as dashed line), if the density is reduced to 2840 kg/m³ (as the west of laxmi ridge), the lower crustal thickness reduces to about 10 km.

The gravity high (H3 and H6, Fig. 5.40) and corresponding magnetic anomalies of the Murray and Saurashtra ridges in the north Arabian Sea as discussed (Fig. 5.33) above are similar to those over the volcanic plugs of Deccan volcanism (Mishra et al., 2001) on the adjoining Indian continent that consist mainly of late Cretaceous alkaline mafic and ultramafic rocks. Hence, ridges in the north Arabian Sea may consist of alkaline complexes caused by the movement of the Indian plate over Reunion hot spot during the late Cretaceous.

(iii) Geodynamic Significance of Crustal Structures: Continental Margins and Laxmi Ridge and Basin

(a) The whole crustal block east of Laxmi Ridge (L7, Fig. 5.40) in the Arabian Sea is different from the block towards the west. It appears that, during the process of Deccan volcanism and rifting of the west coast, the whole crustal block east of Laxmi ridge, including ridge it shelf is modified which may represent structured crust due to volcanism and break up of continents. Even crustal thicknesses and crustal densities observed on either side are distinctly different. This may suggest that the crustal structure under the Laxmi Ridge and to the east of it is the transitional crust. Minshull et al., (2008), based on wide angle seismic data along a profile across the Laxmi Ridge also suggested that Deccan magmatism from the Reunion plume that was responsible for separation of Seychelles has modified the crust in this section. Accordingly there were two phases of extension, viz. extension between the Laxmi ridge and the Indian continent that gave rise to Deccan related magmatism while during the second phase Seychelles separated and the continental margin was developed. The effect of Deccan Volcanism remained confined up to this region towards the north. They have also shown the underplated crust both under the Laxmi ridge and the continenta margin of India that is typical of magmatic provinces due to plumes. This is a familiar situation of volcanic rifted margins world over that are accompanied by two phases of eruption. The dashed line in the Arabian Sea (Fig. 5.40 and 5.41) marks the estimate for the western boundary of the continent–ocean transition crust (COTC) in the Arabian Sea.

Gravity anomalies east of this boundary follows the west cost of India and structural trend of the Indian peninsular Shield , while west of it, the gravity trends are 60-90^0 with that of the transition zone and follows the trends of oceanic crust. Detailed investigations based on seismic and satellite gravity field studies, Collier et al., (2004) have suggested continental crust under Laxmi Ridge and Seychelles as conjugate pairs of rifted margins. They have reported Seaward dipping reflectors on both sides, viz. on southern edge of Laxmi Ridge and close to the foot of the Seychelles Bank (Fig. 5.32(a)) which are definite indicators of rifting. Minshull et al., (2008), based on seismic studies have suggested two stages of extensions, first between the Laxmi ridge and the Indian continent and second stage is related to the break up Seychelles and formation of the continental margin. They have also shown underplated crust of high velocity both under the Laxmi ridge and the continental margin of India that is typical of such volcanic provinces.

(b) The older sea floor spreading magnetic anomalies of the Arabian Sea are largely confined to west of the COTC, where the oldest reported magnetic isochrones 28 belonging to early Eocene is later to the eruption of Deccan Trap. The COTC appears to represent the Indian crust that was down dropped and submerged at time of the plumes related break-up of Madagascar and Seychelles along the west coast of India. The crustal model in the eastern Arabian Sea, east of the Laxmi ridge suggests the extension of Mesozoic sedimentary rocks from the Indian continent.

(c) In the Bay of Bengal, the gravity trends are primarily N-S, being dominated by the trends of Ninety East ridge. Ninety East Ridge presumably formed during the break-up India from Antarctica in the Middle Cretaceous from eruptions of the Kerguelen hot spot as discussed above. Volcanic rocks similar to Ninety East ridge such as the Rajamahal Trap have been reported from eastern part of Indian continent north of this region. The gravity high H24 under Bengal Basin in Bangladesh may, therefore represent an underplated crust which is typical of plume activity.

(d) Modelled crustal thickness (22-23 km) in the Bay of Bengal and structural trend similar to East Coast of India also suggest a transitional crust in this section.

5.6 Conjugate Structures of India and Antarctica: Eastern Margin of India and East Antaretica

In regard to the studies related to Indian plate and Indian continent, its conjugate coastlines and structures with continent to which it was connected before their break up is important. As its connection with Antarctica in Gondwanaland is most important, it is investigated depending on magnetic and gravity anomalies along the coast and contemporary structures on the two continents. Magnetic anomalies are especially important in this regard as direction of remanant magnetization present in rocks can be inferred with some constraints on source configuration which can be used to obtain paleolatitudes as discussed in Section 3.7. Gravity anomalies are useful to check on contiguous structures on two continents and their extensions. As East Coast of India is supposed to be in conjunction with northern margin of east Antarctica (Fig. 5.45, Fedorov et al., 1982; Mishra et al., 1999), we examine here some significant magnetic and gravity anomalies of structures in these two sections. Even in Rodinia reconstruction during Meso-Proterozoic period related to Grenville orogeny, thse two sections of Indian continent and East Antarctica have been shown together (Fig. 5.4). It is apparent from this reconstruction that East Coast of India and Enderbyland and Lambert rift-Amery Ice shelf of East Antarctica were in conjunction in the Gondwanaland reconstruction before their break up. It also shows the position of Kerguelen hotspot which was considered for this break up as discussed in the previous section and has also been responsible for contemporary 90 East Ridge in the Bay of Bengal and Rajamahal and Shylet traps in the eastern part of the Indian continent.

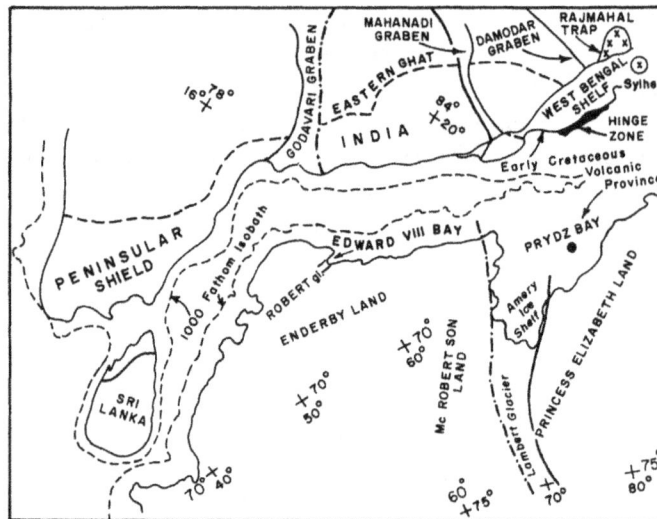

Fig. 5.44 Reconstruction of India and Antarctica before break up showing the east coast of India in conjunction with the northern margin of East Antarctica. Godavari and Mahanadi Rift basins along the east coast of India and Enderly Land and Lambert Rift of East Antarctica being conjugate structures on two continents. Position of Kerguelen hotspot which has been responsible for the break up of two continents is shown tentatively as a dot in Prydz Bay (Fedorov et al., 1982; Mishra et al., 1999).

5.6.1   Gravity and Magnetic Anomalies – Godavari, Mahanadi and Cauvery Basins and Continental Shelves Offshore

East Coast of India is largely characterized by Eastern Ghats comprising rocks of Meso-Proterozoic period (~1.1 Ga) which are much older compared to the period of Gondwanaland (~250 Ma). We have therefore examined sections along the East Coast of India which are contemporary or slightly older to the Gondwanaland. In this connection, Gondwana Rift Basins such as Godavari and Mahanadi Rift basins along the east coast of India and their extension off shore (Fig. 5.45) assume special significance. The continental shelf off east coast of India is relatively narrow of about 20-40 km in the south to about 40-80 km in the north. A bathymetry map of the region as given above (Fig. 5.39) depicts the same.

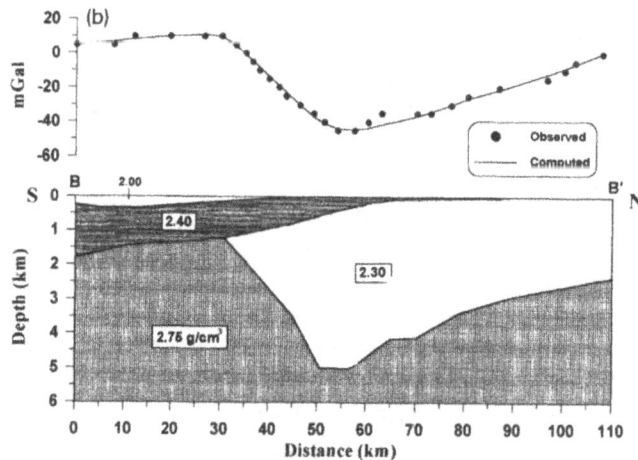

Fig. 5.45 A gravity profile BB' across Godavari coastal basin (Fig. 2.30(a)) and modeled basement structure (Mishra et al., 1999) showing the extension of Gondwana sediments (2.30 g/cm^3) towards the coast intercepted by a coastal basement ridge as inferred in Fig. 5.46(a).

(i) Godavari Rift Basin

Bouguer anomaly map of Godavari basin along with the coastal part is given in Fig. 2.30(a) which shows gravity low in the central part and a high south of it related to a basement depression and uplift, respectively. A gravity profile BB' from that map in coastal part is shown in Fig. 5.45, which also shows a central low increasing to normal values on either sides. It is modeled constraining from the results of a seismic profile (Kaila et al., 1990; Tewari et al., 1996) which shows maximum depth to the basement as about 5 km under the gravity low, which rises to 1.5 km towards the south and 2.5 km towards the north (Mishra et al., 1999). It shows Tertiary sediments with Deccan trap intrusives which has raised the bulk denisity of sediments (2.4 g/cm^3) and Mesozoic sediments (2.30 g/cm^3) overlying the basement indicating the extension of Gondwana sediments in the coastal part.

A total intensity map of the coastal part of this basin is given in Fig. 5.46(a) (Venkata Raju et al., 2002) which shows several magnetic anomalies related to basement and intrusives. The magnetic low (ML) truly represents a basement anomaly, which represents a coastal ridge. A magnetic profile M1 from this map is shown in Fig. 5.46(b), which shows a magnetic low and is modeled by a basement ridge. In order to model it using a basement ridge at appropriate depth, the causative source must have a remnant magnetization of declination = 335° and inclination = -55° which is similar to that for Deccan traps (Table 5.1) and provides a paleo latitudes of about 35° S. This indicates that this region along east coast of India was located in southern hemisphere at the time of the emplacement of these basement dykes and is affected by Deccan trap intrusions.

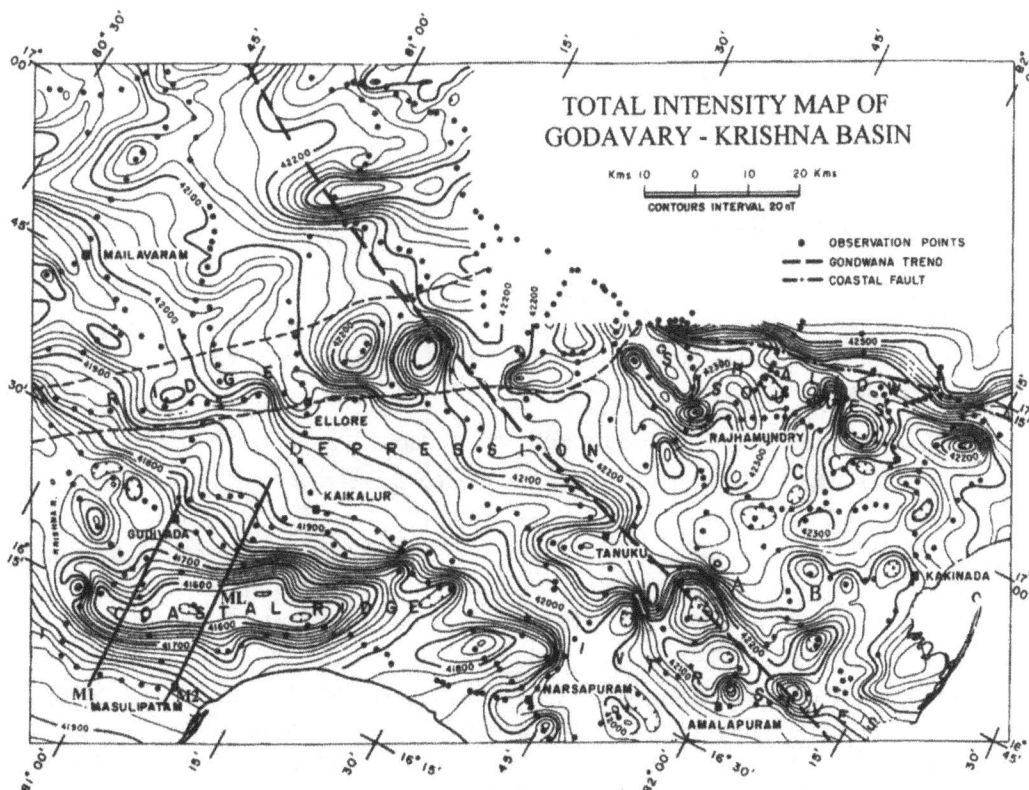

Fig. 5.46(a) Total intensity map of the coastal part of Godavari basin showing a linear magnetic low along the coast indicating a costal basement ridge which intercepts the Gondwana sediments north of it.

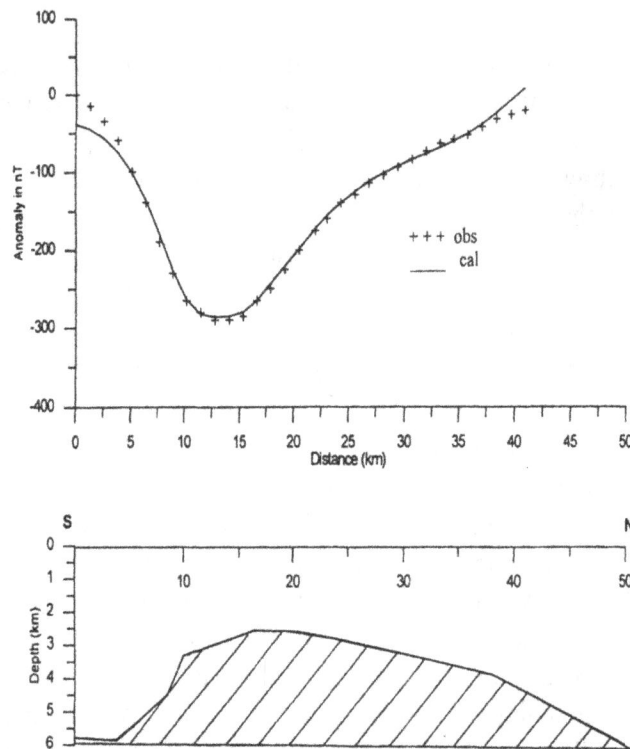

Fig. 5.46(b)   A magnetic profile and computed basement ridge for remnant magnetization of D= 335^0 and I = -55^0 similar to Deccan trap (Table 5.1) which provides a paleolatitude of 35^0 S during intrusion of this dyke (Venkata Raju et al., 2002).

(ii)   Continental Margin offshore Godavari Basin

In order to analyse the breakup of period along east coast of India, it is important to analyse magnetic anomalies along continental margin offshore east coast of India. Fig. 5.47 (Murthy et al., 1994, Venkata Raju et al., 2002) depicts a magnetic high across the continental margin in this section. Its modeling in the form of a continental margin suggests a magnetization vector of declination = 310° and inclination = -67° which is similar to that of Rajamahal trap as given in Table 5.1 that corresponds to paleo latitude of about 50° S. Following two conclusions can be drawn from these results:

(a)   During breakup of India, it was located in southern hemisphere at about 50° S and

(b)   Breakup of India was caused by same source, which gave rise to Rajmahal trap, viz. Kerguelen hotspot.

(c)   Direction of magnetization inferred for basement along the coast suggest Deccan trap activity while that for continental margin indicate older activity similar to Rajmahap trap due to Kerguelen hot spot suggesting that former was confined up to the coast while the latter was responsible for its break up from Antarctica.

As the corresponding feature contiguous to Godavari basin on Antarctica (Fig. 5.44) is Enderby Land, which is mainly occupied by Proterozoic rocks, a comparison between two sections could not be made. However, it may be noted that basement under Robert Glacier in Edward Bay may be a contemporary structure but in absence of any geophysical data, it could not be ascertained for certain. This kind of comparison is attempted in the next section based on geophysical data from the Mahanadi basin, India and Lambert rift, Antarctica that is more conclusive as discussed below.

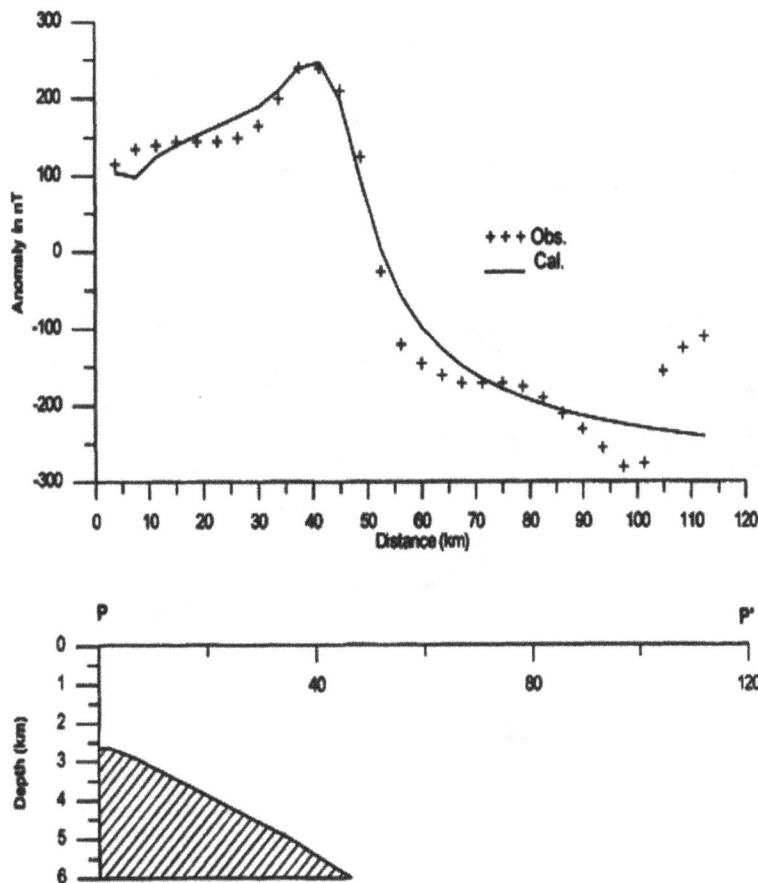

Fig. 5.47 A magnetic profile offshore Godavari basin (Murthy et al., 1994) and modeled continental slope for a remnant magnetization of D =310^0 and I= -67^0 similar to Rajamahal trap (Table 5.1) indicating a paleolatitude of 50^0S during break up of two continents and formation of continental slope in this section (Venkata Raju et al., 2002).

(iii) Mahanadi Rift Basin

As shown in Fig. 5.44, Mahanadi rift basin of Premo-Triassic period is shown in alignment with the contemporary Lambert rift of Antarctica and therefore, they assume special significance for analysis of events leading to the breakup of the two continents.

Fig. 5.48 depicts the Bouguer anomaly map of Mahanadi basin along with the local geology. As typical of rift valleys, it shows linear lows (L1-L5) and shoulder highs (H1-H4) extending up to the coast. Gravity profiles XX' and YY' in the continental and coastal parts are modeled in Fig. 5.49(a) and (b) constrained from seismic profiles in the neighboring section in the coastal part (Kaila et al., 1987; Tewari, 1998). The computed crustal model (Fig. 5.49 (a) and (b)) show thin crust of about 33-34 km, a low density layer (2.65 g/cm^3) in the middle crust and slightly higher density (3.0 g/cm^3) for lower crust as underplated crust which are typical of active continental rift basins formed due to igneous intrusions as discussed in Section 2.7 and demonstrated in Section 2.9. Deep seismic profiles in coastal Mahanadi basin also suggest high velocity in the lower crust (Reddy et al., 2005)

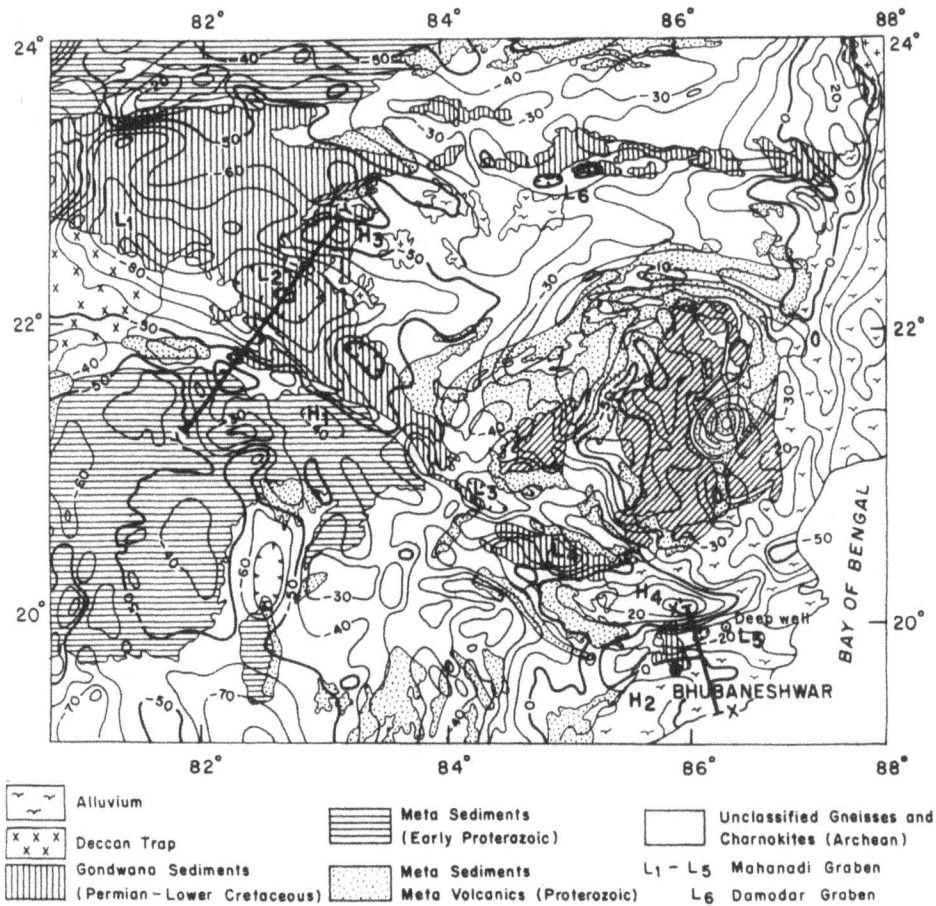

Fig. 5.48 Bouguer anomaly map of Mahanadi Rift basin up to the east coast of India with superimposed geology showing gravity lows, L1-L5 related to the sediments of the basin and gravity high, H1-H4 forming the shoulders.

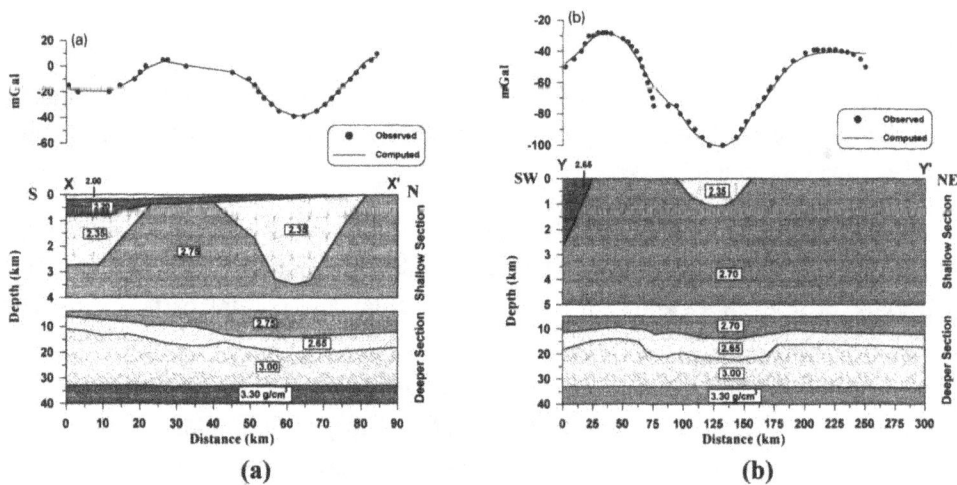

Fig. 5.49 Gravity profile XX' and YY' and modeled crustal structure across Mahanadi basins showing crustal thinning (up warping) and under plated crust (3.0 g/cm³) with a mid crustal low density layer (2. 65 g/ cm³), typical of rift basins formed due to igneous activities referred to as active rift basins (Mishra et al., 1999).

(iv) Continental Shelf off Shore Mahanadi Basin

Airborne total intensity map of continental shelf offshore Mahanadi basin is given in Fig. 5.50. The data recorded at height of about 600 m and flight line spacing of 2 km covers 60 km offshore on the continental shelf. A basement relief model computed from this data using harmonic inversion scheme as described in Section 4.3 constrained from nature of continental shelf and some bore hole data (Baishya and Ratnam, 1981) is given in Fig. 5.51 (Mishra, 1984). It is computed for a total magnetization of 900 nT which suggest mafic rocks as the source rock. The direction of magnetization required to obtain this relief is declination = $15°$ and inclination = $+ 65°$ which represent a reverse magnetization of the direction of magnetization reported for Rajamahal trap and one computed for magnetic anomalies offshore Godavari basin as described above. Such high magnetizations of these rocks suggest that it is caused by volcanic rocks and therefore, the eastern margin of India in this section represents a volcanic passive margin.

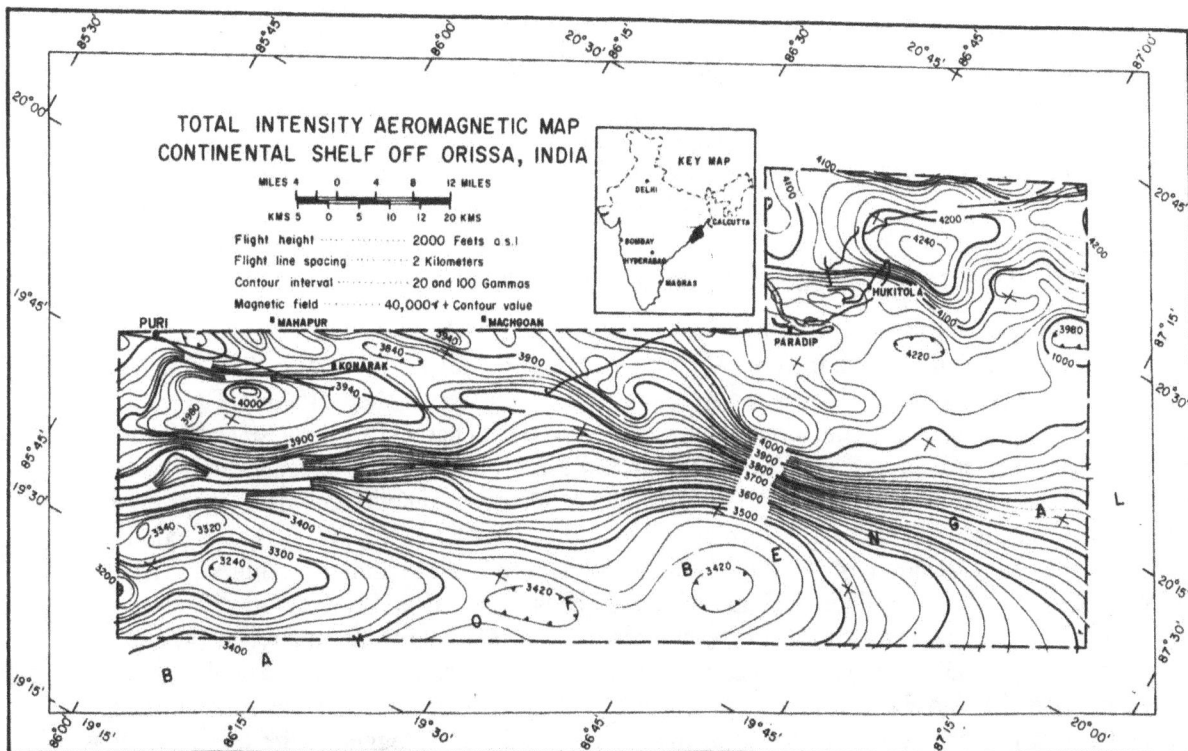

magnetic high and low indicating mafic intrusive rocks along the coast (Mishra et al., 1984).

As described in Section 3.2.5 normal and reverse directions of magnetizations from contemporary rocks suggest stable directions, which can be used for paleomagntic studies. Therefore, the direction of magnetization inferred for the causative sources of magnetic anomalies offshore Mahanadi basin appears to be quite suitable for computation of paleolatitudes, which provides southerly latitude of about $50°$ S during formation of the rocks that is consistent with measurements from Rajamahal trap (Table 5.1). This indicate that this break up must have been caused by the same source, viz. Kerguelen hot spot

Fig. 5.51 Computed model of basement (Continental shelf) from airborne magnetic data for a remnant magnetization of D= 15⁰ and I= +65⁰ which represent a reverse magnetization corresponding to Rajmahal trap (Table 5.1) related to the break up of India and Antarctica in southern hemisphere. In such a case it provides a paleolatitide of 50⁰ S which is consistent with other observations related to their break up (Mishra, 1984).

(v) Continental Shelf Offshore Cauvery Basin in South India

We have discussed magnetic data offshore Godavari and Mahanadi basins, which are located offshore east coast of India in the central and northern parts, respectively. We shall now discuss the nature of magnetic anomalies in the southern part of continental shelf offshore Cauvery basin along the East Coast of India. Cauvery basin on shore is characterized by several horst and Graben type of structures (Fig. 9.41(b)) that indicates compression tectonics and makes it important for hydrocarbon exploration. A total intensity magnetic anomaly map offshore Cauvery basin in South India is presented in Fig. 5.52 (Murthy et al., 2007). This section is in continuation of a major east-west trending tectonic unit on the Indian continent known as Cauvery Shear Zone or Palghat Gap, which is considered as a Proterozoic collision zone (Section 7.9) and reactivated as a rift basin after breakup of India from Antarctica (Mishra and Vijaya Kumar, 2005). The magnetic anomaly map (Fig. 5.52) presents an E-W trending magnetic high between F1 and F2 (~11° – 12° N), which is in continuation of the Cauvery Shear Zone (Palghat Gap) on the continent that also presents E-W trending magnetic high along the coast. The magnetic anomalies north and south of this section depict different trends indicating that the Cauvery Shear Zone appears to extend over the continental shelf. It also shows different trends of magnetic anomalies compared to off shore Mahanandi basin presented in Fig. 5.50. This is in conformity with the effective elastic thickness (Section 4.4), which suggest a higher value for the northern part (10-25 km) compared to the southern part (5 km) (Chand et al., 2001) indicating a weak crust in the southern part that can be attributed to tectonic activities associated with the formation of Palghat Gap (Mishra, et al., 2006). The magnetic high at this magnetic latitude suggests a depression in the continental shelf bounded by E-W faults F1 and F2, which is also indicated by bathymetry in this section (Murthy et al., 2007) and supports the continuation of palghat gap over the continental shelf. The epicenter of Pondicherry earthquake in 2001 (black dot) coincides with the central part of this magnetic high along its axis indicating this area to be seismically active.

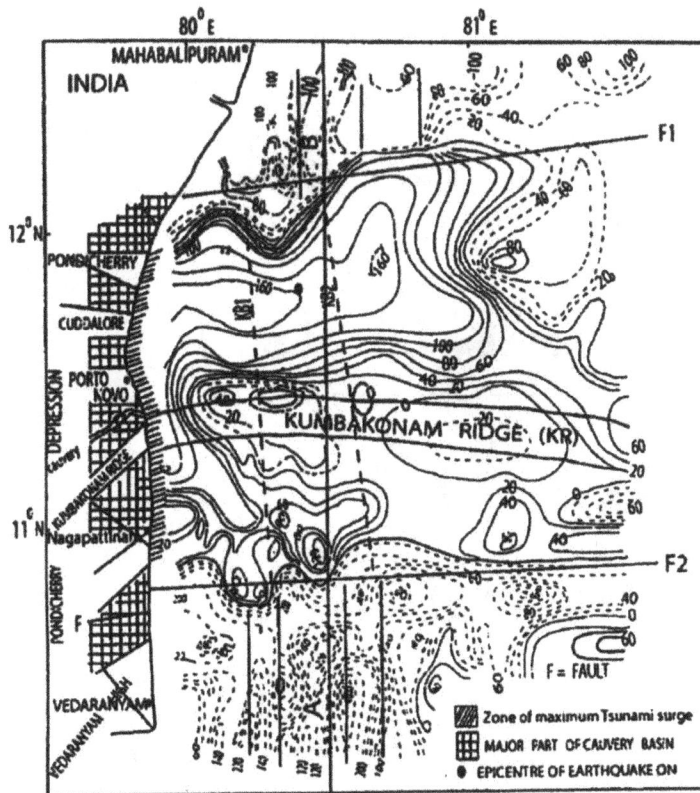

Fig. 5.52 Total intensity map off shore Cauvery basin, east coast of India (Murthy et al., 2007). It show a linear magnetic high in the central part striking E-W similar to that of Palghat gap on the Indian continent indicating its extension over the continental shelf.

5.6.2 Gravity and Magnetic Anomalies – Lambert Rift and Amery Ice-Shelf, East Antarctica

As Amery Ice Shelf and Lambert Rift of Antarctica represent structures contiguous to Mahanadi basin, India (Fig. 5.44), gravity and magnetic anomalies of this section are analysed for a comparison with Indian counterpart. Thick Permian-Triassic sediments (5-6 km) in the Prydz Bay basin based on seismic studies (Stagg, 1985) suggest that they are also contemporary to Mahanadi Gondwana basin. Seismic sections also show seaward dipping reflectors in this section suggesting rifted margin due to plume activity as described above in section 5.3. Long sediment record from Early Permian to Early Cretaceous suggest a NW drainage pattern due to buoyant uplift of SE part of Antarctica (Kent, 1991) which might have been maintained for such a long time due to Kerguelen hot spot that ultimately resulted into the break up of two continents (Charvis et al., 1999). Fig. 5.53(a) is the ice surface contours showing elongated linear nature, typical of rift basins. It also shows location of magnetic profile modeled in Fig. 5.55(a) and (b) and deep seismic sounding profile used to constrain the density models. Free air anomaly map of this region, (Fig. 5.53(b), BMR, 1982) suggests linear gravity high and low corresponding to this unit. Two profiles G1 and G2 from this map are shown in Fig. 5.54(a) and (b), which are modeled constraining it from a seismic profile (Fedorov et al., 1982) in the neighborhood. The crustal models show a thin crust and a high density layer (3.05 g/cm^3, under plated layer) along the Moho. These signatures indicate continental active rift basin similar to those described above under Mahanadi basin across east coast of India.

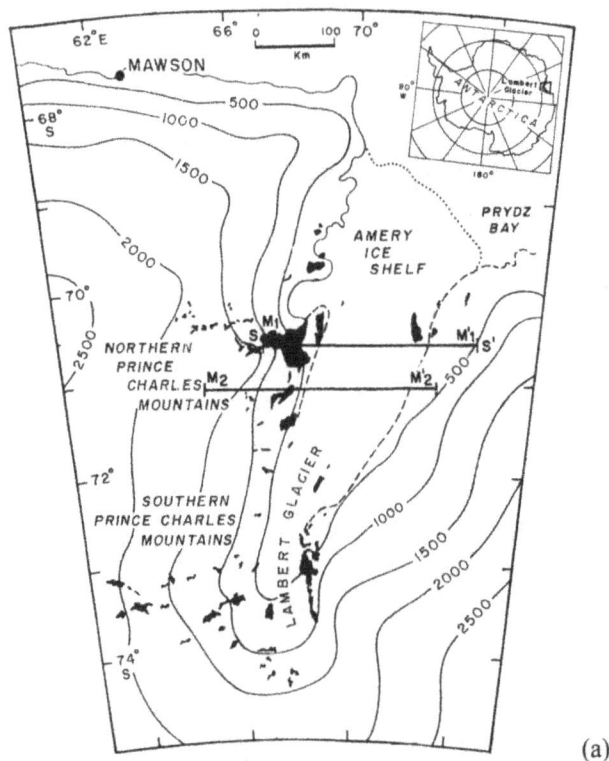

(a)

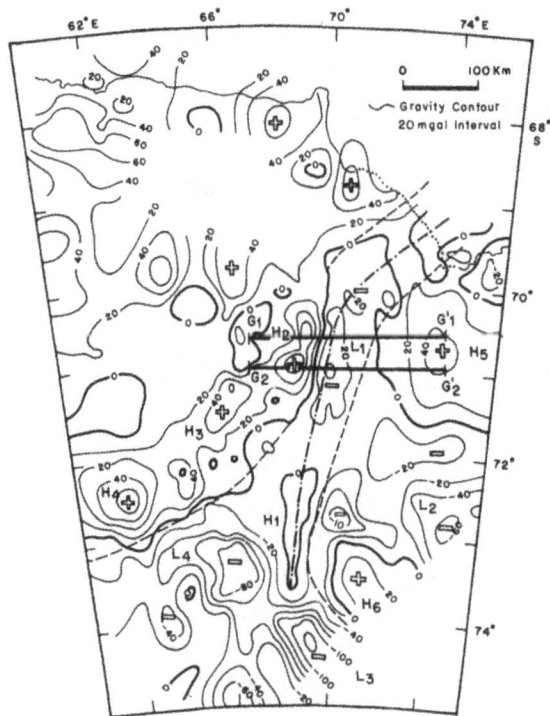

(b)

Fig. 5.53 (a) Lambert rift and Amery ice shelf showing ice surface contours in meters. (Fedorov et al., 1982). M1-M1' and M2-M2' are the location of magnetic profiles modeled in Fig. 5.55. SS' is the location of deep seismic sounding profile used to constrain gravity modeling in Fig. 5.54. (b) Free air anomaly map of the Lambert Rift, Antarctica (BMR, 1982, Mishra et al., 1999).

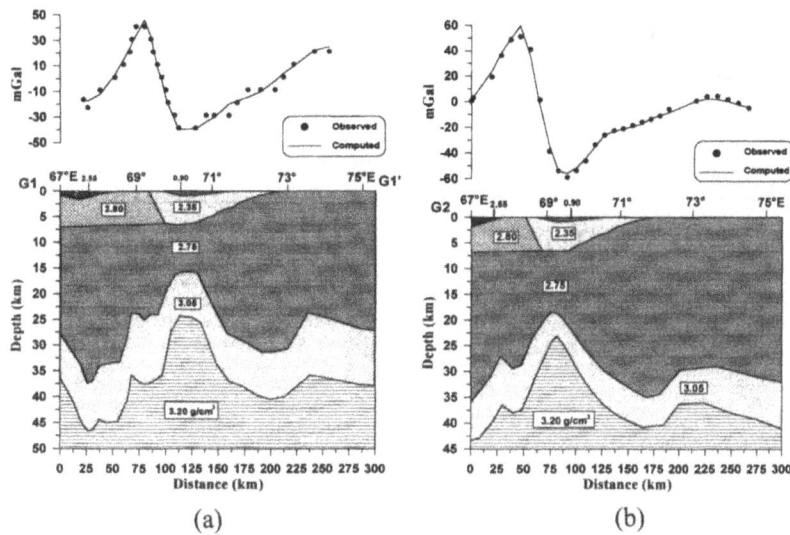

(a)   (b)

Fig. 5.54 Gravity profile and modeled crustal structure under the Lambert Rift, Antarctica showing crustal thinning and under plated crust (3.05 g/cc³) indicating an active rift basin similar to Mahanadi basin along east coast of India (Mishra et al., 1999).

Two magnetic profiles across Lambert Rift, Antarctica along 70.5° and 71° S obtained from Pozdeyev (1994) are presented in Fig. 5.55(a) and (b) and modeled for magnetic sources with basement depth controlled from gravity models. Order of magnetic anomalies indicate mafic intrusive rocks as basement and primarily negative anomaly suggests presence of remnant magnetization as being close to magnetic pole, it should have been a high for induced magnetization. The computed models show high susceptibility and high natural remanent magnetization confirming the nature of causative sources as mafic intrusive bodies indicating active nature of this rift basin as described above. The direction of magnetization for computed models are declination = 150° and inclination = -60° whose declination is same as that obtained for magnetization of continental shelf offshore Mahanadi basin, India and inclination is opposite of it and similar to those reported for Rajmahal Trap in East India (Table 5.1). This inclination provides paleo latitude of 50⁰ S, same as that inferred from the direction of magnetization of rocks off shore Mahanadi basin along the east coast of India.

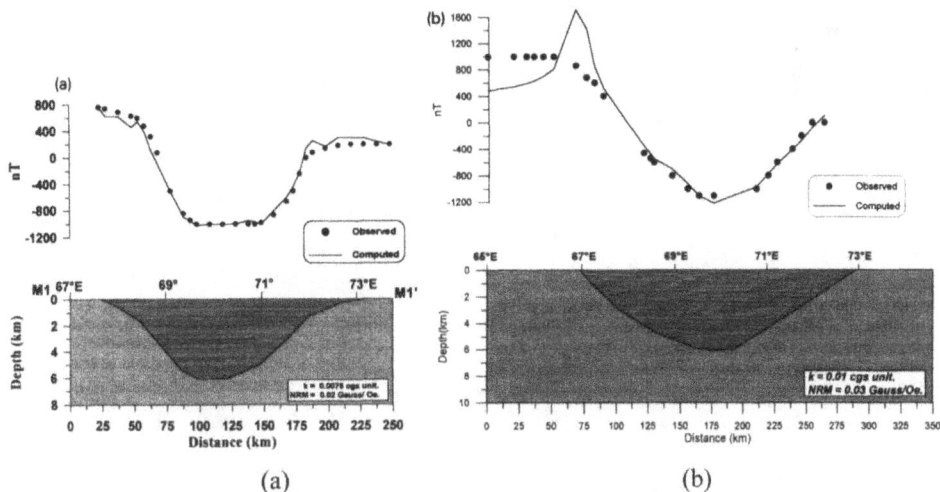

(a)   (b)

Fig. 5.55 Magnetic profile across Lambert Rift, Antarctica (Pozdeyev, 1994) showing large amplitude magnetic anomalies related to mafic intrusives. The basement model provides a remnant magnetization of D = 150⁰ and I = –60⁰ similar to Rajmahal trap (Table 5.1) indicating paleolatitude of 50⁰ S as obtained previously from magnetic anomalies offshore east coast of India.

The data set and results presented above suggest that East coast of India and Enderby Land of Antarctica were conjugate structures and Mahanadi rift basin and Lambert rift on respective continents were connected to each other. Both represents active rift basins which can be attributed to their proximity to Kergulen hotspot (Fig. 5.44). They represented contiguous structures formed due to Kergulen hotspot on two continents prior to the break up.

Collision of Indian and Eurasian Plates and Seismotectonics: Himalayan and Tibetan (Terrains)

6.1 Introduction

The Himalayan-Tibetan orogenic belt is one of the highest and largest systems of fold belts with the highest mountains and plateaus in the world. It is a classical case of continent-ocean convergence culminating into continent-continent collision. It shows all features of such a collision viz. thrusts, suture zones, mountain building, and typical rock types related to them like ophiolites, and granitic batholiths. It has a unique characteristic of being partly continent-continent collision and partly oceanic-continental collision along the Sumatra-Java trench towards the southeast (Fig. 6.5). As shown in Fig. 5.13, the Eurasian plate is formed by parts of Asia and Europe which explains its name. However, in context of the Indian plate, the Asia part of Eurasia consisting of Tibet is more important as this part collided with India to form the Himalayan-Tibetan collision zone.

We have, therefore, referred to Eurasia as Asian plate in our context. First, we have described the general features of the collision between the Indian and the Asian plates, including its general tectonics and satellite gravity data, to obtain first hand information about the whole region. This is followed by a section-wise description of geology/tectonics and various geophysical data sets such as the central, the Eastern, and the western parts to work out details of crustal and lithospheric structures and their role in geodynamic and seismotectonics of this region. In case where only geological periods are referred, readers can find the corresponding approximate ages in million years in Appendix I.

6.1.1 Collision of Indian and Asian Plates

Allegre et al., (1984) have provided a schematic diagram of convergence between the Indian and the Eurasian (Asian) plates since 140 Ma and the development of important tectonic elements due to this convergence. The Indian plate drifted almost from 50°S to the present position during this period and the entire Tethys Ocean between them subducted under the Asian plate due to this convergence. The subduction of the Tethys Ocean started since the breakup of the Indian plate from Gondwanaland and its movement northwards at ~120 Ma (Chapter 5). The Lhasa block of the Asian plate at that time was located at the equator and was pushed northwards to join the main Asian block. Fig. 6.1 (Royer and Patriat, 2002; Avouac, 2007) provides a recent reconstruction since 80 Ma based on magnetic anomalies of the Indian ocean, which suggests movement of the Indian plate almost from 30°S to the present position of ~ 35 °N covering a distance of almost ~ 6000 km during this period. This is almost similar to the drift curve of India based on paleomagnetic results (Fig. 5.1, Mishra, 1965). Fig. 6.1 (Bettineli et al., 2006) shows the convergence rate between two plates based on GPS measurements at the western and the Eastern syntaxis that shows two major changes in the speed at ~ 50 Ma and ~ 40 Ma that may represent the first collision of the western syntaxis (Fig. 6.3) followed by the collision along the entire northern margin of the Indian plate (Fig. 6.4), respectively.

The Indus Tsangpo Suture Zone between them formed when the Indian plate collided with the Asian plate along the whole northern section at about 35-40 Ma. It also formed the core complex of oceanic crustal rocks squeezed upwards at the surface (thrusted) known as ophiolites; and subduction related magmatism north of the suture (Fig. 6.2).

Tapponier and Molnar (1976), Baranowski et al., (1984), and Molnar (1990) have provided schemes of Indian and Asian plates, collision and the under-thrusting of the Indian plate under the Asian plate. The crust under Tibet has thickened due to under-thrusting of the Indian crust under the Asian crust. Subsequently, due to convergence of two blocks and under-thrusting of the Indian plate and its northward movement, a north dipping thrust fault known as the Main Central Thrust (MCT) is formed at about 20 Ma along the northern margin of the Indian plate when the crystalline rocks of Higher Himalaya thrusted over the Paleozoic and the Proterozoic metasediments and subduction related leucogranites intruded in the Indian crust south of the suture zone. With further convergence and compression, another thrust south of the MCT developed, known as Main Boundary Thrust (MBT) at about 12 Ma along which the Paleozoic and the Proterozoic metasediments are thrusted over the Plio-Quaternary Siwalik sediments to form the Lesser Himalaya. Another thrust south of MBT known as the Himalayan Frontal Thrust (HFT) is formed at about 4-2 Ma along which Siwalik sediments of Plio-Quaternary times are emplaced over the recent sediments/alluvium of Ganga basin. Formation of new thrusts can be attributed to a balance of gravitational downward pull of thrusted rocks and compressional force that stops thrusting along the previous thrust. The new thrust is formed due to further compression. However, their shift from north to south can also be attributed to erosion due to monsoon and related uplift as described in Section 6.2.6. These thrusts that form the Himalayan Collision Zone extend as semicircular features from the western to the Eastern margin of the Indian plate as given in Fig. 6.5. Himalayan thrusts, HFT, MBT, and MCT are joined to the Main Himalayan Thrust (MHT) or Himalayan Sole Thrust (HST) at a depth that represents a plane of decollement as described with details in the following sections (Fig. 6.2).

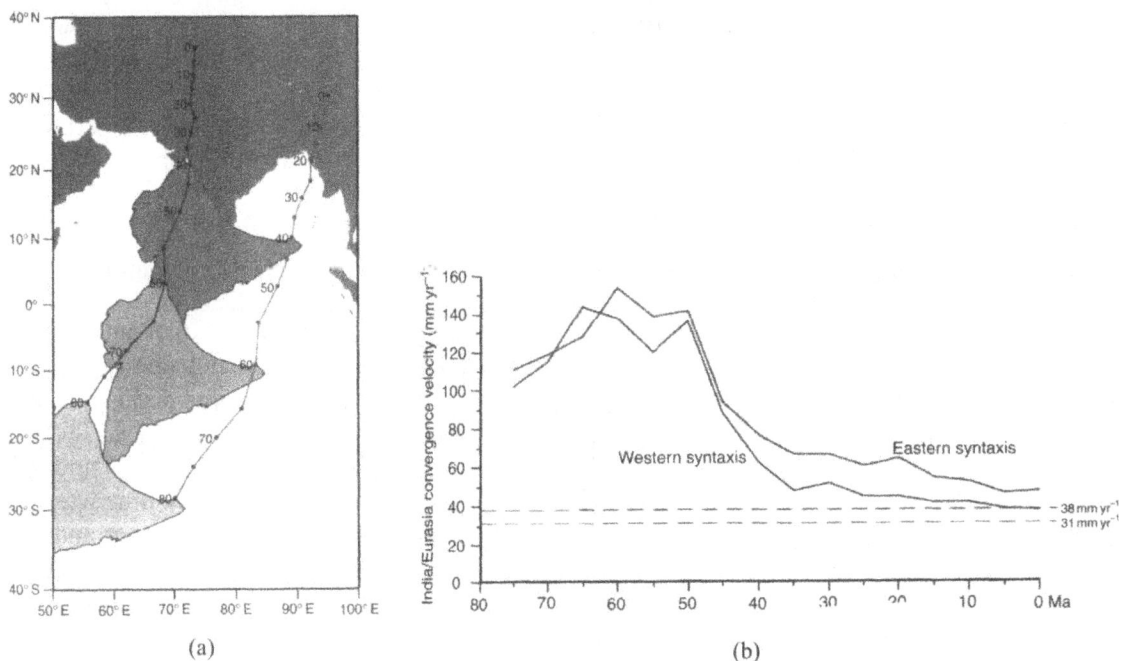

(a)   (b)

Fig. 6.1 A recent reconstruction of India since 80 Ma and (a) India-Eurasia convergence velocity based on magnetic anomalies of Indian Ocean (b) (Royer and Patriat, 2002; Avouac, 2007). Modern velocities from GPS measurements are given as dashed lines (Betinelli et al., 2006). It shows major changes in velocity at ~50 Ma and ~ 40 Ma that might be related to first collision along the western syntaxis and along the entire northern front, respectively.

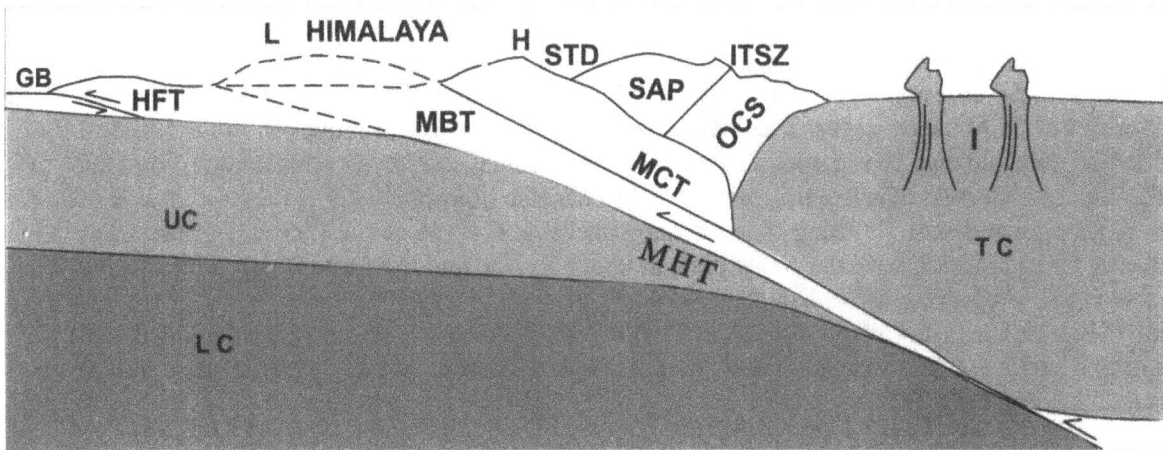

Fig. 6.2 Evolution of Himalayas after collision of India and Tibet at about 50-40 Ma with Indus Tsangpo Suture Zone (ITSZ) as a suture. Due to this convergence, various thrusts are formed on the Indian side during different periods: the Main Central Thrust (MCT) at about 20 Ma, Main Boundary Thrust (MBT) at ~10 Ma, and Himalayan Frontal Thrust (HFT) at ~ 4-2 Ma. All are connected to the Main Himalayan Thrust (MHT) in depth. These thrusts have given rise to the Himalaya, Higher (H) and Lower (L) and Ganga Foreland Basin (GB) and its foredeep along the Himalayan front. On the Tibet side, the crust has thickened and several subduction related intrusives (I) are found similar to the island arc.
OCS-Oceanic crustal rocks mixed with sediments, UC: Upper Indian Crust, LC: Lower Indian Crust, TC: Tibetan Crust, STD: South Tibet Detachment, and SAP: Sediments of Accretionary Prism (Mainly Paleozoic and Mesozoic Periods)

(i) Period of collision

The time of actual collision plays an important role in several aspects of geodynamics of continents such as its influence on climate change, and faunal extinction and is, therefore, essential to determine accurately from different considerations. As discussed in Section 5.3, after the break up of the Indian plate from Gondwanaland during 118-120 Ma at about 48°S latitude, the Indian plate drifted northwards to acquire its present day position. In this process, the Tethys ocean subducted along the southern boundary of Eurasia (Fig. 5.28(a)) till the two collided at about 50-55 Ma (5.28b) when the speed of the Indian plate was reduced to about 11.2 cm/year from about 18.4 cm/year prior to it. It further reduced to 6.4 cm/year at about 40 Ma (Fig. 5.30(b); Molnar and Tapponnier, (1975). An almost similar rate of convergence is also obtained from the recent synthesis of the magnetic anomalies of the Indian ocean (Royar and Patriat, 2002) that also shows rate of convergence velocity changing from about 14 cm/yr to about 8 cm/yr between 50 to 45 Ma and about 5 cm/yr between 55 to 35 Ma. Convergence velocity between two plates finally approaches the present day convergence rate (38 mm/yr, Bettinelli et al., 2006) as obtained from GPS measurements (Fig. 6.1) (Avouac, 2007). The first change in speed at 55 Ma was attributed to collision of the Indian and the Eurasian plates, when part of the Western Syntaxial Bend collided with the Asian plate (Fig. 6.3). Based on the continental drift and rock types in this section, Wadia (1931) suggested that the western part of the Indian plate was first to collide with the Asian plate forming the Western Syntaxial Bend. However, at this stage a part of the Tethys Ocean was still present between the two plates along their Eastern parts (Fig. 6.3). Period of collision along Western Syntaxis has been inferred from dating of ophiolites along the western fold belt, Pakistan, such as the Bela-Muslim Bagh ophiolites (Fig. 6.5) which have also provided a period of 55-50 Ma. Further investigations of sedimentological history and stratigraphy in Himalaya and Tibet also suggested the date of this collision as 50-55 Ma (Beck et al., 1995; Zhu et al., 2005). However, based on detailed sedimentological history and igneous activities in Himalaya and Tibet, Aitchison et al., (2007) suggested that the two might have collided at about 35 Ma, while the first event at 55 Ma might be related to the collision of an intra-oceanic arc of the Tethys Ocean with the Indian plate. The period of 35 Ma approximately coincides with the second stage of reduction in the speed of the Indian plate and may be regarded as the period when actual continent-continent collision took place due to subduction of remaining part of Tethys oceanic lithosphere along the Eastern side of the Indian plate and underthrusting of the Indian plate (Fig. 6.4). Since the Western Syntaxis was locked after its collision with the Asian plate at about 55 Ma, and the Eastern part with oceanic crust was still subducting, the Indian plate would experience an anticlockwise rotation during this period. Fig. 6.3 and 6.4 show the Chitral fault along which the Indian plate collided and subducted under the Asian plate. This eventually formed the present day Indus-Tsangpo Suture Zone (ITSZ, Fig. 6.5).

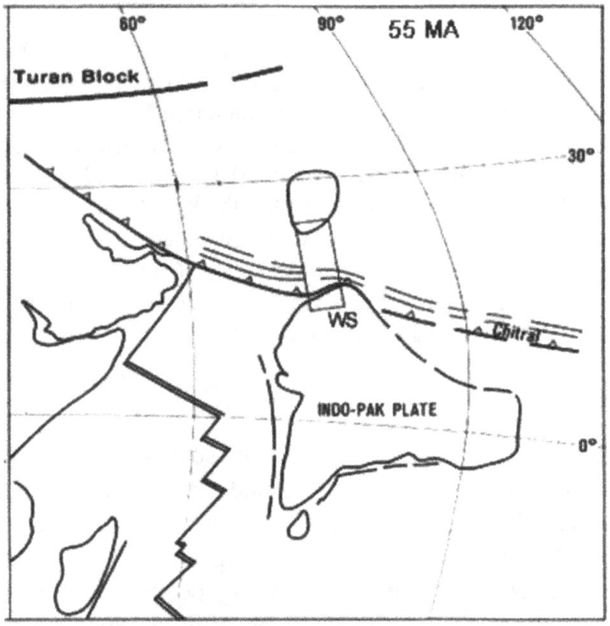

Fig. 6.3 Collision of western India and Tibet along Chitral fault at about 55 Ma showing collision of the western part of India with Tibet (Asia) and part of the unsubducted part of Tethys ocean in the Eastern part. Sea floor spreading magnetic anomalies in the Indian ocean that caused the drifting of the Indian plate are shown in the western part, WS-Western Syntaxis (modified after Bannert, 1992). It conforms to the reconstruction of Patriat and Segoufin (1988) and Klootwijk et al., (1986) (Fig. 5.29).

6.1.2   Important Tectonic Elements of Himalaya, Tibet, and Adjoining Terrains

The collision of the Indian and the Eurasian plates gave rise to several important tectonic elements of a typical continent-continent convergence zone. Fig. 6.5 presents the important tectonic elements of this region and adjoining plates along with some important tectonic elements of the North Indian Shield to show their interrelationship. This figure is just shown to get an overview of the entire region. Tectonics of the individual sections are shown in different sections related to them. The figure shows the Indus-Tsangpo suture zone between the two plates with obducted ophiolitic mélange of rocks implying oceanic crustal rocks. These ophiolites are host rocks for several minerals such as sulphide, chromite, magnetite tin, tungsten and, platinum. The figure shows several thrusts both on the Indian plate and the Tibet block of the Asian plate which developed on account of this collision. Considerable amount of geological information about the Himalaya have been published (Gannser, 1964, 1980; LeFort, 1975; Valdiya, 1984, 1989; Thakur, 1981; Jain et al., 2003; Najman, 2006) which form the basis for future work.

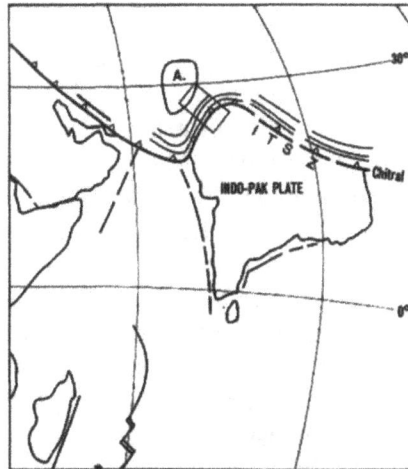

Fig. 6.4 Collision and subduction of India under Tibet along the whole section at about 40 Ma. Sea floor spreading magnetic anomalies in the Indian ocean that caused the drifting of the Indian plate are shown in the western part, ITSZ- Indus Tsangpo Suture Zone (Modified after Bannert, 1992). It conforms to the reconstruction of Patriat and Segoufin (1988) and Klootwijk et al., (1986) (Fig. 5.29).

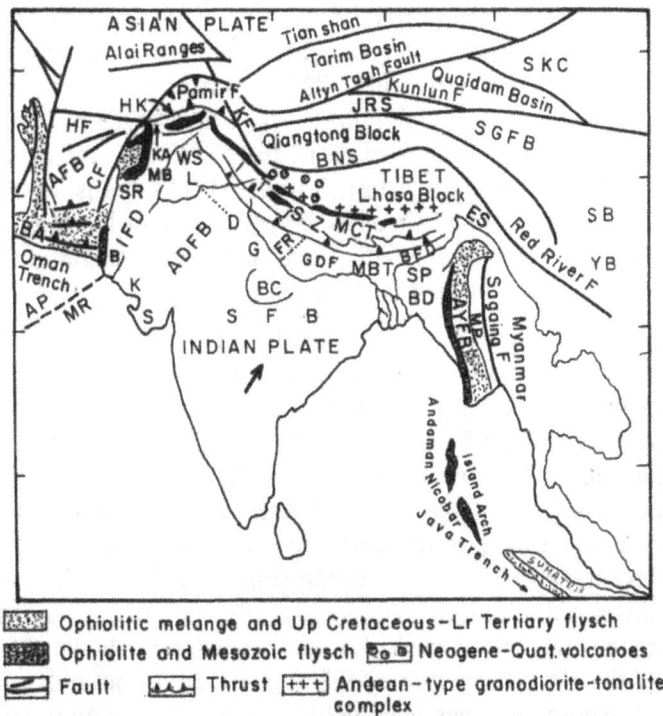

Fig. 6.5 Tectonic map of Himalayas, Tibet and surrounding regions of Asian plate including Afghanistan and Myanmar blocks. Arrow in the Indian continent indicates principal stress direction. Notations as per alphabetical order are as follows: ADFB = Aravalli-Delhi Fold Belt, AFB = Afghanistan Block, AP = Arabian Plate, B = Bela ophiolite, BA = Baluchistan Arc, BNS = Bangong-Nujiang Suture, BC = Bundelkhand craton, BD = Bangladesh, BFD = Brahamputra Foredeep, CF = Chamman Fault, D...L = Delhi-Lahore-Sagodha Ridge, ES= Eastern Syntaxes, G = Ganga, GDF = Gang Fore Deep, HF = Herat Fault, HK=Hindu Kush Ranges, IFD = Indus Fore Deep, ITSZ = Indus-Tsangpo Suture Zone, JRS = Jinshan River Suture Zone, K = Kutch, KA = Kohistan Arc, KF = Karakoram Fault, MP = Mandalay Plateau, MB = Muslim Bagh Ophiolite, MBT = Main Boundary Thrust, MCT = Main Central Thrust, MR = Murray Ridge, S = Saurastra, SB = Sichuan Basin, SFB=Satpura Fore Belt, SGFB= Songpan Ganzi Fold Belt (Sichuan plateau), SKC = Sino-Korean Craton, SP = Shillong Plateau, SR= Sulaiman Range, WS = Western Syntaxes, YB =Yangtze Block (Modified after Valdiya, 1989). Detailed tectonic maps for different parts of this region are provided below in sections dealing with them.

The Murray Ridge in the Arabian Sea and Chamman fault in Pakistan define the Western boundary of the Indian plate with the Arabian plate and the Afganistan-Baluchistan block of the Iranian plate, respectively. West of Murray Ridge is the Arabian plate which is subducting along the Makran coast under the Iranian plate giving rise to ophiolite complex of oceanic crust; Oman trench is to the south of it in the Arabian Sea, and Baluchistan Arc is north of it. Murray Ridge is primarily a non-spreading ridge of magmatic rocks which might be contemporary with Deccan Traps and is characterized by strike slip and normal faults giving rise to several graben type troughs such as Dalrymple Trough (Edwards et al., 2000). Oman trench and associated Makran subduction zone is one of the youngest (10-15 Ma) subduction zones and is separated into the Eastern and the western parts along a strike slip fault called Sonne fault which is presently active and forms the Ormara microplate with the northern part of Murray Ridge as its Eastern boundary (Kukowski et al., 2000). Chamman fault is a transform (strike slip) fault, with several obducted ophiolite complexes along it. The most important being the Bela-Waziristan and Muslim Bagh ophiolites in Pakistan. East of the Chamman fault, there are several thrust zones forming the mountain ranges as Kirthar and Sulaiman ranges that are referred to as the western fold belt, Pakistan. Further east, there is the Indus fore deep with thick sedimentary sequences. The collision began with the emplacement of the Bela and Muslim Bagh ophiolites towards the end of the Cretaceous period during Paleocene (~ 55 Ma, Fig. 6.3). The Bela-Waziristan ophiolites were emplaced on Jurassic and Cretaceous sediments. Bela ophiolite is a late Cretaceous supra subduction zone ophiolite. It consists of an island arc terrain of tholeiites in the north and a basinal terrane of back arc basinal basalt in the south. Alkaline basalts represent oceanic sea mounts. Chamman fault joins the Pamir fault in the north and takes an easterly turn. This turning of all terrain elements from approximately N-S to E-W is called the Western Syntaxis (Fig. 6.5). The Kohistan Arc (KA) lies in between the Indian and the Asian plates and this section is, therefore, characterized by two sutures; one south of it with Indian plate known as Main Mantle Thrust (MMT) that is synonymous with the ITSZ towards the east, and second north of it known as the Shyok Suture or Northern Suture. Towards north, the Kohistan arc underthrusts under the Hindu Kush and Pamir ranges of the Asian plate. Further details of the Western Syntaxis are discussed in Section 6.3.

Tibet, in fact, is a microplate formed due to an agglomeration of several smaller blocks such as the Lhasa block and Qiangtang block (Fig. 6.5) which finally joined the Asian plate before its collision with the Indian plate. Among themselves, they are separated by different suture zones such as the Bangong Nujian suture zone (BNS), Jinshan River Suture (JRS), and Kun Lun fault. The Tarim basin and Tian Shan Ranges are the effect of this collision over the Asian plate. Comparing the collision zone with plate tectonics model (Fig. 5.11), the Indus, Ganga, and Brahmaputra fore deeps on the Indian plate represent the fore land basins synonymous with trenches in case of oceanic subduction zones such as the Andaman-Java trench in the Indian ocean (Fig. 6.5). The Delhi-Lahore ridge in the western part represents a basement ridge caused by a crustal bulge due to the lithospheric flexure of the Indian plate synonymous to the outer rise in case of oceanic trenches as shown in Fig. 5.21(a). This crustal bulge extends all along the Himalayan front (Mishra et al., 2004) but due to the sedimentary cover of Ganga basin, it is not exposed and has been delineated based on gravity studies in the next sections (6.1.6 and 6.1.9). There are also some basement ridges perpendicular to the Himalayan fronts such as the Faizabad ridge which may be an extension of the Bundelkhand Craton (Fig. 6.5). The two important features of northern part of Indian Peninsular Shield are the Satpura Fold Belt and the Aravalli-Delhi Fold Belt, which represent collision zones of Proterozoic Period similar to present day Himalaya (Mishra et al., 2000) as discussed in Chapter 7.

At about 95-100°E longitude, most of the collision tectonic features in this section, take a southerly turn along an axis known as Eastern Syntaxis. East of the Indian plate is the Myanmar-China block of the Asian plate separated by the Sagaing fault. The collision zone in this section is known as the Arakan- Yoma Fold Belt (AYFB) characterized by several thrust faults known as the Schuppen belt occupied by several exposed sections of ophiolite rocks and flysch sediments. It indicates oceanic subduction changing into continental convergence. Further south, the continent-continent convergence changes to ocean-ocean convergence along the Sumatra-Java trench. Andaman-Nicobar islands represent the accretionary prism of this convergence. All along the collision zone in Tibet subduction related volcanic rocks as granodioritic batholiths are exposed along the ITSZ (Fig. 6.5).

6.1.3 Satellite Free Air Anomaly Map

A combined gravity map of the entire region is available only through satellits. We therefore analyzed gravity and geoid fields in conjunction over Himalaya and Tibet including adjoining sections of the Asian and Iranian plates obtained from satellite (http://icgem.gfz-postdam.de/ICGEM/ICGEM.html). Satellite gravity data as free air anomaly and geoid anomaly maps are first presented here and described below to acquire firsthand information about these terranes and to show their importance in such regions. Some preliminary analysis based on their spectral analysis is also made in this section for quick assessment of these data sets while detailed section wise analysis are made latter.

Fig. 6.6 is the free air anomaly map of Himalaya and Tibet and adjoining regions of the Asian and the Iranian plates as derived from CHAMP-GRACE satellite model referred to above. Comparing the free air anomaly map (Fig. 6.6) with tectonics of the region (Fig. 6.5), it is noted that most of the known features are well reflected in this map. Their correct amplitudes can be assessed only after terrain and Bouguer corrections. It is however described here to introduce the original data set that is used subsequently. Moreover, sometimes the anomalies are lost in processing and therefore it is essential to look for them in the original data and compare with the processed data. It is also described here to show the importance of free air anomaly maps. Some of the significant gravity anomalies qualitatively are as follows:

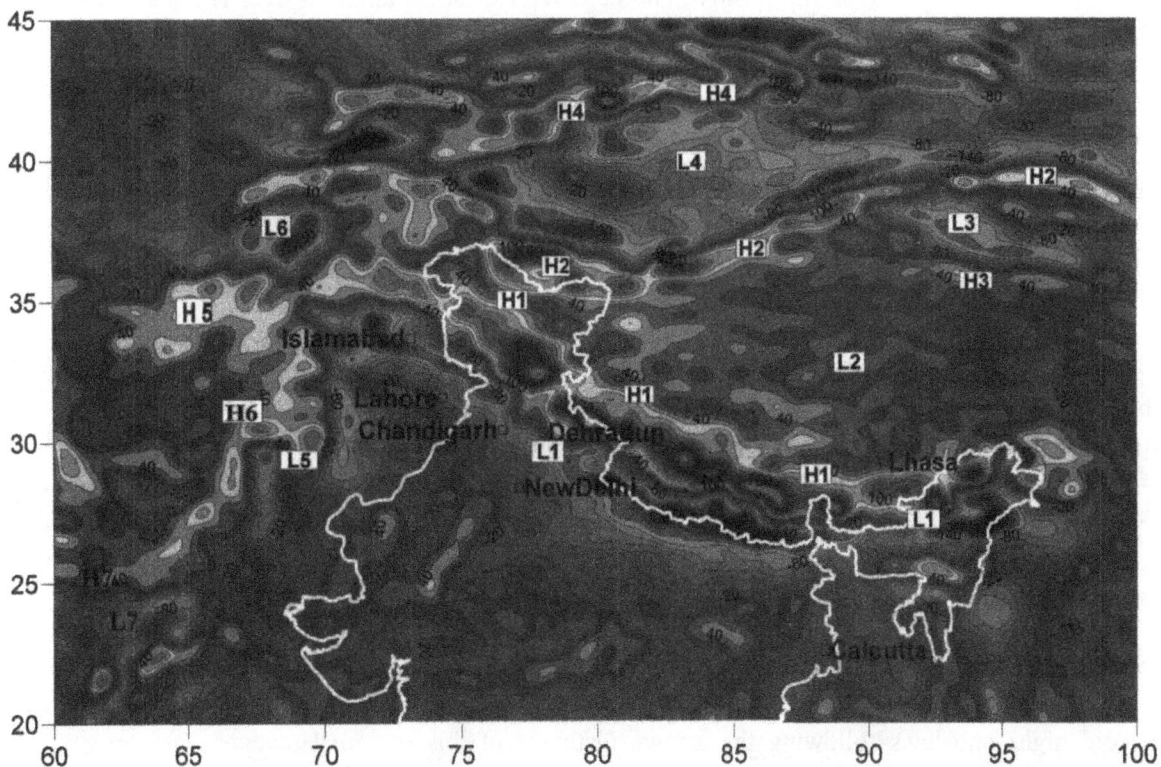

Fig. 6.6 Free air anomaly map of Ganga basin, Himalayas, and Tibet and adjoining regions derived from satellite data. H1-H7 and L1-L7 represents gravity highs and lows respectively discussed in the draft. The white line indicates the boundary of the Indian continent. Thrusts and sutures are largely characterized by gravity highs while basins and plateaus by gravity lows (http://icgem.gfz-postdam.de/ICGEM/ICGEM.html).
[Colour Fig. on Page 775]

(i) Gravity highs are associated with fold and thrust belts of Indian and Asian plates. They are as follows: Himalayan fold and thrust belts and suture zone, ITSZ (H1), Altyn Tagh fault (H2) Kunlun fault (H3), Alai Ranges (H4), Herat fault (H4), Chamman fault (H5). They are caused by high density mafic and ultramafic rocks thrusted along them however, being small wavelength anomalies their amplitudes, can be assessed only after terrain correction. Caporali (2000) and Tiwari et al., (2006, 2008) have shown such gravity highs associated with Himalayan thrusts in limited sections in the Western Himalayan Syntaxis and Sikkim Himalaya.

(ii) Basins are reflected as gravity lows such as Ganga basin (L1), Qaidam basin, (L3), Tarim basin (L4), and Indus basin (L5) etc., that are primarily caused by low density sediments. These are large wavelength features, and their correct gravity anomalies require both terrain and Bouguer corrections.

(iii) The Tibetan plateau is characterized by small amplitude gravity low, L2 with regional field varying from 0 to –20 mGal suggesting a nearly isostatically balanced crust as per Airys crustal root model to high topography in this section. However, gravity highs over Himalaya suggest under-compensated crust in terms of crustal root and high density shallow crustal rocks. This is usually the case in most of the orogenic belts where backside plateaus are nearly isostatically compensated but their margins are under-compensated. Along fore deeps, the basins are usually over-compensated. Ganga, Brahmaputra, and Indus fore deeps along Himalaya show large gravity lows due to low density sediments. Part of it may be related to over-compensation. This is also reflected in isostatic anomaly map (Qureshy and Warsi, 1980; NGRI, 1975; Tiwari and Mishra, 2008, Fig. 7.9) that shows positive and negative isostatic anomalies over Himalaya and Ganga basin, indicating under-and over-compensated states, respectively. Goetze et al., (1991) studied the isostatic balance of the Alps and Andes and also suggested a mass deficit of about 10% smaller than the topographic load for both mountain ranges. This indicates that most of the recent orogenic belts are under-compensated as per the Airys model.

(iv) A gravity low (L6) is observed over the Tadjik basin, but its SE gradient coincides with the Hindu Kush range. This section is also characterized by the Bouguer anomaly low where deep focus earthquakes are concentrated in the subducting Indian lithosphere (Section 6.3.4, Tiwari et al., 2008).

(v) Pairs of gravity highs and lows (H7 and L7) characterise the Oman trench and Baluchistan arc (Fig. 6.6) where the Arabian plate is subducting under the Afganistan block typically for subduction zones. Similar pair of gravity anomalies are also observed in case of continent-continent collision such as H1 and L1 (Fig. 6.6) over Himalayan collision zones. Such paired gravity anomalies are useful in delineating Archean-Proterozoic collision zones as described in Chapter 7.

(vi) Gravity highs, characteristic of the Himalayan thrust zones in the north (H1) are absent towards Eastern Himalaya along the Eastern Syntaxis and Arakan-Yoma Fold Belt.

6.1.4 Satellite Geoid Anomaly Map

Fig. 6.7 is a geoid anomaly map of the same region obtained from the same source. This map also reflects geoid highs and lows following the known tectonics of this region. However, for correct visualization of anomalies, it must be corrected for topography (terrain correction) but still we look for original data as described in case of free air anomalies. As discussed in Chapter 2, geoid anomalies reflect deep seated features better than the gravity field. It is because geoid decreases inverse of the distance while the gravity field decreases square of the distance. Some of the significant features qualitatively are:

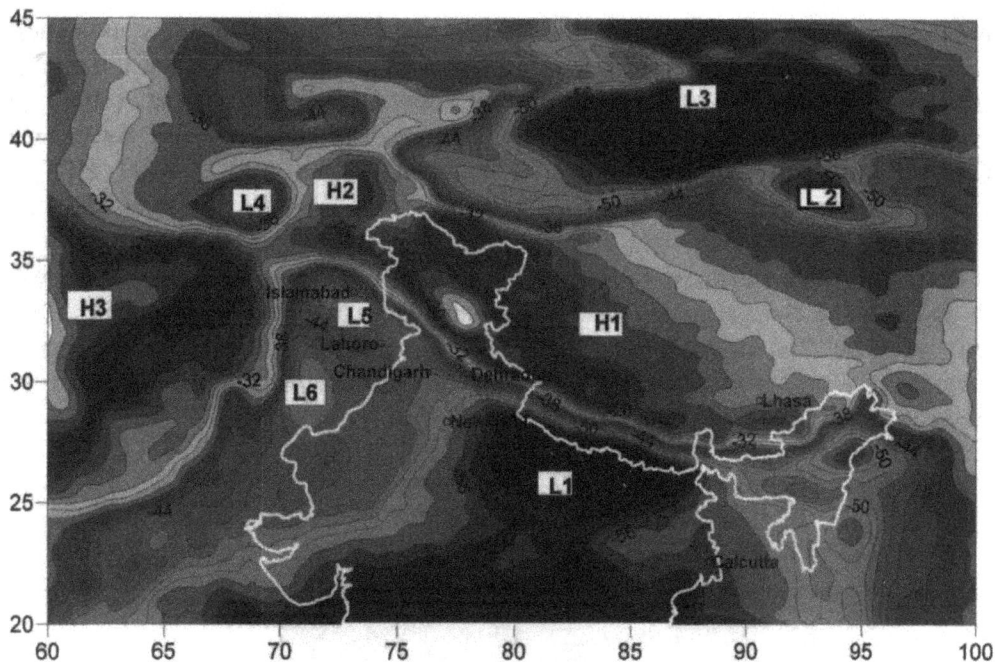

Fig. 6.7 Geoid height map of Ganga basin, Himalayas, and Tibet and adjoining regions. H1-H3 and L1-L6 represent geoid highs and lows in meters as discussed in the draft. High lands are characterized by geoid highs while basins by geoid lows (http://icgem.gfz-postdam.de/ICGEM/ICGEM.html). [Colour Fig. on Page 776]

(i) Geoid highs are observed over Tibet and Himalaya (H1) including its western extension (H2) that represents Pamir and Kohistan arc which suggests the predominance of high density rocks in these sections. Gravity high, H3 occupies the high lands of Afganistan block. A part of these highs must be also due to topographic effect that should be corrected for correct assessment of these anomalies.

(ii) Basins are characterized by geoid lows such as the geoid low, L1 over Ganga basin which extends from the Eastern margin of the Indian plate (Assam shelf) to the western margin (Kashmir-Islamabad basin). Gravity lows, L2 and L3 are observed over Qaidam and Tarim basins, respectively. Similarly, gravity lows, L5 and L6 occupies the Indus basin.

(iii) It is interesting to observe in this case that the Hindu Kush Range is also characterized by a geoid low (L4) while all other ranges show geoid highs. Same is the case with Eastern Himalaya (Burmese arc) that also show gravity lows (L7) indicating predominance of low density rocks in these sections. Both these sections are characterized by deep focus earthquakes (> 100 km) found only in these sctions along the Himalayan arc as discussed in Section 6.1.10.

6.1.5 Bouguer Anomaly Map

Shin et al., (2007) have computed the Bouguer anomaly (Fig. 6.8) over Himalaya and Tibet from the free air anomaly obtained from GRACE satellite using necessary terrain and Bouguer corrections. It shows major gravity low (L1) over Tibet, attributed to crustal thickening due to isostasy flanked by sharp gradients G1 and G2 related to thrusts and suture zones which are reflected as gravity highs in case of free air anomaly. The relative gravity highs, H1is related to Tarim basin due to various intrusives in this section (Yang et al., 2007). It is interesting to note that the gravity highs (H1, H2, H3, H4; Fig. 6.6) of the free air anomaly map related to high density rocks along thrusts and suture zones do not find reflection in the Bouguer anomaly. It might be due to large negative effects of crustal roots caused by isostatic compensation and also due to filtering of small wavelength (high-frequency) component to reduce noise in the data. Therefore, to highlight small wavelength shallow anomalies in orogenic belts where isostatic compensation exist to some degree as thickening of the crust, it might be advisable to analyse free air anomalies or the Airys isostatic anomaly map. Tiwari et al., (2010) have computed gradients in three direction planes, X, Y, and Z from the Bouguer anomaly derived from satellite data (CHAMP and GRACE) and found that the thrust zones such as MCT, MBT, HFT, and BNS get much better reflected in gradients compared to the Bouguer anomaly. However, ITSZ does not find reflection in the gradient maps that might be due to its linear narrow nature. They also computed the Effective Elastic Thickness (Te) from the Bouguer anomaly and topography using wavelet analysis and found it changing from ~ 45 km in the Ganga plains to a minimum of 28 km north of the ITSZ over Tibet that increases northwards to ~ 50 km north of the BNS along a profile (88°E). The minimum value might be related to the section where the subducting lithospheres from two sides might be colliding (Zhao et al., 2010).

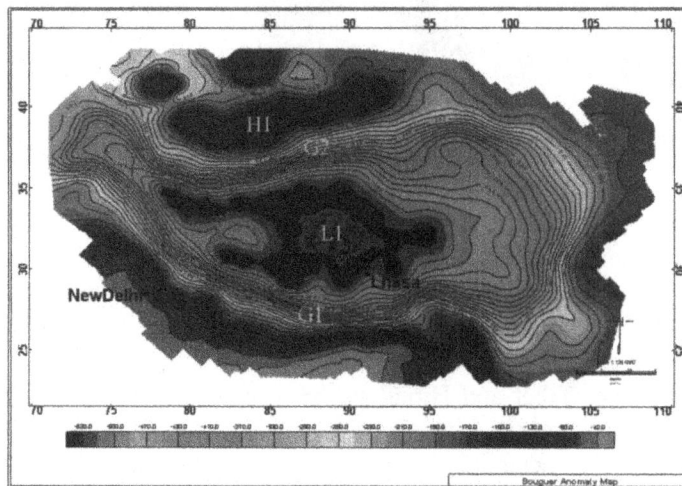

Fig. 6.8 Bouguer anomaly map of Himalayas and Tibet obtained from Satellite free air anomaly map (Shin et al., 2007) showing a major gravity low (L1) over Tibet. Gradients G1 and G2 coincide with Himalayan thrusts and suture zone (ITSZ), and Altyn Tagh fault, respectively. The gravity high, H1 is related to the Tarim basin.
[Colour Fig. on Page 776]

6.1.6   Spectral Analysis and Depth Estimates

Spectral analysis of potential fields (Mishra, and Pedersen, 1982; Fedi et al., 1997) can be used to obtain the average depth of the sources. The corresponding frequency bands can be used to separate the observed fields into different components originating from different levels. Accordingly, the transform of the observed gravity fields (g(k)) for a layered model with relief can be expressed as

$$g(k) = 2\pi G\rho . \Sigma_k \exp - |k| z . \Delta z(k) \qquad \qquad .....(1)$$

where z is the average depth, ρ is the density of the layer, $\Delta z(k)$ is the transform of the relief of the layer, with $|k|$ representing the frequencies given by k = 1, 2 .... n/2, where n is the number of observations.

In case the relief of the layer is less than its average depth and uncorrelated, $\Delta z(k) = 1$ and therefore ln (g (k)) versus k provides straight line segments with slopes equal to the average depth of relief of the layer. However, model studies (Mishra and Pedersen, 1982) have shown that the spectral depths are 10-15% over the estimate of the average depth of the relief, and therefore, they provide the upper bound of depth estimates. In case of two dimensional maps, $k^2 = k_x^2 + k_y^2$, where k_x and k_y are frequencies in x and y directions. Therefore, two dimensional computed spectrum is averaged radially to provide radial spectrum versus combined frequencies, k. In case of multilayered cases, they are separated in different wave bands with respective linear segments that can be used to estimate average depths and separate the observed field in groups by corresponding wave band filters. As equation (1) is valid for gravitational potential, and geoid field (anomaly) is related to gravitational potential, it is also valid for geoid anomaly.

Further, details of spectral analysis of gravity and magnetic fields are given in Chapter 4 of Part I.

(i) Free Air Anomaly Map

Quantitative analysis of such large data sets with several anomalies of different amplitude and wave length, especially for deep seated sources of geodynamic importance, statistical operations in frequency domain including spectral analysis are better suited than time domain methods where individual anomalies are treated separately. Fig. 6.9(a) depicts the spectrum of the free air anomaly (Fig. 6.6) which fits the three layered model by least squares method with linear segments of slopes 66, 26, and 10 km that represent Moho, mid-crustal layer, and shallow sources (basement), respectively. Spectral depths are upper bound on the average depths of the layers in the entire region and in this case, the depth to the Moho varies from about 50 km along the Himalayan front to about 75 km under Tibet. Therefore, an average depth of 66 km under the entire region appears to be quite justified. The second layer between the first and the second segment at a depth of about 26 km represents the lower crust including northern part of the Main Himalayan Thrust as suggested from seismic studies over Tibet described in the next section. Interface at about 10 km depth represents the average depth of upper crustal sources including the southern part of the Main Himalayan Thrust that represents a decollement plane as suggested from INDEPTH profiles as described in section 6.2 (Zhao et al., 1993).

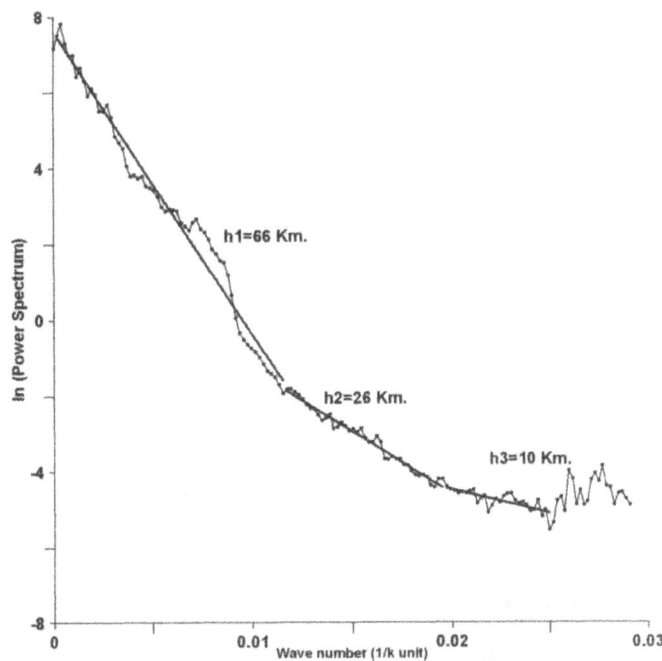

Fig. 6.9(a) Radial Spectrum of the free air anomaly map (Fig. 6.6) showing major sources at three levels with their sources at average depths of about 66, 26 and 10 km representing Moho, lower crust and upper crustal sources including basement and Main Himalayan Thrust.

(ii) Bouguer Anomaly Map

Spectrum versus wave number plot for the digital data of Bouguer anomaly is given in Fig. 6.9(b) (Mishra and Ravi Kumar, 2008) that fits to four linear segments with slopes 320, 122, 60, and 20 km. As discussed below in section 6.2, Priestley et al., (2006), based on S-wave tomography, have suggested the presence of high velocity rocks in depth range from about 100 km to 200-250 km under Tibet, underlain by the low velocity asthenosphere of Tibet where advection takes place. They also suggested thickness of lithosphere varying from 300-320 km under Tibet (Fig. 6.30(b)) to about 120 km along margins towards the west and the east and 120-140 km under the Indian Shield towards the south. Based on tomographic image of the upper mantle beneath Tibet from INDEPTH data (Section 6.2.2) Tilmann et al., (2003) suggested a subvertical high velocity zone from about 100 to 300-400 km depth that is related to subducted part of Indian lithosphere. Based on receiver function analysis, Kumar et al., (2005) have mapped the subducted Indian and Asian lithosphere from about 120 to 270 km that is underlain by Tibetan asthenosphere in the western part. Based on various seismic experiments in Tibet till that time including INDEPTH seismic profile and receiver function analysis (Section 6.2), Kind et al., (2002) have suggested that the subducted Indian and Asian lithospheres have not affected the 410 and 660 km discontinuities indicating that they are confined above this layer. The seismic investigations referred to above are discussed in detail in Section 6.2 while describing seismic studies. It is, therefore, logical to attribute the first two segments of spectral decay as top and bottom of the subducted Indian and Asian lithosphere while the latter two segments represent Moho and top of the lower crust as discussed in the previous case of free air anomaly.

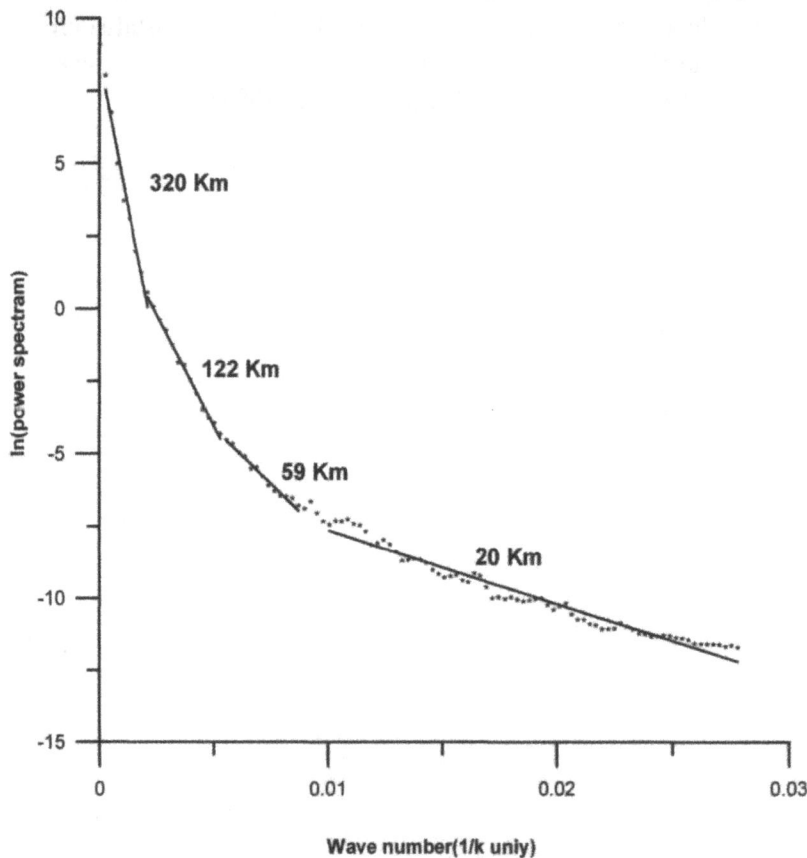

Fig. 6.9(b) Radial spectrum of Bouguer anomaly map (Fig. 6.8) showing major sources at four levels at average depths of about 320, 122, 59 and 20 km representing upper mantle, Moho and lower crustal sources.

(iii) Geoid anomaly Map

Spectrum of the geoid data is given in Fig. 6.9(c) that also fits with four linear segments by the least square methods. These segments depict slopes equal to 328, 132, 72, and 20 km. As discussed above for the spectrum of Bouguer anomaly, the sources between 132-328 km are related to the subducted Indian and Asian lithosphere in the upper mantle, while 70 and 20 km interfaces represent the Moho and the lower crust, respectively. It is interesting that upper mantle sources are reflected only in the spectrum of Bouguer and geoid anomalies, and do not find reflection in the spectrum of the free air anomaly map. This can be attributed to the effect of isostatic compensation of topography taking place in the upper mantle whose effect reduces the negative bias in the free air anomaly but gets reflected in the Bouguer anomaly due to correction for the topography incorporated in it. The purpose of computing spectrum of different data sets and showing similar depth estimates is to test the stability of the computed spectrum and the reliability of results obtained from them. It is known that in case of unknown outputs from a statistical experiment, similar results from different set of samples confirm the stability and reliability of the experiment (Blackman and Tukey, 1958). However, depths of various layers inferred from the spectral decay of three fields match quite well and are within error limits of 10-15%. Due to the inherent ambiguity in potential field, no known methods of depth estimates expected to yield better results than this. This confirms the stability of the computed spectrum and the depth estimated from them.

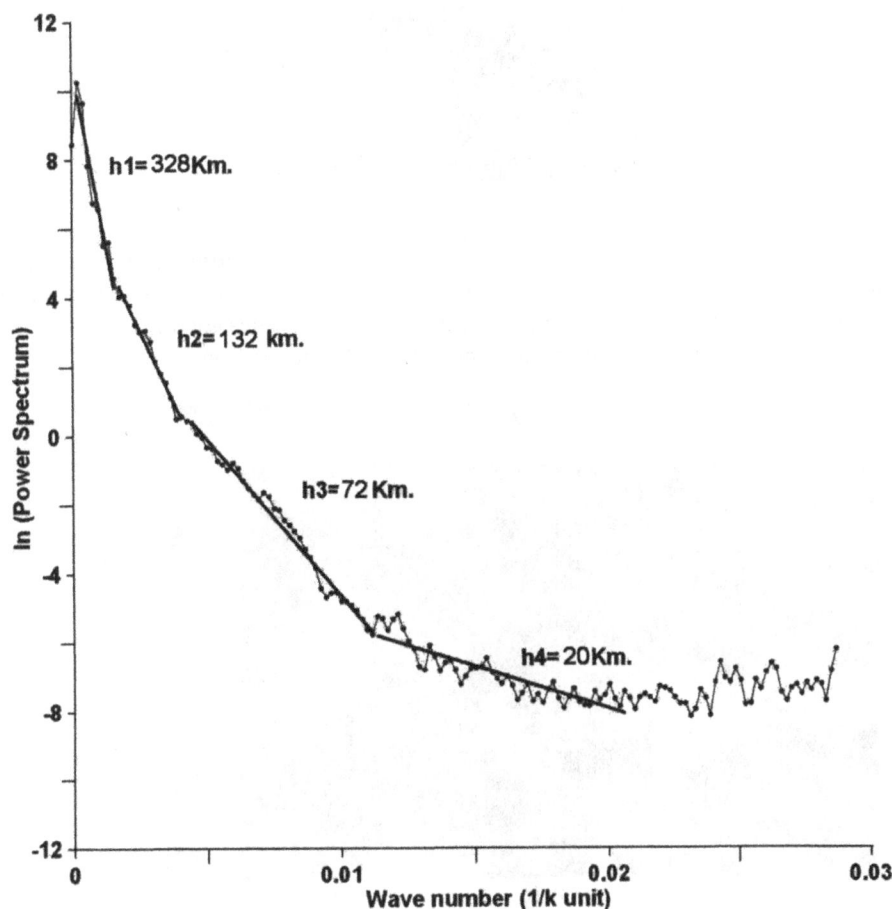

Fig. 6.9(c) Radial Spectrum of geoid data (Fig. 6.7) with sources at average depths of about 328, 132, 72, and 20 km representing the upper mantle, Moho, and lower crustal sources.

6.1.7 Filtered Geoid and Bouguer Anomaly Maps – Regional and Residual Maps

Low and high pass filtered maps of free air and geoid anomaly accentuated several deep seated and shallow sources, respectively almost similar to those described above (Figs 6.6 and 6.7). The most interesting in this regard has been the low pass filtered geoid map that reflected the density characteristics of rocks in the upper mantle. The low pass geoid map for sharp cutoff related to part of the first segment (Fig. 6.9(c)) represents deep seated sources in the upper mantle as given in Fig. 6.10(a) that shows primarily gravity lows (L1) oriented NE-SW (Mishra and Ravi Kumar, 2008). It approximately represents the primary stress direction of the Indian Shield (Gowd et al., 1992) and direction of plate motion perpendicular to the surface tectonics (Fig. 6.5) represented by various thrusts and suture zones. The relative gravity highs, H1 and H2 coincide with the thin section of the lithosphere along the margins of the Tibetan Plateau. The nature and depth estimated from the spectrum for gravity low, L1 with its lowest values under Tibet, suggest that it represents the subducted parts of the Indian and the Asian lithosphere.

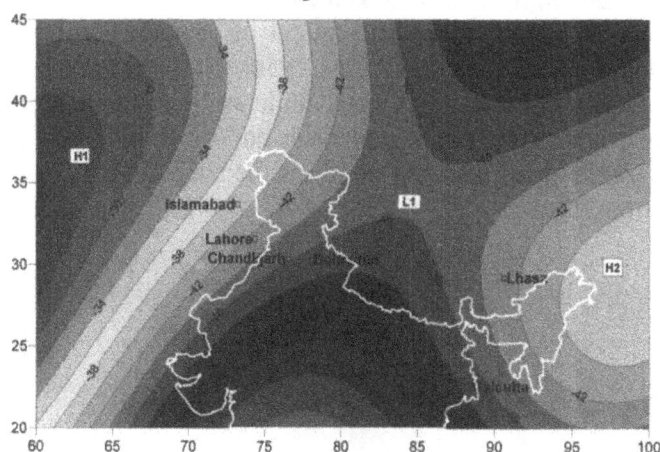

Fig. 6.10(a) Low pass filtered geoid height map related to the first segment of spectrum (Fig. 6.9c) with wave length >3140 km denoting sources in the upper mantle between average depth of 132-328 km. It is primarily a geoid low (L1) with relative highs (H1 and H2) along the margins. These anomalies are oriented NE-SW, in the principal stress direction and motion of the Indian plate. [Colour Fig. on Page 777]

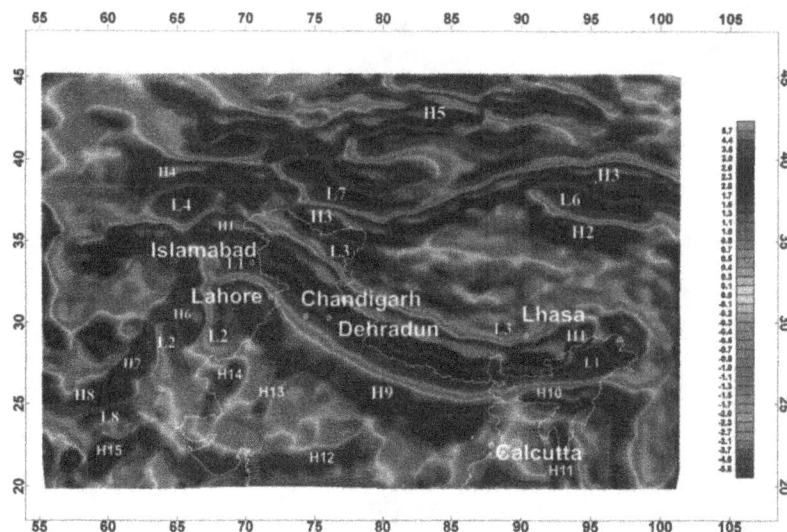

Fig. 6.10(b) High pass filtered geoid height map for wavelengths related to the fourth segment of the spectrum. It shows several highs (H1-H15) and lows (L1-L8) related to shallow/exposed sources).[Colour Fig. on Page 777]

The residual geoid anomaly related to crustal sources represented by the fourth segment of the spectrum (Fig. 6.9(c)) associated with wave number >0.01 is given in Fig. 6.10(b). This map looks like a satellite free air anomaly map (Fig. 6.6, Mishra and Ravi Kumar, 2008) delineating shallow sources which indicates that residual geoid field can be used to infer shallow bodies while regional geoid field delineates deeper sources. It also suggests that the effect of topography is mainly confined to the high frequency (small wavelength) component. Qualitatively, the residual geoid field (Fig. 6.10b) has delineated fold and thrust belts as geoid highs and basins as geoid lows that are as follows :

L1 = Himalayan foredeep from the Assam shelf in the east to the western margin in the west extending southwards as L2 related to Indus fore deep; H1 = Himalayan fold and thrust belt and suture zone north to its extending from the Eastern Syntaxis to the Western Syntaxis (Kohistan arc) and further extending westward related to Herat fault; L3 = Lhasa block with grandioritic intrusives of low density extending from the Eastern margin of Tibet to the western margin (Karakoram Range) and extending westwards as L4 related to Hindu Kush Ranges. While L6 over Qaidam basin suggests low density sediments. The linear belt of gravity highs, H2 and H3 over the Kunlun-Altyn Tagh faults indicate thrusted high density rocks as in the case of Himalayan Fold and Thrust belt (H1). It extends westward as H4 related to Pamir Ranges. Geoid highs H6 and H7 towards the west represents Sulaiman and Kirthar ranges along the Chamman fault suggesting their association with high density rocks such as Muslim Bagh and Bela ophiolites. Similar rocks may be present throughout the section at shallow depths. The set of geoid low and high L8 and H8 typically represents the Oman trench and Baluchistan arc. The gravity anomalies north of the Altyn Taugh fault, such as L7 and H5, are related to Tarim basin and Alai ranges of the Asian block. The geoid highs H9 running almost parallel to the Hiamalayn foredeep (L1) and the Himalayan fold and thrust belt appear to represent the crustal bulge caused by lithospheric flexure of the Indian plate. The Delhi-Lahore-Sargodha ridge (Fig. 6.5) at its western end is the shallow basement manifestation of this bulge as described in the next section (6.1.9). There are several other anomalies that can be related to exposed shallow density structures.

The geoid anomalies, H9 extends eastwards to the Shillong Plateau (H10) and joins with the geoid highs, H11 of Arakan Yoma Fold Belt (Fig. 6.5) and Sagaing fault, to its east. It has also delineated some important tectonic elements of the North Indian Shield that are described in Chapter 7. The geoid highs, H12 coincides with the Satpura Fold Belt (Section 7.6) that also extends to the Shillong Plateau. In certain sections these highs are extending towards the north that represents basement ridges under the Ganga basin. For example, along 70° E (H13) 78° E and 85° E that are referred to as Aravalli-Delhi Fold Belt and it's extension (Section 7.5), Faizabad ridge and Munger-Saharsa ridge (Section 7.4), respectively. Geoid highs, H14 are part of the high density mafic intrusive of Western Rajasthan in the basement (Sections 7.5 and 7.11). The geoid highs, H15 of the North Arabian Sea represents the high density rocks of the Murray ridge that extends in land as the Kirthar range in Pakistan. It also joins with the geoid highs, H12 of the Satpura Fold Belt through Saurastra that shows connection of the high density rocks in this section that may represent Deccan trap as it is extensively found offshore west coast of India (Sections 5.4 and 9.6). It emphasizes the application of residual geoid anomalies for shallow sources that has not been used much for exploration purposes. In fact, it provides an additional data set complimentary to the gravity field and should be utilized for exploration purposes.

The spectrum of the Bouguer Anomaly (Fig. 6.9(b)) suggests a wave number (n) of 0.002 for separation of regional and residual fields due to sources in the upper mantle and crustal sources using low and high pass wave band filter, respectively. Fig. 6.11 is the regional field that shows a gravity low (L1) of about –180 to –360 mGal centered at Tibet indicating low density sources in the upper mantle. The high pass filtered residual field (Fig. 6.12) due to crustal sources also shows a gravity low (L1) over southern part of Tibet and three gravity highs, H1, H2, and H3, south of Himalaya over the northern and NW Ganga basin and Tarim basin, respectively. The gravity low, L1 is related to crustal root due to isostatic compensation of the Tibetan topography while gravity highs, H1 and H2 under the Ganga basin coincide with the section of crustal bulge due to lithospheric flexure, especially the latter (H2) that coincides with the Lahore-Sargodha basement ridge (Fig. 6.5) regarded as the shallow manifestation of such a bulge (Duroy et al., 1989; Mishra et al., 2004). The crustal bulge along the entire Himalayan section is delineated in the next section while describing the Bouguer anomaly of North India (Section 6.1.9). Such crustal bulges are characteristic of all subduction zones referred to as bathymetric high or outer rise in case of oceanic trenches (Turcotte et al., 1978). The gravity high, H3 is attributed to large scale Permian mafic intrusives in the Tarim basin (Yang et al., 2007). This map suggests maximum crustal thickness in South Tibet related to the gravity low, L1 while it decreases northwards as confirmed by various geophysical investigations as described in forthcoming sections.

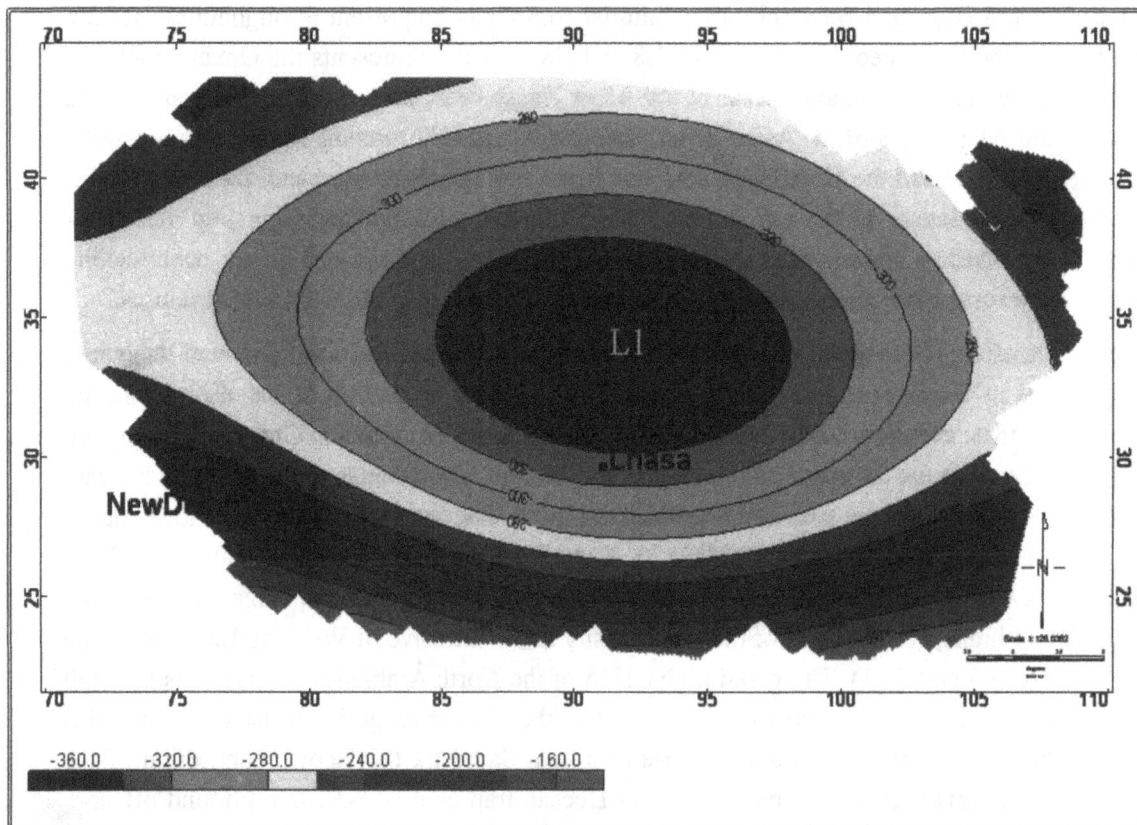

Fig. 6.11 Low pass filtered regional Bouguer anomaly of Tibet showing a major gravity low, L1 in the central part of Tibet related to low density rocks in the upper mantle (120-320 km).

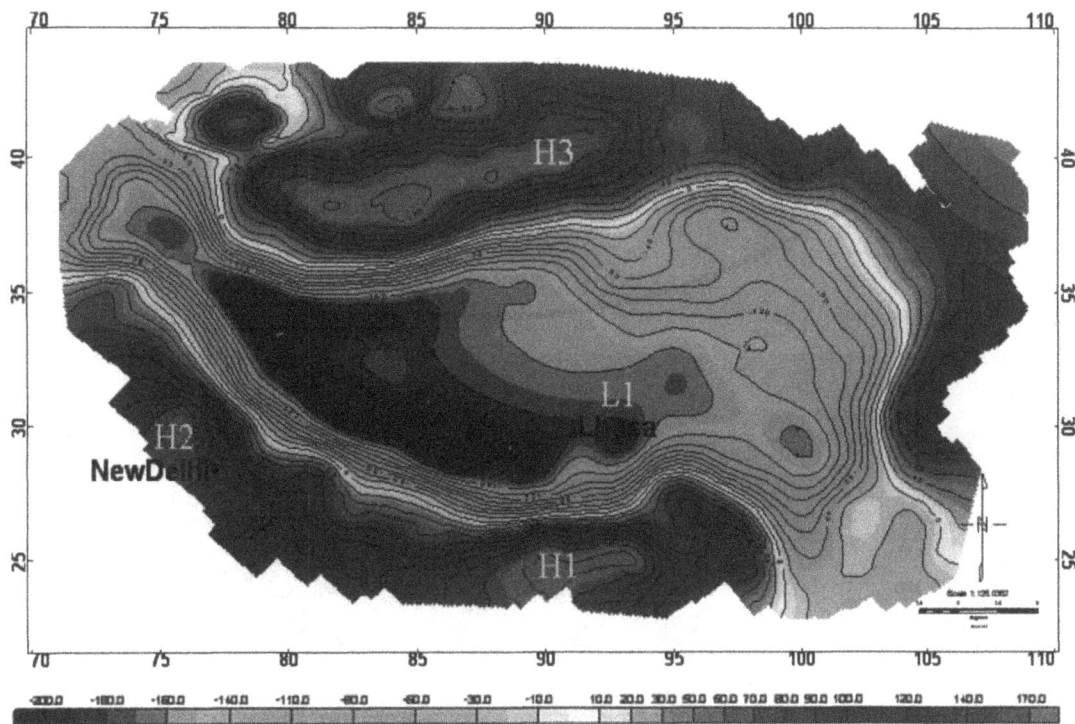

Fig. 6.12 High pass filtered residual Bouguer anomaly of Tibet showing a major gravity low in southern part of Tibet due to crustal thickening and gravity highs, H1 and H2 over Ganga basin and H3 over Tarim basin.

6.1.8 Modelling Regional and Residual Bouguer Anomaly of Tibet along 90°E

A first-hand model of lithosphere and crust under Tibet is obtained by modeling regional and residual satellite gravity fields. However, detailed models of lithospheric and crustal structures in this region are provided in the following sections by integrating the available geophysicsl data sets. The low pass filtered Bouguer anomaly (Fig. 6.11) representing the regional field is modeled along a profile (90°E). The modeled section of the low density body in the upper mantle and the computed and the observed field along the profile are given in Fig. 6.13(a). The depthwise extent of this body (120-300 km) is constrained from the spectral depths (Fig. 6.9(b) and (c)) and density contrast of the body from the match between the observed and the computed fields. This model provides the density of the body that produces the observed regional Bouguer anomaly low as 3250 kg/m^3 while surrounding mantle has a density of 3300 kg/m^3. Seismic receiver function analysis and tomography studies have suggested lithospheric thickness of 260-280 km characterized by high velocity as referred to above and described below in Section 6.2.3. However, the high velocity section extends almost up to 300 km, though with a reduced amplitude. Tseng and Chen (2008) have suggested hydrated subcontinental lithospheric mantle under Tibet due to subduction that would also reduce density due to presence of water and is detectable in the Transition zone based on P-and S-wave velocities. The high pass filtered residual field along 90°E (Fig. 6.12) is given in Fig. 6.13(b) that is modeled due to crustal thickening under Tibet from 30 km up to about 74-83 km as obtained from the spectrum of the fields given above. The adjoining highs over Ganga and Tarim basins are modeled due to crustal bulge up to 30 and 35 km, respectively as shown in the model. This crustal model is quite similar to those provided by Hetenyi et al., (2007) and Avouac (2007) where crustal thickness changes sharply from 38 km under the Ganga plains to 78 km under Himalaya and Tibet, south of the suture zone. The crustal thickening under Himalaya might have taken place as early as 44 Ma, subsequent to the collision of the Indian and the Asian plates (Aikman et al., 2008)

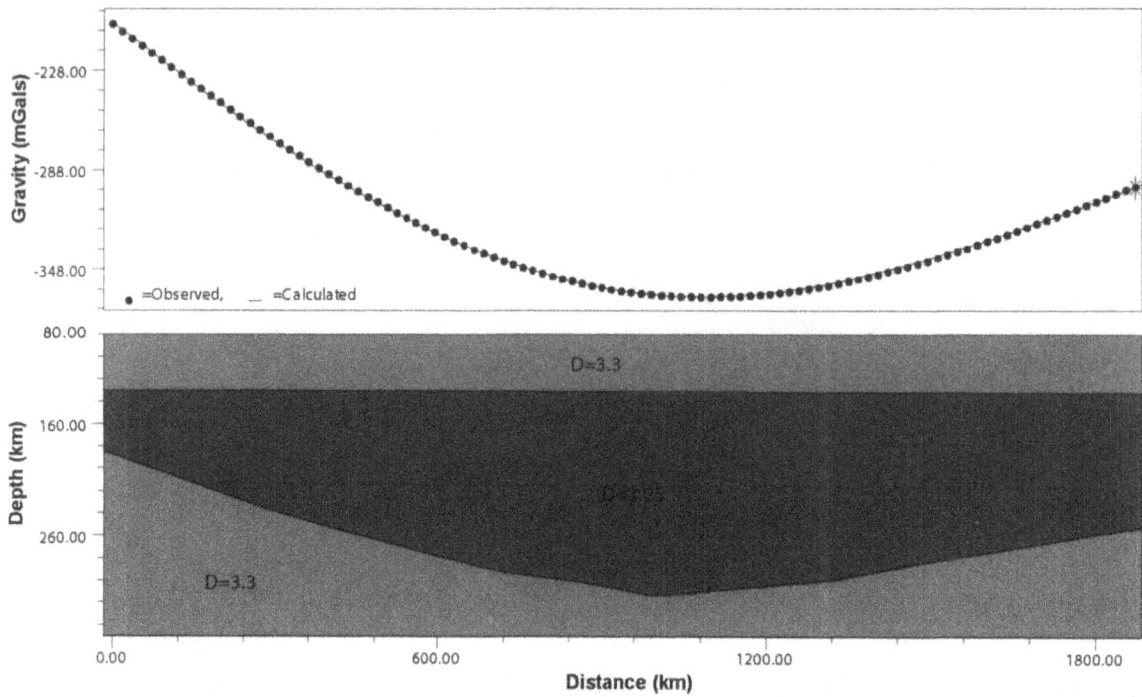

(a)

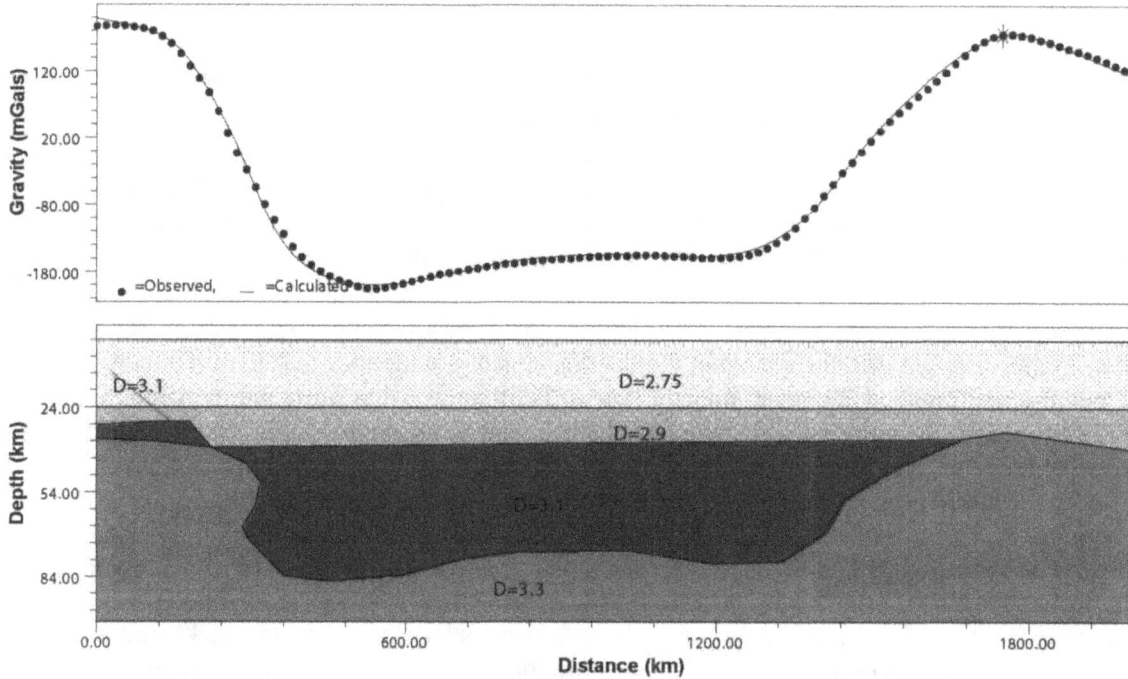

(b)

Fig. 6.13(a) Regional Bouguer anomaly over Tibet along 90°E (Fig. 6.11) and computed field due to low density body (3.25 g/cm³) between 120 to 300 km in the upper mantle showing a good match between the two. **(b)** High pass filtered residual Bouguer anomaly along 90°E (Fig. 6.12). It is modeled due to crustal thickening (38-83 km) under Tibet and crustal bulge up to 35 km under Ganga and Tarim basins. Densities (D) are in g/cm³.

6.1.9 Bouguer Anomaly of Himalaya and Foredeep: Filtered Regional Maps and Lithospheric Flexure

In order to validate the spectral depths obtained from satellite map with respect to ground-based gravity data and to delineate the crustal bulge due to lithospheric flexure that is an important aspect of Himalayan tectonics, we analysed the complete Bouguer anomaly map of India (GSI-NGRI, 2006) for the northern part that includes part of North-Western Himalaya in Kashmir (Fig. 6.14). This map is based on gravity data recorded at about 5 km interval that was processed including terrain and Bouguer corrections providing Bouguer anomaly to an accuracy of ± 1.5 mGal. The western and the Eastern parts of this map basically show gravity highs, H1 and H2 that have been attributed to shallow intrusives along the MBT and the MCT. The gravity lows, L1 and L2 are associated with the sediments of the Ganga basin and crustal thickening under western Tibet and granitic intrusives. The other gravity anomalies of this map are discussed in Chapter 7. As this map covers a significant part of Himalayan section only in the western part, we computed the spectrum of digital data of the western part (72°- 80 °E) as given in the inset of Fig. 6.14 that shows four linear segments with causative sources at depths of 134, 40, 17, and 6 km. Spectrum of Bouguer anomaly of the Western Syntaxis has also provided almost similar depth of 127 km for the first segment in the upper mantle (Fig. 6.55(b), Mishra and Rajasekhar, 2006). This spectrum (Inset, Fig. 6.14) could not provide a segment related to deepest sources suggested by the spectrum of satellite Bouguer and geoid maps due to limited data size and we could not conceive any bigger data set recorded in the Himalaya on the ground. However, the depths provided by the first linear segments in the upper mantle matches quite well within the error limits with those provided by the second segments of spectrum of satellite data. Moreover, the depth of lithosphere (120 km) based on S-wave tomography as given by Priestley et al., (2006) along the western margin of Tibetan Plateau is almost the same. The depths to Moho, lower crust, and basement in this case are 40, 17, and 6 km, respectively that are reduced compared to those observed over Himalaya and Tibet from satellite data, as this data set consists of a significant part of the Indian continent where depths to these interfaces are less.

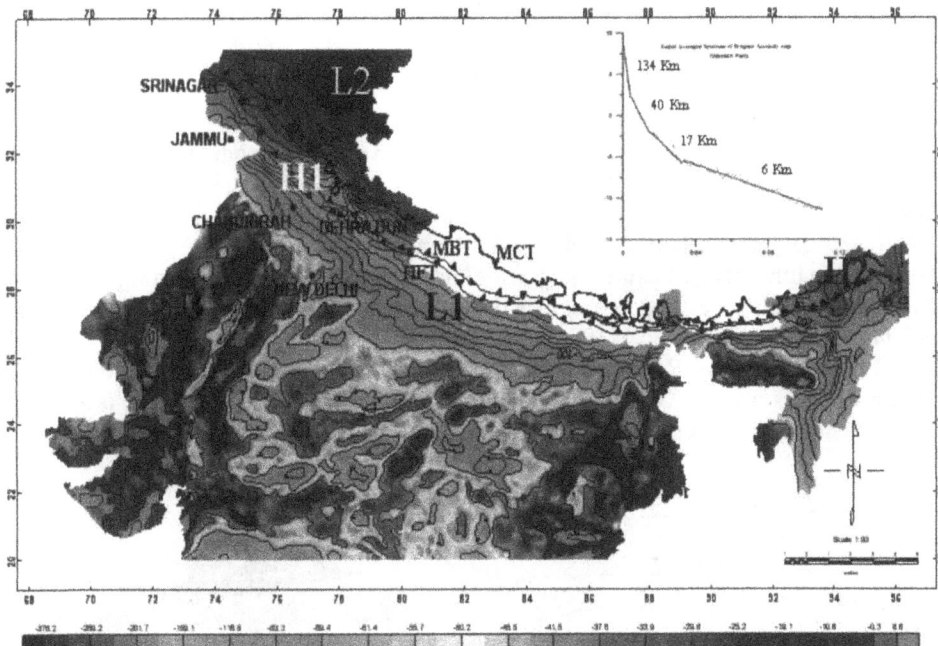

Fig. 6.14 Complete Bouguer anomaly map of North India including western Himalayas based on ground data. Inset shows the spectrum of western part (74-80°E) where data from NW Himalaya is available. It shows causative sources at four levels, 134, 40, 17, and 6 km related to upper mantle, Moho, lower crust, and basement. Depth to Moho, lower crust, and basement is relatively less in this case as most of the data in the present case pertains to the Indian continent. (Colour map is given in Fig. 2.28 and Fig. 7.8(a))

Lithospheric bulges can cause mafic intrusives due to decompression in the upper mantle (Hirano et al., 2006) that may also cause gravity highs. The digital data of Fig. 6.14 is filtered with wave number 0.004 (Fig. 6.15(a)) that represents both upper mantle and lower crustal sources, where the effect of crustal bulge caused due to lithospheric flexure of the Indian plate is quite prominent (H1, H2, H3, and H4). The gravity highs, H1 coincides with the Delhi-Lahore-Sargadha ridge that is the shallow (basement) manifestation of the crustal bulge (Duroy et al., 1989). The effect of lithospheric flexure as lithospheric and crustal bulge (Fig. 6.15(a)) suggests that it affects the total lithosphere including the lithospheric mantle. This crustal bulge is quite significant for seismic activity along the Himalayan front as it produces extension in the upper part and compression in the lower part (Fig. 6.15(b)).

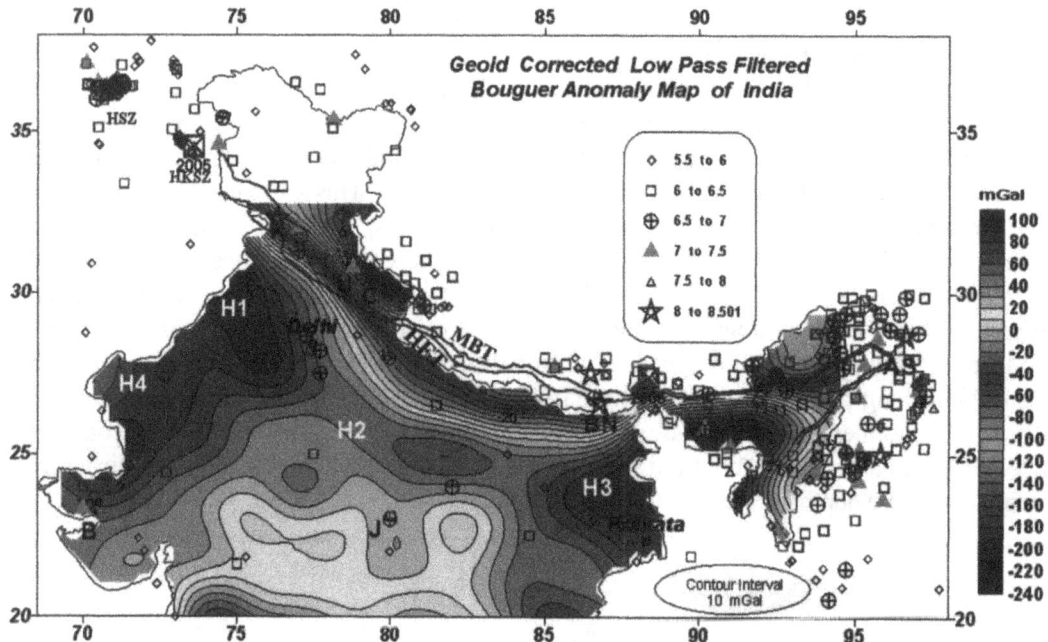

Fig. 6.15(a) Low pass filtered regional Bouguer anomaly of North India showing anomalies from upper mantle and lower crust related to first two segments of the spectrum (Inset, Fig. 6.14). These highs represent crustal upwarp due to lithospheric flexure. It also shows seismic activity along the Himalayan front. HSZ = Hindu Kush Seismic Zone and HKSZ = Hazara-Kashmir Seismic Zone, which also coincide with the gradient between gravity highs (H1-H3) due to lithospheric flexure and gravity low (L1) due to crustal thickening extrapolated towards the NW. Some great and large earthquakes shown by symbols are: K = Kangra (1905), U = Uttarkashi (1991), C = Chamoli (1999), BN = Bihar Nepal (1934), SH = Shillong (1897), AS = Assam (1950), B = Bhuj (2001) and J=Jabalpur (1997) earthquakes. Rectangular cross mark in the HKSZ shows Muzaffarabad (Kashmir) earthquakes of 2005 (Earthquakes from USGS web site). Hazara Kashmir Seismic Zone (HKSZ) is also known as Indus Kohistan Seismic Zone (IKSZ). [Colour Fig. on Page 778]

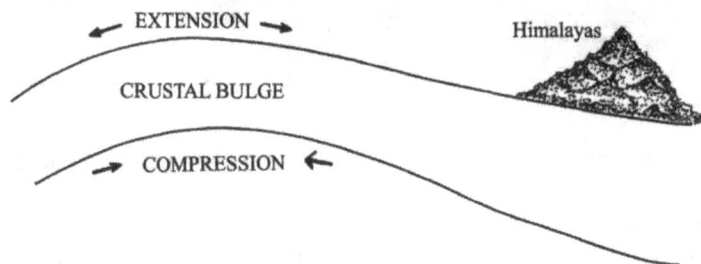

Fig. 6.15(b) Concept of crustal bulge due to lithospheric flexure under Himalayas showing extension in the upper part and compensation in the lower part.

6.1.10 Seismicity of Himalaya and Tibet and Deep Focus Earthquakes

An overview of Himalayan seismicity is given here for a broader perspective while details of some of the major earthquakes of this region, like the Kashmir earthquake of 2005, are discussed under the sections where the the tectonics of those regions have been described (Section 6.34). India's basement rocks flex and slide beneath Himalaya during great earthquakes. With India-Tibet convergence of about 18 mm/yr (Section 6.4), about 1.8 m of potential slip in earthquakes accumulates each century. Less than half of the Himalayan arc is known to have ruptured since 1800 (Fig. 6.16. Bilham et al., 2001). Surface exposures of ruptures, however, have not been found for any of the major earthquakes along the Himalayan arc. Fig. 6.15(a) shows low-pass-filtered Bouguer anomaly map of north India indicating a part of Himalaya with linear gravity highs (H1-H3) in the Himalayan foredeep related to the flexural bulge, which extends NW to coincide with the Delhi-Lahore-Sargodha ridge. This figure also shows the distribution of major earthquakes (> 6 magnitude) along the Himalayan front since 1670, which mainly coincides with the gradient of the gravity highs (H1-H3) due to lithospheric flexure, and gravity low towards the north due to crustal thickening. Some of the important large and great earthquakes reported along this front are Kangra (1905), Bihar-Nepal (1934), Assam (1950), and Shillong (1897) earthquakes; some recent examples are Chamoli (1999) and Uttarkashi (1991) earthquakes, which are plotted on Fig. 6.15(a). The Indus Kohistan Seismic Zone (IKSZ) or Hazara Kashmir Seismic Zone (HKSZ), and Hindu Kush Seismic Zone (HSZ) in Pakistan west of the Indian boundary are also plotted in this map which are seismically the most active zones along the Himalayan front as discussed in Section 6.3.4. These two sections are characterized by thrust types of shallow (crustal) earthquakes that originate in the crust up to a depth range of 50 km, while intermediate and deep focus earthquakes reported from nearby Pamir and Hindu Kush originate mostly from a depth range of 100-300 km. The epicenters of earthquakes in these zones can also be visualized to coincide with the gradient of gravity high due to flexural bulge towards the south and gravity low due to crustal thickening towards the north. The gradient coincides with the Himalayan frontal thrusts.

The epicenters of the Kangra earthquake of 1905 (Mw ~7.8) and Bihar-Nepal earthquake of 1934 (Mw ~ 8.2) are separated by a large distance known as the central seismic gap in this region (Khattri, 1987). Khattri suggested three such seismic gaps along the Himalayan front. However, when paleoseismic activities and their likely rupture zones were plotted (Fig. 6.16), the whole Himalayan front appears to have ruptured during last 1000 years (Bilham et al., 2001; Avouac, 2007) indicating different recurrence periods in different sections of the Himalayan front. One suggestion in this regard has been locking of the subducting Indian plate of Central Himalaya under south Tibet (Schulte-Pelkum et al., 2005; Section 6.2.2) that would increase the recurrence period in this section and may cause large earthquakes (Mw > 8) in future (Bilham and Wallace, 2005). Rajendran and Rajendran (2005) have also suggested a higher recurrence period for Western Nepal that probably is dependent on the nature of underthrusted rock types and ridges (Valdiya, 1994; Raval, 2000; Mishra and Rajasekhar, 2008). There does not appear to be anything common in the four known great Himalayan earthquakes (M ~ 8.0-9.0; from west to east, the 1905 Kangra, the 1934 Bihar, the 1897 Shillong, and the 1950 Assam earthquakes (Kayal, 2010; Kayal, 2008). However, if one considers the role of the crustal the bulge due to lithospheric flexure of the Indian plate as one of the contributory factors towards the Himalayan earthquakes that might be the common factor between all Himalayan great earthquakes whose effects will be different in different sectors due to different rock types. In Section 6.3.5, the extension of the Aravalli Delhi Mobile Belt (ADMB, Section 7.5) to NW Himalayan front has been demonstrated in residual gravity anomaly map and suggested that the section of the Himalayan front where the faults/thrusts of the ADMB and the basement ridge, Jaisalmer-Ganganagar ridge due to the Western Fold Belt (Pakistan) interact with the Western Himalayan front are most seismogenic that has given rise to several large and great earthquakes including Kangra earthquake of 1905. The role of lithospheric flexure in case of seismic activity along the Andaman-Sumatra subduction zone especially for the Sumatra earthquake of 2004 that represents an extension of the Himalayan collision zone in the Bay of Bengal, is demonstrated and described in Section 8.4.3.

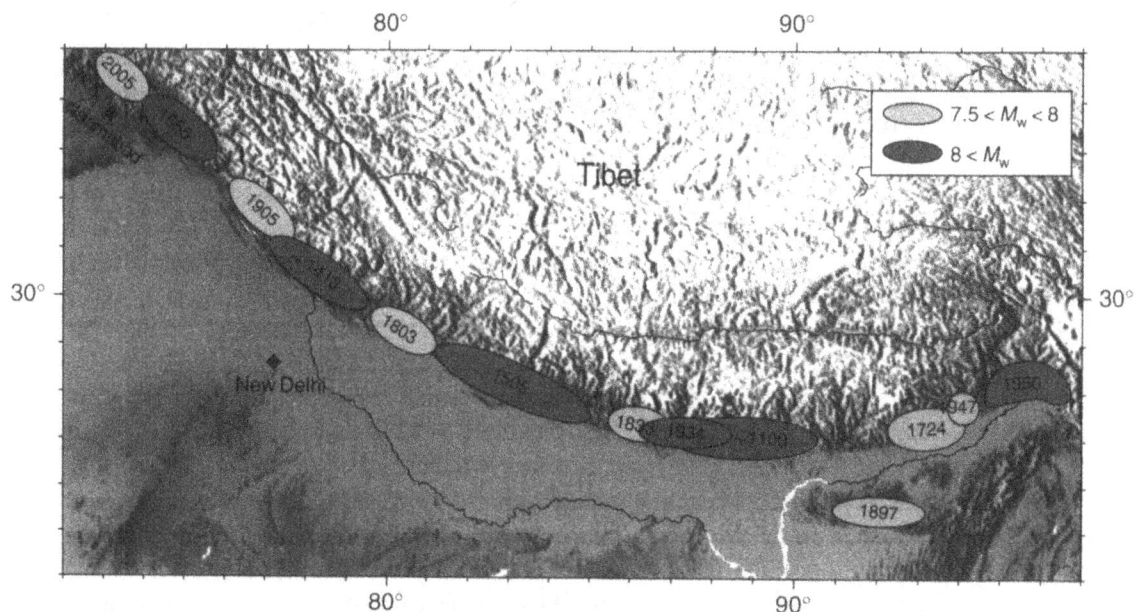

Fig. 6.16 Estimated rupture area of major earthquakes along the Himalayas (Bilham, 2004 and Ambraseys, 2000; Avouac, 2007 and references therein). It shows the different sections of Himalayan front being ruptured during different periods in the last 1000 years. [Colour Fig. on Page 778]

Most of the earthquakes occurring along the Himalayan front are thrust type of earthquakes. The histogram of the number of earthquakes of magnitude greater than five since 1974 along the Himalayan front and their focal depth (Fig. 6.17(a), Mishra and Rajasekhar, 2006) suggests their concentration in the depth range of 5-10 km and 30-35 km. The first level indicates the average level of the Main Himalayan Thrust under the Himalayan front that extends from the Ganga basin across the Himalayan collision zone. Himalayan Frontal Thrusts such as HFT, MBT, and MCT are attached to it in depth (Fig. 6.2). The second level of 30-35 km is the interface separating upper and lower crusts and coincides approximately with the Main Himalayan Thrust in the northern part. This indicates the importance of various rheological interfaces for seismic activity that is important for strain accumulation and failure to give rise to seismic activity. In fact, the Main Himalayan Thrust represents a plane of decollement along which the Indian lithosphere gets detached from the Indian plate and thrusts under the Asian plate.

Monsalve et al., (2006) have also defined bimodal depth distribution for earthquakes in the Himalaya in Eastern Nepal and Southern Tibetan Plateau which are concentrated above the Main Himalayan Thrust in depth range from surface to about 20 km and in the upper mantle from about 60-80 km under the South Tibet. Accordingly, they suggested that in case of Himalaya, the whole crust is brittle while in case of South Tibet, the upper crust and upper mantle may be brittle and lower crust may be ductile in nature (Section 6.2.5). However, it can also viewed as change in the depth of Moho from Himalaya to South Tibet. In both cases, the second level of concentration of earthquakes can be viewed as concentrated in the lower crust along the Moho which in any case represents a zone. In fact, these histograms suggest that the earthquakes originate from all levels up to the upper mantle and these levels of concentrated zones of hypocenters simply indicate their preferred depths of origin where strain concentration is more compared to other levels. Earthquakes originating from the ductile lower crust can be explained based on brittle thin laminated layers sectionwise, that give rise to such seismic activity (Meissner and Kern, 2008).

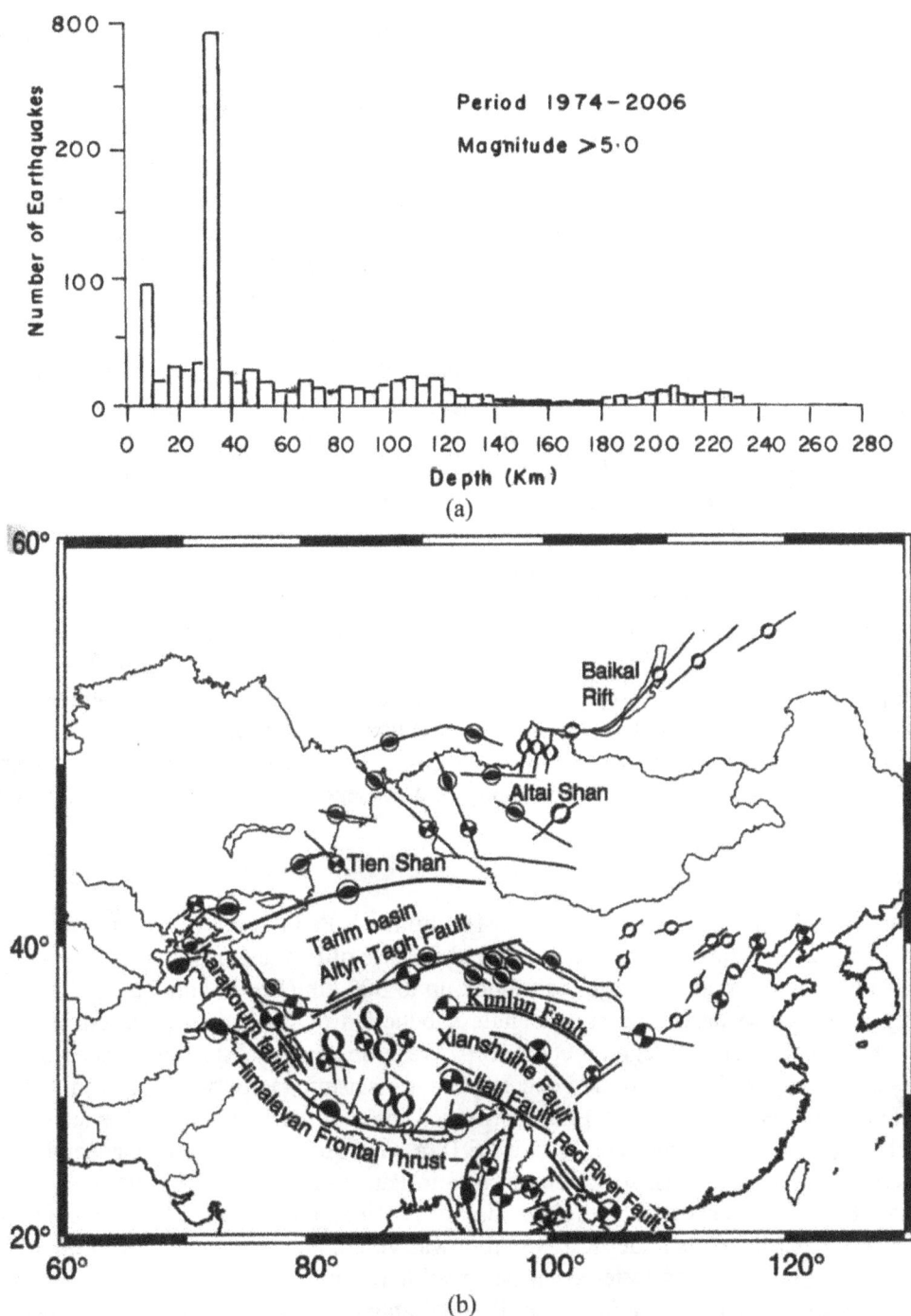

(a)

(b)

Fig. 6.17 (a) Histogram of earthquakes of magnitude greater than 5 since 1974 (USGS website) along the Himalayan front showing their concentration in depth range of 5-10 km and 30-35 km (Mishra and Rajasekhar, 2006). (b) Fault plane solution of some major earthquakes in Himalaya and Tibet showing predominance of thrust type earthquakes in Himalaya while normal type of earthquakes are dominant in Tibet (England and Molnar, 1997).

Deep focus earthquakes originating from the upper mantle in Hindu Kush and Pamir sections can be explained based on the presence of underthrusted cold Indian lithosphere in the upper mantle. Another interesting observation about seismicity in South Tibet relates to normal earthquakes related to extensional tectonics, while those in Himalaya are thrust types. (Fig. 6.17(b) England and Molnar, 1997). This suggests that the thrust type of earthquakes in Himalaya and Tian Shan are caused due to convergence, and normal type of earthquakes in between Tibet and Tarim basin are associated with normal faulting caused by extension. This indicates that the convergence and extension are inter-linked. Historical seismicity shows that slip on MHT is frequently accommodated through M >8 shallow earthquakes but shows a seismic gap in Western Nepal. This can be explained either through a seismic creep on the MHT or long lived elastic strain accumulation as described above. A zone of intense microseismicity indicates a stress build up generated by locking of a seismic creep which occurs along the MHT beneath Higher Himalaya and Tibet (Jouanne et al., 2004).

Priestley et al., (2008) reexamined the seismic data from this region and found that earthquakes occur throughout the crustal thickness of the Indian Shield (35 km), including lower crust that also conforms to the large effective elastic thickness as described in Section 6.2.5. This indicates a temperature of about 500 °C at Moho. Based on the Indian crust underthrusting under Tibet, they concluded that earthquakes in Himalaya and Tibet can occur in (i) upper crust up to isotherm depth of about 350 °C (ii) lower crust of dry granulite facies rocks, and (iii) mantle that is colder than 600 °C. As per the thermal structure under Tibet given by Hetenyi et al., (2007) as described in Section 6.2.5, 350 °C isotherm reaches at a depth of 7-8 km, while 600 °C isotherm reaches 40 km depth. This implies that they are primarily concentrated in upper and lower crusts. Hetenyi et al., (2007) have also shown the concentration of microseismic activity at a depth of 60-80 km under Higher Himalaya where subducting lower crustal rocks change into eclogite. These considerations of ductility and the brittle nature of the crust is related to shallow earthquakes occurring in the crust. 3-dimentional tomographic imaging of seismic velocities and seismic attenuation in Central North Island of New Zealand (Taupo Volcanic Zone) suggest role of fluids in lower crustal earthquakes near continental rifts. It involves the weakening of faults on the periphery of an otherwise dry mafic crust by hot fluids (Reyners et al., 2007).

(i) Deep Focus Earthquakes

There are at least two sections in Himalaya, viz. Hindu Kush-Pamir and Burmese arc, where intermediate to deep focus earthquakes have been reported. In case of Hindu Kush, most of the seismic activity originates below 100 km and is largely confined up to 300 km (Pegler and Das, 1998). They have been largely attributed to faster and higher angle subduction of the Indian plate under Hindu Kush (Kaulakov and Sobolev, 2006; Negredo et al., 2007) resulting into slab breakoff that would be brittle in nature and therefore prone to the occurrences of seismic activity. Based on thermo-mechanical modeling, Negredo et al., (2007) have provided the thermal structure of the subducting slab under the Hindu Kush and Pamir section (Fig. 6.18(a) and (b)) that shows in general lower temperature at depth under Hindu Kush section compared to Pamir section. Hildebrand et al., (2000) have also suggested major differences in the thermal history of Hindu Kush where crustal melting was limited and restricted to 24 Ma, while under Karakoram it was voluminous and continued from 37 to 9 Ma. Van der Hilst (1998) has reported faster shear wave velocities in the upper 100 km under Hindu Kush compared to the Lhasa block indicating lower thermal gradient in case of the former. Based on earthquake data, Lister et al., (2008) have suggested the existence of an elongate boudin type feature in the upper mantle under the Hindu Kush section coinciding with the hypocenters of the deep focus earthquakes (100-300 km) that may represent the remnant of the subducted Indian lithosphere. This might be caused by the stretching of subducted rocks under gravitational effect as subduction is almost vertical in this section. Fast subduction in this section may also imply subduction of high density rocks implying mafic/ultramafic rocks related to the oceanic crust that would cause more stretching.

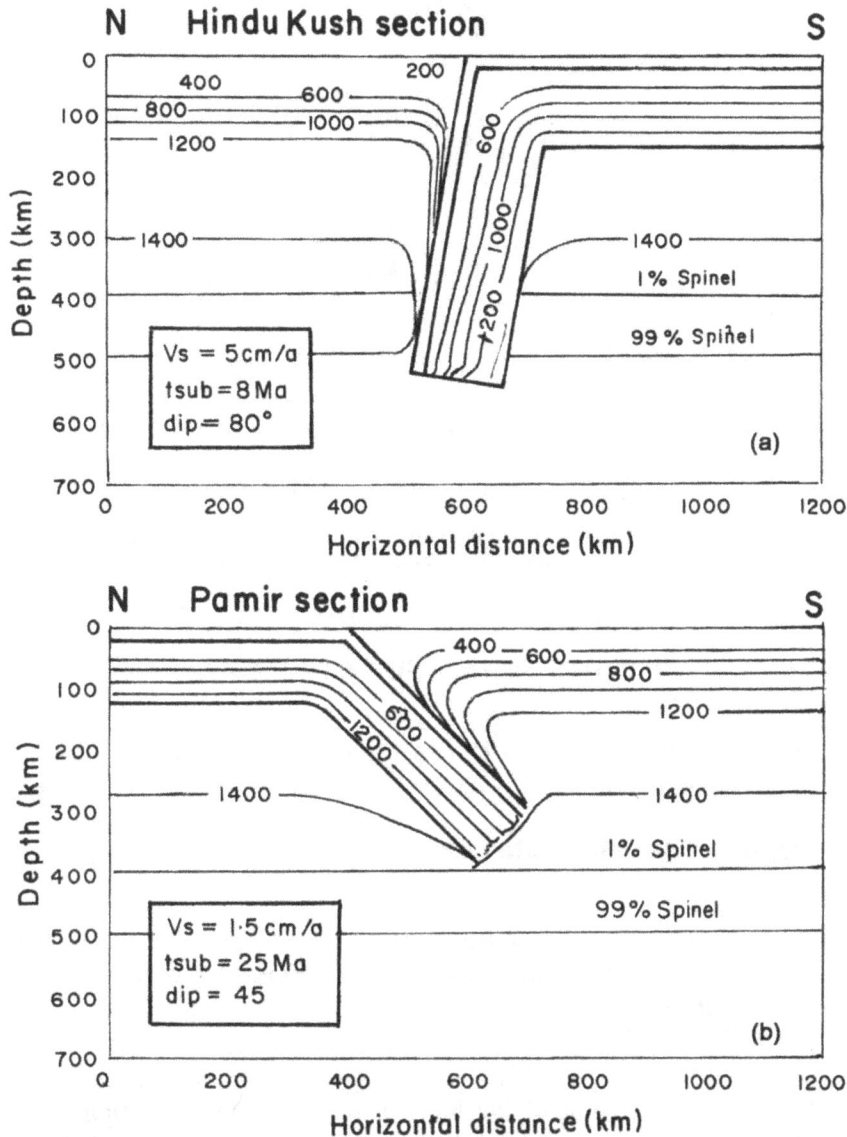

Fig. 6.18 Geotherm in subducting plate under the Hindu Kush (a) and Pamir (b) sections showing lower temperature in subducting slab under Hindu Kush in the upper mantle where rocks can remain brittle causing earthquakes (Negredo et al., 2007).

 Some deep focus earthquakes below 100 km have also been reported from the Burmese arc where subducting slab breaksoff at about 90 km has been postulated (Rao and Kalpana, 2005). They have given the various stages of subduction of Indian plate under the Burmese plate that shows slab bending westwards at 410 km transition zone and breakoff at about 90 km, which would cause thrust type faulting and earthquakes below this level (Fig. 6.19) while those occurring above this level are associated with strike slip. It is interesting to note that gravity highs (Fig. 6.6) that characterize thrusts and suture zones of Himalaya and Tibet are absent from Hindu Kush and Eastern Himalaya indicating the absence of or limited thrusting in these sections. Bettinelli et al., (2008) have suggested seasonal variations of seismicity and geodetic strain in the Himalaya induced by surface hydrology. Accordingly, they observed that the seismicity rate was twice as high in winter compared to summer. They attributed this to high pore pressure at seismogenic depth. Seismotectonics of deep foucus earthquakes of Hindu Kush and Burmese Arc are further dicussed in Section 6.3.4 and 6.2.5.

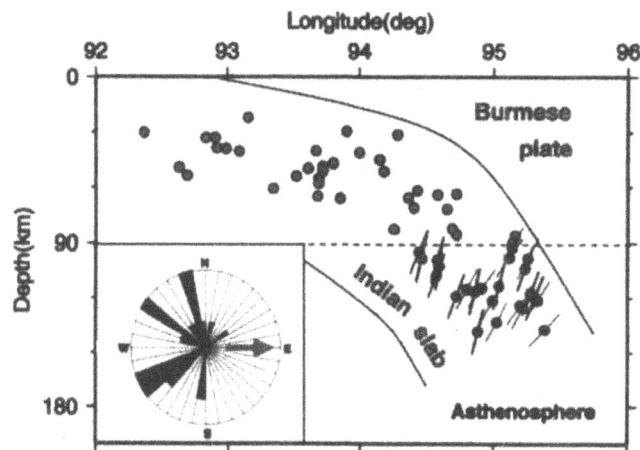

Fig. 6.19 Subduction of the Indian plate under the Burmese plate and hypocenters of earthquakes in Eastern Himalaya showing shallow strike slip (circles) and deeper reverse fault earthquakes (circle and line). Slab breakoff is indicated at about 90 km (Rao and Kalpana, 2005).

6.2   Central and Eastern Himalaya, Burmes Arc and Tibet – Crustal and Lithospheric Structures

This part relates to the section between 80 ° – 90 °E (Fig. 6.5) that primarily consists of Nepal and NE Himalaya, Burmese Arc and Tibetan terrain.

6.2.1   Geomorphology, Geology, and Tectonics

(i)   Geomorphology

Geomorphology and important tectonic elements of this region are presented in Fig. 6.20(a) (Braitenberg et al., 2000) that also shows the location of INDEPTH profiles described in the next sections. It shows Main Boundary Thrust (MBT), almost a semicircular line which is followed by Main Central Thrust (MCT) that shows a more crooked nature. In general, the Lesser Himalaya between MBT and MCT is 1.5-2.0 km high and is characterized by Paleozoic-Proterozoic metasediments, while the section north of the MCT reaches 4-5 km elevation with thrusted crystalline basement rocks in between them. The ITSZ is the suture between the Indian plate and the Tibetan block of the Asian plate towards the north and is characterized by association of ophiolite rocks and melanges. There are Trans-Himalayan batholiths in the Lhasa block just north of ITSZ such as Kailash and Gangdese etc. (Fig. 6.21) that were formed due to the subduction of oceanic lithosphere of Tethys Ocean under Tibet, are granodioritic-tonalitic in composition and are exposed as peaks over the Tibetan Plateau. To the south of the suture zone, there are peaks of leucogranites such as Mt. Everest, which is the highest peak in the world (~ 8848 m). They were formed along MCT due to underthrusting of the Indian lithosphere, therefore contain more of continental crustal rock, and are referred to as leucogranites. Mt. Kailash is one of the Trans-Himalayan batholiths that is one of the highest peaks in that region. Overall, the Tibetan plateau is flat over which these batholiths rise as peaks. Those south of ITSZ belong to post-Miocene period, while north of ITSZ (Fig. 6.21) belong to pre-Miocene period (Gannser, 1964). Mt. Kailash (~ 6700 m) is one of the sacred places of worship for Hindus and has been considered to be the abode of the most worshipped God Lord Shiva (Shanker) and his family of Vedic period (6-5 Ky BP). It has also been revered by Buddhists and Jains. The area around Mt. Kailash-Manasarovar (Fig. 6.20(b)) forms a hydrological divide with river Indus and Satluj flowing NW and Tsangpo towards the east, which are older than the Himalaya, including the Higher Himalaya. Several rivers like Yamuna, Ganga, and Kali, etc., flow southwards from Higher Himalaya indicating that they are post-Miocene after formation of Higher Himalaya. The major rivers of the region are antecedent, meaning that they predate cession or uplift. At least 8 Ma uplift initiated or strengthened the Asian monsoon.

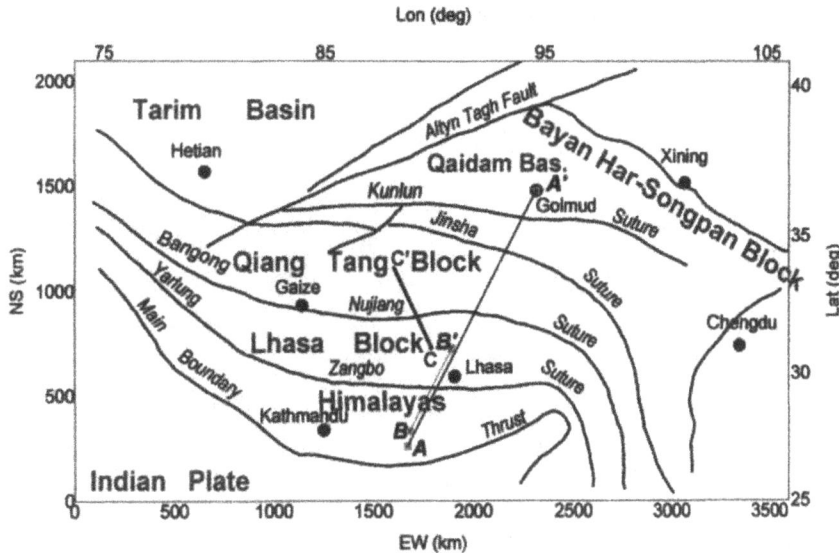

Fig. 6.20(a) Tectonic map of Tibet with important tectonic elements and INDEPTH profiles I and II (BB′) and III (CC′) plotted on it. It also shows Golmud-Yadong (AA′) geotransect described in Section 6.2.5 (Braitenberg et al., 2000).

River Ganga (Fig. 6.20(b)) is also revered by Hindus since the Vedic period. It is, in fact, the life line for one of the most populated land, the Indo-Gangetic plains. This, along with its tributaries, dumps almost two billion tonnes of fertile sediments in the plains, bringing down the slopes of Himalaya causing erosion and uplift. Vedic scholars have also suggested a connection between Lord Shiva at Mt. Kailash and the river Ganga. Legend has it that the Ganga descended from heaven with great speed and Lord Shiva at Mt. Kailash retarted its speed and tamed it for mankind. The Vedic period at approximately 6-5 Ky BP corresponds to just after the last glaciation in this region, known as neoglaciation (8-6 Ky BP) (Shroder Jr. and Bishop, 2000) when large amount of water must have flowed down from the north (heaven) to the south (earth) flooding the entire region. Even the melting of Pleistocene glaciation at the start of Holocene (12-10 Ky BP) must have caused havoc because of a thick sheet of ice of about 1-2 km that had covered the Tibetan plateau during that time (Kuhle, 2007) indicating the evolution of civilization at the Himalayan foothills after the Pleistocene and Holocene glaciations.

Major rivers like Satluj and Indus along the western Himalaya, and Yamuna and Ganga in the central part follow the Himalayan trend suggesting that they are controlled by basement structures related to Himalayan tectonics. The former, originating from Tibet, follow the trends of Western Himalaya and discharge in the Arabian sea. The Yamuna and Ganga, originating from NW Himalaya, flow southwards almost up to the exposed Precambrian rocks of the Indian shield and take an easterly turn along them to discharge in the Bay of Bengal, forming a garland to the Himalaya. Even small changes in the river course appears to be controlled by basement tectonics such as NE trend towards Himalaya before Varanasi, and easterly trend after it. This NE change in the river course is parallel to the Son lineament towards the east and follows the North Narmada Fault, which is a well known basement fault affecting a large section of the north Indian shield as the Narmada-Son lineament (Section 7.6). The basement tectonics in the

Indo-Gangetic plains, which represents a foreland basin of Himalayan collision, are primarily controlled by two factors, viz. (i) precambrian tectonics prior to Himalayan collision, and (ii) tectonics due to Himalayan collision such as crustal bulge caused by lithospheric flexure of the Indian plate as discussed in Section 6.1.6 and 6.1.8.

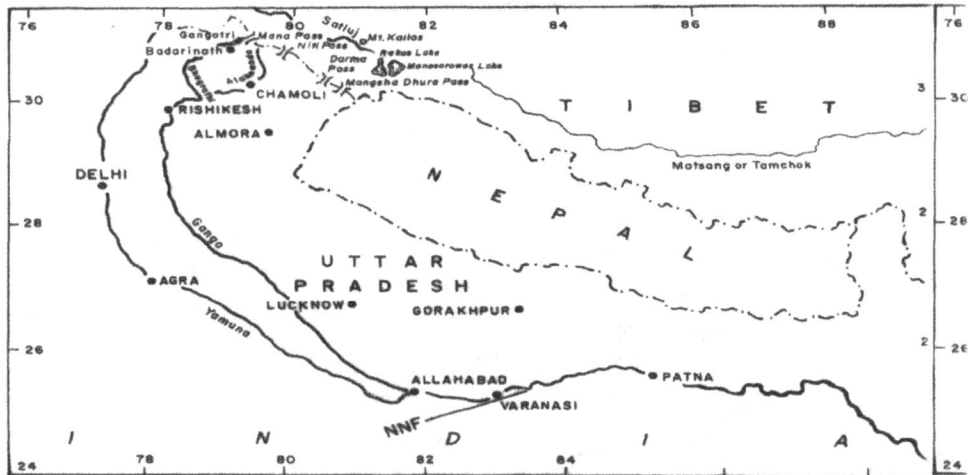

Fig. 6.20(b) Some important rivers of NW Himalaya with Mt. Kailash in Tibet and their connection.

Fig. 6.20(c) Photograph of Mt. Kailash representing a subduction related Pre-Miocene (~30Ma) granite batholith in SW Tibet, north of the suture zone that shows the almost vertical face of such intrusive. Manasarover Lake lies at the base of it. It is one of the largest fresh water lakes at a high altitude. (By courtesy R.K.Math). [Colour Fig. on Page 779]

(ii) Geology and Tectonics

There are five major terrains forming Himalaya, the Tibetan Plateau, and the adjoining Asian Plate. They were initially located at low latitudes separated by the Tethys Ocean which drifted northwards and accreted to the Asian plate separated by suture zones. These terrains as given in Fig. 6.20(a) are (a) Himalayan terrain, (b) Lhasa Terrain, (c) Qiangtang and Songpan-Ganzi Terrain, and (d) Kunlun Terrain-Qaidam basin (Gongjian et al., 1991). The adjoining Asian plate to its north consists of (e) the Tarim basin and the Tian Shan Ranges north of the Altyn Tagh fault.

(a) Himalayan Terrain

A detailed geological map of Himalaya is given in Fig. 6.21 (Sorkhabi and Macfarlane, 1999). Major structures of the Himalaya include the south vergent thrusts and suture zones as has been described above in Section 6.1: (i) The Himalayan Frontal Thrust (HFT) that has thrusted Plio-Quaternary Siwalik sediments southward over recent alluvium (ii) Main Boundary Thrust (MBT) along which Paleozoic and late Precambrian sedimentary rocks are emplaced over the Plio-Quaternary Siwalik sediments, and (iii) MCT that has thrusted crystalline rocks of High Himalaya over Paleozoic and late Precambrian sedimentary sequences (iv) the ITSZ is the suture zone between India and Tibet and is largely characterized by ophiolite rocks that represent remnant of the oceanic crustal rocks (v) North Himalayan normal fault known as South Tibetan Detachment System (STDS), north of the MCT and south of the ITSZ separates the Tethyan sediments that represent trench deposits towards the north from crystalline rocks along MCT towards the south. Uplift along this fault suggests a ductile shear zone at the depth. In contrast to thrusts, STDS is a normal fault. Thrusting in orogenic belts are against the gravitational forces. Therefore, mountains that are formed due to thrusting rise till the compressional forces balance the gravitational forces (Lliboutry, 1999). Subsequently, new thrusts zones are formed that demonstrate the shift of MCT to MBT and to HFT.

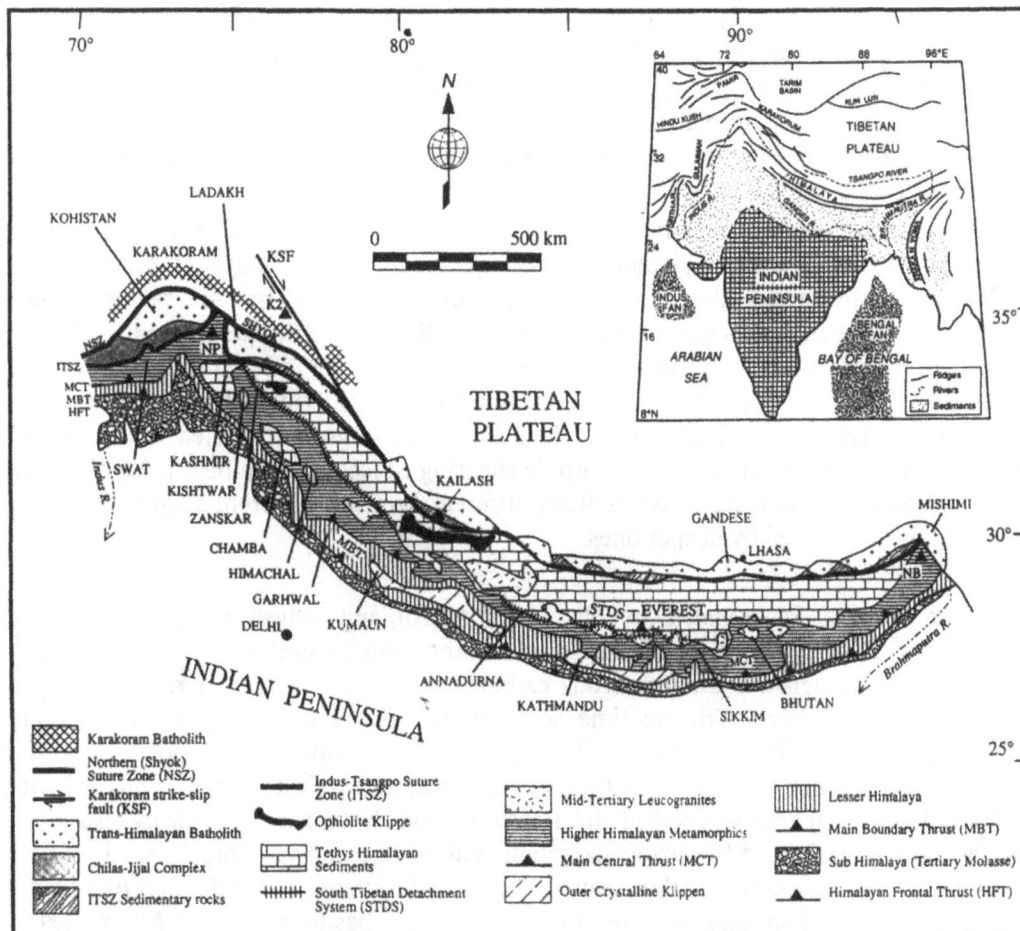

Fig. 6.21 Important geological elements of Himalayas with exposed ophiolites and different units shown prominently. Trans-Himalayan batholiths, ophiolites, and leucogranites along with various thrusts are clearly marked (Sorkhabi and Macfarlane, 1999 based on Gansser 1981 and Windley 1983).

Trans-Himalayan granite batholiths and several Neogene and recent acid volcanic centers are shown north of ITSZ. It also provides important leucogranites as Neogene acid volcanic centers just north of MCT in the Tethyan sediments. Two suture zones in the Western part may be noted. They are known as Main Mantle Thrust (MMT) synonymous to ITSZ towards the east and the Northern suture or Shyok suture separating the Kohistan arc and Karakoram batholith. The ages of leucogranites as crystalline rocks decreases from the west to the east (Guillot et al., 1999), e.g., the Nanga Parbat-Harmosh massif shows an age of 30-35 Ma that decreases eastward as follows: Zanskar (22-20 Ma), Garhwal (22-18 Ma), Everest (16-13 Ma), and Bhutan (13-11 Ma) which also suggests that continent-continent collision first took place towards the western side as discussed in Section 6.1.1 along the Western Syntaxis which progressed towards the east. However, it is quite intriguing that younger leucogranite are not found on the western side, and none younger than 9 Ma along the entire Himalayan belt. This can, however, be tentatively explained due to the shifting of thermal conditions required to form leucogranites successively towards the east as continent-continent collision progressed eastwards and time required in transformation of underthrusted continental crust to leucogranites. Leech (2008) has explained the absence of younger leucograintes on the western side based on interruption of N-S channel flow due to the Karakoram fault. However, it still does not explain the systematic younger ages of leucograintes towards the east.

The Himalayan terrain north of MBT primarily consists of Proterozoic metasediments equivalent to those found in the Indian Shield and crystalline basement rocks. Paleozoic and Mesozoic sediments display Gondwana marine facies. Marine sedimentation ended in the Eocene when the Indian plate collided with the Lhasa block. During intra-continental orogeny, the STD separated North Himalayan terrain toward the north, and higher and Lesser Himalaya towards the south.

High temperature reversed metamorphism has been reported from the shear zones. This implies an upward increase of metamorphic grade across Himalaya (LeFort, 1975). Initially, it was observed along the MCT specially along the Everest section where metamorphic grade increases upwards from chlorite (300 °C) to sillimanite grade (700 °C). Due to poor mineral assemblage along the MBT, it could not be inferred till graphitization was used for this purpose using Raman Micro-Spectroscopy providing a temperature range of 350-530 °C (Bollinger et al., 2004). This has been largely attributed to recumbent isotherms (LeFort, 1975) or post-metamorphic shearing of isograds (Hubbard, 1996). It can also be attributed to thrusting of rocks from different levels at different stages, first from shallower levels followed by deeper ones.

(b) Lhasa Terrain

The Lhasa Terrain is bounded by the Bangong-Nujiang suture towards the north (Fig. 6.20(a)). It largely consists of para-gneisses and granitic gneisses as basement. The cover rock is Ordovician to Carboniferous carbonate and clastic rocks. A rift occurred in the late Permian or Early Triassic time south of the Tsangpo River and subsequently expanded to form the Tethys ocean in Triassic Time. With northward subduction in the Cretaceous of the Tethys oceanic crust along the ITSZ, an active continental margin with trench-arc-basin system developed in the Lhasa Terrain. Suture zone is characterized by ophiolite suite of rocks and Mesozoic sediments with diorite and volcanic rocks. There are two stages of magmatization in this section, 120-80 Ma related to subduction of oceanic crust and extrusion of andesite volcanic rocks in back arc basins and 70-40 Ma related to collision, which are primarily oceanic crustal rocks. The magmatic rock during collision stage is dominated by granodiorite with higher content of SiO_2 and K_2O. Gangdise magmatic arc in Eastern Lhasa block suggest three phases of tectonic activities (i) Late Cretaceous-Tertiary (ii) Paleocene-Eocene, and (iii) Post Eocene (Gongjian et al., 1991).

(c) Qiangtang and Songpan-Gonzi Terrain

This terrain is bounded by the Bangong-Nujiang the suture towards the south and the Jinsha River suture towards the north (Fig. 6.20(a)). It is characterized by a plateau towards the west and basin towards the east. The basement rocks are gneisses and metavolcanics over the plateau while the basin towards the east consists of Ordovician and Silurian unmetamorphosed or slightly metamorphosed sediments which rest on crystalline basement. BNS is characterized by presence of ophiolite rocks which may represent an earlier subduction zone similar to ITSZ (Allegre et al., 1984). The Songpan-Ganzi Terrain north of the Jinsha River suture represents trench-arc-basin system of a convergent margin that is largely covered by Tertiary deposits. Towards south, ophiolites occur. The island arc comprises andesite with Lesser amount of basalt (Gongjian et al., 1991)

(d) Kunlun Terrain

This terrain north of the Kunlun fault (Fig. 6.20(a)) consists of Late Ordovician metasediments and slabs intercalated with limestones, dolomites, silicious, and pillow basalts overlain by late Devonian molasse sandstone intercalated with minor volcanic rocks. These rocks are intruded by late Caledonian Calc-alkaline diorites and granodiorites of island arc type with ages 423-394 Ma (Xiao et al., 1986). The rocks are interpreted to represent a trench- arc-basin system formed during Caledonian convergence. Qaidam basin is occupied by Cenozoic sediments underlain by Caledonian tectonic belt. The tectonic activity might have continued up to late Permian that represents island arc volcanism within the Kunlun terrain. There appears to be a suture separating north and south Kunlun terrains during the late Devonian-Carboniferous periods that is followed by large scale intracontinental magmatism during the late Hercynian (Permo-Triassic) period (Gongjian et al., 1991). The Ultra High Pressure Metamorphic (UHPM) rocks of Qaidam basin of 500-440 Ma and protolith ages of eclogite of 800-750 and 1000 Ma suggest following sequence of events: At about 1000 Ma ago, a number of continents were amalgamated to form Rodinian continent in this area. This part of Rodinia was then rifted at about 800-700 Ma ago to form an oceanic basin with a variety of MORB and ocean island basalts. Closure of this basin produced Neoproterozoic ophiolites and granite gneisses in the north Qaidam mountains. In the early Paleozoic, another Qilian basin formed. Subduction of this oceanic lithosphere formed the island arc volcanic rocks at about 500 Ma ago, and subduction related granites at 470-500 Ma (Yang et al., 2006). Subsequently, continent-continent collision gave rise to ultra high pressure metamorphism.

(e) Tarim Basin and Tian Shan Ranges of the Asian Plate

Tarim basin (Fig. 6.20(a)) is the largest basin in China. It is surrounded by the Tian Shan and Kunlun orogenic belts and forms a foreland basin over the Tarim microplate. It is important for oil and natural gas occurrences and primarily consists of Precambrian and Phanerozoic rocks with a long magmatic history of 2500-260 Ma. Exposed Permian rocks consist of volcano-sedimentary sequences, limestone, and basic-intermediate-acidic volcanic rocks. Bimodel Permian volcanic rocks (248-292 Ma; Yang et al., 2007) are tectonically significant as they indicate a rifting phase during that period. They have also been correlated with a mantle plume. Tian Shan represents an intracontinental mountain belt that formed in response to the India-Asia collision. It consists of high mountain ranges in the west decreasing in altitudes towards the east. It is an active fold and thrust belts with Mesozoic and Cenozoic sedimentary cover that vary in both stratigraphy and structure from east to west. It shows a thin-skinned geometry with decollement at about 6-10 km within the Mesozoic and Cenozoic sedimentary cover similar to MHT under Himalaya (Fig. 6.24). This implies that there are similarities in subsurface structures both on the Indian and the Asian plates, though signatures of collision tectonics are better observed on the Indian plate. Highest parts of Tian Shan are adjacent to areas of active shortening indicating that major uplifts in Tian Shan are young during Cenozoic or Quaternary in age (Burchfiel et al., 1990). The Tian Shan Carboniferous-Permian rift related volcanism in northwestern China represents an overly recognized large igneous province related to a mantle plume (Xia et al., 2008).

(iii) North East Himalaya and Syntaxis

Fig. 6.22 is the geological and tectonic map of the Eastern Syntaxis (Namche Barwa and Siang Syntaxis) along which the Himalayan structures changes from E-W to almost N-S. It shows the extension of ITSZ as the Lohit Thrust or Tiding suture with exposed ultrabasic rocks along it, and the Mishmi Thrust that is synonymous of MBT in the Eastern Himalaya (Ding et al., 2001). The Assam Shelf in the Brahmaputra foredeep south of Pasighat is occupied by alluvium which is surrounded by thrusted Siwalik sediments along the HFT. Siang Syntaxis is occupied by Tertiary and Mesozoic sediments followed by rocks of Lesser and Higher Himalaya represented by Proterozoic metasediments and crystalline rocks, respectively. It also shows the Eastern part of BNS and Guyu thrusts and associated Gangdese arc.

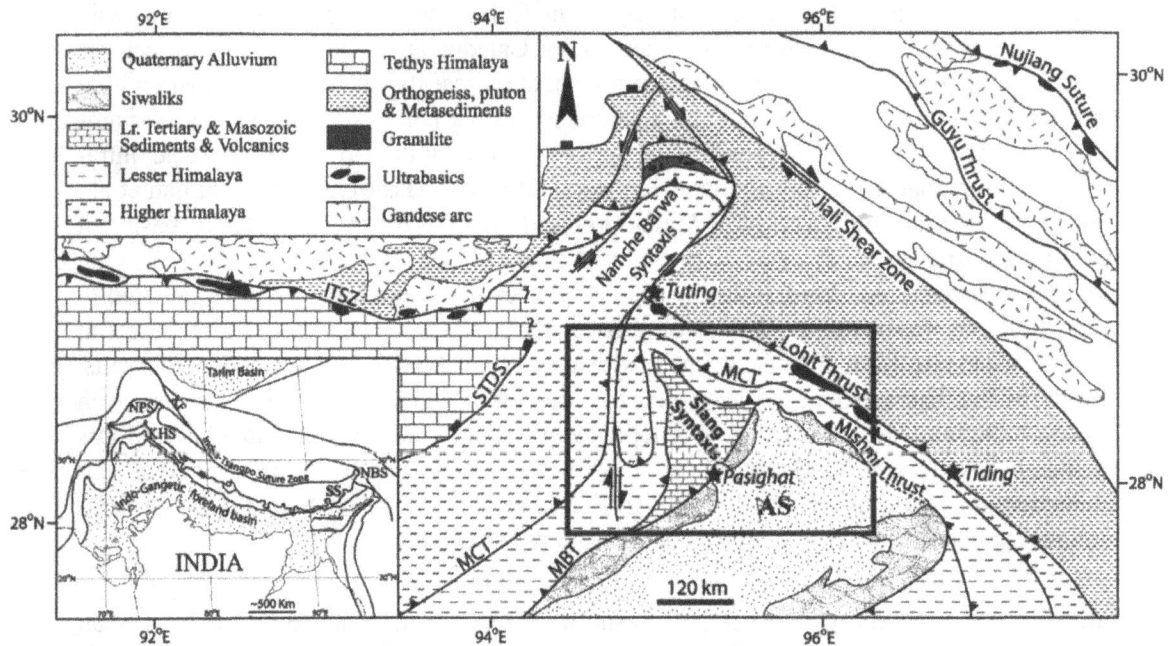

Fig. 6.22 Geological map of the NE syntaxis of Himalaya and surrounding regions. GT = Gangdese thrust; ITSZ = Indus-Tsangpo suture zone; MBT = Main Boundary Thrust; MCT = Main Central Thrust; STDS = South Tibet Detachment System; KF = Karakoram Fault; NBS = Namche Barwa Syntaxis; SS = Sian Syntaxis; NPS = Nanga Parbat Syntaxis; KHS = Kashmir Hazara Syntaxis, As = Assam shelf (Ding et al., 2001).

6.2.2 Seismic Studies: Crustal Structures – INDEPTH Profiles and Other Seismic Studies

Crustal and lithospheric structures are important for geodynamic studies. Methods which are commonly used to obtain details of these structures and their physical properties are seismic, magnetotelluric, and gravity-magnetic methods. Seismic studies are best suited to obtain crustal structures in any region. One of the earliest seismic studies in the Himalaya was carried out by Finetti et al., (1979) across the Western Himalaya extending from the Nanga Parbat to the Alai Ranges. It provided a maximum crustal thickness of about 70 km under the Karakoram Ranges reducing to about 60 km under the Pamir and Alai Ranges. It also provided a low velocity zone at a depth of 30 km under the northern part of the Karakoram Ranges. Kaila et al., (1978, 1984) have reported some seismic profiles across the Western Himalaya towards the Indian plate which suggested Moho varying from 40-78 km under the Kashmir Valley and the Higher Himalaya. However, the most important results from seismic studies were obtained from INDEPTH profiles (Fig. 6.20(a)) as described below.

(i) INDEPTH Profiles

INDEPTH (International Deep Profiling of Tibet and Himalaya) was a multinational and multi-institutional project involving China, U.S.A., Germany, and Canada. Such experiments were undertaken under a collaborative scheme across ITSZ and southern Tibet popularly known as INDEPTH I, II, and III which included seismic wide angle reflection surveys, magnetotelluric measurements and broad band studies. Fig. 6.20(a) (Braitenberg et al., 2000) shows the INDEPTH I and II as Profile BB' superimposed over a detailed tectonic map of Tibet. It also shows INDEPTH III across Bangong-Nujiang suture as Profile CC'. Profile AA' is a geotransect from Golmud to Yadong described in Section 6.2.5. Several other seismic studies like broad band studies, and receiver function analysis etc., were also conducted along these profiles that are discussed in the forth coming sections (Kind et al., 2002).

These measurements and their analysis provided one of the most reliable results about the subsurface structures under this region. Fig. 6.23 (Hauck et al., 1998) shows a detailed layout of INDEPTH I and II located across the ITSZ and INDEPTH III (Fig. 6.20(a)) is located across the Bangong-Nujian Suture Zone (BNS). Fig. 6.23 also shows the important tectonic elements across INDEPTH I and II. It shows in detail the nature of various thrusts, suture zones, and disposition of leucogranites vis-a-vis suture zone and the STD. Detailed tectonics of this region is provided in this figure for readers to comprehend the kind of rock types and structures along and across a typical section of the suture zone in the Himalaya. These profiles are primarily located in the Yadong-Gulu Rift across the ITSZ. In spite of being an overall compressional regime, there are local extensional regions almost perpendicular to the compressional forces giving rise to rift valleys in South Tibet. This conforms to the normal type earthquakes in Tibet (Section 6.10). There is a North Himalayan anticline south of the suture zone with several domal shaped batholiths (Kangmar dome etc.,) while towards the north of it is the Gangdese batholith representing subduction related volcanic rocks. Numbers 1-7 represents different parts of the profile, with sections 1-4 representing INDEPTH I, and 5-7 representing INDEPTH II. Several publications have appeared on the results from these experiments. Some significant ones are as follows, : Zhao et al., (1993); Brown et al., (1996); Hauck et al., (1998), Nelson et al., (1996); Haines et al., (2003); Tilmann et al., (2003); Unsworth et al., (2005); Zhao et al., (2004a, b, 2006); etc., some of which are discussed below.

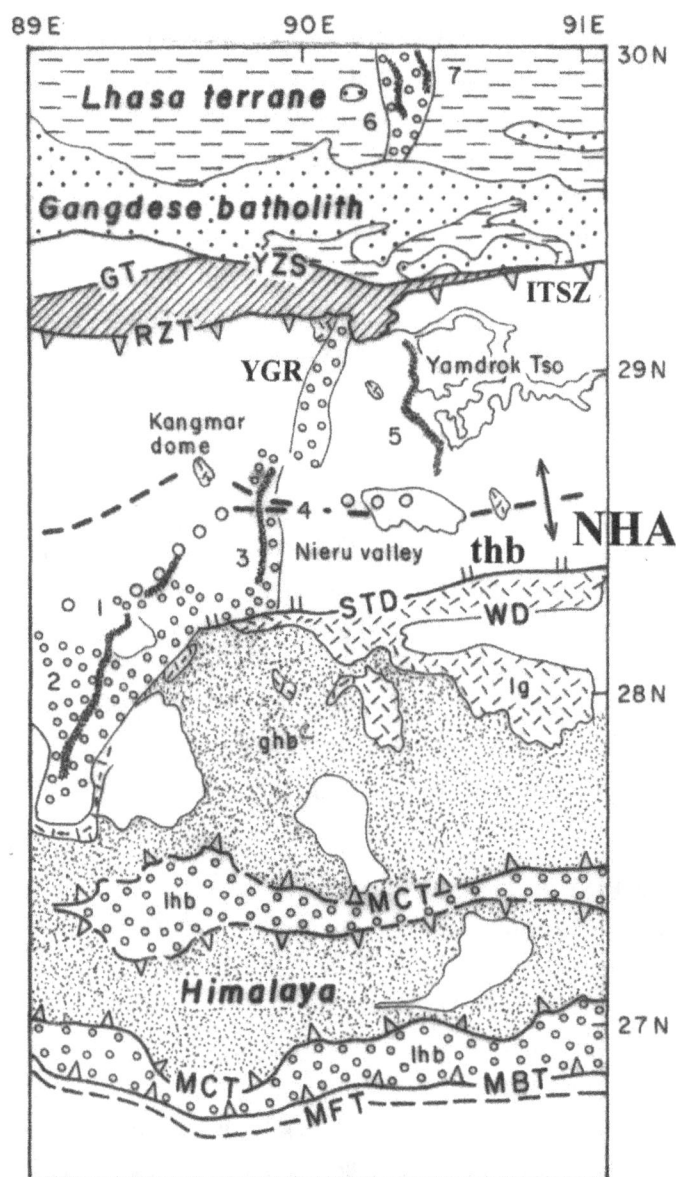

Fig. 6.23 Detailed tectonic and geological map of a section of suture zone (ITSZ) where INDEPTH I and II profiles are located (Bold line with numbers 1-7). Sections 1-4 represents INDEPTH I and 5-7 INDEPTH II. It shows major thrust and suture zones, such as MFT (HFT), MBT, MCT, YSZ (ITSZ). These profiles are primarily located in the Yadong-Gulu rift (YGR) shown by a circle. NHA = North Himalayan Anticline; ghb = Great Himalayan belt (crystalline rocks); lhb = Lesser Himalayan belt (meta sedimentary rocks); lg = Leucogranites; thb = Tethyan belt (Sedimentary strata); RZT = Renbu-Zedong thrust and GT = Gangdese thrust (Hauck et al., 1998).

(a) INDEPTH I and II

Deep seismic reflection data acquired under INDEPTH I along a profile (Fig. 6.23) provides most valuable and reliable information about crustal structure in this region. The interpreted section along INDEPTH I (Hauck et al., 1993; Fig. 6.24) shows Moho at a depth of 75 km under South Tibet and a set of strong reflections in the middle crust which is identified as the MHT that represents a decollement plane to which all Himalayan thrust are connected in depth as shown in this figure. The extrapolation of these interfaces suggest Moho at depth of about 40-45 km under Ganga Basin and basement at a depth of 6 km which eventually joins with other thrusts like MBT, and MCT. Most of the shallow seismic activities are concentrated along junction of the MBT and MHT between Lesser Himalaya and Higher Himalaya. A similar section is also shown by Zhao et al., (1993) that shows Moho at a depth of ~ 80 km under the ITSZ.

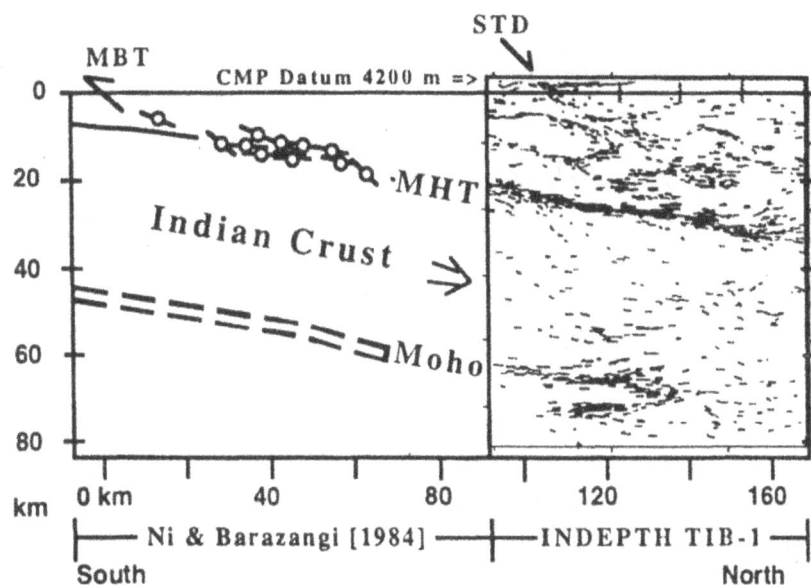

Fig. 6.24 Results from the seismic section INDEPTH I and its extrapolation under the Indian plate. Circles are hypocenters of earthquakes projected on this profile. They mostly lie in a depth range of 8-15 km at the junction of MBT and MCT with MHT (Hauck et al., 1998).

Combining the seismic reflection data from INDEPTH I and II, Hauck et al., (1998) provided details of the crustal model along the whole profile. Fig. 6.25 shows the interpreted crustal section. It is also constrained from Moho obtained from receiver function analysis which shows a thick crust of 70-78 km in this section. The interpreted crustal section shows MHT south of the suture zone that joins the Moho in a ramp like feature. Kangmar Dome is shown to affect the crustal structures in the form of domal uplift of the MCT. South Tibet Detachment (STD) is shown as a normal fault which controls the Tethys Himalaya. Suture zone is bounded by thrust on either sides which form the central core complex of collision tectonics, with Renbu-Zedong thrust towards the south and the Gangdese thrust towards the north. The crust under the Lhasa block shows several domal type reflections which might be related to the convergence of two blocks. Magmatic fluid bodies in the upper crust under the Lhasa block were delineated based on bright spots in the seismic section (Brown et al., 1996) which may represent subduction related granitic magma trapped in the crystalline crust that also provides low velocity conductive layer in the upper crust as discussed below. This is also confirmed from earthquake data from INDEPTH passive source experiment (Kind et al., 2002). These magmatic bodies along this profile are responsible for the Yadong-Gulu Rift as magmatic intrusives are essential elements for the formation of active rift basins as discussed in Section 2.7.

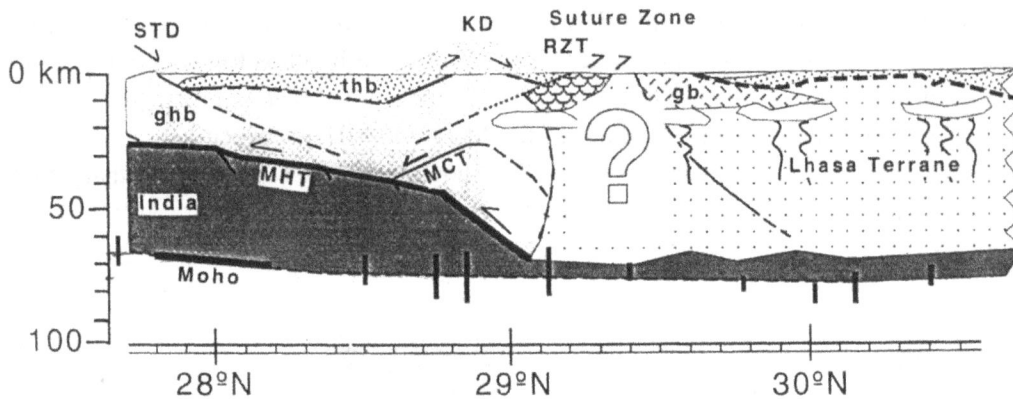

Fig. 6.25 Interpreted crustal section across the suture zone (ITSZ) based on INDEPTH I and II. Abbreviated features are same as given in Fig. 6.23. KD = Kangmar dome (Hauck et al., 1998) RZT = Renbu-Zedong Thrust.

The German component of INDEPTH profiles involved measurements using broad band seismometers and mobile seismic stations. The results from these studies are reported by Kola-Oja and Meissner (2001). They have also reported an upper crust of 35 km and a low velocity channel between 15-20 km depth that may represent conductive bodies inferred from MT measurements as described below. Based on velocity and temperature models they have suggested west-east flow of Tibetan crust as the most likely model for the deformation of Tibetan Plateau.

(b) INDEPTH III

Seismic study along INDEPTH III profile, described by Haines et al., (2003) and Zhao et al., (2004a, b, 2006), is located across BNS suture zone (West line, Fig. 6.20(a)). The interpreted velocity model under Lhasa and Qiangtang blocks and BNS suture zone is given in Fig. 6.26 (Zhao et al., 2004a, b) which shows inversion of structures below 33 km with buckling and upwarping of layers above and below it, that indicates compressional and extensional tectonics, respectively. This indicates that the Lhasa and Qiangtang blocks must have rifted before their collision and subsequently sutured along BNS. It also shows higher velocities of 7.1-7.3 km/sec at the base of the crust. The upper crust shows typical crustal velocities with layers dipping inside from two sides under BNS which is typical of convergence zones. High velocity at the base of the crust indicates mafic rocks which may represent the 10 km thick Jurassic oceanic crustal mélange under BNS, that also represents a rifting phase prior to convergence. The seismic section under the Qiangtang block does not show bright spots related to magmatic fluid bodies as in case of South Tibet along INDEPTH I as described above. Based on the crustal seismic section, Haines et al., (2003) suggested that the convergence along the BNS suture zone is largely accommodated through crustal thickening, and the thickened lower crust might have escaped through ductile flow.

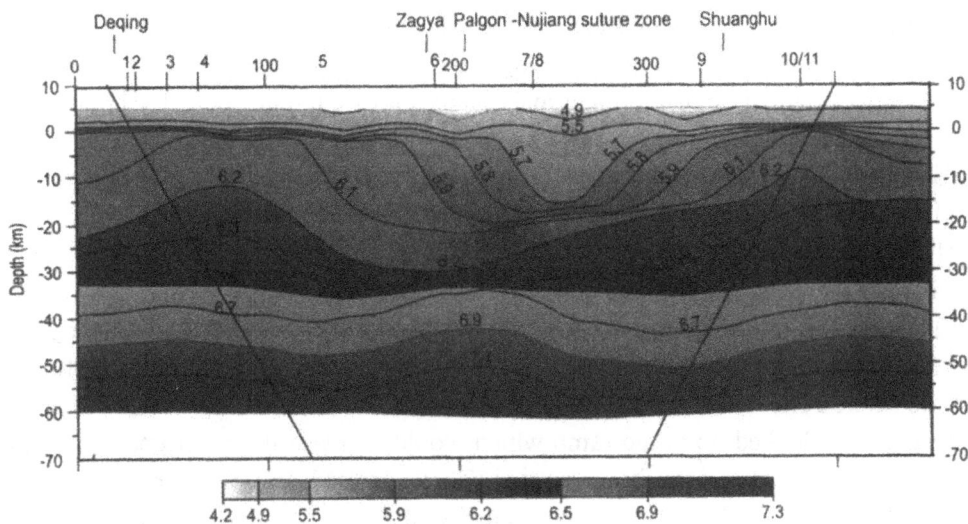

Fig. 6.26 The seismic velocity structural profile across Bangong-Nujiang Suture Zone (BNS) along INDEPTH III West line (Fig. 6.20(a)) showing an underplated crust of high velocity (7.1-7.3 km/sec), upwarping in the lower crust, and buckling in the upper crust (Zhao et al., 2004a, 2004b) indicating both extension and compression related first to rifting, followed by convergence, respectively. [Colour Fig. on Page 779]

(ii) Seismic Studies, Eastern Tibet

Based on deep seismic sounding profiles and gravity studies along two profiles in Eastern Tibet across the Sichuan Plateau and Sichuan basin (Yangtze Craton, Fig. 6.5). Wang et al., 2007 have provided crustal structures in this section. They suggested a crustal thickness of 62 km under Eastern Tibet reducing to 43 km east of it under the Sichuan basin (Yangtze craton) that might have resulted from the convergence of Tibet and Yangtze craton. This suggests a thick crust even under Eastern Tibet where convergence was related between different blocks. They also reported a low velocity layer in the upper crust of Sichuan Plateau (Eastern Tibet) that is absent under the Sichuan basin east of it, which may be synonymous to East-West flow under Tibet taking a southward turn along the Eastern Tibet.

Based on receiver function analysis from 25 broad band temporary stations and one permanent station in SE Tibet, Xu et al., (2007) suggested that crustal thickness varies from 60 km under Eastern Tibet to 46 km under the Sichuan basin, and 40 km under the Yangtze block. They also detected S-wave low velocity intra-continental zone that supported from high heat flow indicated partial melt in the crust which supports intra-crustal flow along the southeast margin of the Tibetan plateau. It is also supported by the high Poison ratio under the southEastern margin of the Tibetan plateau reported by them. Royden et al., (2008) have associated the West-East flow under the Tibetan crust with the trench rollback in the western Pacific and Indonesia. Wang et al., (2008), however, opposed the West-East flow model based on the similarity between crustal and upper mantle fabrics based on seismic studies.

(iii) Broad Band Studies and Integrated Crustal Structure

Based on broad band seismometer recording in Nepal and Tibet, Schulte-Pelkum et al., (2005) mapped the underthrusting Indian plate from receiver function analysis of P-and S-waves under Himalaya and South Tibet which was also constrained from the results of INDEPTH Profiles. They suggested that seismic anisotropy is prevalent above the MHT which is a decollement plane attributed to shear stress causing slip in great earthquakes at shallow depths. They also suggested that high concentration of small earthquakes along the decollement plane under High Himalaya may represent brittle to ductile transition and locked portion of the decollement. The lower part of the subducted Indian crust shows high velocity due to eclogite formation under high pressure characterized by high density and is, therefore, important for gravity modelling. The decollement plane (MHT) varies in depth almost from a few km under Lesser Himalaya to about 40 km under the suture (ITSZ), while Moho varies from about 42 km to 76-78 km in the same section. Maximum changes in depths of these horizons are observed between High and Tethyan Himalaya. In this model, the MHT increases in depth consistently from Lesser Himalaya up to the suture zone, while in the previous model (Fig. 6.25) it increases in depth consistently up to the north of the suture zone where it dips abruptly forming a ramp. In the latter model, the plane of decollement may be locked up at the ramp which would increase the recurrence period, and subsequently great earthquakes may be generated. They claimed an uncertainty of 2 km on inferred depths. Based on high seismic P-wave velocity of 8.4 km/sec in the lower crust under High Himalaya in the Eastern Nepal (Fig. 6.27), Monsalve et al., (2008) suggested the formation of eclogite due to metamorphism. Based on low Vp/Vs in the lower crust, they suggested fluids with high pore pressure that is also suggested by high conductivity in almost the same section (Fig. 6.32(a)). Singh et al., (2006, 2007) have reported E-W anisotropy related to MCT and south of it in NE Himalaya and N-S in the Burmese arc that follow the structural trend. Shapiro et al., (2004) have reported radial anisotropy of Rayleigh and Love waves in the middle and the lower crust across Tibet which they attributed to the thinning of the middle crust by 30% that may be related to deformation of a mechanically weak layer flowing as a channel towards the east.

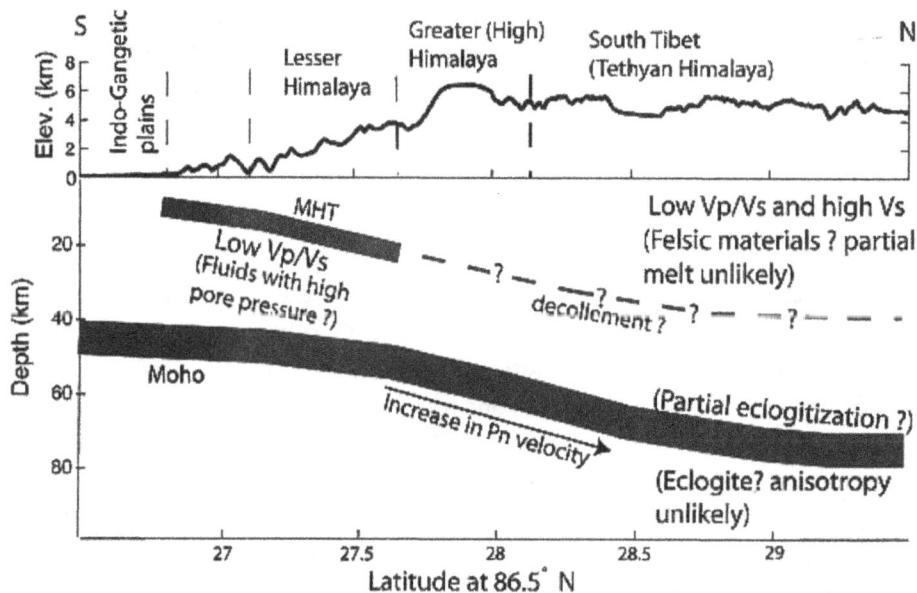

Fig. 6.27 Underthrusted Indian plate under Himalaya and Tibet based on P- and S- wave travel time using local network of broad band stations in central Nepal and south Tibet. Discontinuities are constrained from the results of INDEPTH profile and Shulte-Pelkum et al., (2005) with their thicknesses representing depth uncertainties. It provided high velocity in the lower crust and the upper mantle that are attributed to eclogite formation. Most of the earthquakes occurring in the southern part of High Himalaya are confined to the upper and lower crust defined by the Main Himalayan Thrust (MHT) while those in the northern part of High and Tethyan Himalaya occur mainly in the upper mantle (Monsalve et al., 2008).

(iv) Shallow Seismic Section in Nepal

A shallow seismic section across Southern Nepal (Mines and Geology, Nepal, 2005) is given in Fig. 6.28 and shows 4-6 km of sediments in the southern part south of HFT. The important point is the occurrence of several blind thrusts that are not exposed. These blind thrusts might be seismogenic that cannot be inferred from surface investigations. Further, anticlines and thrusts that are linked in their formation may form structural traps for oil and gas occurrences. Gondwana and Surkhet rocks may act as seal rocks for oil and gas occurrences. Another important point is the occurrence of Gondwana sediments along basement faults and thrusting of Proterozoic metasediments (Vindhyan and others) along the MBT and crystalline rocks along the MCT. Siwalik stratigraphy is shown nicely towards the south.

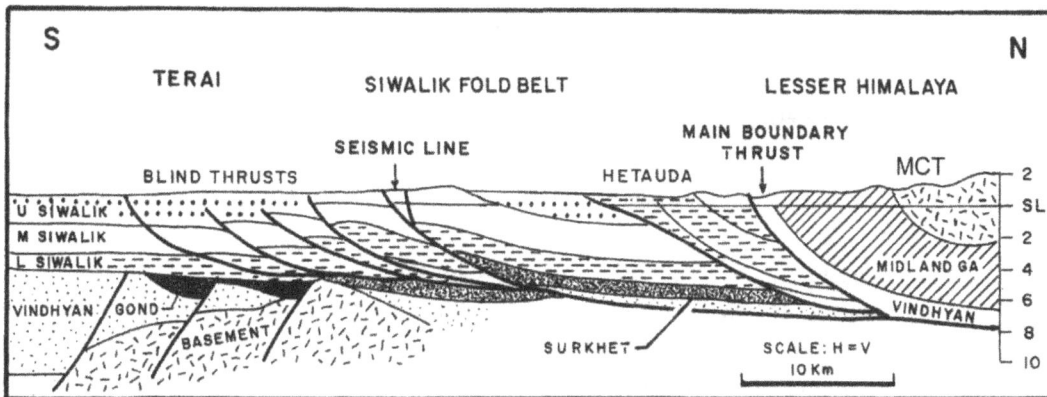

Fig. 6.28 A shallow seismic section in Nepal that shows basement structures and blind thrusts, south of MBT (Mines and Geology, Nepal, 2005, Brocher).

6.2.3 Seismic Studies: Lithospheric Structures

(i) Central Himalaya and Tibet

Travel time tomography has provided different models of subducting Indian Lithosphere at depth from a steep subduction (Replumaz et al., 2004; Li et al., 2006) to a gentle dipping near horizontal underthrusting (Zhou and Murphy, 2005). Based on tomographic imaging of mantle under Tibet and India, Van der Voo et al., (1999) have the suggested that the subducted Indian lithosphere lies in the mantle as cold high velocity body while a part of it slips under Tibet giving rise to plateau uplift and other associated tectonics.

Kosarev et al., (1999) used P-to S-converted tele-seismic based data recorded by temporary broad band networks across Tibet to image the underthrusted Indian lithosphere. They suggested Indian lithosphere (100-200 km) subducting up to about 200-300 km under north of BNS where it plunges to about 400 km. This plunging of Indian lithosphere causes upward thrust under North Tibet providing buoyancy to that part. DeCelles et al., (2002) have also envisaged two slab breakoffs involving Neotethyan oceanic lithosphere at about 45-35 Ma, and Greater Indian lithosphere at about 20-10 Ma, that may imply collision of Indian and Asian plates and period between formation of MCT and MBT, respectively. This implies that thrusting might have been accentuated due to slab breakoff due to buoyancy caused by it. Based on receiver function analysis of INDEPTH II and III data, Kind et al., (2002) suggested that 410 and 660 km discontinuities are well defined under central Tibet indicating the lack of subducting Indian lithosphere in this section. Subsequently, Kumar et al., (2006) mapped the underthrusting Indian and Asian lithosphere based on S-to P-wave converted receiver function analysis under Tibet. They integrated the previous results and provided a composite model as given in Fig. 6.29. In general, the Indian lithosphere dips from a depth of 160 km south of the ITSZ to about 220 km south of BNS suture while Moho varies from 80 to 60 km, respectively. On the Asian side the structures are simple with Moho at a depth of about 60 km and the lithosphere is almost horizontal from central Tibet to Northern Tibet under the Kunlun fault. However, the position of Indian lithosphere under BNS and north of it is not clear in this model that reaches to a depth of 300-400 km in the model given by Kosarev et al., (1999).

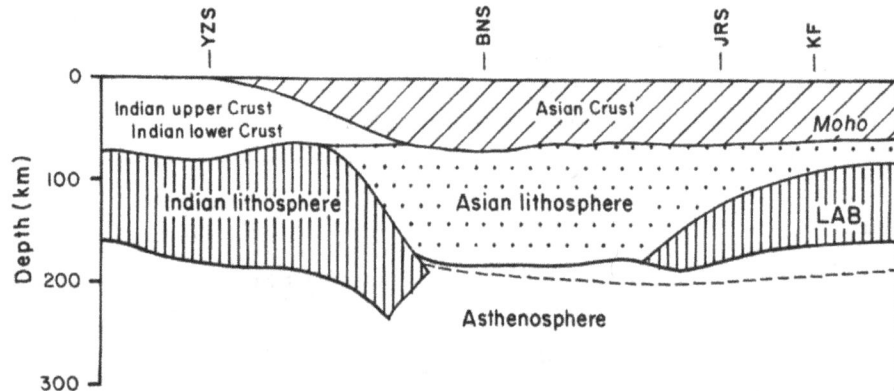

Fig. 6.29 Indian and Asian lithosphere under Tibet based on P-and S-wave receiver functions. The Indian plate subducts up to south of the BNS suture while the Asian lithosphere is confined to its north (Kumar et al., 2006).

Based on tomographic image of the upper mantle under central Tibet from INDEPTH data, Tilmann et al., (2003) suggested subvertical high velocity subducted Indian lithosphere from 100 to about 400 km just before BNS suture zone, compensated by upward magma flow under the Asian mantle towards the north. This may set up convection cells under the Asian mantle causing its erosion and detachment and the warm mantle under north-central Tibet.

P_n tomography over Tibetan Plateau (Liang and Song, 2006) suggested a west-east low velocity zone under Qiangtang block and Songpan Ganzi Fold Belt (SGFB) in northern Tibet that bends southwards with SGFB and extend to Indo-China (Fig. 6.30(a)). This might be analogous to the west to east flow of the Tibetan lower crust which is ductile in nature. The low velocity zone would correlate to the hot and deformable lithosphere in the north, east and S-E. A low velocity zone is noticed under the Yadong-Gulu rift. Based on these low velocity zones, they have suggested a west-east-southeast flow under the Tibetan Plateau as shown in Fig. 6.30(a). They suggested changes in crustal thickness across suture zones and also suggested that the Tibetan plateau south of the Kunlun fault is moving eastward relative to both India and Eurasia. This movement is accommodated through the rotation of material around the Eastern Syntaxis. Based on high resolution triplicate waveforms across apertures of over 1000 km, Chen and Tseng (2007) have shown high P-wave velocity near-bottom of the mantle transition zone and suggested that under thrusted rocks from the Indian shield get detached and are confined to the Transition Zone. Based on high P-wave speed and its absence in S-wave speed, Tseng and Chen (2008) further suggested the occurrence of hydrated subcontinental lithospheric mantle under Tibet caused by subduction of the India plate. This leads to detachment and sinking of cold hydrated lithosphere that may be trapped in the transition zone giving rise to high P-wave velocity. Based on shear wave birefringence, Chen et al., (2010) suggested a rapid increase about 100 km north of the Bangong-Nujiang suture under Qiangtong block in west central Tibet. They attributed it as marking the northern leading edge at the sub-horizontally advancing mantle lithosphere of the Indian plate. This is also confirmed from the results of the finite frequency tomography using both P- and S-waves (Hung et al., 2010) who suggested termination of high wave speed near the advancing Indian mantle front.

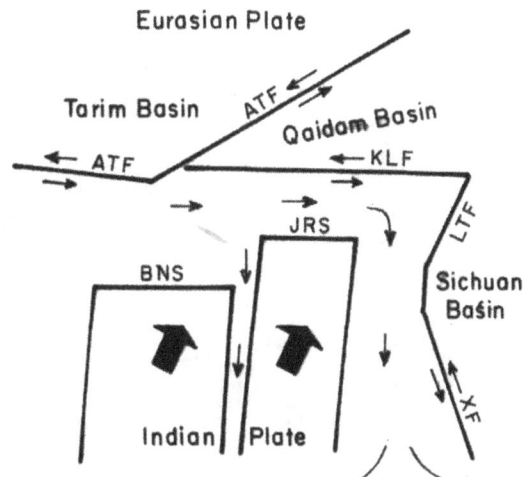

Fig. 6.30(a) A schematic view of the Indian and Tibetan collision zone showing the Eastern part of the Indian lithosphere advancing more northwards along the Yadong-Gulu rift compared to the western and central parts and Tibetan lower crust flowing east and southwards (Liang and Song, 2006).

Based on S-wave form tomography, Priestley et al., (2006, 2008) have provided an upper mantle model up to about 400 km of Eastern Asia including the Tibetan Plateau and adjoining sections. Since variations in S-wave speed are sensitive to temperature, they can be successfully used to differentiate between the lithosphere and the asthenosphere. Priestley et al., suggest ± 9% of perturbation in the velocity compared to the standard PREM model which indicate high velocity from about 100 km up to 260 km under Tibet that may represent underthrusted Indian lithosphere. However, the high velocity zone extends up to about 300 km with reduced perturbation in velocity. The section below this level is characterized by low velocity which may represent the hot asthenosphere under Tibet. They also suggested that the mantle structure up to about 200 km depth follows surface geology and tectonics, while below 200 km depth, the mantle structures follow advection in the upper mantle different from surface tectonics. The lithospheric thickness provided by them varies from ~140 km under the north Indian shield and along western and Eastern margins of the Tibetan Plateau, to about 300 km under the Tibetan Plateau (Fig. 6.30(b); Priestley et al., 2008) that is almost equal to those found under cratons, and has been refered to as craton in making (McKenzie and Priestley, 2008). Fig. 6.30(b) demonstrates two slices of tomography results at depths of 125 and 175 km. Both show high velocity under the Indian shield and Tibet indicating that the lithosphere is thicker than 125 km under the entire region. However, the slice at 175 km depth started showing low velocity under the Indian shield indicating that the lithosphere thickness may be close to 175 km, but in case of Tibet it still shows high velocity indicating that the lithosphere is thicker than this level. The profile at the bottom (Fig. 6.30(b)) shows high velocity under Tibet at a depth of 250-300 km that provides the thickness of lithosphere in this section, while the same under the North Indian Shield is ~ 150 km. The vertical profile with depth (d) provides the standard model of surface wave velocity with depth used for estimating the increase and decrease of the velocity in tomography computations, that show a sudden increase in velocity at ~ 180 km. The lithospheric thickness described above is almost the same as that provided by McKenzie and Priestley (2008) where they have used an additional constraint of occurrences of Kimberlite and suggested a lithospheric thickness of 160-180 km under the South Indian Shield, which may be attributed to erosion and uplift of the Indian lithosphere since the Archean-Proterozoc period. They have also suggested a cold upper mantle in a depth range of 80-90 km where microseismic activity has been reported. Further discussion on the Indian lithosphere can be found in Section 7.11.

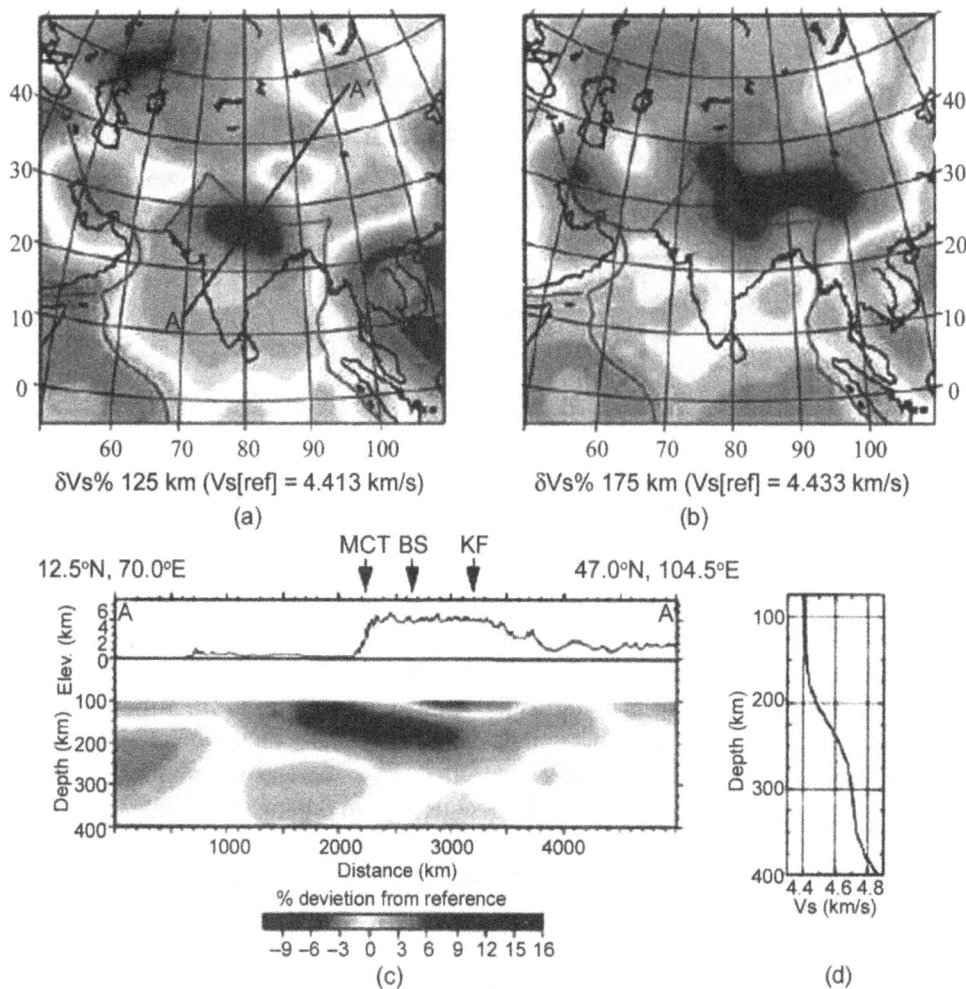

Fig. 6.30(b) Surface wave (Sv) tomography modes at depths of 125 km (a) and 175 km (b). Straight green lines are major tectonic features. (c) shows the depth model along a profile AA' (a) with elevation and tectonic features, the Main Central Thrust (MCT), the Bangong suture (BS) and the Kunlun Fault (KF) marked on it. (d) is the reference model used to find the deviation in Sv velocity (Priestley et al., 2008). [Colour Fig. on Page 780]

Based on surface wave velocities in the Himalayan-Tibetan region and fluid flow driven by topographically induced pressure gradient, Copley and McKenzie (2007) suggested a viscosity of about 10^{20} PaS in southern Tibet and 10^{22} PaS between the Eastern Syntaxis and Sichuan basin. This matches quite well with the flow in the southern Tibet and Burmese arc that might be responsible for the formation of the Eastern Syntaxis.

Based on geological reconstructions and seismic tomography of mantle, Replumaz et al., (2004) suggested that Indian plate is overriding its own sinking mantle and does not seem to under-thrust Tibet beyond ITSZ. According to them, the tomographic images suggest that approximately 1500 km of convergence of Indian plate is absorbed by deformation of the Asian plate. Their model also suggests a gap between the presently underthrusted rocks up to transition zone 410-660 km, and previously subducted and detached Indian lithosphere below transition zone at a depth of about 1000-1700 km. Based on P-wave tomographic imaging, Li et al., (2006) detected high seismic velocity above 410 km discontinuity under Tibet related to the subducted part of Indian lithosphere (1, Fig. 6.30(c)). They also suggested that it may not be connected to the remnant of the Neo Tethys oceanic slab in the lower mantle (2, Fig. 6.30(c)). Based on S-wave form tomography, Lebedev and Van der Hilst (2008) have also suggested high velocity zone under Tibet from about 80 km downwards that extends up to 200 km with almost the same order of increase in velocity, and further up to 330 km with some reduced increase in the velocity.

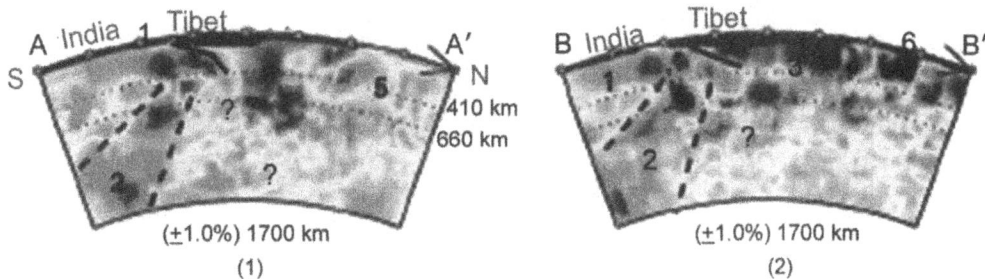

Fig. 6.30(c) Subducted high velocity Indian lithosphere under Tibet, (1) based on tomography while subducted Tethyan lithosphere (2) is detected below the transition zone in the Western Himalaya (AA) but continuous in depth in the Central Himalaya (BB) (Li et al., 2006). [Colour Fig. on Page 780]

(ii) Eastern Himalaya and Syntaxial Bend

Based on P-wave tomography, Li et al., (2008), have provided the extent of subduction of the Indian plate in different sections. Accordingly, it has subducted under the entire Tibetan plateau in the western part while it is confined to south of the BNS in the central part, and Eastern Himalaya in the Eastern part. The image of the P-wave tomography presented by them suggest almost sub horizontal subduction of the Indian lithosphere under the western and the Eastern Himalaya while it is at about of 45° for central Himalaya that may explain the former being seismically more active than the latter. They have also suggested that across the Eastern Himalayan Syntaxis, the lithosphere thickness is 100-110 km that subducts under the Burmese arc in the northern part at an angle of almost 60°, while in the southern part, it is almost vertical which is seismically more active with intermediate-deep focus earthquakes. They also delineated a large low velocity zone in the upper mantle under the Tenchong volcano (25°N, 98°E) that is an active volcano east of the Burmese arc with most recent eruption in 1609 indicating a magma chamber which may be associated with the subduction of the Burmese micro-plate under the Eurasian plate. This is also supported from high conductive anomalies (< 10 Ohm-m) that extend from about 5 km to 15 km and up to 25-30 km with reduced conductivity values (Bai et al., 2001). The same might be connected to a low velocity zone in the upper mantle indicating a magma chamber.

Based on receiver function analysis along two profiles, N-S and E-W (Fig. 6.31) Uma Devi et al., (2010), have provided crustal thickness of about 40-55 km under the Himalayan Foredeep to about 80 km under High Himalaya in the Eastern sector. However, under Lower Himalaya, they show a relatively shallow crust of about 40 km that might be related to crustal bulge due to lithospheric flexure which has been explained in Section 6.1.9. They also suggested almost normal crustal thickness of about 50-55 km under the Burmese arc that reduces to 40 km under the foredeep due to crustal bulge. They also suggested the variation in lithospheric thickness from about 100 km in the foredeep parts to about 170-180 km under the Eastern Himalaya and the Burmese arc that reduces to ~100 km under the foredeep and further increases to ~150 km south of the Shillong plateau, reflecting the effect of lithospheric flexure. This model, based on S-wave receiver function analysis, conforms to global surface wave tomography (Ritzwoller et al., 2002).

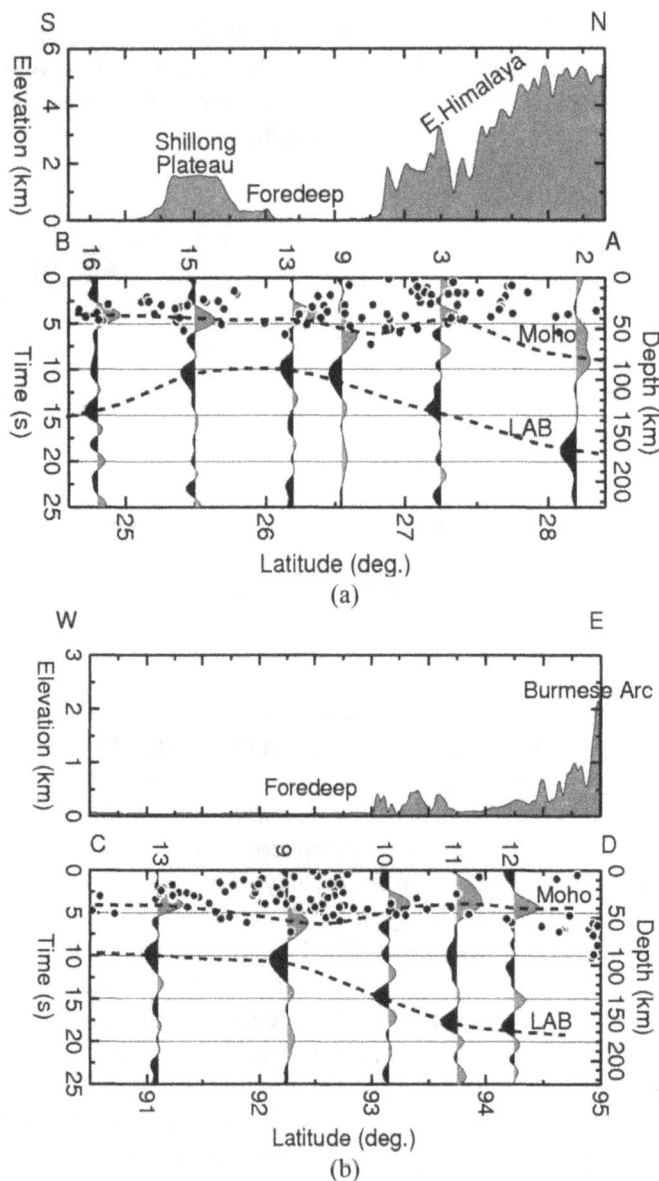

Fig. 6.31 S-wave receiver function analysis across Eastern Himalaya (a) and Burmese arc (b) providing the crustal thickness and lithosphere asthenosphere boundary (LAB). They clearly shows the lithospheric flexure of the Indian plate under foredeep that is also reflected as crustal bulge, though with a reduced amplitude (Uma Devi et al., 2010).

6.2.4 Magnetotelluric Studies

Magnetotelluric (MT) studies involve the measurement of conductivity, which has been quite successful to delineate conductive structures in the crust. High conductivity primarily indicates the presence of fluids or partial melts or fluid magma in the crust. Another explanation usually offered for high conductivity is the presence of graphite with connectivity which can be resolved based on tectonic settings. However, the condition of connectivity is met more easily in the case of fluids as compared to graphite as the latter usually occurs in patches unless its deposits are formed. Magnetotelluric measurements were carried out along INDEPTH profiles, which provided useful information about conductive structures.

Nelson et al., (1996) have described the results of MT studies along INDEPTH I and II profiles and suggested a highly conductive zone as partially molten rocks in the lower curst at a depth of 20-25 km under the ITS (Indus-Tsangpo suture) and southern Tibet, north of the ITS. Unsworth et al., (2005) have suggested high conductive zone in the lower crust north of the ITS (Fig. 6.32(a)) based on magnetotelluric measurements that coincides with the bright spots of seismic reflections confirming the presence of fluids in this section that may represent partial melts and/or aqueous fluids. This conductive zone extends subsurface up to the Kangmar dome that may have been formed due to these partial melts. There is another conductive zone in the upper crust north of the STD. Li et al., (2003) suggested a thick zone of partial melt (> 10 km) or a relatively thin layer of aqueous fluids (100-200 m) or both (Fig. 6.32(b)) which have given rise to a high conductive zone below south Tibet. The features of the collision tectonics in this section got reflected for the first time in seismic sections of INDEPTH I and II as shown in Fig. 6.25 that is superimposed over the conductive section in Fig. 6.32(a). It shows the extension of the Main Himalayan Thrust (MHT) from the Himalayan front up to suture zone (ITS) that shows a linear conductive section along MHT. The conductive section north of the STD under Tethyan Himalayan is largely located above the MHT that might be responsible for the shallow seismic activity originating above the MHT in this section as presence of fluids aid in slippage and seismic activity. Unsworth et al., (2005) have also suggested the extension of this conductive zone for about 1000 km under southern Tibet towards the west under Nepal Himalaya and northwestern Himalaya. However, conductivity values under NW Himalaya are lower than other sections. Unsworth et al., have also delineated a highly resistive body under Higher Himalaya north of the MCT in the lower crust that might be related to subsurface extension of the leucogranites exposed in this section. Lemmonier et al., (1999) have given the electrical structure of central Nepal (Fig. 6.32(c)) that is superimposed on the INDEPTH I (Fig. 6.24) by Hauck et al., (1998). It clearly shows mid crustal conductor at a depth of 10-20 km under Higher Himalaya that coincides with the hypocenters of the earthquakes in this section and may represent the extension of partial melts shown in Fig. 6.32(a). Gokaran et al., (2005) have inferred a conductive zone between MBT and MCT at a depth range of 10-15 km under Sikkim Himalaya just east of Nepal Himalaya. In this section also this zone lies just above the MHT as in the case of Nepal Himalaya.

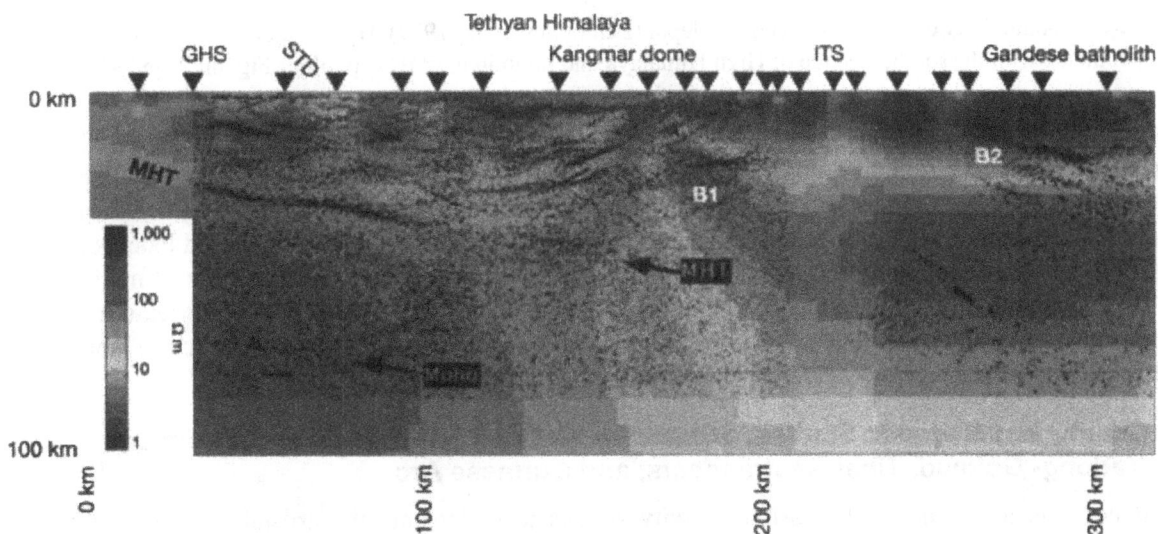

Fig. 6.32(a) Conductivity distribution along INDEPTH Profile with common mid point reflection profile superimposed on it showing high conductivity in the lower crust north of the ITS extending in the upper crust north of the South Tibet Detachment (STD) under the Kangmar dome. It coincides with the seismic bright spots, B1 and B2. It also shows reflections due to Main Himalayan Thrust (MHT) extending subsurface up to Kangmar dome and Moho. GHS: Great Himalayan Sequences (Unsworth et al., 2005). [Colour Fig. on Page 781]

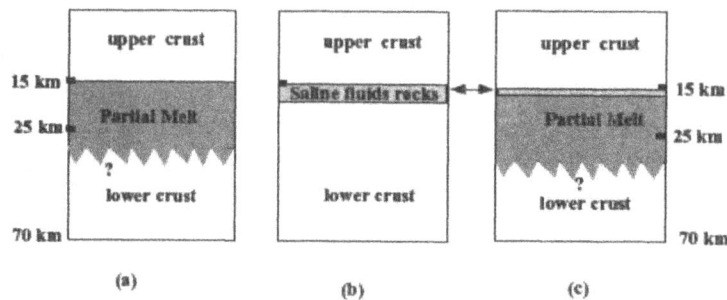

Fig. 6.32(b) Three schematic models of conductivity distribution along INDEPTH profile from magnetotelluric measurements indicating partial melt (a), or saline fluid (b), or both (c) in the lower crust under South Tibet (Li et al., 2003) that extends up to ITSZ. This high conductivity zone coincides with low S-wave velocity and bright spots observed in the INDEPTH section (Brown et al., 1996)

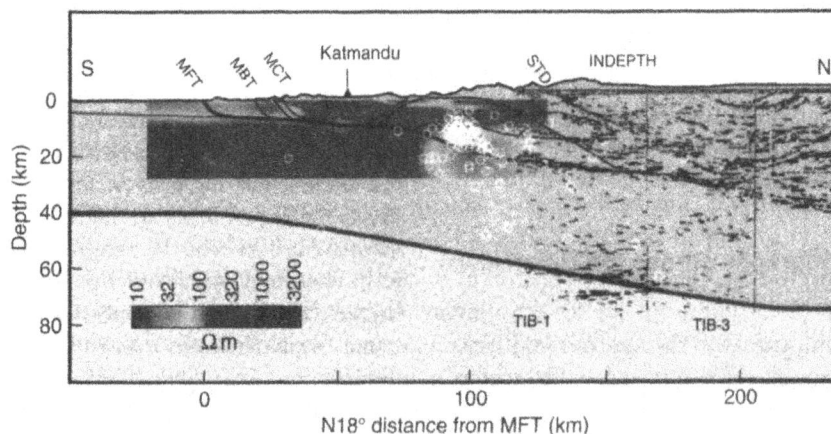

Fig. 6.32(c) Conductivity distribution in central Nepal (Lemonnier et al., 1999). Hypocenters of earthquakes (circle) coincide with the southern slope of High Himalaya and ramp in the MHT. [Colour Fig. on Page 781]

Magnetotelluric measurements over the Altyn Tagh fault (ATF) north of Tibet (Bedrosian et al., 2001) suggest conductive zone (10-100 Ω m) under the fault zone that extends up to 8 km and dips northwards. It is, however, contrary to present day tectonics in this region that are NE vergent thrusts and faults. It is, therefore, suggested to represent relic of some earlier suture, plausibly a mid Paleozoic suture; ATF was activated along this paleosture. They attributed this conductive anomaly to marine sediments and possibly graphite along the paleosuture zone. Based on some conductive feature at depth, they also suggested ATF to be of a lithospheric scale that has activated a Paleozoic suture along which the Tibetan lithosphere has extended eastwards.

6.2.5 Gravity and Magnetic Studies: Structures and EET – Ganga Basin-Sikkim and Yadong- Golmud, Tibet Geotransects, and Burmese Arc

We shall now discuss in detail the various gravity studies to understand the crustal structures and nature of rock types.

(i) Flexural Isostasy and Effective Elastic Thickness

Most of the recent works on gravity studies in Himalaya relates to estimation of effective elastic thickness (EET) based on flexural model. Isostatic compensation is a well accepted physical process, in which the earth responds to any load (surface or subsurface) over the geological time scale and restores the equilibrium. Flexural isostasy implies that the topographic load is supported by flexure of the lithosphere instead of its root as suggested in Airy's model of isostasy. Airy's model of isostasy assumes that the lithosphere does not have any strength and all the loads are supported by crustal root. However, the occurrences of old mountains, especially those belonging to the Archean-Proterozoic period without any crustal root, suggested that the lithosphere in these sections are strong enough to withstand their loads. This strength, however, resides in parts of lithosphere that show elastic properties where stress is proportional to strain up to a certain limit of stress caused by the load. Beyond this limit it breaks, showing brittle characteristics. Thickness of this part of the lithosphere is known as effective elastic thickness (EET) and is a measure of strength of the lithosphere as has been discussed with details in Section 4.4 (Banks et al., 2004). Larger the EET, stronger the crust/lithosphere. As this section of lithosphere breaks due to applied stress and associated strain, this is considered a brittle part of the lithosphere and is important for seismic activity (Burov and Diament, 1995; Jackson, 2002).

Two spectral methods are commonly used to estimate EET. Important points related to them are discussed in Section 4.3. However, for sake of completeness, they are briefly described here and important references cited for those interested in details. In the first approach, the linear transfer functions (admittance between topography and gravity anomaly) are computed and fitted with the admittance values of the elastic plate model, which is referred to as admittance analysis (McKenzie and Bowin, 1976). Another approach is the computation of coherence between topography and gravity anomalies, which assume that short wavelength topography is uncorrelated with Bouguer gravity anomalies, while long wavelength anomalies involved in isostatic compensation, are correlated with topography that would produce high coherence. Theoretical coherence and admittance are computed by the scheme provided by Forsyth (1985), Lowry and Smith, (1994) and Mckenzie and Fairhead (1997), respectively, compared with the computed coherence and admittance for best fit of EET value. In fact, misfit function between the theoretical and computed values of coherence and admittance is computed for different EET values and its value for minimum misfit function provides the actual value of EET.

There have been several attempts to estimate Effective Elastic Thickness (EET) under Himalaya and Tibet. Initial estimate of EET (Te) for Himalaya based on 1-d profile analysis and surface loading was quite high: of the order of 100 km and above (Karner and Watts, 1983) which was reduced to about 40-50 km when Te was estimated based on 1-D coherence analysis (Jin et al., 1994). Elastic plate model based on bending moments however, provided Te of 90 km under Ganga basin and 30-35 km for the Tibetan Plateau (Jin et al., 1996). Cattin et al., (2001) obtained a lower Te of 40-50 km under the Ganga basin compared to those by Lyon and Molner (1983) as 60-70 km and 80 km under the Tibetan Plateau. Rajesh and Mishra (2003) have reported an average value of 50 km along a profile (90°E) extending from the Ganga plains across Himalaya and Tibet. However, based on 2-D analysis, Rajesh et al., (2003) suggested an EET of 35 km under Ganga basin decreasing to 25 km under Himalaya and 30 km under Tibet (Section 4.6.4). Most of the latter works on gravity studies in Himalaya related to the computation of effective elastic thickness and modeling based on it. Jin et al., (1996) modeled gravity data along INDEPTH I profile and its extension towards the north and south up to Qaidam basin and Indo-Gangetic plains, respectively. They modeled this profile for different Te in different sections and suggested a Te = 90 km for Indian plate, Te = 30 km for Himalaya and Tibet, and Te = 50 km for northern part beyond Kunlun fault. With these constraints, the computed field matched well with that of the observed field. The Main Himalayan Thrust (MHT) inferred from INDEPTH matches with that of the top of the subducting Indian crust. Cattin et al., (2001) have suggested over compensation for the foreland and undercompensation of the Higher Himalaya with a steep gravity gradient of the order of 1.3 mGal/km related to steeper Moho. A 10 km wide hinge of the gravity data under South Tibet is attributed to eclogitisation of the subducted Indian crust. They suggested an effective elastic thickness of 40-50 km in the foreland.

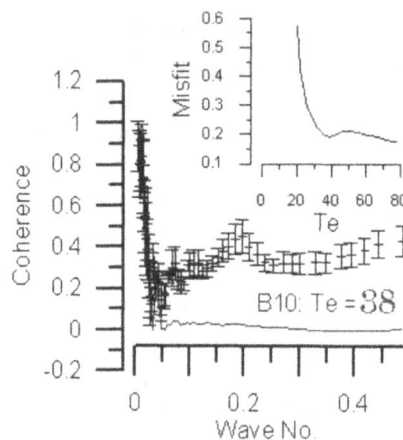

Fig. 6.33 Coherence versus wave number for a 5 × 5 degrees block NW of Delhi based on the free air anomaly and the topography and misfit function showing minimum value for EET = 38 km (Tiwari and Mishra, 2008)

Constraining the crustal thickness from receiver function analysis and INDEPTH profile as described above, Hetenyi et al., (2006) suggested a decrease in Te from 60-80 km under foreland to 20-30 km northwards as it is flexed down beneath Himalaya and Tibet. The lower value of Te for Himalaya and South Tibet can be attributed to the weak crust due to collision of Indian and Eurasian plates and associated high thermal anomaly as indicated by the presence of partially molten rocks discussed above from MT measurements. Hetenyi et al., (2007), based on Airy-type isostasy gravity modeling, suggested that high density eclogites are required beneath the Tibetan Plateau, which has formed due to interaction of water with the lower crust at maximum depth of its descent 70 km (high pressure). Based on thermo-kinematic modeling using radiogenic heat production and heat loss, they provided thermal structure of Indian crust underthrusting Tibet and modeled a gravity profile for plausible density distribution in the crust and upper mantle. They also indicated concentration of microseismic activity in the upper mantle where the underthrusted Indian crust bends under Tibet and changes to eclogite. They suggested a maximum crustal thickness of 75-80 km under the suture zone (ITSZ) and South Tibet.

Based on detailed analysis of Bouguer anomaly and topography, Jordan and Watts (2005) suggested variations of EET from 70 Km in the central region of foreland Ganga basin to 30-50 km towards the east and the west. The Assam shelf along the Eastern Syntaxis in NE India, however, provided the minimum value of 20-30 km in the fore deep part indicating the weakest crust in this section affected both from the Eastern Himalayan and Burmese arc collision tectonics. Rajesh and Mishra (2003) also suggested similar values of 20-23 km for both the Eastern and the Western Himalayan Syntaxsis that are lowest values reported from the Himalayan Fold Belt suggesting weakest sections that are most seismogenic compared to other sections. Tiwari et al., (2006, 2008) have provided EET of 50 and 53 km along profiles across Sikkim Himalaya and the Western Syntaxis, respectively as described in the next sections. These profiles extended from the Ganga plains in the south, to Tibet in the north. High values of EET along these profiles can be explained based on their large sections being located in the Ganga plains. In order to check its variation between these two profiles, we computed EET for a 5 × 5 degree block, NW of Delhi (Tiwari and Mishra, 2008) from the coherence between the free air and the topography that provided an EET of about 38 km for minimum misfit of the coherence function (Fig. 6.33). This shows a high value of 50 km of EET in the central part of the Ganga foredeep and the Himalayan front as computed along Sikkim Profile and lower values in the NW and SE parts that may be attributed to solid basement rocks such as Bundelkhand granite underthrusting in the central part, and disturbed sections like Proterozoic rift basins along the two sides (Mishra and Rajasekhar, 2008). The analysis of Jordan and Watts (2005) also provided qualtitatively almost similar results with higher value in the central part and relatively lower values on either side.

There are so many variations in estimates of Te as given by different workers that it becomes difficult to pick one or the other value. The same is the case in other parts of the world. This diversity of results might be due to two reasons. First, to compute spectrum minimum, the data set required is $5° \times 5°$ and it is difficult to get such a large homogenous data set where signals from different geological provinces do not interfere. This is especially true in a region like Himalaya. The second problem lies with different methodologies followed by different workers for computation especially in the frequency domain and leakage of energy due to confined data set. However, it can broadly be used to define the crustal strength qualitatively; larger the value of Te stronger the crust.

Modeling along some specific profiles, and estimation of EET in Central Himalaya and Tibet are given below.

(ii) (a) Gravity Profiles across Sikkim Himalaya and South Tibet

A gravity profile across Sikkim Himalaya was recorded by Tiwari et al., (2006) at about 1 km spacing (Fig. 6.34(a), 250 stations) that was extended towards north over Tibet based on the gravity anomaly map of China (Sun, 1989). This profile was extended southwards over parts of India and Bangladesh based on the Bouguer anomaly map of these countries (NGRI, 1975 and Rahman et al., 1990). The whole profile is shown in Fig. 6.34(b) that also shows the computed Bouguer anomalies along this profile assuming flexure of semi-infinite (broken) elastic plate of effective elastic thickness of 10, 50, and 90 km which shows minimum misfit for Te = 50 km. Most of the reported figures of effective elastic thickness for Himalaya and Tibet are about 30-40 km as described above. A higher estimate in the present case can be attributed to a large section of profile extending over the Indian continent and Tibet.

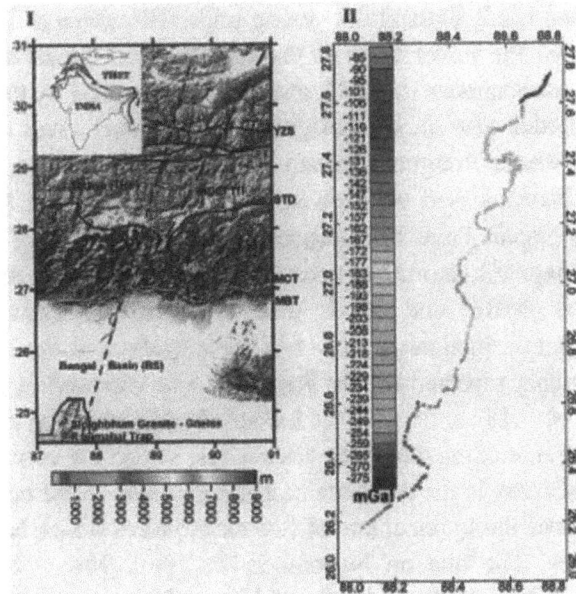

Fig. 6.34(a) Location of gravity and magnetic stations across the Sikkim Himalaya are plotted as red circles on the topographic map of the region. All data (existing and new) are projected along the dashed line. INDEPTH profile is marked by a thick dashed line. All the major thrust faults separating different geological units are also marked. Location of 250 gravity and magnetic stations across the Sikkim Himalaya are plotted as circles on the right hand side. Colour of the circles indicates the range of gravity anomalies (Tiwari et al., 2006).

[Colour Fig. on Page 782]

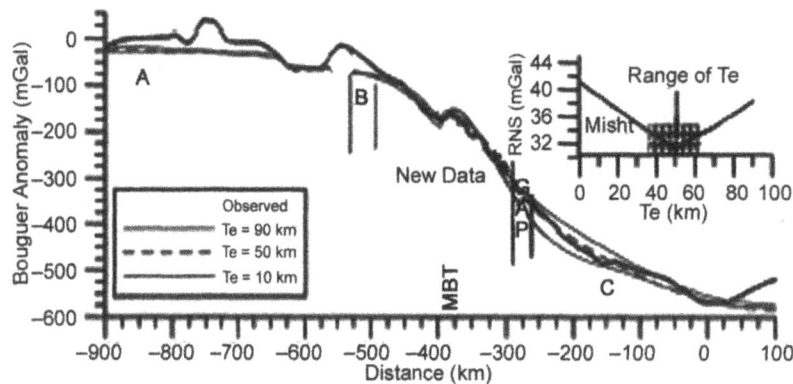

Fig. 6.34(b) Observed and computed Bouguer gravity anomalies along a profile generated from: new recorded data and existing data, (A) from India (NGRI, 1975); (B) from Bangladesh (Rehman et al., 1990); (C) from Tibet (Sun, 1989). Computed Bouguer gravity anomalies are related to flexure of semi-infinite (broken) elastic plate of effective elastic thickness of 50 km as suggested by low rms value plotted in the inset. Bouguer gravity anomalies computed for effective elastic thicknesses of 10 and 90 km (extreme ends) are also plotted for comparison. Distance is plotted from plate break, i.e. X=0 km is considered as plate break (Tiwari et al., 2006).

This gravity profile is modelled constraining from the crustal structure as described above from seismological studies and the INDEPTH I profile that is close by (Fig. 6.34(a)). The crustal model and computed gravity along with the observed field is shown in Fig. 6.35(a) which also shows the Moho obtained from seismic studies under the Indian continent (Krishna and Rao, 2005) and Tibet (Fig. 6.25, Hauck et al., 1998). It shows buckling of the Indian crust and a ramp in MCT that joins the MHT. This ramp may cause locking up of underthrusted Indian lithosphere resulting in larger recurrence period of earthquakes in this section as discussed above in Section 6.1.10 and 6.2.2. Crustal thickening under Himalaya and Tibet is apparently caused by underthrusting of the lower crust of the Indian plate. Small amplitude and short wavelength gravity highs associated with MBT and MCT are caused by thrusted high density rocks along them. This model also shows eclogitization of lower crust of relatively higher density under South Tibet where Bouguer anomaly shows a small relative high forming hinge zone. Monsalve et al., (2008), based on high seismic velocity of 8.4 km/sec under High Himalaya in the Eastern Nepal, have also suggested metamorphism of the underthrusting Indian plate to eclogite. Magnetic anomalies recorded along this profile are primarily related to thrust zones such as MBT and MCT due to intrusives (thrusted) along them (Tiwari et al., 2006). Magnetic inclination of $-64°$ for intrusives at the southern end of the profile (INT) is similar to that reported for the Rajmahal trap exposed in the neighbourhood. Similarly, the inclination of $-21°$ and $-22°$ for Lesser (MBT) and Higher Himalaya (MCT) required to model the corresponding magnetic anomalies, suggest a comparatively southern position for the Indian land mass in the southern hemisphere close to the equator (Tiwari et al., 2006). This figure also shows the hypocenters of two earthquakes which have occurred in this region during recent times. The one on November 19, 1980 (Mw = 6.3) and second on February 14, 2006 (Mw = 5.3), with focal depths of 17 and 37 km, respectively coincide with upper and lower interfaces of the lower crust that is considered to be ductile and would facilitate the slip for earthquakes to occur. The crustal thickness changes from 38-40 km under the Ganga plains to about 75-76 km under South Tibet. Receiver function analysis in central Zagros also suggests abrupt changes in Moho from 45 km along the Main Zagros Thrust similar to MCT as shown above along this profile (Fig. 6.35(a)) wherein the crust of Zagros underthrusts Central Iran. It is similar to the Himalayan Thrust belt suggesting similar processes in this section that connects the Himalayan-Alpine collision zone (Paul et al., 2006). Yin (2010) has compared the two orogens, viz. Himalayan-Tibetan orogen formed due to India-Asia collision and Turkish-Iranian-Caucasus orogen formed due to Arabia-Asia collision, and shown several similarities between them. Accordingly, their interactions have affected the far field Cenozoic deformation in Asia. They have suggested a N-E trending 300-400 km wide and > 1500 km long fault that extends from the Zagros thrust belt in the south to the western Mongolia in the north, and links with the active Tian Shan and Altai Shan intracontinental orogens.

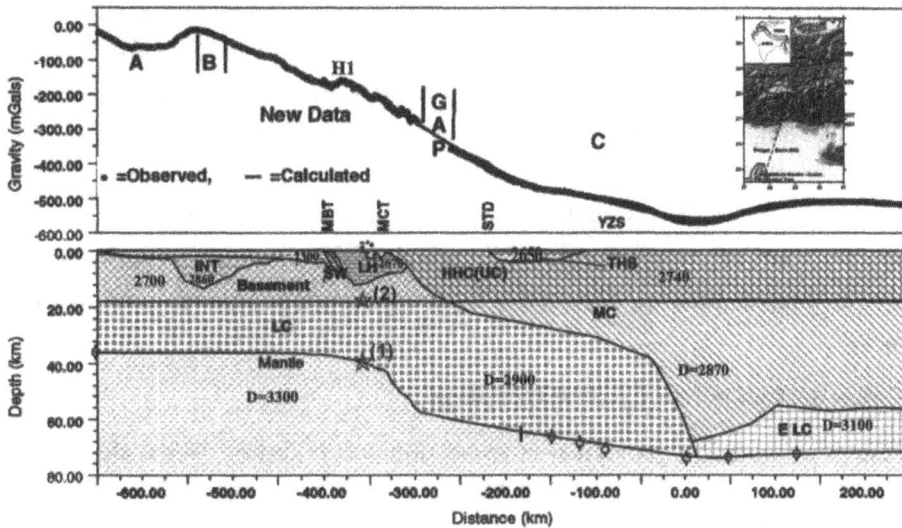

Fig. 6.35(a) Crustal cross-section derived from modeling of gravity data. Also shown are observed and computed gravity anomalies. Densities (D) are given in kg/m³. Moho depths from seismic results are marked by circle and vertical line at Moho boundary for comparison. A, B, C are also marked for data information as mentioned in Fig. 6.34 (Tiwari et al., 2006). Hypocenters of seismic activity on February 14, 2006 (1) and November 19, 1980 (2) are marked by stars.

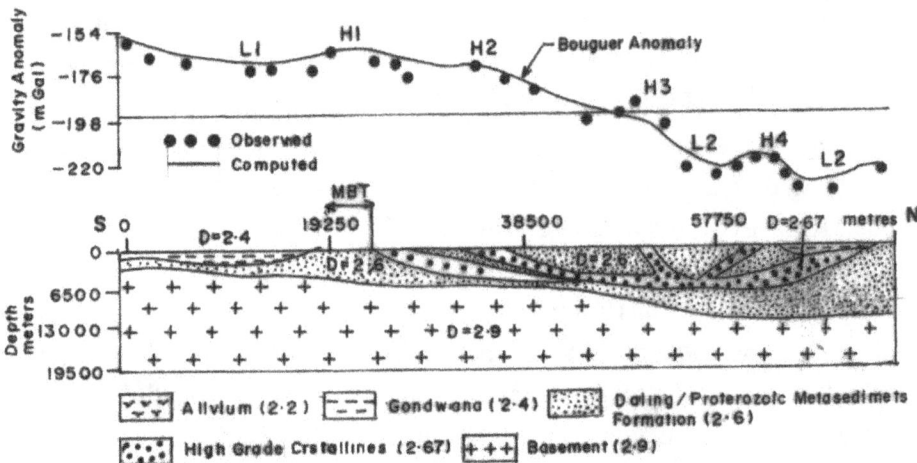

Fig. 6.35(b) A detailed gravity profile from Siliguri to Gangtok and computed geological model (Modified after Lahiri et al., 2008)

(b) Detailed Gravity Profile and Shallow Section across Sikkim Himalaya

A detailed gravity profile recorded at closer interval (Lahiri et al., 2008) from Siliguri to Gangtok (~ 77 km) almost along the same profile described above is presented in Fig. 6.35(b), which shows a general gradient towards the north related to deepening of the basement as crustal thickness remains almost same in this section (Fig. 6.35a). Besides, it shows short wavelength and small amplitude gravity highs, H1-H4 and gravity lows, L1-L3 related to shallow intrusives and thickening of sediments, respectively. The computed model shows several thrusted high density rocks that have given rise to small wavelength gravity highs. The high density rocks may represent high grade crystalline rocks (augen gneiss, granite-mylonite) that occur within the exposed low grade Proterozoic metasediments in this section (Acharyya, 2007). However, as per the geological map very few exposures of these rocks are shown in the map. But from the computed model, it appears to be quite widespread and that has not been reported adequately. Some of these gravity highs may represent blind thrusts as shown in Nepal in Fig. 6.28. The northward gradient suggests increase in the depth to the basement from 3.5 km to 12 km that may represent the plane of decollement. It shows Gondwana sediments in a limited section, thrusted Proterozoic metasediments, and high grade crystalline rocks.

(iii) A Geological Cross Section across Central Himalaya

Fig. 6.36(a) presents a schematic geological cross-section across central Nepal along longitude through Kathmandu, incorporating the exposed thrusts and results of geophysical investigations (Avouac, 2007). It shows the HFT, MBT, and MCT joining in depth to the Main Himalayan Thrust that is a plane of decollement. It also shows a ramp in the MHT under High Himalaya that further extends to South Tibet in depth range from surface at HFT to a maximum depth of about ~ 40 km under South Tibet. It also shows the South Tibet Detachment (STD) as a normal fault and intrusions of leucogranites similar to the Kangmar dome from the partially molten rocks inferred from MT measurements under South Tibet (Fig. 6.32(a)) that might be synonymous with N-S channel flow in the Indian crust. Such a mid-crustal flow has also been inferred from geological studies such as thermobarometry and U-Pb geochronological data under High Himalaya of Nepal (Everest) and South Tibet (Searle et al., 2003). Crust thickens from about 38-40 km under the Ganga basin south of HFT, to about 80 km under South Tibet, as inferred from seismic and gravity studies described above. It also provides details of Tethyan sediments between STD and ITSZ. A N-S trending mid-Miocene dyke swarm in the Tethyan sedimentary rocks of the Indian plate, yields similar Sr-Nd isotopic data as Miocene ductile dyke from north of ITSZ. Dykes on both sides of this suture represent crustal melts derived largely from mid lower crust with emplacement age of 12-9 Ma indicating southward flow of the Asian crustal rocks. It also delimits the age of the southward channel flow at about 12-9 Ma that also coincides with the ages of the leucogranites in this section (King et al., 2007).

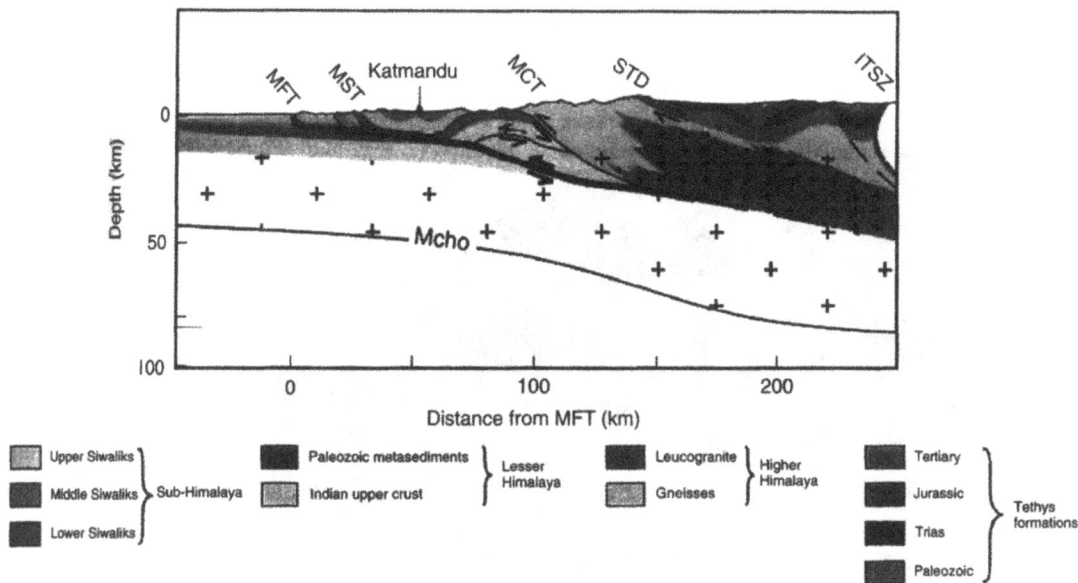

Fig. 6.36(a) A typical geological cross section in central Nepal based on investigations described above. The Leucogranite shown along MCT might have been related to channel flow giving rise to leucogranite intrusive bodies between STD and ITSZ (Avouac, 2007). [Colour Fig. on Page 782]

(iv) Gravity and Magnetic Studies over Tibet: Yadong-Golmud Geotransect

One of the most important expriments over Tibet has been the Integrated Geophysical and Geological Survey along an International Geotransect from Golmud-Yadong, close to the Indian border in Sikkim Himalaya (Fig. 6.36(b)). Besides detailed geological mapping, various geophysical parameters such as gravity, magnetic, seismic, earthquakes data were recorded in a 100 km wide strip and presented as a geotransect (Gongjian et al., 1991). As this figure indicates, this transect passes through all the important terrains of Himalaya and Tibet. This geotransect is about 1300 km long. Some of the significant results along this transect are discussed below.

The sharpest gradient in the gravity field is observed over the Himalayan Belt up to ITSZ as previously shown in Bouguer anomaly map from satellite gravity data (Fig. 6.8) of Himalaya. Contours up to MCT show concave southwards, indicating a residual gravity low in the central part indicating low density intrusives. It changes sign at STD (NHNF) where it becomes concave northwards showing a residual gravity high over the Kangmar dome. The same is the case at the Kangmar thrust. Over the ITSZ, gravity highs of 20-30 mGal related to high density oceanic crustal ophiolite rocks are observed. In this section, linear magnetic anomaly of NW-SE orientation of 100-300 nT due to mafic rocks along thrust and suture zone are observed. Gravity anomalies over the Lhasa block is subdued except at few places that might be caused by mafic/ultramafic intrusives. However, linear magnetic anomalies of 100-200 nT are quite abundant especially over mountains indicating linear thin mafic intrusives, indicating thrusted oceanic crustal rocks due to compression which did not reflect in gravity data.

The Qiangtang terrain also shows subdued gravity anomalies but moderate 100-200 nT magnetic anomalies over mafic intrusives. Towards north, Kunlun Mountains (North Kunlun terrain) shows gravity gradient indicating change in crustal thickness, and linear magnetic anomalies of about 300 nT. Most of the magnetic anomalies along the transect show pairs of magnetic highs and lows, with lows located towards the north indicating induced magnetization. Therefore, these are quite suitable for quantitative interpretation using tabular (dyke) model that can provide information about their depth extent and susceptibilities.

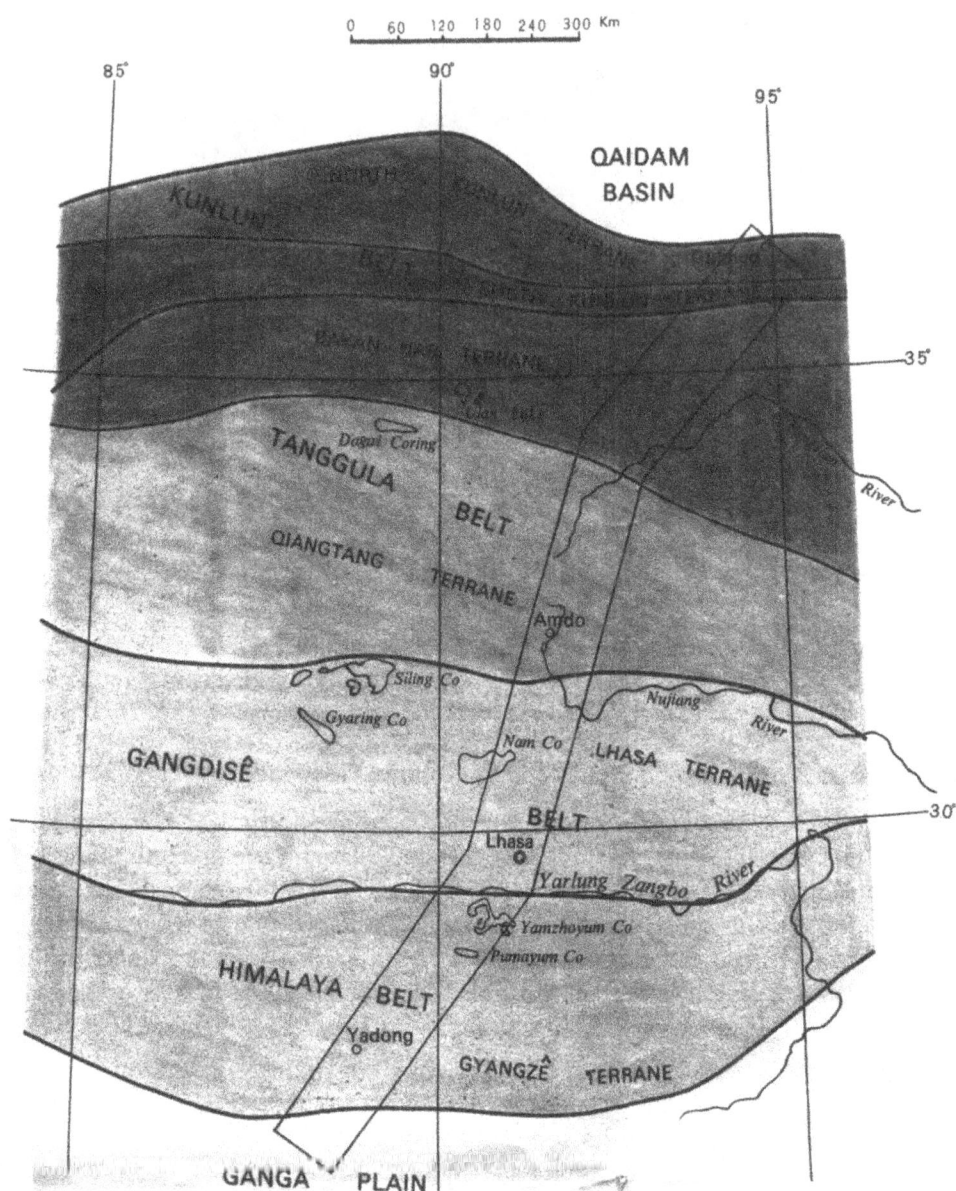

Fig. 6.36(b) Yadong-Golmud geotransect across Northern Himalaya and Tibet with respect to tectonics (Gongjian et al., 1991). This profile is also shown in Fig. 6.20(a) as profile AA'.

Crustal Structure : The crust thickens to 70-75 km (including topography) under Tibet reducing to 51-52 km under Kunlun mountains and 50 km under the Himalayan Belt. Tentative dates for different thrusts have been given (Gongjian et al., 1991) as follows:

Thrust along ITSZ	:	70-40 Ma
NHT (Kangmar thrust)	:	40-20 Ma
MCT	:	~ 20 Ma
MBT	:	~ 10 Ma
NHNF (STD)	:	< 10 Ma

The various tectonic events in Tibet have been summarized as follows: (Gongjian et al., 1991)

1. Breakup of different parts through rifting that drifted apart separated by oceanic basins.

2. It is followed by convergence that involves subduction of the oceanic crust resulting in collision and related tectonics of thrusting and shortening.

3. Intra-continental collision and related tectonics of shortening and thrusting

Heat Flow: Heat flow values reported over Tibet suggest low values in the northern part and high values (90-64 mw/m^2) in the southern part indicating cold and hot lithosphere in the two parts related to old and reworked young crusts, respectively (Gongjian et al., 1991). Thermal conductivity of Higher Himalayan crystalline rocks is suggested to range from 1.5 to 3.6 Wm^{-1}K^{-1} that is considerably higher than that generally found in similar rocks in the cratonic area. This can be attributed to mineralogical changes due to metamorphism (Ray et al., 2007).

Thermal activity: Based on seismic tomography, there have been suggestions (Gongian et al., 1991) that the core-mantle boundary under Tibet is domed and may be hotter than the surroundings. The heat source of this dome might be the deep seated cause for the formation of plateaus.

(v) Integrated Crustal and Lithospheric Structures across Himalaya and Tibet

Jimenez-Munt et al., (2008) integrated all seismic information from central Himalaya and Tibet as described above and some others that are referred to below, and projected them along a profile (Fig. 6.37(a)) to which lithospheric thickness are added as described above. It shows crustal thickness under different terrains from different sources as referred therein. As shown in this figure, Moho changes from 40 km under the Himalayan foreland to about 80 km under the Lhasa terrain, reducing to about 40 km under Qaidam basin and north China block. Crustal thicknesses have primarily changed across thrusts and Qiangtang terrain lies at the center of the S- and N-verging thrusts that shows the extent of the effect of collision between the Indian and the Eurasian plates. The heat flow map presented by them in general shows a higher heat flow in northern Tibet compared to the southern part. They modeled satellite gravity and geoid anomalies and elevation data simultaneously along a profile approximately extending from the Himalayan foreland (23.5°N, 87°E) to north China block (42°N, 100°E, Fig. 6.37(b)). They constrained the crustal thickness from seismic studies (Fig. 6.37(a)) and densities from seismic velocities. Fig. 6.37(b) shows a lithospheric thickness of about 260 km under South Tibet that agrees quite well with that provided by receiver function analysis. It reduces to about 100-160 km under North. Tibet, north of the BNS suture. Accordingly, the crustal thickness changes from 40 km under the Ganga plains to about 80 km under the South Tibet and 60-40 km under the North Tibet. Density values of various layers used to model these fields are given in the caption of Fig. 6.37(b). Variation in the lithospheric thickness is in agreement with high surface heat flow in the northern part and also conforms to the seismic results (Haines et al., 2003 and Tilmann et al., 2003) that suggest asthenospheric upwelling in the northern part as shown in Fig. 6.30. Reduced lithospheric thickness under the Qiangtang terrain is attributed to detachment of underthrusted Asian lithosphere in this section that can be attributed to the interaction of the subducting Indian and Asian lithosphere from two sides. Downward foundering of the subducting Indian lithosphere in the transition zone as suggested from tomography and receiver function analysis (Tilmann et al., 2003; Kosarev et al., 1999) can also be attributed to this interaction. This model suggests the extension of underthrusting Indian lithosphere up to south of BNS (Fig. 6.37 (b)). However, Bendick and Flesch (2007) have suggested its extension up to north of BNS under the Qiangtang terrain based on topography modeling to explain the Qiangtang bulge though the latter approach has been questioned by Klemperer (2008).

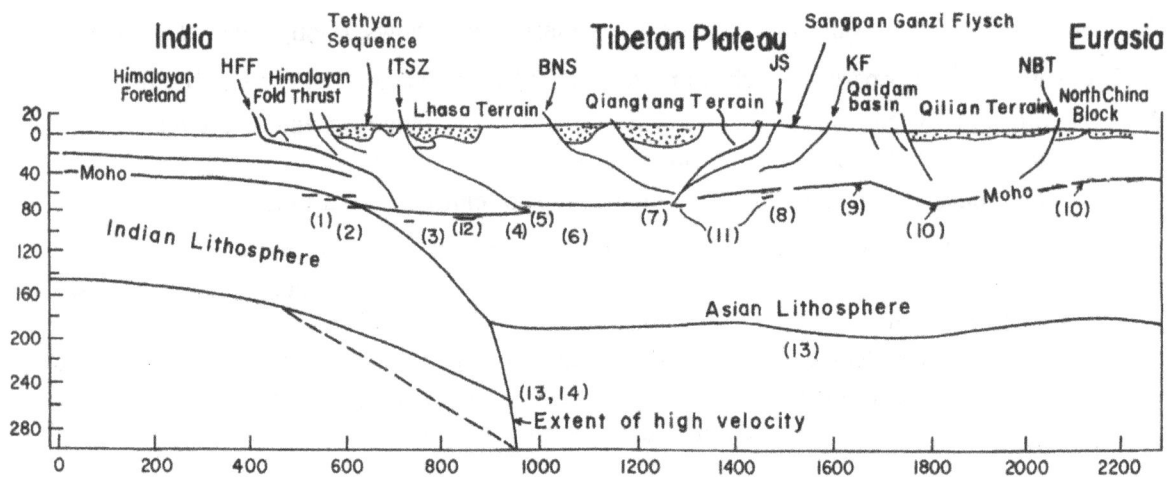

Fig. 6.37(a) Compilation of seismic information from central Himalayas and Tibet with geological details from Jin and Harrison (2000). Seismic references are as follows: (1) black marks (Mitra et al., 2005), (2) (Hauck et al., 1998), (3) (Zhao et al., 2001), (4) (Owens and Zandt, 1997; Meissner et al., 2004), (5) (Haines et al., 2003) (6) (Zhang and Klemperer, 2005), (7) (Owens and Zandt, 1997; Zhao et al., 2001; Haines et al., 2003), (8) (Owens and Zandt, 1997; Zhao et al., 2001; Kind et al., 2002), (9) (Jiang et al., 2006; Zhao et al., 2006), (10) (Rui et al., 1999), (11) (Vergne et al., 2002), (12) (Schulte-Pelkum et al., 2005) (13), (Kumar et al., 2006), (14) (Priestley et al., 2006, 2008). (Modified after Jimenez-Munt et al., 2008)

Jimenz-Munt et al., have also provided the thermal structure of the lithosphere along this profile, which suggests uplift of isotherms in the central part under the Qiangtang terrain due to asthenospheric upwelling causing partial melts in the crust, which would facilitate west to east flow in this section as has been described in Section 6.4. Thickening of the crust under MCT (High Himalaya) and Tethyan and Lhasa block also causes uplift of isotherms under these terrains due to crustal component of the heat flow giving rise to granite melts in the central part of the crust that gave rise to leucogranites. The same flowing through thrusts and fracture zones would give rise to N-S channel flow in the Indian crust has been described in more detail in the next section (Section 6.2.6). This might be responsible for the conductive anomalies in this section as described above, and for the bright spots at depths of 15-18 km on seismic lines (Brown et al., 1996). They also compared temperature and pressure with depth predicted from their model and those derived from seismic data, xenolith and α - β quartz transition for some specific terrains. Based on this study, they suggested a higher temperature at mid-crustal level under the Lhasa and the Qiangtang terrains as compared to those given in the model based on the thinning of the lithosphere (Fig. 6.37(b)) that might be related to some extra heat source in the asthenosphere in the form of magma chamber or plume which is also supported by the various intrusives and underplated crust in this section (Fig. 6.26). It would further facilitate the partial melting of the Tibetan crust and west to east flow in its central part. According to the thermal structure given by Brown et al., 800°C isotherm where lithosphere loses its strength lies at a depth of 72 km under the Himalayan foreland and 30 km under the northern Tibet and Qiangtang terrains, increasing to about 60 km under the Qaidam basin that qualitatively approximately follows the effective elastic thickness in these regions (Jordan and Watts, 2005; Tiwari and Mishra, 2008). Fig. 6.37(b) also shows the hypocenter of earthquakes within 150 km band on each side of the profile (Engdhal et al., 1998) that are mainly concentrated in the crust, especially in the lower crust under Himalaya and in the upper crust under Tibet where isotherms are uplifted indicating that they are primarily related to thermal structure. Some deeper earthquakes occurring in the lower crust and upper mantle are also reported but they are close to major thrusts/sutures such as MCT, BNS, and JS and are related to them. In fact, those under the Himalaya south of MCT tend to concentrate along the interface separating the upper and the lower crusts and the Moho as has been suggested above in case of the Sikkim profile (Fig. 6.35(a)).

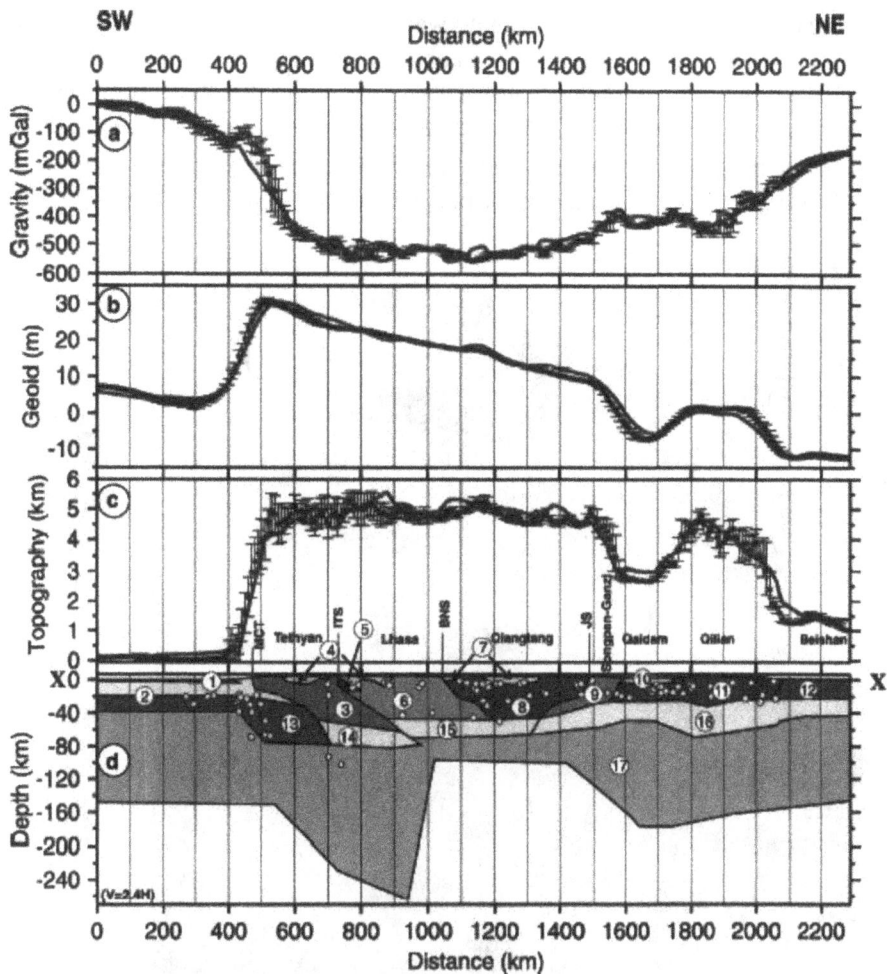

Fig. 6.37(b) Gravity, geoid, and elevation data along profile X (23.5°N, 87°E) X'(42°N, 100°E) and modeled crustal and lithospheric structures based on simultaneous modeling of these data sets constrained from seismic studies. Dots are hypocenters of earthquakes in 150 km wide band on either sides of the profile. Densities of various layers (1-17) are in kg/m³ as follows based on reported seismic velocities and rock types: (1) = 2450; (2) = 2700, 2890; (3) = 2750, 2910; (4) = 2650, 2700; (5) = 2750; (6) = 2700,2930; (7) = 2500,2600; (8) = 2640,3010; (9) = 2600,2850; (10) = 2400; (11) = 2530,2920; (12) = 2550,2880; (13) = 2980; (14) = 3050; (15) = 3050; (16) = 3000; (17) = 3200 (Jimenez-Munt et al., 2008).

(vi) Airborne and Satellite (CHAMP) Magnetic Maps

The Airborne total intensity magnetic map of Tibet (Zhao et al., 2006) has delineated several sets of magnetic anomalies related to collision tectonics with the Indian and Asian plates. There are primarily four types of magnetic anomalies associated with various thrusts and mafic/ultramafic rocks:

(a) There are high amplitude linear magnetic anomalies associated with the ITSZ and the Altyn Tagh faults, which represent mafic rocks thrusted along them due to collision of the respective plates. Magnetic anomalies along ITSZ form liner bands, typical of oceanic crustal rocks in the central core complex of collision zones.

(b) The Lhasa block is characterized by moderate amplitude linear anomalies and may represent mafic intrusives. Some sporadic large amplitude anomalies may indicate concentration of mafic components in these sections.

(c) There are large amplitude linear and moderate wavelength magnetic anomalies in the Tarim basin, which follow the trend of the Altyn Tagh fault and of the Tarim basin. In between, there are sections of low amplitude anomalies and may therefore represent bimodal volcanism of Permian times (240-280 Ma) prevalent in this section (Yang et al., 2007).

(d) Absence of major magnetic anomalies along the BNS suture and Jinshan river suture may be noted, indicating absence of intrusives or oceanic crustal rocks along them.

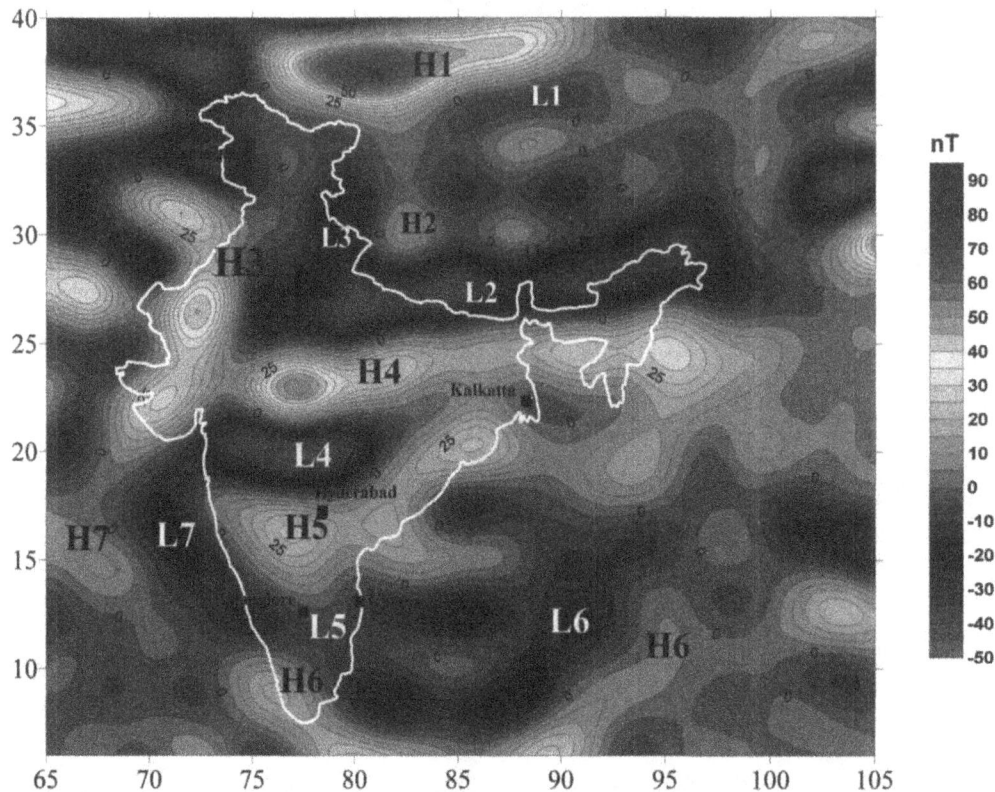

Fig. 6.38 CHAMP satellite magnetic map continued downward to a height of 50 km with magnetic highs, H1-H2 and lows, L1-L2 recorded over Tibet and Himalayas. Other anomalies are related to the Indian continent and are discussed in Section 7.11.4. [Colour Fig. on Page 783]

Satellite magnetic data (CHAMP, Fig. 6.38) shows large wavelength magnetic anomalies which provide pairs of magnetic anomalies H1, L1 and H2, L2 over Himalaya and Tibet, with their gradient coinciding with the Altyn Tagh fault and part of ITSZ and Himalayan thrusts, respectively. This data from CHAMP satellite was recorded at a height of 400 km and continued downwards to 50 km. The CHAMP satellite magnetic map of the Indian continent and adjoining oceans are presented in this figure to check the continuity of the anomalies. However, only the magnetic anomalies related to Himalaya and Tibet are discussed here. High amplitude magnetic high, H1 covering Tarim basin suggests mafic intrusives in this section. The central part of Tibet is magnetically quiet, with a small order of magnetic anomalies as also observed in the airborne magnetic map described above, indicating lack of mafic rocks in this region. Another interesting observation is that magnetic contours do not take a southerly turn along the Eastern Syntaxis following surface tectonics as in the case of gravity anomalies (Figs 6.6, 6.7, and 6.8) indicating that sources of these anomalies are related to deep-seated sources. Magnetic anomaly, H3 extends NE-SW almost from Saurashtra to Kashmir Himalaya in the central part of the Western Syntaxis. As explained in Section 7.11, the gravity highs in the same section are attributed to intrusives from plumes due to the breakup of Gondwanaland, especially the breakup of Africa from the Indian plate. It presumably took place offshore Saurashtra and created several Mesozoic basins in this section, including Saurastra and Kutch as discussed in Chapter 9. The extensions of corresponding magnetic anomalies in the syntaxial bend indicate that this bend itself might have been created due to interaction of respective plumes with the Indian plate. This may also explain the seismically active nature of the Western Syntaxis. Plumes creating such bends in the tectonic plates the world over has been suggested by Vogt (1973) and Vogt et al., (1976). A part of this magnetic high takes a westerly turn, west of New Delhi forming part of the Delhi-Lahore-Sargodha basement ridge that has been considered a shallow manifestation of crustal bulge due to lithospheric flexure of the Indian plate (Section 6.1.9). Satellite magnetic anomalies south of Himalaya over the Indian continent are discussed in Chapter 7 as they are related to the Indian Shield.

As our current interest is primarily long wavelength anomalies, we used CHAMP satellite data for spectral analysis. Spectral method is the ideally suited statistical method for depth estimates when large number of potential field anomalies are present in a map (Section 6.1.6). Spector and Grant (1971), Hahn et al., (1976) and several others have used spectral analysis to analyse magnetic fields, especially airborne magnetic data, to estimate depths to causative sources at different levels. Mishra and Venkatarayudu (1985) used it to estimate the Curie Point Geotherm over India and Himalaya from MAGSAT data (Section 4.6.2). Bilim (2007) has used spectral decay of airborne magnetic field to define Curie Point Geotherm under western Anatolia, validated from geothermal gradients and observed gravity field.

Digital data of CHAMP satellite magnetic is transformed into frequency domain to determine spectral depths (Section 6.1.6) for different wave bands of anomalies. The largest wavelength of this data set provide a depth of about 154 km below surface after subtracting the height of the continued field (50 km) (Fig. 6.39(a)). In order to check the stability of the computed spectrum the whole data set is divided into two blocks, upper (20°N upwards) consisting of Himalaya and Tibet, and lower containing the Indian Peninsular Shield and adjoining oceans. The spectrum for the block of data (20°N upwards) related to Himalaya and Tibet also provided a maximum depth of 159-160 km (Fig. 6.39(b)) for large wavelength magnetic anomalies, which is almost same as the depth for the first segment of spectrum of the total data set (Fig. 6.39(a)) confirming the stability of computed spectrum and depths obtained from them. The second linear segment is related to shallow sources in the upper crust between the surface and 10 km, such as intrusives, and basement. However, magnetic signals are normally confined to above Curie Point Geotherm which in most cases lies in the crust. However, in case of subduction zones, Mahatsente and Ranalli (2004) have provided a thermal structure that shows 300-400°C temperature in case of subducting slab up to 300-400 km, which is well within the Curie point of magnetite (570°C), and can produce magnetic fields especially because these rocks are usually mafic in composition as they are from the oceanic crust or lower crust. Even in the case of a part of the Himalayan orogenic belt in the Hindu Kush and Pamir Ranges, Negredo et al., (2007), based on thermomechanical modeling, have predicted a temperature of 400°C up to a depth of 150-175 km (Fig. 6.18) that is below the curie point of magnetite (575°C) and will produce its magnetic field. Therefore, the magnetic signals recorded in satellite magnetic data from a depth of 150-160 km are attributed to cold subducting Indian lithosphere in the upper mantle, as has been discussed above for gravity anomalies over Himalaya and Tibet. The existence of such a cold slab with high velocity is also inferred from both receiver function and tomography experiments in this region as described above (Fig. 6.30(b) and (c), Section 6.2.3). However, the depth estimated from the spectrum of the CHAMP satellite data could not be verified from direct modeling of the sources at those levels and therefore, cannot be fully relied upon especially because of the absence of magnetization at those levels under normal circumstances.

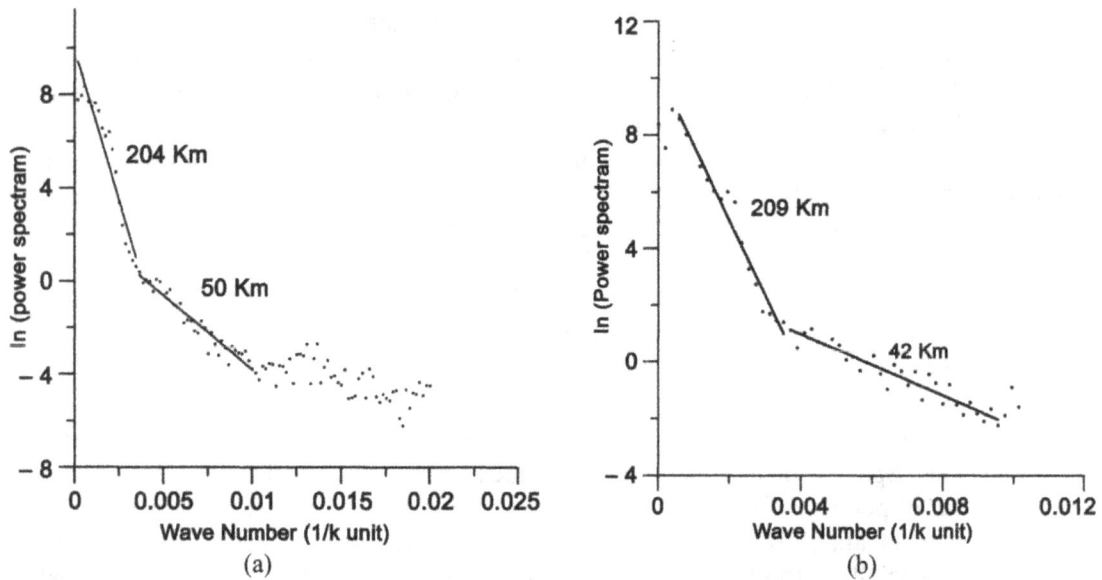

Fig. 6.39 (a) Radially averaged spectrum of the satellite magnetic map with mainly two linear segments with slopes equal to 204 and 50 km, representing sources at a depth of about 154 km and surface as this data pertains to a height of 50 km level. (b) Radially averaged spectrum of the northern part (20°N) of the satellite magnetic map (Fig. 6.38(b)) showing sources at 159 km and approximately surface.

(vii) Magnetic Profiles Across Eastern Syntaxial Bend

Nine magnetic profiles were recorded along profiles, P1-P9 (Fig. 6.40(a)) with a station spacing of about 1 km along roads and tracks in this section. It provided fluctuating magnetic field as the exposed rocks in this section are metamorphosed Proterozoic sediments and intrusives. However, some prominent magnetic anomalies, M1-M4, (Fig. 6.40(b)) with well defined magnetic lows and highs, with lows towards the north indicating induced magnetization, are observed. These anomalies are associated with thrusts and suture zones as shown in Fig. 6.40(a). They are modeled using tabular bodies (Fig. 6.40(b)) that show north dipping bodies as is generally the case for thrusts in the Himalayan collision zone. Their susceptibilities are quite high in the range of $0.6-1.7 \times 10^{-3}$ emu that indicates mafic rocks associated with these thrusts. Most of these mafic intrusives are subsurface as exposed rocks are mainly non-magnetic in nature.

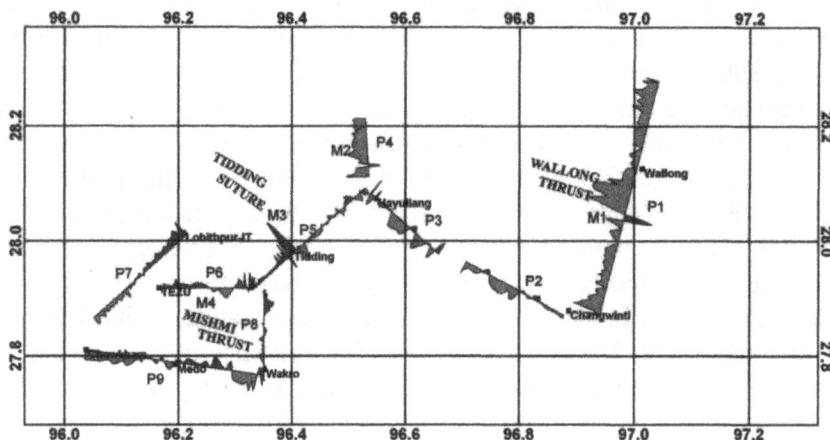

Fig. 6.40(a) Magnetic profiles and important thrusts in the Eastern Syntaxial Bend, India. P1-P9 are magnetic profiles recorded along the roads and M1-M4 are major magnetic anomalies related to the thrusts that are modeled.

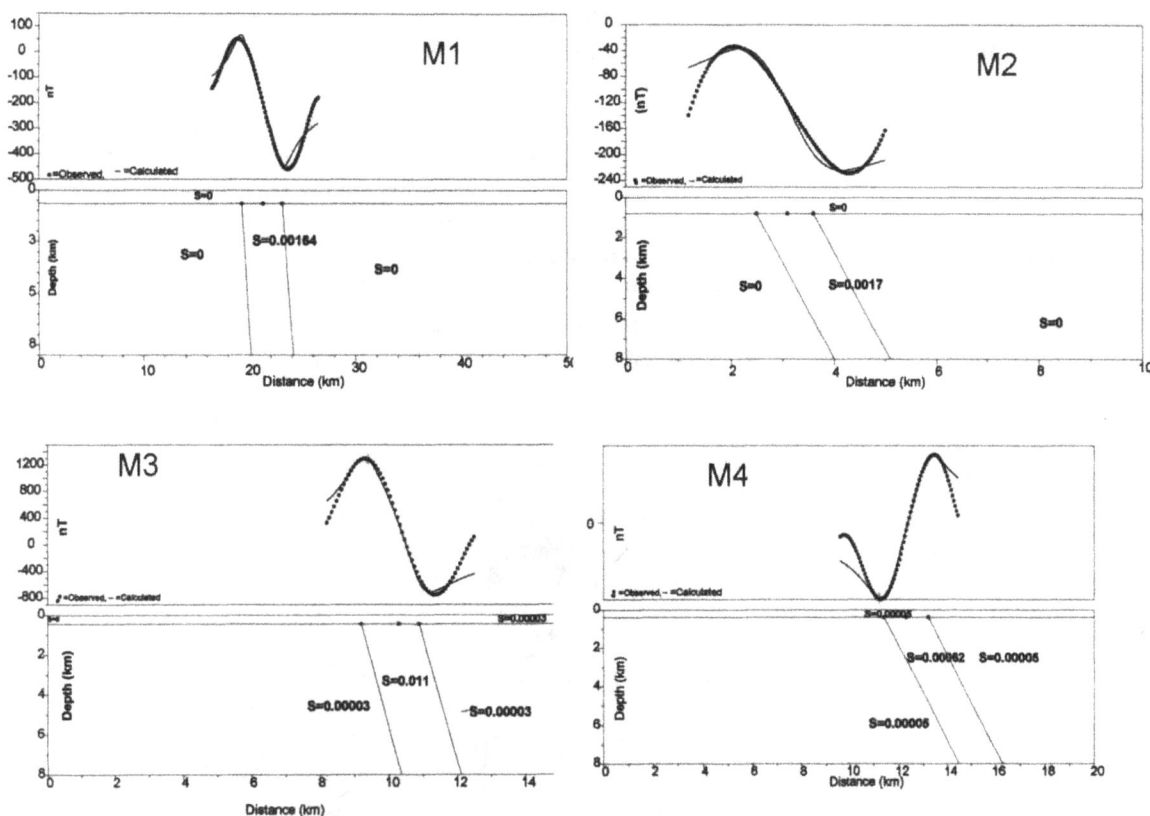

Fig. 6.40(b) Important Magnetic anomalies (M1-M4, Fig. 6.40(a)) and matching computed fields due to modeled bodies showing northward dips with susceptibility of 0.6-1.7×10^{-3} emu indicating mafic/ultramafic rocks.

(viii) Bouguer Anomaly of Burmese Arc, Subduction Model, and Deep Focus Earthquakes

Fig. 6.40(c) is the Bouguer anomaly map of the part of the Burmese (Myanmar) microplate including the Arakan Yoma Fold Belt (AYFB, Fig. 6.5) that was initially recorded for oil exploration by Burma Oil Company (Evans and Crompton, 1946). It shows gravity gradients associated with the AYFB. Another sharp gradient coincides with the Sagaing fault that is characterized by association of high density intrusives. The gravity lows, L1 and highs, H1 are related to the crustal thickening due to subduction across the AYFB, and high density intrusives including ophiolites along the Sagaing fault that is a strike slip fault defining the Eastern margin of the Burmese microplate with the Eurasian plate to its east. Fig. 6.40(d) is a Bouguer anomaly profile along 23.5 °N from East India across the Burmese arc from 92 °E up to 96 °E. The gravity low, L1 east of the AYFB is related to crustal thickening due to isostasy as the in case of the Himalaya and Tibetan Plateau. However, the magnitude and wavelength of this gravity low is much less than those related to the Tibetan Plateau (Fig. 6.8) indicating Lesser crustal thickening (60-62 km) in a limited section as shown in the computed crustal model at the bottom of Fig. 6.40(d). The computed model also shows the part of subducting low density (3.18 g/cm^3) Indian lithosphere that interacts with the Sagaing fault, where deep focus earthquakes have been reported in the lithospheric mantle (Fig. 6.19). The gravity high, H1 (Fig. 6.40(c)) is related to the high density (2.9 g/cm^3) thrusted rocks and other intrusives (Fig. 6.40(d)) along the Sagaing faults. The latter is a strike slip transform fault between the Burmese microplate and the Eurasian plate towards the east, similar to the Chamman fault along the western fold belt (Pakistan, Section 6.3.4). The computed crustal model also shows an underplated crust towards the west under Bangladesh (3.0 g/cm^3) that might have been caused by the interaction of the Kerguelen hotspot during the breakup of India from Antactica, that also gave rise to the Ninety East Ridge as discussed in Sections 5.4 and 7.11. The interaction of the subducting plate with the strike slip Sagaing fault would facilitate slab breakoff in the upper mantle, which would cause deep focus earthquakes. In fact, deep focus earthquakes with hypocenters more than 90 km (Fig. 6.19, Rao and Kalpna 2005) are concentrated along the Sagaing faults where the subducting Indian plate is interacting with the Sagaing strike slip fault, suggesting the importance of this interaction for these earthquakes. Seismic activity to its west is concentrated under the thickened crust below 60 km indicating the importance of collision and subduction along the AYFB for these earthquakes. Shallow crustal earthquakes mainly occur along the western margin of the AYFB, known as Tripura fold belt or Schuppen belt, associated with thrust faults such as Dibang and Naga thrusts.

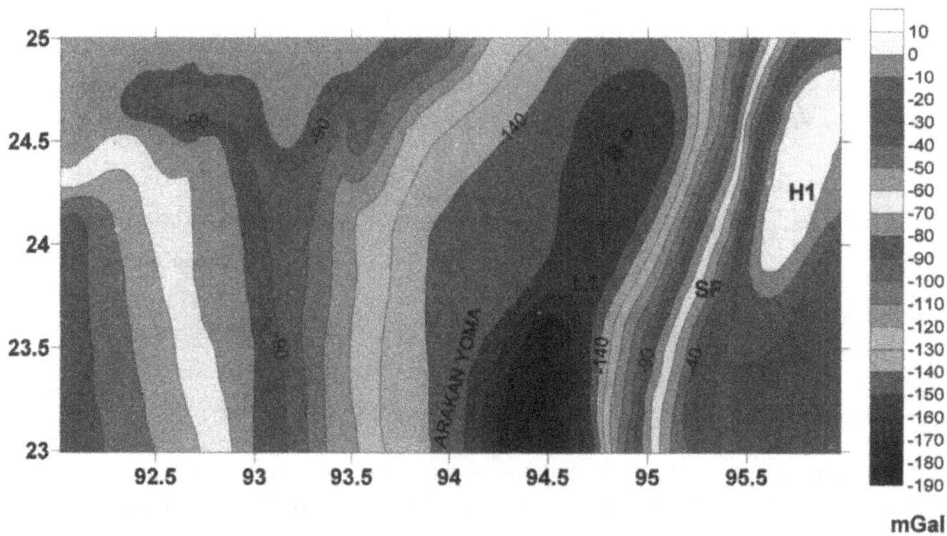

Fig. 6.40(c) Bouguer anomaly map of the Arakan Yoma Fold Belt (the Burmese arc). L1 and H1 denote gravity lows and highs, respectively. SF: Sagaing strike slip fault along the Sagaing range (Fig. 6.5).

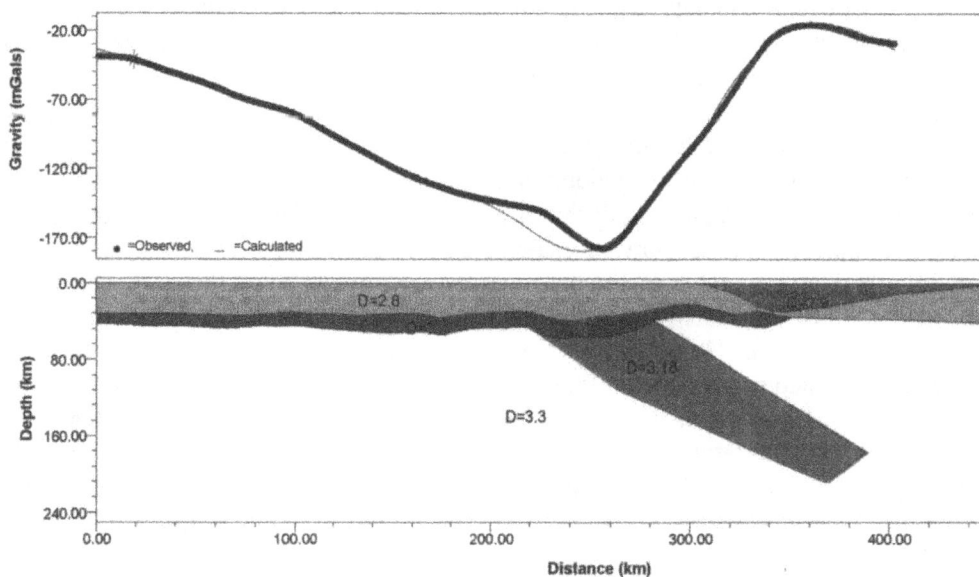

Fig. 6.40(d) Bouguer anomaly profile along 23.5 °N across the Burmese arc, and the computed subduction model of the Indian plate in this section.

(ix) Strength of Himalayan Lithosphere

Lower effective elastic thickness of 30-40 km in collision zones, and bimodal distribution of the hypocenter of earthquakes in the upper crust and upper mantle, suggested a jelly sandwich model of the crust and mantle with a strong upper crust and mantle and ductile lower crust (Fig. 6.41(a), [i] However, Jackson (2002) based on a similar data set from Himalaya and Tibet suggested that most of the continental activities are controlled by the crust, implying instead a strong crust and weak mantle (Fig. 6.41(a), [ii] He also suggested that flexure of the Indian Shield is likely to be a major support of the topography in Himalaya and southern Tibet, and transient lower crustal flow of the type associated with metamorphic core complexes is likely to be controlled by the input of igneous melts and fluid in the lower crust. A similar view was also expressed by Maggi et al., (2000) and Lamb (2002) suggesting a strong crust and weak upper mantle. Maggi et al., (2000) found a correlation between the seismogenic layer and the effective elastic thickness where the strength of the continental lithosphere resides and supports the load (mountain ranges). However, Burov and Watts (2006) favoured the old jelly sandwich model with a ductile lower crust and strong upper mantle (Fig. 6.41(a)). Burov and Watts (2006) have provided Te = 70 km along a profile (~ 80°E) that almost coincides with the crust, and suggest that this whole section is strong and supports the topographic load of Himalaya. However, the large variations in the reported Te makes it difficult to interpret it quantitatively.

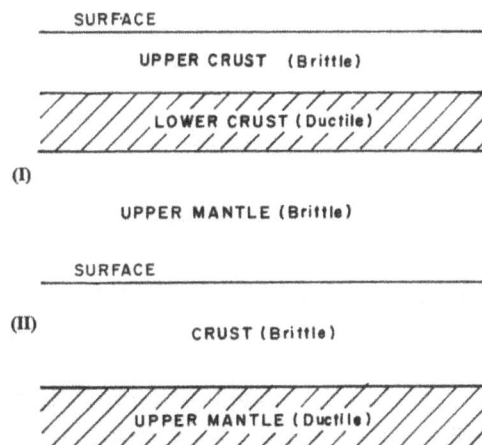

Fig. 6.41(a) Lower crust is ductile in between the brittle upper crust and mantle. This is known as the Jelly sandwich model (I). Upper mantle is ductile under a brittle crust (II).

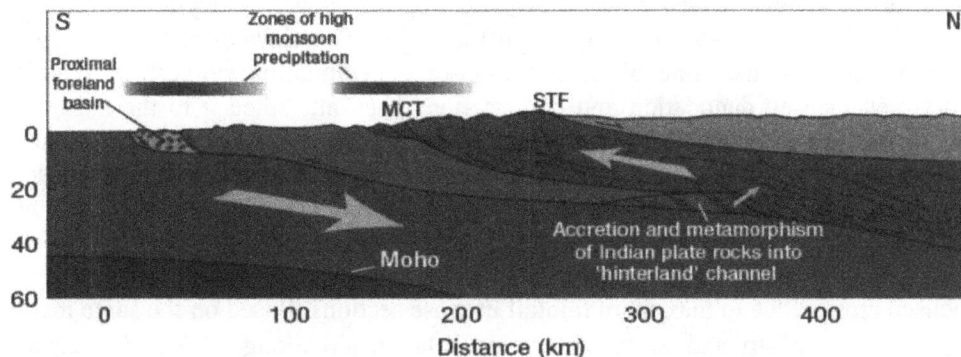

Fig. 6.41(b) A schematic cross-section of the southern margin of the Himalaya-Tibetan orogen that is affected by intense rainfall, coinciding with maximum rainfall, erosion, and uplift along the Main Central Thrust (MCT) and Main Boundary Thrust (MBT) to its south (Clift et al., 2008b). Miocene channel flow is suggested between the MCT and South Tibetan Fault (STF) where High Himalaya is exposed and leucogranites have been reported. Another channel flow during Quaternary period along MBT has also been envisaged. [Colour Fig. on Page 783]

6.2.6 Orography-Climate Induced Uplift and Channel Flow

(i) Central Nepal

Several studies have been made on the effects of orography (climate induced erosional processes) on orogeny. Climate and weather consistently influence the mountains and valleys through erosion over geological period that changes its shape and position. There is a clear relationship between surface topography and climate, which is seen no where better than monsoon in India. But for the Himalaya, water carrying winds would have crossed this region without rains and the Indian continent would have been a big desert like Tibet and Mongolia north of Himalaya. In this regard, the effects of topography on climate is well known, but the reverse is less understood. To understand the effects of climate on topography, one must understand the mountain building processes specially related to neotectonics, and to the reshaping of mountains due to climate. The best example to understand this process is the Himalayan terrain as it is intricately related to monsoon and associated erosion. Besides, in the valleys of Higher Himalaya, wind speeds regularly go as high as 70 km/hour in the afternoon. Willett (1999) created numerical models of mountain belt for wind-driven orography with enhanced precipitation on the side of the mountain chain. He considered two cases: (i) dominant wind in the direction of motion of the subducting plate as it happens in the case of the Himalaya with monsoon, (ii) dominant wind in direction opposite to the motion of subducting plate. Based on numerical models for the two cases, he suggested a broad zone of exhumation mainly in orogen interior in case (i), while in case (ii) the models predicted a focused zone of exhumation along the margin of the orogens. Accordingly, one observes a broad zone of exhumation along MCT and MBT in case of Himalaya. This model also shows how mountains will in this case develop on subducting plate across wind direction. His model also shows crustal thickening and topographic divide between the zone of precipitation and exhumation as is seen between MBT and MCT in case of Himalaya.

As has been stated above, the Himalayan-Tibetan terrain is the result of collision of the Indian and the Eurasion plates at about 50 Ma. Tibet, as part of the Eurasion plate, was amalgamated to it earlier. However, in this process of convergence, there is a shortening accompanied by crustal thickening up to 75-80 km under Tibet, and thrusting to form the Himalayan ranges. Besides material transfer, there also occur processes related to energy transfer in such collision zones. For example, what keeps such large mountains and plateaus in a balanced condition instead of having them falling apart? It has been suggested that the gentle, steady decline in elevation from Tibetan Plateau to the south and east was caused by outward flow of the Tibetan lower crust under Eastern margin of the plateau. Based on field evidences in Marsyandi Valley in central Nepal, Hodges et al., (2004) suggested a zone of rapid uplift in Higher Himalaya during the Quaternary period that coincided with the zone of intense monsoon precipitation suggesting a positive correlation between focused denudation and deformation. They attributed it to the north-south channel flow in the Indian crust. Based on ^{40}Ar/^{39}Ar thermochronological data from detrital muscovites, Wobus et al., (2003) suggested a major discontinuity in cooling ages on two sides of the geomorphic break in central Nepal. Samples to the north are Miocene and younger, while those toward south are Proterozoic to Paleozoic. These observations suggest recent (Pliocene-Holocene) thrusting in central Nepal Himalaya at about 20-30 km south of MCT, which may be related to focused erosion due to maximum rainfall in those sections. Based on the same method, Wobus et al., (2008) constructed an age-elevation relationship along ~1000 m change in elevation (3280-4280 m) profile in central Nepal and suggested a vertical exhumation rate of < 0.1 mm/yr before 10 Ma that increased considerably to ~0.5 mm/yr between 10-7 Ma, which must have been caused by large scale changes in climate. This period coincided with the formation of the MBT and the Lesser Himalaya. Clift et al., (2008a and b) attributed large scale erosion of Himalaya during Holocene to intensified monsoon, and established a correlation between the erosion rate and deformation in Himalaya based on thermochronometric data and stratigraphy records from the South China Sea, Bay of Bengal, and Arabian Sea. It suggested that maximum erosion in Himalaya started at about 23 Ma and continued up to about 10.5 Ma, with a peak at about 15 Ma that slowed up to 3.5 Ma and again accelerated. This must have been connected with large scale changes in climate, such as monsoon. These are also the periods of maximum exhumation when the MCT, the MBT, and the HFT formed. These observations indicated a dynamic coupling between Neogene climate and tectonic activity such as uplift deformation (Fig. 6.41(b), Clift et al., 2008b). Fig. 6.41(b) clearly shows this correlation between channel flow and maximum monsoon precipitation over Higher and Lower Himalaya along the MCT and MBT, respectively, that causes maximum erosion and uplift. The channel flow along the MCT is related to the Miocene period when that section was most active and MCT formed giving rise to leucogranites of that period, while the one along the MBT may be related to the Quaternary period. These observations indicate a major role for the N-S channel flow in the evolution of the Himalayan tectonics including shifting of the thrusts, MCT, MBT, and HFT towards the south. Whipple (2009) has also suggested the influence of climate on the tectonic evolution in case of the Eastern Alps and the St. Elias Range of Alaska.

If erosional channel flow has played a major role in uplift of the Himalaya, its surface expression should be seen in terms of the rate of uplift. In case MCT is the favoured destination of channel flow, there should have been a much higher rate of uplift in this section compared to south. This, however, appears to be true as seen from the elevation along MCT and the hill slopes therein. In case the rate of uplift would not have been more along MCT, the Higher Himalaya should show a mature topography compared to Lesser Himalaya, which is otherwise. Presence of leucogranites has also been considered an indication of channel flow along MCT as shown in Fig. 6.36(a). The concentration of cosmogenic nuclides like Beryllium 10 and aluminium 26 in surface samples, provides their time of exposure which suggests that certain sections in central Nepal near MCT are exposed since much longer than to its south indicating almost threefold higher rate of erosion in these zones of the extruding channel over a time scale of millions of years. Similarly, Carbon-14 dating suggested that erosion had increased in the zone of proposed extrusion along MCT in central Nepal, indicating rapid uplift in this section during the past few million years (Hodges, 2006). Based on study of monsoon rainfall, it has been found that maximum rainfall also coincides with the section of maximum erosion and uplift viz., the zone of extruding channel. These observations have also been confirmed from model studies (Beaumont et al., 2001) that suggest extrusion of lower crustal rocks in regions of high erosion while it travels laterally in case of regions with low erosion rates.

Dietrich and Perron (2006) considered erosion due to various effects of biota which, according to them, are much more pronounced than any other single factor. Accordingly, biota affect climate and climatic conditions affect the erosional processes to a great extent. Using erosion laws that include biotic effects, they have shown that how small biotic processes influence the entire landscape and create distinctive topography. In case their model is to be believed, dip of the various thrusts are controlled by erosional process such as wind, rather than by direction of the subducting plate. Based on $^{40}Ar/^{39}Ar$ experiment from a nearby vertical ~1000 m, age-elevation relationship in Central Nepal, Wobus et al., (2008), suggested slow cooling rate during Early Miocene prior to 10 Ma (< 0.1mm/yr) that accelerated to about 0.5 mm/yr during 10-7 Ma. This change in cooling rate must have been accompanied by changes in climate. Almost at the same time thrusting shifted from MCT to MBT, indicating a relationship between changes in the climate and major tectonics of the region.

Harris (2007) has suggested similar channel flows during Early-Mid Miocene based on exhumed metamorphic assemblages that indicate melt weakening along the south Tibetan Detachment

System and strain softening along MCT. He also indicated Neotectonic extrusion along Quaternary faults south of MCT that is correlated to maximum rainfall during monsoon. This suggests a southward flow of the Indian middle crust, that extrudes along the Himalayan front, that is different from the usual eastward flow of the Tibetan crust. Erosion rate in Himalaya based on cosmogenic isotope studies (^{10}Be and ^{26}Al) of river sediments suggested a rate of 2.7 ± 0.3 mm/year in High Himalaya, which is one of the highest in the world. It reduces to 1.2 ± 0.1 mm/yr and 0.8 ± 0.3 mm/yr along southern margin of Tibet and the foothills of the Himalays, respectively (Vance et al., 2003). This erosion rate is almost the same as the rate of exhumation inferred from apatite fission track ages, indicating that the two are almost balanced over geological periods. However, exposed rocks indicate higher exhumation rate (3-6 fold) that might be related to climatic effects due to erosion and exhumation. The inferred late Miocene onset of the Asian Monsoon System was followed by a decrease in erosional sediment flux and chemical weathering. This might be related to tectonic uplift rates that decreased during Mio-Pliocene and were renewed during the past 1-3 million years (Derry and Lanord, 1997).

Based on thermal-mechanical numerical models, Beaumont et al., (2004) suggested that southward growth of Tibetan Plateau can be explained only due to channel flow in the middle to lower crust, while ductile extrusion of high grade metamorphic rocks along shear zones accounts for the exhumation of the Higher Himalayan sequences. These two processes might be dynamically linked to focused erosion due to monsoon at the edge of the plateau. Searle et al., (2007) have suggested a zone of channel flow in Western Himalaya between MCT and Zanskar Shear Zone that is equivalent of the South Tibetan Detachment Fault of Central Himalaya. They have shown several gneissic domes such as Gianbul, Suru and Chisoti domes, along Dharing and Migar shears that might have been zones of focused denudation and higher rates of uplift related to subsurface channel flow. Quaternary faults in Western Himalaya south of MCT can be investigated for such channel flows during recent times. Mishra and Ravi Kumar (2008) suggested a zone along MCT and MBT in Garhwal Himalaya based on geomorphological changes and residual gravity anomalies as plausible sites for channel flow that needs detailed investigation (Section 6.3.5).

(ii) Residual Bouguer Anomaly of NE India – Eastern Himalayan Front and Burmese Arc

A complete Bouguer anomaly map of Northern India is given in Fig. 6.14. This shows gravity highs, H1 and H2, along the Himalayan front. The gravity high, H2 is located along the Eastern Himalayan front. A detailed Bouguer anomaly map of NE India is given in Fig. 7.30(a) that shows details of this gravity high, H2. The gravity high, H2 (Fig. 6.14) divides the Assam shelf and is almost parallel to the Eastern Syntaxis. It may represent a basement ridge formed due to the crustal bulge caused by the lithospheric flexure similar to the Delhi–Lahore–Sargodha Ridge related to the Western Syntaxis (Section 6.3.4 and 6.3.5).

The gravity highs, H2 (Fig. 6.14) presumably control the flow of rivers like Brahmputra and its tributaries like Lohit, Dibang, and Siang (Fig. 6.42(a)) as they flow from the sides of these highs. Therefore, these highs can be considered to represent recent uplift in this section related to neotectonics that has also been demonstrated in the case of Western Himalaya (Section 6.3.5). It is interesting to observe that these gravity highs coincide with the maximum rainfall in this region (Fig. 6.42(a), Gogoi and Rabha, 2006; Luirei and Bhakuni, 2008) that makes it interesting for N-S channel flow due to orography. This implies uplift due to erosion caused by rainfall.

A large circular gravity low (Fig. 6.42(b)) has been reported from the Eastern Syntaxial bend (27.5°N, 95°E, Maochang et al., 1998; Gahalaut, 2008) that may represent some large feslic intrusive bodies under the alluvium cover extending northwards. This body might have played a definite role in the formation of this syntaxis.

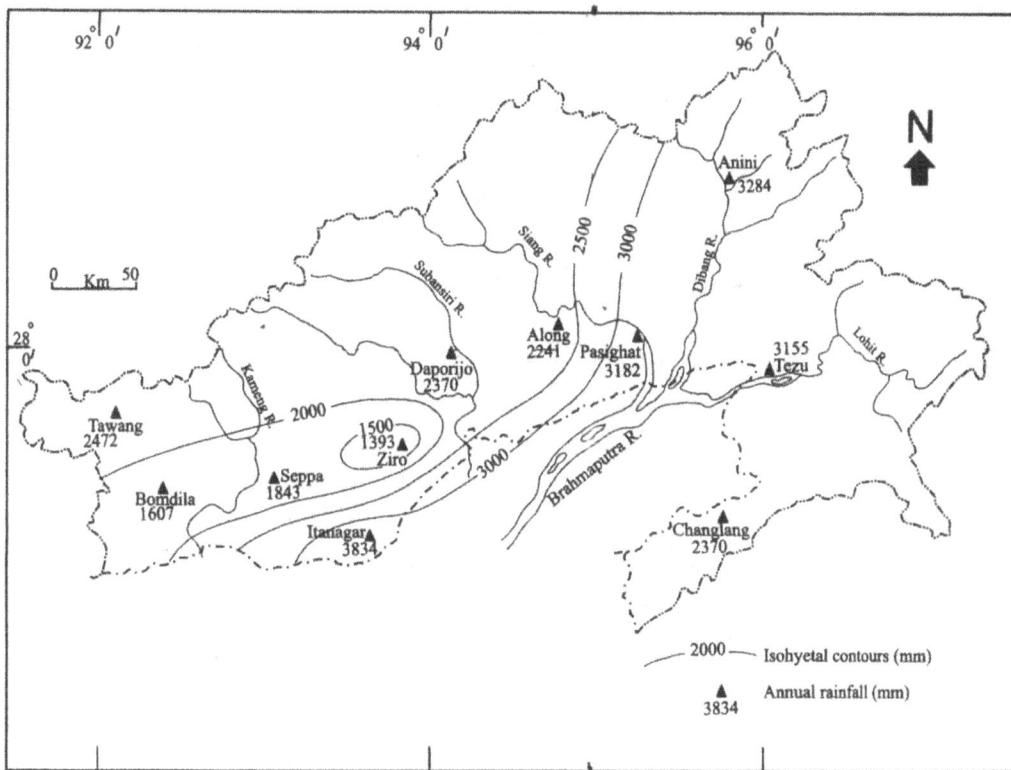

Fig. 6.42(a) Annual rainfall distribution in mm of Arunachal Pradesh and some major rivers of NE India (Gogoi and Rabha, 2006).

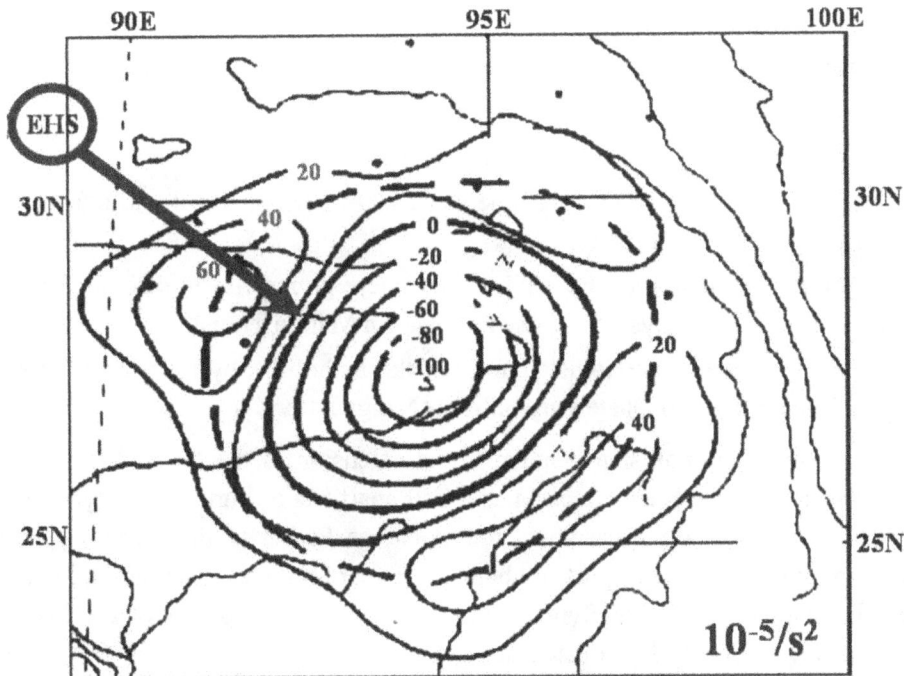

Fig. 6.42(b) A circular gravity anomaly associated with the Eastern syntaxial bend, EHS – Eastern Himalayan Syntaxis (Maochang et al., 1998, Gahalaut, 2008).

6.3   Western Fold Belt (Pakistan) and Syntaxis, and NW Himalaya – Crustal and Lithospheric Structures

6.3.1   Geology and Tectonics

The rocks in Western Syntaxis in general display a higher grade of metamorphism than Central Himalaya, indicating a change in tectonic style in this region. The burial and exhumation history in Western Himalaya are, therefore, different from other parts of the Himalayan orogen (Argles et al., 2003). Fig. 6.43 shows a detailed tectonic map of the western Himalayan Syntaxis showing Kohistan arc in the northern part that is an island arc of late Cretaceous, which accreted to the Asian plate before colliding with the Indian plate. It extends eastward as the Ladakh batholith. Singh et al., (2007) have dated a granodiorite sample near the Shyok suture and a diorite sample close to ITSZ from the Ladahk batholith using SHRIMP-II U-Pb Zircon. They have reported dates of 60.1 ± 0.9 Ma and 58.4 ± 1.0 Ma, respectively, that indicates that Ladakh batholith formed prior to the collision of the Indian and the Eurasian plates. Its contact with the Indian plate is known as the Main Mantle Trust which is synonymous with the Indus-Tsangpo Suture Zone of Central Himalaya (Fig. 6.5). Similarly, the western boundary of the Kohistan arc with the Asian plate forms the Shyok or Northern suture which is synonymous with the Bangong–Nujiang Suture of Tibet displaced by the Karakoram fault (Fig. 6.5). Kohistan arc is a north dipping sequence of rocks that is 30-40 km thick. It consists of the Kohistan batholith in the central part characterized by bimodal volcanics of diorites, granodiorites, and gabbros. Northern part of this arc consists of pillow lavas, island arc basalt, andesite, rhyolite, and Cretaceous sediments. The southern part of the arc consists of a mafic plutonic complex metamorphosed to granulite facies (Chilas complex). Towards the west and east of the Kohistan arc are the Hindu Kush and Karakoram Ranges, respectively. Towards the Hindu Kush Range, the Indian plate-Kohistan arc is underthrusting faster (Koulakov and Sobalev, 2006) compared to other sections giving rise to deep focus earthquakes (100-300 km).

The main events for the evolution of the Western Syntaxis can be summarized as follows (Khan et al., 1989).

1. Formation of island arc during late Jurassic-mid Cretaceous.

2. Closure of the back arc basin, towards north between 100-73 Ma, when it was attached to the Asian plate forming the northern suture.

3. Continued northward subduction of the Tethys lithosphere giving rise to calc-alkaline pluton of 60-40 Ma and closure of Tethys giving rise to the Main Mantle Thrust.

4. Divison of the arc into two parts; Kohistan and Ladakh batholiths, by further movement and intrusion of the Indian plate and Quaternary uplift of Nanga Parbat.

On the western side, Himalayan collision resulted in the extension of the Kabul block along the Chamman fault and formation of the Katawaz basin as a pull apart basin. Further convergence led to southward thrusting of Indian crystalline basement rocks. Nanga Parbat in the Eastern part of the syntaxis is characterized by rapid rates of uplift of about 1 cm/year (Nakata, 1989) and a high rate of surface denudation. The southern part of the Syntaxis yield Paleo-Proterozoic ages of 2.2-2.6 Ga, and represents the basement crystalline rocks thrusted along the MCT (Treloar et al., 2000). However, granite sheets on the NW margin of Nanga Parbat yield dates as young as 26 Ma (George et al., 1993), almost similar to leucogranites in this region.

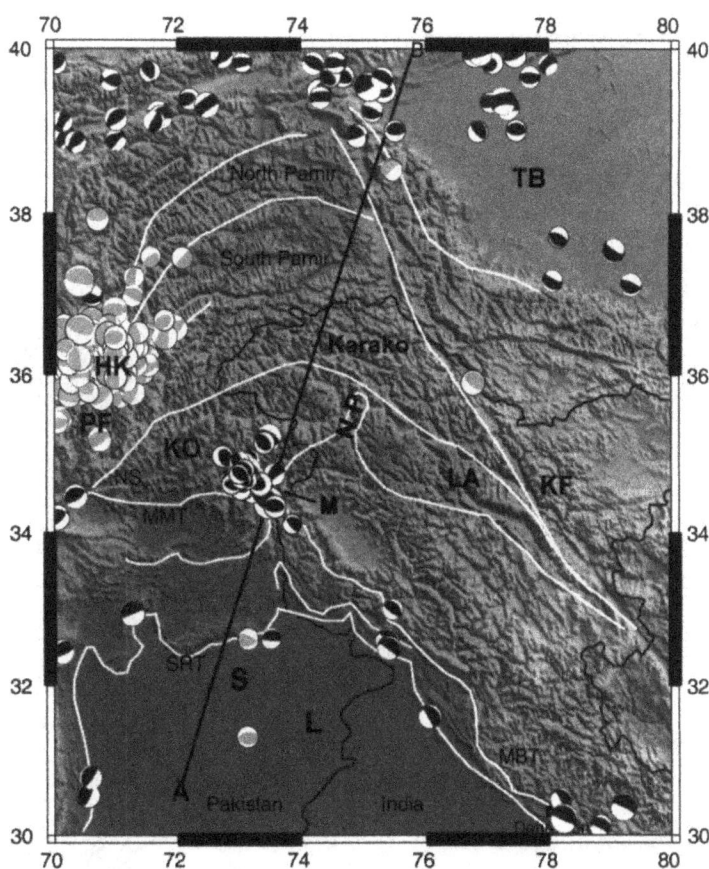

Fig. 6.43 Tectonics of the Western Himalayan Syntaxis with major thrusts and tectonic units marked on the topographic map. LS = Lahore-Sargodha Ridge; SRT = Salt Range Thrust also called Main Frontal Thrust; MBT = Main Boundary Thrust; MMT = Main Mantle Thrust or Indus suture; LA = Ladakh Batholith, KO = Kohistan Arc; NS = Northern Suture; NP= Nanga Parbat; KF = Karakoram Fault; KA = Karakoram Range; HR = Hindu Kush Range; TB = Tarim Basin; L = Lahore; S = Sargodha; PF = Panjshir Fault Epicenters of thrust earthquakes of magnitude > 3 since 1973 are plotted in red and deep focus earthquakes in yellow. M = Muzaffarabad (Kashmir) Earthquake of 2005 (Tiwari et al., 2008). [Colour Fig. on Page 784]

Fig. 6.44 (Karunakaran and Ranga Rao, 1976) shows a detailed geological map of a section of the Western Himalayan front that is reproduced here to familiarize readers with the kind of structures generally encountered along a section of the Western Himalayan front. This map is also reproduced here to explain the gravity map of this section that indicates neotectonic activities as discussed in the next section. It shows the Kangra and Dehradun re-entrants and the Nahan salient between them. Some seismic profiles, K2 and K4, and DS and DN in these re-entrants are discussed in the next section for shallow tectonics. Besides major thrusts, HFT, MBT, and MCT, this map shows several intermediate local thrusts and anticlines and synclines especially in the Kangra re-entrants. The Jawalamukhi thrust south of Kangra is of special interest as the Kangra earthquake (M >8.0) of 1905 (Fig. 6.15(a)) has occurred in this section. In the Nahan salient, where sub-Himalaya is narrow, Tertiary rocks are exposed in imbricated thrust sheets, while in broad sub-Himalayan sections (Kangra and Dehradun re-entrants) alluvium fills wide synclinal valleys (Duns). The geologic evolution of northern India is best recorded in the stratigraphic succession of the Zanskar range (North West Himalaya), which represents the most complete sequence through this ancient continental margin. After the onset of collision between India and Asia close to the Paleocene-Eocene boundary, obduction of the remnants of the neo-Tethys ocean floor on to the Indian margin began, and the latter underwent multi-phase deformation with fold and thrust shortening followed by heating and extension (Gaetani and Garzanti, 1991).

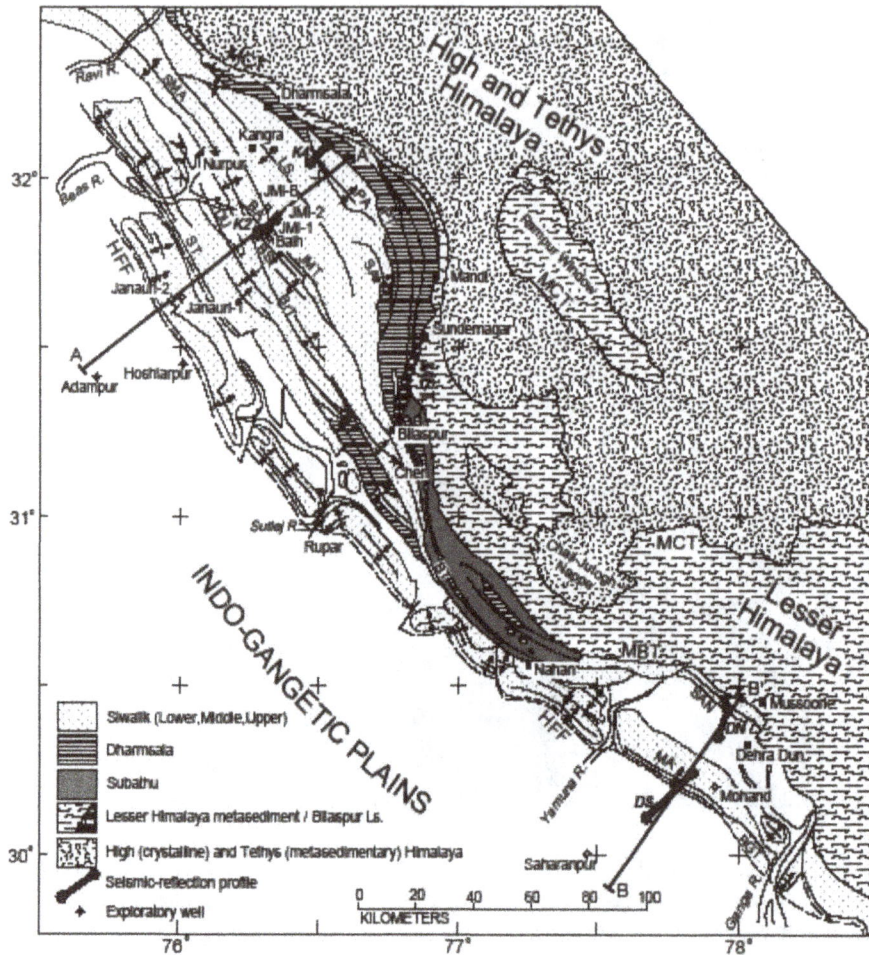

Fig. 6.44 A detailed geologic map of part of the N-W sub-Himalaya showing seismic profiles, and drill holes. Note the great variation in the width of sub-Himalaya (MBT to HFT) due to large part to the sinuous surface trace of the MBT, Structures: BA = Bath acticline; BGT = Bhimgoda thrust; BrT = Barsar thrust; BS = Balaru syncline; BT = Bilaspur thrust; DU = Dumkhar syncline; HFF = Himalayan Frontal fault; JMT = Jwalakukhi thrust; LS = Lambargaon syncline; MBT = Main Boundary thrust; MA = Mohand anticline; MCT = Main Central thrust; PA = Paror anticline; PT = Palampur thrust; SA = Sarkaghat anticline; SAN = Santaurgarh anticline; SMA = Suruin-Mastgarh anticline; ST = Soan thrust. Seismic-reflection profiles: DN = Doon-N; DS = Doon-S; K2 = Kangra-2; K4 = Kangra-4 profile (Karunakaran and Ranga Rao 1976; and Powers et al., 1998).

6.3.2 Seismic Studies

(i) Deep Seismic Studies (DSS) and Receiver Function Analysis

Seismic studies of the Pamir-Hindu Kush region in Western Himalaya are special as they are characterized by relatively deep focus earthquakes in the upper mantle. Tomographic image of the Indian lithosphere using P- and S-wave seismic anomalies (Kaulakov and Sobolev, 2006) suggest the presence of the Indian lithosphere throughout the upper mantle and transition zone, which breaks off at a depth of about 250 km. This breakoff might have occurred at the early stages of collision at about 44-48 Ma (Negredo et al., 2007). After this, the Indian plate continued its northward movement to the present Hindu Kush region where it starts subducting at a fast rate and high angle at about 8 Ma providing a brittle region in the upper mantle, which is prone to seismic activity. The fast rate of subduction indicates the subduction of high density rocks such as the oceanic crustal ophiolitic rocks or volcanic mafic rocks related to the Kohistan arc.

Kumar et al., (2005) have mapped the subducting Indian and Asian plates under the Tian Shan-Karakoram region based on S-wave receiver function analysis presented in Fig. 6.58 along with the density model based on the gravity data along the same profile. Accordingly, Moho lies at a depth of 50-70 km with maximum depth being under the Karakoram-Pamir region. The Indian lithosphere dip from 130 km to 170 km under the Karakoram ranges towards north while Asian plates dips from about 120 km under Tarim basin to 270 km under central Pamir and Karakoram ranges. This indicated that the Asian plate is underthrusting below the Indian plate in this region (Fig. 6.58, Lithospheric Part). A similar model of the Asian lithosphere thrusting under the Indian lithosphere has also been shown by Zhao et al., (2010). They have also shown the respective positions of the Asian and Indian lithospheres along four sections from the Eastern Himalaya upto Western Himalaya, which suggest a crush zone due to collision of two subducting lithospheres from the north and the south in the Eastern part. This section is marked by high temperature and low mantle seismic wave speed. Vinnik et al., (2007), based on joint inversion of P- and S-wave receiver function, also suggested almost similar depths for the underthrusted Indian lithosphere. They suggested fast directions of azimuth anisotropy changing at 160 km parallel to the Himalayan trend in the depth range of 160-220 km indicating the base of Indian lithosphere. They have also reported high S-wave velocity in the section suggesting foundering of lower crust rocks that might be related to interaction of underthrusted rocks with existing lower crustal rocks. Bhukta et al., (2006) and Bhukta and Tewari (2007) reprocessed the deep seismic sounding data (Kaila et al., 1978) along profiles in NW Himalaya (Fig. 6.45) and suggested a detailed crustal model providing a maximum crustal thickness of about 65 km under MMT north of Hazara Syntaxis (Fig. 6.43). They also suggested a thin low velocity layer under the Higher Nanga Parbat (Fig. 6.45).

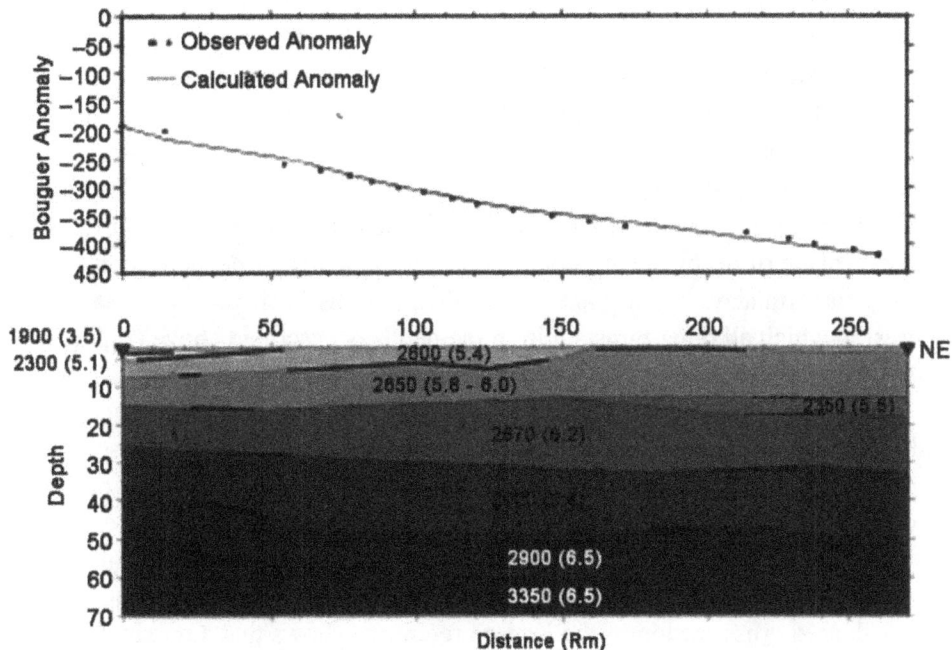

Fig. 6.45 Gravity profiles from Lawrencepur to Astor east of Nanga Parbat across Hazara syntaxis and results of seismic section along the same profile showing a low density and low velocity zone under the Nanga Parbat west of Astor (Bhukta, Personal Communication). This may represent fluid under Nanga Parbat as envisaged based on detailed seismic investigations.

Based on teleseismic receiver function analysis along a profile, Rai et al., (2006) suggested deepening of Moho northwards up to 75 km under the Karakoram fault as has been suggested above from DSS studies. Wittlinger et al., (2004) had suggested further deepening of Moho up to 90 km beneath western Tibet which shallows down to 50-60 km under the Altyn Tagh fault. Oreshin et al., (2008) based on integrated seismic studies such as P and S receiver functions, teleseismic P and S residuals, and shear wave splitting in SKS, suggested a crustal thickness of 65 km in NW Himalaya under Ladakh batholith and Western Tibet. These studies also suggested lack of a low velocity layer in the middle crust indicating absence of fluids or partial melt as has been found from magnetotelluric studies as described below. This can be attributed to lack of such zones along this profile or to large gaps in recording stations in Tethyan Himalaya north of STD. They also delineated high velocity in 200 km range of the upper mantle that was attributed to the subducted Indian mantle lithosphere. Nanga Parbat being an active metamorphic massif, its seismic characterization was done in detail by Meltzer et al., (2001) based on dense seismic array around it. They suggested a sudden drop in microseismicity, with depth in the massif indicating a shallow transition between brittle failure and ductile material. Low seismic velocities are observed at the core of the massif that extend throughout the crust. The main seismic zone and low velocities correlate to high topography, suggesting the role of high temperature and ductile flow in rapid exhumation that suggests the rapid rate of uplift in case of Nanga Parbat, and has indicated the role of channel flow south of the suture zone.

(ii) Seismic Sections and Gravity Anomalies in NW Sub-Himalaya

Fig. 6.44 shows the geology of Kangra and Dehradun re-entrants, and the Nahan Salient. It also shows the layout of seismic profiles K2 and K4 in Kangra re-entrant, and DS and DM in Dehradun re-entrants. Interpreted section up to the basement along profile K2 in Kangra re-entrant and DS in Dehradun re-entrants are given in Fig. 6.46(a) and (b) (Powers et al., 1998). These sections are chosen as they are constrained from deep boreholes along these sections as shown in respective figures. Fig. 6.46(a) shows Bath anticline formed due to thrusts, the prominent being south verging Jawalamukhi thrust between the HFT and the MBT. The Jawalamukhi borehole along this section is the deepest borehole in the Ganga basin drilled by ONGC, Dehradun for oil and gas exploration. Basement is identified as Pre Tertiary Vindhyan sediments of Meso-Proterozoic period of North Indian Shield (Section 7.6) at a depth of about 9-10 km that forms a plane of decollement, along which the Indian crust decouples and the lower part thrusts under the Himalaya. This plane of decollement has been referred to as the Main Himalayan Thrust to which all other thrusts join in the depth as described above (Fig. 6.24). It is interesting to note that the basement (plane of decollement) also shows an upwarp under the anticline that might be related to flexure of the Indian plate. Fig. 6.46(b) shows shallow structures under the Dun valley that is constrained by Mohand bore well. It shows south verging Himalayan Frontal Thrust SW of Mohand. It also shows Pre-Tertiary basement of Vindhyan rocks in the north. In the southern part, the basement is referred to as the Aravalli/Delhi Group of rocks that are exposed in Western India (Section 7.5). The Sub-Himalayan decollement dips 2.5°N beneath the Kangra re-entrant but is steeper by 6° beneath the Dehradun re-entrant (Powers et al., 1998). A balanced cross section of the Kangra re-entrant shows that a maximum of 23 km shortening has occurred since 1.9-1.5 Ma yielding a shortening rate of 14 ± 2 mm/yr, while in Dehradun re-entrant it is at a rate of 6-16 mm/yr. This indicates that about 25% of total India-Eurasia convergence (~ 48 mm/yr) is accommodated in Sub-Himalaya (Powers et al., 1998). This is also supported by GPS studies as discussed in Section 6.4.

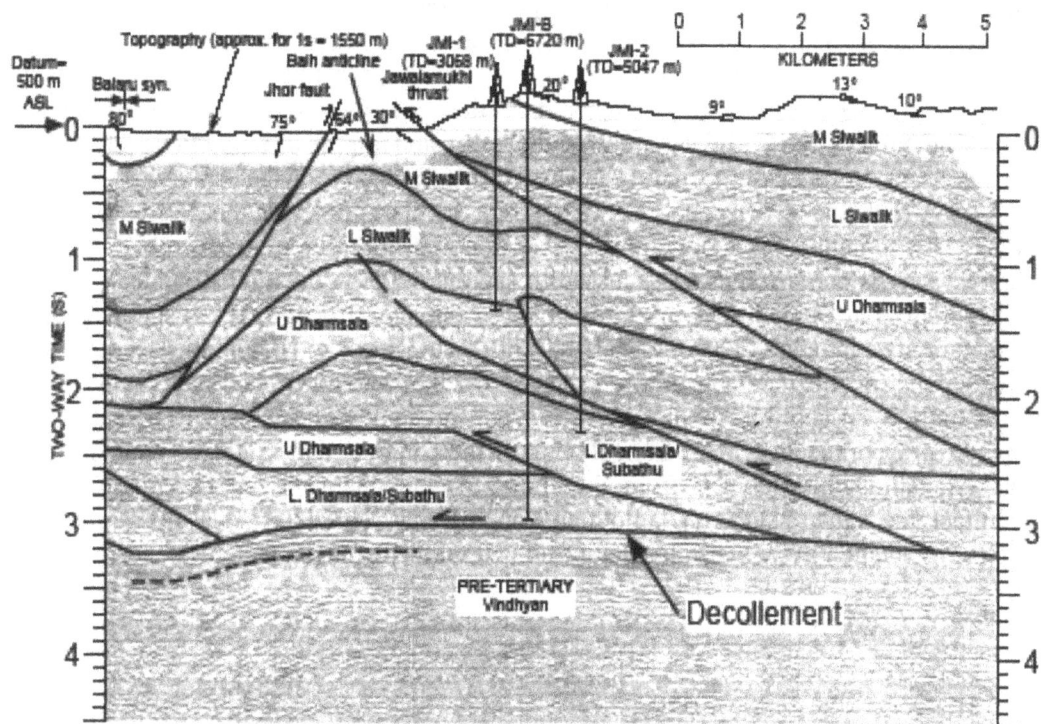

Fig. 6.46(a) Interpreted seismic reflection profile in Kangra re-entrant (Line K2, Fig. 6.44) showing detailed structures of the sedimentary section. It shows the Bath anticline developed due to the Jawalamukhi thrust and Jhor fault. It also shows basement (decollement) upwarp under anticline and downwarp under syncline (Powers et al., 1998).

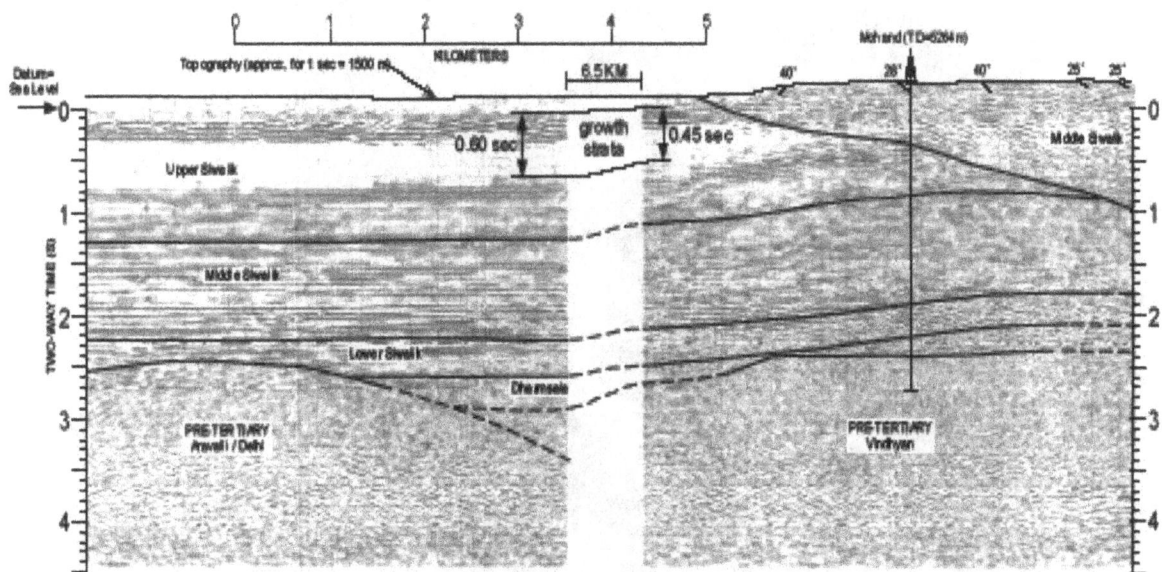

Fig. 6.46(b) Interpreted seismic section in Dun valley (DS, Fig. 6.44) showing a relatively less disturbed section compared to Kangra valley. It shows basement of Aravalli/Delhi Group of rocks in the southern part, while in the northern part, it is Vindhyan rocks (Powers et al., 1998).

A deep seismic section across the southern part of the Kangra re-entrant is given in Fig. 6.47(a) (Tewari, 2007; Reddy, 2010) that shows reflections R1, R2, R3, and Moho. Reflections R1 and R2 with 4.5 and 6.5 seconds TWT may represent the top of the Proterozoic and crystalline basements at depths of 4-5 km and 6-7 km, respectively, that is almost the same as has been inferred from the previous profiles. The Proterozoic basement implies the extension of the Indian shield under the sediments of Ganga basin. In this case, it may be related to the extension of the Aravalli-Delhi fold belt (Fig. 6.5 and 2.26). Moho at 14.5 Seconds TWT suggest a depth of 40-45 km that lies between the HFT and the MBT. Due to undulating topography, the section appears to be quite disturbed. Gravity map of this section vis-à-vis neotectonics is discussed in section 6.3.5. A detailed Bouguer anomaly map of the Ganga foredeep in Dehradun re-entrant is given in Fig. 6.47(b) (Rao, 1973) showing large amplitude negative gravity anomalies along the Mohand Thrust equivalent to HFT in this section. This indicates thick sediments in the foredeep of the order of 6-7 km as shown in Fig. 6.46(b) small amplitude gravity high is descernible to the west of Deoband and west of Muzaffarnagar (H1) extending up to west of Roorkee as evident from flexing of the contours that might be related to the extension of the Aravalli-Delhi fold belt from the Indian Shield characterized by gravity highs (Section 6.3.5). In fact, the Nahan Salient and adjoining Kangara and Dehradun re-entrants might have developed due to the interaction of the Aravalli-Delhi fold belt with Himalayan tectonics.

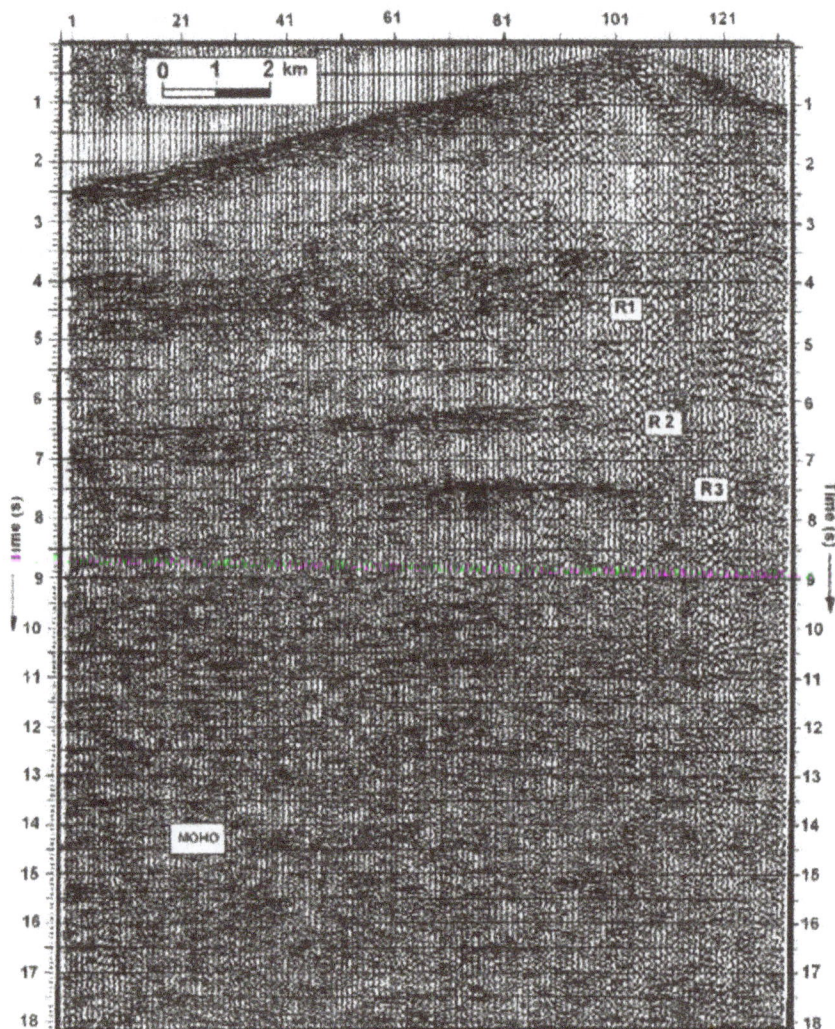

Fig. 6.47(a) A sample seismic record across southern part of Kangra re-entrant suggesting a basement depth of 6-7 km (R2) and Moho depth of 40-45 km between the HFT and the MBT (Reddy, 2010).

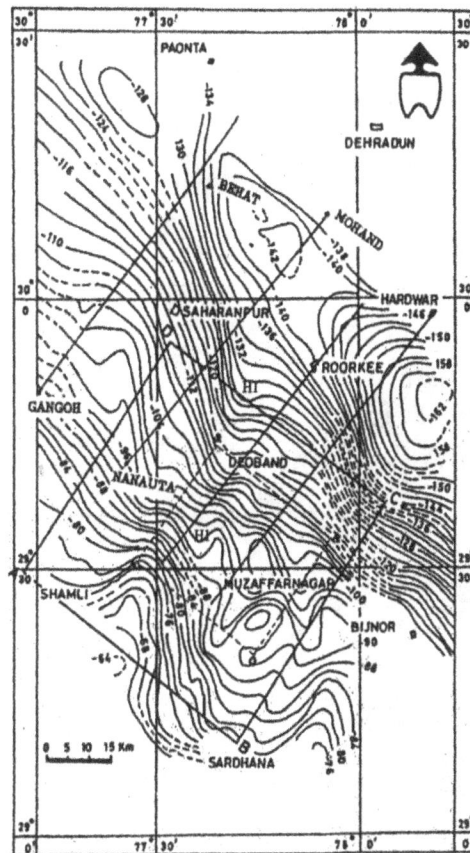

Fig. 6.47(b) Bouguer anomaly map of a part of sub-Himalayan Ganga basin south of Dun valley. It primarily shows gravity lows due to sediments but there are changes in the contour pattern delineating basement structures that can be highlighted by separating residual anomalies (Rao, 1973).

6.3.3 Magnetotelluric (MT) Studies

Magnetotelluric measurements provide conductivity with depths which can be used to identify the rock types and, on that basis, the geodynamics of the region. The first such experiment in this respect was carried out using magnetic arrays that defined a conductor in NW India at a depth of 32 km (Chamalaun et al., 1987) extending to Himalayan foothills near Dehradun. Rao and Prasad (2001) attributed this high conductive zones to serpentinization which usually occurs in subduction zones due to interaction of fluids with ultramafic rocks, peridotite. Long period MT measurements were carried out along a profile Leh to Panamik across ITSZ in NW Himalaya. Fig. 6.48 (Banerjee and Satya Prakash, 2003a) shows the layout of this profile with respect to regional geology and tectonics. It provided a conductive body with a resistivity of 5-10 Ω m under ITSZ and to its north at a depth of 20-25 km, which was attributed to presence of fluids along the underthrusting Indian plate or partial melt at this level (Fig. 6.49, Arora et al., 2007). It might be synonymous with those observed under southern Tibet along INDEPTH profile in central Himalaya as described above in Section 6.2. Magnetotelluric studies in Puga valley, Ladakh located just south of ITSZ suggest highly conductive zone ($\sim 50 \Omega$ m) at a shallow depth of 2 km that correlates with high heat flow, and thereby represents the presence of fluids. Temperature logs indicated high temperature (~ 260 °C) associated with conductive zones that suggests a potential geothermal source of energy in this region (Azeez and Harinarayana, 2007). There are several hot springs along NW Himalayan front of temperature 50-100 °C indicating high thermal gradient in this section and pools of these water may be responsible for high conductivity that can be harnessed for geothermal energy. That also explains high seismic activity in this section (Section 6.3.5).

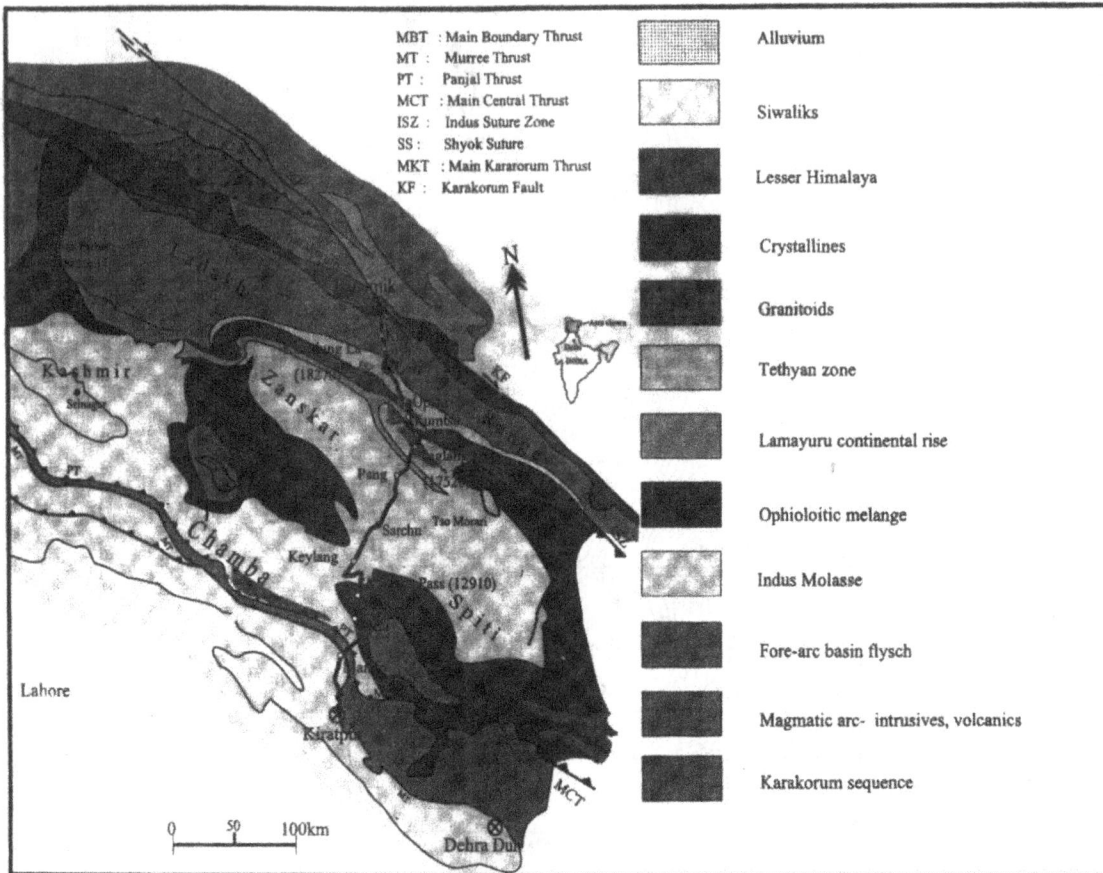

Fig. 6.48 Layout of HIMPROBE profile, the Kiratpur-Leh-Panamik in NW Himalaya superimposed over local geology and tectonics for ready reference (Bannerjee and Satya Prakash, 2003a).

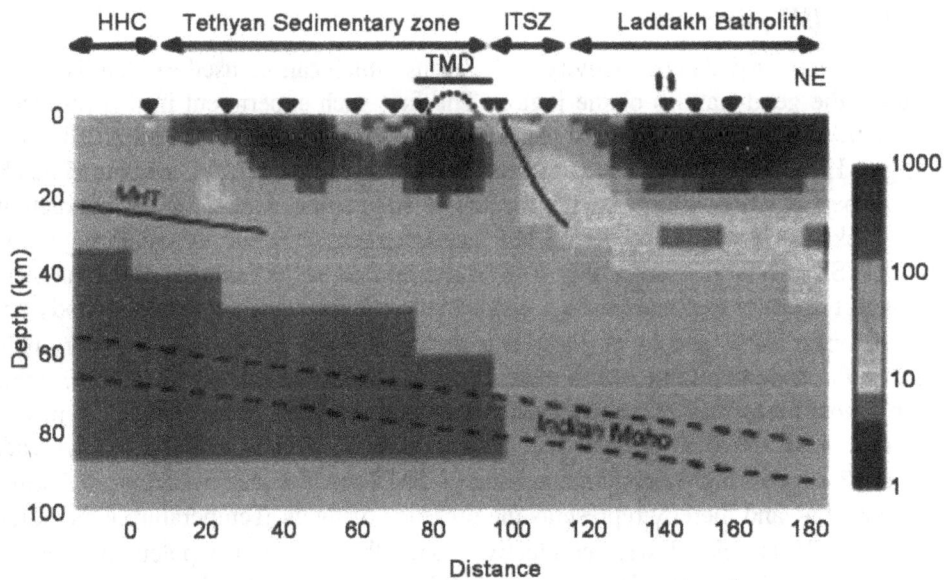

Fig. 6.49 Conductivity profile across suture zone (ITSZ) along the Leh-Panamik profile (Fig. 6.48) showing a conductive body over the subducting Indian plate that lies at a depth of 20-25 km north of the suture zone. (Arora et al., 2007). [Colour Fig. on Page 784]

6.3.4   Gravity Studies and Himprobe: Crustal and Lithospheric Structures and Seismicity – Kashmir Earthquake and Deep Focus Earthquake, Hindu Kush

The earlier measurements of the gravity field by Das et al., (1979) in the NW Himalaya was quite useful as they were used by several workers (Verma and Subrahmanyam, 1984; Verma and Prasad, 1988) to provide first hand information about crustal structure in this region. One of earliest gravity measurements in NW Himalyaya is due to Marussi and Ebblin (1976). Gravity data from this map (Fig. 6.50(a)) along a seismic profile (Finetti et al., 1979) from the Nanga Parbat to Alai Ranges across Karakoram and Pamir Ranges was modeled by Mishra (1982) constraining it from seismic section (Fig. 6.50(a)) that provides a crustal thickness of 58 km under the Nanga Parbat to a maximum of 70-72 km under the Karakoram-Pamir Ranges further reducing to northwards under the Trans-Alai ranges and increasing to 65-66 km under the Alai Ranges due to convergence. This crustal model also provided a low velocity zone in the lower crust at a depth of 35-40 km that may be synonymous with the conductive layer of fluids/partial melt under southern Tibet. Shallow low density bodies under the Karakoram-Pamir ranges are due to low density felsic rocks in these sections. Another gravity profile modeled by him (Mishra, 1982) is from Srinagar to Nanga Parbat-Harmosh massif (Fig. 6.50(b)) that shows a change in crustal thickness from 55-72 km, with a low density layer in the lower crust similar to the earlier profile. The shallow low density body represents the low density Nanga parbat-Harmosh massif.

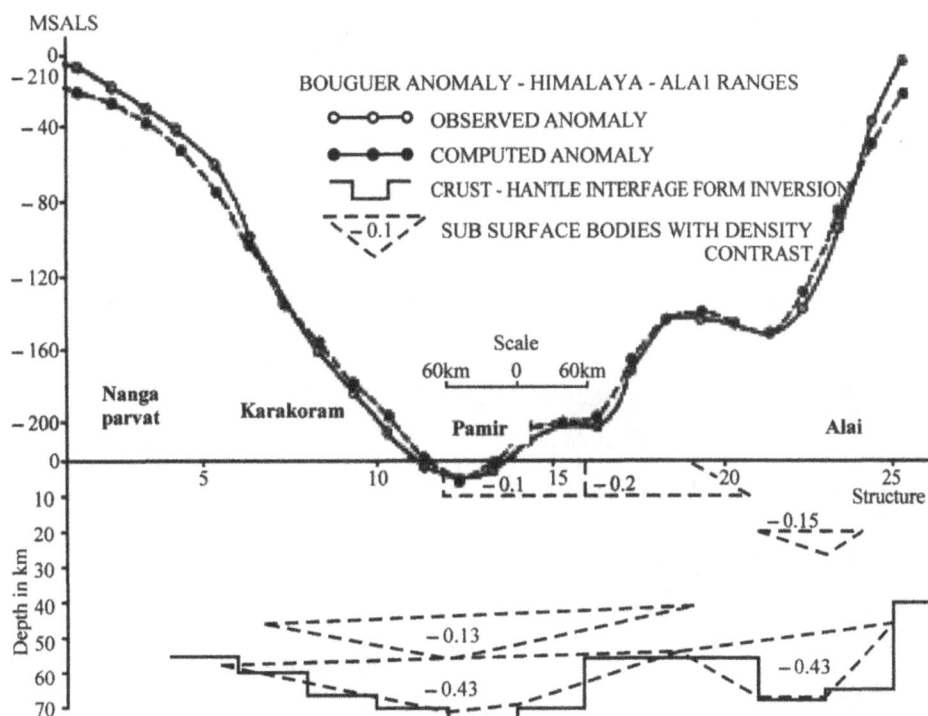

Fig. 6.50(a) The crustal model along the Nanga Parbat-Alai ranges showing crustal thickening under the Karakoram-Pamir section and again under trans-Alai ranges. The low density body at midcrustal level may indicate presence of partial melts or low density bodies that may be the root of the Karakoram-Pamir intrusives (Mishra, 1982).

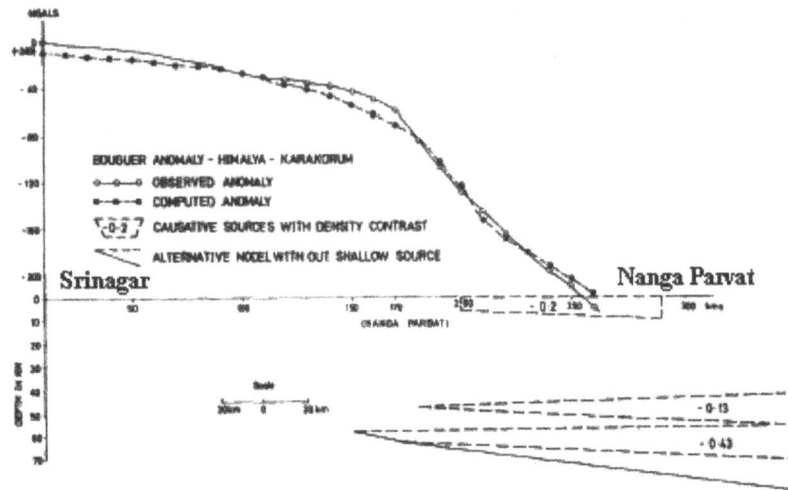

Fig. 6.50(b) A gravity profile from Srinagar (Kashmir) to Nanga Parbat showing crustal thickening, a low density body at the surface representing Naga Parbat massif, and low density layer in the lower crust that may represent fluids as partial melts in this section and is synonymous to low velocity in seismic profiles as described above (Mishra, 1982).

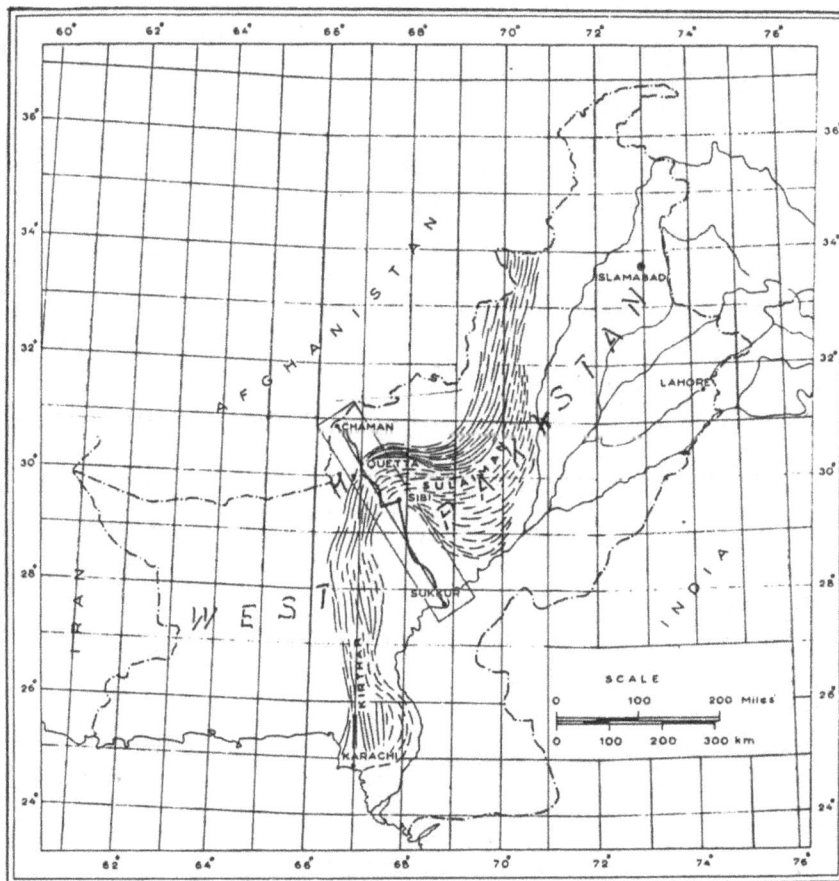

Fig. 6.51(a) Sibi syntaxis of the Western Fold Belt, Pakistan across Kirthar-Sulaiman ranges, Quetta-Chamman Plateau and sibi syntaxial Bend.

(i) Western Fold Belt, Pakistan

Some of the gravity studies reported from Pakistan are described below for a comparison with the observed gravity field in Central and NW Himalaya. The most important tectonic unit in this section is the Chamman fault, which is basically a strike slip fault connecting the western boundary of the Indian plate in the Arabian Sea with those in the northern part (Fig. 6.5). The Sibi Syntaxial bend (Fig. 6.51(a)) just to its east is an important element of the orogenic belt in this section that is the junction of the Kirthar and Sulaiman ranges. A gravity profile across this section (Fig. 6.51(b)) is described by Rahman (1969) and shows a decreasing gravity field over the orogenic belt reaching to a minimum of about 260 mGal over the plateau section of Quetta and chamman. As in case of Central Himalaya and Tibet, the observed gravity shows an opposite correlation with elevation, indicating a thick crust due to isostatic compensation under the Quetta-Chamman section. Modeled crustal section along this profile shows a thick crust of about 60 km in this section for a density contrast of − 0.45 g/cm^3. It may be noted that the maximum elevation, negative gravity anomalies, and crustal thickness are less in this section compared to northern Himalaya and Tibet (Fig. 6.35(a)). The model also shows some high and low density shallow bodies to account for short wavelength and small amplitude anomalies observed over the plateau section, indicating mafic and felsic intrusive bodies. A regional gravity map of the western section along the Chamman fault is given in Fig. 6.52(a) (Guillaume, 1978) that covers Pakistan-Afganistan section of the Western Fold Belt. It shows a small gravity high to the east of Chamman fault, which may represent thrusted high density rocks during collision of the Indian and Afganistan blocks. The gravity highs along this regional profile (Fig. 6.52(b)) are caused by thrusting of high density rocks along the Chamman fault; their bulk densities (3.05-3.10) match those of the ophiolites representing oceanic crustal rocks along the Chamman fault that is east verging. Guillaume (1978) compared it to the Ivrea zone in Italy that is a collision zone in the Alps representing thrusted upper mantle rocks along a strike slip fault similar to the Chamman fault. A detailed gravity anomaly map of Quetta (Fig. 6.51(a)) and Mastung Valleys (Fig. 6.53(a) and (b)) at the western margin of the orogenic belt in this section are published by Mufti and Siddiqui (1961). They normally show circular/semicircular residual gravity lows − 2 to − 12 mGal typical of felsic/granitic intrusives. They might be related to collision related magmatism as has been described in case of South Tibet (Fig. 6.5), just north of the suture zone (ITSZ). Even the regional gravity profile (Fig. 6.52(b)) shows gravity lows typical of granitic intrusives SE of the Chamman fault.

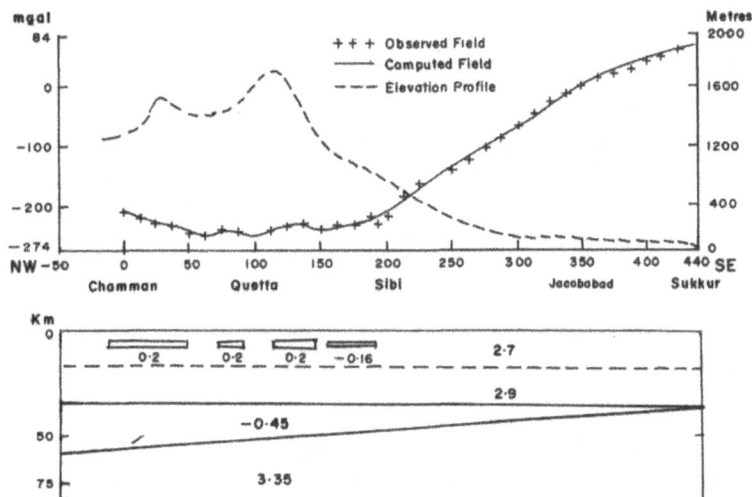

Fig. 6.51(b) A gravity profile across the Sibi Syntaxis (Rahman, 1969) and modeled crustal section with crustal thickness as 65 km and some exposed high and low density bodies indicating intrusives.

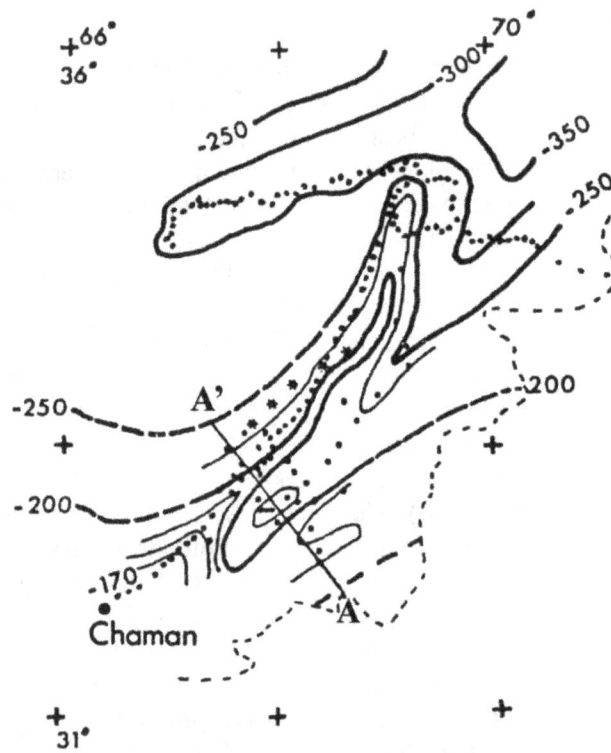

Fig. 6.52(a) Bouguer anomaly map of the Quetta plateau and Chamman fault showing gravity low over the plateau and relative high to its west along the Chamman fault.

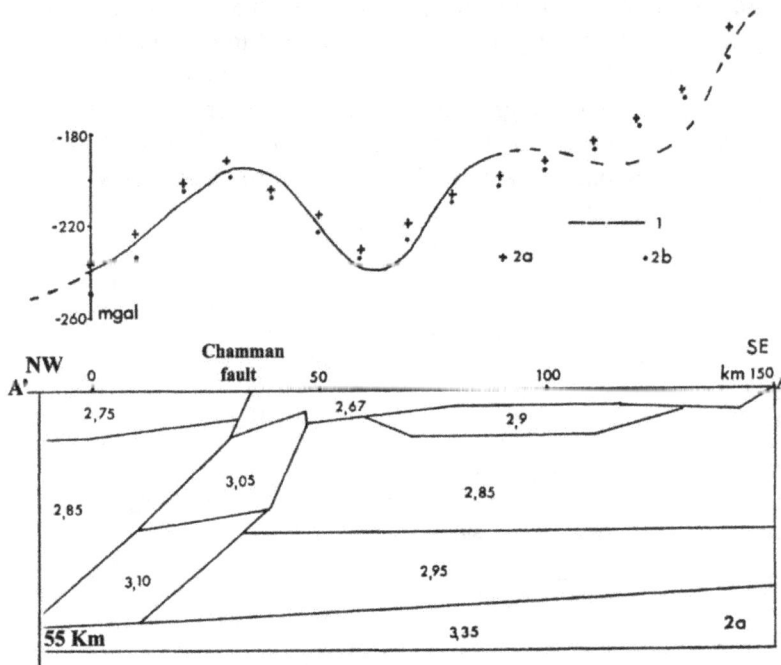

Fig. 6.52(b) Modelling of gravity profile AA' showing the gravity high being caused by east verging thrust of high density rocks that may be ophiolite rocks occuring in this region, (Guillaume, 1978).

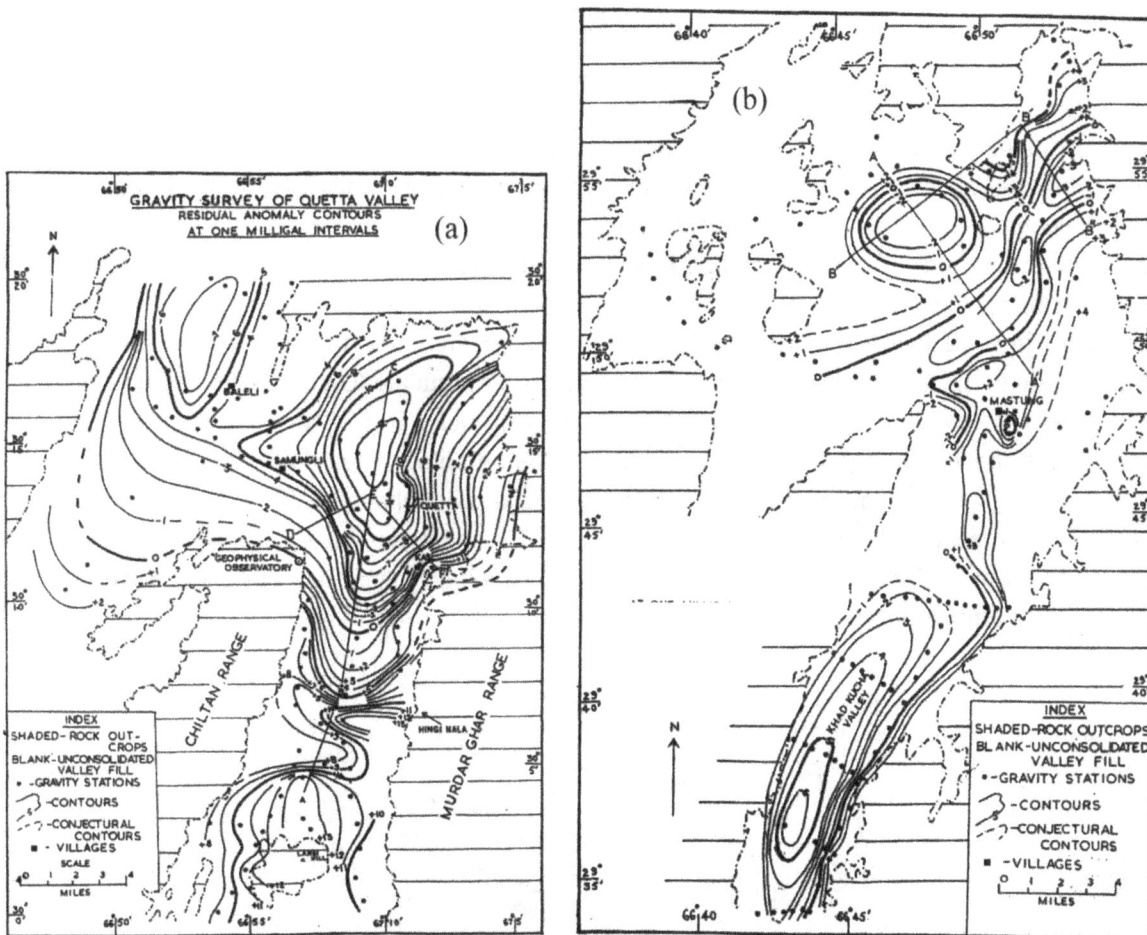

Fig. 6.53 (a) Residual Bouguer anomaly map of Quetta valley that shows typical semicircular gravity highs and lows related to mafic and felsic intrusives associated with collision of the Indian and Eurasian plates in this sector, such as ophiolites and ophiolitic mélanges (Mufti and Siddiqui, 1961). (b) Residual Bouguer anomaly of Mastung valley south of Quetta that also shows similar gravity anomalies as in the Quetta valley (Mufti and Siddiqui, 1961).

The other important feature of the Western Fold Belt is the occurrence of ophiolite rocks such as Muslim Bagh ophiolite along the Chamman fault that represents oceanic crustal rocks (Fig. 6.5). A detailed Bouguer anomaly of this ophiolite section is given by Farah and Zaigham (1979). An E-W geological section (Fig. 6.54(a); Ahmad and Abbas, 1979) across this ophiolite belt is quite revealing. It shows oceanic crustal rocks thrusted on Cretaceous and Jurassic sediments indicating that these rocks must have been thrusted during early Tertiary. This section shows all details of an oceanic crust such as ultramafic tectonite, dolerite dykes, intercalated sediments, and sheeted dykes in the center surrounded by cumulate gabbro and cumulate ultramafics as described in section 5.3.3 (Table 5.2). At least two kinds of igneous activity have been recognized from the Muslim Bagh area, namely the tholeiitic and the alkaline rocks. The former is represented by the tholeiitic group of Bagh complex and dyke swarms, which show MORB like characteristics and are formed near mid-oceanic ridges after the breakup of Gondwanaland. These are mainly considered to belong to the Late Cretaceous that is contemporary to the Deccan trap in Western India that evolved during the breakup of Seychelles from the Indian continent

(Chapter 5). The latter under alkaline group of rocks are represented by alkaline rocks of Bagh complex and Bibai volcanics. These rocks mostly belong to Early Cretaceous and are similar to those formed by a plume hotspot during the breakup of Gondwanaland. Paleomagnetic measurements of basaltic rocks of Bagh complex from the Muslim Bagh area suggest that the direction of magnetization (Inclination = 46.3 and – 53.4°) is almost similar to those reported for late Cretaceous in this area such as the Deccan trap indicating similar relative ages. It also suggests that these rocks formed at about 27-34°S latitude (Yoshida et al., 1992). The Bouguer anomaly of this region (Fig. 6.54(b)) shows gravity highs, since these are mostly high density rocks except in some sections (tectonites) that shows relative gravity lows such as anomaly G1. Maximum gravity highs, G2 are observed over the ophiolite acumulate (Gabbro-ultramafics) that show highest density (3.0-3.3 g/cm^3) among oceanic crustal rocks. It may be noted that sheeted dykes do not show much gravity anomalies due to their limited extent. Another prominent gravity anomaly, G3 is observed over ophiolitic mélange. Rahman (1967) has reported semicircular magnetic (400 nT) and gravity (– 12 mGal) anomalies near Gujranwala in the northern part of the Indus foredeep east of the fold belt, which may represent granitic intrusives similar to those reported from the Quetta-Mastung valleys as described above.

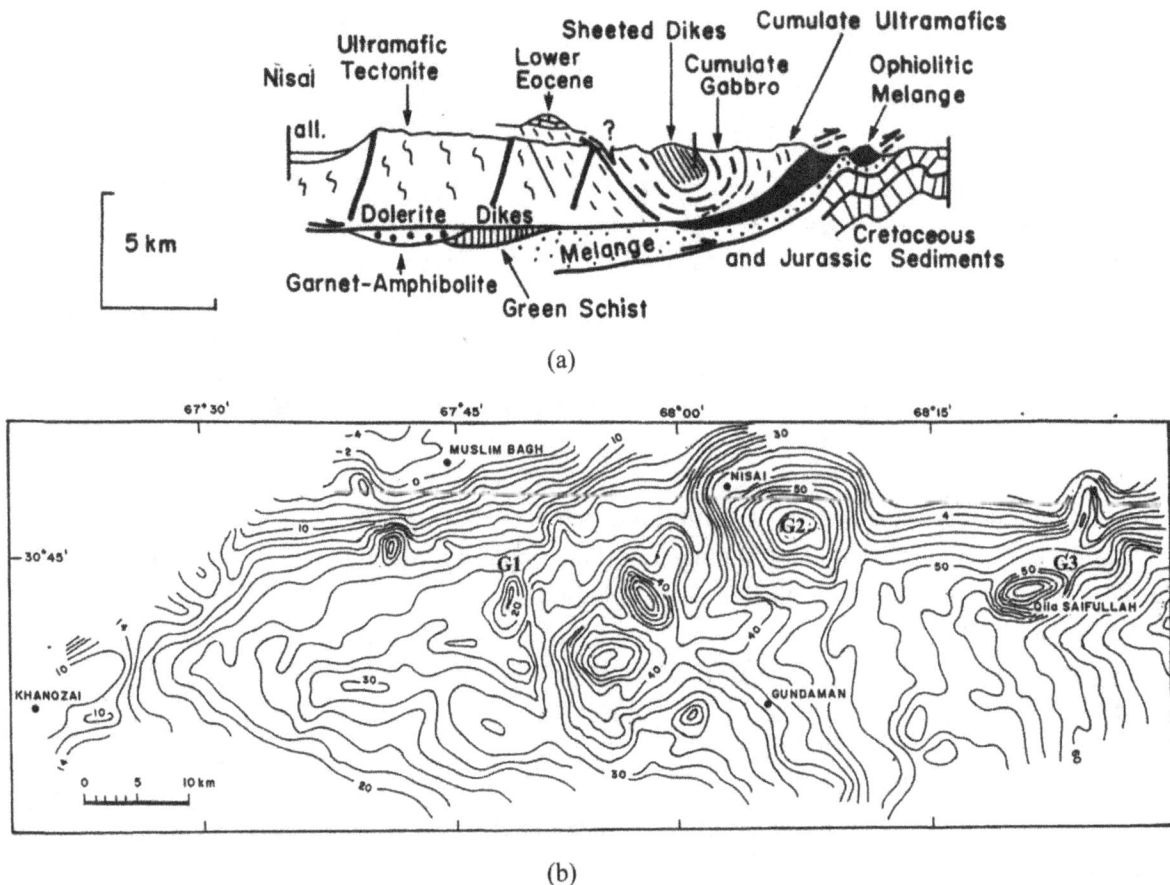

(a)

(b)

Fig. 6.54 (a)A typical cross-section of the Muslim Bagh opiolite (Ahmad and Abbas, 1979) that shows a typical section of the oceanic crust. (b) Bouguer anomaly of Muslim Bagh ophiolite primarily showing gravity highs (Farah and Zaigham, 1979).

Gravity anomaly map of Cholistan east of the Indus River along the Indian border also shows several circular/semicircular gravity highs and lows of about 10-20 mGal amplitude (Mirza, 1973) similar to those found in NW India (H3, Fig. 6.14). These are basement anomalies related to mafic and felsic intrusives, respectively in the basement. The gravity highs due to mafic intrusives in the basement are also confirmed by some large amplitude airborne magnetic anomalies of 100-1000 nT west of Fort Abbas that have been attributed to basement at depths of 4-6 km (BGR, 1992). Even gravity map SE of Karachi (Hyderabad Division), just north of Kutch (Farah and Jafree, 1965), also provides circular/semicircular highs of amplitude 10-20 mGal indicating that the entire section of the Indus foredeep and Western Rajasthan to its east along the Western Himalaya are affected by such mafic intrusives.

(ii) Western Himalayan Syntaxis and Seismicity – Kashmir Earthquake of 2005

A regional Bouguer anomaly map of the Western Syntaxis has been described by Caporali (2000) who also provided an effective elastic thickness of 50 km along the profile extending from the Indian to the Asian plates. He also suggested surface and subsurface folding with a typical wavelength of 100-130 km. Most of the thrusts/faults like MBT, and MCT, have developed close to maxima and minima of this folding where curvature is maximum, and there is a greater likelihood for development of faults. According to him, the upper brittle crust under the Himalaya has a thickness of 20-30 km followed by a ductile channel that extends up to Moho. The upper mantle up to a depth of 100-120 km is again a strong layer, favouring a jelly sandwich model as described above in Section 6.2. Based on the Bouguer anomaly map of the Western Syntaxis presented by him and regional Bouguer anomaly map of the whole region (UNESCO, 1976), a detailed gravity map of a larger section was prepared (Fig. 6.55(a), Mishra and Rajasekhar, 2006). This map also shows the epicenter of major earthquakes (M > 6) that have occurred in this region since 1973. They are primarily concentrated in three zones viz., Indus-Kohistan Seismic Zone (IKSZ) along Hazara Syntaxis also known as the Hazara-Kashmir Seismic Zone (HSZ), and Hindu Kush Seismic Zone (HKSZ). Some seismic activity is also reported from the Pamir ranges over the Asian plate. They are mostly crustal except those in the Hindu Kush Seismic Zone which originate in the upper mantle (100-300 km) as described in section 6.1.10. It is interesting to note that IKSZ coincides with the gravity gradient of high towards the south and low towards the north along MBT and MMT, while HKSZ coincides with a large amplitude gravity low. A spectrum versus wave number plot (Fig. 6.55(b)) suggests 4-5 levels of sources at average depths of 127, 60, 37, 27, and 14 km, representing upper mantle, Moho, lower crust, and upper crustal sources. The first two segments related to upper mantle and Moho are important as they indicate part of the subducted Indian lithosphere and crustal thickening in this section as inferred above in Section 6.1 and 6.2 from other geophysical studies. A residual anomaly map based on wavelength filter with wavelength less than 500 km (Tiwari et al., 2008; Fig. 6.56) provides a set of gravity highs and lows which reflect the tectonics better than the original Bouguer anomaly map. It shows gravity high, H1 related to the Lahore-Sargodha Ridge, which in fact extends southwards up to Delhi as described in Section 6.1. It is associated with the Salt Range Thrust that is equivalent to the Himalayan Frontal Thrust in the NW Himalaya. The gravity high, H2 coincides with the Zanskar Range that is associated with high density rocks. The gravity high, H3 is associated with the Kohistan arc that consists of high density pillow lavas, island basalt, and Chilas volcanics as described in Section 6.3.1 The gravity high, H4 is associated with the Tarim basin, which consists of several igneous intrusives with large magnetic anomalies in CHAMP satellite data (Section 6.2.5) suggesting their mafic/ultramafic nature causing this gravity high. The gravity lows, L2 and L3 coincides with Karakoram and Hindu Kush batholiths, suggesting their felsic composition.

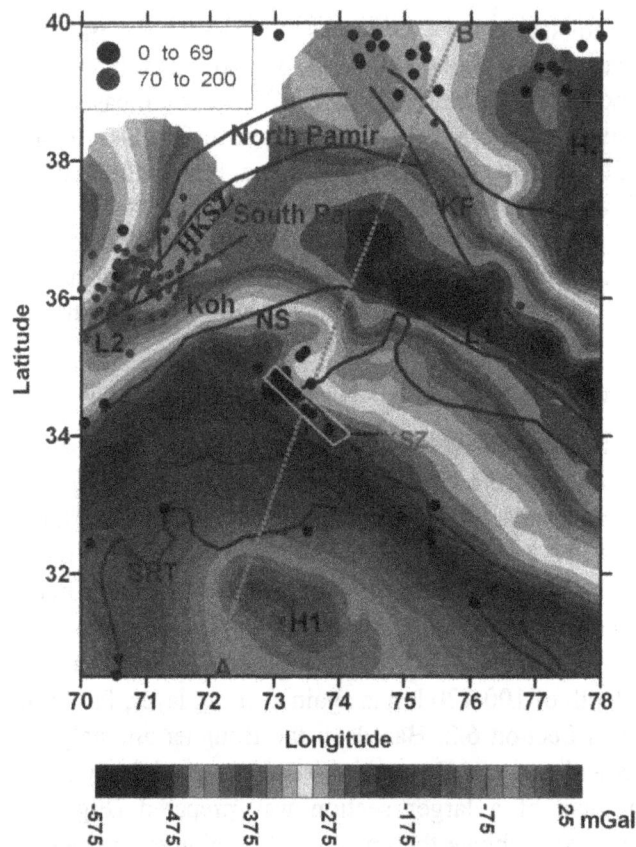

Fig. 6.55(a) Bouguer anomaly map of the Western Syntaxis showing major thrusts and epicenters of thrust earthquakes (black dots) and deep focus earthquakes (red dots). H1 and H2 are gravity highs over thrust belts and L1 and L2 are gravity lows over the Karakoram-Hindu Kush ranges. IKSZ = Indus-Kohistan Seismic Zone and HKSZ = Hindu Kush Seismic Zone (Mishra and Rajasekhar, 2006). Kashmir earthquake of 2005 coincides with the IKSZ (Fig. 6.43). [Colour Fig. on Page 785]

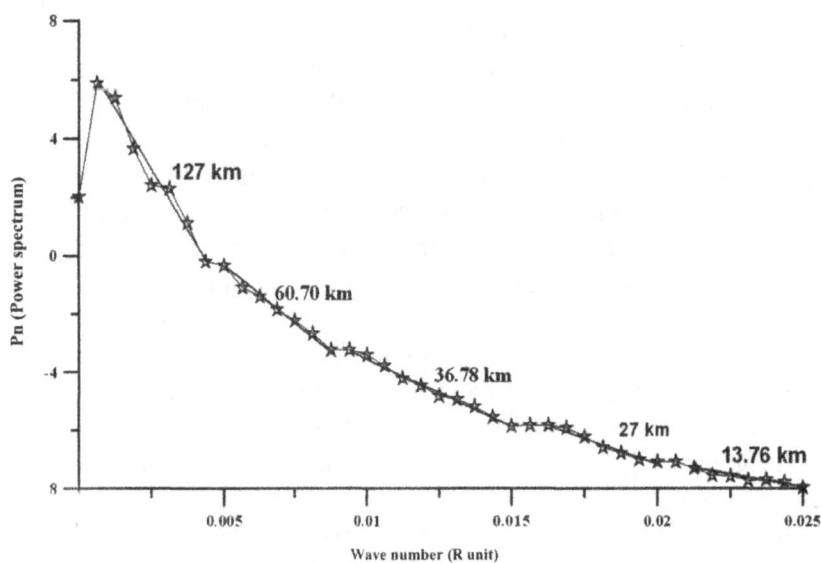

Fig. 6.55(b) Spectrum versus wave number for Bouguer anomaly of the Western Syntaxis and surrounding regions (Mishra and Rajasekhar, 2006).

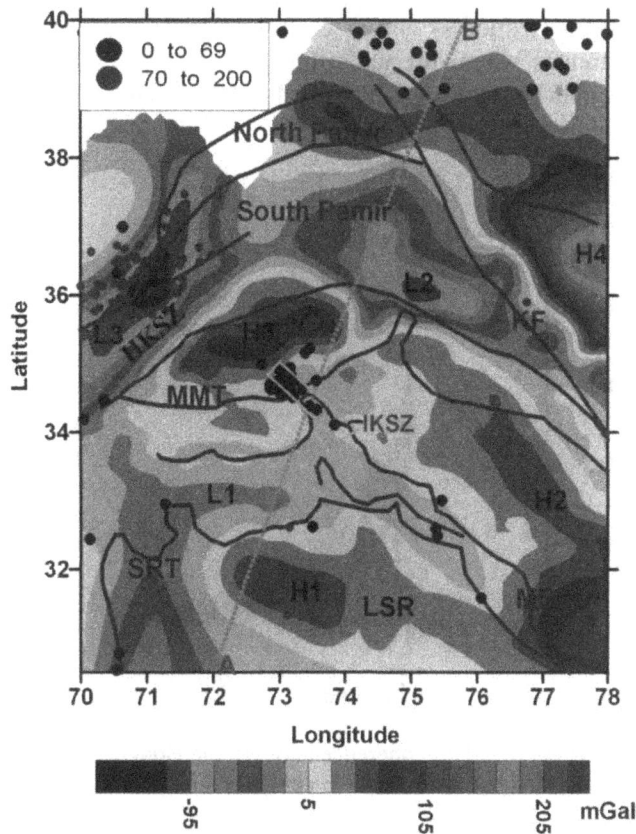

Fig. 6.56 Residual gravity anomaly based on high pass filters of wavelength < 500 km. It shows gravity highs, H1 related to the Lahore-Sargodha Ridge (LSR) and H2 and H3 related to thrust belts. Gravity lows L1, L2, and L3 are related to Siwalik sediments and Karakoram-Hindu Kush ranges. Gravity high H4 is associated with thrust belts of the Asian plate (Pamir ranges) (Tiwari et al., 2008). [Colour Fig. on Page 785]

A profile AB is selected to compute effective elastic thickness Te. Its coherence with topography versus wave number is given in Fig. 6.57 alongwith the modeled curve for Te = 53 km; the two match quite well. Fig. 6.58 shows the regional and the residual gravity fields for low pass filter with wavelength of 500 km and more along the profile AB and the fields corresponding to Airy's model of compensation for elevation plotted in the same figure. The regional and the residual fields match quite well with fields for Airy's model and flexural model for Te = 50 km. It also shows a subsurface model of the lithosphere of the Indian and the Asian plates developed based on the regional and the residual fields for deeper and shallower structures, respectively, described above constrained from the lithospheric structures from receiver function analysis (Kumar et al., 2005) and shallow structures based on known tectonics for bulk densities depending on rock types occurring in this region. This figure also shows the hypocenters of the earthquakes in this region projected along this profile. Most of the hypocenters of shallow earthquakes coincide with the thrust blocks or their projection in the crust, while those of deep focus earthquakes in the upper mantle coincide with the underthrusted slab similar to Wadati-Benioff Zone in the oceanic subduction zones. It also shows the hypocenter of Muzaffarabad-Kashmir earthquake (Mw = 7.6) of October 2005 at a focal depth of 25-26 km, which coincides with the thrusted block between the MCT and the MMT. It also coincides with the interface separating the upper and lower crusts. This model shows the Asian plate underthrusting below the Indian plate. The Indian lithosphere is confined up to 140-160 km while the Asian lithosphere lies between 140-250 km. Maximum crustal thickness is about 65-70 km under the Karakoram-Pamir ranges. This is less than that inferred in the previous models across central Himalaya. The occurrence of lowest gravity field under Pamir ranges where the underthrusted lithosphere from the two sides are interacting with each other, indicate that upper mantle in this section might be serpentinized due to percolation of fluids along underthrusting slab.

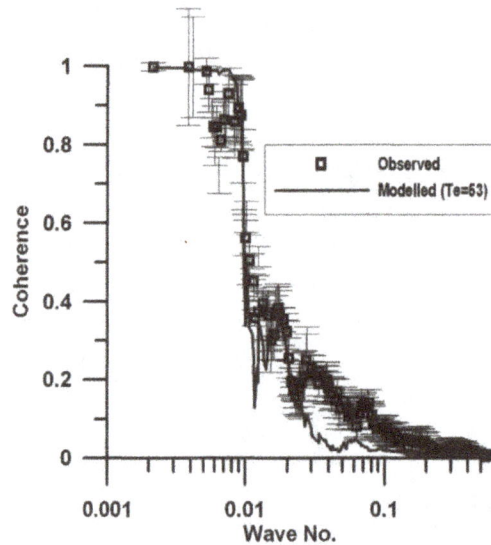

Fig. 6.57 Effective Elastic Thickness (Te) based on coherence between the Bouguer anomaly and topography. The best fit is obtained for Te = 53 km (Tiwari et al., 2008).

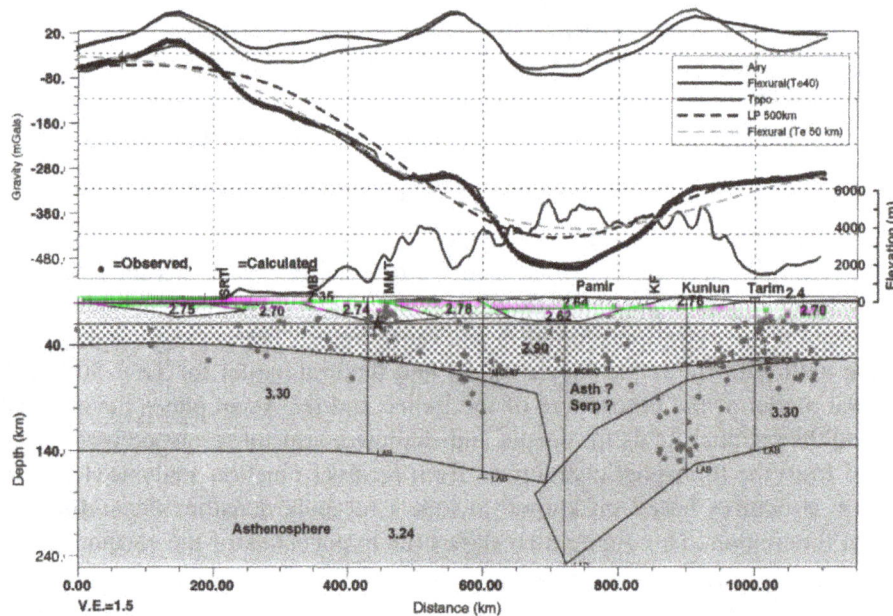

Fig. 6.58 Bouguer anomaly and elevation along profile AA' (Fig. 6.55(a)). Residual anomaly from Fig. 6.56 along the same profile is also plotted with the regional and residual fields for Te = 40 and 50 km. The computed model suggests underthrusting of the Indian and Asian lithospheres, the latter thrusting below the former. It also shows high density rocks associated with thrusts. Hypocenters of thrust earthquakes (Figs. 6.55(a)) projected on this profile coincides with them and their probable extensions depthwise. Hypocenters of some deep focus earthquakes coincide with the underthrusted slab (Tiwari et al., 2008). Cross inicates hypocenter of Kashmir earthquake of 2005. [Colour Fig. on Page 786]

(iii) Satellite Bouguer Anomaly and Deep Focus Earthquakes of Hindu Kush

Intense seismicity to depths of 100-200 km beneath the Hindu Kush is attributed to the fast subduction of the Indian lithosphere in this region with a high angle (Section 6.1.10). Seismic shear wave velocities are significantly faster in this section than those beneath Tibet, where they are limited to the upper crust. Previous geophysical studies indicate elevated thermal conditions and possible crustal melts. U–Pb ages suggest that post India-Asia collision crustal melting beneath Hindu Kush is restricted to 24 Ma whereas in Karakoram, the record is more voluminous and more continuous from ~ 37 to ~ 9 Ma. These observations suggest major differences in thermal histories of these regions where relatively cooler conditions beneath the Hindu Kush are attributed to cold slabs of continental subduction in the upper mantle that cause seismicity in this section (Hildebrand et al., 2000).

Since the Hindu Kush section is significant due to occurrences of deep focus earthquakes (Fig. 6.43), we examined Bouguer and geoid anomalies of this section as obtained from satellite for subsurface density in homogeneity. A detailed tectonic map of this region is given in Fig. 6.59(c) that shows S-type HKSZ located between Chamman-Panjshir and Karakoram faults towards the west and the east and Kohistan arc and Pamir towards the south and the north, respectively. A complete Bouguer anomaly map of Hindu Kush and surrounding sections is computed and given in Fig. 6.59(a) after correcting the free air anomaly (Fig. 6.6) for Bouguer correction and the terrain effect from GTOPO30 topography data. It shows only negative values due to predominance of the effects of changes in the crustal thickness related to isostasy. It primarily shows a gravity low, L1 decreasing towards the north related to high section of mountains such as Kohistan arc, Hindu Kush-Pamir, and Karakoram. The relative gravity high, H1 separates these high regions towards the east from which relatively flat region to the west known as Tadjik basin, which supports their cause due to crustal thickening related to isostasy. Compared to this, the free air anomaly map (Fig. 6.6) has delineated most of the ranges and tectonics of this region as discussed in section 6.1.6. As geoid anomalies reflect deep seated density in homogeneity better than gravity anomalies, we also analyzed the geoid data of Hindu Kush-Pamir section (Fig. 6.7). This map also shows a major geoid low (L4). The Hindu Kush seismic zone (Fig. 6.43) coincides with its Eastern gradient with the geoid highs, H2 related to Kohistan arc and Pamir.

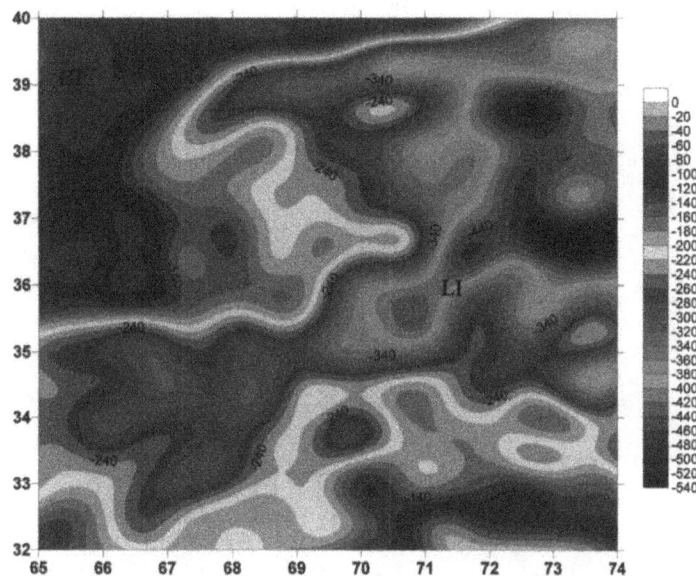

Fig. 6.59(a) Bouguer anomaly map of Hindu Kush and the surrounding region showing large negative values due to thick crust related to isostasy and low density rocks in the upper mantle. [Colour Fig. on Page 786]

Spectral analysis of both Bouguer and geoid anomalies of the Hindu Kush (Fig. 6.59(b)) suggests deep seated sources at depth of ~ 130-137 km that may represent subducted parts of the Indian and Asian lithosphere in this section as discussed above. It conforms to depths inferred from the second segment of the spectrum of large data sets of the same fields from this region (Fig. 6.9(b) and (c) and 6.55(b). The first segments of Fig. 6.9 (b) and (c) related to 320 and 328 km depths can not be delineated from the present spectrum due to limited data size that controls the depth of penetration in spectral analysis. The second segments in the present spectrums (Fig. 6.59(b)) are related to sources in the lower crust up to Moho as discussed above with regard to spectrums in Fig. 6.9(b) and (c). The Spectrum of the geoid data has provided a better estimate of the maximum depth of the Moho while Bouguer anomaly is more representative of sources in the lower crust that might be the characteristic of these data as the geoid data is a better representative of deeper features. As described above, Negredo et al., (2007) have attributed the deep focus earthquakes in this section to the fast subduction at a high angle (Fig. 6.18) and slab breakoff subducting Indian plate. The low density rocks in the upper mantle are present throughout the Himalayan-Tibetan section (Fig. 6.13(a)), but deep focus earthquakes are confined only to the Hindu Kush section. This is attributed to fast subduction and slab breakoff suducting the Indian plate. It would cause low temperature of the subducting slab making it brittle at those depths. In addition, the one unique feature of this section is the interaction of the Panjshir fault that is the northward extension of the Chamman strike fault along the Western Fold Belt (Fig. 6.59(c)) with the subducting Indian plate and slab breakoff can be attributed to this interaction. Fig. 6.59(c) also shows the Karakoram strike slip fault on the Eastern side of the western syntaxis that also interacts with the Hindu Kush Seismic Zone (HKSZ) in the northern part, which would also cause slab breakoff the subducting Indian plate causing deep focus earthquakes. This possibly explains the linear nature of the HKSZ extending from the Panjshir fault towards the west up to the Karakoram fault towards the east. The same is true for the Burmese arc (Fig. 6.40(c) and (d) where the Sagaing strike slip fault interacts with the subducting Indian plate causing slab breakoff and intermediate deep focus earthquakes in that section. Fig. 6.58 also indicates the interaction of the subducting Indian and Asian lithosphere under the Pamir-Hindu Kush section that may also cause slab breakoff subducting lithospheres, facilitating deep focus earthquakes.

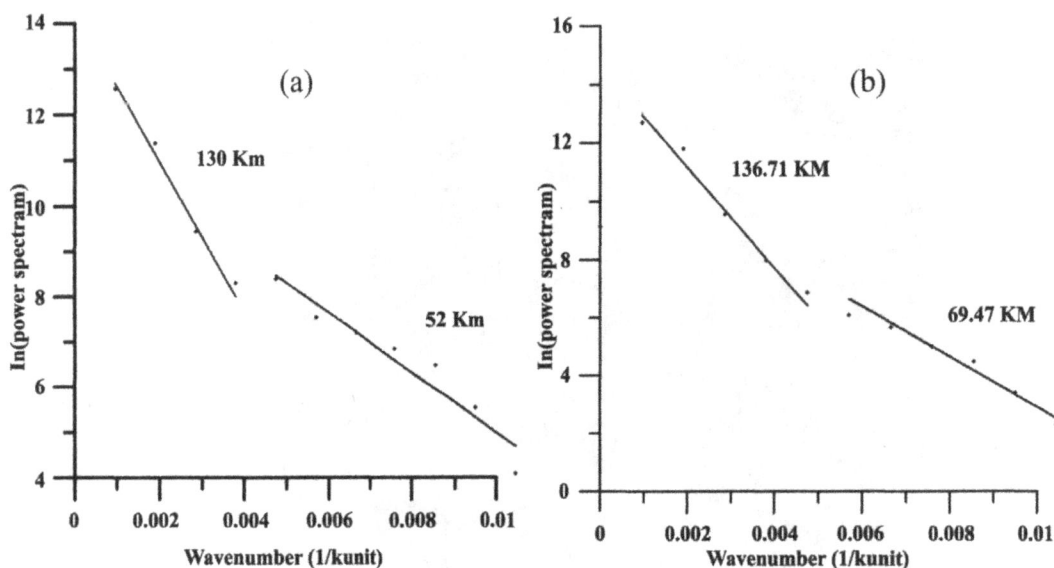

Fig. 6.59(b) Radial Spectrum of Bouguer (a) and (b) geoid anomalies of the Hindu Kush-Pamir section showing deep seated sources in lithospheric mantle related to first segment of the spectrums.

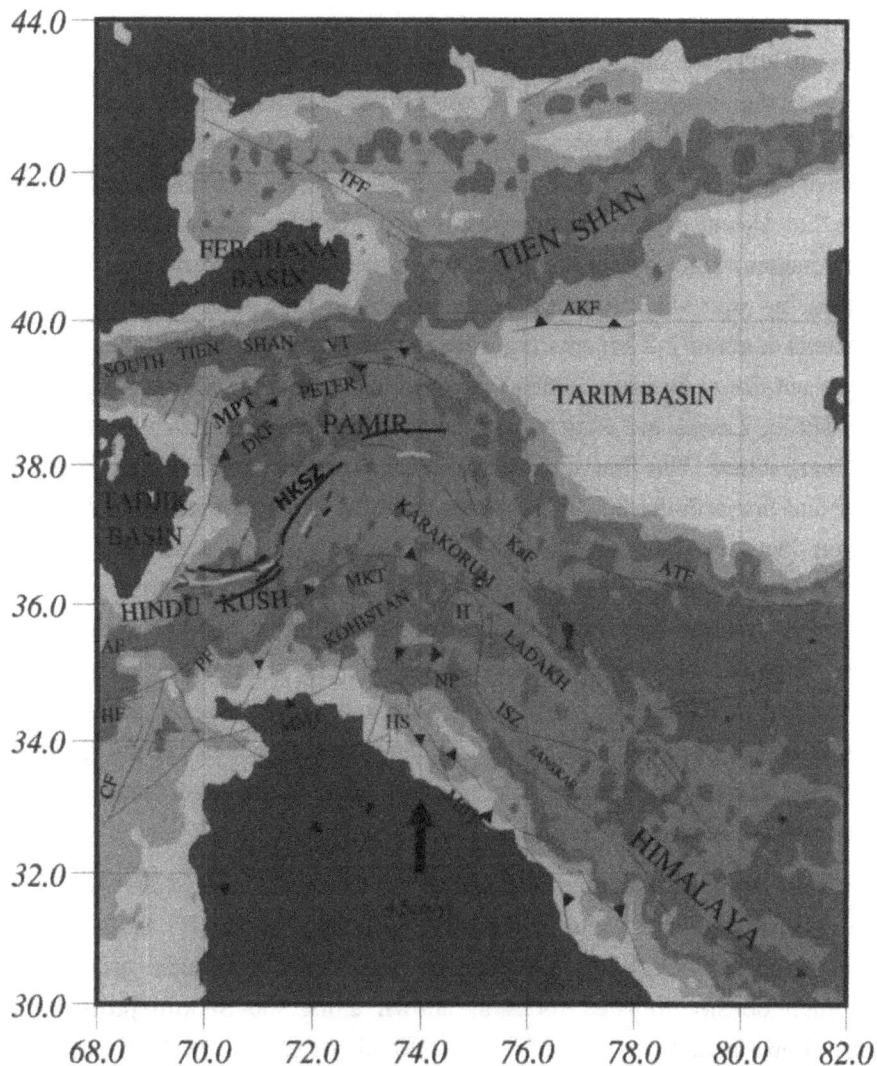

Fig. 6.59(c) Map showing the Pamir-Hindu Kush region. Topography is contoured at 1000 m intervals (0-1000 m green, 1000-2000 m, Yellow etc.). Faults are marked in black VT- Vakhsh; DKF – Darvaz-Karakul Fault; KaF – Karakoram Fault; AF – Andarab Fault; HF – Herat Fault; PF- Panjshir Fault; CF – Chamman Fault; ISZ – Indus Suture Zone; MBT – Main Boundary Thrust; MKT – Main Karakoram Thrust; MMT – Main Mantle Thrust; NP – Nanga Parbat; H – Haramosh; HS – Hazara Syntaxis; ATF – Altyn Taagh Faualt; AKF – Atushi – Keping Fault; TFF – Talas Ferghana Fault HKSZ – Hindu Kush Seismic Zone in S pattern (Focal Depth: red; 100 km; orange: 125 km; yellow: 150 km; green: 175 km) (Pegler and Das, 1998) [Colour Fig. on Page 787]

(iv) North West Himalaya – HIMPROBE

Integrated geological and geophysical studies were undertaken during the 1990s across Trans-Himalayan and Karakoram in Western Himalaya with active support from the Department of Science and Technology, Government of India. It basically consisted of geological mapping and dating of some important rock units (Jain et al., 2003) along with geophysical measurements of gravity field along a profile, from Kiratpur-Leh-Panamik (Fig. 6.48).

This section is mainly occupied by rocks of Lesser and Higher Himalaya associated with MBT and MCT and various intrusive such as Ladakh batholiths and mafic/ultramafic oceanic crustal rocks associated with the suture zones. An important unit in this series is the Tso Morari crystalline which contains dark coloured patches of eclogites that represent Ultra High Pressure (UHP) rocks formed at about 55 Ma (De Sigoyer et al., 2000). The discovery of coesite in eclogites from the Tso Morari crystalline complex in NW Himalaya indicated that it is the only ultra high pressure metamorphic rock in the Himalaya in the Indian territory. Five major types of fluids are identified by micro-thermometry as inclusions in this rock (Sachan et al., 2005). Gravity measurements at about 1-2 km spacing were carried out along this profile, Kiratpur-Leh-Panamik (Banerjee and Satya Prakash, 2003a). This profile is about 550 km long across thrust zones (MBT and MCT), Lesser and Higher Himalaya, and two suture zones in this area, ITSZ and Shyok (Northern) suture. The free air and complete Bouguer anomaly along this profile is given by Banerjee and Satya Prakash (2003a) along with the isostatic anomaly and elevation. A positive correlation between free air anomaly and elevation, and somewhat the opposite correlation with the Bouguer anomaly suggests isostatic compensation to some extent. Isostatic anomaly derived by subtracting the effect of fully compensated crust for Airy's model, provides a relative gravity high in the central part over the Higher Himalaya. This suggests an undercompensated crust in this part or a shallow high density body. Seeing its wavelength and considering that Himalayan orogeny is a young orogeny where activities are still going on, it is expected that it would show an undercompensated crust while to the north under Tibet, it is an overcompensated crust as discussed in Section 6.1.

The Bouguer anomaly along this profile is modeled by Chamoli et al., (2010), and shows a decreasing field towards the north due to crustal thickening up to 75 km under Tethyan sedimentaries and north of them (Fig. 6.60(a)). Several small amplitude gravity highs are modeled due to high density thrusted rocks as shown along the Sikkim profile in Central Himalaya. It also shows a ramp under the Higher Himalaya similar to the Sikkim profile in Central Himalaya (Fig. 6.35(a)). A small gravity high over suture zone (ITSZ) is caused by high density ophiolite rocks while Shyok suture zone reflects a gravity low due to low density that is anomalous. Significantly, Ladakh batholith shows gravity high. Banerjee and Satya Prakash (2003b) have compiled a complete Bouguer anomaly map of NW Himalaya from different sources as given in Fig. 6.60(b). It shows decreasing field from about – 100 mGal along the Ganga foredeep to about – 560 mGal north of the ITSZ. An Airy's isostatic anomaly map computed from this Bouguer anomaly shows a high of about 80 mGal over High Himalaya north of MCT, and lows of almost the same magnitude along Himalayan foredeep and north of the suture zone over the Western Tibet suggesting under-and overcompensated crusts, respectively. Alternatively, they can also imply high and low density rocks at shallow depths, respectively, or a combination of the two with large wavelength features indicating the state of compensation and short wavelength anomalies indicating shallow bodies. The same, however, is observed in most of the present day orogenic belts where frontal part and interior show overcompensated crusts, while central mountain ranges show undercompensation.

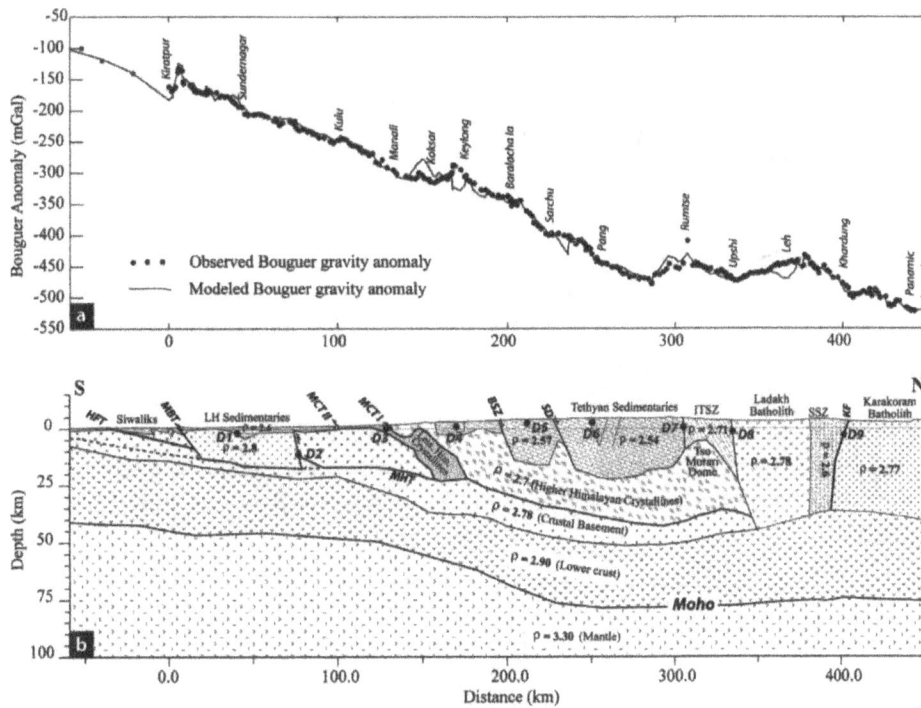

Fig. 6.60 **(a)** Bouguer gravity anomaly along Kiratpur-Panamik profile (Fig. 6.48) and **(b)** modeled crustal section constrained from all available information from this region. MHT: Main Himalayan Thrust, SD: Sarchu detachment, KF: Karakoram fault, STD: South Tibetan Detachment, BSZ: Baralacha la shear zone.
(Chamoli et al., 2010)

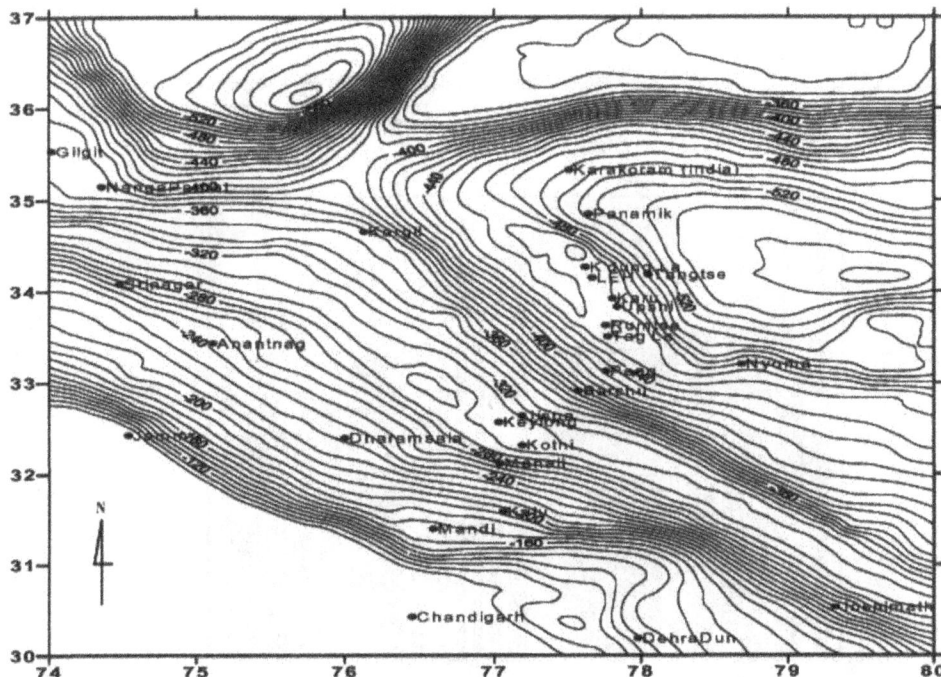

Fig. 6.60(b) A complete Bouguer anomaly map of a part of NW Himalaya compiled from different sources showing gradient overthrusts and suture zones due to high density rocks associated with them, and low over the northern and the Eastern parts due to crustal thickening (Banerjee and Satya Prakash, 2003b and Personal Communication).

Gravity profile across Tso Morari crystalline and suture zones and associated ophiolites (Fig. 6.61(a)) were recorded by Sastry et al., (2004), and were processed for constant density 2.67 g/m^3 and variable density as per exposed rock types for combined Bouguer and terrain correction as given in Fig. 6.61(a). The gravity profile across Tso Morari crystalline (Fig. 6.61(a)) shows almost a constant field for fixed density, but a decreasing field for variable density. As the field across the Himalaya decreases northwards due to crustal thickening, it appears that the latter alternative of variable density shows the correct representation of the field in this case. However, reduction of data with variable density has the problem of selecting the bulk density for exposed rocks and their depthwise extent. The field reduced for variable density also shows gravity high over the ophiolites associated with the ITSZ, viz. Nidar and Zildat ophiolites, while over the Puga formation it shows a small order low that might be related to low density sediments. Tso Morari crystalline does not show any specific anomaly inspite of inclusions of ultra high pressure eclogite rocks in it, indicating the absence of any large body of eclogite associated with it or at subsurface levels. They have given densities and susceptibility of some typical rock types exposed in the region, which are presented in Fig. 6.61(b) and (c) for reference purposes.

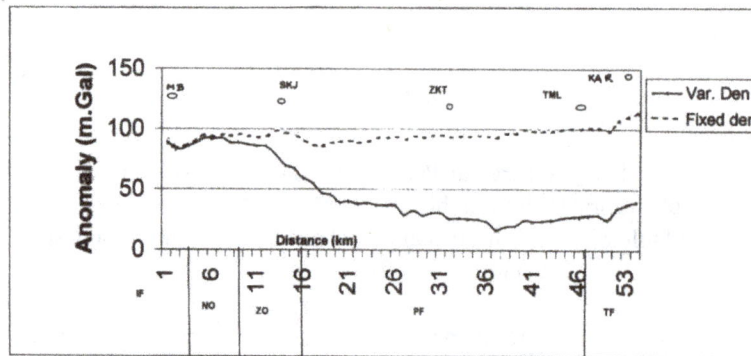

Fig. 6.61(a) A gravity profile across Tso Morai crystalline from Mahe (MB) to Tso Morai (TML). IF = Indus formation; No = Nidar ophilites; Zo = Zildat ophiolites; PF = Puga formation; TF = Taglang la formation (Sastry et al., 2004).

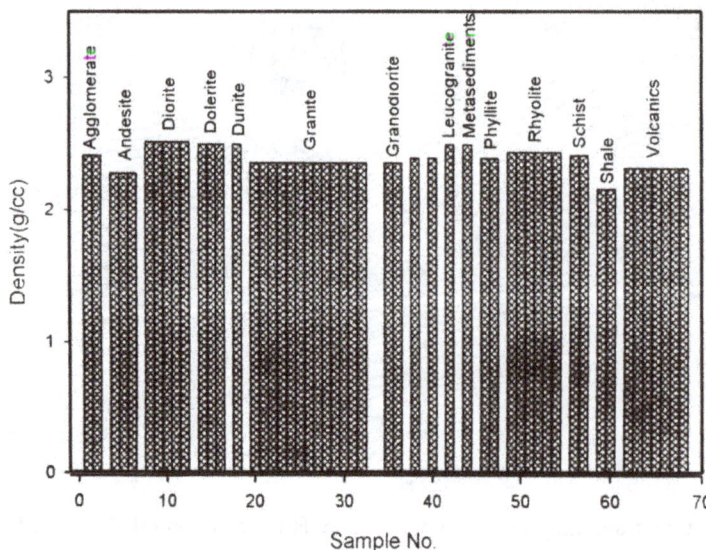

Fig. 6.61(b) Some typical densities of rocks exposed in this sector (Sastry, Personal communication).

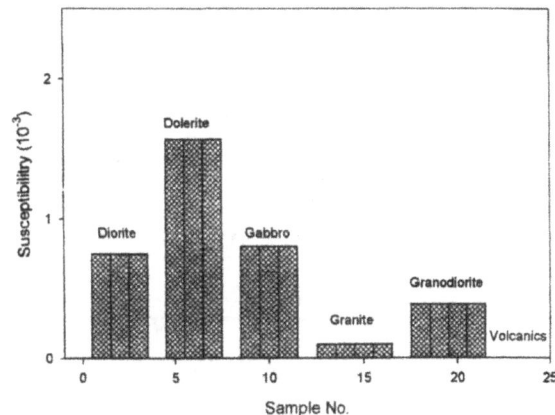

Fig. 6.61(c) Some typical susceptibilities of rocks exposed in this section (Sastry, Personal communication).

(iv)   A Detailed Gravity and Magnetic Profile across MCT

A small (11.7 km long) but detailed gravity profile recorded across the MCT at close station spacing of 20 m has been reported by Seshunarayana et al., (2008). This profile (NE-SW) from Maneri-Pala (A-E, Fig. 6.62(a)) is located in the Garhwal sector of NW Himalaya. This figure also shows the epicenters of various seismic activity (1900-1963) in this section, including the Uttarkashi earthquake of 1991 of magnitude 7.0, and their concentration along the MCT. Fig. 6.62(b) shows the magnetic and the gravity profile with reference to the distance from station A (Maneri). This figure also shows a shallow section whose computed fields matched quite well with the observed field for parameters given in the caption of the figure. The most interesting gravity and magnetic anomalies are observed over the MCT that shows gravity low, and the magnetic highs and lows related to intrusives along it. Crystalline rocks that have been thrusted along the MCT in this section have a lower density giving rise to gravity low. Due to their metamorphic nature, they produce magnetic anomalies. Another interesting observation is the occurrence of several faults or fractures sympathetic to the MCT dipping northwards which are well reflected in magnetic data that is much simpler to record as compared to gravity data in hilly terrain.

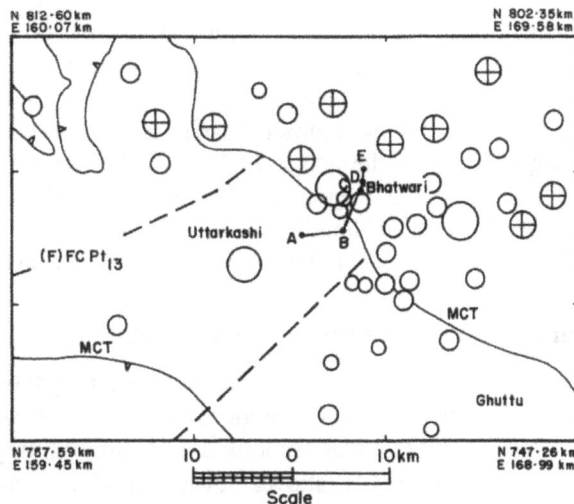

Fig. 6.62(a) Seismotectonic map of study area, showing the location of profile A (Maneri)-E(Pala) (GSI, 2000). N and E indicate North and East. F=Faults, (F) FC = Older folded cover sequence affected by fold-thrust movement, MCT = Main Central Thrust, --- Lineament, B = Kumaltigad, C = Nihargad, D = Papergad. Circles indicate epicenters of earthquakes depending on their magnitudes, largest being the Uttarkashi earthquake of magnitude 7.0. The circle with positive sign indicate hot springs.

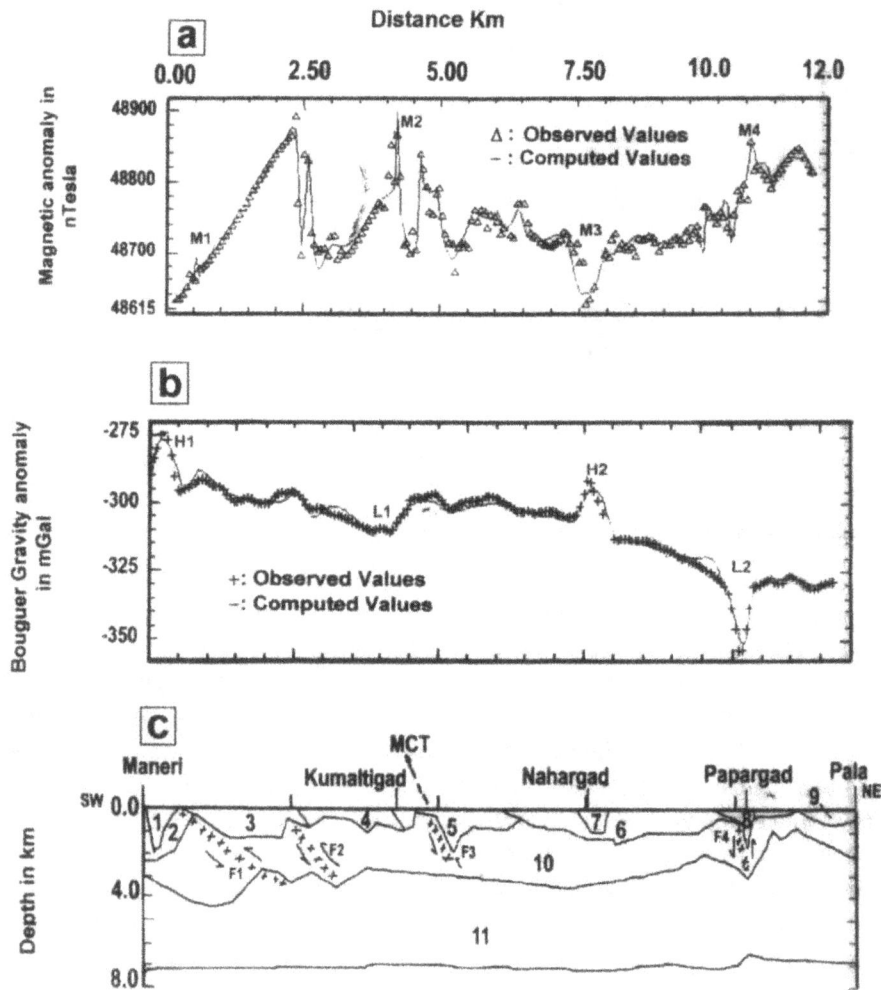

Fig. 6.62(b) Magnetic anomaly data plotted along Maneri-Pala segment; M1, M3 are corresponding to magnetic lows while M2, M4 are magnetic highs (a); Bouguer gravity anomaly data plotted along the Maneri-Pala segment; H1, H2 indicates gravity highs and L1, L2 indicates gravity lows (b); a shallow section based on modeling these data sets and surface tectonics (c). Values of physical parameters used for modeling the fields for different bodies are as follows: 1- (2.80 gm/cc, -0.65 mille cgs), 2- (2.30 gm/cc, -0.10 millle cgs, 3- (2.35 gm/cc, 1.80 mille cgs), 4- (2.10 gm/cc, 0.05 mille cgs), 5- (2.50 gm/cc, -0.55 mille cgs), 6- (2.30 gm/cc, -1.20 mille cgs), 7- (2.75 gm/cc, -1.70 mille cgs, 8- (1.70 gm/cc, 0.90 mille cgs), 9- (2.28 gm/cc, 2.20 mille cgs), 10- (2.60 gm/cc, 0.10 mille cgs), 11- (2.65 gm/cc, 2.00 mille cgs) (Seshunarayana et al., 2008).

(vi) Magnetic Profile – Haridwar to Joshimath across MBT and MCT

This magnetic profile was recorded at 1 km spacing to decipher the magnetic characteristics of thrusts in this section. Due to inhomogeneity of magnetic minerals in the exposed rocks along this profile, recorded data was quite noisy and was smoothed by taking running average of 5 data points as shown in Fig. 6.62(c). This profile shows several moderate amplitude magnetic anomalies marked as M1-M5. M1 coincides with the MBT where Proterozoic metasediments are exposed, while M2 coincides with the Alakhnanda fault along the Alakhnanda River. The magnetic anomaly M3 coincides with the mafic volcanics exposed along this profile. M4 and M5 are related to MCT where crystalline basement rocks are exposed and several seismic activities have been reported (GSI, 2000). These anomalies are modeled using inversion scheme for tabular bodies (Section 4.5) and causative bodies are shown in the same figure under respective anomalies (Tiwari et al., 2002). These models suggest variation in susceptibility from $0.2 - 0.6 \times 10^{-3}$ emu indicating their mafic nature. They all dip northwards, same as the thrusts. Besides, there are some small magnitude magnetic anomalies, such as one corresponding to HFT, indicating that the thrusted rocks in this case are less magnetic in nature. These results suggest that the magnetic method can be successfully used to delineate thrusts and faults in the Himalayan terrain as it is the simplest method to record in the field.

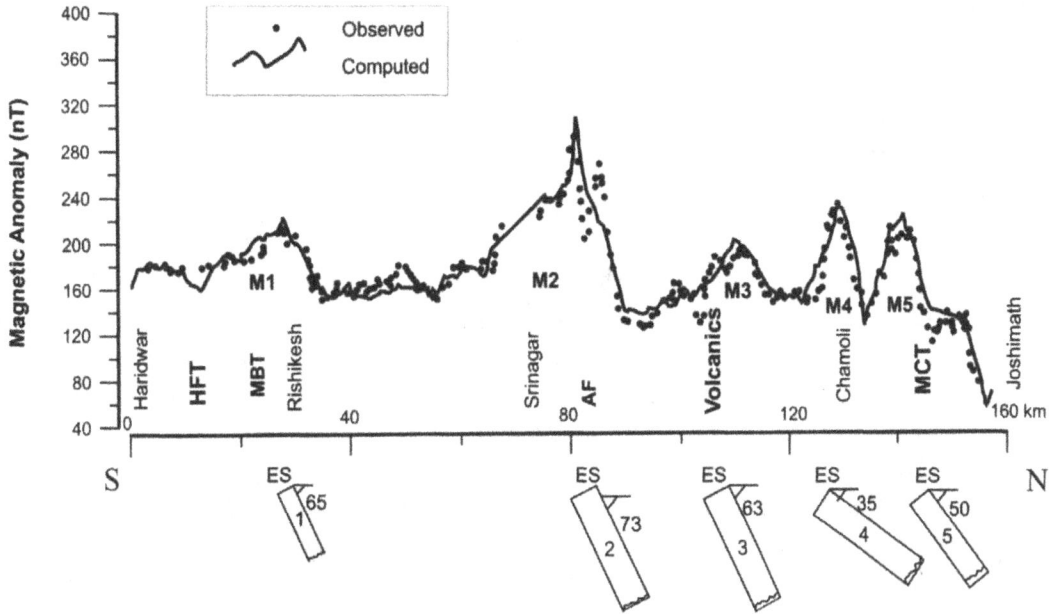

Fig. 6.62(c) A total intensity magnetic profile from Haridwar–Joshimath (Badrinath) showing an averaged observed profile with moderate amplitude magnetic anomalies, M1-M5. Modeled tabular bodies (1-5) with respective susceptibility are: $1 = 0.4 \times 10^{-3}$; $2 = 0.2 \times 10^{-3}$; $3 = 0.4 \times 10^{-3}$; $4 = 0.4 \times 10^{-3}$; $5 = 0.6 \times 10^{-3}$. Dips are shown with respect to Exposed Surface (ES) that is quite undulating changing in elevation from a few hundred meters in the southern part to ~ 3.5 km in the northern part. AF: Alakhnanda Fault. Most of them may represent blind faults (Fig. 6.62(b)) that can be delineated based on magnetic surveys.

6.3.5 Neotectonics and Orography – Gravity Anomalies, Crustal and Basement Structures, Seismicity, Uplifts and River Courses

The activity along the Himalayan front has successively shifted from the north to the south starting from the (MCT) to (MBT) and (HFT) during last 20 Ma. The MBT and the HFT are presently the most active sections characterized by the thrusting of Paleozoic and Proterozoic meta-sediments and Siwalik sediments, respectively. The Siwalik group of rocks along HFT are underlain by gravels and boulders, which indicate recent activities like uplift, and changes in levels in this section. The section between the MBT and the HFT in NW Himalaya is also characterized by Dun structures which indicate broad synclinal longitudinal valleys due to exhumation of the frontal range of Himalaya caused by recent uplifts (Fig. 6.44) that has produced re-entrants such as Kangra and Dehradun re-entrants. The maximum rate of uplift of the order of 3-4 mm/yr in general and 10 mm/yr in Nanga Parbat has been reported in this section (Nakata, 1989). Sites of focused denudation due to monsoon along MBT in Central Nepal leading to maximum rate of uplift have been suggested for north-south channel flow in the Indian crust during recent times, that has been discussed in Section 6.2.6 (Wobus et al., 2003; Harris, 2007). Even a part of the uplift along MCT and Higher Himalaya in central Nepal is attributed to similar processes during Miocene time (Hodges, 2004). Searle et al., (2007) have identified certain domal structures in the Zanskar range in Western Himalayas (Fig. 6.43) that was attributed to channel flow. Willett (1999) and Dietrich and Perron (2006) have shown by numerical modeling that sites with the maximum rate of erosion due to rainfall and biotic effects from humans and animals are also the favoured sites for maximum rate of uplift and flow of lower crustal rocks as channel flow along thrusts in these sections. They have also shown the shift of consecutive thrusts with time towards the erosional front as it is observed in the case of Himalayas. Sedimentological isotopic and mineralogical changes in sediment record between 9-7 Ma have been used as evidence of intensification of the monsoon, implying uplift of Tibetan Plateau above a certain threshold by this time.

As the rate of erosion and uplift are largely controlled by rainfall, it is important to know the periods of maximum rainfall and aridity for neotectonics during Holocene. The climate has been flip-flopping between these two extremes during the Holocene at an approximate periodicity of 1500-2000 yrs. (Tiwari, 2005). The start of Holocene at 12,000 yrs Before Present (BP) marks an arid and warm climate when melting of Pleistocene glaciation took place (Radhakrishna, 1999 a, b) and Tibet was re-inhabited (Dennell, 2008). It is followed by several periods of wet spells and aridity in India during Holocene time. However, the important ones are the deglaciation after Pleistocene glaciation at the start of Holocene. The next major period of glaciation marked in Western Himalayas is the period of Neoglaciation at about 8-6 Ky B.P. (Shroder, Jr. and Bishop, 2000). Accordingly, civilization that developed along major rivers during wet periods migrated when they changed courses or dried up during arid periods. One such case is that of the Vedic civilization during 6-4 Ky B.P. along the banks of river Saraswati in Western India which almost coincides with the period of deglaciation after Neoglaciation. However, this civilization migrated eastwards at about 3.5 Ky B.P. (Lal, 1998) when Western India including the Western Himalayan front experienced an intense arid climate. Based on pollens in sediments it has been suggested that even the Gangotri glacier in Garhwal Himalaya, source of the river Ganga experienced a hot, dry climate during 3.5 Ky B.P (Valdiya, 2002). There is ample evidence of the Vedic Period Civilization known as Harappan culture along the banks of river Saraswati during this period. The most important evidence for this civilization comes from archeological investigations in Harappa, and Mohenjedaro in Pakistan, and Kalibangan, and Siswal in India (Lal, 1998, Wakankar, 1999). It is largely believed that the river Ghaggar in India and Hakra in Pakistan (Fig. 6.63(a), Valdiya, 2002) represent the remnant of that mighty river Saraswati. The existence of such a river has been shown by occurrences of pools of fresh water at a depth of 50-60 m and sometimes as deep as 300 m in Cholistan (Fort Abbas), Pakistan close to present day river Hakra based on airborne electromagnetic surveys (BGR, 1992). They found a copious supply of fresh water located over saline water, and differentiated them based on resistivity with fresh water of higher resistivity (> 20 Ω m) compared to underlying saline water (< 20 Ω m). Some efforts in this direction by the Oil and Natural Gas Corporation in Jaisalmer have yielded positive results and have struck paleowaters in bore wells at a depth of 500-550 m. Fig. 6.63(a) and (b) show the major rivers originating from NW Himalaya and their cross-points along the Himalayan front. The elevation contours superimposed on it (Fig. 6.63(b)) show higher elevation in general in the Nahan salient compared to the Kangra and the Dehradun re-entrants.

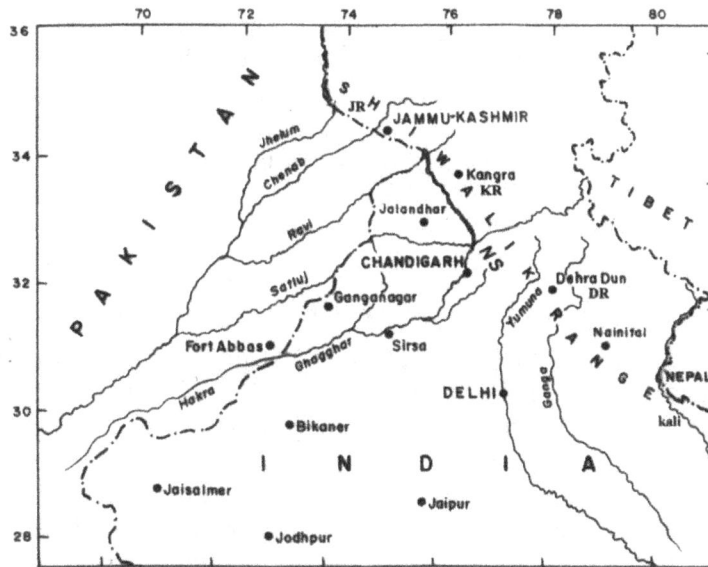

Fig. 6.63(a) Major rivers originating from the NW Himalaya and their intersections with the Himalayan front (Siwalik Range). It also shows the Nahan Salient (NS) in between Kangara re-entrant (KR) and Dehradun re-entrant (DN).

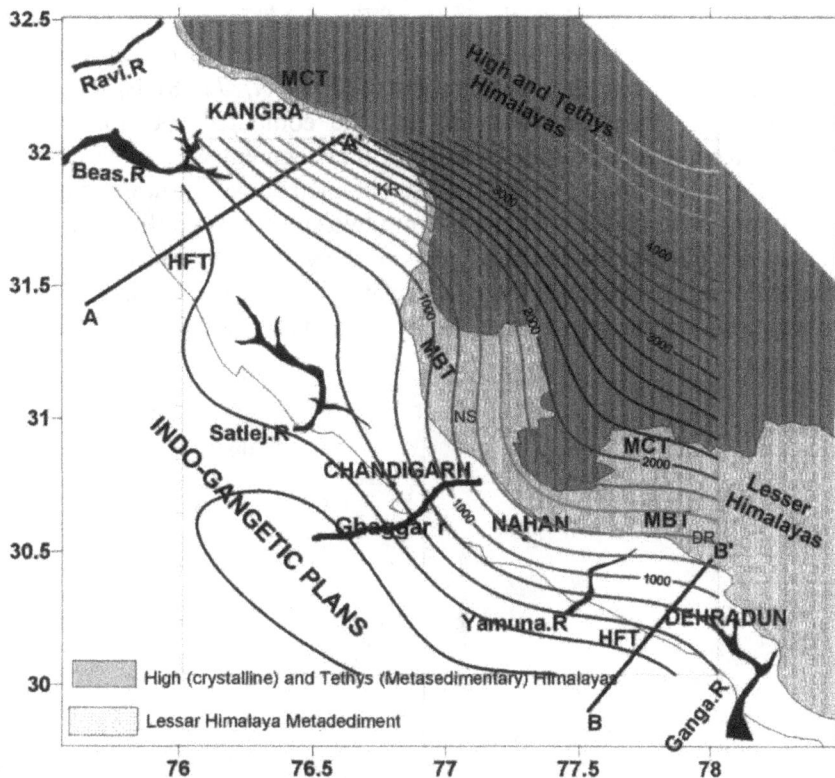

Fig. 6.63(b) Tectonics of the Western Himalayan front with the Nahan salient (NS) and adjoining re-entrants of Kangara (KR) and Dehradun (DR). MCT: Main Central Thrust, MBT: Main Boundary Thrust and HFT: Himalayan Frontal Thrust. It also shows the topography with the highest elevation in the Nahan salient along the front.

Nair et al., (1999) have identified paleochannels in western Rajasthan with wells of paleowaters older than 1.8 and 6.0 Ky B.P. in shallow and deep wells, respectively. Changes in the river courses of Himalayan rivers appear to have taken place several times. As suggested by Clift and Blusztajn (2005) based on isotopic studies of sediments in Indus and Bay of Bengal fans, all the Himalayan rivers including the five rivers of Punjab were discharging eastwards in Bay of Bengal; but due to some tectonic changes these five rivers of Punjab changed their courses five million years ago to discharge in the Arabian Sea. This major change in river courses may indicate the shift from MBT to HFT that would definitely affect the drainage pattern of the region.

(i) Bouguer and Geoid Anomaly Maps of NW India and Himalayan Front

The complete Bouguer anomaly map (Fig. 6.14) of the North Indian Shield including the Himalayan front on the western side shows gravity highs, H1 along the Western Himalayan Front. This map also shows the MBTand the HFT related to the collision of the Indian and the Asian plate. This map also shows major gravity lows L1 and L2 related to low density sediments of the Ganga basin and crustal thickening under the Himalaya, respectively. The gravity highs along the Himalayan front, H1 are masked by the regional gravity lows (L1-L2) that can be better delineated in the residual anomaly map by removing the regional field based on spectral analysis as described above in Section 6.1.6.

Spectral analysis of potential fields (Section 6.1.6) can be directly used to estimate depths to the sources at different levels and separate them in different wave bands as regional and residual fields. The spectrum of the Bouguer anomaly map of the North Indian shield including Himalayan front is computed and given in the inset of Fig. 6.14. It suggests shallow segment with an average depth to the causative sources at 6 km. In order to check the stability of computed spectrum, the data from the western part (20°-34°N, 72°-81°E) of this map is transformed separately in the frequency domain, and the computed spectrum versus wave number is given in Fig. 6.64(a) that provides a straight line segment with slopes equivalent to 134, 33, 17, and 6 km which are almost similar to those obtained above for larger data sets within the permissible error limits. Similar depth estimates from two spectrums confirm the reliability of the computed spectrum and thereby the average depths of the various layers obtained from them. Similar results obtained from the different data sets related to similar experiments confirm the stability of the computations and the results obtained from them (Blackman and Tuckey, 1958).

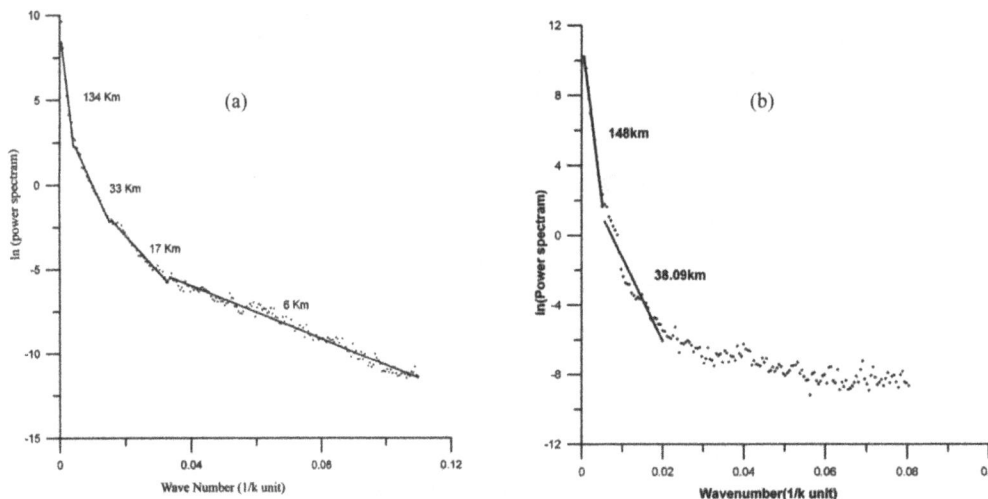

Fig. 6.64 (a) Spectrum of the western part of the Bouguer anomaly map (Fig. 6.14, 20-35°N; 70-81°E) showing linear segments with slopes equivalent to 134, 33, 17, and 6 km. (b) Spectrum of the geoid data for the block shown in Fig. 6. It shows two linear segments with slopes equivalent to 148 and 38 km depth. It shows considerable noise beyond wave number 0.02 that might be due to large data spacing (25-50 km) in satellite measurements.

Another data set that describes density in homogeneity in a region is the geoid undulations (anomalies) described below. This data set for the western part (Fig. 6.7) for almost the same region (20°-34°N, 66°-80°E) as the Bouguer anomaly is transformed in the frequency domain (Fig. 6.64(b)) that provided two linear segments. Their slopes suggest average depths of 148 and 38 km that are almost same for the first two linear segments as shown in Fig. 6.64(a) related to LAB and Moho within the error limit of 10-15% which confirms the stability of these spectrums and results derived from them as suggested above. This spectrum indicates that the geoid field largely represents deep seated sources. Beyond the wave number 0.02, the spectrum of the geoid data shows considerable noise that may be due to large data spacing (25-50 km) in satellite measurements. This indicates that the satellite data primarily reflects deep seated sources and that consistent signals may not be obtained from sources above Moho.

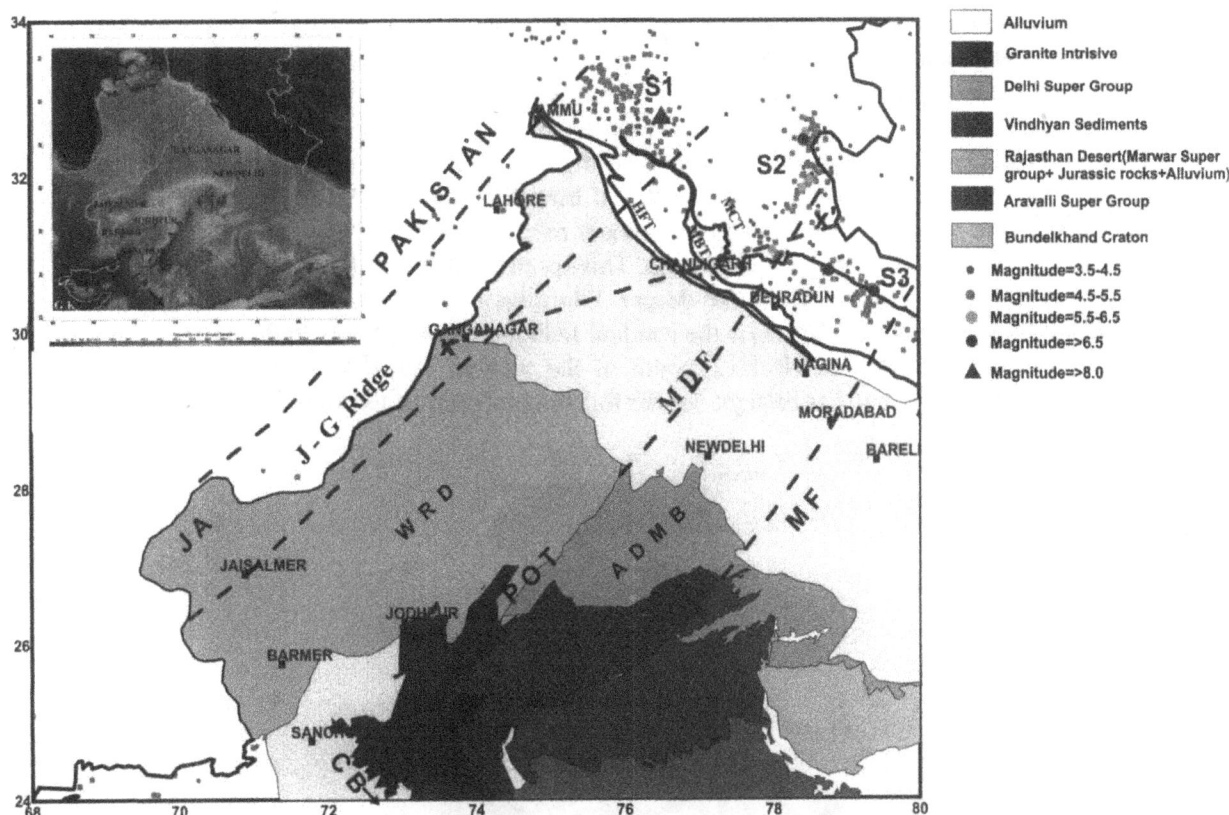

Fig. 6.64(c) A simplified geological map of NW India showing Aravalli Delhi Mobile Belt (ADMB) and Great Boundary Fault (GBF) and Phulad OphioliteThrust (PT) along its eastern and the western margins and their extensions towards the north are shown as Moradabad Fault and Mahendragarh-Dehradun Fault (MDF), respectively. They interact with the Himalayan front along the MBT and seismic activity zones marked as S2 and S3 might be related to this interaction as these faults are west dipping. Jaisalmer-Ganganagar ridge (J-G Ridge) in Western Rajasthan Desert (WRD) is shown as dashed lines that might be related to lithospheric flexure of the Indian plate along the Western Fold Belt (Pakistan) delineated based on the residual Bouguer and geoid anomalies as described below. It interacts with the Himalayan front in seismic activity group S1 where the great Kangra earthquake (filled triangle) of 1905 is located. Thrusts along the Himalayan front are marked as: HFT – Himalayan Frontal Thrust, MBT – Main Boundary Thrust and MCT – Main Central Thrust GK- Gulf of Kachchh and JA – Jaisalmer arch where Mesozoic sediments are exposed while other parts of the Western Rajasthan is covered by alluvium. Jaisalmer, Barmer and Sanchor basins close to these towns are Mesozoic basins that are connected to the Cambay Tertiary basin (CB) leading to the Gulf of Cambay to the SE. Inset is the topography map based on the SRTM data. [Colour Fig. on Page 787]

(ii)   Residual Bouguer and Geoid Anomaly Maps of NW India and Himalayan Front

Fig. 6.64(c) is a simplified geological map of the NW India that also shows important tectonic elements of this region. The most important tectonic element in this section is the Aravalli Delhi Mobile Befft (ADMB) that represent a Meso-Neoproterozoic collision zone consisting of several mafic and ultramafic intrusive like ophiolite, granulite rocks etc. ofhigh density and metasediments (Section 7.5). It is bounded by the Great Boundary Fault (GBF) and Phulad Thrust (POT) related to Phulad ophiolite in southern part of the ADMB against Erinpura granite towards the east and the west, respectively. To the west of the ADMB are exposed granite intrusive (Erinpura granite) and Malani volcanic of Neoproterozoic period that are subduction related magmatism of Proterozoic collision related to the ADMB (Section 7.5). Malani volcanic is quite wide spread and forms the basement of the Marwar supergroup of rocks in Western Rajasthan of also Neoproterozoic period west of the ADMB. Mesozoic sedments of Jurassic period are exposed around Jaisalmer related to the Jaisalmer arch. This map also shows the plausible extension of the margin faults of the ADMB and Jaisalmer-Ganganagar basement ridge based on the residual Bouguer and geoid anomalies as described below.Inset shows the extension of the Jaisalmar, Barmer and Sanchor Mesozoic basins to Cambay Tertiary basin leading to the Fulf of Cambay. It also shows the ADMB and uplands of the Western Rajasthan (Jodhpur) and Punjab (Chandigarh). Residual Bouguer anomaly of NW India including western Himalayan Front is obtained from the observed field (Fig. 6.14) using a high pass filter. The spectrum of the Bouguer anomaly map of NW India (Fig. 6.64(a)) is used to decipher the cut off wave number for this purpose. This spectrum indicates that wave numbers >.03 separates the basement sources from the deeper lithospheric and lower crustal sources that represents the residual field. Fig. 6.65(a) is the residual field that shows gravity highs, H1-H5 along the western Himalayan front and H6-H12, south of the western Himalayan front that requires suitable explanations. This map has brought out the following aspects of tectonics in this region.

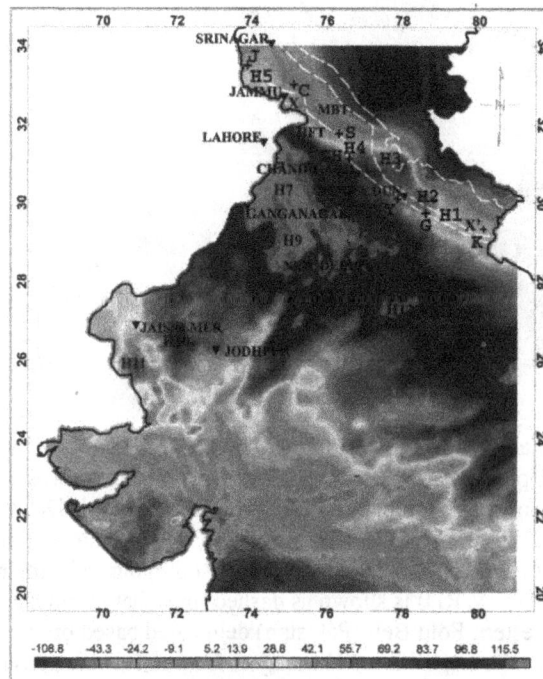

Fig. 6.65(a) Residual anomaly map of Western Himalayas for wavelengths <210 Km showing gravity highs H1-H12 caused by shallow sources. The "+" sign marks the major rivers along the Himalayan front; K = Kali; G = Ganga; Y = Yamuna; GH = Ghagghar; S = Satluj; C = Chenab; J = Jhelum. These rivers flow from the sides of gravity highs, H1, H2, H3, H4, and H5 except Ghaggar which flows from south of gravity high H4.Gravity highs, H6 and H7 represent the crustal bulge due to lithospheric flexure of Indian plate due to Himalayas; and H7 represents part of the Delhi-Lahore-Sargodha basement ridge. Other highs and lows are explained in the text. [Colour Fig. on Page 788]

(c)   The southern part of the gravity high, H7 extends almost up to Jaisalmer and further southwards, almost parallel to the Western Fold Belt (Pakistan) and therefore, it may represent the effect of the lithospheric flexure of the Indian plate due to this fold belt on its western side. This effect is better reflected in the geoid map as described below. It coincides with the Jaisalmer arch where Mesozoic sediments are exposed and related basement ridge, Jaisalmer-Ganganagar that almost separates the Indus basin (Pakistan) and the Rajasthan desert towards the west and the east, respectively (Fig. 6.64(c)). The northern part of the gravity high, H7 extends to the Himalayan front in Kangra reentrant (Fig. 6.63(b) and 6.64(c)) that has plausibly developed due to this interaction and may represent the effect of lithospheric flexure on the western side as described above for the extension of the H7 towards the south. This part of the Himalayan front, viz. Nahan salient and adjoining Kangra and Dehradun reentrants (Fig. 6.63(b) and 6.64(c)) are seismically active sections. It has been suggested that the specific parts of the Himalayan front in this section where the Eastern and the western faults/thrusts of the ADMB, the Great Boundary Fault (GBF) and the Phulad Ophiolite Thrust extending as Moradabad fault and Mahendragarh-Dehradun fault (GSI, 2000), respectively and the basement ridge due to lithospheric flexure on the western side (northwards extension of H7) interact with the Himalayan front are most seismogenic (S1, S2 and S3, Fig. 6.64(c); Mishra et al., 2011)). This interaction has given rise to several major and great earthquakes including Kangra earthquake of 1905. It is interesting to observe that the lithospheric flexure due to NW Himalayan Fold Belt (India) and Western Fold Belt (Pakistan) interact in the Ganganagar-Lahore section where maximum gravity (Fig. 6.65(a)) and geoid (Fig. 6.65(b)) anomalies are observed.

The most seismogenic sections are located west of the GBF and the PT as these are west dipping faults and are almost linear along the extensions of these faults. Among them, the section S1 is more seismogenic compared to the other two sections (S2 and S3, Fig 6.64(c) that might be due to the interaction of the crustal bulge and the basement ridge with the Himalayan front caused by the lithospheric flexure on the western side that would affect the entire lithosphere and is continuously operative. It might be also due to proximity of S1 to the Western Syntaxis where fast subduction is taking place and is one of the tectonically and seismically most active regions (Section 6.3.4).   Intersections of these faults with the Himalayan front are also characterized by straight river courses. For example, the Moradabad and the Dehradun faults are close to river Kali and Yamuna-Ganga, respectively while the basement ridge on the western side occupies the section between the rivers Beas, Ravi and Chenab (Fig. 6.65(a)). This section along the NW Himalayan Front is characterized by several hot springs of temperature 50-100 °C (Section 6.3.3) indicating high thermal gradient that will also reduce the strength of sub-surface rocks facilitating seismic activity.

(d)   The gravity lows in between the gravity highs, H10 and H11 represent Barmer graben that extends NW as gravity lows of the Jaisalmer basin and SE as gravity lows of the Sanchor basin which are hydrocarbon bearing Mesozoic basins. Towards SE, they are connected to the Cambay Tertiary basin that is an oil producing basin where Tertiary sediments are underlain by Deccan trap followed by Mesozoic sediments in certain sections (Section 9.4). The gravity highs, H10 and H11 represent high density mafic rocks forming the shoulders of these rift basins indicating their active nature formed due to mafic intrusive. This indicates that this whole region was active during Mesozoic period forming a large rift basin while during Tertiary period its southern part of Cambay basin was re-activated by Deccan trap eruption through Gulf of Cambay. In this regard, the Mesozoic volcanism that gave rise to these basins might be related to the contemporary break up of Africa from India along its west coast of Saurashtra and Kachchh due to Karoo volcanic (Plume) (Sections 7.11 and 9.4) that might have been initiated through the Gulf of Kachchh (Inset, Fig. 6.64(c)) where maximum thickness of the Mesozoic sediments (~2.5-3 km) including along the coastal parts of Kachchh and Saurashtra with this gulf have been reported (Section 9.2.4 and 9.3.2).

(e)   It is interesting to note that most of the major rivers along the Himalayan front in this section are located in between the gravity highs H1-H5. For example from east to west rivers Kali, Ganga, Yamuna, Satluj, Ravi, Chenab and Jhelum flow from High Himalayas to plains across sides of the gravity highs H1, H2, H3, H4 and H5 (Fig. 6.65(a). These rivers could not be superimposed over this figure as its clarity would have been lost and therefore the cross points of these rivers along Himalayan front are plotted on this figure for ease of comparison. However, river Ghaggar flows from southern flanks of gravity high, H4 that is associated with Punjab upland.

(f)   Geoid undulations also reflect surface/subsurface density in homogeneity similar to the gravity field. As geoid varies inverse of the distance from the source, it reflects deeper (regional) sources better compared to the gravity field that varies inverse of the square of the distance. We, therefore also examined the geoid data of NW India for deep seated density in homogeneity in this section. The observed geoid (Fig 6.7) for $360° \times 360°$ for this region was corrected for a broad regional field of $50° \times 50°$ that is subtracted from it. The resultant geoid anomaly is given in Fig. 6.65(b) that shows several significant anomalies. The most significant in this case are the geoid highs, H1 under Ganga basin related to Delhi-Lahore-Sagodha ridge that is caused by the crustal bulge associated with the lithospheric flexure of the Indian plate along the Himalayan Fold Belt as discussed above for gravity anomalies. As this data set is adopted from the satellite measurements, it extends west wards beyond the limits of the Indian Territory that shows the extension of the geoid high, H1 up to Lahore-Sagodha defining the true extent of this ridge. It also shows the geoid lows, L1 and L2 related to Ganga fore deeps with a small high, H2 in between related to the Nahan Salient. There are some significant geoid anomalies in the western part of the map such as geoid highs and lows, H7 and L4 that are related to the Sulaiman ranges of the Western Fold Belt (Pakistan) and the Indus fore deep, respectively. The geoid highs, H6 over High Himalaya and suture zones are related to its high elevation and associated high density rocks. It is interesting to observe that the geoid highs, H6 and H7 related to the NW Himalayan front and Western Fold Belt (Pakistan), respectively are almost parallel to the geoid highs H1 and H3. This indicates their genetic relationship suggesting the latter representing the effect of basement uplift due to lithospheric flexure of the Indian plate on the western side due to the Western Fold Belt (Pakistan). Lithospheric flexure and related crustal bulge can cause even mafic intrusive due to decompression in the mantle enhancing the gravity and geoid highs (Hirano, 2006).

(g)   Regional and residual gravity anomalies along Ganganagar-Chandigarh profile extended to the MCT (Fig 6.64(a)) adopted from the respective maps are given in Fig. 6.65(c) and (d) that are modeled below these fields. The regional field (Fig. 6.65(c)) shows a sharp gradient beyond Chandigarh that is modeled due to crustal thickening up to 56 km under the MCT for a density of 3.0 and 3.4 g/cm^3 for the lower most crust and the upper mantle, respectively. These density values suggest the presence of eclogite in these sections as has been suggested above for the Central Himalaya (Section 6.2, Fig. 6.27). This model also shows an uplift of Moho by 2-3 km east of Ganganagar and similar depression in the foredeep around Chandigarh that is the effect of lithospheric flexure as described above. Uplift of the lower crust to the north of Chandigarh across the Nahan salient may be related to the interaction of high density rocks of the ADMB with the Himalayan fornt that intersects this profile in this section and has uplifted the Nahan salient as is also described in the next section. Part of it may also be due to Himalayan orogeny caused by convergence and folding. Maximum crustal thickness along this profile to the north of the MCT is ~ 60 km.

(h) Fig 6.65 (d) shows the residual anomaly along this profile and also a total intensity magnetic profile recorded from Ganganagar to Chandigarh. The residual field (Fig. 6.65(d)) is primarily modeled due to the basement ridges and depressions caused by the lithospheric flexure east of Ganganagar while the magnecic field is modeled due to susceptibility variation in the basement. It shows a three layer model with top being Quaternary and Tertiary sediments (2.22 g/cm^3) followed by Proterozoic metasediments (2.55 g/cm^3) that extend from the Indian Peninsular shield and form the basement in this region. It is followed by gneissic basement (2.72 g/cm^3). The high density basement (2.85 g/cm^3) starting to the north of Chandigarh represents the extension of the ADMB and associated rocks such as Malani volcanic under Ganga basin that intersects the profile at this point and extends under Himalaya. This model also shows thrusted rocks along the Himalayan thrusts (HFT, MBT and MCT) and these thrusts join to the plane of decollement (MHT) at subsurface level. It also shows the maximum thickness of sediment of 7-8 km in the Himalayan foredeep that is confirmed from seismic profiles in this section (Section 6.3.2, Powers et al., 1998). The large gravity low north of the MCT is caused by low density metasediments and crystalline basement rocks as shown in the model.

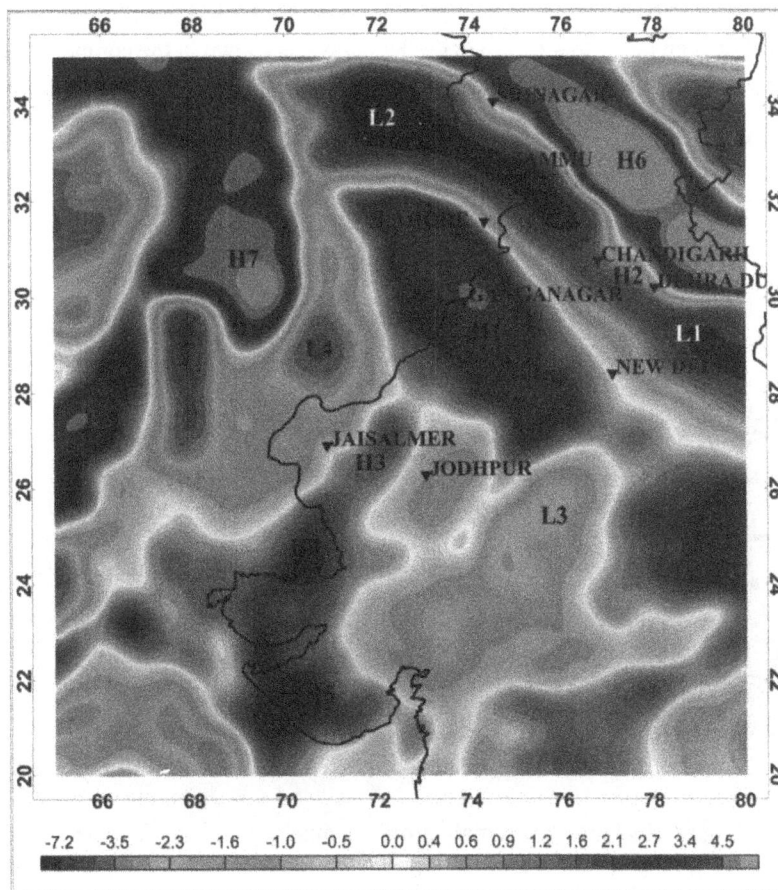

Fig. 6.65(b) Residual Geoid map of NW India western Himalayan front, and adjoining region showing a geoid high, H1 related to the Delhi-Lahore-Sargodha ridge and other highs, H3-H5 that are attributed to the mafic intrusive that has given major rivers along Western Himalayan front. [Colour Fig. on Page 789]

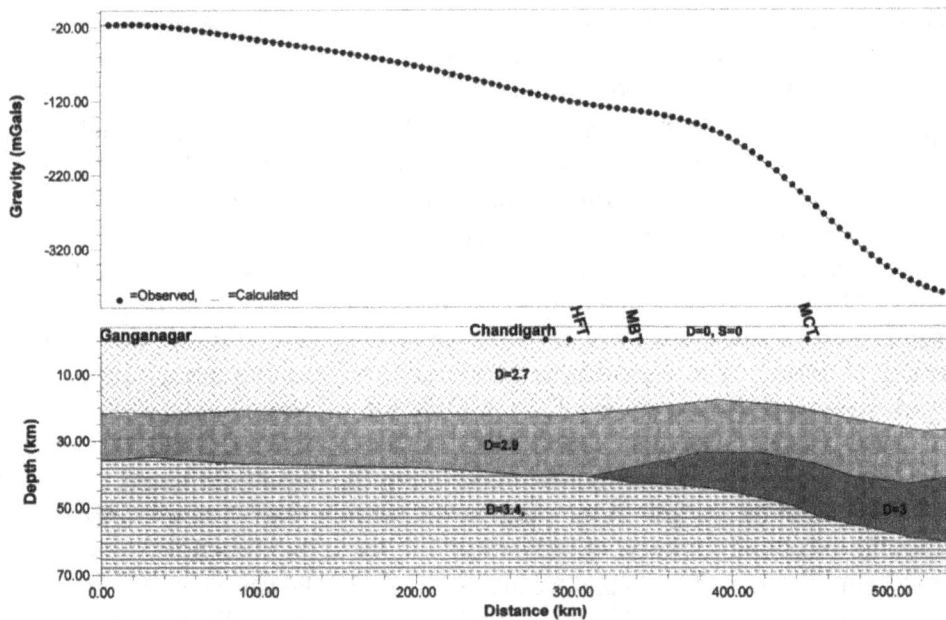

Fig. 6.65(c) A regional gravity anomaly profile from Ganganagar-Chandigarh (Fig. 6.65(a)) based on low pass filter of the Bouguer anomaly map of NW India for wave number < 0.03. It shows a sharp decreasing gradient due to crustal thickening up to ~ 60 km under Himalaya as shown in the computed model for a density of 3.0 and 3.4 g/cm³ of the lower crust and upper mantle, respectively. It also shows a crustal bulge of 2-3 km NE of Ganganagar due to lithospheric flexure of the Indian plate.

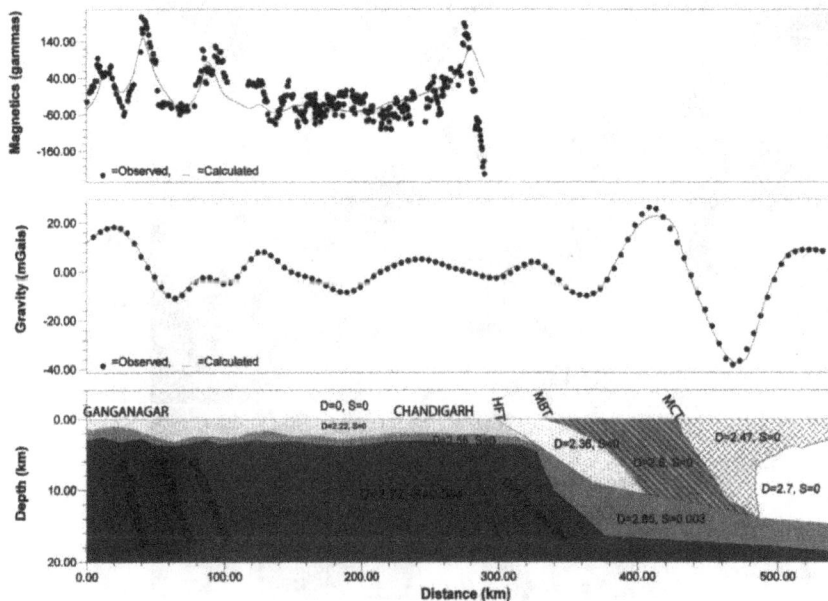

Fig. 6.65(d) The corresponding residual gravity anomaly profile, Ganganagar-Chandigarh (Fig. 6.65(a)) and computed model with densities, D in g/cm³ and susceptibility, S in SI units. It shows a top layer of sediments followed by metasediments of Proterozoic period and basement. The high density body in basement (2.85 g/cm³) represents the Aravalli Delhi mobile belt where it intersects the profile along the Himalayan front that extends under the Himalaya. It also shows the Himalayan thrusts, HFT, MBT and MCT occupied by thrusted rocks joining to the plane of decollement (MHT) at subsurface level.

(iii) Basement Uplift under Nahan Salient

A residual gravity anomaly profile $(X - X')$ across gravity highs centered at the Nahan Salient (H1-H4, Fig. 6.65(a)) with respect to the surrounding region is modeled in Fig. 6.66 to estimate the uplift in this section. Though this profile is almost parallel to the Himalayan front, it is across the gravity highs, H1-H4 as we want to model these gravity highs as basement uplift under the Nahan salient with respect to adjoining re-entrants. As it represents a residual anomaly, it is modeled due to uplift of the basement for suitable densities of the basement and the overlying sediments. Bulk density of the basement is assumed as 2.8 g/cm^3 that is slightly more than that for gneissic basement due to the presence of mafic intrusives as envisaged above. It provides an uplift of about 2-3 km in the Nahan salient with respect to adjoining Kangra and Dehradun re-entrants where a basement depth of 6-7 km has been reported (Fig. 6.46(a) and (b)). It is supported from a seismic reflection profile that also suggested an uplift of 1.5-2.0 km in this section (Baruah et al., 2009). The computed model also suggests the depth to basement as 7.0 km, which almost same as that provided by shallow seismic section (Fig. 6.46(a) and (b)).

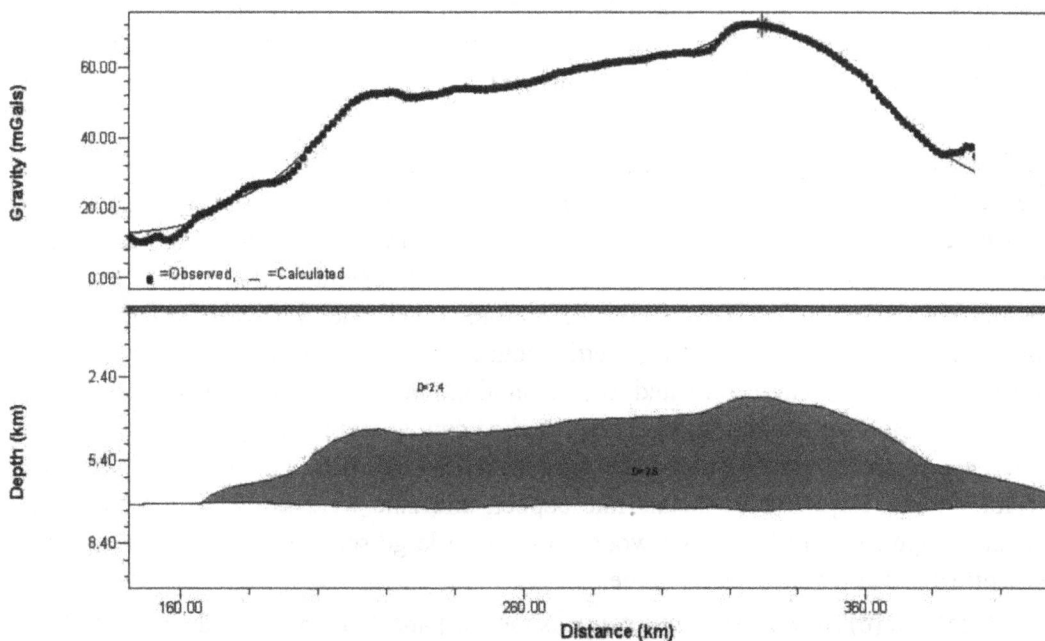

Fig. 6.66 Modeling of gravity highs in the Nahan salient along a profile XX' (Fig. 6.65(a)) showing an uplift of about 2 km and depth to basement as ~7 km.

(iv) Uplift, River Courses, and Lost River Vedic Saraswati

Some significant points in this regard are as follows:

(a) The major rivers along the NW Himalayan front are controlled by gravity highs (H1-H5, Fig. 6.65(a). It suggests that these gravity highs are related to recent uplift. Maximum gravity highs in amplitude and extent are observed over the Nahan Salient, which is connected to the uplift of the Punjab upland in Ganga basin. This uplift in the Nahan Salient along the NW Himalayan front can be attributed to high density sub-surface intrusive. Modeling of the gravity highs related to the Nahan salient suggests a basement uplift of about 2 km under the salient compared to the adjoining Kangra and Dehradun re-entrants.

This uplift has changed the drainage pattern towards the west and the east of this section. This change in drainage pattern has been dated as 6-5 Ma as referred to above (Clift et al., 2008a, Najman et al., 2009).

(b) A part of the above uplift is of recent origin as observed from the higher topography in the Nahan salient compared to adjoining sections (Fig. 6.63). This can be attributed to isostatic rebound due to deglaciations and monsoon activity during Holocene time. Thakur et al., (2007) have shown uplifts south of the Himalayan front along Garhwal Himalaya in the Piedmont zone where sediments of 15-5 Ka, especially at 11.5 and 5 Ka, have been affected by this uplift. Both these periods almost coincide with the period of deglaciation after the Pleistocene and Neoglaciation periods as discussed above (Appendix II) and uplifts during these periods can be caused by isostatic rebound. Uplifts due to isostatic rebound after deglaciation is a well known phenomenon monitored at several places especially in Scandinavian countries where an uplift of 0.9 cm/year has been reported that produced a gravity anomaly attributed to mantle flow (Ekman and Mackinen, 1996). This section of gravity highs in Nahan salient related to recent uplift should be investigated for channel flow that has been advocated for Himalayan orogeny as discussed above (Harris, 2007). Searle et al., (2007) have indicated some signatures of channel flow in the adjoining section of Zanskar range.

(c) Based on isotopic studies in the Indus delta and provenance analysis, Clift et al., (2008b) have suggested that the Holocene erosion of the Lesser Himalaya was triggered by intensified summer monsoon which has caused its maximum uplift during this period due to orography (Hodges, 2004). Erosion during 6-5 Ky BP (Appendix II) is further aided by flow of glaciers and biotic effects due to deforestation and other activities by humans and animals that are important for uplift (Dietrich and Perron, 2006; Zalasiewicz et al., 2008). Among other human activities related to erosion along hill slopes and climate change was smelting of metals for weapon making that required cutting and burning of large amount of woods, which started almost around the same time (5-4 Ky BP). This has been specially true for western Rajasthan as metals like copper, and zinc were available nearby in the Aravalli Ranges. Deforestation and wood burning on large scale would have aided green house effect and arid climate at that time.

(d) Fig. 6.63(a) and (b) show rivers Satluj towards the west and Yamuna towards the east of the Punjab upland which is associated with gravity high H4 observed over the Nahan Salient along the Himalayan front as discussed above. The Punjab upland forms a flat region with a slightly higher elevation than adjoining regions. The river Ghaggar flows from the southern margin of this gravity high (H4, Fig. 6.65(a)) over the Punjab upland which separates it from the catchment area of the Himalayas. If it is true that the river Ghaggar is the proverbial mighty river, Vedic Saraswati (Radhakrishna and Merh 1999; Valdiya, 2002; Rai, 2006), then this recent uplift along MBT and HFT and Punjab upland associated with the gravity high, H4 might be responsible for changes in the geomorphology of this area depriving the river Ghaggar of the perennial source of water. Radhakrishna (1999 b) suggested that this river nurtured the Vedic civilization in Rajasthan and Gujarat during 6-5 Ky BP (Appendix II) which continued up to 3.5-4 Ky BP, when the population migrated towards the east probably due to reduced water supply in this river that might be caused by recent uplift in Nahan salient as described above.

(e) This part of the Himalayan front is characterized by several large magnitude earthquakes (> 5) and hot springs (GSI, 2000), which indicates stressed section due to underthrusting of high density rocks that characterizes this section (gravity highs, H2-H4, Fig. 6.65(a)). The Kangra earthquake of 1905 of magnitude 8.0 with epicenter at about 32°10′ N, 76°20 E is one of the great earthquakes in this section whose epicenter coincides with the western margin of the gravity high, H4. Oil wells in this section are also over pressured (Powers et al., 1998) that also indicate a stressed section where seismic activity is expected.

(f) Sedimentological studies along with luminescence dating in the western Ganga basin has also suggested a dry spell during 6.5–3.6 Ka after a wet spell during 9.6-6.5 ka (Bhosle et al., 2008) that provided sufficient water resources along the various rivers for development of Vedic civilization. However, it reduced slowly during the dry period especially in rivers not connected to Himalayan glaciers, such as Ghaggar, the then river Saraswati. The civilization along this river migrated eastwards. Based on the study of stalagmite records in two caves in SW China using $\delta^{18}0$ record, Hu et al., (2008) have suggested dry spells in Asian monsoon even in that region during 4.8-4.1 and 3.7-3.1 Ka, which almost coincides in general with those suggested for NW India. In fact, excavations along HFT in this section also suggest recent uplift and neo-tectonic activity during Late Quaternary to recent times (Thakur and Pandey, 2007).

6.4 Geodetic Measurements and Deformation Models

Based on study of satellite imagery of large faults in Central and South east Asia, Molnar and Tapponnier (1975) suggested Indenter model for collision of Indian and Asian plates and tectonics related to it. Accordingly, Indo-China, and east of Tibet (Sunda land) moved southward along the strike slip Red River fault due to this collision and Tibet moved eastward along another large strike slip fault, Altyn Tagh fault to the north of it. It is, therefore, suggested that the shortening due to Himalayan collision is accommodated due to movements along these strike slip faults. The lateral extension in E-W direction resulted into N-S graben structures such as Yadong-Gulu rift (Fig. 6.23).

Geodetic measurements using GPS provide a new constraint on the tectonics of a region. One of the earliest results from global GPS measurements have been the convergence of Bangalore at a rate of 36-37 mm/yr (Paul et al., 2001) directed 28°NE. This is significantly slower than 50-47 mm/yr inferred for Indian plate from paleomagnetic measurement and seafloor spreading magnetic anomalies as discussed in Section 6.1, and Nuvel 1A NNR prediction directed 24°NE (Jouanne et al., 2004). Displacement of Bangalore is also consistent with the displacement predicted using revised plate reconstruction proposed by Gordon et al., (1999) (37 mm/yr) with a 38°NE azimuth. This discrepancy between the two has been attributed to the present day shortening of the oceanic lithosphere in the Indian ocean south of Indian continental lithosphere possibly within the Central Indian Ocean Deformation Zone (Figs. 5.35 and 8.5(a)) that is folded and faulted where several large earthquakes have been reported. It extends eastwards and interacts with the Sumatra-Java subduction zone. The epicenter of the great Sumatran earthquake of 2004 that gave rise to the largest Tsunami waves coincides with its junction with this subduction zone (Section 8.4). Further investigations have suggested a convergence of 10-20 mm/yr along the Himalayan arc which varies from west to east suggesting that about half to one-third of the convergence between the Indian and Asian plates is absorbed in the Himalayas due to thrusting and uplift while the remaining part is absorbed in Tibet due to internal deformation (Fig. 6.67(a)). Shortening inferred between the Indian plate and Lhasa (Southern Tibet) is consistent with the east-west extension of the Tibetan Plateau (Banerjee and

Burgmann, 2002; Jade, 2004). The contraction vector across Himalaya changes from N20°E in the northwest Himalaya to N25°W in east Nepal consistent with east-west extension of S. Tibet (Paul et al., 2001). Based on detailed GPS measurements in Western Himalaya, Jade et al., (2004) suggested slow relative motion in Ladakh with a right lateral slip along the Karakoram fault as only 3.4 ± 5 mm/yr. This slow rate of movement is similar to that estimated for Late Holocene time but does not agree with a higher rate of movement of 30-35 mm/yr. This indicates that Tibet behaves more like a fluid than like a solid plate that deforms internally, consistent with West-East flow under the Tibetan Plateau. It has been suggested that the upper crust is decoupled from the lithospheric mantle by partially molten lower crust that indicates that the upper crust deformation must be strongly influenced by mantle lithosphere deformation. An opposite view is expressed by Wright et al., (2004) based on satellite radar interferometry (InSAR) observations who suggested that the deformation in Himalaya and Tibet is distributed throughout the continental crust which appears to behave like putty (Kerr, 2004). They found that the slip on the Karakoram and Altyn Tagh strike slip faults are much less than what it should be to account for the observed convergence, which suggests significant internal deformation in Tibet.

Based on detailed GPS measurements, Calais et al., (2006) suggested that large part of Himalaya Tibet and Asia neither behaves like 'strong and brittle' nor 'weak and viscous' as was previously considered. Instead, it is a combination of both, viz. (i) Internal continuous strain confined to high elevation regions that are quite strong and breaks on collision as suggested by plate-tectonics; and (ii) un-resolvable strain rates over a large part of Tibet that is ductile and deforms like a play-doh on collision. Fig. 6.67(a) (Gahalaut, 2008) shows the GPS vectors in different sections over Tibet that also suggests internal deformation and west to east flow, which takes a southerly turn along Eastern Syntaxis, south of Tibetan Plateau.

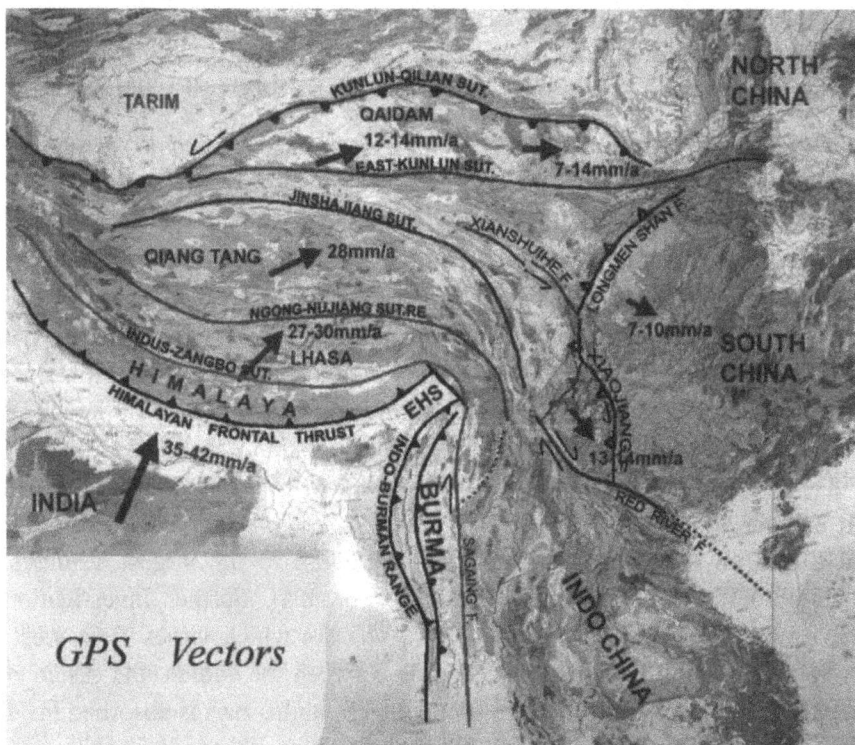

Fig. 6.67(a) GPS vector over Himalayas and Tibet (Gahalaut, 2008). [Colour Fig. on Page 789]

6.4.1 Eastward Extrusion of Tibetan Crust and Sichuan Earthquake of 2008

Based on GPS measurements, Wang et al., (2001) suggested eastward movement of the Tibetan Plateau south of Kunlun fault relative to both the Indian and the Asian plates accommodated through rotation of material along the Eastern Syntaxis. They also suggested a east-southeast-ward movement of North China and South China blocks with a speed of 2-8 and 6-11 mm/year confirming the rotation of the Tibetan crust along the Eastern Syntaxis. The fast direction of shear wave splitting also shows similar changes (Wang et al., 2007), which is consistent with lithospheric extrusion. Zhang et al., (2004) using Global Position System velocities at 533 points in the Tibetan Plateau and its margins, suggested that present day deformation in Tibet can be described as continuous extension in ESE-WNW direction, and shortening in NNE-SSW direction. The interior of the plateau moves eastward and takes a southward turn along Eastern Syntaxial bend. Gan et al., (2007) have provided GPS velocity vector in this region relative to stable Eurasia that suggests the eastward movement of the Tibetan plateau and southward rotation along Eastern Syntaxial bend. They attributed it to the eastward escape of highly plastic upper crustal material driven by a lower crustal viscous channel flow generated by lateral compression and gravitational buoyancy. They also suggested that there is no convergence between the Eastern margin of the Plateau and Sichuan basin to the east indicating southward flow along this margin. Several seismic studies as described in Section 6.2 have also suggested west to east extrusion of the Tibetan crust.

Eastward extrusion of the Tibetan crust has been attributed by Royden et al., (2008) to trench rollback in the western Pacific and Indonesia. The northern Tibet, however, remained low till Eocene/Oligocene time. At this time, large parts of subducted Eurasian lithosphere extruded eastwards. This coincided with the fast eastward shift of the trenches in the western Pacific, Pilippine, and Indonesian oceanic subduction zones. This rapid rollback of trenches in the western pacific continued up to 15-20 Ma when the eastward extrusion of Tibet also halted and material started piling up in Eastern Tibet causing its uplift and east-west extension over Tibet, giving rise to several north-south oriented grabens. At this stage, the eastward extrusion took a southerly turn along the Eastern margin of the Tibetan Plateau and the strong lithosphere of Sichuan basin (Fig. 6.67(b)). However, Kerr (2008), has cautioned that this does not satisfactorily explain the east-west extensional regions observed over Tibet. Wang et al., (2008) based on similarity between crustal and upper mantle fabric in seismic imaging, negated the lower crustal flow model. Royden et al., (2008) have further added that the Sichuan earthquake of magnitude 7.9 on May 12, 2008 occurred beneath the steep margin of the Tibetan plateau at a depth of 10-20 km adjacent to the mechanically strong Sichuan basin which suggests slow uplift consistent with the crustal flow. A geological section (Fig. 6.67(b)) through the epicenter shows intricate thrust sequence emplaced along the Pengguan massif that might be related to west to east extrusion. It is almost similar to thrust earthquakes along the Himalayan front as described in Sections 6.1.10 and 6.2.5 for Muzaffarabad (Kashmir) earthquake of magnitude 7.6 at a focal depth of 20-25 km (October, 2005, Section 6.3.4). Gupta and Rao (2008) have shown similarities in the decay rate of the aftershocks between the two earthquakes separated by thousand of kilometers. Post-seismic displacement studies related to this earthquake may provide some significant information about the Tibetan crust in this section. Based on finite element modeling of topography using Indian plate as an undeformable indenter in a viscous region, Benedick et al., (2008) suggested that some elevated and low relief topography in the northern part of Tibetan Plateau can be attributed to lower crustal flow. They also suggested that in the best fit model the indenter extends north of BNS. Based on a three dimensional numerical model of deformation in a viscous crust, Cook and Royden (2008), developed a model of the plateau that shows several similarities to the Tibetan plateau such as overall morphology, rotation around Eastern syntaxis, and E-W extension that may be related to rapid eastward flow of crustal material in a weak zone.

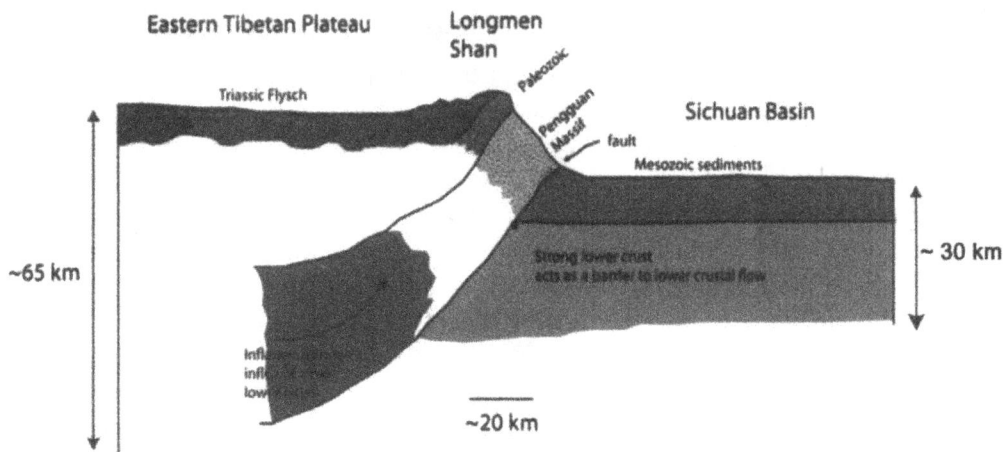

Fig. 6.67(b) A geological cross section showing imbricated thrusts in the epicentral area of the Sichuan earthquakes of May 12, 2008 (Gupta et al., 2008; MIT, 2008). Hypocenter is approximately plotted as black dot under the Eastern thrust.

Based on GPS measurements along the southern margin of the Tibetan plateau, Shen et al., (2005) suggested that the upper crust is composed of several fragmented blocks overlying a mechanically weak lower crust which is deformed due to rotation of upper crustal blocks. There are primarily two basic questions: (i) whether the tectonic deformation is blockwise or broadly distributed; or (ii) the N-S shortening of Tibetan plateau is absorbed by crustal thickening or eastward extrusion. Some believe that the collision zone is composed of a collage of lithospheric blocks and deformation takes place mainly along block boundaries delineated by large scale, rapidly slipping strike slip faults. Since the blocks cannot absorb deformation internally, in this view N-S shortening of Tibetan plateau is accommodated by rapid eastward extrusion. Others, however, believe that crustal strength is reduced by existence of a ductile lower crust making deformation between the upper crust and mantle 'decoupled'. Therefore, northward advancement of the Indian plate results in thickening of the lower crust and broadly distributed deformation. Further, underthrusting of the Indian lithosphere would also result in modifying the thermal structure with relatively high temperature at mid-crustal depths causing ductile flow and decrease in flexural rigidity of the lithosphere. It is, therefore, likely that shortening in Tibet has been absorbed both by crustal thickening and lateral escape (Fig. 6.68).

Deformation in Tibet can therefore take place in following two ways (Vergnolle et al., 2007):

(a) Deformation due to rigid motion of lithospheric blocks with strain concentrated along narrow fault zones.

(b) Viscous flow of a continuously deforming solid where faults play a minor role.

6.5   Summary

6.5.1   Forces and Kinematic Model

Fig. 6.68 (Zhang and Wang, 2007) shows a Kinematic model where compressional forces of the Indian plate are transferred to Tibet across the Himalaya causing deformation in the form of extension of Tibet along strike slip faults, Karakoram, and Altyn Tagh fault taking a southerly trend along Red River fault. However, shortening of Tibet is absorbed in the E-W extension producing rift basins of Tibet related to the eastward flow of the Tibetan crust. There will also be rapid erosion along the Himalayan front due to rainfall and monsoon, causing a southward shift of the front itself. Fig. 6.68 also shows extension and escape of Tibetan lithosphere eastward and internal northward flow up to Altyn Tagh fault where it takes an easterly turn. Briefly, the following are the important forces that have been responsible for structures and tectonic activities in the Himalayas.

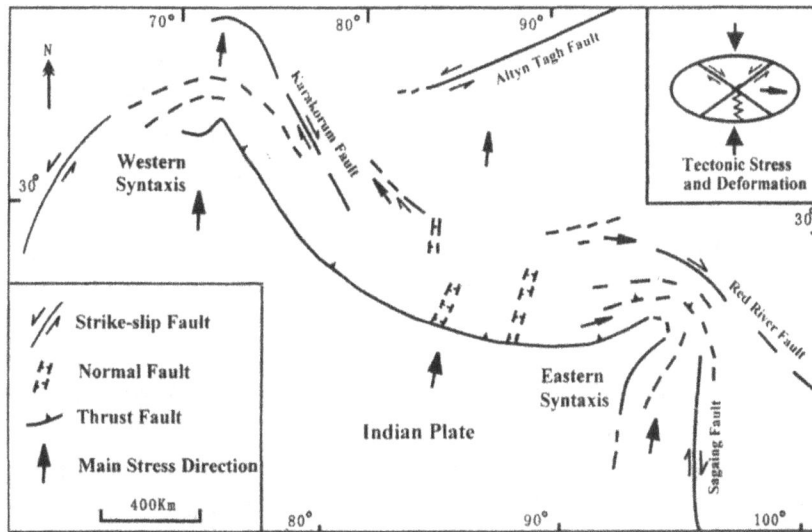

Fig. 6.68 A Kinematic model of tectonics between Indian and Eurasian plate due to compressional forces giving rise to rift basins over Tibet and extrusion of Tibet eastward along two major strike slip faults, viz. Karakoram and Altyn Tagh faults; and southward escape along the Red River and Sagaing faults (Zhang and Wang, 2007).

(a) SW-NE compressional forces due to convergence of the Indian and the Asian plates. This has its origin in the Indian Ocean in the form of sea floor spreading across the Carlsburg ridge and has been responsible for thrusting of the Indian plate under Tibet and related south verging thrusts which have given rise to the Himalayas. Initiation of Tibetan plateau started during Early Cretaceous due to collision of Asia with Lhasa block towards the north where a fold and thrust belt developed between 144-110 Ma. India-Tibet continental collision during 45-35 Ma and underthrusting of the Indian plate led to crustal thickening and isostatic uplift. These compressional forces caused N-S shortening of the Tibetan plateau. The E-W extensional forces over Tibet (Fig. 6.68) and may be related to the eastward flow of Tibetan lower crust that takes a S-ward turn along Eastern margin of Tibet forming Eastern syntaxis and several N-S oriented grabens.

(b) Crustal thickening led to a ductile lower crust that extruded eastward and left lateral Red River fault was initiated at 35 Ma. By late Oligocene (30 Ma), the Tibetan Plateau was relatively low (~ 2 km) and the proto-Himalaya was quite subdued. At about 28 Ma, the Gangdese thrust developed and started moving southward in the from of several south-directed small thrusts. Subsequently, the south verging MCT developed at about 24 Ma and activities continued along it up to ~ 11 Ma when the formation of the MBT was initiated. Subsequently, a broad zone of deformation under the MCT was active during 10-4 Ma giving rise to the classical Himalayan inverted metamorphic sequences. By 9 Ma, the Tibetan plateau had attained its full elevation of 5 km and experienced E-W extension across N-S trending grabens (Harrison et al., 1998).

(c) The Himalayan-Tibetan mountain system significantly influenced Neogene climate. Erosion and uplift related to wind and monsoon known as orography became important during this period. The continuous erosion of the southern section of hill slopes due to wind, monsoon, and other related effects over millions of years would cause uplift and shift of the Himalayan front southwards. These factors would also modify the thermal structure under Himalaya and Tibet that would cause N-S channel flow giving rise to leucogranites of Miocene Times (24-10 Ma). N-S channel flow in the Indian plate under the Himalaya also accounts for southward growth of Tibetan plateau and intrusion of leucogranites of Miocene time (Fig. 6.36(a)).

6.5.2   Crustal and Lithospheric Structures – Geodynamics and Seismotectonics

Geophysical studies as described above have brought out the following significant points about crustal and lithospheric structures in this region due to underthrusting of the Indian plate under Tibet:

(a) There are several thrusts along the Himalayan front that brought up the older rocks overlying the younger formations. Important among them are the MBT and the MCT. They are connected to the Main Himalayan Thrust or Himalayan Sole Thrust at depth. Most of the seismic activity originates from their junction with this thrust. This has made the Himalayan front as one of the seismically most active sections. Several large earthquakes have been reported from this section. Most of these activities originate from the upper crust as discussed in case of Kashmir and Sichuan earthquakes of October, 2005 and May, 2008 (Sections 6.3.4 and 6.4.1), respectively except under the Hindu Kush Range and Burmese arc where intermediate-deep focus earthquakes from the upper mantle have been reported.

(b) Deep focus earthquakes in the Himalaya are limited to the Hindu Kush and Arakan Yoma Fold Belt (Burmese arc). They are attributed to fast high angle subduction and slab breakoff subducting Indian lithosphere. However, the subduction models proposed for these sections are almost similar to those in other parts of the Himalaya where slab breakoff will be difficult to visualize. Both these sections, however, are unique in occurrences of strike slip faults: Chamman-Panjshir fault in case of Hindu Kush section, and Sagaing fault in case of Burmese arc, which interact with the subducting Indian plate in these sections to cause slab breakoff causing deep focus earthquakes.

(c) The present day distribution of seismic activity suggests enhanced seismic activity along the western and the Eastern Himalayan fronts while the central part shows a gap that has been attributed to a large recurrence period of seismic activity in this section. Large recurrence period in the central part may be caused by locking of the subducting Indian lithosphere that may happen at the ramps in the decollement plane (Main Himalayan Thrust). It is observed that ramps, discontinuities and interfaces separating rocks of different physical properties are important for seismic activity in Himalayas and might be favoured sites for strain accumulation. Another important factor for seismicity is the angle of subduction. Based on P-wave tomography Li et al., (2008) have suggested almost sub-horizontal subduction of the Indian crust in the western and Eastern Himalaya while it is about $45°$ in the central Himalaya that makes the former seismically more active than the latter. Another important factor for the Himalayan seismicity is the interaction of the structures and rock types extending from the Indian shield to the Himalayan front. It has been shown that seismically active sections along the NW Himalayan front are located where it interacts with the extensions of the Great Boundary Fault and Phulad ophiolite thrust that occur along the Eastern and the western margins of the ADMB, respectively. Similarly, the intersection of the NW Himalayan front with the Jaisalmer-Ganganagar ridge possibly caused by the flexure of the Indian plate due to Western Fold Belt (Pakistan) shows intense seismicity where the epicenter of the great Kangra earthquake of 1905 is located.

(d) The crust thickens under Himalaya and Tibet to a maximum of about 80 km under S. Tibet (Tethyan and Lhasa blocks) reducing to 60 km under North Tibet due to collision of the Indian and the Asian plates. This thickening of the crust is suggested to take place as systematic increase in the depth of the Main Himalayan Thrust (decollement plane) and Moho from Himalayan foreland to the S. Tibet or as a ramp in these discontinuities under High Himalaya. The crustal thickness in the western part west of the Chamman fault and Burmese arc in the east is less of the order of about 55-60 km where predominantly strike slip motion is reported.

(e) Spectral analysis and modeling of Bouguer anomaly and tomography experiments suggest that the lithospheric thickness changes from about 140 km under the Himalayan foreland to a maximum of about 260-300 km under S. Tibet (Lhasa block) reducing to about 100-160 km under N. Tibet (Qiangtang terrain and north of it). The reduced lithospheric thickness under the Qiangtang terrain is attributed to the detachment of under-thrusted lithosphere attributed to the interaction of the subducting Indian and Asian lithospheres. Detachment of lithosphere in the mantle has also been envisaged under Eastern Himalaya and South Tibet (Singh and Ravi Kumar, 2009). Lithosphere in this section is characterized by low density and high velocity rocks indicating that it may be part of the subducted Indian and Asian lithospheres. Most of the studies suggest the extension of underthrusted Indian lithosphere up to south of the BNS though some studies suggest its extension north of the BNS to account for the regional Qiangtang bulge. Li et al., (2008) based on P-wave tomography have suggested the extent of underthrusted Indian lithosphere differently in different sections, the maximum being in the NW Himalaya where it has covered the entire Tibetan plateau, while it is limited to the Lhasa block south of BNS in the central part, and the Himalayan fold belt in the Eastern Himalayan section. In the central part, it is in conformity with the model given in Fig. 6.37(b) which shows maximum crustal and lithospheric thickness under the Lhasa block south of BNS where it encounters the subducting Asian lithosphere from the north.

(f) Effective elastic thickness computed from the observed gravity field and topography changes from 50-70 km under the Himalayan foreland to about 30 km under the Himalayan Fold Belt and South Tibet indicating a ductile lower crust in this section. High conductivity zones under Central Nepal and S. Tibet at depths of about 15-25 km coincide with the hypocenters of earthquakes along the Main Himalayan Thrust. It may be related to fluids or partial melt in these sections, which is synonymous to ductile lower crust as envisaged above. Such partial melts and low velocity zone in the crust has also been envisaged in central Andes based on 3-D models (Prezzi et al., 2009). A conductive zone at almost the same depth (20-25 km) has also been reported in NW Himalaya. This may be related to subducting fluid along the underthrusting Indian plate. It emphasizes the role of fluids for Himalayan earthquakes.

(g) Reduced lithospheric thickness under the Qiangtang terrain causes uplift of isoterms due to asthenospheric upwelling, giving rise to partial melt as ductile lower crust in this section. It facilitates west to east extrusion of the Tibetan crust due to convergence between the Indian and the Asian plates. Similarly, the thick crust under S. Tibet causes uplift of isotherm due to radiogesnic heat causing ductile lower crust. Some of these melts may flow N-S in the Indian crust. This is referred to as channel flow. Formation of leucogranites (24-10Ma) in the High Himalaya is attributed to the N-S channel flow as referred to above. Gaillard et al., (2004) have shown similar experimentally derived electrical conductivities of hydrous granite melts under suitable pressure-temperature-H_2O conditions for crust derived magmas similar to Miocene leucogranite plutons exposed in High Himalayan ranges.

(h) West-East extrusion of the Tibetan crust is also supported by GPS measurements and has been attributed to trench rolloff in the Western Pacific (Royden et al., 2008). They connected the stopping of this eastward rolloff during 30-20 Ma to the piling of material and uplift and thrusting of the Eastern Tibet that is continuing and causing seismic activity such as the Sichuan earthquake of magnitude 7.9 on May 12, 2008.

(i) Erosion due to monsoon along the Himalayan front causes southward shift of the Himalayan front resulting in southward growth of the Tibetan Plateau that has caused the shift of thrusts with time from MCT southwards as MBT and HFT. This, in turn, may also cause W-E extrusion and escape of the Tibetan lithosphere taking a southward turn along Red River fault.

(j) Residual Gravity highs delineated along NW Himalayan front (HFT), control present day courses of the main rivers and therefore represent recent uplifts. One of them obstructs the connectivity of the river Ghagghar with Himalayan glaciers indicating that this uplift might have obstructed the perennial sources of water to this river that was once a mighty river Saraswati during the Vedic period (6-4 Ka BP). This section representing a recent uplift should be investigated for channel flow. A similar gravity residual high is also delineated along the Eastern Himalaya west of river Brahmaputra coinciding with maximum rainfall where detailed investigation related to channel flow should be undertaken.

(k) Lithospheric flexure of the Indian plate under the load of Himalaya has caused a crustal bulge under its foreland Ganga basin in the northern, western, and the Eastern parts that might have caused mafic intrusives from the upper mantle in these sections. This bulge can also account for enhanced seismic activity in the western and the Eastern sectors where its amplitude is relatively large due to flexure of the Indian plate on two sides.

(l) Based on magnetic anomalies in the Indian Ocean and paleomagnetic measurements on Indian rocks, the present day convergence of the Indian plate is about 4.7-4.8 cm/year. However, GPS measurements suggest a convergence of about 3.8 cm/year between the Indian plate and the Lhasa block. The remaining convergence of about 10 cm/year may be observed in Central Indian Ocean Deformation Zone which is characterized by large scale folding and faulting and seismic activity.The present day convergence between the Indian and the Asian plate is about 38 mm/year that is accomodated as follows:

  (i) Movement on regional thrust fault, such as MCT, and MBT, and formation of Himalayan fold belt.

  (ii) West-East extrusion and deformation and Uplift
    The Himalayan fold belts accounts for about one-third of this convergence (12-14 mm/year) between the Indian and the Asian plates as discussed in section 6.4 based on GPS measurements while other two-thirds is accommodated within the Tibetan plateau due to internal deformation and west to east extrusion. The collision caused at least two major pulses of uplift between 21-17 Ma and second between 11-7 Ma related to Higher and Lesser Himalaya along MCT and MBT, respectively when maximum erosion of the Himalayan front have been reported. Another phase of enhanced erosion at 3.5 Ma coincides with the formation of the HFT.

(m) Spectral analysis of CHAMP satellite magnetic data provides a depth of 155-160 km for large wavelength anomalies, which suggests cold subducted Indian and Asian lithospheres at this depth where temperature maybe less compared to curie point of magnetite ($\sim 570^0$) and would produce magnetic anomalies. However, these depth estimates from the spectrum of the CHAMP data are tentative as they could not be verified using direct modeling of the sources at those levels. Magnetic highs in CHAMP satellite magnetic data passing through the Western Syntaxis may represent mafic basement ridges that might be related to intrusives from plumes related to the breakup of Gondwanaland whose subduction, in turn, may be responsible for the formation of syntaxis and for a large number of seismic activity in this section. Similarly, Siang syntaxis along Eastern Syntaxial bend coincides with a basement uplift related to lithospheric flexure delineated based on residual Bouguer anomaly of this section that might be responsible for its formation. Vogt et al., (1976) have suggested the formation of cusp and increase in seismicity due to subduction of aseismic ridges in several cases the world over.

(n) Most of the faults of the Indo-China plateau east of the Burmese arc such as Sagaing fault and Red River fault (Fig. 6.5) are strike slip faults (Zhu et al., 2009) which may be attributed to S-N and E-W convergence in the Andaman and the South China Sea, west of the Philippine plate, respectively. It has also affected the mountain chains of the Indo-China peninsula (Vietnam, Laos, Thailand). The sub-parallelism of trenches in the Western Pacific and SE and East Asian-Siberian mountain chains are significant suggesting the effect of E-W convergence of the Western Pacific plate on the tectonics of SE and East Asia. The effect of E-W convergence in the Western Pacific is also seen on Bouguer anomaly and fracture systems of South China Fold System that are almost sub-parallel to the trench in the South China Sea. Rollback of trenches in the Western Pacific and Indonesia and their interaction with Eurasia have been attributed for evolution of the Eastern Tibetan plateau (Royden et al., 2008) including Sichuan plateau (Fig. 6.5) where the Sichuan earthquake of May 12, 2008 (Fig. 6.67b; M = 7.9) was located.

(o) The E-W convergence of the Western Pacific plate might even be responsible for the N-S turn of the west-east flow of the Tibetan crust along its Eastern margin (Hodges et al., 2004; Wang et al., 2007) that is also supported from high conductivity in the lower crust along the Himalayan front and south Tibet (Nelson et al., 1996; Unsworth et al., 2005) and resistive strong crust east of it (Bai et al., 2001). It is also supported from deep seismic studies (Wang et al., 2007) that provided a crustal thickness of 62 km under the Sichuan plateau and a low velocity layer in the upper crust synonymous with the high conductive layer referred to above. However, the crustal thickness decreases to 43 km and low velocity layer is absent under the Sichuan basin to the east of Tibet indicating a strong crust subducting under Eastern Tibet under the influence of E-W convergence of the Western Pacific plate. These observations support the role of E-W convergence of the Western Pacific plate on the tectonics of East and SE Asia.

7

Geodynamics of the Indian Continent and Seismotectonics: Isostasy, Archean-Proterozoic Cratons, Collision Zones, Rift Basins, Plumes and Lithosphere, and it's Flexure

Forces that are important on a regional scale of plates and are significant for plate tectonics, mostly operate in the asthenosphere and lithosphere. It is, therefore, important to understand the nature of rock types in this section. The Indian plate primarily consists of the Indian continent and surrounding oceans that are relatively young formed after the breakup of India from Antarctica at about 120-118 Ma. However, the Indian continent is composed of several cratons and mobile belts that belong to the Archean-Proterozoic period. Therefore, though the geodynamics of the Indian plate started from the breakup of Gondwanaland as discussed in Chapter 5 and 6 but its storey is not complete without referring to the cratons and mobile belts of the Indian continent and the processes that formed them. Plate tectonics was largely developed to account for observables in oceans, like sea floor spreading magnetic anomalies and mountain building processes over adjoining continents along subduction zones. It, therefore, accounted for about 200 Ma which are the oldest reported sea floor magnetic anomalies and the age of the ocean. However, continents being much older (Archean-Proterozoic period) required either another theory or an extension of the same plate tectonics theory with some vairantion to explain various processes observed over them. In this regard, the first effort to extend plate tectonics backwards in time during Archean-Proterozoic period was made only after a few years of discovery of plate tectonics based on rock types and structures, especially in regard to Archean greenstone belts (Windley, 1973; Burke et al., 1973, Tarney et al., 1976).

Archean calcalkaline volcanics, tonalities, and granodiorites show similar geochemical characteristics as those found in modern island arcs and continental margins. This suggested processes similar to plate tectonics during Archean-Proterozoic period (Windley, 1976). Greenstone belts have been considered the present day analogue of marginal basins behind the volcanic arc system that shows normal oceanic lithosphere such as the Japan Sea, and the Andaman Sea Karig (1971) suggested that they are formed due to crustal extension caused by periodic splitting of arc systems. The best example of marginal basin is provided by Rocas Verdes in South Chile where a basin opened up along the western margin of South America during Early Cretaceous and developed into a synform by Mid-Cretaceous, bordered by the batholiths of the Pacific margin. It compared well with typical Archean greenstone belts with low grade metamorphism (Tarney et al., 1976).

442

7.1   Archean-Proterozoic Tectonics

Spanning 3.8 to 2.5 Ga the Archean period is largely characterized by folded granitoids and gneisses of 3.5-2.7 Ga alongwith supra-crustal lavas and sediments, including komatitic lavas that form a typical greenstone belt. Komatitic lavas with high percentage of Mgo (upto 29%) are typical of greenstone schist belts world over and are absent from Phanerozoic plate tectonic settings. They require a high temperature of 1800-2000°C in the upper mantle to provide this melt. This suggests a high thermal gradient in the Archean crust. Alternatively they have been brought up by plumes of that period. However, their wide occurrences in most of the green schist belts world over suggest that they are provided by melting of mantle rocks due to the then prevailing high thermal gradients, or that the period experienced plume activity on a much wider scale than subsequent geological periods. Present day dissipation of heat is primarily attributed to plate tectonic processes (70 %) and therefore, in case the primordial earth consisted of material with high temperature and thermal gradient, some form of plate tectonics is required from almost its beginning stage to cool it down to the present stage. A possible model for Hadean tectonics prior to Archean period (4.8-3.8 Ga) can therefore be summed up as creation of komatitic lavas at mid-oceanic ridges and its subduction due to its high density in spite of being buoyant caused by high temperature (Fowler, 2005). Higher thermal gradient implies high heat flux and fast movement and consumption of smaller plates. Subsequently, during Archean period, layered crustal structure had developed that was required to support plate tectonics (Burg and Ford, 1997).

The quartzite-carbonate-mafic-ultramafic association present in greenstone schist belts in gneissic terrian has no present day equivalent but can be regarded as representing a shelf with oceanic crustal rocks like ophiolite sequences. However, during Paleo-Mesoproterozoic period several features related to plate tectonics have been identified world over such as continental platforms, crustal scales brittle behavior such as dyke swarms, rifted continental margins, and linear mobile belts that have been described below while describing Proterozoic collision zones in India. However, some significant differences have also been noticed such as lack of blue schist facies in subduction zones and ophiolites in Archean-proterozoic collision zones. These differences, however, are explained based on high thermal gradient during those periods that would produce high pressure, medium temperature suites such as green schist, and kyanite bearing rocks instead of blue schist facies (Moores and Twiss 1995). It is also suggested that instead of ophiolites, one gets stratiform mafic/ultramafic complexes commonly associated with greenstone schist belt of Archean-Proterozoic periods. However, these differences indicate significant differences in mantle processes compared to the present day activity. The absence of komatiites and occurrences of signatures similar to the present day plate tectonics as described above during subsequent periods suggests that the earth's crust has stabilized during Archean time to give rise to plate tectonic processes as we envisage it today. However, based on paleomagnetic, geochemical and tectono-stratigraphic data, Cawood et al., (2006) have suggested that processes similar to plate tectonics have been operative since 3.1 Ga. This implies that after the stabilization of cratons during Paleo-Mesoarchean period, their movement and collision have been made possible almost similar to present day plate tectonics.

The Appalachian-Caledonide belt of early Paleozoic period is the oldest example that has preserved the clear signatures of plate tectonics. The Appalachian-Caledonian belt (Fig. 7.1) also provides an example of repeated cycles of breakup and collision almost along the same orogenic belts during different geological periods (Whittington, 1973). This belt is affected by at least two periods of deformation during 550-300 Ma (Williams, 1984). Presently the spreading along mid-Atlantic ridge is in progress since about 180-200 Ma. The first stage of collision accounts for the collision of east coast of America with Africa (Mauritanides-Atlas Mountains) and South America while the second stage accounts for Caledonides in Europe. Williams also identified suspect terrains along the eastern margin of Paleozoic North America that accreted to it during closing of Iapetus and opening of North Atlantic oceans. Suspect terrains are sections along margins of a plate that do not conform to general rock types and other characteristics in the adjoining sections of the plate, and are accreted during processes related to plate tectonics. In this regard collision zones generally represent the suspect terrains. Nance et al., (1990) have suggested that several times in the earth's history, continents have joined to form super continents and subsequently rifted to give rise to plate tectonic processes. They suggested that this process appears to be cyclic in nature. The concept of Wilson cycle with almost 100 million years of cycle between collision and rifting is born out of this concept of repeated orogenies along the same region (Cook et al., 1990). Since older greenstone schist belts of Paleo-Mesoarchean time are smaller compared to those of Neoarchean times, it is suggested that during Archean times, smaller plates were formed in larger oceans that grew due to accretion with time. The gneissic cratons were already formed during Paleo-Archean time due to differentiations in large oceans that moved and collided, subsequently forming thrusts and shear zones between them and green-schist-granite complex as back arc or marginal basins. This suggests the lack of rifting phase during Archean times that is prevalent since Paleo-Mesoproterozoic period as demonstrated in the forthcoming sections in case of Indian cratons. Based on these considerations, Ernst (1990, 2009) provided a schematic model of plate tectonics during Archean-Proterozoic period that shows small platelet type features during Archean times developing into larger plates and super continents by Proterozoic time. The Grenville (Fig.5.4) and Pan African (~ 500 Ma) orogenies the world over are regarded to have formed due to plate tectonics type continental collisions forming super continents, during 1000 Ma and 500 Ma, respectively. The concept of super continent is important for their breakup and subsequent movements as they form caps for heat flux to stop from escaping, aiding in their break up. These reconstructions are important to look for conjugate structures and their similarities on adjoining continents. The occurrences of fold belts of Proteozoic-Palaeozoic period that largely represent collision zones of that period along contacts of different continents (Fig. 5.3) in Permian reconstruction suggest that rifting and collisions have been largely taking place almost along the same lines as they represented plane of weaknesses.

Fig. 7.1 Appalachian-Caledonian-Mauritanides belt that was connected during Early Paleozoic period.

The best example of the Archean crustal model is afforded from the Superior Geological Province in Canada based on gravity modeling (Percival et al., 1983) that is also supported by deep seismic reflection studies. The upper crust is granitoids and the lower crust gneiss, with a total crustal thickness of about 35 km, based on synthesis of paleomagnetic, geochemical and tectonostratigraphic data, Cawood (2006), suggested that plate tectonics must have been operative since about 3.1 Ga. Based on detailed geochemical studies, Wyman and Kerrich (2009) suggested a combination of plume and arc magmatism in the Abitibi sub-province that was responsible for the origin of Archean continental lithospheric mantle and its keel in this region. Keels under Archean cratons are considerably well known that have been attributed to form in the continental lithospheric mantle due to interaction of arc related Archean crust with the buoyant residue of the mantle plume. Interaction of plumes with Archean crust has been considered essential for the formation of Komatiitic lavas that contain up to 29% of Mgo and require high temperature of 1800-2000°C in the mantle. The absence of such lavas in present day orogenic belts suggests that the early earth had high thermal gradient and plausibly started from a hot state. Based on detailed geochemical studies of Pilbara super group in Australia, Smithies et al., (2006) suggested that early Archean basaltic proto-continental crust must have started due to plume. Leech (2001) suggested that orogeny proceeds in a cycle from collision and uplift to metamorphism and delamination of the crustal root, leading to tectonic extensional collapse. As described in Chapters 5 and 6, fluid is essential for eclogitization that leads to delamination and tectonic collapse due to its high density. It was therefore proposed that to complete an orogeny, fluid plays an important role. However, in its absence, the orogeny remains incomplete and is referred to as arrested orogenic development (fossil collision zone) that would show crustal thickening, while complete orogenic belts (fully mature) do not show any significant increase in crustal thickness.

7.1.1   Triple Junction

Triple junctions of sutures are an important feature of present day plate tectonics. Triple junctions are points where boundaries of three plates join each other. Boundaries of plates are defined by any of the three elements: ridge, trench (subduction), or transform faults; therefore, combination of any of these can form a triple junction. For example, in case the boundaries of three plates are characterized by ridges, in that case it is refereed to as RRR triple junction. Similarly, in case the common boundaries of three plates are sutures or subduction zones (trenches), it is known as SSS or TTT triple junction (Fig. 7.2). In case of transforms faults, it is known as FFF. There can be many more combinations such as RRT, RRF and TTR. However, some of them are stable while others are unstable. In most cases of present day stable triple junctions, angles between the three sutures are obtuse angles with at least two of them being > 120 °. The best known triple junction of the Indian plate is the Indian Ocean Triple Junction formed by the mid-Oceanic Indian Ridge, South West Indian Ridge and South East Indian Ridge (Fig. 5.13) that can be truly classified as RRR triple junction. The best example of a TTT triple junction is afforded by Izu Trench, Ryukyu Trench, and Japan Trench in central Japan. Fig. 7.2 shows a TTT type triple junction that shows plates B and C are subducting under A, and plate C subducts under A. This is a case where one plate subducts under the remaining two plates and one plate does not subduct at all. Due to relative motion of plates, the triple junction and the sutures may shift (Fowler, 2005; McKenzie and Morgan, 1969) as shown in Fig. 7.3. Based on these characteristics of stable triple junctions, one can attempt to build past histories of plate tectonic processes that are difficult to decipher in case of Archean-Proterozoic terrain. Burke and Dewey (1973) marked several triple junctions the world over, but they were primarily related to tectonic elements instead of paleo-sutures.

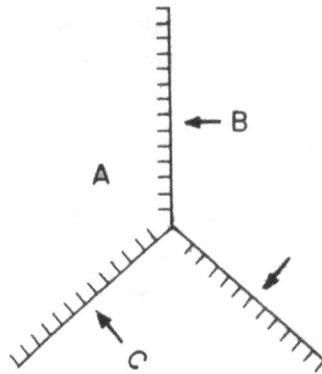

Fig. 7.2 A Trench-Trench-Trench (TTT) or Subduction-Subduction-Subduction (SSS) type triple junction where Plate B subducts under plates A and C, and plate C subducts under plate A. Plate A does not subduct.

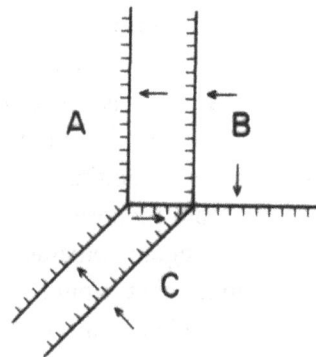

Fig. 7.3 Shifting of sutures at triple junctions are common due to different rates of subduction along these sutures.

7.2   Geophysical Signatures

In a Precambrian terrain, if crustal blocks are separated by a boundary across which there is a marked difference in stratigraphy or tectonic history or a discontinuity in structural trends, such boundaries may be a suture, specially if they are highly sheared. Similarly, there are several geophysical signatures as described below based on the present day collision zones that help in delineating suture zones in Archean-Proterozoic terrains.

(i)   Dipping reflectors from either sides in the crust whose junction, if projected on surface coincides with a sheared zone, and thrusted high velocity lower crustal rocks may occur along it.

(ii)   Dipping reflectors in the upper mantle away from the suture showing trace of subducted rocks.

(iii) Paired gravity anomalies of gravity high observed over Proterozoic terrain representing high density mafic/ultramafic rocks and gravity low over the Archean terrain related to sediments of the foreland basin, are indicative of paleosutures. Another low may be observed on the subducted side of the collision zone due to crustal thickening and felsic intrusives forming a high flanked by lows on either side.

(iv)   Magnetotelluric surveys invariably provide blocks of different conductivities on either side of the Archean-Proterozoic suture. It also provides high conductivity under the paleosutures and at shallow depth on the obducted side (mobile belts) that are related to thrusted rocks from lower crust/upper mantle and associated fluids.

(i) Gravity and Magnetic Anomalies

Gibbs and Thomas (1976) analysed gravity anomalies from several cratons and suggested that highs will be observed over younger Proterozoic rocks as they represent collision zones with thrusted high density rocks, while lows are observed over older Archean cratons. Fountain and Salisbury (1981) also

emphasized the occurrences of paired gravity anomalies over known Archean-Proterozoic collision zones. They provided a schematic model of collision (Fig. 7.4) that shows the pre-and post stages of collision (a and b) and obduction of high density lower crustal rocks providing the gravity high and metasediments at the back of the suture zone giving rise to a gravity low (c) forming paired gravity anomalies. These are primarily regional gravity anomalies useful for geodynamic studies, while corresponding residual anomalies due to shallow and limited size of sources are useful for resource exploration and basement studies as has been discussed in Chapter 9. Thomas (1992) has provided four profiles across well known Proterozoic collision zones and one across Appalachian orogen along the east coast of USA of Early Paleozoic period (Fig. 7.5). It also provides a typical crustal model of the Canadian Shield that shows gravity highs over the thicker and denser crust which is younger than adjoining craton and gravity lows over the Proterozoic metasediments representing green schist facies. It is interesting to observe similar paired gravity anomalies of gravity highs and lows in all five cases as demonstrated in this figure. In case of the Variscan fold belt of the Variscan orogeny of Devonian-Carboniferous period in Germany where the deep continental borehole is located, paired gravity anomalies of high and low (Section 4.7.3, Fig. 4.39(c)) have been observed that have been attributed to metamorphosed basalt in the upper crust and granite intrusive, respectively, based on borehole log. The metamorphosed basalt on one side of a fault may indicate oceanic crust that was involved at the time of this collision to form the fold belt. This observation is quite significant as it is based on direct observation. Mishra (2006) suggested that another large wavelength gravity low due to crustal thickening superimposed with small wavelength gravity highs and lows due to intrusives will be observed over the craton on the other side of the suture zone. This has been discussed in more detail in the next section while describing geophysical signatures observed over the Indian cratons and mobile belts.

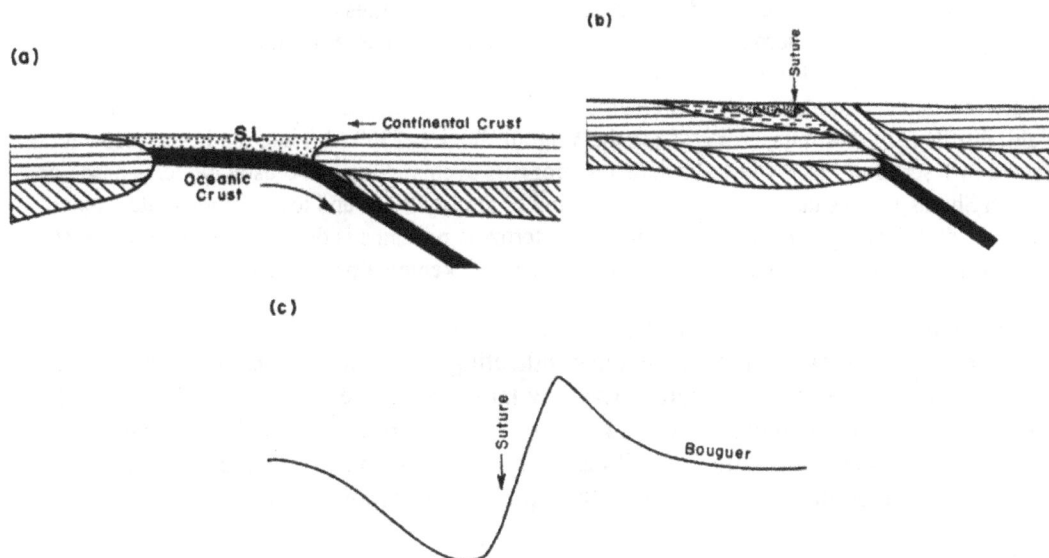

Fig. 7.4 (a) Pre-collision stage between two continents; SL: Sea level (b) postulated geometry after collision; and (c) theoretical Bouguer anomaly for crustal structure in collision zones showing gravity high due to high density thrusted rocks and gravity low due to foreland sediments and crustal thickening (Fountain and Salisbury, 1981).

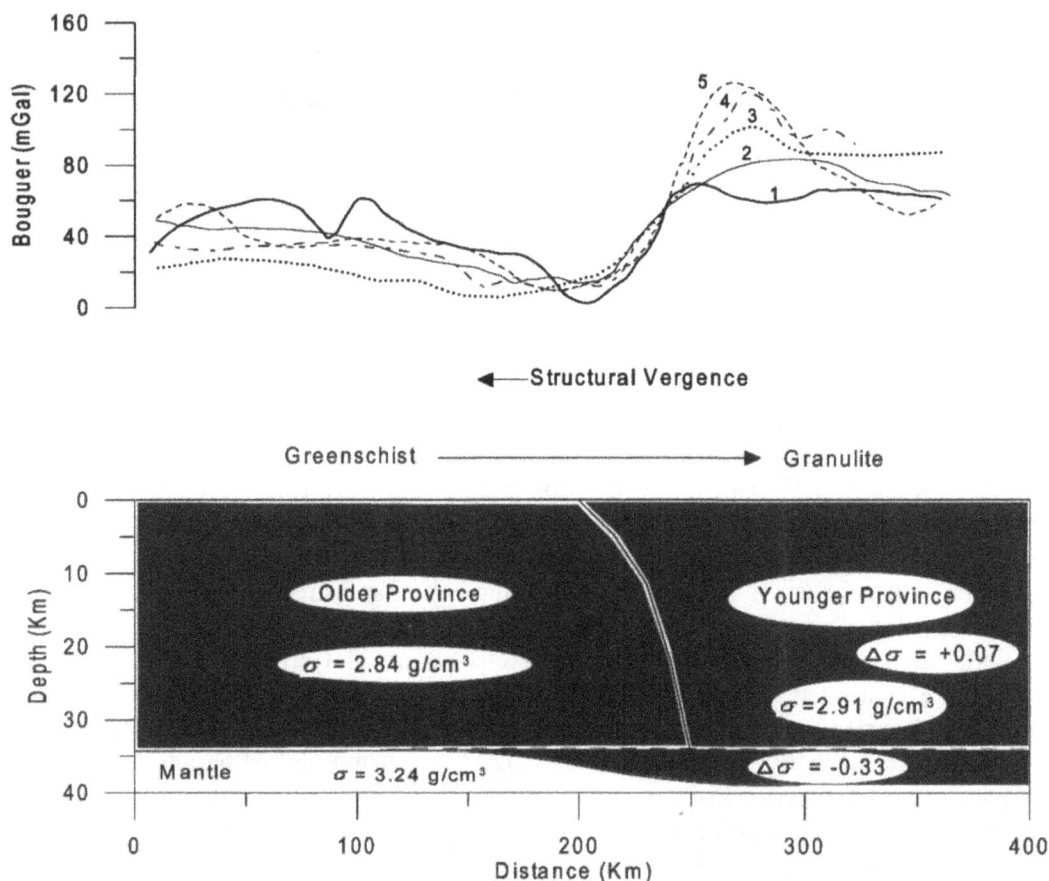

Fig. 7.5 Four gravity profiles across boundaries of various Precambrian structural provinces or terrains and one profile across the Appalchian Orogen. Profile 1 crosses the eastern margin of West African Craton (~ 600 Ma). Profile 2 is the mean curve for five profiles crossing boundaries of the Canadian Shield that juxtapose various Paleo-Neoproterozoic terrains. Profile 3 is the mean curve for nine profiles crossing Appalachian Orogen and may reflect ~ 450 Ma collision. Profile 4 represents average of three profiles from Archean Dharwar Craton across Meso-Neoproterozoic (~ 1500-1000 Ma), Eastern Ghat Mobile Belt of the Indian Shield. Profile 5, in Australia, crosses from the Yilgran Block (>2300 Ma) into the Fraser Orogenic domain affected by a series of Paleo-Meso- Proterozoic orogenic events. A typical crustal model is based on mean Canadian Shield profiles that would produce the kind of gravity highs and lows demonstrated above (1-5) (Thomas, 1992). Gravity highs over the younger Proterozoic province is due to high bulk density of crustal rocks, and low may be related to crustal thickening along the suture.

Magnetic anomalies across an Archean-Proterozoic suture zone show sudden changes in the pattern and the amplitude of magnetic anomalies indicating different terrains on its either sides. Suture/collision zones in general are characterized by large amplitude magnetic anomalies due to their association with mafic rocks. Sometimes, large scale structural trends delineated from gravity and airborne magnetic data over Precambrian terrains may indicate principle stress direction that has affected these terrains during their formation or movement.

(ii)   Seismic Studies

Seismic profiles usually show dipping reflectors on the subducted side of the suture zone due to subducted rocks and high velocity rocks under collision zones due to mafic/ultramafic components of the then oceanic crustal rocks or subsequent intrusives as demonstrated below in case of some Indian Proterozoic collision zones. However, the best seismic signatures are provided by reflections of subducting rocks in the upper mantle as shown in the case of the Abitibi-Grenvile province of the Canadian shield (Cawood et al., 2006) or in the case of Gulf of Bothnia by Riahi and Lund (1994). Seismic surveys also provide crustal thickening on the subducted side along with thrusts associated with high velocity rocks under the collision zone. Most of the significant seismic signatures of Archean-Proterozoic collision are provided by deep seismic profile such as the Consortium for Continental Reflection Profiling (COCORP) in USA (Potter et al., 2007; Cook and Vasudevan, 2006), DEKORP in Europe (Meissner and Krawczyk, 1999; Riahi and Lund, 1994);. BIRPS in U.K (Singh and Mckenzie, 1993), Lithoprobe Project in Canada (Van der Velden and Cook, 1999; Cook and Erdmer, 2005; Clowes et al., 2005, Percival and Helmstaedt, 2006) and Kaapvaal Project in Africa (Geoph. Res. Lett., 2001, 28, 2505-2496). Lithospheric mantle shows considerable heterogeneity and low velocity under mobile belts compared to cratons due to high temperatures (800-900°C). High heat flow has been reported from most of the mobile belts including those from the Indian continent as discussed below. Based on integrated study of seismic, electromagnetic, gravity, and exposed geology along the 1800 km long cross-section of lithosphere across northwestern part of North America covering 4.0 billion years of earth's history, Cook and Erdmer (2005) suggested that the crust and underlying lithosphere largely consist of reworked material and are mostly old, indicating that new accreted rocks of orogenic belts were detached during geological history and were emplaced upon unrelated underlying crust and mantle. This gave rise to the concept of suspect terrains that implies regions where rock types do not match with those from adjoining cratons. Most of the collision zones, whether present day ones like the Himalaya or older ones such as of Archean-Proterozoic period, belong to this class.

(iii) Magnetotelluric investigations

They provide high conductive zones under the collision zones due to deep seated faults/fissures and underlying thrusted mafic/ultramafic, lower crustal/ upper mantle rocks. They also provide blocks of different conductivity across the suture zones. Suture zones, themselves are usually characterized by narrow conductive zones due to presence of fluids along them. Lower crustal rocks have provided high conductivity in Proterozoic orogenic belts in different parts of the world and carbon deposits in them have been considered as one of its primary sources (Jones, 1992). Brasse et al., (2006) have reported the first result of long period EM measurements across Trans European Suture Zone (TESZ) that separates the East European Craton/Baltic Shield (~650 Ma) towards the east from Younger Paleozoic (~650-250 Ma) Central European mobile belts. These measurements delineated three zones with relatively high resistivity under the suture zone compared to adjoining sections. However, resistivity varied considerably along the suture depending on the rock types and presence of fluids. Smirnov and Pedersen (2009) have reported near vertical high conductivity in lower crust and upper mantle (25-60 km) along the Sorgenfrei-Tornquist zone in Europe that separates the Proterozoic Baltic Shield in Sweden to the north from the reactivated part including the Danish basin towards the south.

7.3   Bouguer Anomaly Map of India vis-à-vis Geology and Tectonics

7.3.1 Geology and Tectonics

The observed gravity field is controlled to a great extent by geology and tectonics. Fig. 7.6 is a simplified tectonic map of India. This figure has already been described in Section 2.8 (Fig. 2.26). However, it is reproduced here for the sake of clarity with direction of convergence between different cratons as obtained from the descriptions given below and discussed in Section 7.12. The northern part of India is occupied by the Himalayan Mobile Belt and associated thrusts that have been described in Chapter 6. The Ganga basin to its south is a foreland basin of Himalayan orogeny with thick sediments of ~ 5 km overlying the basement. This a test example of present day collision tectonics that would help in identifying ancient collision zones as described below.

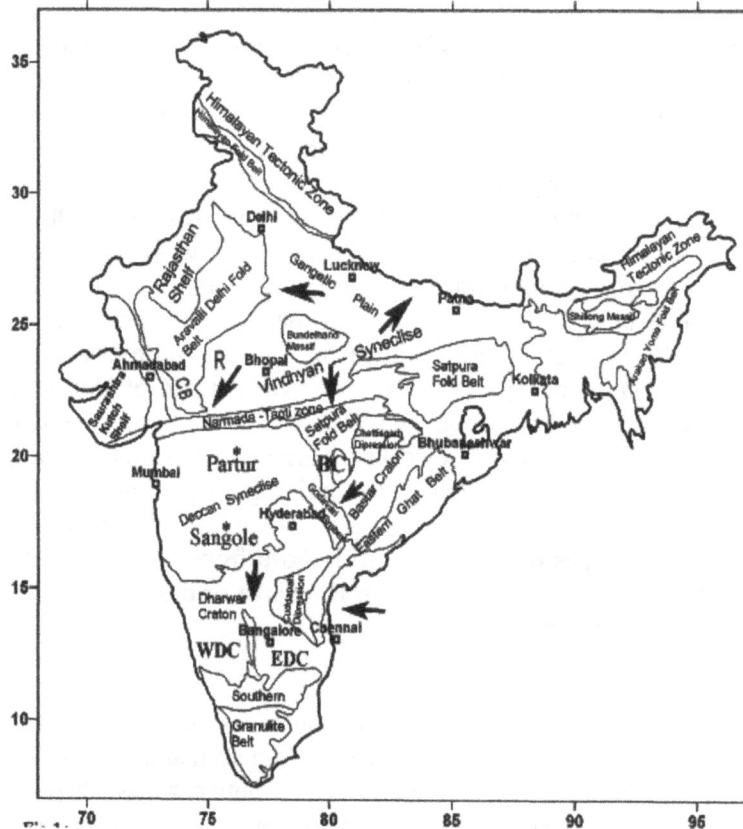

Fig. 7.6 Tectonic map of India showing various cratons and fold (mobile) belts (modified after ONGC, 1968) with arrows indicating direction of convergence during geological periods. The SW-NE directed arrow in the north indicates Cenozoic convergence across the Himalayan Fold Belt, while others indicate Meso-Neoproterozoic convergence between various cratons. A rifting phase during Paleo-Mesoproterozoic period (~ 2.0-1.6 Ga) proceeded by an earlier convergence almost along same tectonic features are also envisaged. BC: Bhandara Craton; CB: Cambay Basin; R:Resultant direction of convergence.

The Aravalli-Delhi and the Satpura Mobile Belts (ADMB and SMB) of the Proterozoic period (Fig. 7.6 and 7.7) occupy the western and the central parts of India respectively with the Ganga and Vindhyan basins lying in between them. The NE-SW trends of the northern part of India (ADMB) changes to NW-SE in the southern part (Dharwar Craton) across the SMB (ENE-WSW). Important geological units of the ADMB and the SMB and of the Bhandara and the Singhbhum cratons south of the SMB and their ages are given in Table 7.1 that also serves a comparison between them. South of 15°N the structural trends are primarily N-S which are abetted against E-W trend of the South Granulite Terrain at about 11-12°N. These structural trends define the major tectonic units. The Eastern and the Western Ghats occupy the two coastlines, the former representing a Proterozoic Mobile Belt. The South Indian Shield is occupied by Archean cratons with Proterozoic and Gondwana basins in between. It is interesting to note that the Archean-Proterozoic Mobile Belts in the tectonics are reflected as highlands in geomorphology that are easily identifiable. Geophysical signatures vis-à-vis geology and tectonics of individual mobile belts and respective models of their evolution are discussed below. Appendix I provides the geological time scale and corresponding ages in million years that can be referred in cases where only geological time scale is referred in the following discussion. Appendix II provides some important activities of recent times that are relevant for social developments.

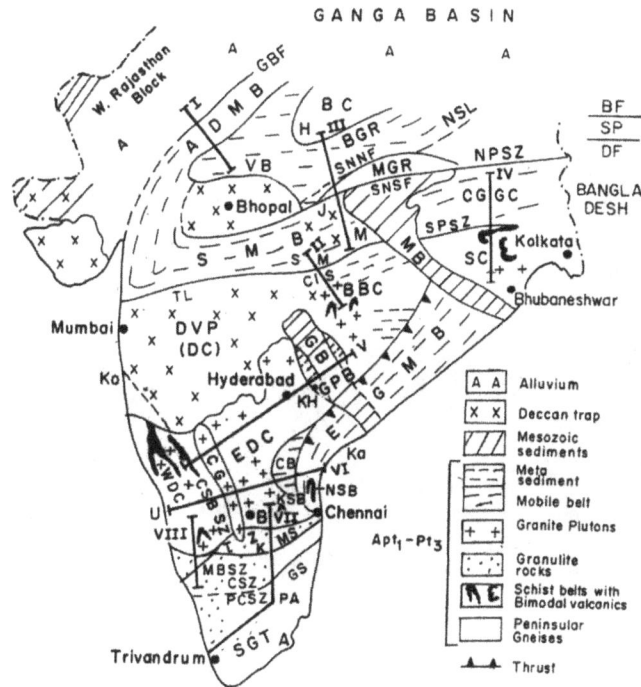

Fig. 7.7 A schematic representation of cratons and fold (Mobile) belts of the Indian Peninsular Shield.
Cratons: BBC = Bhandara-Bastar Craton, BC = Bundelkhand Craton, DC = Dharwar Craton, EDC = Eastern Dharwar Craton, SC = Singhbhum Craton, WDC= Western Dharwar Craton; **Mobile Belts**: ADMB = Aravalli-Delhi Mobile Belt, EGMB = Eastern Ghat Mobile Belt, SMB = Satpura Mobile Belt, CGGC = Chotanagpur Granite Gneiss Complex, GPB = Godavari Proterozoic Belt; **Thrusts/Shear Zones:** BF = Brahamputra Fault, CIS = Central Indian Shear, CSZ = Cauvery Shear Zone, DF = Dauki Fault, GBF = Great Boundary Fault, GS = Gangavalli shear: NE extension of PCSZ, MBSZ = Moyar-Bhavani Shear Zone, MS = Mettur Shear: NE extension of MBSZ, NPSZ = North Purulia Shear Zone, NSL = Narmada-Son Lineament, PCSZ = Palghat-auvery Shear Zone, SNNF = Son Narmada North Fault, SNSF = Son Narmada South Fault, SPSZ = South Purulia Shear Zone, SZ = Shear Zone between WDC and EDC, TL = Tapti Lineament, TZ = Transition Zone; **Geological Provinces/Basins:** BGR = Bijawar Group of Rocks, CB = Cuddapah Basin, CG = Closepet Granite, CSB = Chitradurga Schist Belt, DVP = Deccan Volcanic Province, GB Godavari Basin, KSB = Kolar Schist Belt, MB = Mahandadi Basin, MKG = Mahakoshal Group of Rocks, NSB = Nellore Schist Belt that extends to Khammam (KM) along the GPB, SM = Sausar Meta-sediments, SGT = Southern Granulite Terrain, SP = Shillong Plateau and VB = Vindhyan Basin; **Periods:** APt_1 = Neoarchean-Paleoproterozoic, and Pt_3 = Neoproterozoic; **Profiles:** Profile I = Nagaur-Jhalawar, Profile II = Mungwani- Rajnandgaon, Profile III = Hirapur- Mandla, Profile IV = Along 86^0E between 21^0- 24^0N across CGGC and Singhbhum Craton, Profile V = From Bhandara-Bastar Craton to Dharwar Craton across Godavari Proterozoic Belt, Profile VI = Kavali-Udipi, Profile VII = Kuppam-Kolattur- Palani extended towards the coast, Profile VIII = Along 75.5^0E across TZ and Moyar shear; **Cities:** B = Banglore, BH = Bhuaneshwar, H = Hirapur, J = Jabalpur, Ka = Kavali, K = Kollatur, KH = Khammam, KO = Koyna, M= Mandla, PA= Palni, and U= Udipi.

7.3.2 Complete Bouguer Anomaly Map of India

The terrain corrected Bouguer anomaly map of India prepared based on uniform distribution of data at 5.0 km grid interval representing Complete Bouguer anomaly with accuracy better than ±1.5 mGal (GSI-NGRI, 2006) is given in Fig. 7.8(a) and (b) (Mishra et al., 2008). In order to compute complete Bouguer anomaly, terrain correction is computed and applied to the simple Bouguer anomaly as briefly described below (GSI-NGRI, 2006).

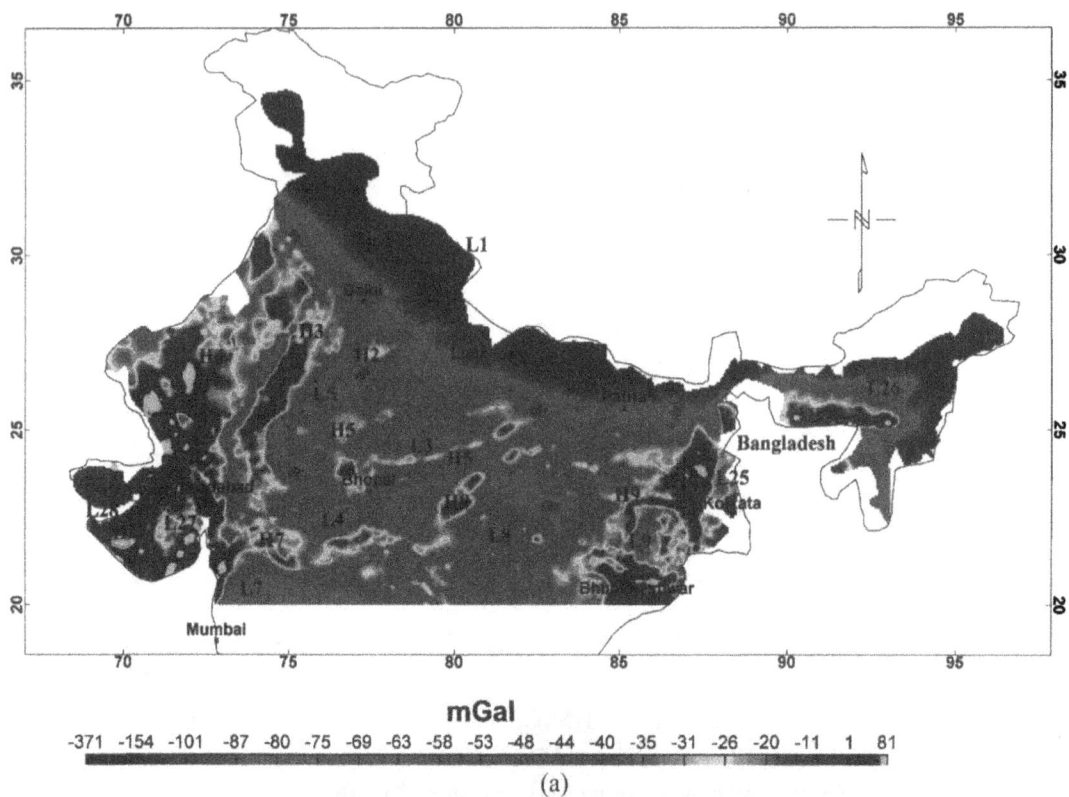

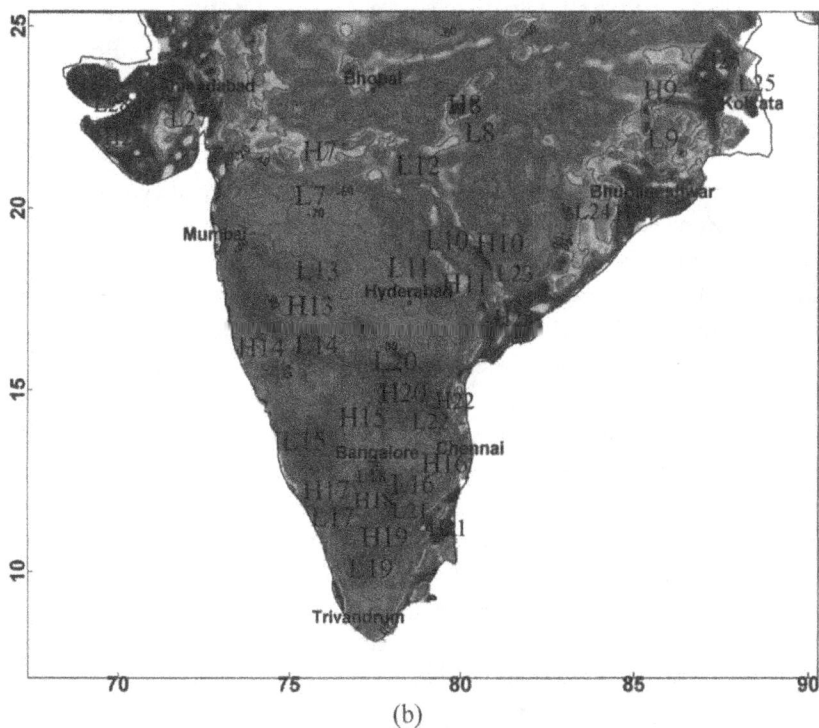

Fig. 7.8 Complete Bouguer anomaly map showing gravity highs and lows marked as H and L, respectively. The following numbers indicate the individual anomalies associated with different tectonic units as described in the text (a) North of 20°N and (b) South of 25°N (GSI-NGRI, 2006; Mishra et at., 2008).
[Colour Fig. on Page 791-791]

Terrain correction accounts for the gravitational effects of surface topography such as hills rising above and valleys lying below the vicinity of the station, which are not covered in the Bouguer correction. This correction mathematically computes the effects of the added/missing masses that are excluded /included in the Bouguer slab calculation. Efficient calculation of terrain corrections requires detailed digital elevation data and computational power. In the present case, terrain correction was made using a global Digital Terrain Model (DTM). The procedure adopted for computation of terrain correction is as follows:

- A station file is prepared with station location and elevation.
- All the stations are projected on same datum of digital elevation data.
- A new digital elevation grid is prepared from the global database. (GEBCO,http://www.gebco.net/,GTOPO-30).
- The station elevation data is extended by 2 degrees (about 200 km) on all sides compared to the station file. Optimization of gridding procedure vis-a-vis grid spacing of both global DTM as well as station spacing is of vital importance to ascertain that the real station coordinates lie on the re-gridded surface.
- Terrain corrections are usually carried out in two parts, the inner and outer zone corrections. The inner-zone corrections require high resolution elevation data to calculate terrain effects up to distance equivalent to Hayford's Zone F of the Hammer's Chart (about 1 km).

For the outer-zone corrections, the terrain effect is computed from a radial distance of 900 m to 200 km around each station, which includes contributions of both topography and bathymetry (along coast). The terrain is represented in terms of prisms of the dimension defined by the grid size, approximately 900 m in the present case. The gravity effect of the mass of each prism is subsequently calculated and added to the Bouguer anomaly of the station according to the procedure outlined by Ehrismann et al., (1966) and Leaman (1998). Finally, the Bouguer anomaly (BA) (terrain corrected) also known as Complete Bouguer anomaly is calculated as follows:

Complete BA = Simple BA + Terrain correction.

Bouguer anomalies (Fig. 7.8(a) and (b)) are mostly negative over Himalayas (L1) and South Indian Shield (L7 and southwards) including Western Ghats, indicating thick crust due to isostatic compensation of topography and low density surface/subsurface rocks. However, they are largely positive over mobile belts (Fig. 7.7) such as the ADMB, the SMB, and the Eastern Ghat Mobile Belt (EGMB) indicating thinner crust under them compared to isostatically balanced crust based on topography. Part of these gravity highs may be caused by crustal high density rocks. Mobile belts being occupied by Proterozoic rocks flanked by cratons will be examined for signatures of Archean-Proterozoic collision zones that are discussed below in detail. The various Bouguer anomalies described below with respect to tectonics and geology (Fig. 7.6 and 7.7) are shown in Fig. 7.8(a) and (b) as H and L indicating gravity highs and lows with following numbers referring to individual anomalies related to different tectonic units. There are twenty eight such anomalies referred to as H1, H2----H28 and L1, L2----L28. However, all of them do not form typical paired gravity anomalies related to Proterozoic collision zones as described below. They are referred in this manner in the figure for the convenience of locating these anomalies.

7.3.3   Isostasy and Effective Elastic Thickness

Isostasy deals with the mode of compensation of the topographic lyad. There are primarily two modes that have been considered in this regard. The earliest being Airy's model compensation where the topographic load is supported by its root just below the load at the Moho that implies absence of crustal strength to support the load. In case of Airy's model of compensation, the crustal thickness would be ~6.7 × topography for a fully compensated crust that depends on the ratio of the density contrast of topography with air ($\sim$2.67 g/cm^3) and density contrast at the Moho between standard lower crust and upper mantle ($\sim$0.4 g/cm^3). This is known as crustal root that supports the topography. As isostasy is achieved on a regional scale, regional free air anomaly as zero by definition suggests Airy's isostatic compensated crust, while its positive and negative values indicate lack of compensation. It also suggests that increasing elevation would cause decrease in the Bouguer anomaly due to crustal roots showing opposite correlation between the two, while positive correlation indicate lack of isostatic compensation. In case of the absence of significant root under a topography for isostatic compensation, the observed gravity lows and highs can be attributed to low and high density rocks in the crust as has been suggested below in several cases in the Indian Shield. This principle has been used to interpret gravity anomalies the world over in combination with seismic studies that provide the estimate of crustal thickness accurately. An interesting situation was encountered in Sierra, USA where the thickest crust was found under northern Sierra with lower elevation while a thin crust was reported from southern Sierra with highest elevation. This reverse correlation was attributed to the loss of the dense eclogite root from southern Sierra (Zandt et al., 2004).

The second mode of compensation relates to flexure of the lithosphere (Section 4.4). In this case, the topographic load is supported by its strength that resides in the part of the crust that is elastic in nature and is known as the effective elastic thickness (EET).

(i) Airy's Isostatic Map of India and Surrounding Oceans

In order to differentiate between the observed Bouguer anomaly over highlands due to isostatic effect and shallow high or low density rocks, the isostatic anomaly map based on Airy's model of isostatic compensation is qualitatively very useful. An isostatic anomaly map of India and adjoining oceans is prepared based on full compensation of topography for a standard crustal thickness of 35 km at sea level (elevation = 0). As in case of terrain correction, a prismatic body of grid size is taken at every station and the crustal thickening (C_t) under this block for Airy's model of compensation is obtained from the regional topography (t) in the block as

$$C_t = [\rho_t / (\rho_m - \rho_t)] \cdot t$$

where ρ_t is the density of the topography (contrast with surrounding air) and ($\rho_m - \rho_c$) is the presumed density contrast between the crust (ρ_c) and the upper mantle (ρ_m).

The gravity field for the prismatic block due to this crustal thickening at the depth of the standard crustal column (T_c) is computed and subtracted from the observed Bouguer anomaly to obtain the isostatic anomaly at that point. In case of a standard crustal column ρ_t = 2.67 g/cm^3 and ($\rho_m - \rho_c$) = (3.3 − 2.9) = 0.4 g/cm^3 and T_c = 35 km. Therefore, in case of a complete compensation based on Airy's model, the crustal thickness would be $\sim$ 6.7 times the regional topography that is added to the standard crustal column to obtain the crustal thickness. In case, the actual crustal thickness is more than the isostatically balanced crust (6.7 × topography), the region is considered overcompensated and it would tend to rise due to buoyancy of the crustal root. However, in case the crustal thickness is less than the isostatically balanced crust, such regions are known as undercompensated and tend to sink due to negative buoyancy forces. However, any other values for the above parameters can be used depending on the information available from a particular region to evaluate the isostatic balance in a region. The isostatic anomaly, therefore removes the effect of isostatic compensation that in most of the orogenic belts is the major source of large wavelength regional anomalies, and highlights the effect of shallow crustal density in homogeneities.

The isostatic anomaly map of India and surrounding oceans computed as above (Fig. 7.9) shows highs over some of the oceanic ridges as free air anomaly map such as Ninety East Ridge, and C-L Ridge suggesting a comparatively thin crust under them, and their association with high density mafic/ultramafic rocks. It also provides highs and lows over ocean basins in the Arabian Sea, Indian Ocean, and Bay of Bengal depending on the thickness of sediments in these basins. Over the continent, it shows highs over topographic highs such as Aravalli, Satpura, Eastern Ghat, and Himalaya; and lows over basins like the Ganga and Godavari basin, indicating that these regions depart from Airy's isostatic model and/or associated with subsurface high and low density rocks. Normally the Proterozoic mountains such as Aravalli, Satpura and Eastern Ghats should have become peniplained over the period due to constant uplift and erosion caused by isostatic adjustments. Fischer (2002), however, attributed the present day high topography of Archean-Proterozoic mountain chains to waning buoyancy of the crustal roots caused by reduced density contrast with respect to surrounding mantle rocks. This density contrast is further reduced in case of the underplated crusts that implies high density ($\sim$ 3.1 g/cm^3) mafic rocks along the Moho, which is generally the case in orogenic belts. The isostatic low and high anomalies indicate over compensated and undercompensated crusts, respectively, and isostatic balance might be due to flexural compensation. As anomalies originating from Moho due to isostasy are characterized by large wave length, the spectral analysis (Section 6.1) can be used to differentiate between the large wave length isostatic anomalies and short wavelength anomalies due to shallow anomalous bodies. Western Ghat reflects near Airy's model of isostatic compensation, whereas the entire section of Himalaya shows positive isostatic anomaly with adjoining lows over the Ganga basin indicating under-and overcompensated crusts, respectively, which is the case in most of the present day orogenic belts like Alps, and Andes indicating that compensation in such cases might be achieved due to flexural compensation.

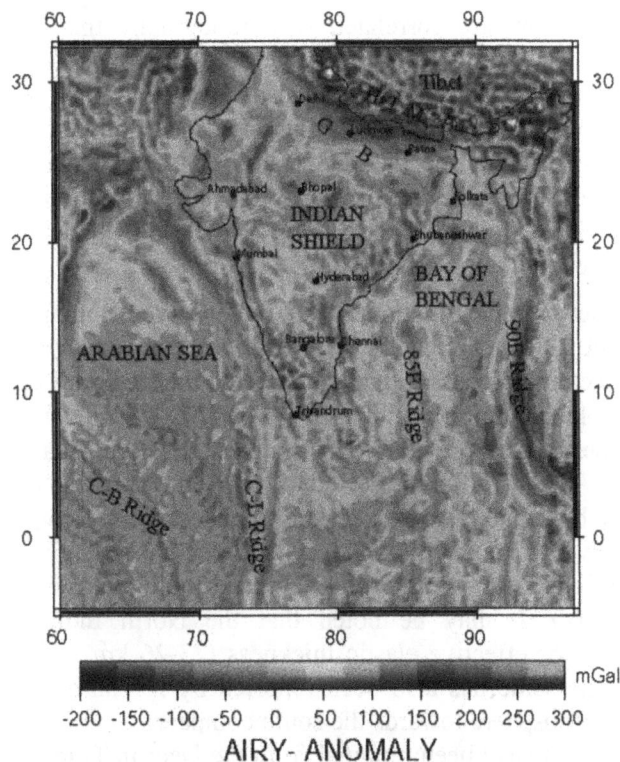

Fig. 7.9 Airy's isostatic anomaly map shows gravity highs over mountain ranges (fold/mobile belts) indicating undercompensation and /or shallow high density rocks. Most significant lows are observed over Tibet, Ganga foredeep and some other basins indicating overcompensation and low density sediments, respectively. Ridges in the oceans show gravity highs indicating high density rocks in the crust associated with them.
Him: Himalaya, GB: Ganga Basin, C-L Ridge: Chagos- Lacadive Ridge, C-B Ridge: Carlsberg Ridge.
[Colour Fig. on Page 792]

(ii) Flexural Isostasy and Effective Elastic Thickness

Flexural isostasy implies that topographic load is supported by flexures of the lithosphere instead of its root as suggested in Airy's model of isostasy. As stated earlier, Airy's model of isostasy assumes that the lithosphere does not have any strength and all the loads are supported by crustal root. However, the existence of old mountains, especially those belonging to Archean-Proterozoic period without any crustal root suggests is that the lithosphere in these sections is strong enough to withstand their loads. This strength, however, resides in parts of lithosphere that show elastic properties where stress is proportional to strain up to a certain limit of stress caused by the load. Beyond this limit, it breaks showing brittle characteristics. Airy type of isostatic compensation broadly agrees with the seismic evidence of the largest thickness of the crust under the highest topography. The part of the lithosphere that behaves elastically over geological time period is termed as Effective Elastic Thickness (EET) and is a measure of strength of the lithosphere (Banks et al., 2004). Larger the EET, stronger the crust/lithosphere. This part of lithosphere behaves elastically and so defines the rheological stratification of lithosphere (Burov and Diament, 1995; Jackson, 2002).

Two spectral methods are commonly used to estimate EET as described in Chapter 4. In the first approach, the linear transfer function (admittance between topography and gravity anomaly) are computed and fitted with the admittance value of the elastic plate model, which is referred to as admittance analysis (McKenzie and Bowin, 1976). The other approach is the computation of coherence between topography and gravity anomalies, which assumes that short wavelength topography is uncorrelated with Bouguer gravity anomalies, while long wavelength anomalies involved in isostatic compensation, are correlated with topography that would produce high coherence. Theoretical coherence and admittance can be computed by the scheme provided by Forsyth (1985) incorporating surface and subsurface loads (Lowry and Smith, 1994; Mckenzie and Fairhead, 1997), and compared with the computed coherence and admittance for best fit of EET value. In fact, misfit function between the theoretical and computed values of coherence and admittance is computed for different EET values and its value for minimum misfit function provides the actual value of EET. McKenzie and Bowin (1976) and Louden and Forsyth (1982) have provided the scheme for computation of the EET.

Based on the coherence between the topography and the observed Bouguer anomaly, one may compute the effective elastic thickness that is a measure of the strength of the underlying lithosphere and in turn, the tectonic processes that has affected that region. Fig. 7.10(a) shows the topography of the Indian Shield and numbered blocks, 1-14 of 5° × 5° with superimposed parts on either side. Mobile (fold) belts like Aravalli-Delhi, Satpura, Eastern Ghat, Dharwar Cratons, Southern Granulite belt (Fig. 7.6 and 7.7) are reflected as high land while the Ganga basin towards the north is reflected as low land. The effective elastic thickness (EET) computed based on the coherence between topography and Bouguer anomaly as outlined above is given in Fig 7.10(b) for the North and South Indian Shield separated by the Satpura Mobile belt with reference to block numbers given in Fig. 7.10(a). It may be noted that the North Indian Shield is usually characterized by higher values of effective elastic thickness (23-26 km) compared to the South Indian Shield (10-16 km). The same has also been reported by Jordan and Watts (2005). This implies, in general, a weaker lithosphere towards the south compared to the north. Minimum value of EET of 8-10 km and 10-15 km have been reported from the Deccan Trap covered region along the west coast of India (Tiwari and Mishra, 1999) and the east coast of India and the SGT (Tiwari and Mishra, 2008). The lower value of EET for the South Indian Shield and along the east and the west coast of India might be the result of breakup of India from Gondwanaland, rifting along the east and the west coasts of India, and their effect inside the continent close to the coastline due to various plumes (Storey, 1995) as described in Chapter 5. The low value of the EET for the South Indian Shield may also be correlated with the Neoarchean-Paleoproterozoic collision of the

Western and the Eastern Dharwar cratons and the Southern Granulite Terrain (Fig. 7.6 and 7.7) and mobile belt type structures of the EDC as discussed in Sections 7.8 and 7.9. Rifts, rifted margin, and collision zones usually provide low values for the effective elastic thickness due to inherent weakness of rocks in these sections caused by related tectonic activities.

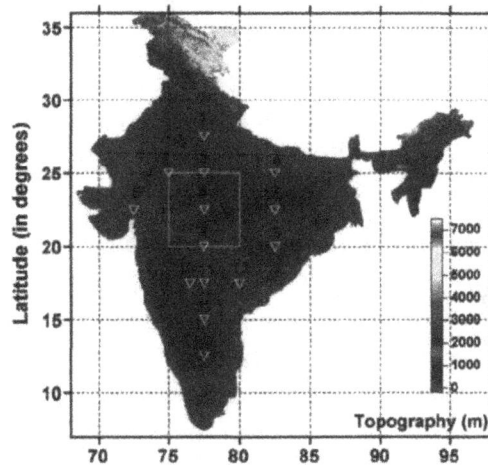

Fig. 7.10(a) Topography with numbered blocks of $2° \times 2°$ superimposing over adjoining blocks

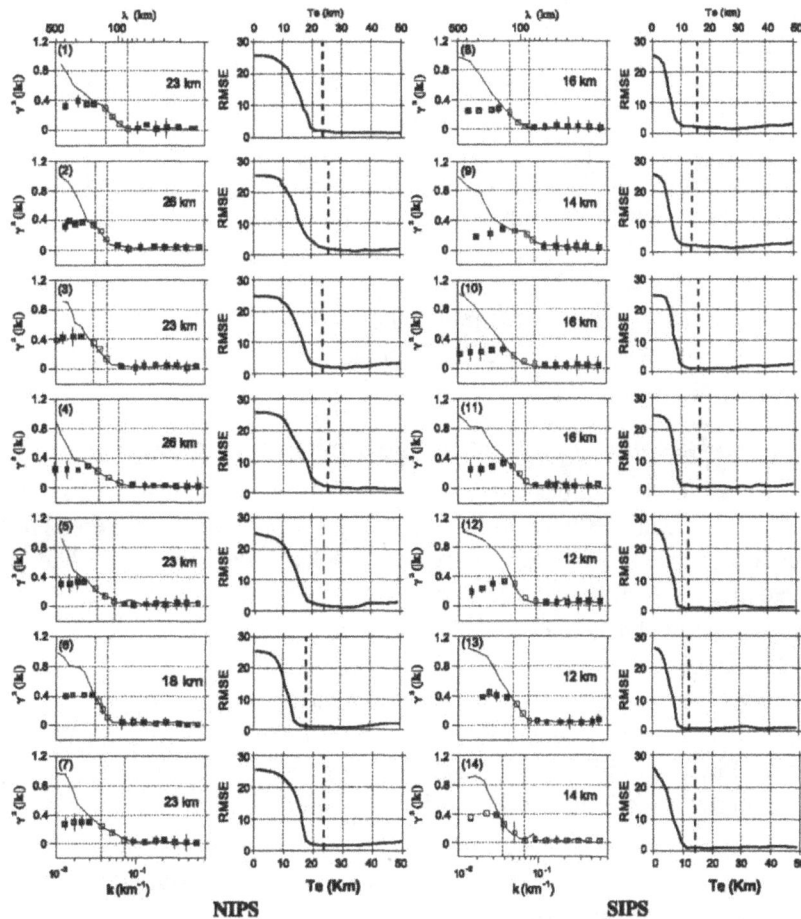

Fig. 7.10(b) Coherence and RMSE with respect to Effective Elastic Thickness (EET) of the North Indian Shield (NIS) and the South Indian shield (SIS) referred to blocks numbers as in Fig. 7.10(a)

7.4   Broad Tectonic Zones of Indian Shield

The northern part of India and its tectonics are largely controlled by collision of the Indian and the Eurasian plates at about 50 Ma with compressional forces in SW-NE direction as has been described in Chapter 6. However, for the sake of completeness, the gravity field observed over the Indian part of the Himalaya (Fig. 7.8(a)) and its relation to the field observed over the Indian Shield south of Ganga basin (Fig. 7.8(b)) are briefly described below. This collision between the Indian and the Asian part of the Eurasian plate has produced the Himalayan Mobile Belt with several thrusts that gave rise to high mountains of Himalaya and the Ganga basin in front of it as a foreland basin. Appendix I provides the approximate geological timescale that can be referred in order to correlate various geological periods in million years.

7.4.1   Cenozoic Tectonics in the Northern Part

(a)   It is characterized by regional gravity lows (L1, Fig. 7.8(a)) of –250 to –350 mGal with increasing elevation indicating isostatic compensation related to crustal thickening caused by the collision of the Indian and the Eurasian plates. However, isostatic anomaly (Fig. 7.9) shows in general a low over Tibet that suggest overcompensated crust in this section, indicating predominance of low density rocks in the lower crust and the upper mantle that has been discussed in Chapter 6.

(b)   Relative short wavelength Bouguer anomaly highs (H1) of 10 to 20 mGal (Fig. 7.8(a)) along Himalayan fold and thrust belt and the Indus Tsangpo Suture Zone (ITSZ) between two plates suggest high density thrusted rocks along various thrusts. These are south verging thrusts indicating SW to NE convergence (Fig. 7.6). It may be noted that these small wavelength gravity highs are better reflected in free air anomaly (Section 6.1.3) compared to the Bouguer anomaly due to the effect of crustal thickening in the latter that produces gravity lows which overshadows the highs due to shallow bodies. However, the gravity low, L1 due to crustal thickening, is better reflected in Bouguer anomaly map compared to free air anomaly suggesting the importance of both maps for specific purposes. The isostatic anomaly map (Fig. 7.9) provides highs over the Himalayan fold and thrust belts, and lows over the foredeep along the Himalayan front and Tibetan plateau on the back side indicating under and overcompensated crusts, respectively in these sections. This, however is true for most of the present day orogenic belts like the Alps, and Andes, that show undercompensated crust under the fold and thrust belts sandwiched between overcompensated crusts in their front and back side, which makes the fold and thrust belts tectonically and seismically active.

7.4.2   Ganga Basin and Basement Model for Hydrocarbon Exploration

(a)   Bouguer anomalies over the Ganga basin are negative with minimum value of – 150 to – 200 m Gal in a plain region (L2, Fig. 7.8(a)) with elevation of a few hundred meters. Therefore, these anomalies are primarily caused by shallow sources of 3-5 km thick sediments. The lowest gravity values are observed along the Himalayan foredeep all along its section in the NE, north, and west related to the Brahmaputra, Ganga, and Indus foredeeps, respectively, indicating the thickest sediments in these sections.

(b)   There are basement ridges and depressions shown by relative gravity highs (H2) and lows (L2), respectively, which in most cases are extensions of structures from the Indian Peninsular Shield towards the north. Gravity highs, H2, for example, represent a basement ridge which is an extension of basement rocks from the Indian Peninsular Shield.

(c)   A basement map under the Ganga basin (Fig. 7.11(a)) was provided based on the airborne magnetic data (Agocs, 1956; Sengupta, 1996). The total intensity magnetic data in this survey was acquired with a terrain clearance of 1500' a.m.s.l (~ 450 m) at profile spacing of 6-24 miles. The accuracy of this data was limited to 5-6 nT. This has delineated some major NE-SW ridges like Faizabad ridge, and Munger-Saharsa ridge (west of Purnea), besides several faults as also supported from the gravity anomaly. It also suggests some depressions along the Himalayan front like Sarda depression (Bareilly), and Gandak depression (Raxul), with maximum thickness of sediment as high as 8 km. Subsequent seismic sections suggest that it has provided 10-20% higher depth estimates indicating maximum sediment thickness as 6-7 km along the Himalayan front as described in Section 6.3.4 based on seismic studies. Such sedimentary thickness maps are very useful for hydrocarbon exploration.

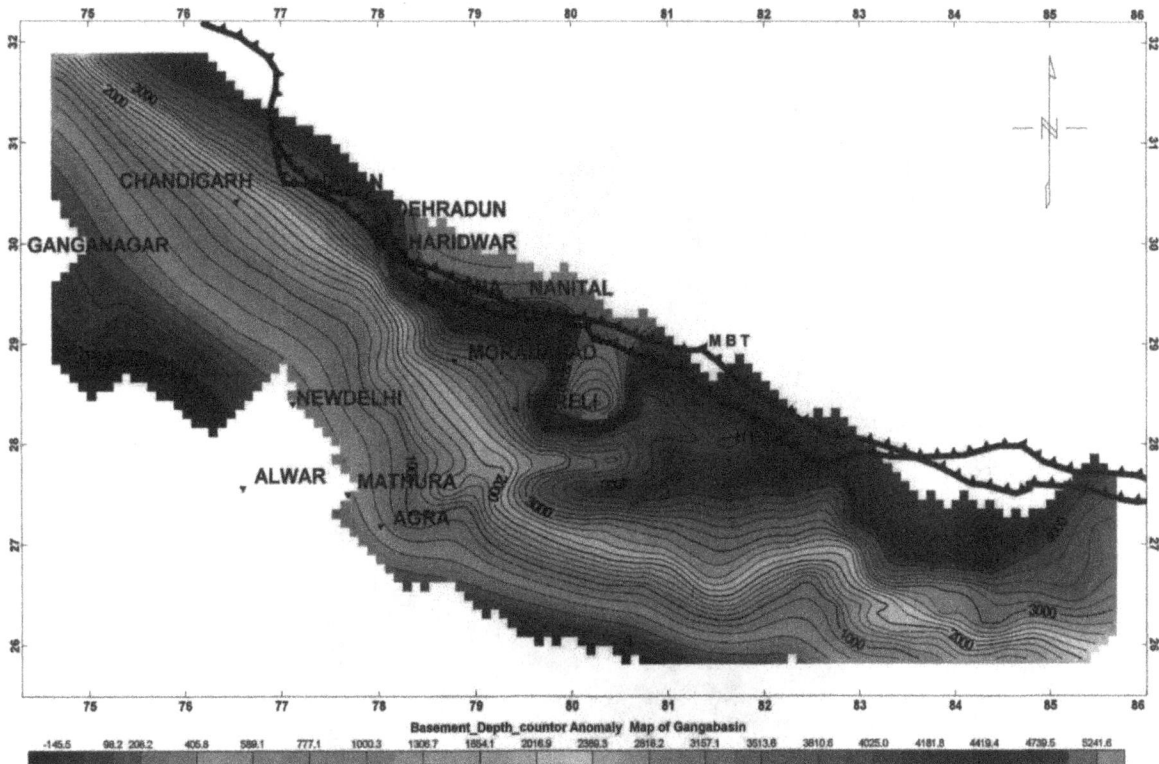

Fig. 7.11(a) Basement under the Ganga basin based on airborne magnetic data (Agocs, 1956; Sengupta, 1996) constrained from seismic and well data (Karunakaran and Rao, 1976). It shows depths to various ridges and depressions in meters with contour intervals of 200 m and depths varying from exposed section in the southern part to ~ 6 km in the northern part along the Himalayan foredeep.
MBT: Main Boundary Thrust and HFT-Himalayan Frontal Thrust. [Colour Fig. on Page 792]

(d) A sample of airborne magnetic data recorded subsequently over the Ganga basin with profile spacing of 2 km at a barometric elevation of 700 m above msl is given in Fig. 7.11(b) (Bahuleyan et al., 1999) that shows two major highs, A2 and A3, seperated by a linear low that coincides with the Moradabad-Haldwani fault. This is a primary fault in the Ganga basin that may represent the extension of the Great Boundary fault (Fig. 7.12(a)) towards the north. The magnetic anomaly, A2 being of higher amplitude (~ 200 nt) suggests a magnetic basement in this section and may represent the extension of the Delhi Group of Proterozoic rocks (Section 7.5) towards the north as it occurs west of the Moradabad fault. Delhi Group of rocks are considerably magnetic as they are characterized by mafic intrusives (Section 7.5). The magnetic anomaly, A3 of lower amplitude indicates a depression with Vindhyan sediment as the basement that occurs east of the Moradabad fault. Some other faults are also shown in this figure that are reflected as changes in the magnetic contours.

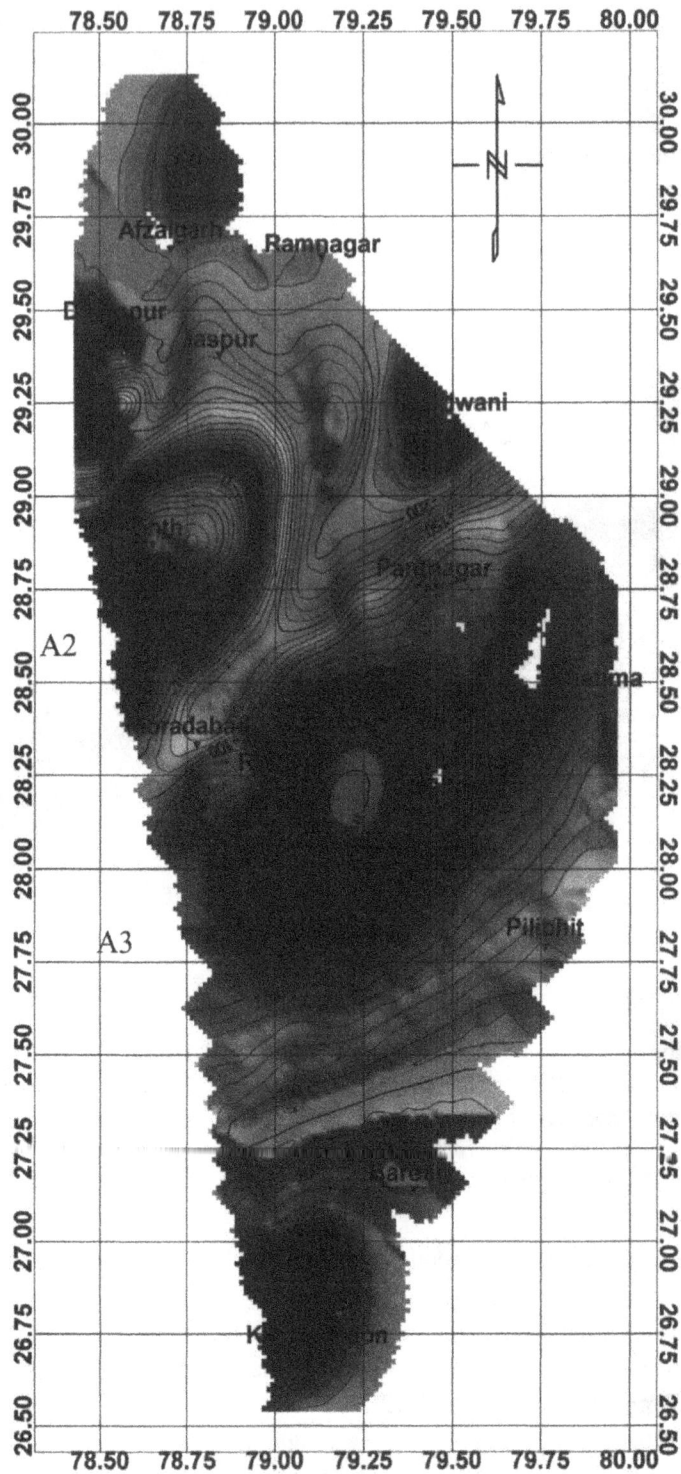

Fig. 7.11(b) A sample of the airborne total intensity map of the Ganga basin flown with 2 km spacing and 700 m as flight barometric altitude. The NE-SW linear magnetic low from Moradabad represents a fault that might be northward extension of thrusts related to the Aravalli-Dehi Mobile Belt. Magnetic highs on either side of it may represent the extension of this belt itself. Contour interval-10 nT (Babhuleyan et al., 1999). TheMoradabad fault is reflected as nosing of contour of a magnetic low.
[Colour Fig. on Page 793]

7.4.3 Archean-Proterozoic Cratons and Collision Zones of Indian Shield

The examination of gravity anomalies of Himalaya that represent an active present day collision zone between the Indian and the Eurasian plates suggests that it is characterized by linear gravity highs due to thrusted high density rocks with adjoining lows caused by crustal thickening towards the north, and sediments of the foreland basin towards the south. The same criteria can be used to decipher Archean-Proterozoic collision zones between cratons. However, with passing time (on a scale of million years), the crustal thickness decreases due to constant erosion and uplift, and so is the gravity low caused by it. Therefore, the ratio of gravity high to low is more (1-2) in case of Archean-Proterozoic collision zones compared to the present day orogenic belts as in the case of Himalaya and Tibet.

The Indian Shield consists of several Archean cratons and intervening Proterozoic fold (mobile) belts (Fig. 7.6 and 7.7). Gravity signatures due to Archean Proterozoic mobile belts (Fig. 7.8(a) and (b) are integrated with other geophysical/geological data sets available from these regions to provide an integrated evolutionary model for them. Based on the structural trends (Fig. 7.6 and 7.7), Indian shield can be divided as North and South Indian Shields separated by the Satpura Mobile Belt that have been described below. A detailed geological map (GSI, 1993) of the North Indian Shield showing important geological units is given in Fig. 7.11(c) that depicts the Aravalli Delhi Mobile Belt (ADMB) and the Satpura Mobile Belt (SMB) and the two appear to be interacting in the western part. The Vindhyan basin is bordered by these mobile belts towards the south and the west with the Ganga basin and Bundelkhand craton exposed towards the north. The South Indian Shield and related cratons are located south of the SMB (Fig. 7.6 and 7.7).

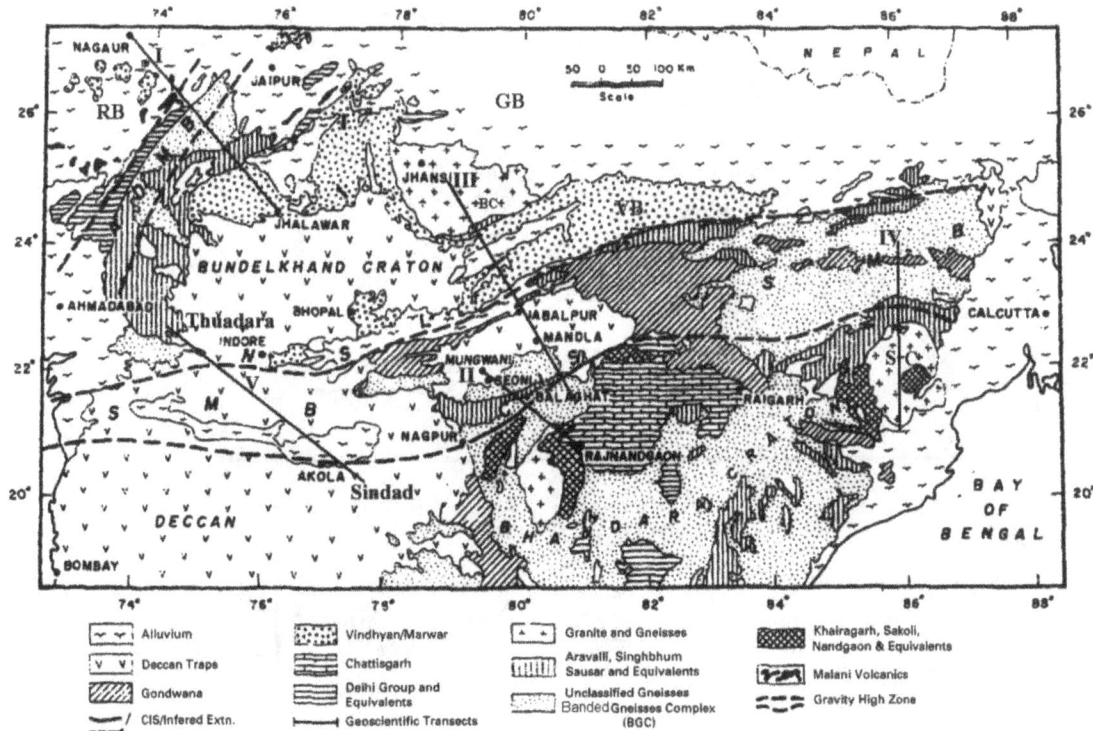

Fig. 7.11(c) Geological map of the North Indian Shield that includes the Aravalli-Delhi Mobile Belt (ADMB), the Satpura Mobile Belt (SMB) and the adjoining regions (GSI, 1993). The Central Indian Shear (CIS) is shown in the central part of the southern margin of the SMB whose northern margin is the Narmada-Son lineament (NSL). The geotransects 'Nagaur-Jhalawar' (I) and 'Mungwani-Rajnandgaon' (II) across the ADMB and the SMB, respectively and zones of gravity 'highs' along them are also shown. Three more profiles III, IV and V across the SMB are also investigated as discussed below. BC: Bundelkhand craton, GB: Ganga Basin, RB: Rajasthan Block, SC: Singhbhum craton, VB: Vindhyan Basin.

7.5   Aravalli-Delhi Mobile Belt (ADMB) and Western Rajasthan

The ADMB forming Aravalli ranges in the Western India is examined below for plausible Proterozoic collision tectonics.

7.5.1   Geology and Tectonics

(a) A detailed geological map of this section is given in Fig. 7.12(a) that also shows the geotransect Naguar-Jhalawar where various geophysical and geological studies were carried out for tectonic studies. This section is characterized by Proterozoic rocks (Table 7.1) with Banded Gneissic Complex (BGC) of Mesoarchean as basement that is separated from the Vindhyan sediments towards the east by the Great Boundary Fault. The BGC consists of metasediments of the Mangalwar complex and Hindoli group of rocks with granite intrusives, and Sandmata complex of lower crustal granulite rocks.

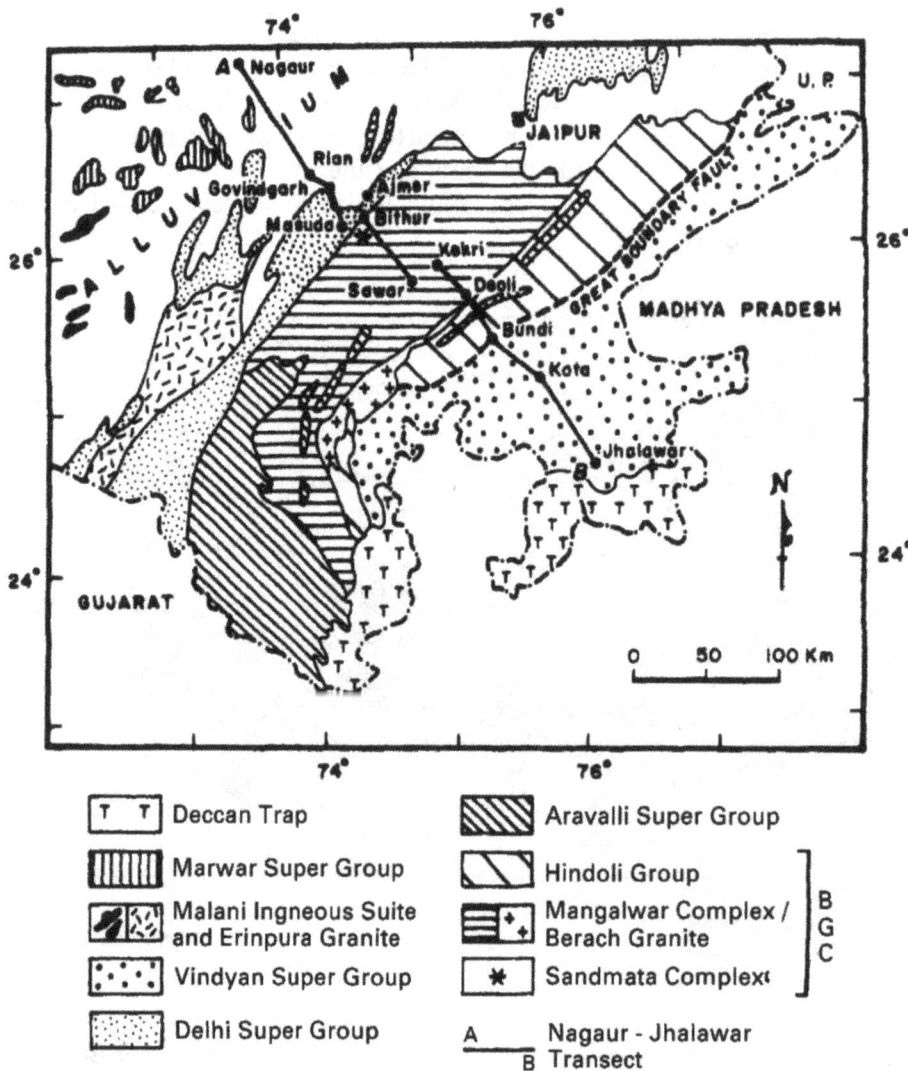

Fig. 7.12(a) Geological map of the ADMB showing the important geological units and the Nagaur-Jhalawar Geotransect (Sinha Roy et al., 1995).

(b) The Hindoli group of rocks consist of volcano-sedimentary sequences and greenstone belts consisting of both mafic and felsic volcanics. The Mangalwar complex is represented by an ensemble of tonalitic gneisses, meta-volcanics (basic and ultrabasic), metasediments and granitic intrusions. The Sandmata complex includes exposed lower crustal granulite rocks and Charnockite bodies. The Aravalli super group of rocks occurring over the Archean basement are exposed south of the transect and consists of metavolcanics and metasediments represented by basic lavas, limestones, quartzites, phyllites, schists, and syenites. The rocks of the Delhi super group occurring on the western side of the ADMB comprise basically of two groups, the western and the eastern. The western group is characterized by basic-felsic volcanics and shallow clastics such as marble, metapelites and calc-schist with conglomerate. The eastern group is free of volcanics and mainly consist of peltic-semipeltic schist with quartzite bands, calc-gneiss, and marble. They are regarded to represent arc-trench sequence towards the east and ophiolites mélange towards the west (Sinha-Roy et al., 1995). The western margin of the Delhi super group in the sourthern part against the Erinpura granite is characterized by the Phulad ophiolote associated with the Phulad ophiolite thrust. At the close of the Proterozoic period, there are magmatic activities such as the Erinpura granite and other granitic bodies and the Malani volcanics. The former is represented by plutonic suite of rocks similar to those found in island arcs while the latter represents bimodal (basic-felsic) volcanics, of calc-alkaline type with rift fills and post-orogenic platform sediments (Marwar group) indicating back arc rift basins and related magmatism (Sinha-Roy et al., 1995). The Marwar super group of rocks occurring at the western end of the transect is flat lying undeformed cover sequence consisting of sandstone, siltstone, clay, dolomite, chert, limestone and evaporites. The ADMB is characterized by two periods of intrusives, viz. Zahazpur granite of 1.7-1.5 Ga that may represent an island arc of that period (Sugden et al., 1990), and Erinpura granite and Malani volcanics of 0.8-0.75Ga west of ADMB (Table 7.1).

(c) Chore and Mohanty (1998) have described a set of bimodal volcanics and coarse clastics in linear rift basins west of the Delhi super group of rocks, which they have identified as back arc basins and associated magmatism developed due to the subduction of Delhi oceanic continental crust under the Rajasthan craton. One of these basins, namely the Punagarh basin depicted hydrothermally altered basaltic rocks of 761 ± 16 Ma (Lente et al., 2009) indicating convergent margin setting during this period. They also suggested U-Pb ages of 800 ± 2 Ma for some granitoid rocks from this section that are related to the post Delhi Erinpora granite (Fig. 7.12 (a)) west of the ADMB. Biswal et al., (1998) have identified a granulite suite in the Delhi super group of rocks characterized by gabbro-norite and basic granulites and similar to the alkaline-calc-alkaline-tholeiitic composition of the present day arc setting. Sinha-Roy (1988) suggested two stages of rifting and convergence along the ADMB related to Aravalli and Delhi group of rocks during Paleo and Mesoproterozoic periods. Sharma (1995) has suggested under-plating due to decoupling of the mantle lithosphere.

Table 7.1 Some impotant rock formations of the ADMB and the SMB and adjoining cratons and their chronological order

Age	ADMB (Sinha Roy, 1988 and Choudhary et al., 1984)	SMB: Central Part of SMB and Bastar Craton (Sarkar et al., 1981 and Jain et al., 1991)	SMB: Eastern Part and Singhbhum Craton (Acharyya et al., 2003)
Neo-Proterozoic	Post Delhi magmatism: Einpura grnite and Malani volcanics: 0.8-0.7 Ga Back arc basins with bimodal volcanics: 0.9-0.8 Ga		
Mesoproterozoic	End of Delhi orogeny: 1.0 Ga Deformation and thrusting of Delhi rocks: 1.1 Ga Delhi rifting and Delhi supergroup of rocks: 1.5 Ga	End of Sausar orogeny: 1.0 Ga Southern-granulite rocks: 1.0Ga Mangikota volcanics: 1.0 Ga Kairagarh volcanics: 1.4 Ga Sausar meta-sediments and gneisses/migmatite complex 1.5 Ga	End of Singhbhum orogeny: 0.9-1.0 Ga Southern granulite belt in CGGC Gangpur granite intrusive: 1.0 Ga Mayurbhanj granite: 1.2 Ga Chankradarpur granite-gneiss: 1.5-1.1 Ga Anorthosite gabbro: 1.5 Ga
Paleoproterozoic	End of Aravalli orogeny: 1.6 Ga Granite of north Delhi fold belt and base metal mineralization: 1.7-1.6 Ga Darwal and Amet granite: 1.9-1.7 Ga Sandmata lower crustal granulite rocks, thrusting: 1.9-1.8 Ga Aravalli rifting and supergroup of rocks: 2.2-2.1 Ga Berach granite: 2.5 Ga	End of Mahakoshal orogeny: 1.6 Ga Dormation of Mahakoshal rocks and northern granulite rocks: 1.6 Ga Sakoli and Nandagon bimodal volcanics of back arc type: 2.2 Ga Dongargarh and Malanjkhand K-granite, Island ar type: 2.3 Ga Granite intrusions: 2.4-1.6 Ga Mahakoshal group of rocks: 2.4 Ga	Ultramafic intrusions northern granulite belt in CGGC: 1.6-1.5 Ga Kohan group: 1.6-1.5 Ga Dalma-Chandeli-Dhanjori volcanics. Similar to back arc basins: 1.7 Ga Dhaibhum stage Chaibasa stage
Archean	Untala and Gingla granite: 2.9 Ga Banded gneissic complex: 3.5 Ga	Unclassified granite and gneisses Amgaon, Sukma etc.,: 3.0-3.5 Ga	Singhbhum granite : 2.95 Ga Older metamorphic group: 3.3 Ga

7.5.2 Geophysical Studies – Nagaur-Jhalawar Geotransect

(a) This geotransect is shown in Fig. 7.11(c) as Profile I. The Bouguer anomaly (Fig. 7.8(a)) shows a linear gravity high, H3 over the ADMB with adjoining linear gravity lows on either side. It also provides paired linear magnetic anomalies in the airborne magnetic map of India (H2, L2; Fig. 3.30) that are typical signatures of Proterozoic collision zones. A detailed Bouguer anomaly map of NW India including ADMB and layout of geotransect from Nagaur-Jhalawar is shown in Fig. 7.12(b). This map adopted from the Bouguer anomaly map of India shows the gravity highs and lows related to the ADMB more clearly. It shows central gravity highs, H of amplitude

20-30 mGal with adjoining gravity lows, L1 towards the west and L2 towards the east coinciding with the Erinpura granite and Vindhyan sediments, respectively. It clearly shows the sets of gravity anomalies typical of Proterozoic collision zones as outlined in Section 7.2 under gravity anomalies that are further analysed below in detail. The striking parallelism between gravity highs, H and lows, L1 and L2 indicate that their sources may have a genetic relationship. The gravity high, H primarily coincides with the BGC and the Aravalli and the Delhi super group of rocks. In some geological maps (Lente et al., 2009) the whole sections of BGC west of the Great Boundary Fault upto east of Ajmer along this transect (Fig. 7.12(a)) is shown as Aravalli super group of rocks indicating the gravity high, H as related to it. The gravity highs corresponding to the ADMB extend to the NW Himalayan front suggesting its northwards extension and intersection of the faults/thrusts associated with the ADMB with the Himalayan front causes seismic activity (Sections 6.1.10 and 6.3.5).

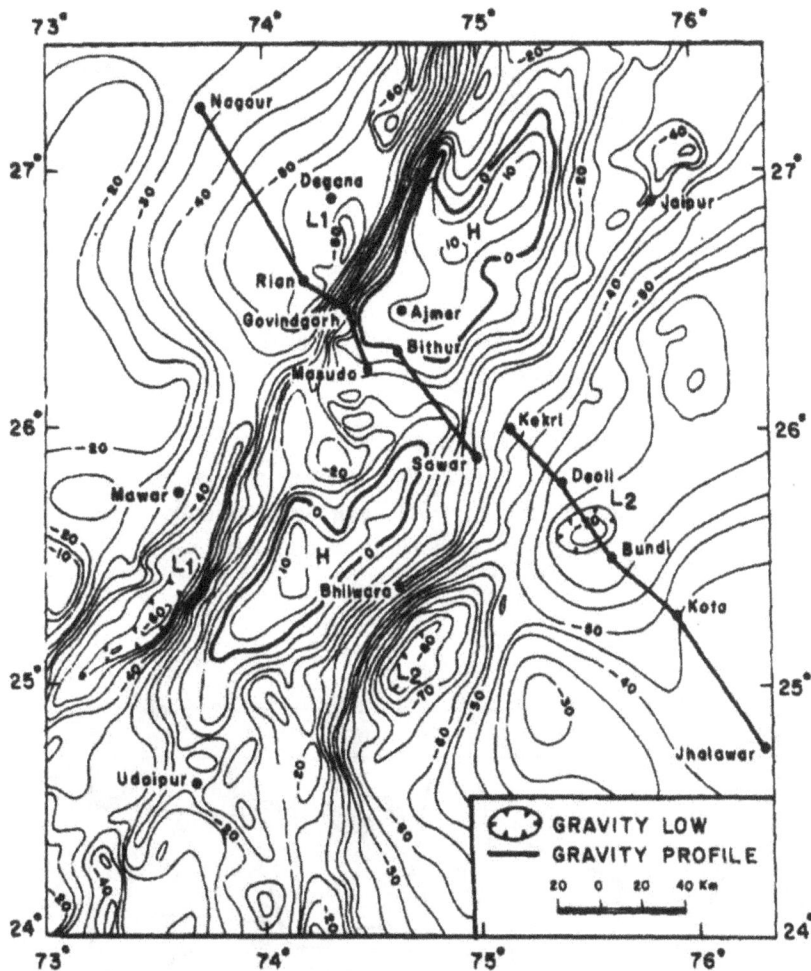

Fig. 7.12(b) Regional Bouguer anomaly map of the ADMB and surrounding region, showing central linear gravity 'highs' (H) and gravity 'lows', L1 and L2 coinciding with the ADMB, felsic intrusives (Erinpura granite), and Vindhyan sediments towards the west and the east, respectively. The geotransect, Nagaur-Jhalawar is also shown.

(b) Line drawing of seismic reflection data along the Nagaur-Jhalawar profile is given in Fig. 7.13(a) (Tewari et al., 1997) that shows domal shaped reflectors in the lower crust under the Delhi Fold Belt and the western part of the BGC. It also shows an east verging crustal scale

Jahazpur thrust (R1) under Hindoli group of rocks and westward dipping reflectors (R2) under the Vindhyan basin east of the ADMB. It also shows crustal thickening under the Delhi group of rocks. The large scale crustal thrusts (F2 and F3) are connected to the domal shaped lower crustal body that indicates its importance in the tectonic development of this region. Tewari et.al., have also provided a slightly higher velocity of 7.3 km/sec for the lower crustal domal shaped body. A shallow fault, F1 separates the Marwar basin towards the west from the Delhi super group of rocks towards the east.

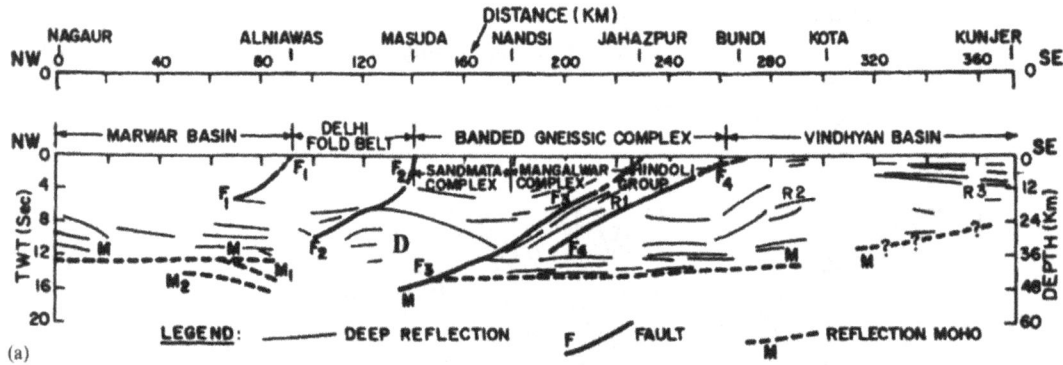

Fig. 7.13(a) Line drawing of the seismic section along the Nagaur-Jhalawar geotransect (Tewari et al., 1995, 1997) showing a thick crust with domal shaped reflectors (D) in the lower crust under the western part and dipping reflections (R1) coinciding with the Jahazpur thrust under Banded Gneissic Complex. The F1-F4 are major faults/thrusts and R1, R2, and R3 are dipping reflectors related to Jahazpur thrust and western margin of the Vindhyan basin.

(c) A magnetotelluric profile over the eastern part of the Bundi-Deoli transect, (Fig. 7.13(b), Gokaran et al., 1995), provided a resistive body (~ 500 Ωm) that represents Jahazpur granite. This section being located over the Jahazpur thrust (F3, Fig. 7.13(a)), the conductive section surrounding Jahazpur granite may indicate the presence of fluids along its contacts.

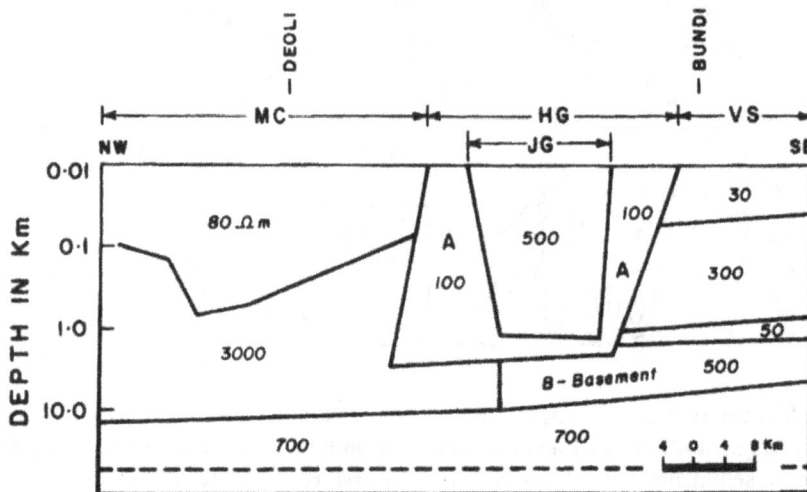

Fig. 7.13(b) Conductivity distribution along the SE part of the transect between Kota and Kekri (Fig. 7.12(a), simplified after Gokaran et al., 1995) in Ωm. It shows high conductivity (A) along margins of Jahazpur granite and a relatively conductive basement (B) below Vindhyan sediments and shallow conductive layers including those over the basement (B) suggest fluid filled weathered layers.

VS: Vindhyan Sediments, HG: Hindoli Group of Rocks, JG: Jahazpur Granite, and MC: Mangalwar Complex.

(d) Fig. 7.14(a) shows specially recorded gravity and magnetic profiles along the geotransect, Nagaur-Jhalawar. It also shows the elevation along this profile. The Bouguer anomaly (~ 10 mGal), free air anomaly (~ 50 mGal), and elevation (~ 400 m) along this geotransect shows an almost parallel nature indicating lack of isostatic compensation and shallower crust than what it should have been for a fully Airy's compensated crust. The total intensity magnetic data shows sharp magnetic anomalies related to the Delhi group of rocks and Sandmata complex representing lower crustal rocks, and Mangalwar complex consisting of volcanics and metasediments. The magnetic anomalies related to the Delhi group of rocks may represent mafic intrusives such as ophiolite rocks. However, as magnetic data is recorded on the ground, it shows sharp fluctuations due to inhomogenous magnetization, typical of metamorphic terrains. Therefore, an averaged magnetic profile of 5 × 5 points is prepared and shown in Fig. 7.14(b) and shows a major magnetic anomaly, M1 in the central part of the ADMB coinciding with the ophiolite and granulite rocks in this section. This can be used for quantitative interpretation. The modeled causative body for this anomaly suggests a west-dipping dyke type of body of high susceptibility that coincides with the contact of the Delhi group of rocks with Sandmata complex of granulite rocks indicating thrusted mafic/ultramafic rocks dipping westwards similar to the known thrusts in this section.

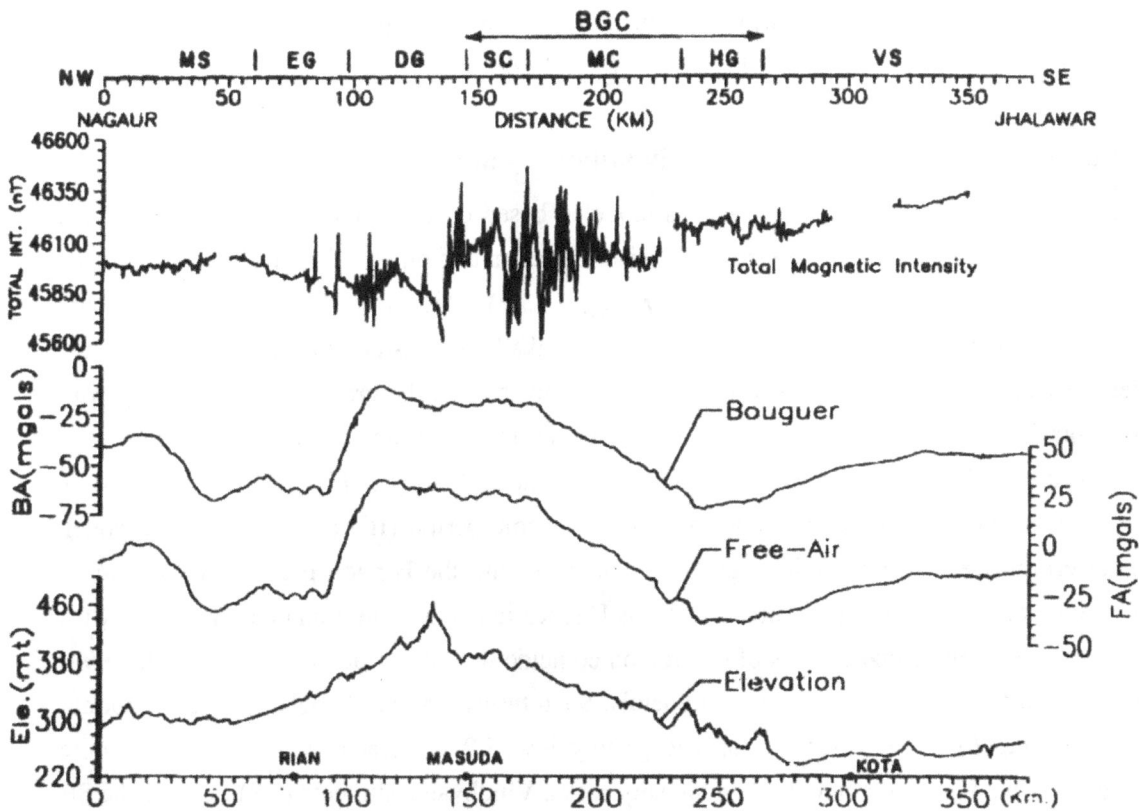

Fig. 7.14(a) Gravity and magnetic profiles and elevations along the Nagaur-Jhalawar geotransect recorded at 0.5 km. It also shows the exposed rock types at the top. Both Bouguer and free air anomaly are highs with increasing elevation suggesting positive correlation between them that indicates lack of isostatic compensation and these anomalies are caused by high density rocks in the crust. The exposed rocks are: MS: Marwar Sediments; EG: Erinpura Granite; DG: Delhi Supergroup, BGC: Banded Gneissic Complex, SC: Sandmata Complex; MC: Mangalwar Complex; HG: Hindoli Group and VS: Vindhyan sediments (Mishra et al., 2000).

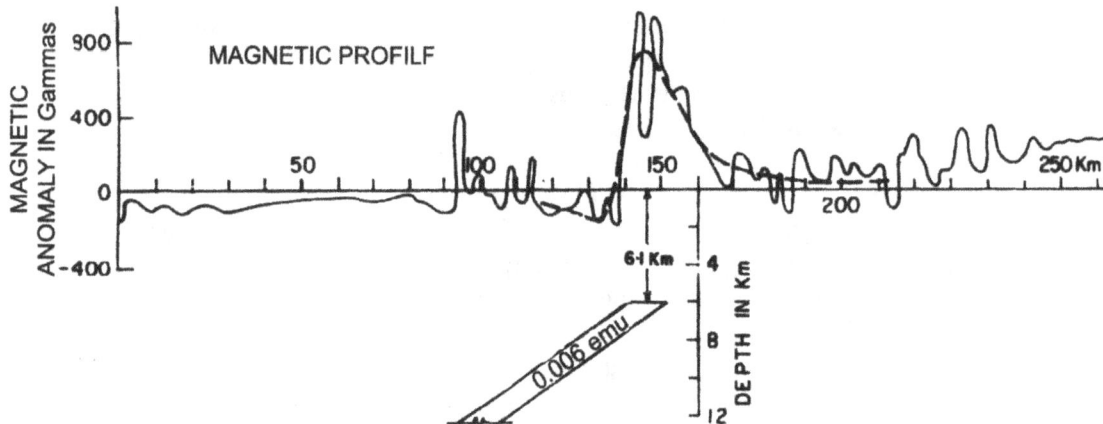

Fig. 7.14(b) Averaged magnetic profile with 5 × 5 data points and magnetic anomaly related to the Delhi thrust modeled using a tabular body of susceptibility 0.006 emu indicating westward dipping mafic intrusive associated with an east verging thrust.

7.5.3 Integrated Crustal Structures and Evolutionary Model

(a) The gravity profile of Nagaur-Jhalawar was modeled based on the comparison of the computed and the observed fields constrained from the seismic section and other information available from this region. Crustal model along this profile (Fig. 7.15(a), Mishra et al., 2000) suggests high density ophiolite rocks along the western thrust (IId) and a high density body (I) in the lower crust under the Delhi group of rocks towards the west responsible for the gravity high, H1 along with some contributions from high density rocks of the Delhi group of rocks, Sandmata and Mangalwar complexes. The linear gravity low west of the ADMB (L1, Fig. 7.15(a)) is caused by the magmatic rocks of Erinpura granite of Neoproterozoic period (III) as shown in this model. The eastern gradient of the gravity high (H1) coincides with the Hindoli group of rocks and an east verging thrust (V, Zahazpur Thrust or Banas Dislocation zone with high density rocks). Most of the base metal mineralized zones of this region coincide with this gradient (Prasad et al., 1999) that has acted as conveyor belt for these minerals. Such thrusts in mobile belts provide important regional setting for mineralized zones. The gravity low, L2 is caused by the Jhazpur granite (2.69 g/cm^3) in contrast to adjoining high density rocks. Vindhyan sediments (VS) towards the SE are underlain by high density rocks (VIb) that form the basement and cause a gravity high at the eastern margin of this transect. This is also indicated by the high conductive basement (Fig. 7.13(b)) in this section.

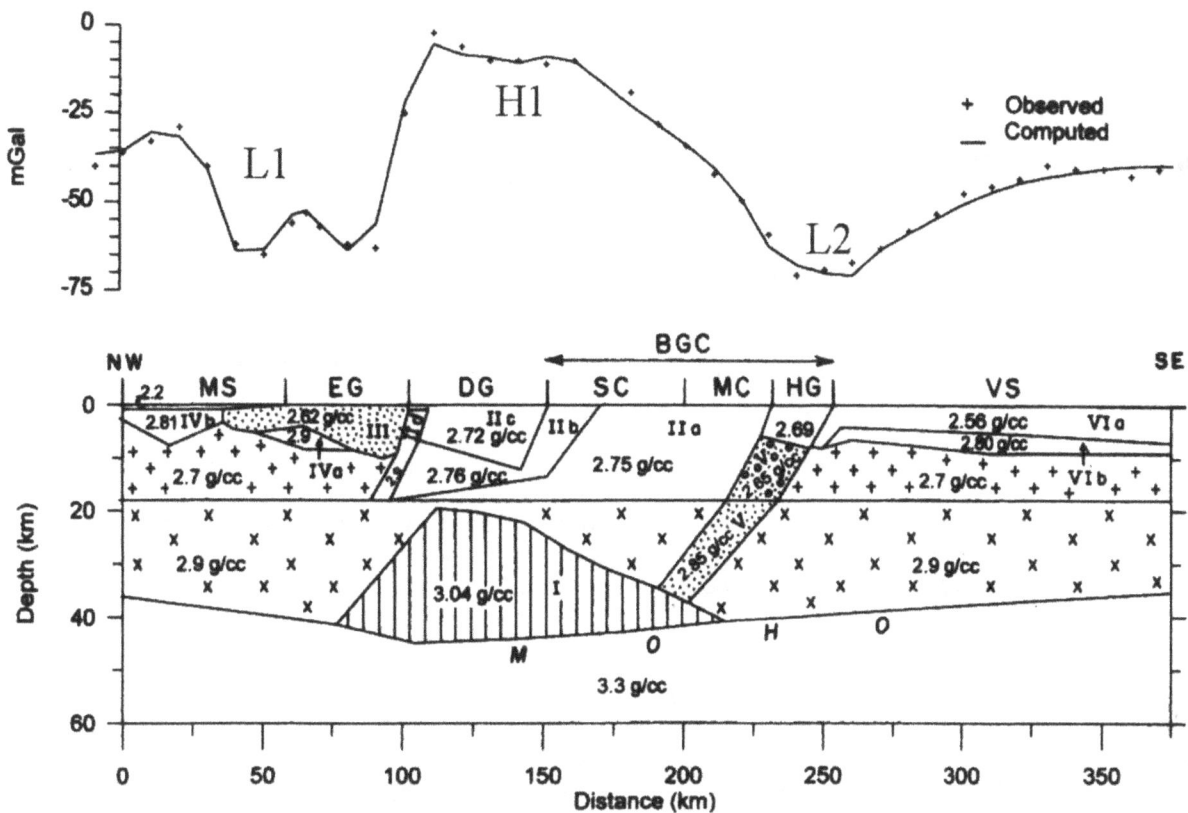

Fig. 7.15(a) The observed and the computed Bouguer anomaly along the Nagaur-Jhalawar geotransect and the crustal section thus derived showing a thick crust and a high density body in the lower crust (I). Densities are in g/cc. Basement of the Vindhyan sediments is formed by a high density body (2.80 g/cc). Causative sources are indicated as I-VI. The exposed rocks are referred to by same letters as given in Fig. 7.14(a). The contact between the Vindhyan sediments (basin) and BGC (HG) is known as the Great Boundary Fault that forms the eastern margin of the Jhazpur thrust (V). Bodies, III, IV a and b are low and high density intrusives on the western side that may be related to E-W convergence (Mishra et al., 2000).

(b) The east verging Delhi thrust (Phulad ophiolite thrust)associated with high density and high susceptibility ophiolites rocks (IId, Fig. 7.15(a)) and Delhi group of rocks of arc-trench sequence and ophiolites mélange towards the west of Mesoproterozoic period (1.5-1.0 Ga, Table 7.1) and low density magmatic intrusives (III, Fig. 7.15(a)) of Erinpura-Malani volcanics of Neoproterozoic period (~ 0.8 Ga) with contemporary back arc settings along the western margin of the ADMB, suggest convergence during Meso-Neoproterozoic period between the Bundelkhand craton and the Rajasthan block. Subduction from east to west is shown by an arrow east of the ADMB in Fig. 7.6. In this case, the Delhi thrust associated with ophiolite rocks (IIId, Fig. 7.15(a)) acted as a suture between the two. Fig. 7.15(b) shows a schematic diagram briefly describing the plate tectonic processes across the ADMB, primarily based on the crustal model (Fig. 7.15(a)) from integrated geophysical study along the Nagaur-Jhalawar geotransect. This cartoon shows the convergence and subduction of the Bundelkhand craton towards the west as subducted slab under the Hindoli group of rocks and backward thrusting of ophiolite and granulite rocks. The Erinpura granite and Malani suite of rocks (MS) west of the ADMB represent subduction related magmatism during Neoproterozoic period. The domal shaped body is shown as underplated crust that might be related to rifting in between the Aravalli and Delhi orogenies during Paleoproterozoic period or to some recent tectonic activity as discussed in (d).

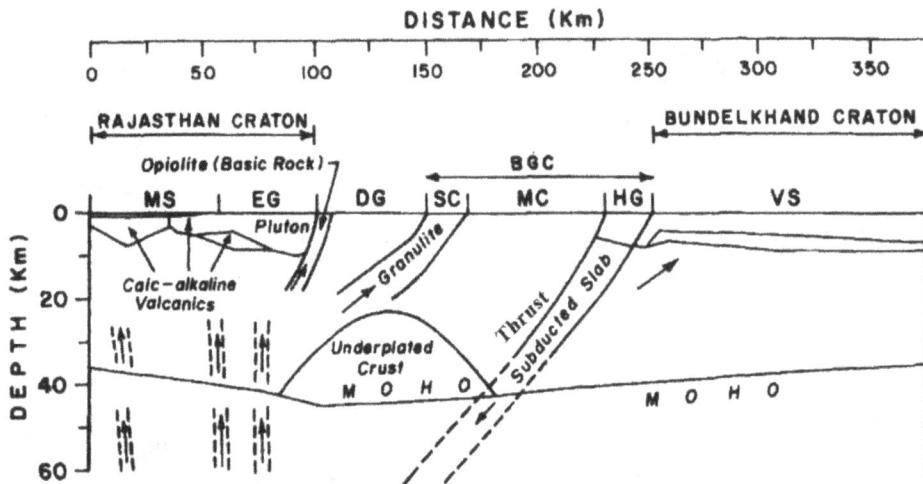

Fig. 7.15(b) A schematic digram artoon of tectonic activities across the ADMB based on crustal model and rock types inferred from integrated geophysical studies (Fig. 7.15(a)) showing obducted and subducted sections along with related intrusives of Erinpura granite and other volcanics on the western side (Mishra et al., 2000).

(c) The Untala and Gingla granite of Neoarchean period (Table 7.1) and the gravity low, L2 towards the east caused by Jahazpur granite of Paleoproterozoic period associated with Jahazpur thrust (V, Fig. 7.15(a)) in the eastern part and Sandmata complex of lower crustal granulitic rocks (Table 7.1), suggest a prior event of convergence and rifting during Paleoproterozoic period (Mishra, 2006). In that case, the Great Boundary Fault (Fig. 7.7) along the eastern margin of the ADMB that merges with the Jahazpur thrust at the depth, acted as a suture between the two blocks. These two phases of convergence were separated by a rifting phase (~ 2.0-1.6 Ga) related to the Aravalli super group of rocks and rocks equivalent to Mahakoshal and Bijawar group of rocks forming the basement of Vindhyan rocks in this Section (VI b, Fig. 7.15(a)) as has been discussed in detail in Section 7.6.2). The Hindoli group of rocks and Mangalwar and Sandmata complexes between the Great Boundary Fault and Delhi super group of rocks grouped under the Banded gneissic complex (Fig. 7.12(a)) have been referred to as belonging to the Aravalli super group of rocks with patches of the BGC in specific sections (Lente et al., 2009). In that case, the Great Boundary Fault representing a Paleoproterozoic suture between the Bundelkhand craton towards the east and the Marwar block (Western Rajasthan) towards the west is perfectly justified as other rocks of the ADMB in this section are younger than the Aravalli super group of rocks and are connected to the Jahazpur thrust in the depth. However, the signature of this event is not as clear as that of later events related to the Delhi group of rocks as described above. East verging nature of Jahazpur and Delhi thrusts suggest convergence in both cases from the east to the west. The absence of ophiolite rocks of Paleoproterozoic period in this section suggested the absence of oceanic crustal rocks during that period. However, the large magnetic anomalies related to mafic rocks of the Mangalwar and Sandmata complex as described above, may in fact represent the same.

(d) Heat flow studies have suggested high heat flow (Roy and Rao, 2000) in the western part of the ADMB. Sen and Sen (1983) have also reported recent uplift especially in the western part of the ADMB. In absence of any recent tectonothermal event, this could not be explained so far. In light of this and the recent uplift, the domal shaped high density body in the lower crust (I, Fig. 7.15(a)) may represent crustal/lithospheric bulge due to flexure of the Indian plate (Fig. 6.15(b)) under the Western Fold Belt (Pakistan) in highly sheared and fractured section of the ADMB that would cause thinning of lithosphere resulting in high heat flow and uplift. This is further explained in Section 7.11 (Fig. 7.60) while describing lithosphere, under the Indian continent.

The gravity highs caused by the crustal bulge due to lithospheric flexure under Himalaya towards the north and the west interact SW of Delhi along the Aravalli belt (Fig. 6.15(b)) enhancing its effect in this section. Lithospheric flexure can cause mantle upwelling and can create melts due to decompression (Hirano et al., 2006). Raval (1995) attributed the high heat flow and recent uplifts to the interaction of Reunion plume with the Indian lithosphere, such as under-plating, in this section. The most important evidence in this regard is the reported high heat flow in the adjoining Cambay basin that is a petroliferous basin supposed to have formed due to Deccan trap intrusions. However, the Deccan trap activity is supposed to cool down during The last 65 Ma, and some neotectonic activity related to Himalayan orogeny such as lithospheric flexure as described above might be responsible for the present day high heat flow in this section.

(e) Western Rajasthan shows semicircular gravity highs (Fig. 7.8(a), H4 = ~ 50 m Gal) of small-medium wavelength that indicate shallow high density rocks. Similar gravity highs have also been reported from eastern Pakistan that can be attributed to mafic intrusives during the breakup of the Indian plate from Africa (Karoo Volcanics) during the Jurassic period that gave rise to several Mesozoic basins (Section 6.3.5) and some mafic dykes of this period in this region. Further details about their causative sources are discussed in Section 7.11.2.

7.6   Satpura Mobile Belt (SMB)

The Satpura Mobile Belt (SMB) in Central India (Fig 7.6, 7.7 and 7.11(e)) forming Satpura ranges extends from the west coast of India up to the Bengal basin in east India, and separates Bundelkhand craton and Vindhyan sediments towards the north from the Dharwar, the Bhandara-Bastar and Singhbhum cratons towards the south. Its northern margin is characterized by the Narmada-Son Lineament (NSL) while southern margin by the Central Indian Shear (CIS) and the Sausar metasediments (Fig. 7.7). Based on gravity and magnetic anomalies, Mishra (1977) had suggested the extension of the Narmada-Son Lineament forming the northern boundary of the SMB to the Shillong Plateau and eastern margin of the Indian plate. The Bouguer anomaly map (Fig. 7.8(a)) shows a linear belt of gravity highs (H7-H9) that extends from the west coast of India up to the Bengal basin in east India, coinciding with the SMB. Accompanying these highs are gravity lows, L7-L9 to their south and the two might be tectonically related to each other as discussed in the forthcoming sections. Some of the geological and geophysical data sets significant for collision tectonics in this region and their evolution are as follows:

7.6.1   Central Part of the SMB: Geophysical Studies – Crustal Structures, Seismicity and Evolutionary model

The important characteristics of the central part of the SMB and the Bhandara craton south of it are discussed below to evaluate Proterozoic collision tectonics.

(i) Geology and Tectonics

(a) A geological map of the SMB in central India is given in 7.11(c) and shows the western part as being primarily covered by the Deccan trap with small exposure of peninsular gneisses that forms the basement. The eastern part is largely occupied by unclassified gneisses known as the Chhotanagpur Granite Gneiss Complex (CGGC). This map also shows the ADMB in the western part and comparative rock types in two sections, with their ages are given in Table 7.1. It also shows the Nagpur-Jhalawar and Mungwani-Rajnandgoan geotransects across the ADMB and the SMB, respectively. A detailed geological map of the central part of the SMB and Bhandara craton to its south is presented in Fig. 7.16, with periods of important tectonic elements given in Table 7.1. It shows the Central Indian Shear (CIS) that extends for almost 500 km from the SE of Nagpur to the ESE of Balaghat and west of Chilpi. It represents a shear zone marked by silicified, brecciated, and mylonite zones showing evidence of ductile deformation, and separates the

NE-SW striking high grade Sausar metasediments and the granulite rocks within the Tirodi gneisses to its north from the N-S striking low grade volcanogenic sequences of Bhandara craton to its south (Fig. 7.16). The Sausar metasediments belong to the Mesoproterozoic period and are represented by quartzite-carbonate sequences with complete absence of volcanic rocks. They are metamorphosed to upper amphibolite to granulite facies and are extensively migmatized. All along the CIS, the granulites are observed as exposed lenses of ca 0.5-7.0 km length and 0.2-2.0 km width within the Tirodi gneisses (Jain et al., 1991).

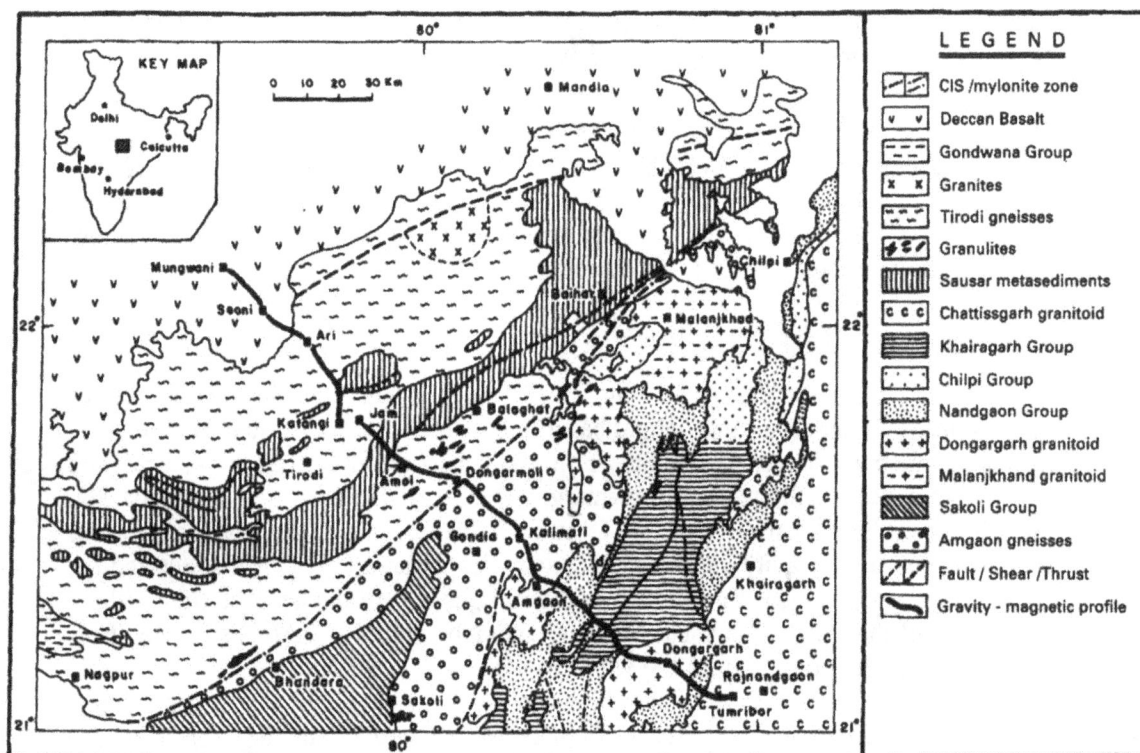

Fig. 7.16 Geological map of the central-western part of the SMB showing Central Indian Shear (CIS) (Jain et al., 1991) separating NE-SW oriented high grade Sausar metasediments almost parallel to the CIS towards the north, from the N-S oriented low grade metavolcanics and metasediments towards the south (Nandgaon-Kairagarh group). Patches of high grade granulite rocks and mylonites are exposed along the CIS. It also shows Mungwani-Rajnandgoan geotransect.

(b) The region south of the CIS is characterized by the Amgaon gneisses which are migmatitic and form the basement of the Sakoli, Nandgaon, Chilpi, and Khairagarh group of meta-volcanics and metasediments of Paleo to Mesoproterozoic period (Table 7.1). The Sakoli and the Nandgaon group of rocks of Paleoproterozoic period are basically represented by phyllites, quartzites, mica schists, banded iron formations (BIFs), and basic and acid volcanics while the Chilpi and the Khairagarh group of rocks represent younger sequences of conglomerates, quartzites, marbles, phyllites and basic volcanics (Fig. 7.16). The volcanics of Sakoli and Nandgaon group of rocks are mainly tholeiitic in character similar to the island arc tholeiitic series to the calc-alkaline series. However, the Khairagarh volcanics show a mixed chemical composition of tholeiitic and calc-alkaline types, which indicates an extensional phase of back arc magmatism (Yedekar et al., 1990). There are some older granitic intrusions in this region such as the Dongargarh and the Malanjkhand granitoids of Paleoproterozoic period which represent calc-alkaline plutons and have intruded into the Amgaon gneisses showing N-S foliation trends. Acharyya and Roy (2000) named this zone, minus the Gondwana (Permian-Triassic) basins as the Central Indian Tectonic Zone and have considered the reactivation of various faults during different geological periods. However, maintaining the age old nomenclature of the SMB or the Satpura Fold Belt has its own advantages.

(ii) Geophysical Studies: Mungwani-Rajnandgaon Geotransect

(a) It is characterized by paired linear gravity anomalies (H8, L8, Fig. 7.8(a) and (b)) and bipolar linear magnetic anomalies (H3, L3; Fig. 3.30) that are typical signatures of Proterozoic collision zones. A detailed Bouguer anomaly map of the central part of the SMB corresponding to the geological map (Fig. 7.16) is given in Fig. 7.17(a) that also shows central gravity highs, H and a large gravity low, L in the SE corner over the Bhandara craton superimposed by several small wavelength gravity highs and lows related to Proterozoic basins in the Bhandara craton. The gravity lows, L1 and L2 are related to Gondwana sediments that represent Permo-Triassic basins that developed subsequently as rift basins and extend eastwards up to the east coast of India as the Gondwana Godavari basin. Formation of these basins have displaced the central gravity high, H towards the south along faults passing through Nagpur and Balaghat that show significant strike slip movement along boundary fault of the Godavari rift basin (Section 7.7). It also shows the CIS and the geotransect, Mungwani to Rajnandgaon (profile II, Fig. 7.11(c)) along which multi disciplinary geophysical surveys were carried out.

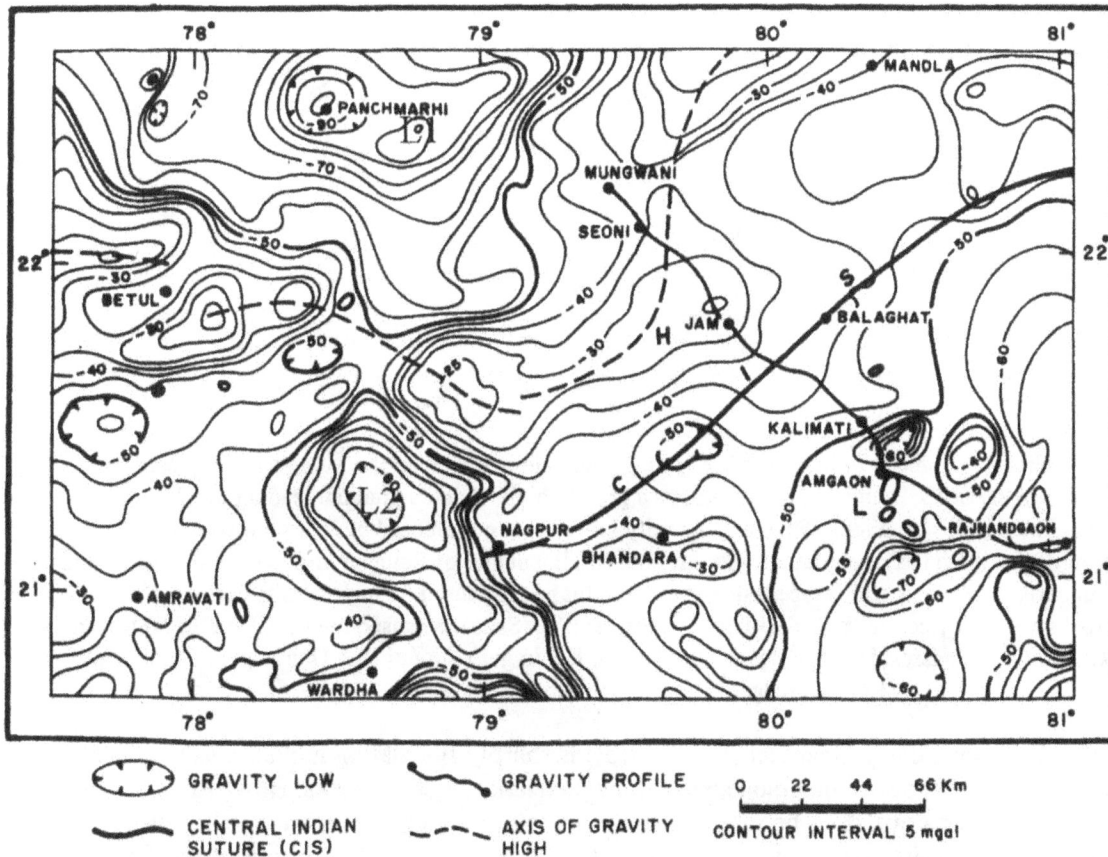

Fig. 7.17(a) Regional Bouguer anomaly map of the SMB and the surrounding region showing CIS separating the gravity 'high' (H) coinciding with the SMB towards the north and a large regional gravity low (L) towards the south. Gravity lows, L1 and L2 are related to Gondwana sediments of the Permo-Triassic period.

(b) The especially recorded gravity and magnetic profile at 1 km interval along the Mungwani-Rajnandgaon geotransect, is given in Fig. 7.17(b). It also shows important geological units at the top of this figure. The gravity data shows highs both in free air and Bouguer anomaly, with increasing elevation that indicate lack of isostatic compensation. The central gravity high, H is therefore caused by high density rocks in the crust. The Bouguer anomaly typically shows a gravity high H, north of the CIS and a continuously decreasing field, L to its south that may form a pair, typical of Proterozoic collision zones, as discussed above. The CIS is characterized by a prominent gravity low, L1 that might be related to the low density body associated with the shear zones.

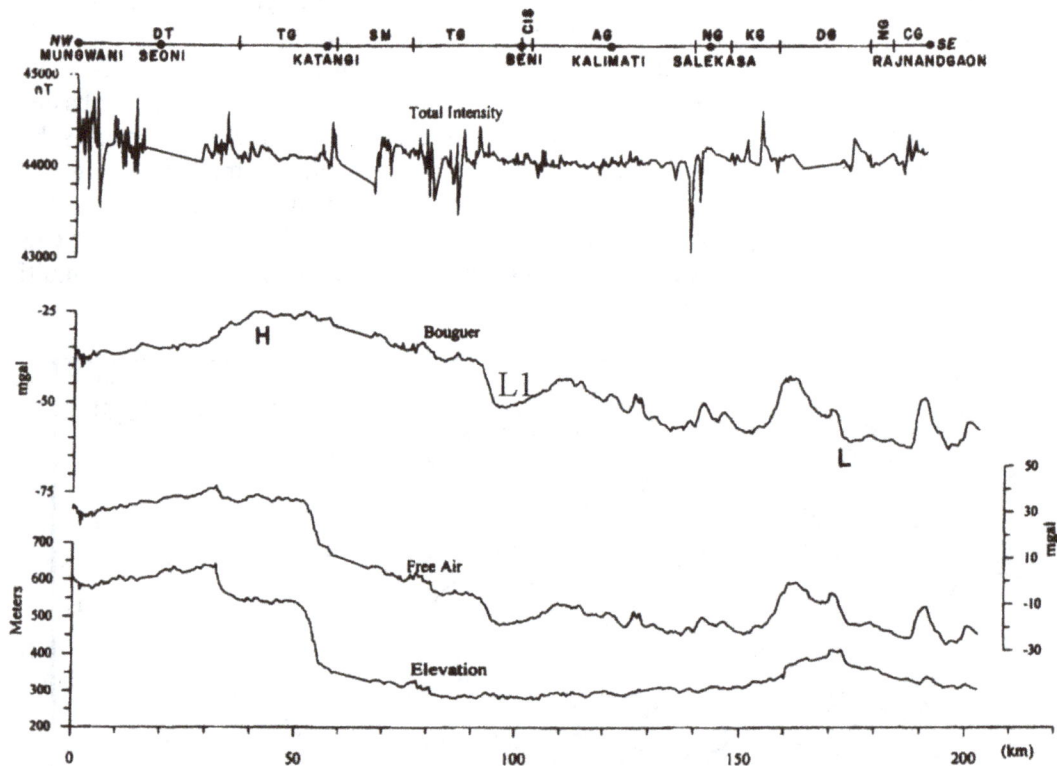

Fig. 7.17(b) Gravity-magnetic profiles, surface geology, and elevation along the Mungwani-Rajanandgaon geotransect across the SMB. Bouguer anomaly shows a gravity high, H and a gradient leading to a low, L. The Central Indian Shear (CIS) coincides with a local gravity low (L1) superimposed over large wavelength gravity gradient towards the south indicating fault/thrust. Almost parallel nature of Bouguer and free air anomalies and elevation suggests lack of isostatic compensation and these highs are caused by high density rocks in the crust. Exposed rocks are; DT: Deccan Trap; TG: Tirodi Gneisses; SM: Sausar Metasediments; CIS: Central Indian Shear; AG: Amgaon Gneisses; NG: Nandgaon Group; KG: Khairagarh Group; DG: Dongargarh Granite; and CG: Chattisgarh Granitoid (Mishra et al., 2000).

(c) Magnetic data along this geotransect (Fig. 7.17(b)) is sharply fluctuating in nature as is typical of metamorphic terrains due to inhomogeneity of magnetization in these rocks. However, it suggests Tirodi gneisses north of CIS to be more magnetic in nature with mafic components compared to Amgaon gneisses to its south indicating different tectonic setup on either side of the CIS. The exposed granulite rocks along the CIS appear to have made the Tirodi gneisses magnetic in nature, which indicates that it has also acted as a thrust. Other significant magnetic anomalies are observed over Nandgaon and Khairagarh group of rocks that represent Proterozoic basins with mafic intrusives (Table 7.1). Further, as this data was recorded on the ground, it has significant amount of noise making it difficult to identify individual anomalies. The same, however, is considerably reduced in airborne magnetic profiles as discussed below.

(d) The airborne total intensity magnetic map (Fig. 3.30) shows thin alternate magnetic highs and lows packed in a compact zone coinciding with the SMB that indicates compressional tectonics. This map also shows major magnetic lows over the Vindhyan sediments flanked by highs related to the Mahakoshal and Bijawar group of rocks, exposed towards the south and the north of the Vindhyan basin along the SMB and the Bundelkhand craton, respectively (Fig. 7.7, 7.11(c)) which also shows gravity highs as has been further discussed in point (iii) and in the next Section 7.6.2.

(e) Fig. 7.18(a) shows a line drawing of a seismic section along this geotransect (Reddy et al., 1997). It shows a fault, F1 along which reflections on either side appear to be displaced. It also shows upwarping of reflections in the lower crust north of the CIS under the SMB and a few reflections in the upper mantle, south of the CIS. Crustal thickness, in general, is about 45-50 km with minimum being north of the CIS and increasing on either side.

(f) The conductivity profile based on magnetotelluric studies (Sarma et al., 1996) shows sharp changes in the conductivity of rocks on either sides of the CIS and high conductive bodies under the CIS and to its north in the upper crust (Fig. 7.18(b)). The high conductive body in the upper crust under the SMB may represent lower crustal rocks as such rocks usually show high conductivity under wet conditions for connectivity that is easily met in orogenic belts due to a fractured matrix.

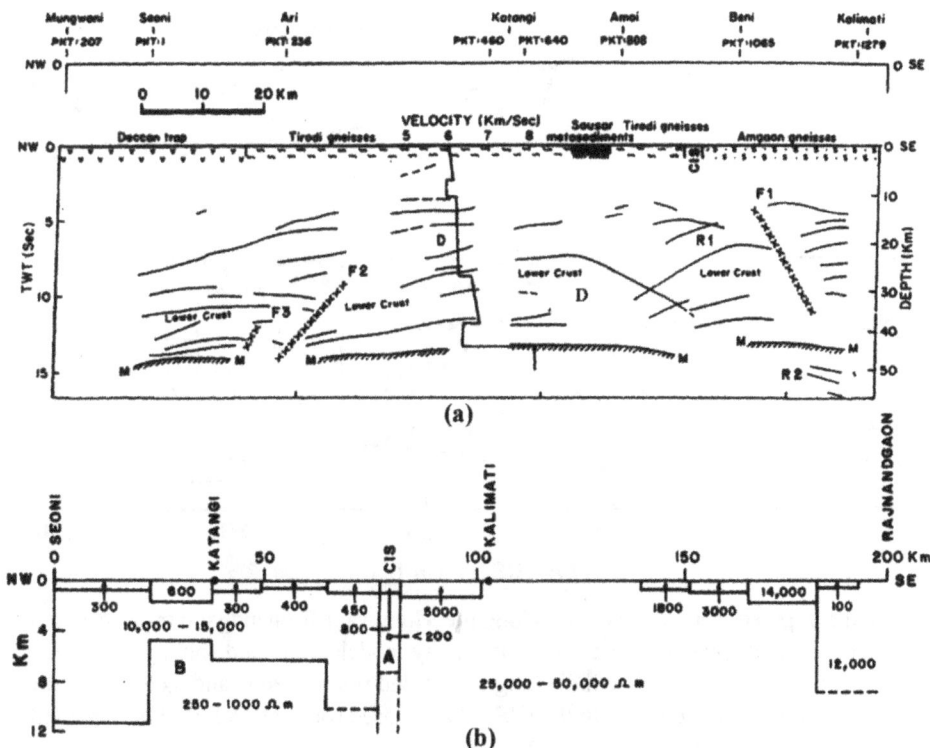

Fig. 7.18 (a) Line drawing of the seismic section along the geotransect from Mungwani to Kalimati. It shows a thick crust (40-45 km) along the whole profile with thin crust towards north and south of the CIS and a deep seated fault (F1) whose projection on surface coincides with the CIS. There are dipping reflectors (R1) just north of the CIS and some below the Moho (R2) south of the CIS. 'D' are the domal shaped reflectors and F2 and F3 are faults under the SMB (Reddy et al., 2000) and (b) Magnetotelluric (MT) sounding along the geotransect, Seoni-Rajanandgaon showing a relatively high conductive crust north of the CIS compared to its south. It also shows a high conductive body under the CIS (A) and another north of it in the upper crust (B), (Sarma et. al., 1996).

(g) Mishra (1992) had suggested high density intrusives in the upper crust related to the gravity high based on the residual field along a profile across central part of the SMB. Modelling of the Bouguer anomaly (Fig. 7.17(b)) along this geotransect (Fig. 7.19(a)) suggested a high density body at a depth of 6-8 km (III, Mishra et al., 2000) coinciding with seismic reflectors and high conductive body as referred to above that suggested lower crustal rocks at shallow depth in this section. This crustal model also suggests a contact (II) separating rocks of different densities in lower crust on either sides of it (Ia and Ib). The projection of this contact at the surface coincides with the CIS. It also shows a low density body under the CIS and several low and high density exposed bodies south of CIS that are related to exposed mafic and felsic intrusives associated with metasediments of Paleo-Meso-Neoproterozoic period like the Nandgaon and Khairagarh group of rocks and Dongargarh granitoid (Table 7.1). Seismic reflections dipping southwards below the Moho south of the CIS represent low density rocks (Id) at the southern end of the profile.

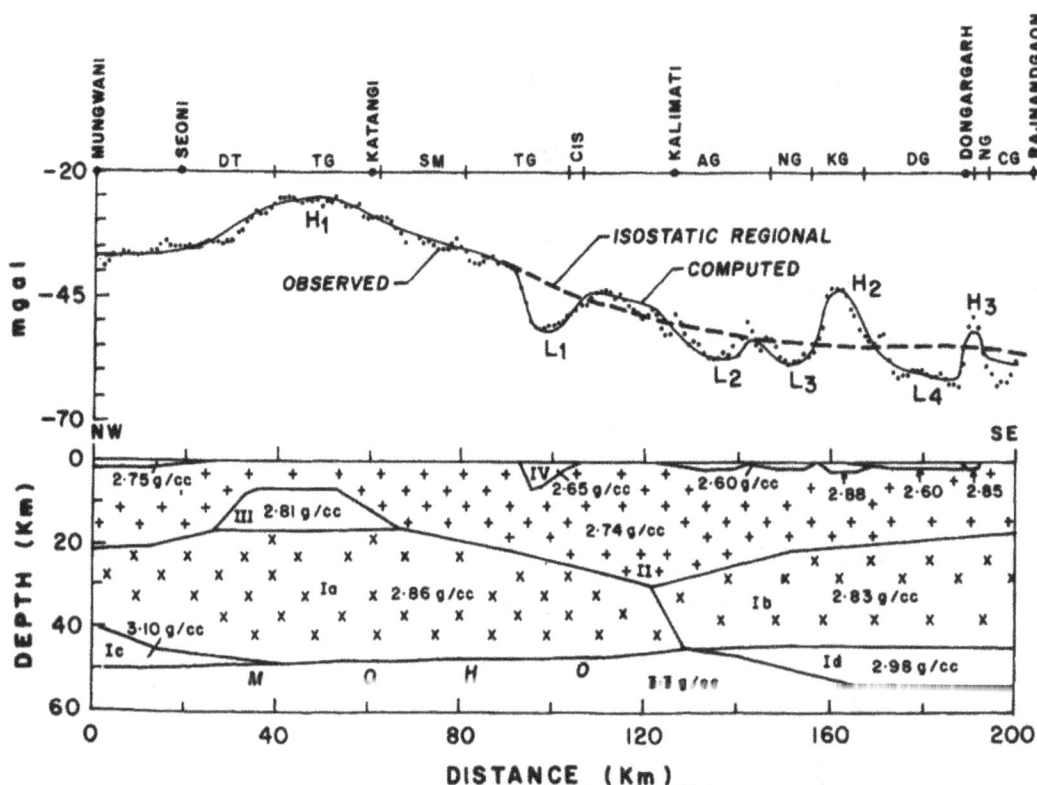

Fig. 7.19(a) Observed and computed Bouguer anomaly along the Mungwani-Rajnandgaon transect and derived crustal section showing an overall thicker crust and different density models north and south of the CIS. CIS is represented by a low density body near the surface indicating fluid filled fracture zone and a contact in the lower crust whose projection on the surface coincides with the CIS. The exposed rocks are referred by same letters as given in Fig. 7.17(b) (Mishra et al., 2000).

(h) Fig. 7.19(b) shows a schematic diagram based on the crustal model (Fig. 7.19(a)) that describes the likely processes involved in the formation of the SMB. It shows the subduction of the northern (Bundelkhand) craton under the southern (Bhandara) craton across the CIS as suture that has caused crustal thickening with low density rocks and subduction related to magmatism on the subducted side (Bhandara craton) in back arc basin type settings. It also shows thrusted lower crustal high density rocks along the CIS under the SMB that indicates the CIS both as the thrust and the suture. These signatures are similar to present day convergence and subductions as in the case of Himalaya, suggesting that the SMB along this profile represents a Proterozoic collision zone between the Bundelkhand craton and the Bhandara craton towards the north and the south, respectively.

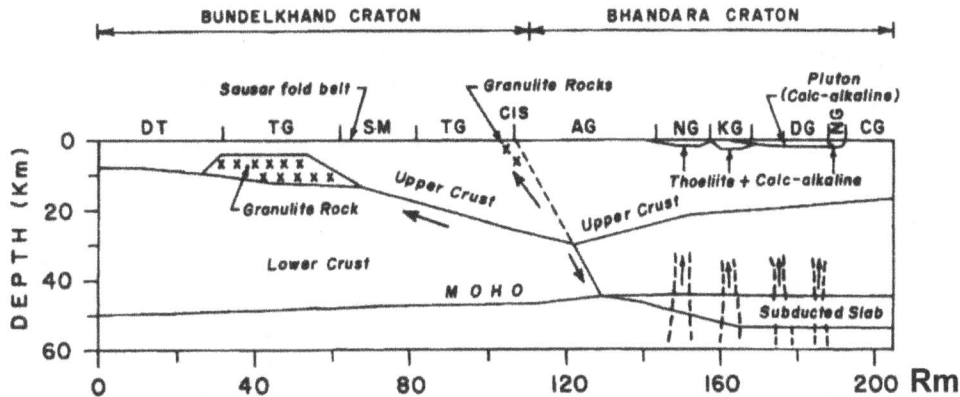

Fig. 7.19(b) A schematic digram of the tectonic activities across the SMB based on the crustal model from integrated geophysical studies showing obducted and subducted sides along with related intrusives with respesct to the CIS (Fig. 7.19(a)) (Mishra et al., 2000).

(iii) Extended Hirapur-Mandla Profile and Jabalpur Earthquake of 1997

(a) Gravity data along another seismic profile Hirapur-Jabalpur-Mandla (Profile III, Fig. 7.7 and 7.11(b)) across the SMB (Kaila et al., 1987; Sain et al., 2000; Murthy et al., 2008) that has been studied extensively was extended towards the north and the south and examined in detail. A detailed geological map of this section along with the Hirapur-Jabalpur-Mandla profile is given in Fig. 7.20 that shows the Bundelkhand craton in the north adjoining the Ganga basin surrounded by Proterozoic Vindhyan sediments whose southern margin is defined by the SMB. Bundelkhand craton in central India primarily represents granitic batholiths of Mesoarchean time (3.3 Ga; Mondal et al., 2002) with several mafic and felsic intrusives almost perpendicular to each other. The NE-SW trending series of shear zones are occupied by quartz reefs while the NW-SE trending swarms of mafic dykes may be related to the opening of Bijawar basin to its south and mark the end of magmatism in Bundelkhand. Singh et al., (2007) have reported both high grade and low grade metamorphics in the Bundelkhand craton known as Bundelkhand gneissic complex and Bundelkhand metasediments and metavolcanics. Vindhyan basin (Fig. 7.20) is a large Meso-Neoproterozoic basin bordered by the SMB towards the south, Bundelkhand craton towards the north, and the ADMB towards the west Itextends northwards under the Ganga basin. It is mostly represented by undisturbed metasediments like sandstones, quartzites, and shales, etc., that are generally undisturbed except along its margin with the SMB and the ADMB. The SMB is defined by the Narmada-Son lineament towards the north and by the CIS towards the south (Fig. 7.11(c)). The Narmada-Son Lineament at the northern margin of the SMB is, in fact represented by two faults referred to as Son Narmada North Fault (SNNF) and Son Narmada South Fault (SNSF) (Fig. 7.20). The section in between these two faults is occupied by alluvium and Mahakhoshal group of rocks of Paleoproterozoic period (Table 7.1) that forms the basement. There are small exposes of the Bijawar group of rocks of Paleoproterozoic period in between the Bundelkhand craton and the Vindhyan basin (Table 7.2) and that may form its basement in the northern part.

Table 7.2 Stratigraphy of Vindhyan Supergroups with stratigraphy and their ages (Modified after Ray, 2006)

Upper Vindhyan Group	Bhander Rewa Kaimur	Bandar Limestone (650-750Ma) Bijaigarh Shale (1100Ma)
Lower Vindhyan Group	Semri	Rohtasgarh Limestone(1600Ma) Rampur Shale (1 599Ma) Deonar Porcellanite (163 lMa) Kaj rahat Limestone (1721 Ma)
Bijawar and Mahakosal Super Groups (Metasediments With Mafic/Ultramafic rocks)		(1 .9-1 .7Ga)
Bundelkhand Granite and Gneisses		(3.3Ga)

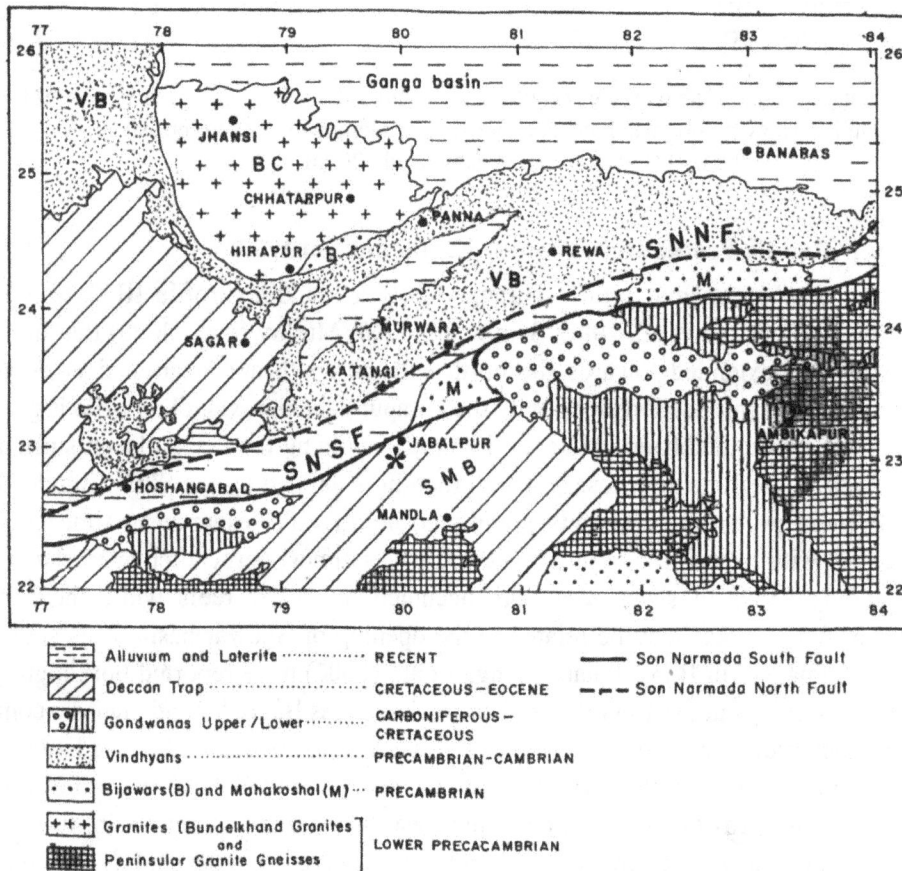

Fig. 7.20 Geology of the central part of the SMB and adjoining region (after GSI, 1993). Abbreviations are as follows: BC = Bundelkhand Craton, VB = Vindhyan Basin SNNF = Son Narmada North Fault, SNSF = Son Narmada South Fault, SMB = Satpura Mobile Belt. Star indicates the epicenter of the Jabalpur earthquake of May 22, 1997 (M = 6.0).

(b) Airborne total intensity map of Bundelkhand craton (Mishra, 1987) shows two sets of perpendicular linear trends, NE-SW and NW-SE, that are related to shear zones with quartz reefs and mafic intrusives, respectively, and may be related to prominent forces that have operated in this region during that period and its effect in perpendicular direction. Gravity, magnetic and airborne magnetic profiles in Central India along Hirapur-Jabalpur-Mandla profile across the Vindhyan basin (Fig. 7.21(a)) encounter both Bijawar and Mahakoshal group of rocks at its northern and southern boundaries, respectively (Mishra, 1992). This profile is, therefore, first discussed below to get acquainted with the nature of these anomalies before describing the maps related to these groups of rocks. This profile shows the gravity high, H3 south of Jabalpur associated with the SMB. It also shows a gravity high, H1 related to the Bijawar group of rocks and a gravity low, L1 in between, related to Vindhyan sediments. The gravity high, H2 is related to the Mahakoshal group of rocks along the northern margin of the SMB. Gravity highs, H1 and H2 associated with the Bijawar and the Mahakoshal group of rocks are special as they are found flanking exposed the Bundelkhand craton and Greater Bundelkhand craton at the northern and the southern margin of the Vindhyan basin. Gravity high, H1 of 30-35 mGal along with the magnetic gradient and anomalies, MG1, MG2, and GM indicate that they are caused by high density and high susceptibility rocks associated with it, suggesting volcano sedimentary rocks with mafic-ultramafic components associated with it that are exposed along the SE margin of the Bundelkhand craton. The gravity high, H2 can be considered to be composed of two gradients G1 and G2 which are typical fault kind of anomalies and are associated with the Son Narmada North and South Faults, respectively (Mishra and Ravikumar, 1998) along this profile that bounds the Mahakoshal group of rocks towards the north and the south respectively. These anomalies, along with corresponding airborne magnetic anomalies M1 and M2, also suggest high bulk density and high susceptibility indicating mafic/ultramafic intrusives associated with the Mahakoshal group of rocks.

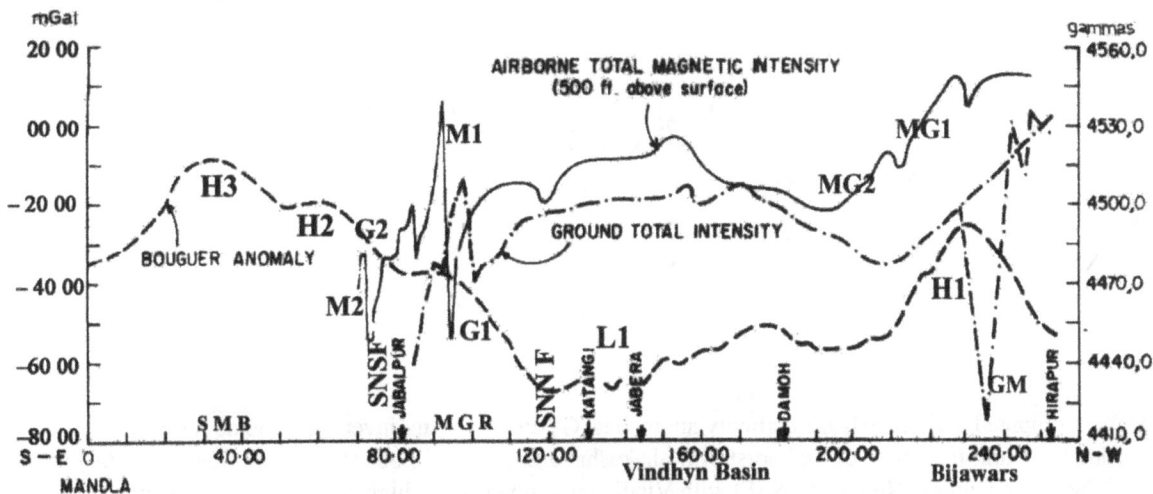

Fig. 7.21(a) A gravity and airborne and ground magnetic profiles, Hirapur-Jabalpur-Mandla across Bijawar and Mahakoshal group of rocks on either side of the Vindhyan basin and northern part of the Satpura Mobile Belt (SMB) (Fig. 7.20). It shows gravity and magnetic anomalies typical of faults, and mafic intrusives, G1, G2 and M1, M2 associated with the Mahakoshal group of rocks, and H1 and MG1, MG2 associated with the Bijawar group of rocks, respectively.
SNNF: Son Narmada North Fault, SNSF: Son Narmada South Fault and MGR: Mahakoshal Group of rocks
(Mishra, 1992).

(c)   Airborne magnetic anomalies observed over the Mahakoshal and Bijawar groups of rocks are quite interesting (Fig. 7.21(a)). The gravity high, H1 is associated with magnetic gradients MG1 and MG2 similar to gravity gradients, G1 and G2 observed over the Mahakoshal group of rocks. The magnetic gradients MG1 and MG2 coinciding with the northern and southern margin of Bijawar rocks are typical fault kind of anomalies, and along with the gravity high, H1, suggest mafic rocks associated with rift kind of structure bounded on two sides by faults. Magnetic anomalies M1 and M2, related to the Mahakoshal group of rocks are paired magnetic anomalies coinciding with the gravity gradient G1 and G2 representing margin faults of Mahakoshal Group of rocks. Paired magnetic anomalies in these magnetic latitudes are related to intrusives suggesting mafic intrusives in this section. Total intensity magnetic field recorded on ground also shows magnetic anomalies similar to airborne magnetic anomalies confirming each other. Lower magnetic intensity plotted in case of ground magnetic data compared to airborne magnetic data is due to base correction, but anomalies are comparable.

(d)   As stated above, the gravity gradients G1 and G2 represent fault kind of anomalies and coincide with the SNNF and the SNSF, respectively. They are modeled using inversion scheme for gravity anomalies due to faults in Fig. 7.21(b) (Mishra and Ravi Kumar, 1998) that has provided opposite dipping faults forming presently a horst kind of structure which is occupied by the Mahakoshal group of rocks. It is also suggested from seismic investigations (Sain et al., 2000) Horst kind of structure along the SMB suggest that it is affected by subsequent compressional forces.

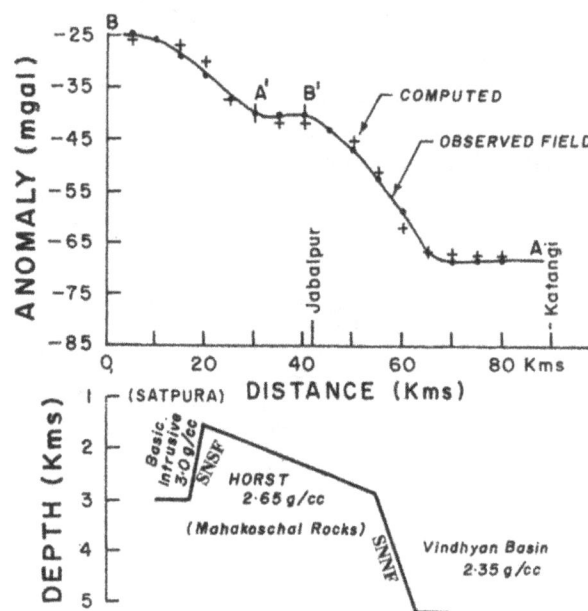

Fig. 7.21(b) Computed fault models from gravity anomalies, G1 and G2 using inversion scheme for gravity anomalies due to a fault. It shows a central horst of Mahakoshal group of rocks bounded by the Son Narmada North and the South Faults (SNNF and SNSF) with Vindhyan sediments and high density mafic intrusives related to the SMB towards the north and the south, respectively.

(e)   The SMB as a whole is seismically active with several earthquakes being recorded in this section (Raval, 1993). This is true for most of the mobile belts as they are tectonically active sections with several faults which are a prerequisite for earthquakes to occur. However, having occurred in the recent past, the Jabalpur earthquake of May 22, 1997 (Fig. 7.20) of moderate magnitude of Mw = 5.7 was studied extensively (Kayal, 2008). Its epicenter south of Jabalpur coincided with the Son Narmada South Fault (Mishra and Gupta, 1998) that is a thrust kind of fault in its present disposition (Fig. 7.21(b)) that might be related to subsequent convergence during the Mesoproterozoic period as discussed above and below in sub-sections (ii and v). The focal depth of about 40 km for this earthquake suggests that this fault (SNSF) extends throughout the crust and its intersection with the Moho might be responsible for this earthquake. Most of the seismic activity of the SMB, including Jabalpur earthquake, is confined to the northern section of the SMB related to the Son Narmada North and South Faults. Its present day active nature may be related to the lithospheric bulge caused by the flexure of the Indian plate under the Himalaya as discussed in section 6.1.9 and above in case of the ADMB and shown in Fig. 7.60. The gravity high due to such a bulge in case of Himalaya extends up to the SMB and has affected the northern part of the SMB making it seismically active due to presence of already existing fault systems from Proterozoic tectonics. The effect of this crustal bulge is seen in the crustal model given in Fig. 7.22(a) that causes extension in the upper crust and compression in the lower crust along the Moho, as shown by arrows in this figure,

(f)   causing thrusting along already existing faults. It is similar to what has been shown in case of crustal model of the Shillong Plateau for the Shillong earthquake of 1897 (Fig. 7.30(b)). Rai et al., (2003) have suggested low mantle heat flow in this section indicating brittle lower crust, which would be amenable to deep earthquakes. The high conductive anomalies under mobile belts imply presence of fluid that reduces their shear strength and makes them prone to seismic activity (Glover and Adam, 2008). Another important event in this section is the Broach earthquake (21.7° N, 73° E) of March 23, 1970 (Chandra, 1977). This region lies at the intersection of eastern margin of Cambay graben and the SMB (Fig. 7.6) making this region susceptiable to seismic activity.

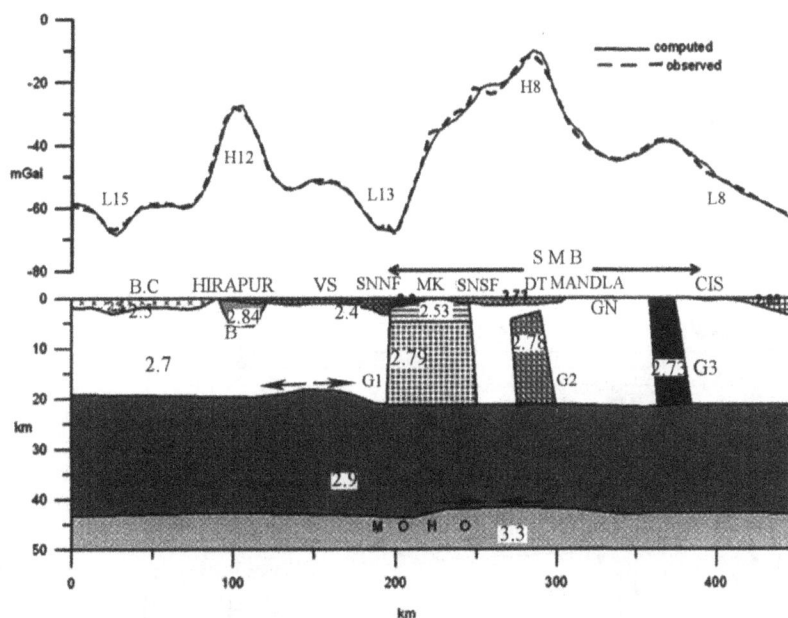

Fig. 7.22(a) Gravity Profile (Hirapur-Mandla) extended to Bundelkhand and Bhandara cratons on either sides (Fig. 20) and computed crustal density model showing exposed geology and important tectonics at the top of the model and causative sources below it. BC: Bundelkhand Craton, VS: Vindhyan Sediments, B: Bijawar Group of rocks exposed in adjoining section, SNNF; Son Narmada North Fault, SNSF: Son Narmada South Fault, MK: Mahakoshal Group of rocks, DT: Deccan Trap, SMB: Satpura Mobile Belt, Gn: Gneisses, CIS: Central Indian Suture. G1, G2 represent subsurface high density intrusives and G3 is exposed lower crustal granulite rocks along the CIS (Arora et al., 2007). Top and bottom arrows indicate extension and compression due to crustal bulge caused by the lithospheric flexure of the Indian plate under Himalaya giving rise to thrust type earthquakes such as Jabalpur earthquake in the lower crust.

(f)   The gravity field along this profile (Hirapur-Jabalpur-Mandla) was extended on either side for about 100 and 150 km towards the north and south based on the Bouguer anomaly map of India (Fig. 7.8 (a) and (b)) to cover parts of Bundelkhand and Bhandara cratons, respectively. The total gravity profile was modeled (Fig. 7.22(a), Arora et al., 2007) constraining from the results of seismic profile, Hirapur-Mandla in the central part (Sain et al., 2000). It provided a relatively shallow crust (41-42 km) under the SMB increasing to about 44-45 km on either side with several high density intrusives (G1-G3). They are associated with the Mahakoshal group of rocks between Son Narmada North and South Faults (Fig. 7.20) along northern margin of the SMB (G1); upper crustal intrusive under SMB (G2) and along Central Indian Shear (G3). The high density intrusive, G1 is supported by high velocity in seismic section as a horst (Sain et al., 2000). Mafic/ultramafic nature of G1 and G2 is also supported by high conductivity (Gokaran et al., 2001). Exposed granulite rocks along the CIS suggest that high density rocks of G3 may represent lower crustal granulite rocks thrusted along it.

(iv) Jabalpur-Raipur Profile and Crustal Structures

(a)   Another profile, Jabalpur-Raipur (IV, Fig. 7.7) was recorded parallel to the previous profile, Mungwani-Rajnandgaon as described above and shown as geotransect (GT) in Fig. 7.31. This profile is important as it traverses through the SMB, CIS, and Chattisgarh basin of the Bhandara craton that is one of the largest Proterozoic basins in the country. The Bouguer anomaly along this profile (Fig. 7.22(b)) shows a large wavelength gravity high, H and a low, L similar to paired gravity anomaly as observed along Mungwani-Rajnandgaon geotransect (Fig. 7.17(b)). The computed crustal model based on the known geology/tectonics and the match between the observed and the computed fields is given in Fig. 7.22(b). Crustal model in this section is shown in two parts: the upper part shows the shallow section, and the lower part shows the deeper section.

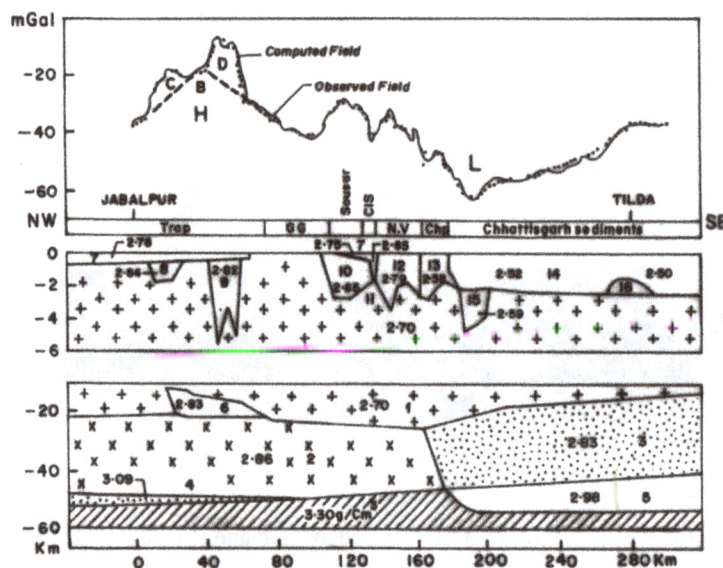

Fig. 7.22(b) Observed and computed Bouguer anomaly along the Jabalpur-Raipur profile (Profile I, Fig. 7.31) with the modeled crustal section. Bodies are numbered 1-16 with their densities shown in g/cm^3. It shows a contact in the lower crust separating blocks (2 and 3) of different densities whose projection on the surface coincides with the CIS and a thick crust south of it and a high density body (6) in the upper crust to its north. Other bodies (7-16) are shallow bodies representing mafic and felsic intrusives. These structures and bodies are similar to those modeled along the geotransect Mungwani-Rajnandgaon (Fig. 7.19(a)) (Mishra et al., 2004).

(b)   The crustal model shows a dipping interface in the lower crust separating rocks of different densities on either side of it. Its projection on the surface coincides with the CIS that is similar to those inferred along Mungwan-Rajnandgaon geotansect (Fig. 7.19(a)). There is a low density body (2.98 g/cm^3) under the Moho in the southern part of the profile and a high density body (2.83 g/cm^3) under the SMB north of the CIS similar to that along the Mungwani-Rajnandgoan profile (Fig. 7.19(a)). There are two high density bodies (2.84 and 2.82 g/cm^3) under the trap in the basement and one north of CIS (2.83 g/cm^3). These bodies are synonymous with high density bodies G1, G2, and G3 (Fig. 7.22(a)) inferred south of Jabalpur along the previous profile.

(c) The first high density body (2.84 g/cm^3) in the northern part of the profile may be related to Mahakoshal group of rocks, while second and third one are related to the intrusive in the upper crust under the SMB north of the CIS, and high grade granulite rocks along the CIS. The chattisgarh basin consist of tuffs of Neoproterozoic period (Basu et al., 2008) suggesting volcanic activity of Meso-Neoproterozoic period that gave rise to the observed gravity high at Tilda (southern end of the profile) and aided the formation of this basin.

(v) Integrated Evolutionary Model of the Central Part of the SMB

(a) There are high grade metasediments of the Sausar group of rocks and granulite rocks north of the CIS with structural trends (NE-SW) being almost parallel to the CIS and low grade island arc type tholeiitic suite of rocks and calc-alkaline magmatism of Mesoproterozoic period (1.5-1.0 Ga) south of the CIS oriented almost perpendicular (N-S) to it. Patches of granulite rocks north of the CIS suggest it to be a thrust and a Paleo suture. Therefore, the high density intrusive in the upper crust (III, Fig. 7.19(a) and, G3 and 6, Fig. 7.22(a) and (b)) north of the CIS coinciding with the central part of the SMB represents part of the thrusted lower crustal granulite rocks that are exposed north of the CIS. The north verging thrusts G1, G2, G3 along the Hirapur-Mandla profile (Fig. 7.22(a)) with high density intrusives and high and low density island arc type magmatism of the Bhandara craton occurring in N-S oriented Proterozoic basins (Fig. 7.16b) suggests collision of Bundelkhand and Bhandara cratons in the central part of the SMB with convergence and subduction from the north to the south during Mesoproterozoic period (1.5-1.0 Ga) as shown in Fig. 7.6 north of the SMB. The low density rocks under the Moho, south of the CIS (Ib, Fig. 7.19(a)), appear to represent remnants of the underthrusted crustal rocks. Fig. 7.19(b) shows a schematic diagram depicting plate tectonic processes across the SMB primarily based on the integrated crustal model (Fig. 7.19(a)) from geophysical studies along the Mungwani-Rajanandgaon geotransect, but also conforms to the crustal models along the other two profiles, Fig. 7.22(a) and(b).

(b) The Dongargarh and Malanjkhand granite plutons of Neoarchean period and Paleoproterozoic magmatic rocks of the Bhandara craton (Sakoli and Nandgaon bimodal volcanics, Table 7.1) suggest a convergence of Paleoproterozoic period (~ 2.5-2.0 Ga) from the north to the south (Fig. 7.6) that gave rise to several granite intrusives of this period in the Bhandara craton (Table 7.1) which was almost contemporary to Aravalli orogeny along the ADMB as described above (Mishra, 2006). Mylonitized contact of Malanjkhand granite of about ~ 2.5 Ga south of the CIS in the Bhandara Craton (Majumdar and Mamtani, 2009) also suggests a convergence phase during Paleoproterozoic period. The Bijawar group of rocks along the Bundelkhand craton during Paleoproterozoic period at the northern margin of the Vindhyan basin and contemporary Mahakoshal group of rocks at the southern margin of the Vindhyan basin with large amount of mafic/ultramafic intrusives, suggest the rifting of the Bundelkhand craton during Paleoproterozoic period (~ 2.0-1.6 Ga). It has also affected the western margin of the Bundelkhand craton at its contact with the ADMB as suggested by gravity highs H2 and extension of H5 described in section 7.6.2. N-S directed convergence and subduction during Meso-Neoproterozoic periods gave rise to back arc basins and associated contemporary magmatism of the Bhandara craton (Table 7.1)

(c) Acharyya and Roy (2000) and Roy and Prasad (2003) have termed this section as the Central Indian Tectonic Zone (CITZ). However, maintaining the old nomenclature of the Satpura Mobile Belt has an advantage of being familiar. Mobile belt signifies a set of geological processes that automatically goes with it. Based on low-medium grade supracrustal belts, gneisses, granitoids, and granulite belts with several crustal scales shear zones and tectonothermal events; they suggested continent-continent collision at ~ 1.5 Ga and S-N subduction in this section. In fact, all these features are typical characteristics of mobile belts world over. Based on directions of reflectors in seismic profiles, Naganjaneyulu and Santosh (2010) suggested both ways subduction, viz. S-N and N-S. However, profile data cannot provide direction of reflectors which will depend to a great extent the on the direction of profiles and processing techniques used to process the data. They have also attributed high conductive body in the upper crust to Deccan trap activity, though such bodies are characteristic of mobile belts the world over and are attributed to upper mantle/lower crustal rocks thrusted during orogeny and presence of fluids. The same has been assigned also in the present case. Exposed Deccan trap does not provide high conductivity as discussed in Section 7.6.3. Further, to understand the direction of convergence and subduction, one must analyse the data from adjoining terrain of basins and cratons instead of mobile belts alone. In the present case, if one looks to Vindhyan and Mahakoshal super group of rocks towards the north of the SMB that represent typical foreland basin and rifted platform deposits, respectively (Section 7.6.2) and the Bhandara craton towards the south with larger crustal thickness and contemporary rift basins with bimodal volcanics perpendicular to the SMB as described above; it becomes amply clear that subduction was from N-S. Moreover, at almost the same time there was convergence and subduction across the Aravalli-Delhi Mobile Belt from E-W that is similar but in opposite direction to the present day convergence across the Himalayan and the Burmese arcs. It would be difficult to conceive S-N subduction across the SMB and E-W subduction across the ADMB at the same time. However, these are the various alternatives and readers can decide for themselves.

7.6.2   Rifting and Platform Deposits: Vindhyan and Bijawar-Mahakoshal Group of Rocks and their Economic Potential – A Plate Tectonics Model

(i) Vindhyan and Bijawar-Mahakoshal Group of Rocks-Plate Tectonics Model

The Vindhyan and Mahakoshal group of rocks north of the SMB with the Greater Bundelkhand craton representing their basement, suggest a rifting phase and platform deposits based on following considerations:

(a) The Vindhyan basin is bounded by the Bundelkhand craton towards the north, the SMB towards the south, and is abutted against the ADMB towards the west (Fig. 7.11(c)). Vindhyan sediments have been encountered in bore wells under the Ganga basin indicating their extension northwards (Fuloria, 1996). The northern part of the SMB is formed by the Mahakoshal group of rocks confined between SNNF and SNSF (Fig. 7.20) consisting primarily of volcano sedimentary sequences of mafic and ultramafic rocks of Paleoproterozoic period (Table 7.1) as described above. Vindhyan sediments are normally undisturbed with a small dip except along its margin along the SMB and the ADMB where they are folded and faulted. They are devoid of any volcanics except the occurrences of porcellanite at their base (Chakravorthi et al., 2007). Towards the north, the Bijawar group of rocks are exposed along the eastern margin of the Bundelkhand craton (Fig. 7.20) that also represents metasediments with mafic and ultramafic intrusives considered contemporary to the Mahakoshal group of rocks (Pant and Banerjee, 1990). The Vindhyan super group (Table 7.2) is composed of four groups of rocks, viz. Semri, Kaimur, Rewa, and Bhander consisting primarily of sandstone, limestone, shale, and carbonate. The first one belongs to the Lower Vindhyan group of volcanogenic sediments, while the latter three comprising primarily of metasediments, represent the Upper Vindhyan group with a major discontinuity at the base of the Kaimur group. Sarangi et al., (2004), Ray et al., (2003), and Rasmussen et al., (2002) have provided an age of about 1.7-1.6 Ga for Semri group of rocks while the Upper Vindhyan groups largely belong to Neoproterozoic period (1.0-0.7 Ga, Kumar et al., 1993; Ray et al., 2003 and Ray, 2006, Table 7.2). The Mahakoshal and Bijawar group of rocks of Paleoproterozoic period (~ 1.9-1.7 Ga) largely consist of mafic and ultramafic rocks that form the basement of the Vindhyan sediments. Stratigraphy of this section is summarized in Table 7.2.

(b) Two most important aspects of plate tectonics are rifting and convergence which explain most of the observations from various basins and orogenic belts. Basins are formed both during rifting and convergence. However, they differ in their characteristics. During rifting, the basins are formed over rifted margins of the cratons while during convergence, foreland basins are formed as the present day Ganga basin associated with the Himalayan fold belt. The basins that are formed during rifting are largely associated with magmatic rocks as magmatism is an essential element for rifting (Section 2.7). These basins are formed over a rifted platform of cratons and consist of large amount of mafic/ultramafic rocks giving rise to volcanogenic sedimentary sequences. However, the foreland basins formed during convergence derive sediments from the adjoining continents and fold belts that have formed due to convergence, as Siwalik sediments of Ganga basin are derived from Himalaya. They are largely free from magmatic intrusion and are undisturbed except along margins of the fold belt. In light of this, the Mahakoshal and Bijawar group of rocks of Paleoproterozoic period along the SMB consisting of mafic/ultamafic intrusives (Mishra and Rajasekhar, 2008) on large scale indicate their formation during the rifting of the Bundelkhand and Bhandara-Bastar cratons that preceeded the meso-Neoproterozoic convergence.

(c) A detailed Bouguer anomaly map of this region north of the CIS is given in Fig. 7.23(a) showing several gravity highs and lows. From the south to the north, the Central Indian Suture (CIS) lies along the southern margin of gravity highs H6 related to the SMB. The SNSF passes through south of the gravity highs, H4 and H5 and forms the southern margin of the Mahakoshal group of rocks (Fig. 7.20) defined by these gravity highs. Vindhyan basin is defined by the gravity lows, L2 due to low density sediments. The gravity highs, H1 partly coincide with the exposed Bijawar group of rocks and can be attributed to similar rocks in the basement that encircle the Bundelkhand craton as gravity highs, H2 and H3 towards the west and the north. Its linear nature indicates rift type of structure. The Bijawar (H1) and Mahakoshal group of rocks (H4 and H5, Fig. 7.23(a)) showing gravity highs along margins of the Bundelkhad craton and greater Bundelkhand craton under the Vindhyan basin, respectively suggest mafic and ultramafic rocks that are associated with the Bijawar and Mahakoshal super groups (Table 7.2, Fig. 7.23(b)).

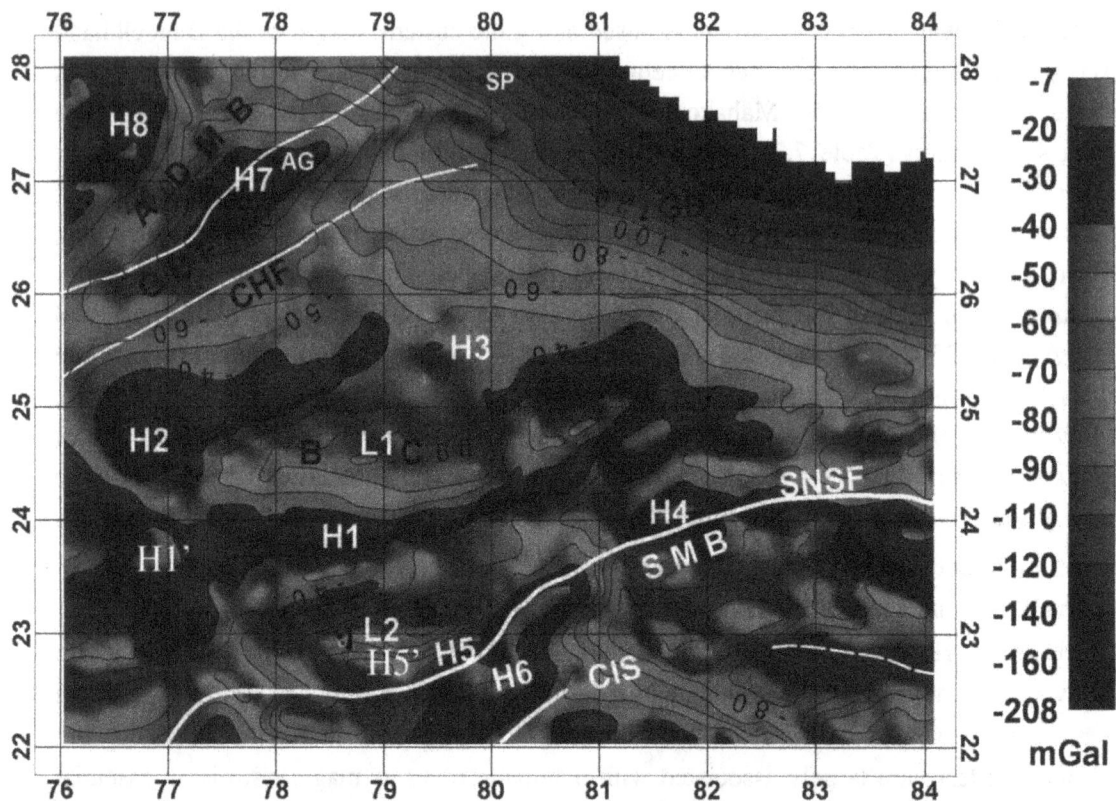

Fig. 7.23(a) Bouguer anomaly map of Bundelkhand craton, Vindhyan basin, and central part of the SMB (Fig. 7.20). Gravity highs and lows important for the present study are indicated by H1-H8 and L1-L2 respectively. Some tectonic elements, important for the present study are schematically shown in limited sections. They are as follows: ADMB: Aravalli-Delhi Mobile Belt, GBF: Great Boundary Fault, CHF: Chambal Fault, GB: Ganga Basin, CIS: Central Indian Shear, BC: Bundelkhand Craton, VB: Vindhyan Basin, SNSF: Son Narmada South Fault. The section between the CIS and the SNSF and their extensions towards east and west is known as Satpura Mobile Belt. The gravity highs due to ADMB (H8) and SMB (H6) are partially presented in this figure. Eastern part of the gravity high, H7, east of the GBF represents Agra (AG)-Shahjahanpur (SH) ridge along the western margin of the Vindhyan basin extending into the Ganga basin. [Colour Fig. on Page 793]

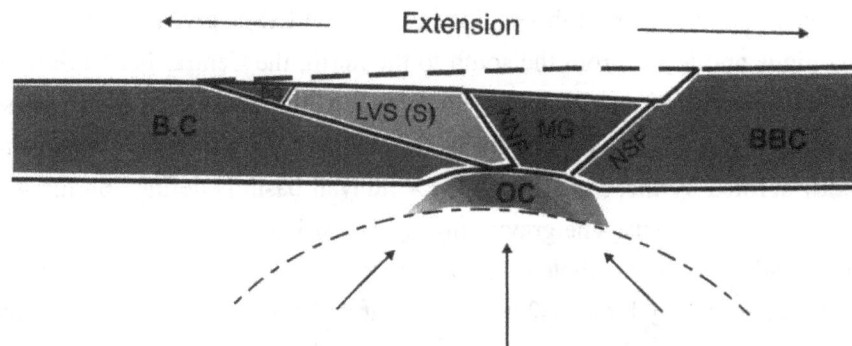

Fig. 7.23(b) A section demonstrating the rifting of Bundelkhand craton (BC) and Bhandara-Bastar craton (BBC) with oceanic crust (OC) in between forming the Mahakoshal (MG), the Bijawar (BG), and the Lower Vindhyan sequences (LVS) of the Semri (S) group of rocks of Paleoproterozoic period over rifted platform of the Bundelkhand craton. NNF: Narmada North Fault, NSF: Narmada South Fault.

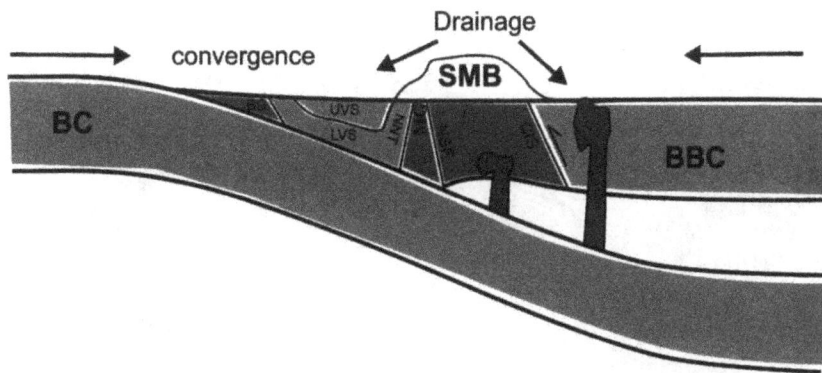

Fig. 7.23(c) A convergence model of the Bundelkhand craton (BC) and the Bhandara-Bastar craton (BBC) along the Satpura Mobile Belt (SMB) showing subduction of the BC under the BBC giving rise to magmatism and thrusting along the Central Indian Suture (CIS). BG: Bijawar group of rocks, MG: Mahakoshal group of rocks, UVS: Upper Vindhyan Sequences, LVS: Lower Vindhyan Sequences, NNF: Narmada North Fault and NSF: Narmada South Fault. Flower structures show subduction related magmatism.

(d) The gravity highs, H7 (Fig. 7.23(a)) at the western margin of the Vindhyan basin along its contact with the ADMB located between the Great Boundary Fault and the Chambal fault towards the east, represent high density rocks in the basement and have been modeled in Fig. 7.15(a) under the Vindhyan sediments along the profile I (VIb). The curvilinear continuity between the gravity highs, H2, H4, H5, and H7 along the SMB and the ADMB indicate that the gravity high, H7 may also represent rocks similar and contemporary to the Mahakoshal or Bijawar group of rocks under the Vindhyan basin along margins of the the greater Bundelkhand craton towards the west. The gravity highs, H5' and H1' connect the two sets of gravity highs and define the limits of the Mahakoshal group of rocks under the Vindhyan sediments forming its basement. Such set of gravity highs (H1-H3) close to continent (craton), and other along the margins (H4, H5 and H7) are typical characteristics of rifted margins as has been reported in several cases like the west coast of Norway (Mjelde et al., 2007), east coast of USA (Talwani and Abreu, 2000), west coast of India (Mishra et al., 2004) as discussed in Section 3.8.1.

(e) The airborne magnetic map of the northern part of the Vindhyan basin recorded at about 500' (~ 150 m) terrain clearance shows three sets of magnetic highs and lows (Zones 1-3) along its contact with the Bundelkhand craton (Fig. 7.24(a)). This section is known as the Panna Diamond belt with volcanic plugs located at Majgama and Hinota, the former being diamondiferous and the only primary source of diamonds in this country. Magnetic lows of these sets of anomalies are located towards the north indicating almost parallel intrusives as dykes in the basement with induced magnetization. It is a typical characteristic of rifted margins showing parallel mafic intrusives as dykes similar to presently being observed in case of the Red Sea Rift (Drake and Girdler, 1964). Modelling of magnetic anomaly of Zone 2 along profile BB' using a tabular body (Fig. 7.24(b)) suggest a magnetic source at a depth of about 700-800 m in the basement of susceptibility 1.0×10^{-3} emu that suggest mafic intrusives in the basement under Vindhyan sediments formed by the Bijawar group of rocks in this section.

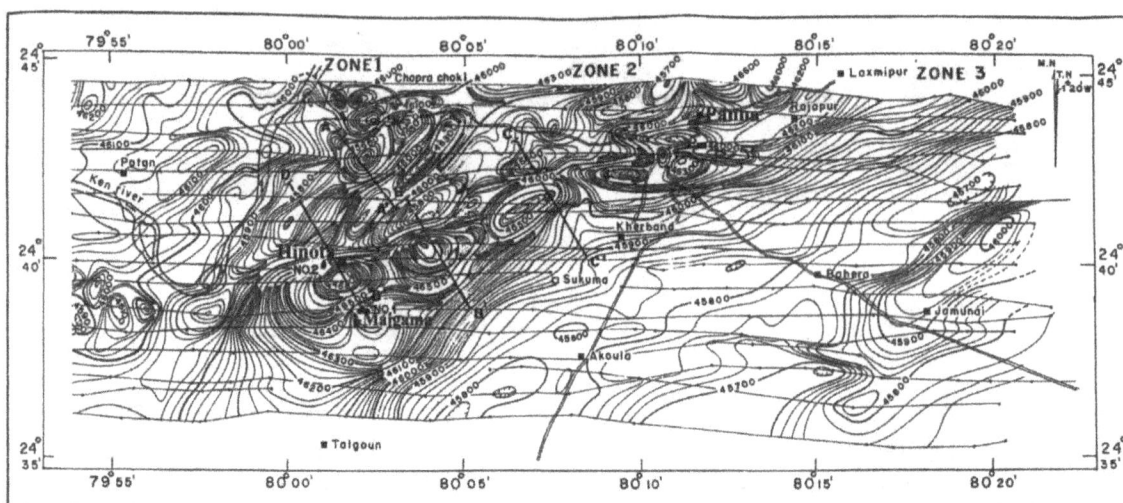

Fig. 7.24(a) Airborne total intensity magnetic map of the Panna diamond belt (SW of Panna, Fig. 7.20) with volcanic plugs, Majgama and Hinota marked on it. E-W oriented flight lines are about 1 km apart flown at altitude of 500' (~ 150 m) above topography. Three pairs of magnetic highs and lows are marked as zones 1, 2 and 3 indicating mafic intrusives in the basement represented by each pair of magnetic high and low. Ken river approximately represents the contact of Vindhyan sediments towards the east and the Bundelkhand Craton towards the west (Mishra 1987).

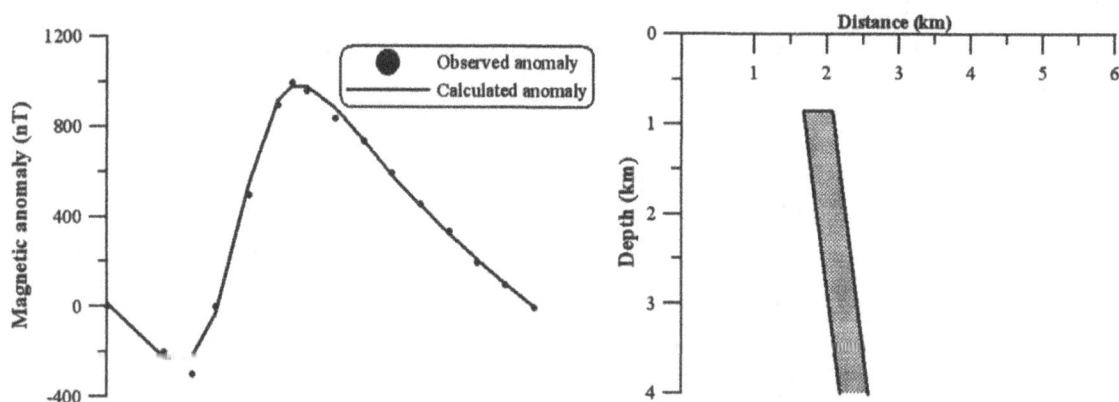

Fig. 7.24(b) Modeling of magnetic anomaly of zone 2 along profile BB' suggesting a depth of 900 m below the plane of observation which implies 750 m below surface for a susceptibility of 1.0×10^{-3} emu that represents mafic intrusions in the basement (shown on right hand side).

(f)   Airborne total intensity magnetic map of Mahakoshal group of rocks south of the Vindhyan basin acquired at a height 5000' (~ 1.53 km) with flight line spacing of 5 km (AMSE, 1995) was digitized and corrected for International Geomagnetic Reference Field (IGRF). The IGRF corrected field is given in Fig. 7.25(a) superimposed over Bouguer anomaly and tectonics which shows large magnetic gradient (MGR) and anomaly south of Sidhi. These anomalies are also paired magnetic anomaly (ML1 and MH1) with magnetic low towards the north indicating induced magnetization, or that magnetization at the time of the formation of these rocks might be similar to the present day field as described in case of airborne magnetic map of Bijawar group of rocks from northern part of the Vindhyan basin (Fig. 7.24(a)). This qualitatively suggests the similarity of mafic intrusives in the two groups of rocks, viz. Bijawar and Mahakoshal. It also shows a gravity high (G1) (H4, Fig. 7.23(a)) coinciding with the magnetic anomalies south of Sidhi. Both are confined between the Son Narmada North and South Faults (Fig. 7.25(a)) over Mahakoshal group of rocks. Another gravity high (G2) towards the north may be noticed, which coincides with the gravity highs, H1 around Bundelkhand craton (Fig. 7.23(a)) that has been attributed to Bijawar group of rocks and mafic intrusives below the Vindhyan sediments in this section (Fig. 7.24(a) and (b)). An airborne magnetic profile along AA' is presented in Fig. 7.25(b), which shows a large magnetic gradient G1 (MGR, Fig. 7.25(a)) that is typical of a fault. The magnetic lows and highs associated with it (L1, H1) are related to intrusives along the fault.

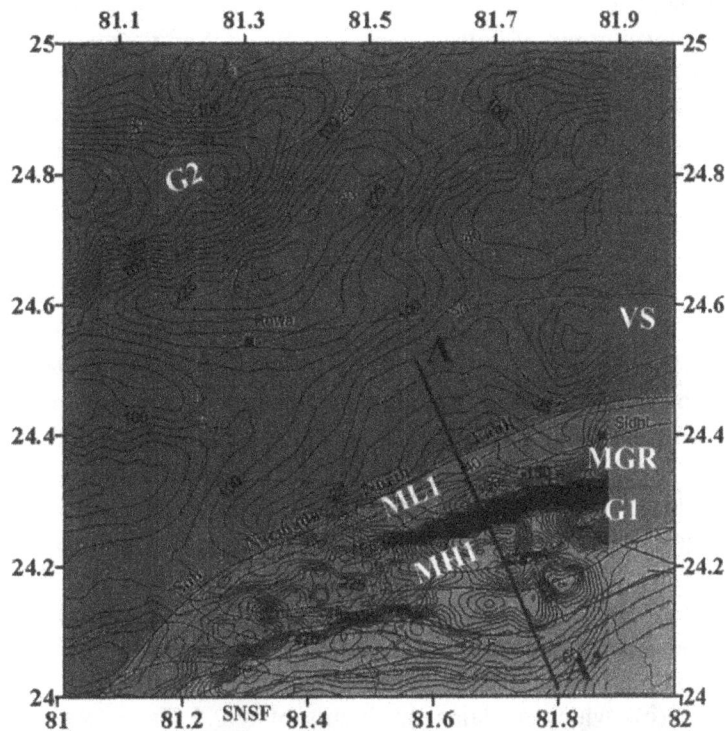

Fig. 7.25(a) Airborne total intensity (100, 75, 50 etc.,) recorded at ~ 750 m above msl and Bouguer anomaly (-40, -50 etc.,) maps of Mahakoshal and part of the Vindhyan basin in central India. The Mahakoshal group of rocks (MGR) between the Son Narmada North and the South (SNSF) Faults shows high amplitude magnetic anomalies (ML1 and MH1) and gravity highs (G1) while gravity highs, G2 in the northern section are related to the Bijawar Group of rocks under the Vindhyan sediments (VS) which are same as H4 and H1 of Fig. 7.23(a), respectively. The gravity high G2 also coincides with magnetic low (< – 25 nT) and a high (> 225 nT) to its south forming a pair due to mafic intrusives in this magnetic latitude corresponding to induced magnetization similar to those related to Mahakoshal group of rocks (ML1 and MH1).

(g) Another common factor between the northern and southern parts of the Vindhyan basin with Bijawar and Mahakoshal Group of rocks as basement, are association of volcanic plugs of Mesoproterozoic period (~ 1.1 Ga) with both of them, that are related to convergent phase subsequent to rifting. The volcanic plugs at the northern margin, known as Majgama and Hinota plugs (Fig. 7.24(a)), and Jungel volcanic plugs in Mahakoshal group of rocks towards the south are associated with large scale regional magnetic anomalies (Fig. 7.24(a), and 7.25(a)) and provide significant magnetic anomalies on ground (Srivastava et al., 1983) indicating their mafic/ultramafic composition.

(h) The gravity highs related to ADMB (H8, Fig. 7.23(a)) starts almost from the Great Boundary Fault (GBF) separating Vindhyan sediments towards the SE and the ADMB towards the NW. However, based on the linear gravity highs (H7, Fig. 7.23(a)) and their similarity with those due to Bijawar and Mahakoshal group of rocks (H1-H3, H4-H5; Fig. 7.23(a)) as described above, Mishra and Rajasekhar (2008) have considered the basement of the Vindhyan sediments, east of the GBF up to the Chambal fault representing rocks equivalent to the Bijawar and Mashakoshal group of rocks that occur along the southern margin of the Bundelkhand craton and are formed during rifting. Even the Bijawar and the Mahakoshal group of rocks towards the east are partially exposed, and partially lie under the Vindhyan sediments as its basement.

(i) These observations suggest that Mahakoshal and Bijawar group of rocks represent the rifting phase of the Bundelkhand craton during Paleo-Mesoproterozoic period (~ 2.0-1.6) and contemporary lower Vindhyan group of volcanogenic sediments (Semri group, ~ 1.7Ga, Sarangi et al., 2004, Ray et al., 2003 and Rasmussen et al., 2002) were deposited on the platform formed due to this rifting (Table 7.2, Fig. 7.23(b)). The volcano sedimentary sequences of Mahakoshal group of rocks and their sectoral nature also indicate that they were deposited along the rifted margin of the craton. The Mahakoshal group of rocks initially occupied a rift basin which was subsequently uplifted as a horst (Fig. 7.21(b)), which may be related to subsequent convergence and collision phase along the SMB. Occurrences of porcellanite in the lower part of the Vindhyan basin also suggest igneous activities during this period, which is a common feature of rifting. Raza et al., (2009) have suggested that the lower Vindhyan volcano sediments succession was deformed and exposed to erosion before the deposition of upper Vindhyan rocks.

(j) The rifting phase of Paleo-Mesoproterozoic period was followed by convergence during Meso-Neoproterozoic period (Section 7.6.1) that caused uplift and deformation of the Mahakoshal group of rocks as a horst (Fig. 7.23(c)) that is located along the margin of the Bundelkhand craton forming a part of the collision zone. This is evidenced from the seismic section of this region (Sain et al., 2000) that shows Mahakoshal group of rocks forming a horst in this section as shown in Fig. 7.21(b). This seismic section has also delineated deformation of the Lower Vindhyan group of rocks (Semri group) along the SMB that was caused by convergence and collision of the Bundelkhand and Bhandara-Bastar cratons. During this convergence, the Upper Vindhyan group of rocks of Neoproterozoic period (~ 1.0-0.7, Kumar et al., 1993, Ray et al., 2002, Ray, 2006) were deposited (Fig. 7.23(c)) which explains the hiatus of ~ 0.6 Ga between two groups of Vindhyan sedimentation (Mishra, 2011). This convergence has also given rise to several subduction-related island arc type mammatic intrusions in Bhandara-Bastar craton as described above.

(k) Both Bijawar and Mahakoshal group of rocks are economically important and that conforms with their formation related to rifted margins of the Bundelkhand craton. Several economic minerals such as base metals are supposed to be associated with the mafic/ultramafic intrusives of these rocks. One of the airborne magnetic anomalies near Gadarwara, SW of Jabalpur (Fig. 7.25(c), Achuta Rao et al., 1993) located under alluvium was examined in detail for its mineral potentiality (Intierra, 2008) suggesting a potential for sulphide mineralization including gold and platinum. There are several such magnetic anomalies in this zone under alluvium (Achuta Rao et al., 1992) similar to the one given in Fig. 7.25(c). This is a typical high amplitude magnetic anomaly of about 600-800 nT observed in this section with magnetic low towards the north as described above for Mahakoshal group of rocks. These characteristics of magnetic anomalies can be used to delineate Mahakoshal group of rocks under alluvium in this section from airborne magnetic data, and followed on the ground for their mineral potentiality. The magnetic anomalies are also associated with small amplitude (5-10 mGal) gravity highs that can be delineated through a detailed gravity survey of this region.

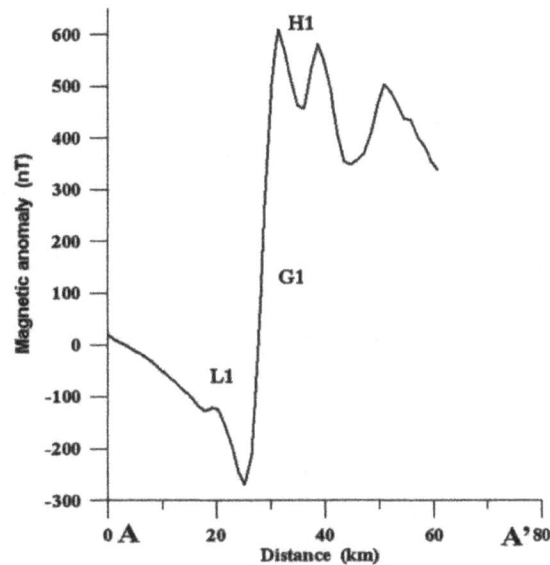

Fig. 7.25(b) A magnetic profile, AA' across Mahakoshal group of rocks showing a sharp gradient (G1) and associated magnetic highs (H1) and lows (Ll) towards the north indicating mafic intrusive.

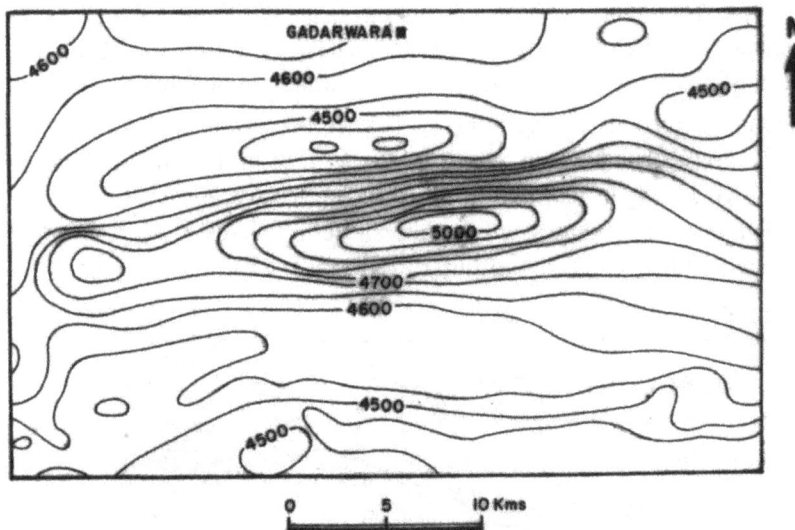

Fig. 7.25(c) Gadarwara airborne magnetic anomaly (Achuta Rao et al., 1993) recorded at 750 m above msl related to the Mahakoshal group of rocks under alluvium SW of Jabalpur associated with sulphide mineralization.

(ii) Basement under Vindhyan Basin and Hydrocarbon Exploration

In order to access the nature of the basement under the Vindhyan basin, the harmonic inversion scheme, as described in Chapter 4 and briefly outlined in Section 6.1.6, is used. Equation (6.1) provides the average depth of layers from spectral decay in case the transform of the relief (h (f)) is assumed as one. However, in case the average depth (h) is provided in this scheme and relief is allowed to vary, one can obtain the depth of the layer at every data point in a plane (Mishra and Pederson, 1982). The Bouguer anomaly of the Vindhyan basin (L2, Fig. 7.23(a)) is digitized at equidistant points of 3.5 km interval and used for spectral analysis in the following steps:

(a)   Digital data of the Bouguer anomaly of the Vindhyan basin (64 × 128) is transformed into the frequency domain for spectral analysis (Section, 6.1.6). Firstly, a regional field related to deep seated sources is removed by subtracting a plane from the observed field that is a usual practice for regional-residual separation from the observed gravity field. The residual gravity field is used to compute the basement with its average depth as 4.0 km that would be less than the maximum depth obtained from the spectral decay (Hahn et al., 1976). The bulk density contrast between Vindhyan sediments and the basement is taken as $2.3 - 2.7 = -0.4$ g/cm^3 that provided 3-Dimensional basement configuration (Mishra et al., 1996) as given in Fig. 7.26(a).

(b)   Fig. 7.26(a) shows NE-SW structural trends with a ridge H1 that corresponds to gravity highs, H1 (Fig. 7.23(a)) related to Bijawar group of rocks, south of the Bundelkhand craton. The basement depression, L1 corresponds to the largest Bouguer anomaly lows, north of the SMB and provides the maximum thickness of Vindhyan sediments as 5-5.5 km. The ridge at the SE margin represents Mahakoshal group of rocks along the SMB associated with the gravity highs, H4 and H5 (Fig. 7.23(a)). The Bijawar and Mahakoshal group of rocks presently occupy horsts (H1) and SE of Jabalpur, respectively as seen from this figure that might have been formed due to effect of convergence across the SMB after the rifting as discussed above (Fig. 7.23(c)). Another ridge, H3 in between H1 and L1 (Fig. 7.26(a)) passes through Damoh and divides the Vindhyan basin in two parts. The importance of the southern section the of the Vindhyan basin for hydrocarbon prospect also suggests that it may represent rifted margin. Subsequent convergence has given rise to structures like Jabera dome (NE of Jabera, Fig. 7.26(a)) and Damoh ridge that are significant for hydrocarbon prospect (Ram et al., 1996). This kind of basement relief map is extremely useful in delineating basement structures and, by proxy, the structures within the overlying sediments useful for explo,ration of hydrocarbons as discussed in Section 4.7 and Chapter 9.

(c)   Some resistivity sounding profiles (Fig. 7.26(b), Das et al., 1990) suggested relatively low resistivity (100-200 Ω m) for Mahakoshal group of rocks indicating volcanogenic sedimentary rocks overlain by resistive Vindhyan sediments and Archean basement (> 500 Ω m), respectively. They also suggested maximum thickness of Vindhyan sediment as 5-5.5 km as given above (Fig. 7.26(a)), underlain by the Mahakoshal group of rocks of 2-3 km.

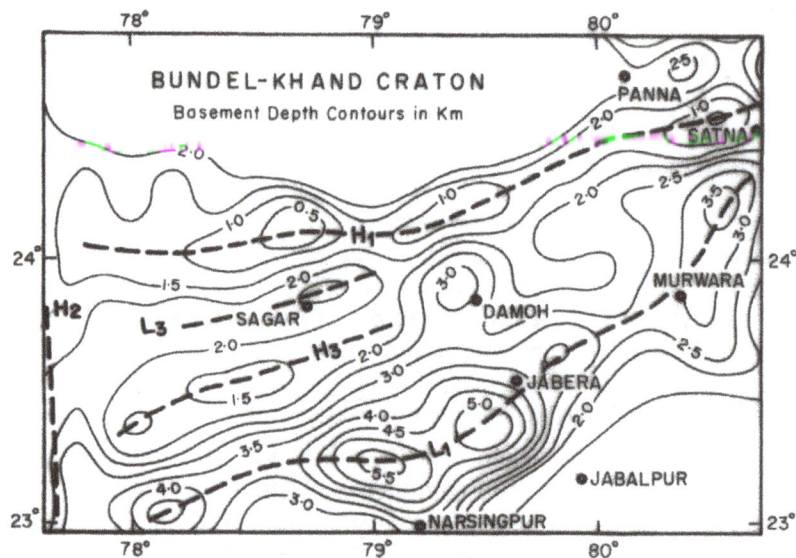

Fig. 7.26(a) The three-dimensional basement relief map computed from the Bouguer anomaly map of the Vindhyan basin for a density contrast of -0.4 g/cm^3 between Vindhyan sediments and the basement, showing ridges and depressions. The Damoh ridge (H3) and Jabera dome are important for hydrocarbon exploration.

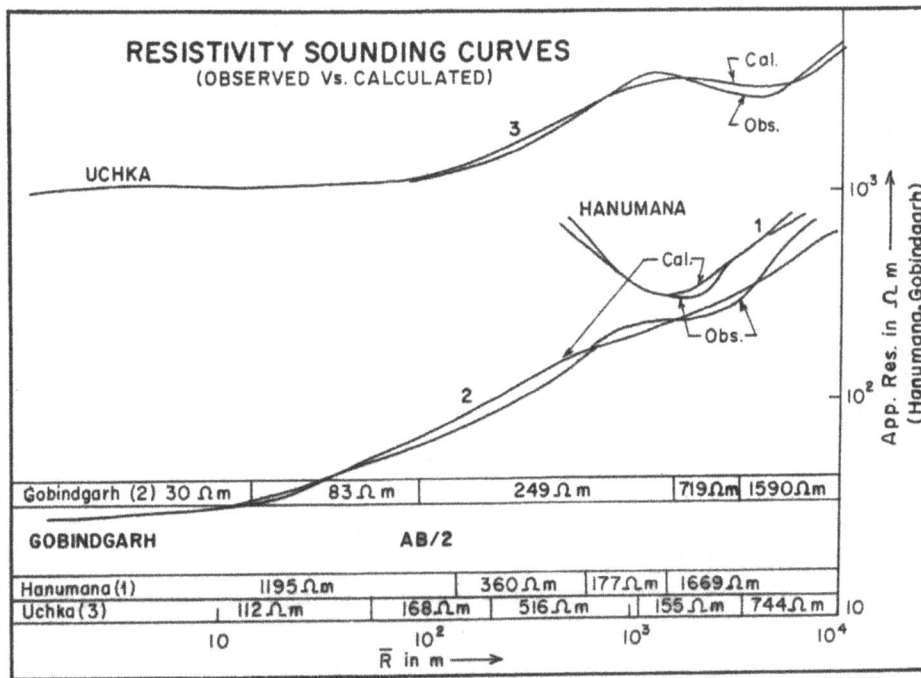

Fig. 7.26(b) Some representative deep resitivity sounding profiles, observed and theoretical, across the Vindhyan basin and the Mahakoshal group of rocks showing top resistivity layer as Vindhyan sediments while intermediate conductive layer (100-200 Ωm) represent Mahakoshal group of rocks bottom resistive layer is. The Archean basement. The maximum thickness of inferred Vindhyan sediments and Mahakoshal rocks are about 5 km and 1-2 km, respectively.

7.6.3 Extension of the SMB towards East and West

(i) Chhotanagpur Granite Gneiss Complex and Singhbum Craton

Some significant points related to the extension of the SMB towards the east and the Proterozoic collisions are as follows:

(a) Extension of Archean-Proterozoic collision zones in the central part of the SMB as described above is examined eastwards over the CGGC and the Singhbhum craton to its south (Fig. 7.7 and 7.11(c)). A detailed geological map of this section is given in Fig. 7.29(a) that shows important tectonic elements of this region including the Singhbhum craton and Shillong plateau. It shows Singhbhum Mobile Belt bounded by South Purulia Shear Zone (SPSZ) towards the north and Singhbhum Shear Zone (SSZ) towards the south that separates CGGC towards the north and the Singhbum Craton towards the south. As the central part of SMB is bounded by the Narmada-Son lineament (NSL) towards the north and the CIS towards the south, with Bhandara craton located to its further south (Fig. 7.11(c)), the CGGC is bounded by the extension of NSL towards the north and South Purulia Shear Zone (SPSZ) towards the south with the Singhbhum craton located to its south. The Singhbhum craton consists of granitic intrusions and several Paleo-Mesoproterozoic mafic/ultra mafic intrusives such as the Dalma-Dhanjori Volcanics (Table 7.1). Banerjee (1982) suggested similarity between intrusives of the Singhbhum craton and island arc type magmatism indicating Proterozoic collision and subduction from north to south. The CGGC is also characterized by narrow belts of ultramafic intrusives of granulite rocks (Acharyya, 2003) of Mesoproterozoic period similar to granulite rocks north of the CIS reported from the central part of the SMB (Table 7.1).

(b) The Bouguer anomaly of the CGGC (H9, Fig. 7.8(a) and (b)) is also characterized by gravity highs and adjoining gravity low, L9 to its south over the Singhbhum craton, similar to gravity high and low, H8 and L8 in the central part of the SMB. To check on the extension of gravity anomalies from the central part of the SMB towards the east and the west, a detailed Bouguer anomaly map of the SMB and Singhbhum craton is given in Fig. 7.27(a) which shows the extension of gravity anomalies of the central part of the SMB (H1, H2 and H3) towards the east (H5-H7). The SPSZ is represented by linear gravity lows marked by L6 similar to CIS in the central part of the SMB (L1, Fig. 7.17(b) and 7.19(a)). Gravity highs, H5 and H6 in the eastern part are related to the granulite rocks of Mesoproterozoic period (Acharyya, 2003). The gravity high H7 is related to the Singhbhum Mobile Belt that includes the Chandil-Dalma volcanics of Mesoproterozoic period (Table 7.1).

(c) The gravity data along a profile 86°E and 21°-24°N (Profile IV, Fig. 7.11(c) and 7.27(a)) across the CGGC and the Singhbhum craton is given in Fig. 7.27(b) that shows gravity anomalies marked by the same notations as shown in Fig. 7.27(a) along with the computed crustal section for a good match between the observed and the computed fields (Rajasekhar and Mishra, 2005). The computed crustal section shows high density granulite rocks (EGR) and upper crustal intrusive, HD north of the SPSZ related to the gravity High, H6 and high density mafic intrusives as Chandil and Dalma volcanics for the gravity high, H7 and crustal thickening to its south related to the gravity low, L7. The gravity high at the southern end of the profile is due to the Iron Ore group of rocks that represents schist belts in this section. This crustal model, showing gravity highs related to lower crustal granulite rocks north of the SPSZ and crustal thickening south to its, therefore indicates SPSZ as a suture between the CGGC and the Singhbhum craton towards the south similar to the CIS in the central part of the SMB related to Mesoproterozoic convergence and subduction from north to south.

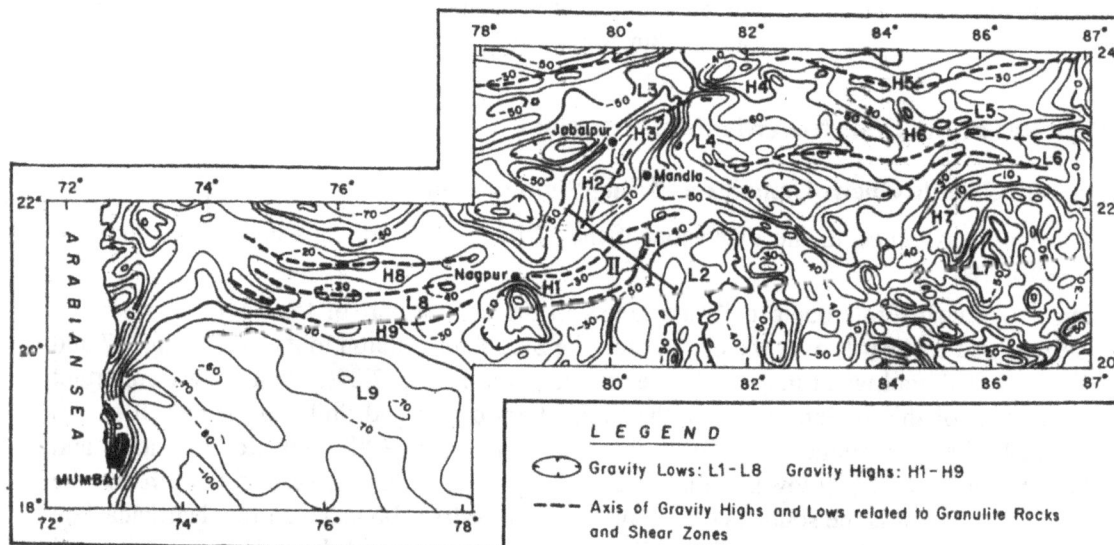

Fig. 7.27(a) Bouguer anomaly map of Central India related to the Satpura Mobile Belt (SMB) and extension of gravity anomalies from central part towards the east, Chottanagpur Granite Gneiss Complex (CGGC) and the west. H1-H9 and L1-L9 represent linear gravity highs and lows related to surface/subsurface high density rocks of the SMB. L1 is related to the Central Indian Shear (CIS), which extends eastwards as L6 related to the South Purulia Shear Zone (SPSZ) north of the Singhbhum craton. The gravity highs H1, H2, H3, and H4 in the central part of SMB extend as H5 and H6 towards the east related to two phases (1.6 Ga and 1.0 Ga) of exposed lower crustal granulite rocks in CGGC. The gravity highs and lows, H7 and L7 are related to the Mesoproterozoic Mobile Belt (Fig. 7.29(a)) with Chandil-Dalma volcanics and crustal thickening, respectively. The gravity highs and lows, H8-H9 and L8 are related to the western part of the SMB. It also shows the geotransect (Profile II) in the central part of the SMB (Fig. 7.19(a)) and Profile I is in the eastern part across the CGGC and the Singhbhum craton.

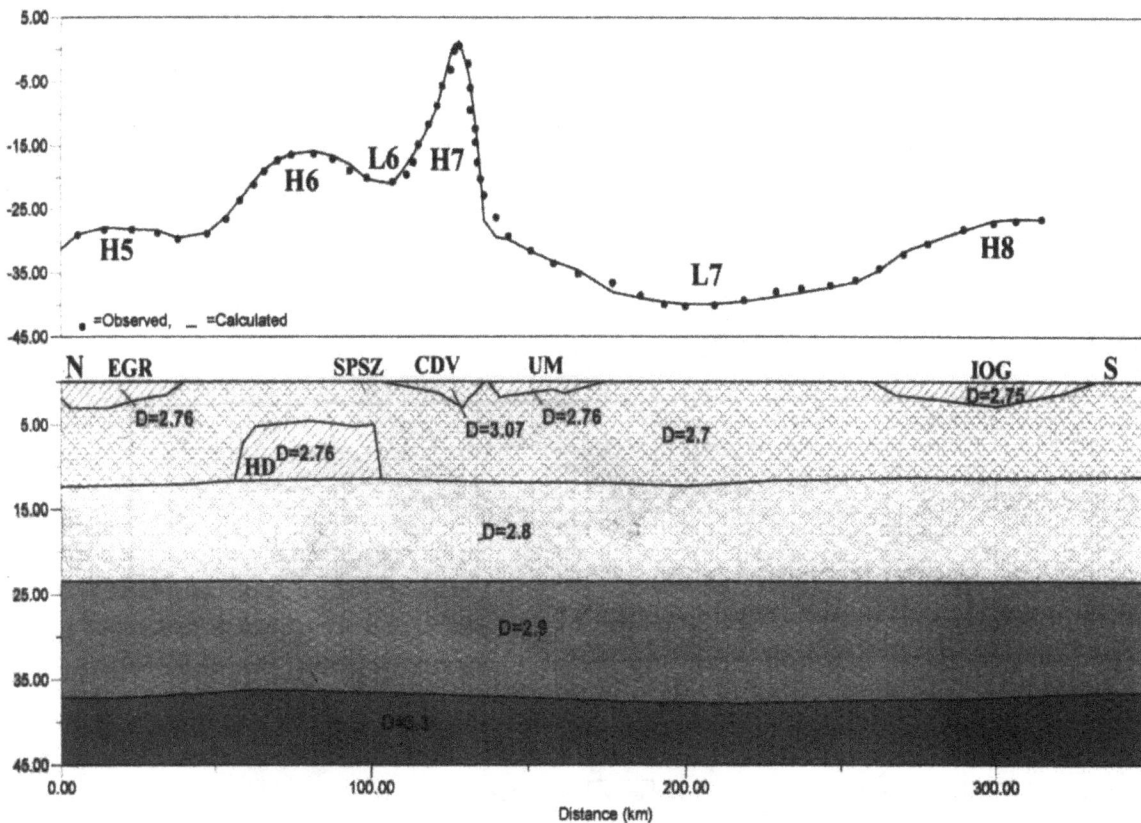

Fig. 7.27(b) Bouguer anomaly and crustal model along a profile 86°E (21-24°N) across CGGC and the Singhbhum craton, Profile IV (Fig. 7.7) and Profile I (Fig. 7.27(a)). Densities (D) of bodies in the model are given in g/cm³. Dots are the observed field, with continuous line showing the computed gravity field. Gravity anomalies are marked as H5-H8 and L6-L7, same as those given in Fig. 7.27(a). Exposed geology along the profiles are denoted as EGR: exposed Granulite rocks, SPSZ: South Purulia Shear Zone, CDV: Chandil Dalma Volcanics, UM: Unclassified Metamorphics, IOG: Iron Ore Group of rocks representing greenstone belt (Rajasekhar and Mishra, 2005).

(d) The Paleoproterozoic intrusives of the Singhbhum craton, such as Singhbhum granite and Chaibasa and Dhaibhum stages (Table 7.1), suggest that there might also be a convergence during Neoarchean-Paleoproterozoic period followed by a rifting phase during Paleo-Mesoproterozoic period as previously described in case of the central part of the SMB.

(e) An airborne magnetic profile from Allahabad to Calcutta (Fig. 7.27(c), Achuta Rao et al., 1970) flown at an altitude of about 2.0 km amsl provides significant magnetic anomalies related to the extension of the Narmada-Son Lineament (NSL, Fig. 7.29(a)) towards east. It shows a major magnetic anomaly 1, that coincides with the NSL in this section and represents a fault kind of magnetic anomaly. Magnetic anomaly, 2 coincides with the northern contact of the CGGC. The magnetic anomaly, 3 represents a fault kind of magnetic anomaly and coincides with the Damodar Gondwana graben cross-section between Dhanbad and Raniganj, known for several coal fields. This magnetic anomaly indicates the mafic basement in this section related to granulite rocks exposed in the adjoining section as described above in (b) associated with the Proterozoic collision in this section.

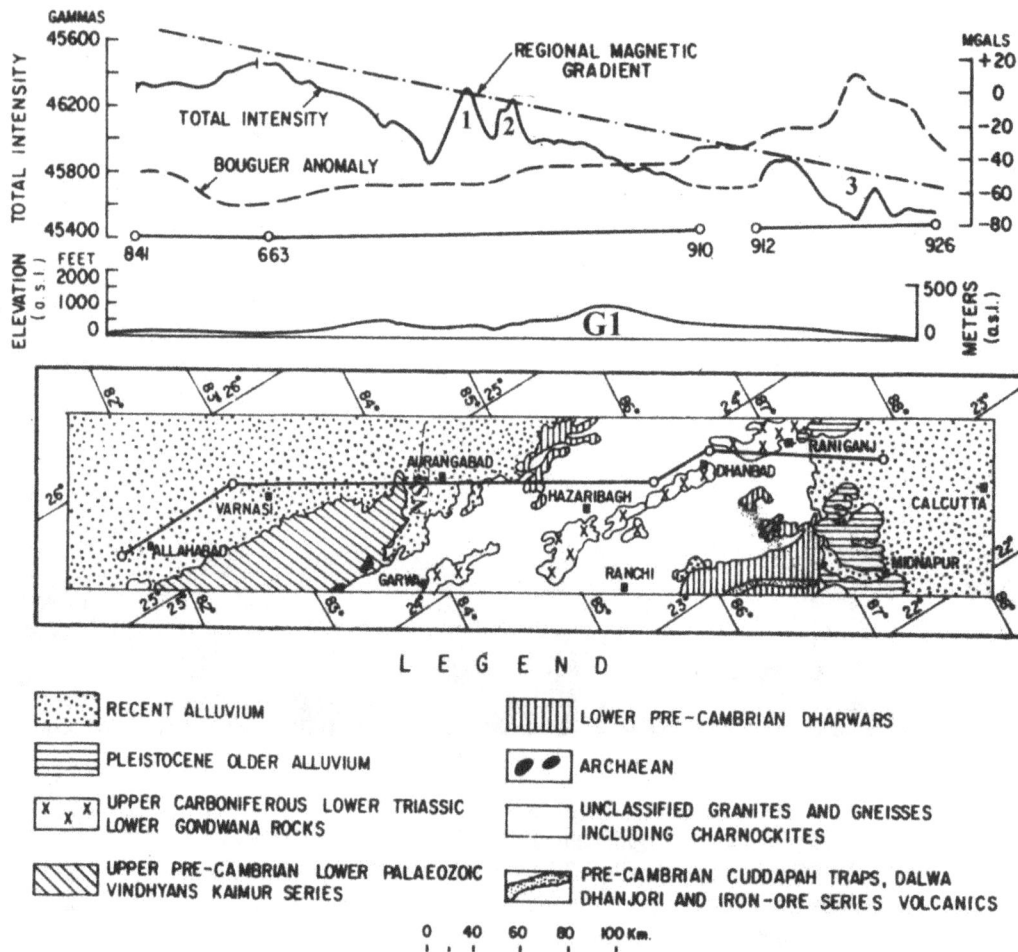

Fig. 7.27(c) Airborne total intensity magnetic profile from Allahabad to Calcutta flown at an altitude about 2.0 km. Flight path over a geological section is shown for ready reference. The Narmada-Son Lineament (NSL) coincides with a fault kind of magnetic anomaly, 1. Bouguer anomaly along the same profile is also shown for comparison.

(ii) Western Part of the SMB and Tapti Lineament

Signatures of Proterozoic collision on the western side of the SMB as in the case of the central and eastern part described above is difficult to visualize as this region is covered by Deccan trap volcanics. However, there are certain features which draw attention to plausible Proterozoic collision in this section also:

(a) The Tapti Lineament extends from the west coast of India upto central India (Fig. 7.7) along the southern margin of the western part of the SMB characterized by gravity gradient between H7 and L7, (Fig. 8(b)) that is typical of thrusts and the two may form paired gravity anomalies related to Proterozoic collision zones as discussed above.

(b) The western section of the SMB is also characterized by the linear gravity high H7 (Fig. 7.8(b)) that is similar to H8 over the central part of the SMB and the two are connected with each other These gravity highs may therefore indicate thrusted high density intrusives in the basement as exposed granulite rocks in the central part of the SMB. Since this section of the SMB is covered by Deccan trap volcanic, surface exposures of older rocks are absent to check them on the ground.

(c) A gravity profile on the western side from Bhopal-Banglore-Trivandrum across the western part of the SMB and the South Indian Shield is given in Fig. 7.28(a) that shows a regional gravity high, H over the SMB and a large low, L to its south over the Eastern Dharwar craton. Besides this large wavelength gravity low, there are small wavelength lows and highs (c-g) that are caused by local sources of low and high density rocks. The gravity high, H is significant in this regard as it coincides with the western part of the SMB and may be caused by thrusted high density rocks in this section as in the case of the central part of the SMB. The large wavelength gravity low, L is modeled due to low density lower crust implying its felsic nature, as has been confirmed from low Poisson's ratio in receiver function analysis (Vinnik et al., 2007). This gravity low can also be modeled due to crustal thickening. However, as it mainly passes through the Eastern Dharwar craton where crustal thickness is almost constant as discussed in Section 7.8, it is modeled due to low density rocks in the lower crust suggested by low Poisson's ratio. But some of the residual gravity anomalies of relatively large wavelength such as d and g, south of Bangalore, may represent crustal thickening as the crust is thick in this section as described below for the Southern Granulite Terrain (Section 7.9).

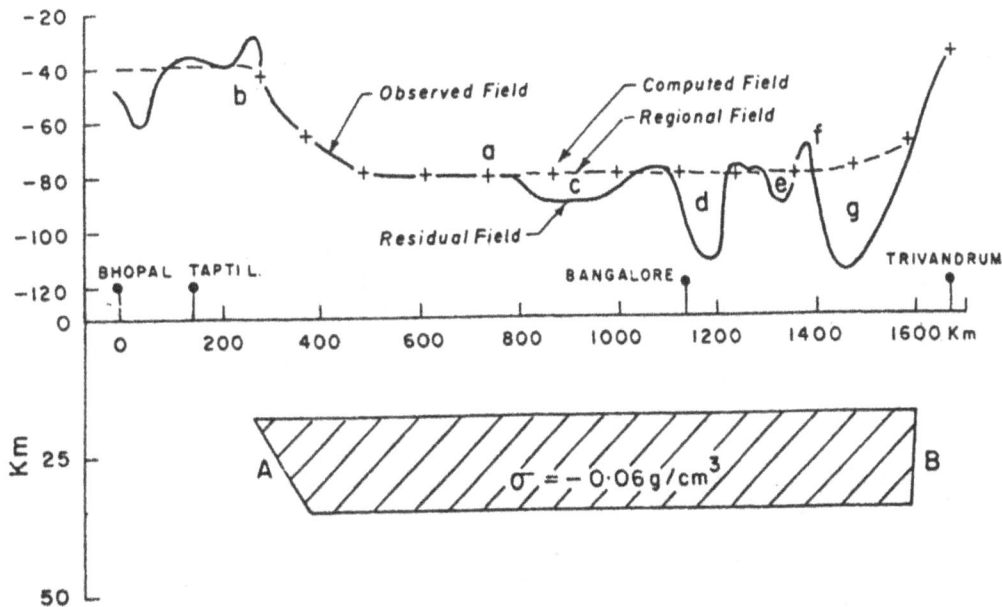

Fig. 7.28(a) A gravity profile (Fig. 7.7) from Bhopal to Trivandrum across the western part of the SMB and the Dharwar craton. The regional gravity anomaly is represented by a large wavelength gravity low a, while b, c, d, e, f and g are residual gravity anomalies caused by shallow localized sources discussed in the text. The regional gravity low (a) is modeled due to low density (-0.6 g/cm^3) of the lower crust (18-35 km) compared to the density of the normal lower crust. The projection of contact, A, on surface coincides with the Tapti Lineament. The largest residual gravity low (g) is observed over the Southern Granulite Terrain, corresponding to L19 (Fig. 7.8(b)).

(d) As this profile (Fig. 7.28(a)) is sufficiently large, the gravity high, H due to the SMB, is not defined properly. Therefore, a gravity profile, Thuadara-Sindad across the western part of the SMB (Fig. 7.11(c)) is given in Fig. 7.28(b) that shows gravity high over the SMB. This profile is modeled constraining it from a seismic profile along the same section (Sridhar et al., 2007; 2009). The modeled crustal section shows a high density body (2.85 g/cm^3) in the upper crust that represents mafic intrusive similar to those modeled under the central part of the SMB (Fig. 7.19(a)). The gravity high at the northern end of the profile is related to the exposed Aravalli group of rocks of density 2.83 g/cm^3.

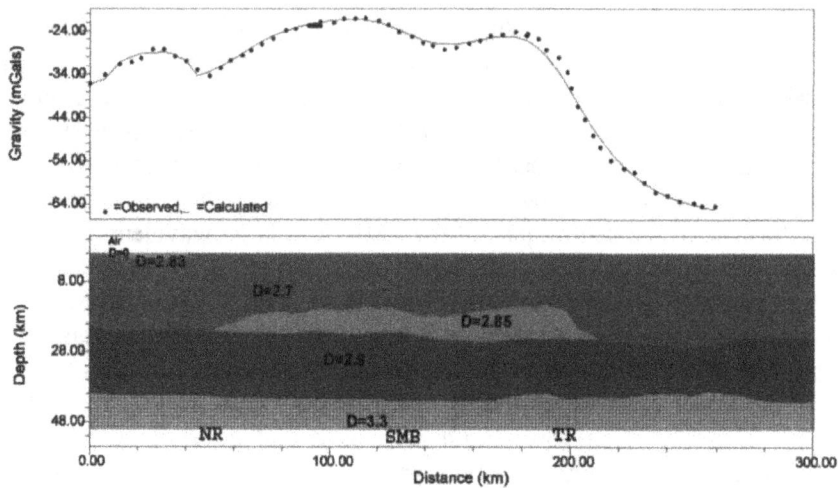

Fig. 7.28(b) Gravity profile across the western part of the SMB from Thuadara to Sindad (Fig. 7.11(c)) showing gravity high over the Satpura Mobile Belt (SMB) due to high density intrusive in the upper crust separated by the Tapti River (TR). NR: Narmada River.

(e) Conductivity structure over the western part of the SMB is given in Fig. 7.28(c) (Rao et al., 1995). It shows a conductive body at a depth of about 4 km under the SMB and along the Tapti and Narmada-Son Lineaments. This conductivity structure is similar to those observed across the central part of the SMB (Fig. 7.18(b)) and is attributed to mafic/ultramafic intrusions as in the previous case. It may represent thrusted lower crustal granulite rocks as in the previous case. Narrow zones of high conductivity under the NSL and the Tapti Lineament may be related to fluid along these lineaments similar to those along the CIS (Fig. 7.18(b)). Patro et al., (2005) have also reported conductive bodies associated with these lineaments and central part of the SMB in the upper crust as described above.

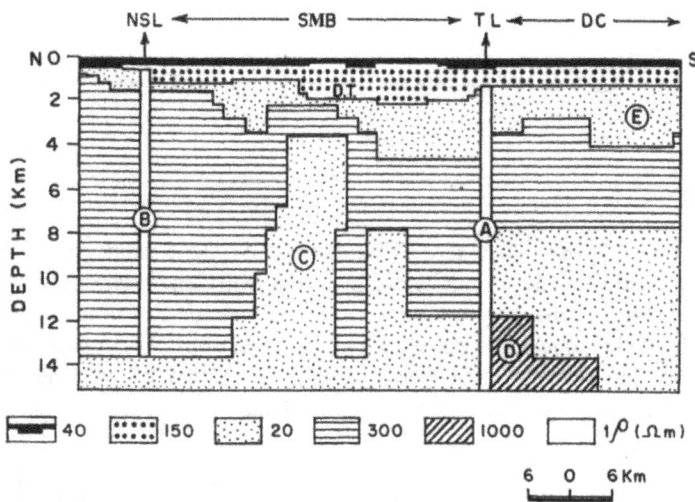

Fig. 7.28(c) Conductivity distribution across the western part of the SMB bounded by the Narmada Son Lineament (NSL) and Tapti Lineament (TL) along a profile (modified after Rao et al., 1995). It shows two highly conductive zones A and B coinciding with TL and NSL. They are similar to the signatures of the CIS (Fig. 7.18(b)) in the central part of the SMB where it is believed to indicate a fluid-filled fracture zone. The section between NSL and TL shows different conductive structure compared to those on either sides. The high-conductivity body under the SMB along this profile lies almost at the same depth as along the previous profile (Fig. 7.18(b)), attributed to thrusted lower crustal rocks.

(f) An airborne magnetic profile from Nagpur to Bhopal (Fig. 7.7) in central India flown at an altitude of about 2 km amsl (Fig. 7.28(d), Achuta Rao et al., 1989) has provided some significant magnetic anomalies related to the extension of the SMB towards the west. There are three sets of magnetic anomalies, H1, L1; H2, L2; and H3, L3 that are typical paired magnetic anomalies observed in low geomagnetic latitudes. The magnetic anomalies, H1, L1 and L2, H2 coincide with the northern margin of the gravity high, GH1 that represents the SMB. Therefore, these magnetic anomalies appear to represent the mafic intrusive rocks of the SMB that have been modeled along previous profiles (Fig. 7.19(a) and 7.22(a)).

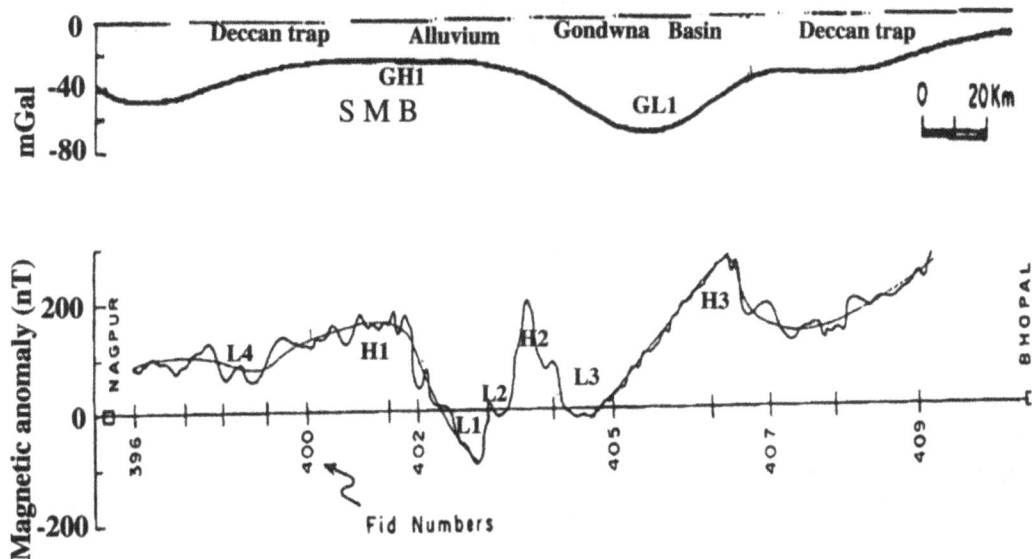

Fig. 7.28(d) Airborne magnetic profile flown at 6500 ft (1970 m amsl) across the Satpura Mobile Belt (SMB) from Nagpur to Bhopal along with Bouguer anomaly and exposed rocks, are shown schematically at the top. Numbers along the profile indicate fiducial numbers used to locate flight path on the ground. Gravity high, GH1 and magnetic anomaly, H1, L1 and H2, L2 are related to the Satpura Mobile Belt (SMB) while gravity low, GL1 and magnetic anomaly, L3 and H3 are related to the Gondwana Satpura basin.

(g) Another important signature of collision and subduction from north to south across the western part of the SMB is observed in the upper mantle conductive structure along a profile from Partur to Sangole (Fig. 7.28(e), Patro and Sarma, 2009) south of the western part of the SMB (Fig. 7.6 and 7.40(a)). It shows an upper mantle conductor consistently dipping southwards at almost 45°. This figure also gives a Bouguer anomaly profile from Sangole to Partur at the bottom showing a regional high increasing towards the SMB that appears to represent a remnant of lod subduction zone as per its geometry giving rise to high conductivity that may be related to Proterozoic subduction and thrusting across this part of the SMB. Thrusting would give rise to regional gravity high towards the collision zone, the SMB in this case.The Bouguer anomaly profile also shows residual gravity high and low, H1 and L1 that are well known Sangole high and Kurduwadi low as discussed in Section 7.8.3. The residual gravity high, H2 is part of the gravity highs due to the SMB (H7, Fig. 7.8(a) and (b)). High conductivity in such cases is attributed to fluid along the thrust. Such consistent dipping body cannot be explained by any other tectonic feature in this section. The high conductive body in the crust just south of Partur may represent thrusted upper mantle/ lower crustal rocks that have been traced throughout the SMB along profiles II-V (Fig. 7.11(c)) as discussed above. Similar conductive and gravity anomalies related to deep seated thrusts have been delineated from other Proterozoic mobile belts such as the ADMB (V, Fig. 7.15(a)).

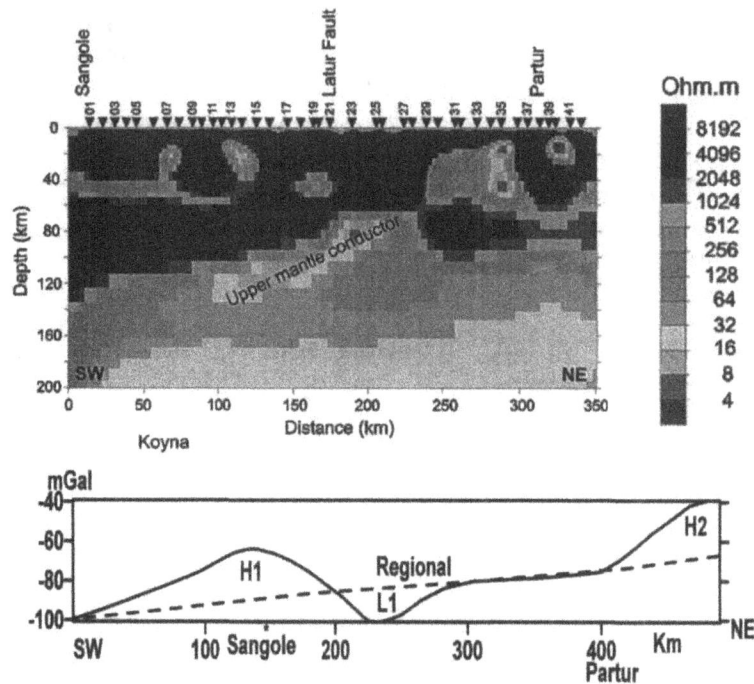

Fig. 7.28(e) Two dimensional geoelectric model based on MT profile Sangola-Partur (Fig. 7.40(a), Patro and Sarma, 2009) showing an upper mantle conductor in the lithospheric mantle with decreasing depth northwards in the crust near the Proterozoic Satpura Mobile Belt (Fig. 7.11(c)). The bottom profile is Bouguer anomaly from SW of Sangole to NE of Partur (Fig. 7.40(a)) showing an increasing regional high towards the SMB indicating high density rocks that may represent a thrust. [Colour Fig. on Page 794]

(h)   The above observations suggest that the western part of the SMB also represents a Proterozoic collision zone between the Bundelkhand craton towards the north and Dharwar craton towards the south with Tapti Lineament being a suture, synonymous to the CIS in the central part, and represents its extension westwards up to the coast. The high density mafic intrusive under the western part of the SMB may represent the thrusted lower crustal rocks in this section as under the central part of the SMB.

(iii) Connection of the SMB and the ADMB and Plausible Convergence Directions

(a)   The gravity anomalies of the ADMB and the SMB are joined in the western part (Fig. 7.8(a)) forming an arcuate shaped collision zone between the Bundelkhand craton in the center, the Rajasthan block towards the west, and Dharwar-Bhandara-Singhbhum cratons towards the south (Fig. 7.7 and 7.11(c)). These observations suggest simultaneous E-W convergence between the Bundelkhand craton and Rajasthan (Marwar block) across the ADMB and N-S convergence between the Bundelkhand craton and the Southern cratons (Dharwar-Bhandara-Singhbhum cratons) across the SMB during Mesoproterozoic period (Fig. 7.6). In fact, Mishra (2006) suggested that there might have been a NE-SW convergence between the Bundelkhand craton towards the NE and Rajasthan block towards the west and the southern cratons towards the south with E-W and the N-S components in the two cases. The NE-SW oriented shear zones with quartz reefs of the Bundelkhand craton indicate primary stress in this direction during the Proterozoic period. The junction of the ADMB and the SMB (H3, H7, Fig. 7.8(b)) SE of Ahmedabad might have formed a triple junction with the Damara mobile belt between the central (Zaire), East African and Kalahari cratons that got separated after rifting of the Gondwana land. However in the absence of the third suture in its present form, this can not be ascertained definitely.

(b) Based on rock types and their ages (Table 7.1), Mishra (2006) also suggested that there might have been a similar phase of convergence during Neoarchaean-Paleoproterozoic period (3.0-2.0 Ga) across the ADMB and the SMB related to Aravalli and Mahakoshal and Singhbhum orogenies in the respective cases that also gave rise to magmatic rocks of this period related to these orogen. It is followed by a rifting of the Bundelkhand craton that gave rise to Bijawar and Mahakoshal group of rocks, with large scale mafic rocks and volcanogenic sequences over the rifted platform of the Bundelkhand craton along the ADMB and the SMB as discussed above. This is followed by the second convergence during Mesoproterozoic period (1.5-1 Ga) that is dicussed above. Rifting and convergence along the same features during different geological periods have been reported from several places the world over as discussed in Section 7.1.

7.6.4 Shillong Plateau and Bengal Basin, India and Bangladesh and Shillong Earthquake

(i) Geology and Bouguer Anomaly

In order to check the extension of the Satpura Mobile Belt orogeny further eastwards gravity data and related geological/tectonic settings of the Shillong plateau (Fig. 7.6), where meso-Neoproterozoic rocks are also exposed and depict a gravity high (H26, Fig. 7.8(a)) similar to the SMB, were examined. A detailed geological/tectonic map of this section along with that of the adjoining SMB (CGGC) is presented in Fig. 7.29(a) (Acharyya, 2003). The Shillong plateau represents a flat highland of 1.0-1.5 km regional elevation flanked by plain lands of the Himalayan foredeep towards the north and the Bengal basin towards the south. Some significant geological and geophysical information related to the Shillong plateau is given below and shows the extensions of the SMB eastwards up to the eastern margin of the Indian plate.

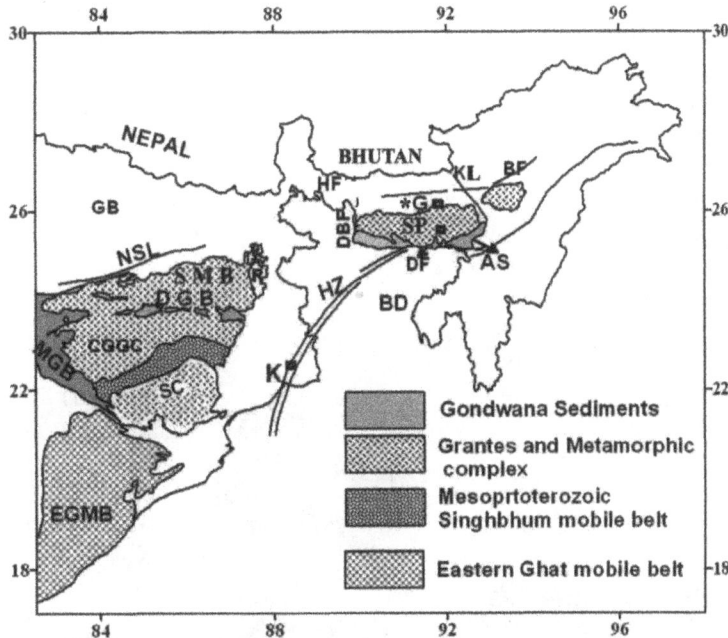

Fig. 7.29(a) Geological map of the Shillong Plateau and adjoining regions (GSI 1993, Acharyya, 2003). Various notations in alphabetical order are as follows: AS: Assam Shelf, BD: Bangladesh, BF: Brahmaputra Fault, K: Kolkata, CGGC: Chota Nagpur Granite Gneiss Complex, DBF: Dhubri Fault, DF: Dauki Fault, DGB: Damodar Gondwana Basin, EGMB: Eastern Ghat Mobile Belt, GB: Ganga Basin, G: Guwahati, HF: Himalayan Fore deep, HZ: Hinge Zone, KL: Kopili Lineament, MGB: Mahanadi Gondwana Basins, M: Mikir hills, NSL: Narmada Son Lineament, R: Rajmahal traps, SC: Singhbhum Craton, SMB: Satpura Mobile Belt, SP: Shillong Plateau, SPSZ: Singhbhum Purulia Shear Zone and SSZ Singhbhum Shear Zone. Location of epicenter of the 1897 Shillong earthquake is denoted by symbol star.

(a) The Shillong Plateau is basically E-W and changes to NE-SW at its eastern margin to form the Shillong-Mikir hills. The two are separated by a major NW-SE oriented Kopili Lineament, which extends on either side to the Main Boundary Thrust and Naga Thrust of Himalayan orogeny towards the NW and the SE, respectively (Fig. 7.29(a)). In fact, the Shillong Plateau lies at the junction of the Himalayan orogeny on three sides viz. north, east, and NE. Basement rocks in the Shillong Plateau consist of gneisses and migmatites with enclaves of amphibolite and high-grade supracrustals overlain by the Shillong group of rocks of Proterozoic metasediments. The oldest member of this group are of Mesoproterozoic period (1530-1550 Ma). Several granite plutons of 885-480 Ma (Mitra and Mitra, 2001) are exposed over the plateau. Similar rock types and metamorphic grades of almost similar ages (~ 1.6 Ga) have also been reported from the Chhotanagpur Granite Gneiss complex (CGGC) (Mitra and Mitra, 2001). The CGGC forms part of the Satpura Mobile Belt as discussed above and represents a Mesoproterozoic collision zone with the Singhbhum craton towards the south. The Shillong Plateau has also been considered as the Shillong-Meghalaya Gneissic Complex (SMGC) with Mesoproterozoic ages of 1596 ± 15 Ma based on well constrained Monazite ages (Chatterjee et al., 2007). They have also reported a Pan African event of 500 Ma and older ages of 1078 ± 31 Ma and 1472 ± 38 Ma from SMGC, considering it as a part of Pan African and Antarctic amalgamation along the east coast of India.

(b) Shillong Massif is associated with several E-W, N-S, NE-SW, and NW-SE oriented faults and lineaments (Fig. 7.29(a)). The Dauki fault along its southern margin is the most prominent one and is an E-W oriented south dipping fault, that is continuous throughout its section extending eastward up to the Haflong thrust related to the Himalayan orogeny towards the east (Srinivasan, 2005; Nag et al., 2001). The Shillong Plateau is also bounded by the Dhubri fault, the Brahmaputra fault, and the Kopili lineament towards the west, the north, and the east, respectively. Evans (1964) has considered the Dauki fault as a tear strike slip fault along which the Shillong Plateau has shifted about 200-250 km eastwards after getting detached from the eastern part of the peninsular shield. The Assam Shelf flanking, the Shillong-Mikir Massif (Fig. 7.29(a)) towards SE consists of Gondwana sediments similar to those found along the east coast of India (Dasgupta and Biswas, 2000) showing its similarity to the eastern Ghats.

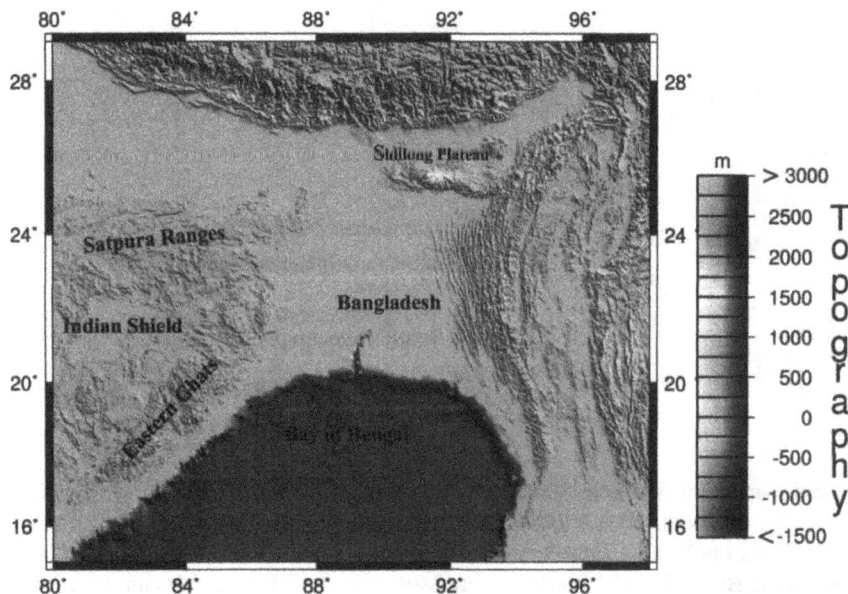

Fig. 7.29(b) Geomorphology of NE India showing Satpura ranges, Eastern Ghats, Shillong Plateau, and their interrelationship. Northern margin of Satpura ranges appears to be joining the Shillong plateau across the Bengal basin.

(c) Bangladesh, south of the Shillong Plateau, is largely occupied by alluvium underlain by thick sediments of about 10 km in general, and about 15 km in the Sylhet trough south of the Shillong Plateau over a mafic crust similar to oceanic crust (Alam et al., 2003). The eastern part of the Bengal Basin in Bangladesh (Fig.7.30) is connected to the East Indian Shield through a Hinge zone across which sudden changes in thickness of sediment and crust from W-E have been reported (Kaila et al., 1992). The Hinge zone oriented NE-SW is at least 500 km long connecting the Shillong Plateau to the east coast of India and offshore near Calcutta (Fig. 7.30(a)).

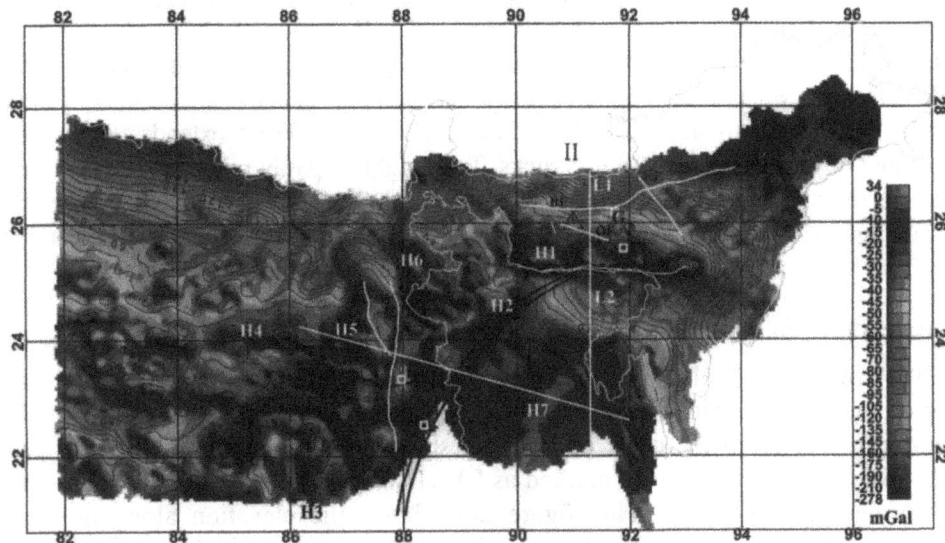

Fig. 7.30(a) Complete Bouguer anomaly map of NE India and Bangladesh. H1-H7 and L1-L2 represent gravity highs and lows. Some important tectonic units and towns are as follows: HZ: Hinge zone, S: Shillong, C: Calcutta, G: Guwahati, BF: Brahmaputra Fault, OF: Oldham Fault, HZ: Hinge Zone, DF: Dauki Fault, KL: Kopili Lineament, RF: Rajmahal Fault, SBF: Sainthia Bahmani Fault. The epicenter of the 1897 Shillong earthquake is shown by a red triangle. Location of Profiles I and II across the Chhotanagpur Granite Gneiss Complex, Bengal basin (India and Bangladesh) and Shillong Plateau are also shown. The deep seismic sounding (DSS) profile covers the central part of Profile I in Bengal basin (India) across the Hinge zone (Rajasekhar and Mishra, 2008). [Colour Fig. on Page 794]

(d) Topographic image of the area (GEBCO, http://www.gebco.net; Fig. 7.29(b)) suggests that the Satpura Ranges bounded by the Narmada-Son Lineament (NSL) towards the north appear to be connected to the Shillong Plateau and the Brahmaputra fault north of it, respectively, as their linear extensions. In Gondwana land reconstruction, Powell et al., (1988) and Harris and Beeson (1993) have suggested the extension of the Satpura Mobile Belt, the Precambrian rocks of NE India (Shillong plateau), and the Eastern Ghat Mobile Belt to the Albany Mobile Belt of Western Australia.

(e) A complete Bouguer anomaly map of the eastern part of the Indian peninsular shield (CGGC), the Shillong Plateau, and Bangladesh (82^0E – 96.5^0E and $21.5\ ^0$N – $28\ ^0$N) (Fig. 7.30(a)) is prepared based on data from the Indian region (GSI-NGRI, 2006) and gravity map of Bangladesh (Rahman et al., 1990). The gravity data from India and Bangladesh are basically regional at about 3-5 km interval along roads and tracks with an overall accuracy of ± 1 mGal. However for the present study related to crustal structures, this accuracy is sufficient where anomalies of order of 5 mGal and above are considered. This map shows a major gravity high over the Shillong Plateau (H1, Fig. 7.30(a)) flanked by gravity lows over the Ganga basin of the Himalayan foredeep (L1) and the Surma Basin (Sylhet trough; L2), related to sediments in these basins. The gradient of gravity high, H1 towards the south coincides with the Dauki fault, and towards the north coincides with the northern margin of the Shillong Plateau (Fig. 7.29(a)). The linear gravity high, H2 coincides with the Hinge zone, which extends from the east coast of India up to the Shillong Plateau. Gravity high H3 oriented NE-SW are related to the Eastern Ghat Mobile Belt, while H4 oriented E-W are related to the Satpura Mobile Belt. The gravity high H3 related to Eastern Ghat Mobile Belt that has been described in detail in the next section, extends towards the north as gravity high, H5 along the eastern part of the CGGC, which interacts with the gravity high, H4 related to the Satpura Mobile Belt in the eastern part of the CGGC. The gravity high, H5, however extends further northwards up to Sikkim-East Nepal border as kinks in the gravity map, which appears to be embedded in the large gravity low due to sediments of the Himalayan foredeep. Some workers (Mukhopadhyay et al., 1986) have attributed H5 and H6 to the Rajmahal Trap due to their proximity to exposed trap rocks. However, if observed carefully, the exposure of Rajamahal traps coincides with the eastern part of gravity high, H5, which extends further westwards. Moreover, the Rajamahal Trap is confined between two faults, (Fig. 7.29 (a)), Rajmahal fault (RF) and Saithia Bahmani faults (SBF) (Narula et al., 2000), while H5 extends even west of it and therefore cannot be attributed solely to trap rocks. This might be partly related to the high density lower crustal rocks of the CGGC as discussed above in Section 7.6.3 (H5, Fig. 7.27(b)). The gravity high, (H7) in southern part of Bangladesh may be manifestation of oceanic crust in this section and volcanic rocks from Kerguelen hotspot during the breakup of the Indian plate from Antarctica.

(ii) Crustal Structure under Shillong Plateau and Bangladesh

Profile II, (Fig. 7.30(a)) across the Shillong Plateau (Fig. 7.30(b), 600 km long) passes through one of the highest values of the gravity field over the Shillong Plateau and extends from the Himalayan foredeep in the northern part up to the coast of Bangladesh towards the south. The major gravity anomalies along this profile (Fig. 7.30(b)) are marked as L1, H1 and L2 representing the anomalies as given in Fig. 7.30(a) by same notation. This figure also shows the elevation along this profile depicting the extent of the Shillong Plateau. This profile shows a gravity high over the Shillong Plateau (H1) flanked by gravity lows related to the Himalayan foredeep (L1) and the Surma basin (L2, Sylhet trough) towards north and south of Shillong Plateau. This profile is modeled using the following constraints:

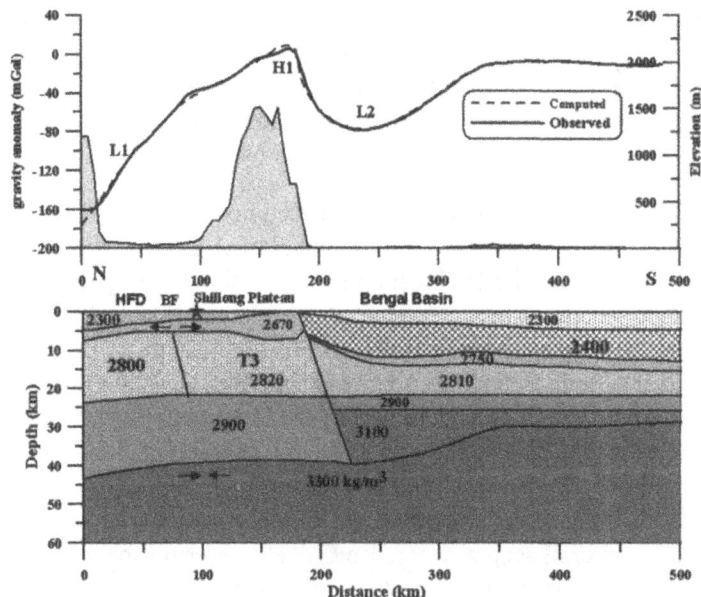

Fig. 7.30(b) Crustal model along profile II across the Shillong Plateau. Densities of bodies are given in kg/m^3. T3 is high density thrusted body in the middle crust. Elevation is plotted in meters. BF – Brahmaputra fault, HFD – Himalayan Fore Deep. Star indicates the approximate location of Shillong earthquake of 1897 projected on to this profile. Arrows indicate direction of forces due to crustal bulge caused by lithospheric flexure under Himalaya towards the north. It indicates compression at Moho and extension at the shallow/basement level.

(a) As the free air anomaly in Bangladesh along the coast is close to zero (Mishra et al., 2004) and elevation in this section is also small, it indicates a compensated crust.

(b) Based on deep seismic profile, in the Bengal basin east of the Hinge zone, a crustal thickness of 29-30 km has been reported (Mall et al., 1999)

(c) Receiver function analysis suggests a crustal thickness of 37-38 under the Shillong Plateau which increases northwards upto 44-45 km under the Himalayan foredeep in the northern part of the profile and 41 km under the Surma basin (Sylhet trough), south of the Shillong Plateau (Mitra et al., 2005).

The gravity high, H1 cannot be explained due to changes in Moho as given above and is, therefore, modeled due to a high-density body (2820 kg/m^3) in the upper crust as intrusives (T3) which is supported from high seismic velocity in receiver function analysis. The density contrast of this intrusive (T3) corresponds to almost 5% increase in S-wave velocity as suggested by receiver function analysis (Mitra et al., 2005) and similar increase in P-wave velocity from local earthquake tomography. The projection on surface of the contacts of the high-density intrusive (T3) approximately coincides with the Dauki and the Brahmaputra faults dipping towards the south. The effect of crustal bulge (shown by arrows) in the Himalayan foredeep due to lithospheric flexure under Himalaya towards the north of the Shillong Plateau (Section 6.9, Fig. 6.15(b)) can be seen in this model. These would introduce compressional and extensional forces (Arrows, Fig. 7.30(b)) at the Moho and basement levels, respectively, and are likely to cause seismic activity. Crustal model along another gravity profile I (Fig. 7.30(a)) is given in Fig. 7.55 that shows changes in crustal thickness across the Hinge zone from 40 km under the Indian shield to 34 km under Bangaladesh, and a thrusted block of high density under the Indian shield almost similar to that deciphered along Profile II across the Shilong plateau.

(iii) Airborne Magnetic Map

The airborne total intensity map of the Shillong plateau and the adjoining Benal basin, Indian shield and Bangaladesh is given in Fig. 7.30(c). The airborne magnetic map is adopted from USGS open file report 97-470H entitled 'Digital Geologic and Geophysical Data of Bangaladesh, (Persists et al., 2001) and published by Rahman et al., (1990). This data set was acquired at a flight altitude of 150' (~ 45.5 m) with flight-lines spaced 2 km apart and tie-lines 5 km apart. Sampling interval was 2 seconds with an over all accuracy of ±1.5 nT. The airborne magnetic data of the Shillong plateau was flown by NGRI (1980) and was carried out at a flight height of 1380 and 2100 m above msl with a line spacing of 1 km (Kayal, 2005), while those for the Indian Bengal basin and adjoining shield is obtained from AMSE (2001) which were flown with flight lines spaced at 5 km apart at flight altitude varying from 5000' to 7000' above msl. Due to varying flight altitude, all the data set were continued to a common height of 7000' (~ 2100 m) and a composite map was prepared and is presented in Fig. 7.30(c) (Rajasekhar, 2010). However, due to gaps in between different data set and mismatch between them three different scales for three sections are given in this figure. However, magnetic highs and lows can be correlated from different maps and tectonics in individual sections can be inferred. The Hinge Zone (Pair of solid lines) is clearly reflected as gradient between lows towards the west and NW and relative highs towards the east that separates the sediments and non-magnetic basement from magnetic basement of mafic rocks, respectively. The magnetic highs in the southern part of Bangaladesh (BH2) may represent the mafic rocks that were responsible for the breakup of India from Antarctica due to the Kerguelen hotspot that also gave rise to 90 East Ridge in Bay of Bengal and Rajmahal and Sylhet traps on the Indian continent (Chapter 5). It may be noted that the Bouguer anomaly map also presented gravity highs in the southeren part of Bangladesh (Fig. 7.30(a)) that is also related to the same event. Towards the west, the Indian Shield is characterized by both E-W (SH) and N-S (EH) magnetic trends that are related to the Satpura Mobile Belt (SMB) and the Eastern Ghat Mobile Belt (EGMB) as described in Sections 7.6 and 7.10, respectively. The magnetic highs due to the SMB (SH) show its extension via the Bengal basin high (BH1) to the Shillong plateau highs (SPH) indicating continuation of these structures as discussed below. The magnetic lows due to the Bengal basin, India (IL1) and Bangladesh (BL2) also show continuation of basement structures that continue further eastwards south of the Shillong plateau separated by a fault known as the Dauki fault (Fig. 7.30(a))

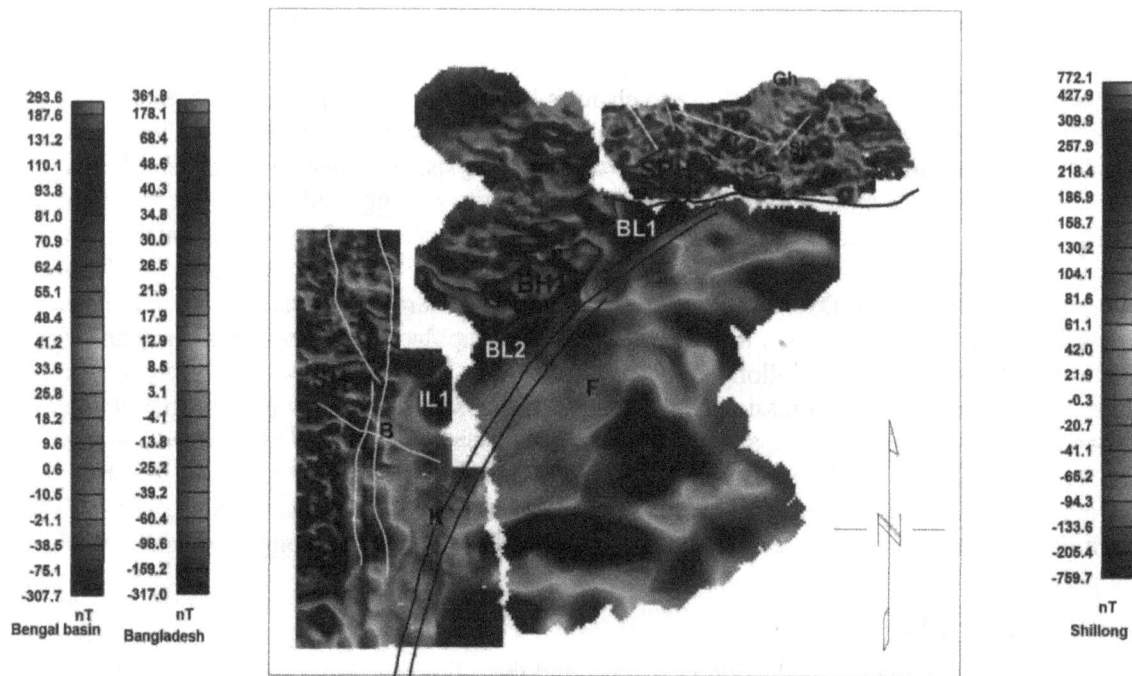

Fig. 7.30(c) Airborne magnetic map of Shillong plateau and adjoining Bengal Basin, Indian Shield and Bangladesh with different scales for the three sections (Rajasekher, 2010). K: Kolkatta; Gh: Gauhati; Sh: Shillong; EH: Eastern Ghat trend magnetic high; SH: Satpura trend magnetic high; BH1: Bangaladesh magnetic high 1; BH2: Bangaladesh magnetic high 2; SPH: Shillong Plateau magnetic high; BL1: Bangaladesh magnetic low 1; BL2: Bangaladesh magnetic low 2; IL1: India Bengal Basin magnetic low 1.
[Colour Fig. on Page 795]

(iv) Shillong Plateau as Plausible Extension of the SMB

(a) Crustal model under the Shillong Plateau deduced from gravity modeling constrained from seismic studies (Fig. 7.30(b)) suggests a crustal thickness of 37-38 km under the Plateau, which increases to about 44-45 km towards the north under Himalayan foredeep and about 41km just south of the Shillong Plateau. Further south, the crustal thickness under Bangladesh decreases to 29-30 km, which is attributed to oceanic crust based on high crustal density supported from high seismic velocity, referred to above. The large crustal thickness (40-41 km) under the Surma basin (Sylhet trough) is primarily attributed to the thick sedimentary column (12-14 km) in this basin.

(b) The crustal model under the Bangladesh show an underplated crust of higher density (3100 kg/m^3) at the base of the crust, which however, is limited to the Dauki fault towards the north and the Hinge zone towards the east. It might have resulted from the eruption of Sylhet-Rajmahal trap of 116 ± 3.5 Ma (Ray et al., 2005) in this section that is considered to have same origin as 90° East Ridge from the Kerguelen hotspot and has been responsible for the breakup of the Indian continent from Antarctica (Mahoney et al., 1983). Underplating is typical of such volcanic provinces.

(c) Crustal model under the Shillong Plateau shows high density (T3, 2820 kg/m^3) thrusted rocks in the middle crust with crustal thickness of about 37-38 km as in the case of the CGGC. High density-high velocity thrusted rocks in the middle crust, in general are characteristic of Proterozoic collision zones. The inferred density of the thrusted block under the Shillong Plateau is 2820 kg/m^3, which is similar to that inferred under the Satpura Mobile Belt, west of the Shillong Plateau (Fig. 7.19(a), 7.22(a) and (b)).

(d) Saha et al., (2007) have reported high conductivity at a depth of 7-8 km under the Shillong Plateau which almost coincides with the high density thrusted block. High conductivity has also been reported for high density thrusted blocks under the Satpura Mobile Belt (Fig. 7.18(b)).

(e) The Shillong Plateau shows geomorphological alignment with the Satpura Mobile Belt (Fig. 7.29b) and similar crustal structure with rock types of almost the same age as the CGGC. This suggests that it might also be a part of the Satpura Mobile Belt and may represent its extension eastward. In adjoining Bangladesh, Ameen et al., (2007) have reported granite gneissic basement rocks from drill holes at depth of 227 m between the CGGC and Shillong Plateau which provided almost similar age of Proterozoic period (1.6 Ga) as oldest gneissic rocks from CGGC and the Shillong Plateau that may represent the connection between the two units viz. the CGGC and the Shillong Plateau. As the crust under Bangladesh is oceanic in nature, the CGGC and the Shillong Plateau might have been a part of the coast of the Proterozoic Indian craton on this side and the SMB extended up to the then plate margin.

(v) Seismotectonics of Shillong Earthquake of 1897: Near Plate Boundary Earthquakes

Based on crustal structures and other details of tectonics related to the Shillong plateau, an effort is made below to understand the seismotectonics of this earthquake.

The Shillong earthquake of 1897 was one of the most devastating earthquakes away from plate boundaries. Oldham (1899) recorded the effects of this earthquake in detail forming, one of the best kept records of earthquakes of that time. Gutenberg (1956) estimated a magnitude of 8.7 to this earthquake with epicenter located at the northern margin of the Plateau (Narula et al., 2000). Based on large destruction and aftershocks in the northern part of the Shillong Plateau, Oldham (1899) ascribed it to a rupture extending from the Dauki fault south of the Shillong Plateau to its northern section. However, subsequent workers suggested that due to its large magnitude and related devastation it must have been caused by north dipping fault related to the Himalayan orogeny in the northern part of the Shillong Plateau (Seeber and Armbruster, 1981).

Based on geodetic measurements after this earthquake reported by Survey of India and present day GPS measurements, Bilham and England (2001) suggested an uplift of 11 m along the northern edge of the Shillong Plateau. Accordingly, they suggested a subsurface reverse fault in the northern part of the Shillong Plateau dipping towards south away from the Himalayan front as the causative fault for this earthquake, and named it 'Oldham fault'. In fact, they suggested the Shillong Plateau as a 'Pop-Up' structure between the south dipping Oldham fault and north dipping Dauki fault towards its north and south, respectively. Based on paleoseismic investigations, Rajendran and Rajendran (2004) suggested Brahmaputra fault (Fig.7.29a) as a causative fault for this earthquake. Recent compilation of micro seismic data supports a south dipping fault that projects on ground surface in Brahmaputra valley along this river (Kayal et al., 2006).

The Shillong Plateau and its adjoining regions are seismically active (Kayal et al., 2006). The Shillong Plateau, being high with an average regional elevation of 1.0 km, should have at least 7.0 km of crustal thickening based on Airy's model of isostatic compensation showing gravity low over it. The observed gravity high however, suggests crustal thinning and some high density rocks within the crust. Thin crust under Shillong Plateau (~ 37-38 km) with respect to its elevation, introduces the isostatic instability resulting in local stress along the boundary faults. A low pass filtered Bouguer anomaly map suggests a gravity high under the Ganga basin (Fig. 6.15(b), Mishra et al., 2004b, Fig. 7.60) due to crustal bulge caused by lithospheric flexure of Indian plate due to Himalayan orogeny.

This gravity high due to crustal bulge shows the highest amplitude over Shillong Plateau and adjoining sections, which can be attributed to its location at the junction of Himalayan orogeny on three sides, viz. north, east, and northeast. As shown in Fig. 7.30(b) (shown by arrows), the crustal bulge would introduce compressional and extensional forces at the Moho and basement levels, respectively which would introduce differential stress in this section of the crust. It is interesting to note that the Brahmaputra fault coincides with the crustal bulge which might have been caused by the latter.

Seismotectonics of the Himalayan collision zone have already been discussed in Chapter 6. Plate boundaries and collision zones are usually characterized by large and great earthquakes due to the large amount of strain caused by thrusting. However, some of the seismic activity (M > 7) occurring close to the plate boundaries may not be termed as truly plate boundary earthquakes but might be connected to some of the plate boundary activities. They can be termed as near plate boundary earthquakes. The Shillong earthquake of 1897 may be one such earthquake in India, the other being the Bhuj earthquake of 2003 as discussed in the next Chapter 8.

This model (Fig. 7.30(b), shows the effect of crustal bulge caused by lithospheric flexure only towards the north. However, the Shillong Plateau is also affected by crustal bulge caused by lithospheric flexure towards east and the two together might be responsible for maximum crustal bulge in this section. The large continued crustal bulge under the Shillong Plateau due to lithospheric flexure under Himalaya might even be responsible for its uplift. This gravity high due to crustal bulge extends northwards upto the southern part of Bhutan, which shows more seismicity compared to the northern part of Bhutan (Velasco et al., 2007). They attributed it to brittle deformation in the crust and strike slip faulting in southern Bhutan. The effect of crustal bulge due to lithospheric flexure on near plate boundary earthquakes appears to be one of the effects of plate boundary activity. Another factor that may affect near plate boundary earthquakes are the connecting faults and lineaments to plate boundary such as the Dauki fault, Brahmaputra fault, and Kopli Lineament in case of the Shillong plateau that may transfer some of the strain across plate boundaries to these sections.

The seismicity of the Shillong Plateau, including the Shillong earthquake of 1897, can therefore be attributed to (i) local stress due to isostatic instability (ii) regional stress due to lithospheric flexure and crustal bulge due to plate tectonic forces causing brittle deformation in the crust, and (iii) contacts of mafic intrusive as stress concentrators connected to the Himalayan front towards north and east through extension of various faults and lineaments to plate boundaries in these directions. These factors have been attributed to large earthquakes in Kachchh including the Bhuj earthquake of January 26, 2001 as discussed in the next Chapter (Mishra et al., 2005). In this respect, it is different from plate boundary earthquakes along the Himalayan front, which are primarily caused by thrusts related to the collision of the Indian and Asian plates.

The Shillong earthquake of 1897 was attributed to the south dipping Oldham fault by Bilham and England (2001) and similar dipping Brahmaputra fault by Rajendran and Rajendran (2004) as discussed above. However, the epicenter of this earthquake as given by Geological Survey of India (Narula et al., 2000) based on the description of Oldham (1899) and shown in Fig. 7.30(b) suggests that it may be associated with south dipping Brahmaputra fault with focal depth as 25-26 km which coincides with the interface separating upper and lower crusts that is usually the favoured sites for strain accumulation. Saha et al., (2007) did not find any signature of the Oldham fault in magnetotelluric measurements though they suggested more MT measurements to resolve it. Their seismological data recorded by the permanent network in the Shillong Plateau suggest hypocenters of present day seismic activity concentrated south of the Brahmaputra faults and the latter dips southwards. The effective elastic thickness computed from Bouguer anomaly and topography in the Ganga basin adjoining Shillong Plateau lies between 23-26 km (Fig. 7.10(b), Rajesh and Mishra, 2004), which suggests that this level almost defines the separation of brittle upper crust and ductile lower crust and that major seismic activity is likely to originate above this level, conforming to the focal depth reported for most of the present day seismic activity in this section (Saha et al., 2007) and also inferred for the Shillong earthquake.

7.7   Godavari Proterozoic Belt (GPB) and Gondwana Basins – Crustal Structures, Seismirity Proterozoic Triple Junction

7.7.1   Godavari, South Rewa-Mahanadi Basins – Geology, and Bouguer Anomaly

(i) Godavari Basin – Geology, Bouguer Anomaly, and Bhadrachalam Earthquake

(a) This section is characterized by Gondwana (Permian-Triassic) sediments in the central part (Raju, 1986) flanked by Meso-Neoproterozoic metasediments of Pakhal-Sullavai basins (Rao, 1987) that are exposed on either sides of the Central Godavari basin (Fig. 7.7). A regional geological map of this region and Profile V across the Godavari Proterozoic belt of Fig. 7.7 is shown in Fig. 7.31 as Profile II. It shows the Godavari Proterozoic belt on either side of the Godavari Gondwana (Permian-Carboniferous period) south of the Satpura Fold (Mobile) belt. Contemporary Gondwana sediments are also found within the Satpura Mobile belt as shown in Fig. 7.11(b) and 7.20 that may be regarded as extensions of the Godavari Gondwana basin towards the north. The Bouguer anomaly, Fig. 7.8(b) shows a central gravity low, L10 related to Gondwana sediments flanked by gravity highs, H10 and H11 that coincide with the Meso-Neoproterozoic metasediments of the Pakhal and Sullivai Group of rocks along the shoulders. A seismic profile across the Chintalpudi sub-basin (section 2.9.1) (Kaila et al., 1990; Pandey and Rao, 2006) has provided high velocity at a depth of about 6 km.

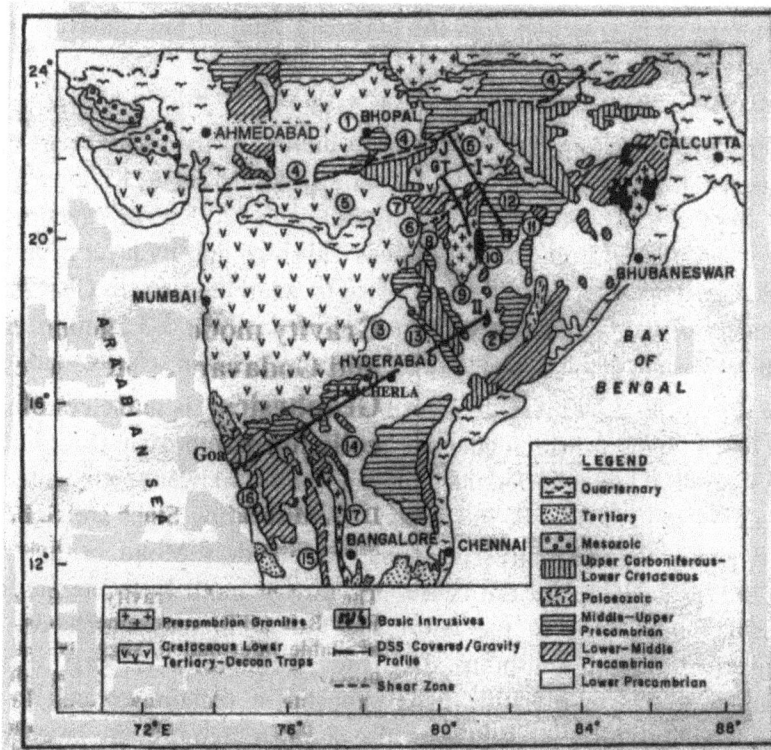

Fig. 7.31 Geological map of central India showing the relationship of the SMB with the Proterozoic Godavari Belt and the South Indian cratons. Geological units discussed in the present study are marked by encircled numbers. 1, Bundelkhand craton; 2, Bhandara-Bastar craton; 3, Dharwar craton; 4, Narmada Son Lineament; 5, Satpura Fold Belt; 6, Cental Indian Shear; 7, Sausar group; 8, Sakoli group; 9, Bhandara-Bastar craton; 10, Basic Volcanics; 11, Sonakhan Schist Belt; 12, Chhattisgarh basin; 13, Godavari Proterozoic Belt; 14, Eastern Dharwar Craton; 15, Shear Zone; 16, Western Dharwar Craton and 17, Closepet granite. The geotransect (GT) across the central part of the SMB (Profile II, Fig. 7.7 and 7.27(a)) and two new profiles, I and II are also shown. Profile I extends from Jabalpur (J) to Raipur (R) across the SMB while Profile II extends from Bastar craton to the west coast of India across the GPB and Dharwar craton.

(b) North of the exposed Proterozoic rocks is the Bhandara-Bastar craton that consists of several basins of Mesoproterozoic period. The contact of Bhandara-Bastar craton with the Proterozoic metasediments of the Godavari Proterozoic Basin consists of the Bhopalpatnam granulite belt of 1.6 Ga. The southern part of the Godavari Proterozoic basin is in contact with the Dharwar craton that is characterized by the Karimnagar granulite belt of 2.4-2.2 Ga at its contact (Santosh et al., 2004). However, the latter (southern part) also shows a thermal event of 1.6 Ga that has been reported from Karimnagar dykes in the vicinity of the granulite belt (Rao et al., 1990) which is contemporary to the Bhopalpatnam granulite belt.

(c) A detailed Bouguer anomaly map superimposed over a simplified geological map of the Godavari basin and surrounding Proterozoic section is given in Fig. 2.30(a) (Mishra et al., 1987) that shows central gravity lows (L1-L3 associated with the Gondwana sediments and flanking highs, H1 and H2 along the shoulders associated with the Proterozoic metasediments as described in Section 2.9.1. Godavari basin, being a rift basin, has been seismically active with several microseismic activities in this region (Mishra et al., 1987). However, the Bhadrachalam earthquake of magnitude 5.3 in April, 1969 was the largest reported earthquake in this region whose epicenter (Fig. 2.30(a)) lies at the intersection of Mailaram basement high with northern boundary fault of the Godavari basin. Mailaram high is part of the Eastern Ghat Mobile Belt (Section 7.10) that is exposed in this section bounded by faults and therefore its intersection with the boundary fault of the Godavari rift basin is prone to seismic activity.

(d) A gravity profile across the Chintalpudi sub-basin (L3, Fig. 2.30(a)) is modeled (Fig. 7.32(a)) constraining it from a seismic profile limited to the central part covered by Gondwana sediments (Kaila et al., 1990). It shows a high density (2.85 g/cm^3) layer at a shallow depth of about 5-6 km under the Godavari sediments (2.35 g/cm^3) synonymous with the high velocity layer inferred from seismic studies. This high density layer is uplifted along the shoulders that may represent high density thrusted rocks of the Eastern Ghat Mobile Belt, forming the basement of the Gondwana sediments which is about 3-3.5 km in this section. Moho is almost at a constant depth of about 41 km as provided from seismic studies.

(e) A gravity profile AA′ (Fig. 2.30(a)) is given in Fig. 7.32(b) that is modeled using multiple causative bodies based on the density model in the Chintalpudi sub-basin (Fig. 7.32(a)). This profile also shows a central low modeled due to low density Gondwana sediments up to 5 km as obtained using the inversion scheme in Chapter 4, and flanking highs are modeled due to high density (2.80 and 2.90 g/cm^3) intrusives along the shoulders. Moho, being at a constant depth as shown in Fig. 7.32(a), will not cause any gravity anomaly in the observed field. These high density bodies towards the north and the south may correspond to mafic intrusives as suggested from the airborne magnetic profile (Fig. 7.36(a)), including the Bhopalpatnam and Karimnagar granulite belts exposed in certain sections as discussed above. Seismic section in Chintalpudi sub-basin also shows a high velocity layer at a shallow depth of 5-6 km that may be synonymous with these high density bodies in the upper crust. Crustal model along Profile BB′ (Fig. 2.30(a)) across Chinnur high is almost same as that along Profile AA′ (Fig. 7.32(b)) except that Chinnur high that divides the basin in to two parts with basement exposed around Chinnur represents a median high, typical of continental rift basins world over. The basin is, therefore, disturbed in this section and flanks of the Chinnur high are important for exploration of coal and hydrocarbons.

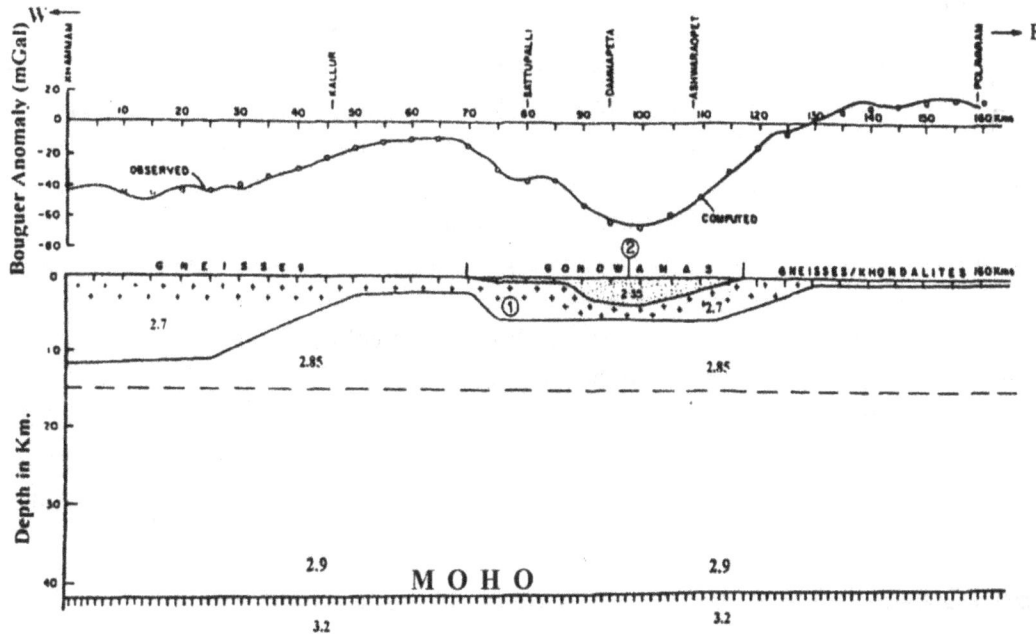

Fig. 7.32(a) An E-W gravity profile across the Chintalpudi sub-basin (L3, Fig. 2.30(a)) and crustal model and its computed field for comparison with the observed field. It shows high density (2.85 g/cm³) layer at a shallow depth in the upper crust that props up along the shoulders. The thickness of the Gondwana sediments is 3-3.5 km in this section.

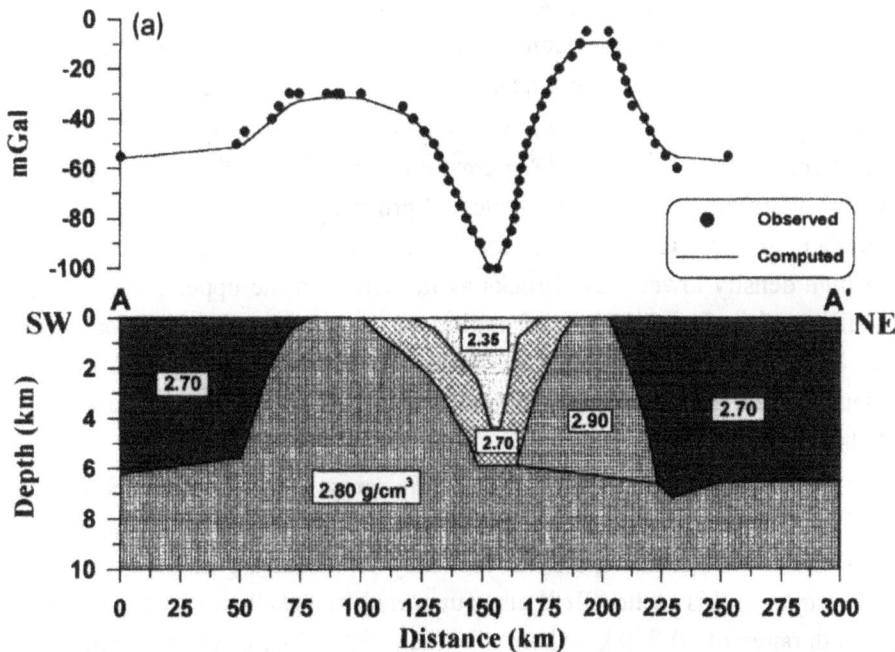

Fig. 7.32(b) Computed density model along profile AA′ (Fig. 2.30(a)) showing low density Gondwana sediments (2.35 g/cm³) in the center and high density (2.80 and 2.90 g/cm³) rocks along the shoulders. The rocks with density 2.70 g/cm³ represent Proterozoic metasediments exposed on either side of the Gondwana sediments (Mishra et al., 1999).

(ii) South Rewa-Mahanadi Basin

Another Gondwana basin extending from the SMB up to the east coast of India in the same manner as the Godavari basin, is the South Rewa-Mahanadi basin that separates the Singhbhum and the Bhandara-Bastar cratons (Fig. 7.7). However, in the absence of any exposed Proterozoic rocks along its flanks, it has not been considered related to Archean-Proterozoic collision. Due to its importance in the breakup of Gondwanaland, it has been described in Chapter 5 (Section 5.6) and crustal structures across this basin are given in Fig. 5.49(a) and (b) that do not show any high density intrusive rocks along the shoulders as in the case of the Godavari basin described above. This indicates that there might not have been any Proterozoic collision along this section, yet there must have been a zone of weakness between the Singhbhum and the Bhandara-Bastar cratons that opened up during the breakup of Gondwanaland to form the South Rewa-Mahanadi basins as discussed in Section 5.6. This zone of weakness could have been created during Archean-Proterozoic collision of cratons along the SMB and the EGMB as discussed in Sections 7.6 and 7.10.

7.7.2   Regional Gravity and Magnetotelluric Profiles across the GPB and Dharwar Craton

(a) A regional gravity profile (profile V, Fig. 7.7; Profile II, Fig. 7.31) and modeled crustal section is given across this region in Fig. 7.33(a) that shows a paired regional gravity anomaly of a regional high (H) over the Godavari Proterozoic Belt and exposed granulite rocks and regional low (L) over adjoining cratons due to crustal thickening. This being a regional profile, the gravity low due to Gondwana sediments (Fig. 7.35(c)) does not find a prominent reflection. However, the regional gravity high, H over the Godavari basin is composed of two highs, B and C, with an intermediate low (A) that is caused by the Gondwana sediments.

(b) The crustal thickening, however, is seen mainly in the western part under the Deccan Volcanic Province with relatively higher density (3.18 g/cm^3) along the Moho that may represent under plated crust which is a common feature under volcanic provinces. Deep seismic sounding profile in this section (Kaila et al., 1981) has also suggested thick crust of 40-41 km. The gravity high H is attributed to a high density lower crustal rocks as intrusives in the upper crust (2.8 g/cm^3) as shown along profile AA′ in Fig. 7.32(b) and similar to those under the SMB as described in Section 7.6 (Fig. 7.19(a)). Short wavelength anomalies, A-F being related to shallow bodies of high and low densities are not modeled and may be related to mafic and felsic intrusives, respectively (Mishra et al., 2002).

(c) A magnetotelluric profile from Goa-Jadcherla (Fig. 7.31) across Dharwar craton SW of the GPB has been reported by Gokaran et al., (2004) that shows a high resistivity lithospheric keel, D (Fig. 7.33(b)) SW of the GPB under the EDC that may represent a granitic batholith as such bodies are quite common in this section. It is also supported by low P-wave velocity residuals in the lithospheric depth range of 60-250 km (Ramesh et al., 1993). The conductive bodies, F and G are associated with the Eastern and the Western Dharwar cratons, respectively, separated by the shear zone that are discussed in the next Section 7.8. Other features are shallow conductivity distribution in the region.

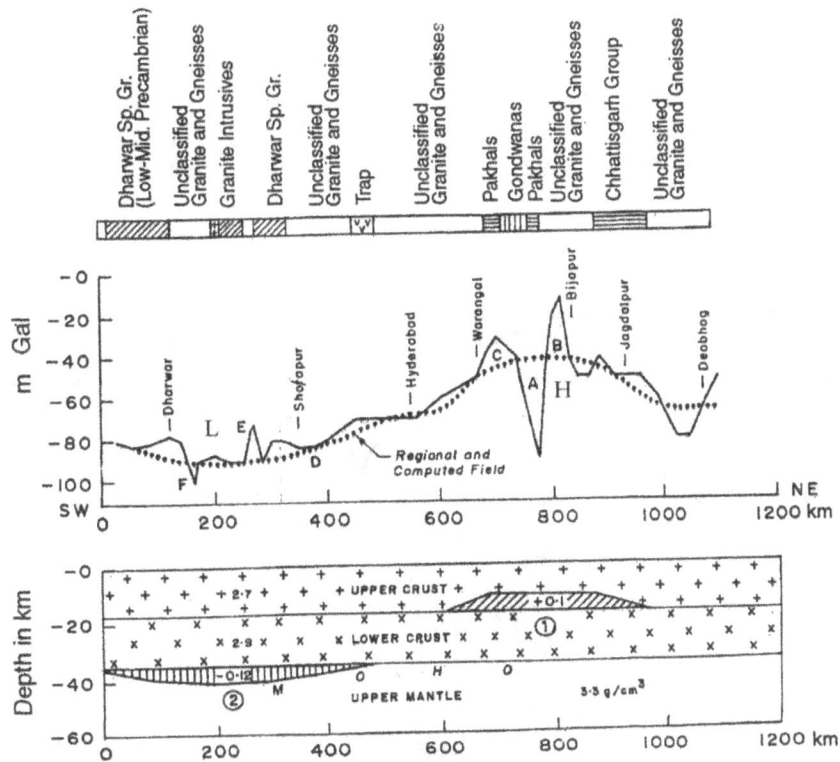

Fig. 7.33(a) Gravity profile II (Fig. 7.31) across the Godavari Proterozoic Belt and computed regional field due to bodies with their density contrast in g/cm³. Body 1 represents high-density rocks under the GPB, and body 2 represents the low-density body along the Moho, indicating a thick crust under the WDC. Top bar shows the exposed geology (Mishra et al., 2004).

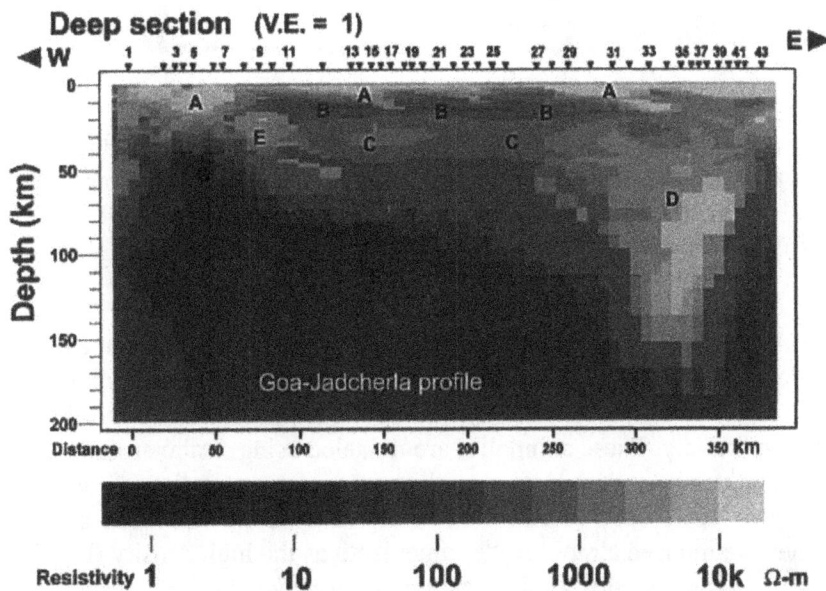

Fig. 7.33(b) Two dimensional geoelectric section along Goa-Jadcherla (Fig. 7.31) MT profile (Gokarn et al., 2004) showing a resistive body, D south west of the GPB that may be related to a granitic pluton while conductive bodies, F and G are related to the Eastern and the Western Dharwar cratons that props up along their contact.

7.7.3   Airborne and Ground Magnetic Maps of Godavari Basin

(a)   Airborne magnetic map of the Godavari basin along with that of the Eastern Ghat Mobile Belt (EGMB) in this section recorded at an altitude of 1.5 km above msl is given in Fig. 7.34(a) (GSI 2001). It shows pairs of magnetic lows and highs, L1, H1 and L2, H2 related to Proterozoic rocks along the southern and northern flanks of the Godavari basin that are also characterized by gravity highs, H1 and H2 (Fig. 2.30(a)) suggesting mafic/ultramafic rocks. These flanks of Godavari basin are also characterized by exposed Karimnagar and Bhopalpatnam granulite rocks, respectively. It is interesting to observe that lows in these anomalies are located towards the north indicating induced magnetization or the magnetization at the time of these intrusions or metamorphism was same as present day earth's magnetic field. The latter is true as the reported direction of magnetization during Mesoproterozoic period (Vindhayan sediments) was almost the same as present day field (Mishra, 1965). It may also be noted that the Eastern Ghat Mobile Belt (EGMB) shows large amplitude magnetic anomalies indicating mafic rocks associated with it.

Fig. 7.34(a) Airborne magnetic map of Godavari basin including that of the central part of the Eastern Ghat Mobile Belt (EGMB) toward the east recorded at a height of 1.5 km above msl (GSI, 2001). It shows pairs of magnetic highs and lows H1, L1 and H2, L2 related to Proterozoic rocks of the Godavari basin indicating mafic rocks. L3 represents Gondwana sediments of Godavari basin. Star indicates Chinnur high (CH) that is a median high in the basin. [Colour Fig. on Page 795]

(b)   A magnetic profile AA′ (Fig. 2.30(a)) based on the total intensity magnetic map of Godavari basin recorded on the ground (Mishra et al., 1987) is given in Fig. 7.34(b). It shows a magnetic low and high related to Godavari Proterozoic belt along the southern and the northern flanks of the Godavari basin, respectively. These anomalies are modeled using a tabular body and intrusive along a fault, respectively, since they depict similar magnetic anomalies as observed in the present case. It is interesting to observe that the two dip in opposite directions similar to a rift kind of structure. They are inferred almost at the same depth as the high density (Fig. 7.32(a) and (b)) and high velocity rocks reported from this section. They may represent subsurface extension of Karimnagar and Bhopalpatnam granulite rocks along the southern and the northern flanks of the Godavari basin, respectively, as basement of Proterozoic rocks.

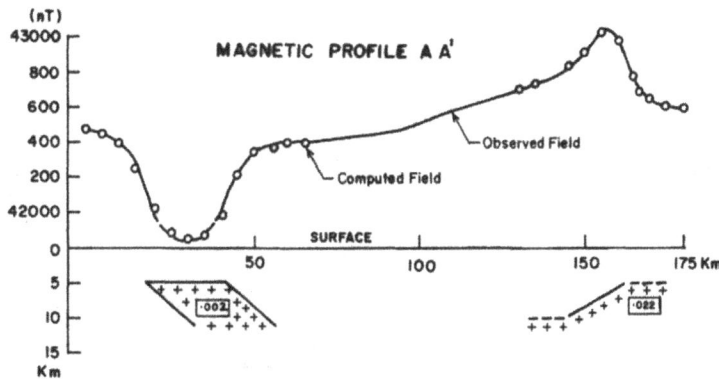

Fig. 7.34(b) Total intensity magnetic profile AA' across Chintalpudi sub-basin that shows prominent magnetic anomalies along the flanks over Proterozoic belts that are modeled due to mafic intrusive bodies of high susceptibilities dipping opposite to each other similar to graben type of structure.

(c) In order to assess the nature of the basement rocks under the Chintalpudi sub-basin (Fig. 7.31) that forms the southern most part of the Godavari basin where it interacts with the EGMB a total intensity map of this subbasin was prepared (Fig. 7.35(a), Mishra et al., 1987). It shows a magnetic low towards the north and high towards the south indicating induced magnetization. A magnetic profile YY' is given in Fig. 7.35(b) that is modeled using the inversion scheme for a dyke model (Chapter 4). The model along with the computed field and its susceptibility is given in this figure. It shows a north dipping body at about 40° at a depth of 6.4 km and susceptibility of 0.001 emu. It conforms with the depth of the high density (Fig. 7.32(a) and (b)) and high velocity body at a depth of about 6 km under the Chintalpudi sub-basin and its high susceptibility suggests mafic body. As this sub-basin lies south of Mailaram high (Fig. 7.31(b)) the basement under this section might be formed by the extension of the EGMB (Fig. 7.7) that is composed of lower crustal rocks and intrusives as described in Section 7.10. It demonstrates the application of magnetic methods to know the nature of rock types and their configuration that connects to the geodynamics of the region.

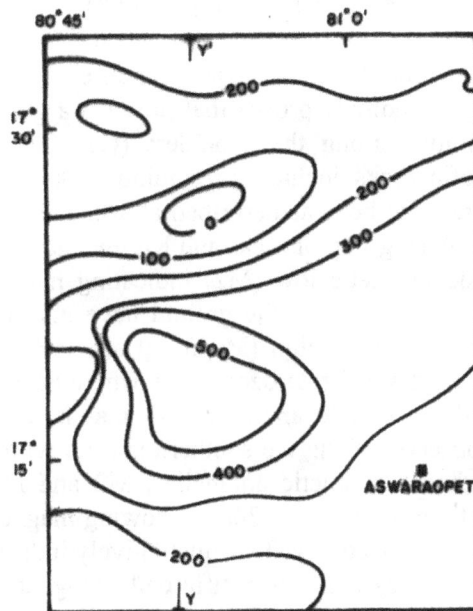

Fig. 7.35(a) Total intensity map of the southern part of the Godavari basin-Chintalpudi sub-basin showing a low towards the north indicating induced magnetization.

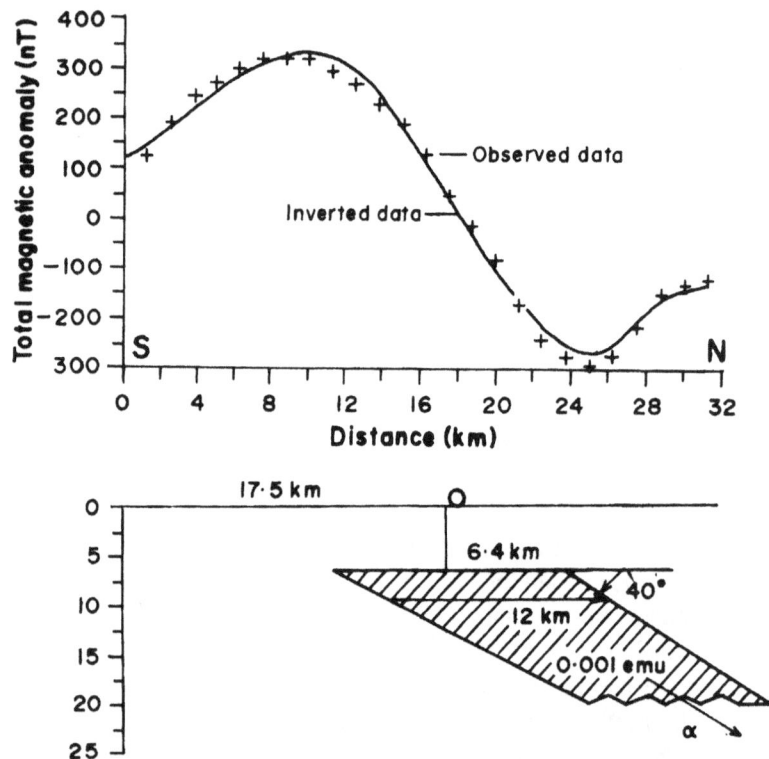

Fig. 7.35(b) Magnetic residual anomaly along a profile YY′ (Fig. 7.35(a)) assuming a regional field of 200 nT. It is modeled for a tabular body at a depth of 6.4 km that is almost the same as the depth of high density rocks (Fig. 7.32(a)). Its susceptibility of 0.001 emu indicates mafic intrusion (Mishra et al., 1987).

7.7.4   Airborne Magnetic Profile from Hyderabad to Puri across the GPB

(a)   A magnetic anomaly of large amplitude ($\sim$ 550 nT, Fig. 7.36(a)) was recorded at a height of about 9000′ ($\sim$ 2.7 km a.m.s.l) in an airborne magnetic profile from Hyderabad-Puri along southern margin of the Proterozoic rocks (Karimnagar Granulite Belt) (Mishra et al., 1971). Fig. 7.36(a) also shows the Bouguer anomaly profile that presents a gravity low over Godavari Gondwana basin (GD, G2) and high along the shoulders (G1, G3) related to Proterozoic metasediments and mafic/ultramafic rocks including granulite rocks on the two sides of the Karimnagar and Bhopalpatnam granulite belts as described above. The gravity high (G1) on the southern side, where Proterozoic Pakhal group of rocks and Karimnagar granulite belt are located, coincides with the large amplitude magnetic low (M2) indicating mafic rocks in this section. Several large amplitude magnetic anomalies on the ground have also been reported from the Proterozoic Pakhal group of rocks in this section (Mishra et al., 1987). The gravity high, G3 related to the northern flank of the Godavari Proterozoic basin where Bhopalpatnam granulite belt is exposed also coincides with the magnetic anomaly, M3 that is a magnetic high while a dominant magnetic low (M2) is observed along the southern margin with a small magnetic high M1 towards the south. These airborne magnetic anomalies, M2 and M3 are similar to those observed along the ground magnetic profile (Fig. 7.34(b)) showing magnetic low and high along the southern and the northern flanks, respectively. This qualitatively indicates that the sources for these magnetic anomalies (M2 and M3) are either differently magnetized or dip in opposite directions that can be clarified only by modeling the respective causative sources as shown in Fig. 7.34(b).

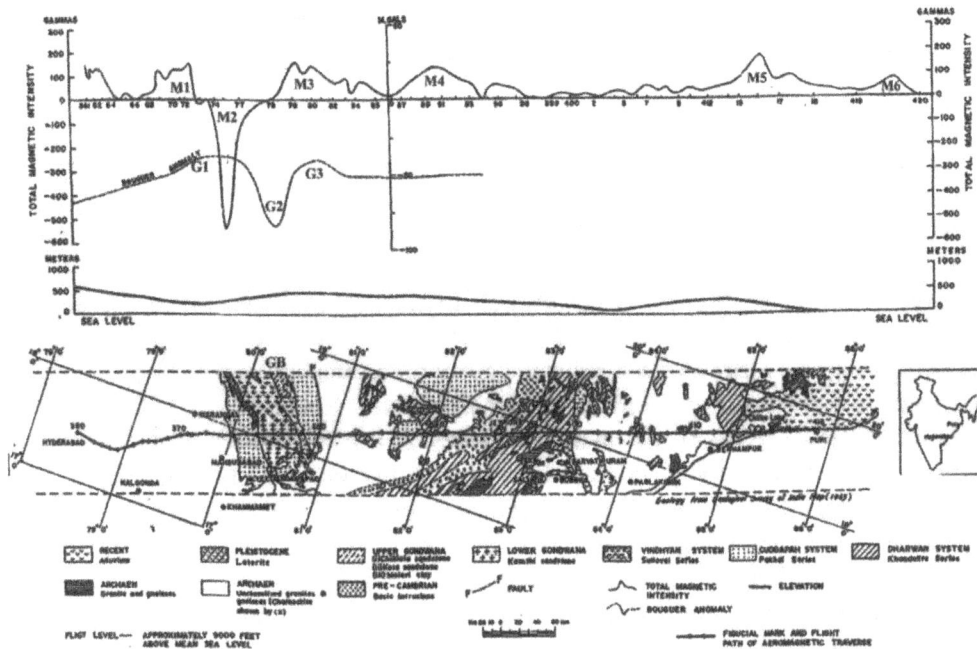

Fig. 7.36(a) Airborne magnetic profile across the GPB from Hyderabad-Puri recorded at a height of 9000'
(~ 2.7 km amsl). The magnetic anomalies, M1, M2, and M3 coincide with the gravity highs, G1 and G3, related
to the Proterozoic rocks along the shoulder of the rift basin. The gravity low, G2 is related to the Gondwana
sediments.

(b) The magnetic low, M2 is modeled using an inversion scheme for tabular dyke model
(Fig. 7.36(b)) that suggests a mafic intrusive of high susceptibility 3.0×10^{-3} emu at a depth of
about 5.3 km below m.s.l dipping almost 60° towards the NE that may represent mafic/ultramafic
intrusive almost at same depth (~ 6 km) where high density (Fig. 7.32(a) and (b)) and high
susceptibility (Fig. 7.34(b)) bodies have been modeled. The remnant magnetization of this body,
inclination = – 40° and declination= 300° required to match the observed and the computed field
is typical of Mesoproterozoic rocks (Vindhyan sediments, Mishra, 1965) in this region indicating
the period of collision and thrusting. It is also similar to those reported for dykes of the same
period in this region (Rao et al., 1990).

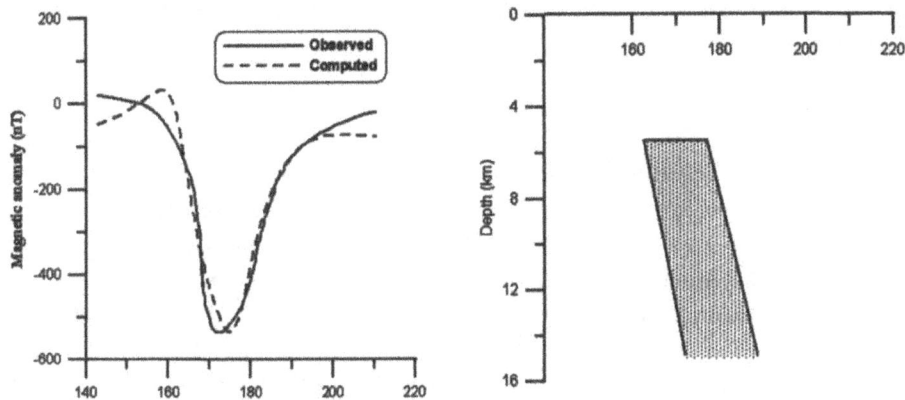

Fig. 7.36(b) Computed model for airborne magnetic anomaly, M2 (Fig. 7.36(a)) observed over the southern
section of the GPB showing a NE dipping tabular body at a depth of about 6 km in the basement for a remnant
magnetization of inclination = – 40° and declination = 300°.

7.7.5   Proterozoic Collision Tectonics across the GPB, Proterozoic Basins, and Triple Junction

(a)   Gravity highs over Proterozoic terrain caused by high density and high susceptibility bodies in the upper crust and exposed lower crustal granulite rocks and gravity lows over Archean craton caused by crustal thickening suggest collision tectonics of Meso-Neoproterozoic period between the Bhandara-Bastar and the Dharwar cratons. It has given rise to this mobile belt with Bhopalpatnam granulite belt of this period and Pakhal-Sullavai metasediments along the GPB. This convergence is approximately NE-SW direction (Fig. 7.6) perpendicular to the strike of the Proterozoic basin that might be related to convergence along the SMB towards its north as discussed above. The resistive lithospheric keel (D, Fig. 7.33(b)), SW of the GPB supported from the low P-wave residual as discussed above may represent part of the subducted crustal rocks that are resting there. Prior to it, there might be another convergence phase during Paleoproterozoic period similar to that along the SMB that is indicated by large mafic/ultramafic intrusives in the basement delineated from the airborne magnetic anomaly as described above and Karimnagar granulite belt of 2.2-2.4 Ga.

(b)   The same model as used in the case of the Lower and Upper Vindhyan group of rocks (Fig. 7.23(b) and (c)) can also be used to explain the formation and the hiatus between the Pakhal and the Sullavai group of rocks of Paleo-Meso and Neoproterozoic periods, respectively, along the Godavari Proterozoic Belt separating the Bhandara-Bastar and Dharwar cratons towards the NE and the SW, respectively (Fig. 7.31). The Pakhal super group of rocks with large scale mafic and ultramafic components formed during the rifting between the Dharwar and the Bhandara-Bastar cratons, while the Sullavai group of rocks formed during convergence after collision and mountain building that explains large scale disturbances of Pakhal metasediments and hiatus between the two.

(c)   The gravity highs due to central and the western parts of the SMB (H8 and H7) and the GPB (H10 and H11, Fig. 7.8(b)) separating the Bundelkhand craton towards the north, the Bhandara-Bastar craton towards the SE, and the Dharwar craton towards the SW form a triple junction of Mesoproterozoic period around Nagpur (L12, Fig. 7.8(b)). This junction is characterized by a circular/semicircular large amplitude gravity low that may be associated with felsic intrusives in the basement. These sutures at this junction make almost ~ 120° from each other, that is one of the characteristics of present day triple junctions of plate boundaries. With the Bundelkhand craton subducting southwards under the Dharwar and the Bhandar-Bastar cratons across the SMB and the Bhandara-Bastar craton subducting under the Dharwar craton across the Godavari Proterozoic Mobile belt, their junction (Fig. 7.7 and 7.8(b)) represents a stable SSS type triple junction during Mesoproterozoic period as shown in Fig. 7. 2.

7.8   Peninsular Shield: Eastern and Western Dharwar Cratons and Deccan Volcanic Province

7.8.1   Geology and Tectonics

(i)   Geology Investigations and Crustal Structures

(a)   The Dharwar craton (Fig. 7.37) consists of two parts as the Eastern and the Western Dharwar cratons (EDC and WDC) separated by a shear zone. It typically consists of a granite-greenstone schist belts with tonalite-Trondhjemite-Granodirite (TTG) gneisses of 3-3.4 Ga as basement. A large section of the Dharwar craton towards the north is covered by the Deccan volcanic province of Late Cretaceous (Fig. 7.6 and 7.7). The EDC is characterized mainly by large linear K-granite plutons such as Closepet granite of Neoarchean-Paleoproterozoic period (2.6-2.5 Ga, Jayananda et al., 2000) and some schist belts such as the Kolar schist belts of Neoarchean-Paleoproterozoic age. This figure also shows the shear zone between the EDC and the Western Dharwar Craton (WDC) and the transition zone between the WDC and the EDC towards the north and the

Southern Granulite Terrain (SGT) towards the south. The granitic intrusions and schist belts of EDC show a curvilinear trend with eastward convexity. The schist belts of EDC such as the Kolar schist belt south of the Cuddapah basin are economically important as they contain gold mineralization. Radhakrishna and Naqvi (1986) have considered the EDC as a mobile belt based on foliations and structures. This figure also shows the SGT and the Cuddapah basin and the Eastern Ghat Mobile Belt that are discussed in the next sections. Chadwick et al., (2000) have considered an oblique convergence from east to west across the shear zone between the EDC and the WDC. Various granitic batholiths of Neoarchean to Paleoproterozoic period of EDC might have formed as subduction related magmatism between the western and the Eastern Dharwar cratons across the shear zone between them (Mishra, 2006).

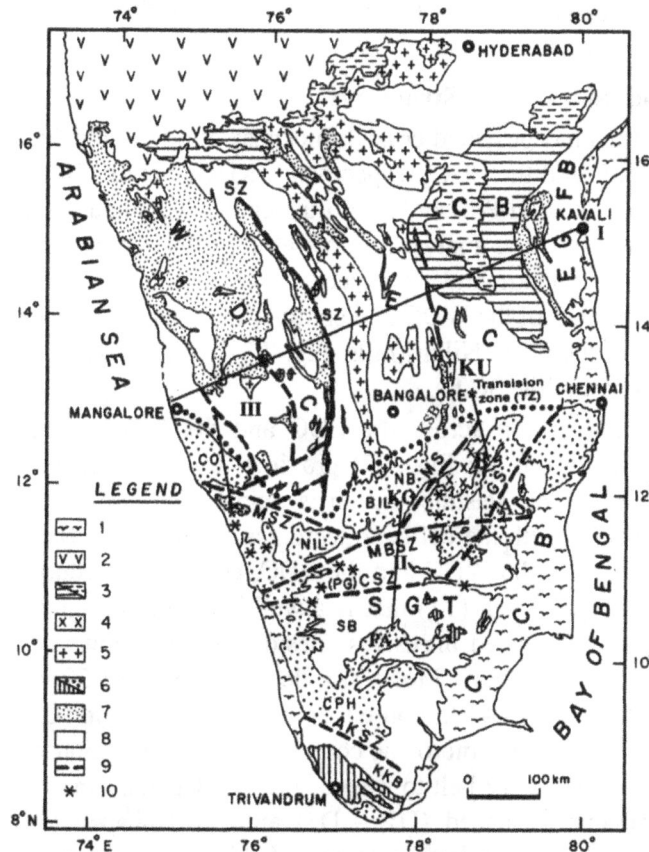

Fig. 7.37 Simplified geological map the South Indian Shield showing the Dharwar craton and the Southern Granulite Terrain (SGT), which is to the south of the Transition zone (TZ). Various abbreviations are as follows: AKSZ: Achankovil Shear Zone, AS: Attur Shear, BIL: Biligirirangan Hills, CB: Cuddapah basin, CCB: Cauvery coastal basin, CO: Coorg Hills, CPH: Cardoman-Palani Hills, CSZ: Cauvery Shear Zones, EDC: Eastern Dharwar Craton, EGFB: Eastern Ghat Fold (mobile) Belt, GS: Gangavalli Shear Zone, KKB: Kerala Khondalite Block, MS: Mettur Shear Zone, MBSZ: Moyar Bhavani shear Zone, NB: Northern Block, NIL: Nilgiri Hills, PCSZ: Palghat Cauvery Shear Zone, PG: Palghat gap, SB: Southern Block, SZ: Shear Zone between EDC and WDC, S: Sangola, WDC: Western Dharwar Craton. Various geological formations and symbols are as follows: 1– Cretaceous– recent sediments, 2– Cretaceous- Eocene, Deccan basalts, 3– Proterozoic metasediments, 4– alkaline complexes and Carbonatites, 5– Granites (Un differentiated), 6– Charnockites-Granulites/Khondalites, 7– Archaean greenstone belts, 8– Gneisses, 9– Major lineaments and 10– Epicenter of some earthquakes associated with shear zones in the SGT. Charnokites-Granulites north of the PCSZ are primarily Archean granulites (~2.5 Ga) while those south of it are primarily Neoproterozoic (~ 0.55 Ga). It also shows Profiles I (Kavali-Udipi, north of Manglore) across the Cuddapah basin and Dharwar craton, Profiles II (Kuppam-Bommidi and Kolattur-Palani) across the eastern part of the SGT and Profile III across the western part of the SGT. KU: Kuppam, B: Bommidi, Ko: Kolattur, PA: Palani.

(b) The WDC is characterized primarily by schist belts (Fig. 7.37) of Meso-Neoarchean period (Dharwar group 2.9-2.6 Ga, Anil Kumar et al., 1996). There are some small clusters of old schist belts of Mesoarchean time (Sargur group, 3.4-3.0 Ga) in the southern part of the WDC. Schist belts primarily consist of metasediments of platform sediments with mafic and granite intrusives along margins such as the Chitradurga schist belt (Naqvi and Hussain, 1972). They usually show gravity lows related to metasediments and felsic intrusives and crustal thickening (Fig. 7.8(b)). Naqvi et al., (2006) and Naqvi and Rana Prathap (2007) have reported adakites from the schist belts of the EDC and WDC, respectively that indicate active continental margin during Neoarchean-Paleoproterozoic period. Based on geochemical characteristics of metagreywackes of the Ranibennur basin in Shimoga Schist Belt (Dharwar group) of the WDC, Hegde and Chavadi (2009) have suggested that this basin formed near a magmatic arc in a continental arc setting.

(ii)   Geophysical Investigations and Crustal Structures

(a) Receiver function analysis has provided crustal thickness of 54-55 and 31-32 km, under the southern part of the WDC and the EDC, respectively (Kiselev et al., 2008). They have also suggested a felsic crust based on Poisson's ratio under the EDC and mafic crust under the WDC. Based on receiver function analysis, Gupta et al., (2003) provided crustal thickness increasing consistently from 35 km under the Deccan Volcanic Province south of Pune to 54-55 km under the southern part of the WDC and south of the transition zone and Moyar Shear under the western part of the SGT (Nilgiri hills, Fig. 7.37), that reduces to 43-45 km further south under the central part of the SGT.

(b) A deep seismic sounding profile across the EDC and the WDC from Kavali to Udipi (Mangalore) (Fig. 7.37) (Profile VI; Fig. 7.7) provided some significant results from this region (Kaila et al., 1979). This seismic section is reproduced here (Fig. 7.38) in order to connect some of its features with the collision tectonics. It shows a thrust type feature (T2) under the shear zone between the WDC and the EDC, east of Chitradurga similar to those along the Eastern Ghat Mobile Belt (T1) that is a well known thrust giving rise to thrusted high velocity rocks and deep seated intrusives. Besides, it also shows high velocity at the surface under the EDC and a shallow Moho at a depth of about 31-32 km (C2) under this thrust (T2) increasing to about 40-41 km on either side of this section (C1 and C3), that are typical characteristics of Archean-Proterozoic collision zone. The section with shallow Moho extends from the Chitradurga schist belt to west of the Cuddapah basin almost for about 120 km. There are several domal shaped (D1 – D3) and bowl shaped (W1-W3) reflectors indicating extensional and compressional tectonics. Schist belts may represent extensional phase while thrusts are in response to compressional tectonics

(c) An airborne regional magnetic profile from Manglore to Madras recorded at a height of 9000' a.m.s.l (~ 2.75 km) is given in Fig. 3.32. It suggests two fault kinds of anomalies along the eastern margin of Closepet granite (E) and over the Chitradurga schist belt (D). It also shows moderate magnetic anomalies towards the east of the Kolar schist belt and west of the Chitradurga schist belt indicating mafic crust in these sections. However, the section between the two does not show any significant magnetic anomaly implying absence of mafic rocks/minerals in this section. Bouguer anomaly along the same profile shows a regional low over the Western Ghat (Kudremukh) with increasing elevation indicating crustal thickening due to isostatic compensation as has also been shown in the seismic section (Fig. 7.38). Airborne magnetic map of this region (Mishra et al., 2006; Fig. 7.47) also shows a NW-SE trend west of Bangalore and NE-SW trend to its east indicating different terrains on either sides of it.

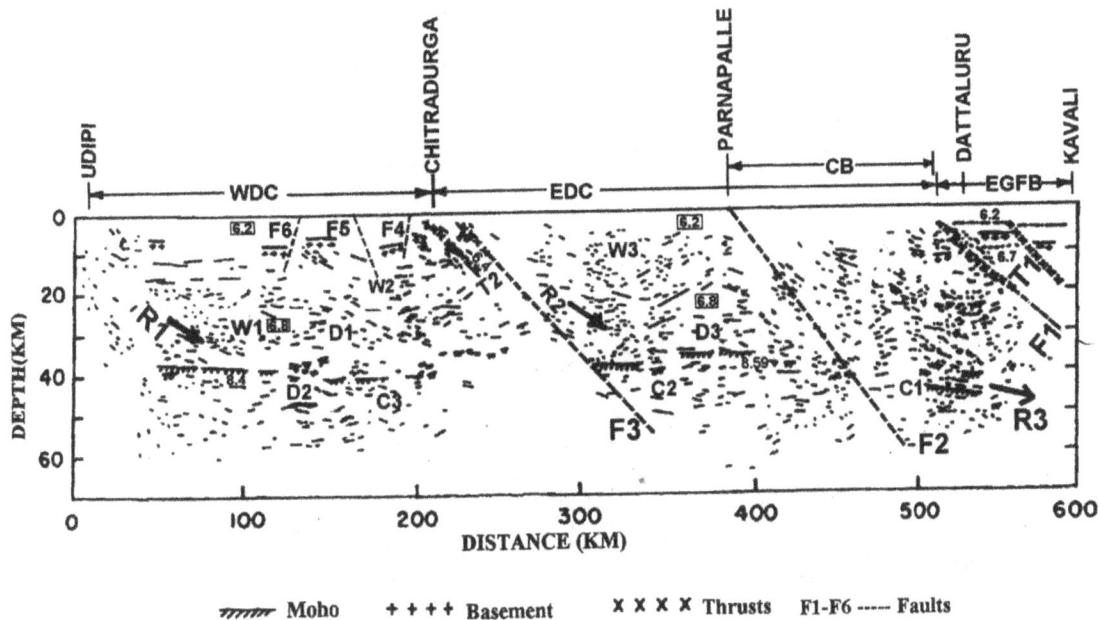

Fig. 7.38 Seismic section along Kavali-Udipi (Mangalore) profile (VI, Fig. 7.7 and 7.37; modified from Kaila et al., 1979). Velocities in km/sec are given in between from Reddy et al., (2000). It shows domal shaped D1, D2, D3 and downwarped W1, W2, W3 reflectors, R1, R2, R3 are eastward dipping reflectors and T1 and T2 and F1-F3 are west verging thrusts and faults. C1 and C3 indicate zones of crustal thickening of 40-41 km while section C2 between faults F2 and F3 shows thin crust of 31-32 km.

(d) Based on heat flow studies, Roy and Rao (2000) suggested that the Archean Peninsular Shield south of the Narmada-Son lineament is characterized by low heat flow (25-50 mW/m^2) compared to the northern part such as the ADMB and the SMB. This may be attributed to the effect of the relatively thick lithosphere in the southern part as has been discussed in the next section. However, the high heat flow and neotectonic activities and thin lithosphere in the northern part has been attributed to crustal bulge due to lithospheric flexure of the Indian plate under the Himalayas as discussed in Section 7.11.

7.8.2 Gravity and Magnetic Fields over a Typical Granite Pluton (Hyderabad) and Earthquakes

(a) The EDC is occupied by a large number of granite plutons (Fig. 7.37), the largest being the closepet granite that extends from the transition zone in the south to the northern boundary of the exposed EDC. To assess the gravity and magnetic fields over a typical granite pluton of the EDC, detailed surveys of these fields were conducted over the exposed Hyderabad granite pluton at station spacing of 0.5 to 1.0 km. A residual Bouguer anomaly map (Fig. 7.39(a)) shows several centers of gravity lows, L1-L5 representing low density rocks in these sections. They may represent individual plutons forming the batholith. There are some gravity highs such as H5 that represent mafic dykes exposed in this section. The gravity highs along the eastern margin represent the gravity highs related to the southern shoulder of the Godavari basin that have been described in Section 7.7.1 A pole reduced total intensity magnetic map (Fig. 7.39(b)) shows a large magnetic low over this batholith with adjoining highs that indicates that this batholith has intruded a relatively mafic rock surrounding it, that is Peninsular gneisses.

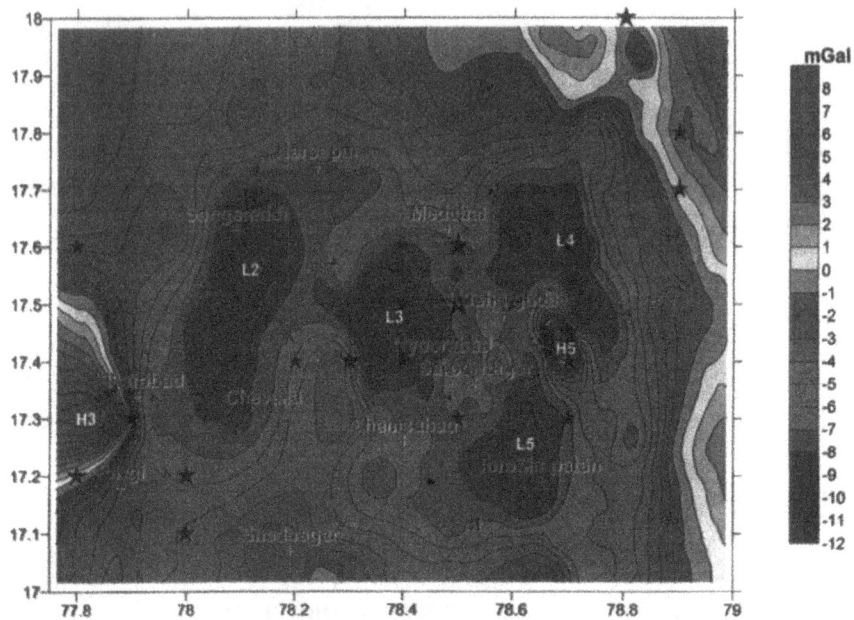

Fig. 7.39(a) Residual Bouguer anomaly (in mGal) of the Hyderabad granite showing lateral extent of various isolated granite plutons and distribution of micro-seismic activity around Hyderabad that ia mainly located at the contacts of the plutons (Singh et al., 2004).

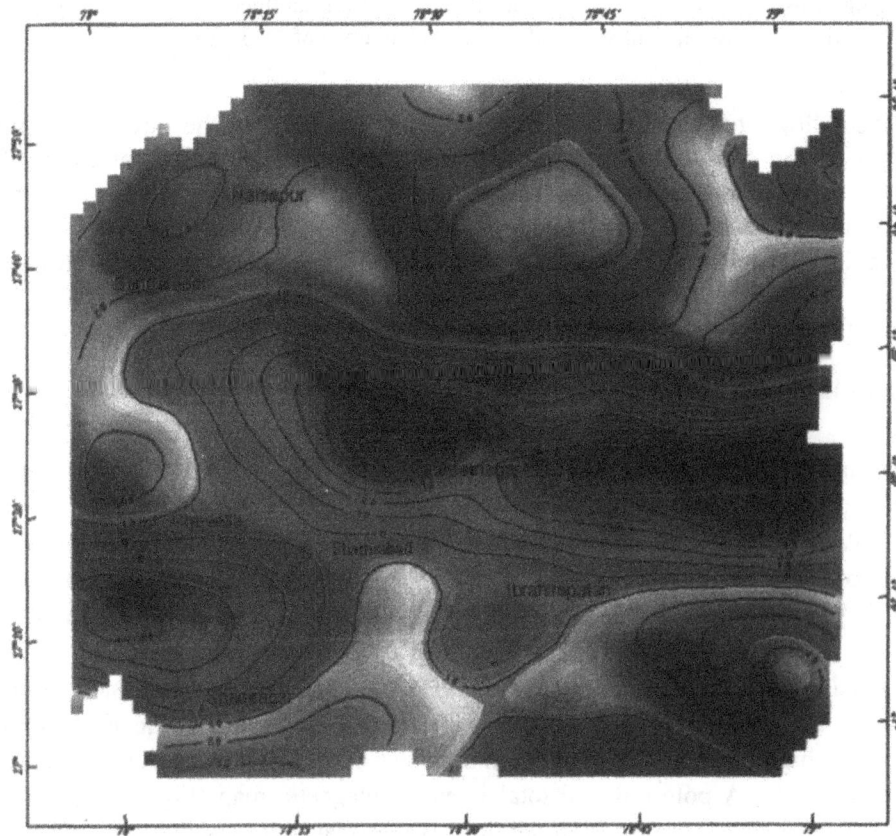

Fig. 7.39(b) Pole-reduced map of the low-pass filtered total intensity magnetic anomaly of the region showing a magnetic low over the Hyderabad granite pluton delineated from the gravity anomalies (Singh et al., 2004).

(b) A 3-dimensional model of the roots of the granite pluton (Fig. 7.39(c)) computed from the spectral analysis of Bouguer anomaly (Section 6.1, Fig. 7.39(a)) for a density contrast of -0.1 g/cm^3 suggests the thickness of batholith as 4-5 km with discrete roots of plutons extending up to 10-11 km. These roots may represent the discrete centers of intrusion forming a batholith at the surface. Roots of plutons are usually conical shapes that explain the semicircular or elliptical nature of batholiths. Fig. 7.39(a) also shows microseismic activity in this region that mainly coincides with the gradient of gravity lows which suggest their concentration along margins of the granite plutons acting as stress concentrators. It shows the importance of intrusives for seismic activity in stable continental regions.

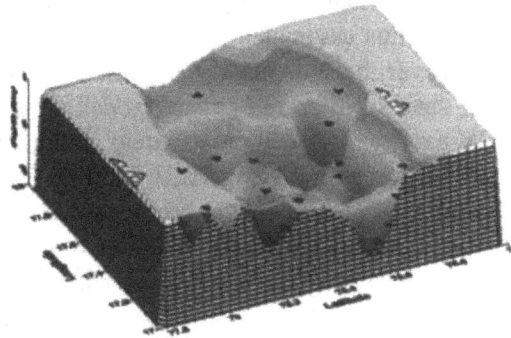

Fig. 7.39(c) Three-dimensional subsurface relief of the Hyderabad granite plutons based on the residual gravity anomaly (Fig. 7.39(a)) showing plutons having individual roots extending up to 10-11 km that are connected at surface to form a composite batholith (Singh et al., 2004).

7.8.3 Deccan Volcanic Province: Bouguer, Airborne Magnetic and Conductivity Anomalies and Seismotectonics – Koyna and Latur Earthquakes

A large section of the Dharwar craton in the western part is occupied by the Deccan Volcanic Provinces (DVP) of Late Cretaceous time (Fig. 7.6 and 7.7). It caused the breakup of Seychelles from India as has been discussed in Section 5.3. The actual line of eruption has been the Chagos-Laccadive Ridge striking the continental shelf of India south of Mumbai. However, it spreads all along this section both in the ocean and the western part of the Indian continent as has been described in Chapter 5. This rock being young has not played any role in the formation of Indian cratons and mobile belts and therefore, has not been discussed in detail in the present context. However, due to its importance in Indian geology, and the basement under it being formed by the Dharwar craton, the Bouguer anomaly and other geophysical data observed over this region are discussed and modeled along a profile to understand the significance of these anomalies. A detailed account of its geochemistry and geodynamics is given in Section 9.2 while describing the Deccan trap province of Saurashtra and subtrappean Mesozoic sediments.

(i) Bouguer Anomaly and Crustal Structures

(a) It is interesting to observe that most of the gravity anomalies, both lows and highs (L13, L14 and H13, and H14, Fig. 7.8(b)) observed in this section, have been attributed to the sources in the basement mainly the extension of felsic and maic intrusives, respectively, associated with the EDC and the WDC exposed along the southern and the eastern margins of the DVP. Normally, being high density rocks, they should have given rise to several gravity highs and lows related to variations in the thickness of the basalt and density inhomogeneity within the basalt. The absence of such anomalies indicate that neither is there any sharp and sudden change in thickness nor much change in the density contast of various rock types forming this province. However, contrary to this several gravity highs have been observed over similar rock types in Saurashtra and Kachchh (H27 and H28, Fig. 7.8(a)) related to volcanic plugs of alkaline complexes associated with Deccan tap eruption in these regions that have been discussed in Chapter 9. This implies that probably the flows, being relatively thin and low density compaed to plugs, do not produce as significant gravity anomalies as the latter that extend throughout the crust and consist of high density rocks.

(b) A detailed Bouguer Anomaly of the Deccan Volcanic Province (DVP) is given in Fig. 7.40(a) and is based on an earlier regional map (Kailasm et al., 1972) and new data acquired along three profiles I-III recorded with a station spacing of about 1 km (Tiwari et al., 2001). It shows two major gravity lows, A and B, referred to as Koyna and Kurduwadi gravity lows, respectively separated by Sangola gravity high, C. These gravity lows have been attributed to sediments associated with rift basins under the DVP (Krishna Brahmam and Negi, 1973). However, to understand the causative sources for these gravity lows, they were modeled along these profiles and crustal model along profile II is presented in Fig. 7.40(b). This is an E-W gravity profile approximately along 18°N across Kurduwadi gravity low that corresponds to the gravity low, L13 in Fig. 7.8(b). It was modeled constrained from a seismic profile almost along the same profile (Kaila et al., 1981).The computed crustal model shows a crustal thickness of about 38 km in the eastern part increasing to 40 km under the Western Ghat that is associated with the regional field consistently decreasing towards the west. It may, therefore, be related to isostasy where a change in elevation of about 200-250 m is noticed. The crustal model also shows relatively high density rocks in the lower crust (3.1 g/cm^3) and a low density body (3.2 g/cm^3) in the upper mantle below the Moho along the coast that represent underplated crust and depleted upper mantle, respectively, which is typical of such volcanic provinces. The Kurduwadi gravity low in the central part of the profile is modeled due to a low density body (2.62 g/cc) of a conical shape that may represent a granitic intrusion.

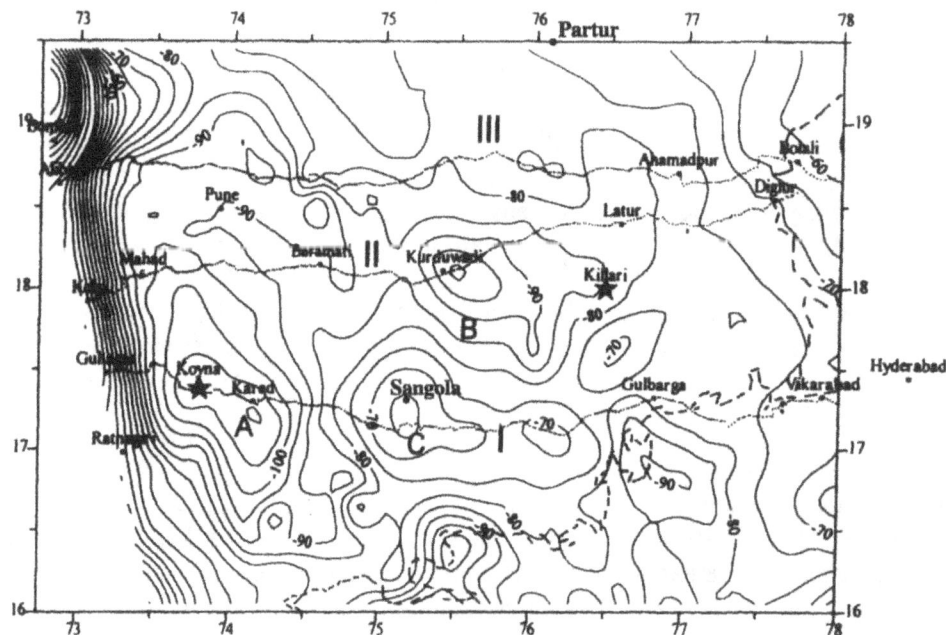

Fig. 7.40(a) Bouguer anomaly map of Deccan Volcanic Province (Kailasam et al., 1972) modified based on specially recorded gravity profiles I, II and III at station spacing of 0.5-1.0 km for delineating crustal structures. Star indicates epicenter of Killari (Latur) earthquake of 1993 and Koyna earthquake of 1967. It shows two major gravity lows, A and B, known as the Koyna and the Kurduwadi gravity lows with an intervening Sangola gravity high, C. (Tiwari et al., 2001).

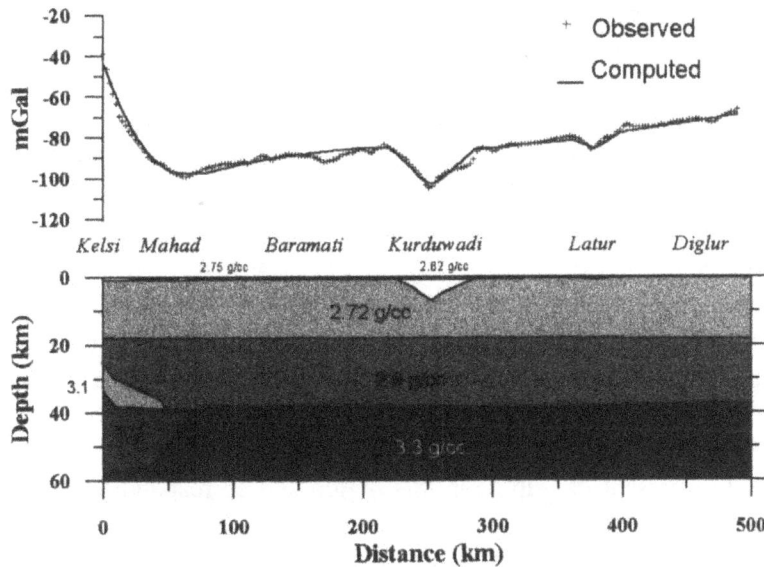

Fig. 7.40(b) Observed Bouguer anomaly along Profile II (~18⁰ N) across the Kurduwadi gravity low and computed gravity field due to a plausible crustal model given below. Densities are given in g/cc (Tiwari et al., 2001).

(ii) Seismotectonics – Koyna and Latur (Killari) Earthquakes

The Deccan Volcanic province has been seismically active with two major earthquakes of Koyna and Latur recorded during recent time that are discussed below:

(a) The Kurduwadi gravity low is a linear low that extends even outside the extent of the exposed Deccan Volcanic Province, south of Gulbarga but is intercepted by an almost perpendicular gravity high south of Killari. Their intersection might be responsible for Killari earthquake (Fig. 7.40(a)) of magnitude 6.3 on September 30, 1993 (Gupta, 1993) at shallow depth of 5-6 km. This gravity high of small amplitude may represent a mafic intrusive such as a dyke and intersections of different intrusives tend to act as stress concentrators and trigger seismic activity. This earthquake is known for its occurrence in a stable continental region and the great devastation that killed approximately 11,000 people. The epicenter of the Koyna earthquake of December 10, 1967 (Fig. 7.40(a)) lies in the Western Ghat that is affected by recent tectonic activity (Gunnell et al., 2003; Valdiya et al., 2001) related to Himalayan orogeny (Mishra and Vijai Kumar, 2005) and is therefore affected by seismic activity. There has been continuous seismic activity in and around Koyna since the impounding of a large reservoir in this region that might have triggered these seismic activities (Gupta et al., 1967; Gupta, 2002). They are mainly confined in the upper crust (< 10 km). Veeraswamy and Raval (2005) suggested that the epicenters of these earthquakes lie on Precambrian tectonic boundaries causing these earthquakes. Bouguer anomaly map indicates an intersection of a N-S trend of Westsern Ghat with a NE-SW trend (deflections of contours) in the epicentral region of Koyna earthquake (Fig. 7.40(a)).

(b) The airborne magnetic map of the Koyna region recorded at a height of 2134 m amsl and 4 km spaced profiles (Fig. 7.40(c), Negi et al., 1983; Nayak et al., 2006) has delineated a magnetic gradient, MG1 that represents a shear/fractured zone that may be the extension of the shear zone between the Eastern and the Western Dharwar cratons as inferred based on gravity highs in the next section. This shear zone extends up to the Koyna River and dam and would therefore facilitate the percolation of water in the hypocentral zone of seismic activity in this region. Another interesting magnetic anomaly is M1 that is a bipolar magnetic anomaly related to a mafic intrusive in the shear zone, which is different from Deccan trap rocks forming the country rock in this section. This intrusive located at the gradient of the bipolar magnetic anomaly coincides with the Koyna river and would therefore also facilitate the percolation of water in this region and also act as stress concentrators for slip to occur. Normally, the Deccan trap covered region shows haphazard magnetic anomalies due to hetrogenous magnetization (sharp changes in susceptibility) of these rocks but intrusives in them can be identified by the definite nature of magnetic anomalies due to their homogenous magnetization especially in airborne magnetic surveys that reduces surface noise. Pandey et al., (2009) have attributed the seismic activity of the Deccan trap covered region to the stress generated from the continous uplift of the crystalline basement and erosion since Precambrian time. The present day uplift in this region and corresponding seismic activity may be related to the eruption and erosion of Deccan trap and its loading and unloading effect as it happens in case of glaciers and their melting (Mishra, 1989).

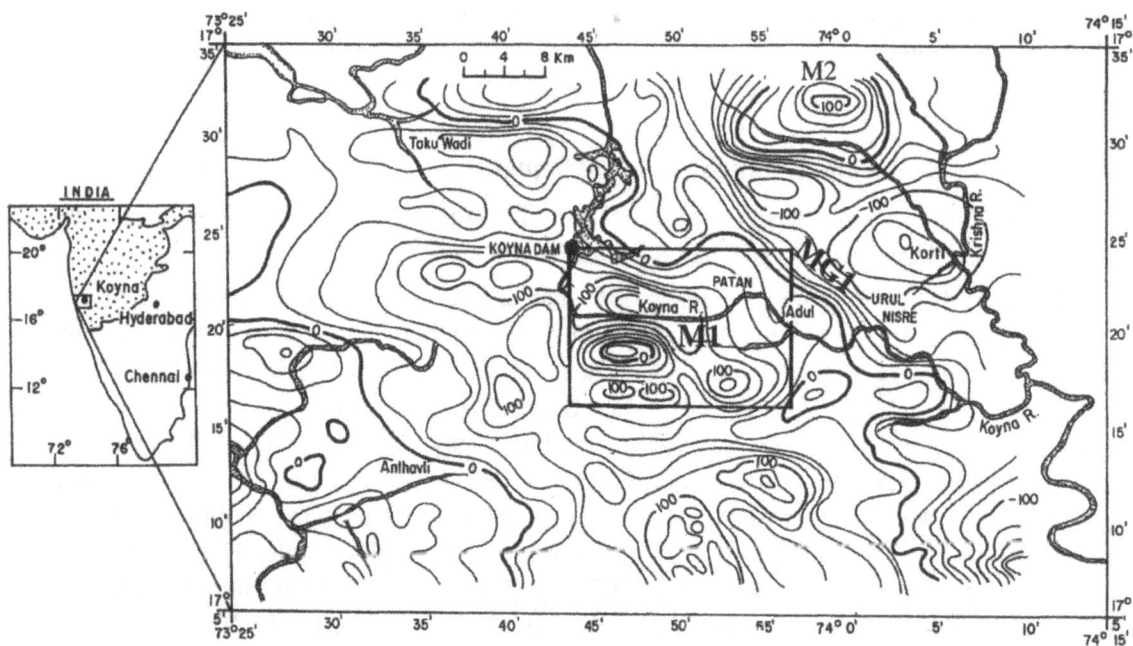

Fig. 7.40(c) Airborne total intensity map of part of Deccan Volcanic province recorded at 2134 m amsl. MG1 is the magnetic gradient coinciding with the Koyna Shear Zone and M1 represents an intrusive mafic body along the Koyna River and M2 is another intrusive on other side of the Shear zone, MG1 (Nayak et al., 2006).

(c) Magnetic intrusion and shear zone in the epicentral region of the Koyna seismically active zone is supported by conductivity anomalies that show relatively high conductivity in the hypocentral region of Koyna seismic activity (Fig. 7.40(d); Guhaghar-Sangole Profile, Patro and Sarma, 2009). Conductivity anomalies show two resistive blocks, R1 and R2 in the crust with an intermediate conductive zone, C1 where the hypocenters of Koyna seismic activity are located. The resistive blocks, R1 and R2 may represent mafic intrusives while the relative conductive zone in between coincides with the Koyna shear zone and may represent the abundance of fluids in this section highlighting the role of fluids in seismic activity of this region. The intermediate conductive zone extends up to the lithospheric mantle. The samples of the crystalline basement rocks from the boreholes in the epicentral zone of Latur-Killari earthquake provided 2% of CO_2 fluid composition (Pandey et al., 2009) that might be responsible for the crustal and the upper mantle conductors in this section.

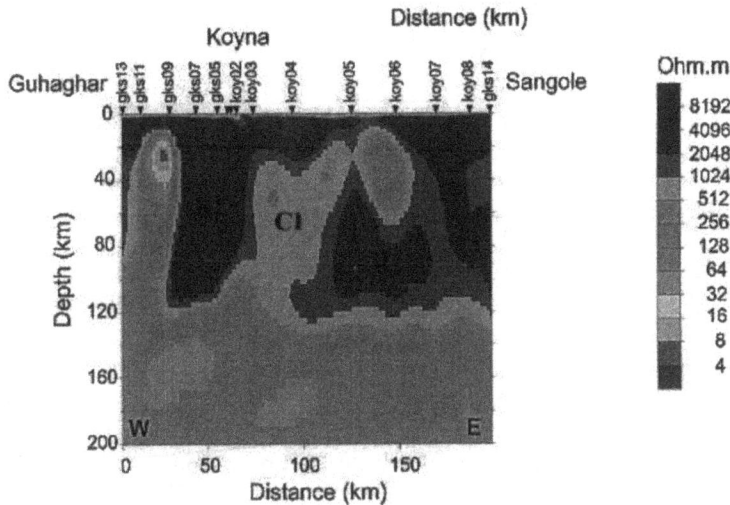

Fig. 7.40(d) A conductivity profile Guhaghar (along the coast)-Koyna-Sangola (Fig. 7.40(a)) shows two resistive blocks (R1 and R2) extending from surface to the lithospheric mantle with a conductive section (C1) in between that corresponds to fluid filled shear zone as cause for Koyna seismic activity. [Colour Fig. on Page 796]

(d) In order to understand the shallow structures in the epicentral region of the Killari earthquake and its causes, a detailed Bouguer anomaly map of this region was prepared based on gravity survey specially conducted for this purpose at 0.5-1.0 km station interval (Fig. 4.2, Mishra et al., 1994). It shows several gravity highs, H1-H3 and lows, L1-L5 of small amplitude that does not find reflection in the regional map. The epicenter of the Killari earthquake coincides with the gradient of the gravity high, H1 and low, L1 that implies a shallow fault in this section. This figure also shows two zones of high conductivity at a depth of 9-10 km inferred from a magnetotelluric profile CC' in the central part marked by vertical lines (Sarma et al.,1994). A gravity profile, AA' is analysed by first removing the regional field from it based on a plane and modeling the residual field for shallow structures (Fig. 7.41(a), Mishra et al., 1998). It shows some high density intrusives below the 300-400 m thick Deccan trap that may represent mafic dykes and act as stress concentrators. Besides, there is a low density layer at a depth of 8-11 km that has intruded in the epicentral zone up to 4-5 km. This layer might be synonymous with the high conductive layer that has intruded in the epicentral zone through faults and fractures. Presence of fluids in the hypocentral zones is known to facilitate slip and cause earthquakes and the same may be true in the present case also (Gupta et al., 1996). Another important observation after this earthquake was the high helium anomaly of 38 ppm against a background level of 0.2 ppm in soil gas samples from a depth of 1.5 m in vicinity of surface ruptures (Rao et al., 1994) suggesting that continous monitoring of such gases in seismically active regions may provide a clue to impending events.

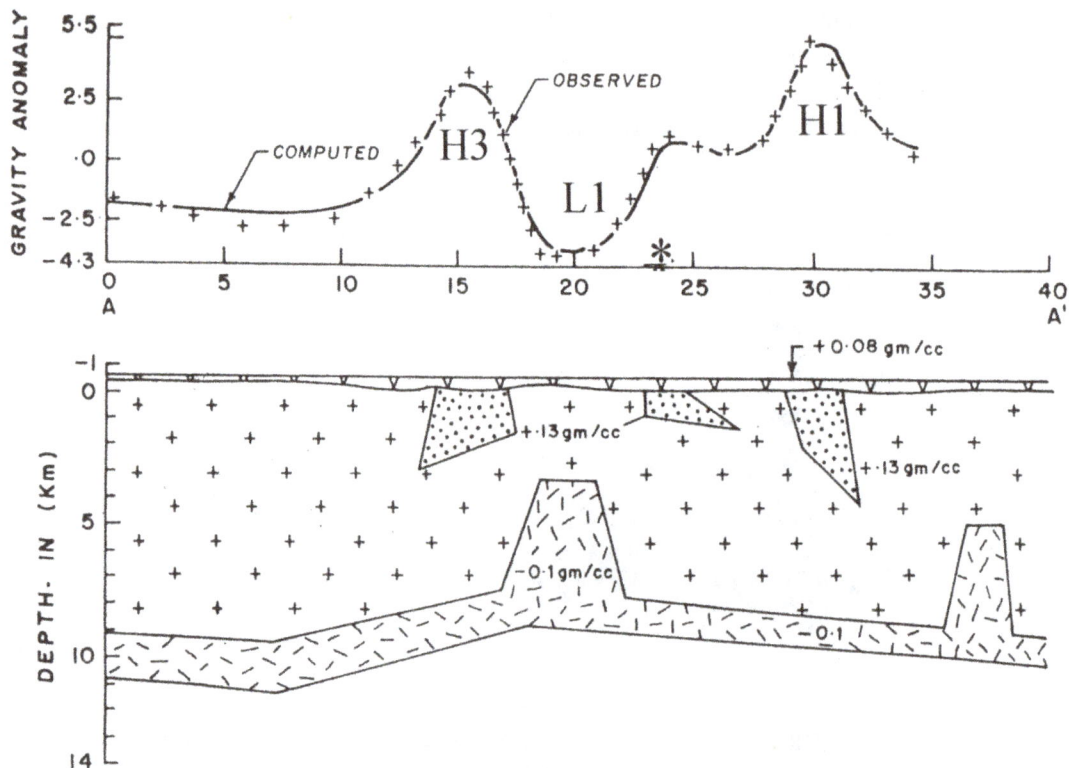

Fig. 7.41(a) The observed residual and the computed gravity fields due to a shallow crustal section along Profile AA′ (Fig. 4.2) showing high density intrusives under the Deccan trap and a low density layer at depth of 9-11 km intruding up to 4-6 km that may be synonymous with high conductive fluid filled zone (Mishra et al., 1998).

(e) Four bore holes drilled in the surface rupture zone of the Killari earthquake up to a maximum of ~ 600 m provided some significant information that is as follows (Gupta et al., 2003). Thickness of the trap in this section is 300-400 m of bulk density 2.72 g/cc and it was associated with a pre-existing fault that might be part of the Kurduwadi Lineament. Heat flow measurements indicated normal heat flow for cratonic parts (43 mWm^{-2}, Roy and Rao, 1999) indicating the absence of contribution from Deccan trap eruption on it. In situ horizontal stress measurements provided values of gradients of S_{Hmax} and S_{Hmin} equal to 39 and 21 MPa/km, respectively (Srirama Rao et al., 1999) that are the same as in other parts of Indian continent like Hyderabad (Gowd et al., 1986), but normally ~ 30% higher compared to other intra-continental regions that may explain for more and higher magnitude Stable Continental Region earthquakes in the Indian continent.

7.8.4   Crustal Structures – Eastern and Western Dharwar Cratons and Madagascar

(a) The gravity anomalies of schist belts of WDC are characterized by large gravity lows and some highs that has been attributed to felsic and mafic bimodal volcanics, respectively, in rift kind of settings (Subrahmanyam and Verma, 1982; Mishra, 1990). The gravity highs, H15 (Fig. 7.8(b)) along the shear zone between the WDC and EDC extend almost from the transition zone (Bangalore) in the south up to the west coast of India near Panaji (H14) towards the north following the trend of geological features in this section like the Closepet granite and schist belts of both the Western and the Eastern Dharwar cratons. This feature also finds reflection in the airborne total intensity map of this region (H6, L6; Fig. 3.30). The gravity highs, H15 (Fig. 7.8(b)) spread over east of the Chitradurga schist belt to the Closepet granite towards the east. There are another set of gravity highs, H13 that are almost parallel to the previous set of gravity highs, H14 that extend from Mumbai to Sangola (Fig. 7.40(a)). This is also reflected in the magnetic anomaly map as H6 and L6 (Fig. 3.30). There is another gravity high H16 along the Kolar Schist Belt (Fig. 7.8(b)) south of the Cuddapah basin that is discussed in detail in Section 9.10.1 with regard to gold mineralization. These gravity highs (H14 and H13) are flanked by gravity lows, L14 and L13 known as the Koyna and the Kurduwadi gravity lows, respectively, that have been described above in regard to Deccan Volcanic Province.

Fig. 7.41(b) Observed gravity field and computed crustal model across the Dharwar Craton along the Kavali-Udipi Profile (Profile VI, Fig. 7.7) and its extension to Madagascar along a profile (Fig. 7.57). The gravity field across Madagascar is adopted from Pilli et al., (1997). It shows three sets of paired regional gravity anomalies; XX′, YY′ and ZZ′ attributed mainly to high density lower crustal rocks along the thrusts and crustal thickening. The short wavelength anomalies L1, L2 and H1, H2, H3 are modeled due to shallow bodies. Densities of various layers and shallow bodies are given in figures as 2.70 g/cm³ etc., which are adopted based on seismic velocities (Mishra and Prajapati, 2003). EGFB: Eastern Ghat Fold (Mobile) Belt.

(b) Several authors (Kaila and Bhatia, 1981; Mishra and Prajapati, 2003; Singh et al., 2004) have modeled gravity highs (H15, Fig. 7.8b) east of Chitradurga constrained from the seismic section (Fig. 7.38). One such model is presented in Fig. 7.41(b) (Mishra and Prajapati, 2003) along the Kavali-Udupi seismic profile (I, Fig. 7.37). This profile is extended westwards by joining gravity data from Madagascar (Pilli et al., 1997) assuming that Madagascar was joined along the west coast of Precambrian India as shown in various reconstructions. The computed crustal model shows a thrust (T2) of high density 2.77 g/cm³ east of Chitradurga extending up to the Closepet granite (2.69 g/cm³). This thrust is almost similar to those modeled under the EGFB (T1) that is a well known thrust formed due to the collision of the Indian continent with Antarctica during Mesoproterozoic period as described in Sections 7.10. Another thrust, T3 of high density 2.78 g/cm^3 is obtained along the east coast of Madagascar. Thrusts are associated with thin crust of 34-35 km that increases to 41-42 km, west of these thrusts. Similar gravity high and adjoining low and a magnetic anomaly, typical of a thrust kind of anomaly have also been reported over the shear zone between the WDC and the EDC along another profile near Gadag, north of this profile (Ramadass et al., 2006). The gravity highs, related to the thrust, T2 between the WDC and the EDC is also reflected in the geoid map of this region as relative geoid high north of Bangalore (H1, Fig. 7.62(a)) in spite of the regional geoid field being low over the entire region, indicating it to be a primary and deep seated structure in the Indian peninsular shield. It is interesting to note that no other gravity anomaly from this region finds reflection in the geoid map due to the prominent regional low in this whole region.

7.8.5   Crustal and Lithospheric Structures under the WDC and Offshore

(a) A detailed Bouguer anomaly and geoid fields obtained by joining continental gravity data with those from satellite over the Arabian sea (Arora et al., 2007) is given in Fig. 7.42(a) and (b). The geoid data is also obtained from the same source as satellite gravity (http://icgem.gfz-potsdam.de/ICGEM/ICGEM.html). Both these data sets are corrected for bathometry and topography to obtain the Bouguer anomaly for the oceanic part and corrected geoid field for continent and adjoining Arabian sea. The field from the coastal part is added while computing fields along the coast in order to reduce the effect of thin crust and other anomalous bodies offshore on the computed field. The lowest gravity and the geoid fields are observed south of Hasan that are modeled in three dimension (Arora et al., 2007) by matching the observed two fields along cross-sections with those computed along the same cross section from crustal and upper mantle sources (Gotze and Lahmeyer, 1988) constrained from the seismic studies along Kavali-Udipi seismic profile in the northern section (Fig. 7.38) and receiver function analysis as discussed above.

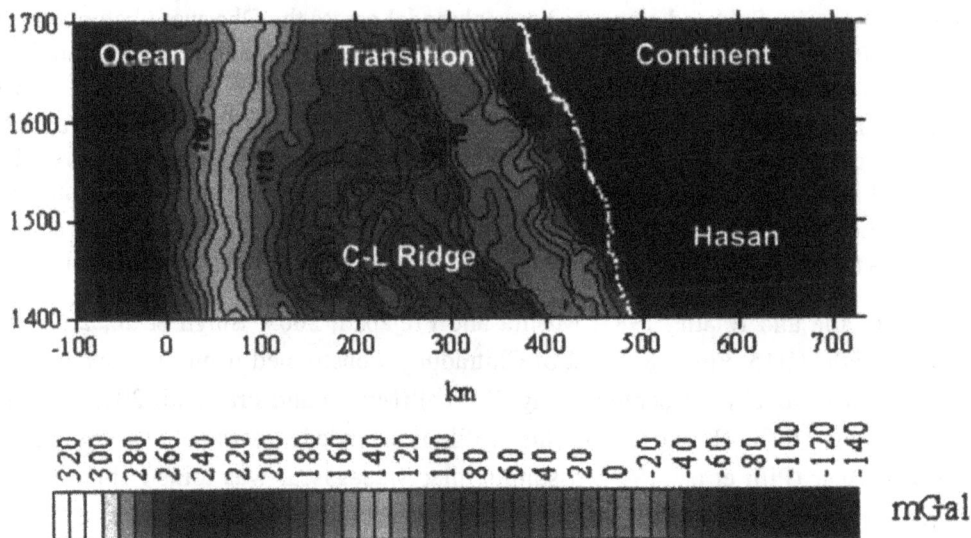

(a)

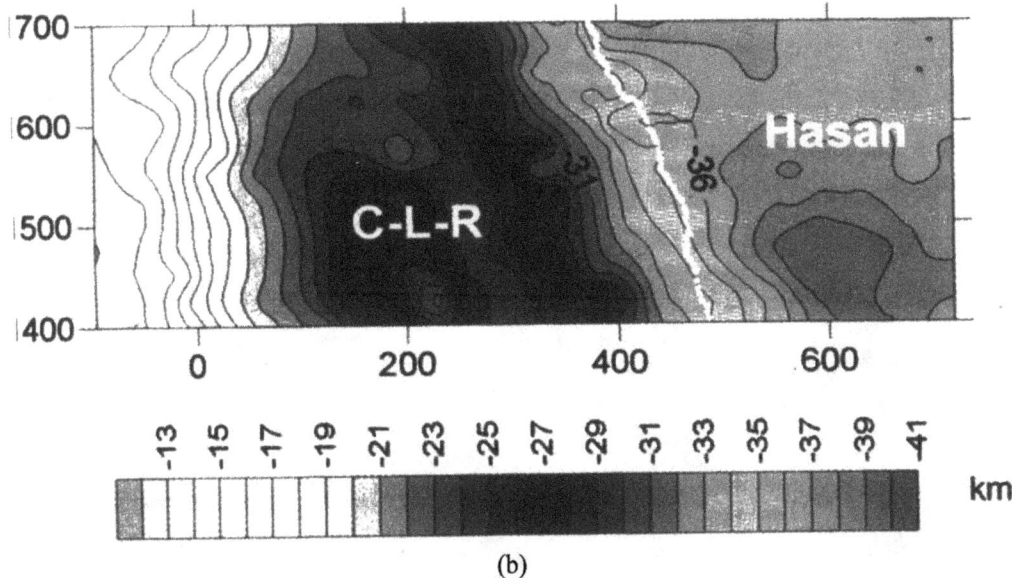

(b)

Fig. 7.42 (a) Bouguer anomaly map of southern part of the Western Dharwar Craton and offshore; white line represents the coastline. The lowest Bouguer anomaly is observed south of Hasan in this section near Sargur schist belt that belongs to an older class of schist belts (~ 3.4 Ga). Zero on the horizontal scale refers to west of the Chagos-Laccadive Ridge almost from where oceanic crust starts. (b) Geoid data for the same area, C-L-R- Chagos-Lacadive Ridge (Arora et al., 2007).

(b) The observed and the computed Bouguer anomaly and geoid fields for plausible crustal and upper mantle models along a profile parallel to 13°N passing through minimum value of the observed Bouguer anomaly, (Fig. 7.42(a)) is given in Fig. 7.43(a) and (b). Fig. 7.43(c) provides a Moho map of the region (Arora et al., 2009) based on several such cross sections both towards the north and south of the section presented in Fig. 7.43(a) and (b). It provides maximum crustal thickness of ~ 50 km south of Hasan that is associated with the minimum Bouguer anomaly that decreases to ~ 41 km towards the north. There are small exposures of older schist belts (Sargur type; ~ 3.4 Ga) east of the lowest Bouguer anomaly and thickest crust indicating crustal root. The density of the lower crust is 2.95 g/cm³ implying a relatively more mafic lower crust as also suggested from receiver function analysis (Kiselev et al., 2008; Gupta et al., 2003). The crust thins to about 10 km towards the ocean (OC) west of the Chagos Lacadive Ridge where a thick crust of ~ 30 km is encountered due to isostatic compensation. The body with density 2.4 g/cm³ represents sediments of this bulk density while 2.7 is the density of trap rocks that formed the continental shelve and Chagos-Lacadive ridge that intruded during the times of Deccan trap eruption (~ 65 Ma) from the Reunion plume. The sections with densities 2.67 and 3.0 g/cm³ towards the ocean represent bathymetry filled with upper crustal rocks that starts from the coast itself but is not visible due to scale and under plated crust under the C-L ridge, respectively. The body with density 2.8 g/cm³ represent a part of the old schist belt (Sargur type) that is exposed in this region. Other densities represent upper and lower crusts and lithospheric mantle under the ocean and the continent. Thickness of lithosphere changes from about 80 km under the Arabian sea to160 km (Fig. 7.42(b)) under the continent that is almost same as obtained from modeling low pass filtered Bouguer anomaly over the Indian continent (Fig. 7.61(a) and (b)). Peschler et al., (2004) who investigated gravity fields of the Pilbara craton, Australia (3.4-3.2 Ga) and the Yilgarn craton, Australia (2.7 Ga) and Abitibi sub-province, Canada (2.7 Ga) also suggested that the mantle root and greenstone belts keels under the Pilbara craton are deeper compared to those under the younger latter two terrains.

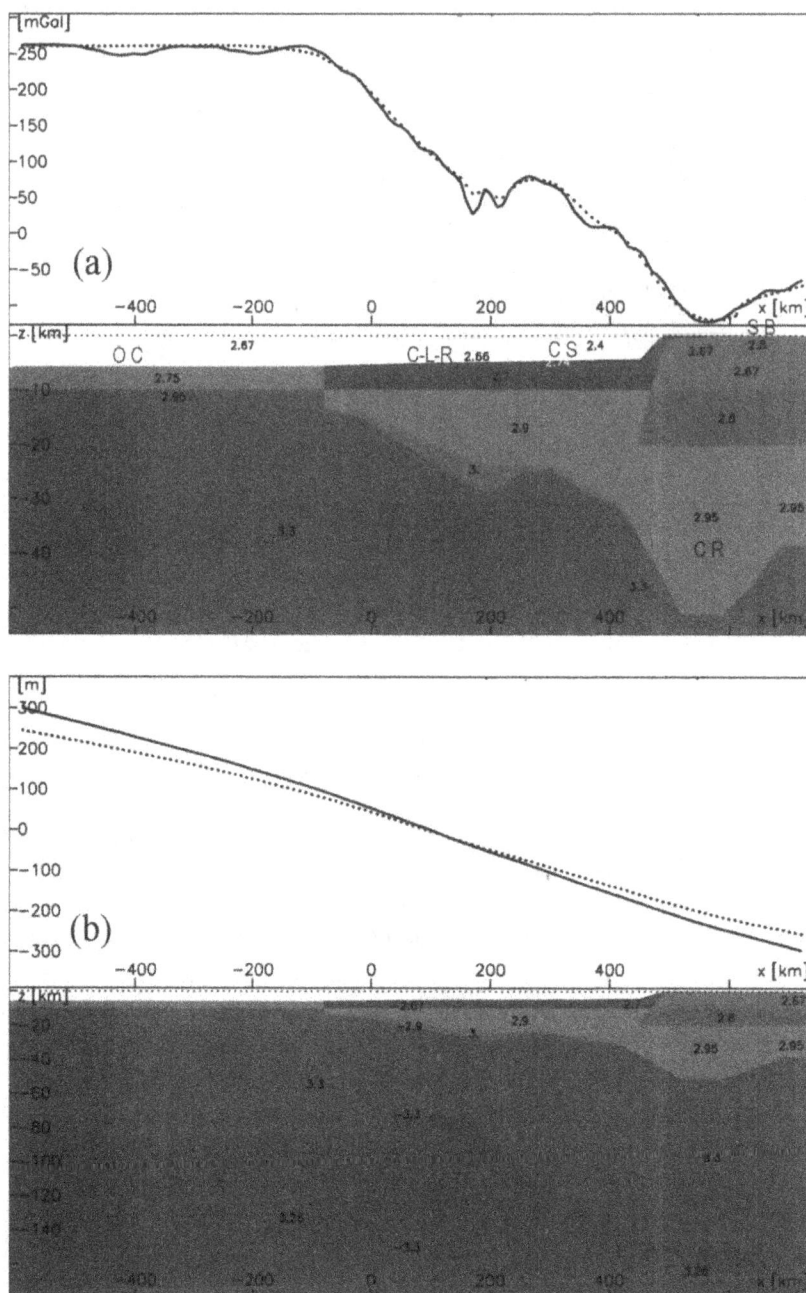

Fig. 7.43(a-b) Crustal and lithosphere density structures along a representative vertical section through lowest Bouguer anomaly south of Hasan based on the computed gravity and geoid fields and their comparison with the observed field. Figure (a) shows shallow section upto ~ 50 km while figure (b) shows full section up to ~ 160 km. They show maximum crustal thickness of about 50 km where old schist belts (Sargur type; ~ 3.4 Ga) are exposed in surrounding region forming crustal root. Parallel sections similar to this one were used to construct the 3D model of the causative sources. Numbers indicate density in g/cm³ that are primarily related to upper and lower crusts and upper mantle for oceanic and continental parts. Some important densities structures are: 2.4: Sediments, 2.7: Deccan trap rocks associated with C-L-R-Chagos-Laccadive Ridge, 2.67: Bathymetry filled with standard crustal rocks, 3.0: Under plated crust and 2.8: Schist belt on continent. Computations are made with reference to standard crustal and upper mantle densities given in figure (b) as − 2.67, − 2.9 and − 3.3, respectively. CS: Continental Shelve, CR: Coastal Root, SB: Schist Belt, OC: Oceanic Crust (Arora et al., 2007).

[Colour Fig. on Page 796-797]

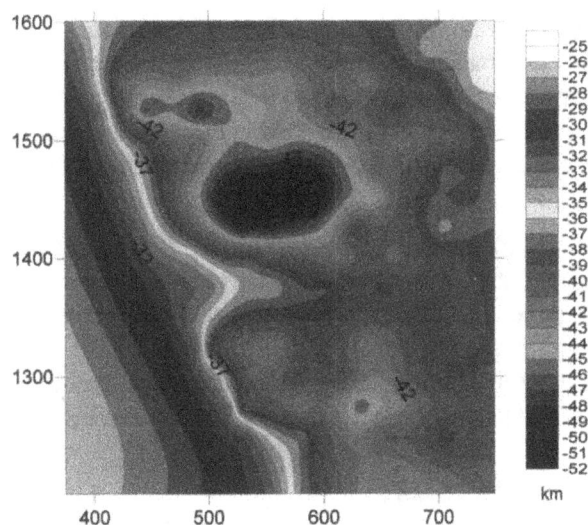

Fig. 7.43(c) A 3-D model of the Moho based on parallel sections as given in Fig. 7.42(a) separated 30' apart based on the comparison between the observed and the computed field showing maximum crustal thickness of 50 km south of Hasan over Sargur schist belt that belongs to older class of Schist belts (~ 3.4Ga) (Arora et al., 2009). [Colour Fig. on Page 797]

7.8.6   Evolutionary Model – Neoarchean Plate Tectonics and Triple Junction

(a)   The shear zone between the EDC and the WDC is characterized by gravity highs (H14 and H15, Fig. 7.8(b)) that spread from the Chitradurga schist belt to the eastern margin of the Closepet granite and is associated with high density rocks. Crustal thickness of 41-42 km along the Kavali-Udipi seismic profile (Fig. 7.38) to a maximum of ~ 50 km under the southern part of WDC has been suggested by gravity modeling that decreases considerably east of the shear zone (31-32 km). These observations suggest that this shear zone forms a suture during Neoarchean time between the two blocks and the EDC subducted under the WDC due to E-W convergence (Fig. 7.65(b)). It has given rise to contemporary granite intrusions and schist belts of the WDC as island arc and back arc or marginal basins with platform type of sediments and mafic rocks similar to oceanic crust. The Closepet granite east of the shear zone (Fig. 7.37) may represent an island arc between the EDC and the WDC formed during convergence. The gravity highs, H13 (Fig. 7.8(b)) extending from Mumbai to Sangola and the Kolar schist belt (Fig. 7.7) are almost parallel to the previous set of gravity highs related to the shear zone (H14 and H15) and geological trends in this region. Crustal thickness also changes almost along this zone showing 31-32 km within this zone and 41-42 km on either sides of it (Fig. 7.38). This indicates that the section between these two sets of gravity highs may represent a diffused collision zone between the EDC and WDC and subduction/underthrusting might have shifted with time from the shear zone to the Kolar schist belt (Fig. 7.37). This, however, cannot be ascertained for certain as differences in rock types on either side of the Kolar schist belt that is the basis for assuming it as an Archean-Proterozoic suture (Krogstad et al., 1989) can be explained based on thrusting from south up to this schist belt (Section 7.9.2). Shifting of sutures and underthrusting are, however, quite common as is presently observed in case of Himalaya where underthrusting has shifted from the Indus Tsangpo Suture Zone to Main Central Thrust, Main Boundary Fault and Himalayan Frontal Thrust during the last 40-45 Ma as discussed in Chapter 6.

(b)   Shift of suture in the present case from the west to the east conforms to the ages of rocks in these sections as schist belts and intrusives west of the Chitradurga schist belt are older, of Neoarchean time (2.9-2.7 Ga), while those to its east are younger, of 2.6-2.5 Ga. This explains the differences in rock types towards the east and the west of the Kolar schist belt as discussed above. The rocks towards the east being part of a thrust zone are mafic/ultramafic in nature as indicated by magnetic anomalies, while those towards the west up to the Chitradurga schist belt represent a reworked crust as part of the collision zone. With passage of time, this collision zone got amalgamated with the EDC which explains the structures similar to mobile belts in the EDC (Radhakrishna and Naqvi, 1986).

(c)   The above convergence and subduction gave rise to a thick crust (~ 50 km) under the older group of schist belts (~ 3.4 Ga, Sargur group) in the southern part of the WDC as crustal root. Based on the thermal, petrologic and geological studies in Slave and Churchill provinces of the Canadian Shield, Canil (2008) and Schmidberger et al., (2007) have suggested that mantle roots under Archean provinces are formed at convergent margins that is also the case for the WDC. It has been even suggested that cratonic roots were formed due to vertical stacking of the subducted plate that soon become strong enough to resist deformation and thereby control the tectonics of that region.

7.9   Southern Granulite Terrain (SGT)

7.9.1   Geology, Tectonics and Geophysical Investigations – Kuppam-Palani Geotransect

Due to complexity of structures in the SGT, a multidisciplinary geotransect from Kuppam-Palani (Fig. 7.37) was carried out. It was planned in two sections from Kuppam-Bommidi and Kolattur-Palani. Some significant results of these investigations are presented below.

(a)   The terrain south of the Eastern and the Western Dharwar cratons after transition zone is known as the Southern Granulite Terrain (Fig. 7.37). It is largely occupied by lower crustal granulite rock and several deep seated intrusives of Neoarchean-Proterozoic period (Bhaskar Rao et al., 2003). Clark et al., (2009) have suggested U-Pb metamorphic age of about 2538 Ma for a charnockite block just north of the PCSZ. It is also characterized by several shear zones and thrusts such as Moyar-Bhavani Shear Zone (MBSZ) and Palghat Cauvery Shear Zone (PCSZ; Fig. 7.37). These two major shear zones enclose the linear Cauvery Shear Zone (CSZ) also known as Palghat gap as it represents a geomorphological low land. The terrain south of the CSZ is also characterized by Pan African event of Neoproterozoic-Cambrian times (Bhasker Rao et al., 2003). Pandey et al., (2005) have provided Sm-Nd model ages of 2.1 Ga for most of the rocks like mafic granulites, and charnockites in the Madurai block south of the CSZ. However, they have also reported Neoproterozoic (~ 800 Ma) and Pan African ages of 532-491 Ma for Rb-Sr model ages. The Grenvillian (1000-800 Ma) and Pan African assemblages of the continents (Fig. 5.3 and 5.4) show the Indian continent in contact with East Antarctica that would have affected the east coast of India and the southern part of the SGT during these orogenies. Based on geological field investigations, Chetty et al., (2003) mapped opposite dipping structures in CSZ and suggested it as a collision zone. Ramakrishnan (2003) has considered the contact between the Dharwar craton and the SGT as a mobile belt (Pandian mobile belt) with Moyar-Bhavani-Attur shear zone as its northern limit while Gopalakrishnan (1996), based on mafic and ultramafic nature of rock types along Mettur shear (Fig. 7.37) north of the Attur shear considered it a suture. Mahadevan (2003), however regarded it as a rifted margin. Raval and Veeraswamy (2007) suggested that the shear zones of the SGT are connected to those from the Dharwar craton to form different tectonic boundaries along which they have been accreted. Santosh et al., (2009) based on the various rock types exposed in the CSZ and the Madurai block, especially the ultra high temperature and high pressure rocks, have suggested a Pacific margin type of setting with southward subduction during Neoproterozoic-Cambrian times that gave rise to a magmatic arc of width > 200 km. This Pacific type orogeny switched over to collision type during Cambrian time at the final phase of Gondwanaland amalgamation. Several good reviews on various aspects of the SGT can be found in Chetty et al., (2006)

(b)   A residual Bouguer anomaly map of the SGT by subtracting the gravity field of geoid low (Fig. 7.62(a)) in the Indian Ocean (Section 7.11.3) from the observed Bouguer anomaly (NGRI, 1975) is given in Fig. 7.44(a) that shows essentially all those anomalies that are marked in Fig. 7.8(b) in this section along with their amplitude. The significant gravity anomalies H16-H18 and L16-L18 (Fig. 7.8(b)) are marked in this figure as H1-H3 and L1-L3 and gravity anomalies H15 and L15 of WDC are marked as H4 and L4 in this figure. Transition zone and Cauvery shear zone are characterized by gravity highs, H1-H3 while adjoining sections show gravity lows, L1-L3. The southern margin of the gravity highs, H1 coincide with the eastern part of the transition zone that extends for about 100 km towards the north indicating that the high density rocks extend northwards along this section of the transition zone. The transition zone in the eastern part is slightly modified based on the southern gradient of the gravity highs, H1. The same however is not true for the western part of the transition zone that coincides with the southern margin of the gravity lows, L4. The gravity highs, H2 and H5 on the western side are related to exposed high density granulite rocks associated with the MBSZ in this section (Fig. 7.37). They appear to form pairs typical of Archean-Proterozoic collision zone that is investigated below. Based on gravity modeling, Mishra and Rao (1993) modeled the gravity highs and lows, H3 and L3 along a specially recorded profile and suggested crustal thickening up to 45 km south of the CSZ and thinning under the CSZ indicating different terrains on either sides of the PCSZ. Krishna Brahmam and Kanungo (1976) attributed most of the gravity lows in southern peninsular shield to granitic batholiths and Verma (1985) and Radhakrishna et al., (2003) have attributed most of the gravity highs of the SGT to high density lower crustal granulite rocks. This figure also shows the lay out of the Kuppam-Palani geotransect that crosses two pairs of gravity anomalies, H1, L1 and H3, L3.

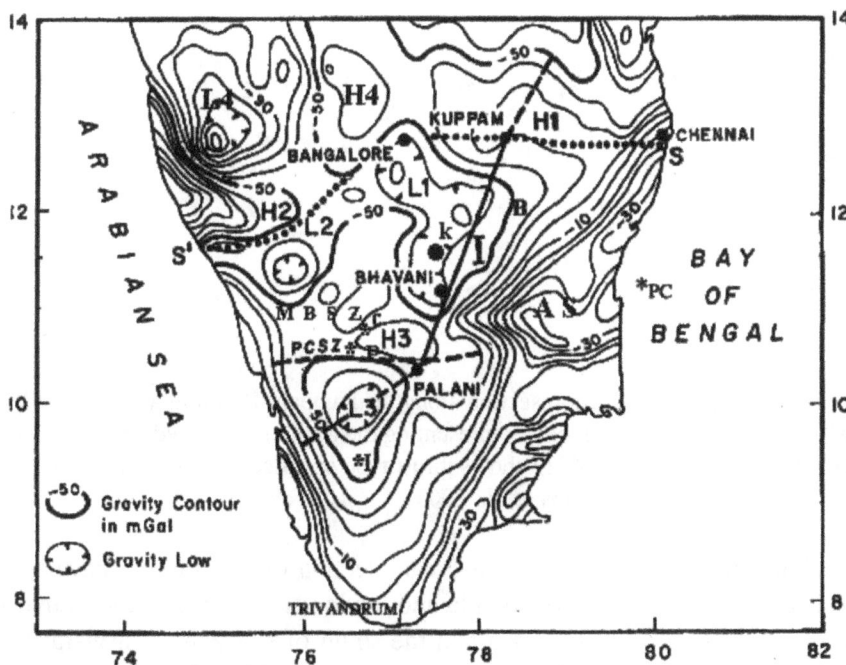

Fig. 7.44(a) Regional Bouguer anomaly map of southern granulite terrain corrected for gravity effect of the Indian Ocean geoid low. H1-H4 and L1-L4 indicate gravity highs and lows. Profile I is the geotransect from Kuppam-Palani, which is extended northwards up to about 13°30′N and southwards upto the coast across the gravity anomalies H1 and L3, respectively which are major gravity anomalies related to SGT. The dotted line SS′ separates gravity highs H1 and H2 towards north and L1, L2 towards south that coincides approximately with the Transition Zone towards the east and Moyar Shear towards the west (Fig. 7.37). Similarly PCSZ separates the gravity high (H3) and low (L3). Extension of the MBSZ as Attur Shear (AS) under Cauvery delta is clearly reflected as nosing of contours. Stars indicate epicenter of Coimbatore (C), Palgat (P), Pondicherry (PC) and Idduki (I) earthquakes, B: Bommidi and K: Kolattur (Mishra and Vijai Kumar, 2005).

(c) To obtain the details of the crustal structures and plausible collision tectonics, a multidisciplinary programme of geophysical and geological investigations was carried out along the Kuppam-Bommidi and Kolattur-Palani geotransect (II, Fig. 7.37 and 7.45(a)) across the northern and the central parts of the SGT. Seismic section for the northern part, Kuppam-Bommidi (Fig. 7.44(b)) shows southwards dipping reflectors with increase in the crustal thickness from about 31 km to 46 km under the Mettur Shear Zone (Reddy et al., 2003; Vijaya Rao et al., 2006) where reflectors are dipping from both sides similar to the collision zones and thrusts. The second part of the transect, Kolattur-Palani suggested thin crust under the CSZ (~ 38 km) that increases on either sides of it up to 42-44 km. It also suggested a low velocity layer in the lower crust (24-30 km) underlain by a high velocity layer representing an underplated crust that is typical of collision zones due to large scale magmatism in these sections and associated oceanic crustal rocks.

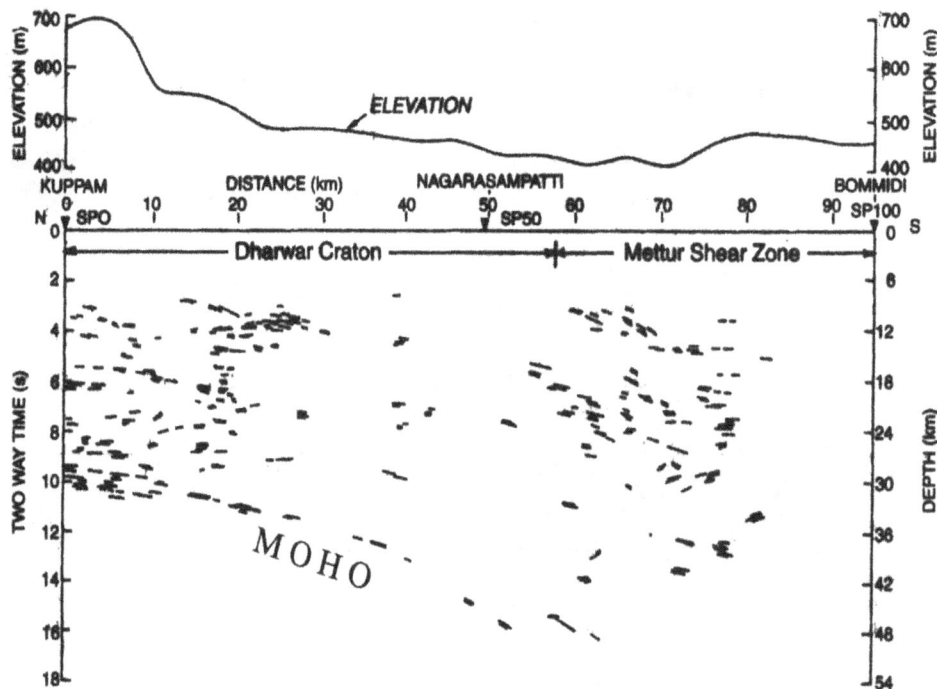

Fig. 7.44(b) Line drawing of the seismic reflection section along the Kuppam-Bommidi (II, Fig. 7.37) profile along with elevation. A deep crustal reflection band representing the Moho is observed under the Eastern Dharwar craton while the Mettur Shear Zone (MSZ) shows thrust type of reflectors similar to collision zones (Reddy et al., 2003)

(d) Magnetotelluric investigations along the Kollatur-Palani section of the transect have provided a relatively high conductive zone in the upper mantle and the lower crust under the Cauvery Shear Zone north of the PCSZ compared to the north and the south of this zone (Harinarayana et al., 2003). Kuppam-Bomidi section (Fig. 7.44(c), Harinarayana et al., 2003, 2006) has also provided high conductivity under the Mettur Shear Zone that is part of the CSZ collision zone. These high conductive zones may be related to mafic/ultramafic rocks in these sections. Ohyama et al., (2008) have suggested the presence of CO_2 rich fluid inclusions in granulite rocks along the PCSZ. Extensive CO_2 metosomatism of the ultramafic rocks generated magnesite deposits around Salem in the Mettur Shear Zone (Santosh et al., 2009) that are manifestions of mafic/ultramafic intrusives from the lower crust and the upper mantle in this section, and would produce high conductivity anomalies.

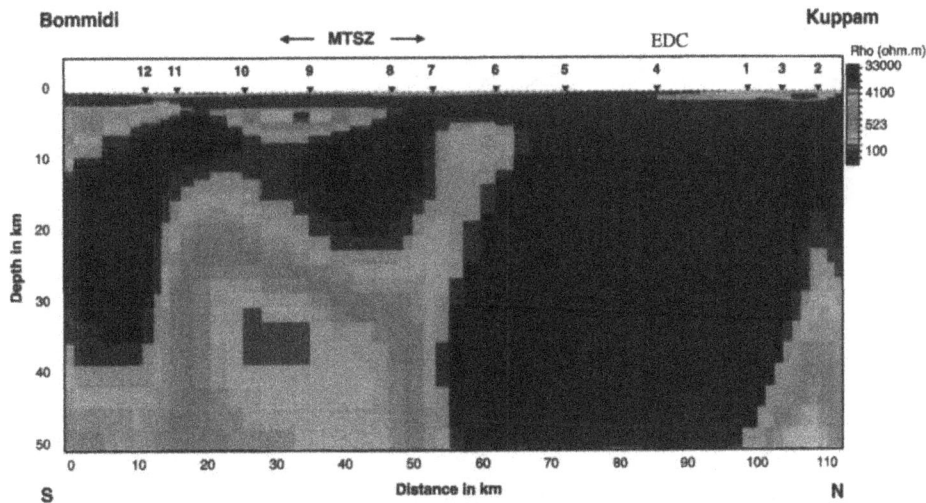

Fig. 7.44(c) 2-D conductivity distribution along the Kuppam-Bommidi geotransect showing high conductivity under the Mettur shear zone (MTSZ) similar to collision zones and a resistive crust under the Eastern Dharwar craton (EDC) (Harinarayana et al., 2003). [Colour Fig. on Page 798]

7.9.2   Bouguer Anomaly, Integrated Crustal Structures and Seismicity

(a)   During the multidisciplinary surveys along the Kuppam-Palani transect, gravity survey was conducted at closer station spacing of 0.5-1.0 km along this transect and 2-3 km spacing along roads and tracks in 100 km wide strip along this transect that provided a detailed Bouguer anomaly map of this region (Fig. 7.45(a), Singh et al., 2003). It shows almost all the essential gravity anomalies that are delineated in the regional map with more details of amplitude etc. The gravity highs and lows, H1, L1 and H4, L4 are placed the same with respect to the transition zone and the PCSZ as in the regional map (Fig. 7.44(a)). It also shows the layout of the transect, Kuppam-Mallapuram (Bommidi) and Kollattur-Palani. This new data combined with some more new data in southern part of the SGT and earlier existing data (NGRI 1975), a detailed Bouguer anomaly map was prepared by Kumar et al., (2009a, Fig. 7.45(b)). This detailed Bouguer anomaly map shows almost all major gravity highs and lows, H1-H4 and L1-L4 of the regional map of the SGT in addition to several small wavelength anomalies that may be related to shallow density inhomogenities. Contours in this map are relatively not so smooth, which might be due to merging of different data sets or may be reflecting surface inhomogeneity. The Achankovil Shear Zone (AKSZ, Fig. 7.37) is a major shear zone in south India that separates the Kerala Kondalite Belt towards the south from Madurai block towards the north, but does not find reflection in the regional maps (Fig. 7.44(a), NGRI, 1975). The same, however, is reflected as kinks (K1) in gravity contours in the present map and finds even better reflection in the map with 5 mGal contour interval (Kumar et al., 2009b) showing a small but a definite anomaly of 5-10 mGal. It also demonstrates the importance of contour interval. Almost subparallel to K1, there are another set of kinks, K2 that may indicate another shear zone with mafic intrusives. This NW-SE trend is not common in the eastern part of the SGT and reminds of Dharwarian trend north of it. It coincides with domal type reflectors in the crust. Further south, these reflectors are sharply dipping towards the south (Prasad et al., 2007). These signatures may represent southward subduction under the Madurai block as discussed below in more details.

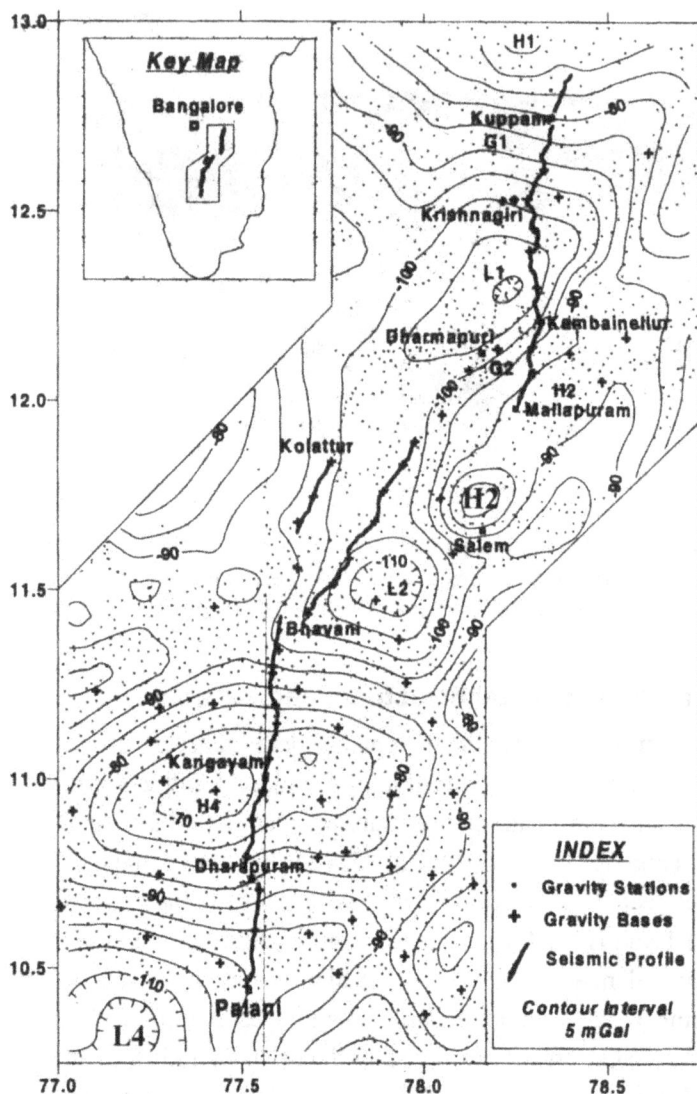

Fig. 7.45(a) Bouguer anomaly map of about 100 km wide strip along the Kuppam-Bommidi (Mallapuram)-Kolattur-Palani geotransect. Dots indicate gravity stations (Singh et al., 2003). Gravity gradients, G1 and G2 indicate the transition zone and the Mettur Shear Zone, the latter is shown as high reflective (Fig. 7.44(b)) and high conductive (Fig. 7.44(c)) zone. The gravity high, H4 is related to the Cauvery Shear Zone.

(b) The specially recorded gravity profile, Kuppam-Palani at 0.5-1.0 km station spacing was extended on either sides based on the regional map of this region (Profile I, Fig. 7.44(a)) that is given in Fig. 7.45(c). It represents Profile VII (Fig. 7.7) extended towards north and SW. It shows two sets of paired regional gravity anomalies. viz., H1, L1 and H3, L3 associated with the transition zone and the CSZ with highs located towards the north and lows towards south of these features along with some short wave length high and low (A and B) related to shallow sources. The gravity high, H1 extends north of the transition zone. This profile is modeled using a four layered model constrained from the seismic profile in its central part (Reddy et al., 2003). It shows four layered crustal model with intermediate low and high density layers 2.65 and 3.0 g/cm^3), the latter may represent under plated crust. Gravity highs, H1 and H3 are modeled due to shallow high density rocks (2.82 and 2.8 g/cm^3) in the upper crust and may represent lower crustal rocks thrusted in this section and gravity lows (L2 and L3) are caused by crustal thickening up to 43-44 km. The high density body related to gravity high, H1 extends north of the transition zone indicating that the transition zone has acted as N-verging thrust in this section, and Kolar schist belt lies at its western margin. This would explain the changes in rock types on either side of this schist belt. Towards the east is the Mg rich andesite rocks from the upper mantle due to thrusting and towards the east is the reworked crust of the EDC that has been the basis for assigning the Kolar schist belt as a Precambrian suture (Krogstad et al., 1989) as described in Section 7.8.1. The computed model (Fig. 7.45(c)) also shows a low and a high density layers that may represent ductile and under plated parts of the crust that are typical of orogenic belts. Most of the reflectors south of CSZ are also dipping southwards (Prasad et al., 2007) indicating north-south convergence. These two signatures in conjunction, viz., high density lower crustal rocks at shallow depth and crustal thickening are indicative of Archean-Proterozoic collision zones that has given rise to various shear zones and high density lower crustal granulite rocks along with deep seated intrusives of anorthosites, and carbonatites, etc., in this region. In this case these anomalies and corresponding model suggests that the CSZ represents a Neoarchean-Paleoproterozoic collision zone depending on ages of the granulite rocks exposed in the northern part of the SGT and transition zone is one of the thrusts related to this collision.

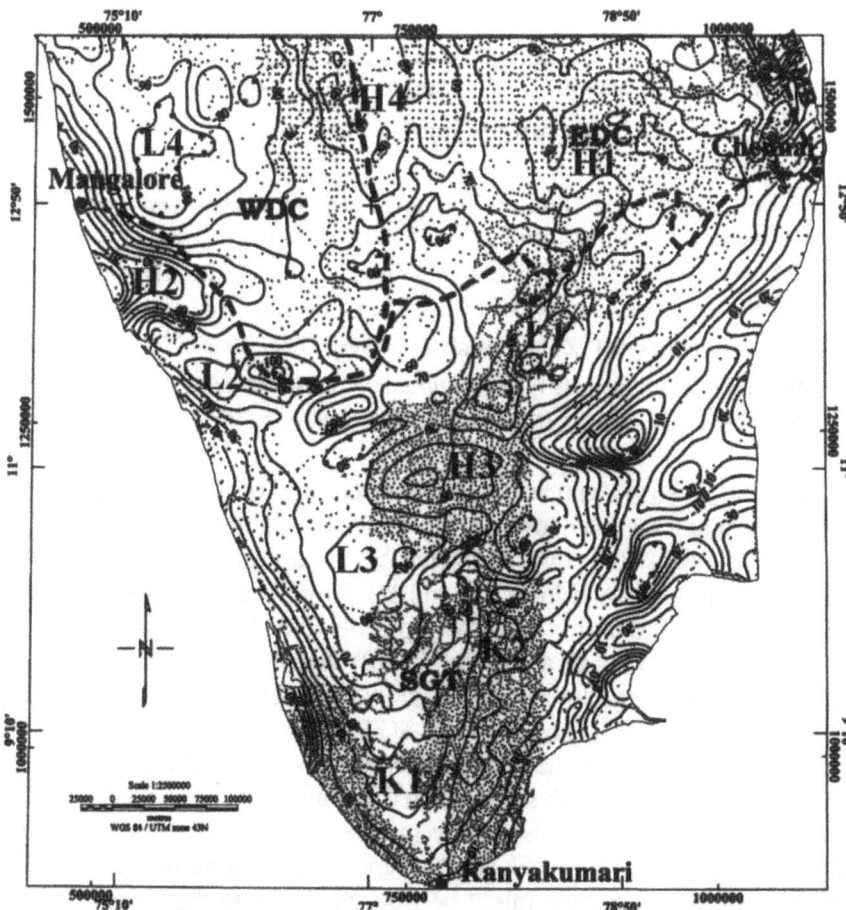

Fig. 7.45(b) A detailed Bouguer anomaly map of the Southern Granulite Terrain (SGT). Dots are gravity stations (Kumar et al., 2009a). It shows some significant gravity highs and lows such as H1-H4 and L1-L4 as given in the regional map (Fig. 7.44(a)). Besides it also shows small anomalies as kinks of contours, K1 and K2 related to Kerala-Khondalite block (Fig. 7.37) and shallow high density intrusive, respectively.

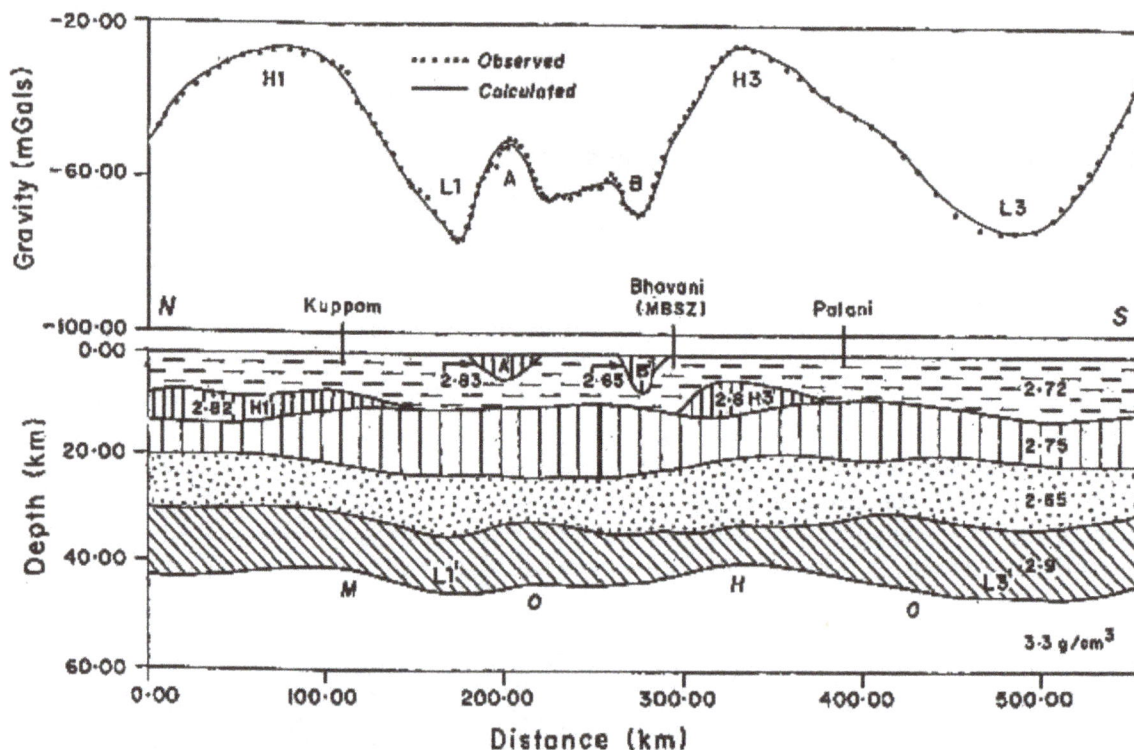

Fig. 7.45(c) The geotransect Kuppam-Palani (Profile VII, Fig. 7.7) and its extension up to about 13°30'N towards north and up to coast towards south (Profile I, Fig. 7.44(a)). It shows two sets of paired gravity anomalies H1, L1 and H3, L3 with highs (H1, H3) related to high density mafic bodies H1$'$ and H3$'$ along the Transition Zone and Cauvery Shear Zone respectively (Fig. 7.37). Lows (L1, L3) are related to crustal thickening towards south of Transition Zone (TZ) and Palghat-Cauvery Shear Zone (PCSZ) (L1$'$, L3$'$). The gravity high A and B are modeled due to high density mafic/ultramafic intrusives (A$'$) and a low-density felsic body (B$'$), respectively (Mishra and Vijai Kumar, 2005).

(c)   The gravity highs, H2 (Fig. 7.44(a)) representing the western part of the transition zone separate the WDC towards the north and the SGT towards the south. It is flanked by two gravity lows L4 towards the north and L2 towards the south separated by the transition zone and Moyar shear, respectively. Both the gravity highs and lows, L4 and L2 shows parallel trends with the gravity high, H3 indicating their inter-relationship. A gravity profile 75.5 °E across the gravity high, H2 and lows, L4 and L2 (III, Fig. 7.37) is given in Fig. 7.45(d) that shows a gravity high (H2) in the central part associated with exposed lower crustal granulite rocks of Coorg hills and crustal thickening (50-52 km) under the WDC (L4) and the SGT (L2) separated by the transition zone and the Moyar shear, respectively. A higher density for the lower part of the crust is adopted in the present case as receiver function analysis suggested a mafic crust under the WDC (Kiselev et al., 2008). These combination of sources for the gravity high and the low suggests that the western part of the Transition Zone-Moyar Shear also represents a suture between the WDC and the SGT, and convergence and subduction might be from the north to the south as above in the case of EDC. However, the gravity highs, H5 being parallel to H2, suggest that the thrust on the western side is formed by the combination of the Moyar and the Bhavani shear zones that are associated with the gravity highs, H2 and H5.

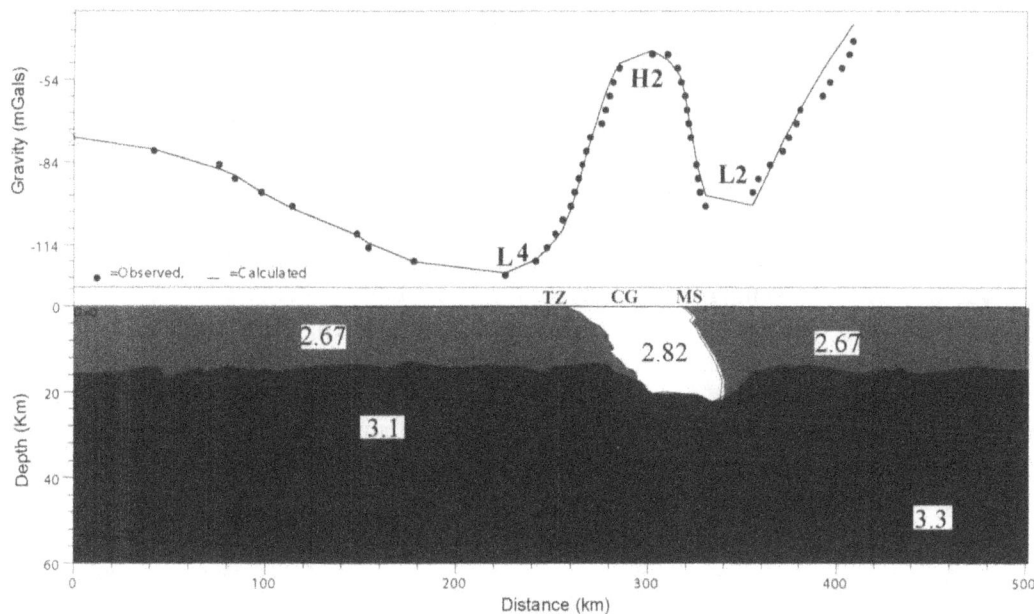

Fig. 7.45(d) A gravity profile along 75.5°E across the WDC and the SGT (Profile VIII, Fig. 7.7) with gravity lows, L4 and L2 and intervening gravity high, H2 located along this profile (Fig. 7.44(a)). The computed model shows crustal thickening up to 50-52 km for the gravity lows and thrusted high density rocks for the gravity high that represents lower crustal granulite rocks of Coorg hills (CG).

(d) Several seismic activities have been reported from the SGT (GSI, 2000). However, the Cauvery Shear Zone (Fig. 7.37) has been seismically most active section in this region. Being bounded by two major shear zones, viz. MBSZ and PCSZ and a deformed zone with several deep seated intrusives, it is highly susceptible to seismic activity. Since historical times, some significant ones are the Coimbatore earthquake of 1900 (Rao, 1992) and the Palghat earthquake of 1975 of magnitude ~ 5.0 (Fig. 7.44(a)). Epicenters of both these earthquakes lie at the eastern margin of the gravity high, H3 along its northern and southern gradient with adjoining lows, respectively, that are related to the Bhavani shear and the PCSZ which are deep seated thrusts (Fig. 7.37). This region is geomorphologically a low land referred to as the Palghat gap that is supposed to have formed due to recent activity since Miocene along the west coast of India associated with Himalayan orogeny (Mishra and Vijai Kumar, 2005) which has made this section seismically active. This part of the Palghat gap is characterized by several deep seated intrusives of granite batholith and alkaline plutonic complexes (GSI, 2000) that may act as stress concentrators. Another important earthquake in this section has been the Pondicherry earthquake of September 25, 2001 of magnitude 5.5 (Fig. 7.44(a)) along the continental shelf off Pondicherry as discussed in Chapter 5 whose epicenter cincides with the extension of the MBSZ and the Attur shear eastwards (Fig. 7.37). In this section, the gravity high, H21 (Fig. 7.8(b)) due to southward extension of the EGMB as discussed in the next section extends westwards bounded by the MBSZ and the PCSZ towards the north and the south, respectively, that suggests the extension of the CSZ eastwards on the continental shelf. The CSZ (Palghat gap) oriented E-W shows signs of recent tectonic activity as discussed above and its interaction with the continental shelf with NE-SW structural trends would be prone to seismic activity. The Pondicherry earthquake is its manifestation.

(e) Besides these moderate magnitude earthquakes, there have been several tremors around the Idukki reservoir that have been attributed to reservoir induced seismicity. In this regard, an earthquake of magnitude 4.5 was reported in June, 1988 that was studied extensively (Rastogi, 1992). These activities are located at the southern gradient of the gravity low, L3 that has been attributed to crustal thickening of 47-48 km (Fig. 7.45(c)). In one of the specially recorded gravity profiles from Trivandrum to Banglore (Fig. 7.37 and 7.44(a)) at close station interval of 0.5-1 km (Fig. 7.46(a)), besides recording regional gravity lows, L3 and L1 of Fig. 7.44(a), a residual gravity anomaly (A) similar to a fault kind of anomaly was recorded in the epicentral zone of Idduki seismic activity. The regional field shown in this figure is related to deep seated sources in the upper mantle. As shown in previous crustal models (Fig. 7.45(c)), the regional gravity low, L3 is modeled due to crustal thickening and gravity low, L1 is modeled due to a low density body that coincides with the southern part of the Closepet granite (L18, Fig. 7.8(b) and 7.7(b)). However, the residual gravity anomaly, A is modeled due to a fault model (Fig. 7.46(b)) that suggest a fault extending up to 9 km for a density contrast of 0.113 g/cm^3 indicating low density intrusives along the fault towards the north. This fault may be an extension of the Periyar fault along Periyar River which is the source of the reservoir and therefore, impounding of reservoir may trigger seismic activity along this fault.

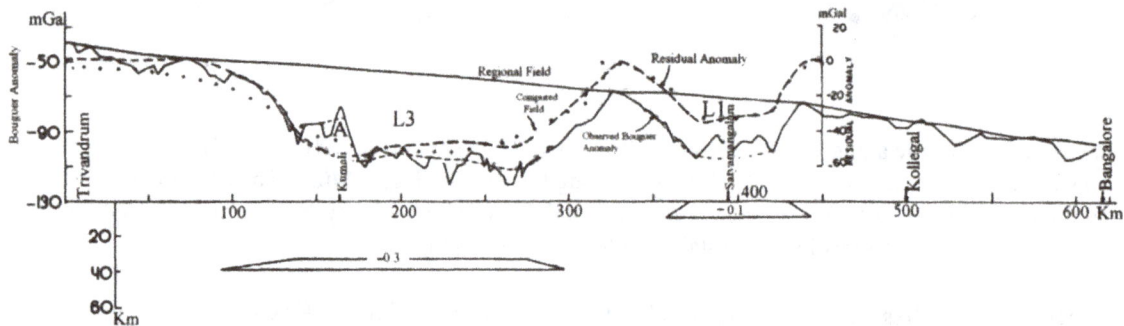

Fig. 7.46(a) A detailed Bouguer anomaly profile from Trivandrum to Banglore showing regional gravity lows, L3 and L1 and a fault kind of anomaly, A around Idduki seismic zone (Mishra et al., 1989).

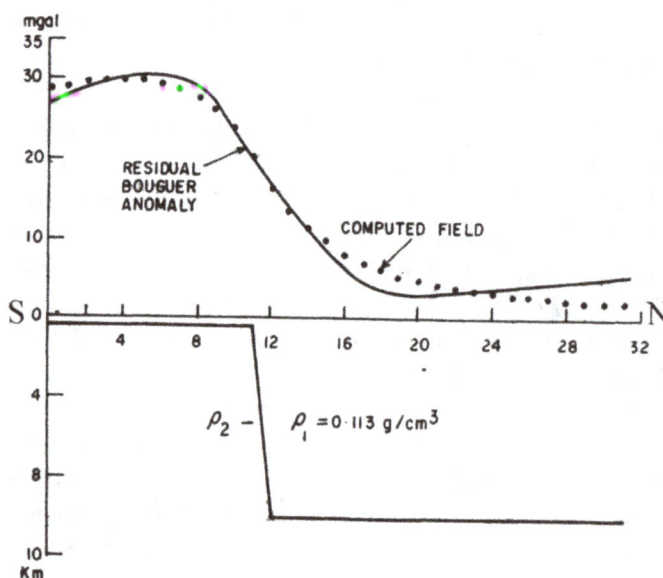

Fig. 7.46(b) Gravity anomaly, A modeled due to a fault coinciding with the Periyar river (Mishra et al., 1989).

7.9.3 Airborne Magnetic Map and Shear Zones

(a) Total intensity magnetic data recorded on the ground along Profile I (Fig. 7.44(a)), Kuppam-Palani (Fig. 7.47(a)) shows a sharp magnetic gradient (M1) across the Mettur Shear (Singh et al., 2003) that coincides with the transition zone in this section indicating rocks of different susceptibility on either sides which may be caused by thrusting of lower crustal rocks along this shear zone as suggested above based on gravity anomalies. The same profile also shows combination of high amplitude magnetic anomalies (M3) associated with the PCSZ that indicates mafic intrusives in this section and a different terrains compared to the northern part. The section of the magnetic profile between M1 and M3 shows different kind of magnetic anomalies compared to the northern and the southern parts indicating different terrains on either sides compared to the central part.

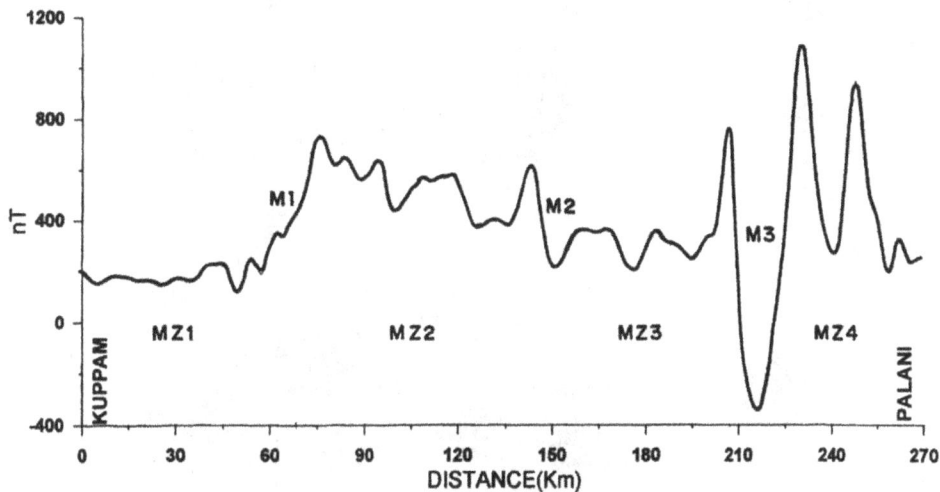

Fig. 7.47(a) Total intensity magnetic anomaly in nT along the Kuppam-Palani profile (Fig. 7.44(a)). It shows a contact type of magnetic anomaly, M1 along the transition zone between the Dharwar craton and the Southern Granulite Terrain that separates terrains of different magnetic characteristics on either sides. M2 and M3 represent mafic intrusions along the Moyar-Bhavani and Palghat-Cauvery shear zones, respectively (Singh et al., 2003) indicating a different terrain south of the PCSZ and mafic intrusions along it.

(b) There are significant curvilinear bipolar magnetic anomalies observed in the airborne total intensity map of the country (H7, L7; H8, L8; Fig. 3.30) related to shear zones of the SGT. Detailed airborne magnetic anomalies observed at a height of 7000' (~ 2.1 km) (Mishra et al., 2006) have suggested that MBSZ and PCSZ joins with the Mettur Shear and Gangavalli Shear zones in the eastern part (Fig. 7.47(b)) to form curvilinear shear zones that may represent thrusts in the Southern Granulite Terrain. Gravity and airborne magnetic profiles from Kolattur to Palani along Profile VII (Fig. 7.7) as shown in Fig. 7.47(b) are presented in Fig. 7.48 and the two data sets are modeled simultaneously to constrain the composite model. The magnetic low, L3 characterizing the CSZ with magnetic highs, H2 and H3 along the MBSZ and PCSZ quantitatively suggest high susceptibility mafic rocks under the CSZ for induced magnetization in low geomagnetic latitude. The crustal model and the density and susceptibility of various layers that constitute the crustal model obtained from a good match between the computed and the observed fields is given in Fig. 7.48. It shows a shallow Moho of 37-38 km under the CSZ increasing to 44-45 km towards north of MBSZ and south of PCSZ. It shows relatively high density and high susceptibility rocks in the upper crust under the CSZ indicating high density mafic rocks. The MBSZ dipping southward and PCSZ dipping northwards are similar to the central core complex of collision zones between two blocks with high density mafic rocks squeezed upwards. The CSZ, therefore represents the collision zone with thrusting along the MBSZ and the PCSZ and also along the transition zone north of the MBSZ. Subsequently, there were intrusives of anorthosites and carbonatites in the collision zone of the CSZ.

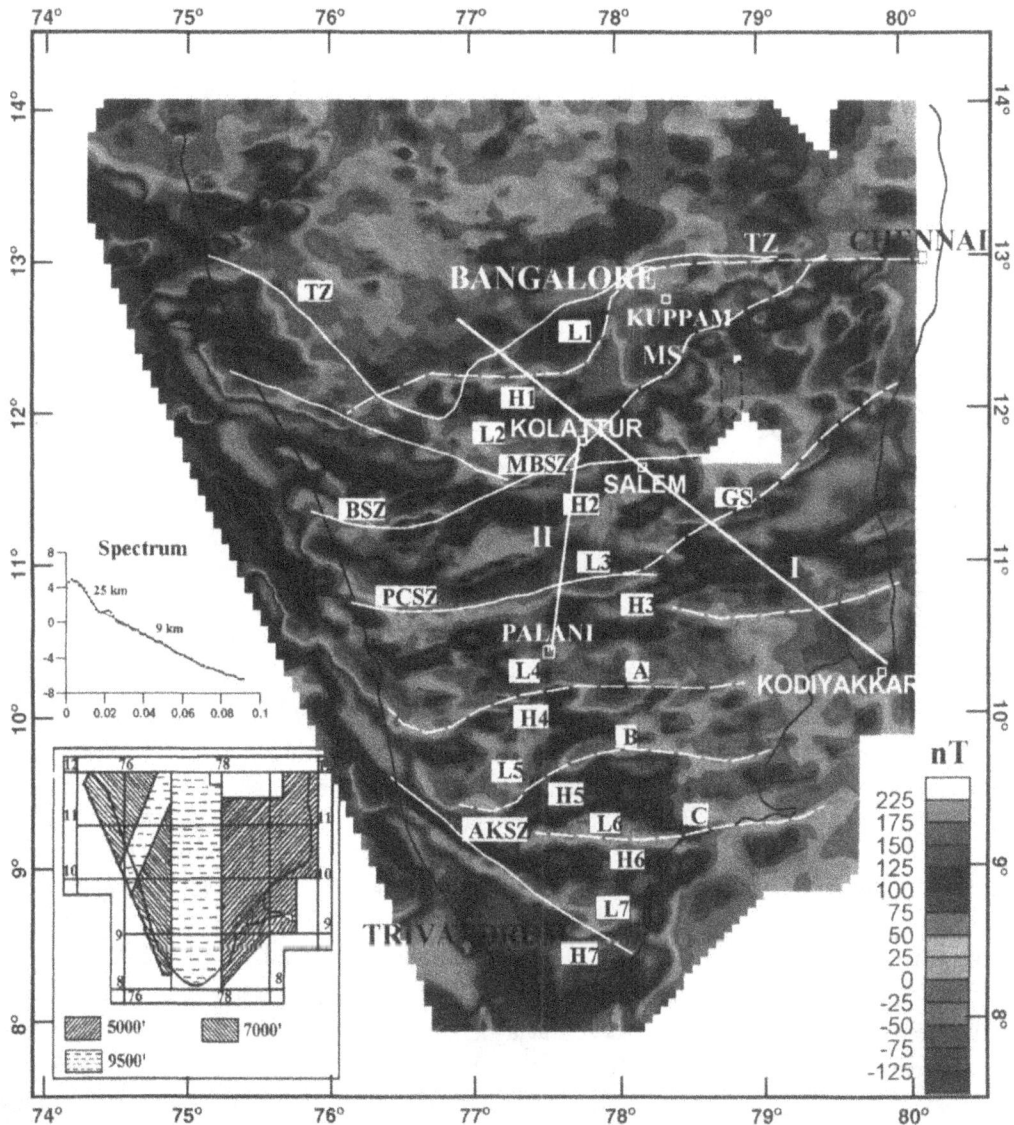

Fig. 7.47(b) Airborne total magnetic intensity map of Southern Granulite Terrain, India, recorded at flight elevations of 5000', 7000' and 9500' above mean sea level (Inset) and have been reduced to a common datum of 7000' (~ 2.1 km) for a composite map. Some data gap in the NE part marked by black line is filled from ground magnetic data continued to same height. H1......H7 and L1.....L7 indicate magnetic highs and lows forming pairs of magnetic anomalies. The inset of spectrum presents its plot versus wavelength, which shows two linear segments with their slopes equivalent to 7 and 23 km depths below surface representing top and bottom of the predominantly magnetic mafic layer, respectively. Abbreviations are the same as given in Fig. 7.37 (Mishra et al., 2006). [Colour Fig. on Page 798]

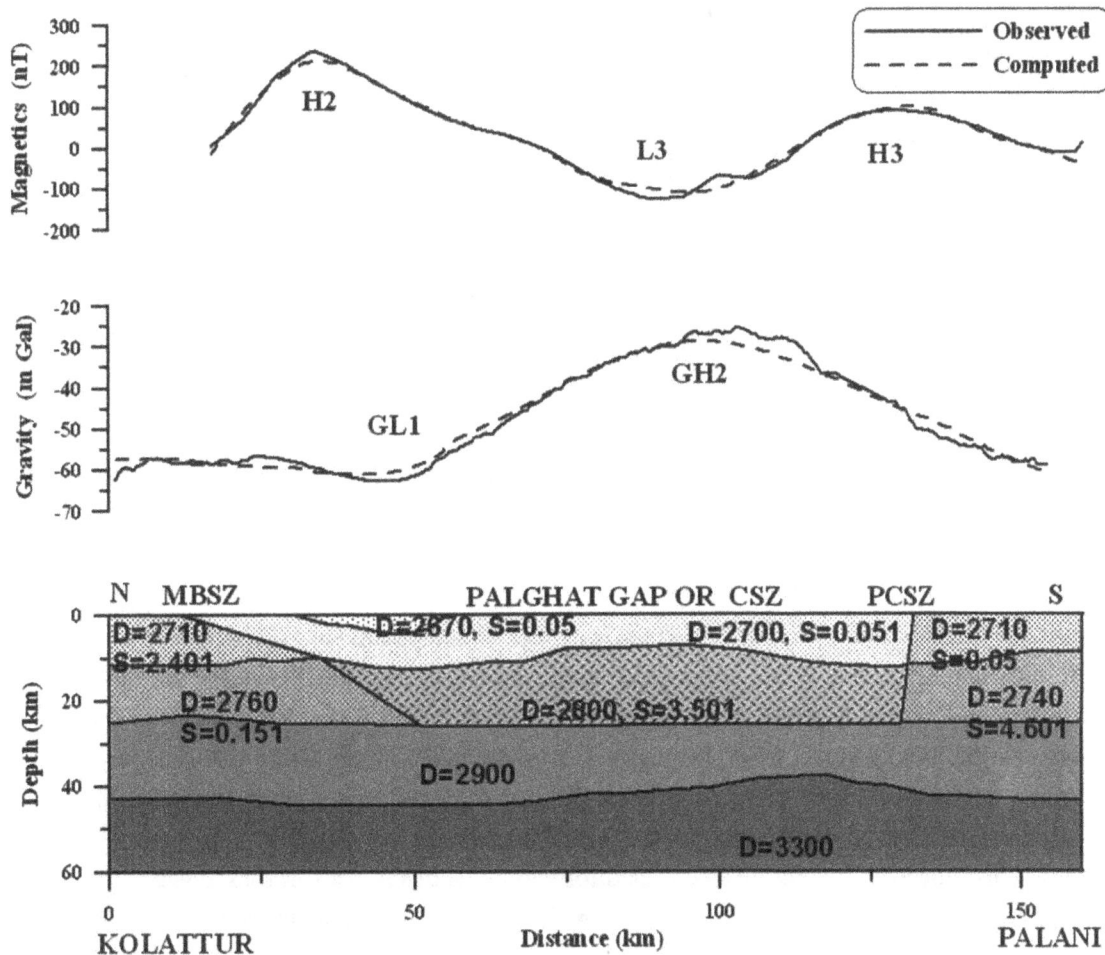

Fig. 7.48 Gravity and air borne magnetic fields along parts of profile Profile I (Kolattur-Palani, Fig. 7.44(a)) and computed crustal model with their physical properties, D denotes density in kg/m^3 and S denotes susceptibility in 10^{-3} CGS units. Susceptibility is primarily limited to first and second layers with high susceptibility rocks in the second layer showing variations across shear zones. It is supported by spectrum of the air borne magnetic field (Fig. 7.47(b)). It shows an upwarping of Moho under Cauvery Shear Zone (Palghat gap) and crustal thickening on either sides towards the north and the south with MBSZ and PCSZ dipping in opposite directions, typically representing central core complex of a collision zone (Mishra et al., 2006).

7.9.4   Evolutionary Model – Neo-Archean Plate Tectonics and Triple Junction

(a)   Gravity highs and thrusted high density rocks along the transition zone and the CSZ with relatively thin crust under the CSZ suggest that the CSZ represents a Neoarchean-Paleoproterozoic collision zone and the EDC and the WDC have subducted under the SGT due to N-S convergence (Fig. 7.6). In this regard, the transition zone – MBSZ-Mettur shear towards the north and the PCSZ – Gangavalli shear towards the south acted like thrusts and sutures. The MBSZ and the PCSZ dipping opposite to each other (Fig. 7.48) enclosing the CSZ with high density and high susceptibility rocks indicating mafic/ultramafic rocks suggest the CSZ as a central core complex of a collision zone. This is supported by several exposed anorthosite and carbonatite complexes in this section such as the Sittampudi anorthositic complex with eclogite and garnet gabbro rocks and minerals with REE composition similar to basaltic oceanic crustal protolith metamorphed in a subduction regime (Sajeev et al., 2008). The present status of the CSZ as Palghat gap of low lying land has been attributed to a rift like structure due to neotectonic activity caused by a crustal upwarp (Fig. 7.45(c), Mishra and Vijai Kumar, 2005) that might have been caused due to present day plate tectonic forces. The neotectonic activity in this section is also suggested by high upper mantle heat flow in this section (Roy et al., 2003). The Pan African and Grenvillian dates related to the Madurai block of the SGT south of the PCSZ may be related to another Pacific type collision during Neoproterozoic period as suggested by Santosh et al., (2009). It can also be related to collision along the southern part of the Eastern Ghat Mobile Belt (EGMB) with East Antarctica as discussed in the next section 7.10. Similar dates have been extensively reported from the exposed sections of the central part of the EGMB.

(b)   The eastern part of the transition zone (H1, Fig. 7.44(a)) and the MBSZ in the western part (H2 and H5) join with the shear zone between the EDC and the WDC (H4, Fig. 7.44(a)) with obtuse angles of about 120° each indicating that their junction may form a Neoarchean Triple Junction of SSS (TTT) type (Fig. 7.2) as previously described in case of the western and the eastern parts of the Satpure Mobile Belt and the Godavari Proterozoic Belt. In this case, the EDC subducted under the WDC and the EDC, the WDC subducted under the SGT, and the SGT did not subduct satisfying the condition of a stable triple junction as shown in Fig. 7.2. Shifting of the thrust/ suture between the EDC and the WDC from the shear zone to the Kolar schist belt eastwards as described above (Section 7.8.6) might have been facilitated by this triple junction as shifting of sutures at triple junction is quite common due to differences in the spead of convergence and subduction. This, however can not be astertained for certain as differences in rock types on either side of the Kolar schist belt can be explained on the basis of thrusting from south along the eastern part of the transition zone while towards the west is the normal reworked crust of the EDC as discussed above in Section 7.9.2.

(c)   The gravity data of the South Indian Shield is corrected for the geoid low of the Indian Ocean and the residual anomaly is obtained by removing regional field using zero free air anomaly (Subba Rao, 1996). This residual anomaly is converted to apparent density map (Singh et al., 2003) through Fourier transformation. It uses the similar scheme as given in equation 6.1 to obtain the depth to the sources and basement relief model. In this case, it is assumed that the observed anomalies are caused by variations in the physical parameter (ρ), instead of relief, h(z) that is considered a plain surface. The apparent density map (Fig. 7.49(a)) highlights the high density rocks along the shear zone, the eastern part of the transition zone and the MBSZ between the WDC, the EDC and the SGT marked as H1, H2 and H4 corresponding to the gravity highs, H1, H2 and H4 as given in Fig. 7.44(a). This map clearly defines the triple junction at Bangalore with axis of high density sections around Bangalore making almost 120° from each other indicating a stable triple junction. The high density rocks along the shear zone between the WDC and the EDC (H4) spread almost from east of the Chitradurga schist belt to the west and the south of the Cuddapah basin (Kolar schist belt) as suggested above based on gravity highs in this section indicating this whole section as a diffused collision zone.

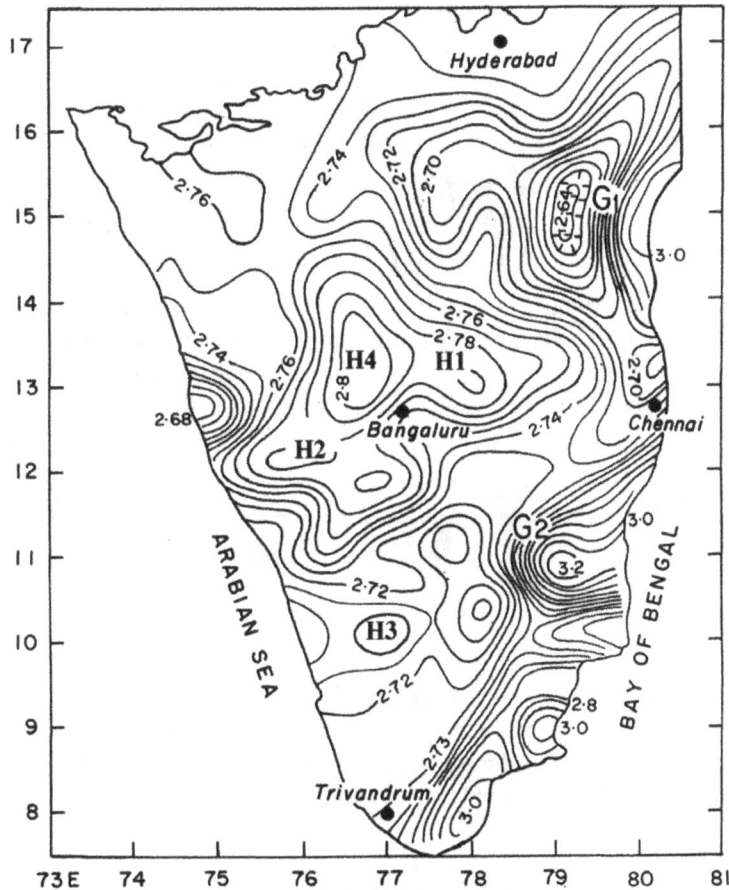

Fig. 7.49(a) Apparent density map based on the deconvolution of the gravity field (Fig. 7.44(a)) to a mean density of 2.75 g/cm³ showing linear high density sections, H1, H2, H3 and H4 representing gravity highs, H1, H2, H3 and H4 (Fig. 7.44(a)) forming a triple junction north of Bangaluru (Singh et al., 2003).

(d) The gravity lows, L1 and L2 and L4 (Fig. 7.44(a)) are interacting in the southern part of the Closepet granite at the triple junction where a major gravity low is observed which suggests that the Closepet granite might have intruded here and spread northwards along the shear zone between the WDC and the EDC that explains gravity highs being observed over the Closepet granite due to subsurface thrusted high density rocks.

(e) Fig. 7.49(b) and c provides a plausible schematic scheme of Neoarchean-Paleoproterozoic collision and its effect in the SGT. Fig. 7.49(b) shows the collision between the Western and the Eastern Dhwarar cratons and the Southern craton, presumably the Madurai block. It shows the direction of convergence from the north to the south in the Dharwar craton that subducts along the MBSZ-Mettur shear which has also acted as a thrust and lower crustal granulite rocks have thrusted along it extending northwards upto the transition zone. In Madurai block, the convergence is from the south to the north that has subducted along the PCSZ-Gangavalli shear forming a Central Core Complex of mafic/ultramafic rocks and different deep seated intrusives (Fig. 7.49(b)). An alternative model of one sided subduction of the Dharwar craton from the north to the south along the MBSZ is presented in Fig. 7.49(c) where the Dharwar craton subducts under the Madurai block and the PCSZ is formed along the Madurai block due to this subduction and collision. In such cases, there may be limited subduction from the other side. This case might be related to Neoproterozoic convergence across the PCSZ that would explain the corresponding dates from the Madurai block (Santosh et al., 2009).

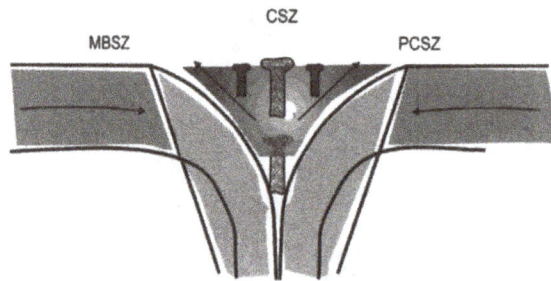

Fig. 7.49(b) A schematic cross section of the CSZ between the MBSZ and the PCSZ where two cratons have collided to form the Central Core Complex of this collision zone where several mafic/ultramafic and other deep seated intrusives are found as shown by the flower structure. Horizontal arrows (N-S and S-N) indicate convergence and subduction and SE-NW and SW-NE arrows indicate thrusting.

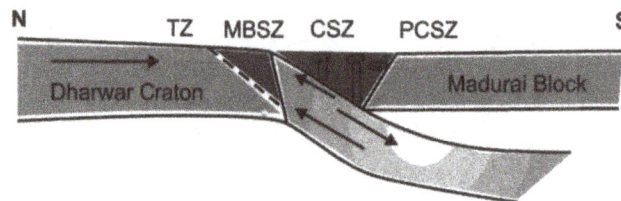

Fig. 7.49(c) An alternative model of collision between the Dharwar craton and the southern craton plausibly Madurai block where only the Dharwar craton is subducting that gave rise to mafic/ultramafic and other deep seated intrusives of the CSZ as subduction related magmatism shown as flowers and the PCSZ is formed as a thrust. Based on Neoproterozoic ages south of the PCSZ, this might be a scenario during Pan African time.

7.10   Eastern Ghat Mobile Belt (EGMB) and Adjoining Cuddapah, and Coastal Godavari Basins

The Eastern Ghat Mobile Belt (EGMB) extends from north of Chennai up to Bhubaneshwar (Fig. 7.6 and 7.7) along the east coast of India and is characterized by linear gravity highs as in case of other mobile belts. There are several Proterozoic basins to its west such as the Cuddapah basin in the southern part, that assume special significance due to its size, shape and mineral potentiality.

7.10.1 Southern Part of the EGMB and Multi-Parameter Geophysical Investigations: Crustal Structures, Seismicity, and Evolutionary Model – Proterozoic Plate and Plume Tectonics

The EGMB extends from Bhubaneshwar in the north to north of Chennai in the south (Fig. 7.7). Some of the significant geological and geophysical signatures of collision tectonics along EGMB are as follows:

(i) Geology and Tectonics

(a)   A detailed geological map of the southern part of the EGMB and the adjoining Cuddapah basin (Fig. 7.37) is given in Fig. 7.50(a). The EGMB is characterized by Peninsular gneisses flanked by recent alluvium and high grade Charnockite and Khondalite rocks towards the east. The Dharwar super group of rocks known as the Nellore-Khammam schist belt with felsic and mafic intrusives occurs on the western side along the western margin of the EGMB. There are several deep seated intrusives such as gabbro, anorthosite, carbonotite, etc., exposed in the EGMB. The Cuddapah basin, west of the EGMB (Fig. 7.37) is primarily characterized by metasediments of Cuddapah super group and Kurnool super group of rocks of Paleo-Neoproterozoic periods (~ 1.9-0.6 Ga), respectively. The Cuddapah super group of Paleoproterozoic period occupies two distinct basins viz. Nallamalai and Papagani sub-basins in the western and the eastern parts of the Cuddapah basin, respectively. However, the Papagani group in the eastern part is older than the Nallamalai group. Mafic basic volcanics as large sills and dykes are exposed along the western margin of the Cuddapah basin in the Papagani sub-basin. French et al., (2008) dated these mafic sills as ~ 1900 Ma and also suggested similar dates from mafic intrusives of the Bastar craton, west of the northern part of the EGMB (Fig. 7.6). This indicates a similar or slightly older dates for the Cuddapah Super group of rocks in the Papagani sub-basin. There are several occurrences of Uranium mineralization in this basin along its western margin (Sharma et al., 1995) and baryte deposit at Mangampeta. The metasediments of the Nallamalai basin are highly disturbed with tight folds and faults. There are some schist belts of Dharwar super group of Neoarchean time towards the the south and west of Cuddapah basin. One such schist belt towards south is the Kolar schist belt that is significant for gold occurrences (Section 9.10.1). It is surrounded by granite intrusives and Peninsular gneisses of the EDC on the western side. This figure also shows the layout of two deep seismic sounding (DSS) profiles and Profile I extends from Kavali (east coast) to Udipi (west coast) and has been described in Section 7.8 (Fig. 7.38).

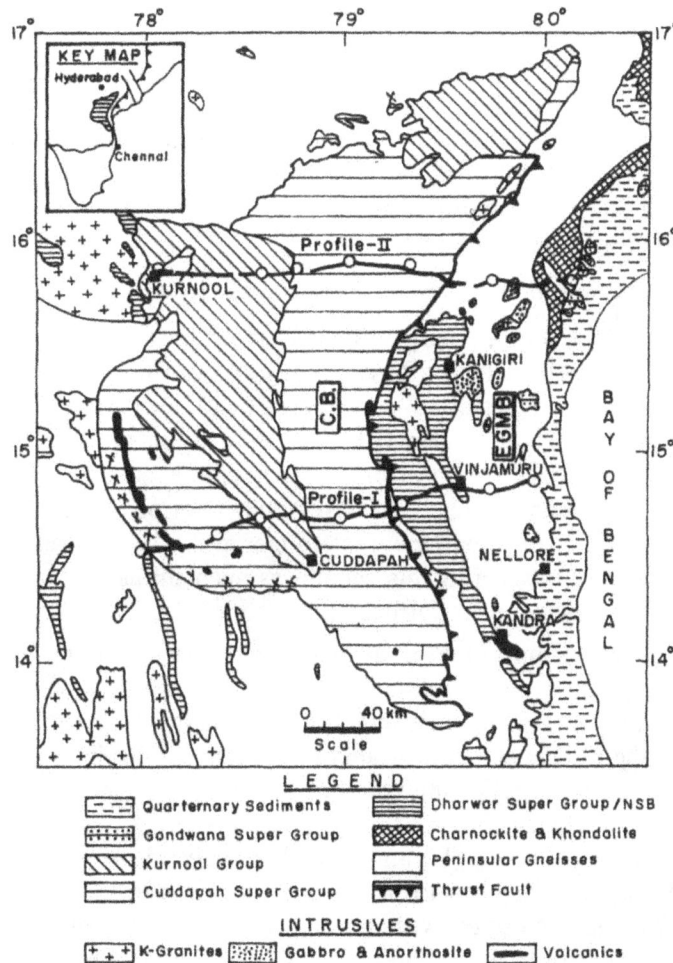

Fig. 7.50(a) Geological map of the Proterozoic Cuddapah basin (CB) and the adjoining Eastern Ghats Mobile Belt (EGMB; GSI, 1998). Their contact zone is characterized by a thrust fault along the eastern margin of the Cuddapah basin and the Nellore Schist Belt (NSB) towards the east. Crosses (x) along the western margin indicate uranium mineralization. The profiles I, used for the density modeling across this region is part of a deep seismic sounding profile from Kavali-Udipi (Fig. 7.38).

(b) High grade charnockite and khondalite rocks of the EGMB and alkaline intrusives have provided wide range of dates from 1.6-1.0 Ga (Mezger and Cosca, 1999). However, the most prominent metamorphic event in the Eastern Ghat Mobile Belt is reported at 1.0-1.1 Ga (Paul et al., 1990). The exposed part of the EGMB extends from the Bhubaneshwar in the north to Nellore in the south, north of Chennai along the east coast of India (Fig. 7.6 and 7.7). It has been divided into several sections based on metamorphism and rock types. However, the most prominent divisions, have been north and south of the Godavari basin (Fig. 7.7). The southern part is characterized by low grade Nellore-Khammam Schist Belt of Neoarchean-Paleoproterozoic period (Fig. 7.50(a)) that is absent from the northern part. Similarly, the high grade metamorphic event of Neoarchean time reported from the northern section is absent from the southern section (Mukhopadhyay and Bask, 2009). Granite intrusives and carbonatite complexes of Mesoproterozoic period (1.5-1.0 Ga) are more predominant in the southern section.

(c) The northern part of the Eastern Ghat Mobile Belt consists of gneisses and migmatites with high grade granulite rocks of Mesoproterozoic period (1.6-1.5 Ga) (Dobmeier and Raith, 2003), which are similar to those reported from the Chhotanagpur Granite Gneiss Complex (Acharyya , 2003) and the Shillong massif as referred to above. In fact, Dobmeier et al., (2006) have provided dates of about 1.6 Ga for magma emplacement and low grade metamorphic overprint of 0.5 Ga related to Pan African event in the southern part of the Eastern Ghat Mobile Belt. Vijaya Kumar and Leelanandam (2008) have identified two stages of rifting and convergence in the southern part of the EGMB during Paleoproterozoic period (2.0-1.6 Ga) and Mesoproterozoic period (1.55 Ga) respeetively. There may be another phase of rifting and convergence related to Grenvillian/Pan African collision (1.0/0.5 Ga). Gravity highs due to the EGMB are caused by exposed rocks of relatively high density (2.8-3.0 g/cm^3) in Fig. 7.50 (a) and (b) along the east coast of India that may be related to exposed charnokite and khondalite.

(ii) Bouguer Anomaly Map and Seismicity: Ongole Seismic Activity

(a) The Eastern Ghat Mobile Belt (EGMB) is largely characterized by gravity highs along the east coast of India (H22-H24, Fig. 7.8(b)) which extends from south of Bhubaneshwar up to north of Chennai. It is interesting to observe that these highs, H22-H24 are accompanied by gravity low L22-L24 towards the west forming paired gravity anomalies that are significant for Proterozoic collision zone. It is also characterized by linear bipolar magnetic anomalies (H4, L4; Fig. 3.30) that is also significant for Proterozoic collision zones. Another set of regional gravity high and low, H20 and L20 (Fig. 7.8(b)) occurs along the western margin of the Cuddapah basin that may be related to mafic and felsic intrusives in this section as such intrusives are exposed in this section. These anomalies are modeled in the forthcoming sections.

(b) A detailed Bouguer anomaly map of the Cuddapah basin and adjoining EGMB (Krishna Brahmam et al., 1986) is separately given in Fig. 7.50(b) which shows major gravity highs over the EGMB (H2-H5) and another gravity high in the western part of the Cuddapah basin (H1) primarily coinciding with the Cuddapah super group of Papagani sub-basin and basic sills. Large amplitude gravity lows (L1 and L4) following the trends of the eastern margin of the Cuddapah basin coincide with the Nallamalai sub-basin of the Cuddapah super group. The western gravity high (H1) is surrounded by gravity lows, (L3 and L6) which coincide with the granite intrusive bodies of the EDC and therefore, appears to be associated with them. There is another large amplitude gravity low L7 cutting across the western margin of the Cuddapah basin that is similar to L3 across the NW margin of this basin and may be caused similarly by granite intrusive bodies which are not exposed. As it is cutting across the western margin of the Cudddapah basin where uranium mineralization is found, this anomaly assumes special importance for further exploration. Some of these gravity anomalies are modeled in the next section. This figure also shows the layout of two (DSS) profiles.

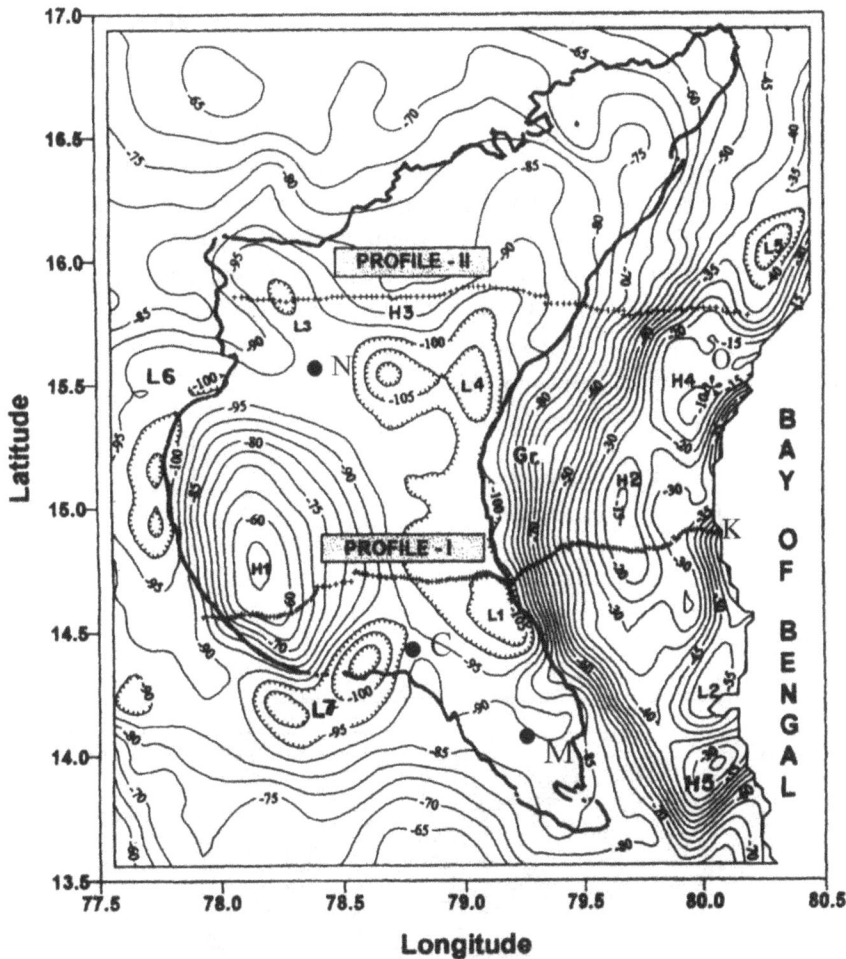

Fig. 7.50(b) Bouguer anomaly map (in mGal) of the Proterozoic Cuddapah basin and the adjoining Eastern Ghats mobile belt (after Krishna Brahmam et al., 1986) showing prominent gravity high (H1) over SW Cuddapah basin and gravity lows, L1 and L4 over the eastern Cuddapah basin. Gravity highs, H2, H4 and H5 and lows, L2 and L5 coincide with the Eastern Ghats mobile belt. Faulted contact zone thrust is characterized by a sharp gravity gradient (Gr). Star indicates epicenter of Ongole seismic zone (O).
Towns shown are; C: Cuddapah, N: Nandyal, M: Mangampeta and K: Kavali.

(c) Several earthquakes have been reported from the southern part of the EGMB and the adjoining Cuddapah basin but the Ongole region has been seismically most active (Krishna Brahmam et al., 1986). There have been three significant earthquakes, on October 12 and 13, 1959 and March, 27, 1967 of magnitude ~ 5.0 in this section. This region lies at the eastern gradient of the gravity highs, H2 and H4 defining the eastern margin of the thrusted block related to the Eastern Ghat Orogeny (Fig. 7.52(a)). Being a deep seated thrust, it can be easily be reactivated by any recent activity. It is also the junction of the N-S and NE-SW trends of the east coast of India towards the south and the north, respectively that may form a node for strain accumalation. In fact, the Ongole section is most seismogenic section along the east coast of India that might be due to this curvature of the Eastern Ghat at this point where stress gets concentrated resulting in seismic activity. Stress in this section may be generated from plate tectonic activity along the Andaman trench. The seismogenic section of Onogle also marks the transition zone between the high and low grade rocks of the EGMB towards the east and the west, respectively. These considerations make it seismically active.

(iii) Gravity, Airborne Magnetic and Magnetotelluric Profiles, and Crustal Structures

(a) An airborne magnetic profile flown at an altitude of ~2.7 km above msl is given in Fig. 7.50(c) along with the Bouguer anomaly and the geological cross section with flight path marked on it (Mishra, 1984). It is close to the DSS profile I (Fig. 7.50(a) and (b)). It shows gravity highs, H1 and H2 related to the Western part of the Cuddapah basin and the EGMB as shown in Fig. 7.50(b). These gravity highs also coincide with the magnetic anomalies A1 and A2-A4, respectively. Magnetic anomaly, A1 is a paired magnetic anomaly related to mafic intrusives in this section while magnetic anomalies, A2-A4 are fluctuating magnetic anomalies that are generally observed over metamorphic rocks that characterize the EGMB.

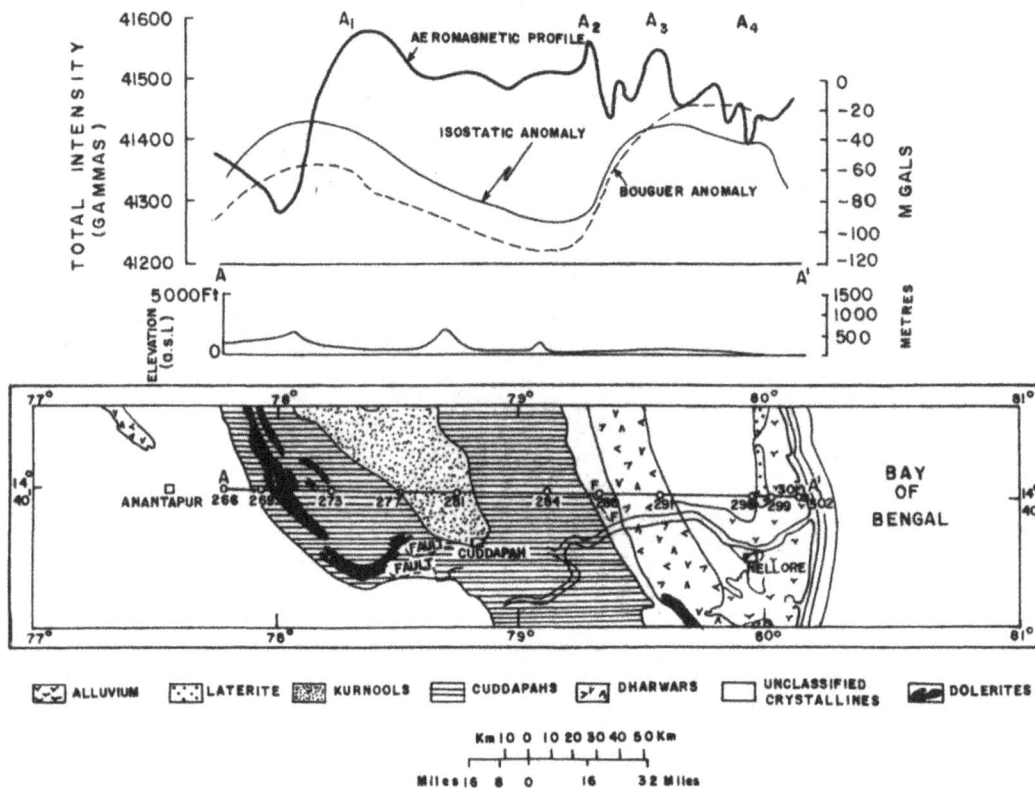

Fig. 7.50(c) Airborne total intensity magnetic profile recorded at 2.7 km above msl and Bouguer anomaly showing magnetic anomalies coinciding with gravity highs related to the EGMB and the western Cuddapah basin that indicate mafic intrusives in these sections. Flight path over a generalized geological section is shown for ready reference.

(b) The gravity highs and low, H1, H2 and L1, related to the EGMB and Cuddapah basin along the DSS Profile I (Fig. 7.50(b)) was modeled by Mishra and Tiwari (1995) and Singh and Mishra (2002). Mishra and Tiwari (1995) modeled the gravity and the airborne magnetic data adopted from Fig. 7.52(a) along Profile I simultaneously and provided a composite crustal model (Fig. 7.51(a)) that shows thrusted high density rocks T1 related to gravity high, H2 and crustal thickening from 35 km along the east coast of India up to 41-42 km under the eastern part of the Cuddapah basin related to the gravity low, L1. This model also shows underplated lower crust (3.05 g/cm^3) under the eastern part of the Cuddapah basin that is typical of orogenic belts. The gravity high H1 and associated magnetic anomaly are related to high density, mafic rocks in the basement under the western part of the Cuddapah basin. This mafic body also shows a high susceptibility with remnant magnetization I = 6° and D = 119° that is similar to Cuddapah dykes and sills (Radhakrishnamurthy et al., 1967). This crustal model under the EGMB and the Cuddapah basin is almost similar to those deduced based alone on gravity data (Fig. 7.41(b)) with some differences in the high density high susceptibility mafic body in the western part of the Cuddapah basin that is better defined in the present model due to inclusion of the airborne magnetic data and its modeling (Fig. 7.51(a)). The thrust, T1 coincides with the density gradient, G1 (Fig. 7.49(a)) that extends southwards up to the southern margin of the east coast of India as density gradient.

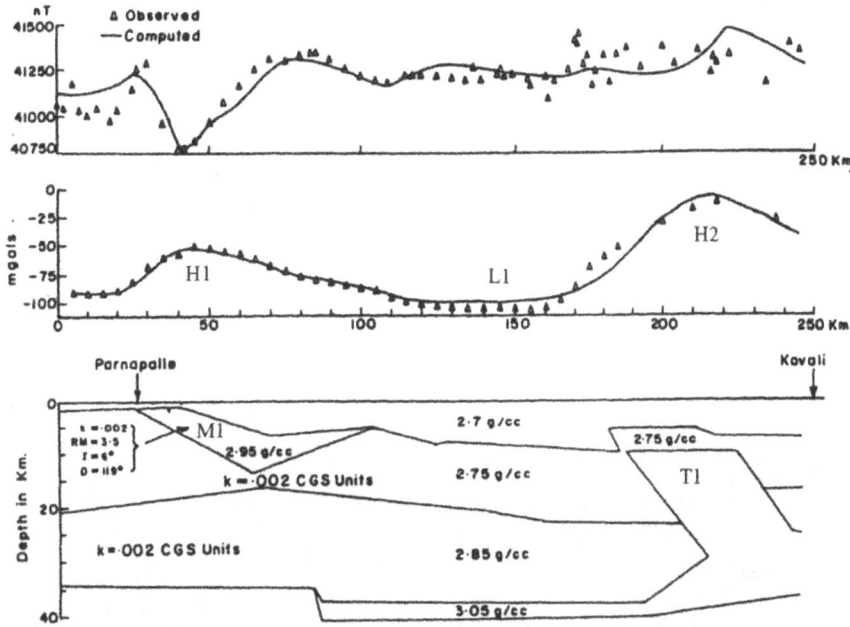

Fig. 7.51(a) Bouguer anomaly (Fig. 7.50(b)) and airborne total intensity (~ 150 m a.g.l, Fig. 7.52(a)) along profile I. The modeled crustal section and the computed fields are also shown with their physical properties mentioned separately for each body. K is susceptibility in c.g.s. units. R.M. is remnant magnetization in A/M and I and D represent the inclination and declination of remnant magnetization. Density is shown in g/cc. High density and high susceptibility body (MI) in the western part represent a mafic/ultramafic lopolith as the basement   (Mishra and Tiwari, 1995).

(c)   The thrusted block (T1) of high density and adjoining crustal thickening suggest collision of the Indian Peninsular Shield presumably with East Antarctica during Mesoproterozoic period as suggested by the similarity of rocks in two sections (Yoshida, 1999) and they were part of Grenvillian formation (Fig. 5.4). The gravity lows, surrounding the gravity high, H2 is caused by felsic intrusives along the pheriphery of the Cuddapah basin. The uranium mineralization reported from the western margin of the Cuddapah basin (Papagani sub-basin) may be related to these felsic intrusives. West verging thrust of high density rocks T1 and crustal thickening under eastern part of the Cuddapah basin suggest east to west convergence and subduction during Mesoproterozoic period (1.5-10 Ga).

(d)   The gravity profile across the EGMB given by Thomas (1992) in Fig. 7.5 is an averaged profile of four profiles reported by Subrahmanyam and Verma (1986). Two of the profiles are located across the Cuddapah basin and the EGMB (L22 and H22, Fig. 7.8(b)) and two across the Bastar craton and the EGMB (L23 and H23, Fig. 7.8(b)). This averaged profile (Fig. 7.51(b)) is modeled on similar lines as the profile across the Cuddapah basin (Fig. 7.51(a)). This profile essentially shows similar gravity anomalies as the previous profile with a high over the EGMB and a low to its west. The crustal model whose computed field matches with those from the observed field (Fig. 7.51(b), Kumar et al., 2004) shows a thrust along the EGMB which has thrusted mid crustal rocks of high density intrusives (2.86 g/cm^3) at the surface. It also shows crustal thickening up to about 40 km, west of the thrust as in the case of previous profile (Fig. 7.51(a)). This provides in general a characteristic crustal model across the EGMB as it represents an average of four profiles across this terrain suggesting similar crustal structures through out the EGMB.

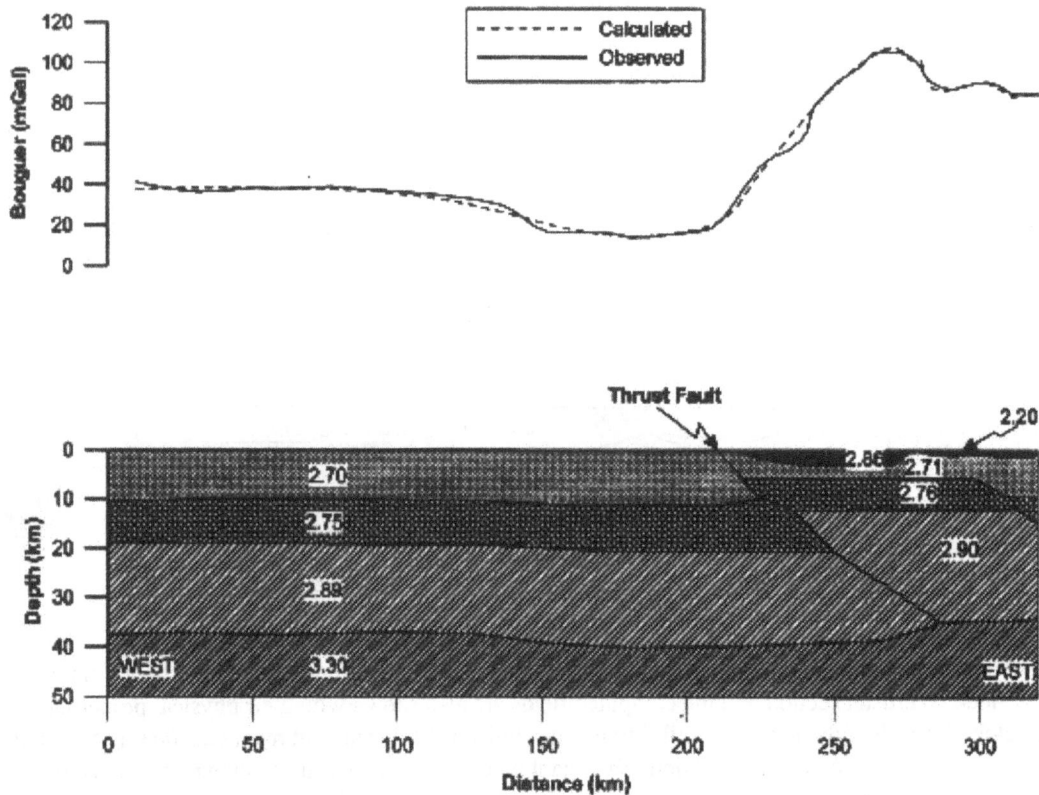

Fig. 7.51(b) Bouguer anomaly and 2-D crustal configuration along a profile that is average of four gravity profiles across the EGMB (Fig. 7.5) and modeled crustal cross section showing an overall thicker crust beneath the adjoining older (Archean) cratons and a thinner crust beneath the younger (Proterozoic) EGMB towards the east. Their contact zone signifies a thrust and welding of two distinct crustal domains through continental collisions. Densities are in g/cm^3 (Kumar et al., 2004).

(e) Magnetotelluric measurements across the EGMB and the Cuddapah basin along Profile I (Fig. 7.50(a)) have provided conductivity distribution (Fig. 7.51(c), Naganjaneyulu and Harinarayana, 2004) in this section, suggesting high conductivity under the EGMB from upper mantle up to a depth of 10-12 km which coincides with the thrusted block. High conductive body associated with thrusted block has also been reported from the Satpura Mobile Belt (Fig. 7.18(b)). The second high conductive body under the Kurnool basin coincides with the mafic body delineated under the western part of the Cuddapah basin (Fig. 7.51(a)), which suggests that this mafic intrusion has modified the crust in this section. The occurrence of fullerence bearing Shungite suite of rocks (Misra et al., 2007) in this section also indicates, that high conductivity may be related to these mafic/ultra mafic intrusives from the mantle.

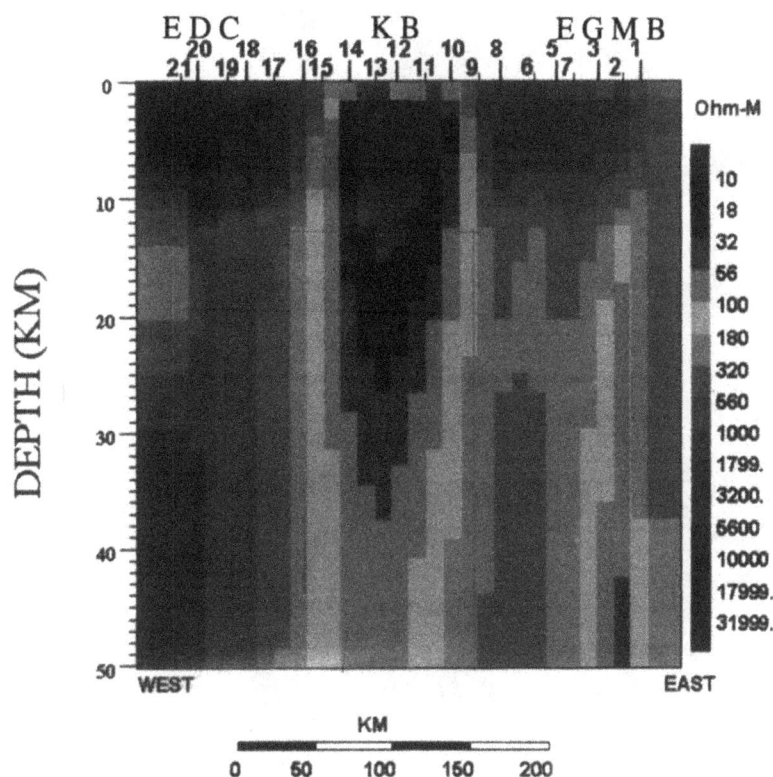

Fig. 7.51(c) 2-D model of conductivity distribution across the EGMB and Cuddapah basin along Profile I (Fig. 7.50(a)). It shows two high conductive bodies related to thrusted blocks of the EGMB and western part of the Cuddapah basin (Kurnool basin, KB) especially the latter that may be related to large scale mafic intrusives in these sections and crust-mantle interaction (Fig. 7.51(a)), while the Eastern Dharwar craton (EDC) shows a resistive crust (Naganjaneyulu and Harinarayana, 2004). [Colour Fig. on Page 799]

(iv) Airborne Magnetic Map of Western Cuddapah Basin and Uranium Mineralization

(a) The airborne total intensity map of the western part of Cuddapah basin recorded at a terrain clearance of about 500' (~ 152 m) with a profile spacing of 1 km (Babu Rao et al., 1987) provided a complex picture of magnetic anomalies due to several mafic intrusives in the Cuddapah basin and metamorphosed sediments. The magnetic data west of the Cuddapah basin was further complicated due to exposed gneisses and various intrusives that included both mafic dykes and granite batholiths. However its image (Babu Rao et al., 1998) shows several predominant NW-SE magnetic trends some of which are correlated with the occurrences of volcanic kimberlite plugs west of the Cuddapah basin and can be used for mineral exploration (Fig. 3.13(b)). The airborne magnetic data of the western part of the Cuddapah basin was corrected for the earth's geomagnetic reference field (IGRF) and filtered for high frequency component (Mishra et al., 1987) that provided a regional magnetic anomaly map. Its image (Fig. 7.52(a)) broadly shows a magnetic low (L1) along the S-W margin of the Cuddapah basin with a magnetic high towards its north and south (H1 and H2). The amplitude and extent of these magnetic anomalies suggest a large mafic body in the basement that has given rise to this set of magnetic anomalies. The magnetic low, L1 is offset by a NE-SW oriented magnetic lineament (ML) that extends across the magnetic low, L1 and defines the eastern limit of the magnetic high, H2. This magnetic lineament coincides with the gravity low, L7 (Fig. 7.50(b)) plausibly due to granite intrusive bodies and is important for uranium mineralization along the western margin of the Cuddapah basin.

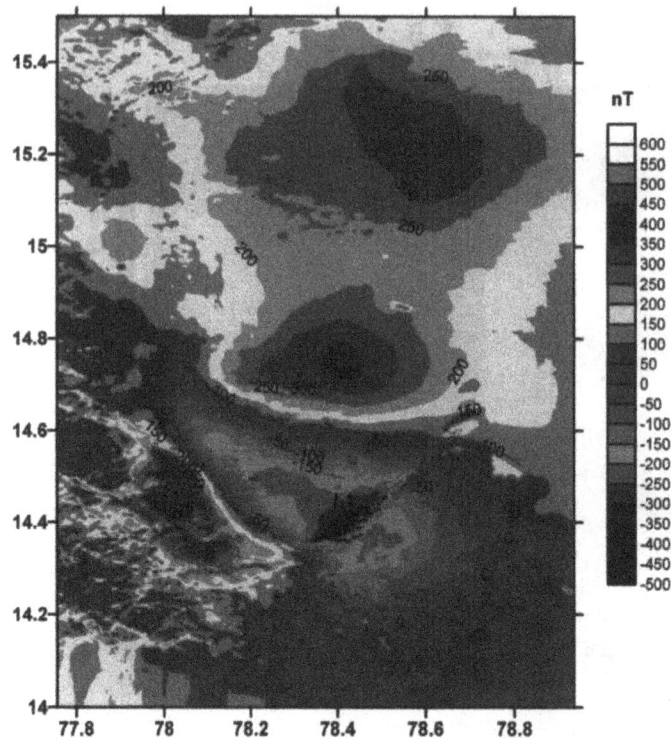

Fig. 7.52(a) Airborne total intensity magnetic anomaly map of Western part of the Cuddapah basin. The IGRF field has been removed from the observed field. It shows a magnetic high (H1) in the northern part and a low (L1) in the southern part along the S-W margins of the Cuddapah basin. Another high (H2) is observed just outside the SW margin of the Cuddapah basin. A magnetic lineament (ML) cuts across the granite batholith, SW of Cuddapah basin and the western margin of the Cuddapah basin that may be important for Uranium mineralization along western margin of the Cuddapah basin. [Colour Fig. on Page 799]

(b) The Cuddapah basin is known for uranium mineralization that occurs along the western margin of the basin. Important host rocks are the strata bound dolomitic limestone and quartzite in the lower part of the Cuddapah super-group (Fig. 7.50(a)) and some structurally controlled Uranium mineralization in the Neoarchean-Paleoproterozoic granitoids along eastern, south western and northern margins of the Cuddapah basin (Sharma et al., 1995). In this regard, the airborne magnetic lineament, ML (Fig. 7.52(a)) is important as it passes through a gravity low related to felsic intrusive body SW of the Cuddapah basin and cuts across the metasediments of the Cuddapah basin in this section where abundant uranium mineralization has been reported. Along the western margin of the Cuddapah basin in Anantpur district around ~ 15^0N, 78^0E, gas leakages have been reported from ground water wells. Resistivity and magnetic surveys suggested that they come through fractures in the mafic sills, and required further detailed exploration to map the structures below the sills exposed in the western part of the basin (Sreedhar Murthy and Sarma, 2009).

(v) 3-D Basement Model of Cuddapah Basin and Mafic Lopolith

(a) The gravity high, H1 observed over the western Cuddapah basin is ideally suited for three dimensional modeling that would provide the depth and nature of the basement rocks. Gravity anomalies usually consist of two components, viz. the regional and the residual fields. The simplest approach to approximate regional field in the observed field is through the use of polynominals of first or second order. The remaining field is known as the residual field that can be used to model the shallow sources like basement etc., In this manner, the residual gravity field for the SW part of the Cuddapah basin is obtained by subtracting a plane from the observed field and this residual field is modeled using 3-Dimensional polyhedra (Barnett, 1976). The polyhedra of high density mafic body that would produce a gravity field almost similar to the residual observed gravity at the surface is given in Fig. 7.52(b), (Mishra et al., 1987). This polyhedra defines a body of density contrast + 0.3 g/cm^3 with its top at a depth of 5 km extending up to 7.5 km in the central part with various apexes along the margins at shallow depths of 1-3 km. This body looks like a mafic/ultramafic lopolith intrusive in the basement that also satisfies the observed magnetic field along a profile (Fig. 7.51(a)).

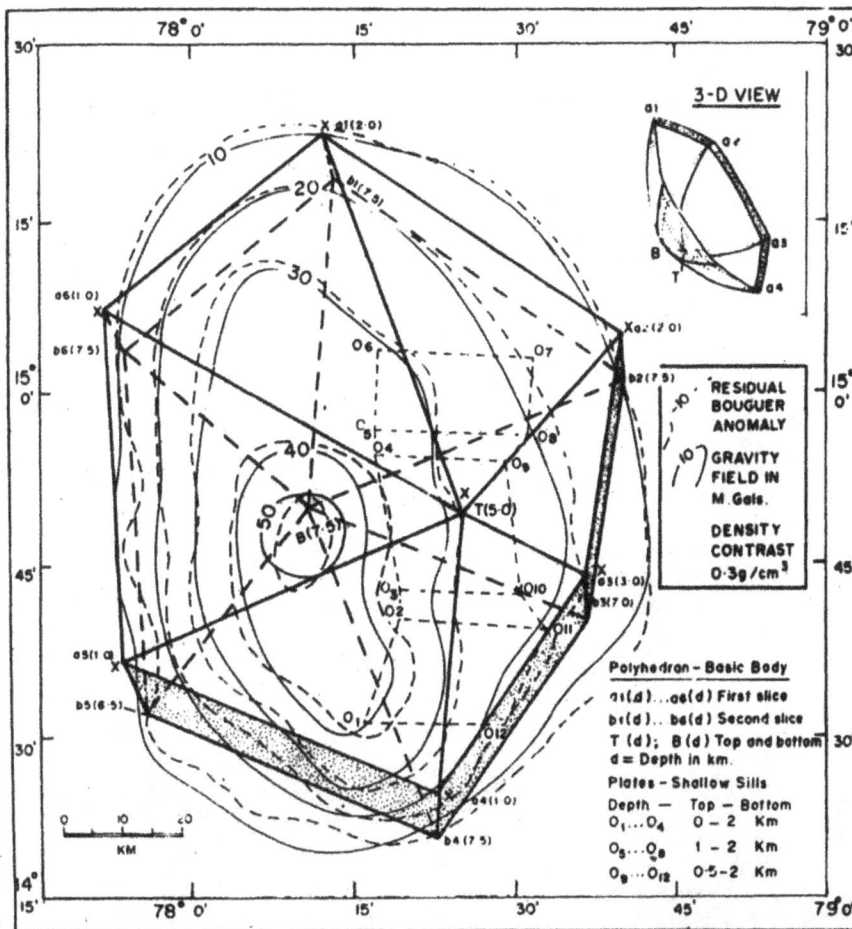

Fig. 7.52(b) Gravity field due to a 3-Dimensional polyhedron with top apices (a$_1$-a$_6$) at depths of 1-3.0 km and bottom apices as b1... b6 at depths of 6.5-7.5 km and its top and bottom lies at depths of 5 and 7.5 km. The computed and the residual observed gravity fields of the western Cuddapah basin are also shown for comparison (Mishra et al., 1987).

(b) A schematic cross section of this lopolith is given in Fig. 7.52(c), and is thickest along the S-W margin of the Cuddapah basin which might have served as conduit for this intrusion and the same might be responsible for the evolution of the western part of this basin. Bhattacharji and Singh (1984) had suggested cycles of thermal event for the evolution of the Cuddappah basin that would give rise to high density ultramafic rocks as basement under this basin. Raval and Veeraswamy (2007) however have attributed it to plume activity.

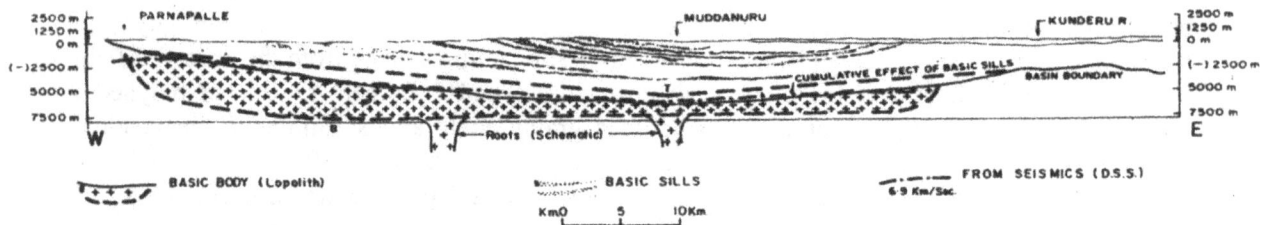

Fig. 7.52(c) A cross-section of the mafic intrusive body from west (W) to East (E) as basement under the western part of the Cuddapah basin in the form of an asymmetrical lopolith.

(vi) Central Part of the EGMB: Coastal Godavari Basin and Hydrocarbon Exploration

(a)   In order to assess the magnetic characteristics of the EGMB, ground magnetic surveys were conducted over a section of the EGMB where it is intersected by the coastal part of the Godavari basin (Fig. 7.7, 7.31). A total intensity map of this region that is covered by alluvium has been given by Mishra et al., (1989) (Fig. 7.53(a)) that shows high amplitude and small wavelength several sporadic magnetic anomalies that is typical of metamorphic terrain which characterizes the EGMB. However, one well defined low was observed along the east coast referred to as Coastal ridge (Fig. 7.53(a)) and a NE-SW magnetic profile (M2, Fig. 7.53(a)) across this anomaly is given in Fig. 7.53(b). As this region lies in a low geomagnetic latitude, a magnetic low defines the characteristics of an almost vertical causative body.

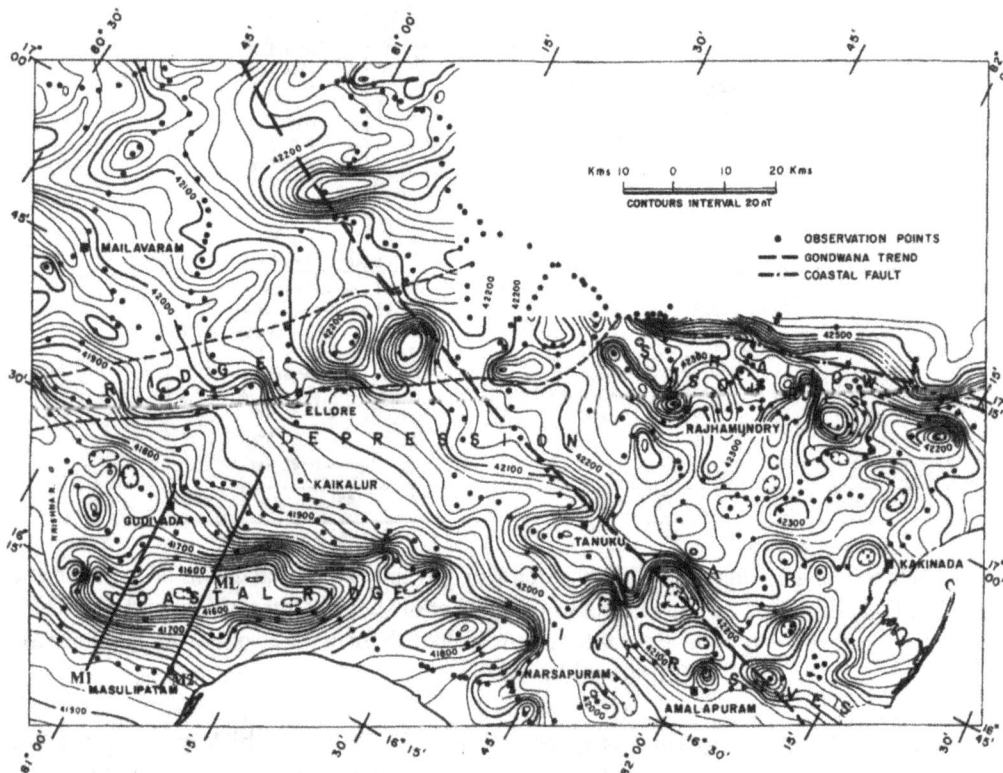

Fig. 7.53(a) Total intensity map of central part of the EGMB as basement under the Coastal Godavari Basin showing sporadic high amplitude anomalies as generally observed over in homogenously magnetized rocks such as trap and metamorphic rocks that have been found in this section as basement in oil wells. A well defined magnetic low (ML) represents the Coastal Ridge (Mishra et al., 1989).

(b) This profile is modeled using the inversion scheme for a tabular body (Section 4.5) and the computed body and its field are given in the same figure. It shows a south dipping body at a depth of about 2.5 km that provides the thickness of sediments in this section. In fact, this coastal Ridge has been important for hydrocarbon exploration in coastal Godavari sub-basin as it formed a barrier to stop the migration of oil towards the sea. It has made the Gudivada depression, north of this ridge (Fig. 7.53(a)) as favourable site for oil occurrence. South-west dipping body is in conformity with the west verging thrusts of the EGMB (Fig. 7.51(a) and (b)). High susceptibility of 0.014 emu suggests mafic rocks that comprises the EGMB as discussed above. It is interesting to observe that similar susceptibility values ares also observed for the basement rocks under the Chintalpudi sub-basin towards its north (Fig. 7.35(b)) suggesting the extension of the EGMB as basement under the southern part of the Godavari basin where it interacted with the Godavari Proterozoic Belt exposed along the flanks of the Godavari basin north of the Chintalpudi sub-basin (Fig. 2.30(a)).

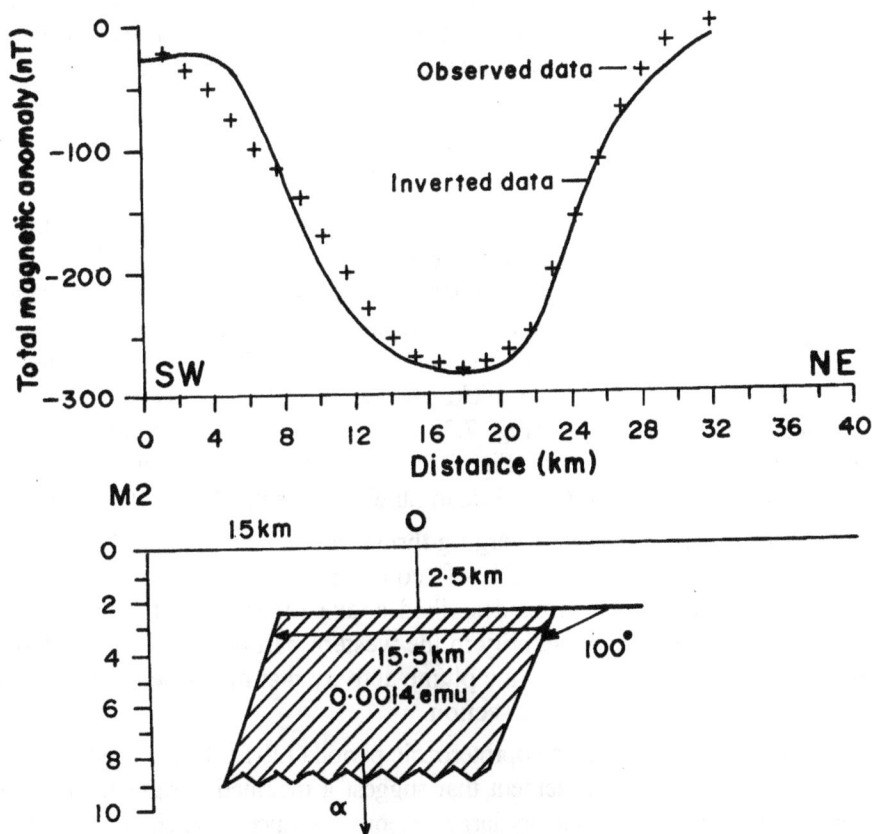

Fig. 7.53(b) Magnetic low due to the coastal ridge modeled along a profile, M2 showing a tabular dyke type intrusive body at a depth of about 2.5 km of susceptibility 0.0014 emu dipping eastwards (Raju et al., 2002) similar to the thrust along the EGMB (Fig. 7.51(a) and (b)).

(c) A gravity profile CC′ from Narsapur to Chintalpudi (Fig. 2.30(a)) across the coastal Godavari basin with the EGMB as the basement is modeled in Fig. 7.53(c) constrained from a seismic profile in this section (Tewari et al., 1996). It shows high density rocks (2.75 g/cm^3) as basement which corresponds to the density of rocks forming the EGMB. Rocks of density 2.30 g/cm^3 represent tertiary sediments that are oil bearing. The shallow basement towards the south represents coastal ridge as shown in Fig. 7.53(a) and (b).

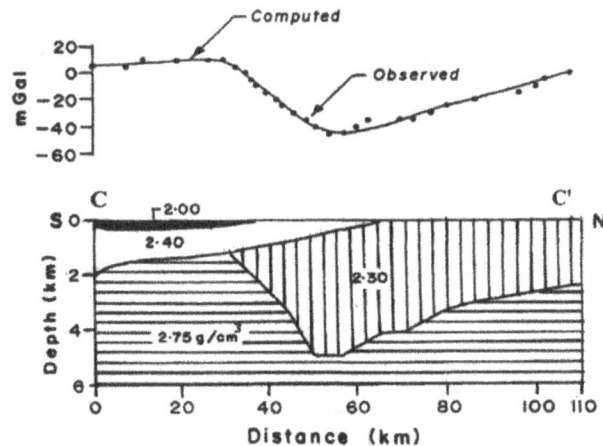

Fig. 7.53(c) A gravity profile CC′ (Narsapur-East of Chintalpudi; Fig. 2.30(a)) across the coastal Godavari basin and modeled shallow section that shows bulk high density (2.75 g/cm^3) rocks for basement equivalent to that for the EGMB that is shallow along the coast representing the coastal ridge.

(vii) Evolutionary Model – Proterozoic Plate Tectonics, Plume and Triple Junction

(a) West verging thrust under the EGMB formed due to collision of the Indian continent and presumably Antarctica giving rise to high grade charnokite and kondalite rocks. Several deep seated intrusives of Mesoproterozoic period indicate west to east convergence and subduction. There have been large scales magmatic intrusions in the Cuddapah basin as suggested above. Mangampeta baryte deposits of the Nallamalai sub-basin along the eastern margin of the Cuddapah basin and associated fullerence bearing shungite suite of rocks (Misra et al., 2007) also indicate volcanic activity. These observations along with crustal thickening, west of the EGMB as suggested above from gravity modeling and seismic studies (Fig. 7.38 and 7.51(a) and (b)), suggest east to west convergence during Mesoproterozoic period (Fig. 7.58(a) and (b)), that is also supported from contemporary direction of convergence in North Indian Shield across the ADMB (Fig. 7.6).

(b) In such a case of E-W convergence, the west verging thrust can be explained at the margin of the obducting Indian craton. The above model of convergence across the EGMB during Mesoproterozoic period conforms with those described above for the ADMB and the SMB in north Indian Shield. Subsequently, plume activity in the western part of the Cuddapah basin gave rise to this part of the basin. It might be the same plume activity that might be responsible for the breakup of the Grenvillean agglomerate in this section.

(c) The Cuddapah super group of rocks of the Paleoproterozoic period (1.9-1.6 Ga) consists of mafic sills and flows and even a mafic lopolith as basement that suggest a magmatic origin for it. Moreover, these magmatic rocks form a part of a plausibly large igneous province (French et al., 2008) that is usually associated with plumes. It is, therefore, suggested that these rocks formed during the rifting phase (Fig. 7.58(a)) that might have been caused by a plume. This rifting might be related to the breakup of the Columbia super continent in this section that has been considered to have existed approximately at the same time (~ 1.9 Ga. Rogers and Santosh, 2009). This rifting phase of Paleoproterozoic period (1.9-1.6 Ga) is followed by convergence (Fig. 7.58(b)) during Meso-Neoproterozoic period that caused deformation of the Nallamalai group of rocks belonging to the Cuddapah super group of rocks along the collision zone of the EGMB. During convergence, the Kurnool super group of rocks of Neoproterozoic period was deposited after collision and the mountain building that supplied these sediments and explains their undisturbed nature being devoid of any magmatic rocks (Fig. 7.58(b)). This explains a hiatus of 0.5-0.6 Ga between the Cuddapah and the Kurnool super group of rocks as suggested previously in case of Lower and Upper Vindhyan group of rocks (Table 7.2).

(d) The Nellore schist belt of the Paleoproterozoic period (~ 2.6-2.5 Ga) in the southern section south of the GPB and Neoarchean high grade metamorphism in the northern section as referred to above suggest that there might have been a prior phase of convergence during Neoarchean to Paleoproterozoic period (~ 2.6-2.0 Ga). It is followed by a rifting phase that gave rise to large scale magmatism of the Cuddapah super group of rocks in Papagani subbasin (~ 1.9 Ga) and Kondapali layered complex of ~ 1.7 Ga in the EGMB along the eastern margin of the Cuddapah basin before the Mesoproterozoic convergence between the Indian cratons and East Antarctica during 1.5-1.0 Ga. Large scale magmatism in the form of large sills, flows, and mafic lopolith in the western part of the Cuddapah basin and contemporary magmatic rocks in the Bastar craton (French et al., 2008) west of the northern part of the EGMB suggest that this magmatism might have been related to plume. During this rifting phase the Cuddapah super group of rocks of Paleo Protepzoic period were deposited (Fig. 7.58(a)). Deformation in the Nallamalai subbasin is related to subsequent convergence during Meso-Neo-Proterzoic period that is described above. The Kurnool super group of rocks of Neo Proteozoic period that are undisturbed and devoid of magmatic rocks must have been formed during final phase of this convergence and collision as basins in such cases are formed after mountain building which supplies sediments to these basins (Fig. 7.58(b)) causing a long hiatus of (0.6-0.5 Ga) between the Cuddapah and Kurnool super group of rocks.

(e) The gravity high, H23 due to high density rocks of the EGMB form a triple junction with the gravity highs, H10 and H11 of the GPB (Fig. 7.8(b)). However, in case the entire EGMB (H22-H4) represented a single suture, it can not be termed as a triple junction. However, if the northern part of the EGMB (H24) related to the Bhandara-Bastar craton and its southern part related to the Dharwar craton and the SGT (H22, H21) formed different sutures with the Antarctica and collided almost at the same time as envisaged above (Fig. 7.6), this junction can be termed as triple junction. It is interesting to observe that the three parts of this junction (H22), viz. the GPB (H10 and H11) and the southern and the northern parts of the EGMB (H24 and H22) make almost 120^0 from each other as in the previous cases of triple junctions discussed above. It may be noted that the gravity high, H23 and magnetic anomaly (Fig. 7.34(c) and 7.35(b)) due to the EGMB extend northwards where the exposed part of the GPB starts and are connected to the gravity highs, H10 and H11 emphasizing their interrelationship. The gravity high, H23 related to this junction extends from the GPB to the gravity high, H22 east of the Cuddapah basin that defines the southern end of the EGMB up to Nellore (Fig. 7.50(a)), which is exposed. The gravity high, H21 that defines its extension southwards is partially located on the continental shelf and partially under the alluvium of the Cauvery basin along the coast forming the basement in this section.

(f) The formations of this triple junction also explains the change in the trend of the EGMB (Fig. 7.6) at this junction (H23), the northern part being NE-SW and southern part N-S and the differences in the metamorphism and rock types between the northern and the southern sections of the EGMB. The southern section of the EGMB south of the GPB is characterized by the low grade Nellore-Khammam schist belt (Fig. 7.8) that formed as back arc marginal basin due to collision of the Dharwar craton with Antarctica along the southern section, which is absent in the northern part. Similarly the high grade metamorphic event of Neoarchean time reported from the northern section is absent from the southern section (Mukhopadhyay and Basak, 2009). Granite intrusives and alkaline complexes of Mesoproterozoic period (~ 1.5-1.1 Ga) are more prevalent in the southern section, south of the GPB compared to the northern section. These observations indicate that the two parts, northern and the southern parts of the EGMB have different collision histories, and are therefore likely to form a triple junction with the GPB. This is a unique case where both ends of the GPB form triple junctions, the northern end with the SMB, and the southern end with the EGMB.

7.10.2 Extension of the EGMB Towards South including Sri Lanka – Crustal Structures

(a) The gravity high due to the EGMB extends even south of Chennai (H21) and there is a corresponding low, L21 to its west related to crustal thickening as H22 and L22, to its north (Fig. 7.8(b)). It is better reflected in the apparent density map given in Fig. 7.49(a) that shows high density rocks (2.8-3.0 g/cm^3) and density gradient (G1 and G2) along the east coast of India extending up to its southern margin. This suggests the extension of the Eastern Ghat orogeny southwards along the entire east coast of India, a part of it being along the continental shelf as the gravity high H21 (Fig. 7.8(b)), extends over it. It appears even to extend in the central part of Sri Lanka as gravity highs H1, (Fig. 7.54(a), Hatharton et al., 1975) related to the Highland Complex of Mesoproterozoic period, almost same as the EGMB. The Highland complex consists of Kadugannawa complex characterized by mafic intrusives of about 1.0 Ga (Brown and Kriegsman, 2003). This map is corrected for the geoid low (Fig. 7.62(a)) in the Indian ocean (Marsh, 1979) to remove the negative bias due to deep seated sources from the mantle. Inset of Fig. 7.54(a) shows the spectrum of this Bouguer anomaly map that suggests three layers at depths of 34, 9 and 4.7 km representing Moho, upper crustal, and basement sources that are almost similar to those found across the east coast of India (Fig. 7.51(a) and (b)).

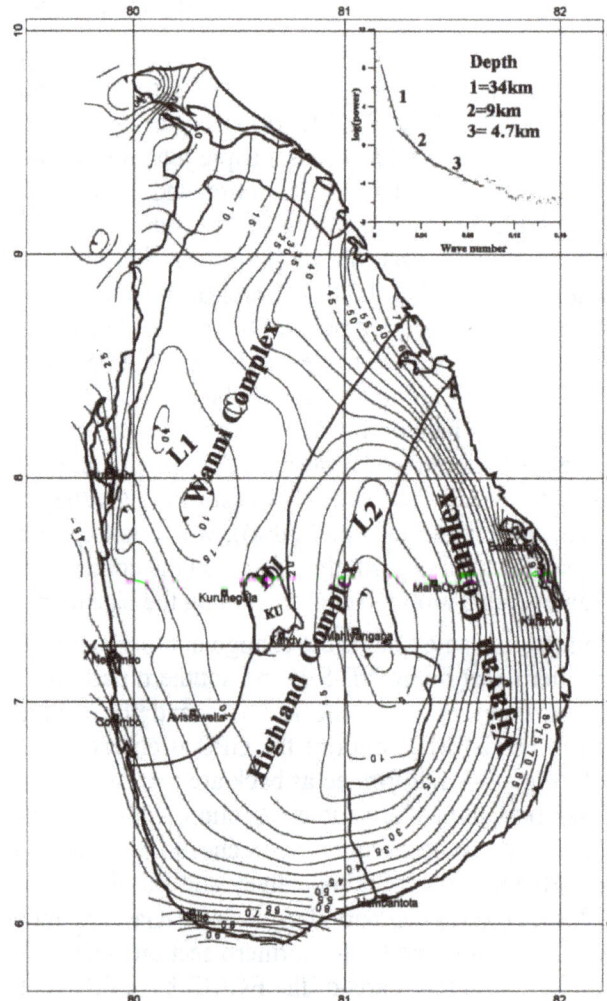

Fig. 7.54(a) Geoid corrected Bouguer anomaly map of Sri Lanka with geological boundaries superimposed on it. XX′ is the profile along which gravity anomaly is modeled as given in Fig. 7.54(b). Spectrum versus wave number of Geoid corrected Bouguer anomaly is given in the inset, which shows three linear segments of slopes corresponding to 34, 9, and 4.7 km which may represent average depths to Moho and sources in the upper crust (Mishra et al., 2006).

(b) An E-W gravity profile (X – X') across the gravity high, H1 (Fig. 7.54(b)) over the highland complex of Sri Lanka suggests high density intrusive similar to that modeled for the EGMB with crustal thickening from about 35 km up to 40-41 km towards the east. Receiver function analysis has also suggested a crustal thickness of 34 km and a low Poissons ratio of 0.25 (Pathak et al., 2006). The crustal thickness along the east coast of India and under the Eastern Dharwar Craton (EDC) is also 32-34 km that is almost same as that along the west coast of Sri Lanka which was supposed to be juxtaposed together and subsequently got separated. The EDC has also shown low Poissons ratio of 0.24 (Gupta et al., 2003) indicating felsic crust as in the case of Sri Lanka indicating a similarity between the two. An average crustal thickness of 35 km based on Bouguer anomaly is also inferred by Tantrigoda and Geekiyanage (2008). Kroner et al., (2003) have suggested Grenville age (1.0 Ga) deformation in Sri Lanka that is also the most predominant age of metamorphism in the EGMB. They suggested active margin settings for Wanni and Vijayan provinces with magmatic arcs of Grenville age. As a suggestion, Sri Lanka might have separated from Gulf of Mannar along east coast of India during the Eastern Ghat orogeny or the offshore Godavari-Cuddapah basin where there might have been more severe effect being part of a triple junction.

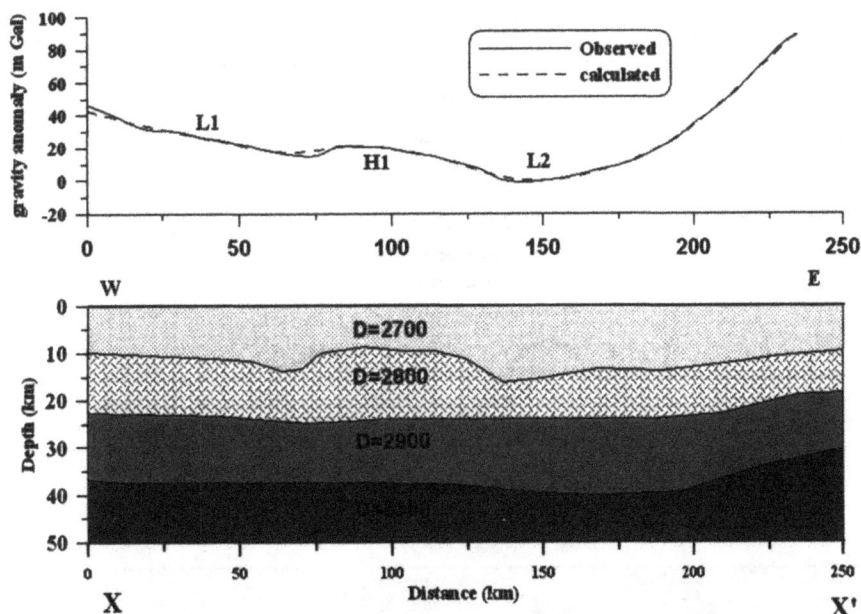

Fig. 7.54(b) Crustal model along Profile XX' (Fig. 7.54(a)) computed using three layered standard crustal model with densities given in kg/m^3. L1 and L2 are flanking lows with a central gravity high (H1). It shows crustal thickening (40-41 km) towards east with high density intrusive related to second layer in the upper crust for gravity high, H1 (Mishra et al., 2006).

7.10.3 Extension of the EGMB towards North – Crustal Structures across Hinge Zone and Bengal Basin, India and Bangladesh

(a) The gravity highs H21-H24 due to the EGMB extends northwards as H25 over the eastern part of the CGGC (Fig. 7.8(a) and (b)). This gravity high (H25) along a profile (I, Fig. 7.30(a)) across the CGGC, Hinge zone and Bengal basin (700 km long) is modeled constraining a part of the profile in West Bengal from deep seismic sounding studies in the central part of the profile (Kaila et al., 1992, Mall et al., 1999). The gravity profile, I (Fig. 7.30(a)) is modeled by extending the seismic sources to either side with some variation to match their computed gravity field with that of the observed field as shown in Fig. 7.55. Gravity high at the western margin of the profile west of the Hinge zone is attributed to the exposed basement and a high density (T2, 2870 kg/m³) body extending from the lower crust up to the upper crust, which is similar to thrusted block as in the case of the EGMB (T1, Fig. 7.51(a) and (b)) including their inferred densities. The relative gravity low in Bangladesh towards the east is modeled due to increase in the sediment thickness. This model also shows the Hinge zone across which the crustal thickness decreases from about 40 km to 34-35 km under the Bengal basin that is considered to represent oceanic crust. It is, therefore, likely that the Hinge zone represents the western margin of Proterozoic India.

(b) Comparison of the crustal model described above across the eastern part of the (CGGC) and Bangladesh (Fig. 7.55) with that across the EGMB (Fig. 7.51(a) and (b)) brings out a very good correlation vis-à-vis high density thrusted lower crustal rocks (T1, Fig. 7.51(a) and T2, Fig. 7.55) and crustal thickness in the two sections. The high density thrusted lower crustal rocks under EGMB have been attributed to Eastern Ghat Orogeny (Singh and Mishra, 2002; Mishra and Tiwari, 1995) related to the collision of Indian cratons with Enderby-land of East Antarctica during Mesoproterozoic period. It, therefore, appears that the Eastern Ghat Orogeny extends northwards beyond the exposed extent of Eastern Ghat and covers the eastern part of the CGGC where it has interacted with the eastward extension of the SMB. The breakup of the Shillong plateau from the Indian Shield along Dauki fault (Evans, 1964) can be attributed to this interaction. The part of the CGGC affected by the Eastern Ghat Orogeny is located with respect to Hinge zone representing the eastern margin of Proterozoic India in the same manner the as the Eastern Ghat Mobile Belt with respect to the present day east coast of India.

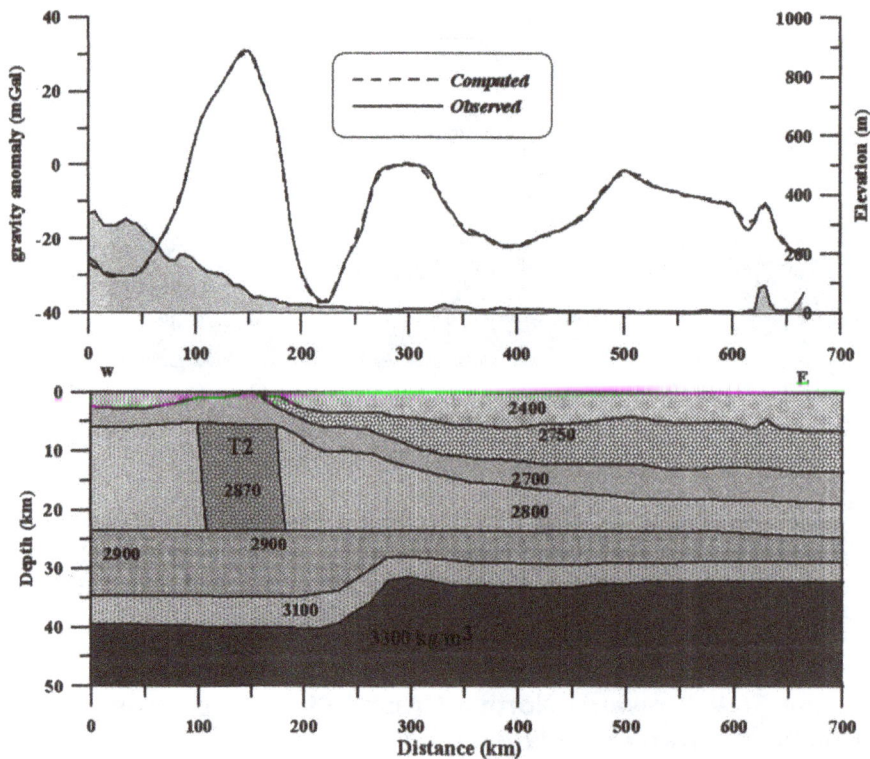

Fig. 7.55 Crustal model along profile I (Fig. 7.30(a)) across the Chhotanagpur Granite Gneiss Complex and Bengal basin. Densities of bodies are given in kg/m³. Thrust, T2 of high density rocks represents thrusted block. Crustal thickness changes from about 40 km under the CGGC (Indian Shield) to about 33 km under the Bengal basin across the Hinge zone. T2 is a high density intrusive in the upper crust similar to T1 along the EGMB (Fig. 7.51(a)).

(c) Chatterjee et al., (2008) have reported 1.55 Ga and Grenvillian metamorphism in the eastern part of the CGGC that is the same as that reported from the EGMB as discussed above. The Hinge zone is supposed to separate the oceanic crust under Bangladesh from the Indian continental crust and therefore, can be considered to represent the extension of the eastern margin of Indian continent before formation of the oceanic crust of Bangladesh. If collision of the Indian Shield with Antarctica gave rise to the EGMB, then this orogeny must extend along the entire eastern margin of the Indian Shield implying its extension towards the north and the south as envisaged above. The gravity high H25 related to it, plausibly extends northwards along the eastern part of the Chhotanagpur Granite Gneiss Complex (H25, Fig. 7.8(b)) and extends further northwards up to the Sikkim-East Nepal border (Fig. 7.30(a)) and may be further extending northwards under the Himalaya as the subducting part of the Indian peninsular shield. The occurrence of Proterozoic alkaline complex of Gorkha-Ampipal in East Nepal along 85° E (Leelanandam et al., 2006) similar to those occurring in Eastern Ghat Mobile Belt further indicates the extension of this mobile belt in Himalaya. It is interesting to note that this section of the Himalayan front is seismogenic and several large and great earthquakes like the Bihar-Nepal earthquake of 1934 (Fig. 6.15(b)) have been reported from this section, which might be attributed to the interaction of Eastern Ghat Mobile Belt with Himalayan orogeny.

7.10.4 Bouguer Anomaly across East Antarctica, Crustal Structures and Rifting, and Convergence

(a) The free air gravity anomaly of Enderbyland and deglaciated topography (Wellman, 1982) shows a good correlation indicating a compensated crust. With average elevation of about 1-1.5 km in this region, crustal thickening of 8-10 km for Airy's model of isostatic compensation is expected. Free air anomaly and deglaciated topography along a N-S profile following the 54°E longitude from the northern tip of the Napier Complex to the central part of Enderbyland (Fig. 7.57(a)) are used to compute Bouguer anomaly profile, which is given in Fig. 7.56. This profile shows gravity high (H1) in the centre followed by gravity lows (L1 and L2) on either sides. The northern gravity low (L1) coincides with intrusive granites of the Napier Complex, while the central gravity high coincides with the high grade rocks of the Napier Complex, and the southern gravity low with rocks of the Rayner Complex.

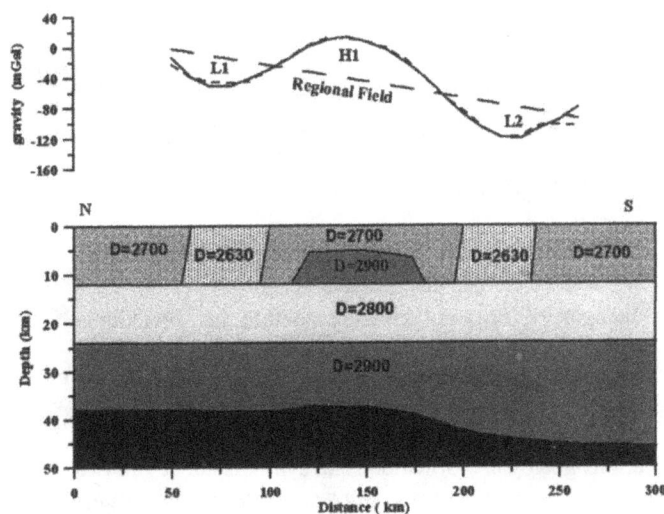

Fig. 7.56 A gravity profile across the Enderby Land and East Antarctica. Location of the profile is approximately shown in Fig. 7.57. The regional field shows thickening of crust towards south with a central gravity high (H1) and flanking lows (L1 and L2). The individual gravity lows and highs are modeled due to low and high density felsic and mafic intrusives, respectively (Mishra et al., 2006).

(b) The modeled crustal section based on this Bouguer anomaly profile is given in Fig. 7.56, which shows a three layered standard crustal model modified based on comparison between computed and observed fields. It shows a crustal thickening from 37 km along the north coast up to 45-46 km inside the continent, as tentatively suggested by the regional field (Fig. 7.56) and isostatic compensation discussed above. Besides changes in crustal thickness indicated by the regional southward dipping field, the local gravity lows (L1 and L2) and high (H1) are interpreted as felsic and mafic intrusives of low (2630 kg/m^3) and high (2900 kg/m^3) densities, respectively in the upper crust. The mafic intrusive in the central part may represent a thrust as in the case of the EGMB (Fig. 7.51(a) and (b)). A gravity profile in the neighbouring region across the Lutzow Holm complex on land also shows a consistently decreasing gravity field from the coast to the interior of the continent related to crustal thickening up to 45-46 km, as supported by seismological studies (Kanao et al., 1994).

(c) The rock types and predominant metamorphic ages of rocks in the Rayner complex are almost the same (1.0 Ga) as those of the EGMB, India and Vijayan Complex of Sri Lanka which were juxtaposed with each other (Rao et al., 1995; Fitzimons, 2000). The gravity anomalies and their sources indicating compressional phases suggest that these continents appear to have joined during the Grenvillean orogeny (1.0-1.0 Ga) and formed the orogenic mobile belt (Fig. 7.57(a)). In this regard, the Antarctica continental Margin Magnetic anomaly (ACMMA) (Golynsky et al., 2002) showing linear magnetic highs offshore Enderbland to Lutzow-Holm Bay and extending further west up to Dronning Maud Land where Africa was attached to Antarctic, assumes special significance. Satellite magnetic data has also shown magnetic highs over the EGMB along the east coast of India and offshore, Sri Lanka, and along the northern margin of Antarctica (CGMW, 2007). These linear magnetic highs can represent mafic rocks related to rifting of the two continents between Paleoproterozoic and Maso Proterozoic convergences as stipulated above.

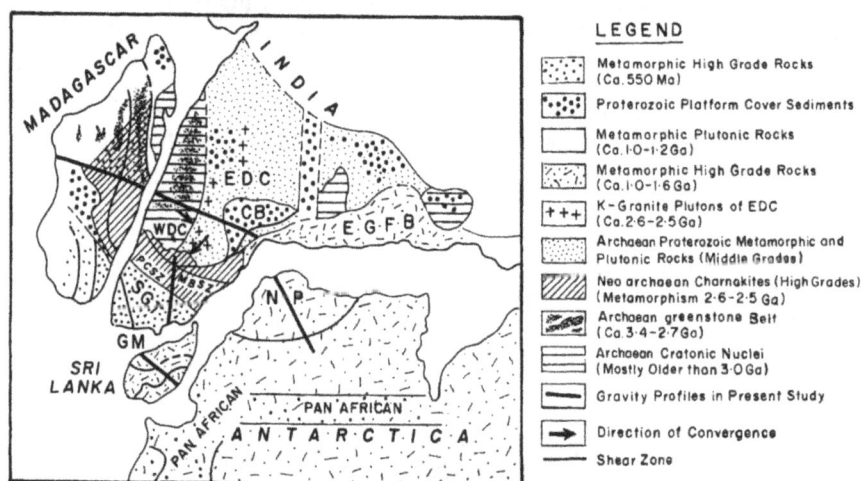

Fig. 7.57(a) Reconstruction of Madagascar, India and Antarctica before their breakup and some common rock types and shear zones (modified after Yoshida et al., 1999). MBSZ and PCSZ represent Moyar-Bhavani and Palghat Cauvery shear zones of Southern Granulite Terrain of India, which appear to extend in Madagascar. The EGFB of India appears to fit with the Napier Complex (NP) of East Antarctica and Sri Lanka, in case Sri Lanka is fitted in Gulf of Mannar (GM). It also shows the layout of gravity profiles across different continents, which have been used in the present study. Various geological formations and symbols are as follows: 1 – Metamorphic High grade rocks (Ca.550 Ma), 2 – Proterozoic platform cover sediments, 3 – Metamorphic Plutonic rocks (Ca, 1.0 Ga – 1.2 Ga), 4 – Metamorphic High grade rocks (Ca. 1.0 Ga – 1.6 Ga), 5 – K-Granite plutons of EDC (Ca. 2.6 Ga – 2.5 Ga), 6 – Archaean Proterozoic Metamorphic and Plutonic rocks (Middle Grades), 7 – Neo archaean Charnockites (high grades) (metamorphism 2.6-2.5 Ga), 8 – Archaean green stone belt (Ca. 3.4-2.7 Ga), 9 – Archaean Cratonic Nuclei (Mostly older than 3.0 Ga), 10 – Gravity profiles in the present study, 11 – Direction of convergence and 12 – Shear zone (Mishra et al., 2006).

(d) Fig. 7.57(a) shows the plausible position of India and East Antarctica during Mesoproterozoic period at the time of formation of the EGMB. Fig. 7.58(a) and (b) show schematic sections of their rifting and convergence during Paleo and Mesoproterozoic periods. The former might have been caused by a plume in the first instance giving rise to large mafic sills and flows and mafic lopolith of the western Cuddapah basin that in fact gave rise to this basin and the Cuddapah super group of rocks were deposited on the rifted platform of Indian cratons as discussed above (Fig. 7.58(a)). Other Proteozoic basins along the EGMB specially those in the Bastar craton also might have formed in the same manner as French et al., (2008) have shown similarity between the mafic sills of Cuddapah basin and Dykes of the Bastar craton as described above. This rifting might be related to the breakup of the Columbia super continent in this section that has been considered to have existed approximately at the same time (~ 1.9 Ga, Rogers and Santosh, 2009). It is followed by the Meso-Neoproterozoic convergence between the Indian cratons and East Antarctica that has been discussed above in detail. This convergence and collision of Indian cratons and East Antarctica (Fig. 7.58(b)) gave rise to high grade rocks of the EGMB due to thrusting. The Kurnool super group of rocks of Cuddapah basin were deposited during this convergence.

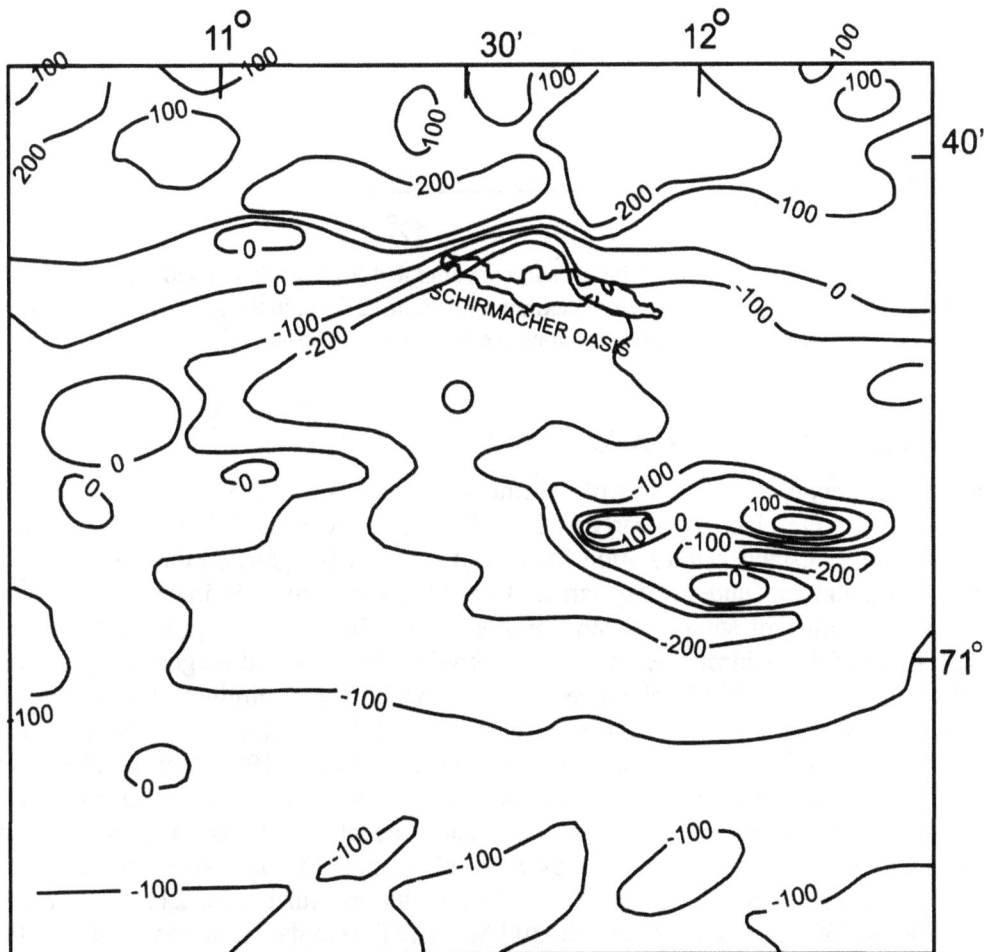

Fig. 7.57(b) Heliborne total intensity residual anomaly map (nT) of Scirmacher Oasis-Wohlthat Mountains that lies south of the surveyed block in Antarctica. N-S profiles, 3 km apart were recorded with a terrain clearance of 325 m. It shows a magnetic gradient north of Schirmacher oasis indicating a fault and moderate amplitude magnetic anomalies in the central-eastern part suggesting mafic intrusive. Elevation changes consistently from 50 m along the northern boundary that is close to the ice shelf, to ~ 1100 m along the southern boundary.

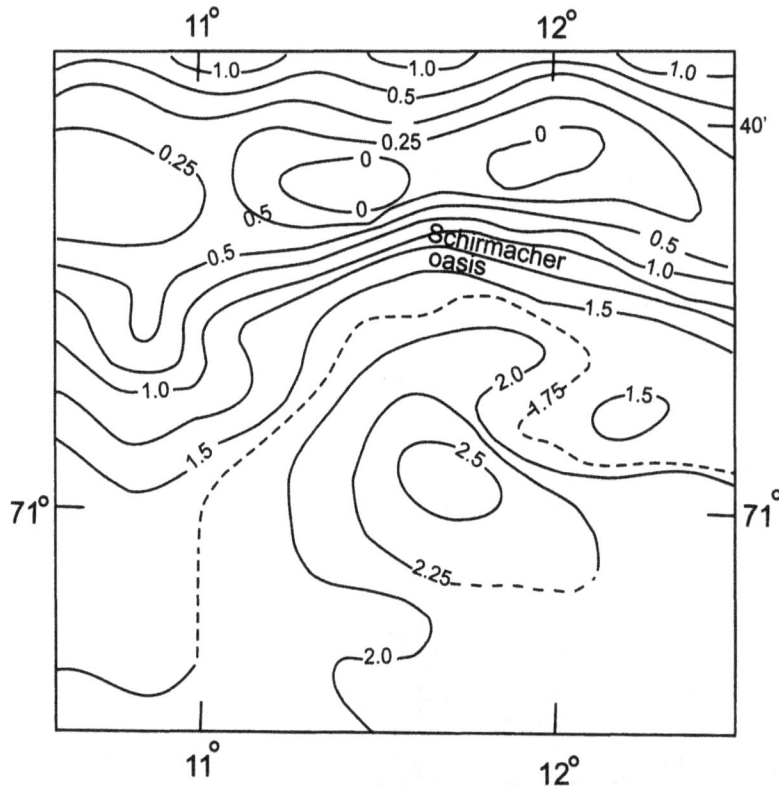

Fig. 7.57(c) Depth to magnetic horizon in km below surface computed from residual magnetic anomaly (Fig. 7.57(b)) through harmonic inversion showing maximum depth of ~2.5 km in the southern part that is exposed in the northern part along the Schirmacher Oasis.

7.10.5 Heliborne Magnetic Survey in Antarctica

In order to assess the magnetic characteristics of basement rocks of Antarctica, magnetic and some airborne magnetic surveys were planned as early as 1987-88 during the seventh Indian expedition to Antarctica. India had already established a station Maitri and Dakshin Gangotri on the ice shelf north of the Schirmacher oasis and was undertaking various expeditions for scientific investigations. First among these was the recording of some magnetic profiles when five magnetic profiles on ground across the northern slope of the Schirmicher oasis were recorded that provided magnetic highs along these profiles (Mittal and Mishra, 1985). However, as these profiles were limited in extent due to logistical problems and did not cross the Schirmacher hills, the full extent of magnetic anomalies could not be recorded and therefore, a heliborne magnetic survey (NGRI, 1989, Verma, 2010) was planned. Again due to logistical problems, a small area was selected to cover Schirmacher hills and to its south upto Wholthat Mountains (85 × 110 sq km) that connects to the Droning Maud Land (Fig. 7.57(b)). Schirmacher hills are characterized by both felsic and mafic intrusives and consist of primarily gneisses and granulite rocks with E-W trending faults on either side. Limited age data suggests gneisses of 860-1050 Ma and basalts of 150-300 Ma. The former date is almost similar to that from the Eastern Ghat Mobile Belt (Section 7.10.1) and the latter one is related to the breakup of Gondwanaland, suggesting convergence and collision between the two during Neoproterozoic period as suggested above.

Magnetic data was acquired using a proton precision magnetometer of o.1-1.0 nT sensitivity for different sample interval and Doppler radar was used for position location. Survey was conducted along the N-S flight lines spaced 3 km apart at a ground clearance of 325 m. The total intensity residual anomaly map after removing a least squares plane is given in Fig. 7.57(b) (NGRI, 1989). This map shows two significant features: first, a magnetic gradient north of the Schirmacher oasis indicating a contact of differing susceptibility, and a set of magnetic highs and lows in the central-eastern part of the map indicating mafic intrusive bodies. Almost the same amplitude of magnetic highs and lows with highs towards the north, qualitatively indicate that the causative sources possess remnant magnetization with inclination as -45^0 almost equivalent to latitude ~ 50-60^0S in southern hemisphere where Gondwanaland was located during the intrusion of these rocks. This period almost coincides with the dates of basalt given above (150-300 Ma). There are some small order anomalies north of the Schirmacher hills over the ice shelf that might be intra basement anomalies. A magnetic basement map of this data set was computed using the harmonic inversion scheme for a magnetization of 300 nT as described in Section 4.3 (Mishra, 1984) and given in Fig. 7.57(c). Basement is exposed along the Schirmacher hills while its maximum depth (2.5 km) is north of Wholthat Mountains. The exposed part of the basement in this map related to Schirmacher hills is shifted towards the north, and is attributed to the effect of remnant magnetization. The magnetic anomaly in the central-eastern part appears to be caused by mafic intrusives at a depth of ~ 1.5 km. A bulk magnetization of 300 nT implies a susceptibility of $\sim 7.5 \times 10^{-3}$ emu that may correspond to gneisses with mafic intrusives.

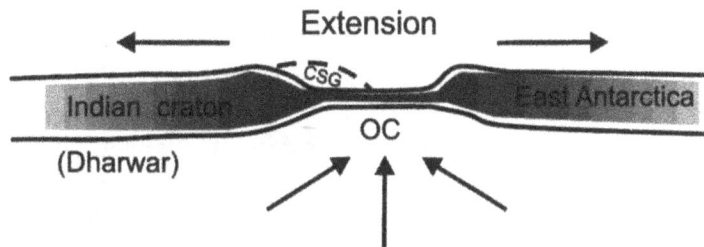

Fig. 7.58(a) A section demonstrating rifting of Indian cratons and East Antarctica with Oceanic crust (OC) in between. The Cuddapah super group (CSG) of Paleo-Mesoproterozoic period is formed over the rifted platform of the Indian (Dharwar) craton.

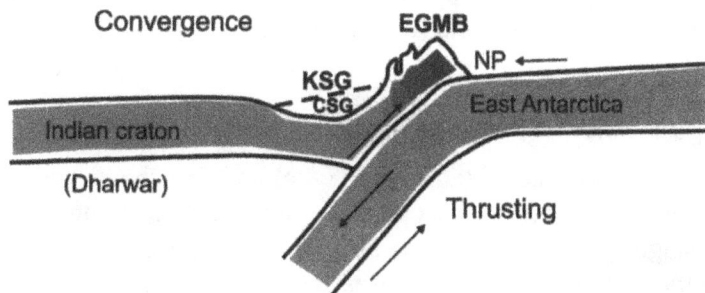

Fig. 7.58(b) Convergence (E-W) of East Antarctica and Indian cratons along the Eastern Ghat Mobile Belt (EGMB) and subduction (NE-SW arrow) causing magmatism and thrusting (SW-NE arrow). The Kurnool super group (KSG) of rocks of Neoproterozoic period are deposited over the Cuddapah super group (CSG) of rocks. Thrusting of rocks has given rise to lower crustal rocks of the EGMB. NP on East Antarctica is the Napier Complex.

7.11 Lithosphere and Asthenosphere under the Indian Continent

The first part of this chapter related to the Archean-Proterozoic cratons and collision zones of the Indian Shield have been primarily dealt with based on the observed geophysical data including gravity (Fig. 7.8(a) and (b)) and its modeling. However, the present day geodynamics of a region are primarily controlled by lithosphere and asthenosphere reflected in large wavelength anomalies, that requires special processing of data where small wavelength anomalies due to shallow sources are suppressed. In this section, the Bouguer anomaly map of India (Fig. 7.8(a) and (b)) is processed for large wavelength features and integrated with the information available from seismic studies to understand the nature of the Indian lithosphere and asthenosphere, which has great implications for present day tectonic activities of the Indian plate including Himalaya and Tibet.

Based on magnetic anomalies in the Indian Ocean, the inferred speed of the Indian plate is about 4.8 cm/yr (Royer and Patriat, 2002) that is one of the fastest among the various plates. The cause for this fast speed of the Indian plate has been a matter of great speculation among the geoscientists. Kumar et al., (2007) have attributed this to the thin lithosphere caused by plumes, while Raval (1993) attributed it to the thick lithosphere driven by plumes and Negi et al., (1987) attributed it to thin lithosphere due to frictional heating. Based on S-wave receiver function analysis, Kumar et al., (2007) have estimated lithospheric thickness of about 100 km, while Priestley et al., (2006, 2008) based on surface wave form tomography, suggested a thickness of about 160 km under the Indian continent increasing to about 320 km under Tibet. Based on the thick lithosphere under Tibet, McKenzie and Priestley (2008) suggested that Tibet may be a craton in the making. Kiselev et al., (2008) suggested crustal thickness of 31 km under Eastern Dharwar Craton (EDC) and 55 km under Western Dharwar Craton (WDC) (Fig. 7.6) and did not notice any change in S-wave velocity between 50 km and 250 km. They also suggested absence of high velocity keel under Indian cratons that characterizes most of the cratons world over. Based on tomographic imaging, Li et al., (2006) suggested a high velocity zone under the Indian continent from 60 km to 200 km that represents the lithosphere. However, the high velocity is further noticed below 600 km (Lebedev and Van der Hilst, 2008) attributed to the subducted Tethyan lithosphere. Priestley et al., (2008) suggested a lithospheric thickness of about 160 km under the Indian Shield that increases to about 200 km under the Himalaya with an upwarp under the Ganga basin. In the light of these controversies about the Indian lithosphere and subducted Tethyan lithosphere, large wavelength gravity anomalies observed over India are analysed to obtain signatures of low density rocks in lithosphere and its consequences. Lithospheric sources are defined by large wavelength gravity anomalies. It is, therefore, essential to know the spectral characteristics of the Bouguer anomaly map of India (Fig. 7.8(a) and (b)) and obtain large wavelength component from it.

7.11.1 Spectrum of Bouguer Anomaly Map of India

As discussed above, Bouguer anomaly map of India (Fig. 7.8(a) and (b)) shows linear gravity highs due to mobile belts and lows due to sedimentary basins. However, the South Indian Shield is characterized by large wavelength gravity lows whose sources are deep seated. To investigate these deep seated sources, a spectral analysis of this map was undertaken (Section 6.1.6). Spectrum of Bouguer anomaly map of India (Fig. 7.8(a) and (b)) and its northern (20°N upwards) and southern parts (20°N downwards) are given in Fig. 7.59 (a), (b), and (c) suggesting long wavelength sources in the upper mantle at a depth of 160-180 km that based on the results of seismic studies as described above, represent the lithosphere-asthenosphere boundary (LAB). The other sources are located at average depths of 36-55 km, 22-26 km, and 6-9 km representing Moho, top of the lower crust, and shallow basement sources. As described above, the depth to Moho varies considerably from Western Dharwar craton (55 km) to the Eastern Dharwar craton (31 km) (Kiselev et al., 2008). Deep seismic sounding and gravity modeling as described above have also suggested crustal thickness varying from about 35-55 km, under the Indian Peninsular Shield. Depths estimated from three spectrums of data drawn from the large data set being within error limits (10-15%), suggest the stability of the computed spectrum and reliability of the results inferred from them (Blackman and Tukey, 1958).

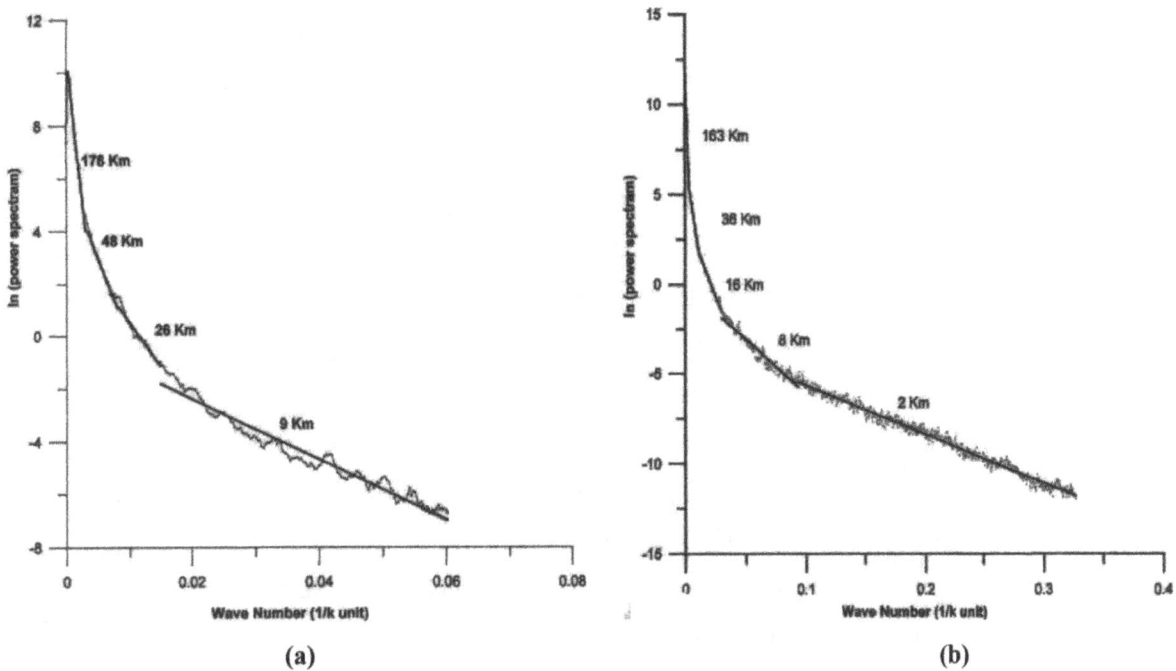

(a)   (b)

Fig. 7.59 (a) Spectrum of the Bouguer anomaly map of India showing linear segments with depths of 178, 48, 26, and 9 km representing lithosphere-asthenosphere boundasry (LAB), Moho, and Conrad discontinuities and upper crustal sources, respectively. (b) Spectrum of the northern part of the Bouguer anomaly of India showing linear segments of depths 163, 36, 16, 8, and 2 km representing almost same sources in this section as given in the caption of Fig. 7.59(a). (Mishra and Ravi kumara, 2008)

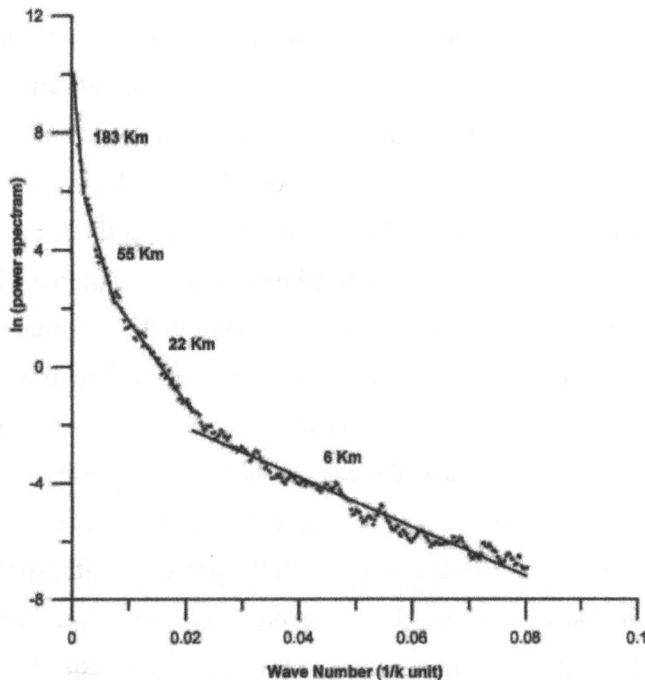

Fig. 7.59(c) Spectrum of the southern part of the Bouguer anomaly map of India showing linear segments with depths 183, 55, 22, and 6 km representing almost same sources as given in the caption of Fig. 7.59(a).

7.11.2 Low Pass Filtered Regional Map and Lithospheric and Crustal Bulges and Seismicity

The low pass filtered Bouguer anomaly for the first segment (Fig. 7.60) related to wave number (n = 0.012) shows large amplitude gravity lows L1 and L2 over the Himalaya and the South Indian Shield, respectively, indicating the dominance of low density rocks in the lithospheric mantle under these sections; while SE and NW sections show gravity highs, H1 and H2, respectively. Tomography images also show high velocity in the central part of India upto a maximum depth of 200 km starting from the Himalayas to South India while NW and SE part shows normal or lower velocity sections (Lebedev and Van der Hilst, 2008). It is well known that the Keruguelen hotspot affected the breakup of India from Antarctica at about 118-120 Ma, the Mahanadi basin along the east coast of India coincides with the center of gravity high H1, and the Lambert rift of Antarctica were conjugate structures (Section 5.6). The Kerguelen plume might have been centered at this junction of Mahanadi and Lambort rift causing lithospheric upwarp or thinning of the lithosphere and intrusives in this section (Mishra et al., 1999). The gravity high, H2 in NW India centered at Saurashtra and Kachchh, may in fact represent signatures of the breakup of Africa from the Indian plate in the form of lithospheric upwarp and high density rocks as this breakup was initiated from offshore Saurashtra on the Indian side and the Somali basin on the African side. Further, there are several Mesozoic basins in NW India, including Saurashtra and Kachchh that might have been produced due to these intrusives. Kachchh itself represents a Mesozoic (Jurassic) rift basin where there are several uplifted parts that might have been affected due to magmatic intrusives related to the breakup of Africa from India and the formation of this rift basin as described in the next chapter (Chapter 8). The airborne total intensity map also shows significant magnetic anomalies similar to those due to volcanic plugs in this section (H9-H11, L9-L11; Fig. 3.30) that extend from Kachchh to Kashmir along the India-Pakistan border. Raval (1989), however, suggested that NW India is affected by the Reunion plume while Kennett and Widiyantoro (1999) had suggested low velocity in upper mantle under NW India that is affected by the Reunion plume which gave rise to the Deccan trap during late Cretaceous. The gravity highs, H3 and H4 in the northern part is caused by flexures of the Indian lithosphere (Section 6.10, Mishra et al., 2004) under the Himalaya towards the north and the west as has been suggested in case of oceanic subduction zones known as outer rise (Turcotte et al., 1978) and described in Section 6.1. The gravity high, H4 extending from Delhi westwards appears to represent the part of the lithospheric bulge whose shallow manifestation occurs in the form of the Delhi-Lahore-Sargodha ridge. The part of gravity high, H1 north of Kolkata over the Shillong plateau is also related to the lithospheric bulge caused by the lithospheric flexure and high density rocks of the Shillong plateau as discussed in Sections 7.6.4 and 6.1.9.

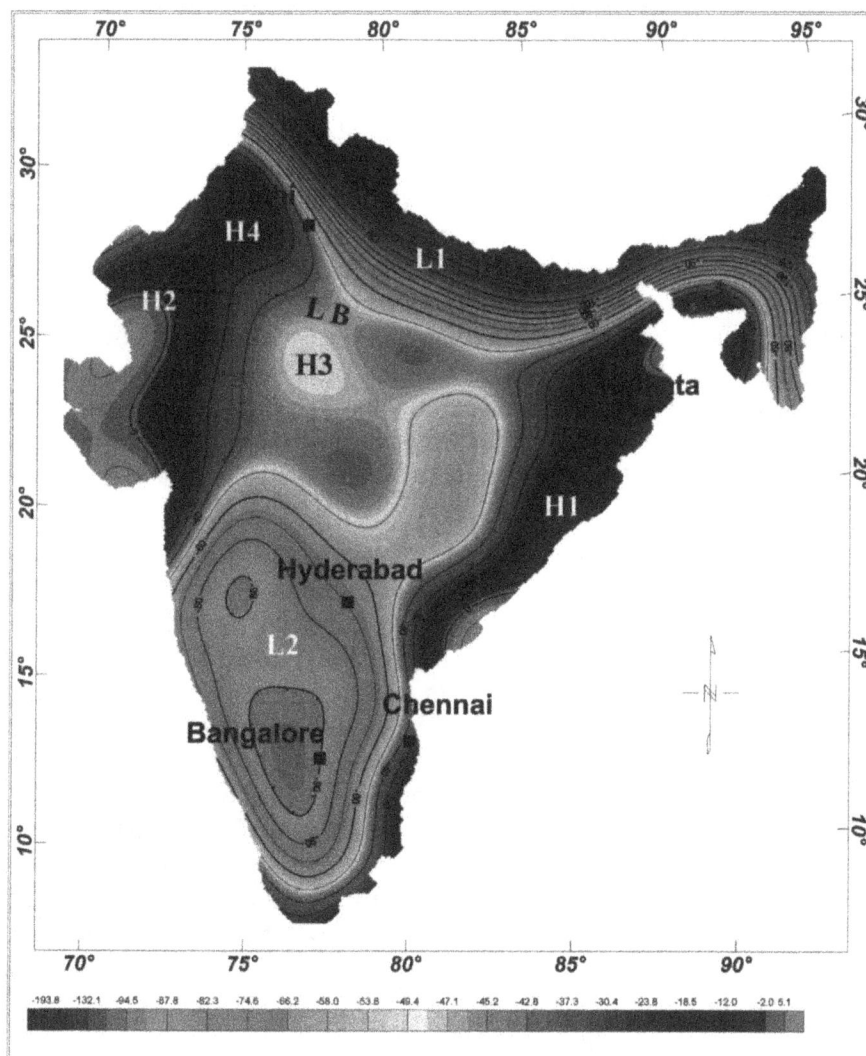

Fig. 7.60 Low pass filtered Bouguer anomaly map for first and second segments of spectrum (7.59a) corresponding to wave number 0.012 and wavelength > 523 km related to sources in the lithospheric mantle showing gravity highs, H3 and H4 over the north Indian Shield related to lithospheric bulge caused by flexure of the Indian plate under Himalaya and lows over the south Indian Shield. Gravity highs, H1 and H2 along the coasts may represent the effects of the breakup of the Indian plate from Gondwanaland and Africa along the east and the west coasts of India, respectively LB: Lithosphsic Bulge. [Colour Fig. on Page 800]

In order to access the effect of the lithospheric bulge due to flexure of the Indian plate on crustal structures, a low pass filtered map of India including third segment of the spectrum (wave number < 0.02) is given in Figs. 6.15(a) and 4.28(b). These maps show the effect of lithospheric flexure more prominently due to effects of both lithospheric and crustal bulges as gravity highs, H1-H4 over the North Indian Shield extending from Himalaya upto the Satpura Mobile (Fold) Belt (SMB) in the south and Eastern India (Shillong Plateau) to the Rajasthan in the west including Aravalli Delhi Mobile (Fold) Belt (ADMB) (Sections 6.1.9 and 6.3.5). Both the mobile belts, viz. the ADMB and the SMB, are reflected as gravity gradients and highs due to associated high density rocks with them as described in Sections 7.5 and 7.6. It also reflects the known tectonics better than the previous map (Fig. 7.60) due to the crustal component in it. This explains the present day uplift in the North Indian shield specially in case of the ADMB and the SMB along old fault systems as discussed in Sections 7.5 and 7.6. It also makes these sections seismically more active as it activates some of the old faults as has been discussed above in case of Jabalpur and Shillong earthquakes (Sections 7.6.1 and 7.6.4). The role of crustal bulge for seismic activity was empsasized by Mishra and Rajasekhar (2006) for seismic activity along the Western Himalayan front (Section 6.1.9) and by Bilham (2004) in the Ganga basin. This crustal bulge introduces compression in the lower crust and extension in the upper crust as shown in Fig. 6.15(b). Therefore, thrust type earthquakes originate in the lower crust due to this factor. The South Indian Shield is primarily characterized by gravity lows, L2 that are centered over the SGT and the Western Dharwar Craton and are related to crustal thickening in these sections. The same flexure, would cause subsidence just south of the SMB that may cause compression in the upper crust and activate pre-existing fault systems. Latur (Killari) and Koyna earthquakes (Section 7.8.3) might have been contributed by it which explains their focus being shallow (< 10 km). The two of the largest earthquakes away from the plate boundary in India, viz. The Shillong earthquake of 1897 (Section 7.6.3) and the Bhuj earthquake of 2001 (Chapter 8) with their hypocenter in the lower crust, are affected by flexure of the Indian plate (Fig. 4.28(b)) that have been classified by the author as Near Plate Boundary earthquakes and are related to E-W oriented faults almost perpendicular to the plate motion.

Two profiles (i) N-S along 77°E and (ii) E-W along 12.5°N passing the through the minimum value of the gravity low over the South Indian Shield in the low pass filtered map for lithosphere (Fig. 7.60) are modeled at average depth obtained from the spectrum. Fig. 7.61(a) is the N-S profile along 77°E that is modeled for a density of 3250 kg/m^3 in the lithospheric mantle below the Moho while surrounding rocks are of density 3300 kg/m^3 associated in general with the upper mantle. It shows a maximum lithospheric thickness of 179-180 km that reduces to 140 km under the lithospheric upwarp of Ganga basin, further increasing northwards to 200 km under the Ganga basin (section 6.2). Based on seismic tomography, Priestley et al., (2008) have shown a high velocity zone up to 150-160 km under NW India increasing to about 200 km under the northern part of the Ganga basin. Their velocity profile also shows a bulge under the Ganga basin that might be synonymous with the lithospheric upwarp up to 140 km shown in the gravity model (Fig. 7.61(a)). The second profile along 12.5°N is modeled in Fig. 7.61(b) that also shows a maximum lithospheric thickness of 179 km reducing to 160 km in the eastern part for a low density body of 3250 kg/m^3.

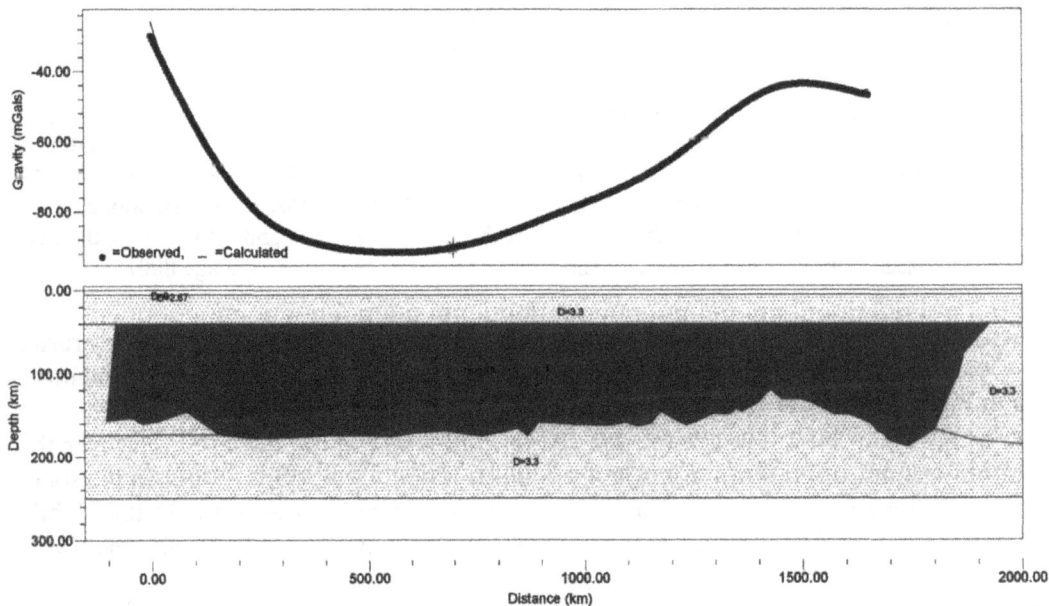

Fig. 7.61(a) A N-S profile along 77°E of low pass filtered map (Fig. 7.60) and model of lithospheric mantle showing low density rocks of 3.25 g/cm^3. This low density body may represent the rocks from the subducted parts of the Indian lithosphere during collision of the Indian and the Asian plates.

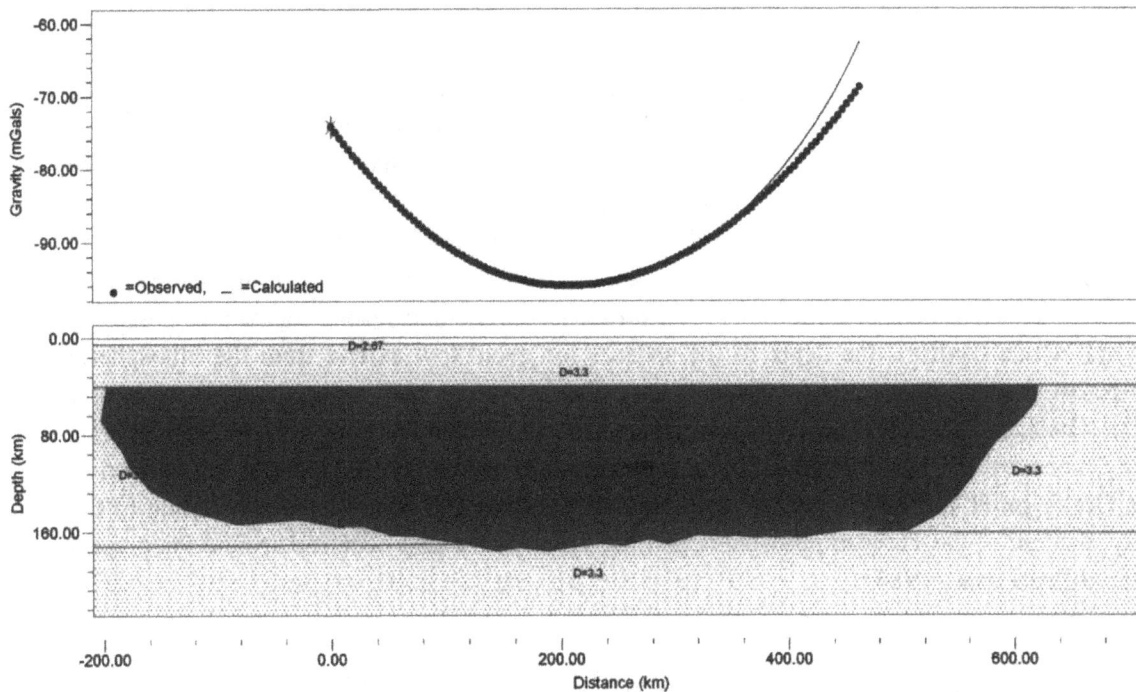

Fig. 7.61(b) An E-W profile along 12.5°N of the low pass filtered map (Fig. 7.60) and model of lithospheric mantle for low density rocks of 3.25 g/cm^3.

7.11.3 Spectral Analysis of Geoid Low of the Indian Ocean

Another data set useful for lithospheric and deep seated sources are geoid data. In fact, geoid data is more sensitive to deep seated sources compared to gravity data as emphasized above in case of the Himalaya (Chapter 6). The occurrence of the largest geoid low in the Indian Ocean (Fig. 7.62(a)) centered at 5°N, 80°E and spreading over a large section of the Indian Ocean and Indian Peninsular Shield (15°S-20°N, 50°-90° E) is analysed to obtain first hand information about the deepest sources of low density in this region. Fig. 7.62(a) is the geoid anomaly of this region in meters obtained from satellite (Reigber et al., 2002) that primarily shows a large low spreading over the entire region. Besides the large wavelength gravity low centered in the Indian Ocean south of Sri Lanka, there are some small wavelength anomalies such as relative gravity highs, H1 associated with the shear zone between the WDC and EDC that represent high density crustal rocks in this section as discussed in Sections 7.8. These anomalies suggest that geoid anomalies can also be used to infer crustal density sources. The only method that can provide first hand estimate of the depth to the causative sources of large wavelength components in the present case is the spectral analysis as discussed in Section 6.1.6. However, as this region consists of both continental and oceanic parts, depth estimate for deeper horizons like the transition zone will be common while depth to shallow horizons like Moho would be different under continents and oceans. Therefore, only the first and second segments of the spectrum can be relied upon. Power spectrum versus wave number for this data set is given in Fig. 7.62(b) that provides the depth estimate for the first segment as 621 km and second segment as 74 km where low density rocks related to the geoid low appears to be concentrated. The first segment appears to be related to the transition zone, while the second one is related to the average depth of lithosphere asthenosphere boundary (LAB) under the continent and the ocean representing mass deficiency in lithosphere and asthenosphere extending below the transition zone. The other two segments represent sources at depths of 22 km and 8 km, which may represent average depth to lower crustal sources under continent and ocean and shallow upper crustal sources, respectively. Since these depth estimates for crustal sources are average for continental and oceanic parts they have not been used for any specific purpose. Moreover, the present study relates to deep sources in the lithosphere and the asthenosphere. Geoid data of this map was divided into two groups lower and upper blocks as 15 °S to 5 °N and 0 ° to 20 °N and 50 °-90 °E and their power spectrums were separately computed that also provided similar depth estimates for the first segment (deepest layer) and second segment as ~ 630 km and ~ 70 km, which confirms the depth to the sources for geoid low extend from the lithospheric mantle to below the transition zone (~ 620-630 km). These sources may represent the low density rocks of the subducted Indian lithosphere below the transition zone that has been inferred from seismic tomography (Lebedev and Van der Hilst, 2008). Concentration of low density rocks in this section of the Indian Ocean geoid low can be attributed to the first collision and sinking and subduction of the break away material of the Tethyan lithosphere and the Lhasa block that occurred around the present day equator (Allegre et al., 1984).

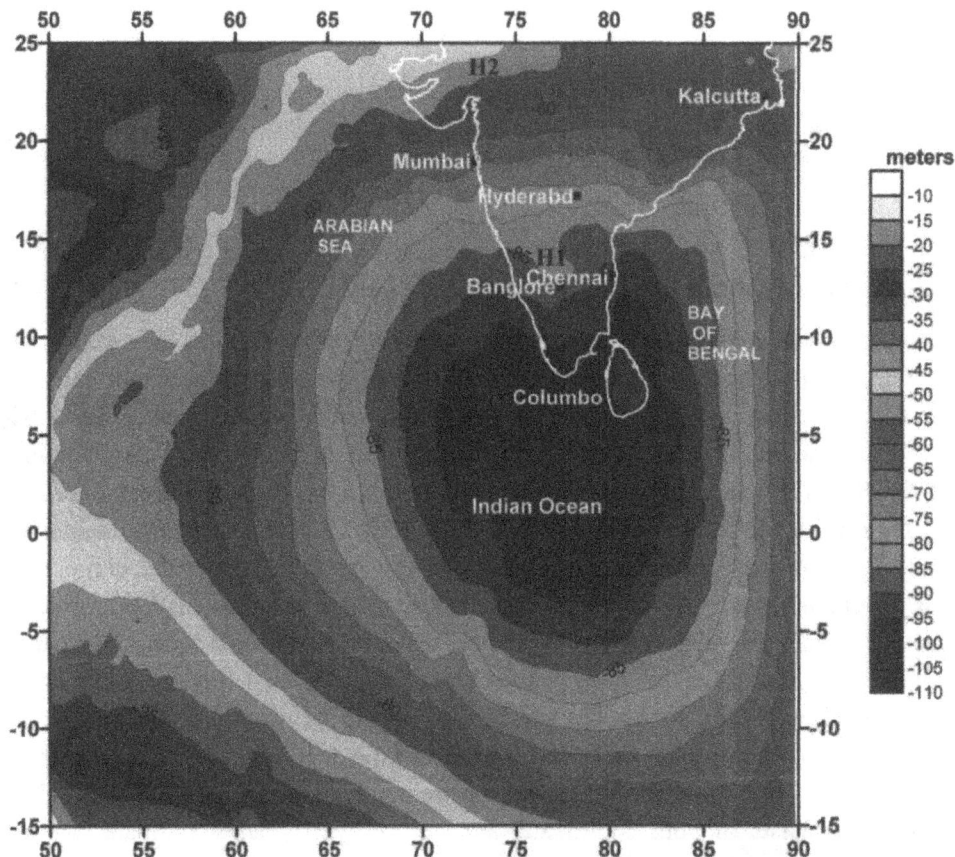

Fig. 7.62(a) Geoid low of the Indian Ocean extending over the Indian Peninsular shield with a relative high, H1 coinciding with parts of the eastern Dharwar craton including the shear zone between the eastern and the western Dharwar cratons and another high, H2 over Kachchh and the Ganga basin that may be a reflection of the lithospheric and crustal bulges due to flexure of the Indian lithosphere. [Colour Fig. on Page 801]

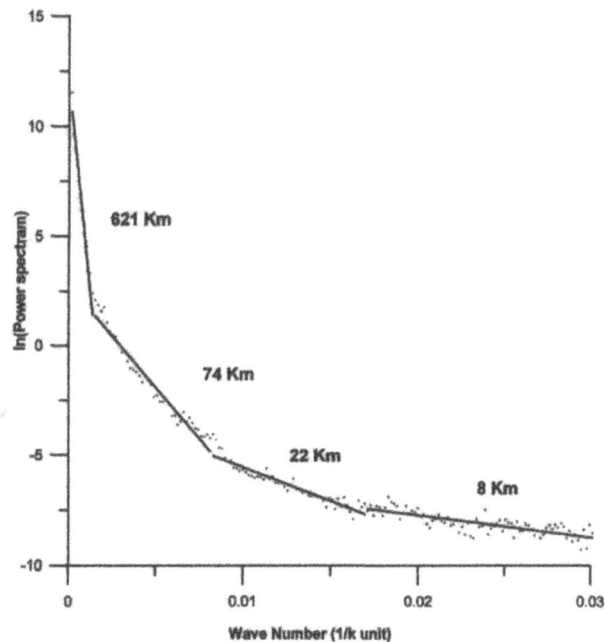

Fig. 7.62(b) Spectrum of the Indian Ocean geoid low showing sources at four levels at respective depths of 621, 74, 22, and 8 km representing asthenospheric sources below transition zone, average of LAB and Moho under ocean and continent and crustal sources, respectively in this region.

7.11.4 CHAMP Satellite and Airborne Magnetic Data Over Indian Peninsular Shield and Spectral Analysis – Moho as Magnetic Discontinuity

As we are interested in deep seated magnetic anomalies in this section, CHAMP satellite data was used for this purpose. Fig. 6.38 is the CHAMP satellite magnetic map of India and Tibet observed at 400 km and continued downwards at 50 km obtained from the website: http://geomag.colorado.edu/lithomod.html. Most of the major magnetic anomalies of this map are correlated to surface geology and tectonics of Indian continent which was also true in case of earlier MAGSAT anomalies (Section 3.6.3, Mishra and Venkatarayudu, 1985) with some differences mainly in the amplitude of these anomalies. Some of the major magnetic anomalies observed over the Indian Shield of this map (Fig. 6.38) are as follows:

(a) The magnetic lows, L3 and highs, H3 coincide with the Aravalli Delhi Mobile Belt and gravity highs, H4 to its west (Fig. 7.8(a)). The ADMB in fact coincides with the gradient of these anomalies which represent a Proterozoic collision zone and consist of mafic rocks including ophiolites and granulite rocks of Paleo-Mesoproterozoic period as discussed above in Section 7.5. The magnetic highs, H3 coincide with the gravity highs, H4 (Fig. 7.8(a)) that have been considered to be caused by shallow intrusives related to the breakup of India due to Karoo and Reunion plumes during Late Jurrassic and Late Cretaceous, respectively (Section 7.11.2). These magnetic anomalies, H3 and L3 extend to the western Himalayan Syntaxis and a part of it extends towards the NW that coincides with the Delhi-Lahore-Sargodha Ridge and might be caused by it. The part of these magnetic anomalies extending to the Western Syntaxis represent mafic intrusives due to plumes and might even be responsible for the formation of this syntaxis as several examples of plumes being responsible for changes in the trends of the subduction zones have been cited by Vogt (1973). Toward the south, the magnetic high, H3 extends to Saurashtra which is occupied by several volcanic plugs of Deccan trap eruption from Reunion plume (Section 9.2).

(b) The gradient of magnetic high, H4 and low, L4 coincides with the Satpura Mobile Belt that represents another Proterozoic collision zone and consists of mafic/ultramafic intrusive (Section 7.6). However, these magnetic anomalies as a whole coincide, respectively, with the Vindhyan basin and the northern part of the Deccan Volcanic Province and may also be related to the basement and the mafic intrusives of the latter.

(c) The magnetic highs, H5 extending from the west to the east coast with NE-SW trend, and along the east coast of India coincide with Dharwar folding and the Eastern Ghat Mobile Belt that also represent a Proterozoic collision zone consisting of mafic/ultramafic intrusives as discussed above in Section 7.10. Its southern margin coincides with the transition zone between the Dharwar craton towards north and Southern Granulite Terrain towards the south. (Fig. 7.37), characterized by mafic intrusives as discussed in Section 7.9.

(d) Magnetic low, L5 and high, H6 coincide with the Southern Granulite Terrain with lower crustal granulite rocks that are predominantly mafic in nature which are represented by these magnetic anomalies. The airborne magnetic map of this region (Fig. 7.47) also presents paired magnetic anomalies with lows towards the north indicating induced magnetization as in the present case of satellite magnetic data. The gradient of magnetic low, L5 and high, H6 coincides with the Palghat Cauvery Shear Zone (PCSZ, Fig. 7.37) that is a major tectonic boundary between the northern and southern blocks of the SGT representing a Proterozoic collision zone (Section 7.9). These magnetic anomalies extend in the Bay of Bengal indicating the extension of the SGT in this section as basement.

(e) Arabian Sea and Bay of Bengal provide some significant magnetic anomalies, H7, L7 and L6 and H6, respectively. The former set (H7 and L7) are caused by the continental shelf and margin occupied by igneous intrusives from the Reunion plume which gave rise to Deccan trap eruption (Section 5.4). The latter (L6 and H6) in Bay of Bengal showing a NE-SW trend appears to represent basement anomalies caused by intrusives related to 85 and 90 East Ridges.

These observations demonstrate the application of satellite magnetic data for deciphering tectonics of a region.

(i) Spectral Analysis of CHAMP Satellite-Airborne Magnetic Data

The CHAMP Satellite-airborne magnetic map (Fig 3.30) over the Indian continent (EMAG2; Maus et al., 2010) shows intermediate-short wavelength anomalies and has reflected most of the tectonics of the Indian Shield as described in Section 3.8.2. This map has been prepared by combining the long wavelength CHAMP satellite data with the available airborne magnetic data from the Indian continent and reduced to 4 km altitude above geoid. In order to estimate the depth to the magnetic sources, digital data of this map (Fig 3.30) was subjected to spectral analysis that is given in Fig 7.63(a). To test the stability of the computed spectrum, spectral plot for the southern parts (25° N southwards) is given in Fig 7.63(b). These spectrums provide the average depths to the magnetic sources. The deepest sources lie at the depth of ~24-25 km as the data pertains to 4 km height above geoid. This defines the average thickness of the magnetic crust where magnetization is reduced sufficiently and cannot be recorded. Curie point isotherm defines the level where magnetization of rocks vanishes totally. However, to record magnetic anomalies from that depth at an altitude of 4 km, there should be a significant magnetization suggesting that thus isotherm would be below this level. It conforms with the normal-low heat flow in the Indian shield (14-20 mWm^{-2} mantle heat flux, Roy and Mareschal, 2011) that suggests Curie point isotherm below the Moho. The second segment of this spectrum at depths of ~ 3-4 km represents the average depth of basement in basins and intrusive in the upper crust as there are several basins with this value as average thickness of sediments such as the Ganga basin and, Godavari basin etc., The third segment primarily represents noise. Maus et al., (2011) have updated this map as EMAG3 by incorporating ship borne magnetic data for the Arabian and Bay of Bengal but it is the same as EMAG2 over the Indian continent.

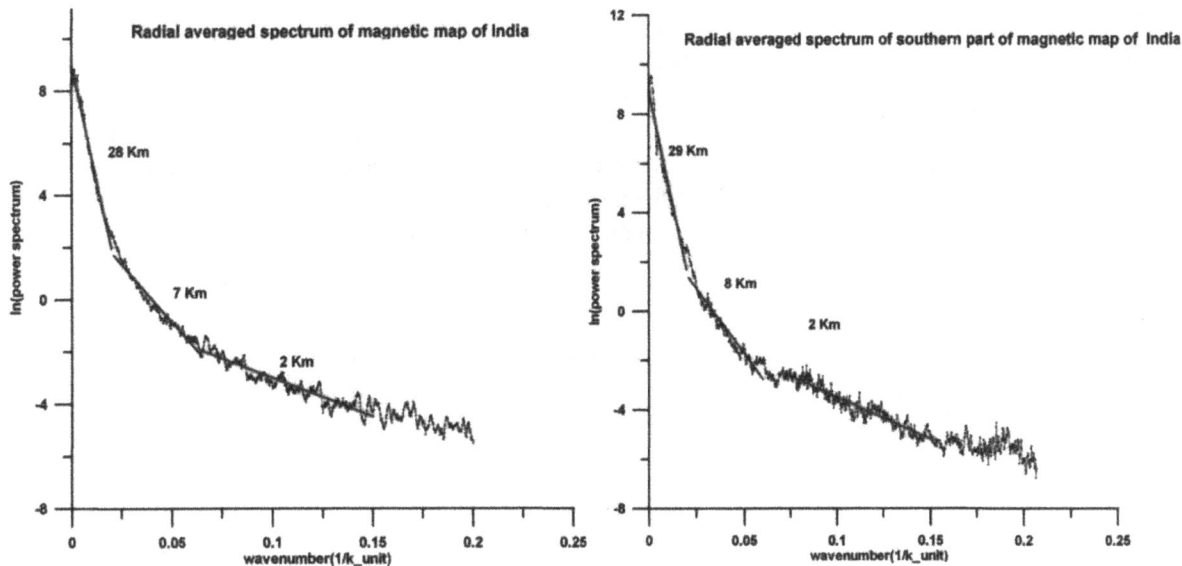

Fig. 7.63 Radially averaged spectrum of the magnetic anomaly map of India reduced at 4 km altitude over geoid (a; EMAG2; Fig 3.30) and its southern part (b; south of 25^0N). They show depths of 24-25 km and 3-4 km. The last segment is primarily noise.

(ii) Spectral Analysis of Airborne Magnetic Profiles and Moho as Magnetic Discontinuity

(a) Three long range high altitude (2.70 km above msl) airborne magnetic profiles are described in this Chapter to access the magnetic characteristics of Dharwar craton, viz. Mangalore to Madras along ~ 13°N parallel (Section 7.8), one across the Cuddapah basin (Fig. 7.50(c)), another also across Cuddapah basin, north of the previous profile (Mishra, 1984). Magnetic anomalies along these profiles have been qualitatively used to ascertain the magnetic nature of the sections where they are located, but could not be used quantitatively because of high altitude and uncertainties in the actual location of the anomalies. However, such profiles are useful to find approximate depths of deep seated sources related to large wavelength features using spectral analysis as discussed in Section 6.1 and demonstrated above for large wavelength gravity and magnetic anomalies.

(b) In order to find the depth to the magnetic sources of different wavelengths, the digital data of magnetic profile from Mangalore-Madras along ~ 13° parallel is transformeds to frequency domain and its ln(amplitude spectrum/k) is given in Fig. 7.64(a) (Mishra, 1984). The factor 1/k is a correction factor for the rectangular shape of the magnetic sources as discussed by Mishra and Pedersen (1982) (Chapter 4). However, as the spectral decay provides an average estimate, it can be used for an approximate estimation even without this correction. This spectrum provides basically two straight line segments, with slopes equal to 35 and 10 km, that imply average depths of about 32-33 km and 7-8 km below the surface. The interface at a depth of 32-33 km coincides with the Moho under the Eastern Dharwar Craton (Section 7.8.4) where largest section of this profile is located.

(c) The two long range airborne magnetic profiles across the Cuddapah basin (Section 7.10.1, Fig. 7.50(c)) that is part of the Eastern Dharwar craton are also transformed to the frequency domain and their ln(amplitude spectrum/k) is given in Fig. 7.64(b) and (c). These spectrums can be approximated by two straight line segments with slopes equal to 44 km and 9 km that imply approximate average depths of 41-42 km and 6-7 km for respective sources. The interface at a depth of 41-42 km coincides with the Moho in the eastern section of the Cuddapah basin (Section 7.10.1).

(d) The depth to sources related to the first segment approximately coincides with the depths to Moho in these sections while the second segment represents sources in the upper crust. In case of the Cuddappah basin, the second segment represents the mafic lopolith forming the basement as described in Section 7.10 (Fig. 7.51(a)). It implies that the Moho approximately coincide with the magnetic crust and represent a magnetic discontinuity similar to density discontinuity. This implies that even at this level there is sufficient magnetization to produce magnetic anomalies at the surface/flight altitude. Curie point isotherm may, therefore lie below it.

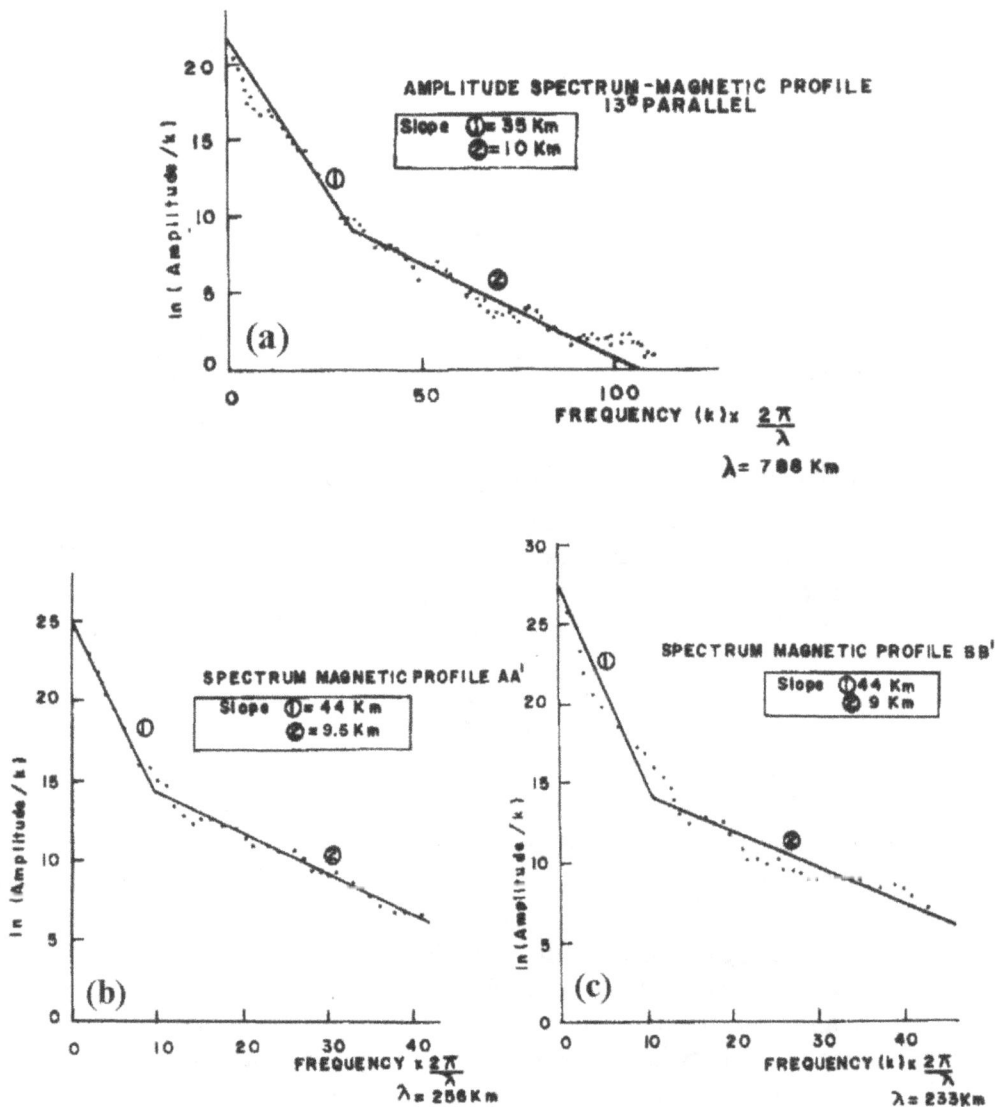

Fig. 7.64 (a) Spectrum of long range, high altitude airborne magnetic profile from Manglore to Madras (Chennai) across Dharwar craton flown at an altitude of 2.7 km above msl (Fig. 3.31) showing linear segments with slopes of 35 and 10 km representing an average Moho and upper crustal magnetic sources. Moho in this case may represent a magnetic discontinuity similar to Curie point geotherm. (b) and (c) Spectrum of long range, high altitude airborne magnetic Profiles across Cuddapah basin (Fig. 7.50(c) and south of it along 15^0 N) flown at altitude of 2.7 km above msl showing two segments of slopes 44 and about 9 km representing approximately the Moho and the basement under the Cuddapah basin. In this case also Moho appears to represent the Curie point geotherm.

7.12   Conclusions: Archean-Proterozoic Convergence Model of Indian Cratons, Rifted Platforms, Plumes, and Basins

The study and observations presented above suggest following important conclusions about Indian cratons and mobile belts.

7.12.1 North Indian Shield: Paleo and Mesoproterozoic Convergence and Rifting

The North Indian Shield consists of the ADMB and the SMB surrounding the Bundelkhand craton in the central part and Rajasthan block (craton) towards the west. The following significant points about Proterozoic plate tectonic have emerged from the above discussions:

(a) Fig. 7.6 depicts the convergence of Indian cratons across various mobile belts as described above. As suggested previously, there have been two phases of convergence and an intermediate rifting phase across the ADMB and the SMB from east to west and north to south during Paleoproterozoic and Mesoproterozoic periods, respectively. The first cycle of convergence and rifting during Neoarchean-Paleoproterozoic period (2.9-1.6 Ga) across the ADMB related to granite intrusives of that period and the Aravalli group of rocks (Table 7.1). It is followed by convergence between the Bundelkhand craton and the Rajasthan block during Meso-Neoproterozoic period (1.5-0.8 Ma) related to the Delhi group of rocks, Neoproterozoic back arc basins, and related magmatism to its west (Table 7.1) (Fig. 7.65(a)).

(b) N-S convergence and rifting during Neoarchean-Paleoproterozoic period (2.9-1.6 Ga) across the central part of the SMB gave rise to various older granite intrusions of Bhandara craton and Mahakoshal and Bijawar group of rocks of Paleoproterozoic period with mafic/ultra mafic intrusives deposited in rifted basins over the rifted platform of the Bundelkhand craton. In fact, the large scale mafic and ultra mafic intrusives might be responsible for the rifting of the Bundelkhand craton. This rifting phase also affected the eastern margin of the ADMB with Bundelkhand craton where rocks equivalent to Mahakoshal and Bijawar Group of rocks have been envisaged. NW-SE oriented mafic dykes of Paleoproterozoic period in the Bundelkhand craton also support this rifting phase during this period. Subsequently, the Lower Vindhyan group of rocks (Semri group, ~1.7-1.6 Ga) of volcanogenic sedimentary sequences were deposited over the rifted platform of the Bundelkhand craton along the SMB and the ADMB during the rifting phase, with Bijawar and Mahakoshal group of rocks as basement (Fig. 7.23(b)). The second cycle of convergence during Meso-Neoproterozoic period along the SMB gave rise to the Sausar and Singhbhum orogenies and contemporary magmatic rocks of Bhandara and Singhbhum cratons (Table 7.1) with the Central Indian Shear (CIS) and the South Purulia Shear Zone (SPSZ) along southern margin of the SMB, respectively, as thrust and suture. The Upper Vindhyan group of rocks (Kaimur, Rewa, and Bhander groups) of Neoproterozoic period (~ 1.0-0.7 Ga) were deposited during the final phases of this convergence as foreland basin after collision and mountain building that supplied these sediments which are largely undisturbed. This explains the long hiatus of 0.6-0.5 Ga between the Upper and the Lower Vindhyan groups.

(c) The SMB and the CIS extends towards the west with the Tapti Lineament as suture separating the Dharwar craton from the Bundelkhand craton towards the north. The best signature of a deep seated thrust south of the western part of the SMB is accorded by an upper mantle conductor dipping about 45° southwards with an accompanying regional gravity high (Fig. 7.28(e)) suggesting high density rocks and fluid along the thrust. It implies that the whole of the SMB was involved in this collision and subduction during Meso-Neoproterozoic period between the Bundelkhand craton towards the north and Dharwar, Bhandara, and Singhbhum cratons towards the south (Fig. 7.65(a)). The latter convergence also caused the uplift and deformation of the Mahakoshal, the Bijawar, and the Aravalli group of rocks and disturbed Vindhyan sediments along their margins with the SMB and the ADMB that produced folding and faulting in these sections.

(d) The gravity highs due to the ADMB and the SMB are joined in NW India forming a curvilinear collision zone. Simultaneous E-W and N-S convergence of Bundelkhand craton across the ADMB and the SMB suggests that primary stress direction might be NE-SW with these two directions as their components (Fig. 7.6). This is also supported from large shear zones with quartz reefs oriented NE-SW in the Bundelkhand craton. The junction of these two mobile belts might have formed a triple junction with mobile belt between the Central and the East African cratons in Africa that got separated after the breakup of India and Africa.

(e) High heat flow and recent uplifts along the ADMB and the SMB suggest neotectonic activities. These are specially prominent along the ADMB. These neotectonic activities are attributed to the crustal bulge caused by the flexure of the Indian plate under Himalaya towards the north and the Western Fold Belt, Pakistan towards the west. These effects are more prominent along the ADMB and the SMB as these sections are affected by prominent thrusts of crustal scale that get reactivated due to neotectonic activity. The anomalous high velocity and high density domal shaped body in the lower crust under the Delhi group of rocks of the ADMB (Fig. 7.13(a) and 7.15(a)) have been attributed to asthenospheric upwelling due to Delhi rifting or remnant of oceanic crust of that time. However, high heat flow and recent uplift in this section indicate that it may be a manifestation of the lithospheric flexure in the form of crustal bulge bounded by crustal scale thrusts during recent times due to Himalayan orogeny.

(f) Simultaneous to the convergence across the SMB during Paleo-Mesoproterozoic period, the Bhandara-Bastar craton converged towards the Dharwar craton across the Godavari Proterozoic Belt (GPB, Fig. 7.6) in almost the same direction of NE-SW that gave rise to the Karimnagar ganulite belt (2.4-2.2 Ga) and Pakhal group of rocks with large scale mafic/ultramafic components were deposited on the rifted platform of the Dharwr and the Bhandara-Bastar cratons. The two convergences across the SMB and the GPB might in fact be related to each other. The Bhopalpatnam granulite belt of 1.6 Ga is related to convergence during the Mesoproterozoic period when the Bundelkhand craton, Bhandara-Bastar craton, and Dharwar craton towards the north, SE and SW collided and subducted under each other. The Sullavai group of rocks of Neoproterozoic period were deposited during the final phase of this convergence that explains the hiatus between the Pakhal and Sullavai group of rocks as in the case of the Upper and Lower Vindhyan sediments as discussed above (Fig. 7.65(a)).

(g) The linear gravity highs due to the Godavari Proterozoic belt and the western and the central parts of the SMB (H7, H8, and H10 and H11, Fig. 7.8(b)) make a triple junction east of Nagpur intersecting at almost 120° from each other forming a SSS type (Fig. 7.2) stable triple junction of Mesoproterozoic period. A schematic section of convergence across the SMB, the ADMB, and the GPB is shown in Fig. 7.65(a) that shows the Bundelkhand craton subducting under the Rajasthan block and Bhandra-Bastar craton due to E-W and N-S directed convergences, respectively, related to NE-SW primary stress direction at that time. It also shows the Bhandra-

Bastar craton subducting under the Dharwar craton across the GPB. This is in accordance with the triple junction (Fig. 7.2) where one plate subducts under the remaining two plates and one plate does not subduct and therefore, they appear to form a stable triple junction. These processes have given rise to the subduction related magmatism on the subducted sides and high grade mafic/ultramafic intrusives in mobile belts (obducted sides).

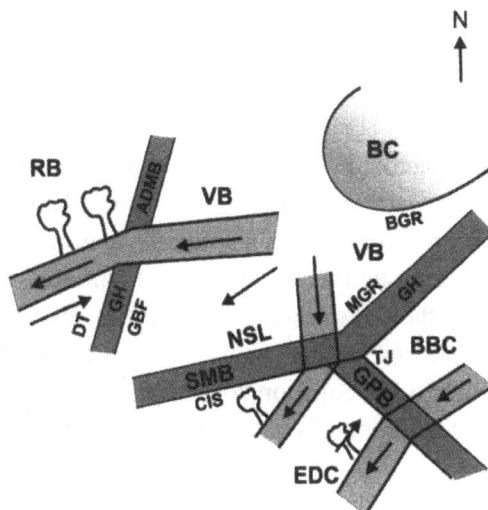

Fig. 7.65(a) A schematic section of convergence during Meso-Neoproterozoic period in central India across the Satpura Mobile Belt (SMB), Aravalli Delhi Mobile Belt (ADMB), and Godavari Proterozoic Belt (GPB). NE-SW arrows with N-S and E-W components indicate direction of convergence while NE-SW and SW-NE arrows indicate direction of subduction and thrusting, respectively. Flowers indicate subduction related magmatism. NSL: Narmada Son Lineament, CIS: Central Indian Suture, GBF: Great Boundary Fault, DT: Delhi Thrust, VB: Vindhyan Basin, EDC: Eastern Dharwar Craton, BC: Bundelkhand Craton, MGR: Mahakoshal Group of Rocks, BGR: Bijawar Group of Rocks, BBC: Bhandara-Bastar craton and TJ: triple Junction. A similar convergence during the Neoarchean-Paleoproterozoic period followed by rifting is also envisaged when MGR, BGR and VB are formed on the rifted platforms of the adjoining cratons.

7.12.2 South Indian Shield: Neoarchean Convergence and Triple Junction

This part of the Indian Shield south of the SMB consist of the Western and the Eastern Dharwar cratons (WDC and EDC) and the Southern Granulite Terrain (SGT) separated by shear and transition zones, respectively, (Fig. 7.37) and the EGMB along the east coast of India (Fig. 7.6 and 7.7). Some significant points related to convergence and collision of South Indian cratons are as follows:

(a) Linear gravity highs along the shear zone between the Western and the Eastern Dharwar cratons, that extend from the eastern margin of the Chitradurga schist belt to Panaji along the west coast of India following the geological trends of this region and differences in rock types across it, suggest that it may represent a suture between them. The section between the Kolar schist belt and the shear zone between the Western and the Eastern Dharwar cratons may represent a diffused collision zone of Neoarchean time where crustal thickness is less (31-32 km) compared to adjoining regions (40-41 km).

(b) Transition zone is characterized by differences in rock types and crustal structures between the Western and the Eastern Dharwar cratons towards the north and the southern granulite terrain towards the south, characterized by gravity highs. The E-W oriented Cauvery Shear Zone (CSZ) south of the transition zone is another tectonic boundary in the SGT characterized by linear gravity highs and thin crust (38 km) that increases to 44-45 km on either sides. It may, therefore, represent a collision zone of Neoarchean to Paleoproterozoic period with contemporary lower crustal granulite rocks and deep seated intrusives exposed in this region and the MBSZ-Metter Shear and the PCSZ-Grenville Shear as thrusts on either sides of it (Fig. 7.37). These two thrusts dip opposite on either sides with CSZ as the central core complex of typical collision zones (Fig. 7.49(b)). The transition zone may represent a thrust related to this collision.

(c)   The eastern part of the transition zone between the SGT and the EDC, the MBSZ between the WDC and the SGT, and the shear zone between the EDC and the WDC are characterized by gravity highs (H16, H17 and H15, Fig. 7.8(b)) that interact with each other at about 120° around Bangalore suggesting an Archean-Proterozoic SSS type triple junction (Fig. 7.65(b)). Fig. 7.65(b) is a schematic section of E-W convergence and subduction of the EDC under the WDC across the shear zone-Closepet granite where thrusting and subduction have given rise to high density rocks under the shear zone-Closepet granite and schist belts with bimodal volcanics of the WDC. Closepet granite may represent an island arc caught between the WDC and the EDC or formed due to subduction related magmatism. It also shows the shifting of the thrust/subduction eastwards (W-E arrow) forming a diffused collision zone responsible for the felsic intrusives and schist belts of the EDC, and mobile belt kind of structures in this section.

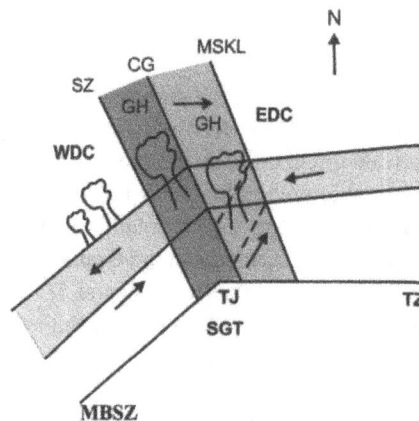

Fig. 7.65(b) A schematic scheme of E-W convergence and collision of the Western and the Eastern Dharwar cratons (WDC and EDC) during Neoarchean-Paleoproterozoic period and subduction (NE-SW arrow) across the shear zone (SZ)-Closepet granite (CG) giving rise to high density rocks of this section and schist belts with bimodal volcanics of the WDC. Opposite arrows (SW-NE) indicate thrusting. Subsequently, the thrust shifted to the Mumbai-Sangola-Kolar lineament (MSKL) shown by the horizontal W-E arrow. Flowers indicate subduction related magmatism. TJ is the triple junction between the shear zone, the eastern part of the transition zone (TZ) and the Moyar Bhavani shear zone (MBSZ) towards the west. GH represents gravity and geoid highs.

(d)   The EGMB represents a collision zone between the Dharwar craton and Antarctica with convergence and subduction from the east to the west during Mesoproterozoic period (~ 1.5-1.0 Ga) that gave rise to thick crust to its west. The disturbed nature of the Nallamalai sub basin in the eastern part of the Cuddapah basin is another indication of collision along the EGMB. This may be related to the Grenville agglomeration. Prior to this convergence, a convergence and rifting phase during the Neoarchean-Paleoproterozoic period (2.9-1.6 Ga) is suggested to account for the Nellore schist belt of the Neoarchean-Paleoproterozoic period and the Kondapalli layered complex (~ 1.7 Ga) in the southern part of the EGMB. The Cuddapah super group of rocks with large scale mafic sills and flows (~ 1.9 Ga) also formed during this rifting phase (Fig. 7.50(a)). In fact, this rifting might have been caused due to these intrusives that might be related to a plume. This rifting may be related to the breakup of the Columbian agglomeration in this section as similar date has been envisaged for this agglomeration (Rogers and Santosh, 2009) and the EGMB was juxtaposed with East Antarctica. Even during subsequent reconstructions of Grenville and Gondwanaland agglomerations, the EGMB is shown juxtaposed with East Antarctica (Rino et al., 2008). The Kurnool super group of rocks of the Cuddapah basin of Neo-Proteozoic period, formed during the final phases of the Meso-Neoproterozoic convergence, collision, and mountain building along the EGMB as envisaged above. This supplied sediments to this basin, which are largely undisturbed (Fig. 7.65(c)). This explains a long hiatus between the Cuddapah and the Kurnool super group of rocks.The gravity highs due to EGMB in the central part of the east coast of India extends northwards up to Himalaya and southwards up to the southern part of the east coast of India and in Sri Lanka with almost similar crustal structures indicating the extension of this orogeny of Meso-Neoproterozoic period in these sections.

(e) The other Proterozoic basins along the EGMB towards the north associated with the Bhandara-Bastar craton such as the Indravati basin also might have formed in the same manner as the Cuddapah basin towards the south. French et al., (2008) have reported similar dates for dyke swarms in the Bastar craton (~ 1.9 Ga) as for mafic sills of the Cuddapah basin and suggested a large igneous province of that time in this region that indicates a plume origin for them, which was responsible for the breakup of the Columbian agglomeration in this section as envisaged above.

(f) The northern and the southern parts of the EGMB and the GPB form a triple junction separated by about 120° each. This explains the changing of the trend and metamorphic history of rocks of the EGMB towards north and south of this junction. The GPB is unique in that its northern end has formed a triple junction with the SMB while the southern end is joined to the EGMB in the same manner. Convergence and subduction across the EGMB is summarized in a schematic section (Fig. 7.65(c)) that shows subduction and thrusting due to E-W convergence which gave rise to high density thrusted rocks of the EGMB and deep seated intrusives of this section, and the Cuddapah basin to its west. It could also be NE-SW directed convergence similar to other cases, as indicated by similar oriented shear zones in the EGMB as described above.

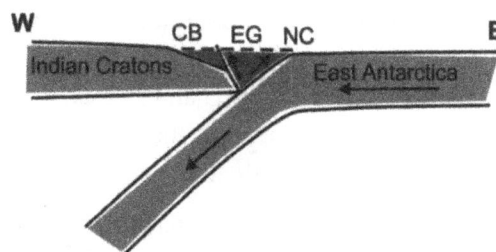

Fig. 7.65(c) A schematic section of the Mesoproterozoic collision between the Indian cratons and East Antarctica along the EGMB that gave rise to this mobile belt and associated intrusives and formed contemporary basins to its west. E-W and NE-SW arrows indicate direction of convergence and subduction, respectively, while opposite arrows (SW-NE and SE-NW) indicate direction of thrusting.
CB: Cuddapah Basin, EGL: Eastern Ghat Mobile Belt, NP: Napier Complex.

7.12.3 Integrated Convergence Model of Indian Cratons and Rifting

(a) It is interesting to observe that the direction of convergence during Paleoproterozoic and Meso-Neoproterozoic period in the north and the south Indian Shields, are in the same direction of E-W and N-S, with NE-SW as primary stress direction that is supported by NW-SE lineaments and fracture zones in the Indian Peninsular Shield. This indicates a consistent pattern of plate tectonic forces during Neoarchean-Proterozoic period. It is, therefore, suggested that primarily the convergence and the rifting between the Indian cratons and the East Antarctica propelled other Indian cratons for similar processes that might have been caused by major driving forces like mantle convection. This is also indicated by the fact that in all the past reconstructions like Columbia, Grenville, and Gondwanaland, the EGMB is shown juxtaposed with East Anarctica. It is further observed that the primary stress direction during Proterozoic period was opposite to that of the present day convergence of the Indian and the Asian plates across the Himalayan collision zone (Fig. 7.6). This also implies the inversion of processes responsible for continental breakup and their drifting, such as mantle convection cells since Proterozoic period to Gondwanaland and subsequently when it rifted apart to give rise to the present day movement of the Indian plate.

(b) Significant points of convergence and rifting of Indian cratons as described above (Fig. 7.6) suggest the formation of an agglomeration of cratons during Neoarchean-Paleoproterozoic period (~ 2.9-2.0 Ga) that rifted into different smaller cratons during Paleoproterozoic period (2.0-1.6) and again started to converge and collide almost along the same lines during 1.5-1.0 Ga that gave rise to various Proterozoic mobile belts (ADMB, SMB, and EGMB) and associated Proteozoic basins. This agglomeration of cratons may be related to the Columbia and Grenville agglomerations as similar dates have also been reported from East Antarctica (Veevers and Saeed, 2009). Repeated cycles of convergence and rifting along the same tectonic elements have been discussed above in Section 7.1 that appears to be applicable in case of Indian cratons. However, the Dharwar craton and SGT had remained stabilized since Neoarchean-Paleoproterozoic period as described above. Formation of super continents is important for their breakup as in case of Grenville and Rodinia formations, when they tend to cap the heat flow from the earth's interior that is important for the breakup of the supercontinents.

(c) Proterozoic basins associated with different mobile belts like ADMB-SMB, and the EGMB show a consistent pattern of older groups like the Semri group and Cuddapah super group, respectively, formed during the rifting phase (2.0-1.6 Ga) while the younger groups like Kaimur, Rewa and Bhander groups of the Vindhyan basin and the Kurnool super group of the Cuddapah basin, respectively formed during the last stage of Meso-Neoproterozoic convergence, collision, and mountain building that supplied the requisite sediments for these basins. Schematic sections of convergence across the SMB, the ADMB, and the EGMB and formation of Proterozoic basins are given in Fig. 7.65(a) and (c) as an example demonstrating the sequence of events responsible for their evolution that also shows the direction of convergence.

(d) Though geochemical analysis of mafic/ultramafic rocks from Bijawar, Mahakoshal, and Pakhal group of rocks have not been attempted to establish their correlation with contemporary rocks from the Cuddapah super group of rocks but they might have been caused by the same plume. Lithologically, however, several similarities between rocks of Cuddapah super group and Pakhal super group have been shown by Rao (1987). French et al., (2008) have shown the spread of the igneous province that affected the Cuddapah super group of rocks to the Bastar craton where Pakhal group of rocks are located and which was involved in rifting with the Bundelkhand craton where the Mahakoshal and Bijawar group of rocks were deposited. Contemporary Proterozoic basins along the SMB-ADMB, the EGMB, and the GPB separated in space formed due to similar processes indicate the importance of plate tectonics for the evolution of Archean-Proterozoic terrains. The plate tectonic model presented here for the formation of Proterozoic basins and long hiatus between Upper and Lower group of rocks can be used to explain similar processes anywhereelse. As indicated above, the rifting during Paleoproterozoic period might be related to the breakup of the Columbia agglomeration that was caused by a large igneous province along and west of the EGMB, which extended from the Cuddapah basin to the Bastar craton. Large igneous provinces are usually caused by plumes that also are responsible for the breakup of supercontinents.

(e) Signatures of platform deposits as foreland basins are absent from Archean collision zones indicating that they may represent the difference between the Proterozoic and the Archean convergence and collision. The other difference lies in their elevation. Proterozoic fold belts are still observed as highlands while those of Archean period are mostly peneplained plateaus with small variations in elevation and amalgamated to the adjoining cratons unless lower crustal granulite rocks are exposed or affected by latter events as in the case of the SGT.

7.12.4 Lithosphere and Asthenosphere under the Indian Plate and it's Rapid Drift

Some of the significant features of the Indian lithosphere which may be responsible for its buoyant nature and rapid rift, are as follows:

(a) Spectral analysis and wave band filtering of Bouguer anomaly and modeling along profiles provide average thickness of the lithosphere as 160-180 km. It also suggests predominance of low density rocks in the lithospheric mantle that provides high velocity in seismic tomography results as discussed above. The low density and high velocity rocks in the lithospheric mantle suggest that it is a part of the cold subducted Indian lithosphere. However, NW and SE sections along coastal margins of the Indian continent show high density rocks in the lithospheric mantle, supported by normal velocity or slightly low velocity in seismic tomography. These high density sections along coasts can be attributed to the lithospheric upwarp or high density intrusives caused by the plumes during the breakup of India from Gondwanaland. The east coast of India is presumably affected by Kerguelen hotspot during the breakup of India from Antarctica at about 118-120 Ma, while the west coast in the northern part off Saurashtra and Kachchh is affected by several plumes e.g., at the time of breakup of Africa from India (Karoo Volcanics), and subsequently by the Reunion plume that gave rise to Deccan trap rocks on the western part of the Indian continent. Lithosphere under most of the cratons are thicker such as in the case of Kaapaval and Zimbabwe cratons where the lithosphere is thick (250-300 km) compared to the Indian cratons (160-180 km). Gravity highs in the northern part of low pass filtered Bouguer anomaly of India along the Himalayan front represent lithospheric upwarp due to flexure of the Indian plate whose shallow manifestation is the Delhi-Lahore-Sargodha Ridge in the western part of the Indian plate where lithospheric thickness reduces to 140 km. It is quite significant that this lithospheric upwarp also find reflection in surface wave tomography as reduction in lithospheric thickness upto 140 km (Priestley et al., 2008).

(b) As the general elevation of the Indian Shield (Karnataka-Deccan Plateau) is only 700-800 m, the large amount of low density rocks in the upper mantle would cause overcompensation, making the whole region buoyant. The low density rocks of the lithospheric mantle and asthenosphere provide buoyancy to the Indian continent that drives the Indian plate with a relatively higher speed as discussed in section 7.11. Moreover, these low density rocks as subducted Tethyan/Indian lithosphere appear to be extending even under the Indian Ocean south of the Indian continent as is apparent from large gravity and geoid lows of the Indian Ocean south of the Indian continent that would cause tilting of the Indian plate towards the north. This tilting would also cause faster underthrusting of the Indian plate along the Himalayan thrusts. Accordingly, the faster drift of the Indian plate would produce high strain and consequently several large and great earthquakes along the Himalayan front (Mishra and Rajasekhar, 2006; Kayal, 2008). The uplift of the southern part of the Indian plate is further accentuated due to the load of the Himalayas and Tibet at its northern margin, as in the case of a loaded beam.

(c) This kind of uplift in recent times of the southern Indian Shield might be responsible for the larger numbers and higher magnitudes of SCR earthquakes in this region compared to any other stable continental regions (Kayal, 2008). This is also evident from higher horizontal stress in the Indian shield compared to other shields, as discussed in Section 7.8.3. In addition, receiver function analysis (Kiselev et al., 2008) suggests the absence of high velocity keel under the Indian Shield (cratons) that is present under most of the cratons world over, which provides further buoyancy to the Indian plate. The keel might have been delaminated due to large movement of the Indian plate since Proterozoic period that was supposed to be located in northern hemisphere (34°N) at that time and drifted from there to the southern hemisphere (60°S) to form Gondwanaland and again back to the northern hemisphere (Fig. 5.1).

Seismotectonics and Geodynamics: Bhuj, New Madrid and Sumatra Earthquakes and Tsunami with Co-Seismic Changes

8.1   Introduction – Grabens and Horsts

Earthquakes are generated due to movement of rocks along faults that are defined by their dip and strike in the usual sense (Fig. 8.1(a)). As faults can be easily identified based on the gradients in the observed gravity and magnetic fields, they play an important role in delineating seismotectonics as demonstrated below based on the case histories related to some specific earthquakes. Depending on the nature of faults, there are different kinds of earthquakes. Broadly, there are three kinds of faults, viz. normal, reverse, and strike slip faults.

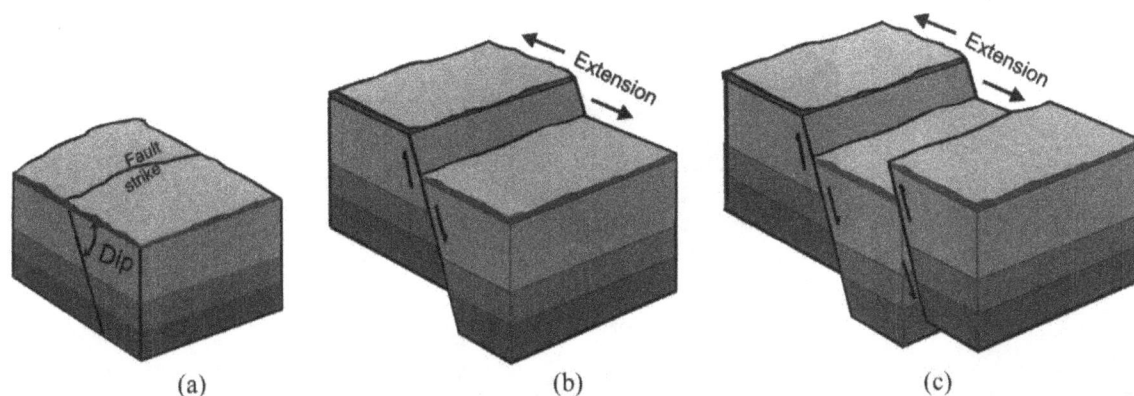

Fig. 8.1 (a) A fault in a block defined by its dip and strike (b) Normal faults with arrows showing direction of movement (c) Graben structure or rift valley with normal faults on both sides of the subsided block

(i) **Normal Fault and Graben:** In this case one block shifts downwards with respect to the adjoining block as shown in Fig. 8.1(b). They are caused by extensional forces (tensile stress). Arrows indicate the direction of movement. In case there are two such faults on either side of a block that has moved downwards and dips inside such that the central block subsides, it is known as a rift valley or a graben (Fig. 8.1(c)). Grabens have a special significance for seismotectonics as several of the stable continental earthquakes are found to occur in old grabens due to reactivation of faults associated with them as in the case of Bhadrachalam earthquake of Godavari valley, India discussed in Section 7.7. These faults are generated due to extension and are generally encountered in sedimentary basins, significant for hydrocarbon exploration. The most important rift basins in India are the Gondwana rift basins such as Godavari and Mahanadi basins of Permian-Triassic period (Section 7.7) and Cambay Tertiary basin (Chapter 9).

(ii) Reverse Fault and Horst: In this caseone block slips over the adjoining block (Fig. 8.2(a)) Reverse Faults are, therefore, opposite in nature to the normal faults. In case there is another fault such that the block in between moves upwards and is raised along faults dipping opposite to each other, it is known as a horst structure (Fig. 8.2(b)). In case the reverse fault is a low dip, it is known as thrust. These structures are formed due to compression and are mostly encountered in orogenic belts such as the Himalayas.

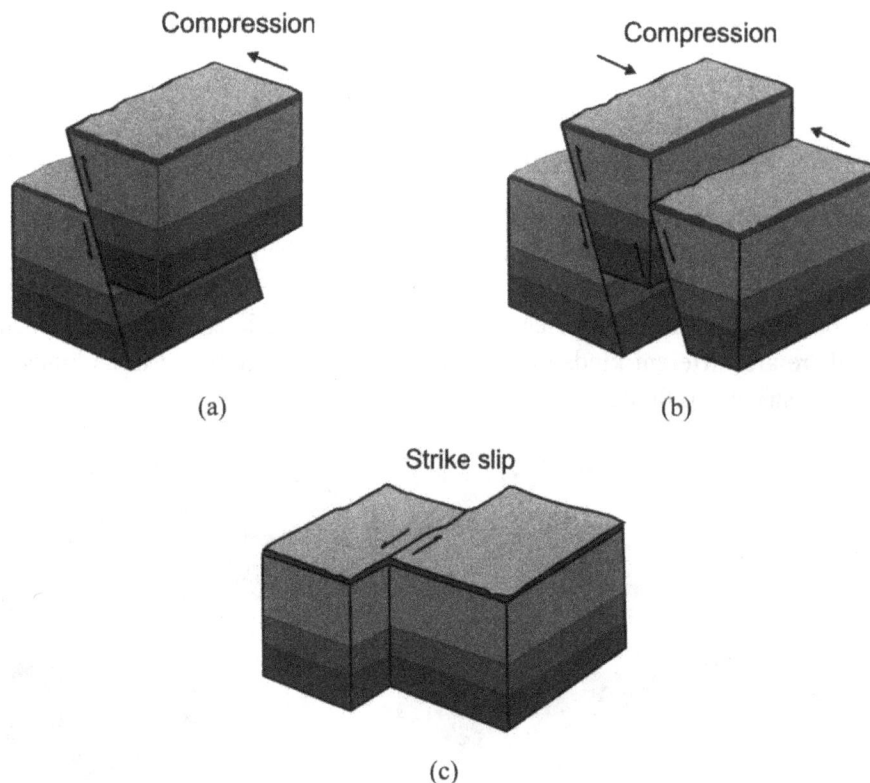

Fig. 8.2 (a) Reverse fault or a thrust with arrows showing direction of movement. (b) Horst structure with two reverse faults on either side of the uplifted block. (c) Strike slip fault with lateral movement in strike direction. Transform faults of plate tectonics belong to this category.

(iii) Strike Slip Fault: In case blocks of rocks slip past each other horizontally, it is known as strike slip faults (Fig. 8.2(c)). In this case, there are no upward or downward movements. Strike slip faults form an important element in plate tectonics as transform faults such as the Chamman and Sagaing faults along the Western fold belt (Pakistan) and the Burmese arc, respectively, as described in Chapter 6 (Sections 6.3 and 6.2). Another important strike slip transform fault is the San Andreas Fault along the West Coast of USA between the Pacific and the North American plates. It is seismically one of the most active and most studied fault systems. There are two types of these faults: right lateral and left lateral strike slip faults. In case of right lateral, the movement is on the right hand side if one stands on the fault, and vice versa in case of left lateral.

8.1.1   Characteristics of Earthquakes

The movement of rocks on faults creates waves that are recorded at seismic stations which locates their position, magnitude and type of faulting associated with them. Primarily, three types of waves are generated: primary or P- waves, secondary waves or S-waves, and surface waves. In P-waves, the motion of the particle is in the same direction as the motion of the waves similar to push and pull and are, therefore, also called compression or longitudinal waves. In case of S-waves, particle motion is perpendicular to the motion of the waves which are therefore, also called transverse or shear waves. Surface waves travel along the surface and consist of a combination of complex motion both rolling as in S-waves and back and forth similar to P-waves. But, due to the combination of motion and large amplitude, they cause maximum damage in any earthquake. These waves are recorded at seismic stations by seismographs and depending on their velocity and time taken by them to travel, their point of origin, known as focal point or focus, and vertical projection on the surface of the earth, known as epicenter, can be determined. P-waves travel faster (6 km/sec) in rocks compared to other waves and reach first followed by S-waves (3.6 km/sec); last to reach a seismic station are surface waves showing large amplitudes. Fig. 8.3(a) shows a typical seismograph where the difference in travel time taken by P and S waves can be ascertained and used to find a the distance between the origin point of the earthquake and the recording stations. However, distance from single station cannot specify the direction from which side the waves are traveling. But, using such travel time interval for P and S waves from several stations to a minimum of three stations, one can pinpoint the exact location where circles drawn from the recording stations intersect each other.

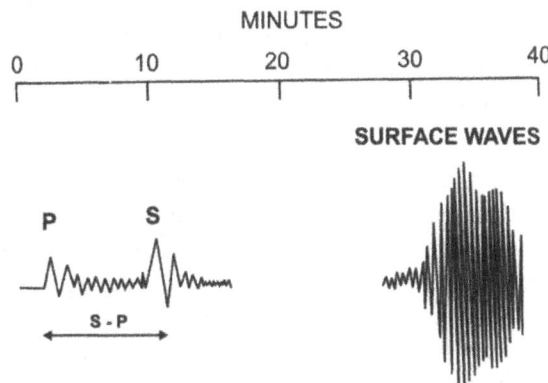

Fig. 8.3(a) A typical seismogram showing P-waves, S-waves, and surface waves.

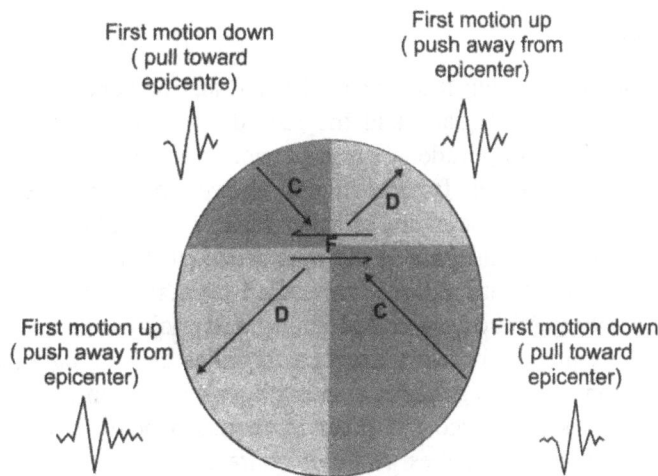

Fig. 8.3(b) First motion on seismogram in different quadrants showing compression and dilatation representing strike slip motion at focal point (F) of the earthquake, also known as hypocenter; C: Compression, D: Dilatation.

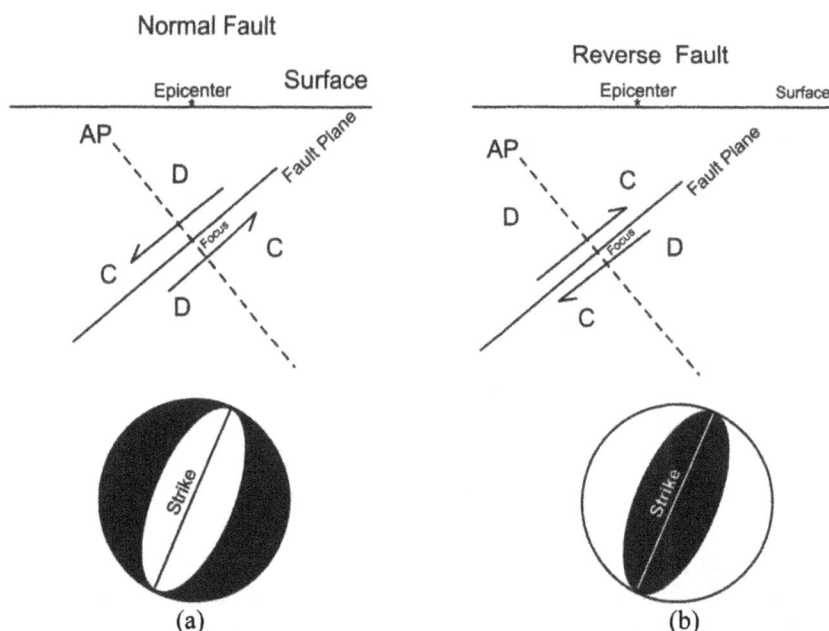

Fig. 8.4 (a) A normal fault showing dilatation between fault plane and auxiliary plane, and compression outside represented as white (blank) ellipse in the central part of a black (filled) circle, with strike direction along major axis and dip direction perpendicular to it. Vertically above the focus is the epicenter. (b) A reverse fault showing compression between fault and auxiliary planes and dilation outside it. It is indicated by a black (filled) ellipse within a white (blank) circle with fault plane along the major axis and dip direction perpendicular to it. Vertically above the focus is the epicenter.

Having determined the point of origin, the next is the magnitude of the earthquakes. Initial attempts in this regard were qualitative depending on the damage caused by an earthquake. Modified Mercalli scale was most popular for this purpose where seismic activity was assigned on scale varying from I-XII with I indicating notice by a few people and XII denoting total damage in the epicentral region. However, subsequently some quantitative scales were developed so as to reduce subjectivity due to individuals. The Richter scale is the most popular scale in this class and uses the amplitude of P- and S-waves at a recording station to estimate the magnitude of an earthquake. The seismograph plot as shown in Fig. 8.4 shows the measure of maximum amplitude, X, and time period, T defining the Richter magnitude, M :

$$M = \log (X/T) + Y$$

where Y depends on the time lag between the arrival of the P and S waves, and accounts for the distance between the epicenter of the earthquake and the recording station. Due to the logarithmic factor, there is a ten fold increase in the amplitude, X (ground motion) corresponding to an increase of one in magnitude. However, due to other factors in the above equation, the energy released corresponds to 32-fold increase for an increase of one in magnitude. This explains the amount of large damage caused by earthquakes with every increase in the magnitude of the earthquake on the Richter scale. Earthquakes of magnitude 5.0-5.9 and 6.0-6.9 are called moderate and a large earthquakes, while those of 7.0-7.9 and 8.0-8.9 magnitude are called strong and major earthquakes, respectively. Earthquakes of magnitude greater than 9.0 are called great earthquakes causing havoc in the epicentral regions. A more precise description of the magnitude of an earthquake is given by computing moment magnitude (M_W) that is found to describe large and great earthquakes better than the Richter scale, which may remain same for moderate earthquakes. Another characteristic of an earthquake is related to the kind of fault that has given rise to a particular earthquake. As discussed above, there are basically three kinds of faults that give rise to earthquakes that are described based on the movement of the adjoining blocks and can be determined based on the fault plane solution. Further details about magnitudes and other characteristics of earthquakes can be obtained from any text book on seismology.

8.1.2 Earthquake Focal Mechanism

Another important aspect of earthquakes is to understand the kind of faults associated with them and the direction of the slip that can be ascertained based on the direction of the first movement of seismographs at seismic stations world over. For example, in case an earthquake initiates at focal point, F (Fig. 8.4(a)), it will produce push and pull movements (dilatation and compression) in different sectors that will be indicated by the first upwards and downwards movements of the seismograph, respectively as shown in Fig. 8.4(a). A similar situation will exist if the movement takes place on the auxiliary plane perpendicular to it. However, the two can be differentiated based on field investigations that would show maximum damage close to the fault plane. If the plot of push and pull (dilatation and compression) are made as black and white quadrants in a sphere, it will look as shown in Fig. 8.5(a) showing the direction of movement along the central axis. This is the case with strike slip motion. In case of normal faults (Fig. 8.4(b)), compression is observed between the fault plane and the auxiliary plane while dilatation is on either sides of it; therefore, it is represented by blank (white) ellipse in a filled (black) sphere with major axis showing the strike of the fault plane and dip direction perpendicular to it. However, in case of reverse faults or thrusts, it is just the opposite (Fig. 8.4(b)) and therefore is represented by filled (black) ellipse in a white circle with major axis in the strike direction and dip direction perpendicular to it (Lillie, 1999).

8.1.3 Type of Earthquakes

Fig. 8.5(a) (Murthy et al., 2010) is the plot of the epicenter of seismic activity (USGS website) in the major parts of the Indian plate classified in four parts, I-IV. It shows the maximum concentration of seismic activity along the Himalayan fold belt (I) and Sunda Arc (IV) that is a continuation of the Himalayan fold belt in the Indian ocean and represents a plate boundary between the Indian and Eurasian plates as described in Chapter 6. These are known as plate boundary or interplate earthquakes. Besides, there is sporadic seismic activity within the Indian plate, both over the continent (II) and ocean (III), known as intraplate or Stable Continental Region (SCR) earthquakes. Plate tectonic theory offers a suitable explanation for plate boundary interplate earthquakes caused by the stress generated due to collision of two plates, and subduction that produces strain causing slip along various faults associated with it. Therefore, most of the large and great earthquakes in the world belong to this category. In case there is a regular slip, small earthquakes are generated. But in case it is locked, and stress and strain are allowed to accumulate, and if the strain produced due to accumulated stress is more than the elastic limit of the rocks, they rupture and large earthquakes are generated. Slip is facilitated due to the presence of fluids along fault planes while locking is facilitated due to changes in the slope of density interfaces such as ramps where stress and strain get accumulated. This is popularly known as the elastic rebound theory. Most of such theories were developed to account for observations of seismic activity along San Andreas Fault system along the west coast of USA that displays strike slip transform fault between the Pacific and north American Plate, and represents one of the most seismically active present-day plate boundaries. Several large and major earthquakes of magnitude >7.0-8.0 have been reported from this region in the last century such as San Francisco, Landers, Loma Prieta earthquakes, on April 18, 1906, June 28, 1992, October 17, 1989, respectively (Robinson, 2002). However, the elastic rebound theory has not been able to explain several observations in case of these earthquakes related to the San Andreas Fault system, the most important being the absence of large drop in the stress field and high heat-flow after the earthquakes that are expected from this theory. It was suggested that this may be accounted for by aftershocks, but they do not occur only in the direction of the main shock but infact occurr in all directions. It is probably nature's way of saving the catastrophe, as so much damage is done with the release of a fraction of stress so one can imagine what would happen if the total accumulated stress is released.

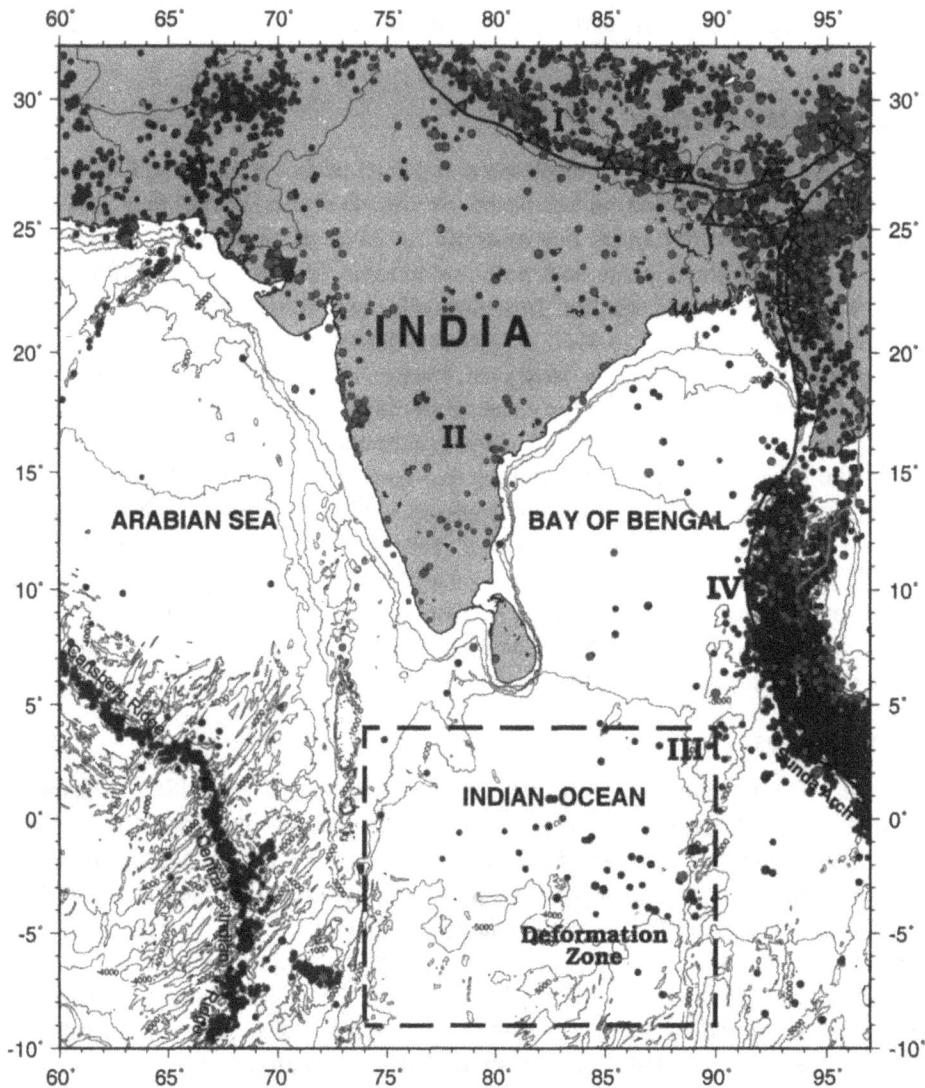

Fig. 8.5(a) Epicenters of seismic activity of the Indian continent and surrounding regions showing maximum concentration along the Himalayan fold belt and along its Western and Eastern extensions that represent interplate earthquakes. Those within the Indian plate are intraplate earthquakes or stable continental region (SCR) earthquakes.

The SCR earthquakes are, however, mysterious in their occurrences and are primarily associated with some preexisting major tectonic elements like faults, lineaments, and grabens. Fig. 8.5(b) (Murthy et al., 2010) shows some well known SCR earthquakes of the Indian continent which are associated with major preexisting tectonic elements such as Narmada-Son Lineament (e.g. Jabalpur earthquake), West coast and continental margins (e.g. Koyna earthquake), Godavari rift valley (e.g. Bhadrachalam earthquake), East coast and continental margin (e.g. Pondicherry-Chennai-Viakhapatnam earthquake), Kurduwadi lineament (e.g. Latur earthquake). Based on seismic activity of the Indian continent (Fig. 8.5(a) and (b)) and tectonics likely to produce intraplate earthquakes, the country is divided into four seismic zones, II-V (Fig. 8.6, NDMA, 2007; BIS, 1986) ranging from lowest to highest seismic risk zones. Some of the large and great earthquakes of the Himalayan plate boundary and most of the SCR earthquakes of the Indian continent are described in Chapters 6 and 7, respectively. However, seismotectonics of the Bhuj and the Sumatra earthquakes of 2001 and

2004 – representing intraplate and interplate earthquakes, respectively – are described below as examples of the two types of earthquakes. Two specific experiments of change in elevation and gravity field after the Bhuj earthquake and temporal changes in the gravity field in Koyna-Warna region (Section 7.8.3) where seismic activity of magnitude < 5 occur regularly are also described to understand co-seismic changes in these parameters.

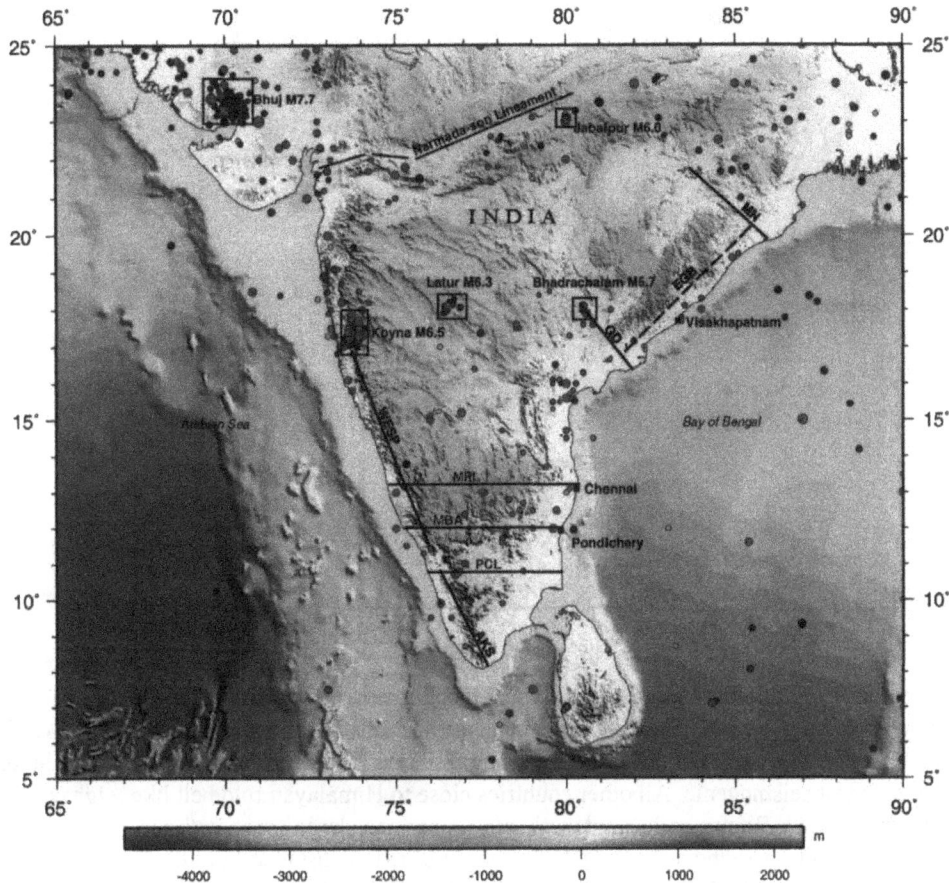

Fig. 8.5(b) Some important intraplate (SCR) earthquakes of Indian continent with their magnitudes showing their connections with known older tectonic elements.

8.1.4   Seismic Zonation Map of India and SAARC Countries

Based on the seismic activity in different parts of the country (Fig. 8.5(a) and (b)), a seismic risk zonation map was prepared (Fig. 8.6) showing zones II to V from low-damage risk to very high-damage risk. The Himalayan collision zone including Andaman Nicobar islands belongs to high and very high damage risk zone. In the continental part, Kachchh and NE India including the Shillong plateau are placed in the same category as Himalayan collision zone. This is quite significant as parts of these regions belong to the intraplate region. It indicates some tectonics specific to these regions that make them seismically active on par with the Himalayan collision zone highlighted in sections 7.64 and 8.2. Seismic Zonation map of SAARC countries is explained in Appendix IV with reference to seismic zonation map of Pakistan and Bangladesh that mainly shows very high to moderate and low seismicity depending on distance from the Himalayan fold belt.

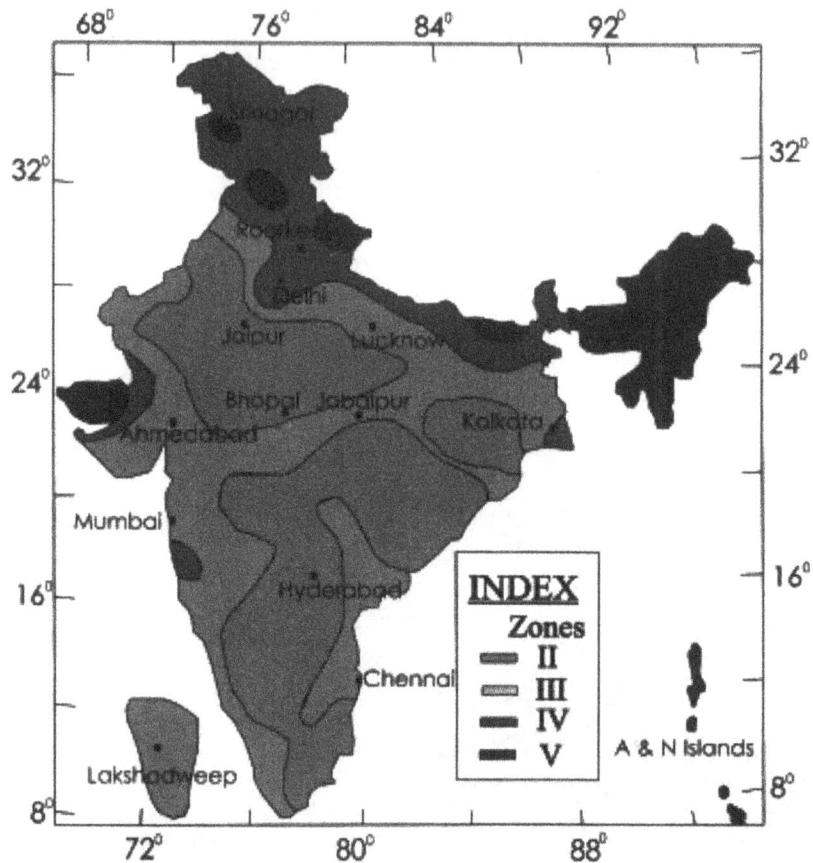

Fig. 8.6 Seismic Zonation map of India (India Meteorological Department, New Delhi) showing seismic zones II-V in increasing order of seismic activity. Some of the Himalayan sections where major/great earthquakes have been reported, Kachchh (Section 8.2), and NE India including Shillong plateau (Section 7.6.4) are included under group V that are most seismogenic. All other countries close to Himalayan fold belt like Afghanistan, Pakistan, Tibet (China), Nepal, Bhutan and Bangladesh, also experience large/great earthquakes and show similar pattern of seismic zoning varying from Zone V to Zone II based on distance from the Himalayan Fold Belt as shown in Appendix IV for Pakistan and Bangladesh. [Colour Fig. on Page 801]

8.2   Bhuj Earthquake of January 26, 2001

The Bhuj earthquake of January 26, 2001, was one of the strong intraplate earthquakes that are unlikely to be produced in stable continental regions and therefore aroused sufficient interest in the international geoscientists for detailed investigations as described below.

8.2.1   Regional Plate Tectonic Settings and Seismicity in Kachchh (Kutch)

The Bhuj earthquake of Jan. 26, 2001 of Mw = ~7.8 with epicenter around 23.4° N and 70.3° E (Fig. 8.7(a)) and focal depth ~24 km (Yogi and Kikuchi, 2001), caused considerable damage both to human beings (dead > 20,000, injured > 150,000) and property (> $ 10 billion). Since its occurrence, it has been considerably debated as being an intraplate or interplate boundary earthquake (Ellis et al., 2001). Though the seismic activity of Kachchh is regarded as intraplate, its connection with the activities at plate boundaries about ~300 km West and NW of Kachchh cannot be ruled out. Considerable amount of seismic activity can be seen (Inset, Fig. 8.7(a)) over the junction of the plates towards the west and the northwest of Kachchh, viz. Indian, Arabian, Iranian, and Eurasian plates along Owen fracture zone, Murray ridge Makran coast, and Kirthar range-Chamman fault evident from the distribution of epicenters since 1668 (USGS, Website). Another microplate (Ormara plate) west of Murray ridge and south of Makran shear zone in north Arabian Sea has been conceived based on detail bathymetry (Kukowski et al., 2000) that would modify the stress distribution in this section, indicating several active plate boundaries NW of Kachchh. Fig. 8.7(a) also shows the epicenters of seismic activity extending from the Western Fold Belt (Pakistan) towards SE up to Kachchh. Further towards the north is the Eurasian plate, which is colliding with the Indian plate along its northern boundary; the latter is subducting under the former giving rise to large mountain chains of Himalayas and generating considerable amount of stresses. Bathymetry of North Arabian Sea suggests that the Kachchh-Lakhpat Lineament (KL, Inset Fig. 8.7(a)) extends westwards up to the Murray ridge. The transcontinental Narmada-Son lineament/fault in central India, south of Kachchh, is considered to have extended up to the Murray ridge (Mishra, 1977). This indicates that there are lineaments/faults that extend from the Indian continent to plate boundaries in North Arabian Sea. This implies that old continental tectonics have affected the young oceanic tectonics, and vice versa in this region.

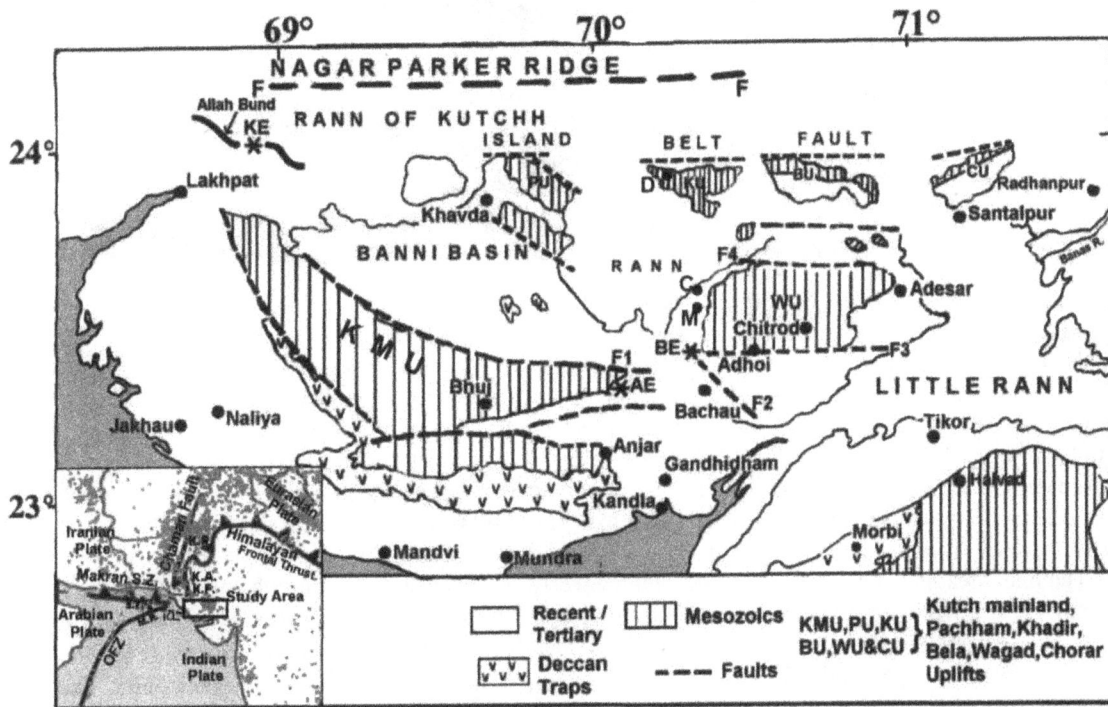

Fig. 8.7(a) Geology and tectonics of Kachchh showing various formations and faults referred to as F1-F4. The Nagar Parker Ridge and Island Belt Fault (IBF) form the northern margin of the basin with several uplifts occupied by Mesozoic sediments south of it. BE, KE and AE are the epicenters of Bhuj, Kachchh, and Anjar earthquakes of 2001, 1819, and 1956, respectively. KMU: Kachchh Mainland Uplift and WU = Wagad uplift; C and M denote Chobari and Manfara villages where large lateral displacement and strike slip movement, after Bhuj earthquake, respectively were reported. The hypocenter of Bhuj earthquake (BE) coincides with the junction of the projection of the North Wagad Fault (F4) dipping southwards (Fig 8.13, 60°) with strike slip fault, F2 in depth range of ~25-26 km that is the focal depth of this earthquake and that has displaced F1 (KMF) northwards (Mishra et al., 2005). Inset shows regional tectonics in this area with different plates and plate boundaries. Grey dots show epicenter of earthquakes since 1968 (USGS, 2001). GF: Gedi fault, KA: Karchi arc, KF: Karachi fault, KL: Kachchh-Lakhpat Lineament, KR: Kirthar range, MR: Murray ridge, OFZ: Owen fracture zone, SF: Sonne fault of Ormara micro plate between OFZ and Makran SZ (shear zone).

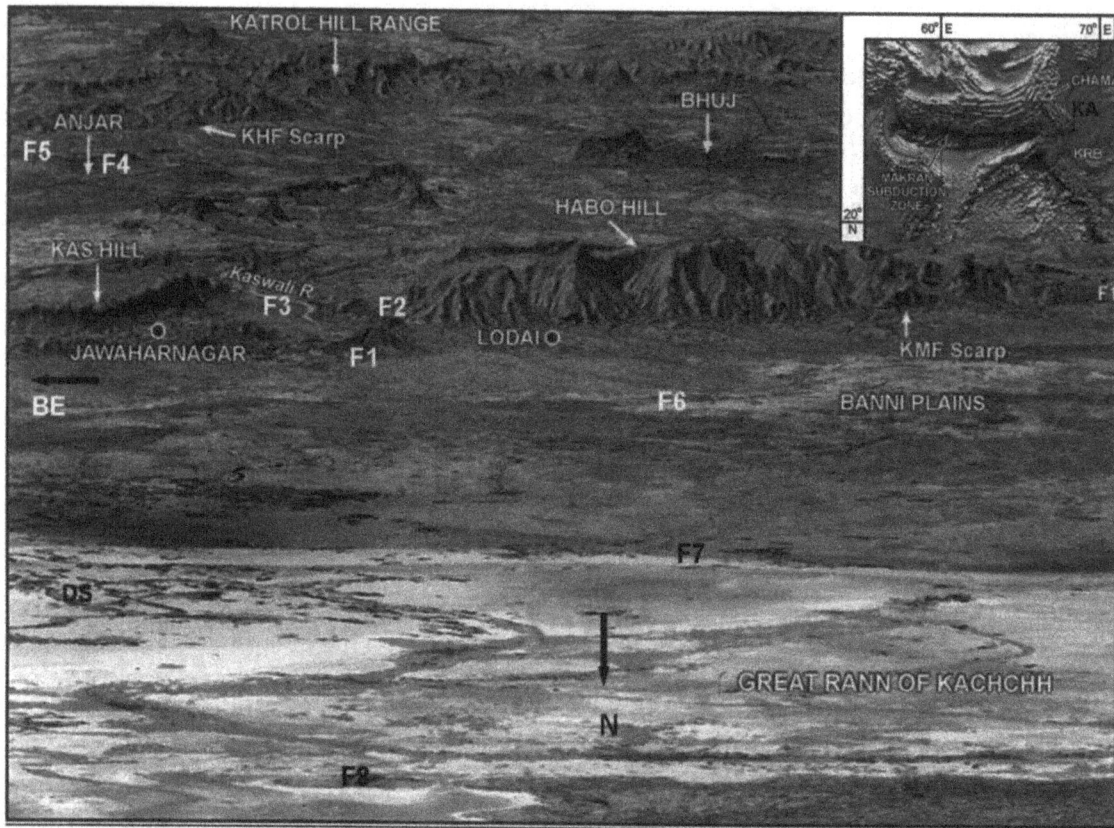

Fig 8.7(b) Topographic image of Kachchh based on Shuttle Radar Topographic Mission (SRTM) data (Courtesy Dr Prabhas Pande, GSI, Personal Communication). It shows some significant scarps and faults. E-W (F1, Kachchh Mainland Fault etc.,) oriented faults are thrust faults that control most of the scarps in this region while NE-SW (F2) and NW-SE (F3) faults along Kaswali River and several others (F4 and F5 etc.,) may be strike slip faults displacing these scarps. Such faults are quite common in Banni Plains and Rann of Kachchh similar to Khavda-Bachau and Chorbari-Manfara faults (Fig 8.7(a)) as apparent from this map. The linear feature, F7, separates the Banni plains from the Great Rann of Kachchh, which shows a highly disturbed section (DS) at its eastern margin and coincides with the uplift axis of the Wagad uplift and Chorbari and Manfara villages (Fig 8.7(a)) towards the East where maximum lateral displacements were reported after the Bhuj earthquake. The linear features, F6 and F8 coincide with the westward extension of the South and North Wagad Faults (F3 and F4, Fig 8.7(a)), respectively that are significant for Bhuj earthquake (BE), located ~37 km east of Jawaharnagar along F6 (F3, Fig 8.7(a)). This clearly indicates that these faults extend for large distances beyond their exposed sections (Fig 8.7(a)) and demonstrates the importance of space imageries for seismotectonics. The airborne magnetic map of Kachchh (Fig 9.29(b)) also reflects NW-SE and NE-SW oriented faults intersecting in the epicentral area of the Bhuj earthquake. Inset shows the Western Fold System in Pakistan with Chamman fault (CHAM) and Karachi Arc (KA) emanating from it and extending NW-SE towards the Kachchh Rift Basin (KRB). This may manifest as NW-SE fault system in Kachchh and suggest its plate boundary connection. [Colour Fig. on Page 802]

Kachchh has experienced several large seismic events in the past (Malik et al., 1999). Notable among them are the Kachchh earthquake of almost similar magnitude in 1819 (Rajendran and Rajendran, 2001) and Anjar earthquake of magnitude 6.0 in 1956 (Chung and Gao, 1995) (Fig. 8.7(a)). They are associated with E-W oriented Island Belt Fault and eastern part of Kachchh Mainland Fault (IBF and F1, Fig. 8.7(a)), respectively, whose extension towards the west almost joins at Lakhpat along the Kori creek. At least three pre-historic earthquakes affecting the Indus Valley Civilization in the Khadir uplift along the IBF (Fig. 8.7(a)) during 2900-1800 B.C. also have been suggested based on archeological excavations (Lal, 1984). Due to the recurrence of large magnitude earthquakes, this region is placed in zone V of the most vulnerable earthquake prone regions of this country on par with the Himalayan collision zone at the northern boundary of the Indian plate (Fig. 8.6). Based on distribution of earthquakes and seismicity of Kachchh, the western boundary of the Indian plate has been considered as a diffused boundary extending up to Kachchh (Stein et al., 2000). Regional tectonics of Kachchh and its surrounding region, and their role in the present earthquake have been widely discussed by several workers (Talwani and Gangopadhyay, 2001; Raval, 2001). However, interaction of regional and local tectonics is important for seismic activity in a particular region. Gravity and magnetic studies have provided valuable insights on the strength of the crust by characterizing the zones of possible weaknesses such as lineaments and faults, which strongly correlate with the siesmicity patterns.

8.2.2 Tectonics of Kachchh (Kutch) Basin – Faults and Radar Topographic Mission (SRTM) Images

Kachchh is an E-W trending Mesozoic rift basin, which is considered to have formed due to the breakup of Africa from the Indian plate during the middle Jurassic and related activities (Biswas, 1987; Mishra, 1999). Important geological and tectonic elements of Kachchh are plotted in Fig. 8.7 (a) and (b). It consists of several uplifted blocks controlled by faults. The Island Belt Fault (IBF) in the north is associated with several small uplifts where Mesozoic sediments are exposed, while the adjoining region between the IBF and the Kachchh Mainland Fault (F1, Fig. 8.7(a)) is covered by sand and salt flats known as the Banni plains and the Rann of Kachchh. The elevated land south of (F1) is the Kachchh Mainland uplift (KMU), which is occupied by Mesozoic sediments. It may represent a median uplift that characterizes most of the rift basins the world over (Section 2.7 and 2.8) as also suggested previously in case of Godavari rift basin in Section 7.7, which may be caused by compression during the close of the basin. Deccan traps of the late Cretaceous period and Cenozoic sediments are exposed (Merh, 1995) in the southern part, which is separated from Mesozoic sediments towards the north (Fig. 8.7(a)) by faults. The Deccan trap in NW India occupies a vast area and represents the volcanic basaltic flows formed due to eruption from the Reunion plume during the late Cretaceous when the Indian plate passed over it (Morgan, 1981). The Kachchh mainland fault (F1) is composed of two components viz. NW-SE trend towards the west and E-W trend towards the east, which are almost parallel to the coastline of the Gulf of Kachchh (Fig. 8.7(a)). The boundaries of various geological formations varying in age from Jurassic to recent south of F1 also follow almost the same trends indicating the importance of these trends in the geological and tectonic evolution of this region. The Wagad uplift located east of KMU is bounded by E-W oriented faults known as the South and North Wagad faults (F3 and F4, Fig. 8.7(a)), and is occupied by the Mesozoic (Middle-Upper Jurassic) and Tertiary sediments. The South and the North Wagad Faults (F3 and F4, Fig 8.7(a)) extend over large distances in Banni plains and Rann of Kachchh (F6 and F8, Fig 8.7(b)). In fact, Middle Jurassic sediments are exposed primarily in the northern part over Wagad uplift and islands like Bela, and Khadir. This indicates that after the deposition of Jurassic sediments, the northern part was uplifted due to compression during the closing phase of the basin that gave rise to these islands. This part of the basin suffered maximum disturbances. Most of these important tectonic elements and faults of the Kachchh basin are reflected as gravity gradients (Chandrasekhar and Mishra, 2002). Some NW-SE oriented faults have also been reported from the northern part such as south and east of Khavda (Fig. 8.7(a)) which are almost in line with the NW-SE strike slip reported by Seeber (2001) at Manfara (M, Fig. 8.7(a)) after the Bhuj earthquake. Kachchh and its surrounding area towards the north (Bannert, 1992) are also dominated by NW-SE and E-W trending lineaments, which extend to the western boundary of the Indian plate. The more important among them are NW-SE oriented Karachi arc (Fig. 8.7(b)) emanating from the Kirthar range of the western boundary of the Indian plate (Bannert, 1992) and E-W oriented Karachi fault (Inset, Fig. 8.7(a) and (b)) which are seismogenic in nature (Kazmi, 1979). Sen et al., (2009) have studied the mantle xenoliths from Kachchh that are formed by reaction of transient carbonatite melts and Iherzolite forming the lithosphere. In terms of isotope ratios, they are mostly similar to the Indian Ocean Ridge basalts and only slightly overlap the field of the Reunion lavas that might be due to their intrusion close to the Indian Ocean Ridge as suggested in Section 9. 2.6.

8.2.3 Seismological and Field Investigations of the Bhuj Earthquake

Field investigations after this earthquake located several sites of mud volcanoes, water fountains, cracks in the ground, and lateral displacements for tens of meters (Rejendran et al., 2001; Wesnousky et al., 2001). However, the causative fault rupture of this earthquake could not be located. This might be due to inaccessibility of the region or the fact that the rupture did not reach up to the surface. Theoretical calculations based on the fault plane solution (Negishi et al., 2002) and modeling of changes in the gravity field and elevation after the earthquake as discussed below, suggest that the causative fault reached a depth of 9-10 km in the basement and therefore represents a blind thrust. Fault plane solution of main and aftershocks suggest E-W and NE-SW and NW-SE oriented thrust faults dipping 40-60° as causative faults respectively for these events (Rastogi et al., 2001; Kayal et al., 2002). The aftershocks are concentrated in an area of 75 × 40 sq km in depth range of 15-38 km (Fig. 8.18) in western part of Wagad uplift and eastern part of the Kachchh Mainland uplift, while the main shock is located almost at the center of this zone at a depth of about 23-24 km. This figure also shows the fault plane mechanism of the main shock that shows E-W oriented fault implying N-S compression. The main shock was followed by a large number of aftershocks and some of them were as large as M_W= 5.6 (Gedi earthquake) quite close to the main shock. Based on detailed analysis of Vp, Vs, and Vp/Vs of aftershocks during 2001-06, Mandal and Chadha (2008) suggested large variations in these parameters in a depth range of 0-34 km. Most of the aftershocks presented E-W oriented fault plane solution with N-S compression but some of the aftershocks presented compression directions as NNE-SSW and NW-SE. They delineated a high velocity zone under the North Wagad Fault suggesting a mafic intrusive body. Based on seismic tomography, Mandal and Pandey (2010) have suggested a reduced crustal thickness of 34 km under the central part of the Kuchchh basin that increases to 42-44 km in the surrounding region. This inference is, however, based on seismic wave velocity modeling from seismic activity and is opposite to what has been inferred from seismic imaging based on active deep seismic sounding and gravity modeling as described below. This might be due to large scale mafic intrusions in this section that would increase the velocity at every level. They have adopted Moho where Vp is 7.75-8.1±0.3 km/sec while velocity of 8.2 km/sec is regarded as upper mantle. In fact, the former can easily pass for an underplated layer while the latter can be regarded as Moho providing a crustal thickness of 42 km. Underplated layer of velocity 7.2-7.5 km/sec has been reported from the neighboring region of Cambay basin (Section 9.4).

8.2.4   Seismic Reflection Imaging

Sarkar et al., (2007) and Reddy (2010) have provided the seismic imaging of the epicentral zone of this earthquake (Fig. 8.8(a)). The Kachchh rift basin basically developed during the Mesozoic period and is therefore characterized by an extensional phase followed by a compressional phase during closure of the basin. The effects of these two major tectonics in this region, viz. extension and thrusting, can be visualized very well from the crustal seismic image. It shows N-verging reflectors, which primarily dip upward towards the north in the upper crust in the direction of the plate motion up to about 10 km depth indicating thrusting. Reflectors below 30 km primarily dip downwards towards the north causing crustal thickening, indicating a tectonic inversion at this depth. Between 20-30 km both type of reflectors intermingle with each other indicating intersection of these two distinct tectonic styles, causing a weak crust. The hypocenter of this earthquake lies in this depth range at 24-25 km as described above. Based on seismic tomography (Kayal et al., 2002; Mishra and Zhao, 2003), presence of fluid in the hypocentral region has been suggested, which also indicates a weak fractured rock matrix in this section. Presence of fluids along faults in Kachchh such as KMF is also suggested by high conductivity that separates resistive blocks (Sastry et al., 2008). A shallow section based on seismic reflections along Profile Mundra-Anjar-Bachau (Fig. 8.7(a)) is given in Fig. 8.8(b) (Gupta et al., 2001) that suggests maximum thickness of Mesozoic sediments as ~ 4 km along the coast of the Gulf of Kachchh, and uplifts and depressions in the basement are fault controlled in comparison to Fig. 8.7(a). Prasad et al., (2010) based on several refraction/wide angle reflection profiles have suggested maximum thickness of Mesozoic sediments along the cost of the Gulf of Kachchh between Mandvi and Mudra that decreases NE to Waged uplift and basement is shallowest NE of Bachau where Wagad uplift starts (Adhoi) and Middle Jurassic sediments are exposed (Fig 8.7(a)).

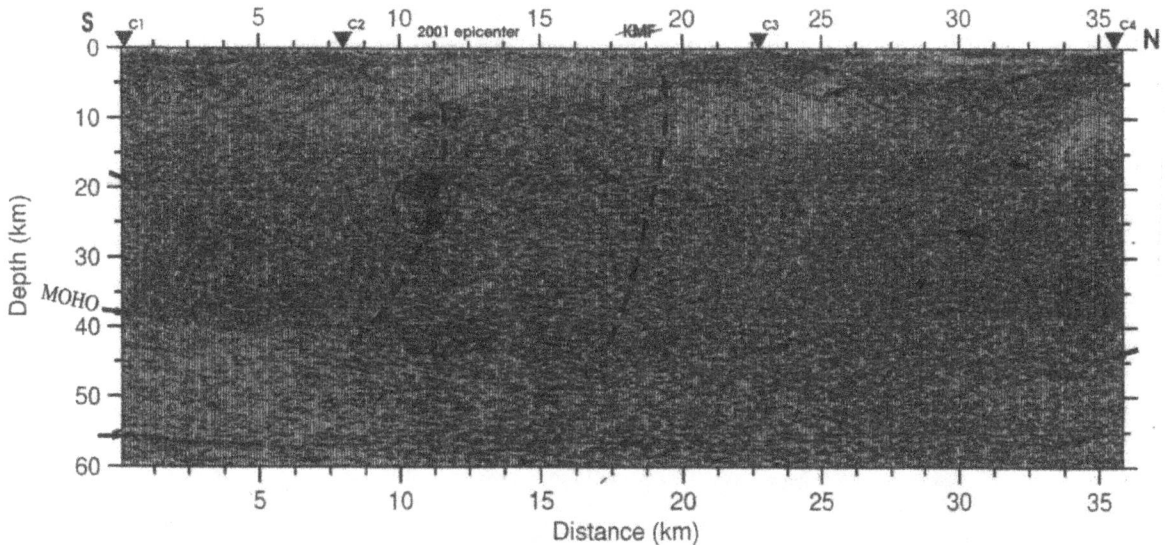

Fig. 8.8(a) Prestack depth migrated seismic section across epicentral zone of Bhuj earthquake showing prominent reflections by arrows. The Kachchh Mainland fault (KMF) and the fault related to Bhuj earthquake is plotted using dashed lines. Arrows in the shallow section up to ~10 km are north verging in direction of the plate motion indicating thrusting. It also shows thickening of the crust under the epicentral zone (Sarkar et al., 2007).

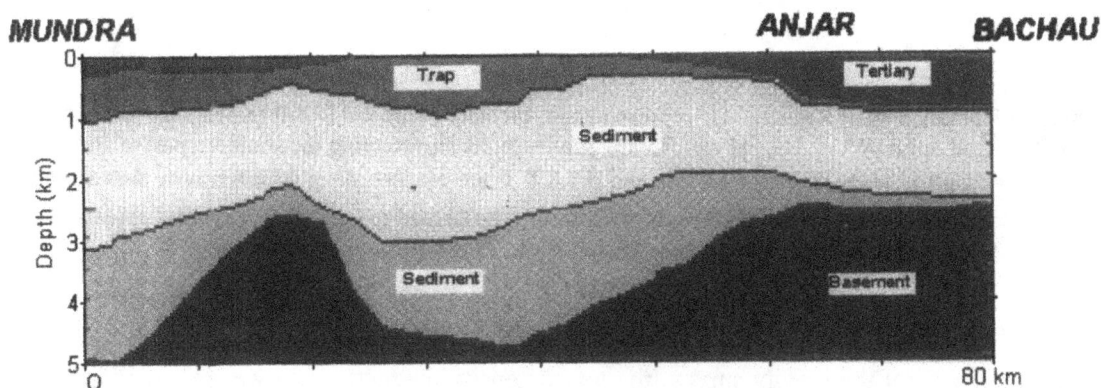

Fig. 8.8(b) Shallow cross section from Mundara-Anjar-Bachau based on seismic reflection profile (Gupta et al., 2001).

8.2.5 Bouguer Anomaly Map of Kachchh (Kutch)

To delineate various trends and structures of this region, a complete Bouguer anomaly map of Kachchh (Fig. 8.9) was prepared incorporating some additional gravity data (250 stations) acquired in the gap area with the existing gravity data (about 1000 stations) (NGRI, 1975). The Bouguer anomaly map (Fig. 8.9) depicts a central gravity high (H1) coinciding with Kachchh Mainland uplift, which extends almost up to Bhuj. Further east there is a gravity low around Bhachau (L1) representing a local Tertiary basin (Chandrasekhar and Mishra, 2002; Mishra and Rajasekhar, 2006; Mishra et al., 2005, 2008). Another gravity low occupies the Wagad uplift (L2, Fig. 8.9) separated by gravity high H2 in between, which is part of the Wagad uplift itself. The other gravity highs H3, H4 and H5 correspond to junctions of Wagad and Bela uplifts, Khadir uplift, and Pachham uplift, respectively (Fig. 8.9). These gravity anomalies (H3-H5) are almost circular/semi-circular gravity highs and appear to be caused by mafic intrusives. Fig. 8.9 also shows two profiles (I and II) which are modeled below for basement and crustal structures. This figure also gives the sites of several wells logged for hydrocarbon exploration and log for one at Sutri (Srinivasan and Khar, 1995) is given in the inset of this figure.

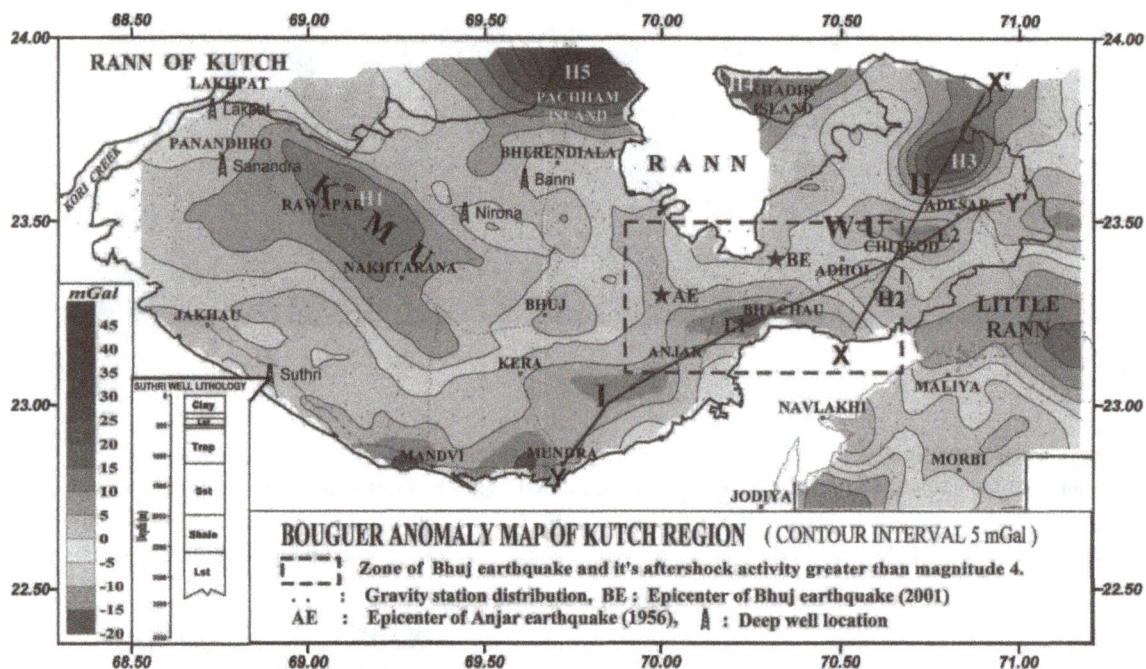

Fig. 8.9 Bouguer anomaly map of Kachchh. H1 represents the Kachchh Mainland Uplift (KMU) and H2 and L2 constitutes the Wagad uplift (WU). H3, H4 and H5 are gravity highs representing the southern part of Bella, Khadir and Pachham Uplifts, respectively. I (YY') and II (XX') are profiles along which gravity data are modeled. BE and AE are epicenters of Bhuj and Anjer earthquake. The dashed red rectangle indicates epicentral zone of Bhuj earthquake and its aftershocks (Mishra et al., 2005). [Colour Fig. on Page 803]

8.2.6 Total Intensity Magnetic Map of Epicentral Area

To supplement the results of the gravity survey, the total intensity magnetic field was recorded along roads and tracks at about 1 km interval in the epicentral area of the Bhuj earthquake and its aftershocks. The total intensity data was acquired using a Proton precision magnetometer with 1 nT accuracy. The total intensity magnetic map after correction for diurnal variation and International Geomagnetic Reference Field (IGRF) (Fig. 8.10) shows a major magnetic low ML1 and a magnetic high MH1 over Wagad uplift and north of it, separated by a large gradient. Since this area is located in a low geomagnetic latitude of 25° N, a basement uplift or a mafic intrusive will be characterized primarily by a large magnetic low accompanied by a small high (Mishra et al, 2005).

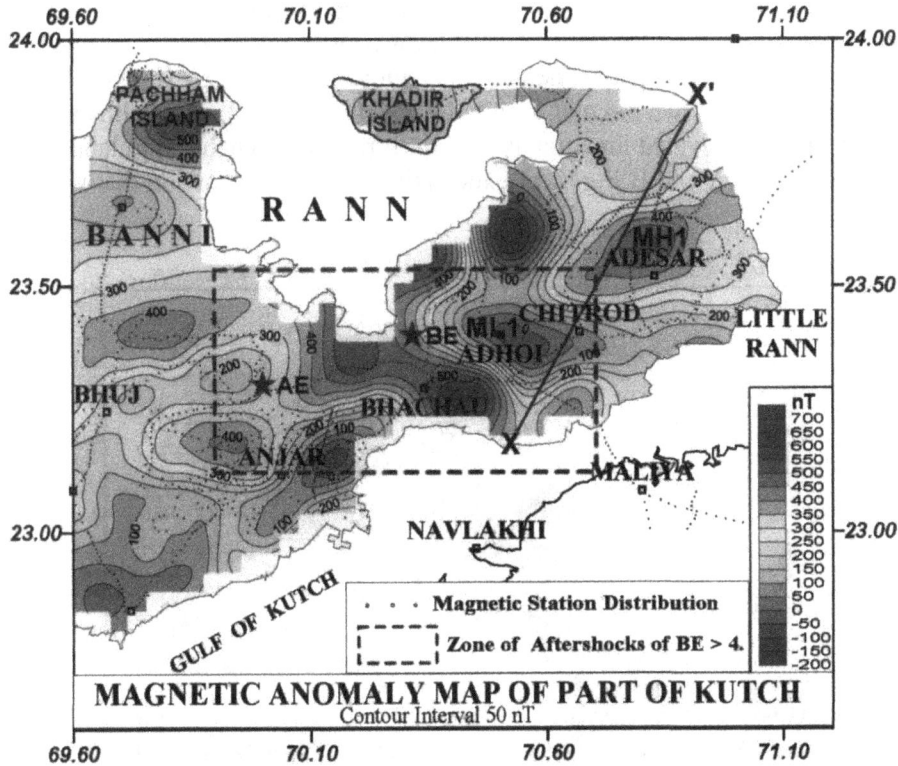

Fig. 8.10 Total intensity magnetic map of the epicentral zone showing some important magnetic anomalies. ML1 and MH1 indicate magnetic lows and highs corresponding to the Wagad uplift. XX' is a profile along which these anomalies are modeled in Fig. 8.14. (Mishra et at., 2005)

8.2.7 Isostasy and Isostatic Regional Gravity Anomaly

Knowledge about the state of isostasy can provide vital information about the nature of vertical forces operating in a region. For example, if the excess mass due to topographic load is overcompensated at depth, then the region tends to rise to maintain the isostatic balance. Similarly, under compensation may cause subsidence. The Free air and Bouguer anomalies have been widely used to infer the state of compensation in a region (Woolard, 1959). The Free air anomaly indicates the state of isostatic compensation; in case of complete equilibrium, its value is zero on a regional scale as the excess of topographic load is balanced by the low density compensating mass. In case of Airy's model of compensation:

$$(\rho_m - \rho_C).T_C = \rho_C.H \qquad\qquad .....(1)$$

where T_C = Thickness of crustal root for compensation

ρ_C = Bulk density of crust

ρ_m = Density of upper mantle and H = Regional Topography.

This equation states that the negative mass of the compensating root cancels the positive mass of the topography. In case of isostatic equilibrium, free air anomalies will be zero or near zero. The corresponding Bouguer anomaly would represent the isostatic regional (Subba Rao, 1996) and its value would indicate the state of isostatic equilibrium. Isostatic regional is part of the regional field related to isostasy and does not account for other deep seated sources that are classically included under the regional field. Its negative values reflect the negative density root of the long wavelength topography indicating overcompensation, while positive values indicate under compensation

(Woolard, 1959, Turcotte and Schubert, 2001) that is directly related to geodynamics causing uplift in the former and subsidence in the latter. The same procedure was used to obtain the regional field in Kachchh from the observed Bouguer anomaly. The isostatic regional (Fig. 8.11, Mishra et al., 2005) so obtained in the present case provides two circular gravity lows, RL1 and RL2, of amplitude –11 and – 13 mGal. They coincide with the elevated topography of the SE part of Kachchh Mainland Uplift and Wagad uplift, respectively (Inset, A and B Fig. 8.11), indicating over compensation that would cause uplift in these sections.

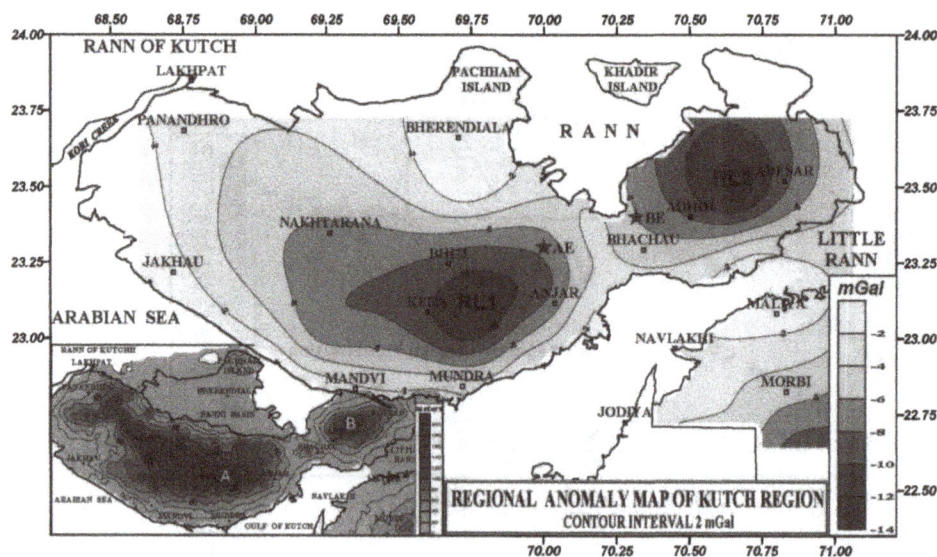

Fig. 8.11 Isostatic regional gravity anomaly map based on Airy's model of isostasy derived from Bouguer anomaly for zero free air anomaly corresponding to regional elevation. RL1 and RL2 denote regional gravity lows corresponding to elevated zones of the eastern part of KMU and WU indicating deep seated mass deficiency. Inset shows the elevation in meters with elevated parts A and B representing KMU and WU, respectively (Mishra et at., 2005).

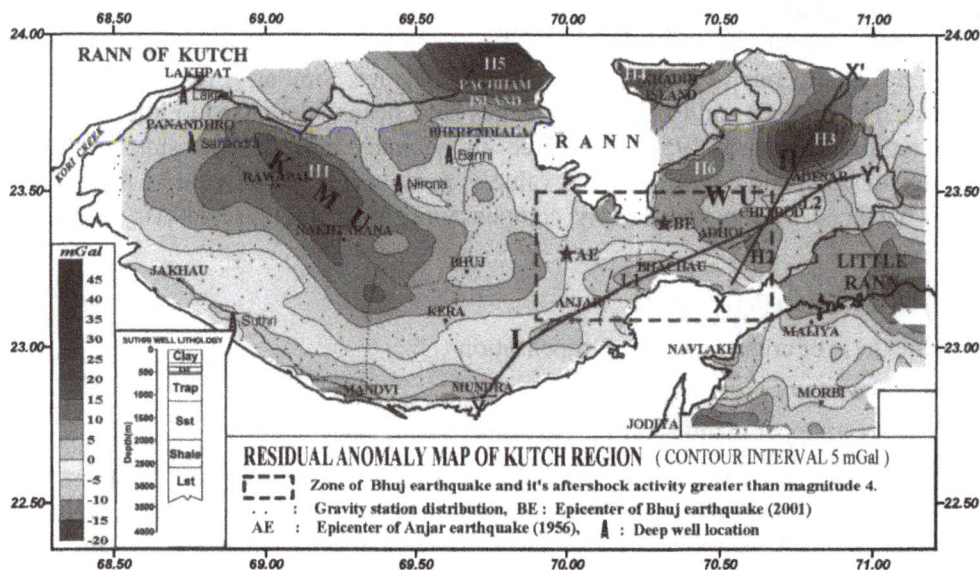

Fig. 8.12 Residual anomaly map with highs (H1-H5) and lows (L1-L5) representing same features as given in Fig. 8.9. Profiles, I (YY') and II (XX') are used to model shallow and deep seated structures.

[Colour Fig. on Page 603]

8.2.8 Residual Gravity Anomaly

Residual Bouguer anomaly (Fig. 8.12) is obtained by subtracting the isostatic regional due to Moho variations (Fig. 8.11) from the observed Bouguer anomaly (Fig. 8.9). It shows an almost similar pattern the observed Bouguer anomaly with reduced amplitude of anomalies. The observed gravity anomalies H1-H5 and L1-L5 as shown in Fig. 8.9 are also marked in Fig. 8.12 by the same notations. The Kachchh Mainland Uplift is characterized by a gravity high H1 with a maximum amplitude of about +20 mGal. It is flanked on either side by lows till it encounters different islands such as Pachham Island etc., towards the north. The gravity high due to Kachchh Mainland Uplift continues eastwards up to the Wagad uplift, which shows a gravity low of about – 6 mGal. The almost circular gravity high, H3 north of Wagad uplift is quite significant as it extends further northwards to include the gravity anomaly due to the Bela uplift. The sources for these anomalies become clear only after modeling as discussed in detail below.

8.2.9 Basement Faults, Structures and Crustal Thickness across Wagad Uplift

(i)   Profile I: YY[1], Mundra-Bachau-Chitrod-Adesar (Fig. 8.9)

Fig. 8.13 shows the residual Bouguer anomaly along this profile, which is obtained by subtracting the isostatic regional (Fig. 8.11) from the observed Bouguer anomaly (Fig. 8.9) as given in Fig. 8.12. The residual Bouguer anomaly is modeled for 2.5 dimensional bodies (Webring, 1985) using the constraints from the seismic studies for the part of the profile up to Bhachau (Fig. 8.8(b)). Some constraints to model gravity data is also obtained from the bore hole litho section at Sutri (Inset; Fig. 8.9) (Srinivasan and Khar, 1995). The bulk density of Tertiary and Mesozoic sediments is obtained from seismic velocity, and reported density logs from the bore hole and laboratory measurements of exposed rocks.

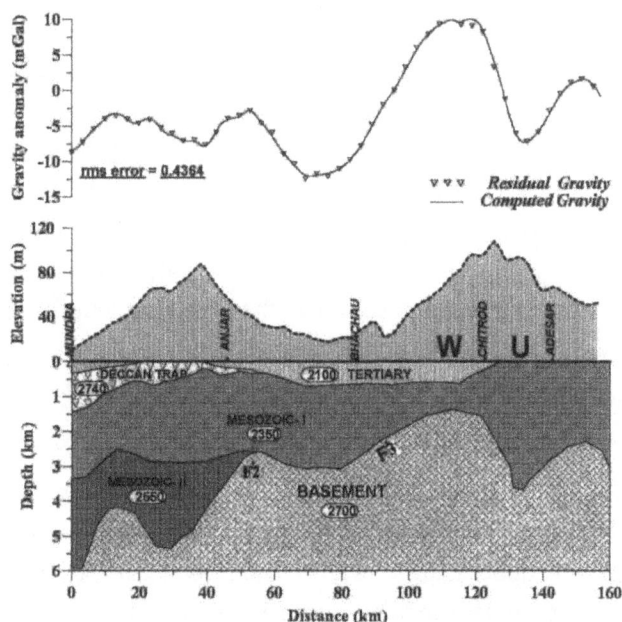

Fig. 8.13 Shallow section up to the basement derived from 2.5-D modeling of Residual gravity anomaly along a profile Mundra-Bachau-Adesar (I, Fig. 8.12) using constraints from seismic section up to Bachau (Fig. 8.8(b)). Densities of different layers are given in kg/m³. The graph shows a basement uplift of 1.0-2.0 km across contacts F'2 and F'3 whose projections on the surface coincide approximately with faults F2 and F3 (Fig. 8.7(a)). The computed and observed fields show r.m.s. error much less (1-2%) than the observed anomaly, indicating a good match between them (Mishra et at., 2005).

There are inherent intrinsic ambiguities in gravity modelling basically due to two unknowns, body configuration including their depths and densities. However, in case one of these unknowns can be constrained from other methods such as seismic sections or bore hole information, the other unknown can be uniquely determined.   Further, if the section derived from gravity modeling provides results within the reasonable geological limits, it can be relied upon for tectonic interpretation. The computed model shows basement uplifts of about 1-2 km across various contacts (F'2 and F'3, Fig. 8.13) whose projection on the surface coincides with F2 and F3, respectively (Fig. 8.7). The projection of F3 on the surface also coincides with the northern limit of the Tertiary sediments south of Chitrod. This implies that the gravity highs are primarily caused by basement uplifts, which are basically fault controlled. Mesozoic sediments in this model are divided in two groups of different densities based on the seismic section showing a lower group with a higher velocity compared to the upper group (Fig. 8.8(b)). This section also shows the thickest sedimentary section towards Mundra along the coast and the basement rises towards the north across ascending thrust faults.

(ii)   Profile II: X- X' (Fig. 8.9)

To map the faults controlling the Wagad uplift in the epicentral area of the Bhuj earthquake and its aftershocks, residual gravity (Fig. 8.12) and magnetic data (Fig. 8.10) along a profile XX' across this zone are modeled (Fig. 8.14) using constraints based on the previous profile as described above. The sharp gravity and magnetic gradients between Chitrod and north of Adesar suggest a contact separating blocks of different densities and susceptibility, which may represent a fault. The basement model computed jointly from the gravity and the magnetic data for 2.5 dimensional bodies (Webring, 1985) and the corresponding density and magnetization distribution is given in Fig. 8.14. In this model Mesozoic sediments are treated as one single unit of bulk density 2350 kg/m^3 due to the predominance of Upper Mesozoic sediments in this region. There are some tertiary sediments towards the north and south of Chitrod with density 2100 kg/m^3. The sediments being non-magnetic in nature, the magnetic field in sedimentary basins primarily originates from susceptibility distribution in the basement. The basement model (Fig. 8.14) shows an increase in susceptibility towards the north. The inclined contact in the basement (F'4) between Chitrod and Adesars separates the high susceptibility basement north of the the Wagad uplift from that of its southern part indicating its mafic nature. The high density (2800 kg/m^3) and high susceptibility body (MI) in the basement may represent a mafic intrusive in the basement that has caused the Bela uplift.

The regional gravity anomaly (Fig. 8.11) along profile XX' (Fig. 8.14) represents a deep seated mass deficiency along the Moho. This regional gravity low (Fig. 8.14) is modeled due to crustal thickening using bulk densities of lower crust and upper mantle as 2900 and 3300 kg/m^3, respectively. It provided a crustal thickening of up to 42-43 km under the Wagad uplift that was subsequently confirmed from seismic studies (Fig. 8.8(a)). The receiver function analysis at Bhuj (Ravikumar et al., 2001) and seismic studies along profile I (Fig. 8.8(a)) suggest crustal thickening from 35 km along the coast up to 45-46 km under the Wagad uplift.

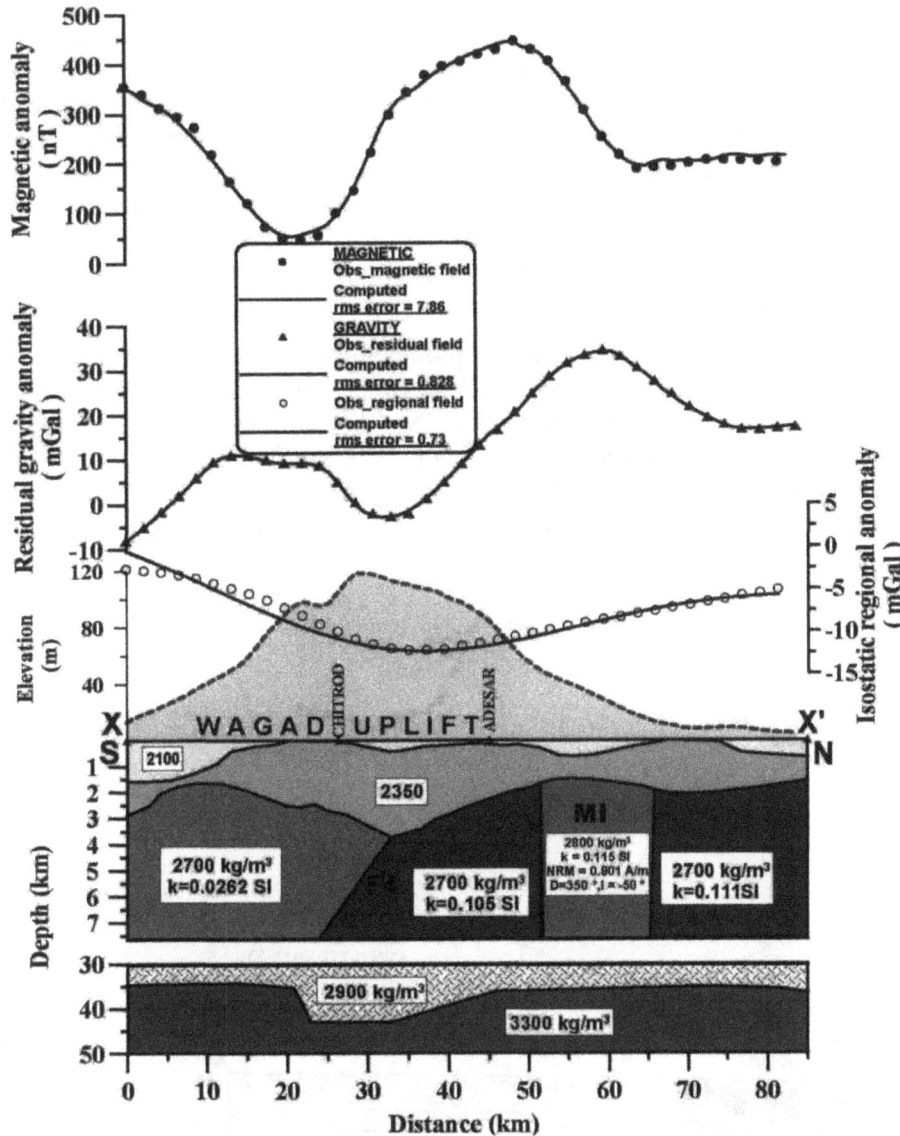

Fig. 8.14 Shallow cross section up to the basement derived from 2.5-D modeling of residual gravity anomaly and total intensity magnetic field along a profile XX' (II, Fig. 8.10). It shows a basement contact (F'4) between Chitrod and Adesar showing variation in susceptibility across it. Its projection on the surface coincides approximately with the north Wagad fault (F4, Fig. 8.7(a)). The regional field (Fig. 8.11) along the same profile is modeled for mass deficiency at Moho of about 8 km thick crust for density contrast of 400 kg/m³ between the lower crust and upper mantle under the Wagad Uplift. MI represents a mafic intrusive similar to a volcanic plug. Susceptibility and density of different bodies are given in SI units and kg/m³, respectively. The computed and the observed fields show an r.m.s. error much less (1-2%) than the observed anomaly indicating a good match between them (Mishra et at., 2005).

8.2.10 Basement Faults and Structures across Kachchh Mainland Uplift

The two profiles described above present the subsurface structures under the Wagad uplift. However, a large part of Kachchh is occupied by Kachchh Mainland Uplift (KMU, Fig. 8.7(a)) and therefore the residual Bouguer anomaly (Fig. 8.12) along two profiles, viz. Mundra-Bherendiala and Sutri to Pachham Island, were modeled to understand the structures in this section.

Fig. 8.15(a) shows the residual Bouguer anomaly and isostatic regional (Fig. 8.11) along the Mundra-Bherendiala profile and modeled for shallow structures and Moho configuration. It shows the KMU as a horst bounded by reverse faults F1 and F3 in the basement, shown by same notation as in Fig. 8.7(a). It may be related to the close of the Kachchh Mesozoic basin representing a median uplift that characterizes rift basins the world over. A basement of density 2.7 g/cm^3 is overlain by Mesozoic sediments (2450 kg/cm^3) and Deccan trap in the southern corner. The maximum thickness of the Mesozoic sediment of ~ 4 km is observed along the coast of the Gulf of Kachchh. It shows a crustal thickening up to ~ 45 km as in the case of the previous profile (Fig. 8.14). In case the basement faults controlling the KMU, F3 and F1 are joined with those controlling the crustal thickening, they form listric faults as shown in Fig. 8.8(a) based on seismic study.

Fig. 8.15(b) is the residual Bouguer anomaly along the profile Sutri-Pachham Island and the modeled shallow section based on it constrained from the bore-well log of the Sutri well for oil exploration (Fig. 8.9). This model also shows a horst kind of structure for KMU bounded by faults F1 and F2 in the basement (2.7 g/cm^3) shown by same notation as Fig. 8.7(a). The maximum thickness of Mesozoic sediments of ~ 4 km is encountered along the coast of the Gulf of Kachchh. The high density body in the northern corner represents the mafic intrusive that has caused thePachham island uplift.

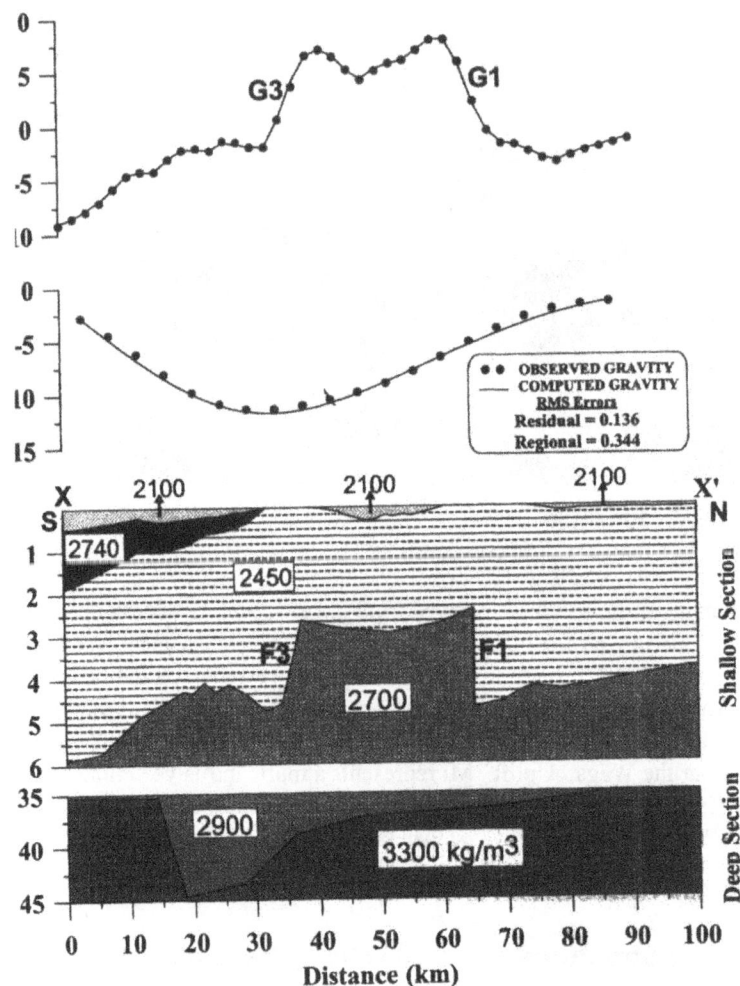

Fig. 8.15(a) Residual Bouguer anomaly and isostatic regional along Mundra-Bherendiala profile and modeled shallow section and Moho configuration, respectively modeled from them. It shows basement uplift of 2-3 km under KMU as a horst and crustal thickening of 35 km along the coast up to ~ 45 km under the KMU.

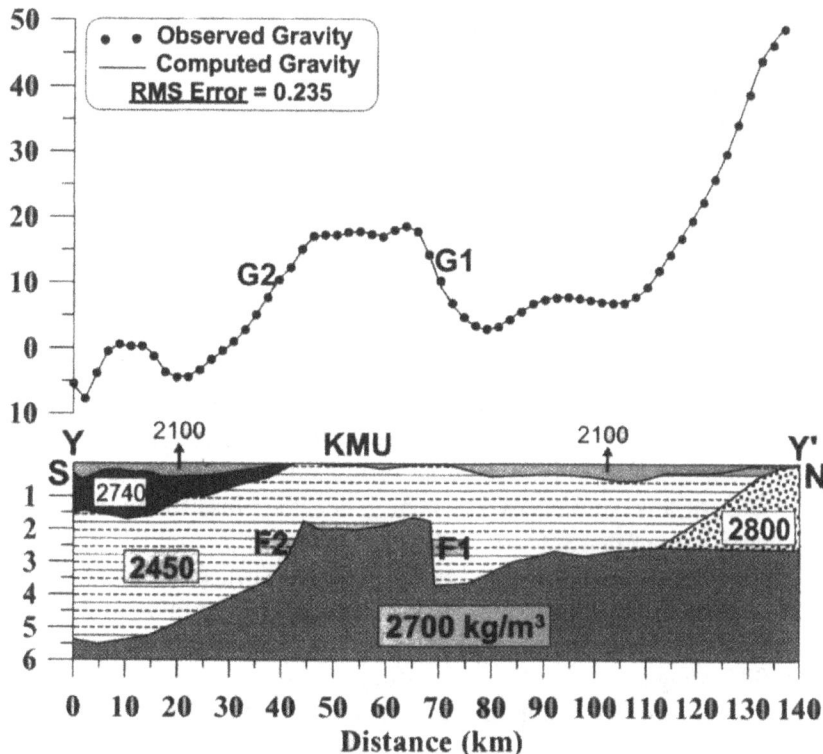

Fig. 8.15(b) Residual Bouguer anomaly and the shallow section modeled from it along the Sutri-Pachham Island showing KMU as a horst bounded by faults F1 and F2. Pachham Island is shown as a mafic intrusive at the northern end of the profile.

8.2.11 Sismotectonics of Bhuj Earthquake and Geodynamics

The gravity modeling along profile I (Fig. 8.9) suggests that the basement is controlled by numerous thrust faults (F'2 and F'3, Fig. 8.13) with the basement uplifted up to 1-2 km towards the north. Their projection on the surface coincides with the south Kachchh Mainland and Wagad faults (F2 and F3, Fig. 8.7(a)). This along with the joint modeling of gravity and magnetic data across Wagad uplift (Profile II, Fig. 8.9) suggests that the basement uplift related to Wagad uplift is bounded by contacts F'3 and F'4 (Fig. 8.13 and 8.14) dipping 50-60° towards the south whose projection on the surface coincides with south and north Wagad faults, respectively (F3 and F4, Fig. 8.7(a)). The basement model (Fig. 8.14) north of Wagad uplift consists of blocks of high density and high susceptibility (MI) which suggest mafic intrusion. The gravity anomaly corresponding to mafic intrusion (MI, Fig. 8.14) is circular in nature (H3, Fig. 8.9) which suggests plug type of igneous complexes as several similar circular/semi circular gravity anomalies due to igneous complexes have been delineated in Saurashtra, south of Kachchh as described in Section 9.2. The various uplifts along, the Island Belt Fault (Fig. 8.7(a)) in the northern part of Kachchh depicting circular/semicircular gravity anomalies (H4, H5; Fig. 8.9), therefore, appear to be associated with mafic intrusives. This was subsequently confirmed from high Vp, Vs, Vp/Vs (Mandal and Chadha, 2008). These intrusives may belong to the Deccan trap activity or to some older activity related to the breakup of India and Africa during the Mesozoic period along the coasts of Saurashtra and Kachchh. This may be especially so because of the occurrences of Mesozoic basins in this section. Role of intrusives as concentrators of regional stress in a region has been highlighted by Hildenbrand et al., (2001).

The fault plane solution of the main shock of the Bhuj earthquake indicates an E-W oriented thrust fault dipping towards south by 40-50° (Yagi and Kikuchi, 2001; Rastogi et al., 2001; Kayal et al., 2002). Based on its reported epicenter and depth and dip of the causative fault, its projection would approximately coincide with contact separating blocks of different susceptibility in the basement under the northern part of the Wagad uplift between Chitrod and Adesar (F'4, Fig. 8.14). The projection of this contact on the surface in part coincides with the north Wagad fault (F4, Fig. 8.7(a)) and therefore might be related to each other in depth. Or, only a part of this fault was reactivated during this earthquake. The seismological studies (Negishi et al., 2002) and modeling of changes in elevation and gravity field before and after this earthquake as described below also suggest that the causative fault for this earthquake is located at a depth of 9-10 km in the basement. As discussed above, several field evidences of liquefaction, lateral spread, sand blows, and mud volcanoes have been reported (Rajendran et al., 2001; Wesnousky et al., 2001) from this area but the fault rupture for the Bhuj earthquake has not been reported so far, which also suggests that it might not have reached the surface. The largest lateral spread and slump, a few meters wide and tens of meters long reported north of Chobari (C, Fig. 8.7(a)) as discussed above, is located close to fault F4 (Fig. 8.7(a)) which in part appears to be the causative fault of this earthquake. Kachchh is located close to the triple junction of three plates Indian, Arabian and Iranian plates towards the east in the north Arabian Sea and western boundary of the Indian plate towards the NW (Inset Fig. 8.7 (a) and (b)). Geological formations of Kachchh and known faults suggest the predominance of E-W and NW-SE trends in this region that lead to plate boundaries as discussed in Section 8.2.1. Based on lineament studies from remote sensing, Singh and Singh (2005) suggested N to NE orientation of stress prior to the Bhuj earthquake, that is direction of the plate motion indicating the importance of plate tectonics forces for this earthquake.

It is interesting to note that the epicenters of the Bhuj earthquake and most of its aftershocks and Anjar earthquake are located at the periphery between the centers of deep seated mass deficiencies at the Moho in the area (RL1 and RL2, Fig. 8.11), which may experience maximum stress. The modeling of these regional gravity lows suggests crustal thickening of 7-8 km for the assumed density contrast of -400 kg/m^3 across the Moho (Fig. 8.14). These gravity lows can be explained by mass deficiency within the crust or at Moho in the upper mantle. Receiver function analysis (Ravikumar et al., 2001) and controlled source seismic suggesting crustal thickness of 35 km along the coast of the Gulf of Kachchh (Mundra), and 44-46 km under Kachchh Mainland Uplift and Wagad uplift as referred to above supports the view that these mass deficiencies are located at the crust-mantle interface. Presence of thick crustal roots creates buoyancy and may lead to accumulation of local stress especially at the periphery of the crustal roots. A positive correlation between seismicity and isostatic anomalies has also been reported from the Iranian plateau (Zamani and Hashemi, 2000). Neotectonic activities in the form of uplift and geomorphological changes have been widely reported from this region (Merh, 1995; Wadia, 1949). Vertical differential stress in Kachchh may be further accentuated due to large scale deposition of sediments in the adjoining north Arabian sea (Clift et al., 2000). The other major earthquakes of Kachchh viz Anjar and Kachchh earthquakes of 1956 and 1819 respectively are also associated with E-W oriented thrust faults F1 and IBF, respectively, (Fig. 8.7(a)) and are located at the pheriphery of isostatic regional low RL1 (Fig. 8.11).

Based on paleoseismicity of this region, Schweig et al. (2003) have suggested that earthquakes in Kachchh relax high ambient stresses that are locally concentrated by rheologic heterogeneities. Various intrusives and changes in susceptibility and density across contacts in the basement as shown in the basement models (Fig. 8.13 and 8.14) will be responsible for heterogeneities, which act as stress concentrators. The NW-SE oriented faults near Khavda (Fig. 8.7(a)) may be extending south wards through Rann of Kachchh as evidenced from the trend of gravity high H2 (Fig. 8.9) forming a knot in the epicentral zone of Bhuj earthquake and its aftershocks. It is supported by the reports of NW-SE strike slip near Manfara after the Bhuj earthquake (Seeber, 2001). Knots of different faults and lineaments tend to facilitate the occurrence of seismic activity. The Wagad uplift where the epicenter of the Bhuj earthquake is located, is bounded by almost parallel E-W oriented thrusts on either sides that may act as a coupled system enhancing the effects of any seismic activity. It is, therefore, inferred that large earthquakes in Kachchh in general and Bhuj earthquake in particular are primarily caused due to (i) coupling of regional stress created by active plate tectonic forces with the local stress generated by crustal buoyant roots (ii) mafic intrusives acting as stress concentrators (iii) cross points of different structural trends as knots (iv) fluids in the hypocentral region and (v) parallel E-W oriented thrusts on either side of the affected blocks such as Wagad uplift in case of Bhuj earthquake. Due to the first factor it can not be treated purely as an intraplate earthquake; to be specific it can be termed as near plate boundary earthquake. Fault plane solutions of most of the aftershocks of the Bhuj earthquake indicate N-S compression, same as the main shock. However, some of them indicate NNE-SSW and NW-SE compression as described in Section 8.2.3. Plate motion of the Indian plate is normally considered to be NNE-SSW and therefore, compression in this direction is related to plate motion while compression in N-S and NW-SE direction, especially the latter, must have a different origin that may be related to activity along the Western Fold Belt (Pakistan). This suggests some contribution from plate boundary on the seismic activity of Kachchh that explains the large number and magnitude of earthquakes in this region compared to other intraplate regions.

The occurrence of the shallowest basement under the Wagad uplift and Islands of Bela, and Khadir where oldest Mesozoic sediments of Middle Jurassic are exposed, suggest that this part of the basin was uplifted after deposition of these sediments during the closing phase of the basin. This section is therefore, highly disturbed causing more seismicity in the northern part of the basin. These uplifts itself might have been triggered by these plug type of intrusives in the form of Pachham, Khader, and Bela islands that provide gravity signatures typical of volcanic plugs. In this light, it can be conjectured that these plugs may belong to Karoo volcanics that was active during Jurassic period and was responsible for the breakup of Africa and India (Sections 5.3 and 7.11), and for the formation of Jurassic basins along the respective coasts of these countries. Due to subsequent deposition of Jurassic-Cretaceous sediments, volcanic rocks related to Karoo volcanic during breakup of Africa and India are not exposed, while Deccan trap is exposed, since it is post sedimentation. Mafic igneous rocks older than the Deccan trap have been reported from Lodhika well by ONGC (Section 9.2.4).

8.3   Source Parameters of the Bhuj Earthquake from Post Seismic Elevation and Gravity Changes

The Bhuj earthquake of January 26, 2001 in Kachchh, India was one of the largest historic intraplate events. Events of this size are commonly associated with a significant amount of ground rupture, which were not found in the present case. However, surface deformation in the form of lateral spreads, sand blows, mud volcanoes, craters, intense liquefaction, and extensive dewatering in the low lying Rann of Kachchh was reported from field investigations (Rajendran et al., 2001, Wesnousky et al., 2001); however, no surface rupture has been found in the epicentral region of the Bhuj earthquake.

Seismotectonics as discussed above and fault plane solutions from teleseismic studies, consistently suggest a steeply south-dipping reverse-slip mechanism for the Bhuj earthquake (Antolik and Dreger, 2003), whereas the lack of surface faulting and the depth distribution of aftershocks (Negishi et al., 2002) suggest that the rupture was deeply buried. Finite fault inversions of teleseismic broadband body waves by Antolik and Dreger (2003) suggest that a substantial slip may have extended to near the surface. The coseismic N35°E displacement of 16 ± 8 mm obtained at Jamnagar ≈ 150 km south of the epicentral area is the only GPS-measured estimate of coseismic surface displacement. The analysis of the motions of historic triangulation monuments of the Great Trigonometric Survey of India last surveyed in 1857 may provide additional constraints on horizontal motions (Jade et al., 2002). Gahalaut (2010) suggested post seismic horizontal velocity vectors almost in the direction of plate motion (N46-50⁰E) that may be attributed to thrusting along the causative fault dipping towards the south. Co-seismic changes in height (from leveling and GPS) and gravity field are presented below to provide a better insight into the source mechanism of the Bhuj earthquake, and to evaluate the sub-surface mass adjustment by intense liquefaction and shallow hydrologic phenomena.

8.3.1   Pre- and Post- Earthquake Measurements of Gravity and Elevation

During 1997-99, a high-resolution gravity survey (NGRI, 2000) was carried out in the Kachchh Basin in connection with hydrocarbon exploration for the Indian oil industry as described above in Section 8.2. Because of severe damage to permanent structures, we could locate only 20 of the gravity base stations along two profiles of the 1997-99 survey (Figure 8.16(a)) after this earthquake in November 2001. Profile I (Mundra-Bachau-Manfara-Chitrod) extends across the epicentral area of the Bhuj earthquake and Profile II (Mundra-Bhuj-Banni Basin) is located to the west of it (Fig. 8.16(a)). Gravity observations were taken in the form of two-fold three-way loops closed within 90 minutes to keep the bias from linear drift of the gravimeter to a minimum. Gravimeter readings were corrected for the earth's tide (Wenzel, 1998). Gravity was measured using LaCoste-Romberg gravimeters having least count of 1 μGal with real measurement precision of around 10 μGal. Station 1 (Fig. 8.16(a)) was taken as reference point for data reduction to determine relative gravity changes between the two surveys. Along with the gravity survey, high precision geodetic survey was also conducted to obtain the changes in elevation after this earthquake in 2001 with respect to those measured in 1997-99 during the previous gravity survey. During the 2001 field compaign, a geodetic survey was carried out using leveling method as was done during 1997-99 so as to be compatible with each other. In addition, during 2001 campaign elevation at these stations was also measured using dual frequency differential GPS for a check on values obtained from the leveling method. Only base stations established during the previous survey (1997-99) were used to reoccupy during the 2001 campaign so as to maintain the required accuracy.

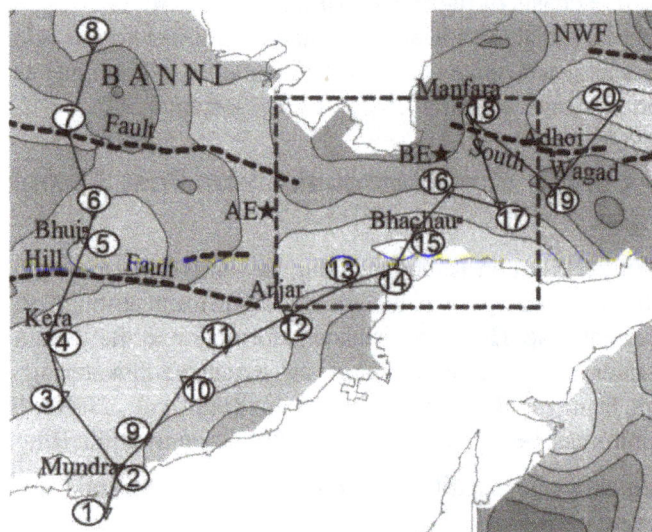

Fig. 8.16(a) Repeat gravity and elevation recording stations marked as 1-20 along Profile I, Mundra-Wagad uplift passing through the epicentral zone of the Bhuj earthquake. Profile II runs from Mundra-Banni Plains with stations marked as 1-8. Station 1 is regarded as the reference station.

To compare GPS heights measured relative to the reference ellipsoid with orthometric heights from leveling, elevations of the geoid above the ellipsoid must be estimated and applied. Thus, the GPS-derived height measurements with respect to the WGS-84 reference ellipsoid were corrected using the global geoid model EGM96 and the local, higher-resolution geoid (Fig. 8.16(b), Chandrasekhar, 2005) computed from the closely spaced (1-2 km) free-air gravity data of NGRI (1975) and Chandrasekhar and Mishra (2002). One standard-deviation uncertainties in the GPS-measured, geoid corrected relative heights are about 0.06 to 0.15 m. The effect of mass anomalies on geoid, and thereby on ellipsoid, is demonstrated in Fig. 8.16(c) that affects the measured height with reference to ellipsoid due to deflection of the vertical which requires to be suitably corrected. Fig. 8.16(b) shows E-W structural trends with a set of geoid highs over the Kachchh Mainland Uplift (Nakhtrana-Bhuj) extending to Maliya and geoid lows over the Wagad uplift (Chitrod) extending to Mundra and Bherendiala similar to the Bouguer anomaly. Epicenters of the Anjar earthquake (AE) and the Bhuj earthquake (BE) and their aftershocks lie on their intersection. The island belt where the epicenter of the Kachchh earthquake (KE) of 1819 was located is characterized by geoid highs delineating mass excess due to high density mafic intrusives as inferred from the Bouguer anomaly. One to one correlation between Bouguer anomaly and local geoid provides an additional data set for confirmation of the gravity anomalies.

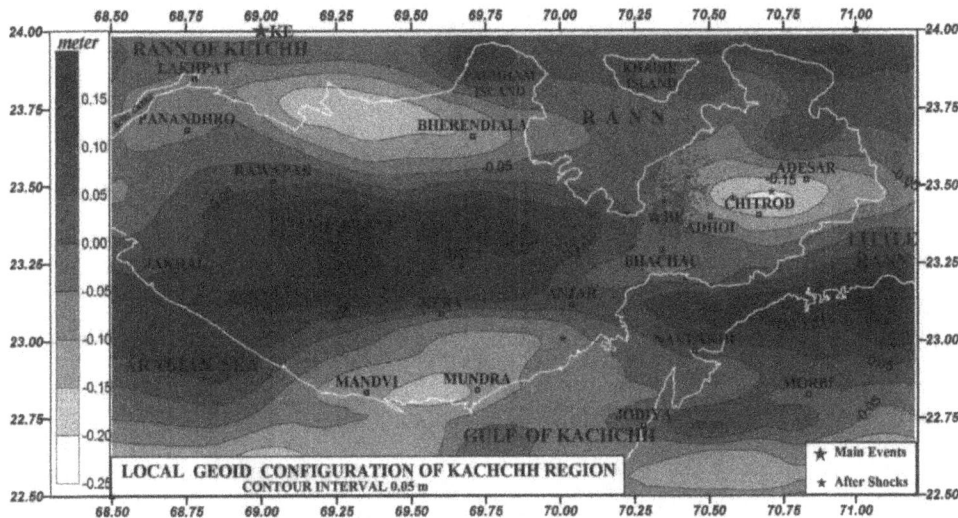

Fig. 8.16(b) High resolution local gravimetric geoid map of Kachchh (Chandarasekhar, 2005) showing geoid high over the Kachchh Mainland Uplift (Nakhtarna-Bhuj) and geoidal low over the Wagad uplift (Chitrod-Adesar) as in case of Bouguer anomaly map (Fig. 8.9) (Chandrasekhar, 2005).

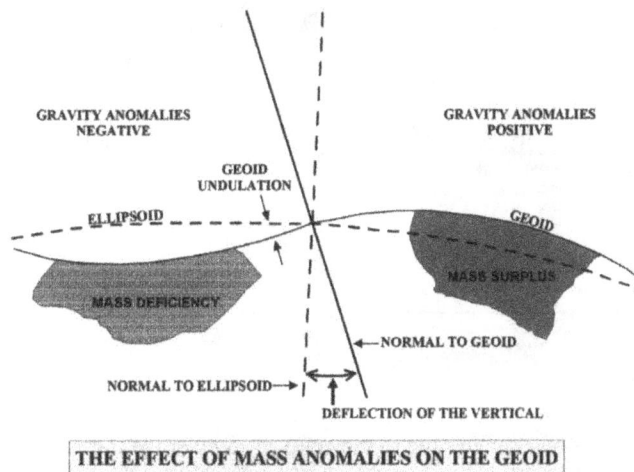

Fig. 8.16(c) The effect of mass excess and deficiency (mass anomalies) on geoid and gravity (Chandrasekhar, 2005).

Fig. 8.17(a) shows the change in elevation (leveling-leveling) from 1997 to 2001 along Profile I (Chandersekhar et al., 2004, Mishra et al., 2008). The data indicates a gradual rise from Mundra to Bachau. Close to the epicenter at Manfara, the uplift increases steeply (1.57 ± 0.46 m) and then decreases to the east on the Wagad highland. Fig. 8.17(a) also shows the co-seismic uplift pattern measured using differential GPS, which shows a maximum change at Manfara of 1.67 ± 0.30 m, and follows the same trend. The observed gravity changes indicate a corresponding peak minimum located at Manfara (-393 ± 18 μGal). As expected, the elevation changes are strongly anti-correlated with the gravity changes. If we assume that the gravity changes are solely due to elevation change of a wide region (Bouguer gravity) we would expect gravity to change − 0.19 μGal for 1 mm of uplift. However, we calculate the full solution of gravity changes due to the dislocation slip (Okubo, 1992) when comparing our source model to the gravity changes. The measurements along Profile II show small changes in elevation, whereas the gravity field shows an increase of up to 311 ± 8 μGal for three sites located on the Banni plains (Fig. 8.17(b)).

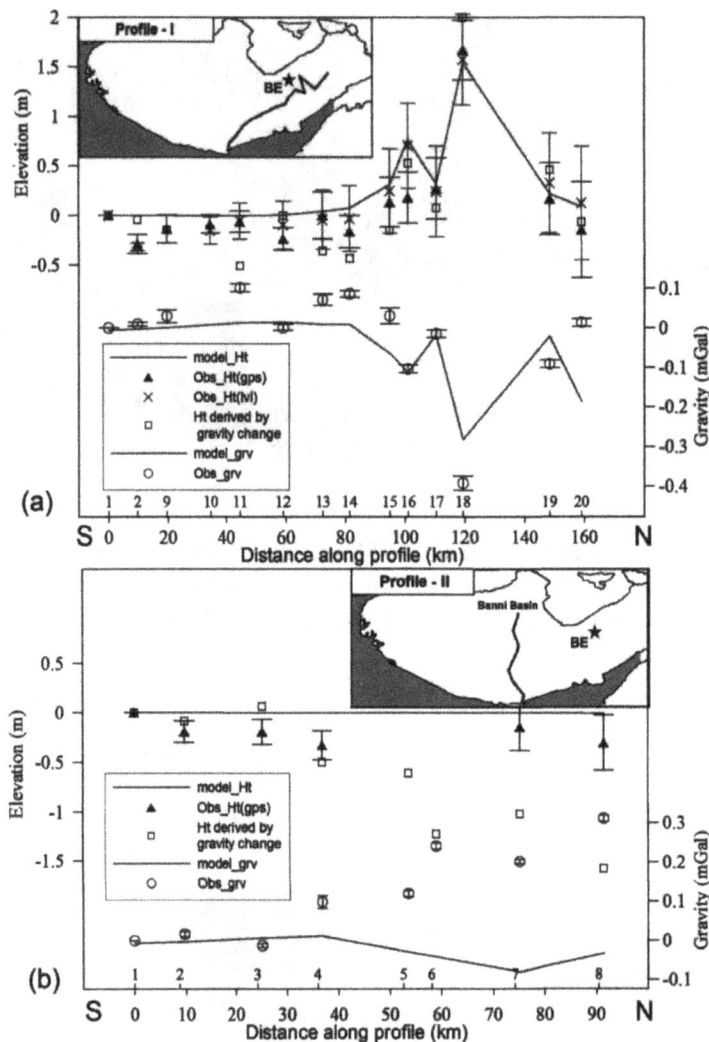

Fig. 8.17 Measured gravity and elevation changes due to the Bhuj earthquake. (a) Gravity and height changes measured along Profile I between 1997-1999 and November of 2001 showing sudden increase and decrease in elevation and gravity changes, respectively in the epicentral area. Error bars represent 1σ error relative to station 1. (b) Gravity and height changes along Profile II (Chandrasekhar et al., 2004).

8.3.2   Co-seismic Dislocation Model

Dislocation models are widely used to deduce the geometry and slip of sub-surface earthquake ruptures. Okada, (1985) and Okubo, (1992) derived analytical expressions for the displacements and gravity changes, respectively, due to faulting on a finite plane in an elastic, isotropic and homogeneous half-space. Inverse methods are used to find a model that minimizes the weighted residual sum of squares $WRSS = (d_{obs} - d_{mod})^T \times cov^{-1} \times (d_{obs} - d_{mod})$, where d_{obs} and d_{mod} are the observed and modeled height changes and cov is the data covariance matrix. Best-fit model is obtained using constrained, nonlinear optimization algorithm (Bürgmann et al., 1997), which allows us to estimate the geometry (parameterized by length, depth, width, dip, strike, and location) and the slip of a single model fault plane. As the data are not able to resolve each of these parameters, we apply additional constraints based on focal mechanism information (Antolik and Dreger, 2003) on the strike (82°) and dip (51°) of the rupture. Fig. 8.17(a) and (b) shows the predicted height and gravity changes for all sites derived from this model. Fig. 8.18 (Chandrasekhar et al., 2004) shows the surface projection of the dislocation model rupture that coincides with the zone of maximum changes in elevation and maximum concentration of aftershocks. It follows the trend of the Wagad uplift and coincides with its western part extending to the Banni plains.

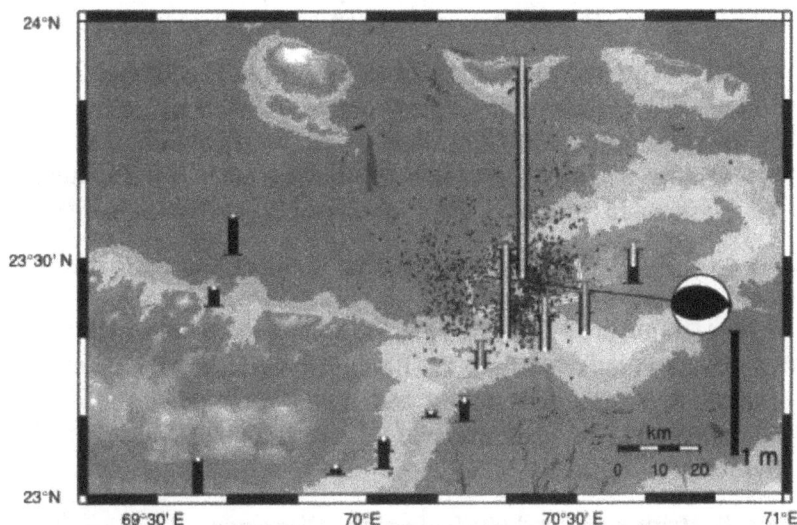

Fig. 8.18 Surface projection (rectangle) of the dislocation model rupture inverted from the leveled elevation changes from the Bhuj earthquake. Red dots indicate aftershock distribution from Negishi et al.,. (2003). The hypocenter location (depth = 22 km) and focal mechanism are from Antolik and Dreger (2003). Color shaded topography is from a 90-m DEM obtained by the shuttle radar topography mission (SRTM). Observed (black wide bars, showing only spirit leveling for stations 11-20 along Profile I) and modeled (yellow narrow bars) elevation changes. The inferred rupture marked by a rectangle on surface that is located near the center of the aftershock distribution (Chandrasekhar 2005).

The small number, low precision, and sparse spatial distribution of the height-change measurements limit the ability to uniquely determine the rupture parameters of the earthquake or to develop more complex (e.g., slip distributed) models of the rupture. Nonetheless, the model we obtain in the inversion is consistent with rupture parameters derived from seismic data (Chandrasekhar et al., 2004). The best-fit uniform-slip dislocation is 27.1 km long, 11.9 km wide, and dips S-ward at the constrained 51° dip angle from 12.5 to 23.2 km depth. Dip slip of 9.1 ± 0.4 m and small right-lateral strike slip of 0.2 ± 0.9 m on this dislocation provide a moment of 1.54×10^{20} Nm, which assuming a rigidity of 45 GPa (Antolik and Dreger, 2003), corresponds to a $M_w = 7.45$ event. This is less than the moment (3.6×10^{20} Nm) inferred from the moment tensor inversion and the finite slip inversion of Antolik and Dreger (2003). The center of the top edge of the model rupture is located at 23.47 °N, 70.37°E. The reduced χ^2 misfit, χ^2_{red} = WRSS/(N-P), where N is the number of data and P is the number of free model parameters of this model is 4.3, suggests that the data are adequately well-fit within their uncertainties. Models terminating below 10 km are preferred, although the inversion is able to find shallower ruptures that fit the data almost as well by adjusting some of the other free model parameters (viz. rupture dimension) in response to the depth constraint. However, models shallower than ~5 km systematically under-fit the uplift of all but stations 17 and 18.

8.3.3 Gravity Changes Due to Faulting and Subsurface Mass Redistribution

The optimized model inverted from the elevation-change data to forward model the associated gravity changes using the solution of Okubo (1991; 1992), it was found that for the sites along Profile I the co-seismic gravity changes are consistent with the model derived from the accompanying elevation changes. The difference between modeled and observed gravity up to 100 μGal for some sites along Profile I could be due to the influence of co-seismic water table rise up to 8 m as reported by Jain (2003) and times of india.com (2004). An increase in water table of 8 m in Kachchh around the epicentral area can contribute up to 40 μGal for an assumed porosity of ~10% for sediments in this area. Additional errors may come from the contribution of ocean-atmospheric loading.

Differences in the observed gravity and modeled field along Profile II of about 275 μGal can be attributed to widespread co-seismic shallow hydrological processes such as extensive de-watering (Bernard et al., 2003) and intense liquefaction (Rajendran et al., 2001), which are widely reported from the Banni plains, and might have caused significant sub-surface mass redistribution. Co-seismic mass change can be estimated from the observed gravity change by using Gauss law: $\Delta m = \left(\dfrac{1}{2}\pi G\right) \oint \Delta g ds$, where G is the gravitational constant and Δg is the gravity change (m/s^2) in a surface area of ds (Hammer, 1945). Applying this to the anomalous mean gravity change of 275 μGal, 2900 t/km^2 of mass was removed from the Banni sub-surface. An area of 1000 km^2 was affected by sand boils, craters, extensive de-watering and liquefaction processes due to the Bhuj earthquake which resulted in removal of total mass of as much as 2.9 Mt in the Banni plains (Chandrasekhar et al., 2004).

8.3.4 Co-seismic Mass Distribution and Hydrological Changes

The Bhuj earthquake took place in a region that was rather poorly instrumented. The lack of nearby mainshock seismic recordings and of a surface rupture precludes a clear picture of the source kinematics. Inversion results are consistent with the geometry and small rupture area suggested in prior studies based on the distribution of aftershocks (Negishi et al., 2002) and waveform inversions (Antolik and Dreger, 2002). This indicates that the Bhuj earthquake was a high-stress drop event on a steep reverse fault in the lower crust that did not rupture near the surface. For comparison, the similarly large magnitude Chi-Chi earthquake (M_w=7.6) of 1999 ruptured a ~ 100 km long and 30 km wide segment of the Chelungpu fault in central Taiwan with an offset of up to 10 m along its prominent surface trace (Johnson et al., 2001).

Although the projected intersection of the rupture plane with the earth's surface does not coincide with any mapped fault of this area, it lies along the westward extension of a fault zone (NWF, Fig. 8.7(a)) that was inferred based on the gravity anomaly of the Kachchh basin (Section 8.2). Even major earthquakes can and do occur on buried faults with no obvious geologic surface expression, which should be considered as a potential earthquake hazard in intraplate regions.

The change in gravity along Profile I follows a linear inverse relationship with the elevation changes; 10 mm of height variation produces ~2 µGal of gravity change. However, the change in gravity along the northern half of Profile II cannot be explained by the height changes alone. Instead, the gravity change appears to be due to mass adjustment by co-seismic fluid ejection related to the intense liquefaction and widespread extensive de-watering in the Banni plains. Using same elevation changes data along Profile I and further subsequent GPS measurements, Chandersekhar et al. (2009) have suggested a visco-elastic flow in the mantle under the epicentral region of this earthquake indicating a weak mantle that was attributed to the effect of Late Cretaceous Deccan plume, which gave rise to trap rocks in western India. The Question that arises is: why are large earthquakes limited to Kachchh, even though the Deccan trap covers a much larger area? In this regard some local tectonics as discussed in Section 8.2.10 must be playing a greater role. Application of changes in the gravity field and elevation for post seismic deformation have also been attempted in other regions such as Turkey (Ergintav et al., 2007).

8.3.5 Comparison of Bhuj Earthquake with New Madrid and Shillong Earthquakes

(i) New Madrid Earthquake of 1811-12 (USA)

In general, Kachchh has experienced more seismic activity compared to other stable continental regions as discussed above. The New Madrid region in Missouri, USA also experiences more seismic activity compared to other stable continental regions. Three major earthquakes of almost the same magnitude (M_W 7.5-8.0) as the Bhuj earthquake on December 16, 1811, January 23, 1812, and February 7, 1812 have been reported from this region (Penick, Jr., 1976). In this respect, the two may be analogous to each other (Ellis et al., 2001). Some important aspect of these two regions in general, and their seismic activity in particular, are as follows (Mishra and Rajasekhar, 2006).

(a) The epicenter of Bhuj earthquake is within 300-400 km of any plate boundary (Inset, Fig. 8.7(a) and (b)) and is connected to the plate boundary towards the west and NW by large lineaments. New Madrid in USA is at least 2000 km away from any known plate boundary. However, there have been large scale extensions, magmatism, and uplift in the nearby Basin and Range province (Section 4.4.2) and Colorado Plateau and Rocky Mountains, respectively from Oligocene to Holocene (Hamilton, 1987; Wernicke, 1987, Roy et al., 2009) that might have affected the New Madrid Seismic Zone as being a rift basin, it represents a weak zone. In fact, there are NE-SW to E-W oriented rivers that connect these sections that may represent lineaments/faults. The reported magmatism in Kachchh relates only to Deccan trap eruptions (~65 Ma) though some limited magmatism during recent times due to lithospheric flexure (Sections 6.1.9 and 7.11.2) caused by Himalayan orogeny cannot be ruled out.

(b) Tomography experiments have suggested the occurrence of a part of the Farallon plate under the Mississippi region where the New Madrid seismic zone is located that subducted about 19 My back along the west coast of the USA. This may explain its plate boundary connection and stress generated due to it.

(c) The New Madrid seismic zone is associated with the Reelfoot basin (Fig. 8.19) that represents a Paleozoic rift basin with several mafic/ultramafic intrusive similar to the Kachchh basin that represents a Mesozoic rift basin with mafic-ultramafic intrusive. However, in both these cases there is a Precambrian ancestry. The Reelfoot basin is associated with the Keweenawan rift of Meso-Protrozoic period (~1100 Ma), and the slip inferred from 1811-1812 earthquakes is consistent with the possible reactivation of ancient rift faults (Ellis et al., 2001). It has been followed by Appalachian orogeny. The Kachchh rift basin is also considered to be associated with Delhi rift of Meso-Proterozoic period followed by Delhi orogeny. Such zones the world over are weak zones characterized by seismic activity.

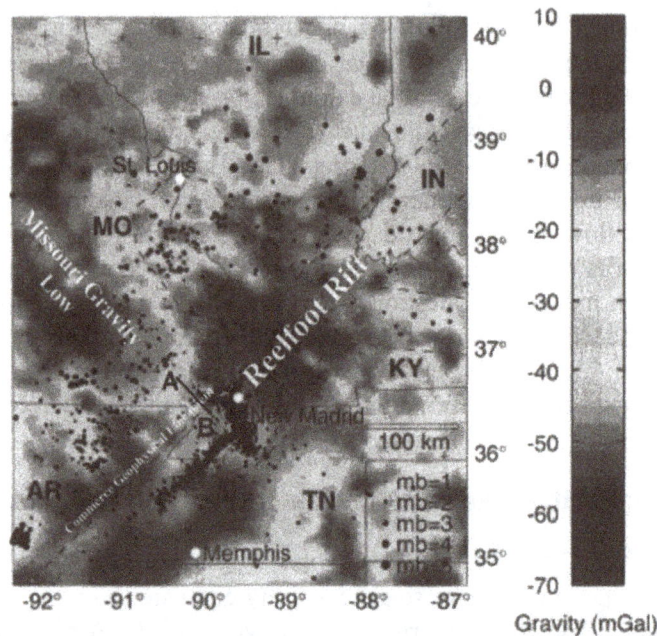

Fig. 8.19 Bouguer anomaly map of the New Madrid Seismic Zone, USA (Hildenbrand et al., 2001) showing a longitudinal gravity high along Reelfoot rift and some tranverse structures such as Missouri gravity low. Seismicity is concentrated at the junction of these structures. [Colour Fig. on Page 804]

(d) The Bouguer anomaly map of the New Madrid Section also shows gravity highs (Fig. 8.19, Hildenbrand, 1985; Hildenbrand et al., 2001) related to mafic/ultra mafic intrusive similar to Kachchh. It also shows a large Missouri gravity low which is extending in the New Madrid seismic zone along the Pascola arch. The intersection of the two appears to be zone of enhanced seismic activity. There is another zone of enhanced seismic activity north of this section and transverse to the rift, indicating the importance of transverse structures and their intersection with longitudinal rift structures for seismic activity in this region. In case of Kachchh, gravity anomalies due to longitudinal structures, H1 and L2 (Fig. 8.9) are intersected by transverse structures, H2 (Fig. 8.9) in the epicentral zone of the Bhuj earthquake that represents a zone of enhanced seismic activity.

(e) Hildenbrand (1985) modeled several gravity and magnetic profiles across this region and suggested an anomalous thick crust up to 40-45 km with a high density body of 3.1 g/cm^3 along Moho similar to an underplated crust. He has also modeled several high density mafic intrusives in the upper crust. As described above, thick crust and high density intrusives with inclined interfaces have also been modeled under the epicentral zone of the Bhuj earthquake (Fig. 8.13 and 8.14).

(f) The main seismic activity was followed by large numbers of aftershocks in the case of both Bhuj and New Madrid earthquakes. This indicates buildup of large stress which cannot be explained as purely intraplate earthquakes. Large scale co-seismic uplift and subsidence have been common to both cases. The absence of foreshocks prior to these large earthquakes in New Madrid and Bhuj has also been a common feature between the two.

(g) Though the New Madrid earthquakes were considered as intraplate earthquakes due to its large distance from plate boundary. But some stress being transmitted from North Atlantic spreading ridge center or San Andreas transform fault that has been highly seismogenic along the western margin of the North American plate cannot be ruled out. Strain rate for

Kachchh appears to be almost same (10^{-7}/year) (Jade et al., 2002) as that for New Madrid seismic zone (Argus and Gorden, 1991), which compares well for plate boundaries.

(h) There are several NW-SE and E-W lineaments in this region that mey be connecting the New Madrid Seismic Zone to the plate boundary along the west coast of USA and other presently active regions such as the Basin and Range province and Colorado plateau. In this regard, the long Missouri River that connects this seismic zone to the west coast of USA (San Andrea Fault System) is worth considering.

(ii) Shillong Earthquake of 1897

The Shillong Plateau in India (Section 7.6.4) which is also not far from the plate boundary (~200-300 km) has experienced more seismic activity compared to normal continental regions similar to Kachchh. The Shillong earthquake of 1897 was one of most devastating earthquakes of presumably high magnitude of 8.3. Some important comparative aspects of the two events are as follows (Mishra and Rajasekhar, 2006):

(a) Both these sections, the Kachchh and Shillong plateaua, are almost equidistant from the western and the eastern boundary of the Indian plate that represents transform faults known as the Chamman and the Sagaing faults, respectively.

(b) The shillong plateau and Kachchh are located over the prominent crustal bulge due to lithospheric flexure of the Indian plate under the Himalayas towards the north and the east and the west, respectively that show highest low pass filtered gravity highs (Fig. 6.15(a) and (b)).

(c) There are several lineaments and faults which extend from plate boundaries to the Shillong plateau that are seismogenic (Section 7.6.4) as in the case of Kachchh.

(d) Both these regions are characterized by E-W thrusts and these two earthquakes of Bhuj and Shillong are associated with similar oriented thrusts (E-W), almost perpendicular to the plate motion. Moreover, in both cases they are related to thrusted blocks of high density bounded by almost parallel front and back thrusts, known as the Wagad uplift and the Shillong plateau.

(f) Gravity data in both cases suggests an uncompensated crust, which implies an isostatically unstable condition. Therefore, it is likely that regional stress related to plate boundary activity and their interaction with local stress caused by buoyant crust and mafic intrusive as stress concentrators are primarily responsible for major seismic activity in both these regions.

(g) Therefore, the seismic activity of these two regions on the Indian plate do not represent truly either intraplate or interplate seismic activity. In fact, they may belong to a different class and can be termed as diffused plate boundary (Stein et al., 2002) earthquakes or near plate boundary earthquakes.

8.4   Sumatra Earthquake of December 26, 2004 and Related Tectonic Settings

8.4.1   Earthquake and Co-Seismic Deformation

A large earthquake on December 26, 2004 was reported from the Andaman-Sumatra section of the Indian-Eurasian plates subduction zone (Zone IV, Fig. 8.5(a)). The details of this earthquake are as follows: magnitude (M_W) > 9.0, focal depth ~ 30 km, epicenter = 3.32°N and 95.85°E offshore Sumatra with nearest town of Banda Aceh about 250 km NNW in the northern part of Sumatra (Fig. 8.20, Mishra and Rajasekhar, 2005). Fig. 8.21 gives a detailed map of the Andaman-Sumatra trench in this section that shows small islands of Simeulue which are close to the epicenter of this earthquake, and the Nias Island coinciding with the flexure indicating uplift along this line. This earthquake had a much larger slip area covering the entire stretch of the Burma micro plate which makes it the second largest recorded earthquake during the last 100 years (USGS website, 2004). It caused a co-seismic uplift of as large as 1.3 m along the NW coast of Sumatra island (Kayanne et al., 2007) and the horizontal displacement of about 14 cm towards the west and 6-11 mm horizontal displacement of the Indian shield eastwards (Catherine et al., 2005). This earthquake was generated by the rupture of the Sunda megathrust over a distance of >1500 km and width of ~150 km with a slip exceeding 20 m (Subarya et al., 2006). GPS measurements in Andaman-Nicobar islands towards the north also suggested similar fault plane characteristics with a co-seismic slip of 3.8-7.9 m (Gahalaut et al., 2006). Large horizontal displacement west to SW varying in magnitude from 10-40 cm with an uplift reaching 16 cm was also reported from the Andaman-Nicobar region (Gahalaut et al., 2008). This indicates the stress direction as the SW which has now relaxed to NE in the plate motion direction (Gahalaut, 2010). The main shock was followed by several aftershocks, which were primarily confined to the Burma microplate extending towards Andaman-Nicobar islands north of the epicentral area. Another earthquake on 23 December, 2004 north of Macquarie Island of magnitude 8.1 at the plate boundary between the Indo-Australian plate and Pacific plate south of New Zealand might be related to the present earthquake as paired earthquake (Mishra and Rajasekhar, 2005). This is important especially because epicenters of both the earthquakes lie on the same plate and both are caused by convergence in NE direction. Such paired earthquakes, one triggering the other, have been reported in some cases (Tibi et al., 2003).

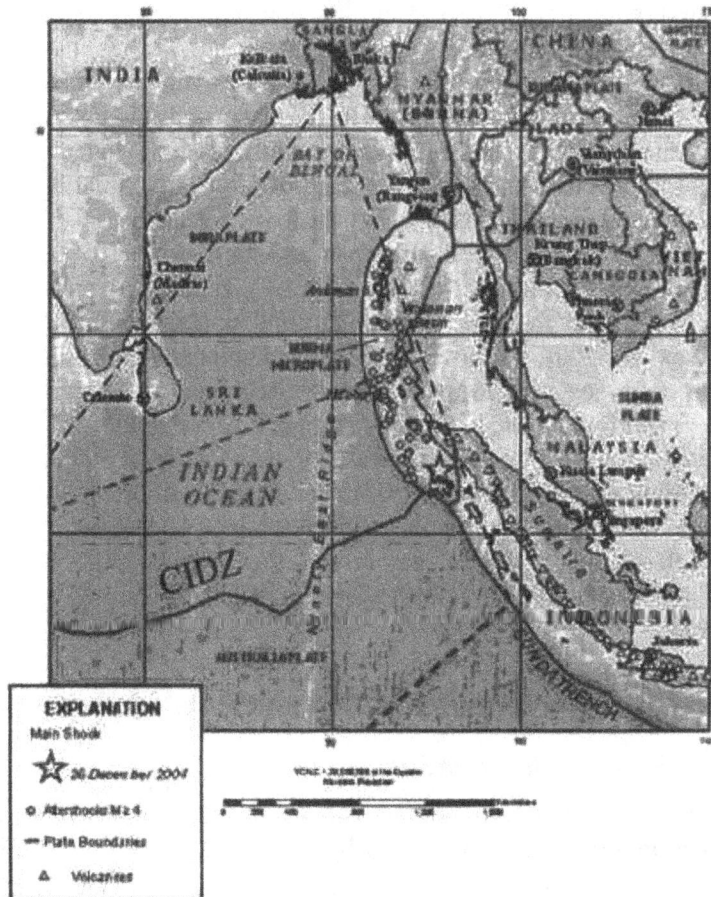

Fig. 8.20 Tectonic setting of North East Indian Ocean and epicentres of earthquake of 26 December 2004 and its aftershocks in the week following main shock (http://earthquake.usgs.gov/activity/past.html). The area between two red-dashed lines in the Indian Ocean is part of the Central Indian Ocean Deformation Zone (CIDZ). Superposed on it is the sedimentary zone in the Bay of Bengal and south of it due to rivers draining into the Bay of Bengal shown by black dashed lines triangle starting from south of Dhaka. The southernmost boundary of sediment cover almost coincides with southern boundary of CIDZ marked by black and red dashed lines respectively. The deformation zone is characterized by large earthquakes, east-west trending folds, thrusts, faults and fractures in sediments. It also shows high heat flow (Verzhbitsky et al., 1998) indicating recent subsurface tectonic activities in this section (Mishra and Rajasekhar, 2005). [Colour Fig. on Page 804]

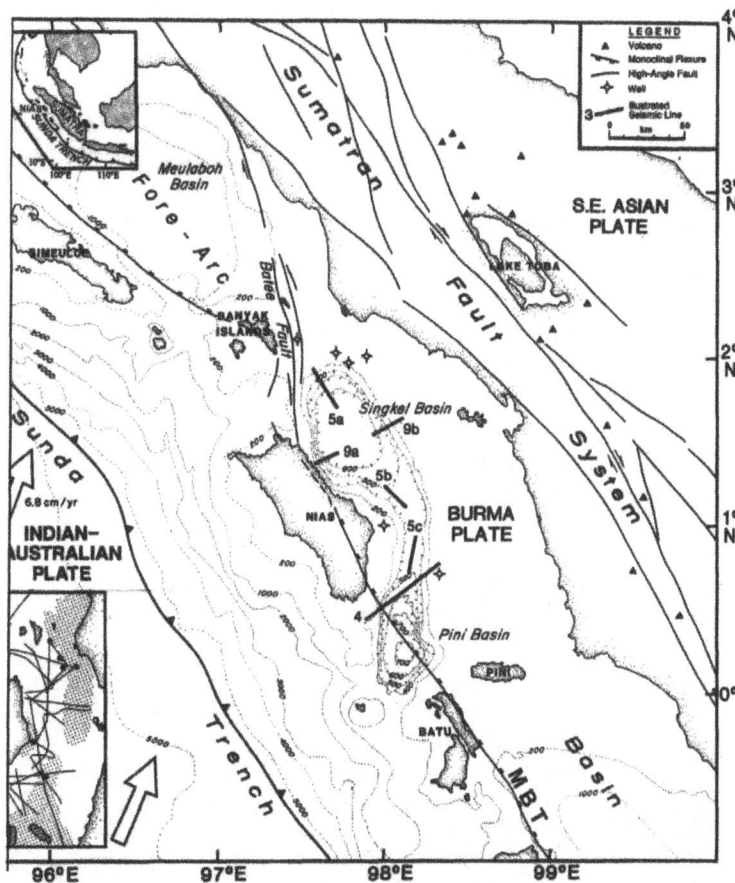

Fig. 8.21 Detailed tectonics of Central Sumatra and offshore showing monoclinal flexure (Matson and Moore, 1992) that coincides with the uplifted islands and epicenter of Sumatra earthquake and its aftershocks (Simeulue island; Fig. 8.20 (a) and (b)). The southward extension of this flexure and uplift almost coincides with the Main Back Thrust (MBT) as proposed by Singh et al., (2010) for tectonics and seismicity of this region. The epicenter of another major earthquake in this region, the Nias earthquake of 2005, also coincides with this flexure.

8.4.2 Associated Tsunami

This earthquake triggered a large tsunami offshore Sumatra at 7:58:53 AM local time on 26 December, 2004 and created havoc in several countries of the Indian Ocean, primarily Indonesia, Thailand, Malaysia, Andaman-Nicobar (India), East Coast of India, Sri Lanka, Somalia, Madagascar, and several small islands in this area (Fig. 8.22). This figure shows the modeled wave height in the Indian ocean with respect to time interval from the occurrence of this earthquake. As shown in this map, the east coast of Africa was affected. Though the wave heights are about 1-2 m in the open sea, the same become at least 10 times more along the coast as described below. This tsunami caused maximum loss in terms of the affected area, leaving millions of people homeless. More than 200,000 human lives were reported to have been lost and millions were injured. It has affected the citizens of more than 50 countries. Tsunamis are created due to displacement of large mass of water at the sea floor that can happen by several means such as volcanism at the sea bed, impact due to asteroids, and earthquakes etc. In the present case, it was caused by uplift of a large rock mass along the rupture zone at the sea bed that was considerably long, wide, and high displacing a large mass of water and triggering one of the largest tsunamis, in history. According to records in the last 60-65 years, at least three large tsunamis have hit the Indian coasts related to earthquakes in the Andaman Sea in 1941, offshore Karachi in 1945, and the present one. The tsunamis occurring in 1941 and 1945 suggest that they can strike even at close intervals specially because they are likely to originate from an earthquake along the plate boundaries in the Arabian Sea, the Indian Ocean, and the Bay of Bengal; or any other activity such as landslides or volcanic eruption at the bottom of these oceans. The second one offshore Karachi in 1945 resulted in a wavefront of almost 11-11.5 m along the coasts of Gujarat, India. Probably the biggest tsunami was also reported from the Indian Ocean related to the Karkatoa volcanic explosion in 1883. This one was so big that it caused about 40 m high waves along the coasts of Indonesia; some of the islands vanished under sea. It affected the entire Indian and Pacific Oceans and even affected the environment in these regions for days and weeks (Winchester, 1883). It was even reported to affect the whole earth, changing its axis of rotation In fact, earthquake related tsunami can cause maximum wave heights of 10-12 m while high speed land slides, high speed volcanic collapse, and volcano land slides are known to cause maximum wave heights of 100-500 m that are much more devastating compared to former. However, the later cases are rare and have been inferred from records of vegetation etc., along hill slopes and some of them might be responsible for mass extinctions in geological time records.

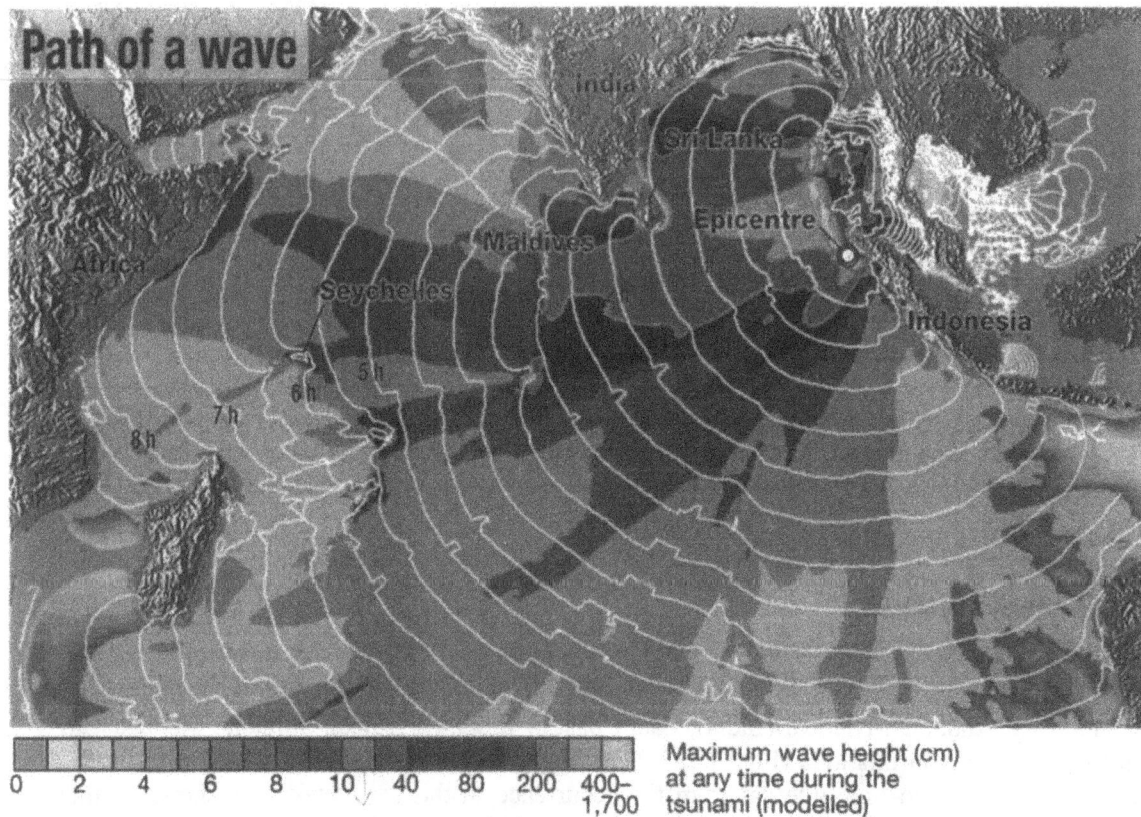

Fig. 8.22 Spread of tsunami waves in meters in the Indian Ocean related to the Sumatra earthquake of December 26, 2004. [Colour Fig. on Page 805]

Such large earthquakes associated with mega thrusts at the ocean bottom cause vertical displacement of large water mass, which subsequently tries to come to equilibrium and triggers tsunami. As tsunamis are generated at the bottom of the sea, they are large wavelength waves in the open sea, traveling with speeds of 700-800 km/h with minimum loss of energy. However, when they encounter shallow sea along the coast, their kinetic energy is converted to potential energy and they grow in height up to tens of meters causing large scale destruction. The most important feature of the present tsunami was its widespread effect towards the west which struck as the far as coasts of Somalia and east Africa. Its effect towards the east was limited and did not affect significantly the west coast of Australia. This can be attributed to the up dip direction of the rupture zone that propelled the water mass towards the west. In fact, the present earthquake consisted of three events occurring within seconds of each other. The primary slip offshore Sumatra is followed by two other slips towards the north of it (Hopkin, 2005). The modelled tsunami suggests maximum amplitude of waves north of the actual epicentre (Schiermeier, 2005) of the main shock which indicates that the amplitude of the primary tsunami might have been enforced due to some secondary tsunami due to northern slips or to landslides triggered by it, especially in the Central Indian Ocean Deformation Zone (Fig. 8.5(a)). The height of waves at the coast will also depend on the bathymetry near the coast along the continental shelf. The east coast of Sri Lanka is almost perpendicular to the the wave motion in this case and therefore, the tsunami has caused maximum damage in this section. The 200m bathymetry of the continental shelf offshore east coast of Sri Lanka extends to Cuddalore Nagapattinam sector along the east coast of India, which might have channelized these waves to cause maximum damage in this part that experienced maximum damage along the east coast of India.

8.4.3 Regional Tectonic Settings

The region offshore Andaman-Nicobar-Sumatra-Java is an active subduction zone and is well known for its high seismicity (IV, Fig. 8.5a). Here, the Indian plate subducts under the Burma micro plate and the Sunda plate with clockwise rotation in NE direction with a speed of 60 mm/year, causing an oblique convergence. It results in high stress generation, which from time to time is released as earthquakes. The tectonics in the epicentral area of the 26 December, 2004 earthquake is further complicated as it is located at the junction of four plates, viz. Indian, Australian, Burma, and Sunda (Fig. 8.20). In the epicentral zone of the present earthquake, stress buildup is further increased due to the presence of a large number of ridges and fractures in the deformation zone, which are subducting under the Sunda plate along with the Indian plate. The boundary between the Indian plate and the Australian plate is a diffused zone (between dashed red lines in Fig. 8.20) and is part of the Central Indian Ocean Deformation Zone. The central continuous red line between dashed red lines (Fig. 8.20) show the boundary between the Indian plate and the Australian plate. In fact, this deformation zone characterized by several large scale folds, faults, and seismic activity defines the Indian and the Australian plates separately. These were otherwise regarded as a single Indo-Australian plate. Another important aspect of tectonics in this region is the presence of thick sediments from the Himalayan rivers draining into the Bay of Bengal along coasts of Bangladesh, which almost covers a triangular zone starting from south of Dhaka extending up to southern diffused boundary of the Indian plate and the Australian plate (dashed black lines in Fig. 8.20). Because of large thickness of sedimentary rocks (8–10 km), crustal thickness in the Bay of Bengal is unusually large (24–25 km) compared to typical oceanic crust as discussed in Chapter 5. The coincidence of southern limit of sediment spread (black dashed line) in Fig. 8.20 and southern limit of CIDZ (dashed red line) suggests the importance of sediments in the tectonics of this region (Mishra and Rajasekhar, 2005). Fig. 8.21 is a detailed tectonics of a part of Sunda trench that shows the monoclinal flexure between the trench and the Sumatra island (Matson and Moore, 1992) that coincides with the small islands like Nias, Simeulue etc., in this section indicating uplift. It is interesting to observe that the epicenters of the main Sumatra earthquake and its aftershocks (Fig. 8.20) and the Nias earthquake of 2005 approximately coincide with this flexure suggesting the importance of flexures in the seismicity of subduction zones. This flexure line coincides with the Main Back Thrust proposed by Singh et al., (2010) as discussed below. The importance of flexure for Himalayan seismicity has already been emphasized in Section 6.1.10 and in Sections 7.6 and 8.2 for some of the intraplate earthquakes in India.

8.4.4   Seismic Imaging, Gravity Anomalies, and Seismic Activity

Based on seismic imaging, Singh et al., (2008) have suggested a broken subducting oceanic crust that is displaced upwards in the overriding plate indicating that the megathrust now lies in the oceanic mantle (Fig. 8.23). This figure also shows several thrusts starting from the subducted oceanic crust up to accretionary wedge. Some other seismic activity of this region also lies in the subducting oceanic plate. It is interesting to observe that the focus of the present earthquake lies almost at the junction of the subducted oceanic plate, continental crust and mantle wedge of different velocities forming a velocity knot similar to lineament knots where stress and strain are likely to accumulate. This earthquake has caused the breakup of the subducting oceanic crust (RZ2) that now lies in the continental lower crust. Singh et al., (2010) have also suggested the importance of backthrusts in seismicity of this region that coincides with the flexure line given in Fig. 8.21.

The Islands of Andaman-Nicobar-Sumatra-Java show relatively curvilinear gravity highs relative to adjoining gravity lows in the ocean following the trends of islands in free air gravity anomaly (Fig. 5.40) that is typical of trenches. It is interesting to note that other ridges of the Indian Ocean such as the Ninety East Ridge, and the Laccadive Ridge etc., which are relatively older and do not represent plate boundaries (Fig. 5.40, Mishra et al., 2004), show highs in free air gravity anomaly. The epicentre of the present earthquake lies within the linear gravity low in free air anomaly representing the Sunda trench, which is the southward extension of the Andaman trench where the main and aftershocks are located (Fig. 8.24). Trench parallel gravity anomalies are significant for seismicity of the trenches. Negative anomalies are correlated with intense seismic activity while positive anomalies correlate with weak seismic activity (Song et al., 2003; Wells et al., 2003) as demonstrated below based on gravity anomaly of Andaman-Sumatra-Java section. A Bouguer anomaly of Andaman-Sumatra-Java sector of Sunda trench is given in Fig. 8.24 along with the tectonics (bottom) showing various fracture zones (Grevemeyer and Tiwari, 2006) that may be subducting under the Sunda trench in this section. The section between the 90 East Ridge and the Investigator fracture zone represents younger (40-60 Ma) oceanic lithosphere compared to those on its west (>80 Ma). It also shows the focal mechanism of major earthquakes of this section that are predominantly thrust related as is widely found along subduction zones. It shows predominantly negative anomaly offshore of Sumatra where large earthquakes are located and positive anomaly offshore of Java where seismic activity is relatively less pronounced. Grevemeyer and Tiwari (2006) modeled the crustal structure along a profile across the epicentral zone of the Sumatra earthquake and Java trench further south (Fig. 8.25(a) and (b)) to compare the two. The two figures show that in case of offshore Sumatra, the coupling zone between the subducting and the overriding plates that ruptures in an earthquake is much larger compared to offshore Java due to oblique subduction of young oceanic crust. As indicated above, the young oceanic crust subducting offshore Sumatra has a higher temperature compared to that subducting offshore Java which would cause the seismogenic coupling zone (>100 °c) at a shallower depth in the former. The bottom of the coupling zone is the junction of the subducting oceanic crust and the mantle of the overriding plate that remains same in both cases. This provides a large area to rupture offshore Sumatra that would cause large earthquakes. This model was validated based on heat flow and seismic studies in this region. The forearc mantle is of higher density (3.3 g/cm^3) than the overriding plate, and produces gravity high offshore Java due to a bulge in the forearc mantle. The crustal density model presented in Fig. 8.25 is almost similar to the seismic model given in Fig. 8.23 showing the focus of the present earthquake at the junction of the subducting plate, and crust and mantle of the overriding plates of different densities. They will form a density knot similar to a lineament knot where the stress and strain are likely to accumulate. These observations suggest that the junctions of different physical parameters tend to accumulate stress and are significant for seismic activity. Co-seismic deformation of this earthquake produced an anomaly of ±15 microGal in GRACE satellite data that may be related to vertical mass displacement in the earth (Han et al., 2006). Satellite gravity may provide a strong tool to study earthquakes in the future as it provides a continuous record of the earth's gravity field and can be used for precursor studies.

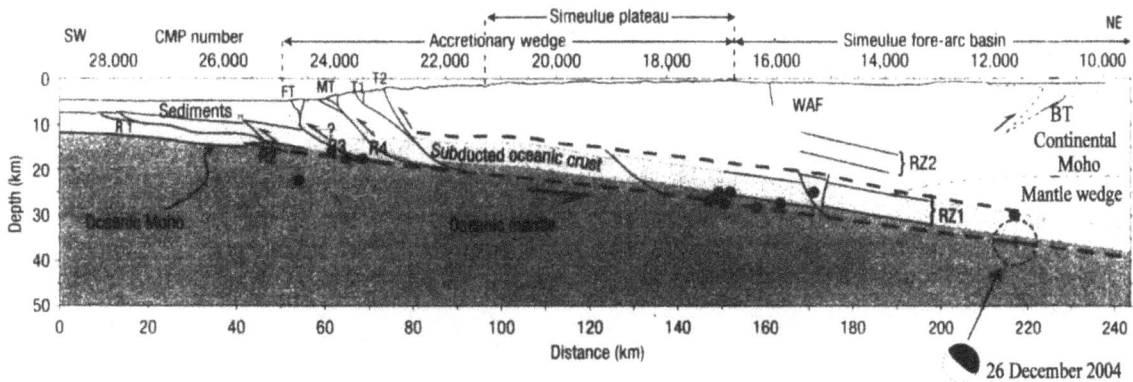

Fig. 8.23 Schematic depth cross section based on seismic reflection profile across epicentral zone of the Sumatra earthquake showing breakaway part of subducting slab displaced upwards (Singh et al., 2008). [Colour Fig. on Page 805]

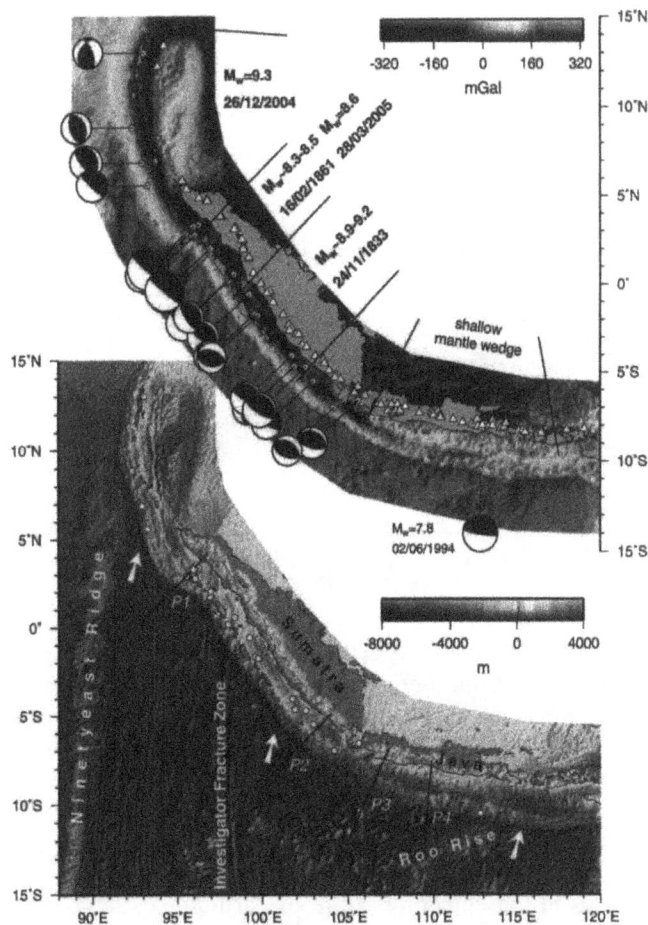

Fig. 8.24 Bouguer anomaly of Sunda trench from Satellite gravity (Grevemeyer and Tiwari, 2006) showing gravity low off Sumatra where most of the seismic activity is located, and gravity high off Java which does not show any major seismic activity. Large thrust earthquakes in this region are from the Harvard University CMT catalog (Mw > 6.5; 1976-2005; Dziewonski et al., 1981). Bottom map shows the tectonics depicting the Ninety East Ridge and Investigator Fracture Zone. The crust between them is younger (~ 40-60 My) compared to the eastern part (~ 60-80 My). [Colour Fig. on Page 806]

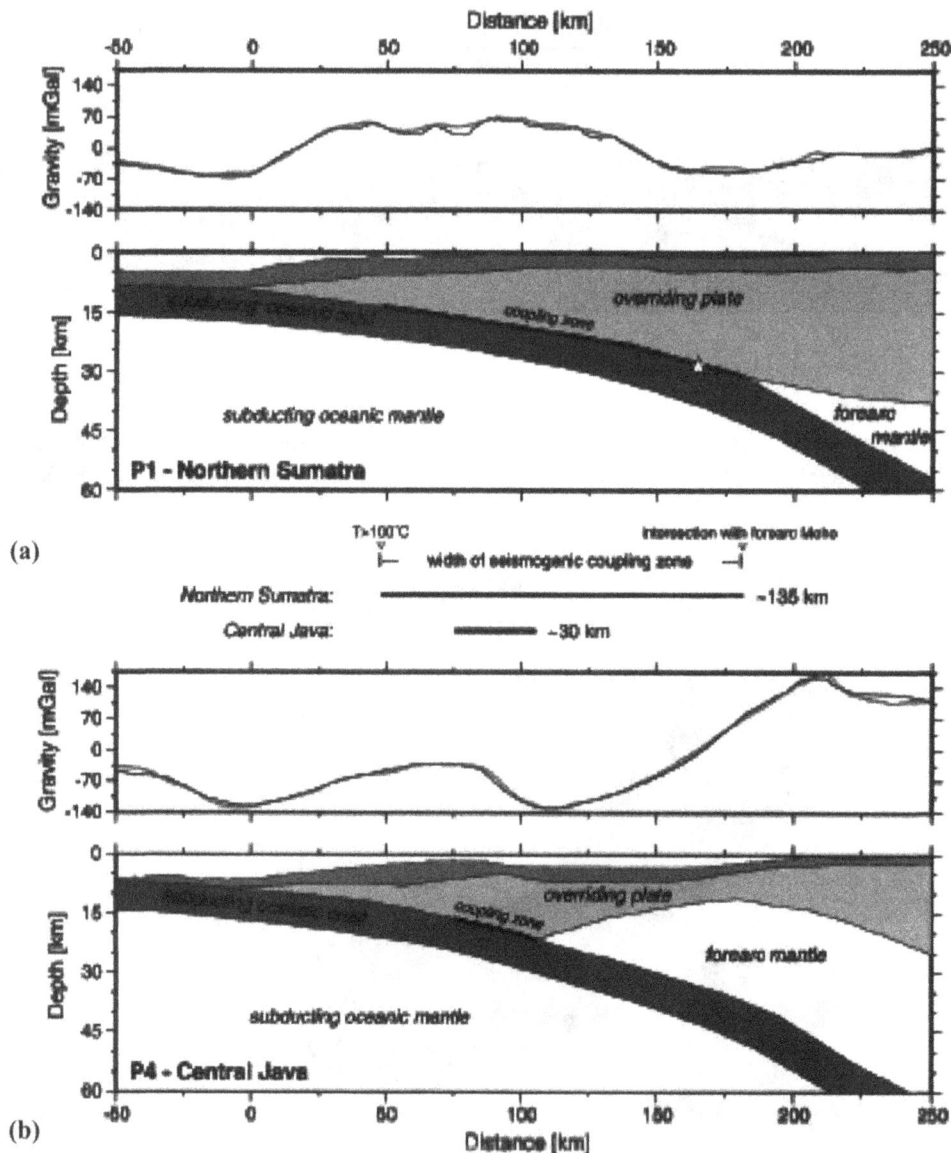

Fig. 8.25 Gravity model (Grevemeyer and Tiwari, 2006) across the Sumatra trench through the epicentral zone of Sumatra earthquake showing a large coupled zone defined by temperature >100°c between subducting oceanic crust and overriding plate that could rupture giving rise to large earthquakes (a). Gravity model across Java trench showing a smaller coupled zone that would produce small magnitude earthquakes (b).

8.5   Co-Seismic Changes, Precursors, and Natural Disaster Management

Some changes in physical properties in the epicentral zone of earthquakes have been reported. These might be useful to understand their mechanism and may eventually be used as precursor studies. Rao et al., (1994) have found an extraordinary high helium anomaly over the surface rupture related to the Latur earthquake of 1993 (Section 7. 8). They reported several tens of high helium anomalies over the rupture zone compared to a regional value of 0.2 ppm that may indicate the fault associated with the earthquake. Gupta et al., (1999) have also reported high helium anomalies in case of the Koyna earthquake of 1967 (Section 7.8) that helped in mapping the surface traces of the causative fault. As fluids in recent years have been found to be associated with the hypocenter of several earthquakes as described in case of Bhuj (section 8.2), Koyna, and Latur earthquakes (Section 7.8). A change in the eletromagnetic field is expected in the epicentral region of earthquakes. Azeez et al., (2009) have reported considerable variations in the telluric and magnetic signals in the frequency range of the MT spectrum during the Bhuj earthquake of 2001 at a site ~ 350 km away from the epicenter of this earthquake. Spectral analysis suggested localized, high amplitude low frequency signals both in time and frequency domains with onset of the earthquake, while the flat spectrum related to natural MT signals were observed before and after the earthquake. Almost similar observations have been reported by Harinarayana et al., (2010) based on magnetotelluric signals observed in Koyna, India where continuous measurements over a long period suggested changes in MT signals during the seismic activity of magnitude M > 4 and < 5 but practically no changes before and after the activity. (Gera et al., 2010) reported a unique atmospheric wave at Vapi on January 25, 2001 close (~ 400 km south) to the epicentral area of the Bhuj earthquake that occurred on January 26, 2001 using a sodar system, which can be trated as precursor to this earthquake. It might be a case of atmospheric gravity wave generated due to thermal anomaly caused prior to an earthquake. This wave has largest amplitude of 480 m and lowest frequency of 70.002 μHz that is quite anomalous and never recorded at any other Indian station or even at the same station (Vapi), which is a permanent sodar station. These physical parameters that have been described above to show co-seismic changes should be further studied to be used for precursor studies. In this class, the satellite gravity offers the best possibility as it is recorded continuously from a satellite and can be processed in time to provide information about temporal gravity changes in a region.

As described above in case of the Bhuj earthquake, temporal gravity and elevation changes being quite common during an earthquake, a detailed experiment related to changes in the gravity field in the seismically active region of Koyna, India is discussed below.

8.5.1 Temporal Variation in Gravity Field in a Seismically Active Region, Koyna (India)

Temporal variations in the local gravity field over a small time period can occur due to deformation of crustal rocks in the form of mass redistribution and change in elevation. The study of variation in the gravity field due to such deformations has been made possible with the advent of high precession gravimeters which measure the variations in gravity field up to microgal levels. Hence, repeat gravity observations using such instruments help to study both aseismic and co-seismic deformations and mass redistribution. Table 2.3 describes the accuracies and details of gravimeters that can be used to record temporal variations in gravity field at fixed locations.

For example, the measure of vertical aseismic deformation in response to loading due to La-Grande-2 reservoir, Quebec, has been studied using repeat gravity observations (Lambert et al., 1986). Variations of the gravity field were recorded and modeled to correlate the gravity change with mass movement due to change in the seasonal groundwater level (Tiwari et al., 2006; Lambert and Beaumont, 1977). Systematic gravity measurements in conjunction with other observations made before and after the Tangshan Earthquake of 1976, revealed dilatational process in the epicentral area (Ruihao and Zhaozhu, 1983). Many works revealed the importance of repeat gravity measurements as described above for the Bhuj earthquake to study the seismogenic deformations (Okubo, 1992, Zhao, 1995).

The Koyna earthquake (M = 6.2) on 10 December, 1967 had shattered the long-held view that peninsular India was seismically stable. This region has attracted attention due to continued seismic activity since the impoundment of Koyna reservoir in 1962, as described in section 7.8.3. An average of 5 to 10 earthquake of M > 4 and several of M > 3 are reported to occur every year in this region (Gupta, 1992). This area, therefore, provides a natural laboratory to study the earthquake processes and their effects on various physical parameters. A local digital seismic network established in this area continuously provides the details of earthquakes occurring in this region. In the present work, gravity observations were made at five permanent stations for more than two months. The objective of this study was to observe changes if any in the gravity field, and investigate whether the observed change is associated with seismic activity. Simultaneously, pore pressure studies (Gupta et al., 2000) were also undertaken to examine the effect of pore pressure changes on the seismicity and the gravity field.

(i) Tectonics

The Koyna region is covered with Deccan basalts considered to have erupted during late Cretaceous (Scetion 9.2). Since this area is totally covered with trap, the underlying geology and tectonics has been a matter of wide speculation. A significant geomorphological observation in this area is the similarity of escarpment trends as observed along the continental divide and the course of Koyna river (N-S) which turns eastward near Koyna (Fig. 8. 26(a)). It is suspected that the Koyna river itself flows along a fault. After the 1967 Koyna earthquake, various geophysical studies (Kailasam et al., 1972, Kaila et al., 1981, Mishra, 1989) were carried out to decipher the major tectonics of this region as described in Section 7.8.3. The Bouguer anomaly map of this area (Fig. 7.40(a)) grossly indicates a sharp gradient, which almost follows the Western Ghats (N-S) up to Koyna where it changes to a NNW-SSE trend, (trend marked in Fig. 8.26a) which is indicative of some structural feature. Subsequent Deep Seismic Sounding studies (Kaila et al., 1981) have delineated a NW-SE basement fault near Koyna coinciding with the Bouguer anomaly pattern. Based on seismicity trends in this region, Gupta et al., (1980) inferred a N-S oriented fault coinciding with 73°45' longitude. However, from the study of composite fault plane solutions of seismic events in this area, Rastogi and Talwani (1980) suggested NW-SE and NNE-SSW as the possible strike directions of the siesmogenic fault(s). It, therefore, appears that the area might contain one or more hidden fault(s) and fractures below the trap.

Based on the empirical relations between the magnitude of an earthquake and the effective distance over which gravity changes are noticeable and precursor time (Scholz et al., 1973) for M > 2 earthquake, five permanent stations (Bopoli, Helwak, Rasoti, Sangamnagar, and Riswad) (Fig. 8.26(a), Tiwari and Mishra, 1997), are established in the Koyna area. The locations of stations were planned such that two stations lie on either side of the inferred fault in this region. One station (station no. 5) was used as a reference station (Fig. 8.26(a)). One of these five stations (station no. 4) could not be occupied after twenty days because of inaccessibility due to heavy rain and was therefore shifted to Sangamnagar (station no. 4A, Fig. 8.26a). At a station at Bopoli, the western side was operated only during the last sixteen days due to logistic problems (Fig. 8.26(a)). Gravity measurements were made using LaCoste Romberg D-type gravimeter (D-116) with electronic read out, which gives a dial reading of approximately 1 microGal.

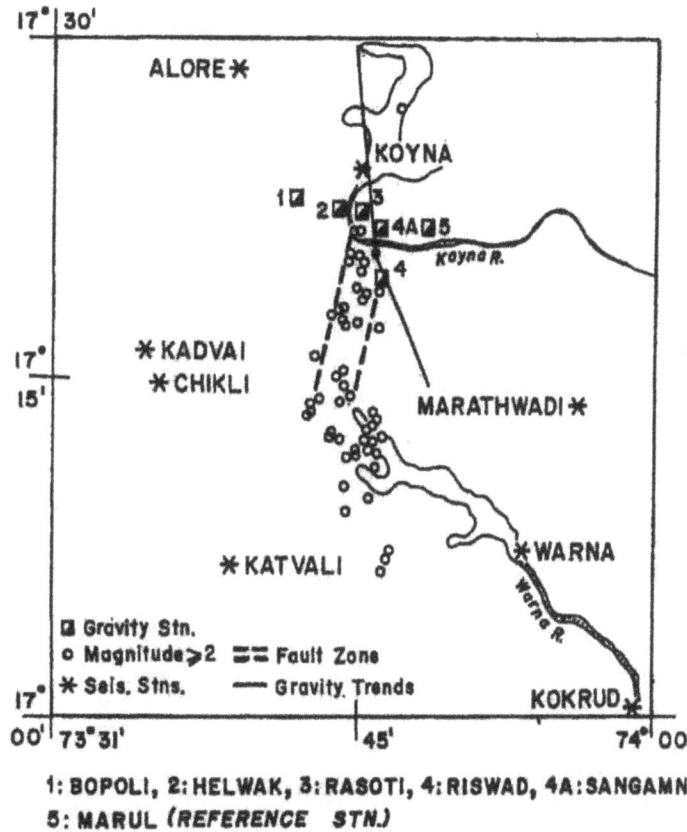

1: BOPOLI, 2: HELWAK, 3: RASOTI, 4: RISWAD, 4A: SANGAMN
5: MARUL *(REFERENCE STN.)*

Fig. 8.26(a) Locations of gravity stations and epicenters of seismic activity of M > 2 during June 2-August 16, 1995.

(ii) Gravity Measurements

The following measures were adopted in order to increase observational precision.

A-B-C-A-B-C survey procedure is followed so that every station is occupied twice in a loop in an interlocking loop system (Fig. 8.26(b)).

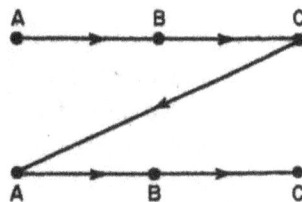

Fig. 8.26(b) Schematic diagram of recording gravity stations in a loop.

1. The road distance between stations is chosen to be almost equal to minimize systematic error due to variable driving time.
2. Each point was re-occupied within one hour so that even a small drift of the instrument can be corrected.
3. During observations the instrument was kept in the same direction at a station in order to eliminate any directional influence on the observations.
4. The entire profile was completed in 3-4 hours, recording at least two readings at every station.

5. The same setting of the gravimeter was used throughout the survey to avoid any error arising from resetting of the instruments.

6. Readings were taken until three consecutive readings were obtained within 1 dial reading.

Tidal correction was made using the standard program GRAVPACK (LRG, 1995) to depict non-tidal variation of gravity. Drift corrections were made assuming linear drift for one hour, which was found quite satisfactory based on static drift recorded both in the field and at NGRI, Hyderabad. Pressure correction was not applied considering that it will be of the order of one microGal as suggested by the supplier. Further, loop error was estimated by taking the average difference between stations in a loop and distributed over all the stations in a loop. The loop closure errors, i.e. A – B + B – C = A – C are usually found to be less than 3 microGal. This combined with other possible random errors may amount to a maximum of about 5 microGal, which is much less than the observed variation of 15 to 30 µGal. The expected observational error was obtained by conducting a similar experiment in a controlled environment at Hyderabad where a random error of 5-10 microGal was noticed between daily observations at a particular station. It is unlikely, therefore, that the recorded variations are related to any observational random error as discussed by Torge and Kanngieser (1979). Finally, the means and standard deviations of average difference between the reference station and other stations, after applying loop closure error, were computed (Table 8.1, Tiwari and Mishra, 1997).

Table 8.1 Means and standard deviations of average differences after applying loop closure errors

Station no.	Mean	Standard deviation
(Fig. 8.27(a))	(mgal)	(microGal)
I	0.577	8.0
2	1.2987	9.0
3	0.2807	6.6
4	0.7092	7.2
4A	8.2532	7.5

Several models have been proposed to explain the temporal variation of the gravity field. These are basically related to fault dislocations or dilatation of a focal body (Okubo, 1992, Zhao, 1995).

(iii) Co-Seismic Changes in Gravity Field

The differences in the gravity field at each of the stations with respect to the reference station (Marul) are computed after accounting for tidal, drift, and loop closure errors. These differences were adjusted to the mean value. The representative plot of variations of adjusted difference at various stations is shown in Fig. 8.27. Significant variations in the relative gravity field are observed (15-30 microGal) which is 2-3 times more than expected random (observational) error (5-10 microGal) and standard deviations (Table 1). The time of occurrences of tremors of $M = \sim 2$ in Koyna area (i.e., north of 17°15', Fig. 8.26(a)) and three tremors of $M = \sim 3$ in Warna area (south of 17°15', Fig. 8.26(a)) (MERI, 1995 a, b) are indicated by arrows in Fig. 8.27. It appears from Fig. 8.27 that relative gravity changes are correlatable from one station to the other and with the occurrence of seismic activities. These changes in gravity field are approximately 15-30 micro Gal. There are approximately twenty seismic events of magnitude 2-3 numbered as A to T in Fig. 8.27. Corresponding to most of these seismic events, variations in gravity field with some differences in amplitude are recorded at different stations as marked in Fig. 8.28. There are three plots of daily variation in the gravity field for almost the full period from 10 June to 17 August, 1995 corresponding to stations at Helwak, Rasoti, and

Riswad/Sangamnagar. At Bopoli, the gravity observations were made only during August, 1995 and therefore the gravity variation plot for this station is confined only to this period. A close examination of these plots (Figure 8.27) suggests change in gravity field correlatable to the seismic tremors marked at the bottom of this figure. Accordingly, the changes in the gravity field in these plots are also named A to T corresponding to the seismic events marked using the same notation.

The observations are plotted in Fig. 8.27

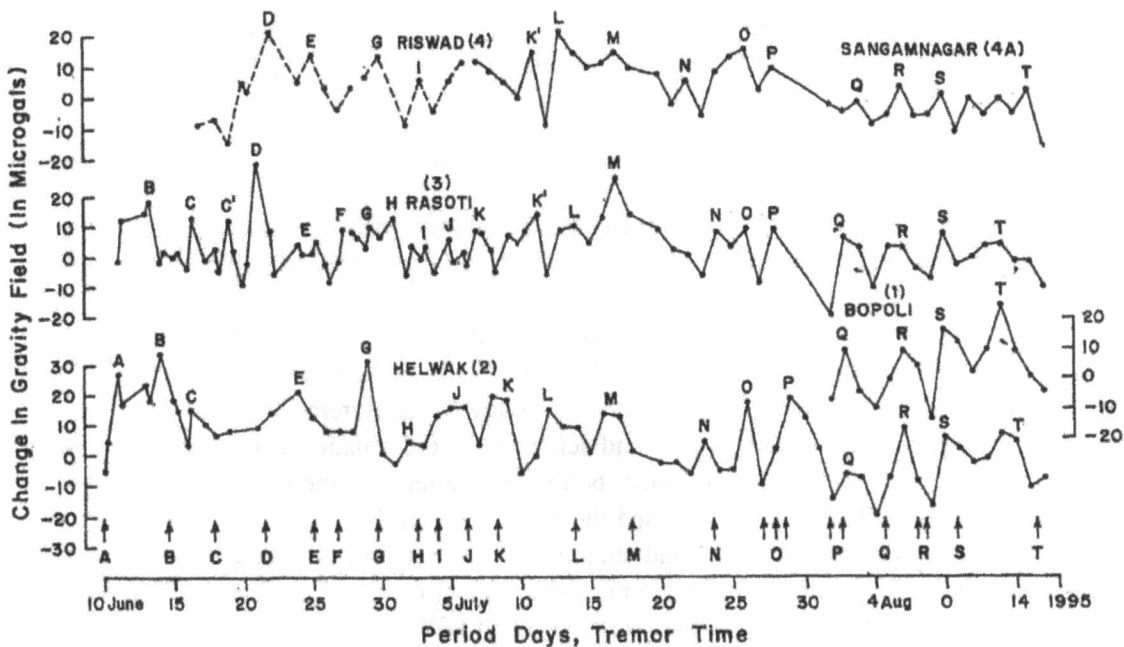

Fig. 8.27 Temporal variation of gravity field from June 10 to August 17, 1995 at fixed stations in Koyna (Fig. 8.26(a)) with reference to fixed station no. 5 (Marul) along with the occurrences of seismic events.

(Tiwari and Mishra, 1997) can be broadly characterized into three groups (Table 8.2) which suggests that:

1. All events (20) are characterized by the relative change in the gravity field at least at one station.

2. Change in gravity field is consistent at all the stations with occurrence of fourteen events.

3. In all the cases, the relative gravity field first increases and then decreases. The seismic events in 18 cases are associated with the decreasing trend in gravity; in only two cases, A and F, which are not very clear, it has taken place when the field is increasing. However, this pattern may change if there is any change at the reference station itself.

4. The changes in relative gravity are more significant at Helwak and BopoJi compared to Sangamnagar, Riswad, and Rasoti. This could be because of their locations on either side of the inferred fault. A similar observation has been noticed by Akin et al., (1991, 1997).

5. There are some cases of change in gravity field at some stations (C' and K') which might be due to some other unknown causes. But, they are not consistent from one station to the other.

6. The amplitude of the variation in the gravity field differs considerably from one station to the other, and is sometimes shifted in time. This might be due to the differences in the time

of observation and the occurrence of seismic tremors, or the location of diferent stations vis-a-vis epicenter of the earthquake.

Table 8.2 Categorization of events

Events occurring during	B, C, D, E, G, H, I, J, K
decrease in gravity field	L, M. N, O. P, Q. R, S, T
Events occurring during increase in gravity field	A, F
Unclear signatures/extra signatures Helwak: Rasoti: Riswad/Sangam Nagar	D,F,I G/ C', K' F,J/ K'

If we consider the models where change in gravity field is accompanied by a change in elevation (2-3 microGal/cm), an elevation change of 8 to 12 cm is required to explain the observed variations in gravity field (15-30 μgal). Frequent changes in elevation of this order over a small time period are very unlikely and therefore this model alone cannot satisfactorily explain the observed changes in the gravity field. The other source of variation is subsurface mass redistributions. However, it is complicated to resolve redistribution of mass with microgravity measurements alone in an area subjected to both change in groundwater regime and seismogenic deformations. In the present case, order of gravity variations are almost the same before and after the monsoon. The effect of commencement of monsoon (21 June) which raised the water level in the reservoir and groundwater level in a well located between station no. 2 and station no.3, has not been noticed in the nature of gravity variations (Fig. 8.27). Further, there is no pumping well in the vicinity of any station which can appreciably change the water table. The study of fluctuation of water level is being carried out in various bore wells (Gupta et al., 2000). Observations of water level change in two wells in this region, one located near gravity reference station and the other in between station no. 2 and station no. 3 show less than 10 cm of relative variation (Radhakrishna, personal communication). This magnitude of change in the water level can produce a variation of hardly 4 to 5 microGal in the gravity field which is much less than the observed variations. Moreover, changes in gravity field due to changes in water level cannot be same at all the stations which are far apart as in the present case. Hence, it implies that the observed gravity changes may not be related to the changes in the water level. Since observed gravity changes cannot be explained either by change in elevation or change in groundwater level, they appear to be caused by the mass redistribution in the epicentral area. The other possibilities to explain the change associated with seismic activity could be dilatancy and dilatancy recovery (Scholz and Kranz, 1974). If dilatancy in fault zone is manifested as opening of multiple pennyshaped vertical cracks, it produces a change in the gravity field which is poorly dependent on change in elevation. This phenomenon can occur in two environments: one, in a horizontal shear stress regime leading to vertical strike slip faulting, and the other in a tensional stress regime causing normal faulting (Sasai, 1986). As most of the minor and major events in this region have strike slip faulting with a small normal component, a decrease in gravity field can be attributed to dilatation which is manifested in the form of opening of cracks. The increase can be attributed to dilatancy recovery. For a better understanding of the gravity changes in seismically active regions and its application as precursor studies, much more gravity, water level, and geodetic observations from different seismically active regions are required.

8.5.2 Natural Disaster Management – Some Suggestions

Natural disaster management includes management of all kinds of disasters such as earthquakes, tsunami, volcanoes, cyclones, and floods. However, here we are concerned only with the first two kinds, viz. due to earthquakes and tsunami. In case of earthquakes, there are two aspects. The first one relates to prediction and forecasting, and the second one to mitigation and hazard preparedness after the earthquake. Forecasting involves the prediction of the region where an earthquake is likely to take place, estimation of its magnitude and time or period of its occurrence. However, the science related to the occurrences of earthquakes such as quantitative estimates of stress, strain caused by it in different rock matrix and their rupture is not yet well understood to predict earthquakes. Lagios et al., (1988) have described the application of microgravity measurements for prediction of earthquakes and volcanic eruption, but their application in actual practice has remained limited due to uncertainty of the occurrence of earthquakes in both location and time. In fact, in case of volcanic eruptions, bore hole gravimetry offers good opportunity as sites for old volcanoes are known where bore hole gravimeters can be installed and monitored for future eruptions. Though some co-seismic changes during the earthquake and prior to it have been recorded as described above, the problem lies in recording these parameters everywhere. This can be gauged from the fact that none of the earthquakes of recent times along the west coast of USA related to San Andreas fault system could be predicted inspite of it being the best studied and instrumented region. However, some claims have been made in this regard to predict seismic activity of moderate magnitude in Koyna (India) based on precursor clusters leading to nucleation (Gupta et al., 2007) where seismic activity is regularly taking place since 1962 after the construction of a dam as described in Section 7.8, which behaves more like a controlled environment. Therefore, the best that can be done in this regard is to differentiate regions based on the probability of the occurrence of the earthquakes that gives rise to seismic zonation maps. The seismic zonation map of India (Fig. 8.6) provides zones marked as II-V from low risk zones to highest risk zones for the occurrence of the earthquakes. It may be noted that the Himalayan fold belt, Kachchh, and the Shillong plateau belong to zone V of the highest risk zone as is also apparent from Fig. 8.5(a) and b that show highest number of seismic activity in these sections. However, it indicates only broad zones that serve only as first order guidelines; their further division in smaller zones known as microzonation maps indicating risk factors is required for better utilization. In this regard, the first step would be assimilation of different geological and geophysical data sets related to seismotectonics and past history of seismic activity in those regions. The seismotectonic Atlas of India (GSI, 2000) and various geophysical data described in chapter 6 and 7 with regard to various seismic activity serve this purpose to some extent. However, integration of such information for most of the seismogenic section is lacking. The next step would require the application of these data to find out the depth extent and dip of the seismogenic lineaments and faults. The most important data sets for this purpose would be the gravity maps of India (Fig. 7.8(a) and (b)) and air borne total intensity magnetic maps available with the AMSE, Geological Survey of India. Air borne magnetic maps are specially suitable to map lineaments due to close spacing of data along profiles. The information from these two maps should be integrated to map the seismogenic lineaments/faults in depth. Gravity maps can also be used to evaluate isostatic conditions in a region that is important for stress buildup as demonstrated above in case of Bhuj (Section 8.2) and Shillong (Section 7.6.4) earthquakes.

Mitigation implies efforts towards reducing the effect of disaster after the earthquake and involves construction of earthquake resistant buildings and preparedness towards rescue and healthcare that require social and governmental efforts. These are beyond the scope of the present book. Several books have been written on this topic such as Ozerde and Jacoby (2006), Coppola (2006), Schneid et al., (2000), Alexander (1993). Mitigation during a tsunami requires timely issue of tsunami warning based on models prepared from the focal mechanism of the related earthquake and rupture parameters (Okal et al., 2009). For example, the Sumatra earthquake of December 26, 2004 created a large tsunami as described above but some other earthquakes in the same region such as the Nias earthquake of magnitude 8.6 on March 28, 2005 created only a local tsunami of about 5m, while another on September 12, 2007 of magnitude 8.4 off southern Sumatra created a very small local tsunami (Srivastava et al., 2007, Okal and Synolakis, 2008). This indicates that the magnitude of earthquakes and the extent and direction of the rupture zone influences the associated tsunami to a large extent. The sumatra earthquake of December 26, 2004, being of higher magnitude by 0.6-0.8 as compared to the other two earthquakes, had an energy content 20-25 times more than the latter ones. Prior warning of tsunami is possible as seismic waves from earthquakes travel faster than to tsunami waves; therefore, the intermediate period between the arrivals of two waves can be used to model tsunami waves and issue advisory about it. Besides tsunami warning issued by tsunami centers world over, the age old method of growing mangroves along the coast or construction of canals slightly away from the coast, as in case of Buckingham canal along the east coast of India (Rao, 2005), remain the best methods to check catastrophic devastation from tsunami waves. These methods also prevent other catastrophic events like cyclones and serve multi-purpose programmes. The Buckingham canal is about 300 km long between Chennai to Nellore (East Coast of India) and was constructed 1-2 km away from the east coast of India in 1878. It gets sea water through pre-existing creeks during high tide on full moon days and has served several purposes like transportation alongwith protecting the neighboring regions from high waves of any kind, including the tsunami associated with the Sumatra earthquake of December 26, 2004.

Resource Exploration and Geodynamics: Hydrocarbons, Groundwater, and Minerals

9.1   Introduction

Resource exploration primarily includes the exploration for hydrocarbons, minerals, and groundwater. However, the strategy of exploration in these cases differs considerably. As the gravity field due to regional sources related to geodynamics display large wavelength (> 100 km) and amplitude (>10 mGal) and those for resource exploration are mostly smaller, it is important to separate the observed field into its regional and residual components. Residual field is used for resource exploration while regional field is used for geodynamics studies. Even in case of resource exploration, the wavelength and amplitude characteristics for hydrocarbon and mineral explorations are different. For example, the structures and gravity anomalies for hydrocarbon exploration may be a few hundred meters to a few kilometers (< 10 km) and amplitude of a few mGal (< 10 mGal). The same for mineral exploration are mostly < 1 km and < 1 mGal. These characteristics can be used to identify respective anomalies and separate them in the regional and the residual fields. There are standard procedures to separate the regional and the residual fields starting from simple graphical methods to mathematical methods of wave length filters and polynominal approximation as described in Section 4.1. A simple method of zero free air anomalies as discussed in Chapter 4 can also be used for this purpose that represent Airy's isostatic compensation. The corresponding Bouguer anomaly represents the effect of isostatic compensation as demonstrated in Section 8.2.6 and 9.2.3. It provides the part of the regional field related to isostatic compensation and its difference with the observed field provides the residual field mostly representing crustal anomaly, which are known as isostatic regional and isostatic residual fields, respectively. This approach, however, might not be useful for mineral exploration due to small order of anomalies involved in it, where even basement anomalies represent the regional field and therefore simple graphical method or appropriate polynominal approximation is used in such cases. However, the zero free air anomaly approach has been found to be very useful for geodynamic studies and oil exploration where Moho configuration is the main source of regional gravity anomaly. Some typical case histories of resource exploration including groundwater are described below where integration of various geophysical methods are attempted to show their efficacies compared to individual methods. Geodynamics for these regions are also inferred from same data sets to show their application for multipurpose exploration schemes.

9.1.1   Hydrocarbon Exploration, Suitable Prospects, Rifted Margins and Seaward Dipping Reflectors

Hydrocarbons are the most important source of energy and are therefore important for development of any country. Present consumption of crude oil in India is 3 mb/day and is likely to grow to 6 mb/day by 2030 while present production is only 0.7 mb/day. At the present day rate of crude oil of $ 100 per barrel, the import of 5.3 mb/day in 2030 would cost ~ $ 190 b/year provided our production remains the same (World Energy Outlook, WEO-2009 at http//www.iea.org/weo/2009.asp). This is a huge cost that requires special efforts towards exploration to mitigate it. Hydrocarbon exploration using gravity surveys are basically indirect to delineate basement structures for detail surveys using seismic method over promising zones. Hydrocarbons are associated with sedimentary basins that are normally characterized by gravity lows. But depending on basement structures, one gets relative gravity highs or lows due to basement uplifts or depression in the basement, respectively. Sedimentary section in a basin generally mimics the basement structures that can be effectively inferred from gravity and the magnetic fields. In fact, three dimensional basement configuration models can be inferred very effectively from airborne magnetic data using harmonic inversion as described in Section 4.3 and demonstrated in Section 4.7. Gravity data can also be used for this purpose but one has to be cautious in this case as gravity anomalies originate from different levels including deeper levels like Moho and in that case only the anomalies from basement should be separated using suitable filters and used for basement computations.

SEDIMENTARY BASIN MAP OF INDIA

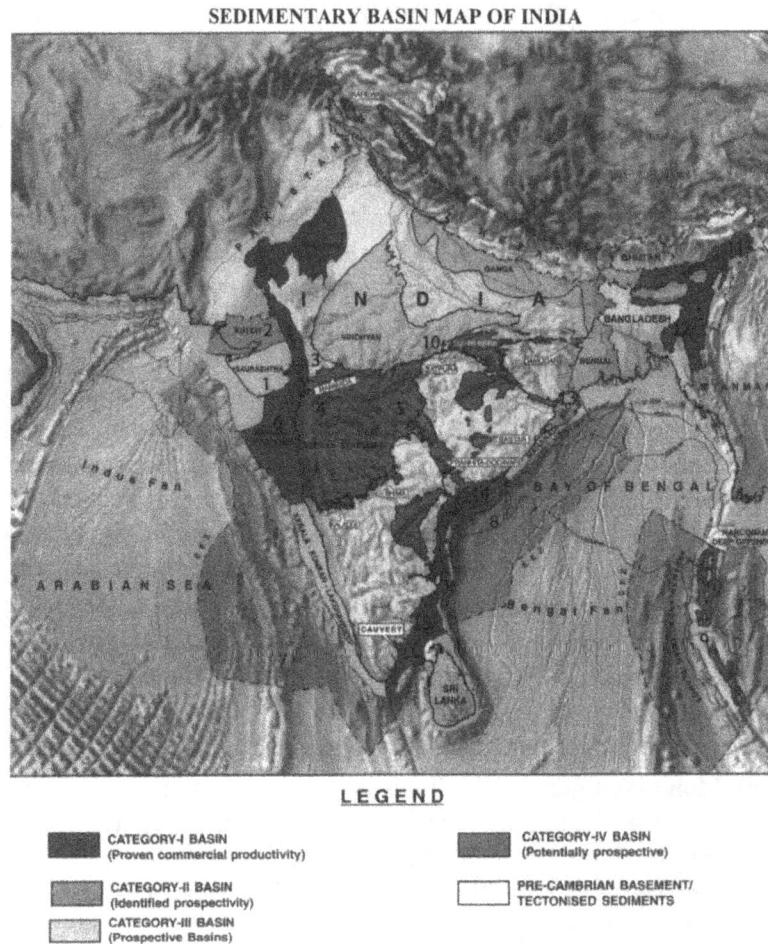

LEGEND

CATEGORY-I BASIN
(Proven commercial productivity)

CATEGORY-II BASIN
(Identified prospectivity)

CATEGORY-III BASIN
(Prospective Basins)

CATEGORY-IV BASIN
(Potentially prospective)

PRE-CAMBRIAN BASEMENT/
TECTONISED SEDIMENTS

Fig. 9.1(a) Major sedimentary basins of India (Srinivasan and Khar, 1995). Numbers 1-15 are regions that are discussed in regard to multi disciplinary geophysical investigations and specially gravity anomalies vis-à-vis hydrocarbon prospects. 1: Kutch, 2: Saurashtra, 3: Cambay basin, 4: Narmada- Tapti Section (NTS), 5: Wardha basin, 6: Bombay high, offshore west coast of India, 7: Laccadive (Lakshadweep) Ridge, 8: Krishna-Godavari basin, offshore east coast of India, 9: Offshore Andaman-Nicobar islands, 10: Vindhyan basin, 11: Upper Assam Shelf, 12: Lower Assam Shelf, 13: Mahanadi basin, 14: Krishna-Godavari basin and 15: Cauvery basin.
[Colour Fig. on Page 807]

Oil and Natural Gas Corporation (Bhandari et al., 1983) has carried out gravity surveys in almost all basins of this country (Fig. 9.1(a), DGH, 2005) primarily to delineate basement structures and thickness of sediments in them. Most of these basins have also been surveyed using seismic methods (Bhandari et al., 1983; Srinivasan and Khar, 1995, DGH, 2005).

There are primarily two kinds of structures suitable for oil accumulation that are known as structural and stratigraphic prospects (Fig. 9.1(b); DGH, 2005). As the name indicates, the former are associated with structures like faults, horst, and rift valleys etc., while latter are related to stratigraphy such as wedge out, pinch out, and build up. Amriti structure that represents a structural prospect in the Vindhyan basin (10, Fig. 9.1(a)) is discussed in detail in Section 9.7.2. Both structural and stratigraphic traps can also occur in conjunction as often happens in case of uplifts as shown in Fig. 5.36(b) along a seismic profile across the Eighty Five East Ridge. Fig. 9.1(c) (DGH, 2005) shows pinchout and turbidite sand in the deep waters of Kerala-Konkan coast that form good stratigraphic prospects in Tertiary sediments. The basement is fractured with almost parallel fracture zones and wedge type of structures and lies at an average depth of ~8-9 km in this section. They are formed due to extension caused by Deccan trap eruption and formation of rifted margin. Top of the basement and following reflectors underneath dip towards the sea that are called seaward dipping reflectors that are different in nature compared to those observed in sediments overlying the basement and are typical of volcanic rifted margins. Volcanic rifted margins show good prospect for oil exploration the world over. As described in Section 5.4, the west coast of India in fact represents a rifted margin formed due to Deccan trap eruption and the basement in this section is represented by Deccan trap that typically shows parallel fracture zones and seaward dipping reflectors. These reflectors are formed due to submarine volcanic eruption of basaltic rocks of high temperature that are chilled in cool water conditions.

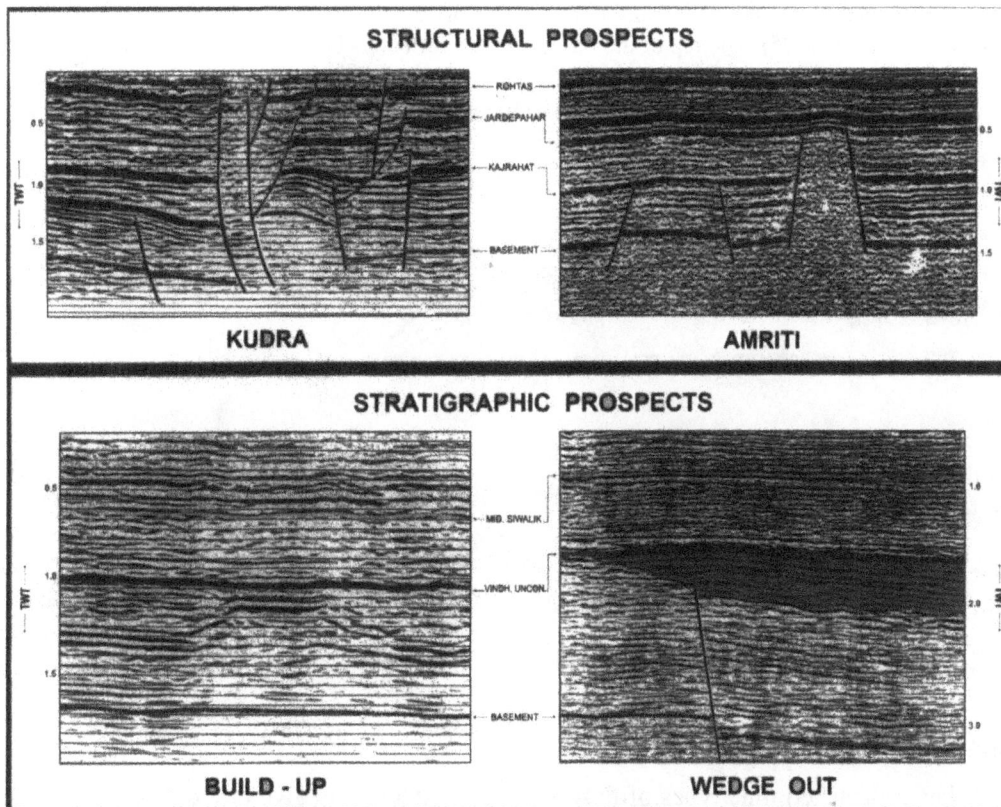

Fig. 9.1(b) Examples of structural and stratigraphic prospects favourable for oil accumulation (DGH, 2005).

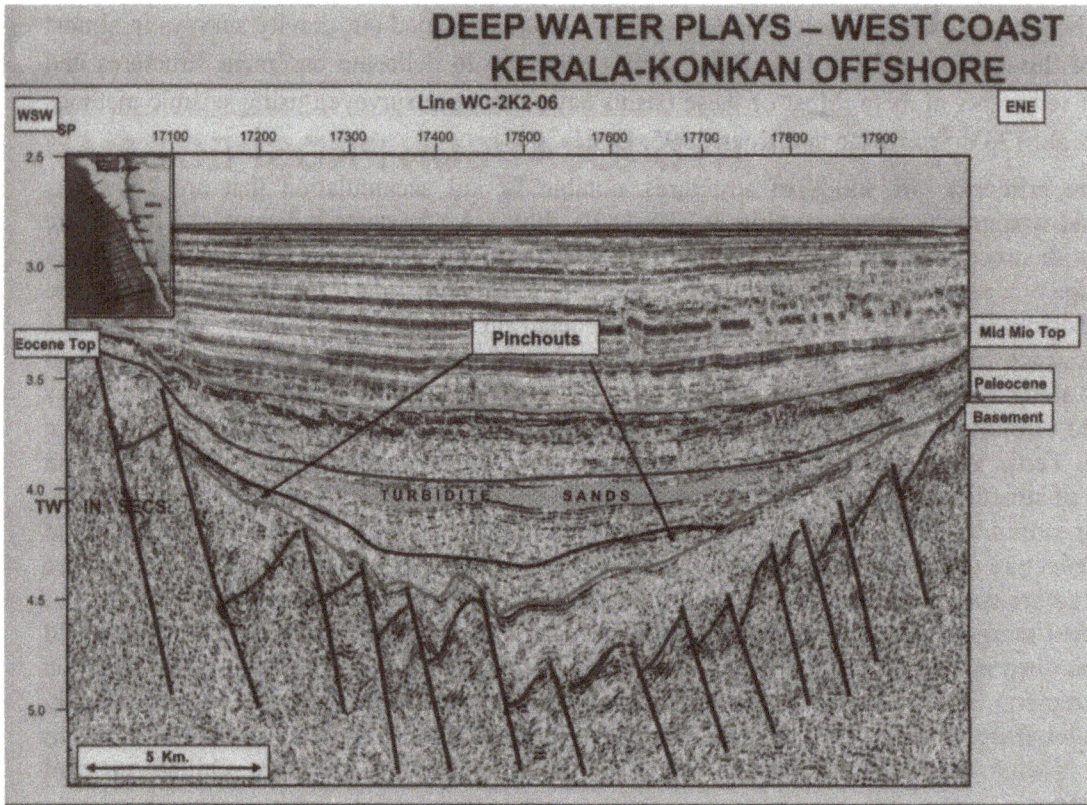

Fig. 9.1(c) Examples of pinchouts and turbidite sand in deep water off the Kerala-Konkan coast that form good stratigraphic prospects (DGH, 2005). The top of the basement and reflectors below it are seaward dipping reflectors that are different from those of sedimentary section above it and typically represent volcanic rifted margins world over. [Colour Fig. on Page 808]

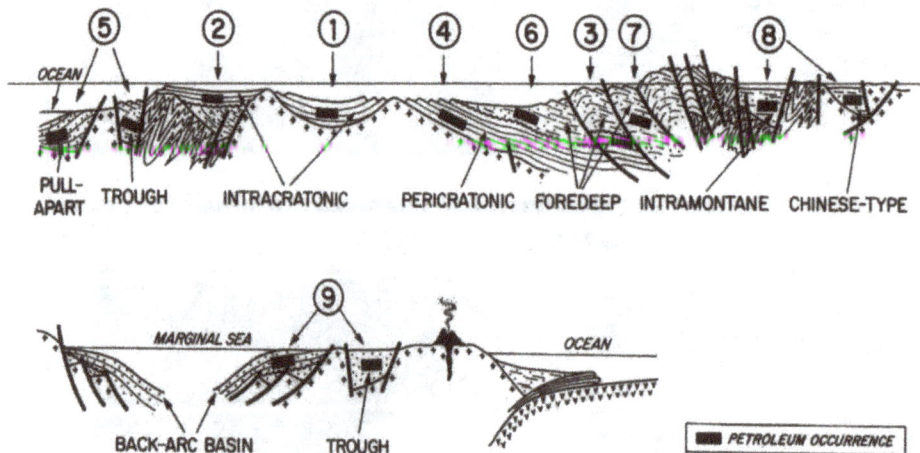

Fig. 9.1(d) Geotectonic classification of basins and their mode of formation based on plate tectonics.

9.1.2 Classification of Indian Basins

Bois et al., (1982) have identified nine types of basins (Fig. 9.1(d)) that evolve due to plate tectonics processes. Based on this classification, Sinha (2004) has classified different Indian basins in these categories as given in Table 9.1.

Table 9.1 Classification of sedimentary basins of India (Sinha, 2004)

Class	Description	Indian Examples
I. PLATFORM OR CRATONIC BASINS		
1	1.Intracratonic basin overlying crystalline	Satpura-South, Rewa-Damodar, Narmada, Pranhita-Godavari, Deccan Syneclise, Bhima-Kaladgi (?), Bastar (?)
2	Intracratonic basin overlying former basin of another category	Rajasthan
3	Pericratonic basins overlain by folded foredeep	Arakan-Yoma Fold Belt
4	Pericratonic basins	Mumbai offshore, Kerala-Konkan, Kutch, Krishna-Godavari, Bengal, Saurashtra, Mahanadi, Vindhyan, Cuddapah, Chattisgarh
II. RAPIDLY SUBSIDING BASINS		
5	Pull-apart basins or associated trough	Cauvery, Cambay
6	Unfolded foredeep or associated trough overlying pericratonic basin	Assam Shelf, Ganga
7	Folded foredeep	Himalayan Foreland
8	Intramontane or Chinese-type basin	Karewa, Spiti-Zanskar
9	Back-arc or association trough	Andaman back-arc

9.1.3 Groundwater Exploration in Hard Rock

Hard rocks having low porosity are devoid of water. In such cases, they usually occur in the top weathered layer or along faults and fracture zones. However, the weathered layer cannot provide perennial source of water and therefore, groundwater investigation in hard rock areas requires delineation of faults and fracture zones that are part of lineaments for this purpose. In trap rocks, as in the case of Saurashtra water may also occur in the intertrappeans in between the flows. However, in these cases also it percolates through faults and fracture zones. As gravity and magnetic methods are quite useful in delineating lineaments, they can be effectively used for groundwater investigations in hard rock areas that are subsequently followed up by more direct methods like resistivity method on ground. In this regard airborne magnetic method is especially useful to delineate lineaments due to close data recording and large area covered. Mishra (1987) and Mishra et al., (1998) have described a large airborne magnetic lineament in the Vindhyan basin that showed good groundwater potential as described in Section 9.6. Kannadasan and Srinivas (2002) have described the application of airborne magnetic map and satellite imagery to delineate groundwater potential zones in Arcot district of Tamil Nadu that is primarily a dry land. They identified 24 deep acquifer zones with good groundwater potential. Rangnai and Ebinger (2008) have shown a very good correlation between magnetic trends and fractures delineated from the airborne magnetric maps and lineaments from Landsat Thematic Mapper (TM) from south central Zimbabwe craton and developed a strategic model for groundwater exploration in this drought prone section of this country. They even delineated small lineaments based on this study that were quite successful in locating sites for hand dug wells and hand pumps. Astier and Paterson (1989) have provided an illustration that shows good yield of groundwater along fault planes and fracture zones though it is far from the run off high land and poor yield from wells that were close by along its slope as demonstrated in Fig. 9.2(a). It shows the importance of lineaments, fractures and fault planes for exploration of groundwater in hard rock areas. Zeil et al., (1991) have demonstrated the application of lineaments for groundwater exploration in Botswana.

The lineaments/faults/shear zones delineated from airborne magnetic maps are followed up on the ground by more direct methods like the resistivity method that is primarily the main geophysical method for groundwater investigations. Water bearing sections being more conductive compared to the surrounding rocks, they can be easily differentiated by the resistivity method. In this method current is sent through the ground using current electrodes and potential differences between potential electrodes are measured. One such combination of current and potential electrodes is shown in Fig. 9.2(b) where C1 and C2 are current electrodes and P1 and P2 are potential electrodes and O is the observation point. Several combinations of current and potential electrodes are suggested in the literature but the Wenner configuration where d1 = d2 = d3 = a is the most popular and widely used configuration. In case C is the current between current electrodes and V is the voltage between potential electrodes, the resistivity is given by R = 2 π .p.V/C., where p is a factor depending on electrode separations and geometry. As the medium is inhomogenous, it is referred to as apparent resistivity. In cae of Wenner configuration where electrode separations are equal to a, it is R = 2 πaV/C, when a is in meters, it is referred to as ohm.m at the central point of electrode configuration. The depth of penetration in the present case is 3a/2 i.e., (separation of current electrodes) / 2 that can be increased by increasing the separation of current electrodes. In case the apparent resistivity is measured at the same point by increasing the electrode separation on either side, it provides the resistivity with respect to depth at the central point and is known as sounding. This book aims to provide some important case histories of groundwater investigations through airborne magnetic and gravity lineaments and therefore, details of electrical methods is beyond the scope of the present book and can be found in any books on Exploration Geophysics such as Telford et al., (1976, 1990), Parasnis (1982).

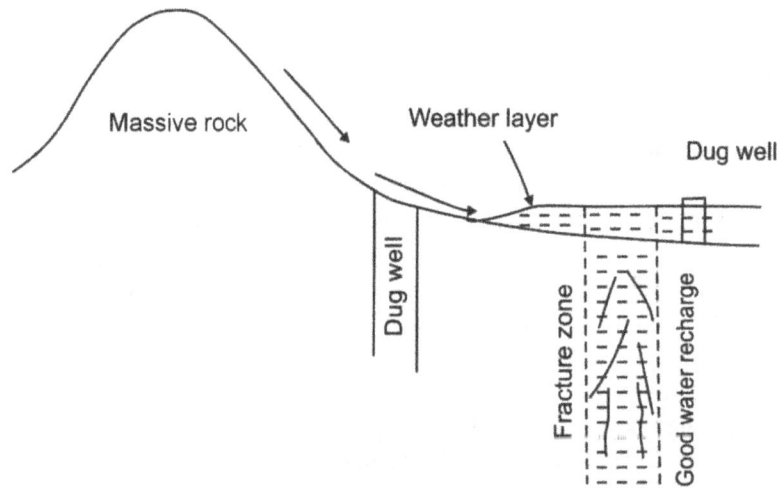

Fig. 9.2(a) Flow of the water runoff from hills that shows better yield from wells located in weathered and fracture zone which are far compared to wells in nearby massive rock.

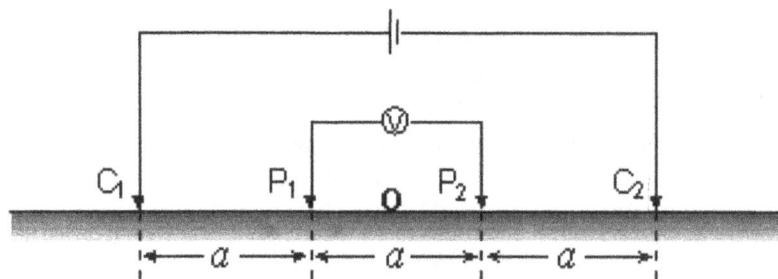

Fig. 9.2(b) Wenner configuration of electrodes for resistivity survey. C1 and C2 are current electrodes and P1 and P2 are potential electrodes at equidistant, a from each other. O is the point of observation.

9.2 Saurashtra, NW India – Hydrocarbon and Groundwater Exploration and Geodynamics

Most of the sedimentary basins in India (Fig. 9.1(a)) have been explored for hydrocarbons by the Oil and Natural Gas Corporation (ONGC) and Oil India Ltd. (OIL). But there are certain sections in the country that have not received adequate attention due to their inaccessibility or problems involved in their exploration using conventional techniques. Conventional seismic techniques, which are the most widely used approaches for oil exploration, have failed to yield the desired results in terrains where the sedimentary formations are covered by a surface layer of basaltic rocks. A large part of western India is occupied by the Deccan trap of Late Cretaceous period (1-6, Fig. 9.1(a)) and Mesozoic sediments in these regions are overlain by trap rocks which are difficult to penetrate by seismic reflection method. About 80-90% of the world's readily accessible hydrocarbon reserves are located in the Middle East and North Sea and come mostly from Mesozoic sedimentary formations. During the Mesozoic period, the worldwide climate was tropical and plankton was abundant in the oceans. Ocean bottoms were stagnant and anoxic and plate tectonic processes facilitated accumulation of thick sediment and produced numerous structural traps where hydrocarbons could accumulate. More than 55% of the world's most petroliferous oil deposits are Mesozoic in age.

Therefore, integrated geophysical exploration programmes were mounted in Deccan trap covered regions of NW India (1-5, Fig. 9.1(a)) to determine the efficacy of these methods to delineate subtrappean Mesozoic sediments in this region. Under this programme gravity, seismic magnetotelluric and electrical methods were simultaneously used to delineate subtrappean Mesozoic sediments. Besides delineating basement structures for hydrocarbon exploration, the same geophysical data were also used for groundwater exploration and geodynamics specific to Saurashtra. The gravity and airborne magnetic maps of Saurashtra are used to delineate structures such as faults, lineaments, and fractures that might be water bearing which were subsequently followed up on the ground by more specific methods like resistivity for confirmation.

9.2.1 Regional Tectonics and Geology

The triangular shape of Saurashtra is the reflection of various system of faults binding on all sides (Fig. 9.3). The Saurashtra peninsula is covered mainly by Mesozoics and Cenozoic rocks. Stratigraphically, the sequence begins with Cretaceous sandstones and shales overlain by Deccan volcanics. These are overlain by Tertiary and Quarternary sediments along the coast.

The west coast of India and adjoining regions have been affected by three major tectonic events during the Mesozoic period after the break up of Gondwanaland, which include (Mishra et al., 2001):

(i) Break up of Africa from India along the West coast of India during middle to late Jurassic. (Besse and Courtillot, 1991)

(ii) Break up of Madagascar along the West coast of India during middle to late Cretaceous (Besse and Courtillot, 1991)

(iii) Break up of Seychelles-Mascarene plateau from the Indian plate (Besse and Courtillot, 1991) and eruption of Deccan Trap from the Reunion hotspot during late Cretaceous (Mc Kenzie and Sclater, 1971)

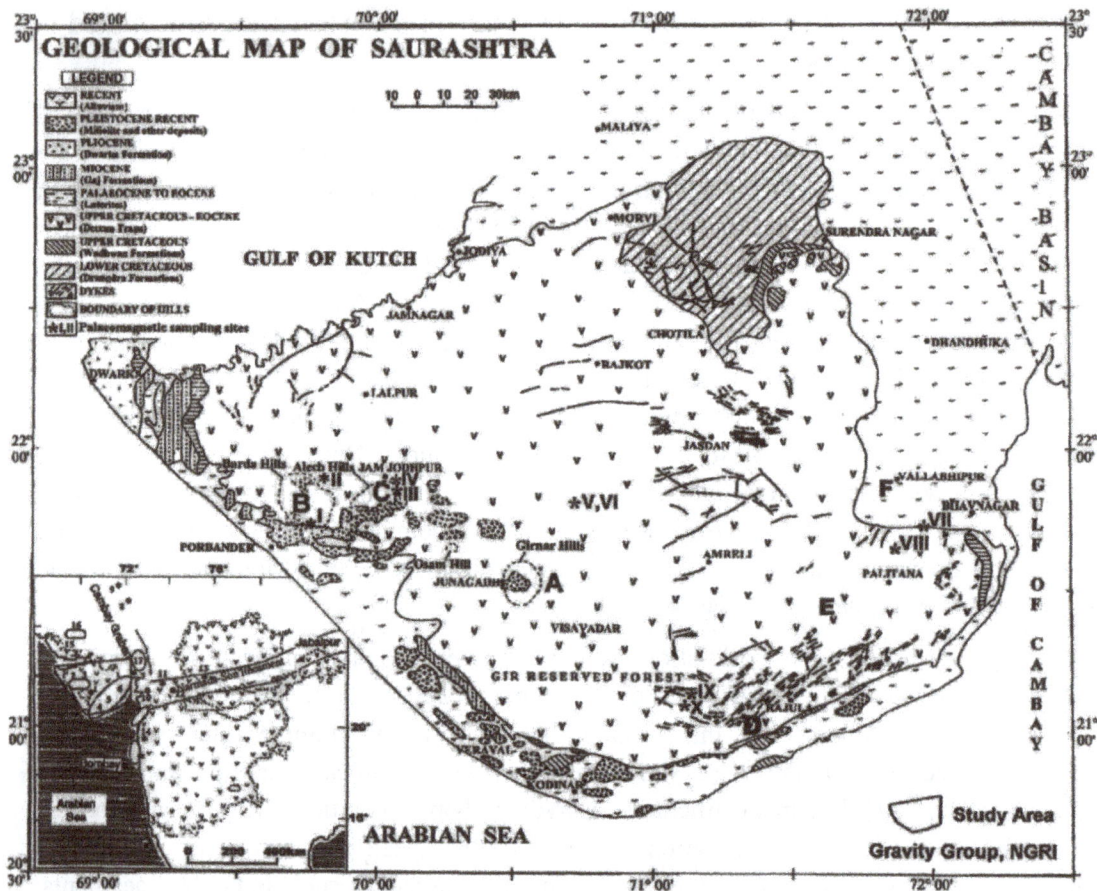

Fig. 9.3 Geological map of Saurashtra (1, Fig. 9.1(a)) with A, B and C representing the Junagadh, the Barda and the Alech volcanic plugs in western Saurashtra and D, E, F represents plugs/stocks in SE Saurashtra. I-X is palaeomagnetic sampling sites. Inset shows alkalic igneous complexes of western and northwestern India marked with stars: 1: Junagadh, 2: Barda, 3: Alech, 4: Rajula, 5: Palitana, 6: Vallabhipur, 7: Kadi, 8: Mundwara, 9: Sarnu-Dandali, 10: Netrang, 11: Phenai Mata, 12: Amba Dongar, 13: Barwaha, 14: Jawhar Nepheline Syenite Dike, Bombay, 15: Alkali Olivine Basalt Lava Flows and Plugs of Central Kutch, 16: Gravity highs in the northern part of Kachchh and 17: Gravity high near Ahmedabad and 8-14 adapted from Basu et al., (1993), (Mishra et al., 2001).

These events have affected the adjoining coastal areas in various ways, the most important being the development of structural trends, rift basins and different kinds of igneous intrusions. The break up of Africa from India along the west coast of India is associated with large scale volcanic eruption over Africa and Antarctica in the form of Karroo volcanics and Ferrar volcanics, respectively (Storey,1995). Some volcanic rocks of Mesozoic period have also been reported from bore wells in Saurashtra (Singh et al., 1997) that might be related to this event as discussed below. Predrift reconstruction of Africa and India shows that plausibly the Karoo rifts (151-159 Ma) (Bosellini, 1992) in India passes through Narmada rift, Cambay rift and Gulf of Kachchh, primarily covering Saurashtra, Kachchh and Western Rajasthan where Mesozoic (Jurassic) sediments are found. It also suggests that Seychelles was located adjoining Saurashtra where Reunion plume was located and was responsible for its break up during late Cretaceous. It gave rise to large volume of volcanic rocks over western India and offshore in the form of Deccan trap. In this reconstruction, Seychelles is close to Saurashtra where the oldest rock (~ 68 Ma) of Deccan trap activity is found as described below while towards the south, younger rocks of this activity are found as the Indian continent moved over this plume during its northward drift (Chapter 5).

The breakup of Madagascar from the west coast of India during 80-90 Ma is related to the Marion hotspot (Morgan 1981). The third event of the breakup of Seychelles and eruption of Deccan flood basalt from Reunion hotspot during late Cretaceous has been most significant and widespread over the Indian continent (Raval, 1989). The Deccan Volcanic Province (DVP) is the largest continental flood basalt volcanic province of the world exposed in western India (Inset, Fig. 9.3). Due to the vast coverage of Deccan trap over the western part of the Indian continent, signatures of the previous events are covered under the Deccan trap and can not be accessed directly in the field. The Saurashtra Peninsula is mostly covered by Deccan Trap (Fig. 9.3) which represents flows of tholeiitic basalt with several intrusions of acidic, alkaline and mafic/ultramafic rocks (Merht, 1995). Deccan Trap of Saurashtra is presumed to be one of the earliest products of magma differentiation due to Reunion hotspot during late Cretaceous as indicated by occurrences of picritic basalt, gabbro and alkaline rocks in this area (Basu et al., 1993) and serpentinization of basic igneous rocks of Girnar hills. Some volcanic plugs have also been reported (A, B, C, Fig. 9.3), out of which the one at Junagadh (A) consists of ring complexes with an assemblage of rock types varying in composition from acidic, alkaline, to mafic/ultramafic. B and C are similar volcanic plugs known as Barda and Alech plugs. Some stocks/plugs are also reported from the south-eastern part of Saurashtra around Vallabhipur, Palitana and Rajula (D, E, F, Fig. 9.3) which are basically alkaline and acidic in composition (Karanth and Sant, 1995). These plugs occupy the western margin of the Gulf of Cambay that is connected to the Cambay rift basin on land.

The most prominent and well studied plug is the Junagadh plug, which represents a layered complex with ring dykes of dolerite and granophyres. Chandra (1999) has given a detailed account of different rock types in this complex, which form three major lithological rock units, namely pyroxenites/peridotite, gabbro and syenite/nepheline syenite. The Barda and Alech plugs in western Saurashtra and Vallabhipur, Palitana and Rajula plugs in SE Saurashtra are primarily composed of acidic and alkaline rocks, respectively. The Barda complex is granitic in composition and characterized by a variety of pyroxenes suggesting an alkaline affinity (De and Bhattacharya, 1971). Based on $^{40}Ar/^{39}Ar$ measurements on tholeiitic basalt from DVP, an age of 66.6-68.5 Ma was suggested for Deccan trap by Duncan and Pyle (1988). Vandamme et al., (1991) have reviewed the reported ages of various rocks from DVP and suggested an age of 65.5 ± 2.5 Ma based on $^{40}Ar/^{39}Ar$ measurements on almost 31 samples from different localities. Another date using ^{187}Re-^{187}Os systematics also suggested almost same period for Deccan trap with further constraints as 65.6 ± 0.3 Ma (Allegre et al., 1999). The igneous intrusive and alkaline complexes of DVP provide an older $^{40}Ar/^{39}Ar$ dates as 68.6 Ma (Basu et al., 1993) for volcanic plugs 8 and 9 (Inset, Fig. 9.3) and 68.6 ± 2.4 Ma for Junagadh plug (Baksi, 1987; A, Fig. 9.3). These dates suggest an older age of 3.0-3.5 Ma for igneous intrusions and alkaline complexes of Deccan trap in NW India compared to main pulse of Deccan trap activity, which may represent incubation period (Basu et al., 1993).

Numerous trap dykes cut the basalt and are seen to show three main directions NE-SW, E-W and NW-SE ranging in thickness from 2-5 meters and rarely up to 15 m (Fig. 9.3) which are dominant structural trends in this area (Mishra et al., 2001). It may be noted that orientation of dyke swarms in north are E-W, almost same as the alkaline complexes of Barda and Alech (Fig. 9.3). There is a patch of Mesozoic sediments in the NE part, which extend below the Deccan trap in certain sections. Several authors have described the magnetic properties of Deccan traps including its remnant magnetization. Sahasrabudhe (1963) has described both normal (335°, − 50°) and reverse (150°, + 55°) magnetizations corresponding to upper and lower Deccan traps respectively. However, subsequently some more reversals were identified in different parts of Deccan trap which have been summarized as N2-R1-N1 polarity Chrons (Vandamme et al., 1991) with bulk of the exposed Deccan trap displaying the middle reverse polarity (R1) and upper normal polarity (N1) at higher levels (Inset, Fig. 9.22).

According to them these polarity Chrons N2-R1-N1 correspond to the magnetic Chrons 30N, 29R, and 29N, respectively, of standard magneto-stratigraphy scale. The reported direction of magnetization for the volcanic rocks of Junagadh plug is D = 137° and I = + 52° for reverse polarity at lower levels and D = 336°, I = – 38° for normal polarity at higher levels (>140 m). Reverse polarity has also been reported for the exposed Deccan trap around Rajkot and other places in Saurashtra (Fig. 9.3), (Verma and Mital, 1972). As referred to above, the exposed volcanic plugs of Saurashtra belonging to Deccan activity show considerable variation in their composition and physical properties. This, in general, is the case with most of the volcanic provinces of the world related to plume activity. However, laboratory measurements provide only the properties of exposed rock samples. Therefore, an integrated programme of exploration using gravity and magnetic methods and laboratory analysis of surface samples for their composition and physical properties is described below.

9.2.2 Lineament Map

Fig. 9.4 shows the major and minor lineaments of Saurashtra prepared from false colour composite thematic maps obtained from NRSA on 1:250,000 scale (Mishra et al., 2001). Shuttle Radar Topography Mission (SRTM) elevation data of the area were also studied during the present work. The SRTM data was processed to create Digital Elevation Models (DEM) and shaded relief maps that can be analyzed by varying the directions and inclination of illumination. The lineaments that are orthogonal to the light will be clearly visible in this process and can be easily detected during interpretation. Thus, the systematic orientation of lineament pattern interpreted from the SRTM shaded relief maps reliably represents the geological structure of the bed rock. The relief maps were used with different illumination directions to extract maximum number of reliable lineaments from the maps. This lineament map shows dense concentration of lineaments along the coast and in areas adjoining it, while the central part shows relatively less number of lineaments. This suggests the marked influence of coastal tectonics in the development of lineaments in this area. The following four major structural trends are clearly visible in the map, which dominate in different regions of Saurashtra.

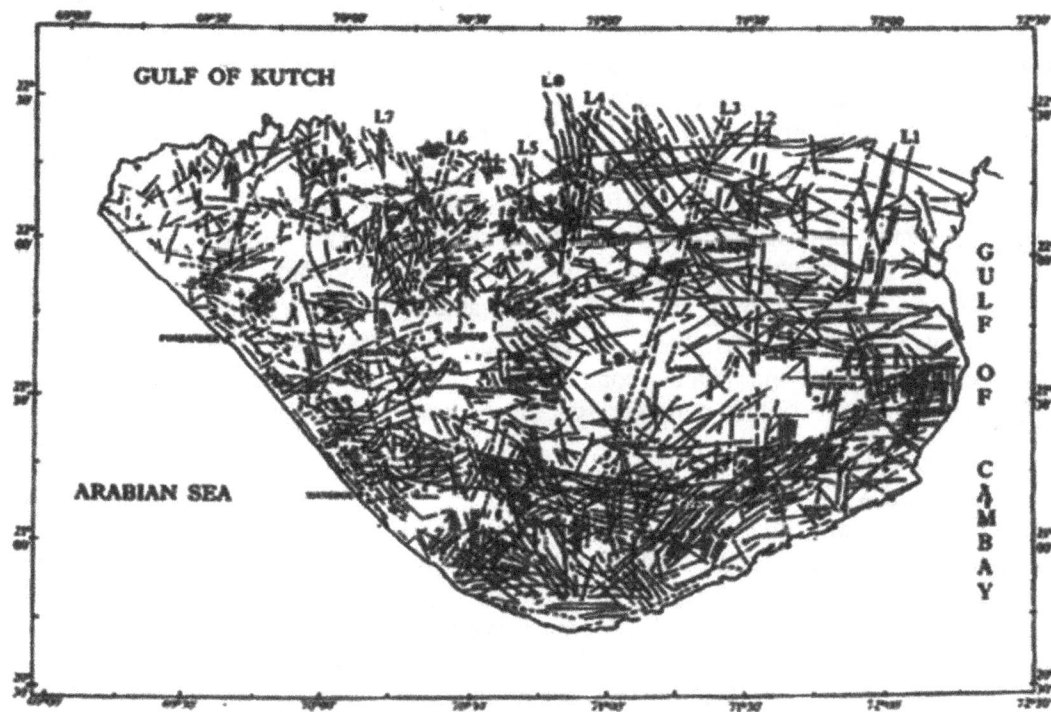

Fig. 9.4 Lineament map of Saurashtra based on false colour, thematic from LandSat-TM data acquired during February 1995. L1-L7 are pairs of NNE-SSW to N-S major lineaments (Mishra et al., 2001).

(i) NE-SW trend is present in almost the whole area. However, it is predominant in the SE part of Saurashtra coinciding with intense dyke swarms of Deccan volcanism and is almost parallel to the coast of Saurashtra with the Gulf of Cambay.

(ii) ENE-WSW to E-W trend is another dominant trend over the entire Saurashtra peninsula. The ENE-WSW trend in the southern section represents the trend of the Narmada-Son Lineament, which is a Precambrian trend in this region

(iii) NW-SE trend, parallel to the west coast of Saurashtra, is predominant in the coastal part and in adjoining areas. It indicates their interrelationship with the evolution of the coastline.

(iv) NNE-SSW to N-S trend is predominant in the eastern and central part, which might have been influenced by the Gulf of Cambay and the Cambay rift basin located east of Saurashtra. A few lineaments, of this trend appear in the northwestern part also but are not very prominent. They are visible in pairs (L1 to L7) as linear zones with contrasting tonal and texture characteristics.

The central part of Saurashtra is made up of an undulating plain broken by hills and considerably dissected by various rivers that flow out in all directions. The intrusive rocks of Girnar rise to 1117 m of elevation while Barda and Alech rise to 637 and about 300 meters respectively. An elevated strip of ground connecting Rajkot and Girnar forms the major water divide of the Saurashtra. The present day drainage pattern in Saurashtra is radial, suggesting a domal uplift around Jasdon coinciding with the Jasdon plateau.

9.2.3 Multi-parameter Geophysical Studies

Gravity, seismic, magnetotelluric and resistivity studies were carried out in Saurashtra for the delineation of Mesozoic sediments beneath the Deccan trap cover for hydrocarbon exploration (NGRI, 1998). These data sets are also utilised for groundwater investigations and geodynamics of this region.

(i) Gravity Studies – Bouguer Anomaly Map and Seismicity

The Bouguer anomaly map of Saurashtra based on ~2 km data spacing is presented in Fig. 9.5 (Mishra et al., 2001) which shows a number of significant features. It was processed for a standard crustal density of 2.67 g/cm^3 and station elevation was obtained from geodetic leveling survey that provided Bouguer anomaly to an accuracy of 0.5 mGal. Gravity 'highs' of 40-60 mGal corresponding to volcanic plugs of Junagadh, Barda and Alech in western Saurashtra are marked as A, B, and C respectively as these plugs are shown in Fig. 9.3. The gravity anomaly A is circular in nature and coincides with the Junagadh plug, while the gravity anomalies B and C are individually circular but are together oriented E-W indicating a fracture zone occupied by these volcanic plugs. A major gravity low, L1 is observed over the Jasdon plateau. Some other significant features of this map are as follows:

(i) Three similar circular gravity highs of almost the same amplitude are also delineated in the SE part of Saurashtra near Vallabhipur, Palitana, and Rajula marked as D, E, and F respectively. These highs are individually circular in nature but together are aligned in NE-SW direction indicating a large fracture/fault zone, which may form the western margin of the Cambay rift basin (Mishra et al., 1998). They coincide with the exposed volcanic plugs/stocks, which are mainly alkaline in nature.

(ii) A broad gravity low (L1, Fig.9.5) over the Jasdon plateau in eastern Saurashtra is delineated, a part of which might be due to isostatic compensation.

(iii) There are other small wavelength gravity highs (H1) and lows (L1 and L2), especially in the northern part of Saurashtra. These lows being represented by moderate to small wavelength and amplitude might be caused by subtrappean shallow sources and the high (H1) may represent a subsurface plug similar to others in this region.

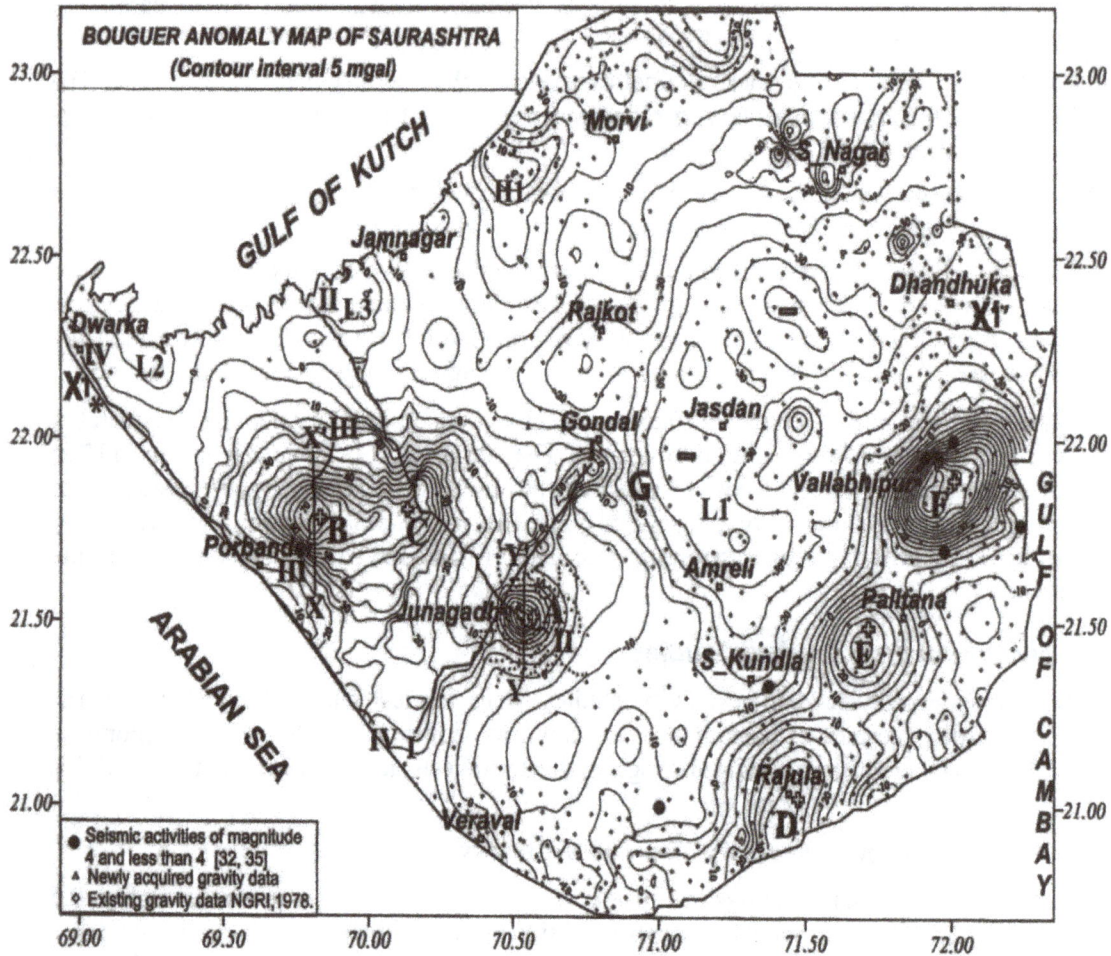

Fig. 9.5 Bouguer anomaly map of Saurashtra based on station spacing of ~ 2 km showing gravity highs A, B, C, D, E and F corresponding to volcanic plugs/stocks and gravity low, L1 is related to Jasdon plateau and L2-L3 and H1 are gravity lows and high, respectively along the north coast of Saurashtra with the Gulf of Kutch. Profiles I-IV are specially recorded profiles with 0.5 km station spacing to define the gravity anomalies of volcanic plugs A, B, C and demarcate any other anomaly due to similar bodies. X1 - X1' is a profile used to demonstrate the effects of isostasy in Fig. 9.6. XX' and YY' are profiles along which the gravity and magnetic fields are modeled in Fig. 9.28(a) and (b) (Mishra et al., 2001). G is the gravity gradient along the western margin of the Jasdon plateau.

Besides these major gravity anomalies, most of the major structural trends observed on lineament map (Fig. 9.4) are also present in the Bouguer anomaly map (Fig.9.5). These are prevalent in different parts of Saurashtra such as (i) NE-SW trend in the SE part (ii) E-W trend in the western Saurashtra (iii) NW-SE trend along the west and the north coast of Saurashtra and (iv) N-S to NNW-SSE trend in the central part. It is interesting to note that the linear gravity gradient G coincides in part with the N-S lineament L4 inferred from satellite data (Fig. 9.4) which are indicative of fractures/fault zones.

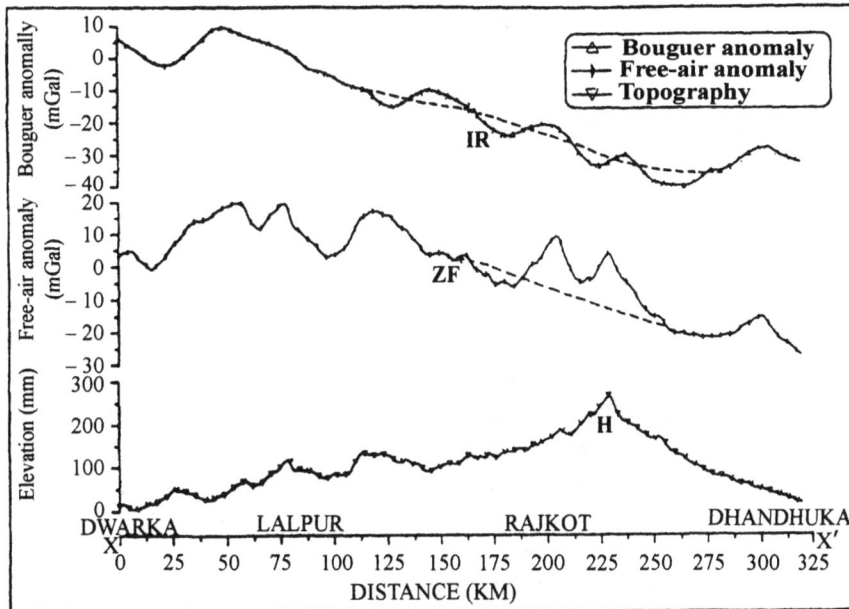

Fig. 9.6 Free air anomaly, Bouguer anomaly and elevation along a profile X1-X1´ (Fig. 9.5) showing opposite correlation between elevation and Bouguer anomalies over Jasdon plateau (H) that indicates Airy's isostatic compensation but negative free air anomaly indicates partial compensation or some mass deficiency in the crust (Mishra et al., 2001). The Bouguer anomaly (~ –15 mGal) corresponding to zero free air anomaly (ZF) represents isostatic regional (IR) related to isostatic compensation. Similarly, the Bouguer anomaly corresponding to zero free air anomalies in the entire region provides the isostatic regional map for the region.

An east-west profile X1-X1´ across Saurashtra showing elevation, free air anomaly and Bouguer anomaly is given in Fig. 9.6. There is a good correlation between free air and Bouguer anomaly over the Jasdon plateau, which are opposite to the regional elevation. This indicates isostatic compensation for this plateau. The zero value of free air anomaly indicates isostatic compensation, and the corresponding Bouguer anomaly represents the isostatic regional as described above. The Bouguer anomaly corresponding to zero free air anomaly (marked on the profile in Fig.9.6) indicates the effect of isostatic compensation in the present case, which is approximately –15 mgal. The elevation profile shows a higher elevation of approx. 300 m over the Jasdon plateau (H), which implies a crustal thickening of approximately 2 km for isostatic compensation. A crustal thickening of 2-3 km is also suggested by seismic investigations (Kaila et al., 1980), which show a crustal thickness of 38 km towards the west and 40-41 km towards the east under the Jasdon plateau. This crustal thickening produces an anomaly of –15 to – 20 mGal as discussed above as effect of isostatic compensation while the observed gravity low is approximately – 35 mGal. Therefore, the difference between the observed and the effect of isostatic compensation viz. approximately –15 to – 20 mGal is produced by other subsurface low density sources such as felic intrusives that might be responsible for the formation of the Jasdon plateau.

The profile X1-X1´ passes through the lineaments L1 to L7 on the lineament map (Fig. 9.4). On projecting these positions on to the Bouguer anomaly profile, one can observe that the lineaments L3, L5 and L7 coincide with gravity lows, while L6, L4 and L1 coincide with falling trends of gravity. This means that these lineaments are associated with gravity low axis. Eaton and Watkins (1970) clearly demonstrated that the course of the buried channel and its axis coincide with gravity troughs. This map (Fig. 9.5) also shows the epicenters of some seismic activities of magnitude ≤ 4 which are associated with the contacts of volcanic plugs. The Junagadh and Vallabhipur plugs (A and F, Fig. 9.5) are significant in this regard. As suggested in Chapter 8 in connection with the Bhuj earthquake, intrusives play an important role as stress concentrators. It has also been suggested in Chapter 8 that overcompensated sections of the crust are prone to seismic activity due to stress related to uplift caused by it that is also applicable in this case as the magnitude of Jasdon gravity low, L1 (Fig. 9.5) is more than required for isostatic compensation as described above.

(ii) Isostatic Regional and Residual Maps

As described above, the zero free air anomaly indicates Airy's model of isostatic compensation. Bouguer anomaly at those points provides the estimate of the effect of isostatic compensation that can be regarded as isostatic regional anomaly. The same procedure is applied to the Bouguer anomaly map of Saurashtra and isostatic regional is obtained as shown in Fig. 9.7. It shows gravity low over the Jasdan plateau indicating crustal thickening under this region. This field, when subtracted from the observed field, provides the isostatic residual field (Fig. 9.8) that shows almost all the essential elements of the observed field with reduced amplitude including Jasdon low (L1, Fig. 9.5). It also shows the gravity highs, A-F due to volcanic plugs (Fig. 9.5) as described above. The gravity lows along the north coast of Saurashtra are medium wavelength and medium amplitude anomalies that are significant for shallow sources and are significant for hydrocarbon exploration as described below. The regional field (Fig. 9.7) is primarily used for the geodynamic studies as discussed in the next section while the residual anomalies (Fig. 9.8) are used to compute basement model and subtrappean Mesozoic sediments.

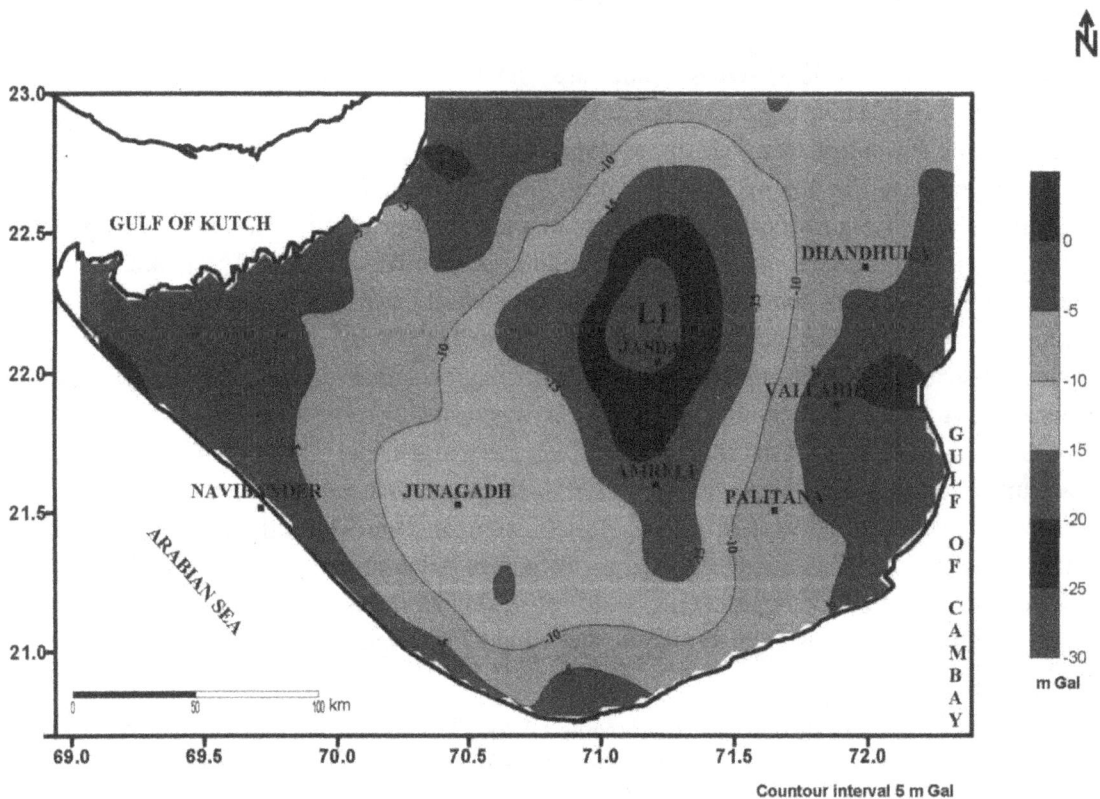

Fig. 9.7 Isostatic regional gravity anomaly based on zero free air anomalies showing a gravity low over the Jasdon plateau that is related to isostatic compensation (Singh et al., 2003).

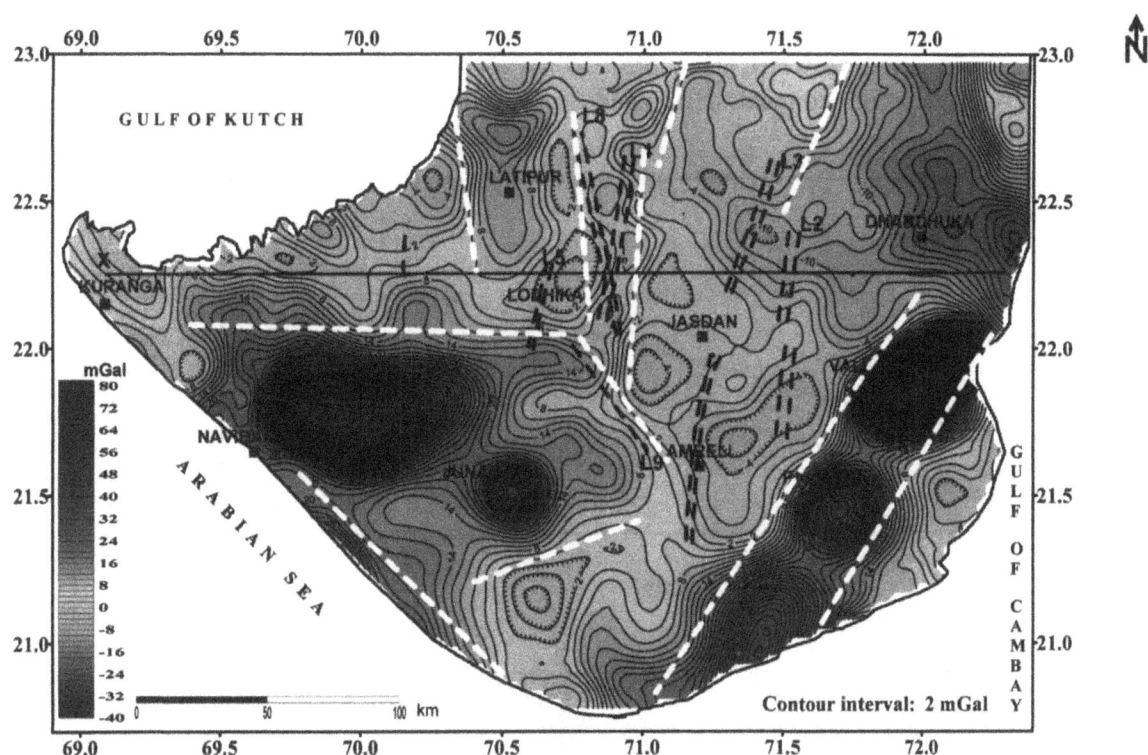

Fig. 9.8 Residual gravity anomaly after subtracting the isostatic regional (Fig. 9.7) from the observed field (Fig. 9.5) that shows several gravity highs and lows related to shallow sources. Dashed lines indicate some major lineaments (NGRI, 1998; Singh and Arora, 2008). [Colour Fig. on Page 808]

(iii) Seismic Method

Traditionally, seismic reflection imaging has been the most powerful geophysical tool for exploration of hydrocarbons. However, the vast sheet of volcanic cover acts as a shield and the conventional reflection techniques for the study of the sub-trappean Mesozoic sediments (Withers et al., 1994) becomes ineffective. Sub-basalt seismic imaging is difficult because of (i) low velocity sediments under high velocity basalts causing problems in identification of critically refracted waves from sediments, (ii) multiples generated in interbedded units of basalt and in the offshore between the top of basalt and the sea floor, (iii) energy scattering from and absorption by the near surface heterogeneities, (iv) wave mode conversion at the top of the basalt (Dhananjay Kumar, et al., 2004). However, during recent years, modeling and inversion of low frequency wide angle reflection/refraction data have proved to be promising on a regional scale (Sain et al., 2002; Sain and Kaila, 1996 and Jarchow et al., 1994). On the other hand, processing based approaches like radon transform, multi component pre-stack migration, can improve the subsurface image (Dhananjay Kumar, et al., 2004).

However, in certain favorable geological situations, a low velocity layer (LVL) is inferred from a "skip" in the first arrivals (Whiteley and Greenhalgh, 1979; Tewari et al., 1995). The presence of a shadow zone or "skip" in the refraction first arrival data is the main response signature in seismic interpretation techniques that has been used in identifying and modelling the subtrappean Mesozoic sediments (Fig. 9.9). The thicker the sedimentary column, the more will be width of the "skip". However, if the thickness of overlying trap is significant and the sediments are thin it becomes difficult and ambiguous to use the "skip" for estimation of sediment thickness. On the other hand, with a thin or moderately thick trap layer overlying a thick sedimentary column the "skip" provides valuable information in the estimation of parameters related to the LVL. Thus, using this response it is possible to demarcate the zones of thick/thin traps or thick/thin sediments (Singh and Arora, 2008).

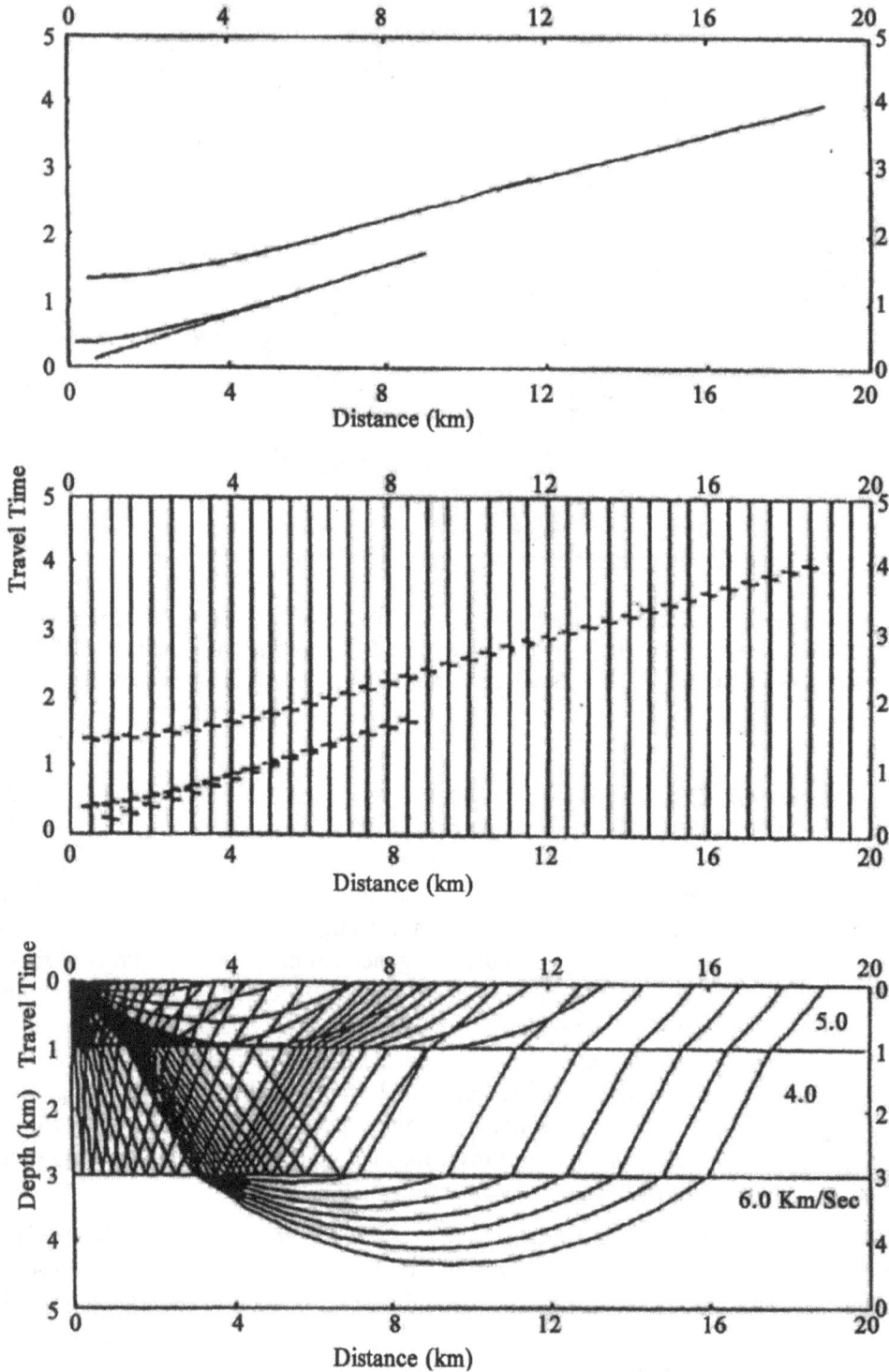

Fig. 9.9 Concept of Seismic Skip in seismic data from Saurashtra that arises due to low velocity sediment under high velocity trap rocks (NGRI, 1998, Singh and Arora, 2008).

(iv) Electrical Methods

The presence of a large resistivity contrast between basalts-sediments-basements provides an appropriate condition for the use of electrical and electromagnetic methods. During recent years magnetotelluric (MT) method has also been extensively used to delineate sediments below basalts (Morrison et al., 1996; Hautot et al., 2000; Withers et al., 1994). In this situation, MT and deep electrical sounding (DRS) responses provide well defined H-type sounding curves for thick sedimentary layers sandwiched between a thick trap cover and the basement, Fig. 9.10 and 9.11 respectively. As the thickness of the trap cover decreases, the apparent resistivity curves gradually assume a shape tending to resemble response of a two-layer structure with the conducting layer on the top whose apparent resistivity tends to assume values closer to the resistivity of sediments. On the other hand, when the sediment thickness becomes smaller, again the sounding curves tend to assume the shape of a two-layer structure with a top conducting layer but with the apparent resistivity of the top layer assuming values closer to resistivity of the traps (Singh and Arora, 2008). With this pattern of the general character of MT and DRS responses in the trap-sediment-basement layered sequence, an examination of the data and its modelling can help identify areas of thick or thin traps and areas of thin or thick sediments. In case of thin sediments, the MT method may not resolve the sediments properly due to the problem of equivalence (Withers et al., 1994). Joint inversion of wide angle seismic reflection and refraction data and MT data has the efficacy to resolve the low velocity conducting sediments below high velocity resistive flood basalts (Morrison et al., 1996; Jones, 1998; Warren, 1996; Hering et al., 1995; Manglik and Verma, 1998).

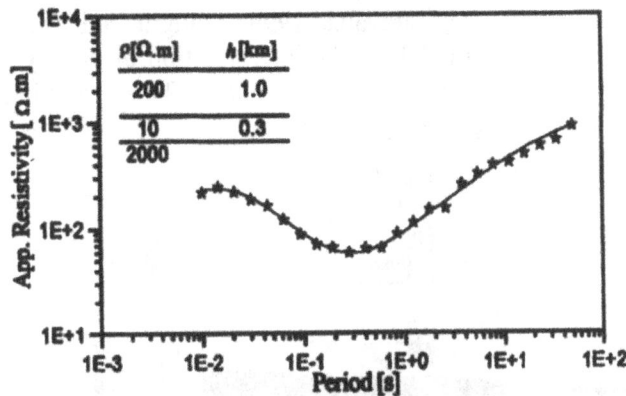

Fig. 9.10 Typical curves for MT interpretation (Manglik and Verma, 1998) showing three layer model with an intermediate conductive layer that may signify Mesozoic sediments.

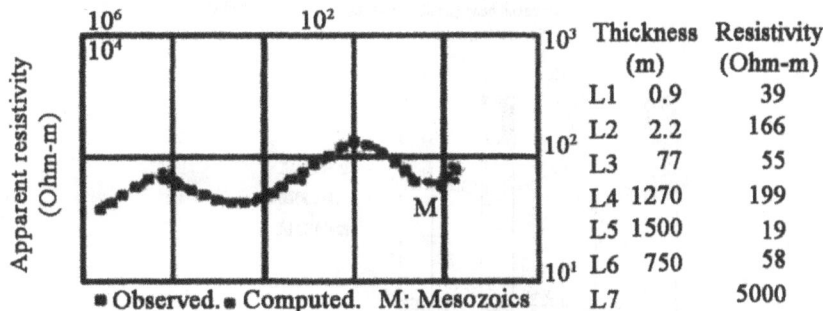

Fig. 9.11 Typical curves for deep resistivity sounding interpretation (Satpal et al., 2006) with low resistivity Mesozoic sediments (M).

9.2.4   Hydrocarbon Exploration – Integrated Approach and Subtrappean Mesozoic Sediment Thickness Map

Gravity lows indicate low density rocks that signify sediments and are important for hydrocarbon exploration. The Bouguer and the residual gravity anomaly maps (Fig. 9.5 and 9.8) show distinctly two types of gravity lows, (i) large wavelength and amplitude gravity low, L1 related to Jasdon plateau that is primarily associated with the isostatic compensation as discussed above and therefore significant for geodynamic studies and (ii) medium-small wavelength and amplitude gravity lows, L2-L3 along the northern coast of Saurashtra (Fig. 9.5) with the Gulf of Kachchh that may be significant for hydrocarbon exploration and are therefore modeled and discussed below. In regard to hydrocarbon exploration, two parametric wells were drilled by the Oil and Natural Gas Corporation (ONGC) at Lodhika and Dhandhuka (Fig. 9.8) within the gravity low, L1 (Fig. 9.5) related to Jasdon plateau. However, as discussed above, this gravity low with large wavelength and amplitude is caused by deep seated sources in response to isostasy instead of low density shallow sediments that is the target for hydrocarbon exploration. Due to information available from these wells, test profiles of seismic, magnetotelluric, and gravity were recorded along a profile joining these wells.

(i)   Seismic, Magnetotelluric, and Electrical Profiles at Lodhika Borehole

A 45 km N-S profile provide constraints on the geometries of the four layers in this area across the Lodhika well shown in a modified form in Fig. 9.12 (Sain et al., 2002; Dixit et al., 2000), viz. the top volcanic layer, the second sedimentary layer (Mesozoic Sediments) and a basalt flow below the sediments, followed by the basement rocks. This is further augmented by joint inversion of MT and DRS data near the well, which also corroborate borehole information, shown in Fig. 9.13 (NGRI, 1998; Harinarayana, 1999, Manglik and Verma, 1998). These constraints serve to define the initial model for the interpretation of gravity data, which is the fundamental tool used to define the layer geometries spatially. This example provides a typical example of the advantages of the integrated approach.

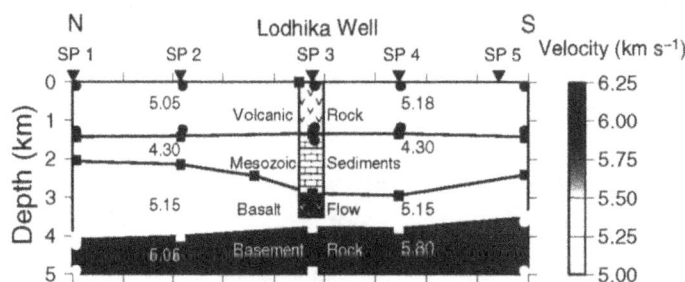

Fig. 9.12 Interpretation of seismic data over Lodhika well. SP are shot points and velocities are given in km/sec (Dixit et al., 2000) showing Mesozoic sediments between two layers of igneous intrusive.

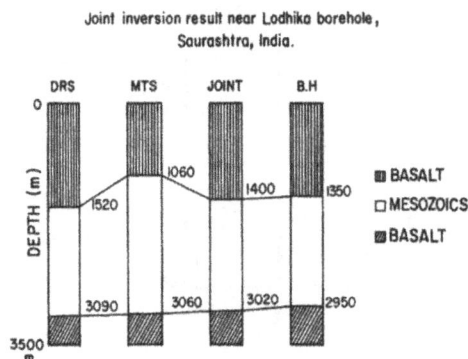

Fig. 9.13 Interpreted section from MT and DRS data and their joint inversion at Lodhika well and borehole data for comparison showing the efficacy of joint inversion (Manglik and Verma, 1998).

(ii) Lodhika- Dhandhuka Gravity Profile

The two borewells at Lodhika and Dhandhuka of ONGC in this region provided constraints on the thickness of trap, Mesozoic sediment and depth to the basement at these points to model a gravity profile across these two wells. The residual gravity anomaly (Fig. 9.8) between Lodhika and Dhandhuka bore wells was modeled constraining the depths from the borehole information for shallow section at these points. The bulk densities were adopted from the likely rock types and borehole density logs and checked vis-à-vis seismic velocities. The computed crustal model is given in Fig. 9.14 that shows upper shallow section up to 6 km and deeper section from 10- 30 km. The deeper section is an artificial body that is placed to account for the regional gradient as apparent from the gravity profile as this gradient can not be produced due to shallow sources. This indicates that the isostatic regional could not remove the total regional field and a part of it remained in the residual field that has to be accounted by a deeper source as also discussed in Section 9.2.3. It shows shallow basement and thining of the sub-trappean Mesozoic sediment towards the east (Dhandhuka). It is overlain by Deccan trap that is exposed in this region. It is also underlain by another layer of volcanic rocks (2.84 g/cm^3) that was also obtained in the bore hole at Lodhika under lying the Mesozoic sediment which might be related to the breakup of India and Africa during Jurassic period as described above in Section 9.1.1 and has great significance for geodynamics. As mentioned above, the low density body in the lower crust is an artificial body to match the regional gradient along this profile that can not be accounted by shallow sources. It is modeled in the form of an actual deep seated body in Section 9.4.2 (Fig. 9.33) along a long profile as modeling of the regional field along a small profile is not justified.

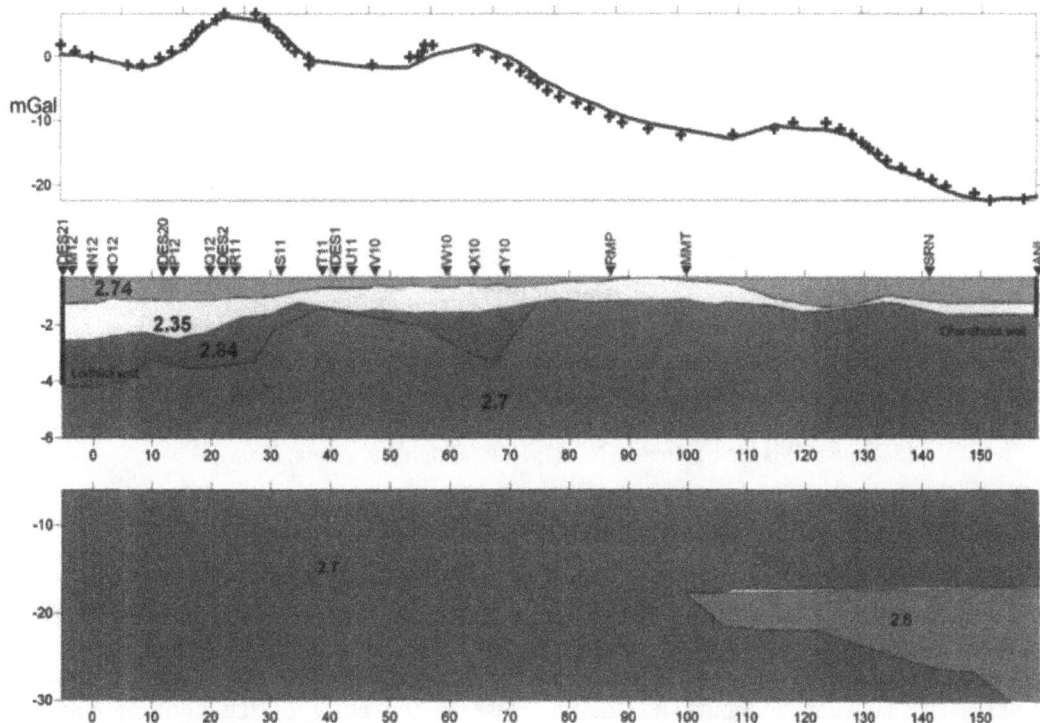

Fig. 9.14 Intepreted section along the Lodhika-Dhandhuka residual gravity anomaly (Fig. 9.8). Shallow section is constrained from borehole data. The lower crustal low density body (2.6 g/cm^3) is an artificial body to account for the decreasing gradient in the observed field that is caused by some deeper features. Numbers indicate densities in g/cm^3. Top layer (2.74 g/cm^3) is Deccan trap followed by Mesozoic sediments (2.35 g/cm^3) and second layer of basaltic intrusion (2.84 g/cm^3) (NGRI, 1998; Singh and Arora, 2008).

(iii) Kurunga-Latipur Profile

As described above based on medium-small wavelength and amplitude, the gravity lows (L2 and L3) observed along the north coast of Saurashtra are significant for exploration of hydrocarbons. Therefore, the residual gravity field (Fig. 9.8) along a profile from Kurunga to Latipur was modeled (Fig. 9.15) constrained from the results of seismic and magnetotelluric measurements along the same profile. Modeling of the gravity field has the advantage of integrating all the available information from a region when the model is constrained using those information. The computed model (Fig. 9.15) shows a top layer of Deccan trap underlain by Mesozoic sediments (2.35 g/cm^3) followed by basement. It shows maximum thickness of Mesozoic sediments of ~2.5 km along the Gulf of Kachchh.

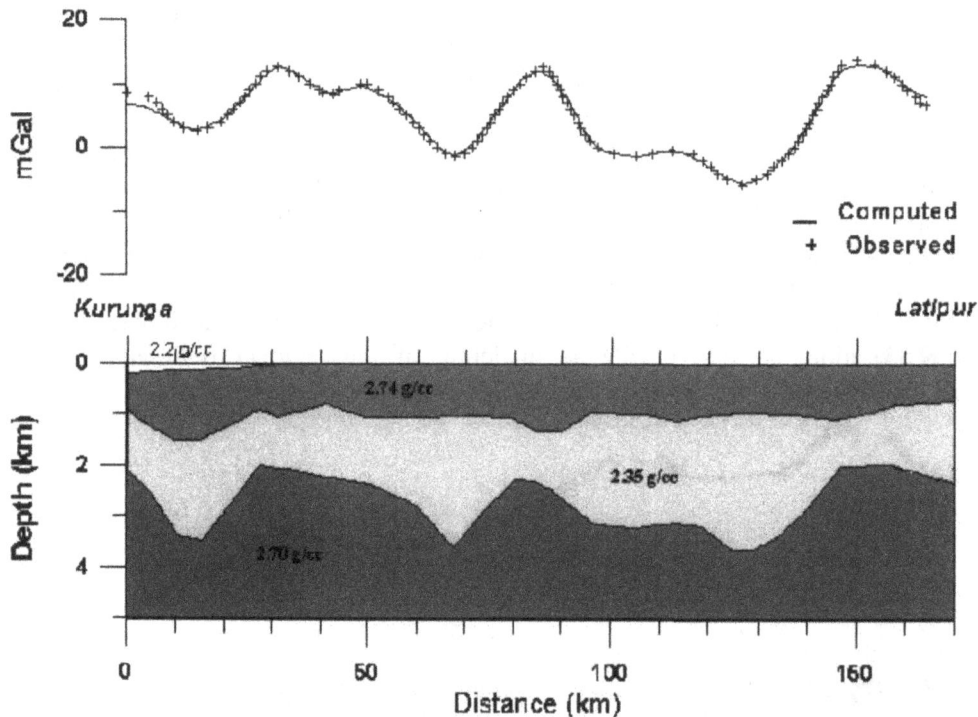

Fig. 9.15 Shallow section based on residual gravity anomalies along the Kurunga-Latipur profile (Fig. 9.8) along the northern coast of Saurashtra with the Gulf of Kachchh. The layer of density 2.35 g/cm^3 represent subtrappean Mesozoic sediments while top layer (2.74 g/cm^3) is Deccan trap and bottom layer (2.70 g/cm^3) is the basement (NGRI, 1998; Singh and Arora, 2008).

(iv) Subtrappean Mesozoic Sediment Thickness Map

Several such 2D sections across major gravity and MT signatures are modelled with inputs from DRS and seismic velocity information and the geometry of the basement structure is derived. The sediment thickness map (Fig. 9.16) reveals that the thickness of the Mesozoic sediments increases from the south to the north and the most significant patch of sediments (~3 km) is interpreted to lie between Jodhpur and Jodiya, close to the coast of the Gulf of Kacchh corroborated by all data sets, viz. gravity, seismic, magnetotelluric, and electrical. Geochemical prospecting has also indicated high concentration of methane, ethane and hydrocarbons in soil samples suggesting it as a prospective area for hydrocarbon occurrence (Devlina Mani et al., 2010). This is a crucial finding vis-à-vis a preconceived notion that the maximum thickness of the Mesozoic sediments is expected below the gravity low related to the Jasdon plateau (L1, Fig. 9.5) where Lodhika and Dhandhuka borewells were located.

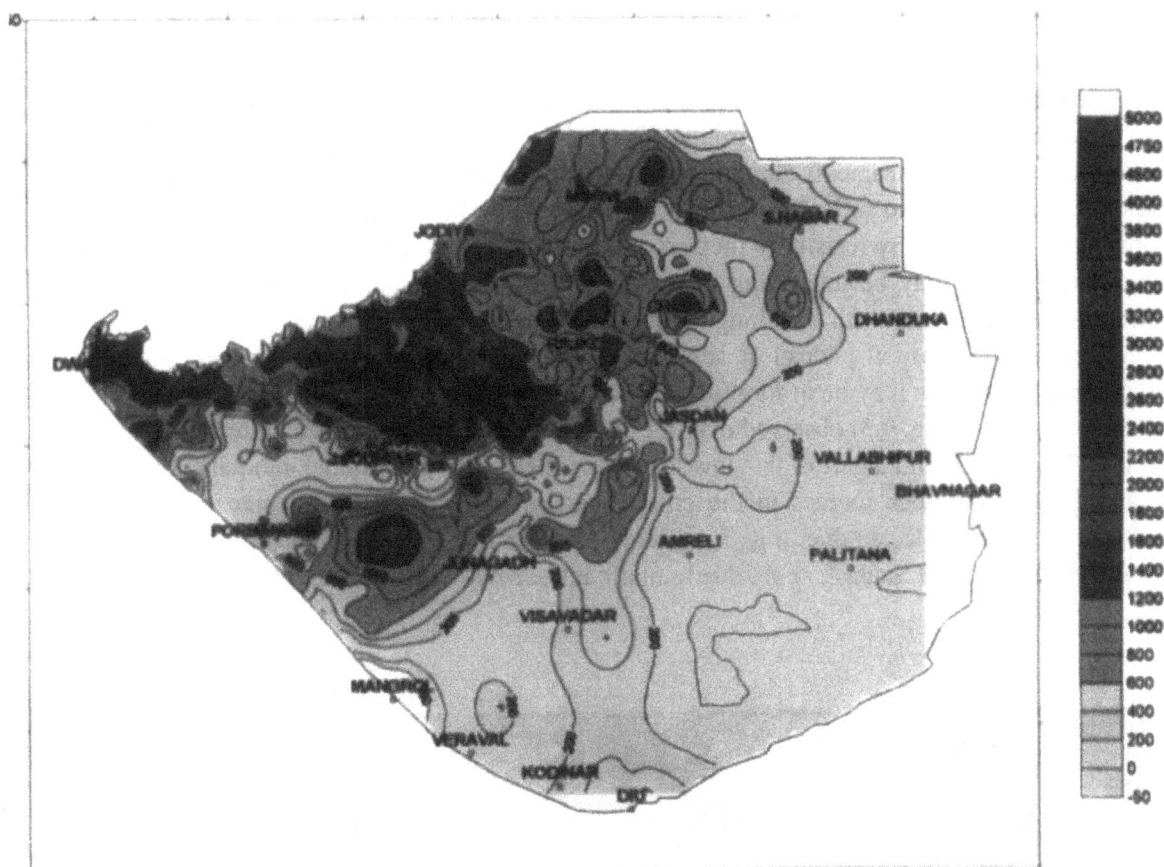

Fig. 9.16 Subtrappean Mesozoic sediment thickness map based on integration of various geophysical data showing maximum thickness of ~3.0 km in the northern part of Saurashtra (NGRI, 1998, Singh and Arora, 2008).

9.2.5   Groundwater Exploration : Airborne Magnetic, Gravity and Landsat Lineaments and Airborne Magnetic Map of Cental Botswana – Karoo Volcanics

Saurashtra is a drought prone area and being primarily occupied by trap rocks, the groundwater is scarce. Gravity and magnetic lineaments indicate change or break in structures that are significant for groundwater investigations. The lineaments being thin and usually large, airborne magnetic surveys are found to be extremely useful due to close data spacing and large covered areas. This is demonstrated below from the airborne magnetic map of Saurashtra and central India in the next section. Gravity maps can also be used to delineate lineaments but only in specific instances. In fact these are known as geophysical lineaments in contrast to imagery lineaments that are delineated from space imageries. Geophysical lineaments are always preferred for any subsurface exploration programmes as they also contain information depthwise.

(i)   **Airborne Magnetic Map, Jasdon Plateau, and Groundwater Potential**

A regional airborne total intensity magnetic map is given in Fig. 9.17 (Misra, 2008) recorded along 2 km spaced profiles. It shows in general magnetic highs in the western and the SE parts related to volcanic plugs and lows in the central part related to the Jasdon plateau. They are separated by sharp gradients on either sides of the Jasdon plateau indicating that it is fault controlled specially towards the NW, SW and SE. These gradients approximately coincide with the gravity gradients along the NW, SW, and SE margin of the Jasdon plateau (Fig. 9.5). They also partially coincide with the imagery lineaments L1 and L4 (Fig. 9.4). Besides these faults, there are some more lineaments inferred from the airborne magnetic map (Fig. 9.18, Misra, 2008) that should be investigated on ground for their groundwater potentiality using resistivity method. One of these lineaments along the western margin of the Jasdon plateau (L4, Fig. 9.4) east of Rajkot that is also reflected in the airborne magnetic lineament (Fig. 9.17 and 9.18) is followed up on ground by resistivity method as described below that indicated good potential for groundwater. It may be interesting to note that the airborne magnetic low due to Jasdon plateau coincides exactly with the Bouguer anomaly low (L1, Fig. 9.5) due to Jasdon plateau suggesting that the Jasdon plateau at subsurface level is occupied by low density and low susceptibility rocks that may be in form of a felsic intrusive. It is also confirmed by the fact that the gravity low, L1 (Fig. 9.5) observed over the Jasdon plateau can not be accounted by isostasy (Section 9.1.3) alone and a part of this low that gets reflected in the residual anomaly (Fig. 9.8) is caused by some other sources in the crust. This low density felsic intrusive might have produced the Jasdon plateau due to its buoyancy. The airborne magnetic highs in the NW, the SW and the SE parts of Saurashtra are due to mafic intrusive as volcanic plugs similar to gravity highs (Fig. 9.5).

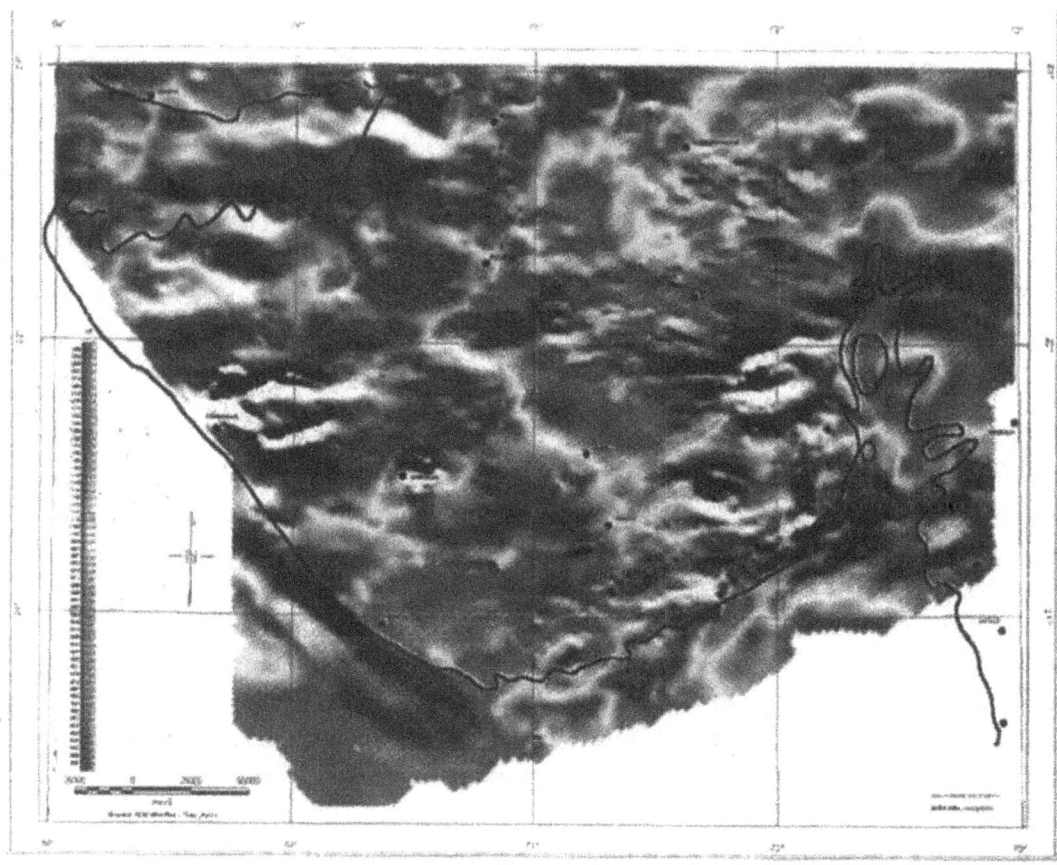

Fig. 9.17 Airborne total intensity map of Saurashtra (Misra, 2008) showing magnetic highs in the western and the SE part related to volcanic plugs (Fig. 9.3) and magnetic low over the Jasdon plateau that coincides with the gravity low (L1, Fig. 9.5) and might be related to subsurface felsic intrusive which might be responsible for the plateau uplift in this part of Saurashtra.

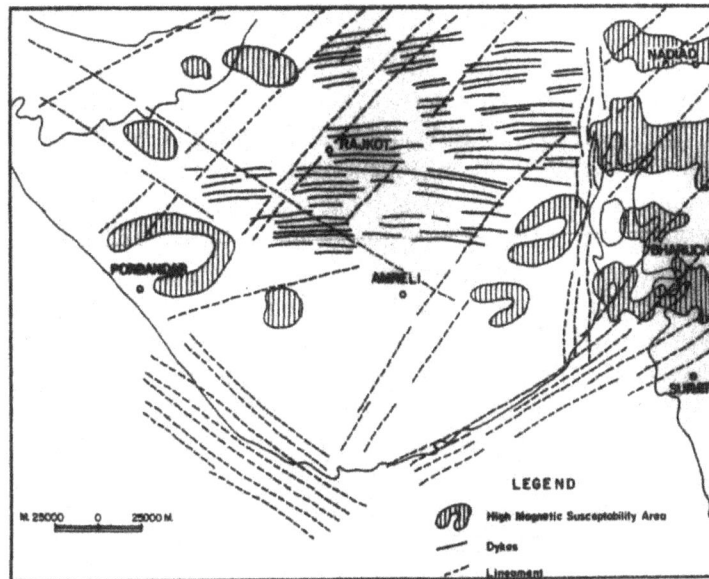

Fig. 9.18 Major magnetic lineaments from airborne magnetic map of Saurashtra (Misra, 2008)

(ii) Gravity and Landsat Lineaments and Groundwater Potential

The Bouguer anomaly map (Fig. 9.5) shows some of the lineaments delineated from imageries such as L1 and L4 (Fig. 9.4), along the NE, SE and the NW margins of the Jasdon plateau that are also reflected in the airborne total intensity map as described above. The lineament, L4 (Fig. 9.4) east of Rajkot was selected for groundwater investigations using resistivity method on ground as this lineament is reflected in all three maps, viz. lineament, airborne magnetic and Bouguer anomaly maps. A typical deep resistivity sounding across L4 is presented in Fig. 9.19 that fits to three layer models with resistivities of 210, 750, 75, and 550 ohm-m at depth of 0, 10, 18, and 458 m, respectively. This indicates that the conductive layer with a resestivity of 75 ohm-m at a depth of 18 m is water bearing and should be drilled for groundwater potential. Similar results were reported from most of the lineaments (Rao, 2007).

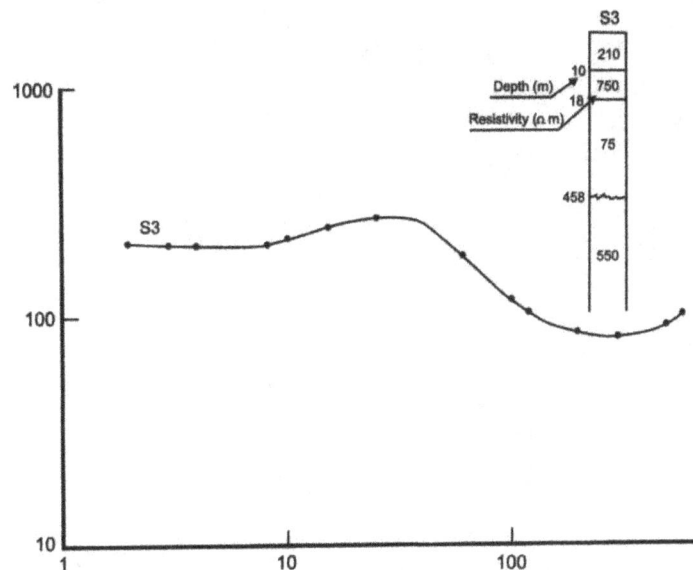

Fig. 9.19 Resistivity sounding across lineament L4 indicating third layer of resistivity of 75 ohm-m at a depth of 18 m to be conductive and favorable for groundwater potential (NGRI, 1998; Rao 2007).

(iii) High Resolution Airborne Magnetic Map of Central Botswana- Karoo Volcanics

A high resolution airborne magnetic map of central Botswana covered with Karoo volcanic similar to those found in Saurashtra is given in Fig. 9.20 (Reeves, 2005, 2010) that shows several interesting features and lineaments suggesting the application of modern day high resolution airborne magnetic maps compared to regional maps (Fig. 9.17). In fact, this map has been extremely useful to delineate several groundwater potential zones in this drought prone region (Reeves, 2010). Zeil et al., (1991) have used airborne magnetic map of SE Botswana to delineate lineaments that were found to be useful for groundwater exploration. They have shown that satellite imageries could not reflect the lineaments due to sand cover that were delineated based on airborne magnetic images and followed up on ground by using resistivity method which yielded good groundwater potential. A high resolution heliborne map of Indian Peninsular Shield has also reflected several magnetic lineaments (Fig. 9.60b) that were not visible on normal airborne magnetic maps. Such high resolution airborne magnetic maps can be used for multiple purposes like exploration of groundwater, minerals and hydrocarbons, in case of sedimentary basins. This high resolution airborne magnetic map of central Botswana shows several NW-SE magnetic lineaments in the central part related to the Karoo volcanic that are abutted against a WNW-ESE magnetic high in the SE corner. It is interesting to observe that both these major trends of Botswana correspond to NW-SE trend of the Indian Shield, and WNW-ESE trend of the Narmada-Son Lineament, and Satpura Mobile Belt indicating a similarity between the major structural trends of the two continents. These observations suggest that these trends formed/reactivated when these continents were joined together before their break up. Karoo volcanic in Africa relates to the breakup of Africa and India during the Jurassic period. The second horizon of igneous intrusion obtained from the Lodhika borehole (Section 9.1.4) in Saurashtra and shown in Fig. 9.14 may be contemporary to the Karoo volcanic. However, it can not be ascertained for certain unless these rocks are dated.

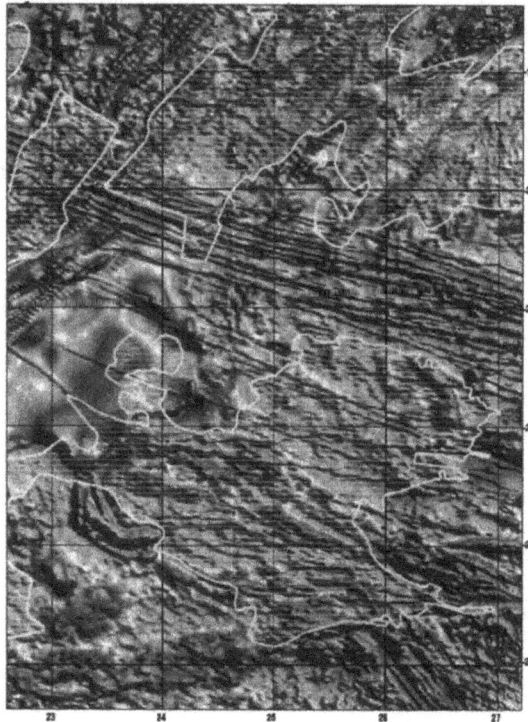

Fig. 9.20 High resolution airborne magnetic map of central Botswana covered by Karoo Volcanics (Reeves, 2010) that shows several lineaments and structures which can be used for multiple exploration programmes in this region (Reeves, 2010). It shows major NE-SE lineaments that can be explored for groundwater potential. [Colour Fig. on Page 809]

9.2.6   Geodynamic Studies – Moho Configuration and Volcanic Plugs with Palaeomagnetic and Geochemical Studies

Geodynamics of a region also influences the exploration strategy. Therefore, the geodynamics of Saurashtra based on the same data sets that were acquired for the hydrocarbon exploration and some additional data acquired for this purpose are discussed below.

(i)   Regional Gravity Field and the Moho Configuration

The regional gravity field representing deep seated sources is primarily used for geodynamic studies. Therefore, this field along a deep seismic sounding profile, Navibandar-Palitana and east of it is modeled due to Moho configuration (Fig. 9.21, Singh et al., 2003) using seismic constraints (Kaila et al., 1980) that suggest crustal thickness varying from 35 km along the coast to 42 km under the Jasdon plateau near Amreli. The computed density Moho approximately coincides with the Moho provided by the seismics for a density contrast of + 0.15 g/cm^3 implying a density of of 3.15 g/cm^3 for the thickened part of the crust for an upper mantle standard density of 3.3 g/cm^3. This implies an underplated crust in this region that is typical of volcanic provinces world over.

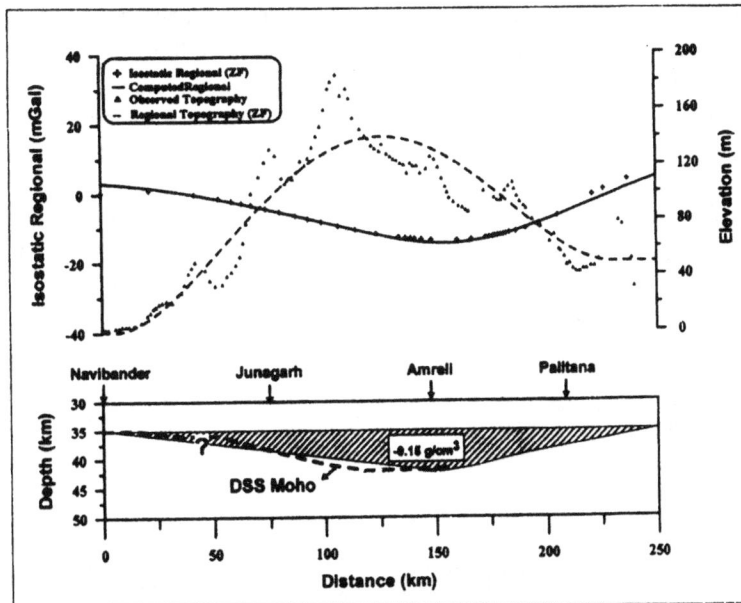

Fig. 9.21 Isostatic regional gravity profile, Navibander-Palitana (Fig. 9.7) and modeled Moho configuration and its comparison with seismic Moho showing 35 km thick crust along the western coast at Navibander increasing to ~ 42 km at Amreli under the Jasdon plateau. It shows a density of 3.15g/cm^3 for thickened part of the crust indicating underplated crust (Singh et al., 2003).

(ii)   Ground Magnetic Survey and Volcanic Plugs

As described above the circular/semicircular gravity anomalies of Saurashtra represent volcanic plugs of Deccan trap origin. Volcanic plugs originate from upper mantle and intrude almost vertically through openings in the crust. Therefore, they are of great significance for geodynamics. As volcanic plug consist of mafic rocks, that are magnetic in nature; magnetic method is most suitable to map them depthwise. These volcanic plugs are reflected in the airborne total intensity map of India (H11, L11; Fig. 3.30) that are spread in the NW part to SE part of Saurashtra. To define these volcanic plugs separately ground magnetic survey was carried out along roads and tracks at approximately 1.5-2.0 km interval using proton precision magnetometers with an accuracy of 1 gamma in parts of Saurashtra that are covered by volcanic plugs. The recorded data is corrected for diurnal variation and the IGRF and the resulting total intensity map is shown in Fig. 9.22. As the exposed rock is Deccan trap, there is considerable variation in the magnetic field due to lateral inhomogeneity of the magnetic minerals. However, the total intensity magnetic map (Fig. 9.22) depicts well defined sets of magnetic "lows" and "highs" marked as A, B, C, D, E and F corresponding to gravity 'highs' of volcanic plugs marked by the same symbols in Fig. 9.5. This clearly indicates that the magnetic characteristics of volcanic plugs are different from the Deccan trap encompassing them. Normally, a body magnetized in the present day geomagnetic latitudes for this area (~30^0N) will produce a strong magnetic 'low' accompanied by a small high towards the north due to induced magnetization and the amplitude of magnetic high will be approximately 50% to that of the magnetic low. However, in the present context, the nature of magnetic anomalies shows large magnetic low accompanied by a magnetic 'high' towards the south suggesting the presence of a strong remanent magnetization corresponding to their period of evolution namely the Deccan trap volcanism. For example the Junagadh plug show a magnetic 'low' of approximately –500 nT towards the north with an accompanying 'high' of 200 nT towards the south. To confirm the effect of remnant magnetization, a map of total magnetic field reduced to pole was prepared (Fig. 9.23) which still shows pairs of magnetic high and lows over the plugs but they are in reverse order compared to the total intensity map. In case of induced magnetization, the field reduced to pole should show a simple high over these plugs. Thus, the presence of +ve and –ve centers in opposite order in pole reduced field confirms the role of remnant magnetization in producing these anomalies. This map (Fig. 9.23) also shows NE-SW trend in the S-E part of Saurashtra which corresponds to exposed basic dykes and volcanic plugs inferred on the basis of the gravity anomalies in this region. The other prominent trends is E-W in the western Saurashtra, which correspond to volcanic plugs of Junagadh, Barda and Alech.

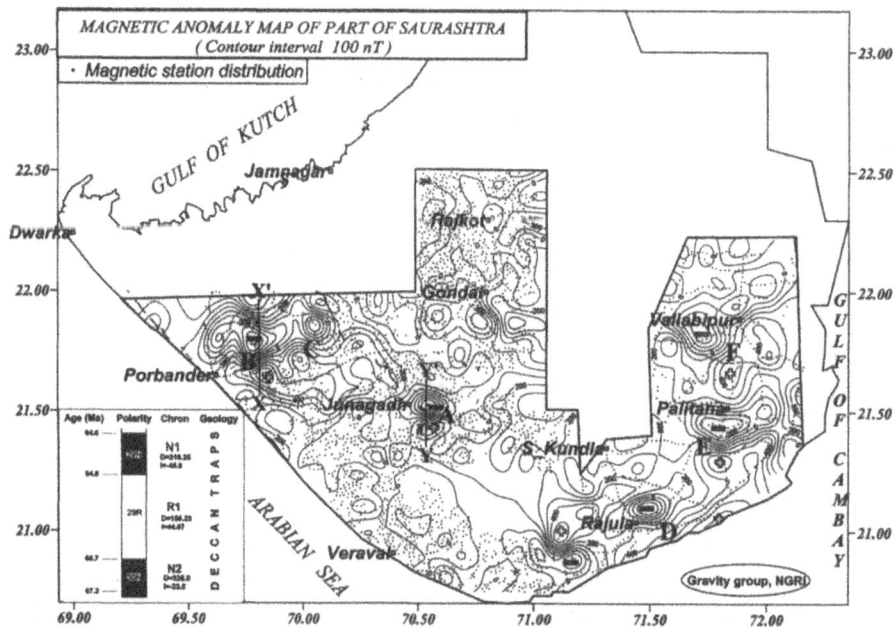

Fig. 9.22 Total intensity magnetic map of Saurashtra showing pairs of magnetic highs and lows corresponding to volcanic plugs/stocks A-F as shown in Fig. 9.3. Inset shows the average directions of magnetization reported for Deccan trap and discussed in the text. XX' and YY' are profiles used for modelling the magnetic fields observed over the plugs (Mishra et al., 2001).

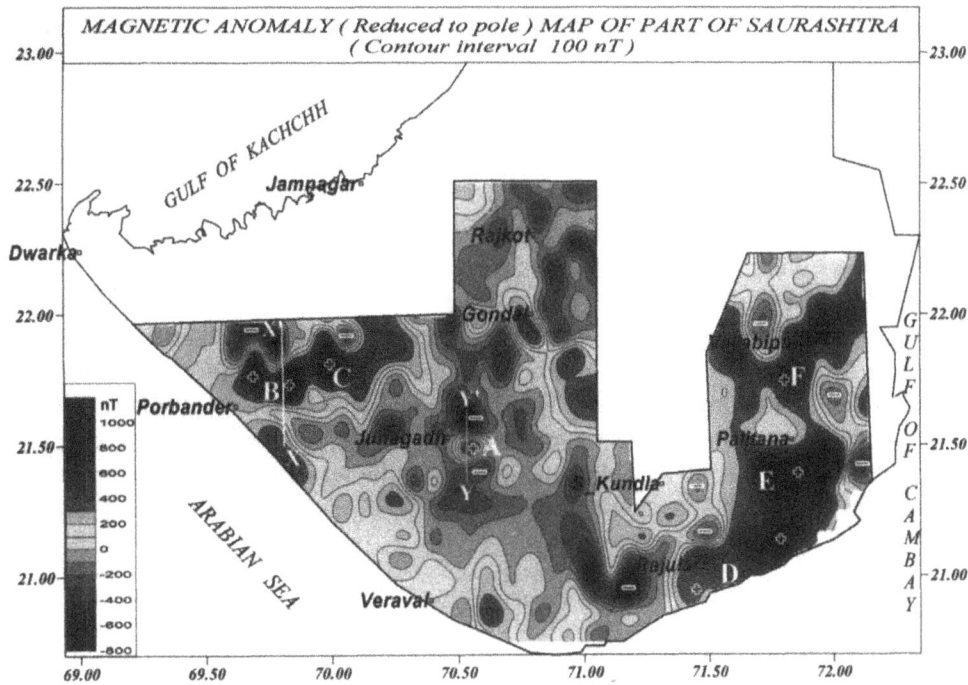

Fig. 9.23 Pole reduced magnetic map of Saurashtra showing pairs of magnetic highs and lows corresponding to magnetic lows and highs respectively of volcanic plugs/stocks of Fig. 9.22 (Chandrasekhar and Mishra et al., 2002).

(iii) Palaeomagnetic Studies

To model the magnetic anomalies discussed above, magnetic properties including palaeomagnetic directions of the exposed rocks surrounding these volcanic plugs were measured in the laboratory. Basic principles of remnant magnetization and paleomagnetic studies are described in Section 3.2. Fifty oriented block samples from 10 sites (I-X, Fig. 9.3) of standard dimensions were used for this study. Natural Remnant Magnetic (NRM) directions and intensity of these samples were measured using Schonstedt spinner magnetometer (Model DSM-2) and susceptibility was investigated using hysteresis and susceptibility apparatus (Likhite and Radhakrishnamurty, 1965). Alternating Field (AF) and thermal demagnetisation of these specimens were carried out to know the primary magnetic direction in them using AF demagnetizer (Creer, 1959) and a Schonstedt Thermal Demagnetizer (Model TSD-1). Hysteresis properties were studied using an electromagnet that can generate DC fields up to 1.5 T. Susceptibility variation with temperature (K – T) in the low temperature region was investigated by cooling the specimens in liquid nitrogen. The K – T observations reveal Multi-Domain (MD) and Cation-Deficient (CD) mixed domain state of grains of the magnetic mineral in these rocks. These properties reveal stable and reliable remnant magnetism in these rocks.

The mean remnant magnetic directions of the volcanic plugs and Deccan trap of the Saurashtra obtained from these measurements and some reported earlier (Fig. 9.24(a)) along with the computed Virtual Geomagnetic Pole (VGP) positions and corresponding statistical parameters are summarized in Table-9.2 (Chandrasekhar et al., 2002). The corresponding VGP for the mean magnetic directions of volcanic plugs and the Deccan traps of Saurashtra (Table-9.2) are plotted on the synthetic Apparent Polar Wander Path (APWP) for India from mid-Cretaceous to the present (Besse and Courtillot, 1991) (Fig. 9.24(b)) along with VGP's corresponding to the magnetic Chrons N2-R1-N1 of the Deccan trap, which are well constrained in age as described above (Vandamme et al., 1991). The VGP for the mean direction of magnetization of the Saurashtra is located close to the VGP for Chron N2 with intersecting circles of confidence. This indicates that the eruption of these volcanic plugs and the Deccan trap of Saurashtra correspond to the initial phases of eruption of the DVP. This is also in agreement with the older ages reported for the volcanic plugs of this area as discussed above. The mean direction of magnetization for Deccan trap of Saurashtra (Table-9.2) provides a palaeolatitude (λm) of 33° S for the Indian plate indicating its position in the southern hemisphere during eruption of these rocks. Table-9.2 shows that in most of the cases for rocks from Saurashtra, the inclination of the remnant magnetization is greater than 50° and average is 48°. Even in case of other volcanic plugs from northern part of DVP such as Barmer napheline syenite, Phenai Mata syenite (9, Inset, Fig. 9.2), Ambadongar carbonatite (Poornachandra Rao et al., 2000), the inclination of remnant magnetization varies from 50°-60°. This is in contrast with the reported inclination of remnant magnitization for rocks from southern part of the DVP, which in most of the cases is less than 50° (Vadamme et al., 1991). This indicates that India was occupying a more southerly position while eruption of the northern part of the DVP in Saurashtra compared to its southern part.

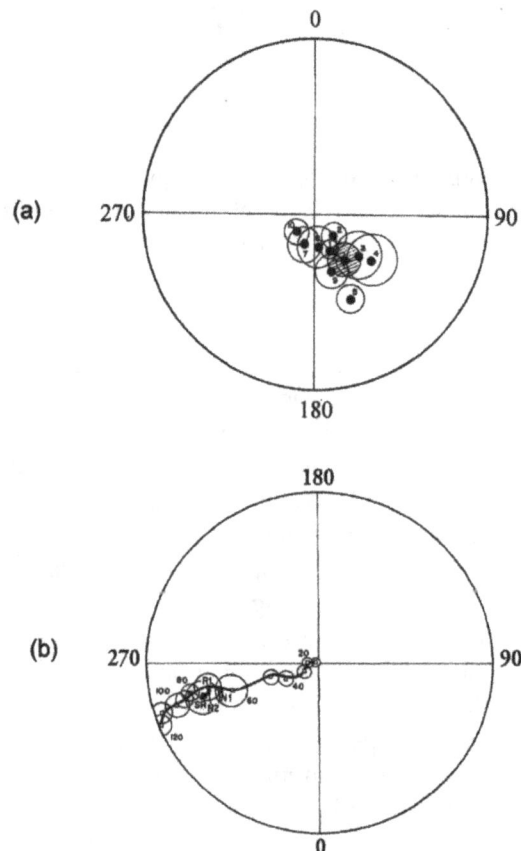

Fig. 9.24 (a) Stereographic plot of site mean magnetic directions of plugs studied from Saurashtra. 1-5 and 7-10 correspond to sites I-V and VII-X (Fig. 9.3). The circles denote confidence limits. Mean of I-X site is shown with horns and hatching. (b) Apparent Polar Wander Path for the Indian plate from mid Cretaceous, (120 Ma) onwards upto 20 Ma (Creer, 1959). Squares represent the apparent pole position and circles around them represent 95% confidence level. VGP's derived from the magnetic directions N2, R1 and N1 of the Deccan traps belonging to Chrons 30N, 29R and 29N, respectively and for the mean direction of magnetization of Deccan trap from Saurashtra (SR) is shown as dots with circles around them indicating 95% confidence level (Chandrasekhar and Mishra et al., 2002).

Table 9.2 Summary of Palaeomagnetic Results of Exposed Rocks from Saurashtra

Locality	Elevation (m)	Sites	Dm °	Im°	K	α_{95}^0	λp^o N	Lp^o W	Age/ References
Deccan Traps (Rajkot)	60-120	3	141	+ 42	-	-	29.9	67.6	[Mital et al., 1976]
Junagadh Plug (A, Fig. 1) Reverse Normal	60-110 140-753	4 14	137 336	+52 -39	- 43.0	- 5.0	22.1 40.4	71.2 79.5	68.6 Ma [Verma and Mital, 1972]
Deccan traps (Saurashtra) Sites I-X, Fig.1	<120	7	147	+56	22.3	11.2	24.2	80.1	Present study
Mean Direction	-	28	326	-48	67.1	8.5	30.0	74.3	-

Dm, Im = Mean Declination and Inclination

K = Precision Parameter = $\frac{N-1}{N-R}$; where, N = Number of samples, R =Length of unit magnetic vector.

α_{95}^0 = Radius of circle of confidence at 95% probability level Fisher, 1953

$\lambda p, Lp$ = Latitude and Longitude of the Virtual Geomagnetic Pole (VGP) (Fisher, 1953)

(iv) Geochemical Studies of Volcanic Plugs

Geochemical studies of major, trace and rare earth elements (REE) of 16 samples representing granophyre and alkaline rocks from the six plugs of western Saurashtra (A, B, C, Fig. 9.3) and southeastern Saurashtra (D, E, F, Fig. 9.3), respectively, were analysed. The range of values for various elements for both the groups of rocks are given in Table-9.3 (Chandrasekhar et al., 2002). The analysed granophyre and alkaline rocks show considerable variation in major and trace elements. The silica and total alkali plot of granophyres show tholeiite composition and range from dacite, trachydacite and rhyolite in composition (Fig. 9.25). The alkaline rocks represented by syenites mostly from plugs in SE Saurashtra exhibit trachyte composition. All the rock types show a calc-alkaline differentiation trend.

Table 9.3 Geochemical Data of Granophyres and Alkaline Rocks

Wt (%)	(1) GRANOPHYRES		(2) ALKALINE ROCKS	
	MINIMUM	MAXIMUM	MINIMUM	MAXIMUM
SiO_2	66.82	72.78	65.79	66.38
TiO_2	0.33	0.80	0.42	0.45
Al_2O_3	11.96	13.74	13.09	14.16
$Fe_2O_3^T$	3.49	5.77	5.18	5.85
CaO	0.60	2.76	0.96	1.46
MgO	0.03	0.82	0.20	0.28
Na_2O	2.80	5.12	5.11	5.77

Table 9.3 *Contd...*

K$_2$O	4.15	5.69	4.72	5.50
P$_2$O$_5$	0.01	0.18	0.04	0.06
MnO	0.11	0.32	0.21	0.32
Sc(ppm)	2.00	11.00	<1.00	3.0
Rb	122.00	209.00	140.00	169.0
Sr	30.00	128.00	25.00	37.0
Y	50.00	106.00	57.00	94.0
Zr	155.00	1486.00	277.00	440.0
Nb	19.00	86.00	179.00	381.0
Ba	595.00	1047.00	537.00	774.0
Hf	7.00	44.00	8.00	14.0
Ta	1.00	5.00	35.00	72.0
La	34.90	61.00	76.40	113.1
Ce	71.80	127.60	168.30	214.2
Pr	8.92	15.17	15.50	20.0
Nd	33.10	51.40	47.90	70.0
Sm	7.12	11.10	8.53	11.6
Eu	1.03	1.93	1.88	2.27
Gd	2.34	3.75	7.44	10.1
Dy	6.29	15.50	9.43	13.90
Er	4.12	13.90	4.98	8.01
Yb	4.42	14.90	5.92	8.84
Lu	0.90	2.35	1.02	1.28
ΣREE	182.7	313.99	352.32	479.80

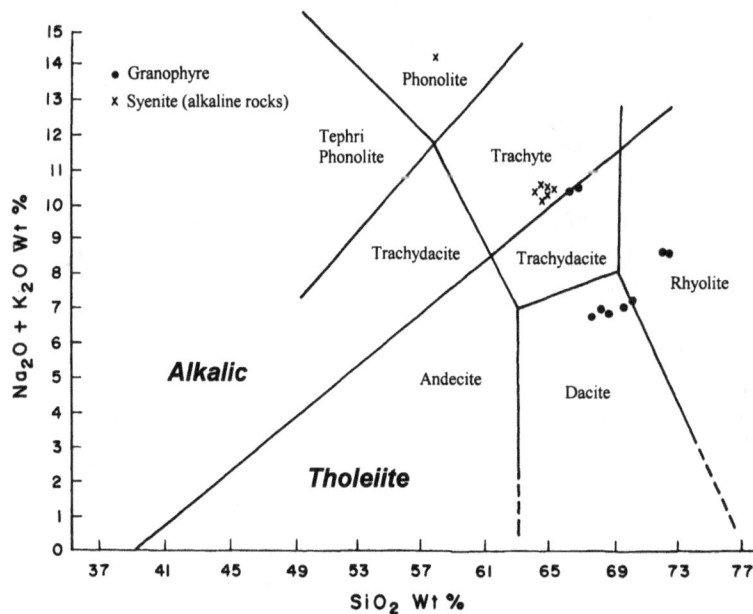

Fig. 9.25 The total alkali-silica diagram. Granophyres show oversaturated nature of silica (Chandrasekhar and Mishra et at., 2002). Demarcation lines are after LeBas et al., (2002). Demarcation line between alkalic and tholeiite field is as given by (MacDonald and Katsura, 1964).

The A, B and C plugs of western Saurashtra show comparatively higher concentration of Zn, Yb, Nb, Ba and lesser concentration of Cr, Ni, Sr, Hf, Rb and Zr compared to the plugs D, E and F in SE Saurashtra. The A, B and C plugs also show lower concentration of REE (183-314 ppm) and a strong negative Europium (Eu) anomaly (Eu/Eu* = 0.39-0.59) and higher enrichment of heavy rare earth elements compared to plugs D, E and F with ΣREE = 352-480 ppm and Eu/Eu* = 0.59-0.79 (Fig. 9.26(a)). A conspicuous enrichment of Light Rare Earth Elements (LREE) suggests that plugs of Saurashtra evolved by olivine and pyroxene fractionation of source magma. One alkaline rock from Junagadh plug (A, Fig. 9.3) of western Saurashtra show distinctly very low contents of Ni, Y, Ba and REE and silica (X, Fig. 9.26(b)) compared to alkaline rocks of SE Saurashtra (Y, Fig. 9.26(b)). This suggests differences in their genesis and evolutionary history of Junagadh plug in western Saurashtra compared to those in SE Saurashtra that might be due to association of latter in SE Saurashtra with Cambay rift basin.

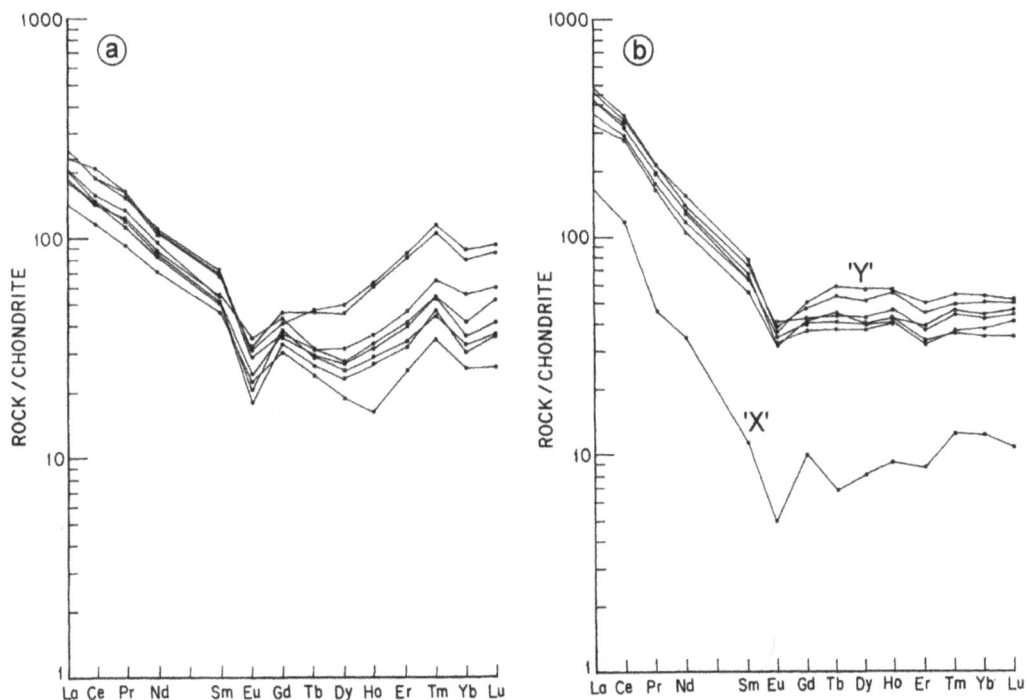

Fig. 9.26 Rare earth element patterns. Chondrite normalised values are after Sun and McDonough (1989). (a) Granophyres from plugs A, B and C (b) Alkaline rocks showing low contents of Ni, Y, Ba and REE and Silica in a sample from Junagadh plug (X) compared to those from plugs D, E and F (Y) (Chandrasekhar and Mishra et al., 2002).

Apart from these known six plugs, some more plugs/intrusive bodies have been reported from this area. One such body is the Essexite body (Sethna and Javeri, 2000), which occurs as a large stock-like intrusion in horizontal flow basalts. Chemically it is characterized by enrichment in large ionic lithosphere elements, high field strength elements and rare earth elements as compared to the other tholeiitic magma predominantly encountered in the Deccan volcanic province. It may represent a separate melt derived as a 5 to 10% partial melt at shallower depth within the upper mantle (Sethna and Javeri, 2000). The present study and earlier studies (Chandra, 1999) suggest that the layered gabbro and cumulates in plugs of western Saurashtra are considered to be a result of fractional differentiation of tholeiitic/olivine tholeiitic magma in a shallow level undisturbed and slowly cooling condition of a magma chamber. The alkaline plugs of SE Saurashtra may have generated by either deriving magma from mantle source through fractionation at a deeper level under high pressure condition or magma generated in its own regime in the lower mantle. The intrusive history of the Mundwara complex (8, Inset, Fig. 9.3) (Basu et al., 1993) which lies in the northern part of the Cambay rift basin provide the established sequence of fractional crystallization of alkaline olivine basalt magma. The parent liquid composition is modelled by different degrees of batch melting of a parent mantle peridotite. The assumed starting peridotite composition is in the garnet- peridotite mineral facies and thus may reflect a greater depth of derivation of the partial melt for the alkaline complexes associated with Cambay rift basin including those in SE Saurashtra. The observed within plate granite tectonic setting (Fig. 9.27(a) and (b)) for the granophyres of plugs A, B and C of western Saurashtra suggest good examples of continental flood basalt provinces, which are generally attributed to mantle plume activity and an attenuated continental crust (Chandrasekhar et al., 2002).

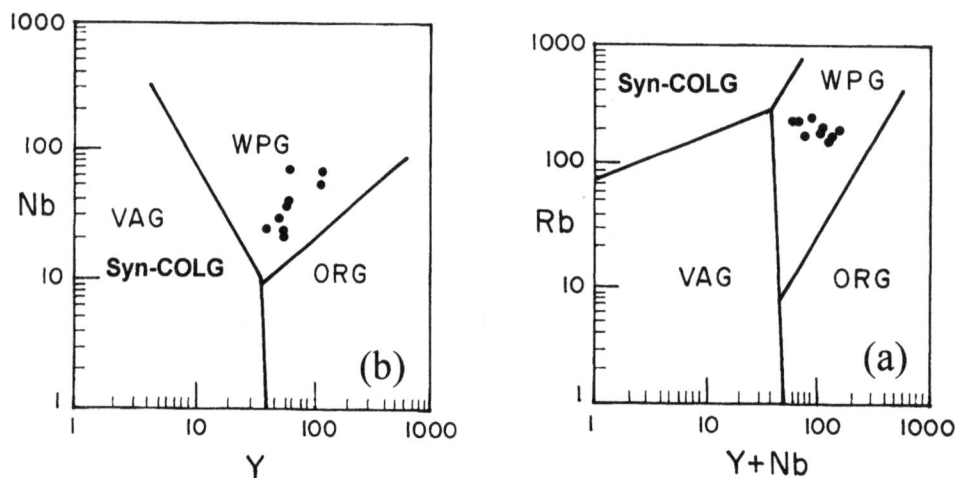

Fig. 9.27 (a) Nb-Y and (b) Rb-(Y+Nb) discrimination diagram for granophyre rocks. Demarcation fields are after Pearce et al., (1984) and Forster et al., (1997). WPG: within plate granite; ORG: ocean ridge granite; VAG: volcanic arc granite; Syn-COLG: Syntectonic collisional granites (Chandrasekhar and Mishra et al., 2002).

(v) Modeling of Gravity and Magnetic Anomalies over the Volcanic Plugs

The amplitude and the nature of gravity and magnetic anomalies over the six volcanic plugs discussed above appear to be almost same and therefore the causative sources are also likely to be similar. Therefore, two profiles XX' and YY' (Fig. 9.5 and 9.22) across the Barda and the Junagadh plugs, respectively, are chosen for detailed modeling. The gravity and the magnetic profiles (Fig. 9.28(a) and (b)) show gravity highs and magnetic low-high pairs over these plugs. The average density of exposed rock types found in these plugs varies from 2650 to 3200 kg/m³ for acidic and mafic/ultramafic rocks with an average density of 2920 kg/m³. The average density of upper crust is taken as 2700 kg/m³ while that of lower crust is 2900 kg/m³. Therefore, the density contrast, producing gravity anomalies in these cases, is mainly confined to the upper crust, which explains the observed gravity anomalies. As the gravity and magnetic anomalies are observed over the same sources the magnetic field is computed for the body defined by gravity anomalies for the magnetic characteristics of rocks as reported above from laboratory measurements. The gravity and the magnetic fields are computed using the expression for 2 ½ dimensional bodies (Webring, 1986), which modify the expression for the gravity and magnetic fields due to 2 dimensional bodies (Talwani et al., 1959) to account for the limited extent in the horizontal direction (Rasmussen and Pedersen, 1979). The bulk characteristics of the rocks are modified to provide a good match between the observed and the computed fields. The computed field and the subsurface model along profiles XX' and YY' are given in Fig. 9.28(a) and (b), which also provides its density and magnetic characteristics. The density contrast ($\Delta\sigma$) of the volcanic plugs with surrounding rocks is approximately +180 to +200 kg/m^3, which extends across the upper crust implying that the bulk density of the volcanic plugs are approximately 2880 and 2900 kg/m^3. The plugs, however, may be extending further below into the lower crust, but they will not produce any measurable anomaly due to lack of density contrast in that part of the crust. The susceptibility is 3.14×10^{-2} S.I. Units with a Koenigsberger ratio of 0.56, which is within the limits of laboratory measurements for the basic and the ultrabasic rocks from these plugs as discussed above. However, the bulk direction of remnant magnetization, which explains the observed field is the normal magnetization of D = 337° and 340° and I = – 38° and – 50°. The reverse direction of magnetization measured from the exposed rock sample reported above cannot explain the observed magnetic anomaly that might be confined only in the upper exposed part while the direction of magnetization inferred from modeling of the magnetric anomalies represent the bulk direction of magnetization of the volcanic plugs.

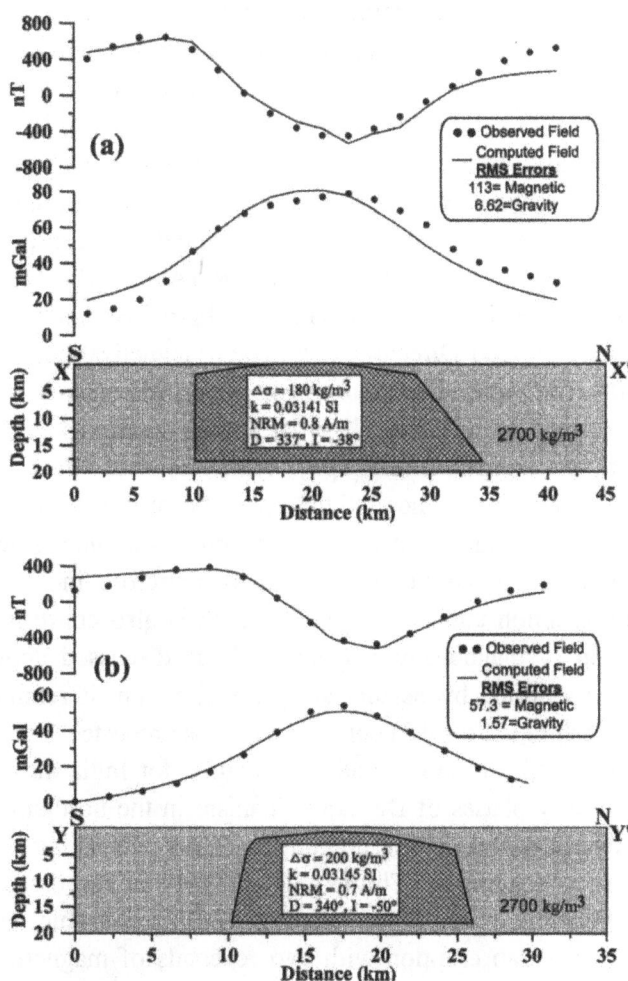

Fig. 9.28 Gravity and magnetic fields along profiles (a) XX' and (b) YY' across the Barda and Junagadh plugs (Fig. 9.5 and 9.22) and the computed fields for the models given at the bottom of the figures and their density and magnetic properties. $\Delta\sigma$ is the density contrast with respect to surrounding rocks (NGRI, 1998; Chandrasekhar and Mishra et al., 2002).

(vi) Geodynamics Model of Saurashtra

Saurashtra provides a unique opportunity to study various aspects of a large volcanic province and its effect on underlying crust and lithospheric mantle. Crustal thickness in Saurashtra changes from 35 km along the west coast to 42 km under the Jasdon plateau towards the east with a high density underplated crust that is typical of volcanic provinces. The gravity low of the Jasdon plateau cannot be accounted for by crustal thickness alone, indicating low density crustal rocks (Fig. 9.14) and is also characterized by magnetic lows (Fig. 9.17) indicating low susceptibility rocks. The subsurface low density and low susceptibility rocks under the Jasdon plateau indicate felsic intrusives that might be responsible for its uplift forming a plateau. Saurashtra has also experienced some seismic activity in recent times (Fig. 9.5) that is mainly confined along the margins of the volcanic plugs that serve as stress concentrators. Uncompensated gravity low of the Jasdon plateau causes overcompensation and related stress. These two factors in combination might be responsible for present day seismic activity in this region as has also been discussed in the case of the Bhuj earthquake of Kachchh in Chapter 8.

The volcanic plugs of Saurashtra and the adjoining region generally represent igneous intrusions and alkaline complexes with widely varying composition from mafic/ultramafic rocks in case of the Junagadh plug (A) to acidic rocks in case of Barda and Alech (B and C) in western Saurashtra and alkaline rocks in case of plugs in SE Saurashtra (D, E, F, Fig. 9.3). Various dates referred to above suggest them to be around 68.5 Ma, while the main Deccan trap activity took place around 65 Ma. This indicates them to be about 3.5 Ma older than the main Deccan trap activity, which may be regarded as incubation time. Modeling of gravity and magnetic anomalies due to volcanic plugs of Junagadh and Barda (A and B, Fig. 9.3) provide the following bulk properties of rocks: Density: 2880 and 2900 kg/m^3, Susceptibility: 3.14×10^{-2} SI Units, Koenigsberger ratio: 0.56, and Direction of remnant magnetization: D = 337° and 340° and I = −38° and −50°. The bulk direction of magnetization shows a normal polarity (340°, − 50°) which may correspond to either N2 or N1 corresponding to magnetic Chrons 30N or 29N respectively on the magnetostratigraphy scale reported from the Deccan trap (Vandamme et al., 1991). However, as most of the magnetic observations are taken at lower levels it is likely to correspond to the normal magnetization of the older flows underlying the volcanic plugs which may belong to Chron 30N. This study, therefore, indicates the presence of earliest phases of Deccan eruption under these plugs which is also confirmed from the VGP position for mean direction of magnetization (Table-9.2) as discussed above and shown in Fig. 9.24(b). This is also supported by usually steeper inclination of remnant magnetization (+50°) for rocks from Saurashtra (Table 9.2) compared to those reported from the southern part of DVP. Steeper inclinations indicate more southerly position for India during the eruption of these rocks and therefore, early phases of Deccan volcanism in the northern part compared to the southern part. The subsurface rocks corresponding to Chron 30N along with exposed rocks from the Junagadh plug corresponding to Chrons 29R and 29N as discussed above similar to those given for DVP (Vandamme et al., 1991) suggests that this volcanic plug constitutes the rocks of total span of Deccan trap eruption with two reversals of magnetic field during this period. A Curie temperature of 500-580 °C suggests the presence of magnetite and titanomagnetite in mafic/ultramafic rocks of these plugs. The volcanic plugs of the SE part of Saurashtra are related to fracture/faults, which form the western margin of the Gulf of Cambay and the Cambay rift basin.

The geochemical data of granophyres from the plugs of western Saurashtra suggest that all of them formed within plate setting and show a continental tectonic setting. The presence of alkaline rocks in volcanic plugs of Saurashtra suggest deep seated products and their origin due to a calc-alkaline differentiation of alkaline magma generated by high pressure fractional crystallization of mantle source. As discussed earlier most of the alkaline plugs follow the deep seated Archaean fractures, which confirm their origin to be deep seated. Further, the greater depth origin of alkaline rocks from the volcanic plugs in the SE part of Saurashtra is also confirmed by the fact that they form the western margin of the Gulf of Cambay and the Cambay rift basin (Mishra et al., 1998). The alkaline complexes in SE Saurashtra along the western margin of the Cambay basin and other alkaline complexes of the Cambay basin such as Mundwara complex discussed above (Inset, Fig. 9.3) may, therefore, appear to represent the early product from parent magma. Partial melt models indicate that the mantle beneath the Deccan Volcanic Province began to melt as early as 68.5 Ma at a greater depth to form the alkaline parent magma while by 65 Ma, the mantle had melted extensively by a larger degree of partial melting at a shallow level, perhaps in the stability field of spinel peridotite, to generate the voluminous tholeiite magma for the main phase of Deccan trap eruption. These rocks appear to be the products of basaltic differentiate or erupted independently from three different magmas of gabbroic, rhyolitic or syenitic composition. The large range of variation in the chemical composition of rocks from the Deccan trap province suggest that this volcanic eruption from the Re-Union plume might have taken place close to the mid-oceanic ridge that was really so when this plume was located close to the Central Indian Ocean Ridge. Due to this eruption the this ridge is supposed to have jumped to the Arabian sea as described in Section 5.4. This is also evident from low value of the effective elastic thickness of region covered by Deccan Volcanic Province on the continent and Chagos-Laccadive ridge in the Arabian sea (Table 4.1). Moreover, the Deccan Volcanic Province is associated with several continental rift basins over the Indian continent like Cambay rift, and Narmada rift that would also modify the magmatic signatures in the neighbouring region resulting in variations in the chemical composition of rocks of this province. Variation in the chemical composition of the Deccan Volcanic Province and some other characteristics have made some scientists suggest that it may represent rift related volcanism from the upper mantle (Chandrasekharam et al., 1999; Mahoney et al., 2000).

9.3   Kachchh – Mesozoic Sediments and Basement Model

Kachchh located north of Saurashtra, is also occupied by Mesozopic sediments (Fig. 8.7). Its geodynamics and tectonics along with various geophysical data have been described in great detail in Chapter 8.

9.3.1 Bouguer Anomaly and Basement Model

The Bouguer anomaly and the regional and the residual gravity anomalies of Kachchh are given in Fig. 8.9, 8.11, and 8.12, respectively. Its Bouguer anomaly is further reproduced in Fig. 9.31 (a) as part of the gravity map of NW India for a comprehensive picture that also shows several semicircular/circular gravity anomalies related to plugs of Saurashtra and Kachchh. Inspite of the exposed Mesozoic sediments associated with the Kachchh Mainland Uplift, it also shows gravity highs, H1 (Fig. 8.9) that indicate that these gravity highs are related to basement structures that might be important for hydrocarbon exploration. The basement structures computed from the combined analysis of gravity, seismic and electrical methods (Fig 8.13, 8.14 and 8.15(a) and (b)) show the thickest Mesozoic sediments of ~ 4.0 km along the southern coast of Kachchh with the Gulf of Kachchh. They also show several basement faults that have caused basement uplift towards the north which have produced several highlands in Kachchh similar to horsts that are significant for hydrocarbon prospect.

9.3.2 Mesozoic Sediment Thickness Map and Hydrocarbon Prospect

Based on the basement models computed along several profiles in Kachchh through integration of gravity, seismic and magnetotelluric measurements, sediment thickness map was prepared that is presented in Fig. 9.29(a) (NGRI, 2000; Singh and Arora, 2008). This map shows maximum thickness (~ 4 km) of Mesozoic sediments in the southern part of Kachchh along the coast of the Gulf of Kachchh. Based on seismic profiles, Prasad et al., (2010) have also suggested maximum thickness of Mesozoic sediments along the coast with the Gulf of Kachchh between Mandvi and Mundra (Fig. 8.7) that might be due to the uplift of the northern part of the Kachchh Mesozoic basin due to intrusives of that period as proposed in Section 8.2.10 where oldest Mesozoic sediments of middle Jurassic are exposed. It may be mentioned here that in Saurashtra, the maximum thickness of Mesozoic sediments of ~3.0 km (Fig. 9.15 and 9.16) has also been encountered close to the Gulf of Kachchh that makes northern section of Saurashtra, Gulf of Kachchh and southern part of the Kachchh an important region for exploration of hydrocarbons. Pandey et al. (2010) have reported seismic imaging of Paleogene sediments of the Kachchh shelf and along with some drilling data suggested sea level fluctuations that has caused transgression and regression type of sedimentation that was dominated by carbonate type of sediments during this period. They delineated some stratigraphic details of Paleogene sediments that have significant resource potential.

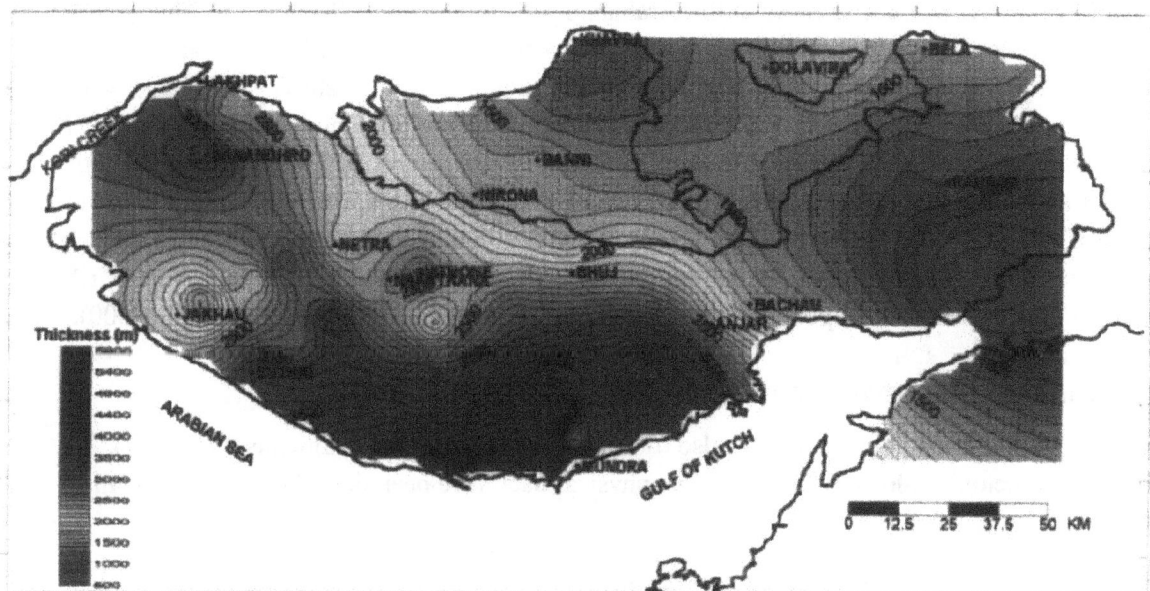

Fig. 9.29(a) Thickness of Mesozoic sediments in Kachchh (2, Fig. 9.1(a)) based on integrated investigations using gravity, seismic, electrical and magnetotelluric methods with maximum thickness of ~ 4.5 km of Mesozoic sediment along the southern coast of Kachchh with the Gulf of Kachchh (NGRI, 2000; Singh and Arora, 2008).
[Colour Fig. on Page 810]

9.3.3 Airborne Magnetic Map and Groundwater Potential

An airborne magnetic image of Kachchh is presented in Fig. 9.29(b) (DGH, 2005) that shows several magnetic highs due to uplifts and islands (Fig. 8.7) indicating their association with mafic intrusives. This map shows mainly two sets of linear magnetic highs and lows, MH1, ML1 and MH2, ML2 that form pairs due to magnetic sources in this latitude and may represent linear dyke type of intrusive. In this regard, the magnetic highs and lows, MH1 and ML1 coinciding with the Pachham island in the north and Wagad uplift in the center, are highly significant as the sharp gradient between them may represent a major fracture zone occupied by mafic rocks that may be even responsible for their uplift. Large numbers of linear magnetic anomalies of high amplitude indicate mafic intrusive in the basement. Their magnetic lows (ML1) being located north of the magnetic highs as observed in case of Saurashtra (Fig. 9.22) suggest remnant magnetization related to southern hemisphere. This gradient would be significant for groundwater potential and should be followed up on the ground by resistivity method as in case of Saurashtra to establish suitable groundwater occurrences. Kachchh is an arid region and therefore any source of groundwater would be of great importance to society. It is interesting to note that the epicenter of the Bhuj earthquake of 2001 (Chapter 8, Fig. 8.7) lies at the intersection of these two pairs of anomalies, indicating a weak zone.

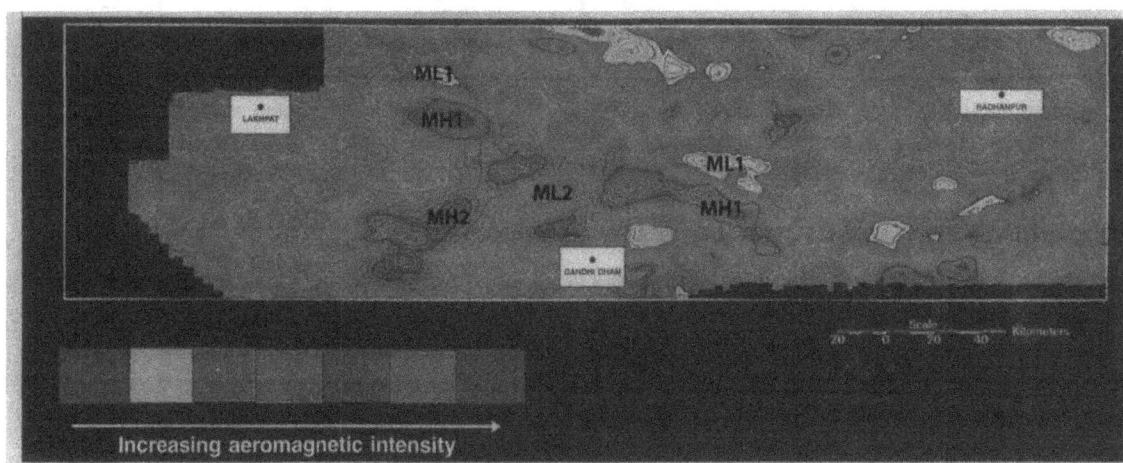

Fig. 9.29(b) Airborne total intensity map of Kachchh showing several pairs of magnetic highs and lows indicating mafic intrusions that might be responsible for various uplifts and islands of Kachchh (Fig. 8.7). There is a significant pair of linear magnetic trends, MH1 and ML1 that coincide with the Pachham island and Wagad uplift (DGH, 2005). [Colour Fig. on Page 810]

9.4   Cambay Basin and its Extension in Western Rajasthan (Barmer and Jaisalmer Basins) – Tertiary and Mesozoic Rift Basins and Geodynamics

The Cambay basin (Fig. 9.30) is a rift basin formed due to the Reunion plume during late Cretaceous (Biswas, 1987). It is one of the major petroliferous basins of India in the NW part of the Indian platform (Avasthi et al., 1971; Raju, 1983). The basin (3, Fig. 9.1(a)) is flanked by the Aravalli-Delhi fold belt on the east and northeast, by the Saurashtra peninsula on the west, and by the Deccan plateau on the southeast. The Cambay basin is regarded as one of the three major marginal rift basins, Kachchh, Cambay and Narmada from the north to the south. These basins originated sequentially between the Early Jurassic and Tertiary, during India's northward migration, after the breakup of Gondwanaland (Biswas, 1982). Based on regional gravity (Tiwari et al., 2001) and tomography study (Kennet and Widiyantoro, 1999), it is suggested that the west coast of India around Mumbai and the Cambay basin shows a low density and low velocity upper mantle, which is attributed to the trace of the Reunion plume.

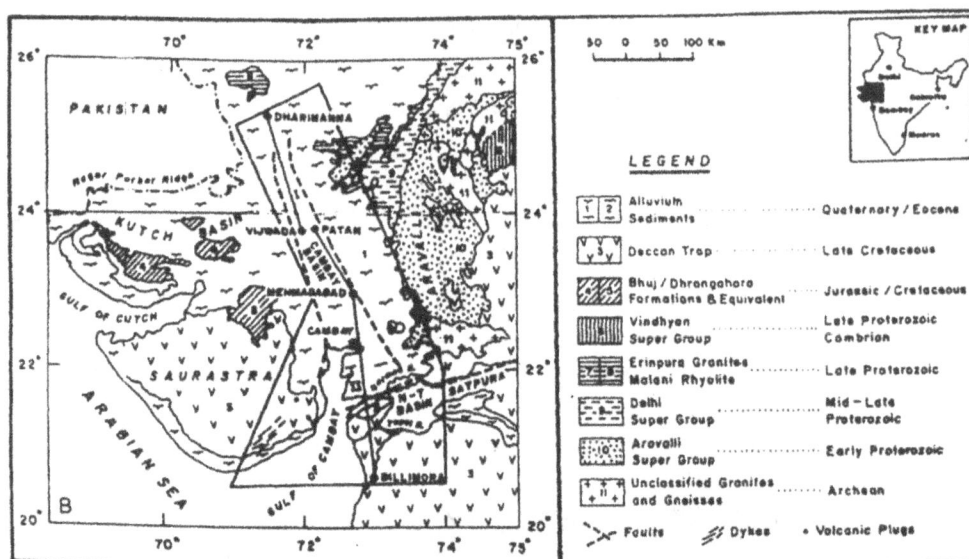

Fig. 9.30 A regional geological map of the Cambay basin (3, Fig. 9.1(a)) marked by five sided straight lines and surrounding regions showing the profile Dharimanna- Billimora. The southern part of this block north of Billimora is the E-W oriented Narmada-Tapti (NT) basin where the two are interacting with each other.

9.4.1 Bouguer Anomaly and Crustal and Basement Models

The Bouguer anomaly of Cambay basin is presented in Fig. 9.31(a) as part of the Bouguer anomaly of the NW India (Mishra et al., 2008; Singh and Mishra, 2000) showing gravity highs, H1 related to cambay besin. They extend NW and merge into lows due to sediments of this basin. The occurrences of Deccan trap below Tertiary sediments in this basin (Fig. 9.32(b)) and in surrounding regions as described above suggest it to be an active rift basin formed due to Deccan trap eruption in this section. The gravity anomalies of Cambay basin extend NW in Barmer (L1) and Jaisalmer (L2) basins (Fig. 9.31(b)) that are primarily Mesozoic basins. Gravity lows of these basins are accompanied by gravity highs, H1and H2 and H3 and H4 along their shoulders that is typical of rift basins (Sections 2.7 and 2.9). These Mesozoic basins also find reflection in the airborne total intensity map (Fig. 9.31(c)). Total intensity magnetic maps of this area were prepared based on profile (flight lines) spacing of 4 miles (~ 6.2 km) flown at an altitude of 1700 ft (~ 512 m) above mean sea level (Agocs, 1956a). These maps presented several magnetic anomalies of hundreds of gamma similar to those produced by mafic intrusive in the basement. The total intensity map could not be digitized due to sharp gradient and high amplitude of these magnetic anomalies. However, the residual magnetic anomaly maps based on 4 miles (~ 6.2 km) template, equal to the flight line spacing and approximate depth to the basement showed moderate amplitude magnetic anomalies section wise. These residual magnetic anomaly maps were digitized and merged together to form an image of the same as presented in Fig. 9.31(c) to compare with the gravity anomaly map of this region (Fig. 9.31(b)). This map (Fig. 9.31(c)) shows magnetic highs and lows, MH1, ML1 and ML2 and MH2, ML3, MH3 related to Barmer and Jaisalmer basins, respectively. Magnetic field being bipolar, magnetic sources provide pairs of magnetic highs and lows in this magnetic latitude and are therefore, spread over a lager area compared to gravity anomalies. Further in case of rift valleys, there are three sources close by, viz. basement under the basin part and shoulders on both side and therefore signals from them get mixed up and one gets sets of highs and lows of magnetic anomalies over a larger area as in the present case compared to the gravity anomalies of these basins. These magnetic anomalies of Barmer and Jaisalmer basins (Fig. 9.31(c)) also show NW-SE trend as observed in case of the Bouguer anomaly.

Besides these definite anomalies, there are gravity and magnetic trends NE-SW marked in Fig. 9.31(b) and c that coincides with the crustal and basement uplifts due to lithospheric flexure of the

Indian plate on the western side (Western Fold Belt, Pakistan) similar to those observed under Ganga basin due to Himalayan Fold Belt (Sections 6.1.10, 6.3.5). It is referred to as Jaisalmer Arch in the tectonics of this region and runs almost parallel to the boundary between India and Pakistan and separates plain lands of the Indus basin towards the west from the Rajasthan desert towards the east. This basement ridge extends to the Himalayan front giving rise to Kangra reentrant and its intersection with the Himalayan front is seismogenic giving rise to several seismic activity including Kangra earthquake of 1905 (Sections 6.1.10 and 6.3.5). Magnetic and gravity anomalies in case of NE-SW trend are low amplitude highs and lows due to depression and uplift of the basement. The gravity anomalies associated with the shoulders of the Barmer rift basin are of high amplitude (H1 and H2; Fig. 9.31(b)) while magnetic anomalies in both cases are of high amplitude as described above that are reduced to moderate amplitude in residual anomaly map (Fig. 9.31(c)) indicating them to be active Mesozoic rift basins formed due to mafic intrusion. These observations suggest the existence of a large rift basin in this region during Mesozoic period that got reactivated in the southern part during Tertiary period due to Deccan trap eruption forming Cambay Tertiary rift basin. Therefore, there appears to have been at least two episodes of volcanic eruption in this region, viz. during Mesozoic period followed by Deccan trap. The former might be related to the contemporary event of the break up of Africa from India due to Karoo volcanic. The second horizon of basic rocks encountered in Lodhika oil bore well (Section 9.2.4) may represent this magmatic intrusion.

(i) Crustal and Basement Models along Profiles across Barmer and Cambay Basins

A NE-SW profile (XX') adopted from the Bouguer anomaly map of the Barmer basin (Fig. 9.31(b)) was modeled for the crustal and basement structures. As there are no seismic constraints available in the region, it is modeled constraining from the results of crustal structures under the Cambay basin as described below based on a deep seismic profile and gravity modeling and general crustal structures for rift basins (Section 2.7 and 2.9). A tentative model of crustal structures is shown in Fig. 9.32(a) that shows Quaternary and Tertiary sediments (2.0 g/cm^3) at the top followed by Mesozoic sediments (2.4 g/cm^3) and high density basement (2.8 g/cm^3) forming the shoulders of the Mesozoic Barmer basin. High density basement indicates mafic basement due to large scale intrusion. It also shows upwelling of Moho up to 27-28 km under the basin that is typical of rift basins (Section 2.7 and 2.9). An alternative model of almost similar shallow structures and up warped part of Moho as under plated crust of density 3.1 g/cm^3 with constant Moho at depth of ~ 32 km was also worked out (Mishra et al., 2011) but in absence of any constraints from seismic in this part, it was considered to keep the model simple as presented in Fig. 9.31(b).

A N-S gravity profile from Dharimanna to Billimora (Fig. 9.30 and 9.31(a)) along the axis of the basin constrained from seismic profile along the same section was modeled. In fact, this long profile was made up of two seismic profiles from Dharimanna-Mehmadabad and Mehmadabad-Billimora (Kaila et al., 1981, 1990) across the Cambay and Narmada-Tapti (NT, Fig. 9.30) basins, respectively that were joined together to model the gravity data along this long profile. The gravity profile and the computed crustal model are given in Fig. 9.32(b) (Mishra et al., 1998) that was referred to as Global Geoscience Transect. It suggests mafic rocks (2.74 g/cm^3) as the basement under Tertiary sediments (2.3g/cm^3) that represent part of the exposed Deccan trap in Saurashtra and high density (3.1g/cm^3) layer along the Moho indicating underplated crust that has also been inferred under Saurashtra (Fig. 9.21) and is typical of volcanic provinces. It also shows certain pockets of Mesozoic sediments below the Deccan trap especially in the southern part and a low velocity layer (2.6g/cm^3) in the upper crust with a shallow Moho (~31 km) that are typical of rift basins. This profile extends further south across the Narmada basin along the east coast of the Gulf of Cambay that also shows thinning of the crust up to ~31 km indicating that this also may represent an E-W oriented rift basin and the two rift basins interact in this section. Seismic profile (Kaila et al., 1981) even shows further reduced crustal thickness of ~21 km. However, it can not account for the observed gravity field and moreover such reduced crustal thickness over the continent is logically not possible. Subsequently, the same seismic data is re-processed by Dixit et al., (2010) who provided the depth to Moho as ~ 40 km under the Narmada-Tapti section. However, this appears to be on higher side as it represents a rift basin and is located along the coast where crustal thickness is likely to be less or normal (~ 34 km).

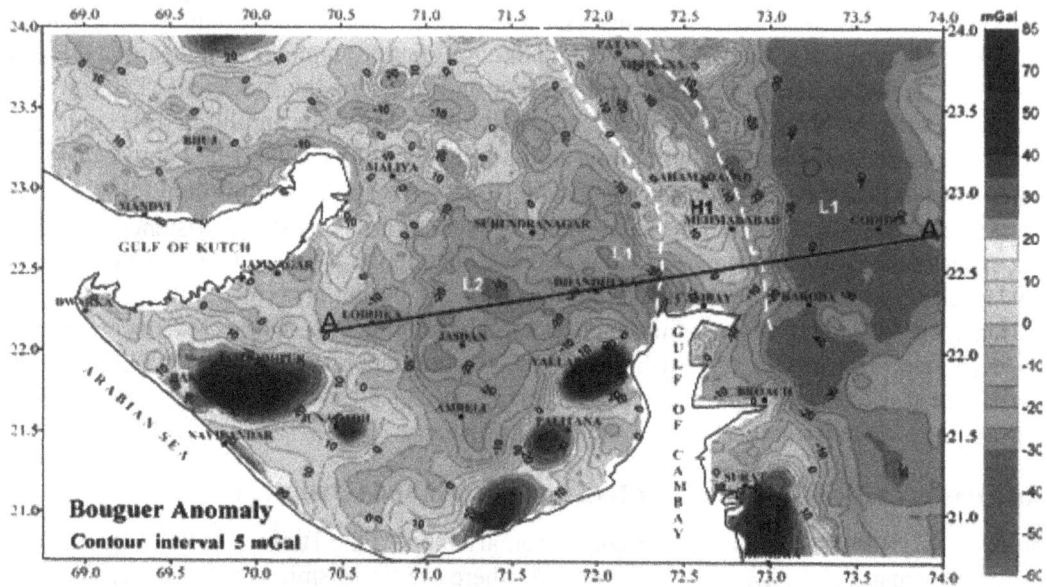

Fig. 9.31(a) Bouguer anomaly map of Cambay basin (H1) and adjoining regions. Semi circular gravity highs are due to volcanic plugs of Saurashtra and Kachchh (Mishra et al., 2008, Singh and Mishra, 2000). AA' is a long profile modeled in Fig. 9.33. [Colour Fig. on Page 811]

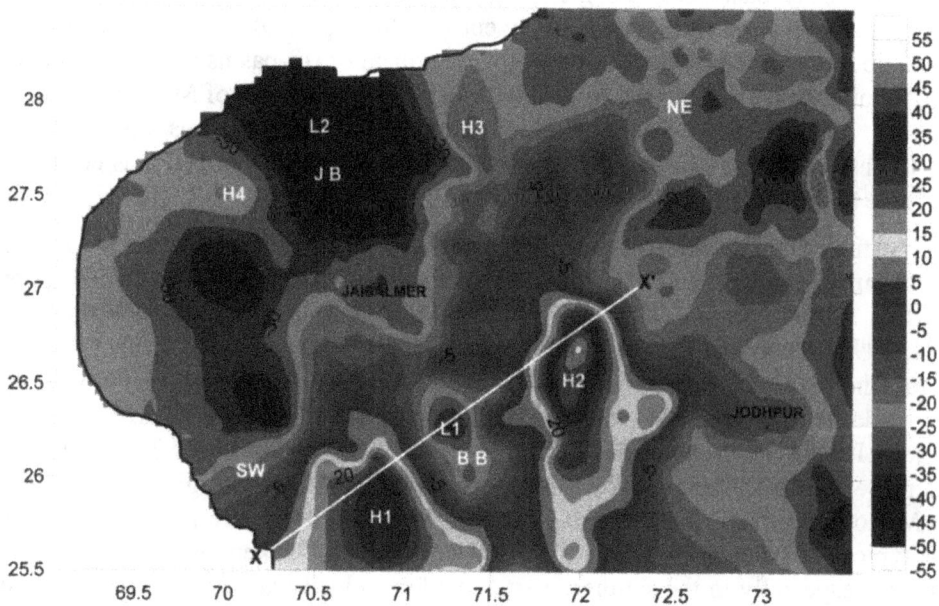

Fig. 9.31(b) Bouguer anomaly map of the Western Rajasthan showing gravity lows due to Barmer (L1) and Jaisalmer (L2) Basins and shoulder highs, H1 and H2 and H3, and H4, respectively. NE-SW gravity trend separates the two basins and appears to be of latter origin representing crustal and basement uplift due to lithoshheric flexure of the Indian plate on the western side (Western Fold Belt, Pakistan) known as Jaisalmer arch. BB – Barmer Basin; JB – Jaisalmer Basin. [Colour Fig. on Page 811]

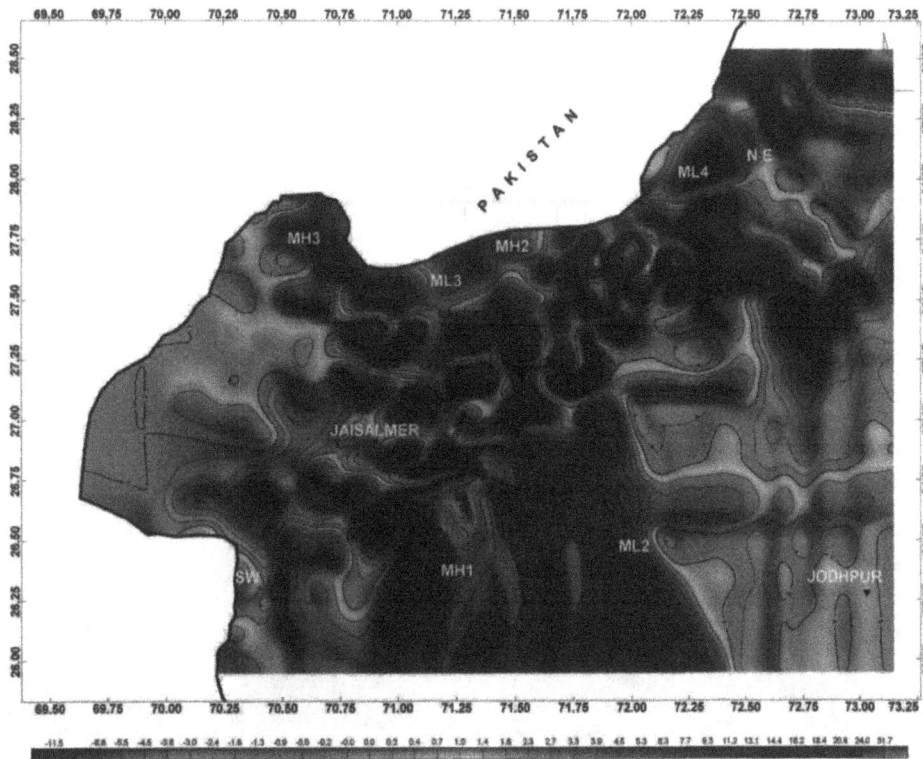

Fig. 9.31(c) Residual airborne magnetic map of parts of the Western Rajasthan based on data recorded along flight line spacing of 4 miles (~ 6.2 km) at an altitude of 1700 ft (~ 512 m) above mean sea level (Agocs, 1956a) showing magnetic anomalies due to Barmer (ML1, MH1 and ML2) and Jaisalmer (MH2, MH3, and ML3) basins. NE-SW magnetic trend is similar to the gravity trend in Fig. 9.31(b). [Colour Fig. on Page 812]

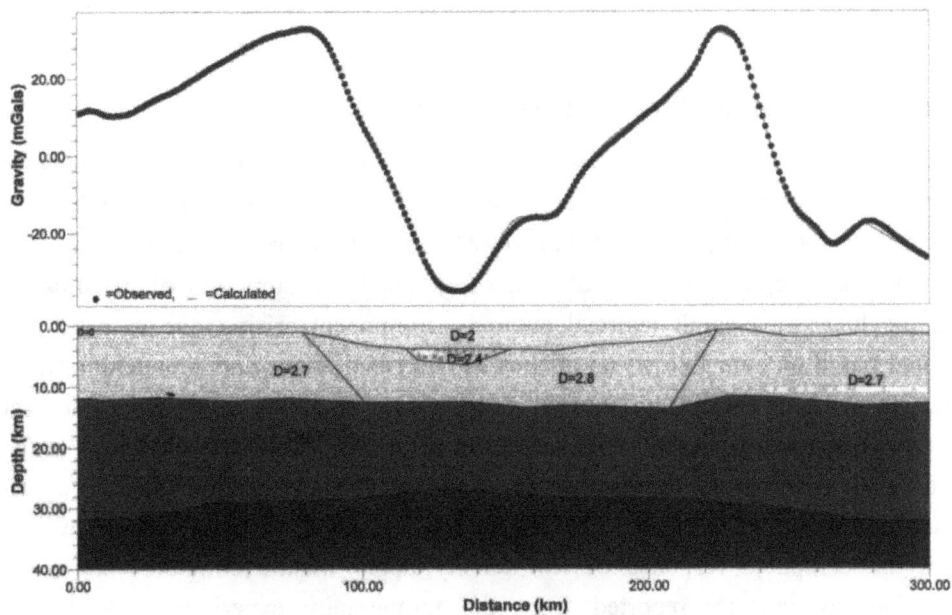

Fig. 9.32(a) Crustal model along a Profile XX' (Fig. 9.31b) across gravity anomalies of the Barmer rift basin showing Quaternary and Tertiary sediments (2.0 g/cm^3) underlain by Mesozoic sediments (2.4 g/cm^3) and high density basement (2.8 g/cm^3) . It also shows up warping of Moho that is typical of rift valleys.

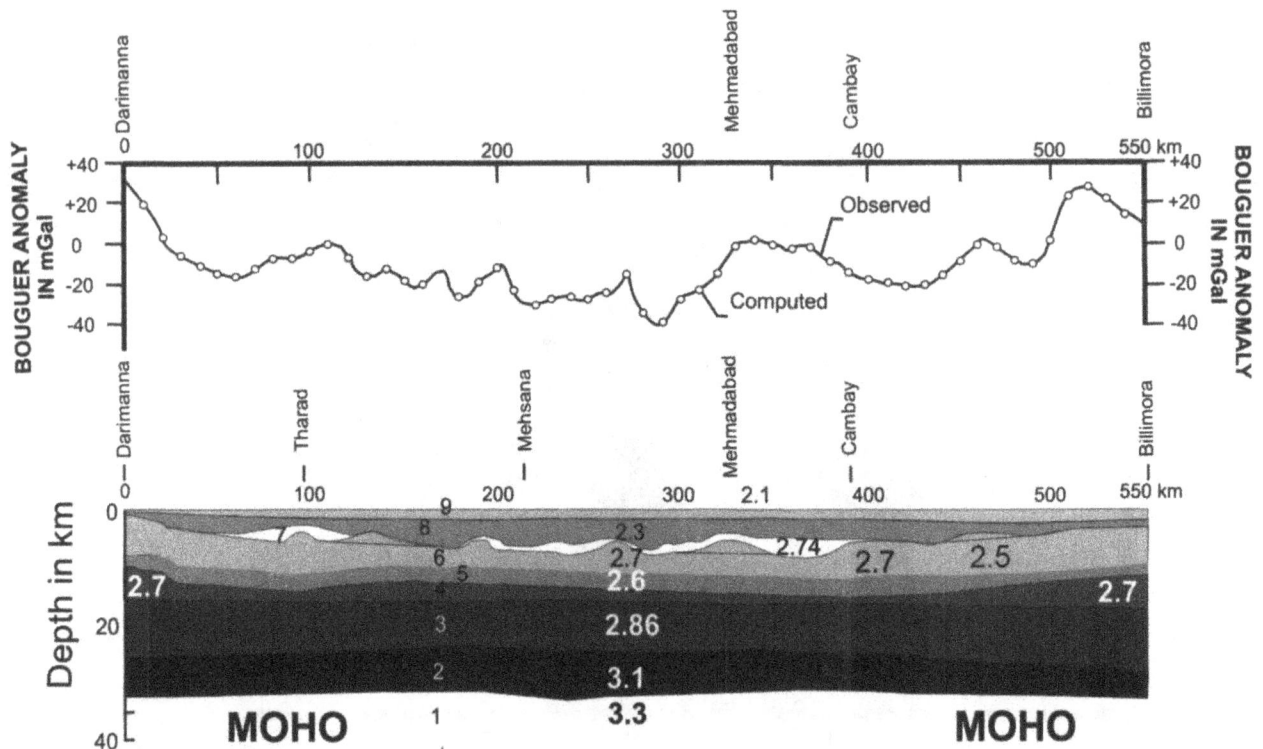

Fig. 9.32(b) Gravity profile from Dharimanna-Billimora (Fig. 9.30) and computed crustal model that shows in general a thin crust of 31-32 km and a high density, 3.1 g/cm^3 underplated crust and a low density layer (2.6 g/cm^3) in the upper crust which are typical of active rift basins. Shallow section shows Tertiary sediments (2.3 g/cm^3) underlain by high density Deccan trap (2.74 g/cm^3) and Mesozoic sediments (2.5g/cm^3)(@.5 g/cm^3) in certain sections (Mishra et al., 1998; Mishra, 2002).

9.4.2   Crustal and Upper Mantle Models across Cambay Basin

Another gravity profile AA' (Fig. 9.31(a)) across the Cambay basin was also modeled to obtain the transverse cross sections of the crustal and lithospheric structures. This profile passes through the bore holes at Lodhika and Dhandhuka (Fig. 9.8 and 9.12) which were used to constrain the gravity model. Since the Bouguer anomaly represents the total sum of gravitational attraction of all subsurface causative sources, the long wavelength regional anomaly component arising due to deep seated sources was separated based on zero free air anomalies as described above and wavelength filtering (Singh and Mishra, 2000). Using the constraints from deep seismic sounding results over Saurashtra, Cambay, and Aravalli (Kaila and Krishna, 1992, Tewari et al., 1997), the long wavelength regional low was attributed to the presence of the low density sub-crustal upper mantle material centered beneath the Cambay basin as shown in Fig. 9.33 that shows shallow and deeper structures in two parts. This model shows the computed response of low-density upper mantle structure that may be synonymous with the low velocity reported from the tomography experiments (Kennet and Widiyantoro, 1999). The most significant aspects of this model are:

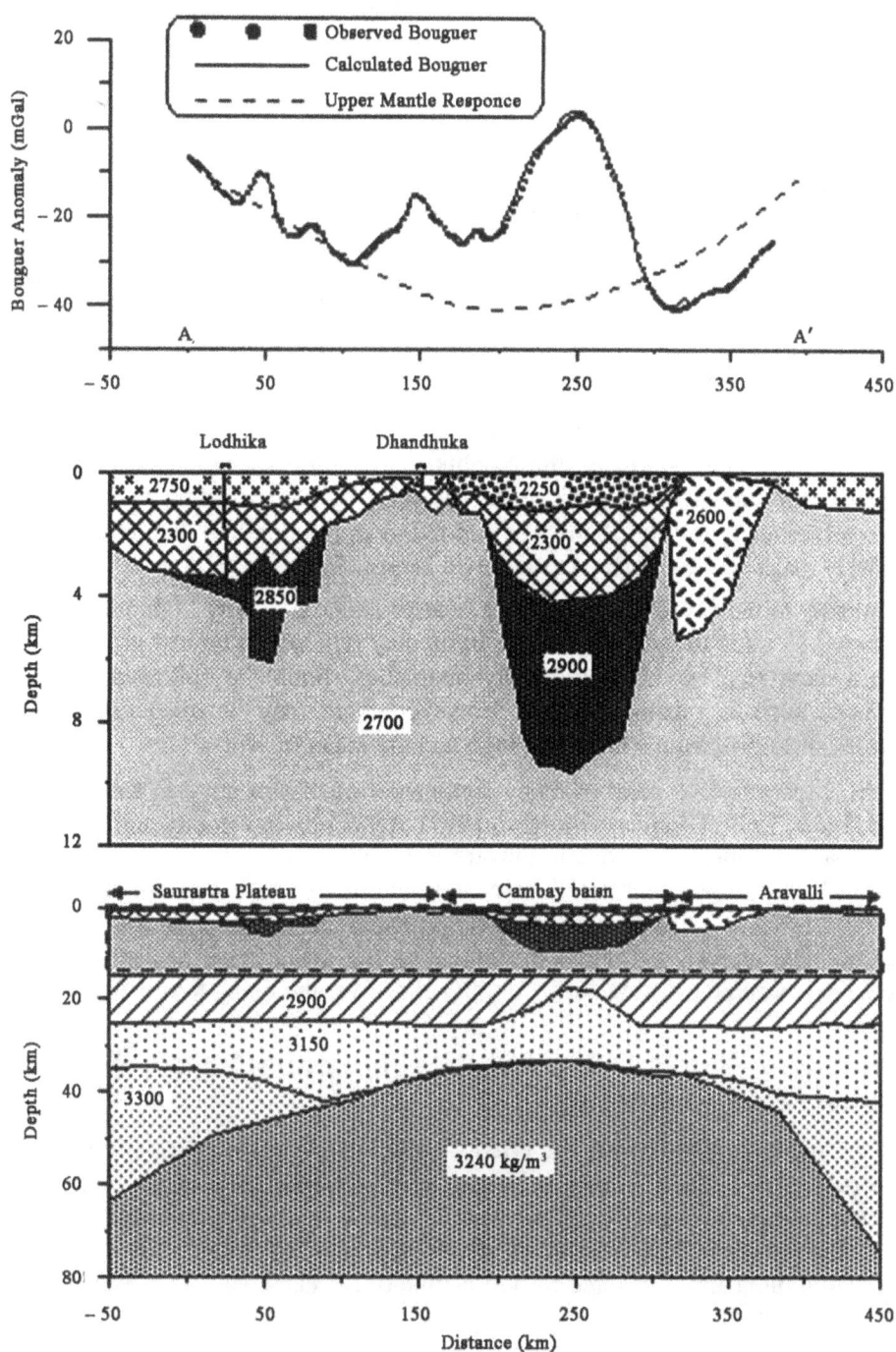

Fig. 9.33 A gravity profile, AA' (Fig. 9.31) transverse to the Cambay basin and computed crustal and upper mantle density model. The regional field is modeled through the deeper structures, viz. Moho and upper mantle structures and then the observed Bouguer anomaly is modeled through combination of both shallow and deeper structures. This profile passes through Lodhika and Dhandhuka boreholes in the western part where borehole information are used to constrain the model. In cambay basin, the shallow section shows Tertiary sediments (2250 and 2300 kg/m³) under lain by Deccan trap (2900 kg/m³). The deeper section shows a pillow type cushion of low density (3240 kg/m³) in the upper mantle and thin and under plated crust that are typical of a rift basins world over (Mishra et al., 2008, Singh and Mishra, 2000).

(i)   Presence of low-density upper mantle structure having a density of 3240 kg/m^3. The top surface of this structure lies at a depth of about 35 km below the Moho. It has a width of about 200 km at the top and begins to broaden at a depth around 55 km. The structure is not confined only to the rift but extends in the surrounding region. It shows asymmetry with reference to basin axis. This kind of structure has been termed as pillow under the rift basins in other parts of the world.

(ii)   The central gravity high of the Cambay basin is explained partly due to the high-density (3150 kg/m^3) thick underplated layer in the lower crust and partly due to high-density trap rock beneath the basin. The top surface of the lower crustal high-density body is constrained from deep seismic sounding and is kept at about 25 km. It is confined mostly within the rift zone. Underplated crust is typical of active rift basins world over as they are associated with large scale volcanism.

(iii)   The shallow section under the Cambay basin shows Tertiary sediments (2250 kg/m^3 and 2300 kg/m^3) underlain by Deccan trap (2900 kg/m^3) as the basement at a depth of 4 km that is almost same as given in Fig. 9.30(b). Under the Lodhika well, however, the layer with density 2300 kg/m^3 represents the Mesozoic sediments underlain by Deccan trap at the surface and underlying second horizon of trap (2850 g/cm^3) is found in the borehole that represents older volcanics. Tertiary rocks of the Cambay basin are oil bearing in this case.

(iv)   The gravity anomaly of the Cambay basin with a relative residual gravity high due to basement high and a regional low due to sediments of the basin may represent a typical gravity signature of oil bearing structures. This kind of gravity anomalies should be followed up by more advanced methods such as seismic surveys. However, there may be other combination of gravity anomalies depending on the basement high and thickness of sediments.

These features (i-iii) appear to be typical of many continental rift basins such as Kenya and other rift basins (Ravat and Braile, 1999; Olsen and Morgan, 1995). Reduction in velocity and density of the upper mantle material has been interpreted due to upwarp in the thermal boundary layer (lithosphere-asthenosphere boundary) and presence of small percentage of partial melt beneath the Kenya rift (Ravat and Braile, 1999). In view of the above, it is quite probable that the inferred low density body in the upper mantle and high density underplated material in the lower crust beneath the Cambay region might have evolved during the passage of the Indian plate over the rising Reunion plume.

9.5   Deccan Syneclise (Narmada-Tapti Section) in Central India – Mesozoic Sediments and Basement Model

The western part of the Narmada basin along the coast (4, Fig. 9.1(a)) is oil bearing and several oil fields exist in this section as is evident from a schematic section based on several seismic profiles (Fig. 9.34, ONGC, 2009) which shows oil window from 2-3 km in Tertiary sediments. The western part of the E-W trending Narmada-Tapti section (4, Fig. 9.1(a) and NT, Fig. 9.30) constitutes the Deccan syneclise of central India and is covered by a thick pile of Deccan lava flows. Evidences of the presence of Mesozoic sediments occur throughout the western part of the Narmada valley as both inliers and outliers which appear to have been deposited during Cretaceous times in a depression between the Satpura and the Vindhyan range (Murthy and Sharad, 1981). These sediments were uplifted, faulted, intruded, and finally covered by the Deccan basalts (Bhattacharji et al., 1996). The presence of C1-C5 hydrocarbons in the adsorbed soil gases in samples collected from parts of the Deccan Syneclise indicates that hydrocarbon generation has taken place in the basin and gases are derived from a thermogenic source. This evidence may open new vistas for commercial discovery of oil/gas in the Mesozoic formations below the Deccan Traps (Vishnu Vardhan et al., 2008 and references therein). The combination of gravity, seismics, DRS and MT techniques were applied in this region in a phase-wise manner, starting with the western most part (NGRI, 2003).

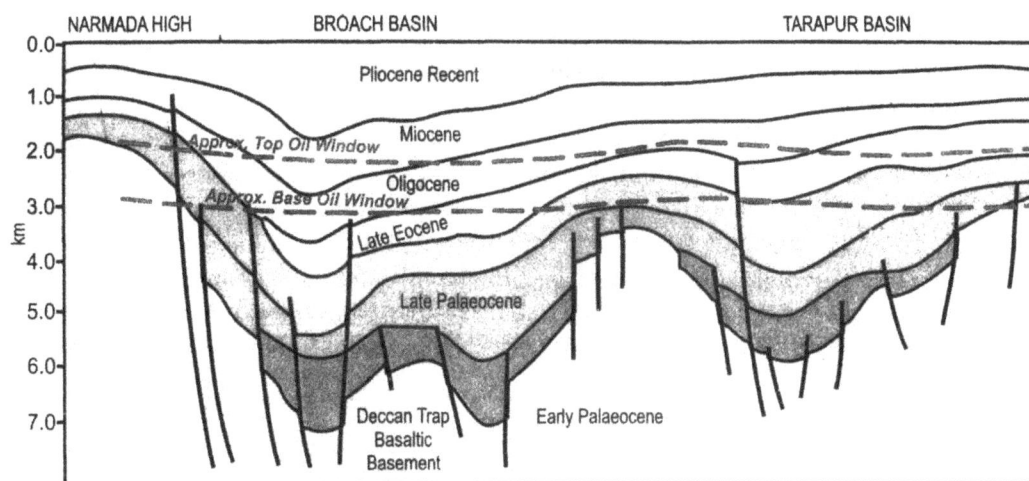

Fig. 9.34 A schematic seismic section across Broach depression in Narmada-Tapti (NT) basin (4, Fig. 9.1(a) and Fig. 9.30) close to the west coast of India where Tertiary sediments are exposed and is oil bearing between 2-3 km depth (ONGC, 2009).

9.5.1 Residual Gravity Anomaly Map

The residual gravity anomaly after subtracting the regional field based on zero free air anomalies from the observed field is given in Fig. 9.35 (NGRI, 2003, Singh and Arora, 2008) that also shows the two profiles modeled below. This map shows two sets of linear gravity highs, Dhule-Akola and through Khandwa that may represent mafic intrusive related to the Satpura Mobile Belt (Section 7.6). The gravity lows in between these sets of gravity highs may be related to subtrappean Mesozoic sediments.

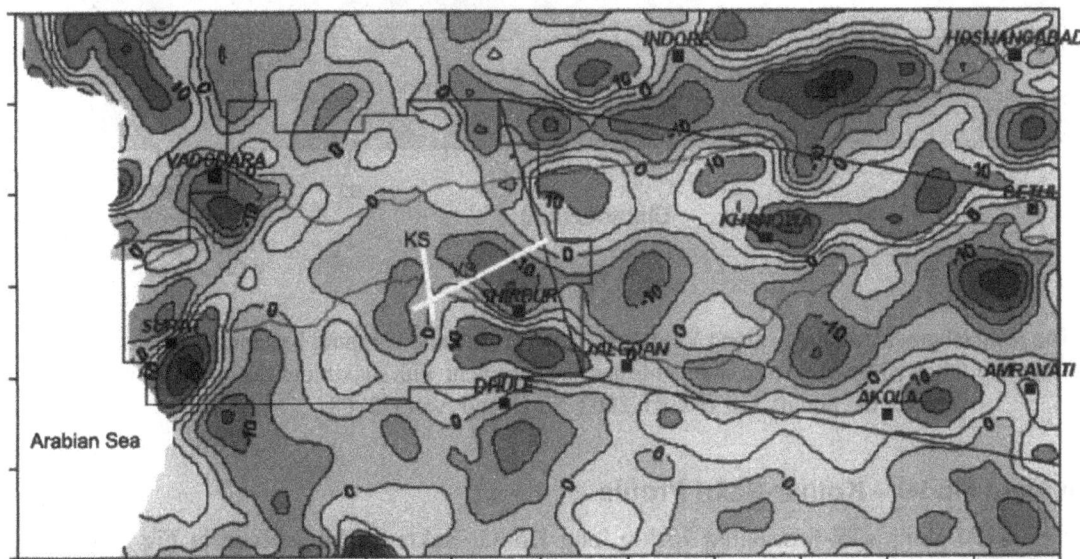

Fig. 9.35 Residual Bouguer anomaly map of the western part of Narmada-Tapti Section (4, Fig. 9.1(a)) referred to as the Deccan Synclise of western part of Central India. It shows two sets of E-W oriented linear gravity highs with an intermediate set of gravity lows that might be related to subtrappean Mesozoic sediments. The gravity low south of Badodra represents Broach depression (Fig. 9.34). White lines indicate profiles that are modeled, VS: Velda- Sendhwa profile and KS: Kothar- Sakri profile (NGRI, 2003, Singh and Arora, 2008).

[Colour Fig. on Page 812]

9.5.2   Basement Model – Velda-Sandhwa Profile

Quantitative modelling of the residual gravity field along the Velda-Sendhwa section (Vs Fig. 9.35) using constraints from other geophysical techniques is shown in Fig. 9.36(a). The first gravity high centred at Shahada is due to the combined contributions from the shallowing of the basement rocks, thinning of the sediments and increase of Trap thickness. The gravity high in the northern part of the profile towards Sendhwa is a typical example of geophysical interpretation through the integration of seismic, MT and gravity results, where the deepening of the basement is discernible from seismic data and the thickening of both basalts and sediments are based on the inversion of MT and DRS data. The observed gravity anomalies then serve to resolve the geometry of the layers spatially, yielding the forms of the sedimentary basins and zones of thick sediments (NGRI, 2003; Singh and Arora, 2008).

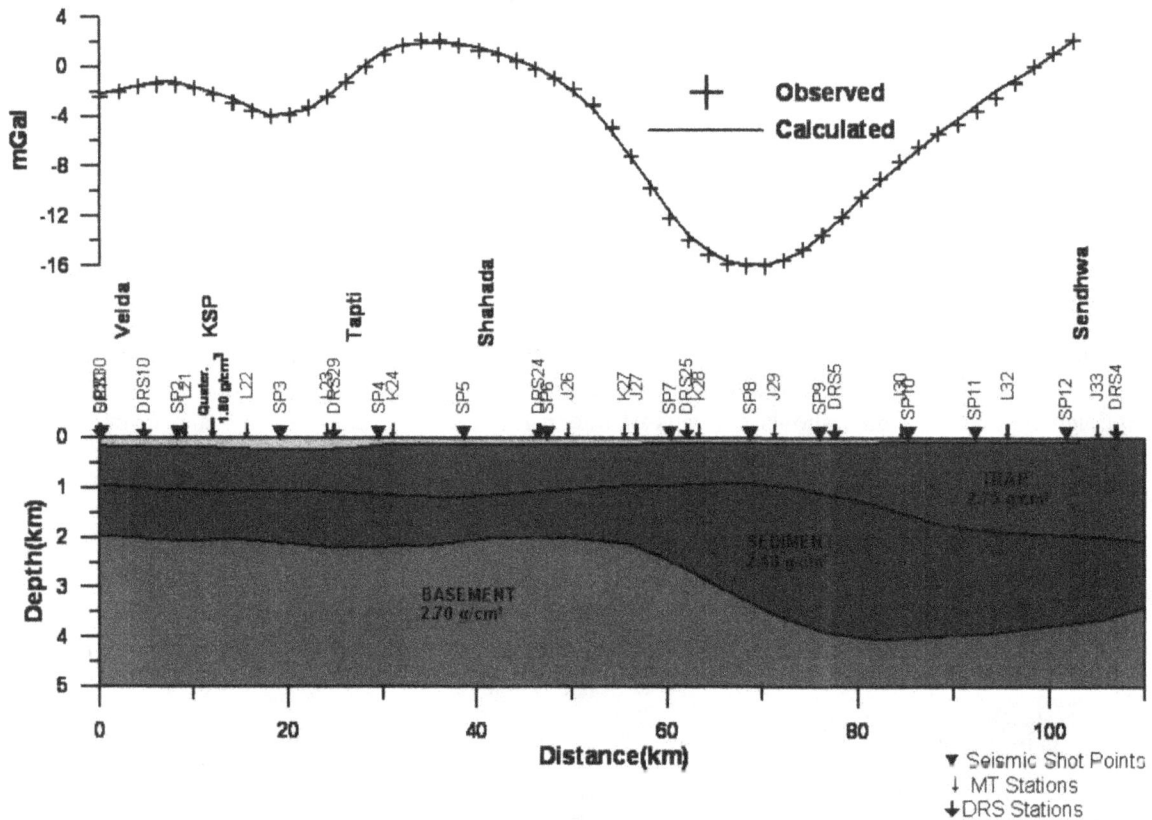

Fig. 9.36(a) The residual gravity anomaly along Velda- Sendhwa profile and computd basement model showing Deccan trap at the top followed by Mesozoic sediments and the basement with maximum thickness of sediment under the gravity low (NGRI, 2003, Singh and Arora, 2008).

9.5.3   Basement Model – Kothar-Sakri Profile

The subsurface model along a N-S oriented 75 km long Kothar-Sakri profile, shown in Fig. 9.36(b) (Singh and Arora, 2008) aims to elucidate another example of integrated interpretation. On the basis of experience during all the previous studies it has been found that the sediment-basement interfaces derived from MT (lower dashed line) and seismic (upper dashed line) methods consistently provide the upper and lower bounds of the depths; these estimates guide the geometry of this interface during gravity modeling. Remaining mismatch of observed and computed gravity fields is adjusted by introducing geologically appropriate mass variations (NGRI, 2003; Singh and Arora, 2008).

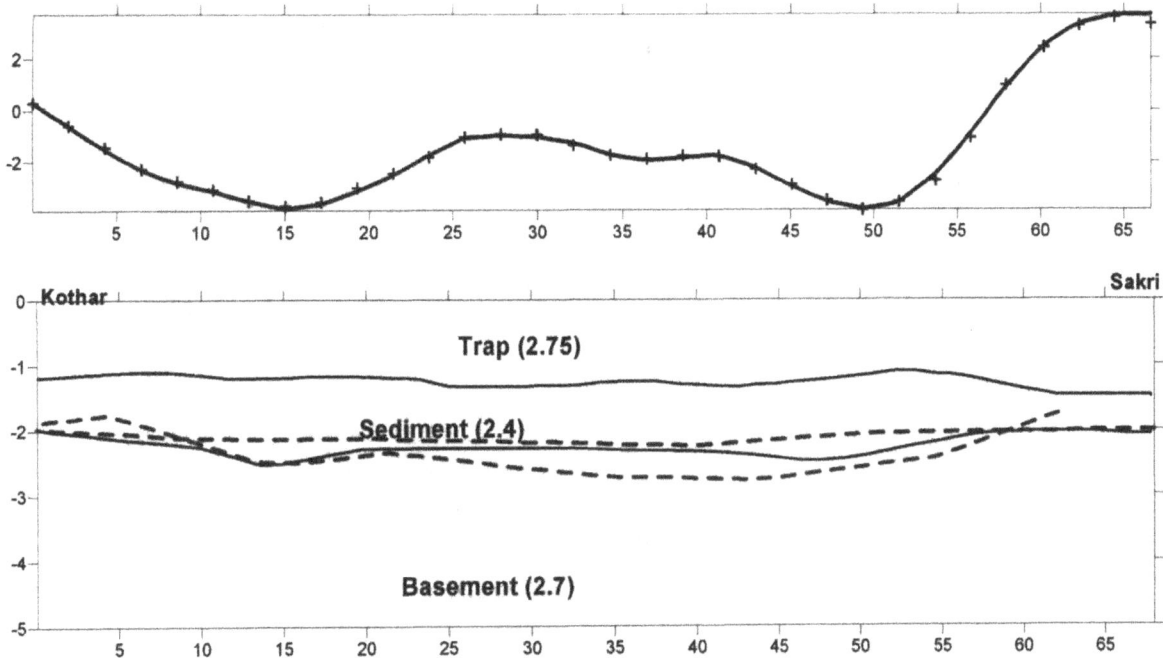

Fig. 9.36(b) The residual gravity anomaly along the Kothar-Sakri profile and computed basement model that is compared with the MT basement (lower dashed line) and seismic basement (upper dashed line). Maximum Mesozoic sediments (> 2 km) are found under two gravity lows separated by an uplift of basement in between (NGRI, 2003, Singh and Arora, 2008).

9.5.4   Subtrappean Mesozoic Sediment Thickness Map

The computed subtrappean Mesozoic sediment thickness map based on multidisciplinary geophysical investigations along several profiles in this region is given in Fig. 9.37(a) that has brought out two prominent subtrappean Mesozoic basins along the central part of the Narmada-Tapti rift zone where the subtrappean sediments attain maximum thickness of about 2.5 km near Sirpur. The trend of these results suggests the possibility of eastward extension of these basins (NGRI, 2003; Singh and Arora, 2008).

9.5.5   Nagpur- Wardha Section

This section known as Wardha basin (5, Fig. 9.1(a)) is located in the eastern part of the Deccan Synclise in central India. As the thickness of sediments in the previous map (Fig. 9.37(a)) showed an increase eastwards, the eastern most part of Deccan Synclise in central India was investigated for subtrappean Mesozoic sediments. This region is characterized by a semicircular gravity low (Fig. 7.8(b)) indicating the presence of sediments below the trap. Moreover, it forms the junction of the Godavari Gondwana basin with the Satpura Mobile Belt that makes it tectonically important. Based on several sections of the basement inferred from magnetotelluric profiles, a basement model (Fig. 9.37(b); DGH, 2005) was inferred showing four depocenters of ~ 4 km depth around Katol under a thin cover of trap rocks (~ 300 m) that is quite suitable for oil occurrences and also favorable for drilling due to thin trap cover.

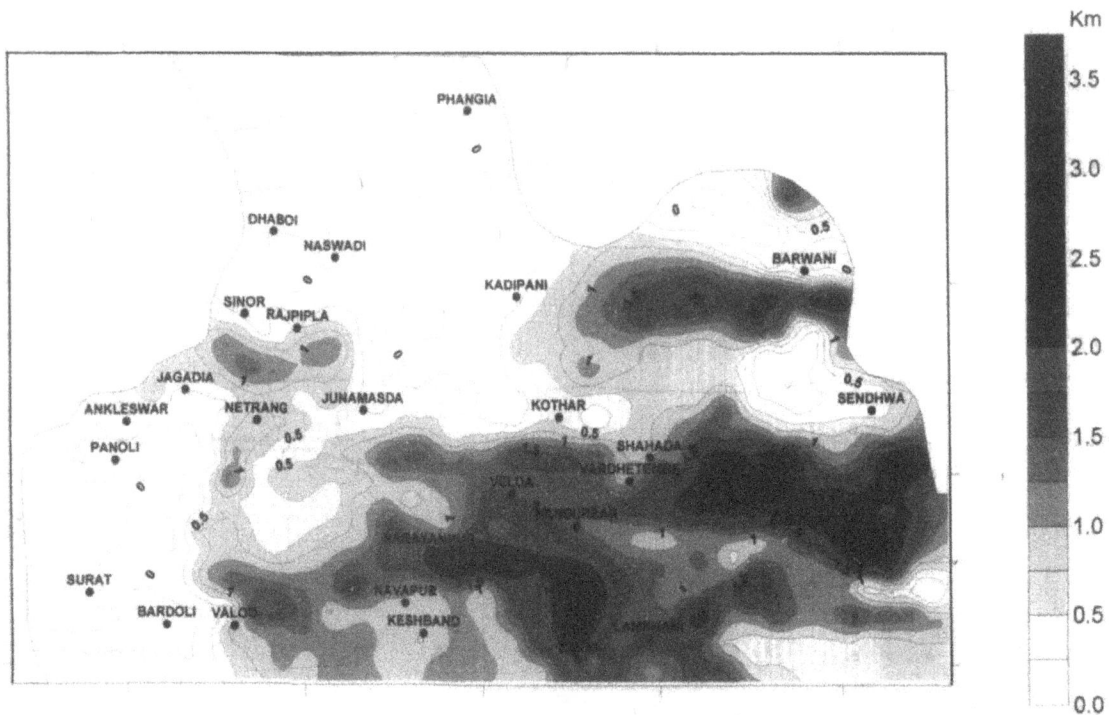

Fig. 9.37(a) Sedimentary thickness map of Narmada-Tapti section (Western part of Deccan Sunclise in Central India) based on integration of all investigations, viz. gravity, seismic, electrical and magnetotelluric along several profiles showing maximum thickness of Mesozoic sediments as ~ 3 km in the eastern part (Saver) and Narayanpur-Sikri in the central part (NGRI, 2003, Singh and Arora, 2008). [Colour Fig. on Page 813]

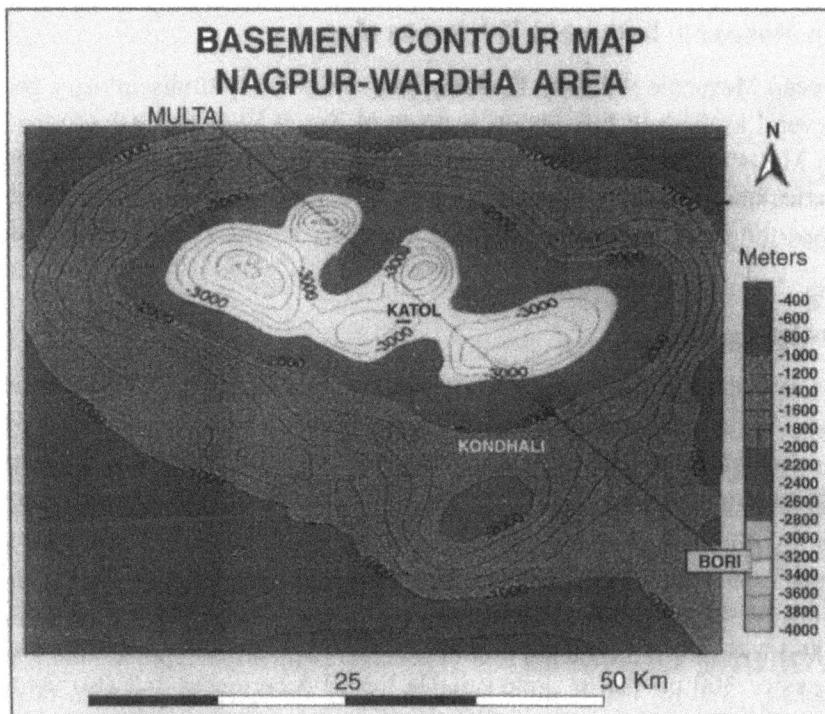

Fig. 9.37(b) Basement contour map of Wardha basin (5, Fig. 9.1a) showing maximum depth of ~ 4 km with subtrappean Meszoic sediment underlain by 200-300m of Deccan trap (DGH, 2005).

9.6   Offshore Exploration – Hydrocarbon Prospect

The tectonics and formation of the continental shelves offshore the west and the east coasts of India have been discussed in Chapter 5. Regional free air and Bouguer anomaly maps of these sections are given in Fig. 5.40 and 5.41, respectively and their application to geodynamics is demonstrated in Section 5.5. The east and the west coasts of India represent passive continental margins which evolved through extension and rifting since the break up of Gondwanaland during early Jurrasic (Section 5.3). The earliest to separate was Africa giving rise to the continental shelf offshore west coast of India followed by the break up of India from Antarctica at ~ 120 Ma that gave rise to the continental shelf offshore east coast of India. Offshore west coast of India was further modified due to the breakup of Madagascar and Seychelles at ~ 88 Ma and ~ 65 Ma, respectively. During the long period of drift from Antarctica to the present day position of India, several coastal basins (Fig. 9.1(a)) with considerable amount of sediments developed both offshore west and east coasts of India. They have been surveyed for oil exploration including for gas hydrates (Avasthi et al., 2008; Sain and Ojha, 2008; Chandra, 2000; DGH, 2005). However, it is beyond the scope of the present book to describe the hydrocarbon exploration offshore in detail, but a few typical cases are presented below.

9.6.1 Offshore West Coast of India – Seismic Section and Bouguer Anomaly

Offshore west coast of India is characterized by Deccan trap activity at Cretaceous-Tertiary boundary (~ 65 Ma) that had covered most of the earlier events. This makes it difficult to explore Mesozoic prospects as described above in case of Saurashtra. However, Tertiary formations that overlie the Deccan trap are represented mainly by Paleocene clastics, Eocene and Miocene carbonates and post Miocene clastics that are quite promising. One of the largest oil field of Bombay high exist in this section. Fig. 5.33(a) and (b) gives the tectonics of the continental shelf offshore west coast of India that shows horst and rift system that are favourable structures for oil accumulation. Fig. 5.33(b) shows Lakshadweep depression with fault controlled Cretaceous and Tertiary basins that may be a prospective target for oil exploration. It shows seaward dipping reflectors suggesting volcanic rifted margin that are important for oil exploration. Important basins along the west coast of India from north are Kutch-Saurashtra offshore basins, Bombay offshore basins, and Kerala-Konkan basins (Fig. 9.1(a)). A seismic section across the Kerala-Konkan offshore is given in Fig. 9.1(c) that shows stratigraphic traps favourable for hydrocarbon exploration. It also shows seaward dipping reflectors as usually found along volcanic rifted margins.

(i) Bouguer Anomaly

A detailed satellite derived Bouguer anomaly map of the continental shelf off the west coast of India is given in Fig. 9.38 (DGH, 2005) that shows linear trends of gravity highs and lows almost parallel to the coast. Though this map is derived from satellite data, it is specially processed for short-medium wavelength basement anomalies that are suitable for oil exploration, while similar maps given in Section 5.5 contain mainly long wavelength anomalies and are suitable for geodynamics study. The most important tectonic feature in this section is the Kori-Comorin ridge that is represented by the linear gravity highs closest to the coast and runs almost parallel to the west coast of India. Towards the west of the gravity highs due to this ridge, there are at least another two trends of gravity highs that may represent the Pratap ridge and Laccadive ridge (Fig. 5.33(a)) beyond which oceanic crust is encountered. These ridges are primarily formed by the Deccan trap and have provided several basement uplifts and depressions forming horst and rift structures (Fig. 5.33(a) and (b)) that are suitable as oil traps. In fact, rifted margins are important for hydrocarbon exploration due to their association with volcanic activity that is important for basin formation due to subsidence and provides the requisite heat for the formation of hydrocarbons. Besides Tertiary basins, Mesozoic sequences are expected offshore Saurashtra and Kachchh that are present on shore as described above. A carbonate build up offshore Saurashtra is shown in Fig. 9.39(a) (Chandra, 2000) that forms a suitable oil trap.

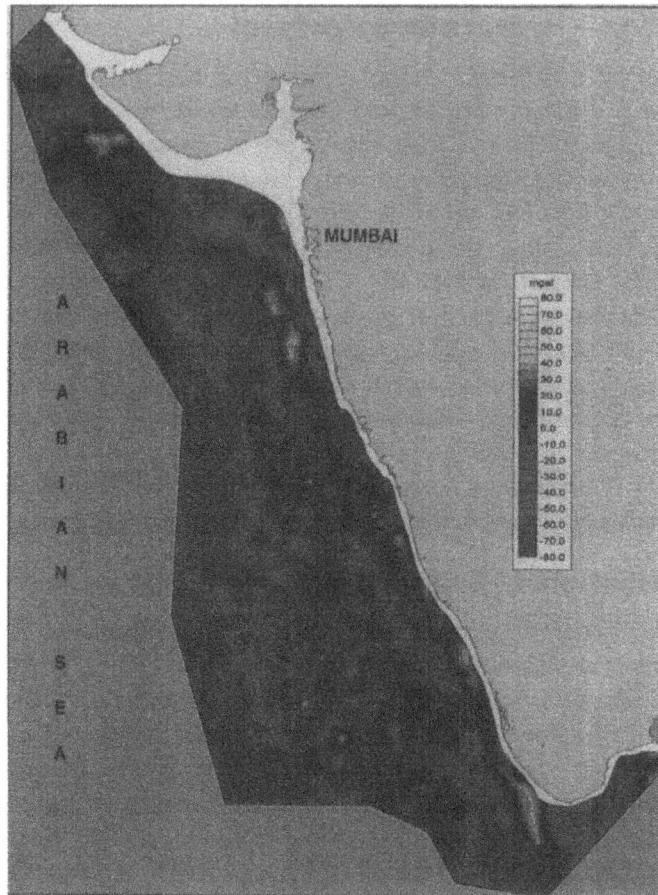

Fig. 9.38 Satellite derived Bouguer anomaly map offshore west coast of India (6, 7, Fig. 9.1a) showing linear gravity highs related to magmatic intrusion (Deccan trap) associated with the formation of rifted margin (DGH, 2005). [Colour Fig. on Page 813]

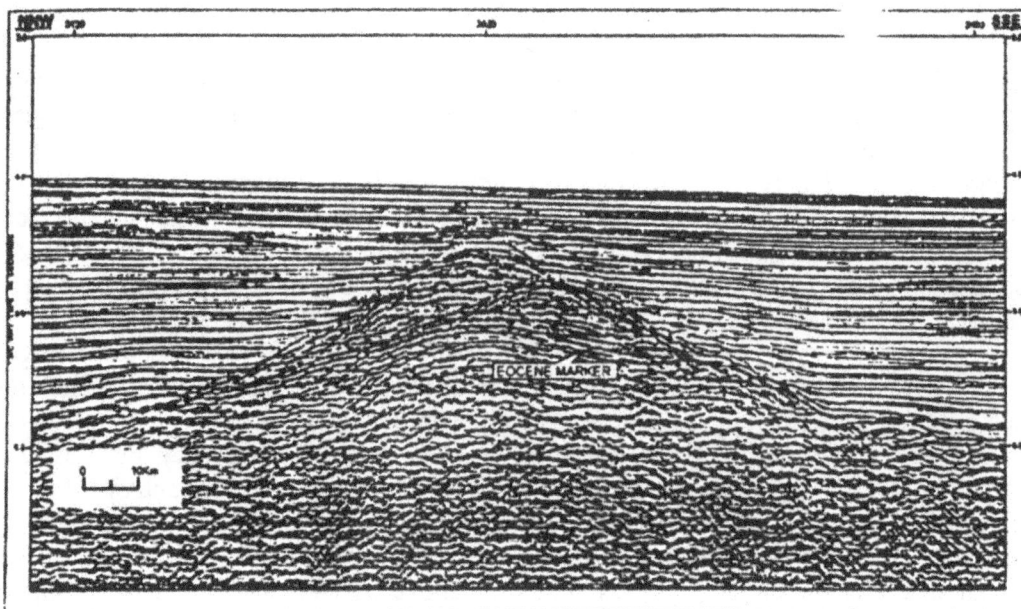

Fig. 9.39(a) Carbonate build up and stratigraphic wedge out in offshore Saurashtra basin (Chandra, 2000)

(ii) Bombay High – Bouguer Anomaly and Plausible Sources

Bombay high offshore Mumbai represents a basement high and offers one of the major oil producing structures in the world. The Bouguer anomaly of this structure is interesting and revealing (Fig. 9.39(b); Sar, 2008) showing a gravity low surrounded by gravity highs. Sar (2008) attributed it to the absence of Deccan trap rocks over it that is prevalent in the surrounding regions. It can also be attributed to thick Tertiary sediments over it that are oil bearing and Mesozoic sediments under it that would also produce gravity lows. Another interesting hypothesis is the presence of a meteoric crater named Shiva crater at this place (Chatterjee, 2010) that was responsible for the extinction of dinosaurs. Metoric craters also produce a similar anomaly of a central low surrounded by ring of highs as discussed in the Chapter 10 with regard to the Lonar crater. However, in the present case the surrounding highs are widespread and therefore, in this respect it is different from gravity anomalies due to craters. Therefore, in all likelihood this gravity low is caused by a thick pile of sediments underneath. This kind of gravity lows surrounded by highs should be explored for their oil potential offshore west coast of India especially in the northern part.

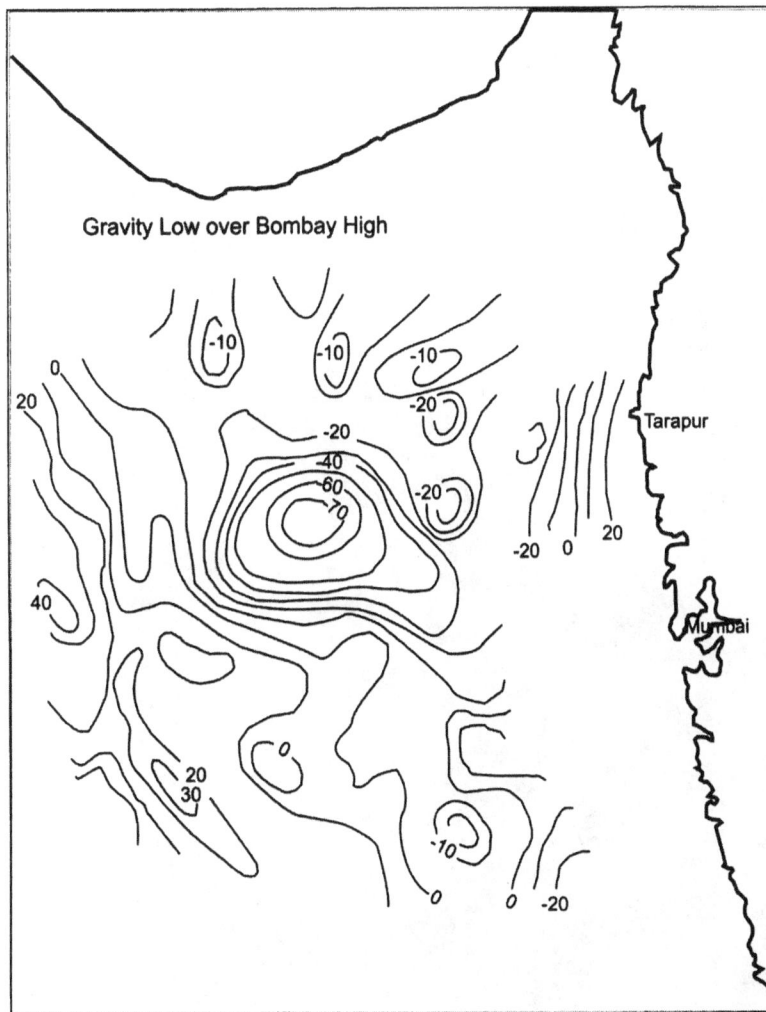

Fig. 9.39(b) Bouguer gravity anomaly map of Bombay High (6, Fig. 9.1(a)) showing a grvity low. It is one of the major oil producing fields (Sar, 2008).

9.6.2   Offshore and Onshore East Coast of India – Seismic Section and Bouguer Anomaly

As per available exploration data, the resource potential of east coast basins is ~ 7.0 billion tons (DGH, 2005). Seismic profiles across offshore East Coast of India are given in Fig. 5.36 (b) and (c) that show thick pile of sediments. Fig. 5.36(b) is across Eighty Five East ridge that shows uplifts and wedge out that are favourable sites for oil accumulation. It has affected the Cretaceous sediments that indicates that it is of latter period. Fig. 5.36(c) is a seismic section offshore Godavari-Krishna basins that show tha nature of continental margin and bright events that are significant for hydrocarbon exploration. It shows thick pile of sediments (~ 8 km) up to Cretaceous top and there are Mesozoic sediments below it as basement has not been encountered so far.

(i)   Bouguer Anomaly

A satellite based Bouguer anomaly off the east coast of India is given in Fig. 9.40(a) (DGH, 2005). This is also a high resolution map suitable for basement structures as described above for similar map offshore west coast of India. In this case also, the Bouguer anomaly shows two sets of linear gravity highs that are almost parallel to the coast. They represent the mafic volcanic rocks of the break up of India with Antarctica plausible due to the Kerguelen hotspot as discussed in Chapter 5. In between these bands of gravity highs, there are gravity lows plausibly representing the basin formed due to this eruption.

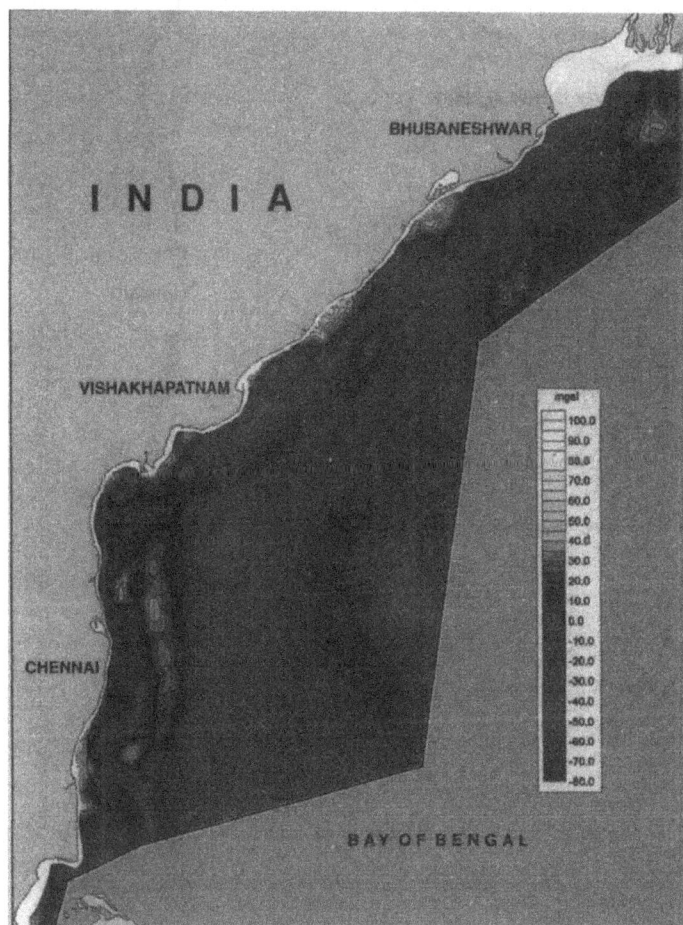

Fig. 9.40(a) Satellite derived Bouguer anomaly map offshore the east coast of India (8, Fig. 9.1(a)) showing linear gravity highs almost parallel to the coast line related to magmatic intrusion during the break up of India from Antarctica (DGH, 2005). [Colour Fig. on Page 814]

(ii) Gas Occurrences offshore Godavari Basin

The most important east coast basins from north to south are the Mahanadi basin, Krishna-Godavari (K-G) basin and Cauvery basin (13-15, Fig. 9.1(a)). They contain almost 4-8 km of sediments from Gondwana to Tertiary and oil window occurs in depth range of 2-3 km. The principle source rocks occur in the coal and carbonaceous shales of Cretaceous and Paleocene age. Dry gas is expected at depths of ~ 6 km. A typical section in K-G basin is shown in Fig. 9.40(b) (Chandra, 2000) that shows listric faults, and rollover anticlines that are favorable structures for oil occurrence. A seismic section (Fig. 9.41(a)) across Dhirubhai-23 by Reliance Industries Ltd. in deep water offshore Krishna-Godavari basin shows a rollover anticline that provides one of the largest gas fields in the world at a water depth of ~ 700 m and target depth of ~ 2000 m with hydrocarbon column of 84 m (DGH, 2005).

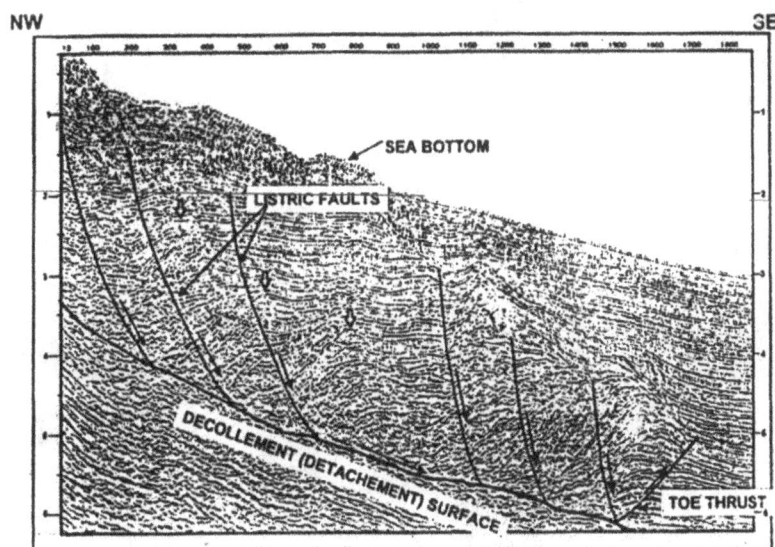

Fig. 9.40(b) Decollment surface, rollover anticline, toe thrust and listric fault etc., in offshore K-G basin (8, Fig. 9.1(a)) (Chandra, 2000).

(iii) On Shore Coastal Basins

As described above, the Mahanadi, Godavari and Cauvery basins (13-15, Fig. 9.1(a)) are three important basins along the east coast of India. General tectonics of the Mahanadi and the Godavari basins are described in Sections 5.6 and 7.7. The coastal basins in these sections consist of ridges and depressions that are important for oil exploration as briefly described in those sections. The coastal cauvery basin and its offshore part also consist of ridges and depressions that are important for oil exploration. Based on several seismic profiles and some well data, a section across the Cauvery basin on shore and offshore is given in Fig. 9.41(b) that shows alternating ridges and depressions (Rama Rao et al., 2010) typical of collision tectonics. They might have formed due to collision of India and Antarctica during the formation of Gondwanaland or prior to it (Section 7.10). The alternating presence of sandstones and limestones makes it a favourable prospect for oil. Geodynamically, the most significant point is the volcanics in the offshore part just over the basement that suggest the presence of volcanics throughout along the east coast of India from offshore Bengal, Mahanadi and Godavari basins extending to the offshore Cauvery basin. Though it has not been dated, but based on the above presumption it can be regarded to have originated from the Kerguelen hotspot during the breakup of India from Antarctica as in the case of other basins along the east coast of India.

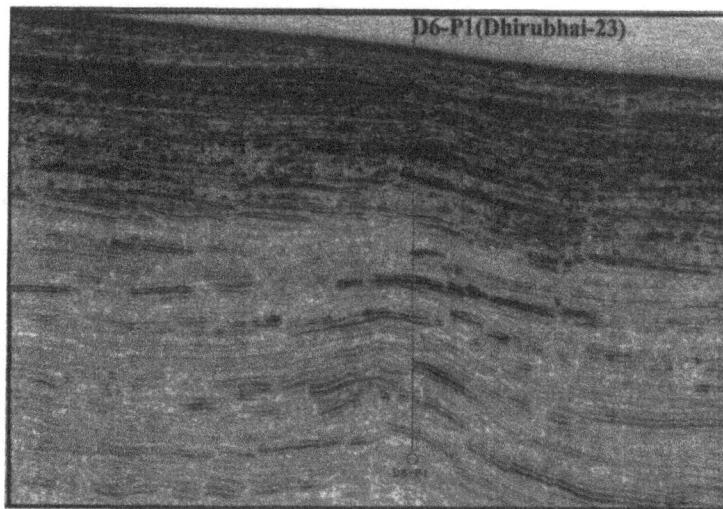

Fig. 9.41(a) Seismic section through discovery well, Dhirubhai-23 in deep waters offshore Krishna-Godavari basin showing rollover anticline giving rise to one of the largest gas field (DGH, 2005). [Colour Fig. on Page 814]

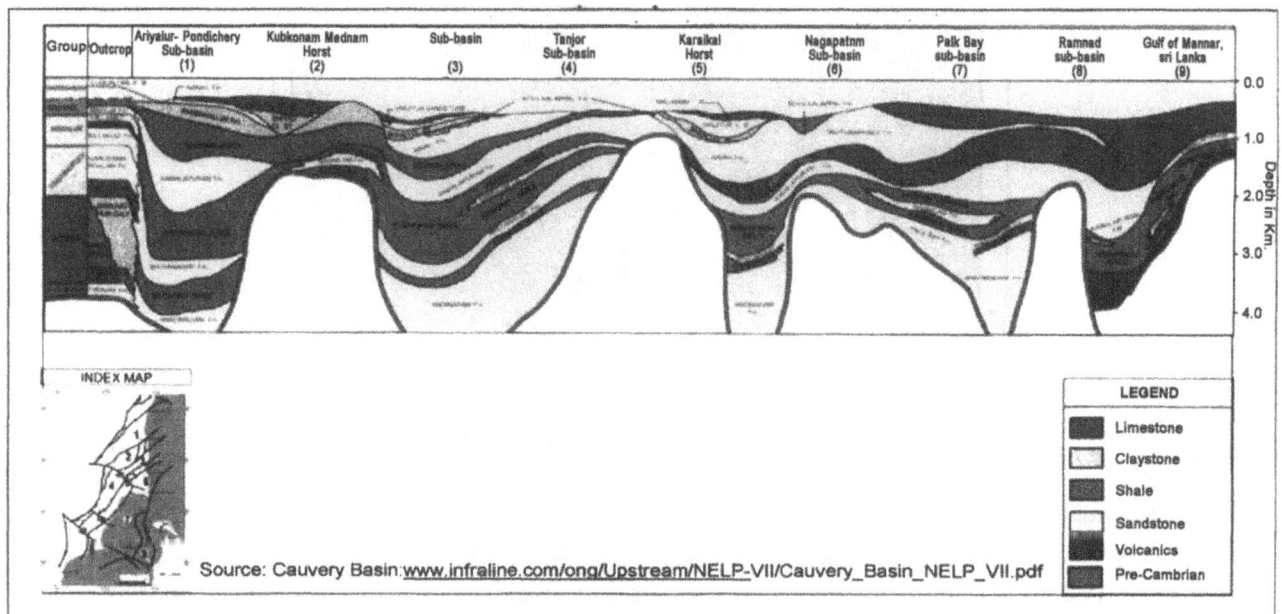

Fig. 9.41(b) A schematic geological section across Cauvery basin onshore and offshore showing basement ridges and depressions (Rama Rao et al., 2010) [Colour Fig. on Page 815]

9.6.3 Offshore Andaman Islands – Seismic Section and Bouguer Anomaly

Andaman-Nicobar islands (9, Fig. 9.1(a)) along with Sumatra and Indinesia represent an active plate margin where the Indo-Australian plate towards the west is colliding and subducting under the Eurasian plate towards the east and is one of the seismically most active margins where great earthquakes have occurred as described in Section 8.4.4. A seismic section offshore Andaman islands is given in Fig. 5.36(d) that shows faulted and folded section and shale diapirism, typical of convergence and collision tectonics. It clearly shows faulting of Miocene and Oligocene sections indicating that these structures are of the latter period. This section also shows bottom simulating reflectors related to gas hydrates. This section being an extension of the Arakan-Yoma fold belt along

Burmese arc towards the north where several oil fields have been discovered, assumes special significance for oil exploration. Fig. 9.41(c) (DGH, 2005) shows satellite based Bouguer anomaly map of Andaman-Nicobar islands and offshore region specially prepared for basement studies. It shows major linear gravity lows where these islands are located, and major gravity highs on either sides of it. The linear band of gravity low represents the trench of this subduction and is one of the largest and oldest reported gravity low in the world that was first noted by Vening Meinesz in 1931 during a marine survey when plate tectonics was unknown. As this band of gravity lows represent the trench, the islands of Andaman-Nicobar form the accretionary prism of thick sediments. The gravity highs east of it represent the marginal sea with the oceanic crust that would produce gravity highs due to high density oceanic crust. The gravity lows embedded in these gravity highs may represent marginal basins that require detailed investigation for their oil potentiality. The gravity highs towards the east are part of Ninety East Ridge that formed during the breakup of India from Antarctica as discussed in detail in Section 5.4. A part of this gravity high close to the trench may also represent the crustal upwarp known as the outer rise in front of oceanic trenches caused due to lithospheric flexure as described in Section 8.4.

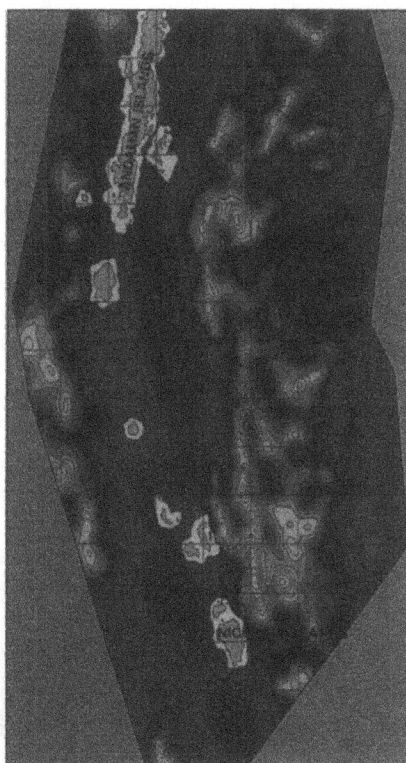

Fig. 9.41(c) Satellite derived Bouguer anomaly map offshore Andaman-Nicobar islands (9, Fig. 9.1(a)) showing a curvilinear gravity low related to trench between the Indo-Australian and Eurasian plates suggesting islands to be part of accretionary prism. The gravity highs towards the west represent the Andaman marginal sea with oceanic crust and some centers of volcanic eruption as arc with basins in between to be explored for hydrocarbons (DGH, 2005). [Colour Fig. on Page 815]

9.7   Vindhyan Basin, Central India and Singhbhum District – Groundwater Potential and Hydrocarbon Prospect

Several attempts were made to expedite the exploration for groundwater using airborne geophysical surveys (Paterson and Bosschart, 1987). Delineation of faults and fractures will aid considerably in groundwater exploration since they control the movement of groundwater. However, the airborne total intensity maps prepared previously for geological mapping or mineral exploration can also be reexamined to aid in groundwater exploration. Several success stories have been reported by delineating potential zones of groundwater using airborne magnetic maps (Astier and Paterson, 1989; Bromley et al., 1994).

Linear features such as fractures and faults, being limited in width, require very closely spaced data, which is usually not possible to acquire on the ground unless the area is already known and limited in extent. It is therefore likely that they are missed in surveys on routine ground checks in unexplored areas. In an airborne survey, the data is acquired at very close interval of a few meters along the profiles and, therefore, long linear features can be mapped effectively from profiles across the strike direction. Once they are accurately mapped, they can easily be followed up on the ground and their groundwater potentiality can be assessed by recording few electrical resistivity profiles across them.

9.7.1   Airborne Total Intensity Map of Vindhyan Basin – Magnetic Lineaments

The intracratonic Vindhyan Basin (10, Fig. 9.1(a)) is bounded by Son-Narmada Lineament towards southeast and the Great Boundary Fault towards northwest as has been described in great detail in Section 7.6.2. The basin represents a sequence of Upper Proterozoic sediments, which are represented by sandstones, limestones and shales, which are largely undisturbed (Fig. 7.20). The Son Valley basin is characterized by gentle synclinal dips plunging towards southwest and simple out crop pattern, except in the southern margin where the sediments are highly tectonised near the Son-Narmada Lineament (Das et al., 1990). In the S-E portion around Katangi (Fig. 9.42(a)), the sediments show considerable disturbance reflected in the form of folds and faults forming alternating basin and dome structures (Choubey, 1971 and Pandey, 1973). One such structure is the Jabera dome which is oval shaped and elongated in ENE-WSW direction where older rocks are exposed in the center surrounded by younger rocks (Fig. 9.42(b)). Such domes have special significance due to associated basement uplifts that are suitable for oil exploration. In case of the Jabera dome, Oil and Natural Gas Corporation has carried out exploration for oil (Ram et al., 1996).

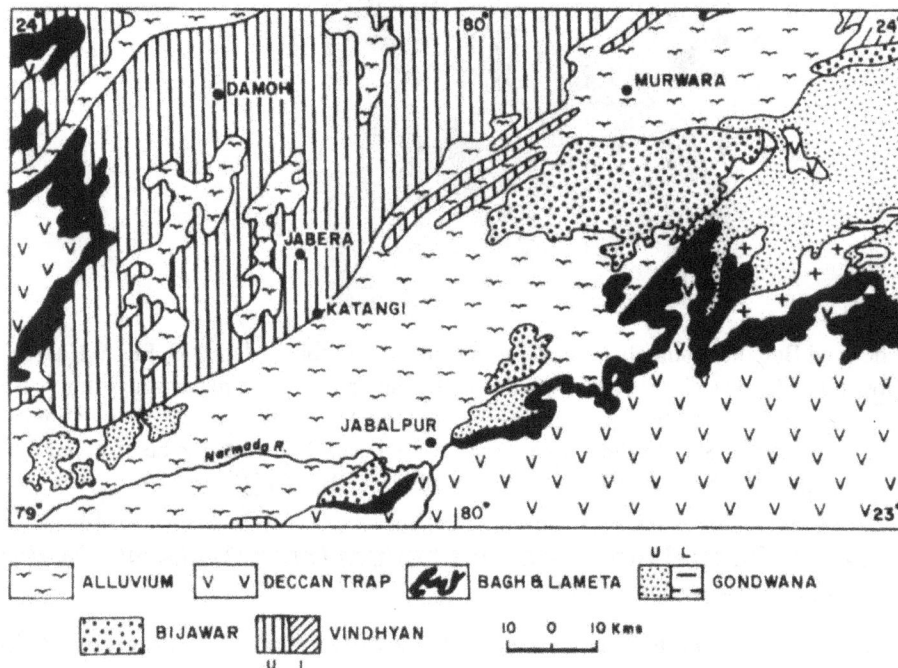

Fig. 9.42(a) Regional geology of the southern part of the central part of the Vindhyan Basin (10, Fig. 9.1(a)).

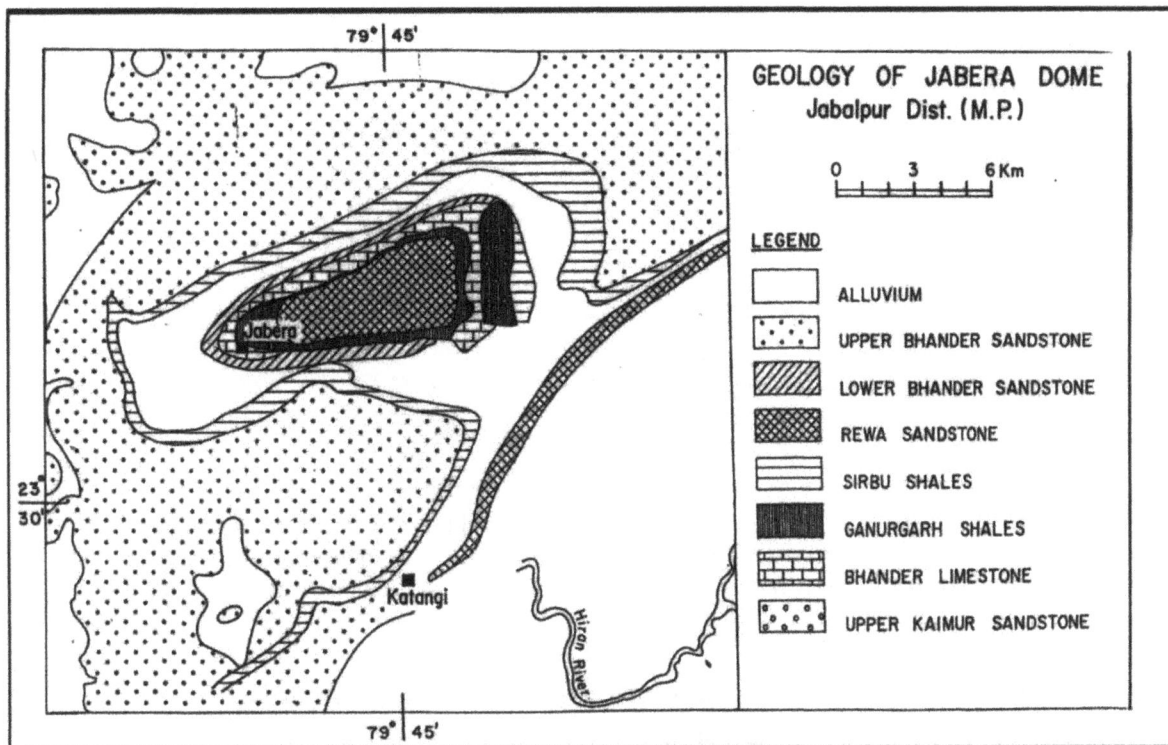

Fig. 9.42(b) Geology of Jabera dome in the southern part of the Vindhyan basin.

The lineament map based on landsat imagery indicates two sets of prominent lineaments, one in ENE-WSW direction and the other in NW-SE in the Jabera-Damoh area (Dey et al., 1993). The former represents the Son-Narmada trend while the latter represents the Precambrian trends of the Indian Peninsular Shield. The airborne total intensity magnetic map of the Vindhyan basin is given in Fig. 9.43(a) (Mishra and Hari Narain, 1977). These measurements were carried out at an altitude of 500 ft (~ 152 m) at a profile spacing of 1 km. The total intensity magnetic map of the Vindhyan basin has reflected several magnetic trends/lineaments and basement topography at a depth of 3-5 km (Mishra, 1987). The most significant magnetic trend in the entire region is the magnetic lineament from Murwara to Tejgarh (Fig. 9.43(a)) with several magnetic anomaly closures of 200-400 nT aligned along the lineament ENE-WSW, that represent the Narmada-Son trend in this section. The nature of these magnetic anomalies suggests that it represents a fracture zone filled with basic rocks at places. Similarly, another set of moderate magnetic anomalies around Nahata, Damoh, and Basa that are perpendicular to the Murwara-Tejgarh lineament representing NW-SE trend and seem to represent basic intrusive rocks at shallow depth. While conducting the ground checks, basaltic rocks were found under soil cover at a location south of Damoh, which may represent the inliers of Deccan Trap exposed west of Basa. Magnetic anomalies in this region are highly reflective as they occur in non-magnetic envoronment of Vindhyan basin. The high amplitude short wavelength magnetic anomalies NW and SE of Katangi coincide with the exposed Deccan Trap. The linear magnetic anomalies in the SE corner of the map represent a part of the Narmada-Son Lineament.

However, the domal structure NE and E of Jabera discussed above (Fig. 9.42(b)) does not produce any major anomaly in the total intensity map. The N-S linear in the center of the map appears to be a defect in data caused by navigational problems that are common in airborne magnetic surveys and are known as herring bone. The area surrounding Jabera and Damoh including Murwara-Tejgarh magnetic lineament is followed up on the ground using gravity, magnetic, and resistivity surveys that are described below.

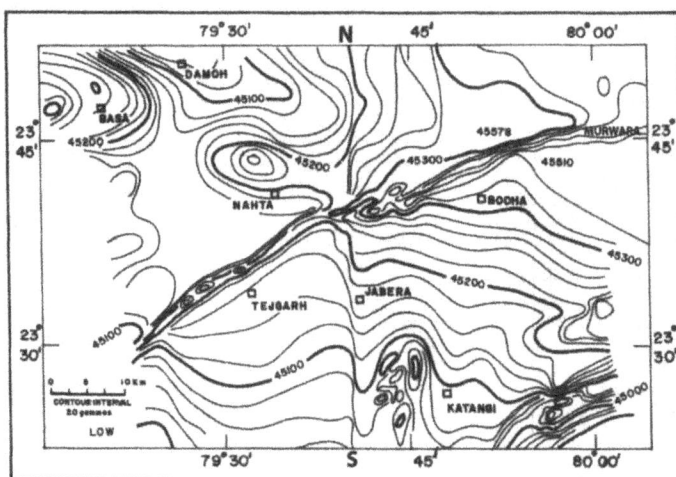

Fig. 9.43(a) Airborne total intensity map of a part of the Vindhyan basin showing a prominent magnetic lineament, Murwara-Tejgarh that is explored for groundwater potentiality. The high amplitude magnetic anomalies NE and SE of Katangi are due to inliers of Deccan trap. Flight altitude: 150 m and Profile spacing: 1.6 km (Mishra, 1987).

9.7.2 Jabera Dome – Bouguer Anomaly and Hydrocarbon Prospect

A detailed Bouguer anomaly of Jabera dome and surrounding area based on data recorded on ground at station spacing of 0.5-1.0 km is given in Fig. 9.43(b) that shows small amplitude gravity high and low, J1 and J2, respectively east of Jabera that are related to Jabera dome. These anomalies do not find reflection in the regional Bouguer anomaly map of the Vindhyan basin (Fig. 7.23(a)). The regional Bouguer anomaly map reflects only the major gravity low, L (Fig. 9.43(b)) that is related to the Vindhyan sediments. This region has been important for oil exploration (Ram et al., 1996) where gas occurrences have been reported. Therefore, the small wavelength and amplitude gravity high and low (J1, J2) assume importance in this regard especially because they occur in a large syncline of the Vindhyan basin. The gravity high, J1 represents a typical case of basement uplift (anticline) in a regional gravity low representing thick sediments that are suitable structures for hydrocarbon occurrences.

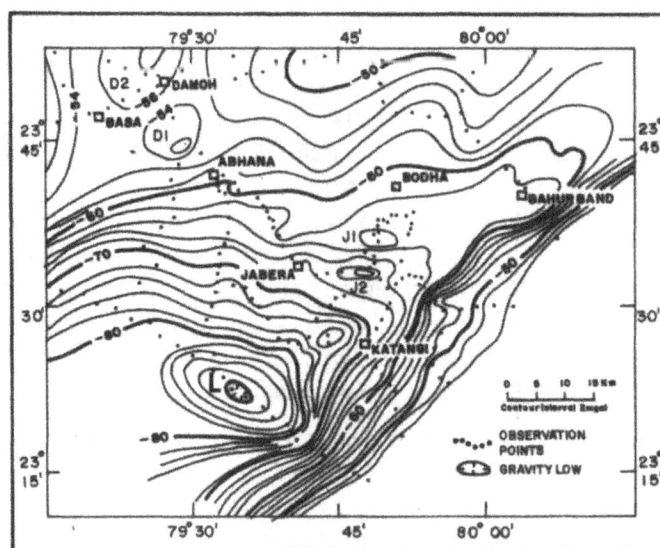

Fig. 9.43(b) A detailed Bouguer anomaly map of Jabera dome and surrounding regions. Gravity anomalies J1 and J2 are gravity high and low that is significant for oil exploration (Mishra et al., 1998).

A schematic regional seismic section is shown in Fig. 9.43(c) from Jabera to Mirzapur that shows ONGC wells at Jabera and Mirzapur showing various Vindhyan formations and one Amriti prospect in Rewa. This figure also shows the gravity profile depicting a regional high and a very significant low over the Amriti prospect. This gravity low appears to be caused by a low density intrusive up to a depth of 1-2 km (NGRI, 2003) which may form an oil trap in the surrounding sedimentary section. The regional gravity high is caused by the margin of the basin.

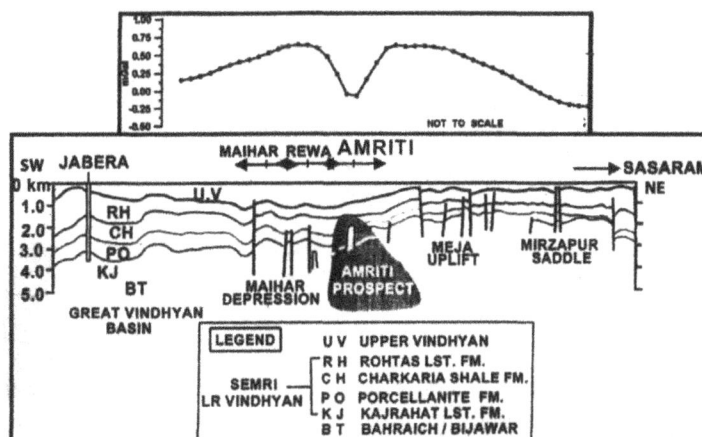

Fig. 9.43(c) A schematic section of the Vindhyan basin from Jabera (SW) to Mirzapur (NE) based on seismic profiles and borehole information and corresponding gravity anomaly with a gravity low over the Amriti prospect.

9.7.3 Ground Magnetic and Resistivity Studies

Ground magnetic and resistivity studies along five profiles were carried out across the Murwara-Tejgarh magnetic lineament (Fig. 9.43(a)). These profiles revealed high magnetic anomalies of 600-800 gamma (Fig. 9.44(a)) and low resistivity (Fig. 9.45(a)) over the lineament. This suggests that there are basic intrusives along the lineament and the region in close proximity of the lineament is more saturated with water than the surrounding regions. The magnetic profiles were modeled to find the depth to the magnetic body responsible for the anomalies. The depth to the top of the body along five profiles was obtained to be ~ 60-80 meters. Modeling along one of the profile is shown in Fig. 9.44(b) that provides a depth of ~ 60 m that appears to be the depth to top of mafic intrusive along this magnetic lineament. There is a possibility that the wells drilled up to this depth may encounter weathered/fractured rock below which hard massive rock is expected.

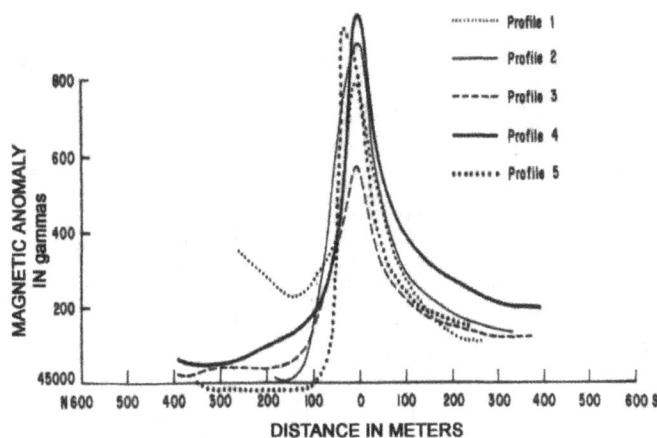

Fig. 9.44(a) Total intensity magnetic profiles recorded on ground across Murwara- Tejgarh airborne magnetic lineament showing a magnetic high of 600-800 nT over the lineament (Mishra et al., 1998).

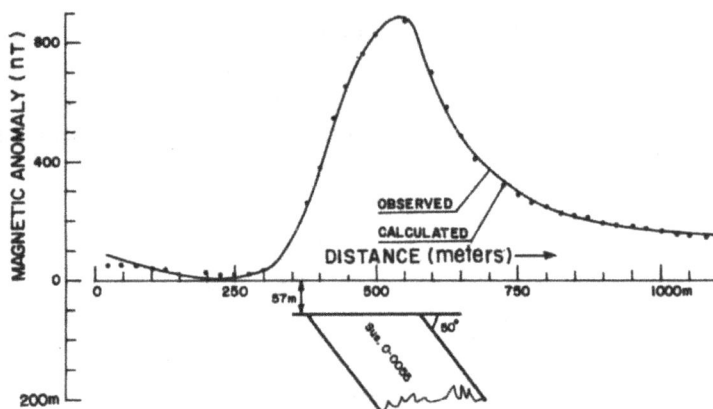

Fig. 9.44(b) Modeling of one of the ground magnetic profiles across Murwara-Tejgarh lineament showing a depth of ~ 57 m below surface.

9.7.4   Resistivity Soundings and Groundwater Potential

Resistivity soundings were carried out at three sites each on profiles 1, 2, and 3 with orientation perpendicular to the profiles. Out of three soundings across each profile, one station, S2 was located at the center of the lineament while other two, S1 and S3 were on either side away from the lineament. Typical resistivity sounding curves across one of the profiles are given in Fig. 9.45(b) that shows minimum resistivity at the station, S2 that is located close to the lineament. The resistivity sounding curves are interpreted using an inversion technique due to Jupp and Vozoff (1975) and Interpex 1D sounding inversion package. The results of these soundings are summarized in Table 9.4. The lower orders of the resistivity (8-10 ohm-m) are found at the second site (S2) across each profile that corresponds to weathered/fractured rock saturated with water. The yield of wells close to the lineament is higher compared to those drilled away on the northern and southern side from it. Thus, the airborne magnetic lineament from Murwara to Tejgarh represents a fracture zone, which is occupied by basic intrusive rocks at places, and is conductive compared to the surrounding regions due to the presence of groundwater along it.

Table 9.4 Results of resistivity soundings at sites S1, S2 and S3 along Profiles P1-P3 perpendicular to Murwara-Tejgarh lineament. Site, S2 lies over the lineament while S1 and S3 are away from it on either side.

Profile No.	Sounding No.	ρ_1 h1	ρ_2 h2	ρ_3 h3	ρ_4
P1	S1	22.4 0.6	7.8 7.4	138	
	S2	38.4 0.3	10.2 9.1	60.0 25.0	9999
	S3	346 0.5	104 1.0	418 13.0	9999
P2	S1	22.5 0.6	9.7 12.5	9999	
	S2	10.0 5.8	26.3 19.5	80.0 30.0	9999
	S3	110.0 1.3	165.0 13.0	9999	
P3	S1	860.0 0.9	200.0 13.8	108.0 46.7	9999
	S2	24.0 0.6	10.5 8.7	44.0 45.0	9999
	S3	32.0 0.5	11.0 6.4	85.8 32.0	9999

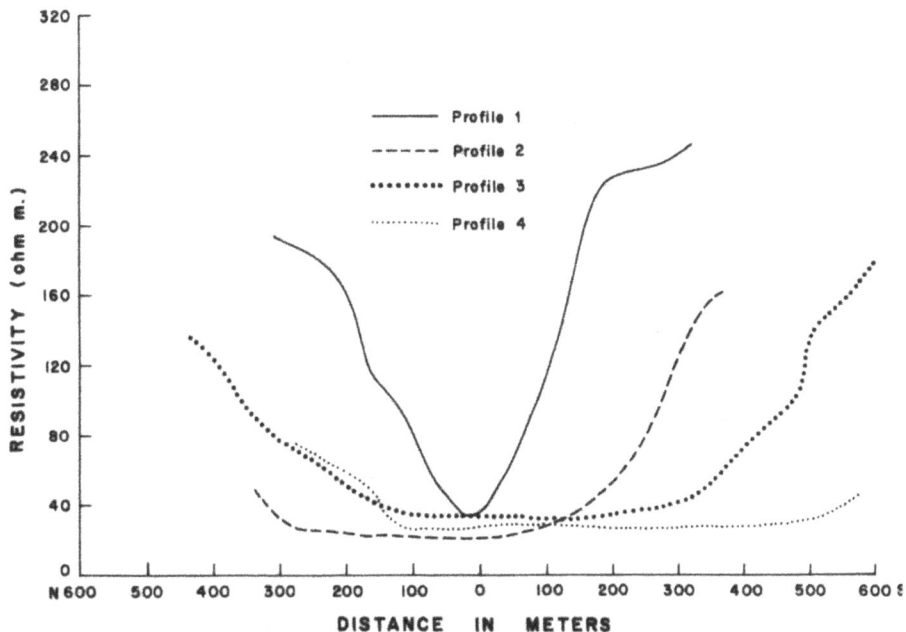

Fig. 9.45(a) Resistivity profiles across the magnetic lineament, Murwara-Tejgarh showing resistivity lows along the lineament indicating favourable groundwater conditions (Mishra et al., 1998).

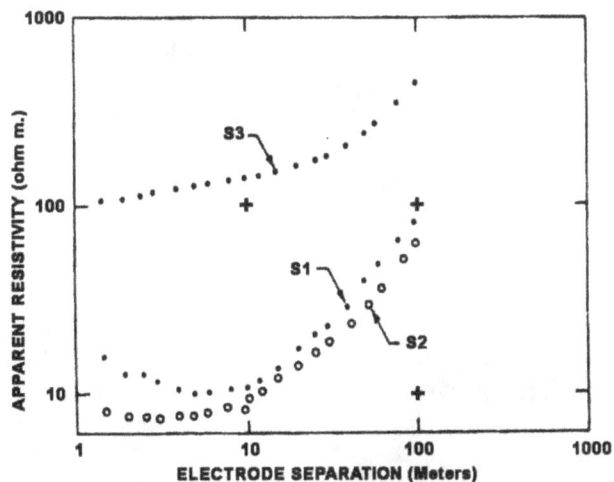

Fig. 9.45(b) Typical resistivity sounding profiles at three sites, S1, S2 and S3 with S2 being over the lineament and S1 and S3 being on either side of it. It shows three layer model as given in Table 9.4 (Mishra et al., 1998).

9.7.5   Singhbhum District, Jharkhand, India - Airborne Magnetic Trends and Groundwater Potential

Singhbhum district is a hardrock region with exposed gneisses where groundwater is scarce. Chandra and Reddy (1987) and Astier and Paterson (1989) have reported airborne magnetic lineaments that were followed up on the ground by different geophysical methods (Fig. 9.46). The sections close to the lineaments show high conductivity (low resistivity) in resistivity and electromagnetic measurements and magnetic gradient in ground magnetic survey indicating fault or shear zones. The two inferences together indicated faults or shear zones filled with water where boreholes were located that yielded good supply of water.

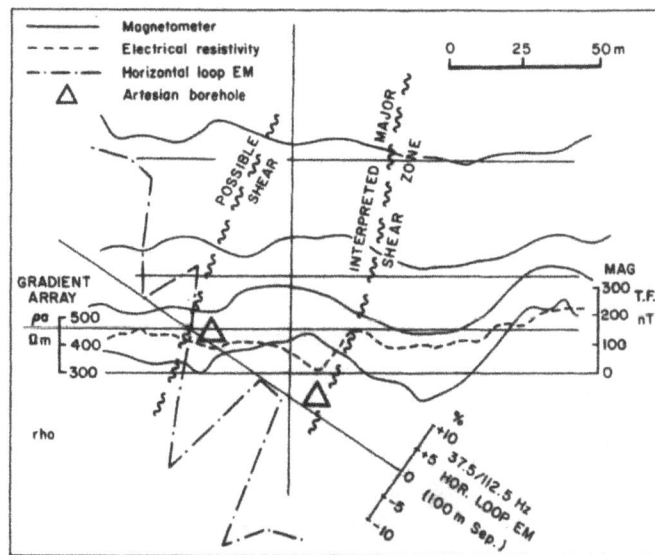

Fig. 9.46 Magnetic, resistivity and EM profiles across airborne magnetic lineaments (major shear zones) at Kudada, Singhbhum district, Bihar, India showing good groundwater potential close to the shear zones marked by triangles (Chandra and Reedy, 1987; Astier and Paterson, 1989).

9.8   Upper and Lower Assam Shelf – Hydrocarbon Prospect and Geodynamics

Oldest oil occurrences in this country come from this region (11, Fig. 9.1(a)) and earliest well drilled in this regard is Digboi oil well by Burma Oil Company. Geologically, it is a complex region being affected by Himalayan orogeny (Chapter 6). A regional tectonics map is given in Fig. 9. 47(a) that shows the area in the Eastern Himalayan Syntaxes between the Mishmi Hills and Mishmi thrust towards the north of it, Himalayan Frontal Thrust towards the east and Naga thrust towards the west. This area forms the Himalayan foredeep towards the east.

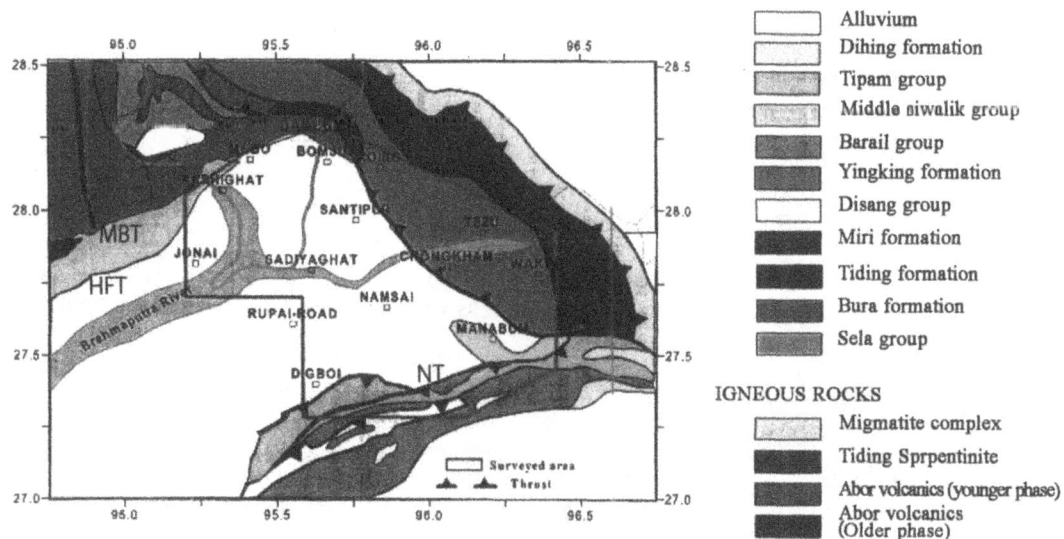

Fig. 9.47(a) A generalized tectonic map of upper Assam shelf (11, Fig. 9.1(a)). Most important elements are Main Boundary and Himalayan Frontal Thrusts (MBT and HFT) west of Pasighat and Misimi thrust west of Santipur and Naga thrust (NT) south of Manabum that are related to Eastern Syntaxes of Himalayan orogeny.
[Colour Fig. on Page 816]

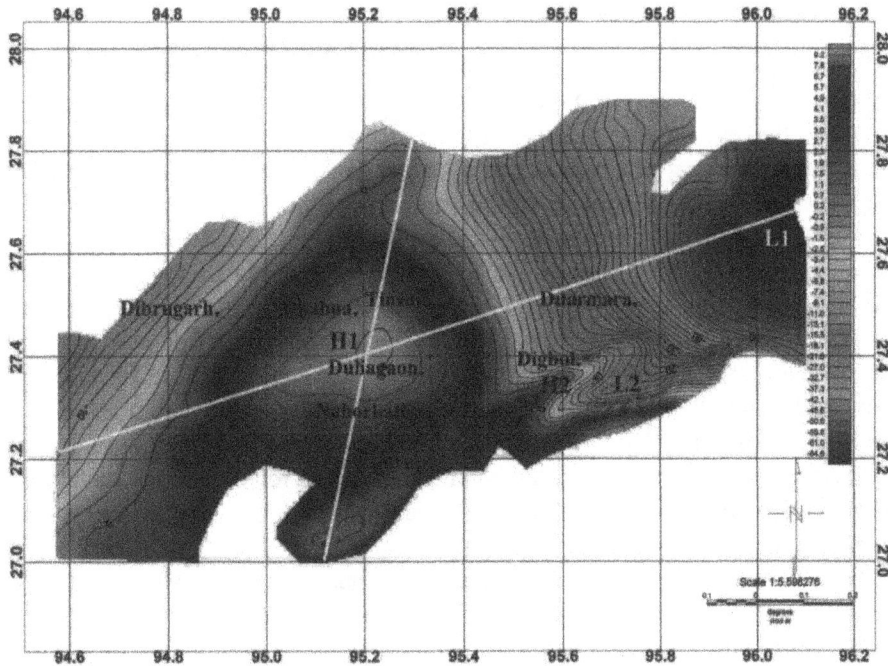

Fig. 9.47(b) Bouguer Anomaly map of Assam Shelf showing a gravity high in the central part and low along Himalayan fore deep. A constant regional field of –200 mGal is removed from the observed field (GSI, 1983, Volume II). [Colour Fig. on Page 816]

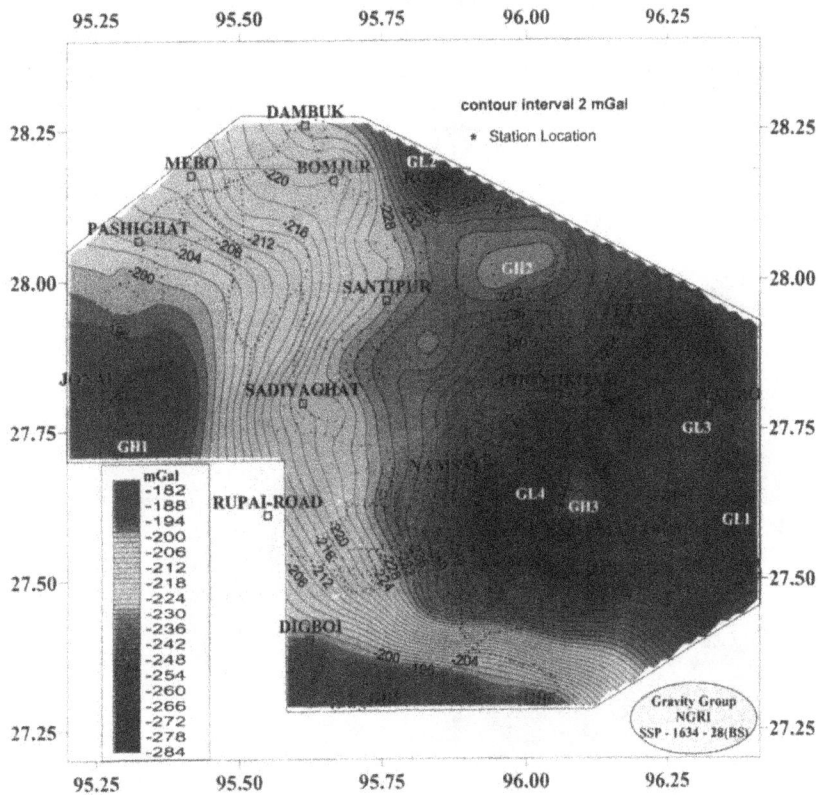

Fig. 9.47(c) Complete Bouguer Anomaly of the Upper Assam Shelf. H: Gravity Highs and L: Gravity lows. (NGRI, 2000; Ghosh and siddiquee 2009). [Colour Fig. on Page 817]

9.8.1   Gravity Anomalies of the Upper Assam Shelf

A regional Bouguer Anomaly of Upper Assam Shelf (11, Fig. 9.1(a)) is given in Fig. 9.47(b) (GSI, 1983, VII) that shows a relative gravity high (H1) in the central part and large gravity lows (L1) towards the NE related to the lithospheric flexure and sediments of the foredeep, respectively. The gravity anomaly by the Naga thrust is reflected as gravity highs (H2) along it followed by a low (L2) caused by thrusted basement rocks and sediments of the foredeep, respectively. The Digboi oil field is found along this thrust that might have provided the oil trap in this section. The gravity high due to lithospheric flexure in the center of the Assam Shelf is caused by underthrusting of the Indian plate towards the north, the NE, and the east synonymous with the gravity highs along the western Himalayan front, Delhi-Lahore Ridge as described in Sections 6.1.9 and 7.11.2. A complete Bouguer anomaly of the Upper Assam Shelf (11, Fig. 9.1(a)) with additional data in this section after terrain correction (Fig. 9.47(c)) was provided by NGRI (2000) that shows a major relative gravity high towards the SW and lowest gravity anomaly in the eastern part indicating a basement uplift and depression, respectively. However, the observed Bouguer anomaly is considerably affected in this region due to crustal thickening by Himalayan orogeny that requires the separation of the regional and the residual fields before the basement anomalies can be assessed. Therefore, a regional field (Fig. 9.48(a)) was separated from the observed field based on finite element method as discussed in Chapter 4 that is similar to polynomial approximation. It may be noted that in this case regional and residual field can not be separated based on zero free anomaly as proposed above in other cases due to its absence in the entire region and free air anomaly assumes large negative value due to crustal thickening. It shows consistently decreasing field towards the north east that indicates crustal thickening in this direction due to Himalayan orogeny. The corresponding residual gravity field is given in Fig. 9.48(b) that shows several highs and lows related to basement uplift and depressions, respectively. Highest gravity anomalies are observed in the eastern and the western parts (GH1 and GH2) where basement rocks are uplifted due to thrusting along the HFT and Mishmi thrusts, respectively. The gravity low GL4 in the central part is plausibly caused by thick sediments in this section. It also shows various profiles that were modeled for depth estimates. The gravity high, GH3 is quite significant as it represents a basement uplift in a large basin.

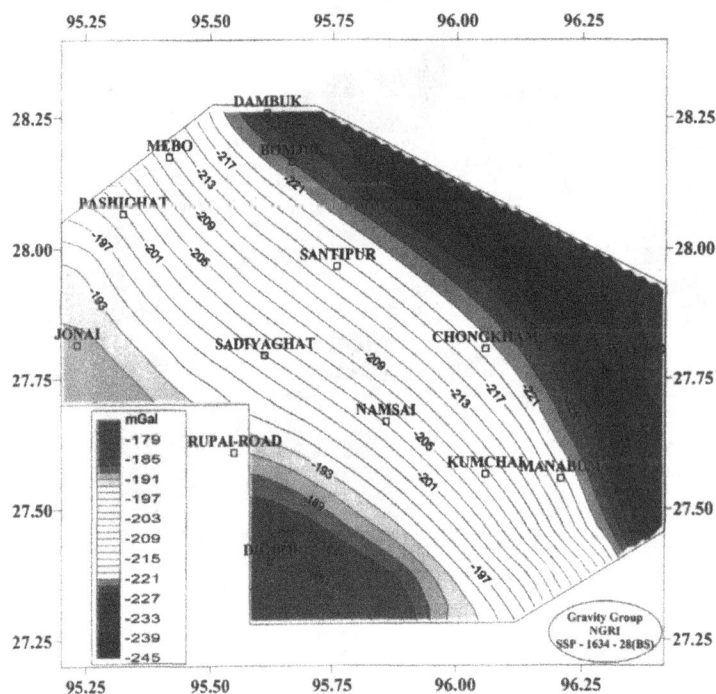

Fig. 9.48(a) Regional anomaly corresponding to a first order polynomial approximation from Bouguer Anomaly showing consistent gravity low towards the east due to crustal thickening. [Colour Fig. on Page 817]

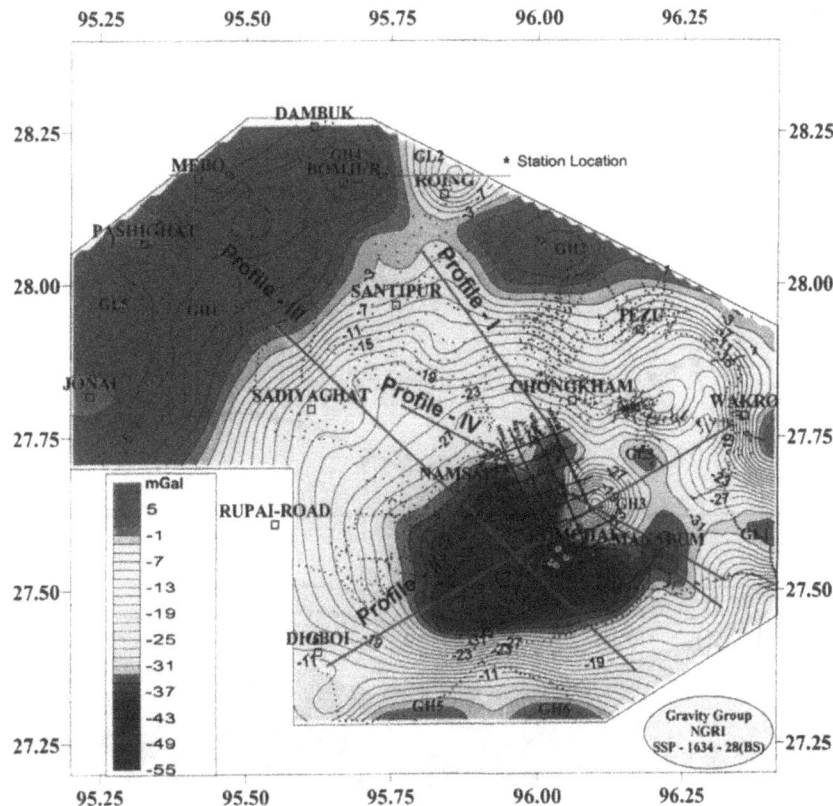

Fig. 9.48(b) Residual gravity anomaly after subtracting the regional field (Fig. 9.48(a)) from the observed field (Fig. 9.47(c)) showing gravity highs related to HFT and Misimi thrust on the western and the eastern sides, respectively. The gravity high, H4 is quite interesting as it represents a basement uplift in a large low due to a basin.
[Colour Fig. on Page 818]

9.8.2 Basement Model and Spectral Analysis

Modeling of gravity anomalies along various profiles provided a maximum depth to basement as ~ 6-7.0 km and minimum as 4.0 km in the eastern and the central parts, respectively and the top alluvium extends up to ~ 2.0 km (NGRI, 2000). Modeling of both gravity and magnetic data along an E-W profile (Jonai-Wakro, Fig. 9.47(c)) is presented in Fig. 9.49(a) that shows horst kind of structures in the eastern and the western parts related to the HFT and the Mishmi thrusts, respectively (Fig. 9.47(a)) and Assam valley in between with thickest sedimentary column of ~ 6 km. Density and susceptibility of different blocks in the basement are given in the figure which suggests that the thrust from the east and the west have affected the basement even in the Assam valley that looks like a rift valley. Ghosh and Siddiquee (2009) have reproduced the Bouguer anomaly (Fig. 9.47(c)) and provided the spectrum of the observed gravity field along several profiles and spectrum along one such N-S profile passing through GH3 (Fig. 9.47(c)) is shown in Fig. 9.49(b). This profile also shows the gravity field that shows the gravity high, GH3 in the center located in a regional low that indicates basement uplift in a large basin and is quite significant for hydrocarbon exploration. The spectrums provided by them invariably shows two layers at depths of ~ 6-7 km and ~ 2 km that implies maximum depth to the basement and bottom of the alluvium layer that forms major density discontinuities (Fig. 9.49(b)). Fig. 9.50a shows a seismic section in this region (Ghosh and Siddiquee, 2009) that also provides a basement depth of ~ 7 km. The basement high in this section is thrust controlled that has affected the entire sedimentary section and can form an oil trap in case other conditions are favorable.

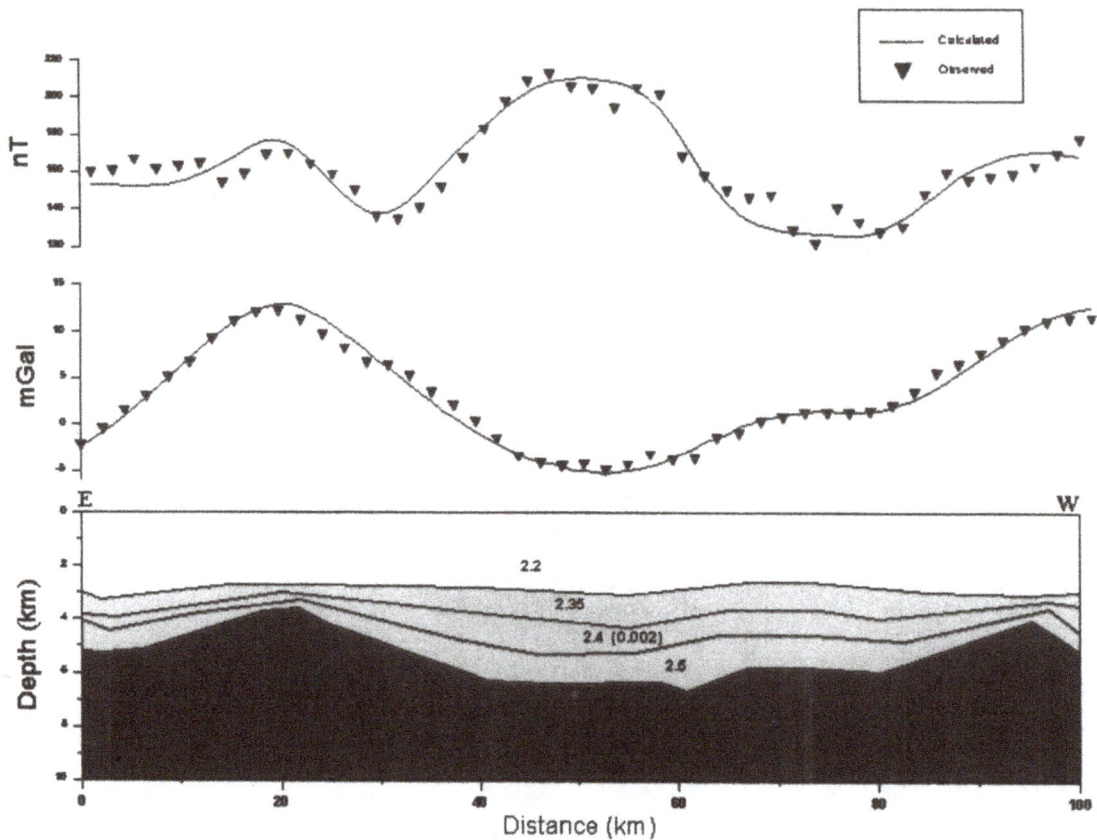

Fig. 9.49(a) Basement model along an E-W profile of residual gravity anomaly showing horst kind of structures due to gravity highs, H1 and H2 related to the HFT and Misimi thrust and gravity lows due to Assam valley in between with a thick sediment of ~ 6 km (NGRI, 2000). [Colour Fig. on Page 818]

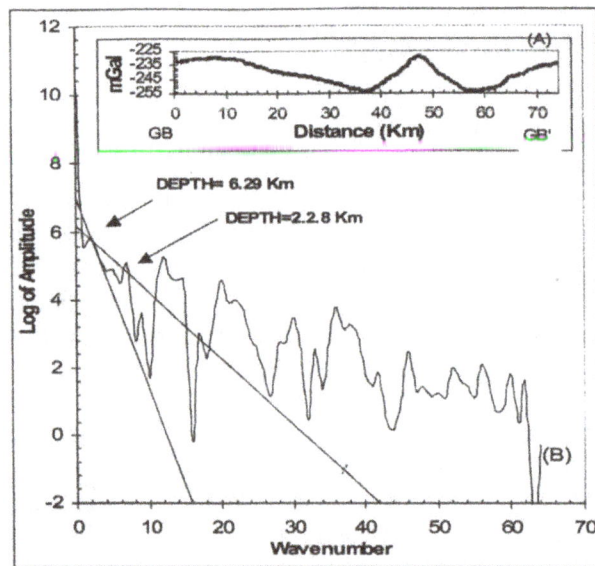

Fig. 9.49(b) Spectrum of the Bouguer anomaly along an N-S profile passing through the gravity high, H4 (Fig. 9.47c). The gravity profile is also shown with a high in the center related to the basement uplift and regional low related to the sediments of the basin. Spectrum shows a two layer model with average depths of ~ 6.3 km and ~ 2.3 km (Ghosh and Siddiquee, 2009).

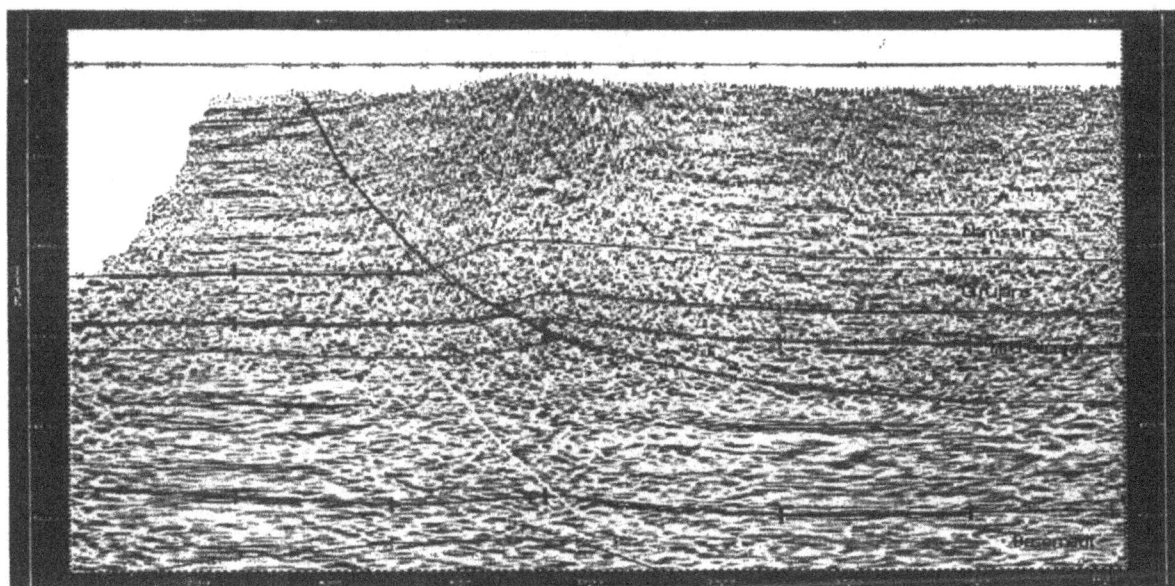

Fig. 9.50(a) An interpreted seismic section across Upper Assam Shelf (Fig. 9.47(a)) showing basement uplift along a thrust that has affected the total sedimentary section. (Ghosh and Siddiquee, 2009)

9.8.3 Geological Section and Oil Occurrences

Fig. 9.50(b) (Pahari et al., 2008) shows a section across Upper Assam Shelf based on borehole data which suggest the importance of basement uplift and anticline structures in sedimentary section for oil reservoirs in this region that occur in Barail, Kopili and Sylhet/Tura formations of the Schuppen belt since Late Miocene. The importance of basement uplifts based on gravity highs in a regional gravity low have been highlighted above. These basement highs might have been caused by the crustal up warp due to the lithospheric flexure of the Indian plate as discussed in Section 7.11 and shown by gravity highs, H1 (Fig. 9.47(b)).

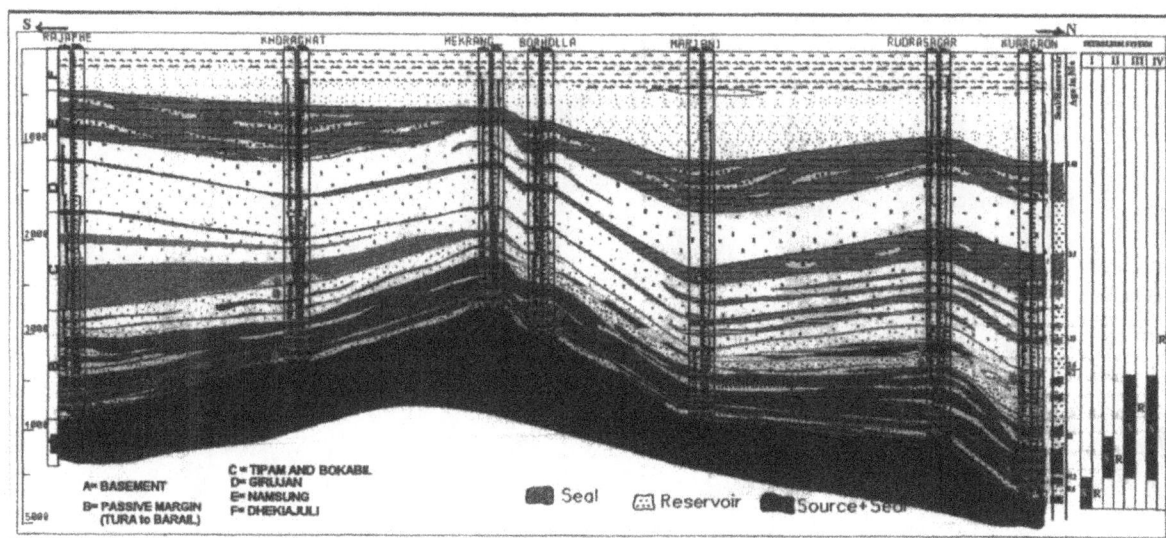

Fig. 9.50(b) Geological cross section across Upper Assam Self showing existing petroleum system (Pahari et al., 2008)

9.8.4   Lower Assam Shelf (Southern Part) – Hydrocarbon Prospect and Geodynamics

As discussed above the Upper Assam self is important for oil exploration as it is a part of the fold and thrust belt along the Eastern Fold Belt of the Burmese arc (Section 6.2). In fact the whole of the NE India is significant for oil exploration and several of the oil fields have been discovered in this region. Fig. 9.50(c) (Singh et al., 1996) shows the main thrusts, Dibang and Naga thrusts known as Schuppen belt due to occurrences of several thrusts in this section. It also shows some of the structures along Naga Thrust that are discovered based on integrated geophysical exploration over decades and are significant for oil exploration. A geological section across Lower Assam shelf (12, Fig. 9.1(a)) extending from Mikir Hills (Fig. 7.30(a)) to Naga Thrust in Southern part of Assam east of Bangladesh is given in Fig. 9.50(d) (Singh et al., 1996) that shows horst and graben structures west of the Naga Thrust which are important for oil exploration. These horst and graben structures are formed due to collision of the Indian and Eurasian plates with Burmese microplate in between. The maximum thickness of sediment is ~ 4 km but it could be more also in different sections. The basement is overlain by Sylhat trap rocks that represent volcanics during the break up India and Antarctica (Section 7.6.4). Oil occurs in Tipam and Surma sandstones.

Fig. 9.50(c) Thrusts along Eastern Fold Belt of Burmese arc and Schuppen belt. It also shows some significant structures for oil exploration along Naga Thrust (Singh et al., 1996).

A typical Bouguer anomaly of a section of this region around Naga thrust is given in Fig. 9.50(e) that shows alternating linear highs and lows with compact highs and broad lows that are typical of collision tectonics. In comparison to Fig. 9.50(d), they can be attributed to horsts and grabens, respectively. They typically represent accretionary prism of plate tectonics (Section 5.2). It is represented by linear bands on surface of Miocene and recent Quaternary sediments caused by horst and graben tectonics, respectively. In gravity map (Fig. 9.50(e)), they are represented by gravity highs and lows, respectively.

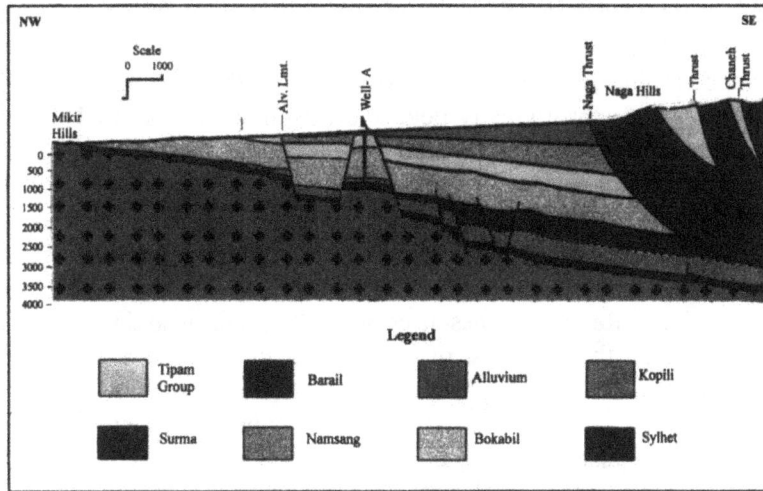

Fig. 9.50(d) A seismo-geological section across Lower Assam Shelf (Dhansiri valley, Singh et al., 1996)) showing horst and graben structures in sedimentary section, typical of collision tectonics.
[Colour Fig. on Page 819]

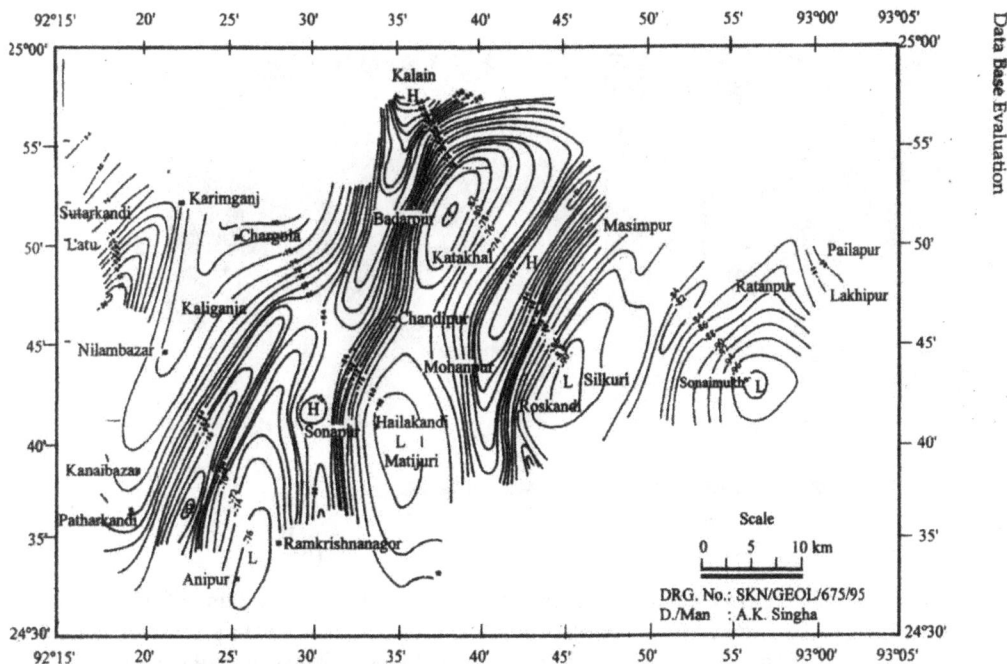

Fig. 9.50(e) A Bouguer anomaly map of Lower Assam (Southern Part, Cachar, Singh et al., 1996)) showing bands of gravity highs and lows typical of horst and graben structures of accretionary prism.

9.9   Groundwater budgeting using GRACE Satellite Gravity Data

GRACE (Gravity Recovery and Climate Experiment) satellite launched in 2002 (Tapley et al., 2009) measures small variations in the earth's gravity field which are used to infer total stored water that is the sum of groundwater, soil moisture, surface water, snow ice, and biomass. Budgeting of glaciers based on temporal variation in the gravity field measured from Satellite has been attempted in Greenland, Antarctica etc., (Chen et al., 2008). The present rate of ice mass loss from Antarctica is estimated at 81 ± 17 km^3/yr. Seasonal variation in the gravity field measured from GRACE has been recently used to monitor the relative depletion of groundwater in a region (Tiwari et al., 2009; Rodell et al., 2009). Temporal variation in gravity field at a place responds to fluctuations in the groundwater level besides anthropogenic changes due to human beings that can be evaluated after applying corrections to the observed field. The GRACE Satellite mission provides monthly and global gravity field solutions at scale of a few hundred kilometer and greater in the form of spherical harmonic coefficients. Tiwari et al., (2009) used coefficients to a maximum of 60° (http://podaac.jpl.nasa.gov/grace) for the period April, 2002 to June, 2008 to compute monthly mass changes in South Asia related to groundwater changes. The anthropogenic contribution is obtained by subtracting the water storage estimates predicted by land surface models. The land surface model used by them for this purpose is version 4.0 of the Community Land Model (CLM) maintained by the National Center for Atmospheric Research (Oleson et al., 2008). The resulting trend is shown in Fig. 9.51(a) and (b). Fig. 9.51(a) shows the total mass of groundwater stored while Fig. 9.51(b) shows the changes in groundwater level after subtracting the anthropogenic contribution from the CLM from the estimated total mass. This figure shows both maximum storage and maximum depletion of groundwater in North India that is one of the most populated and irrigated regions of the world where irrigation has been growing constantly since 1960 after the Green Revolution. Based on an almost similar approach, Rodell et al., (2009) suggested maximum depletion of groundwater from NW India suggesting that the entire North India is suffering from this water depletion. They also suggested minimum depletion of groundwater from the South Indian Shield that may be attributed to this region being a hard rock area. The two centers of recharge of groundwater along the west and the east coasts (Fig. 9.51(b)) are centers of discharge of the Narmada and Godavari rivers, respectively. Dams constructed on them might be responsible for recharge in these regions.

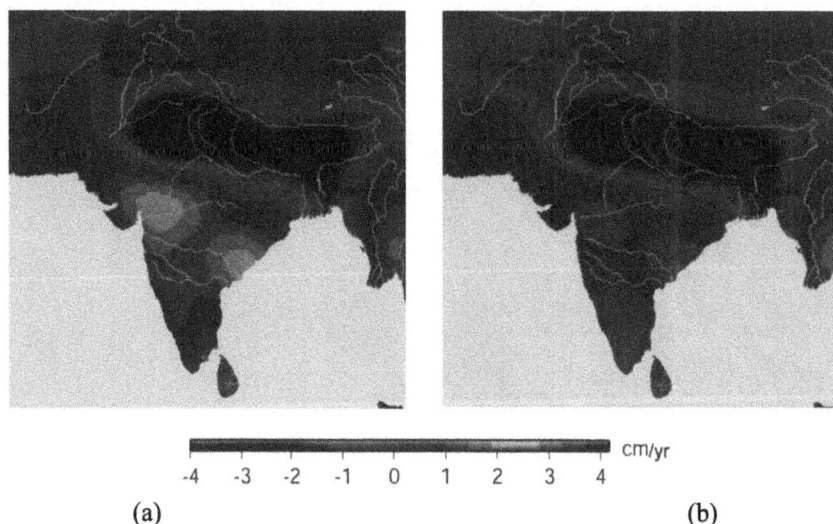

(a)   (b)

Fig. 9.51 (a) Rate of change of terrestrial water storage in cm/yr of water thickness obtained from GRACE gravity solutions. White lines show major rivers. (Tiwari et al., 2009) (b) After subtracting from the naturally occurring water storage variability predicted from the CLM hydrological model from Fig. 9.46(a). This provides the anthropogenically caused groundwater loss that is seen most in the north under the Ganga basin which also happens to be most densely populated region. Maximum exploitation of groundwater has taken place during last decade for agriculture in this section (Tiwari et al., 2009). [Colour Fig. on Page 819]

Another application of satellite gravity in estimating changes such as sea level changes significant for environmental studies is discussed in Section 2.8.3.

9.10   Mineral Exploration and Geodynamics

There are two aspects of gravity and magnetic surveys for mineral exploration, viz. direct and indirect investigations. In case they are associated with mafic/ultramafic rocks, they show slightly higher density compared to normal crustal rocks and therefore can be delineated using gravity method. Mafic associations also produce magnetic anomalies and therefore magnetic methods can also used to explore them. Direct investigations relate to exploration for heavier and lighter minerals compared to average density of the crust (2.67 g/cm^3) which would provide gravity high or low, respectively. Indirect investigations relate to delineation of faults, thrusts, shear zones, and magnetic intrusions, which are host to several mineral occurrences. Direct investigations have been found to be useful in case of minerals like iron ore, chromite, baryte, etc., while for base metal it is used to delineate associated faults, thrusts, etc. The latter, however, are primarily explored using electrical methods. The electrical methods are based on conductive properties of the minerals and therefore, it is primarily used for massive sulfide ores, iron oxides, graphite, vein deposits etc.

The most common electrical methods used for this purpose are, resistivity and self potential methods and electromagnetic and induced polarization methods. The resistivity method is explained briefly in Section 9.1.3 with regard to groundwater exploration. The same principle is used here also where conductive anomalies are due to mineralized zones instead of water. However, mineralized zones can create their own potential and therefore the potential difference in these cases can also be measured without any current electrodes and simply using potential electrodes (Fig. 9.2(b)) and a volt meter that is known as self potential method. The electromagnetic and induced polarization methods are more involved both in field operations and data processing. Briefly, the electromagnetic method is based on the principle of propagation of time varying electromagnetic field in and over the earth through a cable or coil with A.C. current that induces secondary electric currents in the subsurface conductors that oppose the primary field. The secondary field in turn generates both electric and the magnetic fields that can be measured at the surface and interpreted in terms of subsurface conductive bodies. In practice, the magnetic field is measured. The measurement can be made either by the conductive method of electrode contact or by coil induction. The inductive method is more common in electromagnetic prospecting where amplitude of in phase and out of phase components with respect to primary fields is measured. Measurements can be made both in time domain and frequency domain. In the first case, time varying field on the earth's surface is created for a specific interval of time and induced secondary field is measured at specific time interval as it decays, known as windows or channels. As the secondary field is measured after the primary field is switched off, it makes the measurement easy. First window records the signal from the shallow sources, while subsequent windows record signals from deeper sources. Slower decay rate indicates conductive bodies. The airborne version of time domain E.M. system is discussed in Chapter 3.0 and the principle of operation is same in both the cases. Two configurations are mainly used in this case. The first is profiling mode where a small transmitter and receiver are moved along the profile and measurements are made. The second configuration involves laying of a central transmitting loop. A receiving coil is located at the loop center. Several field units based on these principles, along with processing and interpretation softwares, are available in the market and can be used for this purpose.

In frequency domain methods, the signal is recorded for different frequency bands of the transmitting field. Larger the frequency, smaller the depth of penetration. It consists of a small multi-frequency vertical dipole (horizontal loop) transmitter coil and a receiver coil (placed horizontally) at some distance where in phase and out of phase (Quadrature components) of the vertical magnetic fields at different frequencies are measured. The mid-point of the transmitting and receiving coil is the measurement point. The measurement of the real and the imaginary components of the secondary field is carried out by means of a compensator and their values are generally stated as a percentage of the primary field. Depth of penetration is usually claimed to be one half of the coil spacing. Several makes of these instruments along with interpretation softwares are available in the market that can be readily used.

The induced polarization method is similar to resistivity or self potential method where the setup is almost similar, but measurements are made after switching off the current. Once the D.C. current is switched off, potential across the potential electrodes first drops sharply but afterwards decays slowly that can be measured at specific time interval giving rise to time domain I.P. Similarly, there is a frequency domain I.P. where the potential drop with respect to a.c. frequencies is measured. It is measured as dV/V where dV is the secondary voltage after some time t and V is the primary voltage and expressed as millivolt per volt or percent of the primary field. It is referred to as polarizability. It can also be expressed as apparant chargebility similar to apparent resistivity that is defined as the area between the decay curve of the potential and the time window for which the measurement is made and is expressed as millivolt second per volt. There are several make of I.P. units in the market along with the necessary softwares. For further details of these methods readers may refer any text book on Exploration Geophysics such as Telford et al., (1990), Lowrie (1997) etc.,

There are numerous geophysical investigations carried out for mineral exploration in this country, primarily by Geological Survey of India and several review series/books have been published by them (GSI, 1983, 1999, 2002 ; Suguna Tulasi et al., 2003; Ramakrishna, 2006, etc.,) that can be consulted by those interested. Some selected examples are discussed below to highlight the importance of integrated exploration.

9.10.1 Base Metal and Gold Mineralization

As discussed in Chapter 7, mobile belts are most common sections suitable for base metal mineralization that are related to thrusting commonly associated with mobile belts. Geophysical anomalies related to some typical case histories and importance of integrated geophysical investigations are discussed below.

(i) Delhi Group of Rocks, Kayar Mineralized Zone, Ajmer, Rajasthan

This belt is associated with the Delhi group of rocks in the Aravalli Delhi Mobile Belt (Section 7.5) that is a well known fold belt of Proterozoic period where east verging thrusts are common. It is known for occurrences of copper, lead, and zinc. We therefore planned integrated geophysical profiles across some of the known occurrences as test profiles to understand the nature of geophysical anomalies observed over them and most successful method of exploration in this terrain (NGRI, 1999).

Fig. 9.52(a) shows the results of various geophysical methods along one of the profiles. On the western side there are good gravity (G1, ~ 1 mGal) and magnetic (M1, ~ 200 nT) anomalies that are typical of fault/thrust that may be east verging as such thrusts are common in this mobile belt. There is a corresponding S.P. anomaly (SP1) that is slightly displaced eastwards with respect to gravity and magnetic anomalies but coincides with the Zn and Pb occurrences, O1. The displacement of gravity and magnetic anomalies can be explained as they are related to the western side of the thrust while its estern side is mineralized giving rise to the S.P. anomaly. There is no significant electromagnetic response at any of the frequencies for this occurrence indicating that the S.P., gravity and the magnetic methods are successful in the present case. The electromagnetic response, E1 coincides with a small gravity rise but there is no corresponding magnetic response or any known good occurrences of Zn and Pb. They may represent some unknown occurrence that requires to be tested. Another electromagnetic anomaly, E2 is quite consistent at at all frequencies and has given rise to increase in the gravity field (G2) and the magnetic fields (M2) similar to fault/thrust and approximately coincide with occurrence, O2. Some displacement in these anomalies can be interpreted due to shift in the thrust and the mineralized zone in this section.

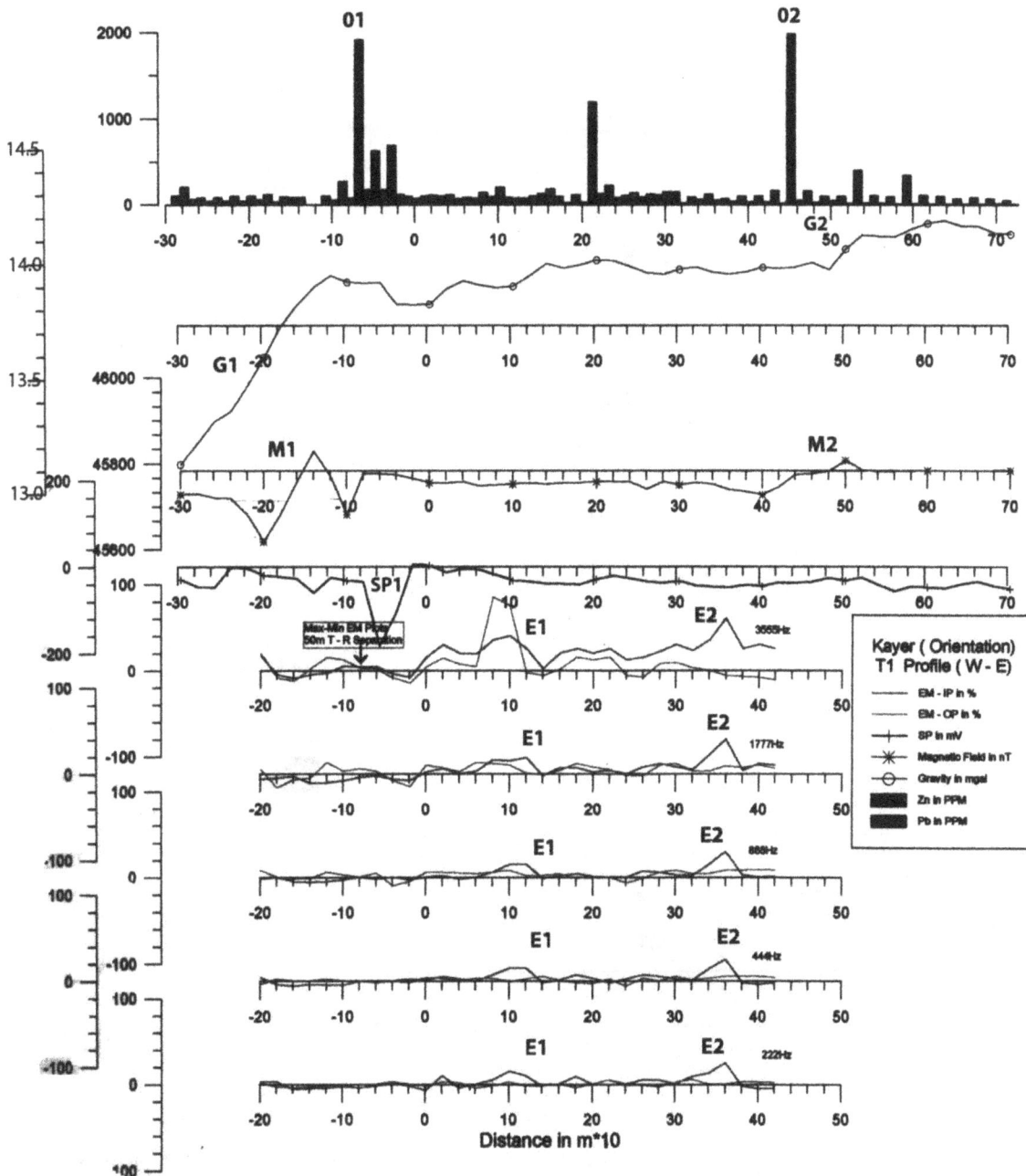

Fig. 9.52(a) Integrated geophysical exploration along a W-E profile in Kayar mineralized zone in the Delhi Mobile Belt (NGRI, 1999).

The results of geophysical investigations along another profile in this region are given in Fig. 9.52(b) that shows time domain E.M. measurements. This figure shows a gravity gradient, G1 and a gravity low, G2 and a corresponding magnetic anomaly, M1 that coincides with the broad E.M. anomaly, EM1 in all channels and are related to the occurrence O1. The gravity anomaly, G1 indicates a fault/thrust while other anomalies indicte the mineralized zone associated with it. In this case, there is no S.P. anomaly that was quite prominent in the previous case.

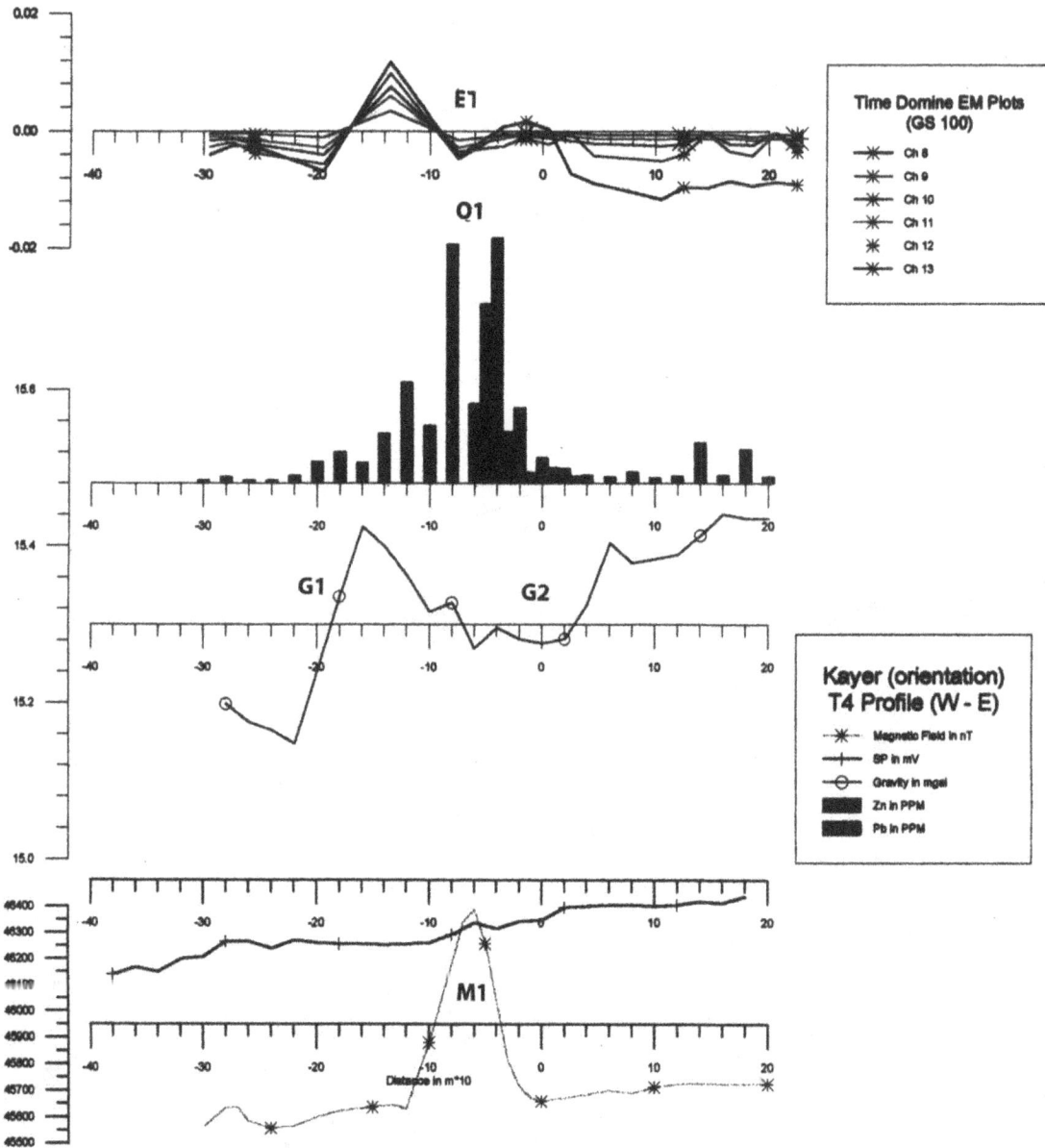

Fig. 9.52(b) Integrated geophysical exploration along a W-E profile in Kayar mineralized zone in Delhi Mobile Belt (NGRI, 1999).

Fig. 9.52(c) is the residual gravity anomaly in a plane over the Khetri base metal mineralized belt in Rajasthan (Ramakrishna, 2006). It occurs in the regional setting of the Aravalli Delhi Mobile Belt which represents a Meso-Proterozoic collision zone as discussed above and in Section 7.5 and provides favorable sites for mineralization due to associated east verging thrusts. It shows a residual gravity anomaly of 0.3 mGal and a significant magnetic anomaly indicating its association with mafic intrusive along east verging thrust.

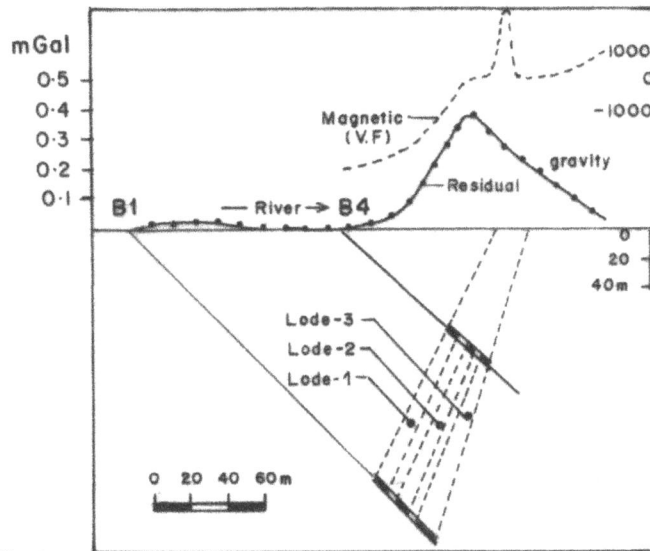

Fig. 9.52(c) Residual gravity and magnetic profiles along boreholes B1-B4 in Banwas block, Khetri copper belt, (Ramakrishna, 2006).

(ii) Mahakoshal and Bijawar Group of Rocks, Central India

As described in Section 7.6, the Mahakoshal and Bijawar group of rocks formed at the margin of the Bundelkhand craton with mafic/ultramafic intrusive represent rifted margin of Paleo Proterozoic period (Section 7.6.2; Mishra and Rajasekhar, 2008) and are suitable host rock for mineralization. Mahakoshal group of rocks are exposed along the North and the South Narmada faults south of Vindhyan basin in patches and is characterized by regional gravity highs (H4, Fig. 7.23(a)) along the northern margin of the Satpura Mobile Belt (Section 7.6). Gravity high at Gadarwara supported by airborne magnetic anomaly provided a good SP anomaly (Achuta Rao et al., 1994) indicating association of the base metal with it. Subsequent follow up by Intierra, Australia (2007), through several bore hole sampling and geochemical analysis, indicated good prospect for copper, gold and nickel associated with the Gadarwara anomaly. There are several other patches of exposed Mahakoshal group of rocks in this section SW of Jabalpur (Fig. 7.20) showing gravity highs (Fig. 7.23(a)) and high amplitude airborne magnetic anomalies (AMSE, 1995). Some of them even occur under thin cover of alluvium, which should be followed up in detail on the ground using integrated geophysical and geochemical exploration. The airborne magnetic maps are readily available from Airborne Mineral Survey and Exploration (G.S.I.), Banglore that should be investigated along with the gravity data in the first instance.

The Bijwar group of rocks on the northern side of the Vindhyan basin is equivalent to Mahakoshal group of rocks and supposedly represents rifted margin platform deposit with several mafic/ultramafic intrusive of the same time (Section, 7.6.2; Mishra and Rajasekhar, 2008). Results of an integrated geophysical exploration in the Bijawar group of rocks (Gurahar Pahar) in the Sidhi district, M.P. (Fig. 9.53(a), Chandra et al., 1999) have provided significant geophysical anomalies of S.P., low resistivity, and high chargeability in I.P. measurements indicating a conductive body that suggest sulphide mineralization. The same is also reflected in contour maps of apparent resistivity and chargeability (Fig. 9.53(a)). Absence of magnetic anomaly indicates the absence of any association of mafic minerals.

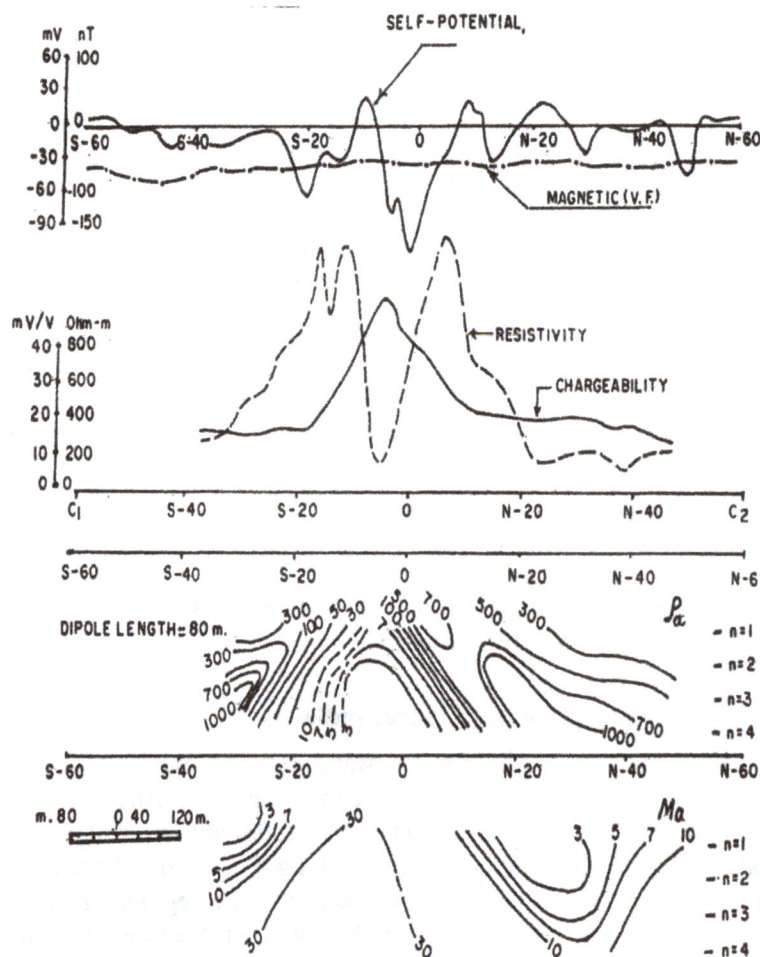

Fig. 9.53(a) S.P., magnetic and I.P./resistivity (gradient and dipole-dipole arrays) profiles in Gurahar Pahar area in Sidhi district, M.P. (Chandra et al., 1999)

(iii) Bijawar Basin, Jhansi District, U.P. – Airborne Magnetic and E.M. Surveys

Due to importance of Bijawar basin in central India for exploration of base metal mineralization, airborne magnetic and electromagnetic surveys were carried out in this section (NGRI, 1975; Mishra et al., 1978). The survey was conducted with a terrain clearance of 120 m along N-S flight lines at spacing of 0.5 km. The E.M. system was a time domain INPUT system (Section 3.6, Gupta sarma et al., 1976). A detailed geological map of the region is given in Fig. 9.53(b) that shows Bundelkhand granite and gneisses as basement towards the north and low-medium grade metamorphic rocks of Bijawar group of Paleo-Proterozoic period. The Bijawar group is largely represented by (i) Sonrai formation of shales, carbonate rocks with greenschists or basalts, and (ii) iron formations of Banded ferruginous quartzites. The Sonrai carbonate formation of the Bijawar group of rocks has shown the prospect of lead, zinc and copper mineralization. This map also shows E.M. anomalies as good, medium, and poor conductors based on the decay rate and their presence in different channels as described in Section 3.6.2. Airborne total intensity map of this region is presented in Fig. 9.53(c) that shows a large magnetic anomaly in the central part over iron formations. Quantitative interpretation of INPUT anomalies are difficult and therefore an empirical approach was desiged to interpret these anomalies. Conductivity-thickness parameter (σt) was computed using the formula for thin sheet (Verma, 1975):

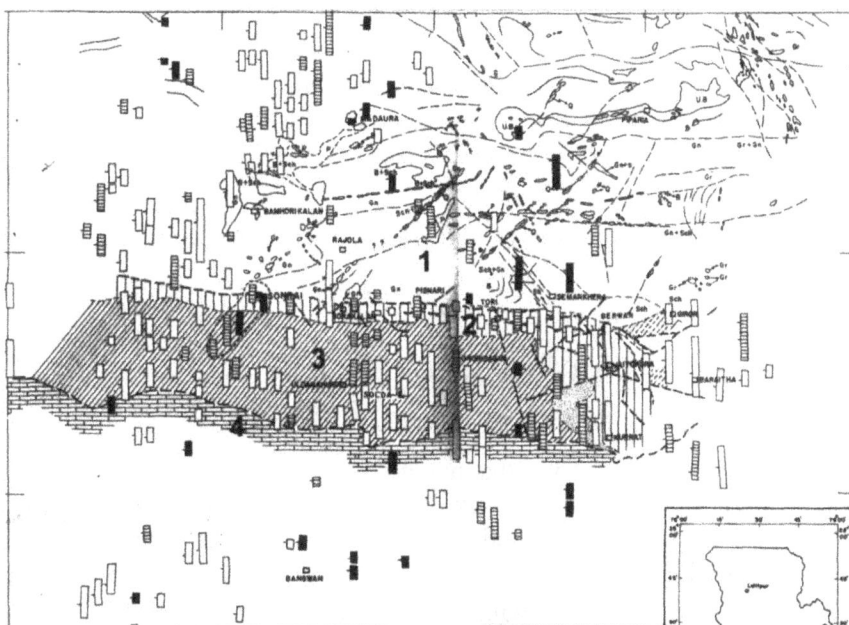

Fig. 9.53(b) Geological map of the Bijawar basin in central India showing Bundelkhand granite as basement rock in the north (1), Bijawar Group of rocks as iron and Sonrai formations (2 and 3) and Vindhyan sediments (4). It also shows good (filled rectangle), medium (rectangle with horizontal lines) and poor (empty rectangles) conductors based on decay pattern of E.M. anomalies.

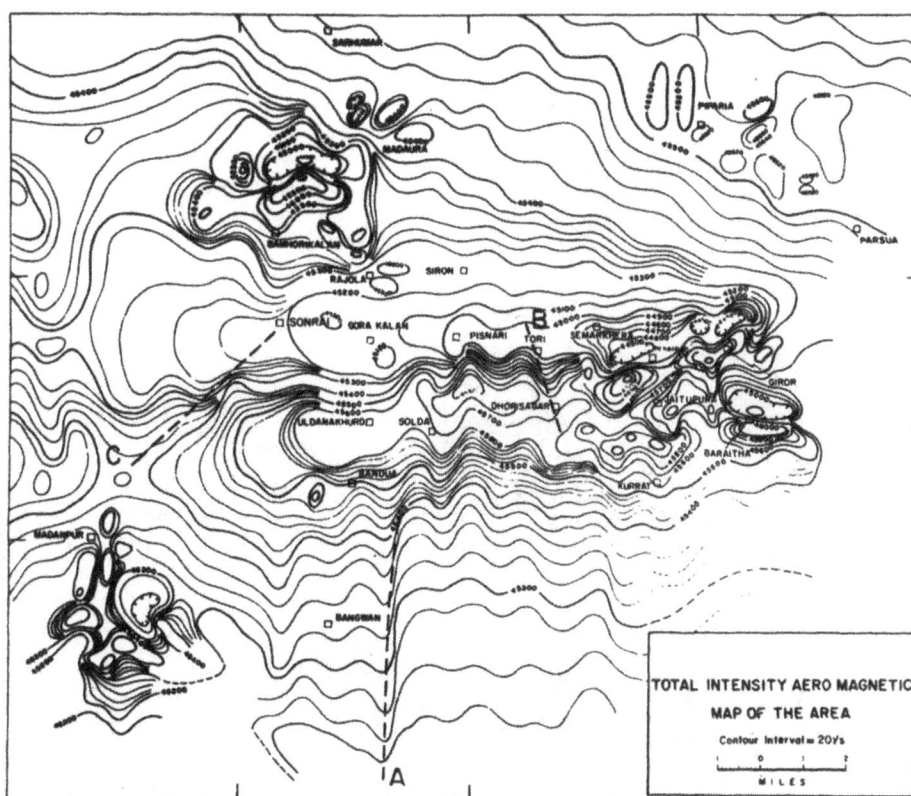

Fig. 9.53(c) Airborne total intensity map of the Bijawar basin large amplitude magnetic anomalies related to iron formations and some significant gradients. B and C represent major magnetic gradients while A is a minor gradient but could also be due to problems in data acquisition.

$$\tau = \sigma \, \mu \, t \, (\rho^2 + h^2)^{1/2}$$

where τ = time parameter; σ = conductivity; t = thickness (meters); μ = Permeability of free space ($4\pi \cdot 10^{-7}$ Henry/ meter); ρ = transmitter-receiver distance (85 meters); h= height of the aircraft in meters. The time parameter (τ) is given as follows (Palacky and West, 1973; Verma, 1975, Mishra et al., 1978):

If Ae_1 and Ae_2 are signals recorded at time t_1 and t_2, the ratio of the two assuming exponential decay is given by:

$$Ae_1 / Ae_2 = R = e^{\Delta t/\tau}$$

$$\text{or} \qquad\qquad = \Delta t \, (\log R)^{-1}$$

where $\Delta t = t_2 - t_1$; time difference of the two recorded signals and $R = Ae_1/Ae_2$

A is a instrument constant and σ t computed in this manner does not represent its actual value and therefore, referred to as apparant conductivity-thickness product. This parameter was computed using the amplitude of signal at channel 4 (C_4) and 2 (C_2) and the computed conductivity thickness product is given in Fig. 9.53(d). It shows two major sections of high conductivity-thickness product marked as B' and C' that also coincides with significant magnetic gradients (B' and C', Fig. 9.53(c)) and good conductors marked in Fig. 9.53(b) based on decay pattern and (C_4/C_2) (Section 3.6.2) indicating contacts/faults that may be mineralized in this case.

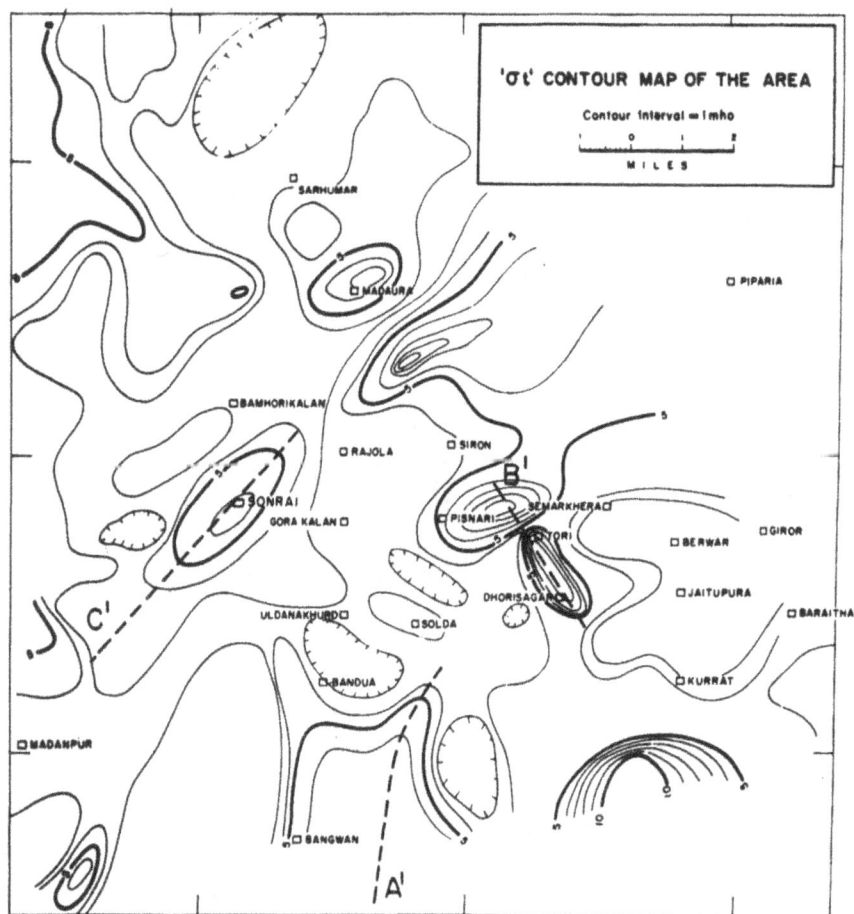

Fig. 9.53(d) Conductivity-thickness product map showing zones of different conductivities. B' and C' represent conductive sections.

(iv) Gold Mineralization

(a) Kolar Schist Belt, India

Gold mineralizations occur in the schist belt of Karnataka, India. In this regard, the Kolar Schist belt is the most famous as it hosts one of the oldest known gold mineralizations in the world. Sometime back an integrated aerogeophysical exploration programme followed by ground follow up work was launched by Geological Survey of India that provided some significant results. Reddy et al., (2002) have provided the airborne magnetic and electromagnetic (EM) anomalies (Fig. 9.54(a)) observed over one of the schist belts of the Sargur group in Western Dharwar Craton (Section 7.8). This schist belt belongs to an older class of schist belts in this region (>3000 Ma). This belt consists of lower granulite facies rocks consisting of talk-tremolite-actinolite-chlorite schist that are host rocks for gold mineralization along with sulphides, mainly pyrite, pyrrhotite and arsenopyrite. Fig. 9.54(a) shows a five channel EM anomaly associated with a magnetic gradient indicating a contact/fault. The five channel EM anomaly indicates good sulphide mineralization (Chapter 3). When followed up on the ground this anomaly provided chargeability and resistivity highs within a broad resistivity, EM and SP lows indicating sulphide/auriferous quartz veins associated with a broad fractured zone. Geophysical logging confirmed the presence of mineralized zones with auriferous quartz veins.

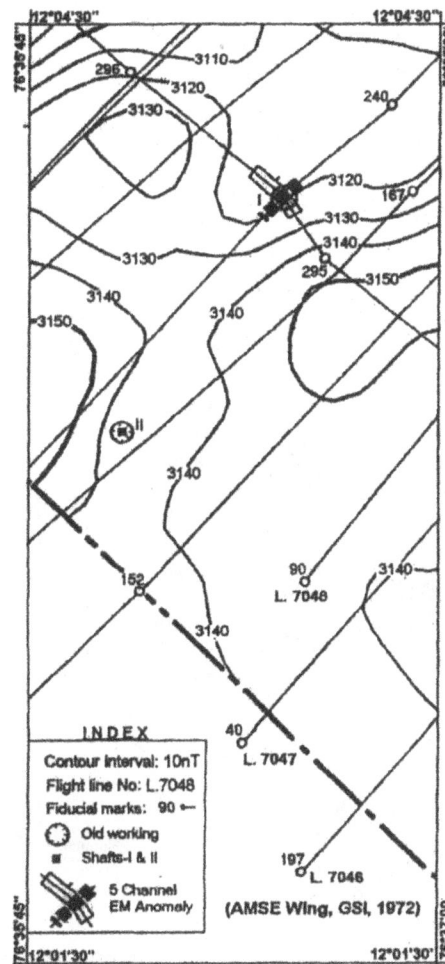

Fig. 9.54(a) Aeromagnetic and electromagnetic (5 Channel) anomaly over Volageri-Ambale schist belt of older group of schist belt (Sargur type) in Western Dharwar Craton (Reddy et al., 2002).

The most important section for gold mineralization in this region is Kolar schist belt where an integrated geophysical exploration programme has yielded some significant results. This Schist belt belongs to the younger group of the schist belts (~ 2500 Ma). The geological map of the Kolar Schist belt is given in Fig. 9.54(b) (Dasu, 2002) that shows granite plutons on the western side and metabasalt and gabbro and dolerite dykes of the schist belt on the eastern side traversed by several faults. It is interesting to observe that the Kolar gold field is located where several NW-SE faults are intersecting the schist belt that might be significant for hydrothermal solutions to percolate and ore concentration. Dasu (2002) has reported that the aeroradiometric K-anomaly associated with the aeromagnetic anomalies related to mafic volcanic rocks were indicators of gold mineralization. This might be due to occurrences of K-granite plutons associated with the Precambrian sutures in Dharwar Craton and Kolar Schist belt has been considered as a Neo Archean-Paleo Proterozoic suture as discussed in Section 7.8. The Bouguer anomaly map of India (GSI-NGRI, 2006) shows a gravity anomaly related to the Kolar schist belt that is reprocessed and the Bouguer anomaly map of this section (Fig. 9.54(c)) shows linear gravity highs associated with metabasalt and lows with the granite plutons that is a typical characteristic of Archean-Proteozoic collision zones (Section 7.2.). The Kolar gold field is located in the central part of this gravity high. The gravity high is intersected by several NW-SE oriented gravity trends that may represent the faults of this section (Fig. 9.54(b)). A major trend changes the course of the gravity high at ~ $13^0 2'$ that appears to represent a major fault that requires to be investigated for mineralization purposes. These sets of gravity anomalies are similar to those observed over deep continental borehole in Germany and surrounding region related to Variscon orogeny (Section 4.7) where gravity highs are related to metabasalt and lows to granite intrusives separated by a fault. The metabasalt extends almost up to 10 km in the upper crust in case of deep borehole in Germany that may represent oceanic crust of that time. The same can be argued even in the present case that explains the presence of mineralized zones in these rocks.

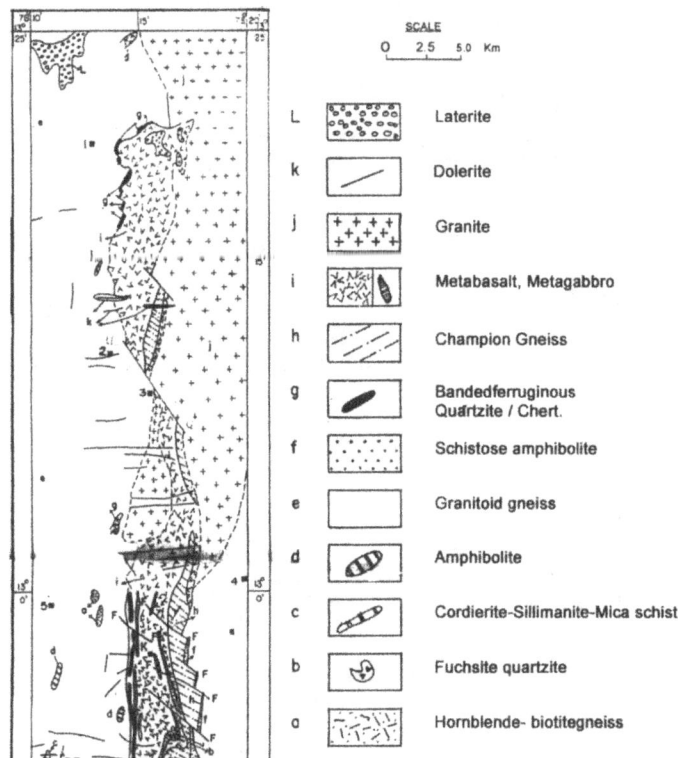

Fig. 9.54(b) Geological map of a part of Kolar Schist belt. K: Kolar Gold Field. (Dasu, 2002)

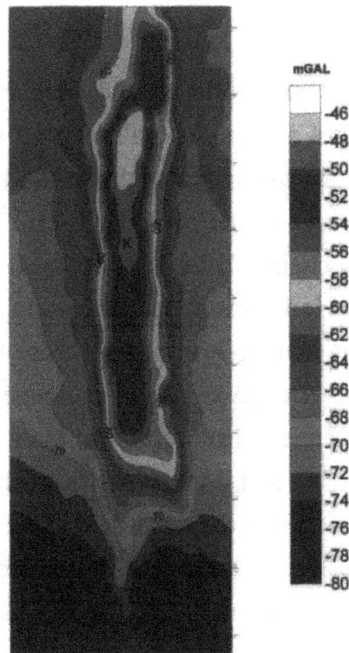

Fig. 9.54(c) Bouguer Anomaly map of a part of the Kolar Schist belt showing linear gravity highs over the schist belt and lows over the granite batholith towards the east (GSI – NGRI, 2006).
[Colour Fig. on Page 820]

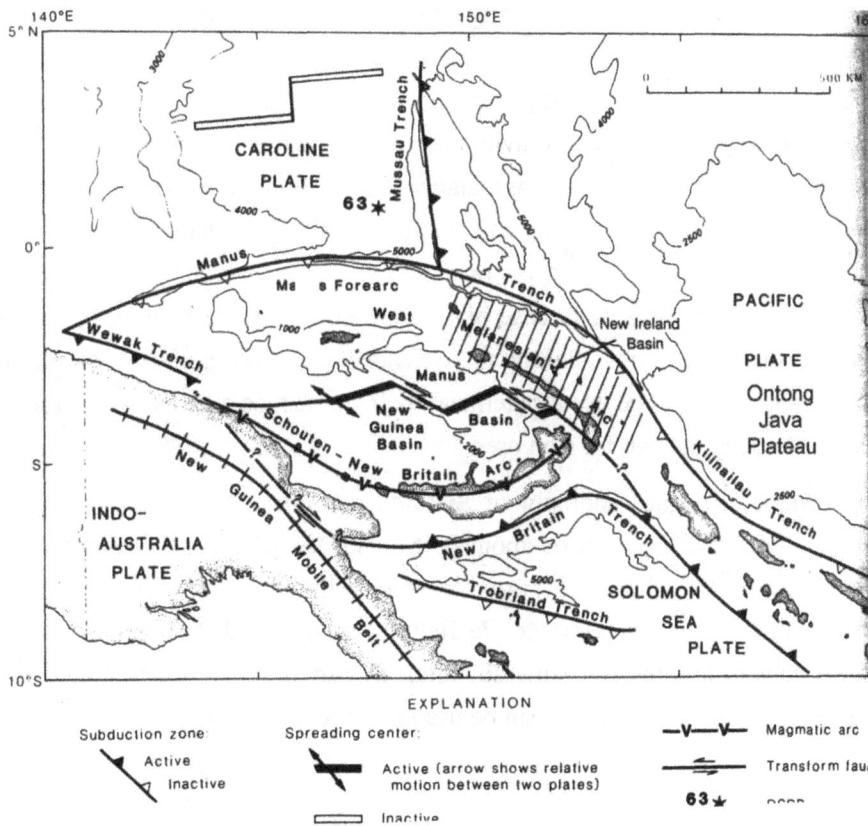

Fig. 9.54(d) Melanesian arc in New Ireland basin between Pacific plate and Indo-Australian plate where recent gold deposits of Pleistocene-Holocene period are found (Marlow et al., 1992).

(b) New Ireland Basin, Papua New Guinea

The association of gold deposits with collision and subduction zones is clear from the occurrence of one of the youngest and largest gold deposit of New Ireland basin, Papua New Guinea (Fig. 9.54(d)) that are associated with the subduction related volcanism and caldera of Pliocene and Pleistocene period in South Pacific (Marlow et al., 1992). Gold deposits occur over Taber-Fenni chain of islands in Melansian arc formed due to collision and subduction of the Pacific and Indo-Australian plate during Eocene to Early Oligiocene. However, due to the presence of Ontang-Java plateau close to the subduction zone, there was subduction and arc reversal that gave rise to prolonged period of igneous activity almost starting from middle Miocene to Holocene. The main rock types in this section are basanite, tephrite, transitional basalt and felsic rocks as trachyte and quartz trachyte. The latter felsic group bears the epithermal gold deposits that are the youngest group of rocks formed during Pleistocene to Holocene and deposits are formed due to near surface heat source, high initial gas and sulfur content in the magma close to the sea water causing episodic boiling and release of gases for gold concentration. These are typical conditions met in oceanic collision and subduction zones

9.10.2 Ferrous Minerals – Iron ore, Chromite and Mangnese Deposits

Ferrous minerals like iron, magnetite, chromite, manganese etc., show higher density and therefore, can be explored using gravity method. Iron ores are very easily delineated by airborne magnetic data due to high magnetization of these rocks. They normally produce very high amplitude magnetic anomalies (~ 5000 – 10,000 nT) and significant gravity anomalies. Magnetic anomalies largely depend on the association of magnetite. Therefore, its association in Banded Magnetite Quartzite that is normally the host rock for iron ores produces magnetic anomalies of ~ 1000 nT as shown in Fig. 3.32 for Chitradurga schist belt. However, in case it is concentrated as iron ore, it would produce much higher anomalies of thousands of gamma as in case of iron ore bodies in the Kudremukh Schist belt (NGRI, 1967) though both these schist belts belong to Archean-Proterozoic period of the same class. Airborne magnetic maps can, therefore, be used to delineat their subsurface extensions without drilling all around and can serve as quality control for iron ores.

Fig. 9.55(a) is the residual Bouguer anomaly of Sukinda chromite belt in Orissa (Bagchi and Banerjee, 1983) which shows a gravity high of about 0.45 mGal trending S - W. Similarly, a well proven manganese ore deposit in Balaghat (M.P.) provides gravity high of about 1 mGal (Fig. 9.55(b), Dash et al., 1983). It is related to the Satpura Mobile Belt (Section 7.6) that is another Proterozoic collision zone providing favorable sites for mineralization. It also provides a prominent magnetic anomaly, which along with the gravity anomaly, can be used for its exploration.

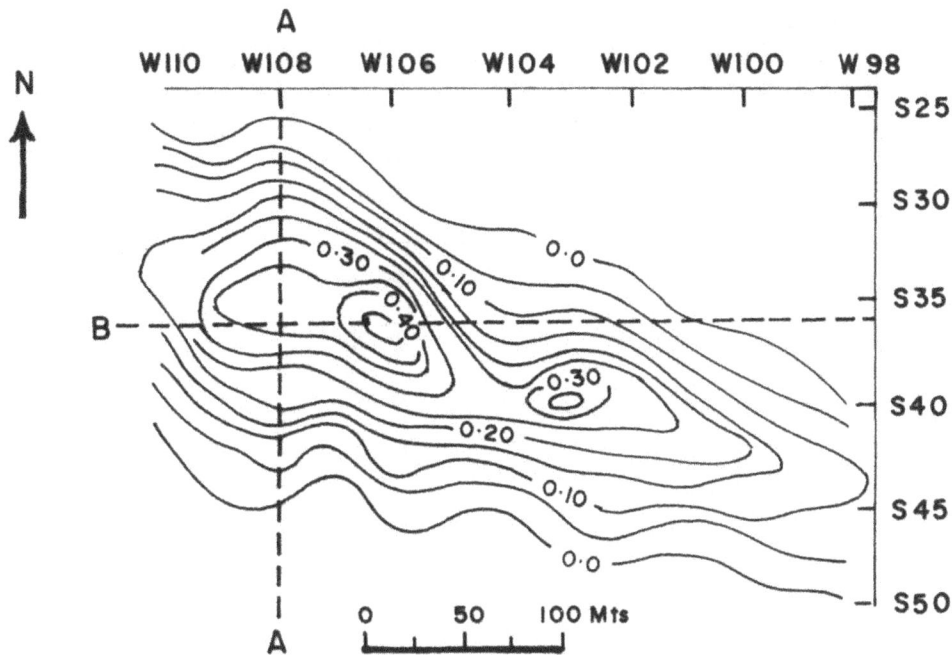

Fig. 9.55(a) Residual Bouguer anomaly of Sukinda Cromite belt (Bagchi and Banerjee, 1983).

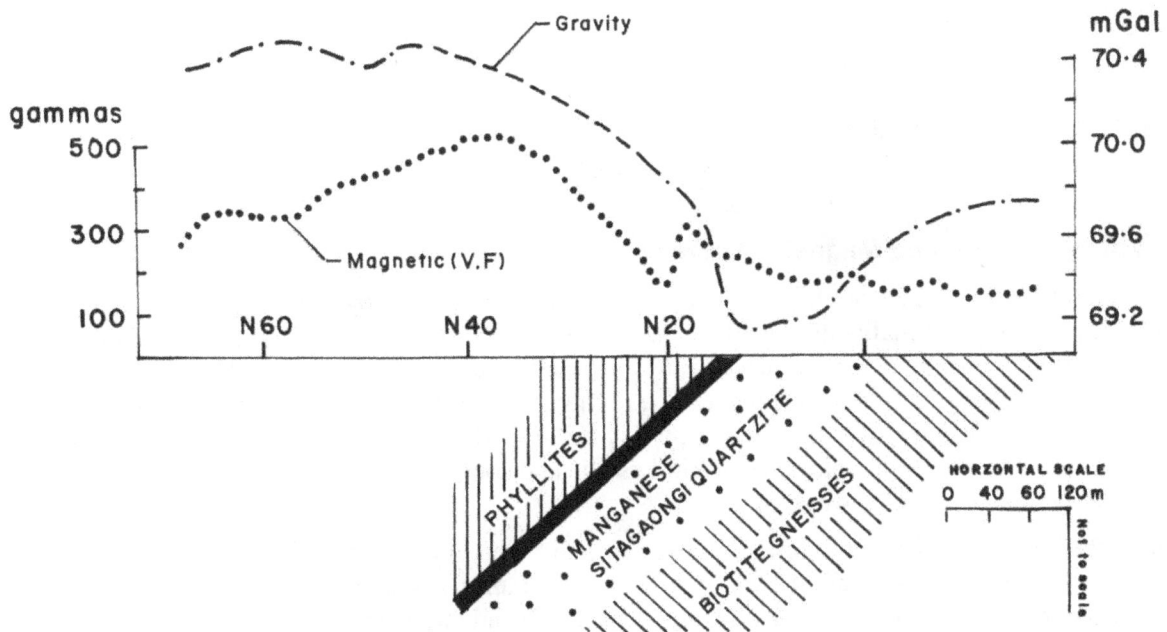

Fig. 9.55(b) Gravity and magnetic anomalies over a manganese ore body at Balaghat, M.P. (Dash et al., 1983).

9.10.3 Non-Metallic Minerals

An example of gravity anomaly over the Mangampeta Baryte deposit in the Cuddapah basin is given in Fig. 9.56 which shows a gravity high of about 1-2 mGal (Bose and Vaidyanathan, 1979). It occurs in the Nallamalai basin associated with regional gravity low related to crustal thickening which has been the attributed to collision of the Indian Shield with Antarctica during Meso-Proterozoic period (Section 7.10). These observations indicate that the Proteozoic collision zones (Chapter 7) are favorable sites for mineral investigations.

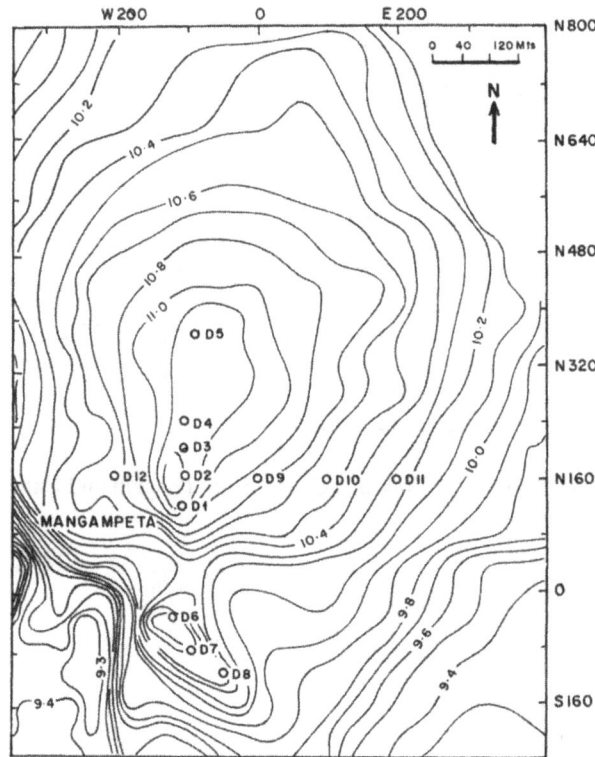

Fig. 9.56 Bouguer anomaly of Mangampeta Baryte deposit of Caddapah basin showing a small high encircling the deposit (Bose and Vaidynathan, 1979).

9.10.4 Volcanic Plugs and Magnetic Anomalies

As discussed in Section 9.2.3 and 9.2.6, volcanic plugs are characterized by high amplitude gravity anomaly due to presence of mafic/ultramafic rocks. Kimberlite group of volcanic plugs are important as sometimes they are diamond bearing, which can be explored using gravity method. However, these plugs do not produce large amplitude anomalies due to their small size and weathering at the top, but observed magnetic anomalies over them are quite significant. Macnae (1979, 1995) have reported magnetic anomalies from several volcanic plugs in Lesotho and southern Australia that have been used to discover several volcanic plugs world over.

As described under the Satpura Mobile Belt (Section 7.6), there are volcanic plugs in the northern part of the Vindhyan basin and one of them at Majhgawan, Panna is diamond bearing. This region shows a moderate amplitude regional gravity high (8-10 mGal) and large amplitude airborne magnetic (Fig. 7.24(a)) and ground magnetic anomalies (Fig. 9.57(a), Mishra, 1987) indicating that the presence of mafic/ultramafic rocks in the basement below Vindhyan sediments and volcanic plugs might be an offshoot from these rocks. It also shows high resistivity. The gravity anomaly over the exposed pipe provided a different picture due to weathering at the top as discussed above. However, the magnetic anomaly is quite prominent at the top of the volcanic plug (450-500 nT) (Fig. 9.57(b), Sharma and Nandi, 1964). The magnetic anomaly in vertical intensity over Majhgawan pipe was modeled by Sarma et al., (1999) who suggested multiple intrusions for different order and type of magnetic anomalies observed over this pipe. However, different order and type of anomalies can also originate from variation in susceptibility due to hetrogeniety of mafic minerals that usually characterizes intrusive rocks. It can also be caused by a different scale of weathering in different parts of the plug. Srivatsava et al., (1983) have also reported a large amplitude magnetic anomaly over Jungel volcanic plug associated with the Mahakoshal Group of rocks along the southern margin of the Vindhyan basin (Section 7.6.2). These observations suggest that the magnetic method is more successful in delineating volcanic plugs.

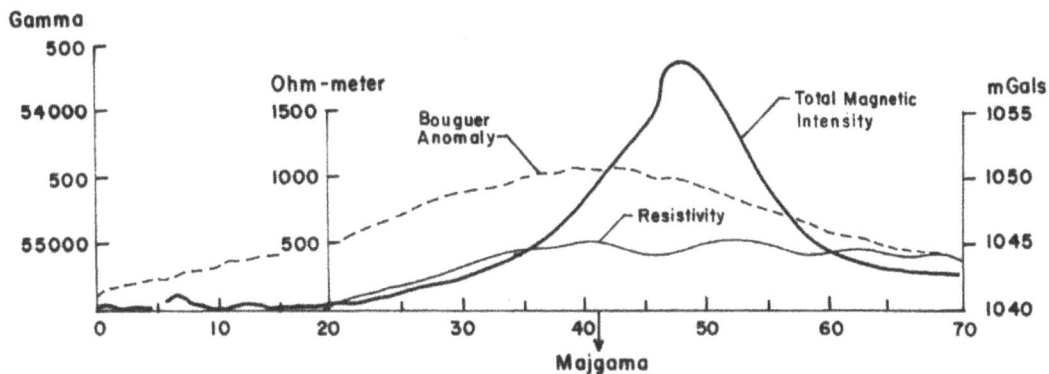

Fig. 9.57(a) Regional Gravity, magnetic and resistivity profiles over the Panna Diamond belt showing regional basement anomalies across northern margin of Vindhyan basin (Mishra, 1987). They may represent mafic intrusions in the basement. Volcanic plugs of Panna may be related to these intrusives.

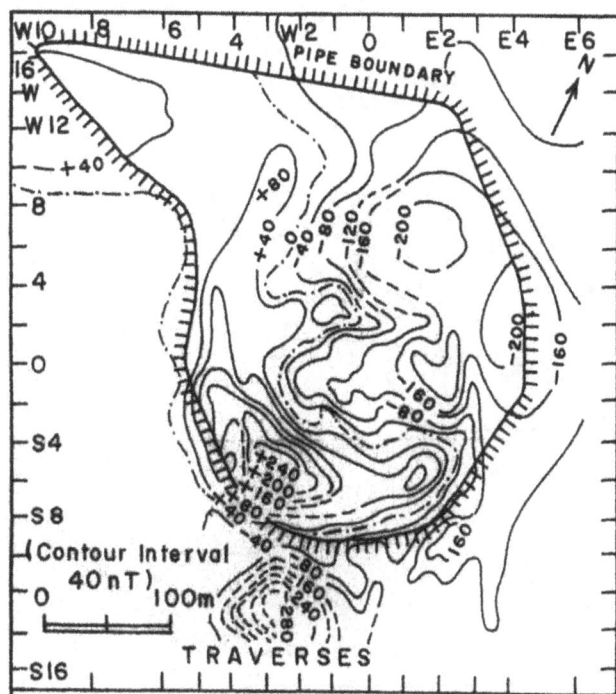

Fig. 9.57(b) Vertical magnetic intensity over Majhgawan volcanic pipe (Sharma and Nandi, 1964) showing a low towards north and high towards south indicating induced magnetization.

The volcanic plug at Wajrakarur in Andhra Pradesh has provided a gravity high of about 0.8 mGal (Fig. 9.58, Subrahmanyam et al., 1991) which is quite significant. Guptasarma et al., (1989) have reported a volcanic plug within a cluster of seven volcanic plugs known previously close to the Wajrakarur pipe (Fig. 9.59). They have shown good EM response for high frequencies from the top weathered layer that is easily recognizable compared to the response from the surrounding region. This response defined the boundary of this plug. The total intensity magnetic map recorded on ground (Fig. 9.60(a)) also shows linear features close to some of the known pipes that may represent shear or fracture zones which might have acted as conduit to these volcanic plugs. The occurrence of mylonitized granite close to these pipes also suggested shear zones.

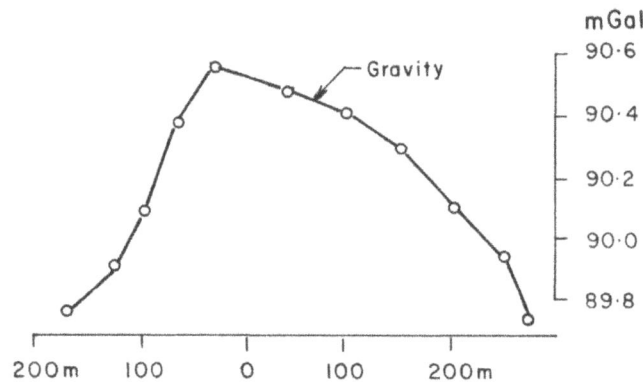

Fig. 9.58 Gravity field over pipe 2, Wajrakarur area, (A.P.) (Subrahmanyam et al., 1991).

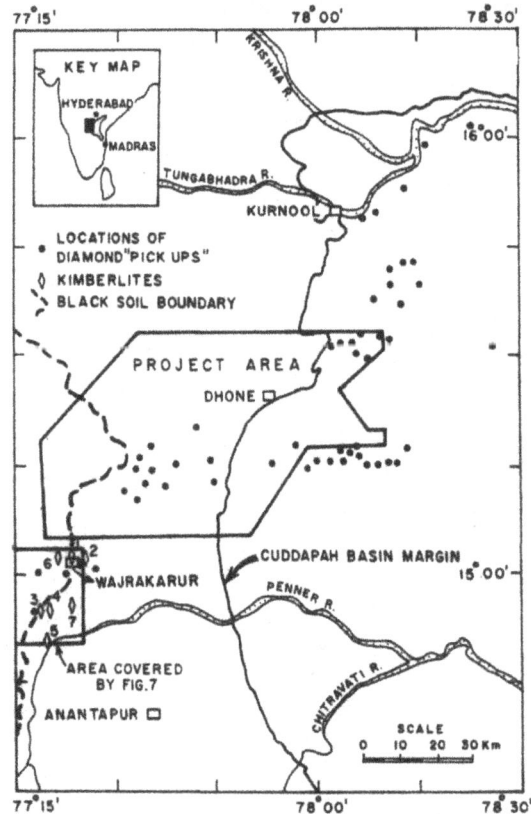

Fig. 9.59 Known Kimberlites, diamond finds (Sakuntala and Krishna Brahmam, 1984) along western margin of the Cuddapah basin.

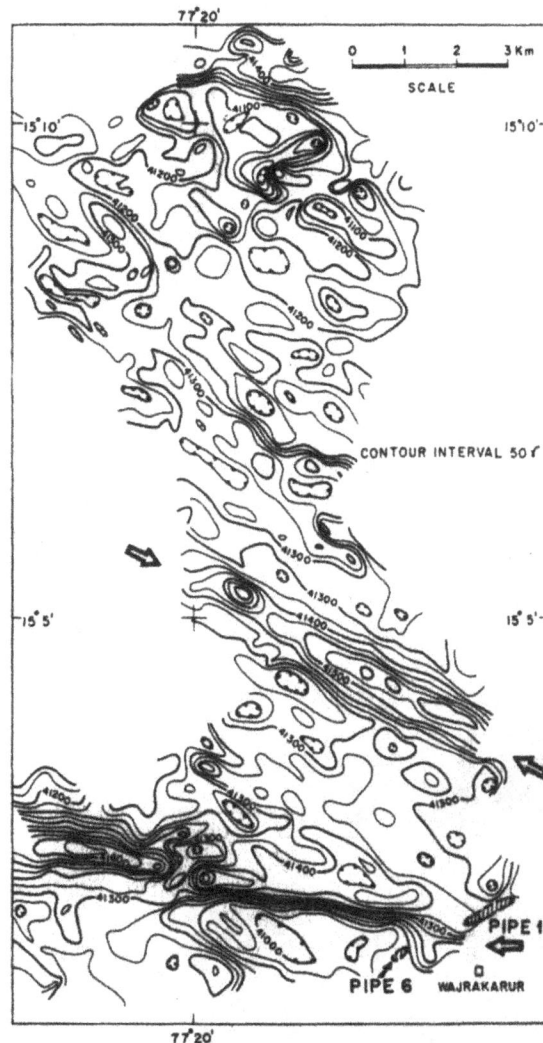

Fig. 9.60(a) Total intensity magnetic map recorded on ground showing volcanic pipes, 1 and 6 associated with magnetic lineaments. There are other magnetic lineaments in this region such as the one shown by open arrowheads that should be explored for volcanic pipes.

These examples have clearly brought out the order of gravity anomalies for mineral investigations that are usually less than 1 mGal in most of the cases and therefore require special gravimeters (microGal) and special precautions to conduct these surveys as discussed in Section 2.3.4 under 'High Resolution Surveys'.

9.10.5 High Resolution Heliborne Magnetic Surveys – Volcanic Plugs and Groundwater Potential

The airborne total intensity map of the Peninsular shield west of the Cuddapah basin (Sections 7.10) where the volcanic pipes of Wajrakarur are located (Fig. 3.13(b)) shows several magnetic lineaments oriented NW-SE and ENE-WSW. These are Precambrian trends in this region related to Dharwarian and Satpura trends, respectively, and may represent shear zones and intrusives. This data was acquired at a ground clearance of ~ 150 m along N-S flight lines 1 km apart. The intersections of two trends are especially significant for exploration purposes. These airborne magnetic lineaments should be followed up on the ground both for exploration of volcanic plugs and groundwater. However, due to large flight altitude and large flight line spacing, lineaments are not that well reflected due to the small order of anomalies associated with them. A high resolution helicopter borne magnetic survey was conducted south of the Wajrakarur area for multi–purpose exploration programmes and for delineating a cluster of volcanic plugs in the Kalyandurg region (Fig. 9.60(b), Ram Babu and Rama Rao, 2008). This survey was conducted at flight altitude of 40 m above ground surface along flight lines 50 m apart that has delineated several magnetic lineaments mostly oriented NW-SE and E-W or ENE-WSW. These are two most prominent structural trends in the Peninsular shield. Group of known volcanic plugs coincide with the NW-SE lineaments, same as in the case of Wajrakarur group of plugs (Section 3.4.4). It demonstrates the application high resolution heliborne surveys to delineate small dimension bodies. The same lineaments can also be investigated for groundwater potential.

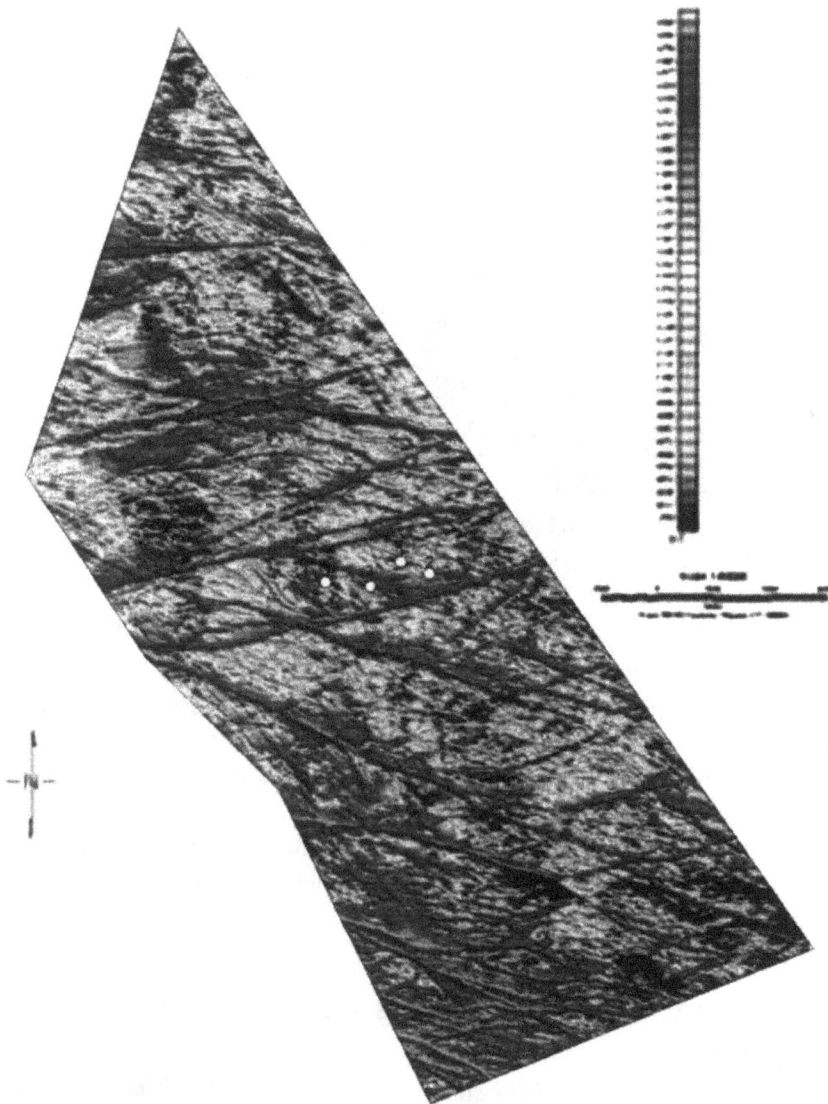

Fig. 9.60(b) High resolution airborne magnetic image of the Kalyandurg area in Anantpur district west of the Cuddapah basin and south of Wajrakarur (Fig. 9.59) based on helicopter borne survey. It shows Kimberlite pipes (white dots) coinciding with magnetic lineaments that are nicely reflected in this map and such lineaments can also be used for groundwater investigations in hard rock areas. Sensor height: 40 m above ground surface, and profile spacing: 50 m (Ram Babu and Rama Rao, 2008). [Colour Fig. on Page 820]

9.10.6 Uranium Mineralization – Integrated Airborne and Ground Geophysical Surveys

Radioactive counters have been primarily used for their exploration. However, the depth of penetration being limited for counters, they cannot be used for exploration of subsurface uranium occurrences. Uranium is generally found in association with sulphides, and graphites etc., that are structurally controlled as faults, fractures etc. Therefore, the geophysical methods that are used to delineate these structures and associated minerals can be used for their exploration. As described above, the gravity and the magnetic methods are extremely useful to delineate these structures while electrical and electromagnetic methods are useful to demarcate sulphide mineralization which in turn can be utilized for exploration of associated uranium mineralization.

(i) Carbonatite Complexes- Rare Earth Elements and Uranium Mineralization

Carbonatite Complexes are plug type of bodies originating from the upper mantle and are therefore, characterized by ultramafic and mafic rocks of high density and susceptibility that would produce significant gravity and magnetic anomalies. As described in Section 8.2 and 9.2, such bodies produce circular/semi-circular gravity and magnetic anomalies of high amplitude that can be easily delineated. Carbonatite complexes assume importance due to their association with rare earth elements like Neobium etc. In case of carbonatite complexes with nepheline syenite, they are hosts of radioactive minerals, and several uranium mineralizations have been discovered in their association.

The Shillong plateau (Section 7.6.4) consists of several carbonatite plugs specially in its N-E part and Mikir Hills. The airborne magnetic map of this section (Fig. 7.30(c)) shows several high amplitude magnetic anomalies. However, due to the small scale of this map, the anomalies due to small bodies could not be correlated in this map. The carbonatite complex in Sung valley in the NE part of the Shillong plateau is a well known plug that could be identified from a large scale airborne magnetic map of Fig. 7.30(c). It provided a large amplitude magnetic low and a small high towards the north (Ram Babu and Rama Rao, 2008). As it has been stated in Section 3.7, disposition of the magnetic low and high depends to a large extent on the magnetization vector. In case of induced magnetization in the present day earth's magnetic field, vertical bodies for the magnetic latitude of this region would show a low towards the north indicating that this body posseses remnant magnetization. Using same pattern of magnetic anomalies a new carbonatite complex was delineated at Jasra (26^0 N, 92^0 30' E) that also shows a large low towards the south and a small high towards the north (Fig. 9.61(a), Achuta Rao et al., 1996). Subsequently, several other carbonatite complexes were also delineated in this region using similarity of magnetic anomalies with those from the Sung Valley and Jasra. As these plugs are deep seated intrusives, they are usually associated with rift basins and rifted margins and these carbonatite plugs of Shillong plateau might have originated during the breakup of Gondwanaland as indicated by the direction of magnetization in them. This is likely as Rajmahal and Shylet traps related to that breakup are exposed in adjoining regions.

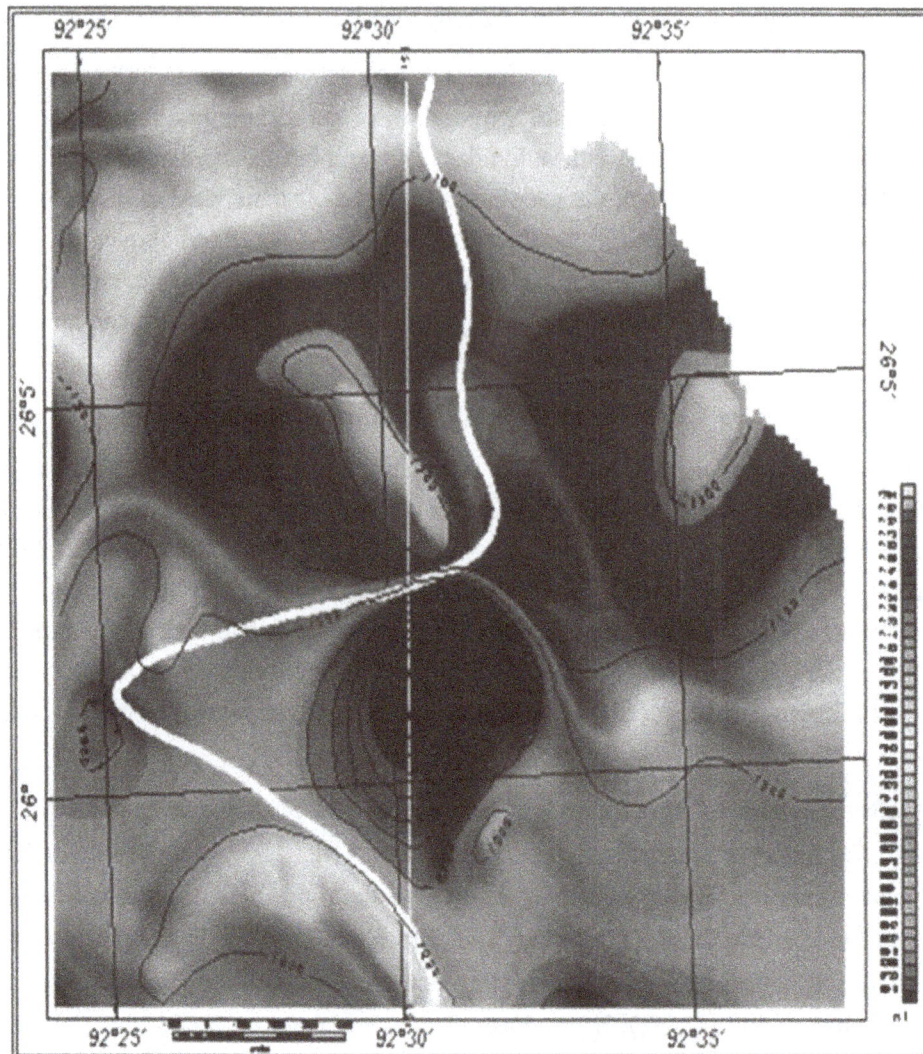

Fig. 9.61(a) Airborne magnetic anomaly image of the Jasra carbonatite complex and a profile across it in the NE part of the Shillong Plateau showing magnetic low and a high towards the north indicating remnant magnetization. Fight altitude: 2100 m above msl and line spacing: 1 km (Ram Babu and Rama Rao, 2008). [Colour Fig. on Page 821]

The importance of carbonatite complexes for uranium mineralization can be seen from Fig. 9.61(b) that shows a carbonatite complex with high radiometric anomaly where both high amplitude airborne magnetic and aeroradiometric anomalies were recorded (Chawla et al., 1993). Ground follow up work showed the area to be occupied by quartzite, pyroxenite and syenite, and uranium and thorium anomalies were detected. Secondary occurrences of radioactive elements along shear zones, faults, lineaments can also be delineated from joint analysis of airborne magnetic and airborne spectrometer surveys. As stated above, airborne magnetic maps are best data sets to delineate these features, its joint analysis with radiometric maps can be directly used for this purpose. One such example is the Singhbhum shear zone (Section 7.6.3) that has shown nice curvilinear gravity high and significant magnetic anomalies along the Satpura Mobile Belt (Section 3.8.2) typical of Proterozoic collision zones where uranium deposits have been reported (Anand and Rajaram, 2004).

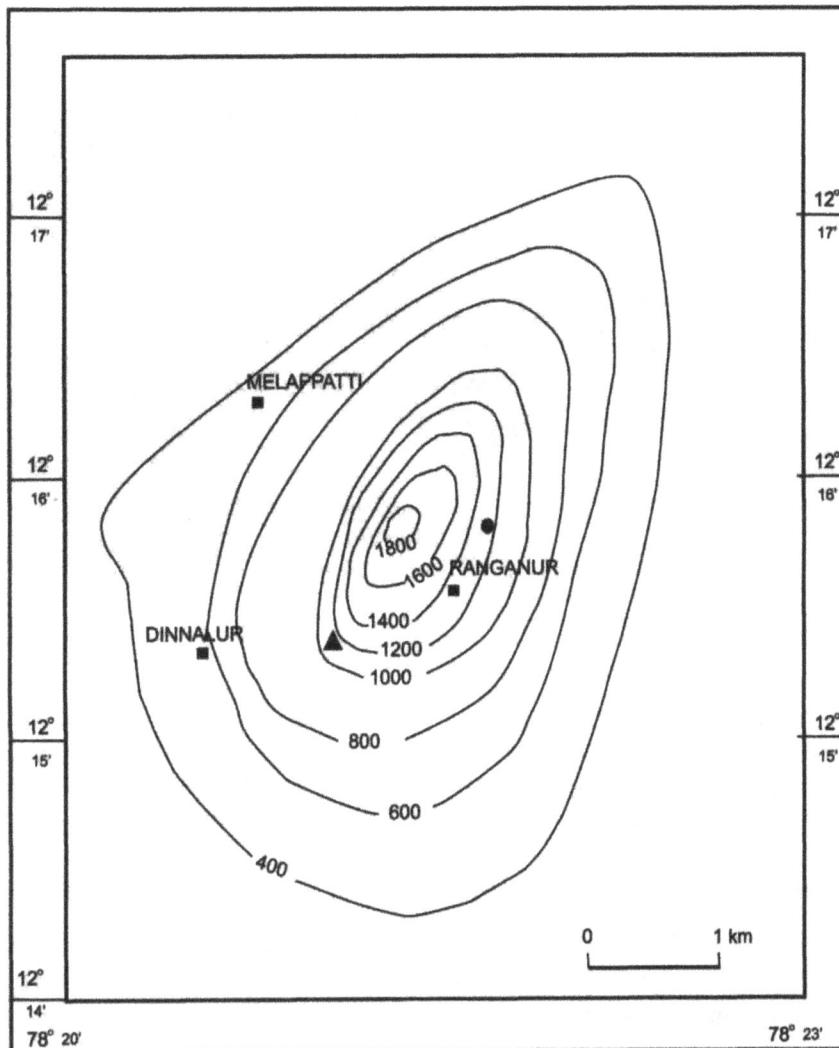

Fig. 9.61(b) Integrated thematic map of a carbonatite-syenite complex at Ranganur, Dharmapuri district, Tamil Nadu, India. Contours represent radioactivity in cps. Filled triangle and filled dot indicate uranium and thorium occurrences, respectively (Chawla et al., 1993).

(ii) Airborne Spectrometer Surveys – Mahakoshal Basin, Central India

Chaturvedi et al., (2008) have described the processing of recorded 256 channel data to first reduce the noise and to convert in four channel count rates related to potassium, uranium, thorium and total count, and present them in ternary images along with the images of other data sets like airborne magnetic, and gravity that provides an integrated picture of geophysical data. Fig. 9.62 shows a ternary image of K, Th, and U superimposed over Bouguer anomaly contours and geology of Mahakoshal basin in Central India (Section 7.6.2; M, Fig. 7.20) that shows a band of high uranium concentration in the northern part coinciding with gravity highs. This gravity high coincides with a large magnetic anomaly (Fig. 7.25(a) and (b)) that represents a fault with mafic/ultramafic intrusive and characterizes rifted platform deposits, which are significant for mineralization (Section 7.6.2). Such maps are highly useful for integration of different data sets. There are strong surface indications of mineralization in association with these rocks as discussed below and above in Section 9.10.1.

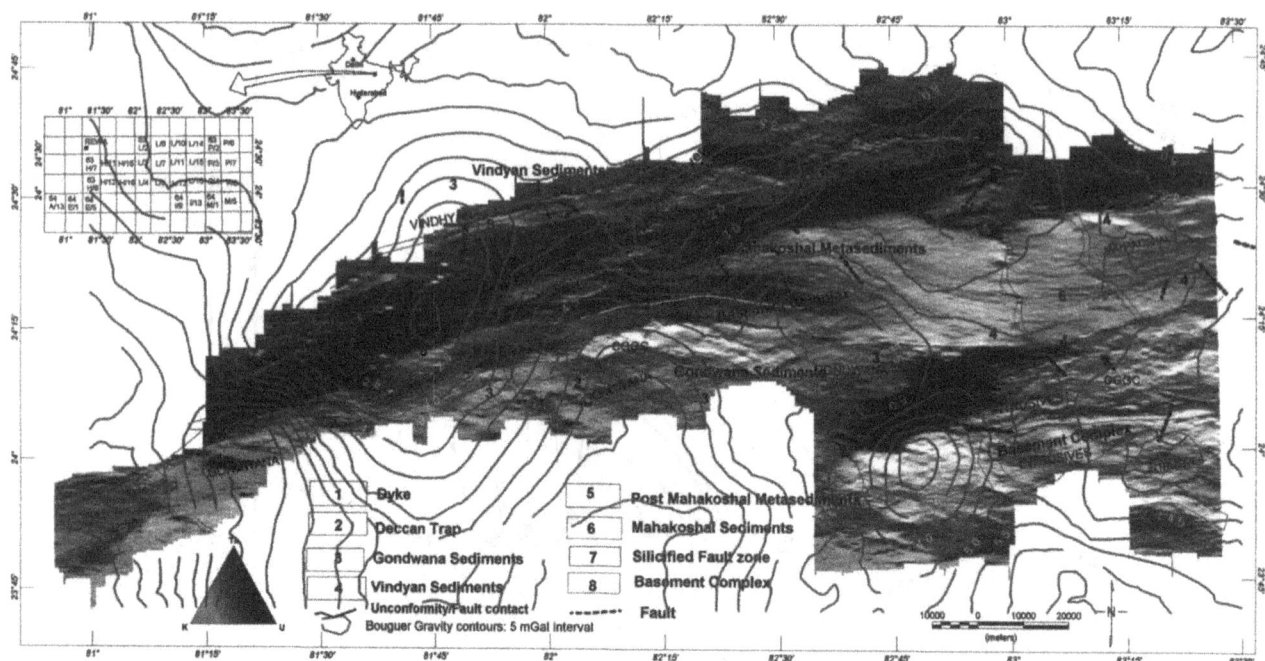

Fig. 9.62 Ternary image of K, Th and U with superimposed Bouguer anomaly and geology of Mahakoshal basin, Central India ((M, Fig. 7.20) (Chaturvedi et al., 2008). It shows linear bands of uranium anomaly in the northern part coinciding with gravity highs indicating mafic/ultramafic intrusive that characterizes Mahakoshal group of rocks (Section 7.6.2). [Colour Fig. on Page 821]

(iii) Integrated Geophysical Exploration for Uranium and Base Metals – Gondwana Basin, Satpura

Markandeyulu et al., (2009) have described an integrated exploration programme for exploration of uranium mineralization in the Gondwana basin of Satpura (Fig. 7.20, South of Hosangabad) that has shown promise for such mineralization. This region lies at the contact of the Upper and Lower Denwa formations of Upper Gondwanas. A pole reduced total intensity magnetic image of this section is given in Fig. 9.63(a) that shows a band of magnetic highs and lows towards the north and the south, respectively. Initially, the total intensity map showed low towards the north and high towards the south, that are reversed in the pole reduced field which usually happens for magnetic anomalies observed in low geomagnetic latitudes of this country. The junction of the magnetic highs and lows usually represents a contact of rocks of different susceptibility caused by faulting and therefore, such contacts are important for mineralization. As seen from this map, the Euler deconvolution solutions are mostly lying at the junction of the gravity highs and lows indicating their importance for location of the magnetic sources and most of them are located at depth of < 20 m. Those authors have also made some faults based on changes in the magnetic trend at the contact of the gravity highs and lows and their intersection is significant for mineralization. These magnetic anomalies are observed almost parallel to the Son-Narmada South Fault (SNSF, Fig. 7.20) and are in line with the Mahakoshal group of rocks towards the NE of the Gondwana basin of Satpura (M, Fig. 7.20) that also show similar magnetic anomalies of lows towards the north and highs towards the south (Fig. 7.25(a) and (b)). It is therefore likely that magnetic anomalies of the present survey block represent the extension of the Mahakoshal group of rocks towards the SW under Gondwana sediments, forming its basement in this section.

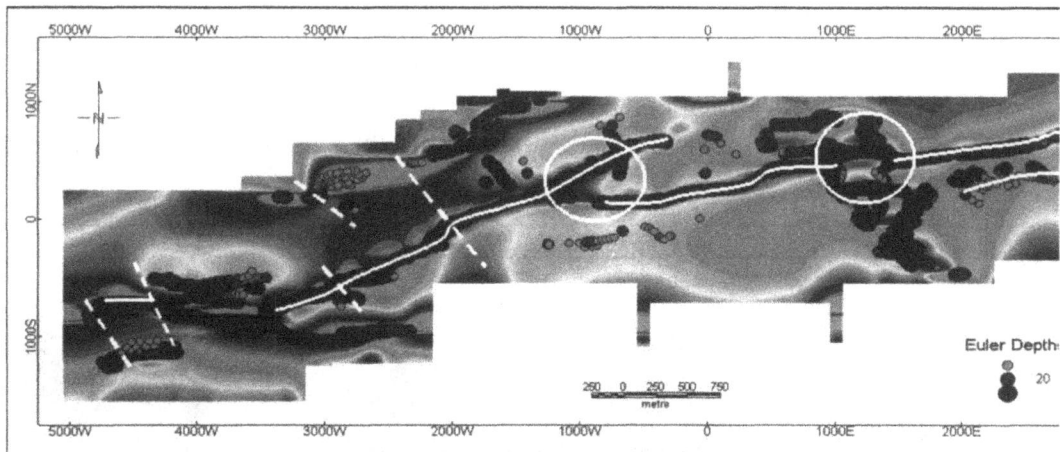

Fig. 9.63(a) Total intensity image of the northern part of the Gondwana basin, Satpura reduced to pole with Euler deconvolution solutions, showing sources coinciding with the linear magnetic trend along the junction between magnetic highs and lows defining a contact or a fault with mafic intrusive. Some faults are inferred based on changes in this linear magnetic trend and their intersections with the magnetic trend are important for mineral exploration (Markandeyulu et al., 2009). [Colour Fig. on Page 822]

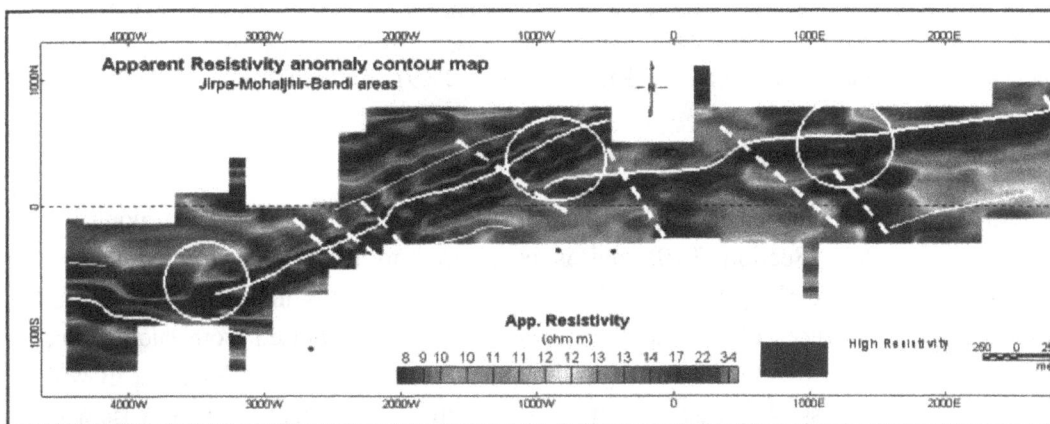

Fig. 9.63(b) Apparent resistivity image of the same area as given in Fig. 9.63(a) showing high resistivity zone along the magnetic linear trend indicating mafic intrusive with low porosity or associated quartz veins (Markandeyulu et al., 2009). [Colour Fig. on Page 822]

The same area is followed up by resistivity and induced polarization methods. Resistivity and chargeability distribution is given in Fig. 9.63(b) and (c) that show zones of high resistivity and high chargeability coinciding with the linear magnetic trend and sources along the contact of the magnetic highs and lows. High chargeability indicates the association of sulfide mineralization, while high resistivity indicates that there are intrusives with lower porosity. High magnetic and gravity anomalies indicate mafic/ultramafic intrusive as discussed above, and represent rifted platform (Section 7.6.2) that are important for mineralization. These anomalies also support their association with the Mahakoshal group of rocks as these rocks are known as host rocks for sulphide and uranium mineralization as discussed above in Sections 9.10.1. These discussions make the Mahakoshal group of rocks important for mineral exploration where planned exploration strategy should be undertaken. In this regard, the alluvium patch between the exposed part of the Mahakoshal group of rocks (M, Fig. 7.20) and Gondwana basin, Satpura (Fig. 7.20) assume special significance as several gravity (GSI-NGRI, 2006) and airborne magnetic anomalies (AMSE, 2001) have been observed in this section, indicating the presence of these rocks at a subsurface level. This section should be the target for future detailed geophysical surveys.

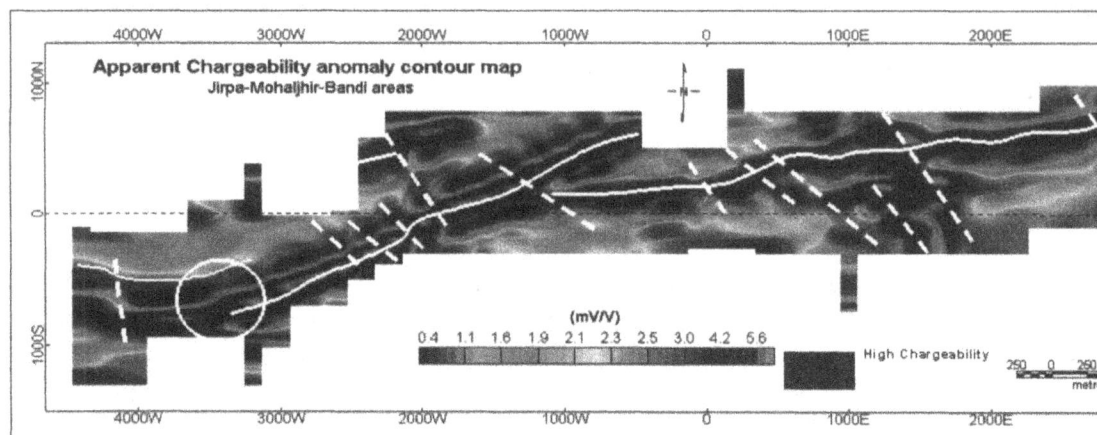

Fig. 9.63(c) Apparent chargheability image based on induced polarization survey of the same block as given in Fig. 9.63(a). It shows high chargeability coinciding with the linear magnetic trend indicating association of sulphide mineralization (Markandeyulu et al., 2009). [Colour Fig. on Page 822]

(iv) Uranium Mineralization – Indravati Basin, Bastar Craton

Indravati basin is a Proterozoic basin in the Bhandara-Bastar craton (2, Fig. 7.31) along the Eastern Ghat Mobile Belt (Section 7.10) and is considered to have been formed due to Mesoproterozoic convergence along the EGMB. Such basins are always important for mineral exploration. Several occurrences of radioactive minerals have been reported from such basins such as the Cuddapah basin (Section 7.10) that is another Proterozoic basin along the EGMB formed due to similar process. The eastern part of this basin at its contact with the basement gneisses has shown several occurrences of radioactive minerals similar to the Cuddapah basin. Results of gravity and magnetic surveys in this section of the basin (Patra et al., 2009) have helped in understanding the genesis of this mineralized section. Total intensity magnetic contour map superimposed over the Bouguer anomaly image (Fig. 9.64) has shown three anomalous zones. Zone I shows high magnetic and gravity anomalies, while Zones II and III show gravity and magnetic lows. These anomalies indicate mafic and fracture zones with felsic intrusive, respectively, that are typical characteristics of supracrustals in Proterozoic terrain as discussed in the case of the Western and Eastern Dharwar Cratons (Section 7.8). Felsic intrusive (Zone II and III) are especially significant for radioactive minerals and should be investigated in detail for prospective mineral occurrences. Bore hole data also suggests fracture zones in the basement with mafic and felsic intrusive. These observations suggests that the circular/semicircular gravity lows are targets for uranium mineralization as they represent felsic intrusives that are the host rock for radioactive minerals.

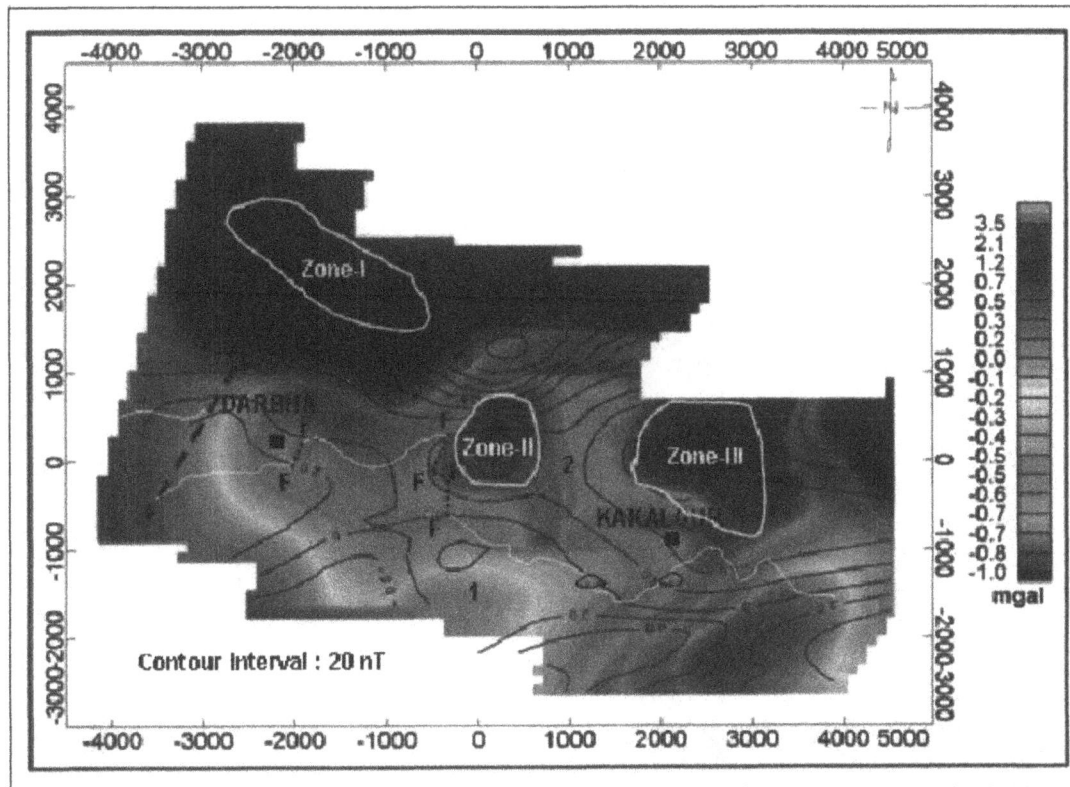

Fig. 9.64 Total intensity magnetic contours superimposed over Bouguer anomaly image indicating magnetic and gravity highs (Zone I) related to mafic intrusive and magnetic and gravity lows (Zone II and III) related to fracture zone with felsic intrusive (Patra et al., 2009). [Colour Fig. on Page 823]

(iii) Airborne Gamma Ray Spectrometer Survey and Paleochannels – Coastal Godavari Basin

Coastal parts are known to contain heavy minerals associated with sand as placer deposits that usually consist of ilmenite, garnet, zircon, monazite and sillimanite. They are generally found as placer deposits concentrated along paleochannels, sand bars, and strand lines etc. Being thorium mineral, monazite is useful as a radioactive element and is therefore, explored along coasts. Coasts along Kerala and Orissa are known examples in this regard. Coastal Godavari basin (Fig. 7.31(b)) consists of several ridges and depressions that are significant for oil exploration. The coastal basin also consists of several shallow structures of recent origin such as paleochannels, and sand bars etc., where the heavy minerals may be concentrated. Rao et al., (2010) have described an airborne gamma ray spectrometer survey for this purpose. The survey was conducted at a height of ~120 m along ~ 1 km spaced E-W oriented flight lines. The thorium image (Fig. 9.65) provides concentration of thorium elements up to 300 ppm. Significantly, several zones show greater than 30 ppm that are important for exploration purposes. It reflects Godavari and Krishna rivers passing through Rajamundary and Kolluru, respectively, and the paleochannels associated with them. Maximum concentration is seen in the northern part that is related to exposed granite, gneisses and charnockite rocks of the Eastern Ghat. However, there are several linear zones of concentration near the coast that are subparallel to the coastline. They appear to be related to paleochannels and are significant for exploration of thorium. The comparison of this map with that of the total count suggested that most of the anomalies are due to thorium. The major long linear trend along the coast covering both Godavari and Krishna basins coincides with the coastal basement ridge (Fig. 7.53(a)) that might be responsible for its formation. Besides exploration of thorium, it provides a strong tool to map paleochannels in coastal regions that are also usefull to explore other minerals which get concentrated along paleochannels such as base metals.

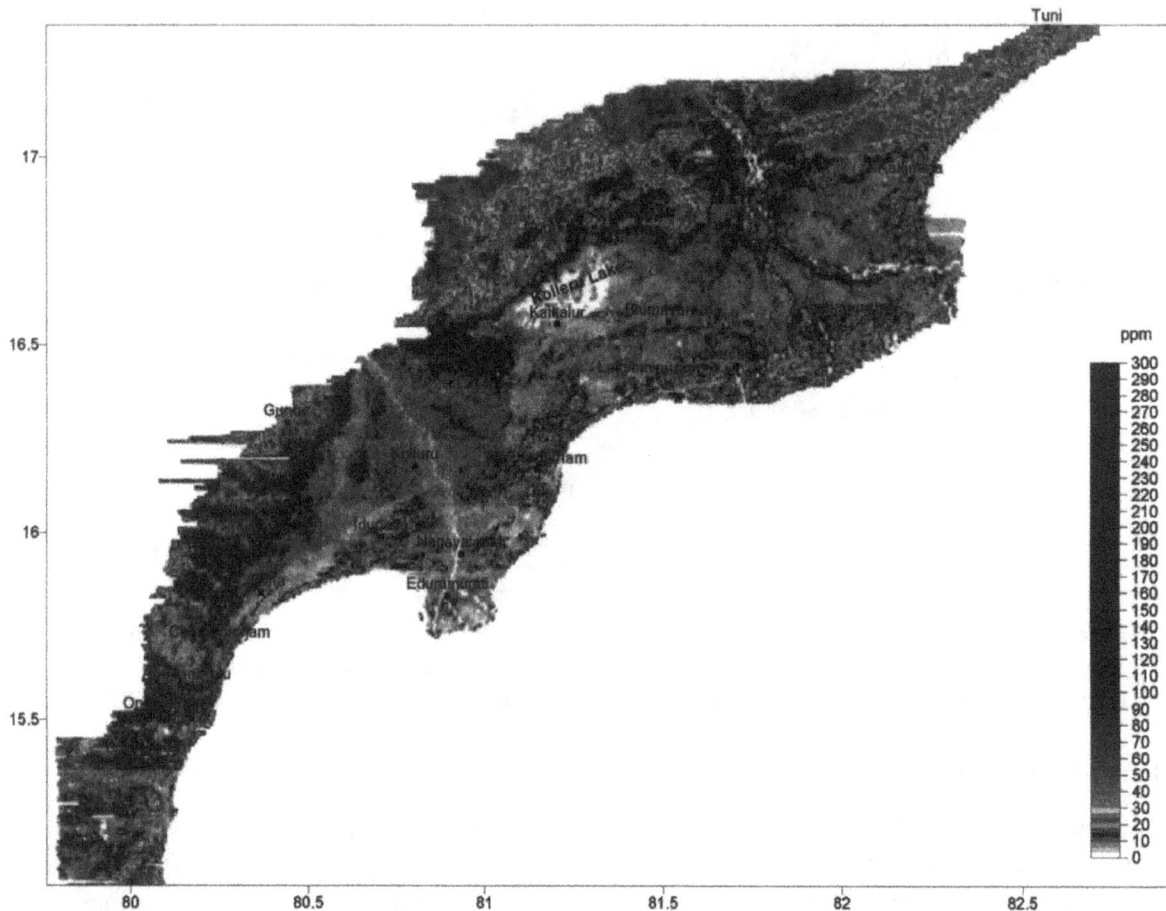

Fig. 9.65 Thorium image of coastal Godavari-Krishna basins based on airborne gamma spectrometer survey showing paleochannels of linear bands of blue and green colors range of thorium. Flight altitude: ~ 120 m and flight line (E-W) spacing: ~ 1 km (Rao et al., 2010). [Colour Fig. on Page 823]

Some Typical Environmental and Engineering Studies: Near Surface Geophysics

10.1 Introduction

Though the approach to exploration for environmental and engineering problems are the same as those used for mineral and groundwater exploration, since target depth is small in the former, survey specifications are different in these cases and are classified under 'Near Surface Geophysics'. Target depth in these cases is usually a few meters to tens of meters (<100 m); therefore, station spacing is maintained at ~1-10 m depending on depth to the target and area to be surveyed. The order of gravity anomalies in such cases are also small (0.01-1 mGal) necessitating high resolution surveys as discussed in Chapter 2, which require high quality geodetic surveys for station elevation. The same is the case with electrical and seismic surveys where electrode separation and geophone separation, respectively are maintained up to a few meters (1-10 m) depending on the depth to the target and size of the area to be surveyed. In most of the cases profiles across the specific target are recorded but in case of detailed investigations, grid surveys can also be undertaken at the same station spacing. Sometimes even depth wise measurements at same points are carried out as in the case of electrical soundings and loggings. In most of the engineering site investigations, knowledge of faults, fractures, and lineaments are essential to avoid such zones for heavy construction. This can be most effectively inferred from airborne magnetic surveys as described in Chapter 9 for groundwater exploration in hard rock areas, and mineral exploration. It may not be possible to undertake airborne magnetic surveys for these investigations due to limited interest but in case such maps are available even on a regional scale, they should be examined for these features.

10.2 Environmental Studies

These studies are related to those aspects of geophysics that are directly or indirectly related to our understanding of the environment. The two important problems related to the environment are (i) ascertaining the quality of aquifers based on resistivity distribution depthwise; and (ii) detection of cavities in mines or heavy construction sites as discussed below. Glaciers budgeting or groundwater budgeting in a region are other aspects that can be studied based on satellite gravity as discussed in Chapter 2 and Section 9.9. On similar lines, geophysical methods can be employed for several other applications in environmental studies. An application of satellite gravity in estimating sea level changes significant for environmental studies is discussed in Section 2.8.3.

10.2.1 Quality of Aquifers

One of the major environmental problems along the coast and in mega cities is the identification of good quality aquifers especially for separating brackish from potable water. It can be achieved to a certain extent by resistivity method. Being saline, brackish water is more conductive than fresh water.

Therefore, vertical resistivity sounding can be effectively employed to separate good quality aquifers from brackish water levels. Fig. 10.1(a) (Saha et al., 2002) demonstrates this application in West Bengal where brackish water and fresh water are mixed in various regions separated at different depth levels. Based on vertical electrical soundings, resistivities at different depths were obtained and aquifers were classified according to these values. Saline/ brackish water zones provided resistivity < 6.0 ohm.m while saturated clay/silt provided a value of 6-8 ohm.m. The best aquifers related to saturated medium/coarse grain sand provided a value of >15.0 ohm.m. Based on these values, good aquifers can be identified and drilled up to that level. Knowledge of clay level is also important as it stops the mixing of fresh with saline water. In coastal and desert regions, airborne electromagnetic methods can be successfully used to expedite the search for fresh water levels. Pakistan had employed the helicopter borne electromagnetic method in collaboration with the Geological Survey of Germany to identify the paleochannels of fresh water in Cholistan to the west of Rajasthan that was found below a saline water zone of high conductivity and was considered to represent the course of the Vedic river Saraswati. A similar approach may yield positive results in Western Rajasthan. Delineation of a paleochannel based on simple resistivity survey in Western Rajasthan (Mishra and Mallick, 2005) has been described in Section 10.7.1. Saha et al., (2002) have used a similar approach to differentiate fresh groundwater zones from saline water zones in Calcutta.

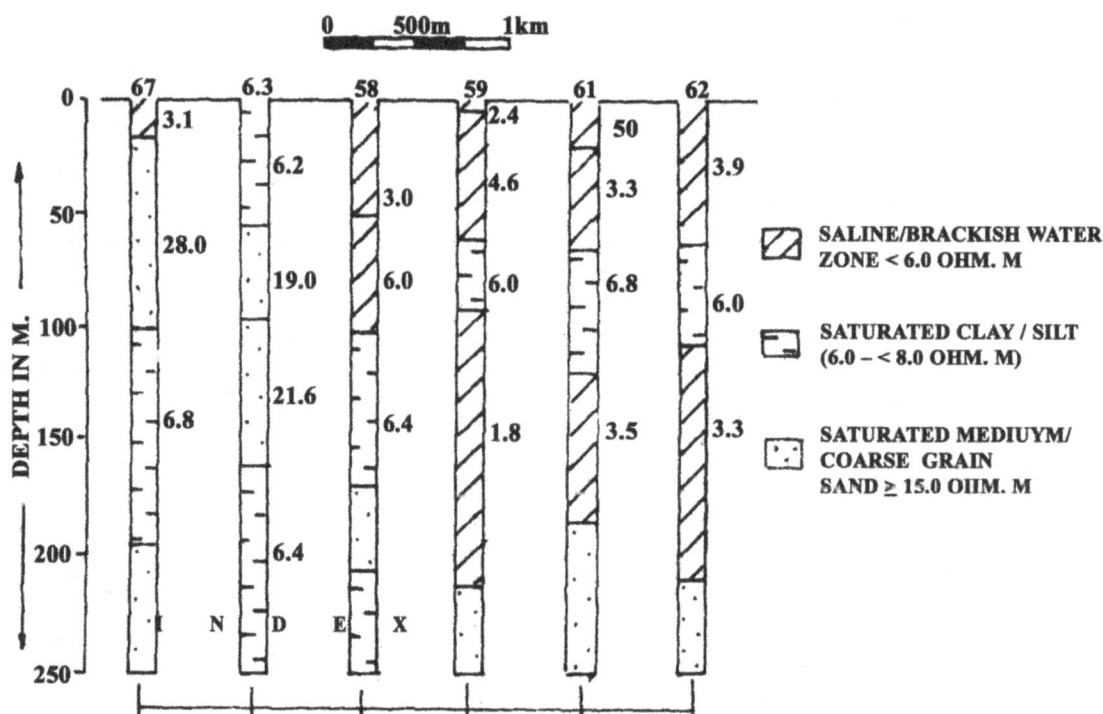

Fig. 10.1(a) Vertical distribution of resistivity with interpreted section of subsurface geology showing lowest values over the saline water zone (Saha et al., 2002)

One of the major problems in certain regions, especially in Andhra Pradesh, India is the presence of fluoride contamination in ground water that causes deformity of bones. Mondal et al., (2010) have suggested that resistivity and induced polarization surveys (Section 9.10) can help in identifying aquifers that are contaminated by fluoride. Accordingly, low resistivity and low chargebility are indicative of high fluoride content in groundwater, while low resistivity and high chargeability are characteristics of fresh water aquifers.

10.2.2 Cavity Detection – Coal Mines

Quite often, cavities occur in mines, especially in coal mines, due to slumping of rocks or discarded mines after excavation. They present future hazards unless they are detected and timely precautions taken. Gravity measurements have been found to be quite useful in this regard. However, anomalies due to such voids are very small (~100-200 microGal) and therefore, high resolution gravity surveys as discussed in Chapter 2 are required for their detection. Resistivity method can also be used effectively in case the cavities are filled with water.

As these cavities usually occur at shallow depth, gravity surveys are conducted at station interval of a few meters (<10 m). A Bouguer anomaly map (Biswas et al., 1999) over a subsided zone in Raniganj coal field, West Bengal provides a gravity low of –300 microGal (0.3 mGal). However, it does not define the void as well as its vertical gradient (Fig. 10.1(b)) that shows a lower gradient of ~200 microGal/meter compared to its surroundings. As gradient maps enhance the anomalies due to shallow sources, and the anomalies due to deeper sources are attenuated, they act like residual maps where the shallow sources are reflected better.

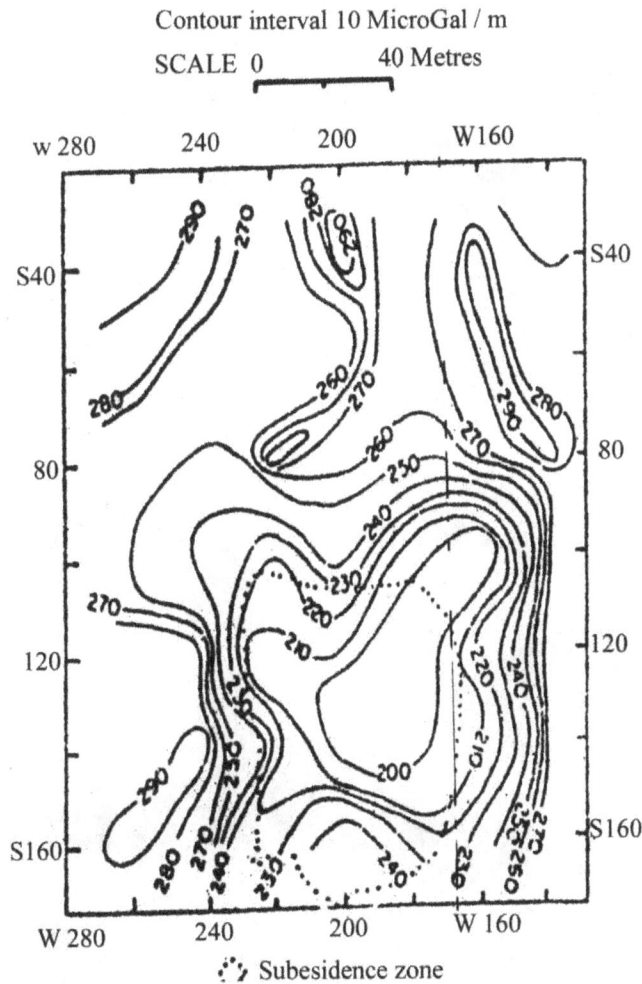

Fig. 10.1(b) Vertical gravity gradient map indicating a lower gradient over a subsidence zone in Raniganj coal field, West Bengal (Biswas et al., 1999).

10.2.3 Mass Changes in Coal Mines – 4-D Gravimetry

Fig. 10.2(a), (b) and (c) (Rahber and Tiwari, 2010) shows the gravity field of a coal mine as it progresses with time. A variation in the gravity field with respect to time is referred to as 4-D gravity accounting 3-D for the causative sources. Fig. 10(a) represents the field due to excavations of 100 pillars of unit dimensions at depth of 25 m that shows gravity low of ~ – 47 microGal assuming a density of 1.5 g/cm^3 for coal. In case it is filled with slurry of density 2.0 g/cm^3, it produces a positive anomaly of ~34 microGal over the section filled with slurry (Fig. 10.2(b)). In case the same mine is inundated with water up to 2.0 m, it would produce an anomaly of ~8-10 microGal that can be easily detected in a high resolution survey (Section 2.3.4).

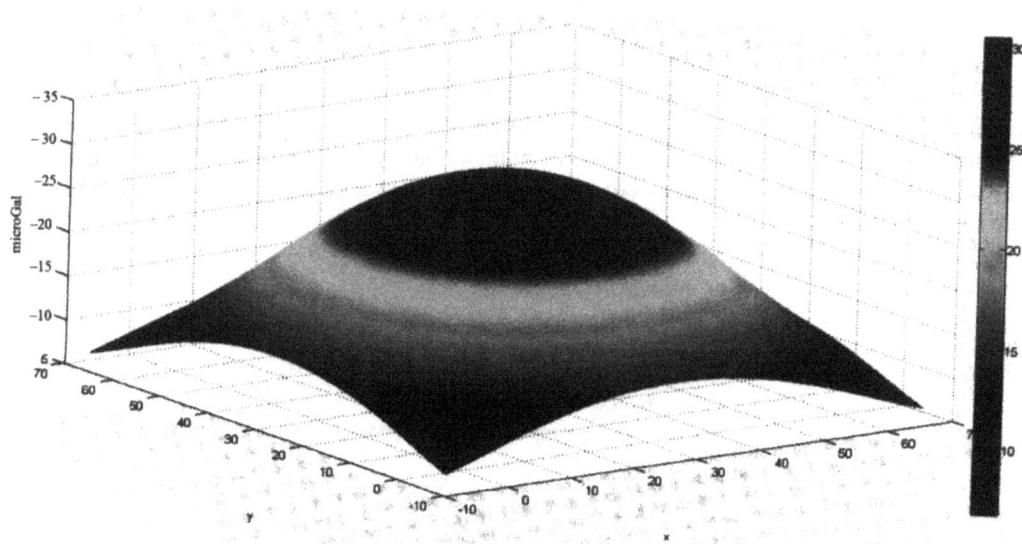

Fig. 10.2(a) Gravity field due to excavation of coal up to 25 pillars of unit dimensions at depth of 25 m. It is a negative field decreasing upwards as –20, –25, –30 mGal due to void of density contrast –1.5 g/cm^3 (density of coal; Rahber and Tiwari, 2010).

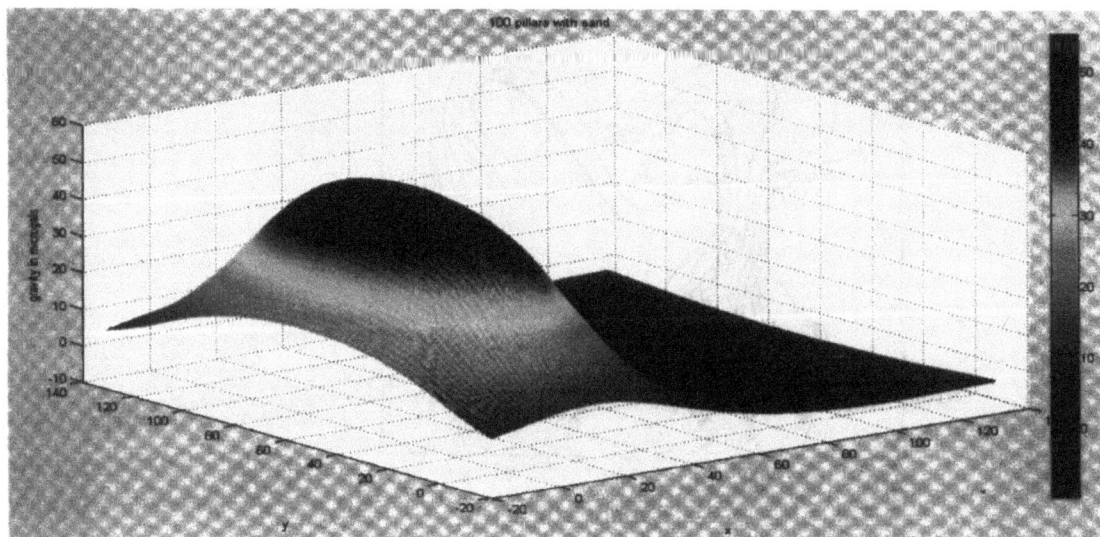

Fig. 10.2(b) Gravity field due to partly stowing slurry of density 2.0 g/cm^3 at a depth of 25 m that shows positive gravity anomaly over stowed part due to high density of slurry compared to coal (Rahber and Tiwari, 2010).

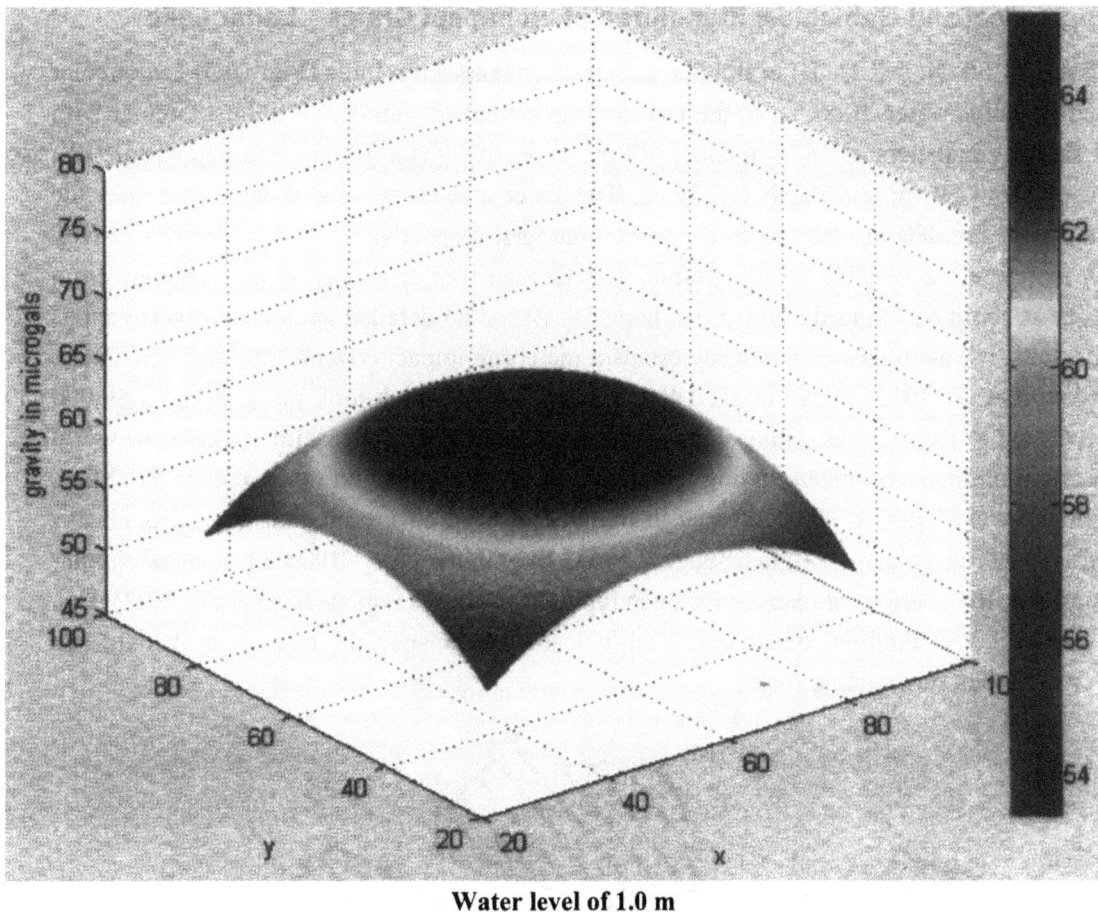

Water level of 1.0 m

Fig. 10.2(c) Gravity field due to water filled mine up to a water level of 1.0 m at a depth of 25 m that shows positive gravity field due to higher water density as compared to void (Rahber and Tiwari, 2010).

10.3 Impact Craters – Lonar Lake in Deccan Trap Province

Impact craters are formed due to impact of meteorites on the surface of the earth. Their identification is important as they provide a connection of the earth with other celestial bodies in the universe. They are also associated with several phenomena in the geological past such as mass extinction and climate change as discussed in Chapter 3. Even the initiation of life on earth is attributed to those that are supposed to have impacted the surface of the earth in large numbers during the Archean-Proterozoic period. The largest of these craters is found on Antarctica that is ~ 300 miles wide lies, under ice cover, and dates back to ~ 250 Ma. It was supposedly responsible for the Permian-Triassic mass extinction (SpaceRef.com, Ralph Von Frese and his team of Ohio State University). It is located in the Wilkis land, East Antarctica, south of Australia that might have even initiated the rifting causing the breakup of the Gondwana supercontinent. Being hidden under ice, this crater was discovered based on its gravity signature of a circular gravity high in GRACE satellite data due to denser material of meteorites. It was also confirmed from airborne radar images. Its effect can be gauged from the fact that the Chicxulub crater in the Yucatan peninsula, considered to be responsible for the extinction of dinosaurs, is only 6 miles wide.

10.3.1 Geophysical and Geological Signatures of an Impact Crater – Lonar Lake

The Lonar lake (19° 58' N; 76° 31' E) in Buldhana district, Maharashtra, India (Fig. 10.3) is a circular lake occupied by saline water. It occurs in the Deccan trap covered region and is almost circular with longest and shortest diameters as 1875 m and 1787 m, respectively (La Touche and Christie, 1912); a raised rim of about 30 m; and depth of 135 m. The inner and outer rims of the crater coincide approximately with the 480 m and 580 m elevation contours, respectively. It is presently filled by water and is therefore inaccessible on the ground. The Deccan trap originated due to eruption from Reunion plume at about 65 Ma as described in Chapter 9. Based on detailed geological investigations and drilling results, it was suggested that it represents a meteorite impact crater (Dube and Sen Gupta, 1984; Fredriksson et al., 1973; Fudali et al., 1980; Nayak, 1972), which took place at about 50,000 years ago according to fission track dating. This is the only impact crater in basaltic rocks in the world and is, therefore, important for understanding the effects of impact craters on other planets like Mars, and the Moon. However, due to its occurrence in the Deccan trap covered region, there has always been a doubt about its volcanic origin (Bucher, 1963; Subrahmanyam, 1985) as several circular volcanic plugs exposed at different places have been reported from this region (Basu et al., 1993). Due to erosion, circular lakes could be created over such volcanic plugs, specially over those that are less resistive to erosion than surrounding rocks.

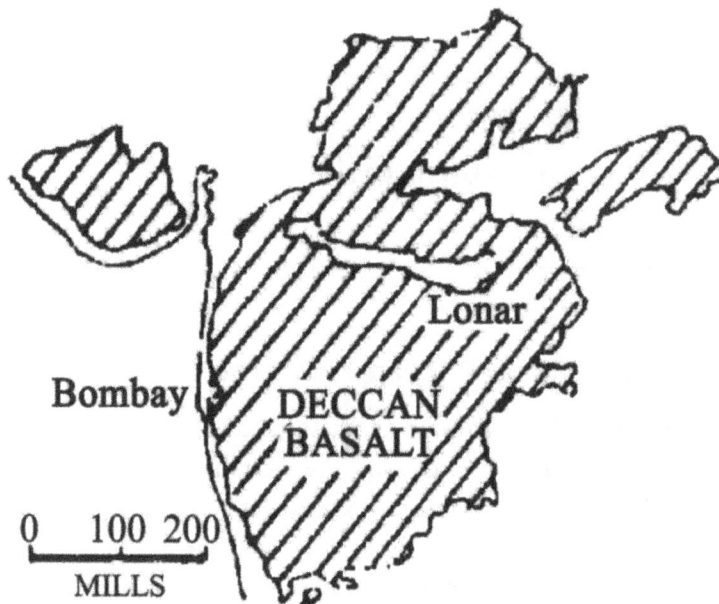

Fig. 10.3 Location map of Lonar in Deccan Trap province.

10.3.2 Gravity and Magnetic Anomalies over Lonar Lake, India – An Impact Crater

Geophysical methods, especially gravity and magnetic methods have been found to be extremely useful to delineate the sites and other details of impact craters the world over (Pilkington and Grieve, 1992). A circular bowl-shaped negative gravity anomaly is observed over simple craters (diameter up to 4 km on the earth). The amplitude of negative gravity anomaly associated with impact craters increases with the crater diameter (Pilkington and Grieve, 1992). The general character of magnetic anomalies associated with impact craters is more complex than gravity signature due to greater variation in magnetic properties of rocks and direction of magnetization. Petrography and petrochemical studies have provided details of basalt occurring around the crater and its shocked facies have already been described (Ghosh and Bhaduri, 2003). The Lonar crater and its environs exhibit six flows of Deccan trap basalt. Based on petrochemistry, it is interpreted that the country rock prior to the extraterrestrial impact was slightly over-saturated tholeiitic basalt belonging to a tectonic setting of normal continental active shield volcano. The basalt from this area had a distinct chemical signature, which differed from the main Deccan basalt with respect to iron and magnesia; the total alkali content showed relatively high magnesia (Ghosh and Bhaduri, 2003).

Meteorite impacts modify the gravity and the magnetic field at the site of impact, which after the impact, shows different signatures as compared to the surrounding rocks. This is especially true about the magnetic field in a basalt province, as the impact will raise the temperature beyond the Curie point of magnetite (550–600^0C) and remagnetize it in the present-day earth's magnetic field. This process modifies the remnant magnetization of igneous provinces corresponding to the period they initially intruded/formed.

10.3.3 Gravity and Magnetic Anomalies

The gravity and magnetic fields observed over Lonar lake were investigated to check on the nature of this crater. The gravity and magnetic anomalies of Lonar lake were obtained using a underwater gravimeter and a vertical intensity magnetometer over a suitable boat (Fudali and Subrahmanyam, 1983). These observations provided almost circular gravity and magnetic anomalies over the lake of almost -2.25 mGal and 550 nT, respectively (Fig. 10.4 and 105(b)). The terrain-corrected gravity anomaly is reproduced in Fig. 10.4, which shows a circular gravity low with $+ 0.3$ mGal contours almost coinciding with 480 m elevation contour, which defines the inner rim of the crater. Circular gravity and magnetic anomalies are observed in case of impact craters and in some cases of volcanic plugs, but their amplitudes and signs differ considerably with a few mGal (<10 mGal) negative gravity anomalies reported for impact craters, and positive $+ 60$-80 mGal anomaly over volcanic plugs, especially those reported from the Deccan trap in Sections 8.2 and 8.9.2 (Chandrasekhar et al., 2002). Magnetic anomalies observed over volcanic plugs of Deccan trap activity are also of a higher order (800–1000 nT) compared to those observed in the present case of the Lonar lake (300 nT, Fig. 10.5(b)). Further, the former shows negative magnetic anomalies, implying magnetic lows due to remnant magnetization corresponding to Deccan trap activity, while the latter provides a magnetic high corresponding to the vertical component of the present-day earth's magnetic field indicating re-magnetization after impact. The higher amplitude of gravity and magnetic anomalies due to volcanic plugs compared to those due to impact craters can be attributed to the depth extent of volcanic plugs and higher contrasts in the physical parameters, viz. density and susceptibility, which produce these anomalies. The nature and amplitude of gravity and magnetic anomalies recorded over Lonar Lake, therefore, tentatively suggest that it represents an impact crater. This can be confirmed from the modelling of the sources causing these anomalies.

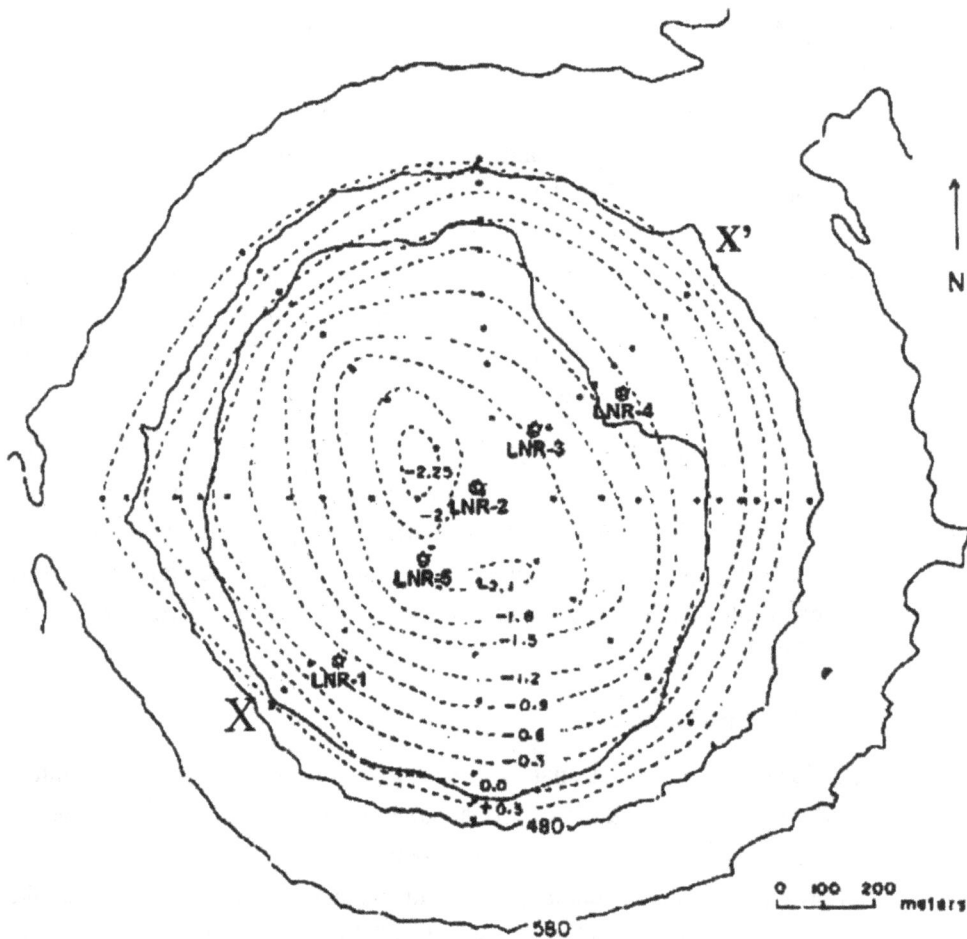

Fig. 10.4 Gravity anomaly with elevation contours (solid lines) superimposed on it. 480 m and 580 m represent the inner and the outer rim of the crater (Fudali and subrahmanyamm. 1983).

10.3.4 Modelling of Gravity and Magnetic Anomalies of Lonar Lake

Fig. 10.5(a) shows a schematic section of Lonar crater and geology based on surface exposures and diamond drill holes LNR1, LNR2, LNR3 (Fredriksson et al., 1973). Fig. 10.5(b) shows gravity and magnetic anomalies along a SW-NE profile XX' (Fig. 10.4), which are modeled for a cross-section given at the bottom of the figure (Rajasekhar and Mishra, 2005). The initial cross-section is made based on information from five boreholes up to 300-400 m shown along this profile (LNR1–LNR5). The initial model is modified by computing and comparing the computed and the observed gravity fields. The gravity and magnetic fields were computed using 2.5-dimensional bodies, implying limited strike direction, and modified using generalized inversion scheme (Webring, 1986). The densities of these layers are assigned from measurements in the laboratory of samples from bore holes (Subrahmanyam, 1985) and density of the massive basalt in surrounding area is taken as 2.75 g/cm^3 based on reported density. It basically shows three layers corresponding to (1) silt layer in the lake of bulk density 2.0 g/cm^3, (2) brecciated zone of bulk density 2.65 g/cm^3, and (3) high density fragmented rocks of density 2.75g/cm^3. The surrounding solid basalt can be assigned a bulk density 2.75 g/cm^3 (Fig. 10.5(b)), as suggested by previous workers for unweathered massive Deccan basalt.

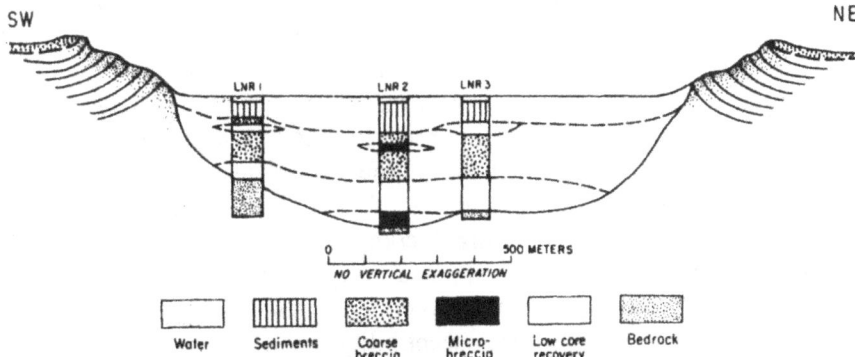

Schematic section of Lonar Crater, geology based on surface exposures and diamond drill holes LNR 1, 2, and 3.

Fig. 10.5(a) A schematic section of Lonar crater and geology based on surface exposures and drill holes shown in the figure (Fudali and subrahmanyam, 1983).

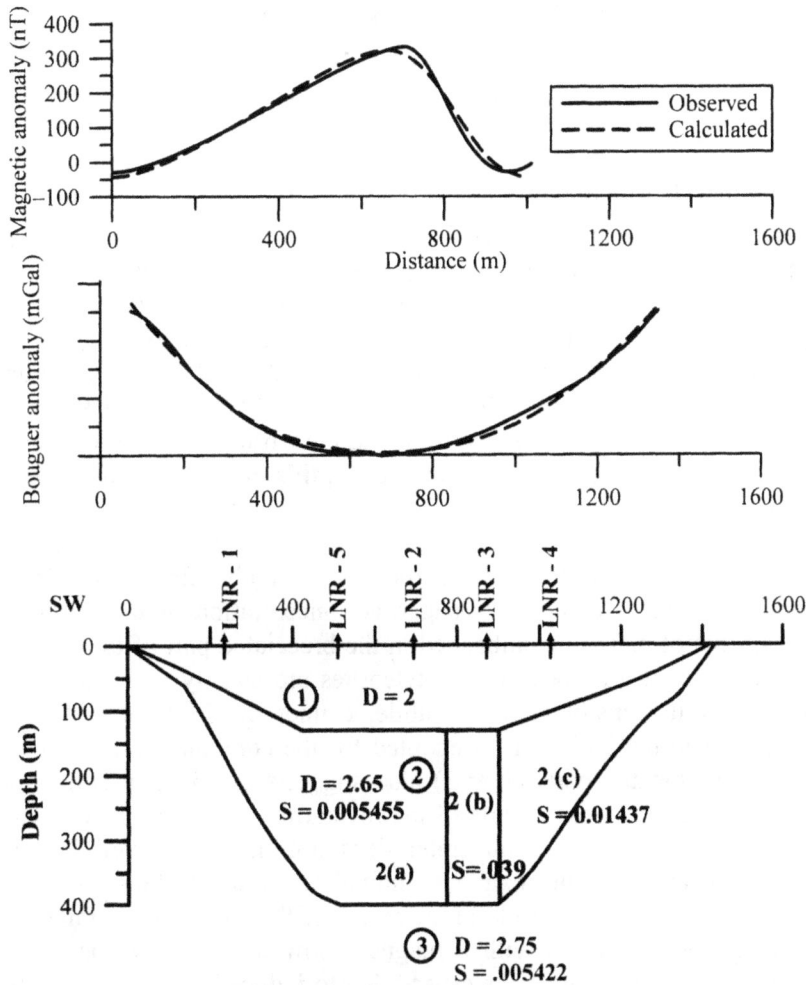

Fig. 10.5(b) A SW-NE magnetic and gravitry profile (XX', Fig. 10.4) across Lonar Lake and modeled subsurface layers of density (D, g/cm^3) and susceptibility (S, SI units) given inside the layers. Three layers, 1-3 represent, 1. silt in the lake, 2a and c. breciated part and 3. Fragmented basalt. The body, 2b is a mafic body of high susceptibility introduced to match the observed magnetic field and may represent parts of the fallen meteorites.

Layer (3) relates to high density fragmented rocks. Being heavier, these rocks fall first after the impact and occupy the bottom of the crater, with brecciated low density parts over them. The magnetic field from this cross-section is computed by assigning susceptibilities similar to those reported for the Deccan trap in adjoining regions (Rao and Bhalla, 1984; Vandamme et al., 1991). However, when the same susceptibility distribution could not produce the observed magnetic field, a dyke type of body numbered as 2b in Fig. 10.5(b) is introduced, which shows the same density as layers (2a and c) but divides them into three parts of different susceptibilities. The model shows a high susceptibility rock/body (2b, Fig. 10.5b) under the magnetic 'high' compared to those surrounding it, which provides a good match between observed and the computed fields. In general, the changes in density are quite subtle compared to susceptibility and therefore, it is quite often possible that density, especially bulk density, may remain the same while susceptibility within the same rock type, specially in case of basalts, changes considerably. In fact, this property of the Deccan trap explains the sudden variations in magnetic field in magnetic surveys over the Deccan trap, even when the gravity field shows a smoothly varying field.

Modelling along the profile XX' is based on a scheme for a 2.5-dimensional body. However, gravity anomaly (Fig. 10.4) suggests that it is a truly three-dimensional body of basinal type with major density contrast along the interface separating the brecciated part and the unaffected Deccan basalt. We, therefore, employed the harmonic inversion scheme as described in Chapters 4 and 6 (Hahn et al., 1976; Mishra and Pederson, 1982) to model the three-dimensional configuration of parts of the Deccan basalt affected by this impact. This method states that the Fourier transform of the observed field is related to the transform of the causative body through a filter function whose inverse transform provides the configuration of the source.

With a density contrast of -0.24g/cm^3, Fig. 10.6 provides a basinal structure with zero contour, almost coinciding with zero contour of gravity anomaly (Fig. 10.4) and suggests maximum depth of 500 m to the unaffected basalt. This depth is with reference to 480 m of elevation contour, which defines the inner rim of the crater. Modelling of gravity and magnetic anomalies in the previous section suggests that an impact crater of approximate dimension $1875 \times 787 \text{ m}^2$ affects up to a depth of 450-500 m from the inner rim of the crater. This amounts to 550-650 m from the surface, which is almost 0.3-0.35 times the surface extent, as is generally found the world over in case of impact craters. It also suggests that the layers are magnetized in present-day earth's magnetic field, thereby indicating that due to impact and related rise in temperature, the original magnetic field of the Deccan trap is modified, which in unaffected parts corresponds to remnant magnetization in the southern hemisphere where India was located during eruption of these rocks. The body (2b) (Fig. 5(b)) with high susceptibility compared to the surrounding rocks suggests concentration of magnetite in this part, which may represent fragments of meteorite embedded in the brecciated part of the Lonar lake as the ejecta cloud later settled along with the rock chips. Meteorites are usually more mafic compared to normal basalt. The variations in density is more subtle, cannot be differentiated in the present computation, and are included in the bulk density adopted for the computation of gravity field. The higher bulk density of the fragmented part (layer 3) also suggests that it may contain some high-density rocks at places, which may represent parts of the meteorite. It may be mentioned here that a satellite crater about 300 m north of the main crater does not show any well-defined magnetic anomaly. Instead, it shows sharply varying magnetic anomalies typical of those observed over the Deccan trap. This suggests that the magnetic field in the region of the main Lonar crater, which shows a pair of well-defined magnetic anomalies with a magnetic low toward north as expected in the present-day earth magnetic field in this region, is caused due to induced magnetization. It is inferred that quantitative modeling of gravity and magnetic anomalies, especially the latter, of impact craters in basalt provinces provides significant information. Since most celestial bodies are composed of basaltic rocks, the satellite gravity and magnetic field recorded over Mars and the Moon can be analyzed in this manner, to differentiate volcanic plugs and impact craters and delineate their subsurface structures.

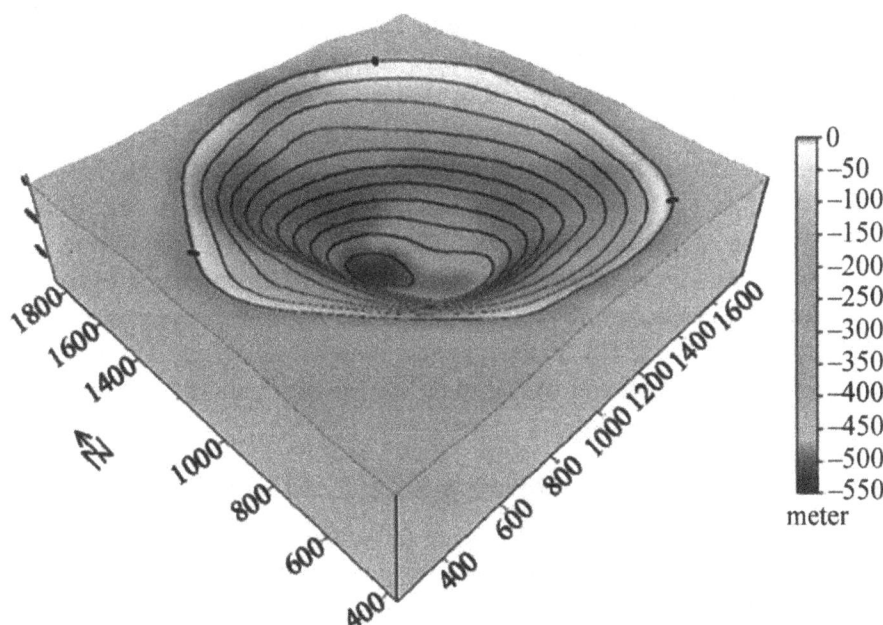

Fig. 10.6 A 3-D view of the impact crater computed from the observed gravity field.

10.4   Effect of Solar Eclipse on the Earths Gravity Field – Allias Effect

Solar eclipses provide an unique opportunity for the study of celestial phenomena such as different parts of the sun like corona and chromospheres, its atmosphere, and their interaction with the earth. The effect of solar eclipses on earth's gravity field was first noted by French, Maurice Allias in 1954, and is known as Allia's effect. While experimenting with a Foucault pendulum, Maurice Allias noted a 13.5° of rotation during a solar eclipse. A basic Foucault pendulum is simply a weight on a wire that moves under the earth's gravitational field when released. Allias further noticed it during another solar eclipse in 1959 (NASA, 2010). He explained it due to certain anisotropy characteristics of space and was awarded several awards in France for this discovery. He was also awarded Nobel Prize in 1988 for economics for his contribution to the theory of markets and efficient utilization of resources which he claimed was inspired from his discovery of Allias effect. Mishra and Rao (1997) noted some peculiar changes in gravimeter during a solar eclipse in India as discussed below. Wang et al., (2000) and Yang and Wang (2002), based on their experiments in China, further reported that the earth's gravity field is affected during solar eclipse. However, there are also certain reports contrary to it when no effect of solar eclipse was noticed on the earth's gravity field (NASA, 2010). In case it is due to anisotropy of the space as suggested by Maurice Alias, it may not act in the same manner over different parts of the globe. This plausibly explains its anomalous behavior. Therefore, scientists are divided on the Alias Effect and its cause that can only be sorted out through more observations throughout the world and its plausible occurrences in specific cases.

10.4.1 Experiment during a Solar Eclipse in India

The solar eclipse on 24 October 1995 starting from sunrise at Iran and ending at sunset at the Pacific Ocean provided a 46 km wide strip for approximately 1800 km in India from Nem Ka Thana (Western Rajasthan) to Diamond Harbour (West Bengal) where the total solar eclipse was observed for some time between 7:22 am to 10:30 am (Bhattacharya, 1995). This solar eclipse was unique due to several scientific experiments which provided several interesting results (Sapra et al., 1997). During the period of this solar eclipse, temporal variation of the earth's gravity field was recorded at Dhoraji (22°, 44', 70° 27' Saurashtra; Fig. 9.1(a)). This region falls within approximately 80% of the total eclipse within. Temporal variation in the gravity field was recorded at this place continuously for approximately 12 hours before and after the eclipse. The variation in the gravity field is recorded using a Lacoste-Romberg gravimeter of 1 microGal accuracy. The temporal variation in the gravity field recorded at a station can be broadly classified as: (1) Very large period (100-10,000 years) variations related to the mantle processes, sea level changes, glacial rebound etc., (2) Large period (10-100 years) variations due to core-mantle interaction, plate boundary deformation, etc., (3) Medium period (days to years) variations due to earthquakes, volcanoes, etc., (4) Short period (hours to days) variations caused by drift of the gravimeter, and ocean tides. (5) Shorter period (hours) variations due to the sudden changes in the atmosphere such as pressure and temperature. (6) Shortest period (seconds to minutes) high frequency noises which are sharp and sudden.

10.4.2 Temporal Variations in Gravity Field During Solar Eclipse on 24 October, 1995

As mentioned above, these variations are largely characterized by their wavelength which is used to classify them in different groups. In the 12-hour record which we made, we can neither record and identify the variations classified under groups (1), (2), and (3). The fourth group of variations shows a cycle of 24 h and is represented by smooth cyclic changes as given in Fig. 10.7(a) which represents the tidal variation on 24 October, 1995 at Dhoraji. It is the group (5) kind of variations which are of present interest and can be recorded in our gravimeter. The variations under group (6) are easily identifiable as they are sharp changes due to sudden jerk or some motion in the vicinity. Some experiments of recording gravitational field during solar eclipses (Tomaschek, 1955; Slichter et al., 1965) were conducted previously to find out the tidal effects and gravitational shielding (Majoranna effect) during that period.

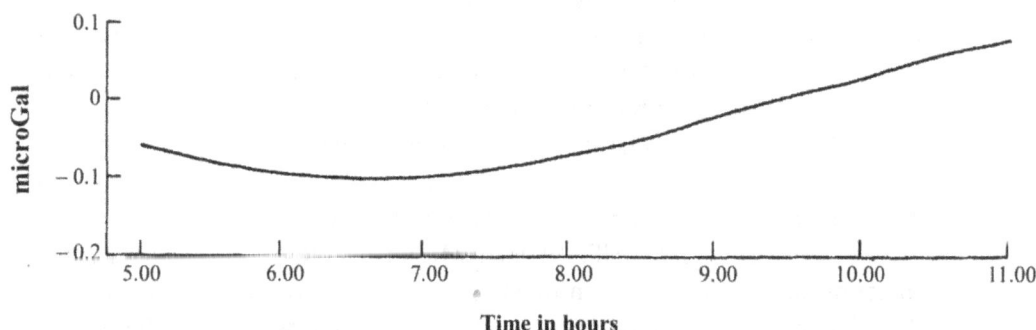

Fig. 10.7(a) Tidal effect on 24 October, 1995 at Dhoraji (Saurashtra).

Normal example of type (4) variations in gravity of short period (hours to days) variations caused by drift of the gravimeter, ocean tides as opposed to type (5) (A, Fig. 10.7(b)) of shorter period (hours) variations due to the sudden changes in the atmosphere such as pressure and temperature can be easily visualized from the daily records of variation in the gravity field. The tidal and the recorded variations of the gravity field on 24 October 1995 for a period from 5:00 am to 11:00 am are shown in Fig 10.7(a) and (b), respectively. Both the graphs show a smooth variation due to short period features like tidal and the drift of the gravimeter over which a shorter period feature of 10-12 microGal (10^{-8} cm/s^2) between 6:30 and 7:30 am (A, Fig. 10.7(b)) is superimposed. This variation can neither be classified under short-period variations due to a tidal effect or drift of the gravimeter under high frequency noise which have special patterns. Therefore, this variation is highly significant as it occurs with the onset of solar eclipse. It may represent sudden changes in the earth's atmosphere with the onset of the eclipse. Exact nature of the changes in the atmosphere is difficult to visualize. However, it could be due to a decrease in the different kinds of radiation such as gamma rays, X-rays, and radio intensity which are found to decrease with the onset of the eclipse or simply due to changes in the atmospheric pressure. Therefore, to understand its exact nature and mechanism, more planned experiments of this kind should be carried out during solar eclipses throughout the world.

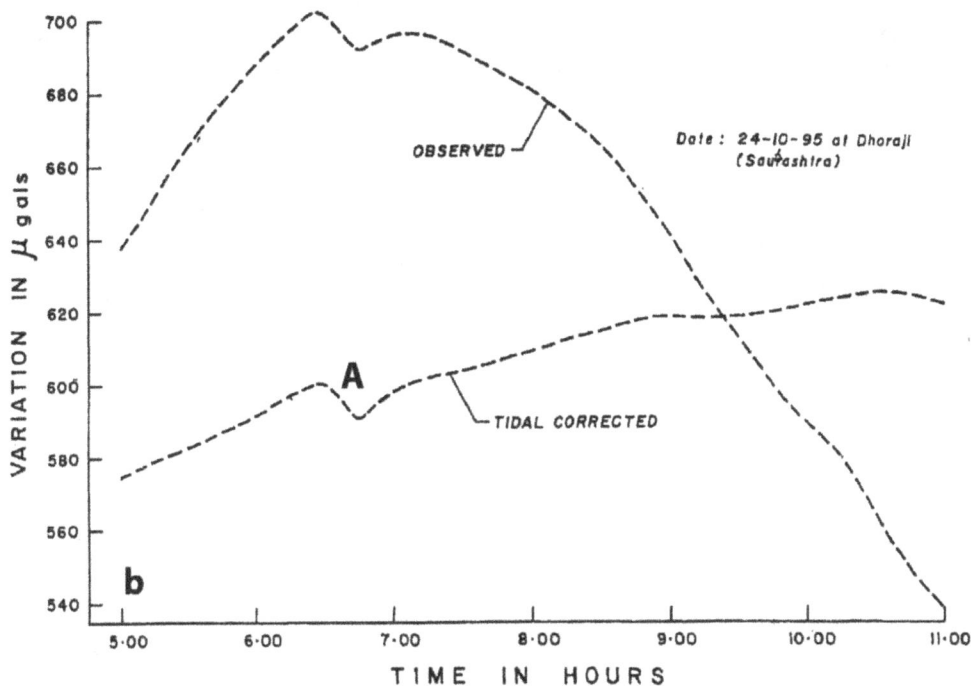

Fig. 10.7(b) Observed (top line uncorrected) and tidal corrected gravity field at Dhoraji on 24 October 1995. The onset of solar eclipse was between approximately 6:30 and 7:00 am local time. The horizontal axis is time in hours (5-11 am), and the gravity scale of the vertical axis indicates values in μGal (10^{-8} cm/s^2).

10.5 Effects of Lightning on the Earth's Magnetic Field

Lightning discharges are negatively charged particles with high current of 10,000 to 100,000 amps in the form of d.c. pulses, and therefore modify the magnetization of rocks where they strike the ground. Several laboratory and field studies (Graham, 1962; and Sukhija et al., 1978) have been made on the rocks subjected to lightning and the following general conclusions have been drawn regarding its effect on the magnetization of rocks;

1. Lightning re-magnetizes the rocks in the vicinity where it strikes usually in a direction different from the natural magnetization of rocks. However, the area affected by lightning is usually small, of a few tens of meters.

2. In case of re-magnetization due to lightning, the magnetic vectors of the rocks from the surrounding region should show concentric contours around the point where lightning has struck.

3. Laboratory studies indicate scattered direction of magnetization in the samples taken from the outcrop affected by lightning.

4. Intensity of magnetization induced by lightning is very large i.e., of the order of 10^{-1} to 10^{-3} CGS units.

5. Model studies have indicated that direction of magnetization depends on the direction of incident lightning discharges.

6. Alternating current cleaning of samples is sometimes effective in restoring the original remnant magnetization.

On March 23, 1997, at approximately 4.00 p.m. thunderbolt lightning struck at Kompalle village and near HMT factory at the outskirts of Hyderabad (Fig. 10.8). Two persons died, one at each site, and three were hospitalized due to severe burn injuries. On March 25, 1997, magnetic surveys were conducted along a few profiles at each of these sites and the results were as follows (Rao et al., 1997).

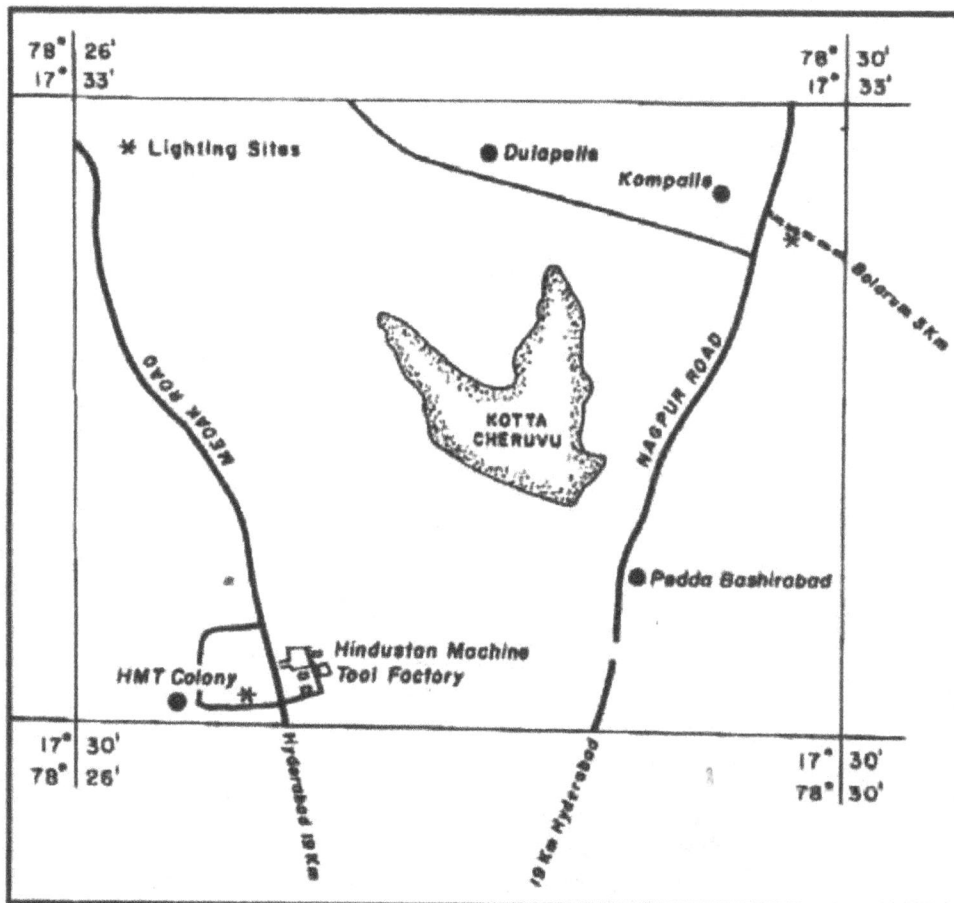

Fig. 10.8 Sites of lightning at Kompally village and HMT factory.

10.5.1 Site at Kompalle Village, Hyderabad

This site is a plain area covered with soil and several small holes were visible at the site where the lightning had struck. It is said by the local people that lightning struck at this place from NW to SE which appears to be its strike direction. As the area is soil covered, rock samples could not be obtained for laboratory measurements. However, taking a central point in the affected zone, two profiles of a few hundred meters each were laid perpendicular to each other and total intensity magnetic field was recorded at every 10 m interval using a Geometrix Nuclear Magnetometer of 1 nT accuracy. These profiles are drawn in Fig. 10.8. Both these profiles show nice consistent magnetic anomalies near the site of lightning, but with a difference as discussed below.

The NW-SE profile which is the strike direction of lightning shows a concentric anomaly of 75 nT spread over almost 150 m (Fig.10.9) beyond which there are sudden fluctuations of magnetic field typical of granites and gneisses. As the concentric "high" is located over the region of lightning and is of different nature than adjoining data, it is reasonable to presume that it is caused by the re-magnetization of rocks due to lightning. This new magnetization appears to be quite consistent and moderate in intensity. The NE-SW profile initially shows a gradient. However, at the point of lightning, there is a magnetic high of approximately 100 nT over and above this gradient. The gradient as such indicates some deeper feature, but the magnetic high over the point of lightning appears to be caused by the magnetization due to lightning. It is interesting to note that along both the profiles the nature and order of magnetic anomaly is same. However, its spread along the strike direction is more (150 m) than perpendicular to it (30 m), which is expected as per the general norm.

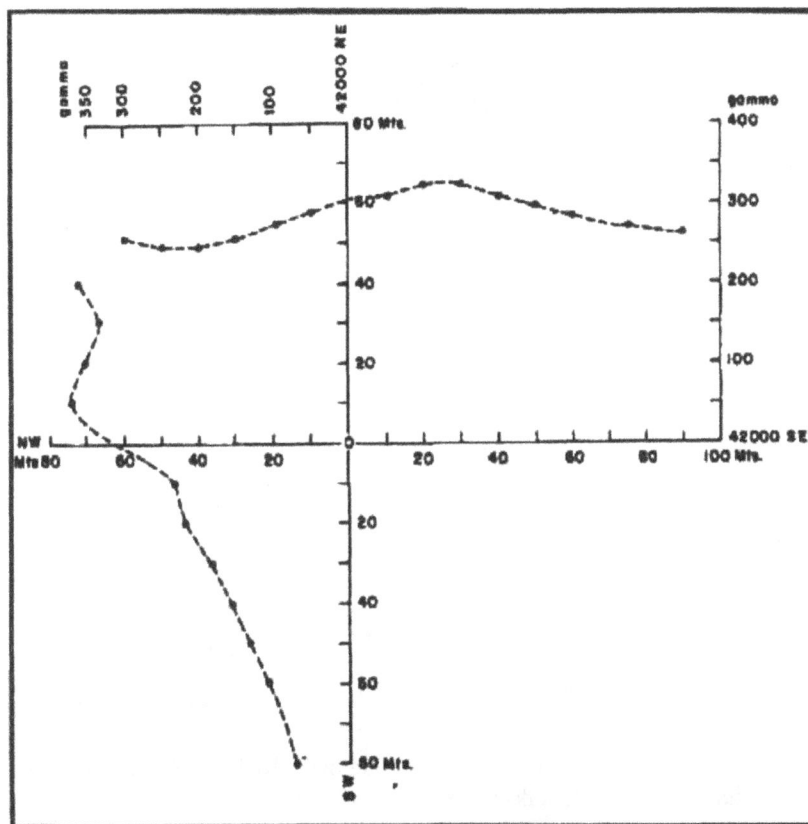

Fig. 10.9 Total intensity magnetic profiles in two perpendicular directions across the site of lightning at Kompalle village showing a magnetic high of ~75-100 nT over the site of lightning.

10.5.2 Site near HMT Factory, Hyderabad

This region is also covered in soil with a few rocks exposed in the surrounding region. The lightning in this area appears to have struck from the east. A few magnetic profiles are recorded on either side of the site of lightning, presented in Fig. 10.10. The central N-S profile shows a magnetic high of 75 to 100 nT over the site of lightning. There is a large magnetic anomaly at the southern end of the profile which coincides with the exposures of gneisses and appears to be caused by them. The central magnetic anomaly (high) at the site of lightning decreases towards the east and the west with a shift towards the north in the profiles recorded west of the central profile which is the strike direction. It may be noted that the decrease in the amplitude of the magnetic anomaly is more towards the east and almost vanishes after 15-20 m east of the central profile. This suggests that lightning has struck from the west and is confined to the central profile. This magnetic high is confined within 20-25 m along N-S and 70-80 m along E-W (strike direction) over the site of lightning and therefore appears to be caused by the re-magnetization of rocks due to lightning.

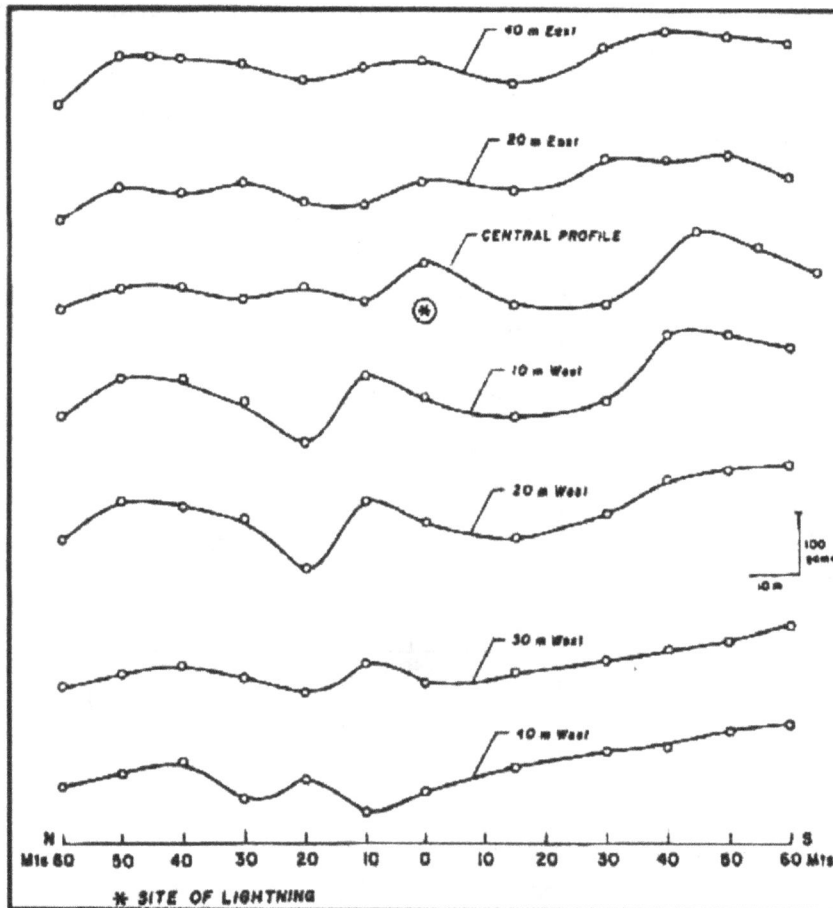

Fig 10.10 Total intensity magnetic profile at the site of lightning near HMT factory showing consistent magnetic highs that decreases away from the site.

10.5.3 Grid Survey at Kompalle Village

As the magnetic anomalies recorded at the site of lightning at Kompalle village are more definite and consistent, a more detailed magnetic survey was planned at 10 m grid interval over an area of 60 × 160 sq.m covering the anomalies observed along the two profiles reported above (Fig. 10.9). The magnetic 'high' referred to in the NE-SW profile (Fig.10.9) persists in all the profiles up to 40 m SE of the site of lightning beyond which it does not exist. Fig.10.11 presents the total intensity magnetic anomaly map which delineated an elongated concentric magnetic high of 150-170 nT close to the site of lightning. The concentric nature of the magnetic anomaly elongated along the strike direction of lightning limited to 30-40 m of the approximate site of lightning suggests that this magnetic high is likely to be caused by the re-magnetization due to lightning. It is interesting to note that, in the present case, the magnetization caused due to lightning is quite consistent and moderate instead of being inconsistent and high as reported previously.

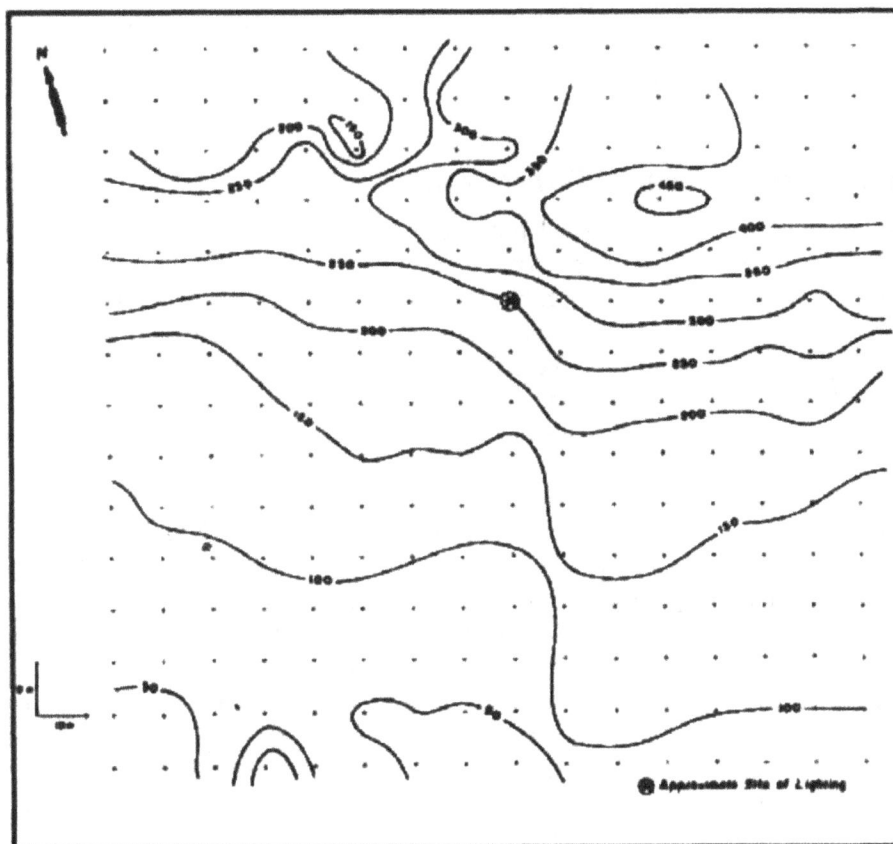

Fig. 10.11 Total intensity map surrounding the site of lightning at Kompalle village showing an elongated concentric magnetic anomaly of~200 nT over the zone of lightning.

10.5.4 Effects of Lightning on Magnetization of Rocks

(i)   Re-magnetization caused by lightning can be consistent over ten to hundred meters depending on the strike direction.

(ii) Intensity of magnetization can be moderate i.e., of 100-150 nT.

(iii) Site of lightning can be located by magnetic surveys over a closed grid of 10-20 m in the regions where they strike.

10.6   Engineering Site Investigations

Solid bedrock topography is important for heavy constructions like dams, power plants, and airports. Its knowledge before hand is important for effective planning and cost estimation. Engineers often require to know (i) disposition of faults and shear zones (ii) thickness of soil cover (iii) nature of bed rock and its topography before planning any heavy construction in a region. These parameters can be effectively evaluated through combinations of geophysical methods that have been discussed below through specific case histories. Most often, the seismic and electrical methods are employed for bed rock topography due to their accuracy for such investigations. However, faults fracture and shear zones can be better evaluated by gravity and magnetic maps, especially airborne magnetic maps in case such maps are available or accessible. Winter et al., (1994) have described in detail the importance of geology, tectonics, and geodynamics for the selection of a suitable site for a hydroelectric project and related tunnel in Kistwar valley in Himalaya at an elevation of 1225 m.

There are several case histories related to engineering sites investigations using geophysical methods. Some of the typical ones are discussed below. Varadarajan and Seshunarayana et al., (1985) based on underwater resistivity survey and seismic refraction survey provided suitable sites for tunneling for Manibhadra dam on Mahanadi in Orissa.

10.6.1 Geophysical Investigations for a Power Project

Wadhwa et al., (2010) have described integrated geophysical investigations for hydroelectric projects in the Himalayas across river Alakhnanda, a tributary to river Ganga. Being in Himalayas it presents a sharp varying topography that causes specific problems for geophysical investigations. Continuous seismic refraction, cross-hole seismic, and resistivity surveys are described for delineating bedrock topography and rock types based on seismic velocities. This project involved construction of a diversion dam of height 55 m, and 14 km long, underground water conductor system, desilting chambers, underground power house, and surface switch yard.

(i)   Continuous Seismic Refraction Survey

Continuous seismic refraction surveys provide the seismic velocities that characterize rock types. Profiles were laid both on land and in water. Figs 10.12(a) and (b) give typical depth sections on land and in water, respectively. Fig 10.12(a) shows three subsurface layers with P-wave velocities as 300-600 m/s, 800-1100 m/s and 3000-4000 m/s that suggest loose boulder bed, compact/partially saturated boulder bed, and quartzite at different depths, as given in these figures. The quality of quartzite along the traverse is inferior up to 35 km (3000 m/s) while beyond that it is good quality (4000 m/s). The rock levels ranged from elevation 1221.4 m at 32.5 km to 1230.2 m that provided a variation in the depth of bedrock (quartzite) of ~10 m in 115 m. Fig. 10.12(b) shows a depth section in water that also suggest three layers with velocities 2700-3000 m/s, 1600 m/s and 4000-5000 m/s representing boulder bed/weathered rock, saturated bed and bedrock, respectively. The quartzite bed underwater appears to be of good quality with higher seismic velocity and its depth ranged between 10-22 m.

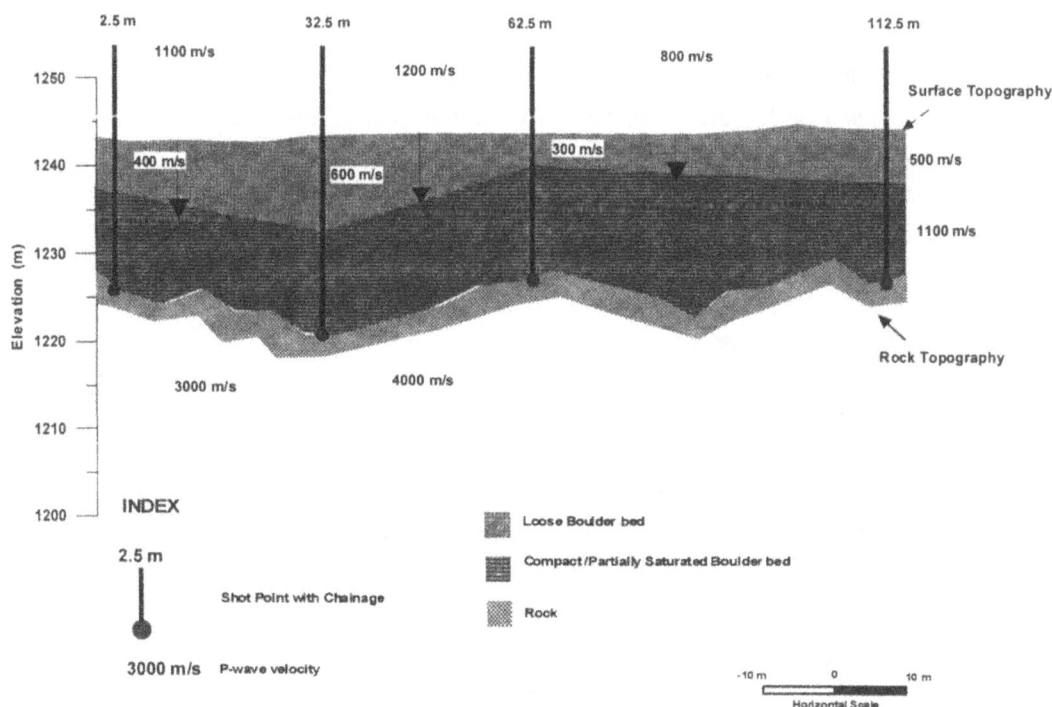

Fig. 10.12(a) Depth section along a land profile based on continuous seismic refraction survey.

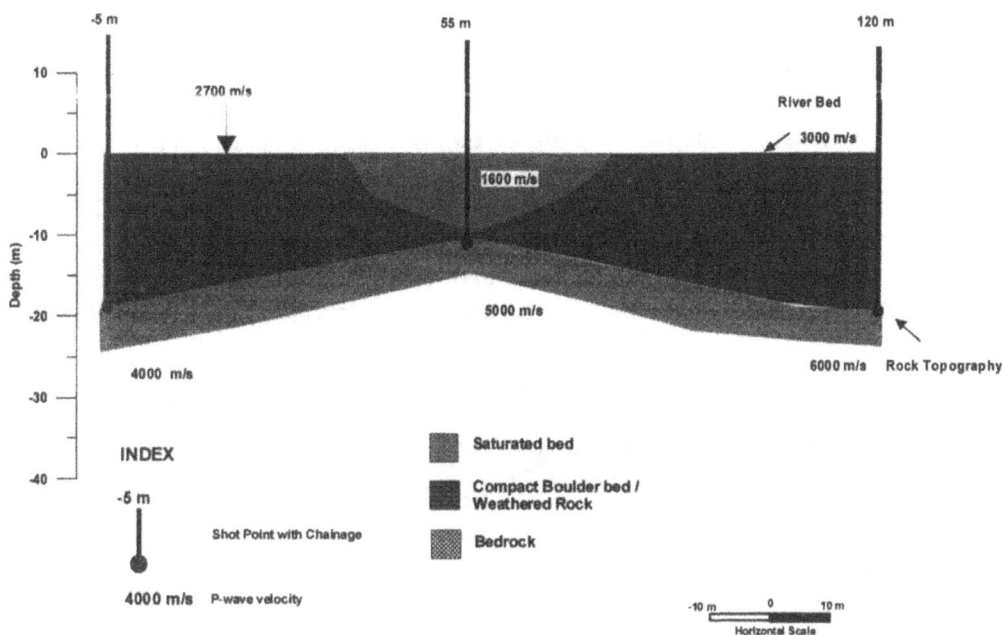

Fig. 10.12(b) Depth section along an underwater profile based on continuous seismic refraction survey.

(ii) Cross-hole Seismic Studies

Cross-hole seismic velocities are better representatives of depthwise rock types as the source and receiver are maintained at the same depth in respective holes that provides bulk velocity in the particular rock formation. Normally, two receivers in holes are used in mutually perpendicular direction from the source. In this particular case the logging interval was kept ~1.5 m and they were logged to a depth of 36 m. A typical plot of P-wave and S-wave velocities with depth is given in Fig. 10.13 (a) and (b), which also shows the inferred rock types with depth. The over burden with smaller velocities is limited to a depth of 6 m followed by shale/slate with higher velocities. This example demonstrates the application of seismic studies for engineering structures.

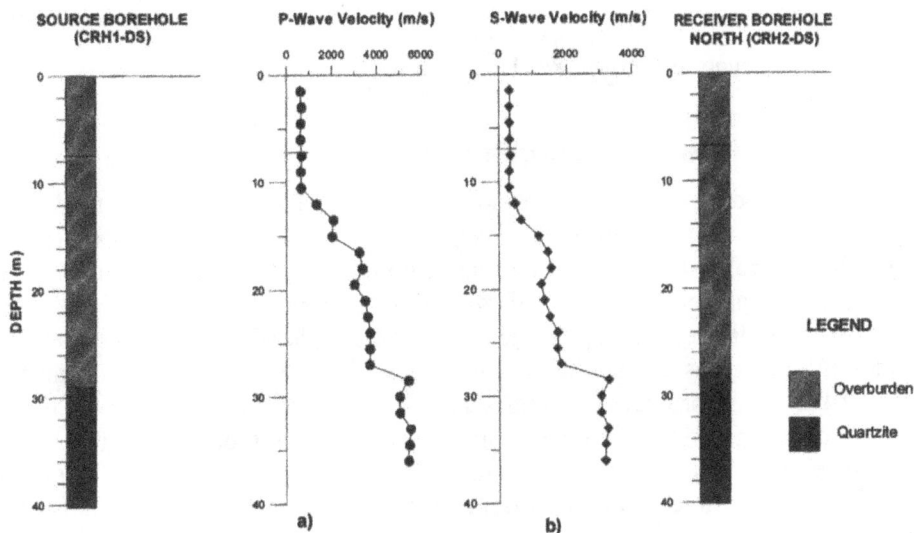

Fig. 10.13 Wave velocities with depth in north direction along with source and receiver hole logs,
(a) Compressional waves and (b) Shear waves.

(iii) Resistivity Survey

Resistivity survey was primarily carried out to delineate conductive zones for earthing of transmission towers in order to minimize contact losses. With lower value of the apparent resistivity, there will be better conduction reducing the transmission losses. The Wenner electrode configuration with 25 m electrodes at 2 m interval at the switch yard site was used. Resistivity imaging was conducted by Scintrex Automative Resistivity Imaging System with 100 Watt output power equipment and processed the data using RES2DINV software package. The pseudo section and the inversion model is given in Fig. 10.14 that shows the conductive section at the end of the profile while the central part is resistive. This should be avoided for transmission towers.

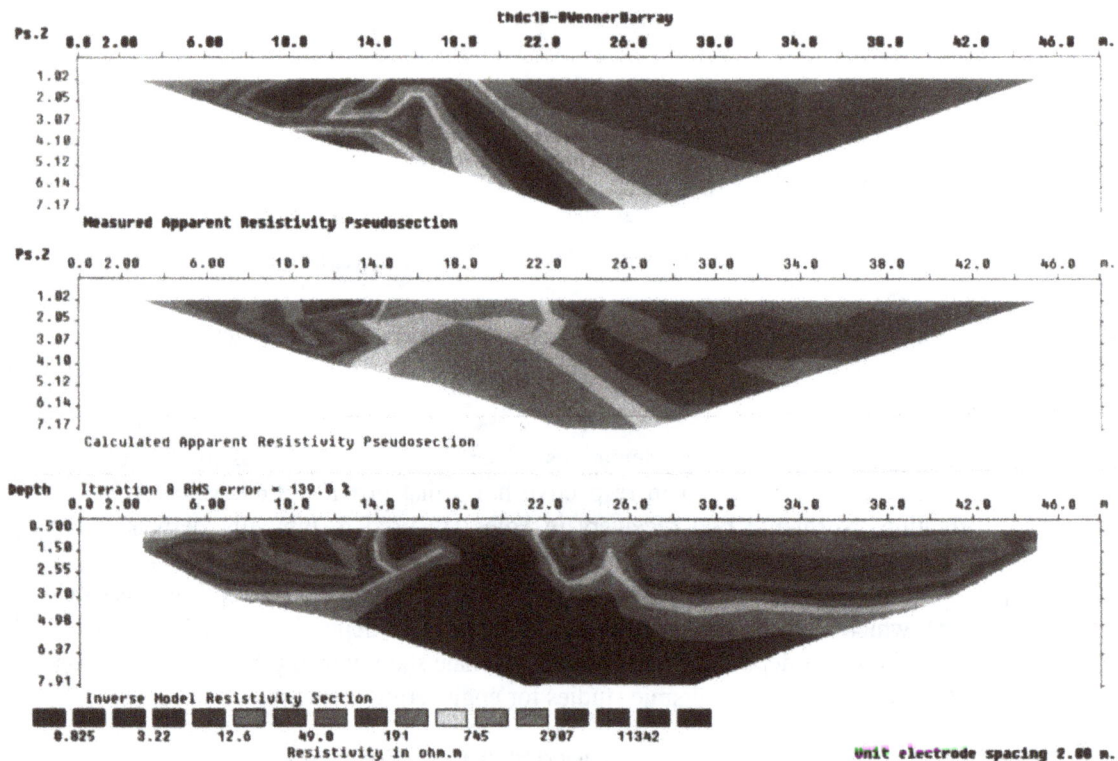

Fig. 10.14 Apparent resistivity pseudo-sections and final model through inversion. [Colour Fig. on Page 824]

10.6.2 Nuclear Power Plant, Kakrapar, Surat, Gujarat

Resistivity logging provides good estimates of resistivity of subsurface formations that helps to identify the quality and nature of rock formations required for heavy constructions. Resistivity measurements using different probes like single point, short normal (16 in), long normal (64 in) and lateral can be used to provide apparent resistivity of formations surrounding the bore hole (Kamble et al., 2010). Similarly, the acoustic or sonic logging provides compressional and shear wave velocities that can be used to identify rock types around the bore hole being logged. It uses dual transmitter and dual receiver array and the sonde provides slowness that is reciprocal of velocity, which in turn provides the velocity of the compressional and the shear waves. A typical plot of various resistivities and P-wave and S-wave velocities from a bore hole up to 100 m depth at the Kakrapar, Nuclear Power Plant are given in Fig. 10.15; their interpretation in terms of rock formations are given in Table 10.1 (Kamble et al., 2010).

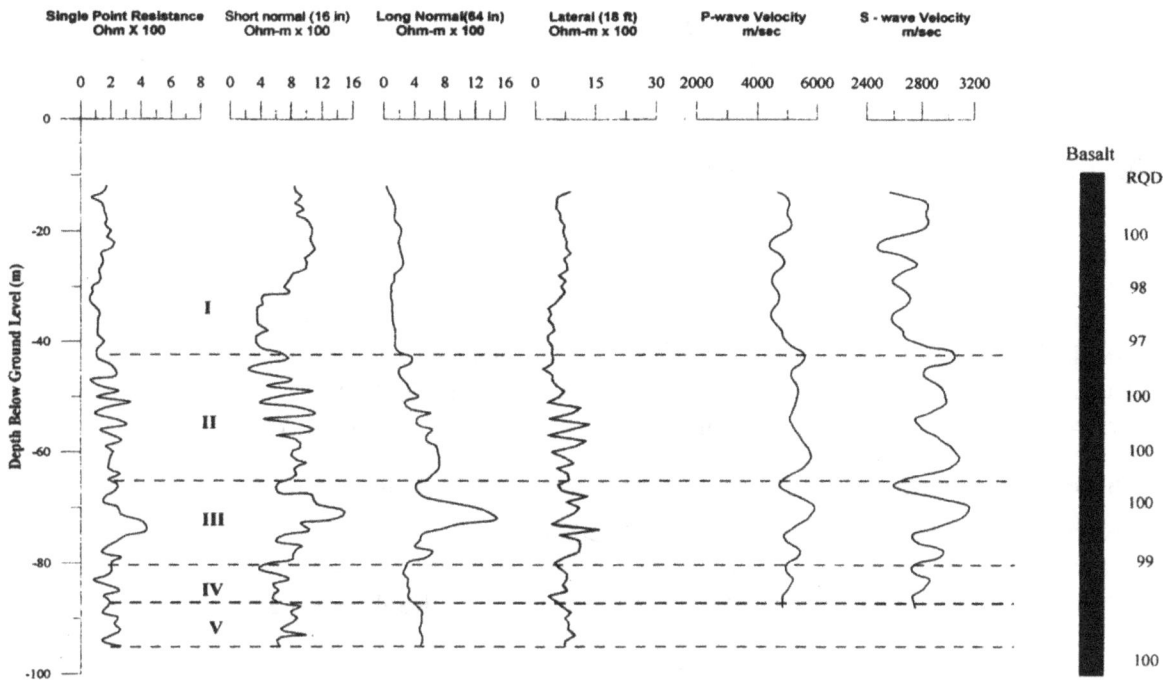

Fig. 10.15 A typical plot of electrical resistivity for various combinations of electrodes and acoustic log at Kakrapar atomic power project, Gujarat.

Table 10.1 Results of Electrical Resistivity and Acoustic Logging At KAPP

Zone below Ground Level(rn)	Nature of Foundation	Resistivity (ohm-rn)	VP (mlsec)	Vs (rnfsec)	Poissons Ratio	Ed × 10⁵ (Kg/crn²)	Gd × 10⁵ (Kg/crn²)
12-42	Slightly Weathered Basalt	100-240	4204-5862	2275-3213	0.23-0 29	4.69-7.58	1.81-2.38
42-65	Slightly Weathered Basalt	200-730	4838-5976	2689-3172	0.27-0.30	4.22-7.50	2.25-2.65
65-80	Sound Basalt	400-1200	5132-6036	2968-3043	0 28-0.30	4 22-7.50	2 25-2.65
80-87	Slightly Weathered Basalt	250-360	4618-5862	2592-2668	0 27-0.30	4.88-7.47	1.92-2.03
87-95	Sound Basalt	440-5 10	4726-5301	2704-2905	0 25-0 28	5 25-6.20	2 09-2 41

It is apparent from Table 10.1 that solid basalt occurs in depth ranges of 65-80 m and 87-95 m that shows high resistivity and high seismic velocity and provides a solid foundation for this nuclear plant. This table also provides Poissions Ratio, dynamic modulus of elasticity (Ed) and shear modulus (Gd) computed from the compressional wave (Vp) and shear wave velocities and density (ρ) as follows: These quantities are also used to classify rock types and their quality. Table 10.1 demonstrates higher values of Poisson's ratio, Ed and Gd for solid basalt providing suitable foundation for large scale heavy engineering constructions.

$$N = (0.5Vp^2 - Vs^2)/(Vp^2 - Vs^2)$$
$$Ed = Vp^2 (1 + \sigma)(1 - 2\sigma)/(1 - \sigma)$$
$$Gd = \rho Vs^2$$

10.7   Archeological Site Investigations

In most of the archeological surveys, one looks for past settlements and articles associated with them such as potteries, walls, kilns, and hearths. While baking to higher temperatures, these articles acquire remnant magnetization and are therefore amenable to magnetic surveys. During baking or drying they loose their moisture content and therefore are more resistive than their surroundings. Therefore, resistivity method is the other geophysical method that can be employed for their successful detection. Both these methods are comparatively cheaper and easy to apply in the field. Most of the archeological sites are located at shallow depths and station intervals are usually small i.e., of ~1-5 m.

10.7.1 Paleochannels and Settlements of Harappa period in Kalibangan, Rajasthan

The Harappa civilization was a part of the Vedic Civilization (6-3.5 Ky BP) whose remains have been found at several places in Western India and adjoining Pakistan (Lal, 1998, and Thapar, 1971). The most important sites are Harappa, and Mohanjedaro in Pakistan and Kalibangan, and Sirsa in India (Fig. 10.16, Valdiya, 2002). The cause for its disappearance or migration is poorly recorded. However, the most popular belief is the disappearance of river Vedic Saraswati along the banks of which this civiliazation flourished. River Ghaggar in India and Hakra in Pakistan that flows with considerably reduced water supply are presently considered remnants of the once mighty river Vedic Saraswati. It is therefore believed that due to uplift along the Himalayan front, either the river Vedic Saraswati was cut off from its perennial sources of water from Himalayan glaciers (Section 6.3.5; Mishra et al., 2008) or its main source was diverted eastwards (Valdiya, 2002).

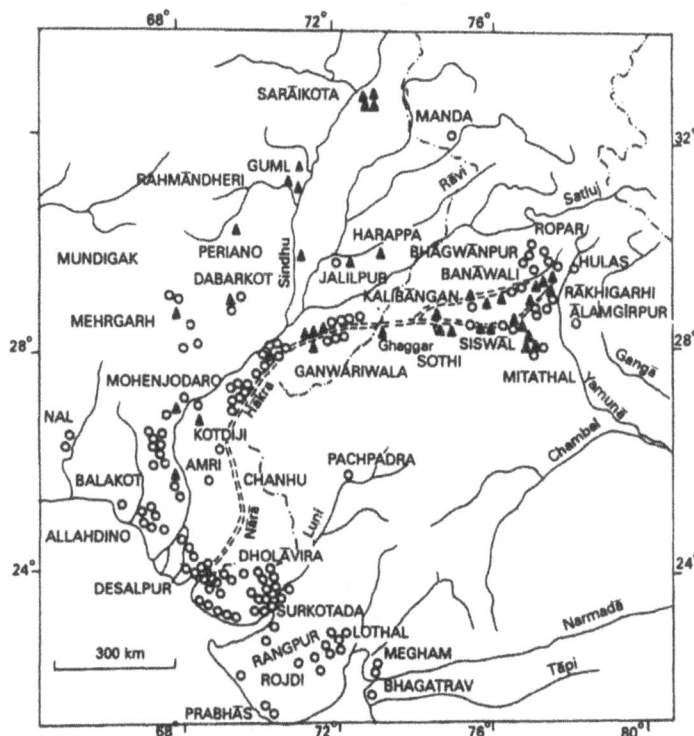

Fig. 10.16 Early and middle Harappan sites in north western India and eastern Pakistan showing a concentration of settlements along river Ghaggar-Hakra-Nara (Valdiya, 2002).

The Holocene period has experienced periods of de-glaciation at 12 Ky BP (Radhakrishna, 1999) after Pleistocene glaciation and at 6 Ky BP after Neoglaciation (8-6 Ky (Shroder, Jr and Bishop, 2000) when large rivers discharged water from Himalayan glaciers into the gangatic plains; one of them being Vedic river Saraswati. However, subsequently during 3.9-3.6 Ky BP there was warm and arid climate when the population along this river migrated eastwards. Several attempts have been made to delineate paleochannels in this region that might be related to river Saraswati. The most important in this regard is the discovery of fresh water at depth of 50-55 m in Cholistan, (Pakistan) based on airborne electromagnetic surveys. Paleochannels have been identified in this region based on dating well waters for periods of 1.8 and 6 Ky B.P that might be related to Harappa and Pre-Harappan civilizations in this region (Nair et al., 1999). Since electromagnetic surveys are most successful in identifying paleochannels, simple electrical method of resistivity measurements at surface was initially carried out to investigate paleochanels in Kalibangan (Mishra and Mallick, 2008) that was one of the biggest settlements of Harappan civilization on the Indian side (Fig. 10.16). Some Pre-Harappan settlements have also been identified in this region by the Archeological Survey of India (ASI).

First, some magnetic and resistivity profiles were recorded over known excavated settlements in Kalibangan. These profiles suggested that magnetic signals were quite weak of order of 1 nT which are difficult to separate from noise in unknown regions, but resistivity values were quite significant. Therefore, resistivity surveys were conducted to delineate shallow settlements in this region. As the settlement in this region is buried under a thin cover of soil and sand, resistivity survey was conducted over a grid using Wennar configuration with electrode separation of 2 and 3 m in unknown area adjacent, to known sites as suggested by the ASI. Due to constant electrode separation, resistance values were plotted directly at the central point of the electrode configuration and values contoured as shown in Fig. 10.17. As this figure shows, there is a channel type of feature in the central part with low value of resistance (L1) and high resistance features (12-17 ohm, H1 & H2) along the flanks of the channel. The sections with low resistance indicate paleochannels as they contain more moisture compared to surrounding regions, while high resistance sections may represent settlements of human based constructions like houses, and burial ground. A similar low resistivity (L2) feature is also observed in the western part with high resistance (H3 & H4) features along the flanks that may represent another settlement. This one being deeper may represent early-Harappan civilization as both occur in the same region at different levels. This investigation suggests that simple low cost resistivity surveys can be used to delineate paleochannels in this region, some of which may contain potable water that can be used for societal needs.

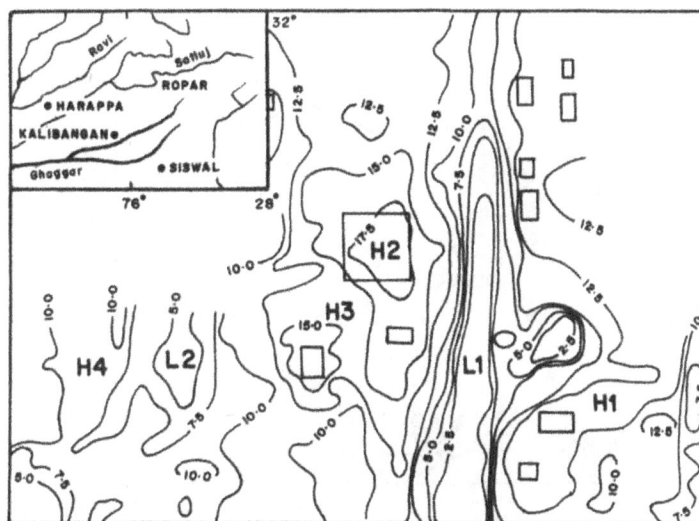

Fig. 10.17 Equiresistivity map of part of Harappan site in Kalibanghan (Rajasthan) at contour interval of 2.5 ohm. The linear resistivity low, L1 indicates a paleochannel while adjoining highs, H1 and H2 indicate settlements along this channel. L2 and H3 and H4 are another set of anomalies that may indicate the same. Inset shows Kalibangan at the bank of the present day river Ghagghar.

10.7.2 Mahastupa, Nelakondapally, Khammam District, A.P., India

The village Nelakondapally is located 23 km SW of Khammam town (Fig. 10.18). Actual sites are shown as 1, 2, 3. It was occupied by Buddhist monks during Lord Buddha's period and a Mahastupa of that time attracts tourists to this part of the country. It was expected that a large settlement of that time must exist nearby. A total intensity magnetic and resistivity profile across Mahastupa was recorded at station interval of 5 m (Fig. 10.19; Achuta Rao et al., 1989) to understand first the nature of anomalies observed over these kind of structures. This figure shows magnetic anomalies, H1, L1, and H2 and corresponding resistivity anomalies are 2 and 3. A Stupa is made up of baked bricks that can produce anomalies of order of 100-200 nT. It is, therefore likely that the magnetic anomalies H1, L1 are caused by the Mahastupa and the corresponding resistivity anomaly 3 is related to it. The large magnetic and resistivity anomalies, H2 and 2, respectively appear to be caused by some geological features. Subsequent search indicated a dolerite dyke in this section that is responsible for large magnetic and resistivity anomalies. They also presented a total intensity map of Mahastupa (Fig. 10.20) that very well defines the boundary of the stupa. This map also shows a high amplitude magnetic anomaly in the southern part that is related to dolerite dyke as indicated above. Similar magnetic and resistivity surveys in the neighborhood delineated several sites where walls and earthen pots of that period were found (Achuta Rao et al., 1989).

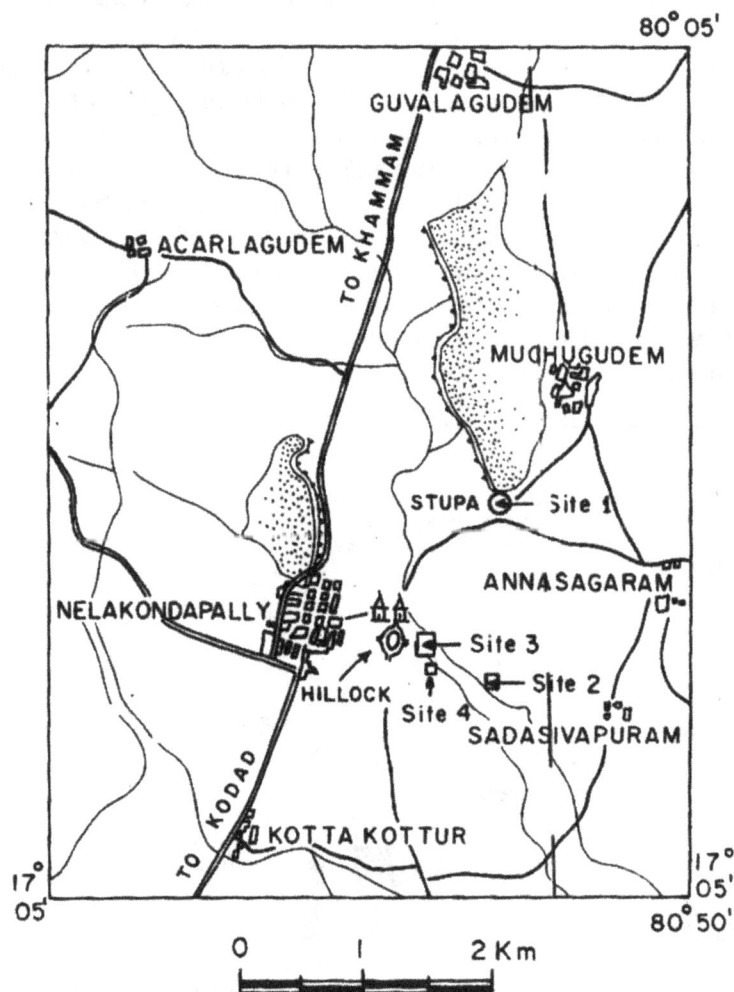

Fig. 10.18 Location map of Nelakondapally in Khammam district of Andhra Pradesh.

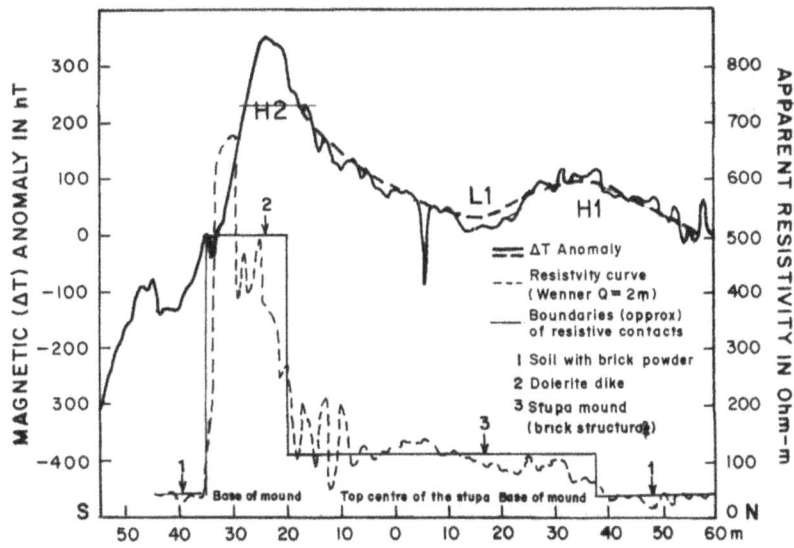

Fig. 10.19 Magnetic and resistivity profiles across Mahastupa.

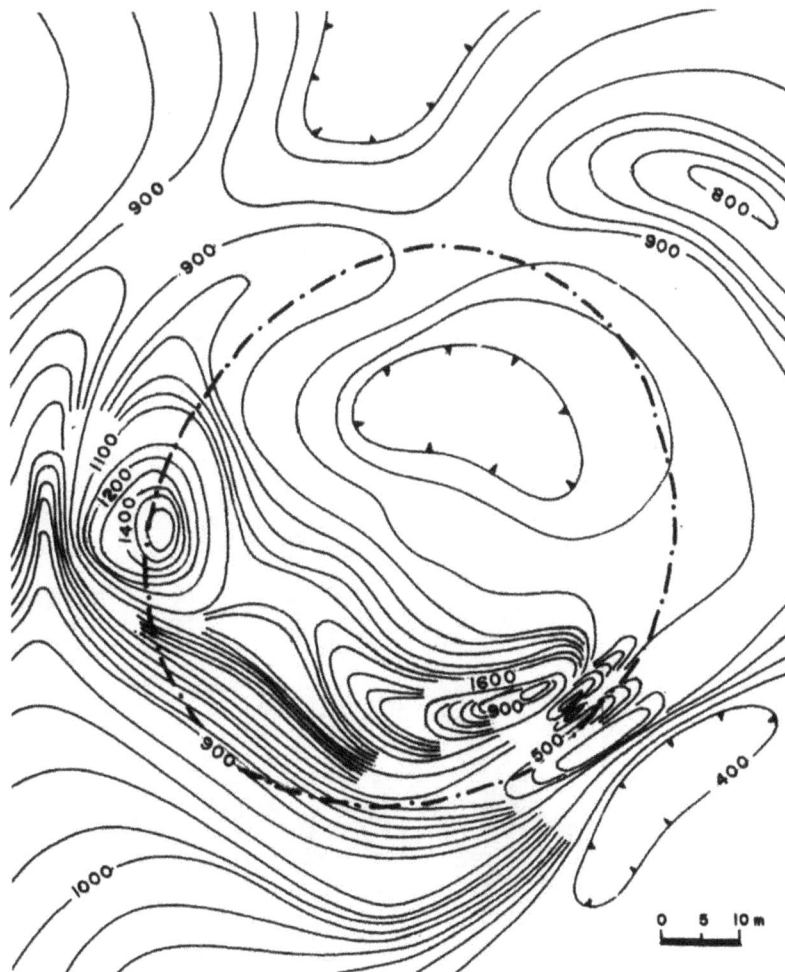

Fig. 10.20 Total intensity map of Mahastupa delineating its circular structure.

Appendix I

Geological Time Scale

Appendix I Geological Time Scale* table

Era	Period/Epoch	Approximate Age (Ma)
Cenozoic (Recent Life)	**Quaternary**	
	Holocene (Most recent)	0.01
	Pleistocene (Recent)	0.01-1.7
	(Alluvium and River	
	Deltas)	
	Tertiary	
	Pliocene	5
	Miocene	25
	Oligocene	34
	Eocene	54
	Paleocene	65
Mesozoic (Gondwana India) (Intermediate life)	**Cretaceous**	145
	Jurassic	205
	Triassic	250
Paleozoic (Ancient life)	**Permian**	295
	Canboniferous (abundance of coal)	355
	Pennsylvanian	325
	Mississippian	355
	(American)	
	Devonian	415
	Silurian	440
	Ordovician	495
	Cambrian	545
Preambrian (Cynobacteria)	**Proterozoic**	
	NeoProterozoic	1000
	Meso Proterozoic	1600
	Paleo Proterozoic	2500
	Archean	
	Neo Archean	2800
	Meso Archean	3200
	Paleo Archean	3600
	Eo Archean	
Hadean		4000
Origin of Earth		4600

Further detailed subdivisions can be obtained from (Ogg et al., 2009).

Appendix II

Some Important Events of Recent Periods (Pleistocene and Holocene)

Appendix II Some Important Events of Recent Periods (Pleistocene and Holocene)

Time	Important Events
1000 years	Industrial Revolution, Scientific Developments, World Wars and Anthropogenic Related Environment Pollution
10,000 years (Holocene)	Middle Ages, Effects of deglaciations after Pleistocene glaciation (~ 10 Ka BC) and Mid-Holocene (~ 6-5 Ka BC) glaciation in form of large rivers especially from Himalaya. Start of civilisations, Scientific thoughts and Various religions and languages like Pali, Sanskrit etc. Pre-Vedic (Purana) & Vedic Periods (3.5-1.7 Ka BC- Indus Valley Civilization and periods of Rama and Krishna) in India may be connected to periods of deglaciations as given above referred to as Pralaya (total devastation due to floods) in Vedas.
1,00,000 years	Cave Man and Modern Homo Sapiens
One Million years (Pleistocene)	Early Homo Sapiens, Neanderthals and Early Homo Erectus

Heat Flow of India

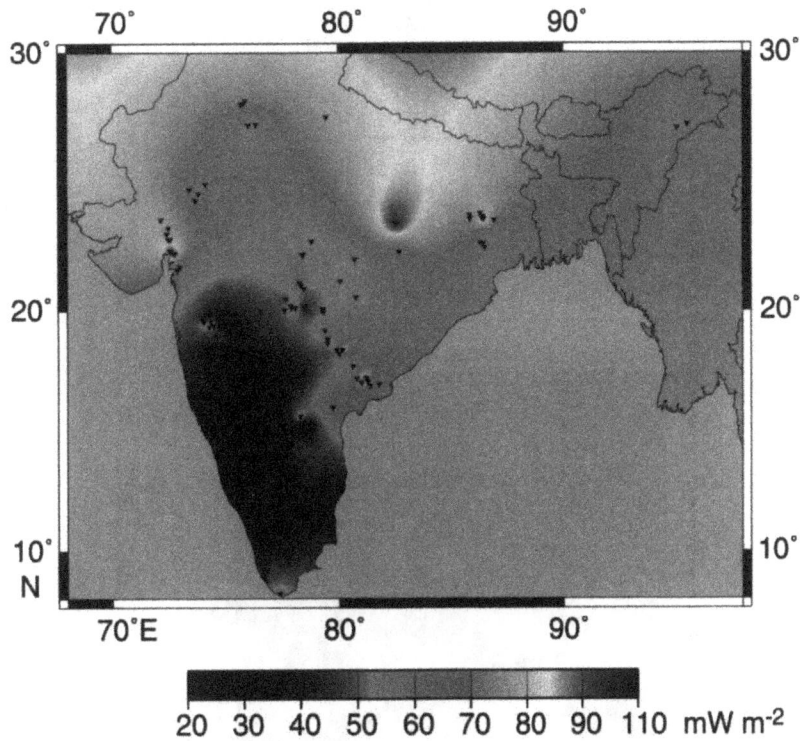

Appendix III Heat flow map of the Indian Shield showing high heat flow over the northern shield in general compared to southern shield (Dharwar craton). Heat flow sites are shown by filled triangles (Roy 2010).
[Colour Fig. on Page 824]

Appendix IV

Seismic Zonation Map of SAARC Countries – Pakistan and Bangladesh

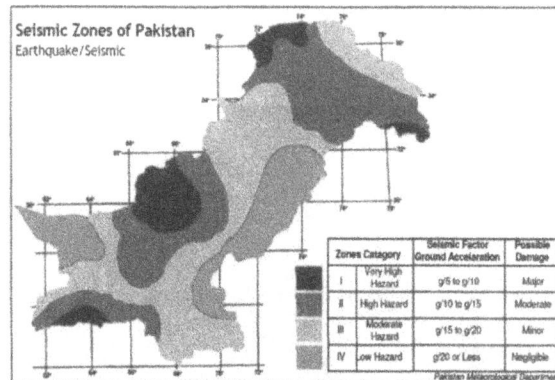

Pakistan

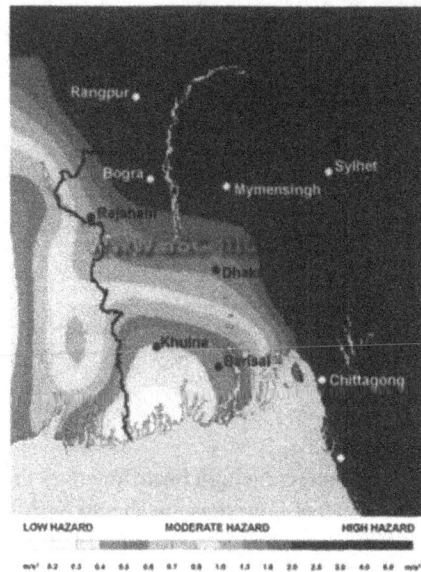

Bangladesh

Appendix IV Seismic zonation map of SAARC countries – Pakistan and Bangladesh. SAARC countries being close to plate boundaries are prone to large seismic activity. These maps show maximum seismically prone sections close to Himalayan Fold belt decreasing away from it. In case of Bangaladesh, the Shillong plateau north of it in India shows maximum seismically active zone as shown in Fig 8.6 and discussed in Section 7.6.4. The other SAARC country like Nepal lies in the Himalayan Fold belt and therefore is characterized by very high and high seismic zones in northern and southern part, respectively while Bhutan's western and the eastern part lies in high to moderate seismic zones and central part has reported low seismic activity. Afghanistan shows very high to high seismic activity in the eastern part that decreases to moderate zones in the central part depending on the distance from the Himalayan Fold belt and again high in the western part west of Herat due to its plate boundary with Iran (By Courtesy Dr O.P.Mishra, SAARC Disaster Management Center, New Delhi).

760

Colour Photographs of All Chapters

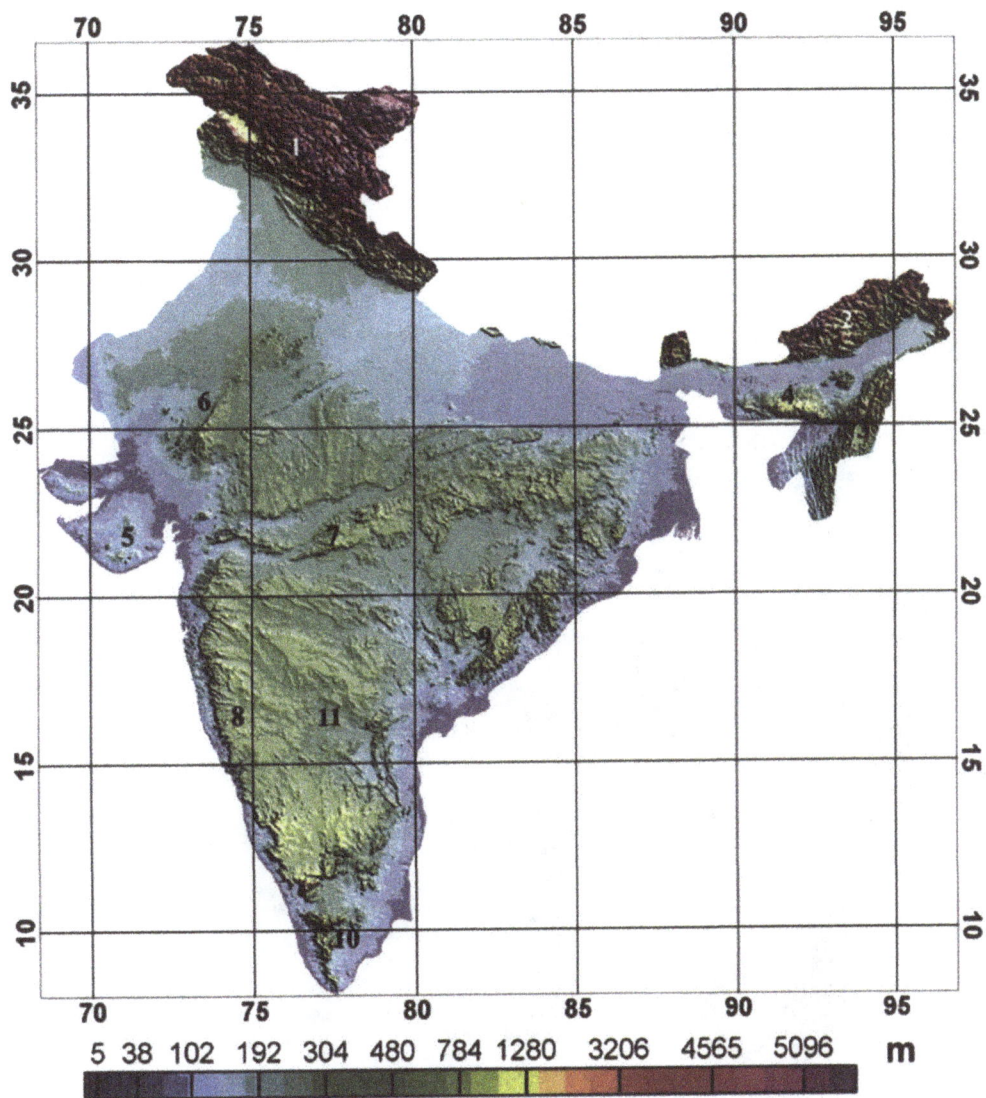

Elevation Map of India

Fig 2.27(a) An elevation map of India. Numbers (1-11) represent important units of the country as follows. (1) Western Himalayas (2) Eastern Himalayas (3) Arkan Yoma mountains (4) Shillong plateau (5) Saurashtra – Kutch shelf (6) Aravalli – Delhi mountains (7) Satpura mountains (8) Western Ghats (9) Eastern Ghats (10) Southern Granulite Terrain (11) Dharwar Craton (Peninsular Shield).

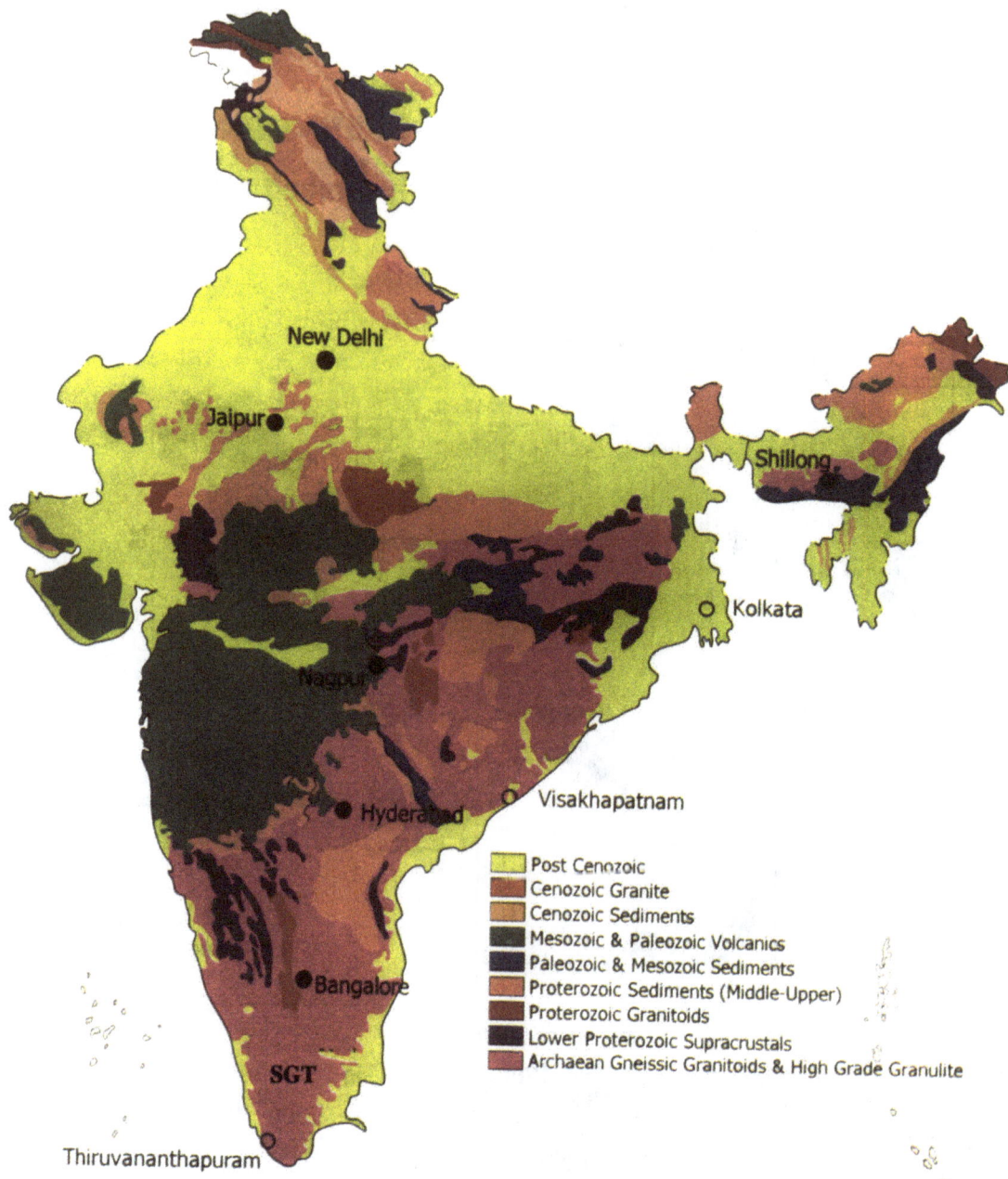

Fig. 2.27(b) A simplified geological map of India (GSI,1993).

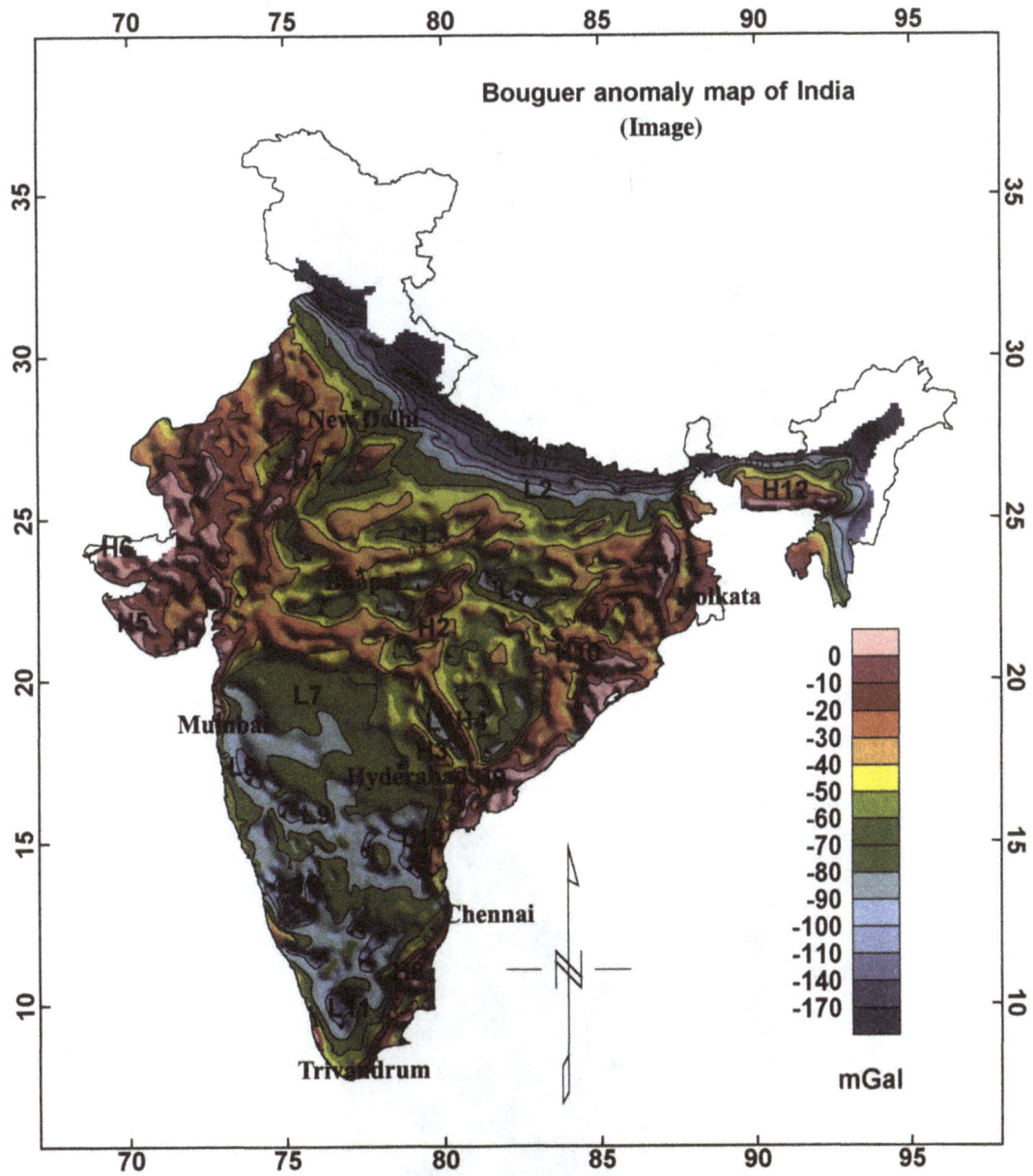

Fig 2.28 Image of the Bouguer anomaly map of India illuminated from the north. H and L denote gravity highs and lows (Mishra et. al., 2008).

Sea Level Trends from Topex-Poseidon (1993-1998)

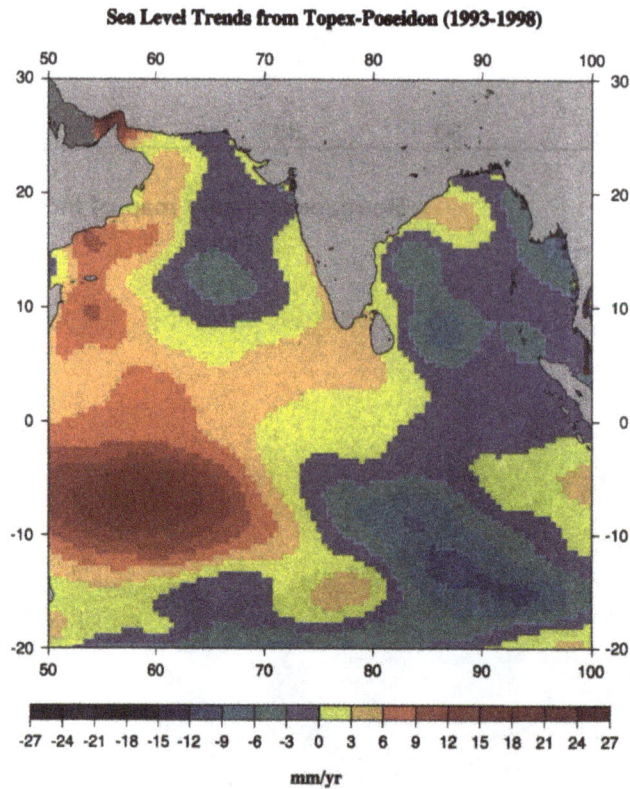

-27 -24 -21 -18 -15 -12 -9 -6 -3   0   3   6   9 12 15 18 21 24 27

mm/yr

Fig 2.29 Sea Level Trends for period 1993-98, derived from Topex / Poseidon satellite data (Tiwari et al., 2004).

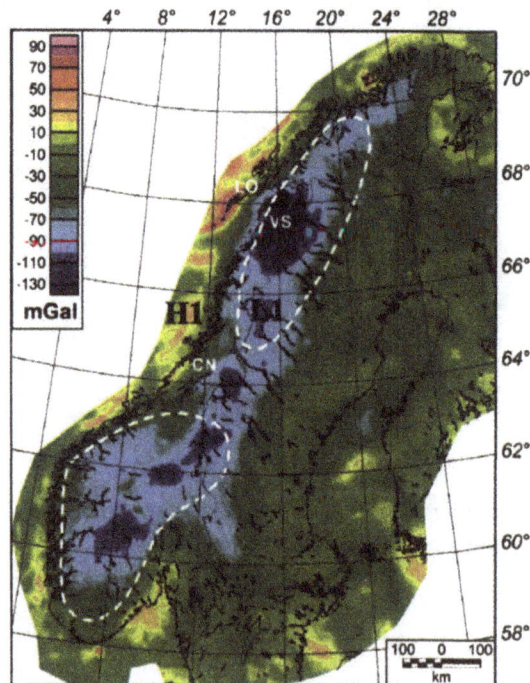

Fig 2.42 Bouguer anomaly map of Fennoscandia providing linear gravity lows (L1) over Caledonian orogeny and small high west of it along the coast (Ebbing and Olesen, 2005).

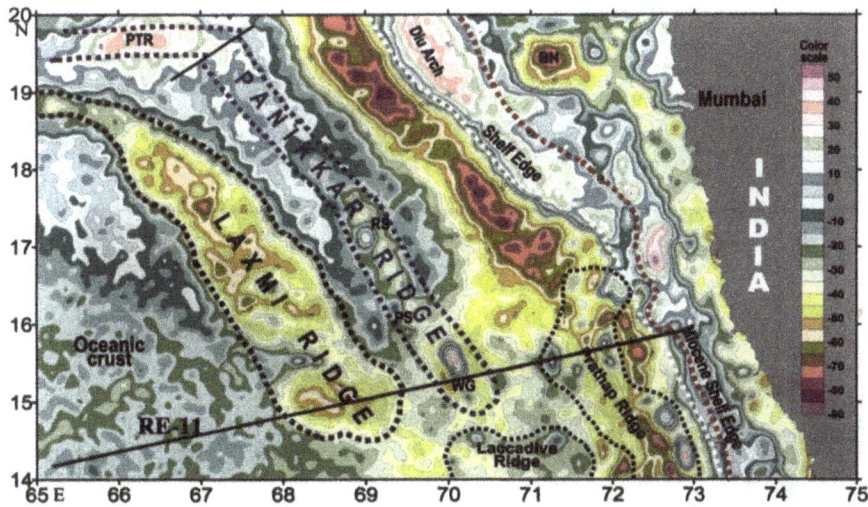

Fig. 3.24(a) Map of satellite free-air gravity anomalies [Sandwell and Smith, 1997] and interpreted geomorphic features. PTR, Palitana Ridge and BH, Bombay High structure (Krishna et al ., 2006).

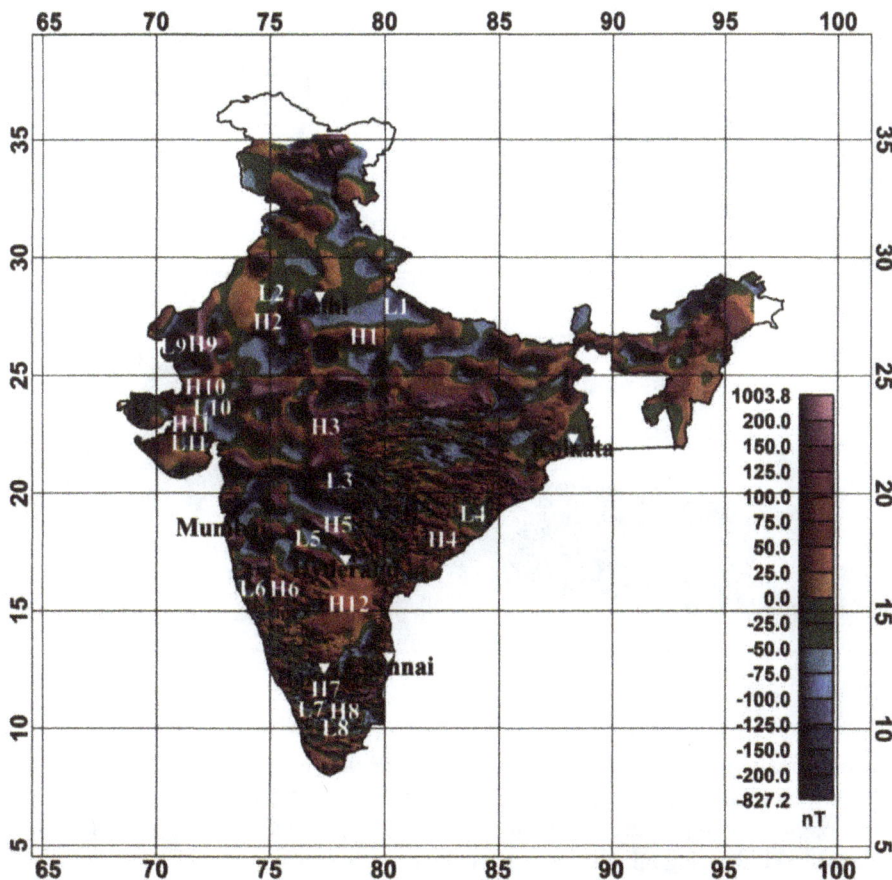

Fig.3.30 EMAG 2: A 2 arc-minute (3.7 km) resolution Satellite-Airborne magnetic map of India compiled from airborne magnetic data from different sources acquired for World Digital Magnetic Anomaly Map (Korhonen et al, 2007) of Commission of the World Geological Map (CWGM, http://ccgm.free.fr/) and long wavelength CHAMP lithospheric field model MF6 EMAG2 (Maus et al., 2008). All data sets are merged at 4 km altitude. H and L denote magnetic highs and lows (Maus et al., 2010).

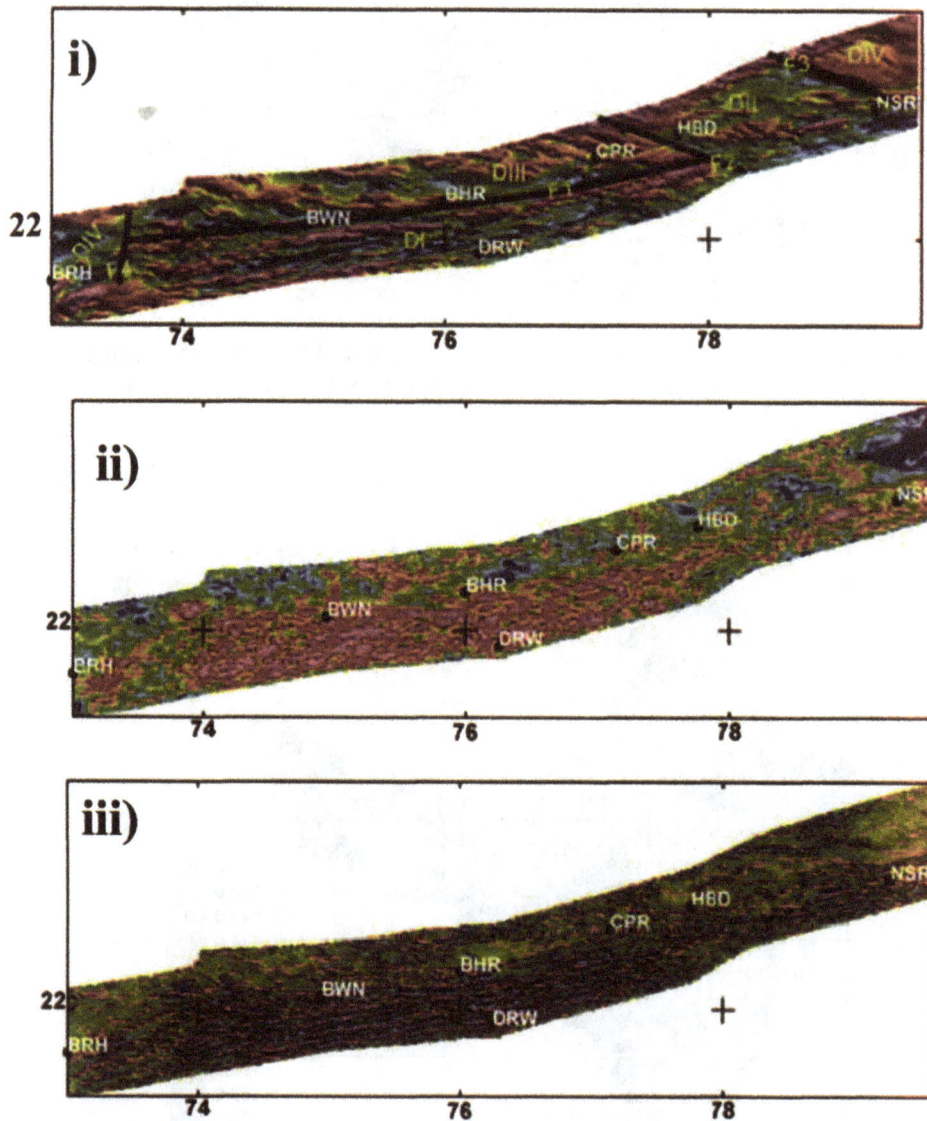

Fig. 4.16 (b) Total field anomaly map and its transformations. Red depicts highs and blue lows. (i) Total field aeromagnetic anomalies with superposed interpreted major faults. F1 indicates Barauni Suktha Fault (BSF). (ii) Analytic signal map, maxima (highs) are located over the magnetic sources south of BSF. (iii) Second vertical derivative of total field, showing minima over magnetic sources south of BSF. HBD – Hoshangabad, NSR - Narsingpur, BRH - Baruch, BWN-Barwani, BHR-Bharwaha, CPR-Chhipaner, DRW-Dorwa. (Anand and Rajaram, 2004).

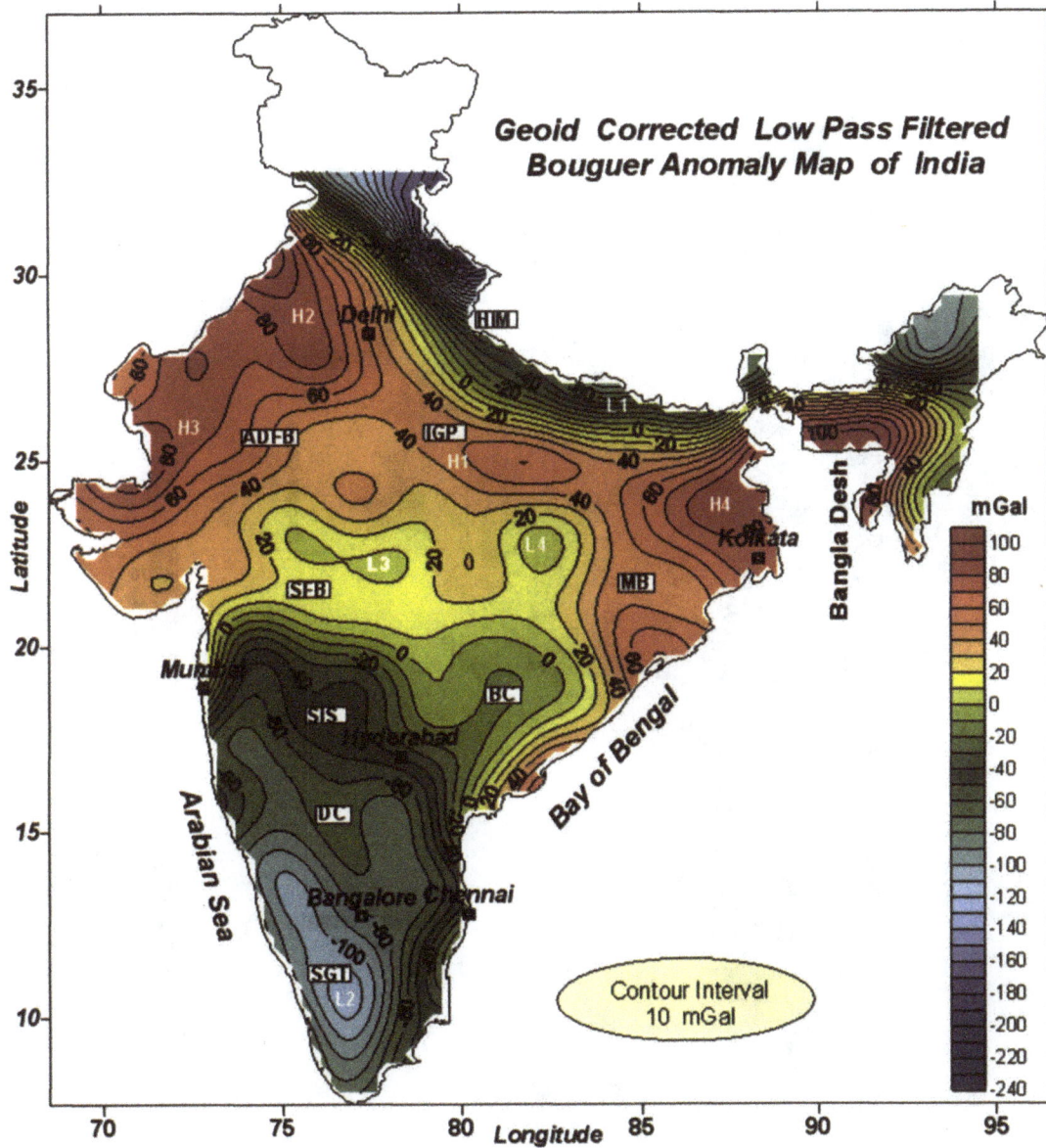

Fig. 4.28 (b) Low pass filtered regional Bouguer anomaly map of India with wave number < 0.02 (first segment of spectrum, Fig. 4.28(a)). L1-L2 are two major gravity lows suggesting crustal thickening in these sections and H1-H4 are gravity highs related to crustal bulge associated with lithospheric flexure due to Himalayas. ADFB=Aravalli Delhi Fold Belt, BC = Bastar Craton, DC=Dharwar Craton, HIM-Himalayas, IGP = Indo Gangetic Plains, MB = Mahanadi Basin, SFB = Satpura Fold Belt, SGT = Southern Granulite Terrain, SIS = south Indian shield (Mishra et al., 2004).

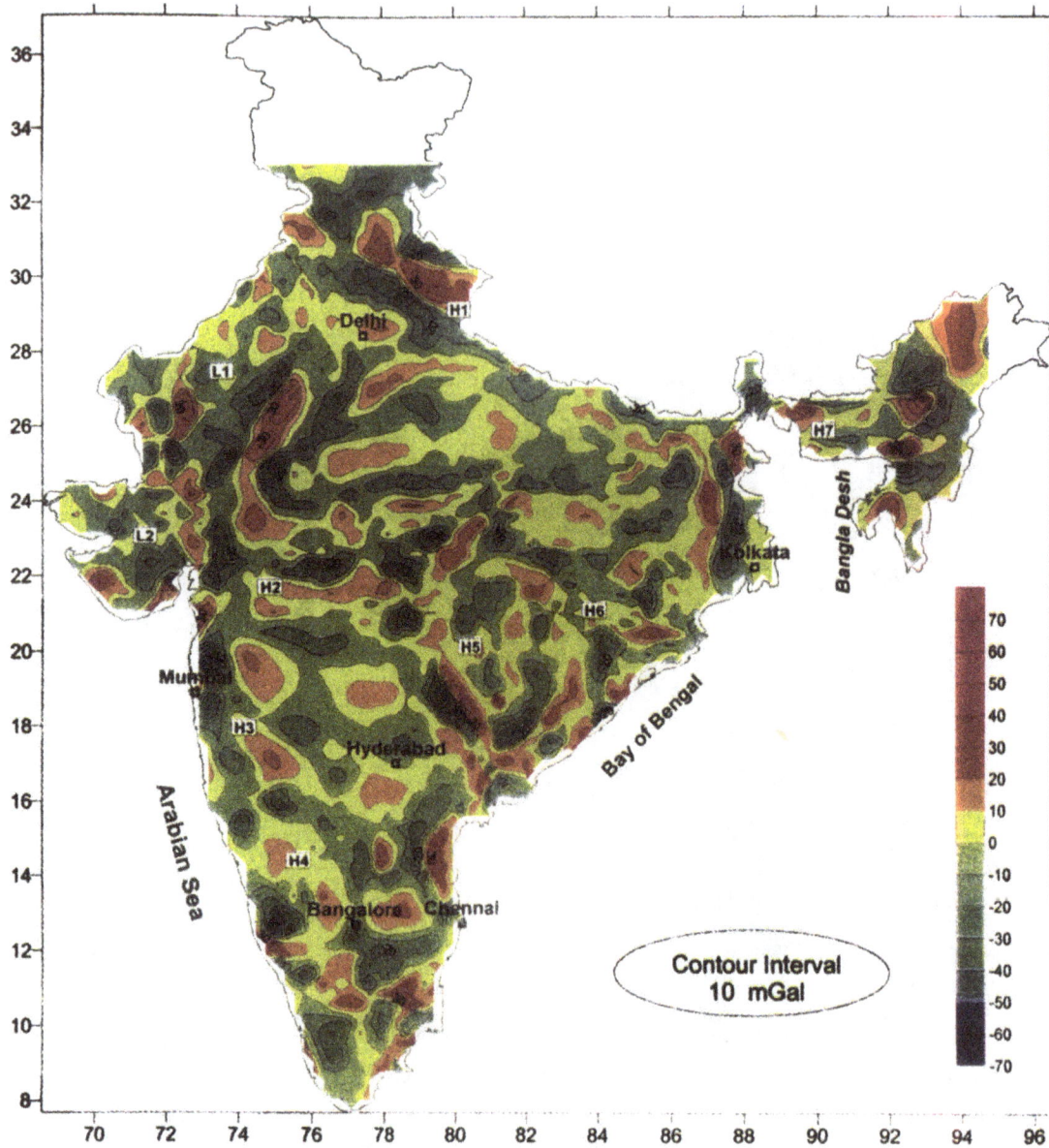

Fig. 4.29 (a) High pass filtered residual Bouguer anomaly map of India (Fig. 2.28). It shows several gravity highs and lows due to shallow and exposed sources, which can be used to delineate these bodies (Mishra et al., 2004).

Fig 5.3 Reconstruction of Pangea during Permian period (~255 Ma) showing the super continents of Laurasia and Gondwanaland (Modified after Scotese et al., 1988, ; http:// www.scotese.com). Numbers 1-8 indicate fold belts (mountains) of ancient times (Proteozoic-Paleozoic period) as follows: - Quachita Ranges, 2- Appalachians, 3- Caledonides, 4- Mauritanides, 5-Eastern Ghats, 6- Napier-Prince Charles Mountains, 7- Hamersley Range, 8- Urals.

Fig 5.17 Black smokers at East Pacific Rise photographed using a submersible vehicle. Such black smokers are treasure trove of minerals (sulphides and gold deposits) and biological species in spite of absence of sunlight, high temperature and hazardous gases in these sections. (Sleep and Woolery, 1978; Malahof, 1983, Macdonald et al., 1980; Kious and Tilling, 1999 that was photographed by Dudley Foster from RISE experiment and courtesy of William R. Normack, USGS as quoted by Kious and Telling).

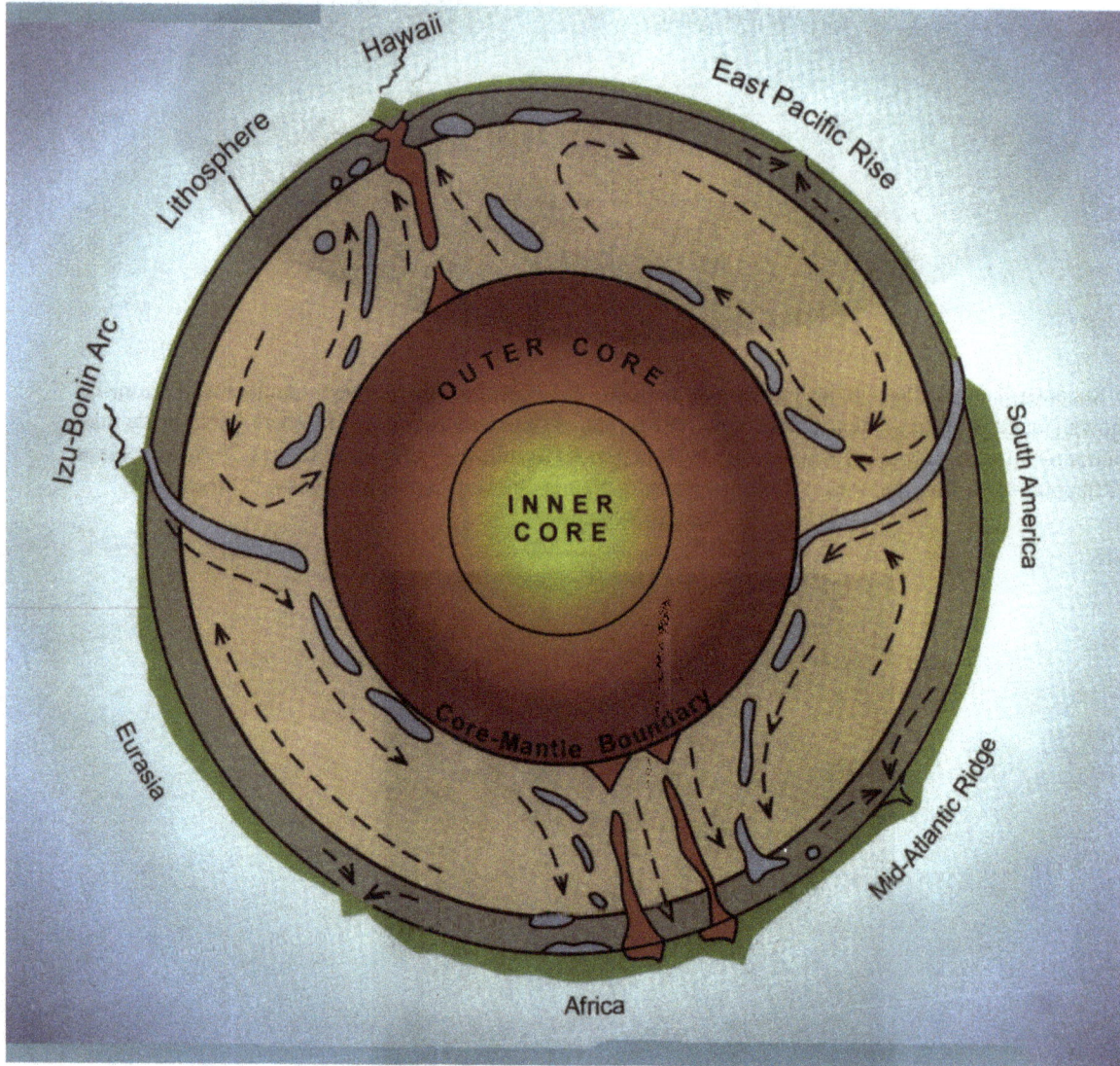

(a)

Fig 5.23 (a) A schematic model showing relationship between subducting slabs which reaches up to base of the mantle and rises as plumes which sets convection cells in asthenosphere that drives the plates and rises at mid-oceanic ridges. (Davies and Richards, 1992, Pirajno, 2007).

DEEP WATER – WEST COAST

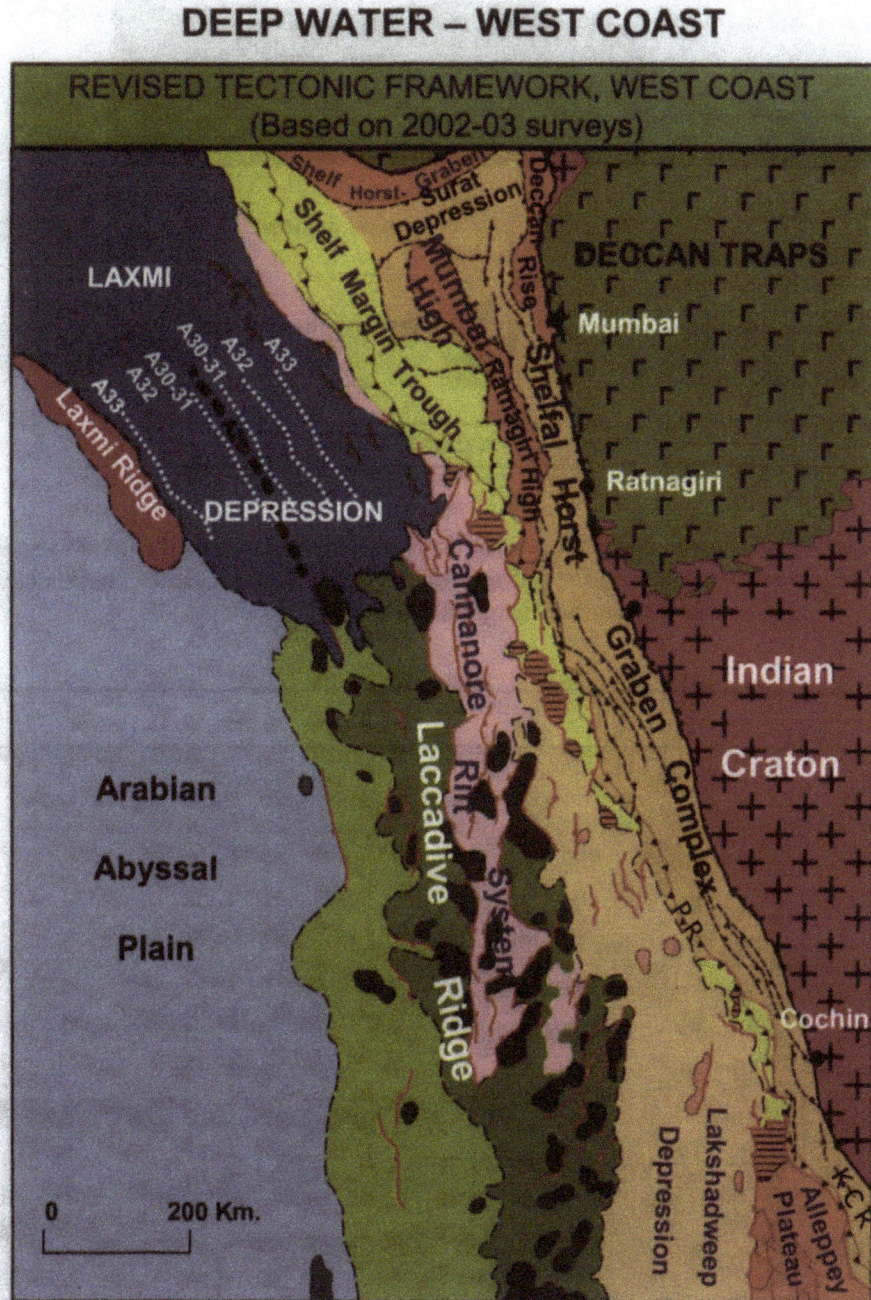

Fig 5.32(b) Continental shelf offshore west coast of India showing horst and rift structure caused by Deccan trap eruption shown in green colour occupied by Lacadive ridge. It shows sea floor spreading magnetic anomalies in the northern part in the Laxmi basin (DGH, 2005). KCR- Kori-Comorin Ridge and PR- Pratap Ridge (DGH, 2005).

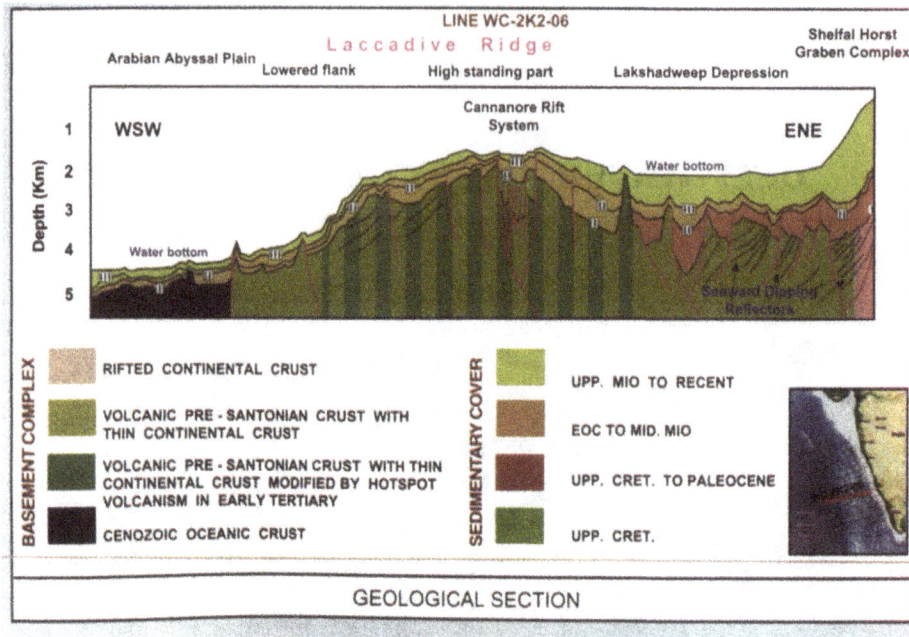

Fig 5.33(b) A schematic geological section across Lacadive ridge offshore west coast of India showing rifted margin formed due to Deccan trap volcanic intrusive and seaward dipping reflectors caused by them indicating extension and rifted margin (DGH, 2005).

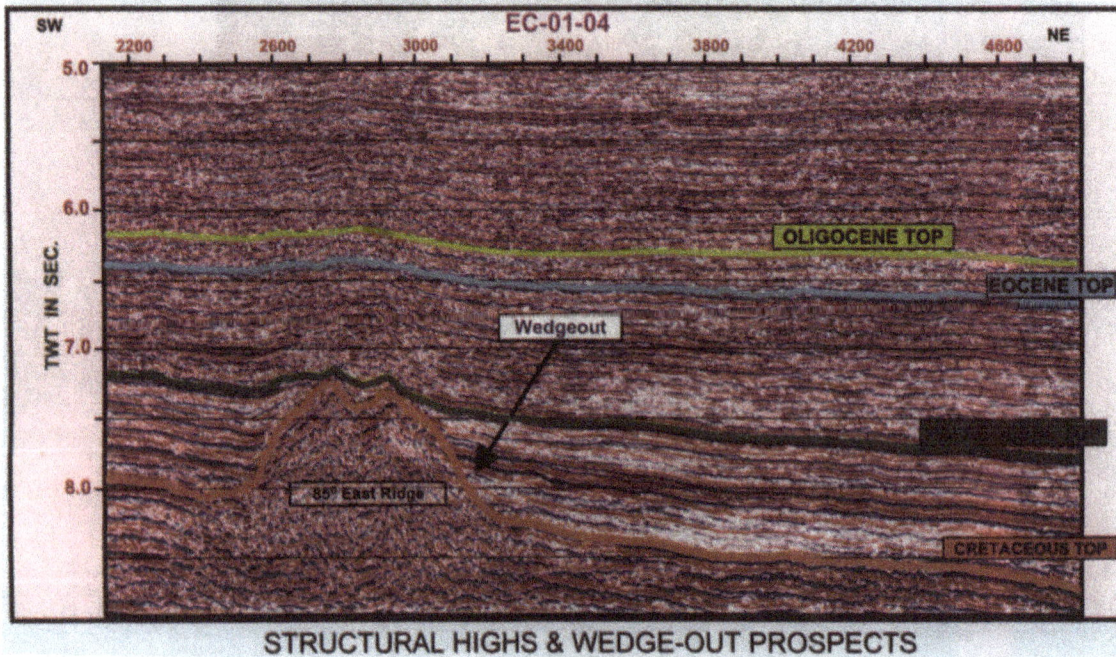

Fig 5.36(b) A seismic section across Eighty Five East Ridge showing a depth of ~7-8 km and the stratigraphy of Tertiary sediments in this section. It also shows the structural high and the stratigraphic wedge out as suitable sites for oil traps (DGH, 2005).

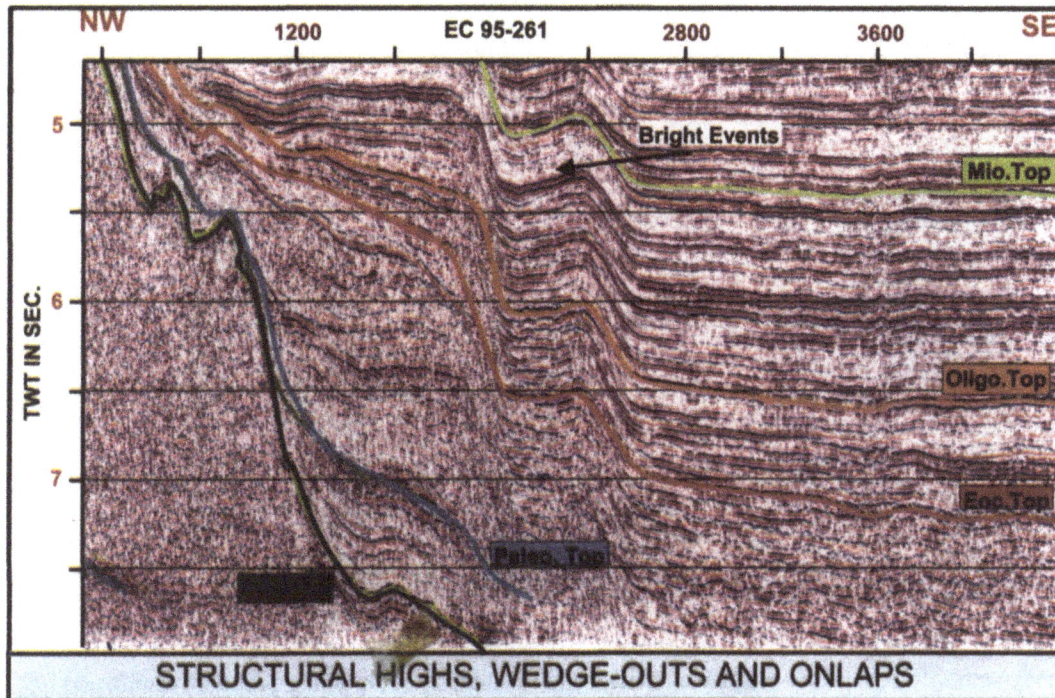

Fig 5.36(c) A seismic section offshore Godavari-Krishna basin along the east coast of India showing Tertiary stratigraphy and a sharply dipping continental margin and structures favorable for oil exploration (DGH, 2005).

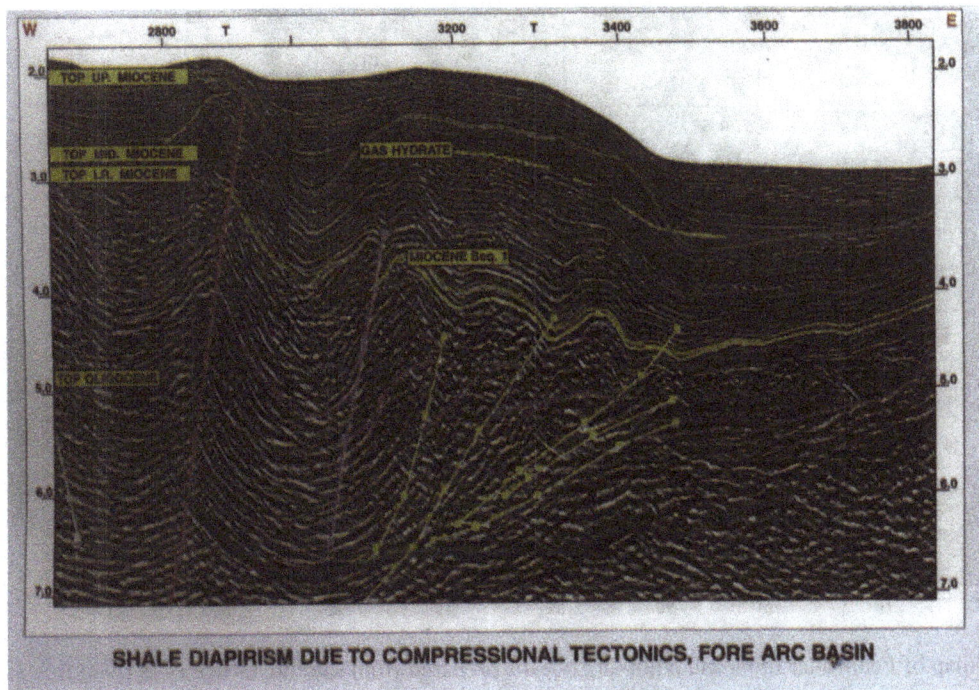

Fig 5.36(d) A seismic section offshore Andaman islands showing a highly disturbed section with several faults, folds and shale diapirism that are typical of compressional tectonics near subduction zones.

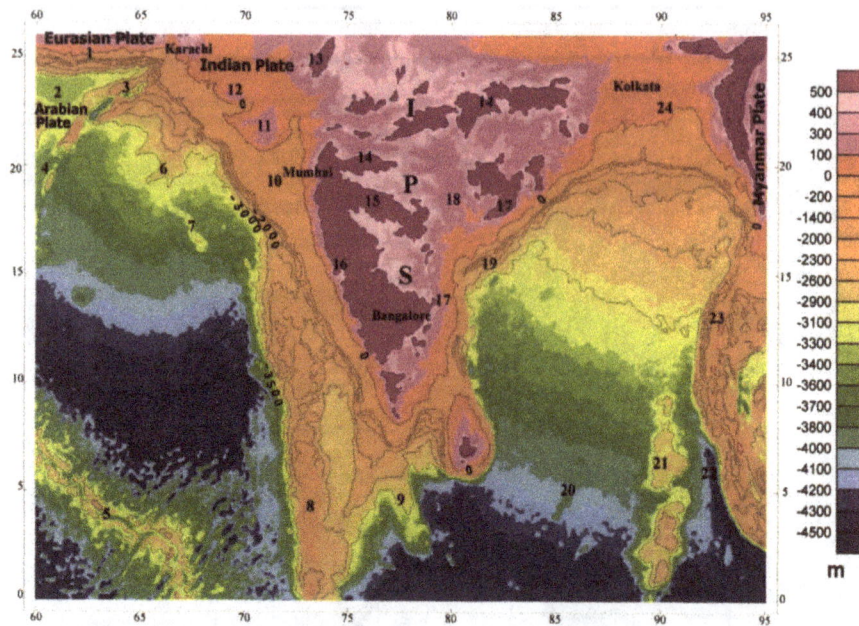

Fig 5.39 Bathymetry of Arabian Sea and Bay of Bengal and topography of adjoining continent (Smith and Sandwell, 1994). Annotated features marked as 1-24 are explained in Table 5.3; IPS= Indian Peninsular Shield (Mishra et al., 2004).

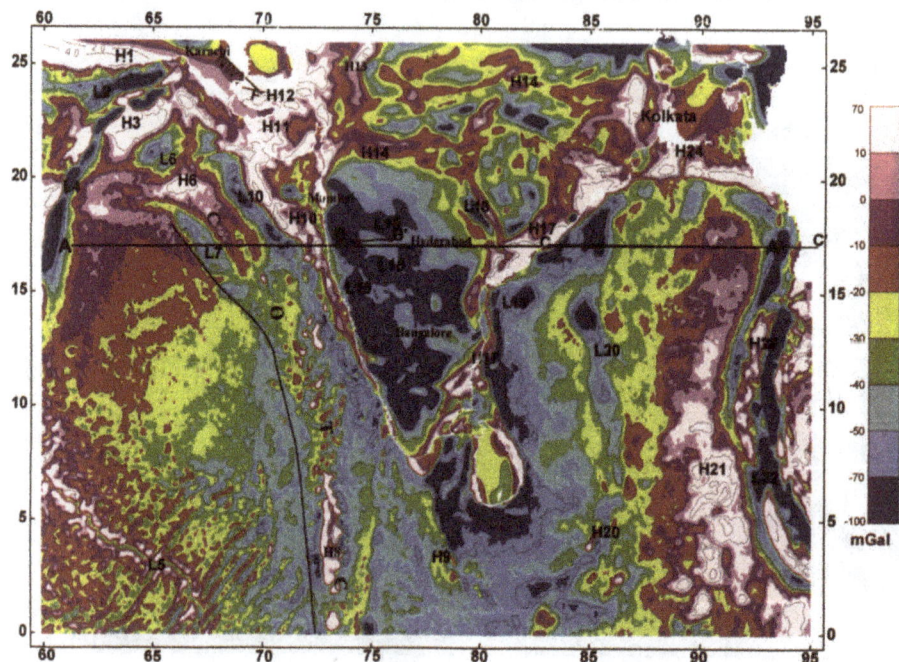

Fig 5.40 Gravity map of Peninsular India and adjoining oceans prepared from satellite altimetry derived free air anomalies over oceans (Sandwell and smith, 1997) and surface Bouguer anomalies over India and adjoining countries (India, NGRI, 1975; Bangladesh; Rahman et al., 1990; Pakistan and Sri Lanka; ESCAP, 1976). Important gravity highs (H) and lows (L) are annotated for the features listed in Fig. 5.39 and explained in Table 5.3. Also shown is the profile AA', modeled in Fig.5.43. BB' and CC' represent the extent of the seismic profiles over the continent and the Bay of Bengal (Fig. 5.42b and c) used to constrain the gravity model. The extent of continent-ocean transition crust (COTC) in the eastern Arabian Sea is indicated by a line (Mishra et al., 2004).

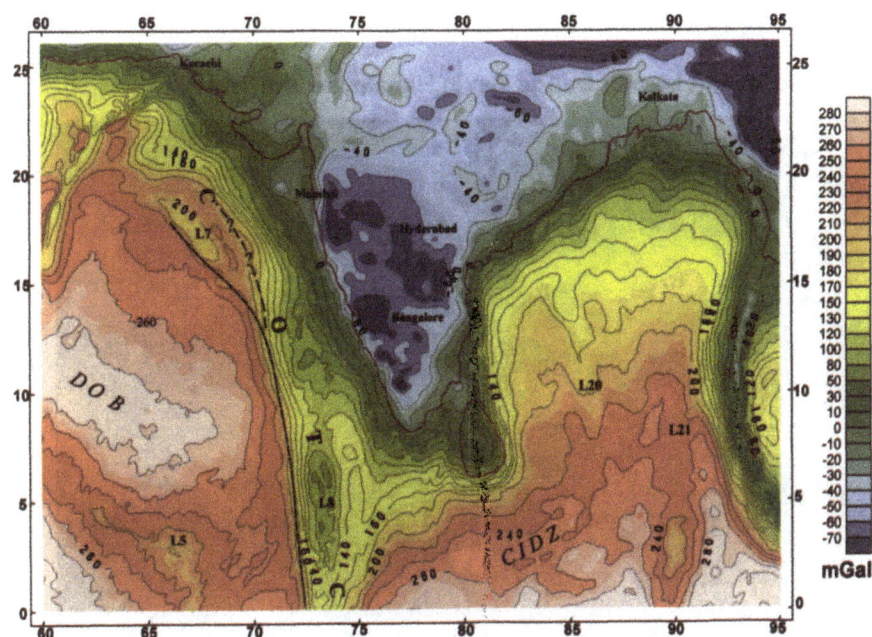

Fig 5.41 Simple Bouguer anomaly maps of the Indian Peninsular Shield and adjoining oceans prepared from the free air anomaly from satellite tracking over continent and altimetry over oceans. The dashed line in the Arabian Sea emphasizes the change in the trends on either side, indicating the COTC. The Bouguer gravity field increases consistently from relative gravity low over the continent and oceanic ridges (L5, L7, L8, L20 and L21) to relative gravity high over the deep ocean basins (DOB) indicating the effect of isostatic compensation. Central Indian Deformation Zone (CIDF) is indicated by trend of relative Bouguer high which does not find reflection in the free air anomaly maps (Mishra et al., 2004) and can be delineated based on this map.

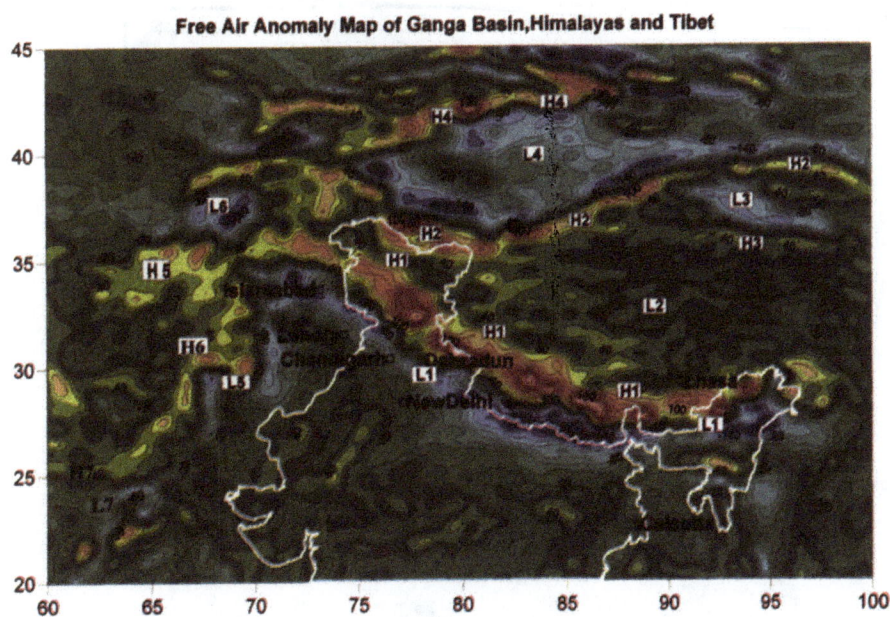

Fig. 6.6 Free air anomaly map of Ganga basin, Himalayas and Tibet and adjoining regions derived from satellite data. H1-H7 and L1-L7 represents gravity highs and lows respectively discussed in the draft. White line indicate boundary of the Indian continent. Thrusts and sutures are largely characterized by gravity highs while basins and plateaus by gravity lows (http://icgem.gfz-postdam.de/ICGEM/ICGEM.html).

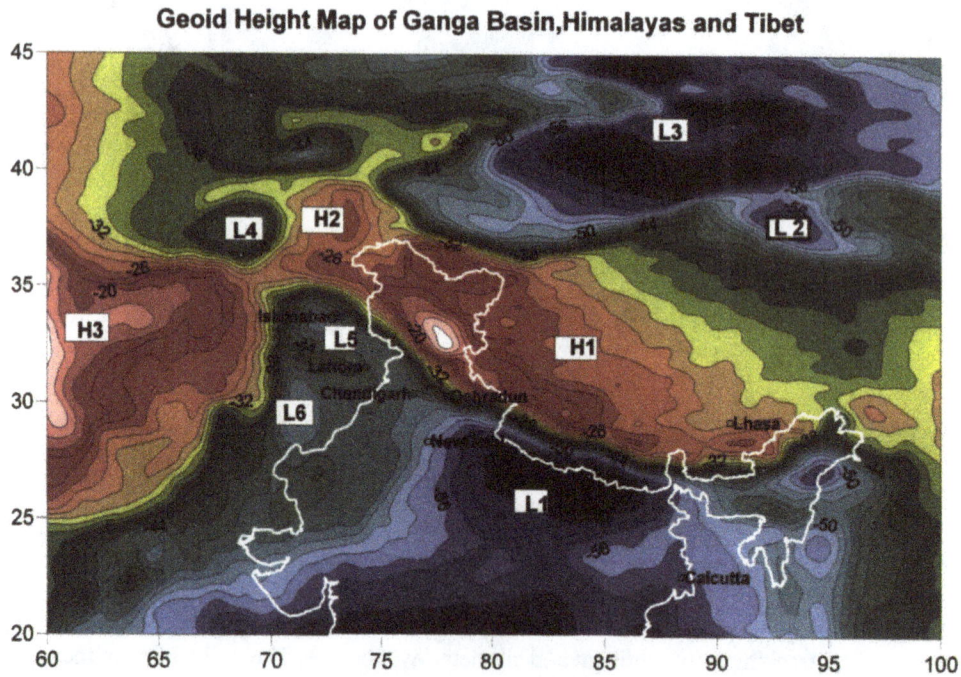

Fig. 6.7 Geoid height map of Ganga basin, Himalayas and Tibet and adjoining regions. H1-H3 and L1-L6 represent geoid highs and lows in meters as discussed in the draft. High lands ares characterized by geoid highs while basins by geoid lows (http://icgem.gfz-postdam.de/ICGEM/ICGEM.html).

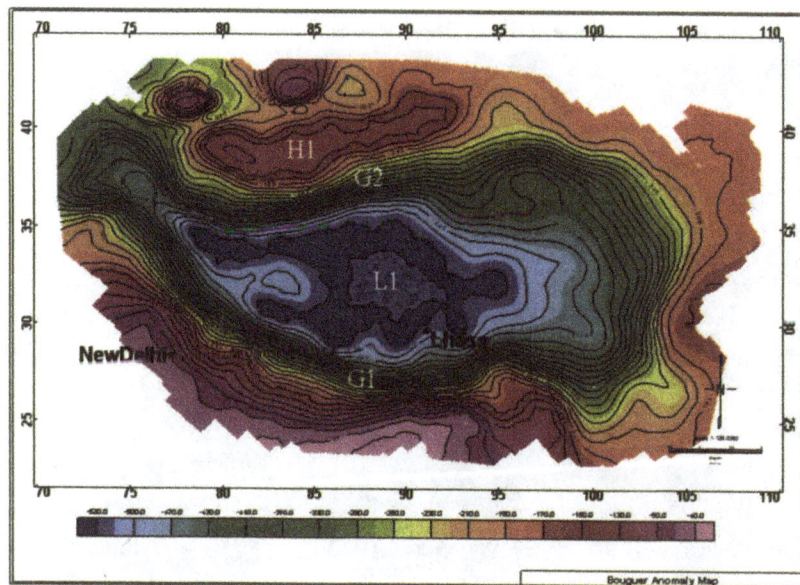

Fig. 6.8 Bouguer anomaly map of Himalayas and Tibet obtained from Satellite free air anomaly map (Shin et al., 2007) showing a major gravity low (L1) over Tibet. Gradients G1 and G2 coincide with Himalayan thrusts and suture zone (ITSZ), and Altyn Tagh fault, respectively. The gravity high, H1 is related to the Tarim basin.

Low Pass Filtered Geoid Height Map Of Ganga Basin ,Himalayas and Tibet

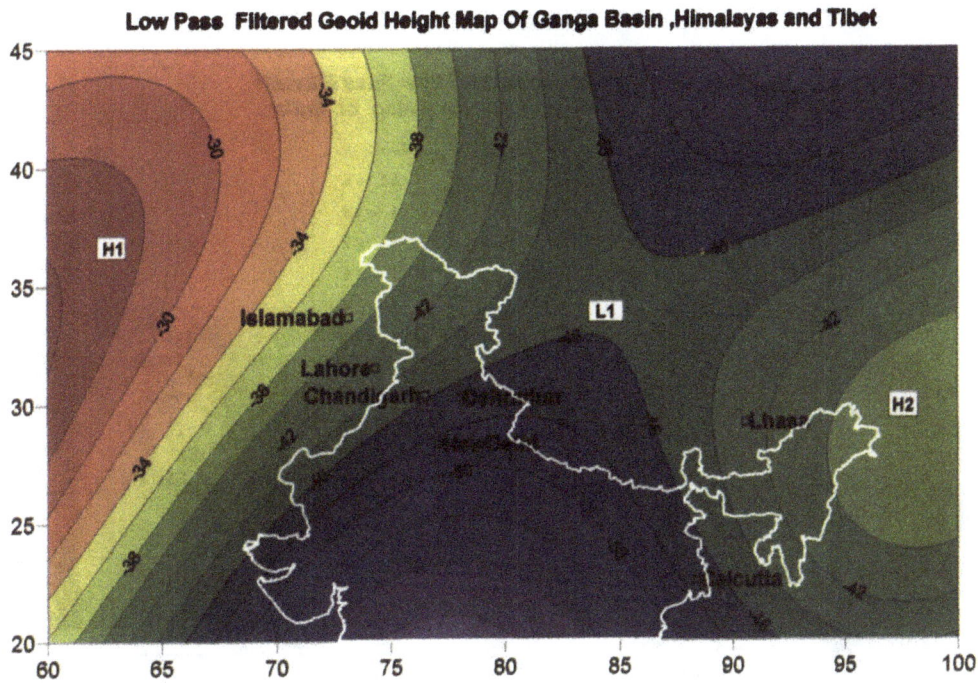

Fig. 6.10(a) Low pass filtered geoid height map related to first segment of spectrum (Fig. 6.9c) with wave length >3140 km denoting sources in the upper mantle between average depth of 132-328 km. It is primarily a geoid low (L1) with relative highs (H1 and H2) along the margins. These anomalies are oriented NE-SW, in the principal stress direction and motion of the Indian plate.

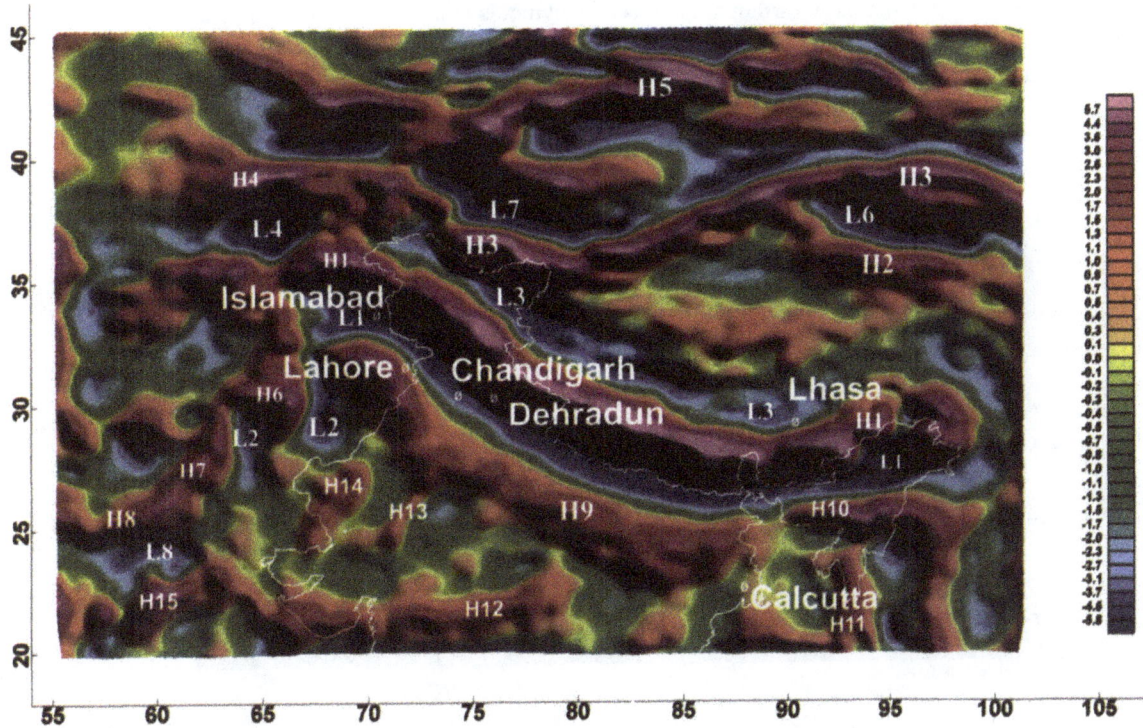

Fig. 6.10(b) High pass filtered geoid height map for wave lengths related to fourth segment of the spectrum. It shows several highs (H1-H15) and lows (L1-L8) related to shallow/exposed sources.

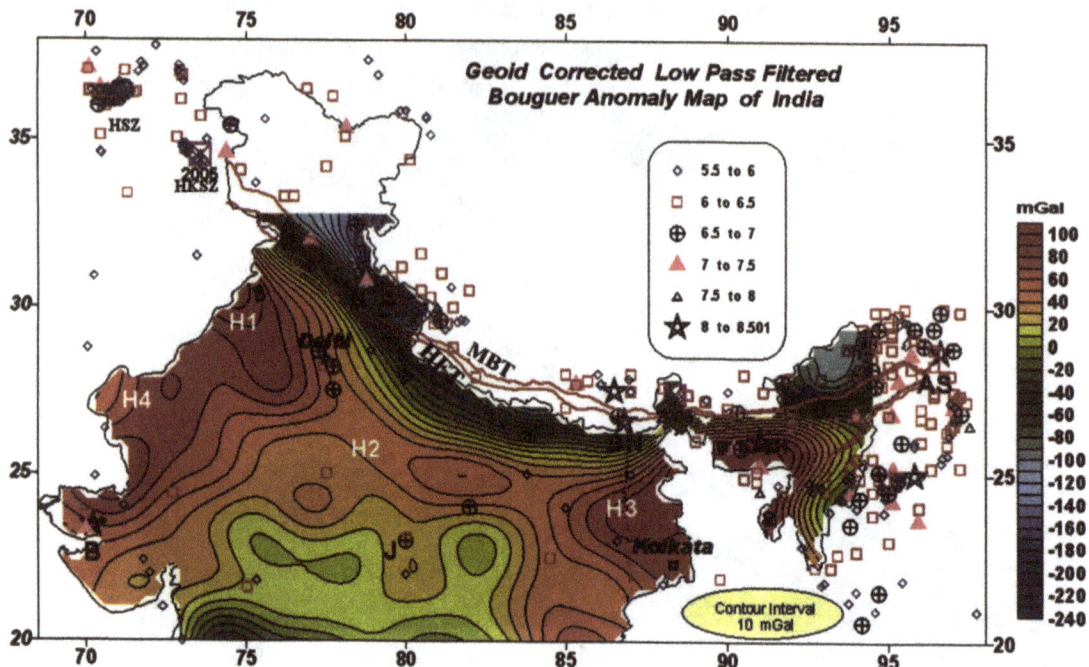

Fig. 6.15(a) Low pass filtered regional Bouguer anomaly of North India showing anomalies from upper mantle and lower crust related to first two segments of the spectrum (Inset, Fig. 6.14). These highs represent crustal upwarp due to lithospheric flexure. It also shows seismic activity along the Himalayan front. HSZ = Hindukush Seismic Zone and HKSZ = Hazara-Kashmir Seismic Zone, which also coincide with the gradient between gravity highs (H1-H3) due to lithospheric flexure and gravity low (L1) due to crustal thickening extrapolated towards the NW. Some great and large earthquakes shown by symbols are: K = Kangra (1905), U = Uttarkashi (1991), C = Chamoli (1999), BN = Bihar Nepal (1934), SH = Shillong (1897), AS = Assam (1950), B = Bhuj (2001) and J=Jabalpur (1997) earthquakes. Rectangular cross mark in shows the HKSZ Muzaffarabad (Kashmir) earthquakes (Earthquakes from USGS web site).

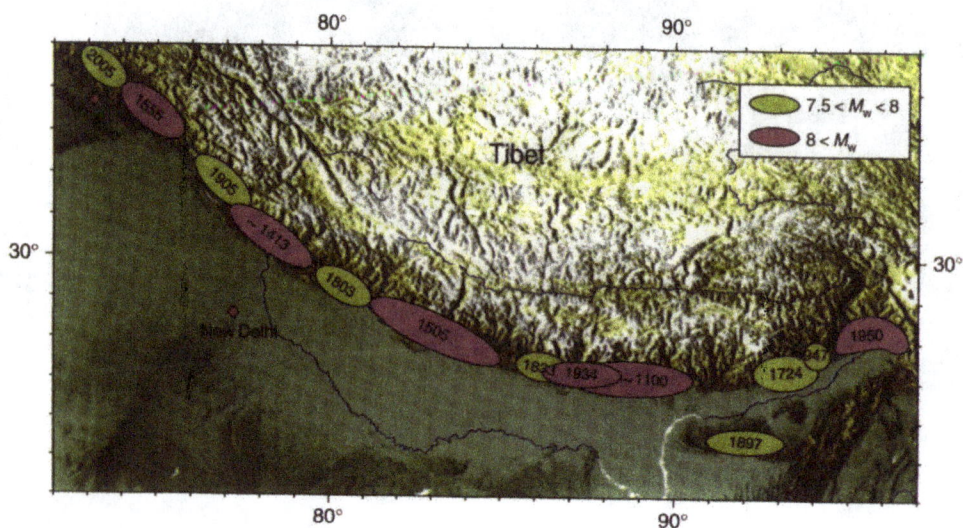

Fig. 6.16 Estimated rupture area of major earthquakes along Himalayas (Bilham, 2004 and Ambraseys, 2000; Avouac, 2007 and references therein). It shows the different section of Himalayan front being ruptured during different periods in last 1000 years.

Fig 6.20(c) Photograph of Mt. Kailash representing a subduction related Pre-Miocene (~30Ma) granite batholith in SW Tibet, north of the suture zone that shows almost vertical face of such intrusive. Manasarover Lake lies at the base of it that is one of the largest fresh water lakes at high altitude and may represent the remnant of the Tethys Sea with modified water column due to deglatciation and rain
(By courtesy R.K.Math).

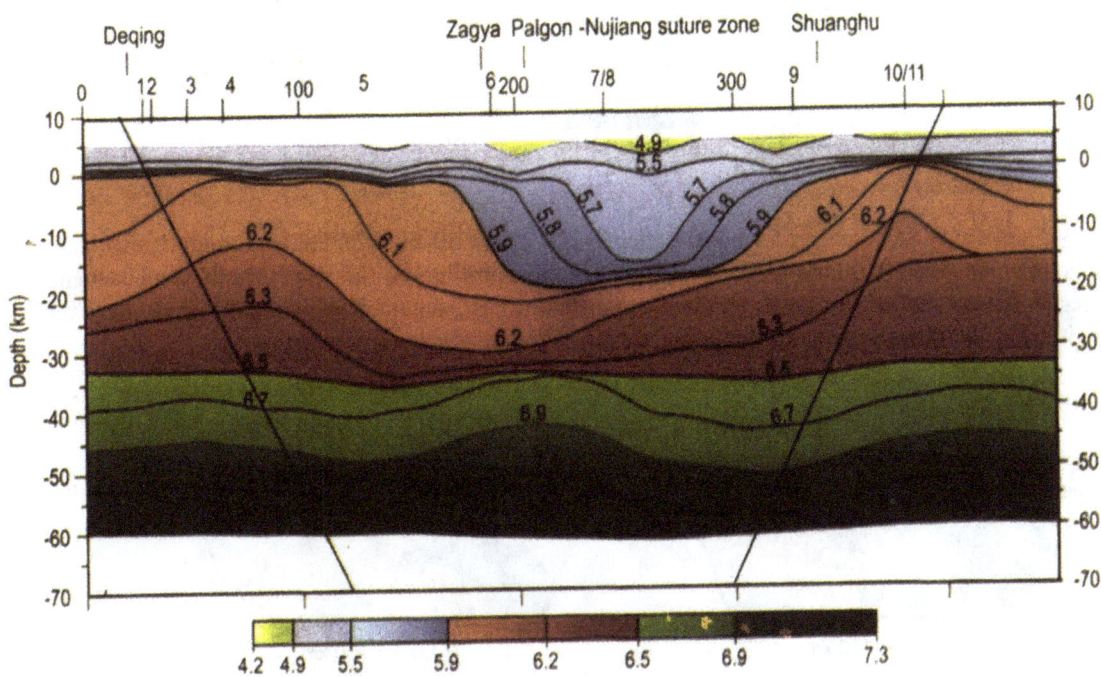

Fig. 6.26 Seismic velocity structural profile across Bangong-Nujiang Suture Zone (BNS) along INDEPTH III West line (Fig. 6.20a) showing an underplated crust of high velocity (7.1-7.3 km/sec) and upwarping in the lower crust and buckling in the upper crust (Zhao et al., 2004a, 2004b) indicating both extension and compression related first to rifting followed by convergence, respectively.

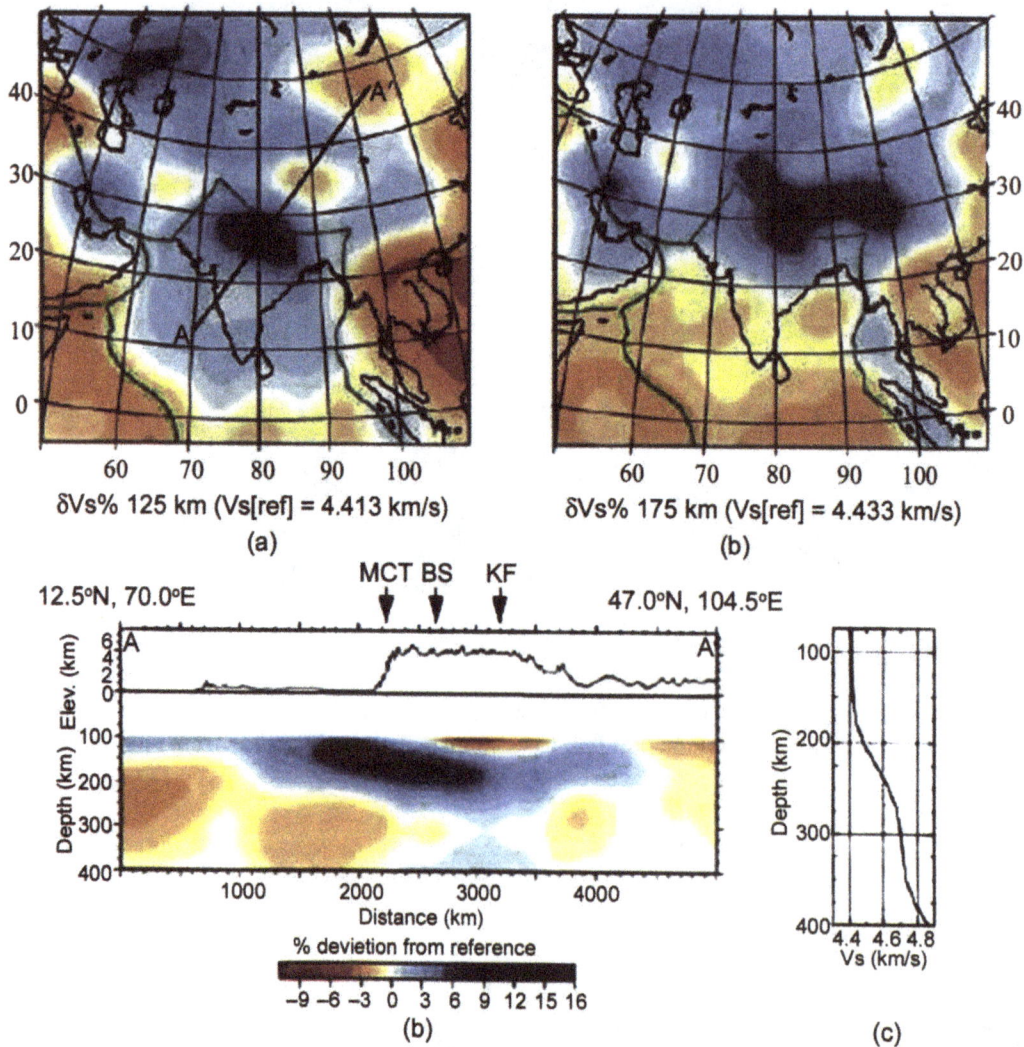

δVs% 125 km (Vs[ref] = 4.413 km/s)
(a)

δVs% 175 km (Vs[ref] = 4.433 km/s)
(b)

MCT BS   KF

12.5°N, 70.0°E   47.0°N, 104.5°E

% devietion from reference

−9 −6 −3 0 3 6 9 12 15 16

(b)   (c)

Fig. 6.30(b) Surface wave (Sv) tomography modes at depths of 125 km (**a**) and 175 km (**b**). Straight green lines are major tectonic features. (**c**) shows the depth model along a profile AA' (a) with elevation and tectonic features, the Main Central Thrust (MCT), the Bangong suture (BS) and the Kunlun Fault (KF) marked on it. (**d**) is the reference model used to find the deviation in Sv velocity (Priestley et al., 2008).

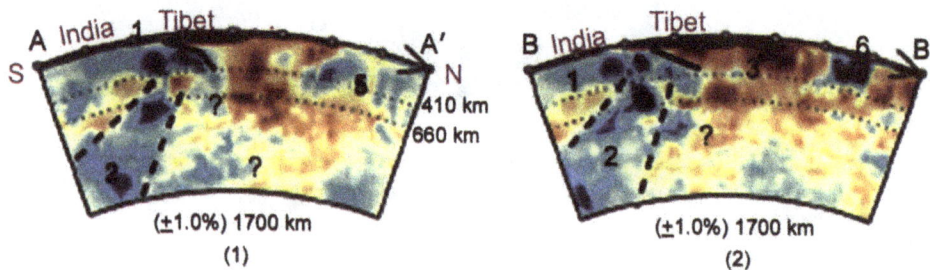

(±1.0%) 1700 km
(1)

(±1.0%) 1700 km
(2)

Fig. 6.30(c) Subducted high velocity Indian lithosphere under Tibet, (1) based on tomography while subducted Tethyan lithosphere (2) is detected below the transition zone in the Western Himalaya (AA') but continuous in depth in the Central Himalaya (BB') (Li et al., 2006).

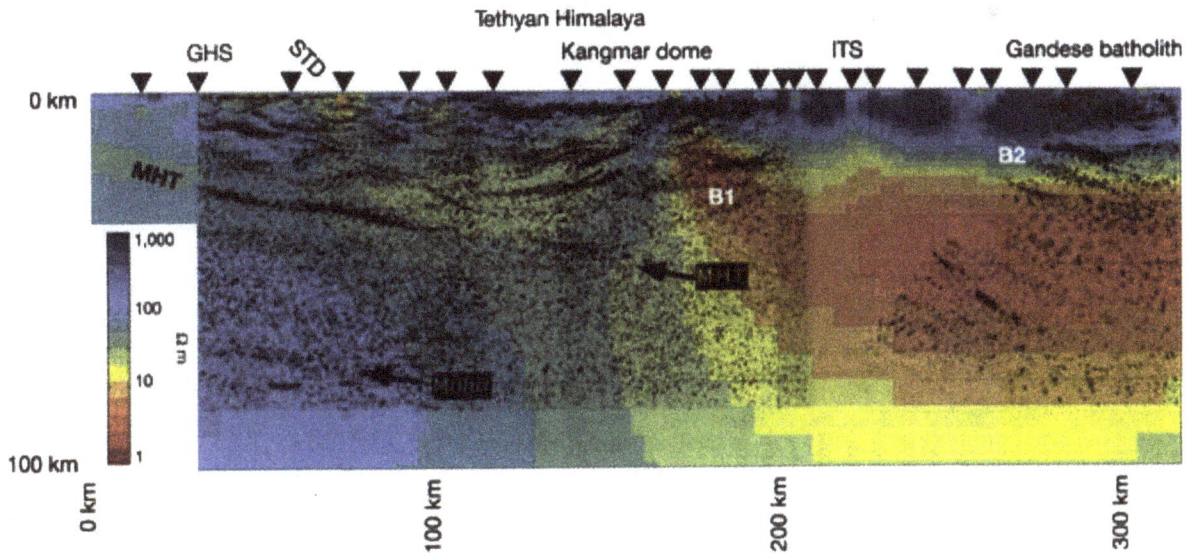

Fig. 6.32(a) Conductivity distribution along INDEPTH Profile with common mid point reflection profile superimposed on it showing high conductivity in the lower crust north of the ITS extending in the upper crust north of the South Tibet Detachment (STD) under the Kangmar dome. It coincides with the seismic bright spots, B1 and B2. It also shows reflections due to Main Himalayan Thrust (MHT) extending subsurface up to Kangmar dome and Moho. GHS- Great Himalayan Sequences.

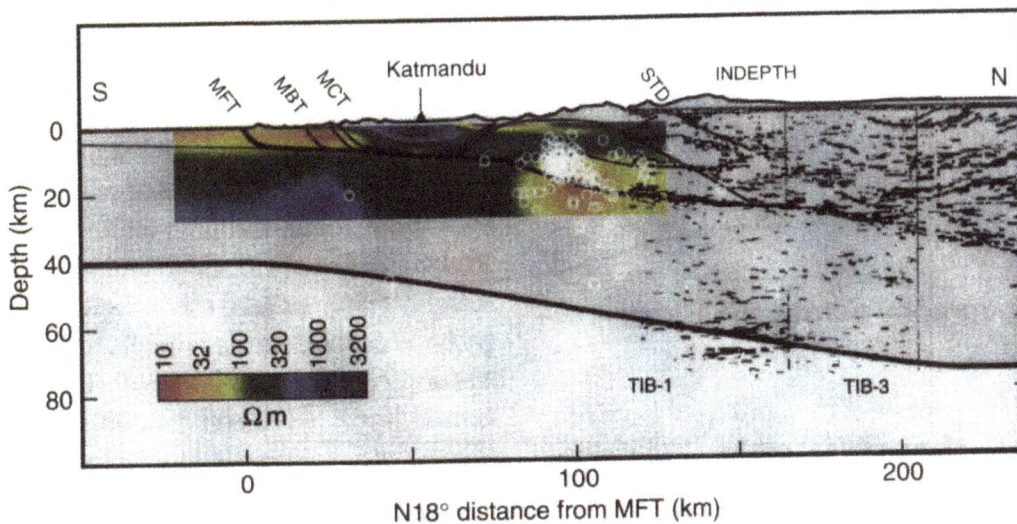

Fig. 6.32(c) Conductivity distribution in central Nepal (Lemonnier et al., 1999) Hypocenters of earthquakes (circle) coincide with the southern slope of High Himalaya and ramp in the MHT.

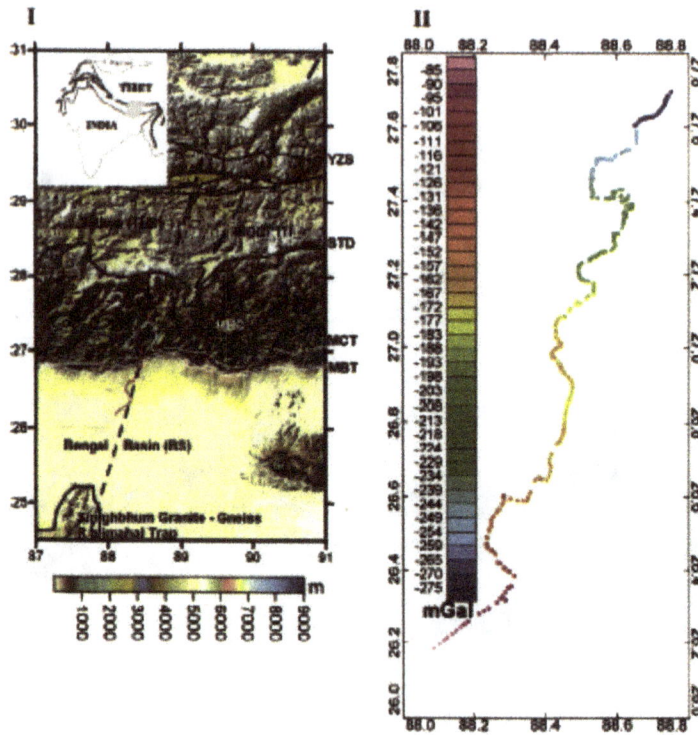

Fig. 6.34(a) Location of gravity and magnetic stations across the Sikkim Himalaya are plotted as red circles on the topographic map of the region. All data (existing data and new data) are projected along dashed line. INDEPTH profile is marked by thick dashed line. All the major thrust faults separating different geological units are also marked. Location of 250 gravity and magnetic stations across the Sikkim Himalaya are plotted as circle on the right hand side. Colour of the circles indicates the range of gravity anomalies (Tiwari et al., 2006).

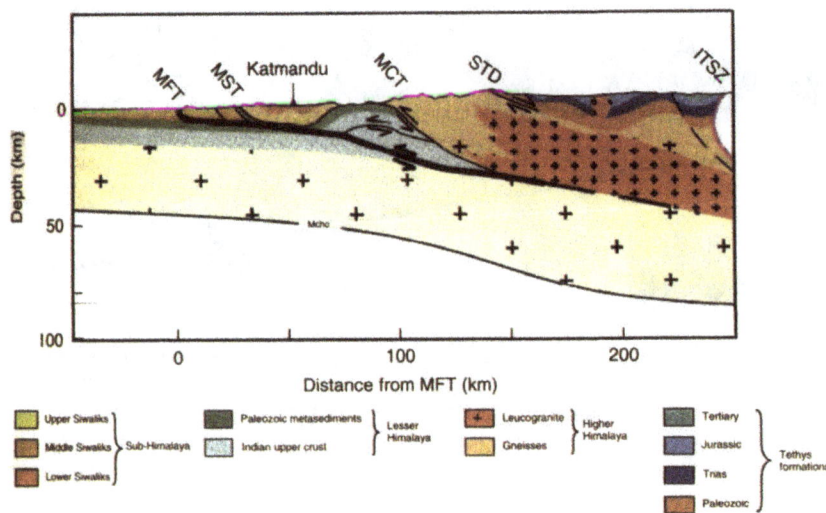

Fig. 6.36(a) A typical geological cross section in central Nepal based on investigations described above. Leucogranite along MCT is shown that might have been related to channel flow giving rise to leucogranite intrusive bodies between STD and ITSZ (Avouac, 2007).

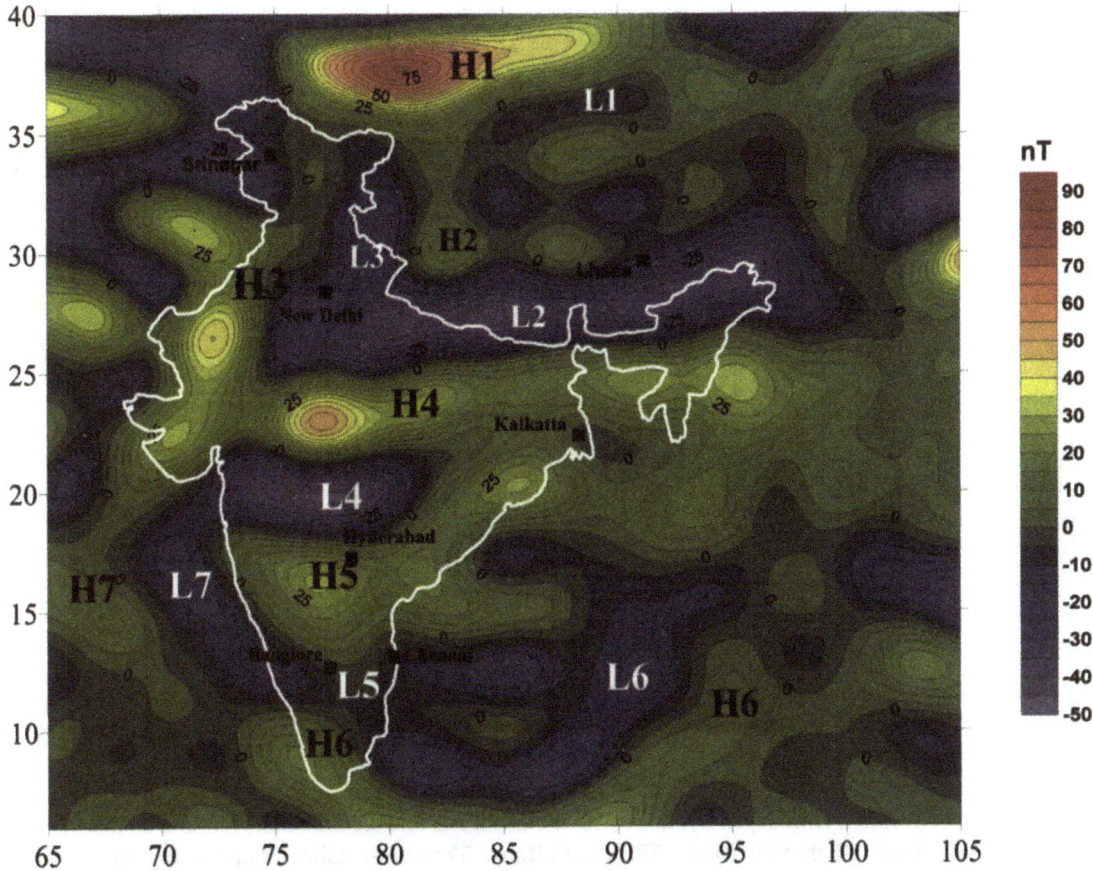

Fig. 6.38 CHAMP satellite magnetic map continued downward to a height of 50 km with magnetic highs, H1-H2 and lows, L1-L2 recorded over Tibet and Himalayas. Other anomalies are related to the Indian continent that are discussed in Section 7.11.4.

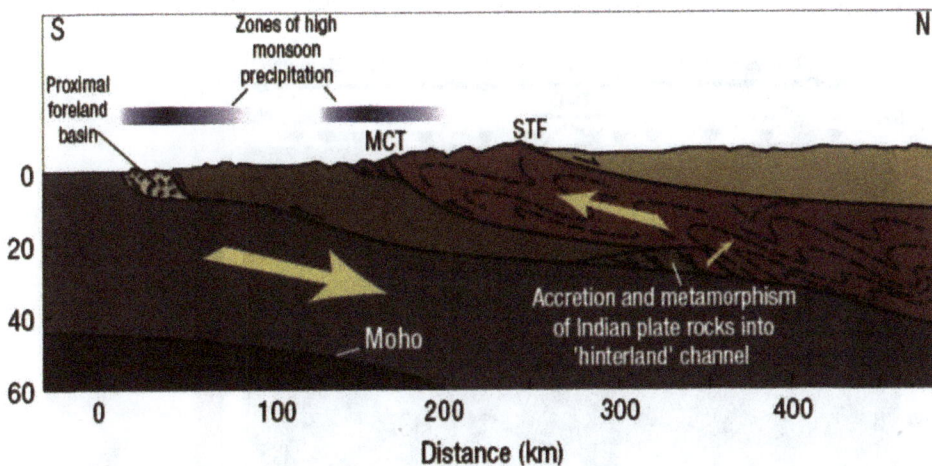

Fig. 6.41(b) A schematic cross section of the southern margin of Himalaya-Tibetan orogen that is affected by intense rainfall coinciding with maximum rainfall erosion and uplift along the Main Central Thrust (MCT) and Main Boundary Thrust (MBT) south of it (Clift et al.2008b). Miocene channel flow is suggested between the MCT and South Tibetan Fault (STF) where high Himalaya is exposed and leucogranites have been reported. Another channel flow during Quaternary period along MBT has also been envisaged.

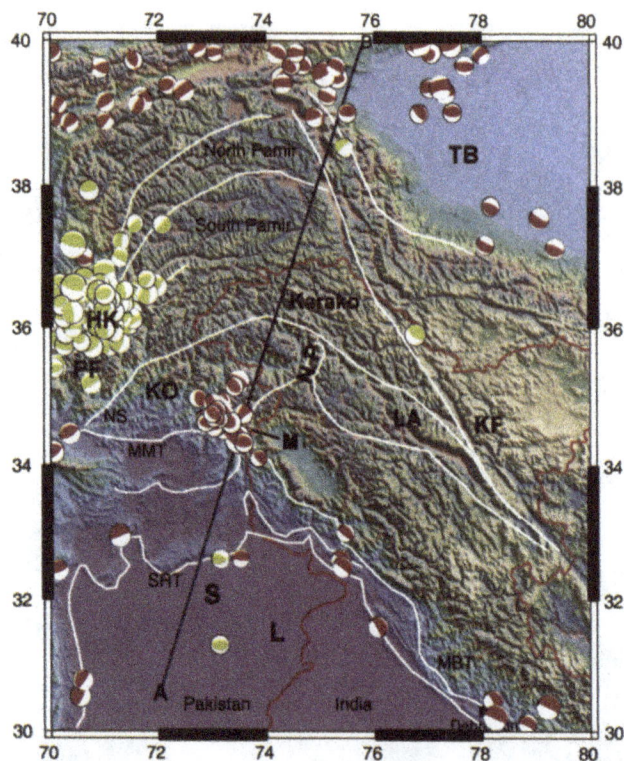

Fig. 6.43 Tectonics of Western Himalayan Syntaxis with major thrusts and tectonic units marked on the topographic map. LSR = Lahore-Sargodha Ridge. SRT = Salt Range Thrust also called Main Frontal Thrust; MBT = Main Boundary Thrust, MMT = Main Mantle Thrust or Indus suture, LA = Ladakh Batholith, KO = Kohistan Arc; NS = Northern Suture; NP= Nanga Parbat; KF = Karakoram Fault; KA = Karakoram Range; HR = Hindukush Range; TB = Tarim Basin; L = Lahore; S = Sargodha. PF- Panjshir Fault Epicenters of thrust earthquakes of magnitude > 3 since 1973 are plotted in red and deep focus earthquake in yellow. M = Muzaffarabad (Kashmir) Earthquake of 2005 (Tiwari et al., 2008).

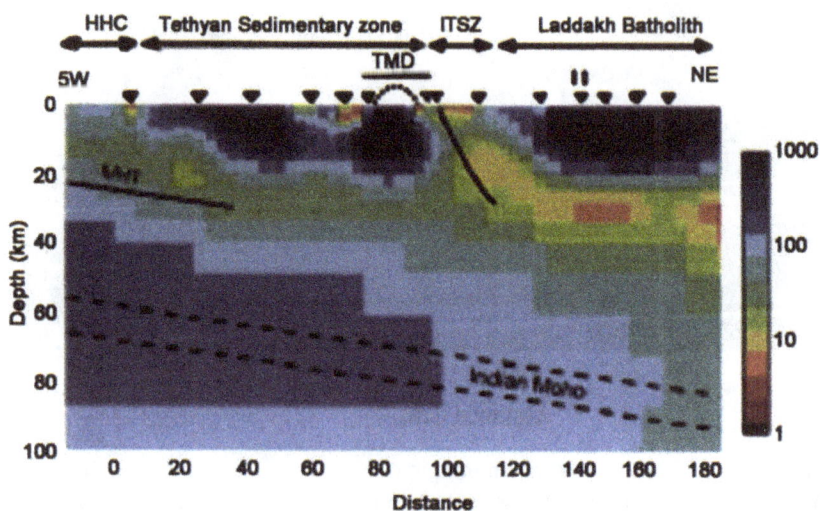

Fig. 6.49 Conductivity profile across suture zone (ITSZ) along Leh-Panamik profile (Fig. 6.48) showing a conductive body over the subducting Indian plate that lies at a depth of 20-25 km north of the suture zone. (Arora et al., 2007).

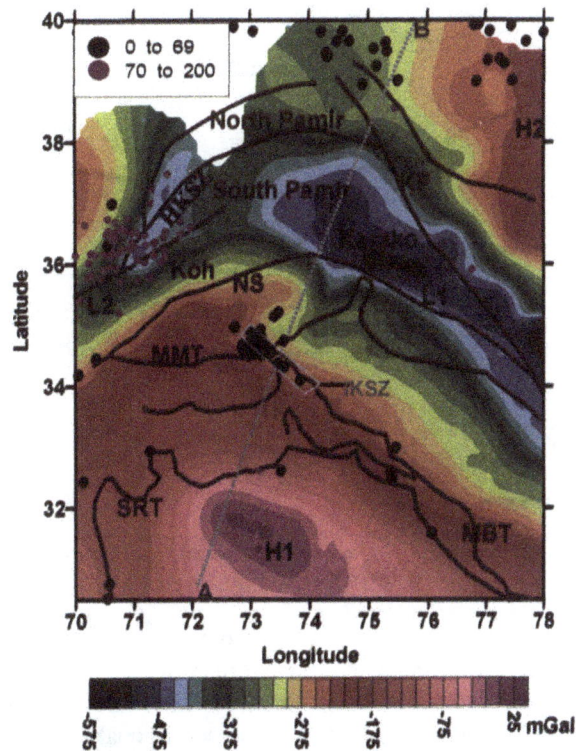

Fig. 6.55 (a) Bouguer anomaly map of Western Syntaxis showing major thrusts and epicenters of thrust earthquakes (black dots) and deep focus earthquakes (red dots). H1 and H2 are gravity highs over thrust belts and L1 and L2 are gravity lows over Karakoram-Hindukush ranges. IKSZ = Indus-Kohistan Seismic Zone and HKSZ = Hindukush Seismic Zone (Mishra and Rajasekhar, 2006).

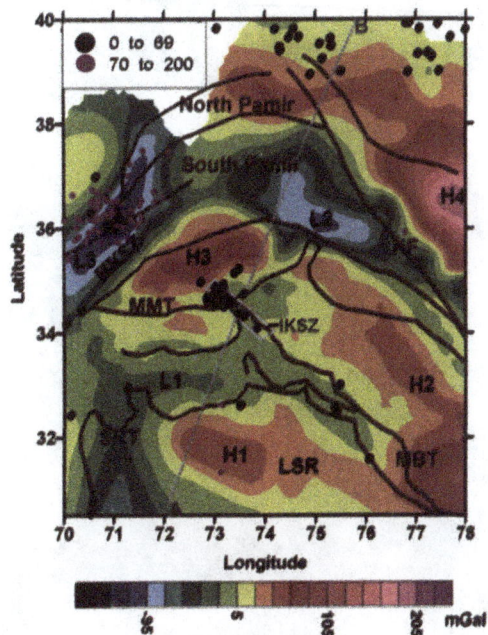

Fig. 6.56 Residual gravity anomaly based on high pass filters of wavelength <500 km. It shows gravity highs, H1 related to Lahore-Sargodha ridge (LSR) and H2 and H3 related to thrust belts. Gravity lows L1, L2 and L3 are related to Siwalik sediments and Karakoram-Hindu Kush ranges. Gravity high H4 is associated with thrust belts of Asian plate (Pamir ranges) (Tiwari et al., 2008).

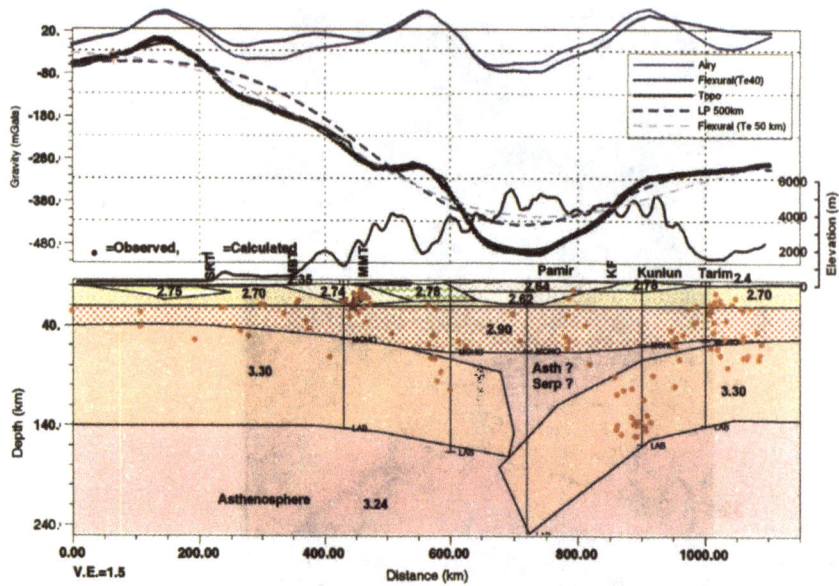

Fig. 6.58 Bouguer anomaly and elevation along profile AA' (Fig. 6.55(a)). Residual anomaly from Fig. 6.56 along the same profile is also plotted with the regional and residual fields for Te = 40 and 50 km. The computed model suggests underthrusting of the Indian and Asian lithospheres, the latter thrusting below the former. It also shows high density rocks associated with thrusts. Hypocenters of thrust earthquakes (Figs. 6.55(a)) projected on this profile coincides with them and their probable extensions depthwise. Hypocenters of some deep focus earthquakes coincide with the underthrusted slab (Tiwari et al., 2008). Cross inicates hypocenter of Kashmir earthquake of 2005.

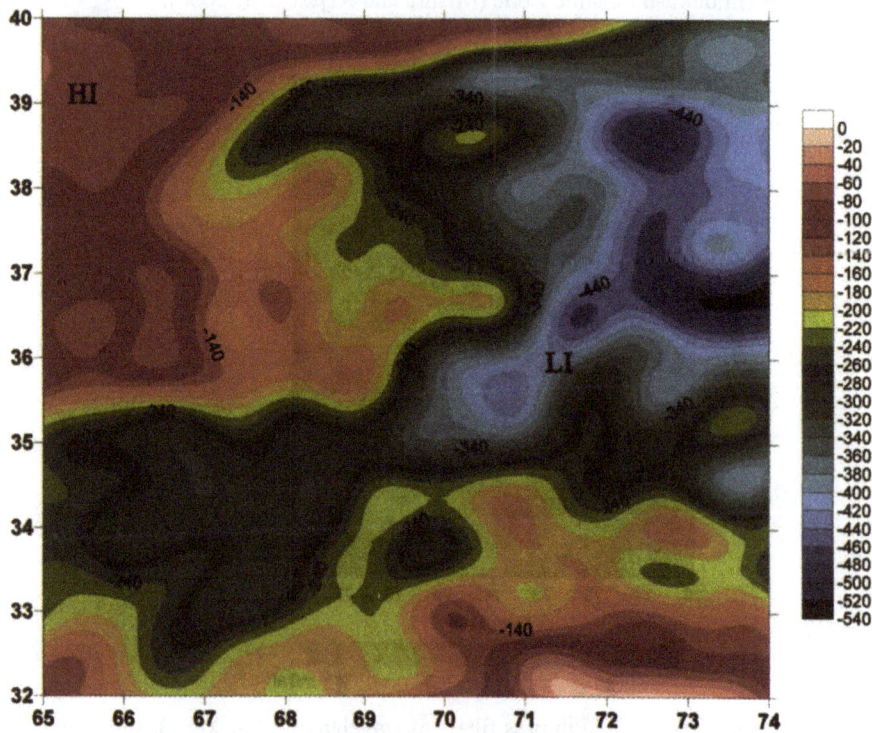

Fig. 6.59(a) Bouguer anomaly map of Hindu Kush and surrounding region showing large negative values due to thick crust related to isostasy. and low density rocks in the upper mantle.

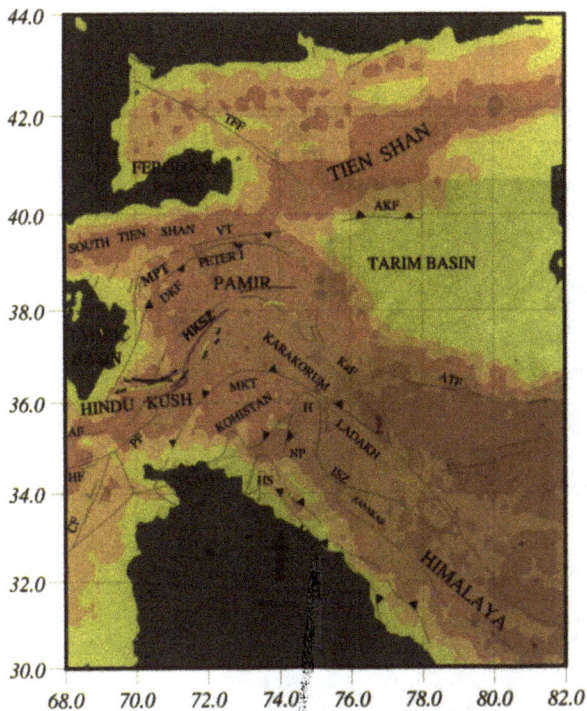

Fig. 6.59(c) Map showing the Pamir-Hindu Kush region. Topography is contoured at 1000 m intervals (0-1000 m green, 1000-2000m, Yellow etc). Faults are ;marked in black VT- Vakhsh; DKF – Darvaz-Karakul Fault; KaF – Karakoram Fault; AF – Andarab Fault; HF – Herat Fault; PF-Panjshir Fault; CF – Chamman Fault; ISZ – Indus Suture Zone; MBT – Main Boundary Thrust; MKT – Main Karakoram Thrust; MMT – Main Mantle Thrust; NP – Nanga Parbat; H – Haramosh; HS – Hazara Syntaxis; ATF – Altyn Taagh Faualt; AKF – Atushi – Keping Fault; TFF – Talas Ferghana Fault HKSZ – Hindu Kush Seismic Zone in S pattern (Focal Depth: red; 100 km; orange: 125 km; yellow: 150 km; green: 175 km) (Pegler and Das, 1998)

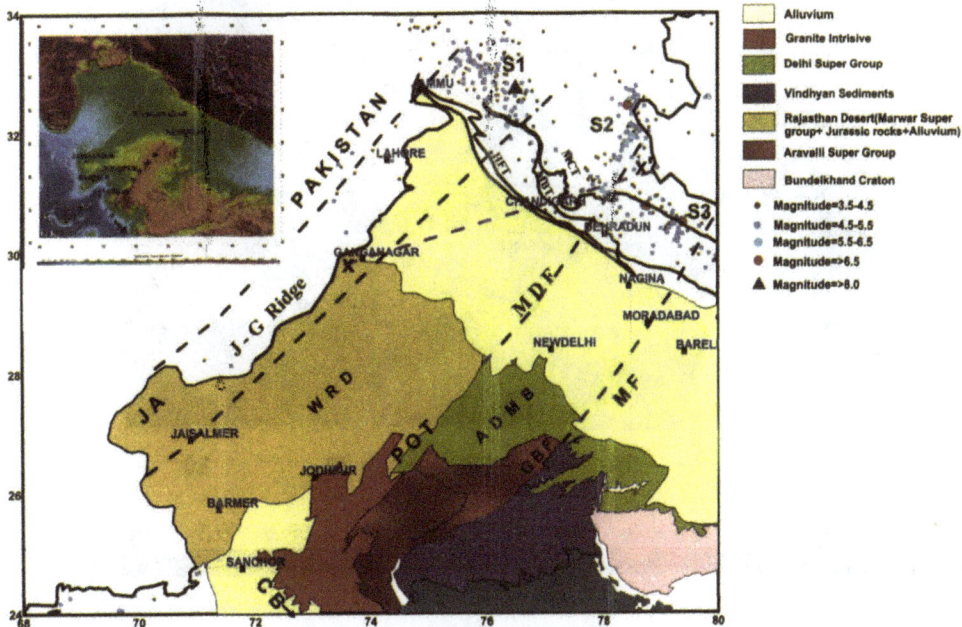

Fig. 6.64(c) A simplified geological map of NW India showing Aravalli Delhi Mobile Belt (ADMB) and Great Boundary Fault (GBF) and Phulad OphioliteThrust (PT) along its eastern and the western margins and their extensions towards the north are shown as Moradabad Fault and Mahendragarh-Dehradun Fault (MDF), respectively. They interact with the Himalayan front along the MBT and seismic activity zones marked as S2 and S3 might be related to this interaction as these faults are west dipping. Jaisalmer-Ganganagar ridge in Western Rajasthan Desert (WRD) is shown as dashed lines that might be related to lithospheric flexure of the Indian plate along the Western Fold Belt (Pakistan) delineated based on the residual Bouguer and geoid anomalies as described below. It interacts with the Himalayan front in seismic activity group S1 where the great Kangra earthquake (filled triangle) of 1905 is located. Thrusts along the Himalayan front are marked as:
HFT – Himalayan Frontal Thrust, MBT – Main Boundary Thrust and MCT – Main Central Thrust.

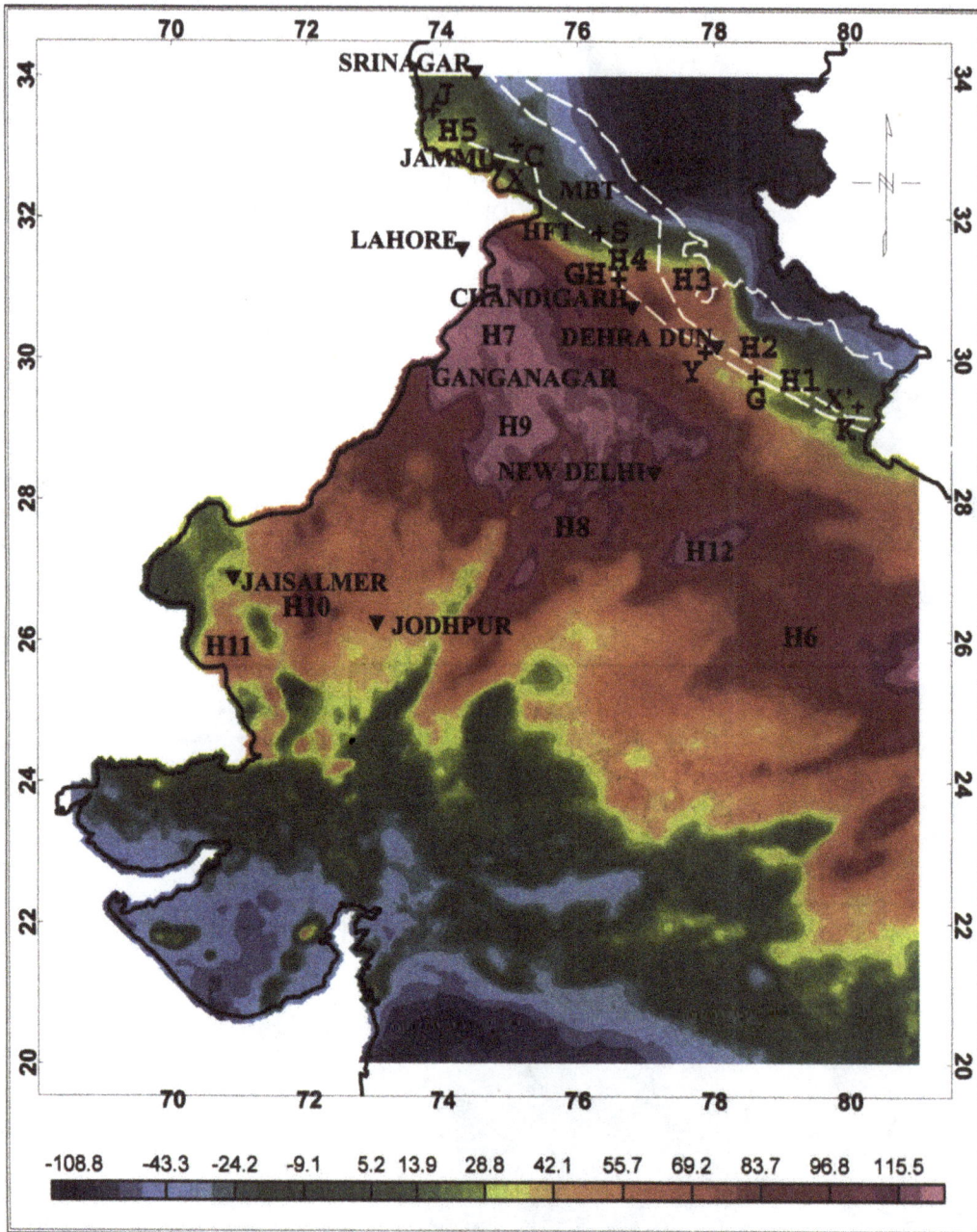

Fig. 6.65(a) Residual anomaly map of Western Himalayas for wavelengths <210 Km showing gravity highs H1-H12 caused by shallow sources. "+" sign marks the major rivers along Himalayan front; K=Kali; G=Ganga; Y=Yamuna; GH=Ghaggar; S=Satluj; C=Chenab; J=Jhelum. These rivers flow from the sides of gravity highs, H1, H2, H3, H4, & H5 except Ghaggar that flows from south of gravity high H4.Gravity highs, H6 and H7 represent crustal bulge due to lithospheric flexture of Indian plate due to Himalayas and H7 represent part of Delhi-Lahore-Sargodha basement ridge. Other highs and lows are explained in the text.

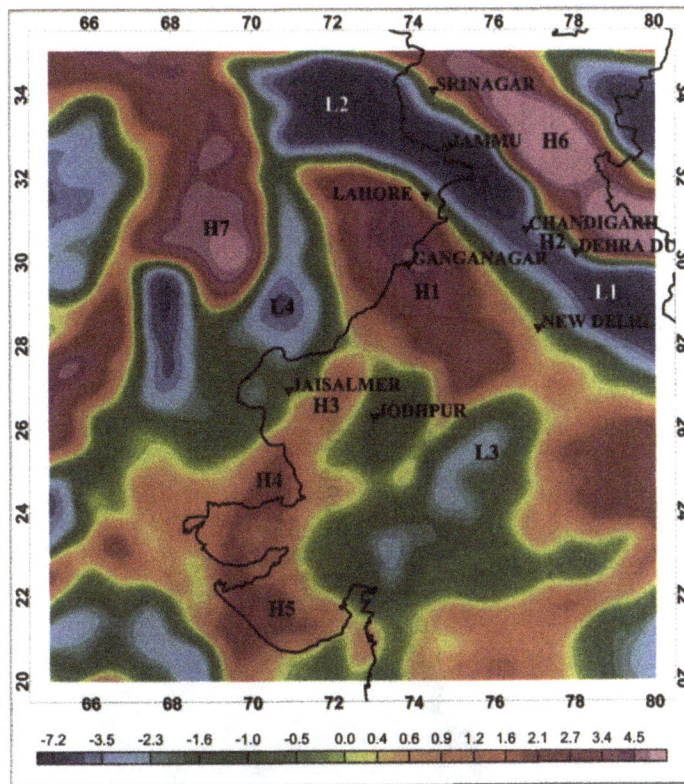

Fig. 6.65(b) Residual Geoid map of NW India and western Himalayan front and adjoining region showing a geoid high, H1 related to Delhi-Lahore-Sargodha ridge and another highs, H3-H5 that are attributed to mafic intrusive that has giveMajor rivers along Western Himalayan front.

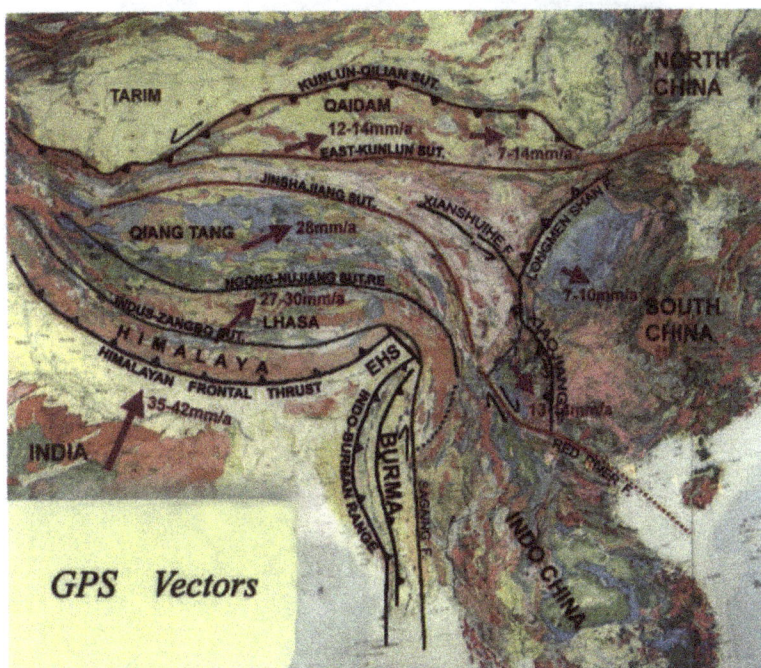

Fig. 6.67(a) GPS vector over Himalayas and Tibet (Gahalaut, 2008).

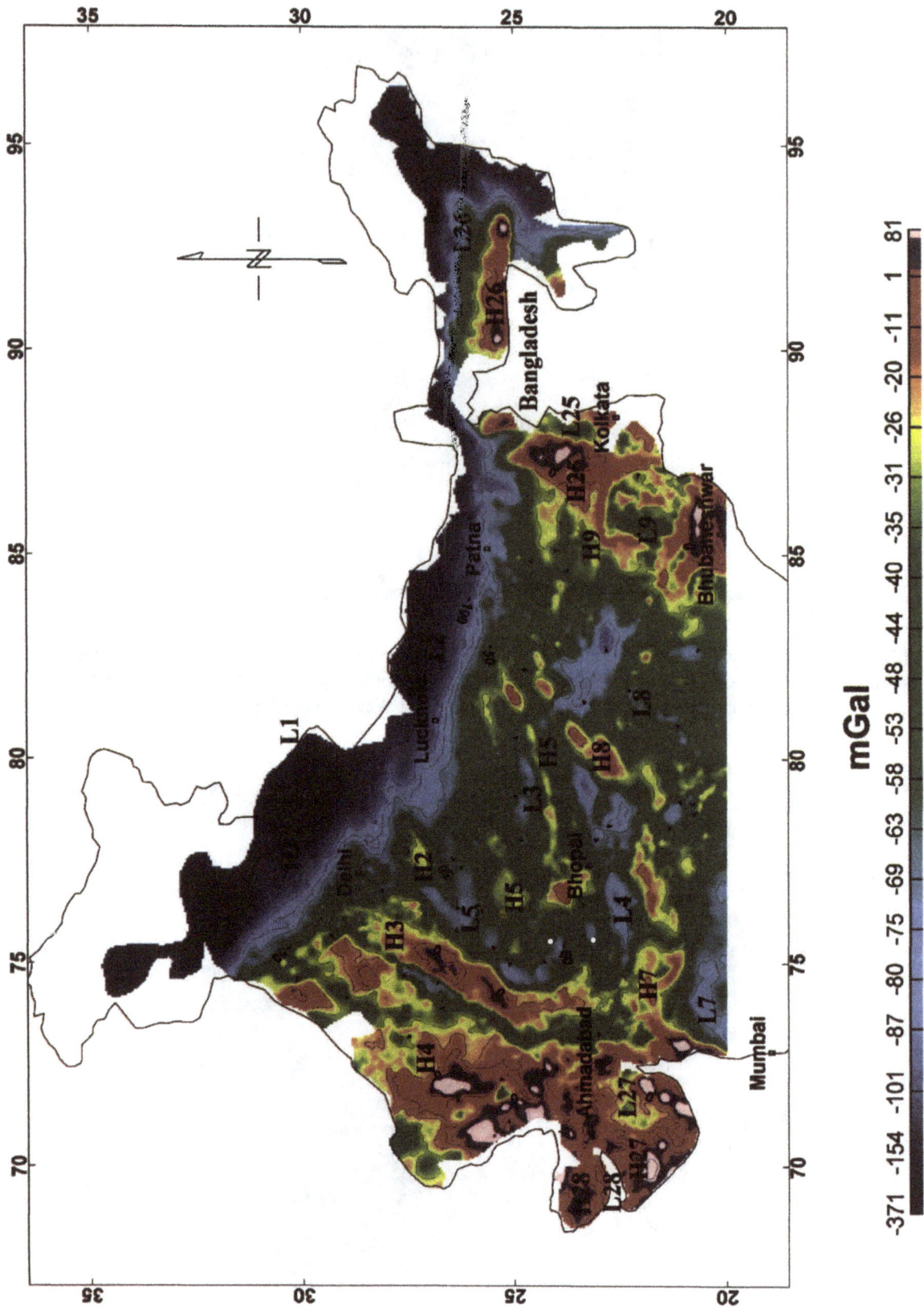

(a)

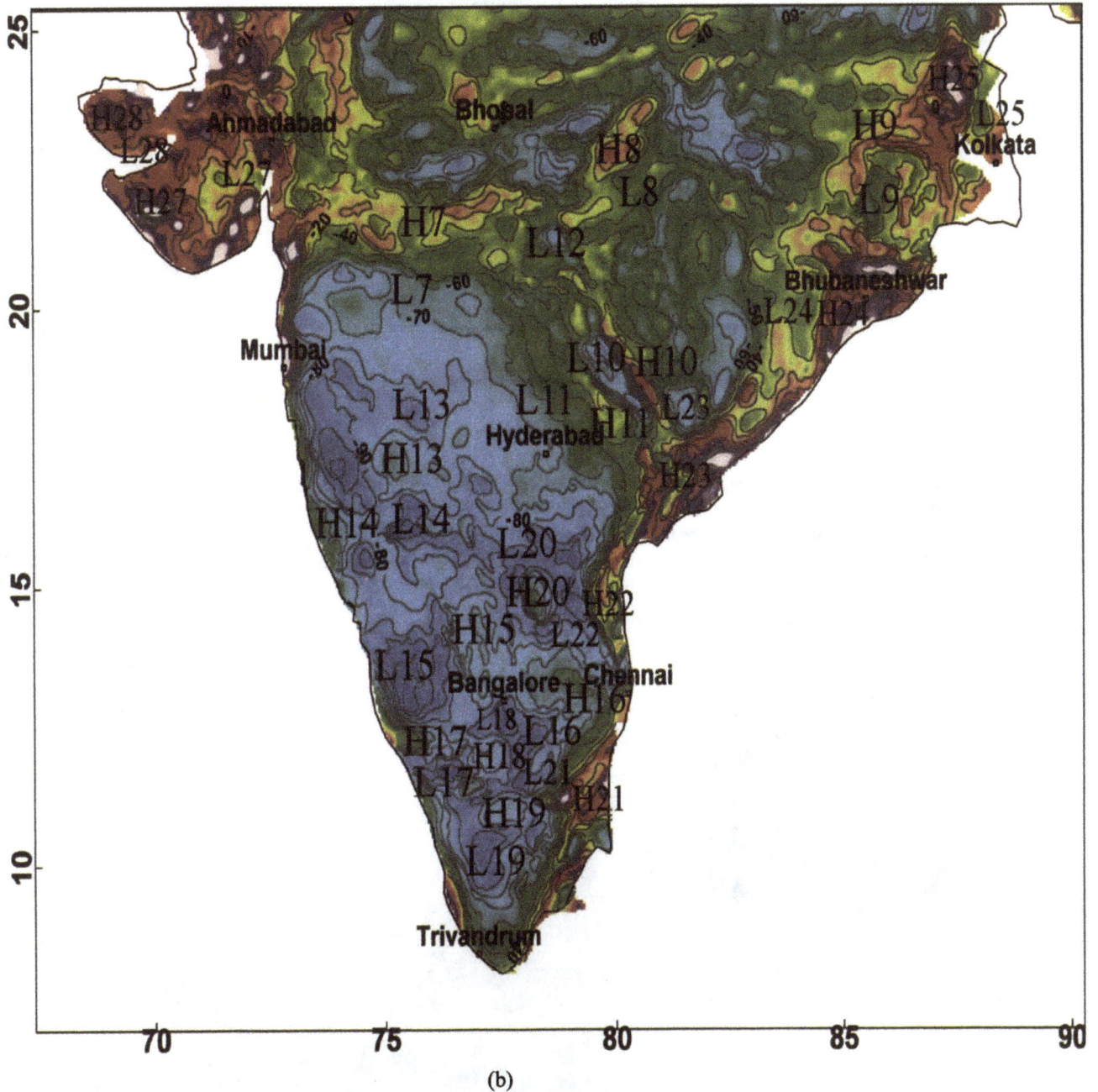

Fig. 7.8 Complete Bouguer anomaly map showing gravity highs and lows marked as H and L, respectively and the following numbers indicate the individual anomalies associated with different tectonic units as described in the text (a) North of 20°N and (b) South of 25°N (GSI-NGRI, 2006 mishra et.at., 2008).

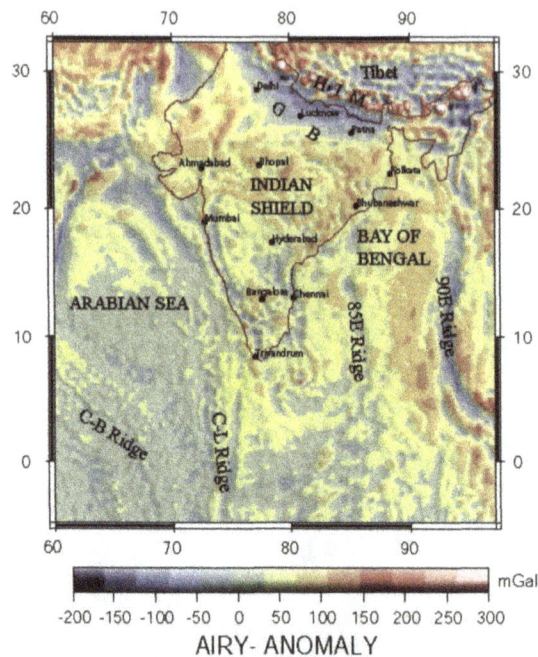

Fig. 7.9 Airy's isostatic anomaly map shows gravity highs over mountain ranges (fold/mobile belts) indicating under compensation and /or shallow high density rocks. Most significant lows are observed over Tibet, Ganga fore deep and some other basins indicating over compensation and low density sediments, respectively. Ridges in the oceans show gravity highs indicating high density rocks in the crust associated with them. Him- Himalaya, GB- Ganga Basin, C-L Ridge- Chagos- Lacadive Ridge, C-B Ridge- Carlsberg Ridge.

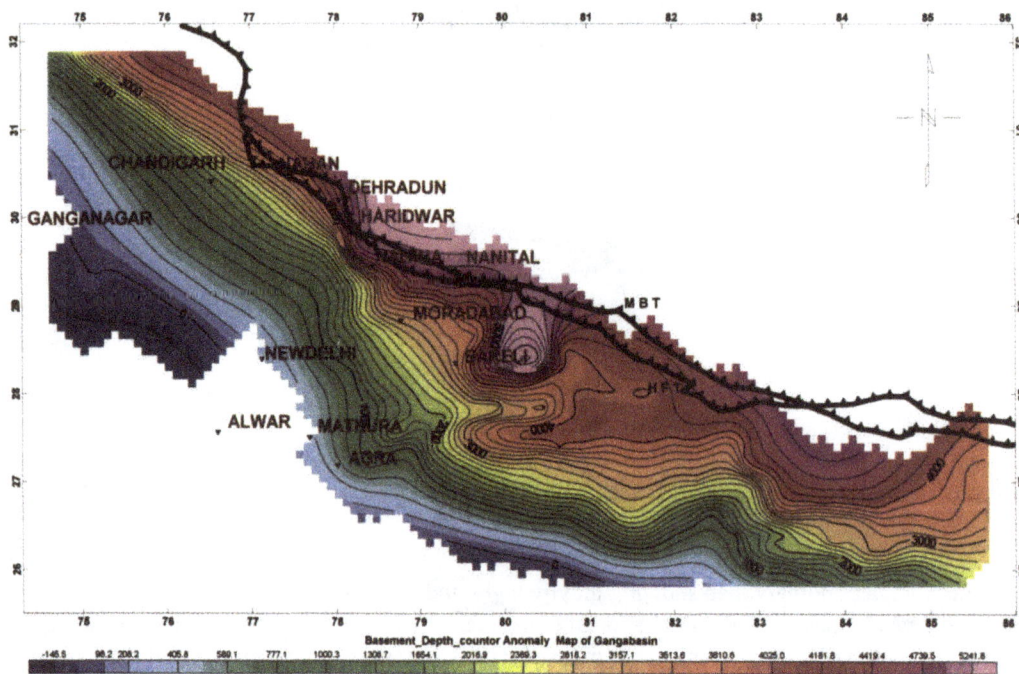

Fig. 7.11(a) Basement under the Ganga basin based on airborne magnetic data (Agocs, 1956; Sengupta, 1996) constrained from seismic and well data (Karunakaran and Rao, 1976). It shows depths to various ridges and depressions in meters with contour intervals of 200 m and depths varying from exposed section in the southern part to ~ 6 km in the northern part along the Himalayan foredeep.
MBT: Main Boundary Thrust and HFT-Himalayan Frontal Thrust.

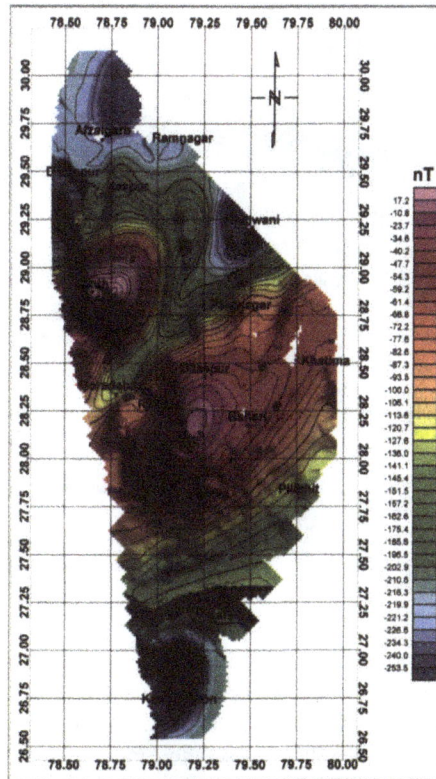

Fig. 7.11(b) A sample of the airborne total intensity map of the Ganga basin flown with 2 km spacing and 700 m as flight barometric altitude. The NE-SW linear magnetic low from Moradabad represents a fault that might be northward extension of thrusts related to the Aravalli-Dehi Mobile Belt. Magnetic highs on either side of it may represent the extension of this belt itself. Contour interval-10 nT (Babhuleyan et al., 1999). TheMoradabad fault is reflected as nosing of contour of a magnetic low.

Fig. 7.23(a) Bouguer anomaly map of Bundelkhand craton, Vindhyan basin and central part of the SMB (Fig. 7.20). Gravity highs and lows important for present study are indicated by H1-H8 and L1-L2 respectively. Some tectonic elements, important for the present study are schematically shown in limited sections. They are as follows: ADMB – Aravalli-Delhi Mobile Belt, GBF – Great Boundary Fault, CHF – Chambal Fault, GB – Ganga Basin, CIS – Central Indian Shear, BC – Bundelkhand Craton, VB – Vindhyan Basin, SNSF – Son Narmada South Fault. The section between the CIS and the SNSF and their extensions towards east and west is known as Satpura Mobile Belt. The gravity highs due to ADMB (H8) and SMB (H6) are partially presented in this figure. Eastern part of the gravity high, H7, east of the GBF represents Agra (AG) – Shahjahanpur (SH) ridge along the western margin of Vindhyan basin extending into Ganga basin.

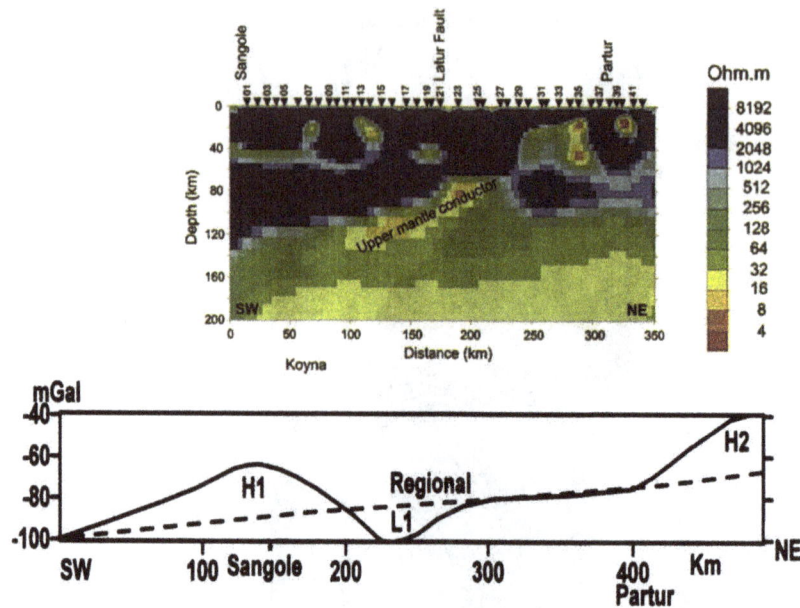

Fig. 7.28(e) Two dimensional geoelectric model based on MT profile Sangola-Partur (Fig. 7.40a, Patro and Sarma, 2009) showing an upper mantle conductor in lithospheric mantle with decreasing depth north wards in the crust where Proterozoic Satpura Mobile Belt is nearby (Fig. 7.11c). The bottom profile is Bouguer anomaly from SW of Sangole to NE of Partur (Fig. 7.40a) showing an increasing regional high towards the SMB indicating high density rocks that may represent a thrust with fluids.

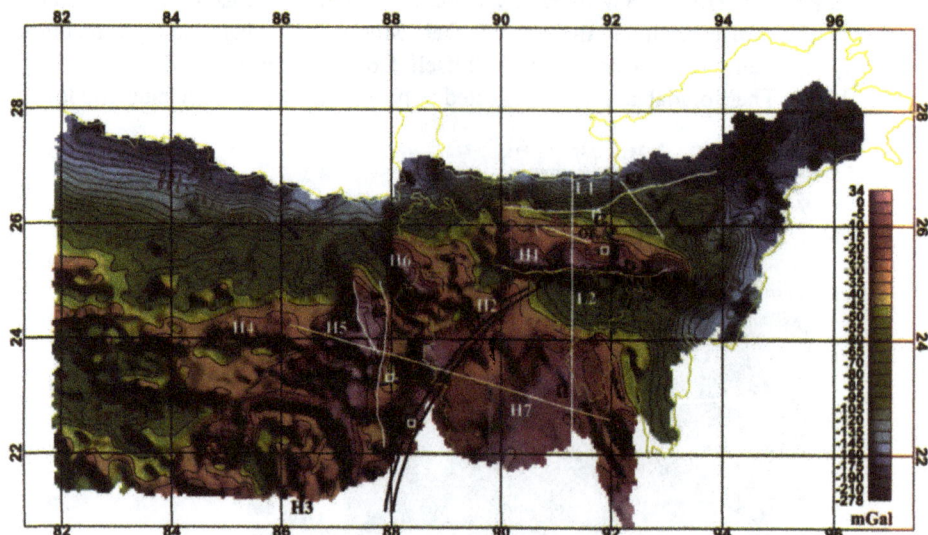

Fig. 7.30(a) Complete Bouguer anomaly map of NE India and Bangladesh. H1-H7 and L1-L2 represents gravity highs and lows. Some important tectonic units and towns are as follows: HZ – Hinge zone, S – Shillong, C – Calcutta, G – Guwahati, BF – Brahmaputra Fault, OF – Oldham Fault, HZ – Hinge Zone, DF – Dauki Fault, KL – Kopili Lineament, RF – Rajmahal Fault, SBF – Sainthia Bahmani Fault. Epicenter of 1897 Shillong earthquake is shown by red triangle. Location of Profiles I and II across Chhotanagpur Granite Gneiss Complex, Bengal basin (India and Bangladesh) and Shillong Plateau are also shown. The deep seismic sounding (DSS) profile covers the central part of Profile I in Bengal basin (India) across the Hinge zone (Rajasekhar and Mishra, 2008).

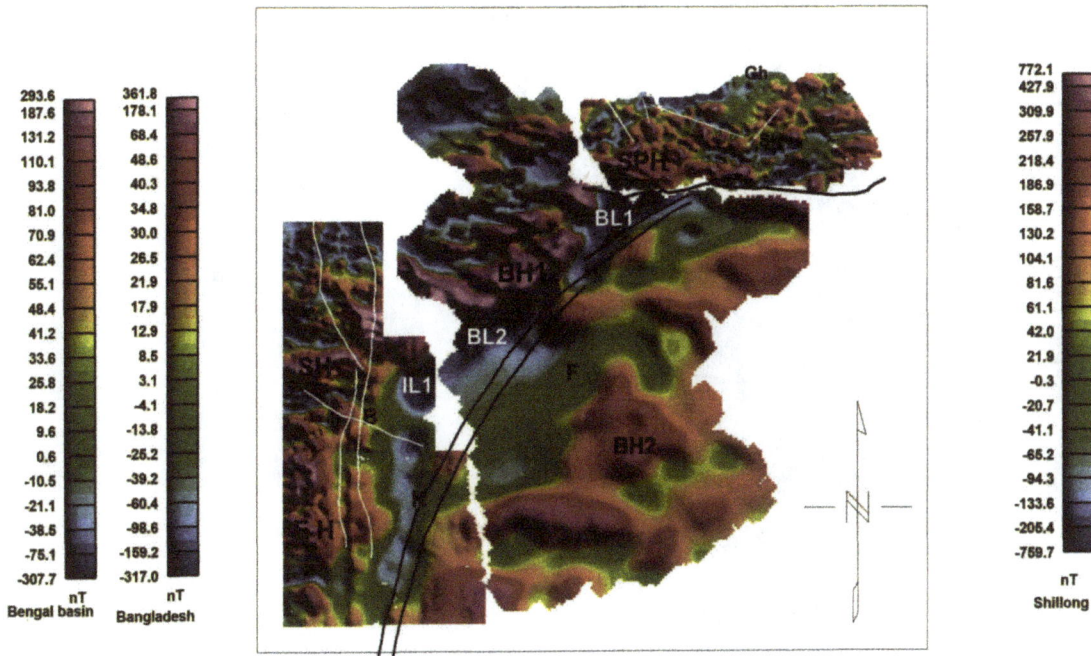

Fig. 7.30(c) Airborne magnetic map of Shillong plateau and adjoining Bengal Basin, Indian Shield and Bangaladesh with different scales for the three sections (Rajasekher, 2010). K-Kolkatta; Gh- Gauhati; Sh-Shillong; EH-Eastern Ghat trend magnetic high; SH- Satpura trend magnetic high; BH1- Bangaladesh magnetic high 1; BH2- Bangaladesh magnetic high 2; SPH- Shillong Plateau magnetic high; BL1- Bangaladesh magnetic low 1; BL2- Bangaladesh magnetic low 2; IL1- India Bengal Basin magnetic low 1.

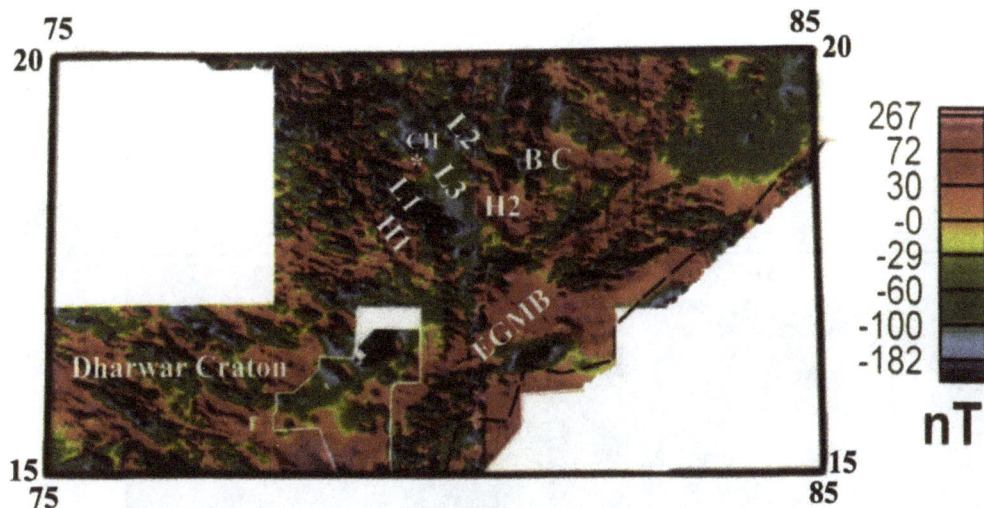

Fig. 7.34(a) Airborne magnetic map of Godavari basin including that of the central part of the Eastern Ghat Mobile Belt (EGMB) toward the east recorded at a height of 1.5 km above msl (GSI, 2001). It shows pairs of magnetic highs and lows H1, L1 and H2, L2 related to Proterozoic rocks of the Godavari basin indicating mafic rocks. L3 represents Gondwana sediments of Godavari basin. Star indicates Chinnur high (CH) that is a median high in the basin.

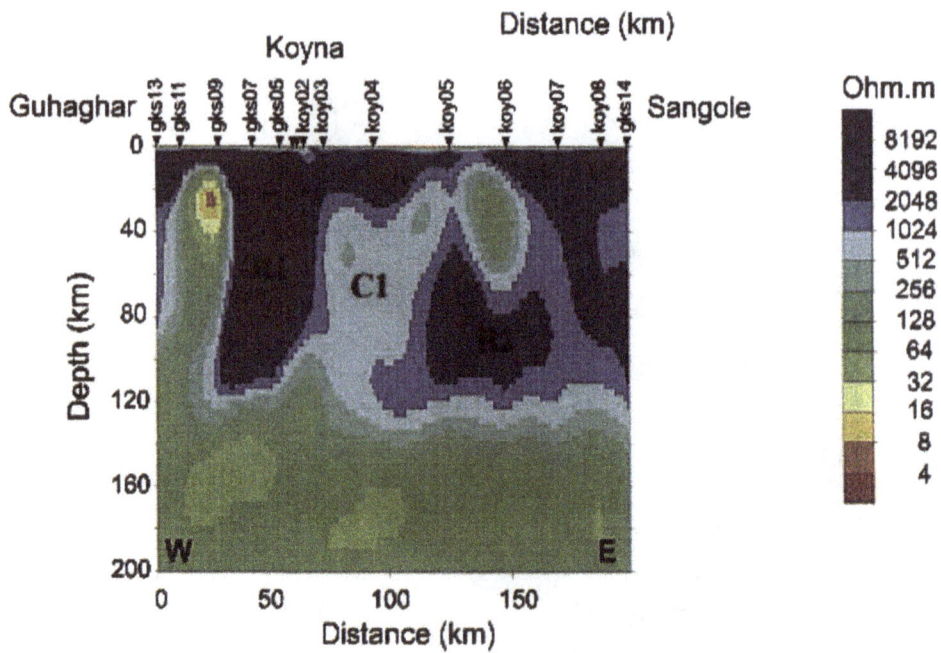

Fig. 7.40(d) A conductivity profile Guhaghar (along the coast)-Koyna-Sangola (Fig. 7.40a) shows two resistive blocks (R1 and R2) extending from surface to the lithospheric mantle with a conductive section (C1) in between that corresponds to fluid filled shear zone as cause for Koyna seismic activity.

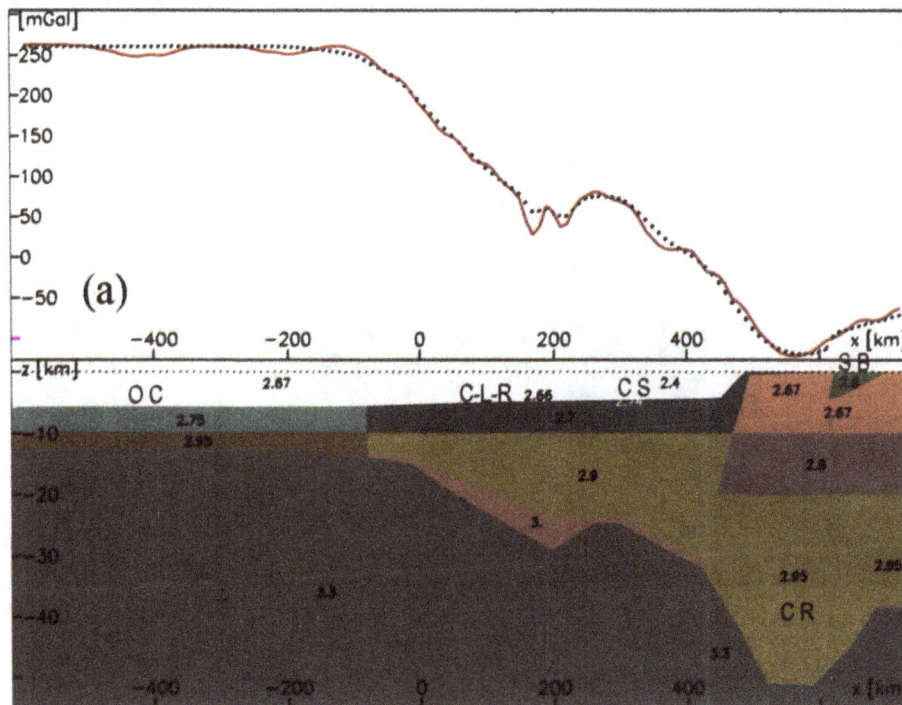

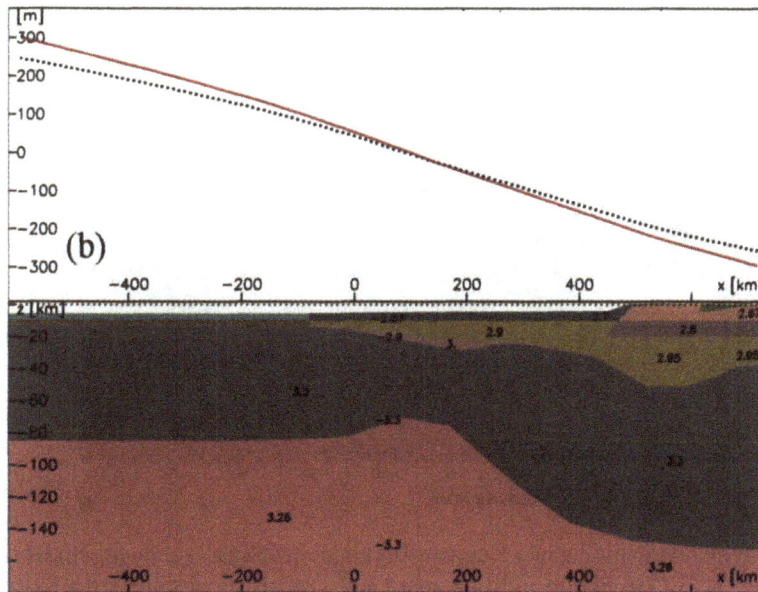

Fig. 7.43(a-b) Crustal and lithosphere density structures along a representative vertical section throughlowest Bouguer anomaly south of Hasan based on the computed gravity and geoid fields and their comparison with the observed field. Figure (a) shows shallow section upto ~ 50 km while figure (b) shows full section upto ~ 160 km. They show maximum crustal thickness of about 50 km where old schist belts (Sargur type; ~3.4 Ga) are exposed in surrounding region forming crustal root. Parallel sections similar to this one were used to construct the 3D model of the causative sources. Numbers indicate density in g/cm³ that are primarily related to upper and lower crusts and upper mantle for oceanic and continental parts. Some important densities structures are: 2.4- Sediments, 2.7- Deccan trap rocks associated with C-L-R-Chagos-Laccadive Ridge, 2.67- Bathymetry filled with standard crustal rocks, 3.0- Under plated crust and 2.8- Schist belt on continent. Computations are made with reference to standard crustal and upper mantle densities given in figure (b) as -2.67, -2.9 and -3.3, respectively. CS- Continental Shelve, CR- Coastal Root, SB– Schist Belt, OC-Oceanic Crust (Arora et al., 2007).

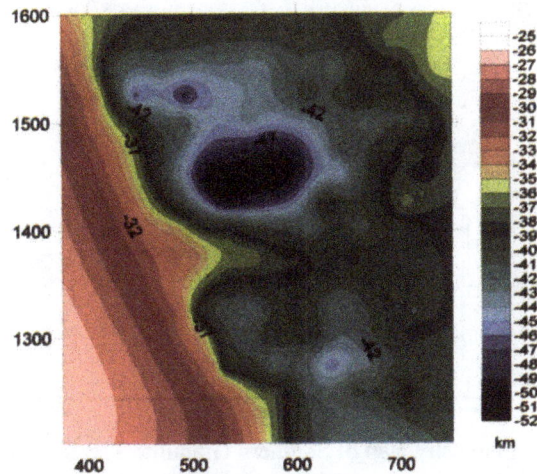

Fig. 7.43(c) A 3-D model of the Moho based on parallel sections as given in Fig. 7.42a separated 30' apart based on the comparison between the observed and the computed field showing maximum crustal thickness of 50 km south of Hasan over Sargur schist belt that belongs to older class of Schist belts (~ 3.4Ga) (Arora et al., 2009).

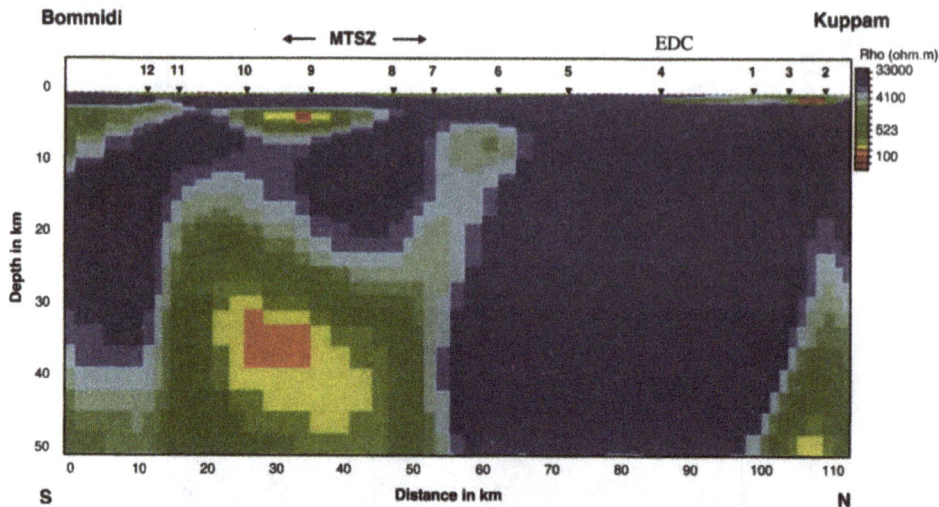

Fig. 7.44(c) 2-D conductivity distribution along Kuppam-Bommidi geotransect showing high conductivity under the Mettur shear zone (MTSZ) similar to collision zones and a resistive crust under the Eastern Dharwar craton (EDC) (Harinarayana et al., 2003).

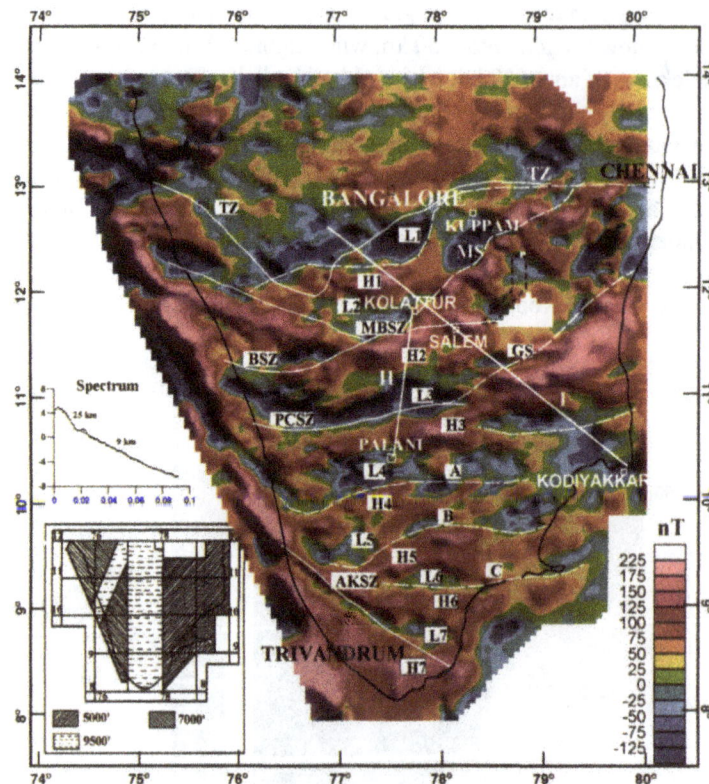

Fig. 7.47(b) Airborne total magnetic intensity map of Southern Granulite Terrain, India, recorded at flight elevations of 5000', 7000' and 9500' above mean sea level (Inset) and have been reduced to a common datum of 7000' (~2.1 km) for a composite map. Some data gap in the NE part marked by black line is filled from ground magnetic data continued to same height. H1......H7 and L1.....L7 indicate magnetic highs and lows forming pairs of magnetic anomalies. The inset of spectrum presents its plot versus wavelength, which shows two linear segments with their slopes equivalent of 7 and 23 km depths below surface representing top and bottom of predominantly magnetic mafic layer, respectively. The latter may represent Curie point geotherm (570^0C). Abbreviations are same as given in Fig. 7.37 (Mishra et al., 2006).

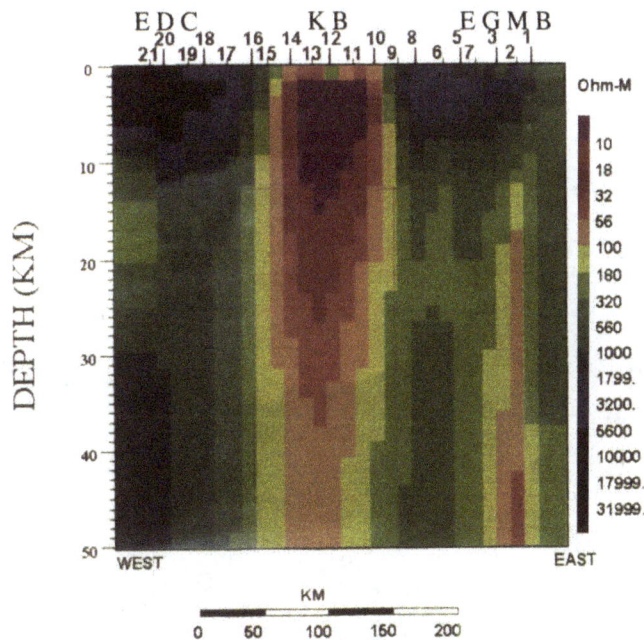

Fig. 7.51(c) 2-D model of conductivity distribution across the EGMB and Cuddapah basin along Profile I (Fig. 7.50(a)). It shows two high conductive bodies related to thrusted blocks of the EGMB and western part of the Cuddapah basin (Kurnool basin, KB) especially the latter that may be related to large scale mafic intrusives in these sections and crust-mantle interaction (Fig. 7.51(a)), while the Eastern Dharwar craton (EDC) shows a resistive crust (Naganjaneyulu and Harinarayana, 2004).

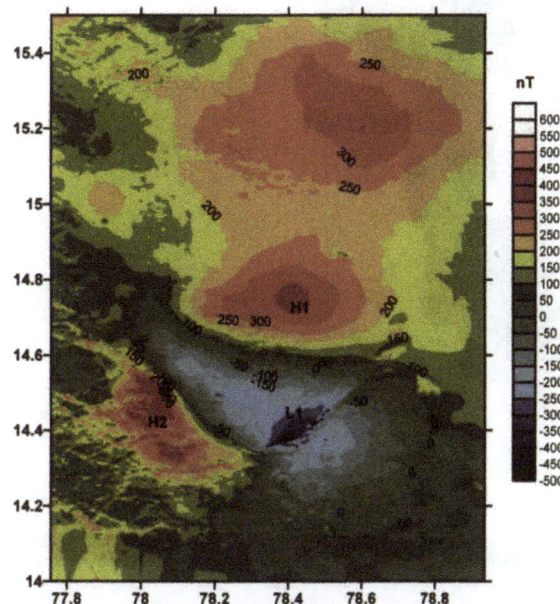

Fig. 7.52(a) Airborne total intensity magnetic anomaly map of Western part of Cuddapah basin. The IGRF field has been removed from the observed field. It shows a magnetic high (H1) in the northern part and a low (L1) in the southern part along the S-W margins of the Cuddapah basin. Another high (H2) is observed just outside the SW margin of the Cuddapah basin. A magnetic lineament (ML) cuts across the granite batholith, SW of Cuddapah basin and the western margin of the Cuddapah basin that may be important for Uranium mineralization along western margin of the Cuddapah basin.

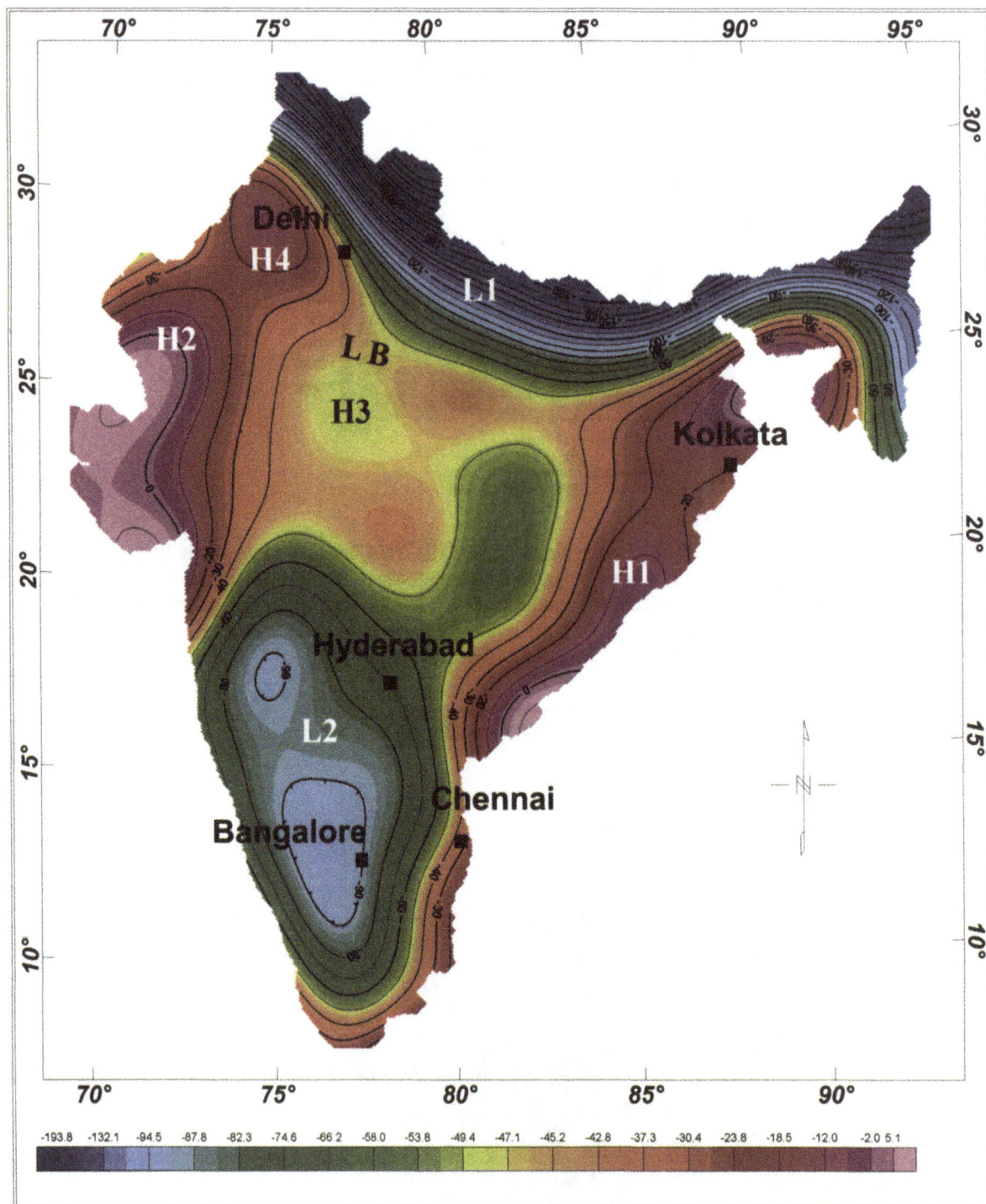

Fig. 7.60 Low pass filtered Bouguer anomaly map for first and second segments of spectrum (7.59a) corresponding to wave number .012 and wavelength >523 km related to sources in the lithospheric mantle showing gravity highs, H3 and H4 over the north Indian Shield related to lithospheric bulge caused by flexure of the Indian plate under Himalaya and lows over the south Indian Shield. Gravity highs, H1 and H2 along the coasts may represent the effects of the break up of the Indian plate from Gondwanaland and Africa along the east and the west coasts of India, respectively.

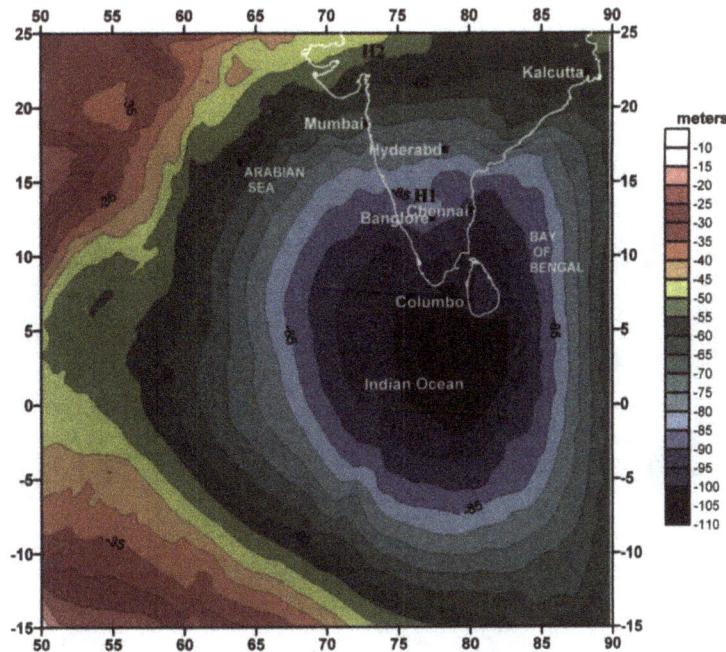

Fig. 7.62(a) Geoid low of the Indian Ocean extending over the Indian Peninsular shield with a relative high, H1 coinciding with parts of the eastern Dharwar craton including the shear zone between the eastern and the western Dharwar cratons and another high, H2 over Kachchh and the Ganga basin that may be a reflection of the lithospheric and crustal bulges due to flexure of the Indian lithosphere.

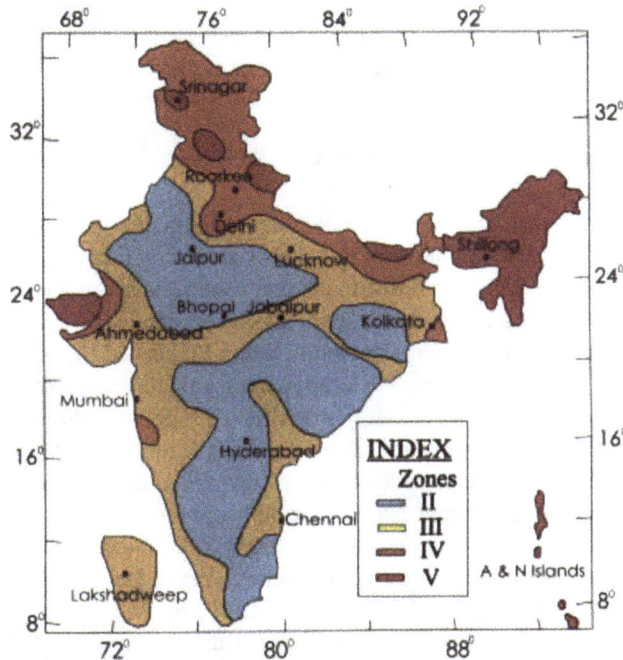

Fig. 8.6 Seismic Zonation map of India (India Meteorological Department, New Delhi) showing seismic zones II-V in increasing order of seismic activity. Some of the Himalayan sections where major/great earthquakes have been reported and Kutch (Section 8.2) and NE India including Shillong plateau (Section 7.6.4) are included under group V that are most seismogenic. All other countries close to Himalayan fold belt like Afghanistan, Pakistan, Tibet (China) Nepal, Bhutan and Bangladesh also experience large/great earthquakes and show similar pattern of seismic zoning varying from Zone V to Zone II based on distance from the Himalayan Fold Belt as shown in Appendix IV for Pakistan and Bangladesh.

Fig 8.7(b) Topographic image of Kachhah based on Shuttle Radar Topographic Mission (SRTM) data (By courtesy Dr Prabhas Pande, GSI, personal communication). It shows some significant scarps and faults. E-W (F1, Kutch Mainland Fault etc.) oriented faults are thrust faults that control most of the scarps in this region while NE-SW (F2) and NW-SE (F3) faults along Kaswali River and several others (F4 & F5 etc.) may be strike slip faults displacing these scarps. Such faults are quite common in Banni Plains and Rann of Kachchh similar to Khavda-Bachau and Chorbari-Manfara faults (Fig 8.7a) as apparent from this map. The linear feature, F7 separates the Banni plains from the Great Rann of Kachchh that shows a highly disturbed section (DS) at its eastern margin and coincides with the uplift axis of the Wagad uplift and Chorbari and Manfara villages (Fig 8.7a) towards the east where maximum lateral displacements were reported after Bhuj earthquake. The linear features, F6 and F8 coincide with the westward extension of the South and North Wagad Faults (F3 & F4, Fig 8.7a), respectively that are significant for Bhuj earthquake (BE), located ~37 km east of Jawaharnagar along F3 (Fig 8.7a). It clearly indicates that these faults extend for large distances beyond their exposed sections (Fig 8.7a) and demonstrates the importance of space imageries for seismotectonics. The airborne magnetic map of Kachchh (Fig 9.29b) also reflects NW-SE and NE-SW oriented faults intersecting in the epicentral area of the Bhuj earthquake. Inset shows the Western Fold System in Pakistan with Chamman fault (CHAM) and Karachi Arc (KA) emanating from it and extending NW-SE towards the Kachchh Rift Basin (KRB) that may manifest as NW-SE fault system in Kachchh and suggests its plate boundary connection.

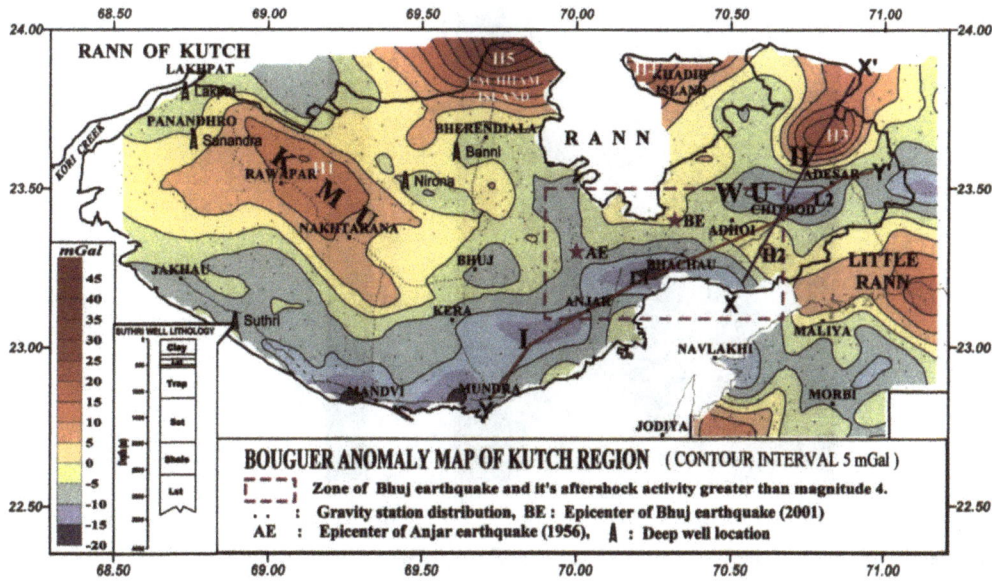

Fig. 8.9 Bouguer anomaly map of Kachchh. H1 represents the Kachchh Mainland Uplift (KMU) and H2 and L2 constitutes the Wagad uplift (WU). H3, H4 and H5 are gravity highs representing southern part of Bella, Khadir and Pachham Uplifts, respectively. I (YY') and II (XX') are profiles along which gravity data are modeled. BE and AE are epicenters of Bhuj and Anjer earthquake. Dashed red rectangle indicates epicentral zone of Bhuj earthquake and it's after shocks. (Mishra et al., 2005)

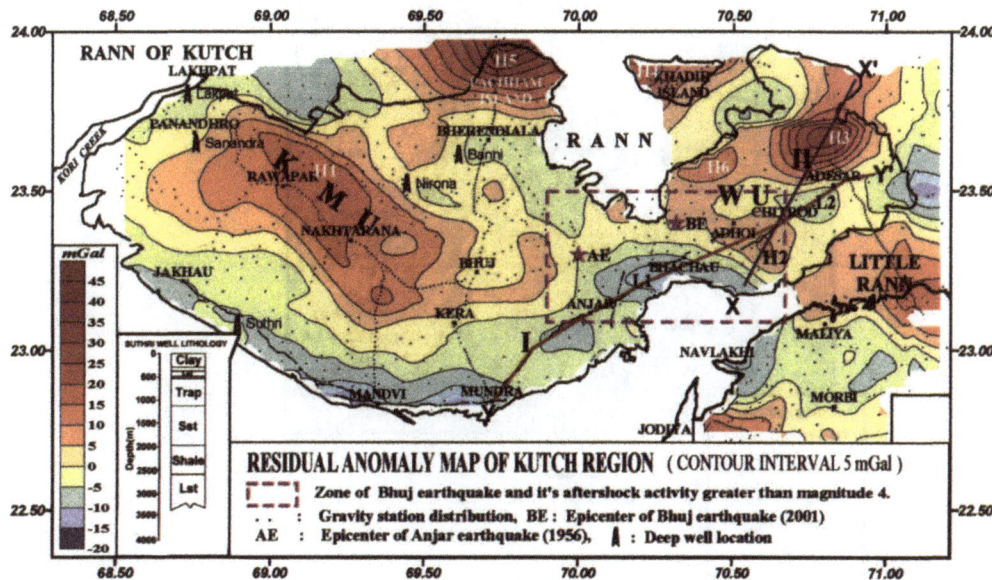

Fig. 8.12 Residual anomaly map with highs (H1-H5) and lows (L1-L5), representing same features as given in Fig. 8.9. Profiles, I (YY') and II (XX') are used to model shallow and deep seated structures.

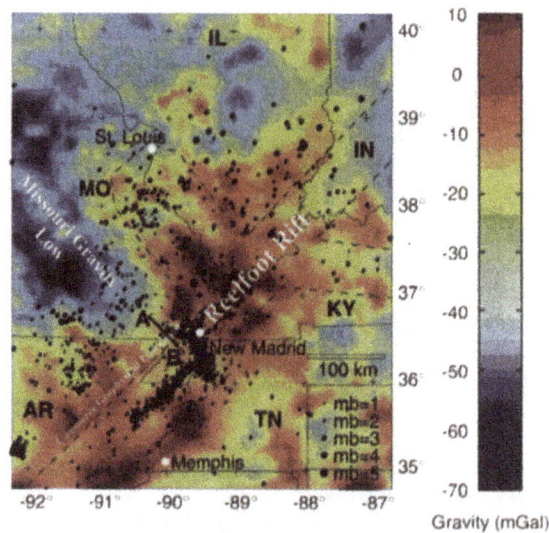

Fig. 8.19 Bouguer anomaly map of the New Madrid Seismic Zone, USA (Hildenbrand et al., 2001) showing a longitudinal gravity high along Reelfoot rift and some tranverse structures such as Missouri gravity low. Seismicity is concentrated at the junction of these structures.

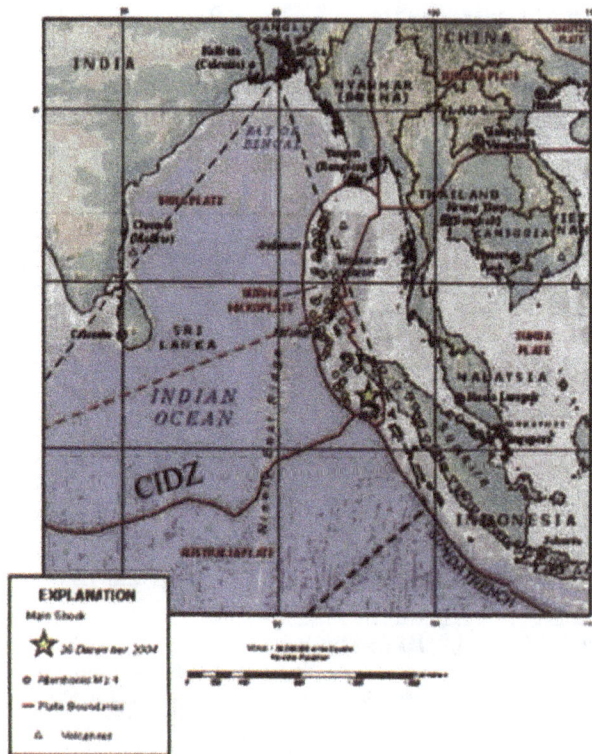

Fig. 8.20(a) Tectonic setting of North East Indian Ocean and epicentres of earthquake of 26 December 2004 and its aftershocks in the week following main shock (http://earthquake.usgs.gov/activity/past.html). The area between two red-dashed lines in the Indian Ocean is part of the Central Indian Ocean Deformation Zone (CIDZ). Superposed on it is the sedimentary zone in the Bay of Bengal and south of it due to rivers draining into the Bay of Bengal shown by black dashed lines triangle starting from south of Dhaka. The southernmost boundary of sediment cover almost coincides with southern boundary of CIDZ marked by black and red dashed lines respectively. The deformation zone is characterized by large earthquakes, east-west trending folds, thrusts, faults and fractures in sediments. It also shows high heat flow (Verzhbitsky et al., 1998) indicating recent subsurface tectonic activities in this section. (Mishra and Rajasekhar, 2005)

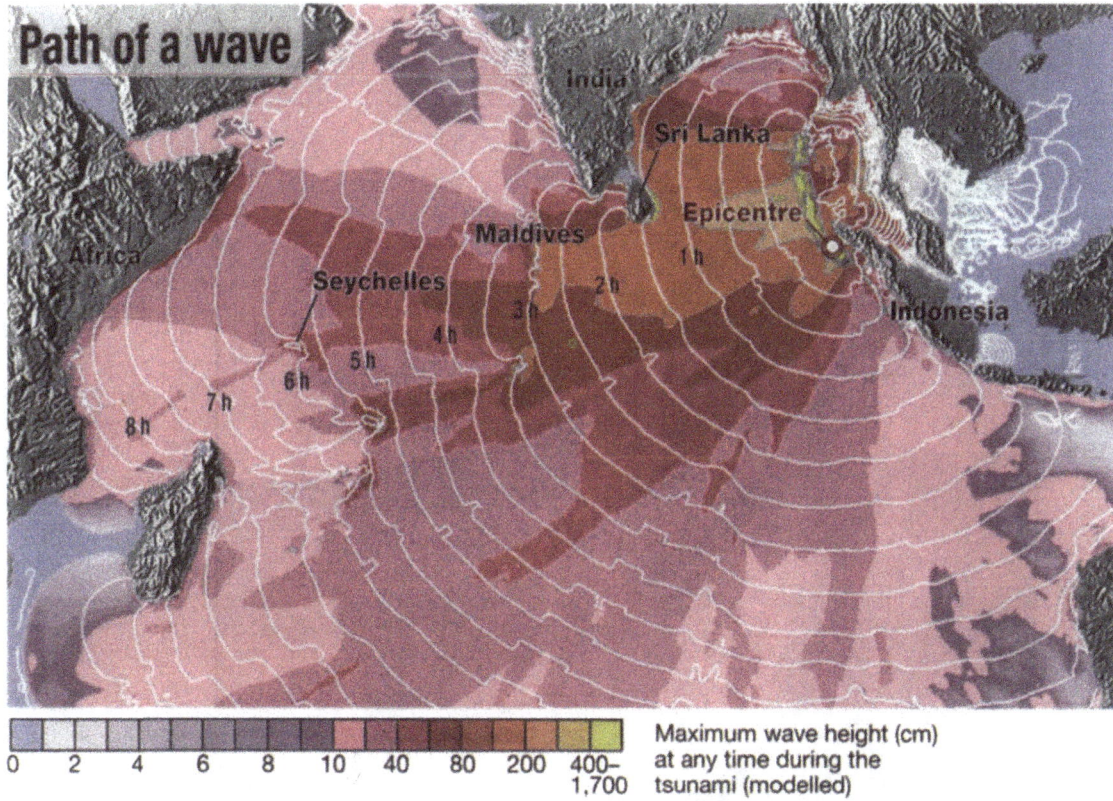

Fig. 8.22 Spread of Tsunami waves in meters in the Indian Ocean related to Sumatra earthquake of December 26, 2004.

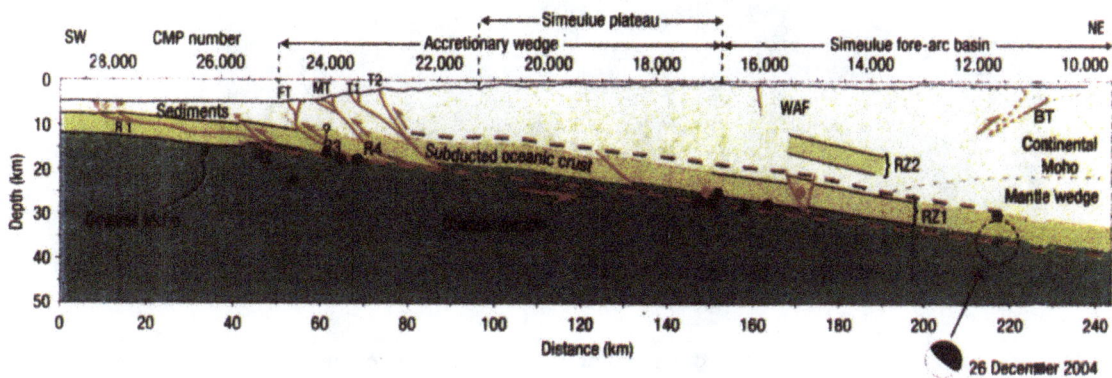

Fig. 8.23 Schematic depth cross section based on seismic reflection profile across epicentral zone of the Sumatra earthquake showing breakaway part of subducting slab displaced upwards (Singh et al., 2008).

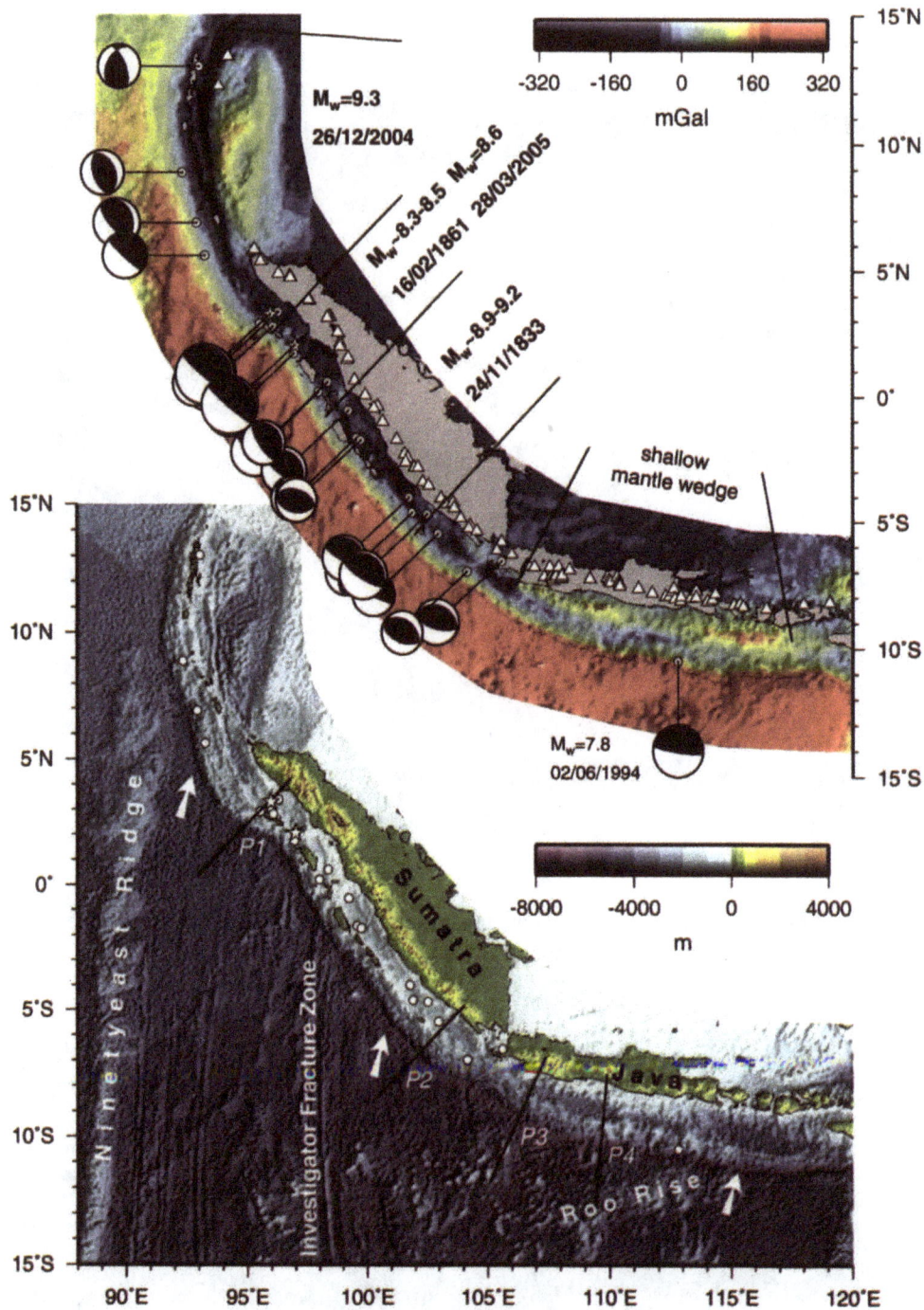

Fig. 8.24 Bouguer anomaly of Sunda trench from Satellite gravity (Grevemeyer and Tiwari, 2006) showing gravity low off Sumatra where most of the seismic activity are located and gravity high off Java which does not show any major seismic activity. Large thrust earthquakes in this region are from the Harvard University CMT catalog (Mw>6.5; 1976-2005; Dziewonski et al., 1981). Bottom map shows the tectonics depicting the Ninety East Ridge and Investigator Fracture Zone and crust between them is younger (~40-60 My) compared to the eastern part (~60-80 My).

SEDIMENTARY BASIN MAP OF INDIA

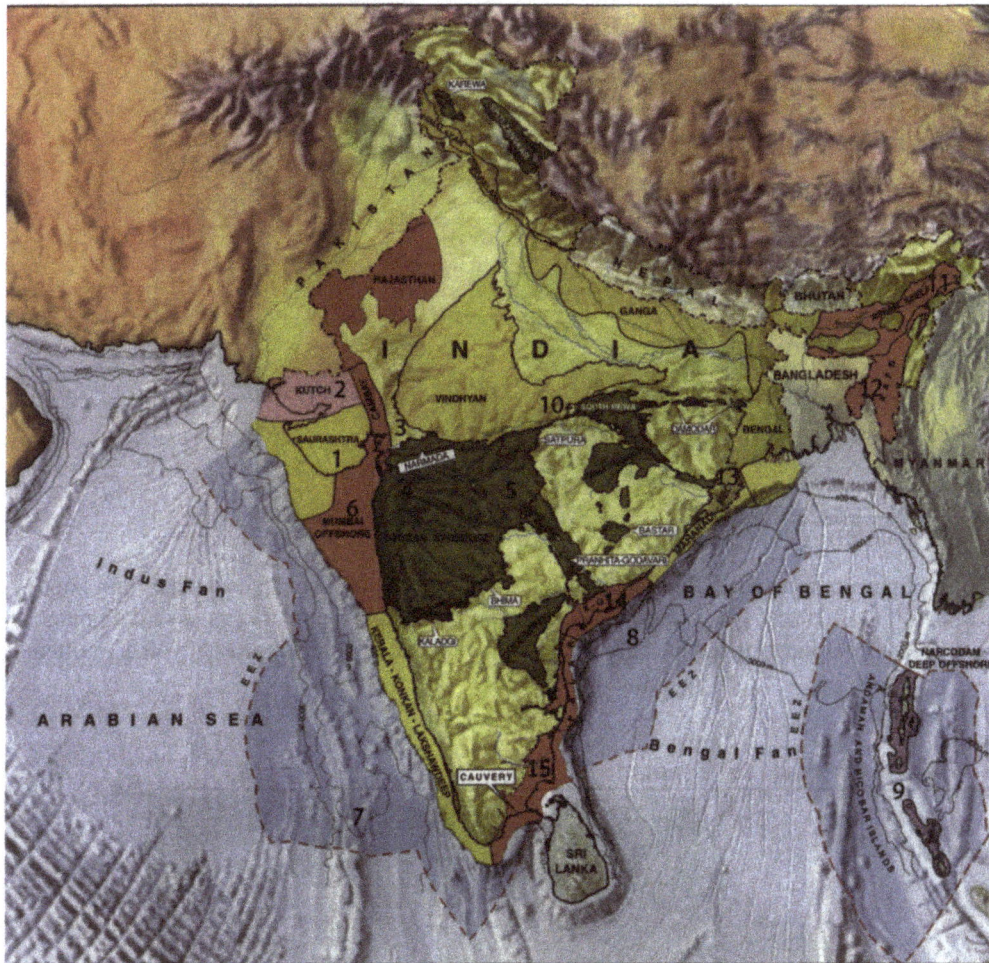

Fig 9.1(a) Major sedimentary basins of India (Srinivasan and Khar, 1995). Numbers 1-11 are regions that are discussed in regard to multi disciplinary geophysical investigations and specially gravity anomalies vis-à-vis hydrocarbon prospects. 1- Kutch, 2- Saurastra, 3- Cambay basin, 4- Narmada- Tapti Section (NTS), 5- Wardha basin, 6- Bombay high, offshore west coast of India, 7- Laccadive (Lakshadweep) Ridge, 8- Krishna-Godavari basin, offshore east coast of India, 9- Offshore Andaman-Nicobar islands, 10- Vindhyan basin and 11- Upper Assam Shelf. 12- Lower Assam Shelf, 13- Mahanadi basin, 14- Krishna-Godavari basin and 15- Cauvery basin.

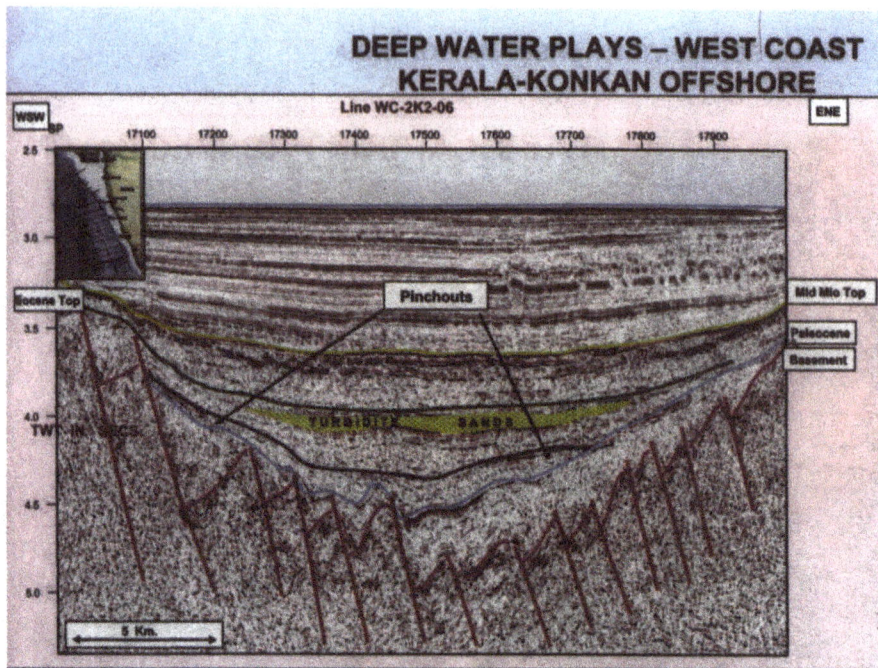

Fig 9.1(c) Examples of pinchouts and turbidite sand in deep water off Kerala-Konkan coast that form good stratigraphic prospects (DGH, 2005). The top of the basement and reflectors below it are seaward dipping reflectors that are different from those of sedimentary section above it and typically represent volcanic rifted margins world over.

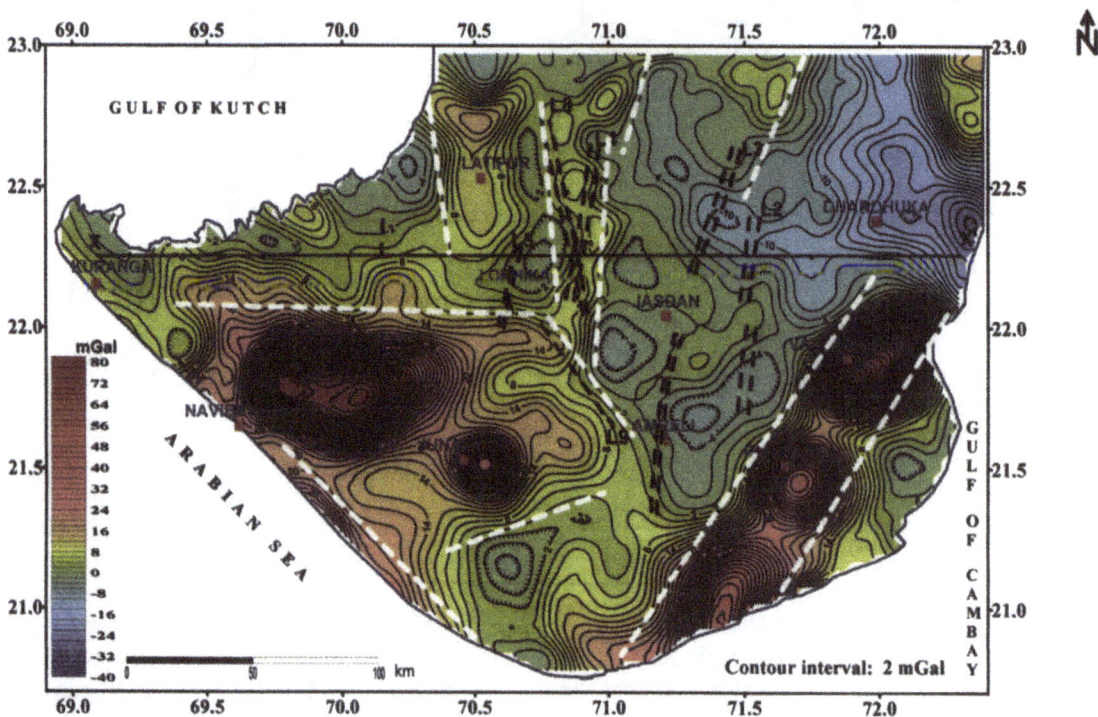

Fig 9.8 Residual gravity anomaly after subtracting the isostatic regional (Fig 9.7) from the observed field (Fig 9.5) that shows several gravity highs and lows related to shallow sources. Dashed lines indicate some major lineaments (NGRI, 1998; Singh and Arora, 2008).

Fig 9.20 High resolution airborne magnetic map of central Botswana covered by Karoo Volcanics (Reeves, 2010) that shows several lineaments and structures which can be used for multiple exploration programmes in this region. (Reeves, 2010)

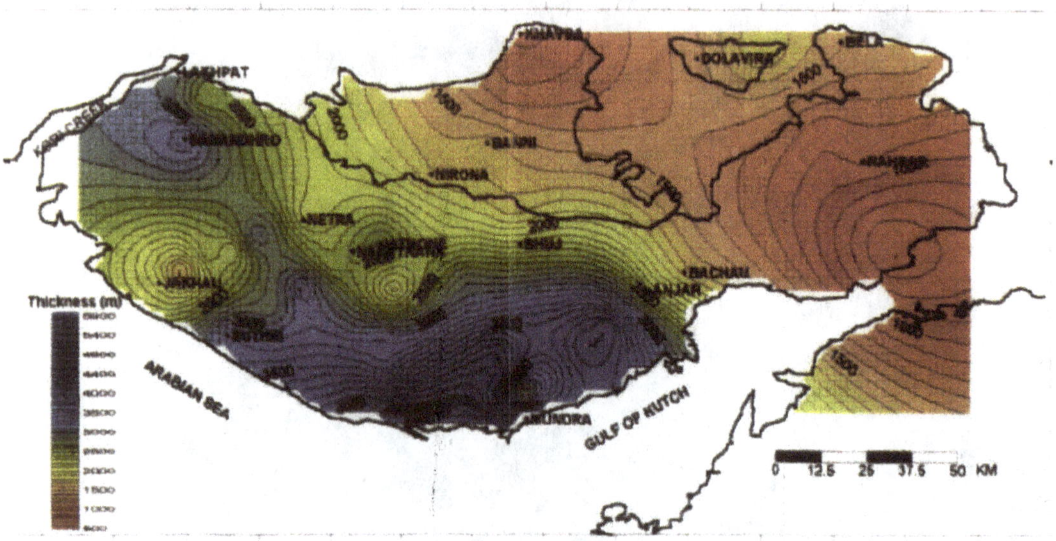

Fig 9.29(a) Thickness of Mesozoic sediments in Kutch (2, Fig 9.1a) based on integrated investigations using gravity, seismic, electrical and magnetotelluric methods with maximum thickness of ~4.5 km of Mesozoic sediment along the southern coast of Kutch with the Gulf of Kutch (NGRI, 2000; Singh and Arora, 2008).

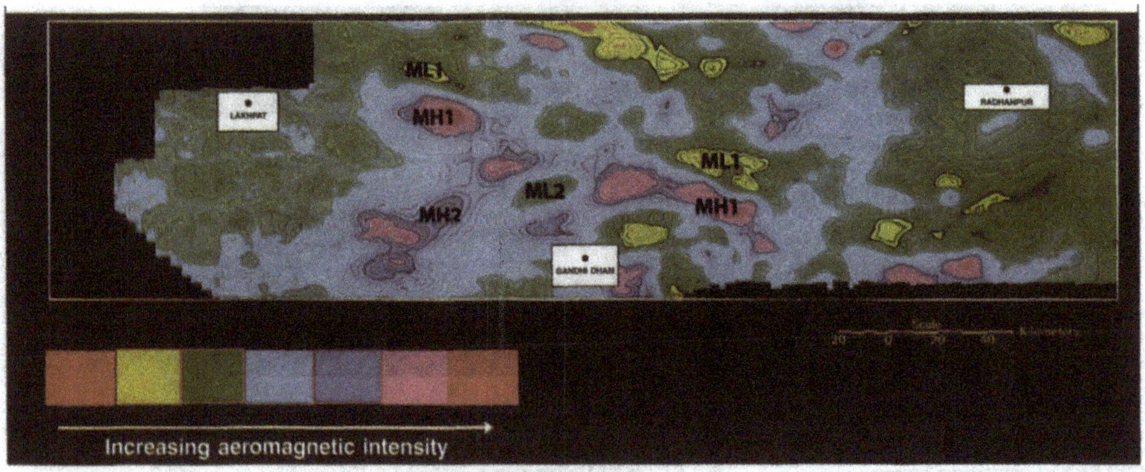

Fig 9.29(b) Airborne total intensity map of Kutch showing several pairs of magnetic highs and lows indicating mafic intrusions that might be responsible for various uplifts and islands of Kutch (Fig 8.7). There is a significant pair of linear magnetic trends, MH1 and ML1 that coincide with the Pachham island and Wagad uplift (DGH, 2005).

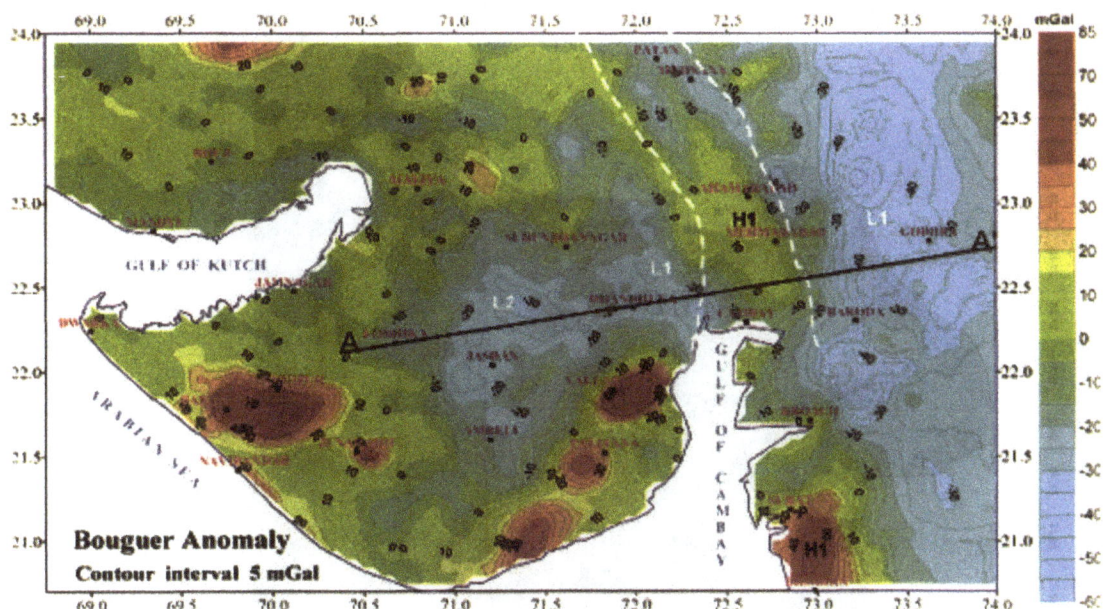

Fig 9.31(a) Bouguer anomaly map of Cambay basin (H1) and adjoining regions. Semi circular gravity highs are due to volcanic plugs of Saurastra and Kutch (Mishra et al., 2008, Singh and Mishra, 2000). AA' is a long profile modeled in Fig 9.33.

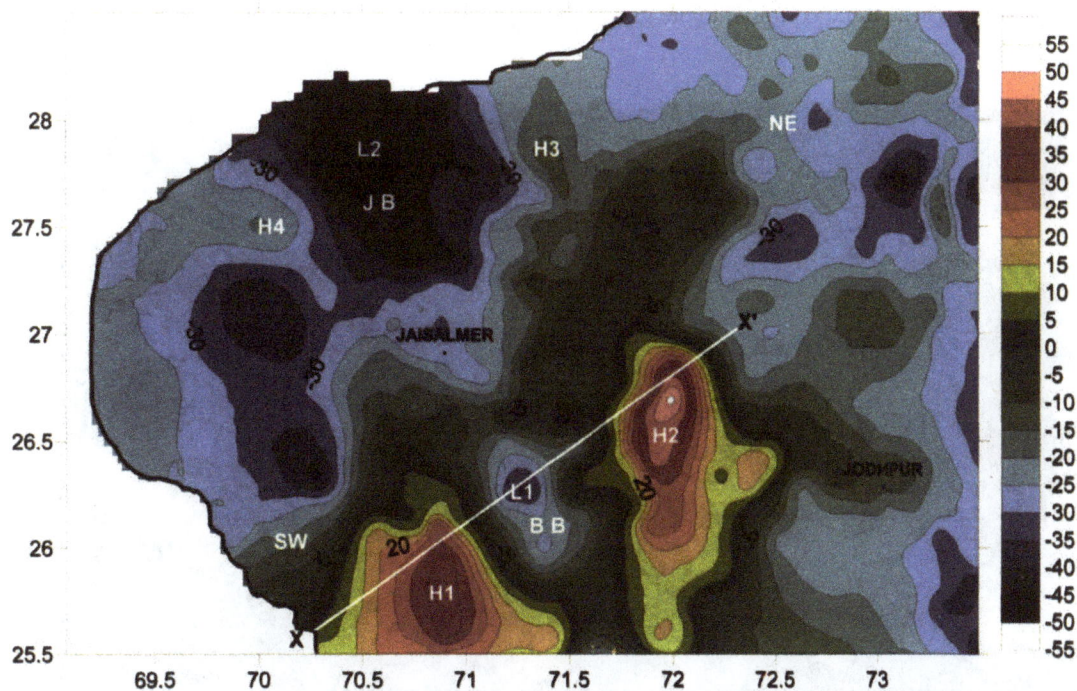

Fig. 9.31(b) Bouguer anomaly map of the Western Rajasthan showing gravity lows due to Barmer (L1) and Jaisalmer (L2) Basins and shoulder highs, H1 and H2 and H3, and H4, respectively. NE-SW gravity trend separates the two basins and appears to be of latter origin representing crustal and basement uplift due to lithoshheric flexure of the Indian plate on the western side (Western Fold Belt, Pakistan) known as Jaisalmer arch. BB – Barmer Basin; JB – Jaisalmer Basin.

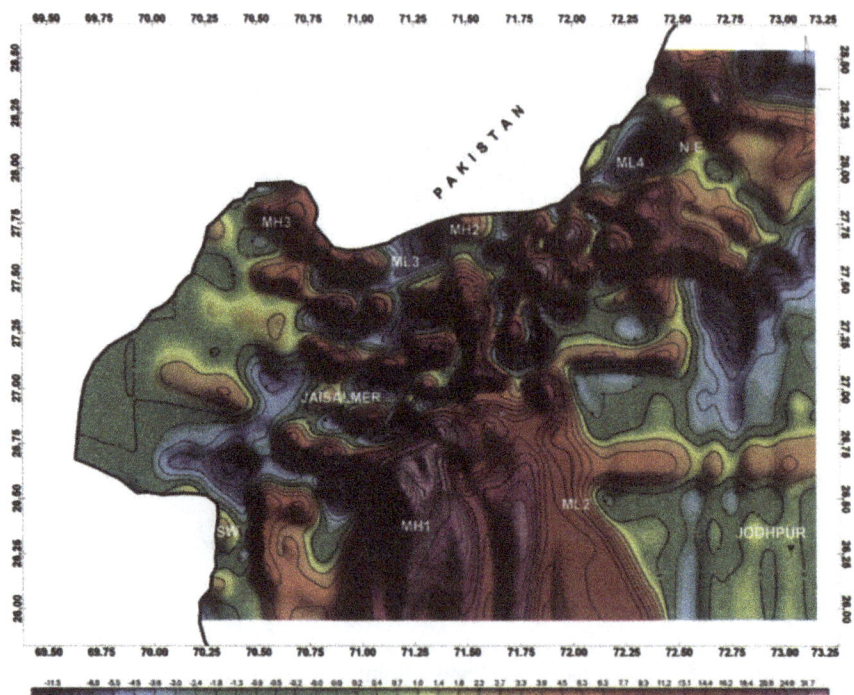

Fig. 9.31(c) Residual airborne magnetic map of parts of the Western Rajasthan based on data recorded along flight line spacing of 4 miles (~ 6.2 km) at an altitude of 1700 ft (~ 512 m) above mean sea level (Agocs, 1956a) showing magnetic anomalies due to Barmer (ML1, MH1 and ML2) and Jaisalmer (MH2, MH3, and ML33) basins. NE-SW magnetic trend is similar to the gravity trend in Fig. 9.31(b) related to Jaisalmer arch.

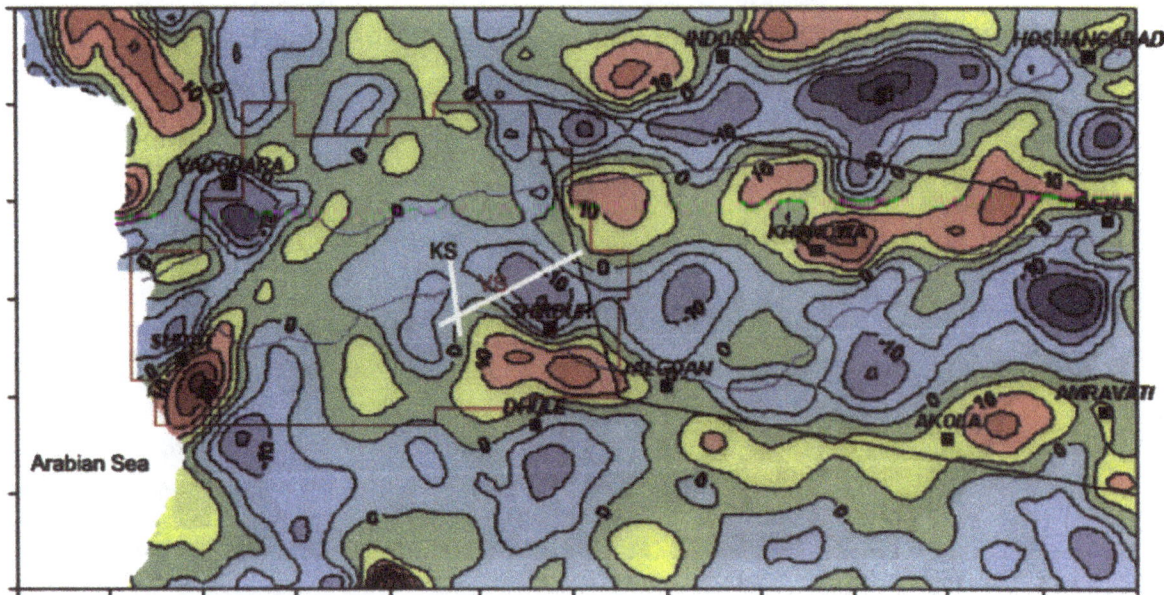

Fig 9.35 Residual Bouguer anomaly map of the western part of Narmada- Tapti Section (4, Fig 9.1a) referred to as Deccan Synclise of western part of Central India. It shows two sets of E-W oriented linear gravity highs with an intermediate set of gravity lows that might be related to subtrappean Mesozoic sediments. The gravity low south of Badodra represents Broach depression (Fig 9.34). White lines indicate profiles that are modeled , VS- Velda- Sendhwa profile and KS- Kothar- Sakri profile (NGRI, 2003, Singh and Arora, 2008).

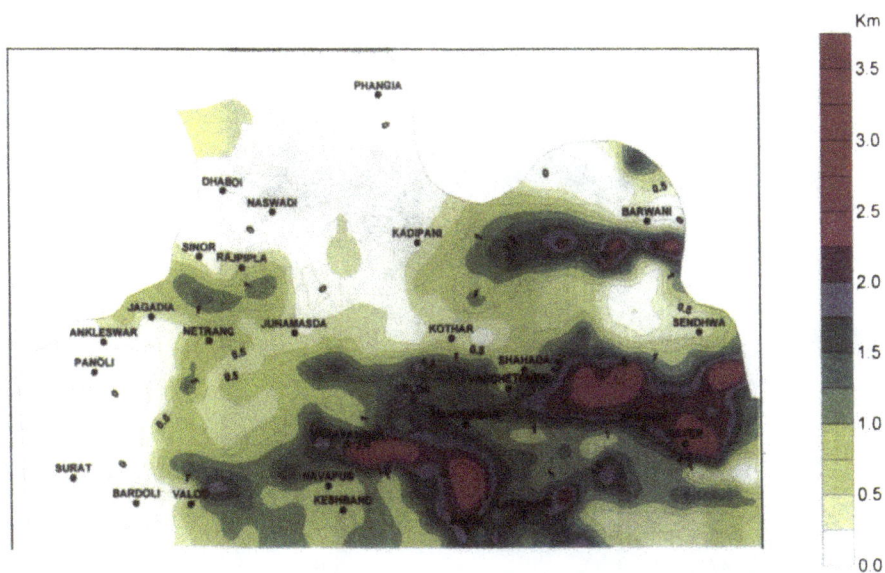

Fig 9.37(a) Sedimentary thickness map of Narmada- Tapti section (Western part of Deccan Sunclise in Central India) based on integration of all investigations, viz. gravity, seismic, electrical and magnetotelluric along several profiles showing maximum thickness of Mesozoic sediments as ~3 km in the eastern part (Saver) and Narayanpur-Sikri in the central part (NGRI, 2003, Singh and Arora, 2008).

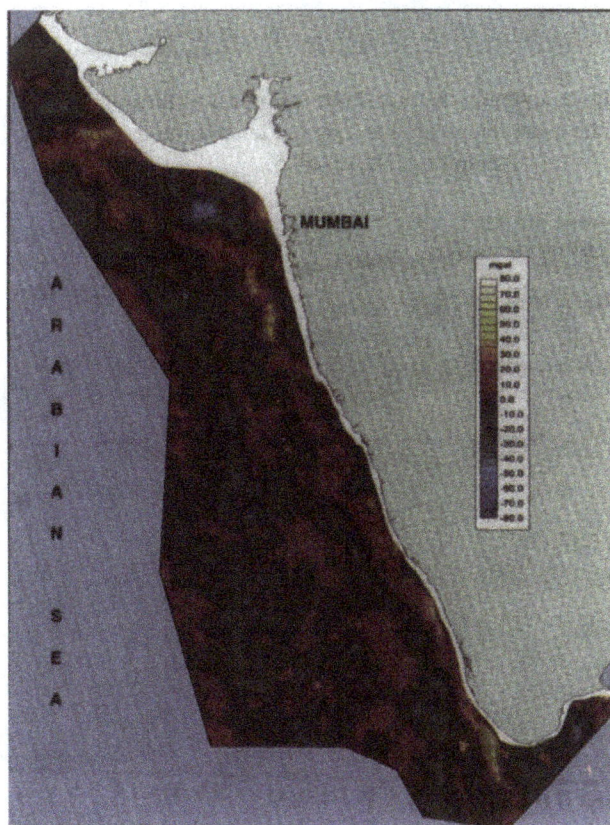

Fig 9.38 Satellite derived Bouguer anomaly map offshore west coast of India (6, 7, Fig 9.1a) showing linear gravity highs related to magmatic intrusion (Deccan trap) associated with the formation of rifted margin (DGH, 2005).

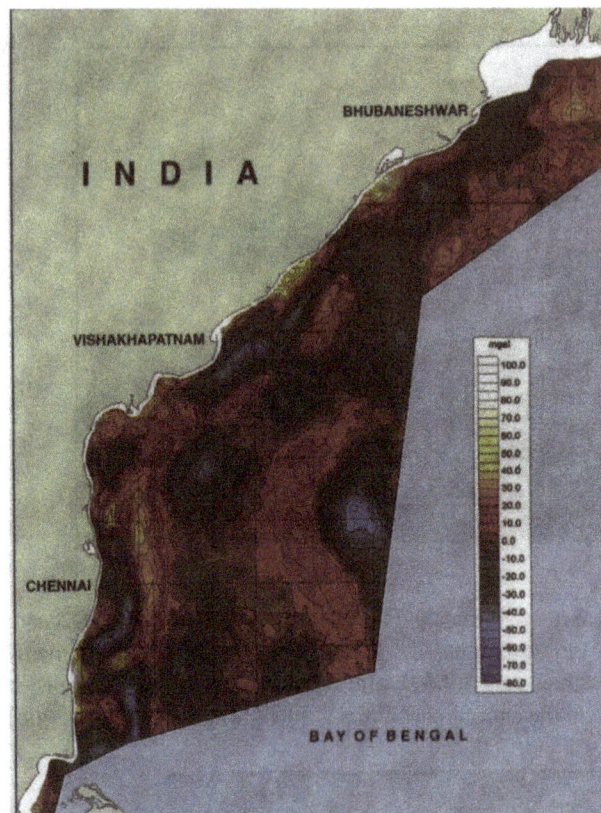

Fig 9.40(a) Satellite derived Bouguer anomaly map offshore east coast of India (8, Fig 9.1a) showing linear gravity highs almost parallel to the coast line related to magmatic intrusion during break up of India from Antarctica (DGH, 2005).

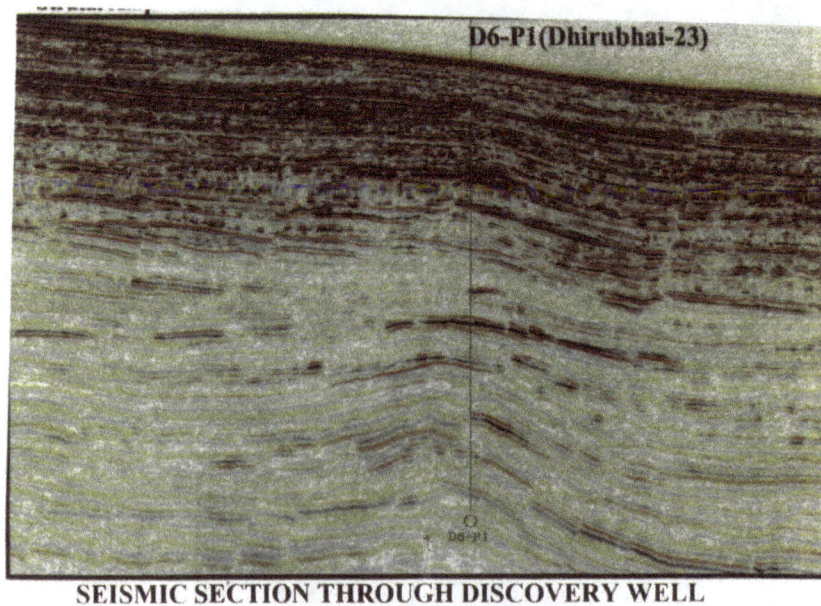

Fig 9.41(a) Seismic section through discovery well, Dhirubhai-23 in deep waters offshore Krishna-Godavari basin showing rollover anticline giving rise to one of the largest gas field (DGH, 2005).

Fig 9.41(b) A schematic geological section across Cauvery basin onshore and offshore showing basement ridges and depressions (Rama Rao et al., 2010)

Fig. 9.41(c) Satellite derived Bouguer anomaly map offshore Andaman-Nicobar islands (9, Fig. 9.1(a)) showing a curvilinear gravity low related to trench between the Indo-Australian and Eurasian plates suggesting islands to be part of accretionary prism. The gravity highs towards the west represent the Andaman marginal sea with oceanic crust and some centers of volcanic eruption as arc with basins in between to be explored for hydrocarbons (DGH, 2005).

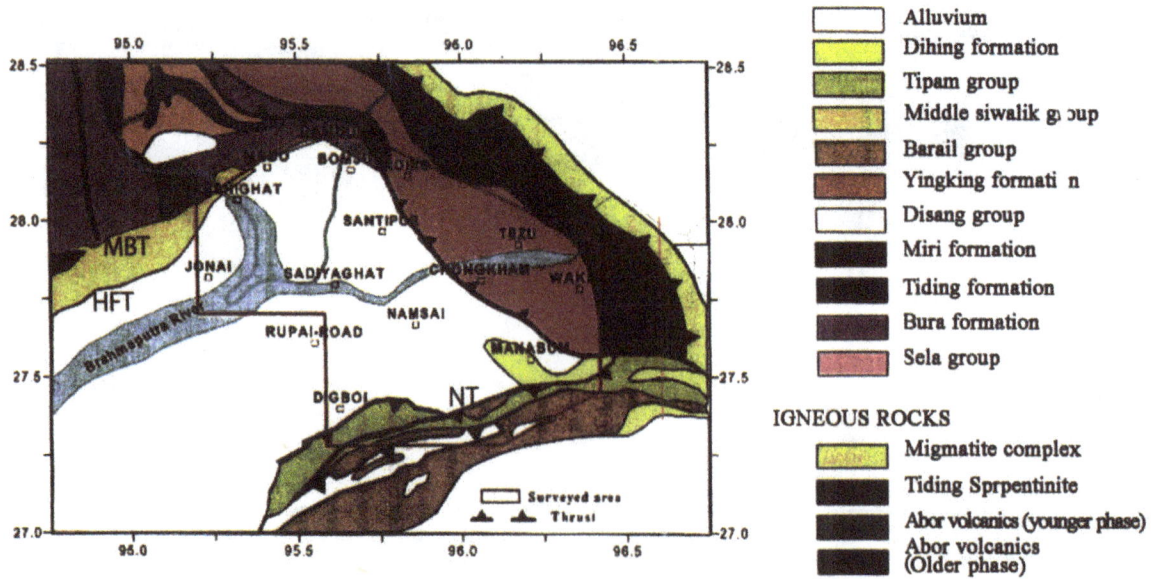

Fig. 9.47(a) A generalized tectonic map of upper Assam shelf (11, Fig. 9.1(a)). Most important elements are Main Boundary and Himalayan Frontal Thrusts (MBT and HFT) west of Pasighat and Misimi thrust west of Santipur and Naga thrust (NT) south of Manabum that are related to Eastern Syntaxes of Himalayan orogeny.

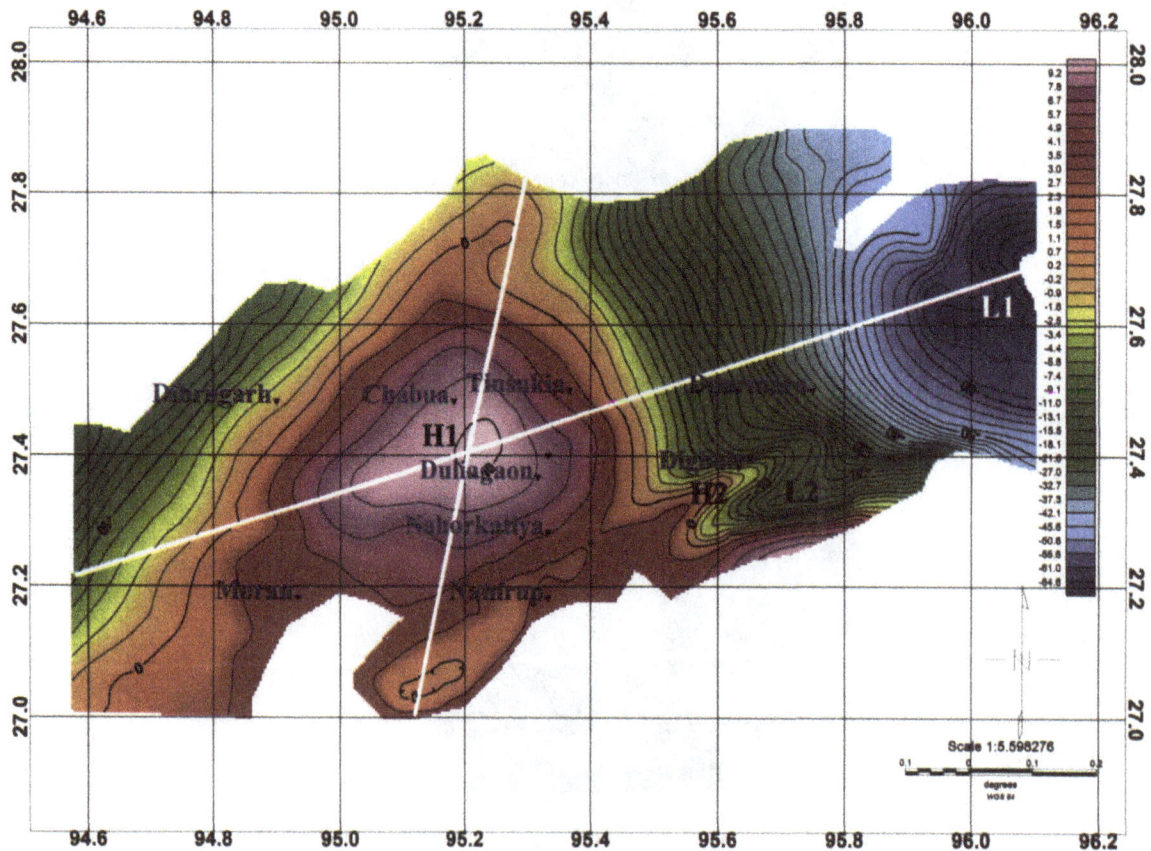

Fig. 9.47(b) Bouguer Anomaly map of Assam Shelf showing a gravity high in the central part and low along Himalayan fore deep. A constant regional field of –200 mGal is removed from the observed field (GSI, 1983, Volume II).

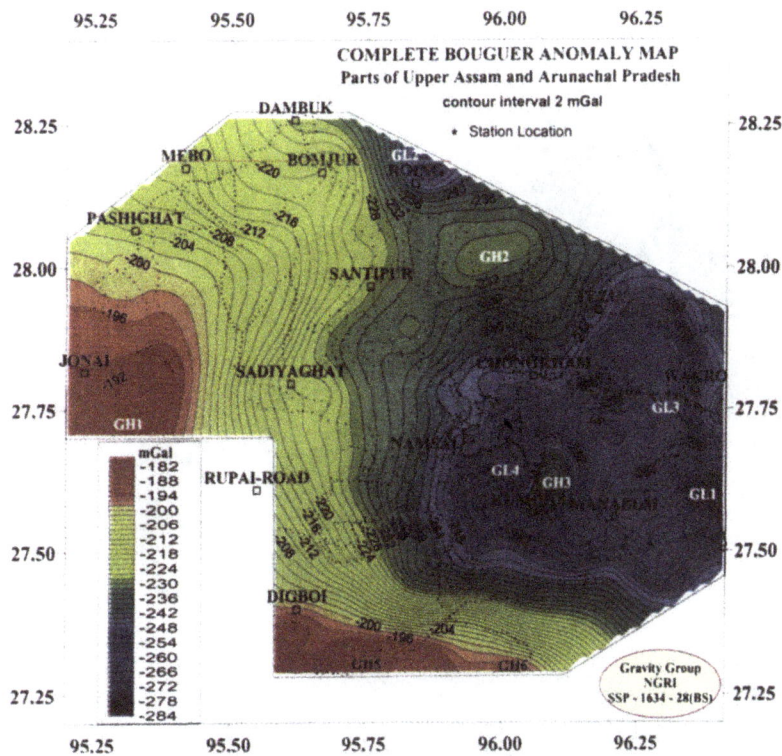

Fig 9.47(c) Complete Bouguer Anomaly of Upper Assam Shelf. H- Gravity Highs and L- Gravity lows.
(NGRI, 2000; Ghosh and siddiquee 2009)

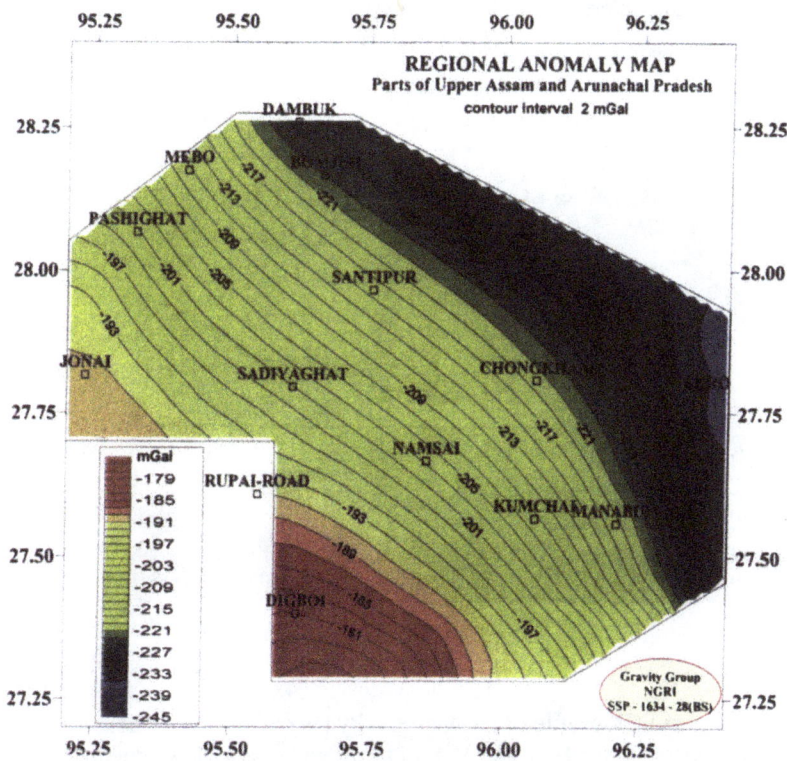

Fig 9.48(a) Regional anomaly corresponding to a first order polynomial approximation from Bouguer Anomaly
showing consistent gravity low towards the east due to crustal thickening.

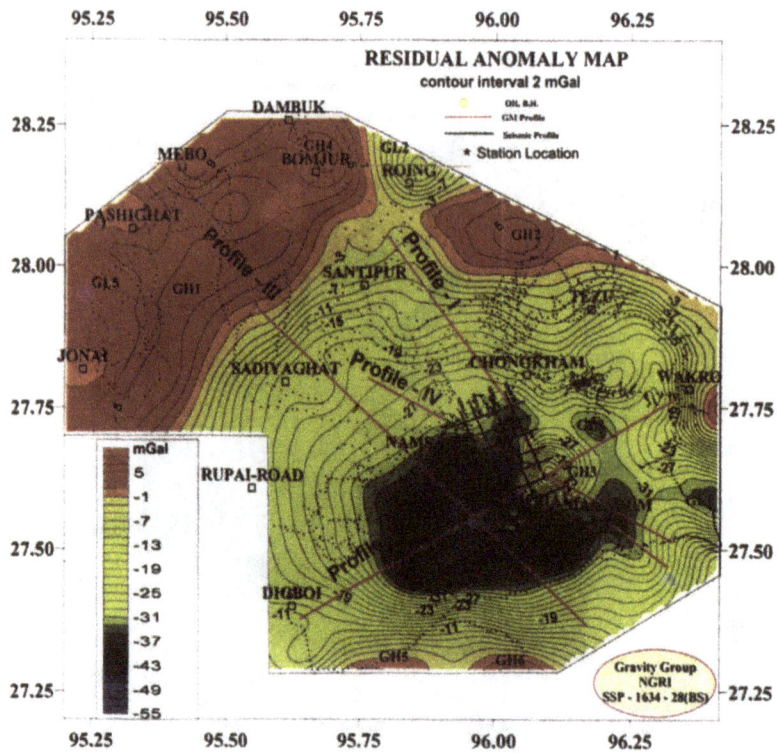

Fig. 9.48(b) Residual gravity anomaly after subtracting the regional field (Fig. 9.48(a)) from the observed field (Fig. 9.47(c)) showing gravity highs related to HFT and Misimi thrust on the western and the eastern sides, respectively. The gravity high, H4 is quite interesting as it represents a basement uplift in a large low due to a basin.

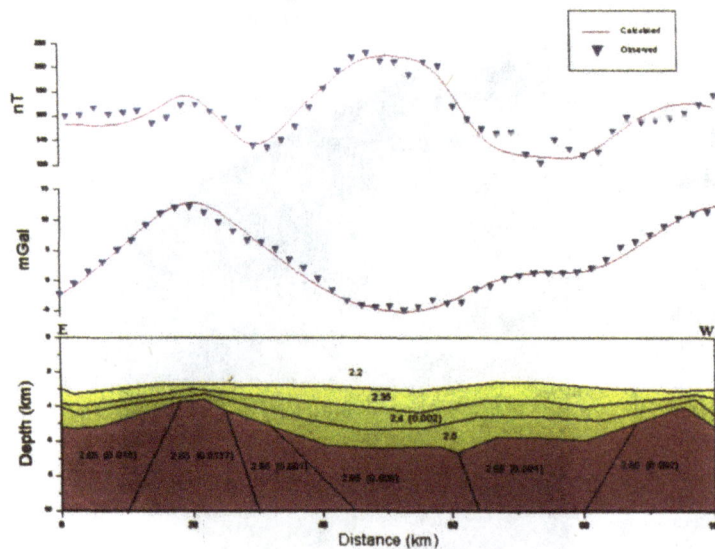

Fig 9.49(a) Basement model along an E-W profile of residual gravity anomaly showing horst kind of structures due to gravity highs, H1 and H2 related to the HFT and Misimi thrust and gravity lows due to Assam valley in between with a thick sediment of ~6 km (NGRI, 2000).

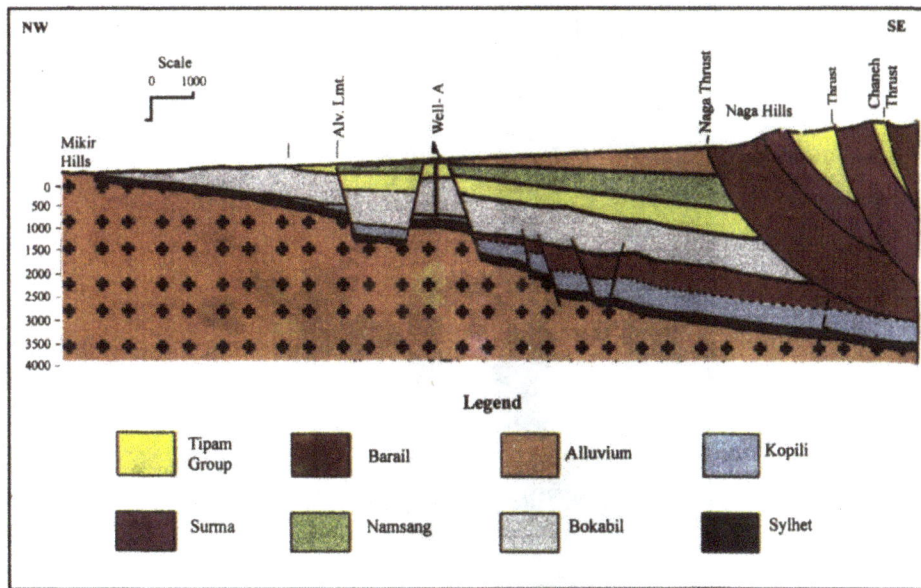

Fig 9.50(d) A seismo-geological section across Lower Assam Shelf (Dhansiri valley, Singh et al., 1996)) showing horst and graben structures in sedimentary section, typical of collision tectonics.

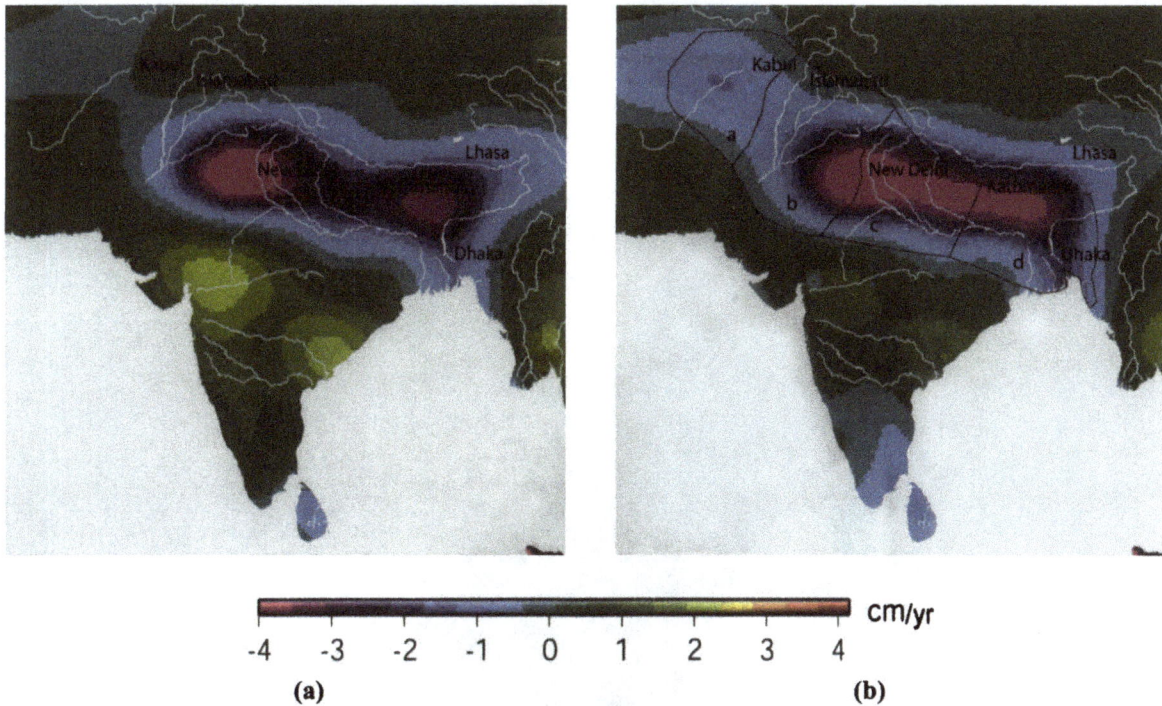

Fig 9.51 (a) Rate of change of terrestrial water storage in cm/yr of water thickness obtained from GRACE gravity solutions. White lines show major rivers. (Tiwari et al., 2009) **(b)** After subtracting from the naturally occurring water storage variability predicted from the CLM hydrological model from Fig. 9.46a. This provides the anthropogenically caused groundwater loss that is seen maximum in the north under Ganga basin which also happens to be most densely populated region and maximum exploitation of groundwater has taken place during last decade for agriculture in this section (Tiwari et al., 2009).

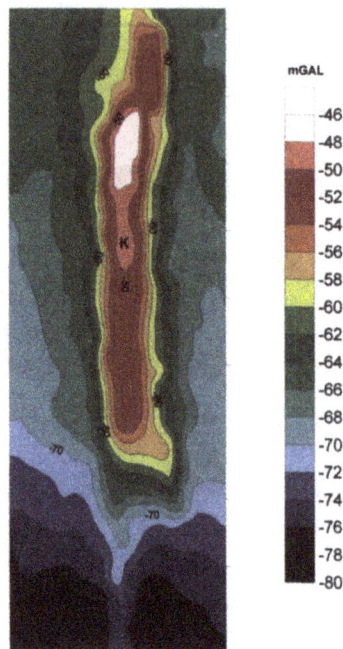

Fig. 9.54(c) Bouguer Anomaly map of a part of the Kolar Schist belt showing linear gravity highs over the schist belt and lows over the granite batholith towards the east (GSI – NGRI, 2006).

Fig 9.60(b)   High resolution airborne magnetic image of Kalyandurg area in Anantpur district west of Cuddapah basin and south of Wajrakarur (Fig 9.59) based on helicopter borne survey. It shows Kimberlite pipes (white dots) coinciding with magnetic lineaments that are nicely reflected in this map and such lineaments can also be used for groundwater investigations in hard rock areas. Sensor height- 40 m above ground surface and profile spacing- 50 m (Ram Babu and Rama Rao, 2008).

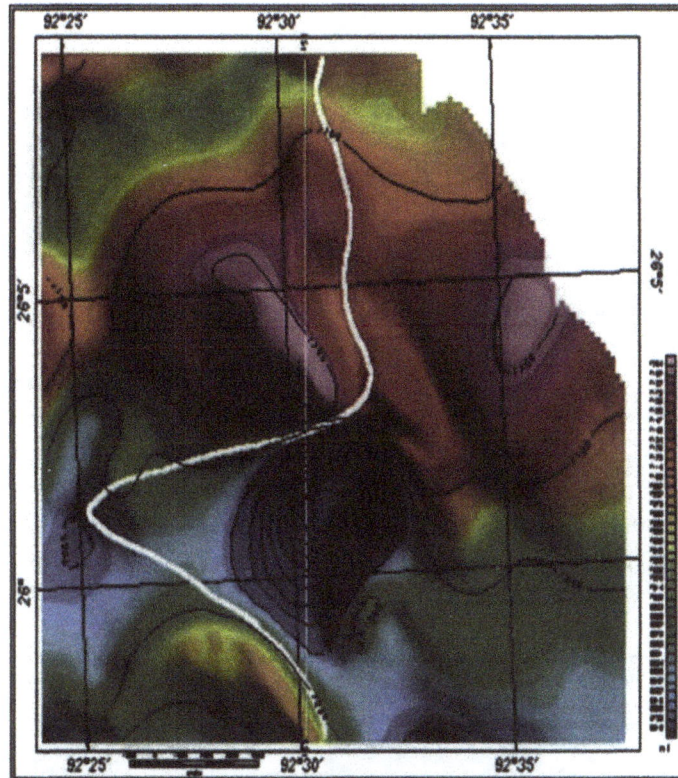

Fig 9.61(a) Airborne magnetic anomaly image of Jasra carbonatite complex and a profile across it in the NE part of Shillong Plateau showing magnetic low and a high towards the north indicating remanent magnetization. Fight altitude- 2100 m above msl and line spacing- 1 km (Ram Babu and Rama Rao, 2008).

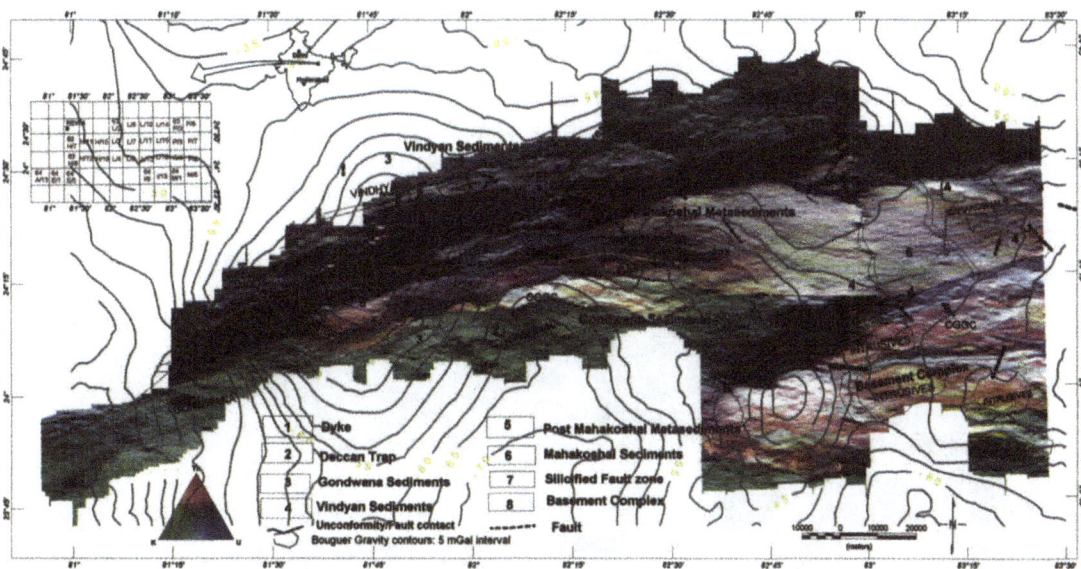

Fig 9.62 Ternary image of K, Th and U with superimposed Bouguer anomaly and geology of Mahakoshal basin, Central India ((M, Fig 7.20) (Chaturvedi et al., 2008). It shows linear bands of uranium anomaly in the northern part coinciding with gravity highs indicating mafic/ultramafic intrusive that characterizes Mahakoshal group of rocks (Section 7.6.2).

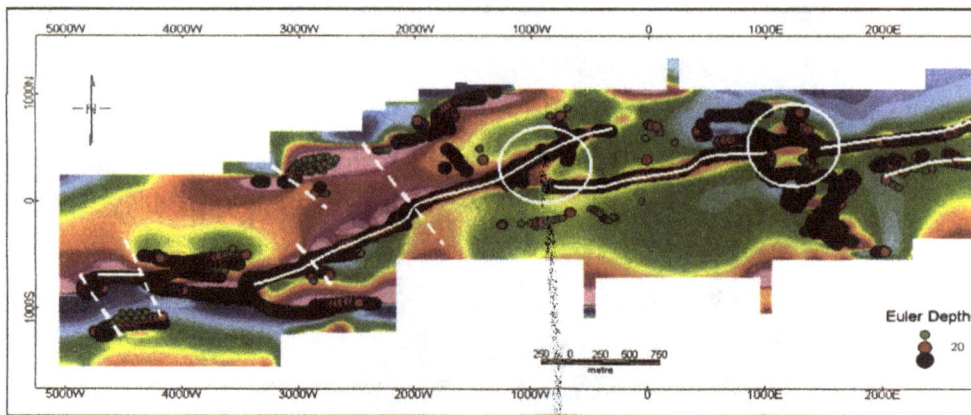

Fig 9.63(a) Total intensity image of the northern part of the Gondwana basin, Satpura reduced to pole with Euler deconvolution solutions showing sources coinciding with the linear magnetic trend along the junction between magnetic highs and lows defining a contact or a fault with mafic intrusive. Some faults are inferred based on changes in this linear magnetic trend and their intersections with the magnetic trend are important for mineral exploration (Markandeyulu et al., 2009).

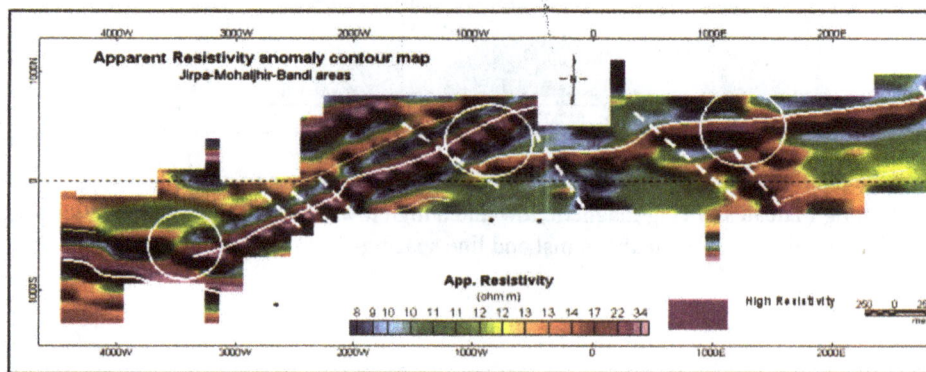

Fig 9.63(b) Apparent resistivity image of the same area as given in Fig 9.63a showing high resistivity zone along the magnetic linear trend indicating mafic intrusive with low porosity or associated quartz veins (Markandeyulu et al., 2009).

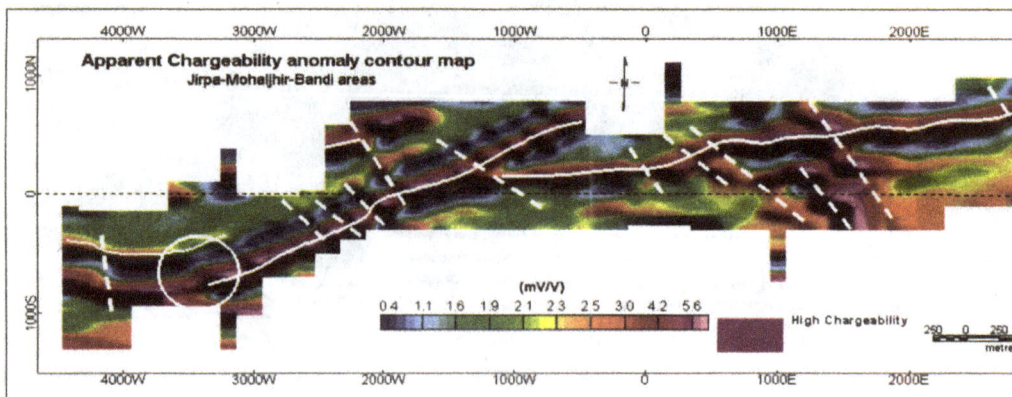

Fig 9.63(c) Apparent chargheability image based on induced polarization survey of the same block as given in Fig 9.63a. It shows high chargeability coinciding with the linear magnetic trend indicating association of sulphide mineralization (Markandeyulu et al., 2009).

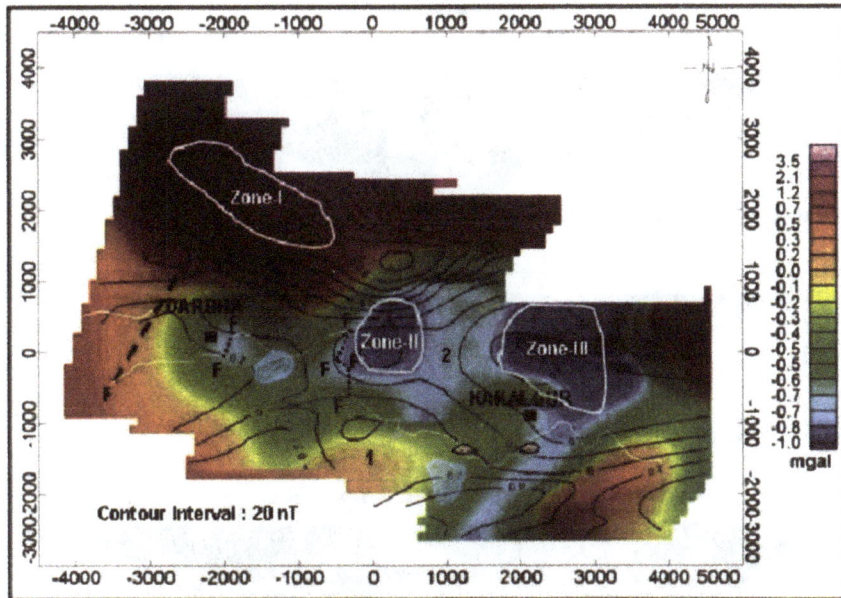

Fig 9.64 Total intensity magnetic contours superimposed over Bouguer anomaly image indicating magnetic and gravity highs (Zone I) related to mafic intrusive and magnetic and gravity lows (Zone II & III) related to fracture zone with felsic intrusive (Patra et al., 2009).

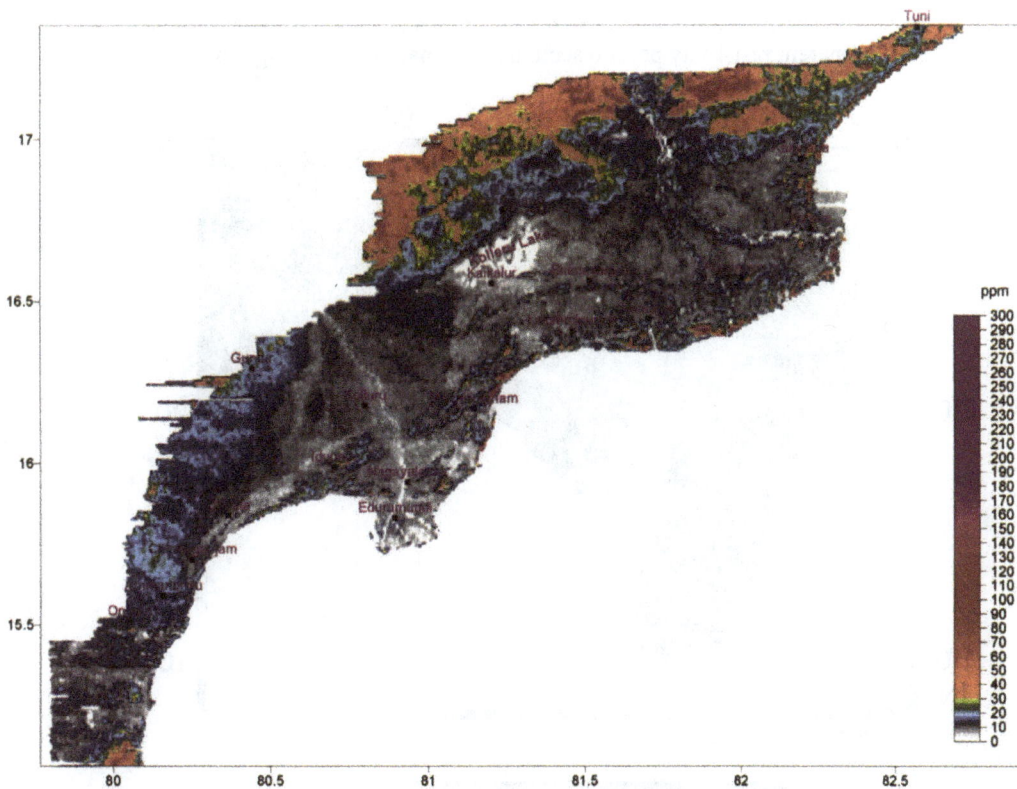

Fig. 9.65 Thorium image of coastal Godavari-Krishna basins based on airborne gamma spectrometer survey showing paleochannels of linear bands of blue and green colors range of thorium. Flight altitude: ~ 120 m and flight line (E-W) spacing. ~ 1 km (Rao et al., 2010).

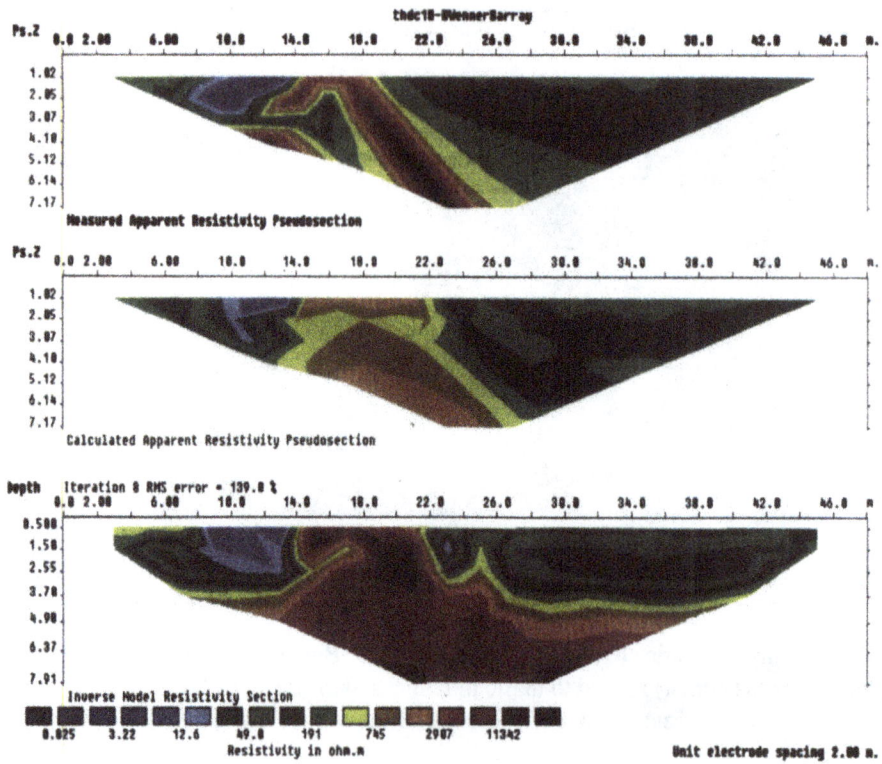

Fig 10.14 Apparent resistivity pseudo sections and final model through inversion.

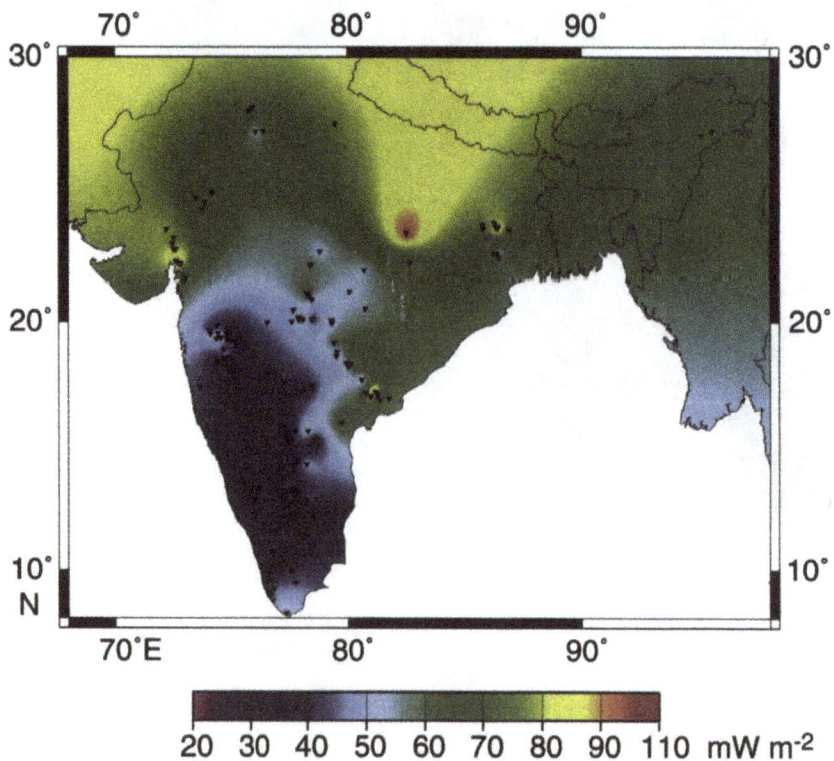

Appendix III Heat flow map of the Indian Shield showing high heat flow over the northern shield in general compared to southern shield (Dharwar craton). Heat flow sites are shown by filled triangles (Roy 2010).

Bibliography

Acharyya, S.K. and Roy, A., 2000, Tectonothermal history of the Central Indian Tectonic Zone and reactivation of major fault/shear zones, J. Geol. Soc. India, 55, 239-256.

Acharyya, S.K., 2003, A plate tectonic model for Proterozoic crustal evolution of Central Indian Tectonic Zone. Gondwana Geol. Magz., 7, 9-31.

Acharyya, S.K., 2007, Evolution of the Himalayan paleogene foreland basin, influence of its litho-packet on the formation of thrust-related domes and windows in the Eastern Himalaya – A review, Journal of Asian Earth Sciences, 31, 1-17.

Achuta Rao, D. and Sanker Narayan, P.V., 1981, Structural control of emplacement of kimberlite pipes at Panna - a suggestion from aeromagnetics, Geoexploration, 19, 207-228.

Achuta Rao, D. Ram Babu, H.V. and Sivakumar Sinha, G.D.J., 1993, Aeromagnetic anomaly: A guide to the location of the possible mid-continent rift related intrusive with sulphide mineralization in Narmada-son lineament belt, Abstract Volume, IGU Convention, 66-67.

Achuta Rao, D., Agarwal, P.K., Sarma, B.S.P., Sanker Narayana, P.V. and Hari Narain, 1970, Analysis of airborne magnetic profile Allahabad-Calcutta, Geophy. Res. Bull, 83-96

Achuta Rao, D., Babu Rao, V., Raju, D.Ch. and Sivakumarsinha, G.D.J., 1989, Analysis of a long range aeromagnetic profile from Nagpur-Bhopal, Proc. Seminar on Advances in Geophysical Research in India, Feb. 8-10, 1989, Indian Geoph. Union, 342-345.

Achuta Rao, D., Ram Babu, H.V. and Sivakumar Sinha, G.D.J., 1992, Crustal structure associated with Gondwana graben across the Narmada-Son lineament in India, an inference from aeromagnetics, Tectonophysics, 212, 163-172.

Achuta Rao, D., Ram Babu, H.V., Rama Rao, Ch., 1989, Geophysical discovery of Archeological sites at Nelakondapally in Andhra Pradesh, NGRI Tech. Rep. No: NGRI-88-EXP-52.

Achuta Rao, D., Ram Babu, H.V., Sinha, G.D.J.S. 1994, Aeromagnetic anomaly: A guide to the location of the possible mid continent rift related intrusive with sulphide mineralization in Narmada-Son lineament belt. Abstract Volume, Ind. Geoph. Union, Hyderabad.

Achuta Rao, D., Sivakumar Sinha, G.D.J., Ram Babu, H.V., 1996, Modelling of aeromagnetic anomaly and its implications on age of emplacement of ultramafic-mafic alkaline complex, at Jasra, Assam, Current Science, 71, 58-60.

825

Afonso, J. C., Ranalli, G. and Fernandez, M., 2007, Density structure and buoyancy of the oceanic lithosphere revisited, Geophys. Res. Letters, 34, L 10302, 1-5.

Agarwal, B.N.P. and Lal, T., 1972, Application of frequency analysis in two dimensional gravity interpretation, Geoexploration, 10, 91-100.

Agarwal, R.G. and Kanasewich, E.R., 1971. Automatic trend analysis and interpretation of potential field data, Geophysics, 36, 330-348.

Agcos, W.B., 1956, Aeromagnetic survey in the Indo-Gangetic Plains, O.N.G.C., Unpublished report.

Agocs, W.B., 1956a, Report on airborne magnetometer survey in Western Rajasthan for Oil and Natural Gas Commission, Ministry of Natural Resources and Scientific Research, Sponsored by Government of Canada under Colombo Plan.

Agocs, W.B., 1958, Results of airborne magnetometer profile from Browns Ville to Guatemala City, Geophysics, 23, 726-737.

Ahmad, Z. and Abbas, S.G., 1979, The Muslim Bagh ophiolites, In: Geodynamics of Pakistan, (eds) A. Farah and K.A. DeJong, Geol. Survey of Pakistan, Quetta, 243-249.

Aikman, A. B., Harrison, T. M., Lin, D., 2008, Evidence for early (>44Ma) Himalayan crustal thickening,Tethyan Himalaya, southeastern Tibet, Earth & Planet. Sci. Letters, 274,14-23.

AISPEI Regional Assembly in Asia. Aug.1-3, (Abstract),

Aitchison, J.C., Ali, J.R. and Davis, A.M., 2007, When and where did India and Asia collide?, J. Geoph. Res., 112, B05423, 1-19.

Akin, D., Demirel, H. and Gerstenecker, C., 1991, Repeated gravity observations in the Mudurnu valley, Turkey, Proc. of the International Conference on Earthquake Prediction-state of the art, Strasbourg, 396-402.

Akin, D., Demirel, H., Friedrich, J. and Gerstenecker, C.,1997, Determination of vertical strain changes using repeat gravity observations, Personal Communication.

Alam, M., Joseph, R., Curray, M., Rahman, L., Chowdhury, R. and Gani, M., 2003, An overview of the sedimentary geology of the Bengal Basin in relation to the regional tectonic framework and basin-fill history. Sedimentary Geology, 155, 179-208.

Alexander, D., 1993, Natural Disasters, UCL Press, London and Research Press, New Delhi, 1-632.

Alldredge, L.R., Van Vooris, G.D. and Davis, T.M., 1963, A magnetic profile around the world, J. Geoph. Res., 68, 3679-3692.

Allegre, C.J. et al., 1984, Structure and evolution of the Himalaya-Tibet orogenic belt, Nature, 307, 17-22.

Allegre, C.J., Birk, J.L., Capmes F. and Courtillot, V., 1999, Age of the Deccan traps using ^{187}Re-^{187}Os systematics, Earth and Planet. Sci. Lett, 170, 197-204.

Allen Osborne Associates, 1998, Rascal operating manual and Turbosurvey® Software-2, Allen Osborne Associates Inc., USA,

Alsdorf, D. and Nelson, D., 1999, Tibetan satellite magnetic low: Evidence for widespread melt in the Tibetan crust? Geology, 27, 943-946.

Ambraseys, N., 2000, Reasppraisal of north-Indian earthquakes at the turn of the 20th century, Current Science, 79, 1237-1250.

Ameen, S.M.M., Wilde, S.A., Kabir, Z., Akon, E., Choudhury, K.R. and Khan, S.H., 2007, Paleoproterozoic granitoids in the basement of Bangaladesh: a piece of the Indian Shield or an exotic fragment of the Gondwana jigsaw, Gondwana Research, 12, 380-387.

Amos, M. J. and Featherstone,W.E., 2006, Comparisons of recent global geopotential models with terrestrial gravity filed data over Newzealand and Australia, Geomatics Research Australasia, Online.

AMSE, 1995, Airborne total intensity magnetic map of 63 H; Airborne Mineral Surveys and Exploration Wing (GSI), Bangalore.

AMSE, 2001, Airborne total intensity magnetic map of 63 H; Airborne Mineral Surveys and Exploration Wing (GSI), Bangalore.

AMSE, 2002, Airborne Mineral Surveys and Exploration Wing of Geological Survey of India, Banglore, Special Publication, 75, Cover Page.

Anand, S.P. and Rajaram, M., 2004, Crustal structure of Narmada-Son lineament, an aeromagnetic perspective, Earth Planets Space, 56, e9-e12.

Anand, S.P. and Rajaram, M., 2004, Identification of uranium deposits through analysis of aeromagnetic data over Singhbhum, Exploration and Research for Atomic Minerals, 15, 121-126.

ANCORP Working Group, 2003.

Andersen, D.B. and Knudsen, P., 2001, Global marine gravity field from ERS-1 and Geosat geodetic mission altimetry, J. of Geoph. Res., 103 (C4), 8129.

Anil Kumar, Bhaskar Rao, Y.J., Sivaraman, T.V. and Gopalan, K., 1996, Sm-Nd ages of Archean meta-volcanics of the Dharwar craton, South India. Precambrian Res., 80, 205-216.

Antolik, M. and D. Dreger, 2003, Rupture process of the 26 January 2001 Mw 7.6 Bhuj, India, earthquake from teleseismic broadband data, Bull. Seismol. Soc. Am., 93, 1235-1248.

Argles, T., Foster, G., Whittington, A., Harris, N. and George, M., 2003, Isotope studies reveal a complete Himalayan section in the Nanga Parbat syntaxis, Geological Society of America, Geology 31, 1109-1112.

Argus, D.F. and Gorden, R.G., 1991, Geology, 19, 1085-1088.

Arora, B.R., Unsworth, M.J. and Rawat, G., 2007, Deep resistivity structure of the northwest Indian Himalaya and its tectonic implications, Geophys. Res. Letters, 34, L04307, doi:10.1029GL029165, 1-4.

Arora, K., Rajasekhar, R.P., Mishra, D.C., 2007, Density models of crust under Dharwar-granite greenstone terrain and the Satpura mobile belt: Archean-Proterozoic Analogue of plate tectonics. Mem. Gondwana Res., 10, 217-226.

Arora, K., Tiwari, V.M., Singh, B. and Mishra, D.C. and Grevemeyer, I., 2009, Three dimensional regional lithospheric structure and tectonic evolution of the western continental margin of India, under publication.

Arur, M.G., Bains, P.S. and Lal, J., 1983, Anomaly map of 'Z' component of Indian subcontinent from magnetic satellite data, presented at 18th General Assembly, IUGG, at Hamburg, 15-27 August.

Astier, J.L. and Paterson, N.R., 1989, Hydrogeological interest of aeromagnetic maps in crystalline and metamorphic areas, (ed.) G.D. Garland, Proc. of Exploration, 87, Ontario Geological Survey, Special Volume, 3, 732-745.

Athavale, R.N., Radhakrishnamurthy, C. and Sahasrabudhe, P.W., 1963, Paleomagnetism of some Indian rocks, Geoph. J. R. Astro. Soc., 7, 304-313.

Athavale, R.N., Rao, G.V.S.P. and Rao, M.S., 1980, Paleomagnetic results from two basic volcanic formations in the Western Himalayas and a Phanerozoic polar wander curve for India, Geophys. J. R. Astron. Soc., 60, 419-433.

Avasthi, D.N., Ramkotaiah, G., Varadarajan, S., Rao, N.D.J. and Behl, G.N., 1971, Study of the Deccan traps of Cambay basin by geophysical methods, Bull. Volcanol., 35, 743-749.

Avasthi, D.N., Singh, K. and Pandey, B.N., 2008, Geophysics in petroleum exploration in India: past, present and future, Mem. Geol. Soc. of India, 68, 213-236.

Avouac, J.P., 2007, Dynamic processes in extensional and compressional settings – mountain building: from earthquakes to geological deformation, Treatise on Geophysics, Crust and Lithosphere Dynamics, 6, (Ed) A.B. Watts, Elsevier B.V., The Neherlands, 377-439.

Azeez, A.K.K. and Harinarayana, T., 2007, Magnetotelluric evidence of potential geothermal resource in Puga, Ladakh, NW Himalaya, Current Science, 93, 323-329.

Azeez, K.K.A., Manoj, C., Veeraswamy, K. and Harinarayana, T., 2009, Co-Seismic E. M. signals in magnetotelluric measurements- a case study during Bhuj earthquake (26th January, 2001), India, Earth Planets Space, 61, 973-981.

Babu Rao, V., Atchuta Rao, D., Rama Rao, Ch., Sarma, B.S.P., Bhaskara Rao, D.S., Veeraswamy, K. and Sarma, M.R.L., 1987, Some salient results of interpretation of aeromagnetic data over Cuddapah basin and adjoining terrain, South India, 'Purana Basins of India', Geol. Soc. Ind., Mem. 6, 295-312.

Babu Rao, V., Sreedhar Murthy, Y. and Govindarajan, K., 1998, Contours to Images-Part II: Aeromagnetic image of western part of Cuddapah basin and adjoining crystallines- A case study, Jour. of Geophysics, 195-203.

Backus, G. and Gilbert, F., 1970, Uniqueness in the inversion of inaccurate gross earth data, Phil. Trans. R. Soc., London, 266 A, 123-192.

Bagchi, A. and Banerjee, B., 1983, Geophysical investigations in Sukinda – Dhenkanal-Keonjhar belt, Cuttack district, Orissa (India). Geol. Sur. of India, Special Publ. No. 2, pp.221-235.

Bahuleyan, K., Sarma, A.U.S., Venkataramani, M.S., Hari Rao, B., Borah, N.M., Sastry, P.S.S. and Singh, Y.N., 1999, Proc. of Golden Jubilee Seminar on Exploration Geophysics in India, Geol. Surv. India, Special Publication No. 49, 413-430.

Bai, D., Meju, M.A., Liao, Z., 2001, Magnetotelluric images of deep crustal structure of the Rehai geothermal field near Tengchong, Southern China, Geoph. J. Int., 147, 677-687.

Baksi, A. k., 1987, Critical evaluation of the age of the Deccan trap, India: implications for flood basalt volcanism and faunal extinctions, Geology, 15, 147-150.

Bandyopadhyay, A., 2004. Sedimentation, tectonics and palaeoenvironment in eastern Kachchh, Gujarat, J. Geol. Soc. Ind., 63, 171-182.

Banerjee, P. and Burgmann, R., 2002, Convergence across the northwest Himalaya from GPS measurements, Geophys. Res. Lett., 29 (13), 1652, doi:10.1029/2002GL015184.

Banerjee, P. and Satyaprakash, 2003a, Crustal configuration in northwestern Himalaya from gravity measurements along Kiratpur-Leh-Panamik Transect, J. Geol. Soc., India, 61, 529-539.

Banerjee, P. and Satyaprakash, 2003b, Crustal configuration across the north-western Himalaya as inferred from gravity and GPS aided geoid undulation studies, Journal of the Virtual Explorer, 12, 93-106.

Banerjee, P., 1998, Gravity measurements and terrain corrections using a digital terrain model in NW Himalaya, Computers and Geosciences, 24, 1009-1020.

Banerjee, P.K., 1982. Stratigraphy, petrology and geochemistry of some Precambrian basic volcanic and associated rocks of Singhbhum district, Bihar and Mayurbhanj and Keonjhar districts, Orissa, Memoir Geological Survey of India, 111, 54.

Banks, R.J, Parker, R.L. and Huestis, 1977, Isostatic compensation on a continental scale, local versus regional mechanism, Geophys. J.R. Astron. Soc., 51, 431-452.

Banks, R.J., Francis, S.C. and Hipkin, R., 2004, Effects of loads in the upper crust on the estimates of the elastic thickness of the lithosphere, Geophys. Journal International, 145, 291-299.

Bannert, D., 1992. The structural development of the western fold belt, Geologisches Jahrbuck Series, B 80, 3-60.

Bansal, A.R., Dimri, V.P. and Vidya Sagar, G., 2006, Depth estimation from gravity data using the maximum entropy method (MEM) and the multitaper method (MTM), Pure & Appl. Geophys., 163, 1417-1434.

Baranowski, J., Armbuster, J., Seeber, L. and Molnar, P., 1984, Focal depths and fault plane solutions of earthquakes and active tectonics in the Himalaya, J. Geophys. Res., 89, 6918-928.

Barnett, C.T., 1976, Theoretical modeling of the magnetic and gravitational fields of an arbitrary shaped three dimensional body, Geophysics, 41, 1353-1364.

Barton, C. E., 1983, Analysis of paleomagnetic time-series techniques and applications, Geophysical Surveys, 5, 335-368.

Barton, P.J., 1986, The relationship between seismic velocity and density in the continental crust-useful constraint, Geophys. J. R. astr. Soc., 87, 195-208.

Baruah, M.K., Joshi, G., Mukherjee, B. and Verma, K., 2009, Tectono- stratigraphic insights from structural balancing in the sub Himalaya zone near Solan, H.P., India and implications in hydrocarbon exploration, Presented in 46 Annual Convention of Indian Geophysical Union at Wadia Institute of Himalayan Geology, October 5-7, 2009.

Basu, A., Patranabis-Deb, S., Schieber, J. and Dhang, P.C., 2008, Stratigraphic position of the ~1000 Ma Sukhinda Tuff (Chhattisgarh Supergroup, India) and the 500 Ma question, Precambrian Research, 167, 383-388.

Basu, A.R., Renne, P.R., Dasgupta, D.K., Teichmann, F. and Poreda, R.J., 1993, Early and late alkali igneous pulses and a high - ^{3}He plume origin for the Deccan flood basalts, Science, 261, 902-906.

Beaumont, C., 1978, The evolution of sedimentary basins on a viscoelastic lithosphere: Theory and examples, Geophys. J. Roy. Astr. Soc., 55, 471-497.

Beaumont, C., 1981, Foreland basin, Geoph. J. Roy. Astr. Soc., 65, 291-329.

Beaumont, C., Jamieson, R.A., Nguyen, M.H. and Medvedev, S., 2004, Crustal channel flows: 1, Numerical models with applications to the tectonics of the Himalayan-Tibetan orogen. Journal of Geophysical Research, 109(B06406), doi:10.1029/2003JB002809.

Beaumont, C., Jamieson, R.A., Nguyen, M.N. and Lee, B., 2001, Himalayan tectonics explained by extrusion of a low-viscosity crustal channel coupled to focused surface denudation, Nature, 414, 738-741.

Bechtel, T.D., Forsyth, D.W. and Swain, C.J., 1987, Mechanisms of isostatic compensation in the vicinity of the East African Rift, Kenya, Geoph. J.R. astron. Soc., 90, 445-465.

Bechtel, T.D., Forsyth, D.W., Sharpton, V.L. and Grieve, R.A.F., 1990, Variations in effective elastic thickness of the North American lithosphere, Nature, 343, 636-638.

Beck, R.A., Burbank, D.W. and Sercombe, W.J. et al., 1995, Stratigraphic evidence for an early collision between north west India and Asia, Nature, 373, 55-58.

Bedrosian, P.A., 2001, Structure of the Altyn Tagh Fault and Daxue Shan from magnetotelluric surveys: Implications for faulting associated with the rise of the Tibetan Plateau, Tectonics, 20, 4, 474-486.

Behn, M. D., Hirtz, G., Keleman, P. B., 2007, Trench parallel anisotropy produced by foundering of arc lower crust, Science, 317, 108-111.

Bell, R.E., 1997, Gravity gradiometry resurfaces, the Leading Edge, 16, 55-59.

Bendat, J.S. and Peirsol, 1993, Engineering applications of correlation and spectral analysis, 2nd ed. John Wiley & Sons, New York.

Bendick, R. and Flesch, L., 2007, Reconcilling lithospheric deformation and lower crustal flow beneath central Tibet, Geology, 35, 895-898, doi: 10.1130/G23714A.1.

Bendick, R., McKenzie, D. and Etienne, J., 2008, Topography associated with crustal flow in continental collisions, with application to Tibet, Geophys. J. Int., 175, 375-385.

Bentley, C. R., 1991, Configuration and structure of the subgalcial crust, In: The Geology of Antarctica (Ed) R J. Tingey, Clarendon press, Oxford, 335-364.

Berger, A., 1988, Milankovitch theory and climate, Rev. Geophys. 26, 624-657.

Bernard, P. et al, 2003. Observing earthquake related dewatering using MISR/Terra satellite data, EOS Trans. AGU, 84, 37&43.

Besse, J. and Courtillot, V. 1998. Paleogeographic maps of the continents bordering the Indian Ocean, since the early Jurassic, J. Geophys. Res., 93, 11791-11808.

Besse, J. and Courtillot, V., 1991, Revised and synthetic apparent polar wander paths of African, Eurasian, North American and Indian plates and true polar wander since 200 Ma, J. Geophys. Res, 96, 4029-4050.

Bettinelli, P., Avouac, J.P., Flouzat, M. et al., 2006, Plate motion of India and interseismic strain in the Nepal Himalaya from GPS and DORIS measurements, J. of Geodesy, 80, 567-589.

Bettinelli, P., Avouac, J.P., Flouzat, M., Bollinger, L., Ramillien, G., Rajoure, S. and Sapkota, S., 2008, Seasonal variations of seismicity and geodetic strain in the Himalaya induced by surface hydrology, Earth & Planet. Sci. Letters, 266, 332-344.

BGR., 1992, Heliborne survey of Cholistan by Bundesanstalft fur Geowissenschaften, Germany.

Bhalla, M.S., Rao, G.V.S.P., Hansraj, A., Rao, M.S. and Rao, N.T.V.P., 1979, Paleomagnetic results of Indian rocks – A review, Geoph. Res. Bull, 17, 159-272.

Bhandari, L.L., Venkatachala, B.S., Kumar, R., Swamy, S.N., Garga, P. and Srivastava, D.C. ,1983, Petroliferous basins of India. Petroleum Asia Journal, 6, 1-189.

Bhaskar Rao, Y.J., Janardhan, A.S., Vijaya Kumar, T., Narayana, B.L., Dayal, A.M., Taylor, B.N. and Chetty, T.R.K., 2003, Sm-Nd model ages and Rb-Sr isotopic systematics of charnockites and gneisses across the Cauvery shear zone, Southern India: implications for the Archaean-Neoproterozoic terranes boundary in the southern granulie terrain, Memoir Geological Society of India, 50, 297-317.

Bhatt, S.C. and Hussain, A., 2008, Structural history and fold analysis of basement rocks around Kuraicha and adjoining areas, Bundelkhand Massif, Central India, 72, 331-347.

Bhattacharji, J.C., 1973, Discussion on reduction of gravity data in India, Geophys. Res. Bull., 11, 153-155.

Bhattacharji, S. and Singh, R.N., 1984, Thermo-mechanical structures of the southern part of the Indian Shield and its relevance o Precambrian basin evolution, Tectonophysics, 105, 103-120.

Bhattacharji, S., Chatterjee, N., Wampler, J.M., Nayak, P.N. and Deshmukh, S.S., 1996 Indian intraplate and continental margin rifting, lithospheric extension, and mantle upwelling in Deccan flood basalt volcanism near the K/T boundary: evidence from mafic dike swarms. Journal of Geology, 104, 379-398.

Bhattacharya, B.K., 1964, Magnetic anomalies due to prism shaped bodies with arbitrary polarization, Geophysics, 29, 517-531.

Bhattacharya, B.K., 1966, Continuous spectrum of total magnetic field anomaly due to rectangular prismatic body, Geophysics, 31, 97-121.

Bhattacharya, B.K., 1980, A generalized multi body model for inversion of magnetic anomalies, Geophysics, 45, 255-270.

Bhattacharya, G.C.,Chaubey, A.K., Murty, G.P.S., Srinivas, K., Sarma, K.V.L.N.S., Subrahmanyam, V. and Krishna, K.S., 1994, Evidence for seafloor spreding in the Laxmi basin, northeastern Arabian Sea, Earth and Planetary Science Letters, 125, 211-220.

Bhattarcharya, J.C., 1995, Current Science, 69, 486-488.

Bhimsankeram, V.L.S., 1975, Paleomagnetism in India – A review, Geophys. Res. Bull, 13, 13-28.

Bhosle, B., Prakash, B., Awasthi, A.K., Singh, S. and Khan, M.S.H., 2008, Role of extensional tectonics and climatic changes in geomorphological, pedological and sedimentary evolution of the western gangetic plain (Himalayan Foreland Basin), India, Himalayan Geology, 29,1, 1-24.

Bhukta, S.K. and Tewari, H.C., 2007, Crustal seismic structure in Jammu and Kashmir region, India, J. of Geophys. Soc. India, 69, 755-764.

Bhukta, S.K., Sain, K. and Tewari, H.C., 2006, Crustal structure along the Lawrencepur-Astor profile in the northwest Himalaya, Pure and Applied Geophysics, 163, 1257-1277.

Bilham, R and Wallace, K, 2005, Future Mw>8 earthquakes in the Himalaya: Implications from 26 Dec 2004 Mw = 9.0 earthquake on India's eastern plate margin, Geological Survey of India Special Publication, 85, 1-14.

Bilham, R, 2004, Earthquakes in India and the Himalaya: Tectonics, geodesy and history, Annals of Geophysics, 47, 839-858.

Bilham, R. and England, P., 2001, Plateau "pop-up" in the great 1897, Assam earthquake, Nature, 410, 806-809.

Bilham, R., Bendick, R. and Wallace, K., 2003, Proc. Indian. Acad. Sci. (Earth Planet. Sci.), 112, 315-329.

Bilham, R., Gaur, V.K. and Molnar, P., 2001, Himalayan seismic hazard, Science, 293, 1442-1444.

Bilim, F., 2007, Investigations into the tectonic lineaments and thermal structure of Kutahya-Denizli region, western Anatolia, from using aeromagnetic, gravity and seismological data, Physics of the Earth and Planetary Interiors, 165, 135-146.

BIS, 1986. Seismic zoning map of India, Published by Bureau of Indian Standards, New Delhi.

Biswal, T.K., Gyani, K.C., Pathasarathy, R., Pant, D.R., 1998. Tectonic implication of geochemistry of gabbro-norite-basic granulite suite in the Proteroxoic Delhi super groop, Rajasthan, India, J. Geol. Soc. Inda. 52, 721-732.

Biswas, S. K. and K. N. Khattri, 2002, A geological study of earthquake in Kachchh, Gujarat, India, J. Geol. Soc. Ind., 60, 131-142.

Biswas, S.K., 1982, Rift basins in western margins of India and their hydrocarbon prospects with special reference to Kutch basin, American Association of Petroleum Geology, ,66, 1497-1513.

Biswas, S.K., 1987, Regional tectonic framework, structure and evolution of the western marginal basins of India, Tectonophysics, 135, 307-327.

Biswas, U.K., Ghatak, T.K., Ghosh, D.C., Naskar, D.C., Chakroborthy, P.K., and Choudhary, D.K., 1999, Proc. of Golden Jubilee Seminar on Exploration Geophysics in India, Geol. Surv. India, Special Publication No. 49, 471-479.

Blackett, P.M.S., 1961, Comparison of ancient climates with the ancient latitudes deduced from rock magnetic measurements, Proc. of Royal Society, A263, 1-30.

Blackett, P.M.S., Clegg, J.A. and Stubbs, P.H.S., 1960, An analysis of rock magnetic data, Proc. of Royal Society, A256, 291-322.

Blackman, K.B. and Tuckey, B.W., 1958, The measurement of power spectra, Dover Publications, New York, 1-190.

Blackman, R.B. and Tukey, B.W., 1958, The measurement of power spectra, Dover Publications, New York, 1-190.

Blakely, R.J., 1996, Potential theory in gravity and magnetic applications, Cambridge University Press, Cambridge, U.L.

Blakely, R.J., Brocher, T.M. and Wells, R.E., 2009, subduction zone magnetic anomalies and implications for hydrated forearc mantle, Geology, 33, 445-448.

Blakely, R.J., Brocher, T.M., Wells, R.E., 2005, Subduction zone magnetic anomalies and implications for hydrated forearc mantle. Geology, 33, 445-448.

Blakely, R.J., Langenheim, V.E., Ponce, D.A. and Dixon, G.L., 2000a, Aeromagnetic survey of the Amargosa desert, Nevada and California: a tool for understanding near surface geologoy and hydrology, U.S. Geol. Survey, Open File Report 00-0188, http://pubs.usgs.gov/open-file/ofoo-188/.

Blakely, R.J., Wells, R.E., Tolan, T.L., Beeson, M.H. and Trehu, A.M., 2000b, New aero magnetic data reveal large strike slip faults in the northern Willamette Valley, Oregan, Geological Society of Am. Bull., 112, 1225-1233.

BMR, 1982, Faye gravity anomalies, Prince Charls Mountains, Bureau of model Resources Canberra, Australia, Antarctica Map no: 25/09/28.

Bois, C., Bouche, P., and Pelet, R., 1982, Global geologic history and distribution of hydrocarbon reserves, Am. Assoc. of Pet. Geol. Bull., 66, 1248-1272.

Bollinger, L., Avouac, J.P., Beyssac, O. et al., 2004, Thermal structure and exhumation history of lesser Himalaya in Central Nepal, Tectonics, 23, (doi:10.1029/2003TC001564).

Bose, R.N. and Vaidyanathan, N.C., 1979, Gravity surveys for barites around Mangampeta, Cuddapah district, A.P. Geol. Soc. Ind., 20, 540-547.

Bosellini, A., 1992, Continental margins of Somalia, structural evolution and sequence stratigraphy, The Geology and Geophysics of Continental Margins (Ed.) J.S.Watkins, F.Zhiqiang and K.J. McMillan, AAPG Memoir, 53, 185-205.

Bosum, W., Casten, U., Fielberg, F.C., Heyde, I. and Soffel, H.C., 1997, Three dimensionl interpretation of the KTB gravity and magnetic anomalies, J. Geoph. Res., 102, B8, 18, 307-321.

Bosum, W., Casten, U.,Feiberg, F., Gotze, H.J., Gobasy, M., Hyde, I., Neubauer, F.M., Rottger, B. and Soffel, H., 1993a, Gravity and magnetic structural models of the KTB area, (Ed) Emmerman, R,, Lauterjung, J. and Umsonst, T., KTB Report 93-2, 319-322.

Bosum, W., Rottger, B., and Schmidt, H., 1993b, Detailed interpretation of magnetic anomalies in the KTB area in connection with borehole magnetic anomalies, (Ed) Emmerman, R,, Lauterjung, J. and Umsonst, T., KTB Report 93-2, 323-325.

Bott, M.H.P. and Hutton, M.A., 1970, A matrix method for interpreting oceanic magnetic anomalies, Geophys. J. Royal. astro. Soc., 20, 149-157.

Bott, M.H.P., 1963, Inverse methods in the interpretation of magnetic and gravity anomalies, In: Methods in computational physics, (Ed.) B.Alder et al., 13, Academic Press.

Bott, M.H.P., 1975, Structure and evolution of north Scottish shelf, the Faeroe block and the intervening region, In: A.W. Woodland (editor), Petroleum and the continental shelf of north west Europe, Geology, Applied Science Publ., England, 1, 105-113.

Bott, M.H.P., 1982, Interior of the earth: its structure, constitution and evolution, II edition, Edward Arnold, London, 1-403.

Bowin, C.O., 1985, Global gravity maps and structure of the earth, In: W.J. Hinze (editor), Utility of Regional Gravity and Magnetic Anomaly Maps, Soc. of Explor. Geophysics, Tulsa, Oklahoma, 88-101

Braitenberg, C., Ebbing, J. and Götze, H.J., 2002, Inverse modeling of elastic thickness by convolution method-the eastern Alps as a case example, Earth & Planetary Sci. Letters, 202, 387-404.

Braitenberg, C., Zadro, M., Fang, J., Yang, Y. and Hsu, H.T., 2000, The gravity and isostatic Moho undulations in Qinghai-tibet plateau, Journal of Geodynamics, 30, 489-505.

Brasse et al., 2006, Probing electrical conductivity of the Trans-European Suture Zone, EOS, 87, 281-287.

Briggs, I. C., 1974, Machine contouring using minimum curvature, Geophysics, 39, 39-48.

Bromley, J., Mannstrom, B., Nisca, D. and Jamtlid. A., 1994, Airborne geophysics, application to a groundwater study in Botswana, Groundwater, 32, 79-90.

Brown, I., Kriegsman, L.M., 2003, Proterozoic crustal evolution of southern most India and Sri Lanka. In: Yoshida, M., Windley, B.F., Dasgupta, S. (Eds.). Proterozoic East Gondwana Super Continent Assembly and Break up. Geological Society, London, Special publication, 206, 169-202.

Brown, L.D., Zhao, W., Nelson, K.D., Hauck, M., Alsdorf, D, Ross, A., Cogan, M., Clark, M., Liu, X. and Che, J., 1996, Bright spot, structure and magmatism in southern Tibet from INDEPTH seismic reflection profiling, Science, 274, 1688-1693.

Brozena, J.M., 1991, GPS and airborne gravimetry: Recent progress and future plans, Bull. Geod, 65, 116-121.

Brudzinski, M. R., Thurbar, C. H., Hacker, B. R., Engdahl, E.R., 2007, Global prevalence of double Benioff zones, Science, 316, 1468-1472.

Brune, J.N. and Singh, D.D., Continent like crustal thickness beneath the Bay of Bengal sediments, Bull. Seismol. Soc. Am., 76, 191-203.

Bucher, W. A., 1963, Crypto explosion structure caused from without or within the earth 'Astrobleme or Geoleme'. Am. J. Sci., 261, 577–649.

Buck, W.R., 1988, Flexural rotation of normal faults, Tectonics, 7, 959-973.

Buck, W.R., 1991, Modes of continental lithospheric extension, J. Geophys. Res., 96, 20161-178.

Buffett, B. A., 2000, Earth's core and geodynamo, Science, 288, 2007-20012.

Buffett, B.A., 2010, Tidal dissipation and the strength of the earths internal magnetic field, Nature, 468, 952-954.

Bullard, E.C., 1949, The magnetic field with in the earth, Proc. R. Soc. London, A197, 433-453.

Bullen, K.E., 1975, Earth's density, Chapman and Hall, London, 1-420.

Burchfiel, B.C. et al., 1990, Crustal shortening on the margins of the Tien Shan, Xinjiang, China, In: Tectonic Studies of Asia and the Pacific Rim (eds) W.G. Ernst and R.G. Coleman, Bellwether Publishing Ltd. for the Geol. Soc. of Am., 142-177.

Burchfiel, B.C., 1990, The continental crust (ed) Mooraes, E.M. Shaping the Earth: Tectonics of continents and oceans, Scientific American, W.H. Freeman and Company, 5-21.

Burg, J-P. and Ford, M., 1997, Orogeny through time: an overview, Geological Society, London, Special Publication No. 121, 1-17.

Bürgmann, R., Segall, P., Lisowski, M. and Svare, J., 1997. Postseismic strain following the 1989 Loma Prieta earthquake from GPS and leveling measurements, J. Geophys. Res., B, Solid Earth and Planets, 102, 4933-4955,

Burk, K. and Dewey, J. F., 1973, Plume generated triple junctions: Key indicators in applying plate tectonics to old rocks, J. Geolo., 81, 406-433.

Burke, K. and Dewey, J.F., 1973, An outline of Precambrian plate development, In: Tarling, D.H and Runcorn, S.K. (Eds.). Implications of Continental Drift to the Earth Sciences. Academic Press, London, 1035-1045.

Burke, K., Dewey, J.F. and Kidd, W.S.F., 1976, Dominance of horizontal movements, arc and microcontinental collisions during the later Permobile regime, The Early History of the Earth, John Wiley and Sons, 113-129.

Burov, E. and Diament, M., 1995, The effective elastic thickness (Te) of continental lithosphere: what does it really mean? J. Geophys. Res., 100, 3905-3927.

Burov, E.B. and Watts, A.B., 2006, The long-term strength of continental lithosphere: "jelly sandwich' or 'crème brulee'?, GSA Today, 16, 1, doi:10.1130/1052-5173(2006)016<4:TLTSO>2.0.CO:2, 4-10.

Cabanes, C., Cazaneve, A., Le Provost, C., 2001, Sea level rise during the past forty years determined from Satellite and insitu observations, Science, 294, 840-842.

Calais, E., Dong, L., Wang, M., Shen, Z. and Vergnolle, M., 2006, Continental deformation in Asia from a combined GPS solution, Geophys. Res., Letters, 33, L24319, doi:10.1029/2006GL028433, 1-6.

Campbell, I. H. and Davies, G. F., 2006, Do mantle plumes exist?, Episodes, 29, 162-168.

Campbell, I.H. and Griffiths, R.W., 1990, Implications of mantle plume structure for the evolution of flood basalts, Earth and Planetary Science Letters, 99, 79-93.

Campbell, W.H., 1997, Introduction to geomagnetic fields, Cambridge University Press, U.K., 1-290.

Cande, S.C. and Kent, D.V., 1995, Revised calibration of the geomagnetic polarity time scale for the late Cretaceous and Cenozoic, J. Geophys. Res., 100, 6093-6095.

Canil, D., 2008. Canada's craton: A bottoms-up view, GSA Today, 18, 4-10.

Caporali, A., 2000, The gravity field of the Karakoram mountain range and surrounding areas, Khan, M.A., Treloar, P.J., Searle, M.P. and Jan, M.Q. (eds) Tectonics of the Nanga Parbat Syntaxis and the Western Himalaya, Geological Society of London, Special Publications, 170, 7-23.

CASE 1985, (Canadian American Seamount Expedition, 1985), Hydrothermal vents on the axis Seamount of the Juan de Fuca Ridge, Nature, 313, 212-214.

Catherine, J.K., Gahalaut, V.K., Sahu, V.K., 2005, Constraints on rupture of the December 26, 2004, Sumatra earthquake from far- field GPS observations, Earth and Planetary Science Letters, 237,673-679.

Cathles III, L.M., 1975, The viscosity of earth's mantle, Princeton, New Jersey, Princeton University Press.

Cattin, R., Martelet, G., Hentry, P., Avouac, J.P., Diament, M. and Shakya, T.R., 2001, Gravity anomalies, crustal structure and thermo-mechanical support of the Himalaya of Central Nepal, Geophys., J. Int., 147, 381-392.

Cawood, P.A., Knoner, A. and Pisarevsky, S., 2006, Precambrian plate tectonics: Criteria and evidence, GSA today, 16, 4-10.

Cenki, B. and Kriegsman, L.M., 2005, Tectonics of the Neoproterozoic southern ganulite terrain, South India, Precambrian Research, 138, 37-56.

CGMW, 2007, Magnetic anomaly map (Scale 1:50,000,000) of the world, Published by Commission for the Geological Map of the World and Printed by Geological Survey of Finland at Lonnberg Promo, Helsinki, Finland. ISBN 978-952-217-000-2.

Chadwick, B., Vasudev, V.N. and Hedge, G.V., 2000, The Dharwar craton, Southern India, interpreted as the result of late Archean oblique convergence. Precambrian Res., 99, 91-111.

Chakraborti, R., Basu, A.R. and Chakraborti, H., 2007, Trace element and Nd-isotope evidence for sediment sources in the mid-Proterozoic Vindhyan basin, Central India. Precambrian Research, 159, 260-274.

Chamalaun, F.H., Prasad, S.N., Lilley, F.E.M., Srivastava, B.J., Singh, B.P. and Arora, B.R., 1987, On the interpretation of the distinctive pattern of geomagnetic induction observed in Northwest India, Tectonophysics, 140, 247-255.

Chamoli, A., Pandey, A.K., Dimri, V.P. and Banerjee, P., 2010, Crustal configuration of the northwest Himalaya based on modeling of gravity data, Pure. Appl. Geophys., DOI 10.1007/s00024-010-0149-2.

Champion, D.E., Lanphere, M.A. and Kuntz, M.A., 1988, Evidence for a new geomagnetic reversal from lava flows in Idaho: Discussion of short polairty reversals in the Brunhes and Late Matuyama polarity Chrones, J. Geophys. Res., 93, 11667-11680.

Chand, S. and Subrahmanyam, C., 2003, Rifting between India and Madagascar-mechanism and isostasy, Earth & Planet. Science Letters, 210, 317-332.

Chand, S., Radhakrishna, M. and Subrahmanyam, C., 2001, India-East Antarctica conjugate margins, rift-shear tectonic setting inferred from gravity and bathymetry data, Earth & Planet. Sci. Letters, 185, 225-236.

Chandra, A., 2000, Deep water prospects in Indian offshore, Presidential Address, Seminar on Exploration Geophysics organized by AEG at Goa, November 8-10, 2000.

Chandra, P., Singh, R.B., Bhattacharya, D., Verma, C.N., Lahiri, A.K. and Mazumdar, K., 1999, Controlled source audio frequency magnetotelluric and time domain spectral induced polarization studies in Gurahar Pahar gold prospect, Sidhi district, M.P., GSI, Special Publication No. 49, 45-52.

Chandra, P.C. and Reddy, P.H.P., 1987, Geophysical surveys in hard rock and sedimentary terrains-Status, Findings and Recommendations: Govt. of India, Ministry of Water Resources, Central Ground Water Board; Report No. 29, Project on Groundwater Studies in Kasai and Subarnarekna River Basins.

Chandra, R. 1999, Geochemistry and petrogenesis of the layered sequence in Girnar Ijolitic Series (GIS) India: The role of differentiation and allied factors, In: Magmatism in Diverse Tectonic Settings, Eds. R.K. Srivastava and R. Chandra , Oxford & IBH Publishing Co., New Delhi, 155-194.

Chandra, U., 1977, Earthquakes of Peninsular India- A seismotectonic study, Bull, Seis. Soc. of America, 67, 1387-1413.

Chandrasekhar, D.V, 2005, Some significant gravity and magnetic anomalies of Sauratra and Kachchh: Associated structures and their implications on the seismicity, A Ph.D. thesis submitted to the University of Osmania, Hyderabad, India.

Chandrasekhar, D.V., Burgmann, R. Reddy, C.D., Sunil, P.S., Schmidt, D.A., 2009, Weak mantle in NW India probed by geodetic measurements following the 2001 Bhuj earthquake, Earth and Planetary Science Letters, 280, 229-235.

Chandrasekhar, D.V., Mishra, D.C., Singh, B., Vijaya Kumar, V. and Burgmann, R., Changes in the gravity field and elevation due to Bhuj earthquake of January 26, 2001, 2004, Geoph. Res. Lett., 31, L19608, 1-4.

Chandrasekhar , D. V. , Mishra , D. C , Rao , G.V.S.P , Rao J.M. , 2002 , Gravity and magnetic signatures of volcanic pipes of Saurashtra, India and their bulk physical properties vis-a-vis paleomagnetic and geochemical studies of exposed rocks, Earth & Planet. Science Letters, 201, 277-292.

Chandrasekhar, D.V. and Mishra, D.C., 2002. Some geodynamic aspects of Kachchh basin and seismicity: an insight from gravity studies, Current Science, 83, 492-498.

Chandrasekhar, D.V., Singh, B., Firozishah, Md. and Mishra, D.C., 2005, Analysis of gravity and magnetic anomalies of Kachchh rift basin, India and its comparison with the New Madrid seismic zone, USA, Current Science, 88, 1601-1607.

Chandrasekharam, D., Mahoney, J.J., Sheth, H.C., and Duncan, R.A., 1999, Elemental and Nd-Sr-Pb geochemistry of flows and dikes from the Tapti rift, Deccan flood Basalt province, India, In: Verma, S.P. (ed.) Rift related volcanism: Geology, Geochemistry and Geophysics, Journal of Volcanology and Geothermal research, 93, 111-123.

Chapman, M.E., 1979, Techniques for interpretation of Geoid anomalies, J. Geophys. Res., 84, 3793-3801.

Charvis, P. and Operto, S., 1999, Structure of the Kerguelen Cretaceous volcanic provinces (Southern Indian ocean) from wide angle seismic data, J. Geodynamics, 28,51-71.

Chatterjee, N., Crowley, J.L. and Ghose, N.C., 2008, Geochronology of the 1.55 Ga Bengal anorthosite and Grenvillian metamorphism in the Chotanagpur gneissic complex, eastern India, Precambrian Research, 161, 303-316.

Chatterjee, N., Mazumdar, A.C., Bhattacharya, A. and Saikia, R.R., 2007, Mesoproterozoic granulites of the Shillong-Meghalaya plateau, Evidence of westward continuation of the Prydz Bay Pan-African suture into Northeastern India. Precambrian Research, 152, 1-26.

Chatterjee, S., 2010, Shiva crater, Personal communication.

Chaturvedi, A.K., Kak, S.N., Tiku, K.L., Maithani, P.B. and Chaki, A., 2008, Airborne geophysical surveys for uranium exploration in India- past, present and future, Mem. Geol. Surv. of India, 68, 309-336.

Chaubey, A. K., Bhattacharya, G.C. , Murty , G. P. S., Srinivas, K., Ramprasad, T. and Gopal Rao, D., 1998, Early Tertiary Seafloor spreading magnetic anomalies and paleo propagators in the northern Arabian sea, Earth & Planet. Science. Letters, 154, 41-52.

Chaubey, A.K., 2007, Personal Communication.

Chauhan, D.S. and Sisodia, M.S., 1989, Phosphorites of Rajasthan, Mem. Geol. Soc. India, 13, 9-22.

Chawla, A.S., Kak, S.N. and Dwivedy, K.K., 1993, An integrated approach for atomic mineral exploration: some case studies, Exploration and Research for Atomic Minerals, 6, 127-137.

Chen, J., A.D. Del Genio, B.E. Carlson, and M.G. Bosilovich, 2008: The spatiotemporal structure of twentieth-century climate variations in observations and reanalyses. Part I: Long-term trend. J. Climate, 21, 2611-2633, doi:10.1175/2007JCLI2011.1.

Chen, W. and Tseng, T., 2007, Small 660-km seismic discontinuity beneath Tibet implies resting ground for detached lithosphere, J. Geophys., Res., 112, B05309, doi:10.1029/2006JB004607, 1-15.

Chen, W.P., Martin, M., Tseng, T.L., Nowack, R.L., Hung, S.H., Huang, B.S., 2010, Shear wave birefringence and current configuration of converging lithosphere under Tibet, Earth and Planetary Science Letters, 295, 297-304.

Chetty, T.R.K., 2001, The Eastern Ghats Mobile Belt, India: A collage of juxtaposed terranes (?), Gondwana Research, 4, 319-328.

Chetty, T.R.K., Bhaskar Rao, Y.J. and Narayana B.L., 2003, A structural cross section along Krishnagiri-Palani Corridor, Southern Granulite Terrain of India, Memoir Geological Society of India, 50, 255-277.

Chetty, T.R.K., Fitzsimons, I .C.W., Brown, L.D., Dimri, V.P. and Santosh, M., 2006, Crustal structure and tectonic evolution of the southern granulite terrain, India. Gondwana Research, 10, 1-206.

Chinnery, M. A., 1961, The deformation of the ground around surface faults, Bull. Seismol. Soc. Am., 51, 355-372.

Chore, S.A. and Mohanty, M., 1998, Stratigraphy and tectonic setting of the trans Aravalli, Neo-Proterozoic volcano sedimentary sequence in Rajasthan. Jour. Geol. Soc. India, 51, 57-68.

Choubey, A.K., Gopala Rao, D., Srinivas, K., Ram Prasad, T., Ramana, M.V., Subrahmanyam, V., 2002, Analyses of multi channel seismic reflection, gravity and magnetic data along a regional profile across the Central-Western continental margin of India, Marine Geology, 182, 303-323.

Choubey, V.D. 1971, Narmada-Son lineament, India, Nature, 232, 38-40.

Choudhary, A.K., Gopalan, K. and Sastry, C.A., 1984, Present status of the geochronology of the Precambrian rocks of Rajasthan. Tectonophysics, 105, 131-140.

Christensen, E.F., Luhr, H., Hulot, G., Haagmans, R. and Purucker, M., 2009, Geomagnetic Research from space, EOS, 90, 213-214.

Chung, W.Y. and Gao, H., 1995. Source parameters of the Anjar earthquake of July, 21, 1956. India and its seismotectonic implications for the Kachchh rift basin, Tectonophysics, 242, 281-292.

Clark, C., Collins, A.S. timms, N.E., Kinny, P.D., Chetty, T.R.K. and Santhosh, M., 2009, SHRIMP U-Pb age constraints on magmatism and high-grade metamorphism in the Salem block, southern India, Gondwana Research 16, 27-36.

Clark, D.A., 1999, Magnetic petrology of igneous intrusions: implications for exploration and magnetic interpretation, Exploration Geophysics, 30, 5-26.

Clarke, L., 1985, Groundwater abstraction from basement complex areas of Africa, Q. J. Eng-Geol, London, 18, 25-34.

Clift, P. D., Hodges, K.V., Heslop, D., Hannigan, R., Van Long, H. and Calves, G., 2008b, Correlation of Himalayan exhumation rates and Asian monsoon intensity, Nature Geoscience, 1, 875-880.

Clift, P., Shimizu, N. and Layne, G., 2000. Fifty five million years of Tibetan evolution recorded in the Indus fan, EOS Trans., AGU 81, 277.

Clift, P.D et al., 2008a, Holocene erosion of lesser Himalaya triggered by intensified monsoon. Geology, 36, 79-82.

Clift, P.D. and Blusztajn , 2005, Reorganization of the Western Himalayan river system after five million years ago. Nature, 438, 1001-1003.

Cloetingh, S. A. P. L. and Topo-Europe working group, 2007, Topo-Europe: The geosciences of coupled deep earth surface provinces, Global and Planetary change, 58, 1-118.

Cloetingh, S., Cornu, T., Ziegler, P.A. and Beckman, F., 2006, Neotectonics and intraplate continental topography of the northern Alpine Foreland, Earth Science Reviews, 74, 127-196.

Closs, H., Hari Narain and Grade, S. C., 1974, Continental margins of India, In: The Geology of Continental Margins, (ed) C. A. Burk and C. L. Drake, Springer Verlag, 629-639.

Clowes, R.M., Hammer, P.T.C., Gabriela, F. V. and Welford, J.K., 2005, Lithospheric structure in northwesren Canada from Lithoprobe seismic refraction and related studies: a synthesis, Can. J. Earth Sci., 42, 1277-1293.

Cochran, J.R., 1979, An analysis of isostasy in the World's Ocean2, Mid Ocean ridge crusts, J. Geoph. Res., 84, 4713-4729.

Cochran, J.R., 1980, Some remarks on isostasy and the long term behavior of the continental lithosphere, Earth & Planet. Sci. Lett., 46, 266-274.

Cogbill, A.H., 1990, Gravity terrain correction calculated using digital elevation models, Geophysics, 55, 102-106.

Coggan, J.H., 1976, Magnetic and gravity anomalies of a polyhedra, Geoexploration, 14, 93-105.

Coles, R., 1976, A flexible iterative magnetic anomaly interpretation technique using multiple rectangular prisms, Geoexploration, 14, 125-141.

Colley, J.W. and Tukey, J.W., 1965, An algorithm for the machine calculation of complex Fourier series, Math. Comp., 19, 297-301.

Condie, K.C., 2001, Mantle plumes and their record in earths history, Cambridge University Press, New York, USA, 1-306.

Cone, J., 1991, Fire Under the Sea, William Morrow & Company, Inc., New York, 1-285.

Cook, F.A. and Erdmer, P., 2005, An 1800 km cross section of the lithosphere through the northwestern North American plate: lessons from 4.0 billion years of earths history, Can. J. Earth Sci. 42, 1295-1311.

Cook, F.A. and Vasudevan, K., 2006, Reprocessing and enhanced interpretation of the initial COCORP Southern Appalachians traverse, Tectonophysics, 420, 161-174.

Cook, F.A., Brown, L.D. and Oliver, J.E., 1990, The Southern Appalachians and the Growth of Continents, In: Shaping the Earth Tectonics of Continents and Oceans (ed) E.M. Moores, Scientific American, Inc., 139-155.

Cook, K.L. and Royden, L.H., 2008, The role of crustal strength variations in shaping orogenic plateaus, with application to Tibet, Journal of Geophysical Research, 113, B08407, doi:10.1029/2007/B005457, 1-18.

Cooper, G.R.G, 2006, Interpreting potential field data using continuous wavelet transforms of their horizontal derivatives, 32, 984-992.

Copley, A. and McKenzie, D., 2007, Models of crustal flow in the India-Asia collision zone, Geophys. J. Int., 169, 683-698.

Coppola, D.P., 2006, Introduction to international disaster management, Butterworth-Heinemann, Oxford, U.K.

Cordell, L. and Taylor, P.T., 1971, Investigation of magnetization and density of a North Atlantic Seamount using Poisson's theorem, Geophysics, 36, 919-937.

Cordell, L., 1978, Regional geophysical setting of the Rio Grande rift, Geol. Soc. Am. Bull., 89, 1073-1090.

Corliss, J. B., 1989, Submarine hot springs again, Origins of Life, 19, and 534-35.

Courtillot, V. and Besse, J., 1987, Magnetic field reversals: Polar wander and core mantle coupling, Science, 237, 1140-1148.

Courtillot, V., Gallet, Y., Le Mouel, Fluteau, F., Genevey, A., 2007, Are there connections between the Earth's magnetic field and climate, Earth and Planetary Science Letters, 253, 328-339.

Crain, I. K., 1970, Computer interpolation and contouring of two dimensional data, Geoexploration, 8, 71-86.

Crain, I. K., Crain, P. L. and Plaunt, M. S., 1969, Long period Fourier spectrum of geomagnetic reversals, Nature, 223, 282

Creer, K.M., 1959, A.C. Demagnetization of unstable Triassic Keuper marls from SW England, Geophys. J. Roy. Astr. Soc, 2, 262-275.

Creer, K.M., Irving, E. and Runcorn, S.K., 1954, The direction of the geomagnetic field in remote epochs in Great Britain, Phil. Trans. R. Soc. Lond, A250, 144-156.

CSIR, 1994, Coloured Bouguer anomaly map of Central India, CSIR News, Published by Council of Scientific and Industrial Research.

Curray, J. R., Emmel, F. J. , Moore, D. G, Raitt, R.W., 1982 , Structure, tectonics and geological history of the north eastern Indian ocean, In: The Ocean Basins and Margins(eds) Nairn, A. E. M., Stehli, F. G., Plenum, New York, 399-450.

Currie, C. A. and Hyndman, R. D., 2006, Thermal structure of subduction zone back arcs, J. Geophys. Res., 111, B08404, 1-22.

Dante, C., 2008, Canada's craton: A bottoms-up view, GSA Today, 18, doi: 10.1130/GSAT01806A.1, 4-10.

Das, D., Mehra, G., Rao, K.G.C., Roy, A.L. and Narayana, M.S., 1979, Bouguer, free air and magnetic anomalies over north western Himalaya, Himalayan Geology Seminar, Section III, Oil and Natural Gas Resources, Geol. Surv. India, Misc. Publ. 41, 141-148.

Das., L.K., Mishra, D.C., Ghosh, D. and Banerjee, B., 1990, Geomorphotectonics of the Basement in a part of Upper Son Valley of the Vindhyan Basin, Journal Geological Society of India, 35, 445-448.

Dasgupta, A. and Biswas, A., 2000, Geology of Assam, Geological Society of India, Bangalore.

Dash, B.R., Venkteshwarlu, P.D., Ghosh, D., Ghatak, S.K. and Singh, R.B., 1983, Geophysical surveys for manganese in Central India. Geol. Sur. India, Spl. Publ., No. 2 (1), pp.237-246.

Dasu, S.P.V., 2002, Significance of multisensor aerogeophysical data in parts of Kolar Schist belt, Karnatka, Geol. Surv. Ind. Spl. Pub., 75, 187-192.

Davies, G. F. and Richards, M. A., 1992, Mantle convection, Journal of Geology, 100, 151-206.

Davy, B.W., 1990, The influence of subducting plate buoyancy on subduction of the Hikurangi-Chatham Plateau beneath the north island, New Zealand, Geology and Geophysics of continental Margins (eds.) J.S. Watkins, F. Zhiqiang and K.J. McMillean, Am. Assoc. of Petroleum Geologists, 53, 75-91.

Dc Mcts, C., Gordon, R.G., Argus, D.F. and Stein, S., 1990, Geophys. J. Int., 101, 425.

De Sigoyer, J., Chavagnac, V., Blichert-Toft, J., Villa, I., Luais, B., Guillot, S., Cosca, M., Mascle, G., 2000, Dating the Indian continental subduction and collisional thickening in the north-west Himalaya: multichronology of the Tso Morari eclogites, Geology, 28, 487-490.

De, A. and Bhattacharya, D., 1971, Phase petrology with special reference to pyroxenes of the acid igneous complex of Barda Hills, western Saurashtra (Gujarat), Bull. Volcanologique, 85, 907-929.

DeCelles, P.G., Robinson, D.M. and Zandt, G., 2002, Implications of shortening in the Himalayan fold-thrust belt for uplift of the Tibetan Plateau, Tectonics, 21, 1062, doi:10.1029/2001 TC001322, 12-25.

Delescluse, M. and Chamot-Rooke, N., 2007, Instantaneous deformation and kinematics of the India-Australia Plate, Geophys. J. Int. 168, 818-842.

Dennell, R.W., 2008, Human migration and occupation of Eurasia, Episodes, 31, 207-210.

Depaolo, D.J. and Manga, M., 2003, Deep origin of hotspots – the mantle plume model, Science, 300, 920-921.

Derry, L.A. France-Lanord, C., 1997, Himalayan weathering and erosion fluxes: Climate and Tectonic Controls, Tectonic Uplift and Climate Change, edited by William F. Ruddiman, Plenum Press, New York, 289-312.

Devleena Mani, Patil, D.J. and Dayal, A.M., 2010, Evaluation of hydrocarbon prospects using new surface geochemical data with constraints from geological and geophysical observations in Saurastra basin, Comm. to Marine and Petroleum Prospecting.

Dey, B.K., Maithani, A., Mitra, D.S. and Agarwal, R., 1993, Photogeological and landsat analysis of Son Valley Vindhyan between Hosangabad and Katni, Technical Report, ONGC, Dehradun.

DGH, 2005, Petroleum Exploration and Production Activities, India, 2004-2005, Directorate General of Hydrocarbons, New Delhi.

Dhananjay, Kumar, Ravi, Bastia and Debajyoti, Guha, 2004, Prospect hunting below Deccan basalt: imaging challenges and solutions. First Break, 22, 35-39.

Dhar, P.C. and Singh R.P. 1993, Evolution of Cambay Graben, In: Rifted Basins and Aulacogens – a geological and geophysical approach. Ed: S.M. Casshyap., Gyanodaya Prakashan, 268-280.

Dickinson, W. R., 1977, Tectono stratigraphy evolution of subduction – controlled sedimentary assemblages, In: Island Arcs, Deep Sea Trenches and Back Arc Basins (ed) M. Talwani and Walter C. Pitonan III , Am , Geophys. Union, D. C., Maurice Ewing Sr, 1; 33-40.

Dietrich, W.E. and Perron, J.T., 2006, The search for a topographic signature of life, Nature, 439/doi:10.1038/nature04452, 411-418.

Dimri, V.P., 2005, Some applications fo fractal theory in potential fields, Himalayan Geloogy, 26, 161-164.

Ding, L., Zhong, D., Yin, A. Kapp, P., Harrison, T. M., 2001, Cenozoic structural and metamorphic evolution of the eastern Himalayan syntaxis (Namche Barwa), Earth & Planet. Sci. Letters, 192, 423-438.

Dixit, M.M., Satyavani, N., Sarkar, D., Khare, P. and Reddy, P.R., 2000, Velocity Inversion in the Lodhika area, Saurashtra Peninsula, Western India. First Break, v. 18, pp. 499-504

Dixit, M.M., Tewari, H.C., Rao, C.V., 2010, Two dimensional velocity model of the crust beneath the Cambay basin, India from refraction and wide angle reflection data, Geophys. J. Int., 181, 635-652.

Dixon,T.H and Parke, M.E., 1983, Bathymetry estimate in the southern oceans from Sea- sat altimetry, Nature, 304, 406-411.

Dobmeier, C.J. and Raith, M., 2003, Crustal architecture and evolution of the Eastern Ghats Belt and adjacent regions of India, In: Yoshida, M., Windley, B.F., Dasgupta, S. (Eds.), Proterozoic East Gondwana: Super continent Assembly and Break up, Geological Society, London, Special publications, 206, 145-168.

Dobmeier, C.J., Lutke, S., Hammerschmidt, K. and Mezger, K., 2006 Emplacement and deformation of the Vinukonda meta-granite (Eastern Ghats, India: implications for the geological evolution of peninsular India and for Rodinia reconstructions, Precambrian Research 146, 165-178.

Dobrin, M.B. and Savit, C.H., 1988, Introduction to geophysical prospecting (4th edition), McGraw-Hill, New York.

Donovan, T.J., Forgey, R.L. and Roberts, A.A., 1979, Aeromagnetic detection of diagenetic magnetite over oil fields, Bull. Am. Assoc. of Petroleum Geologists, 63, 245-248.

Dorman, L.M. and Lewis, B.T.R., 1970, Experimental isostasy, theory of the determination of the Earth's isostatic response to a continental load, J. of Geoph. Res., 75, 3357-3386.

Drake, C.L. and Girdler, R.W., 1964, A geophysical study of the Red sea, Geophys. Jour. Royal Astron. Soc., 8, 473-495.

Duncan, R.A. and Pyle, D.G., 1988, Rapid eruption of the Deccan flood basalts at the Cretaceous/Tertiary boundary, Nature, 333, 841-843.

Durheim, R.J. and Cooper, G.R.J., 1998, EULDEP: A program for the Euler deconvolution of magnetic and gravity data, Computers and Geosciences, 24, 545-550.

Duroy, Y., Farah, A. and Lillie, R.J., 1989, Subsurface densities and lithospheric flexure of the Himalayan fore land in Pakistan, In: Malinconica Jr. L.L. and Lillie, R.J., (Eds.). Tectonics of the Western Himalayas, Geol. Soc. Am., Special Paper, 232, 203-216.

Dziak, R. P, Bohnenstiehl, D. R., Cowen, J. P., Baker, E. T., Rubein, K. H., Haxel, J. H., Fowler, M. J; 2007, Rapid dike emplacement leads to eruptions and hydro thermal plume release during seafloor spreading events, Geology, 35, 597-582.

Dziewonski, A.M. and Anderson, D.L., 1981, Preliminary reference earth model, Physics of the Earth and Planetary Interiors, 25, 297-356.

Eaton, G.P. and Watkins, J.S., 1970, The use of seismic refraction and gravity methods in hydrogeological investigations, Proc. Canadian Centannial Conference on Mining and Groundwater Geophysics, Ottawa.

Ebbing, J. and Olesen, O., 2005, The northern and southern Scandes-structural differences revealed by an analysis of gravity anomalies, the geoid and regional isostasy, Tectonophysics, 411, 73-87.

Ebbing, J., Braitenberg, C. and Götze, H.J., 2006, The lithospheric density structure of the eastern Alps, Tectonophysics, 414, 145-155.

Ebbing, J., Braitenberg, C., Götze, H.J., 2001, Forward and inverse modelling of gravity revealing insight into crustal structures of the eastern Alps, Tectonophysics, 337, 191-208.

Ebinger, C.J. and Casey, M., 2001, Continental break up in magmatic province: An Ethiopian example, Geology, 29, 527-530.

Ebinger, C.J. and Sleep, N.H., 1998, Cenozoic magmatism through out east resulting from the impact of a single plume, Nature, 395, 788-791.

Ebinger, C.J., Bechtel, T.D., Forsyth, D.W. and Bowin, C.O., 1989, Effective elastic plate thickness beneath the east African and Afar plateaus and dynamic compensation of the uplifts, J. Geophys. Res., 94, 2883-2901.

Echtler, H., Lüschen, E. and Meyer, G., 1994, Lower crustal thinning in the upper Rhinegraben: Implications for recent rifting, Tectonics, 13, 342-353.

Edwards, R.A., Minshull, T.A. and White, R.S., 2000, Extension across the Indian-Arabian plate boundary: the Murray Ridge, Geophy. J. Int., 142, 461-477.

Egan, S.S., 1992, The flexural isostatic response of the lithosphere to extensional tectonics, Tectonophysics, 202, 291-308.

Ehrismann, W., Muller, G., Rosenbach, O. and Sperlich, N., 1966, Topographic reduction of gravity measurements by the aid of digital computer, Bollettino di Geofisica teorica ed. Applicata, 8, 29.

Ekman, M. and Mäkinen, J., 1996, Recent post glacial rebound, gravity change and mantle flow in Fennoscandia, Geophys. J. Int., 126, 229-234.

Elkins-Tanton, L.T., 2005, Continental magmatism caused by lithospheric delamination, In: Foulger et al. (eds) Plates, Plumes and paradigms, Boulder, Colorado, Geol. Soc. Am. Special Paper, 388, 449-461.

Ellis, M., Gomberg, J. and Shweig, E., 2001. Indian earthquake may serve as analog for New Madrid earthquake, EOS Tras., AGU 82, 345, 350-351.

Elsasser, W.M., 1946, Induction effects in terrestrial magnetism, 1 Theory, Phys. Rev., 69, 106-116.

Engdhal, E. R., Van der Hilst, R. Buland, R., 1998, Global teleseismic earthquake relocation with improved travel times and procedures for depth determination, Bull.Seismol. Soc. Am., 88, 722-743.

England, P. and Molnar, P., 1997, Active deformation of Asia: from Kinematics to dynamics, Science, 278, 647-650.

Ergintav, S., Dogan, U., Gerstenecker, C., Cakmak, R., Belgen, A., Demirel, H., Aydin, C. and Reilinger, R., 2007, A snapshot (2003-2005) of the #-D postseismic deformation for the 1999, Mw = 7.4 Izmit earthquake in the Marmara region, Turkey by first results of joint gravity and GPS monitoring, Journal of Geodynamics, 44, 1-18.

Erickson, J., 2001, The living earthquakes, Eruptions and other geologic cataclysms, Checkmark Books, New York.

Eriksson, P.G., Mazumder, R., Catuneanu, O., Bumby, A.J. and Llondo, B.O., 2006, Precambrian continental free board and geological evolution: A time perspective, Earth Science Reviews, EARTH-01446, 1-40.

Ernst, W.G., 1990, The Dynamic planet, The Columbia University Press, New York, 1-280

Ernst, W.G., 2009, Archean plate tectonics, rise of Proterozoic supercontinentality and onset of regional episodic stagnant-lid behavior, Gondwana Research, 15, 243-253.

ESCAP, 1976, Bouguer anomaly map of Western ESCAP region, Published by United Nations Economic and Social Commission for Asia and pacific.

Evans, D.A.D., 2006, Proterozoic low orbital obliquity and axial dipolar geomagnetic field from evaporite Paleolatitudes, Nature, 444, 51-55.

Evans, M.E. and Heller, F., 2003, Environmental magnetism, Academic press, USA, 1-299.

Evans, P. and Crompton, W., 1946. Geological factors in gravity interpretation by evidence from India and Burma. Quarterly Journal of Geological Society of London 102, 211-249.

Evans, P., 1964, The tectonic framework of Assam, Journal of Geological Society of India, 5, 80-96.

Farah, A. and Jafree, S.A.R., 1965, Regional gravity survey of Tatta District, Hyderabad Division, West Pakistan, The Geological Survey of Pakistan, XV, Part 2, 6-9.

Farah, A. and Zaigham, N. A., 1979, Gravity anomalies of the Ophiolite complex of Khanozai-Muslim Bagh-Qila Saifullah Area, Zhob District, Baluchistan, Geodynamics of Pakistan, (eds) A. Farah and K. A. DeJong, Geological Survey of Pakistan, Quetta, 251-262.

Fedi, M. and Rapolla, 1999, 3D inversion of gravity and magnetic data with depth resolution, Geophysics, 64, 452-460.

Fedi, M. Quarta, T. and Santis, A.D., 1997, Inherent power law behavior of magnetic field power spectra from a Specter and Grant ensemble, Geophysics, 62, 1143-50.

Fedorov, L. V, and Ravich, M.G., 1982, Geologic comparison of south eastern Peninsular India and Sri Lanka with a part of East Antarctica (Enderby land , Mac Robertson Land and Princess Elisabeth Land) In : Antarctic Geoscience, (ed) C. Craddock, Union of Wisconsin Press, Madison, 73-78.

Fedorov, L. V., Grikurov, G. E., Kurinin, R. G. and Masolov, V. N., 1982, Crustal structure of the Lambert Glacier from geophysical data, In : Antarctic Geosciences, (ed) C. Craddock, Univ. of Wisconsin press, Madison, 931-936.

Finetti, I., Giorgetti, F. and Poretti, G., 1979, The Pakistani segment of the DSS profile Nanga Parbat-Karakul (1974-75), Boll. Geof. Teor. App. 21, 159-171,

Fischer, K.M., 2002, Waning buoyancy in the crustal roots of old mountains, Nature, 417, 933-936.

Fischer, R. L., Jantsch, M. J. and Comer, R. L., 1982, General bathymetric chart of the oceans (GEBCO), Can. Hydro-graphic. Serv., Ottawa, Canada.

Fisher, R.A. 1953, Dispersion on a sphere, Proc. R. Soc. London Ser. A, 217, 295-305.

Fitzimons, I.C.W., 2000, Grenville-age basement provinces in East Antarctica: evidence for three separate collisional orogens, Geology, 28, 879-882.

Flynn, J. and Krause, D., 2000, Monsters of Madagascar, National Geography, 198, 44-57.

Forster, H.J., Tischendrof, G. and Trumbull, R.B., 1997, An evaluation of the Rb vs (Y+Nb) discrimination diagram to infer tectonic setting of silicic igneous rocks, Lithos, 40, 261-293.

Forsyth, D.W., 1985, Subsurface loading and estimate of flexural rigidity of continental lithosphere, J. Geophys. Res., 90, 12623-12632.

Foulger, G.R. and Natland, J.H., 2003, Is "Hotspot" volcanism a consequence of plate tectonics? Science, 300, 921-922.

Foulger, G.R., Natland, J.H. and Anderson, D.L., 2005, Genesis of the Iceland melt anomaly by plate tectonic processes, In: Foulger et al. (eds) Plates, Plumes and paradigms, Boulder, Colorado, Geol. Soc. Am. Special Paper, 388, 595-625.

Fountain, D.M. and Salisbury, M.H., 1981, Exposed cross section through the continental crust: Implications for crustal structure, petrology and evolution. Earth Planet. Sci. Letts., .56, 263-277.

Fowler, C.M.R., 2005, The Solid Earth: An Introduction to Global Geophysics, Cambridge University Press, U.K., 1-685.

Fredriksson, K., Dube, A., Milton, D. J. and Balasundaram, M. S., 1973, Lonar lake, India: An impact crater in basalt. Science, 180, 862–864.

French, J.E., Heaman, L.M., Chacko, T., Srivastava, R.K., 2008. 1891- 1883 Ma Southern Bastar-Cuddapah mafic igneous event, India: A newly recognized large igneous province. Precambrian Research 160, 308-322.

Fu, Lee-Leung and Cazenave, Anny, 2001, Satellite altimetry and earth sciences: a hand book of techniques, Academic Press, 1-463.

Fudali, R. F. and Subrahmanyam, B., 1983, Gravity Reconnaissance at Lonar Crater, Maharashtra, GSI Special Publication Series No. 2, 1, 83–87.

Fudali, R. F., Milton, D. J., Fredriksson, K. and Dube, A., 1980, Morphology of the Lonar crater, India: Companions and implications. The Moon and the Planets, 23, 493–515.

Fullea, J., Fernandez, M., Zeyan, H. and Verges, J., 2007, A rapid method to map the crustal and lithospheric thickness using elevation, geoid anomaly and thermal analysis, Applications to the Gibraltar arc system, Atlas mountains and adjacent zones, Tectonophysics, 430, 97-117.

Fullerton, L.G., Sagar, W. W. and Bandschumacher, D.W., 1989, Late Jurassic-Early Cretaceous evolution of the eastern India ocean adjacent to North-East Africa, J. Geophys. Res., 94, 2937-2953.

Fuloria, R. C., 1994, Geology and hydrocarbon prospects of Mahanadi basin, India, In: Second Seminar on petroliferous basins of India, (ed) S. K. Biswas, ONGC, Dehradun, 355-369.

Fuloria, R.C., 1996, Geology and hydrocarbon prospects of Vindhyan sediments in Ganga valley, Mem. Geol. Soc. of India, 36, 235-256.

Furness, P., 1994, A physical approach to computing magnetic fields, Geophys. Prosp., 42, 405-416.

Gahalaut, V.K. et al., 2008, GPS measurements of postseismic deformation in the Andaman-Nicobar region following the giant 2004 Sumatra-Andaman earthquake, J. Geophy. Res., 113, B08401, doi: 10.1029/2007JB005511.

Gahalaut, V.K. et al., 2010, Post seismic deformation in Andaman-Nicobar region and Kutch in Gujarat, India, Personal Communication.

Gahalaut, V.K., Nagarajan, B., Catherine, J.K., Kumar, S., 2006, Constraints on 2004 Sumatra-Andaman earthquake rupture from GPS measurements in Andaman-Nicobar islands, Earth and Planetary Science Letters, 242, 365-374.

Gaillard, F., Scaillet, B. and Pichavant, M., 2004, Evidence for present-day leucogranite pluton growth in Tibet, Geological Society of America, Geology 32, 9, 801-804; doi:10.1130/G20577.1.

Gaina, C., Muller, R.D., Brown, B., Ishihara, T., 2007, Break up and early Sea floor spreading between India and Antarctica, Geophys. J. Int., 170, 151-169.

Gan, W., Zhang, P., Shen, Z., Niu, Z., Wang, M., Wan, Y., Zhou, D. and Cheng, J., 2007. Present-day crustal motion within the Tibetan Plateau inferred from GPS measurements, J. Geophys., Res., 112, B08416, doi:10.1029/2005/JB004120, 1-14.

Gannser, A., 1964, Geology of the Himalaya, Wiley Interscience, London, 1-289.

Gannser, A., 1980, The significance of the Himalayan suture zone, In: The Alpine-Himalayan Region (ed) J.M. Tater, Tectonophysics, 62, 37-52.

Gansser, A., 1981, The geodynamic history of the Himalaya, in Gupta, H.K., and Delany, F.M., eds., Zagros, Hindukush, Himalaya, geodynamic evolution: American Geophysical Union Geodynamic Series, 3, 111-121.

Gao, H., Wang, J. and Zhao, P., 1996, The updated Kriging Varience and optimal sample design, Math. Geol., 28, 295-313.

Garzanti, E. and Gaetani, M., 1991, Multicyclic history of the Northern India continental margin (Northwestern Himalaya), The American Asso. of Petroleum Geologist Bulletin, 75, 1427-1446.

Gay Jr., S.P., 1963, Standard curves for interpretation of magnetic anomalies over long tabular bodies, Geophysics, 28, 161-200.

George, M.T., Harris, N.B.W. and Butler, R.W.H., 1993, The tectonic implications of contrasting granite magmagtism between the Kohistan island arc and the Nanga Parbat – Haramosh Hassif, Pakistan Himalaya, In: Treloar, P.J. and Searle, M.P. (eds) Himalayan Tectonics, Geol. Soc., London Spl. Publ., 74, 173-191.

Gera, B.S., Gera, N. and Dutta, H.N., 2011, Unique atmospheric wave: precursor to the 26 January 2001 Bhuj, India earthquake, international Journal of Remote Sensing, TRES-TCN-2010-0001.R2.

Gesch, D., Verdan, K. L. and Greenle, S. K., 1999, EOS Trans., 80, 69-70.

Ghosh, G.K. and Siddiquee, F., 2009, Spectral analysis of gravity data- An integrated approach towards hydrocarbon exploration in geologically complex and logistically difficult area of Manabum, Jour. of Geophysics, 30, 37-42.

Ghosh, S. and Bhaduri, S. K., 2003, Petrography and petrochemistry of impact melts from Lonar lake, Buldana district, Maharashtra, India. Indian Minerals, **57**, 1–26.

Gibbs, R.A. and Thomas, M.D., 1976, Gravity signature of fossil plate boundaries in the Canadian Shield. Nature, 262, 199-200.

Glover, P.W.J. and Adam, A., 2008, Correlation between crustal high conductivity zones and seismic activity and the role of carbon during shear deformation, J. of Gephysical Research, 113, B12210, doi: 10.1029/2008/JB005804.

GM-SYS, 2000, Gravity/ magnetic modeling software version 4.6, Northwest Geophysical Assoc., Inc., USA.

Goetze, H-J., Meurers, B., Schmidt, S. and Steinhauser, P., 1991, On the isostatic state of the eastern Alps and the central Andes; A statistical comparison, Geological Society of America, Special Paper 265, 279-290.

Gogoi, B.N. and Rabha, D., 2006, Ground water exploration in Arunachal Pradesh. In: National Seiminar on Land Resource Management for Sustainable Development, pp. 16-17, organized by State Landuse Board, RWD, Unpubl.

Gokaran, S.G., Gupta, G. and Rao, C.K., 2004, Geoelectric structure of the Dharwar craton from magnetotelluric studies: Archean suture identified along the Chitradurga-Gadag schist belt, Geophys. J. Int., 158, 712-728.

Gokaran, S.G., Gupta, G., Rao, C.K.S., Selvaraj. C., 2002, Electrical structure across Indus Tsangpo suture and shyok suture zone in NW Himalaya using magnetotelluric studies, Geophys. Res. Lett., 29-1-4.

Gokaran, S.G., Gupta, G., Walia, D. and Dutta, S., 2005, Magnetotelluric studies in Sikkim Himalaya, DST News Letter (DCS), 15, 16-17.

Gokaran, S.G., Rao, C.K. and Singh, B.P., 1995, Crustal structure in southeast Rajasthan using magnetotelluric techniques Mem. Geol. Soc. India, 31, 373-381.

Gokaran, S.G., Rao, C.K., Gupta, G., Singh, B.P. and Yamashita, M., 2001, Deep crustal structure in Central India using magnetotelluric studies. Geophys. J. Int., 144, 685-694.

Gokarn, S.G., Gupta, G., Walia, D., Sanabam, S.S. and Hazarika, N., 2008, Deep geoelectric structure over the lower Brahmaputra valley and Shillong Plateau, NE India using magnetotellurics, Geophys. J. Int., 173, 92-104.

Goldflam, P., Menzel, H. and Szelwis, R., 1977, Interpretation of gravity data with a periodical structure, Geoexploration, 15, 155-161.

Golynsky, A.V., Alyavdin, S.V., Masolov, V.N., Tscherinov, A.S. and Volnukhin, V.S., 2002, The composite magnetic anomaly map of the East Antartica Tectonophysics, 347, 109-120.

Gomez-Ortiz, D. and Agarwal, B.N.P., 2005, 3 DINVER N: MATLAB program to invert the gravity anomaly over a 3-D horizontal density interface by Parker-Oldenburgs algorithm, Computers and Geosciences, 31, 513-529.

Gongjian, W., Xuchang, X. and Tingdong, L., 1991, Global Geoscience Transect 3: Yadong to Golmud Transect Qinghai-Tibet Plateau, China, American Geophys. Union, Publication No. 189, 1-32.

Gopalakrishnan, K., 2003, An overview of Southern Granulite Terrain, India- constraints in reconstruction of Precambrian assembly of Gondwanaland, Mem. Geol. Soc. of India, 50, 47-78.

Gordon, R. G., DeMets, C. and Royer, J. Y., 1998, Nature, 395, 370–374.

Gordon, R.G., Argus, D.F. and Heflin, M.B., 1999, Revised estimate of the angular velocity of India relative to Eurasis (abstract), EOS Trans. Am. Geophys. Union, 80 (46), Fall Meet. Suppl., F273.

Götze, H.J. and Lahmeyer, B., 1988, Application of three dimensional interactive modelling in gravity and magnetics, Geophysics, 56, 1096-1108.

Götze, H.J. and Li, X., 1996, Topography and geoid effects on gravity anomalies in mountainous areas as inferred from the gravity field of the Central Andes, Phys. and Chem. of Earth, 21, 295-297.

Gowd, T.N. and Srirama Rao, S.V., 1992, Tectonic stress field in the Indian Subcontinent, Journal of Geophysical Research, 97, 11879-11888.

Gowd, T.N., Srirama Rao, S.V., Chary, K.B. and Rummel, F., 1986 , In situ stress measurements carried out for the first time in India by hydraulic fracturing method, Proc. Indian Acad. Sci. (Earth Planet. Sci.), 95, 311-319.

Graham, K.W.T., 1962, The remagnetization of surface outcrop by lightning currents, Geophys. J.. 6, 29-40.

Grant, F.S. and West, W.D., 1965, Interpretation theory in applied geophysics, McGraw Hill Book Co., New York.

Grant, F.S., 1957, A problem in the analysis of geophysical data, Geophysics, 22, 309-344.

Grant, F.S., 1984/85b, Aeromagnetics, geology and ore environments, II, magnetite in igneous, sedimentary and metamorphic rocks: an overview, Geoexploration, 23, 335-362.

Grant, F.S., 1984/1985a, Aeromagnetics, geology and ore environments, I, magnetite in igneous, sediments and metamorphic rocks, an overview, Geoexploration, 23, 303-333.

Gregg, P. M., Lin, J., Behn, M. D., Montesi, L. G. J., 2007, Spreading rate dependence of gravity anomalies along oceanic transform faults, Nature, 448, 183-187.

Greggory, R.W., 1986, The geology of plate margins, Geol. Soc. of America, Map and Chart Series, Mc-59.

Grevemeyer, I., Tiwari, V.M., 2006, Overriding plate controls spatial distribution of mega-thrust earthquakes in the Sunda-Andaman subduction zone, Earth's Planetary Sci. Letters, 251, 199-208.

Groten, E. and Becker, M., 1995, Methods and experiences of high precision gravimetry as a tool for crustal movement detection, J. Geodynamics, 19(2), 141-157.

Grow, J. A. and Bowin, C. O., 1975, Evidence for high density crust and mantle beneath the Chile trench due to descending lithosphere, J. Geophys. Res., 80, 1449-58.

GSI, 1983, Special Publications Series, No. 2, Volume I & II, Published by Geological Survey of India, Calcutta.

GSI, 1993, Geological Map of India, scale 1: 5 Million, Published by Geological Survey of India, Calcutta.

GSI, 1995: The geological map of India published by Geological Survey of India, Calcutta.

GSI, 1999, Proceedings of Golden Jubilee Seminar on Exploration Geophysics in India, GSI, Special Publication No. 49, 1-509.

GSI, 2000, Seismotectonic atlas of India and its environs, Published by Geological Survey of India, Calcutta, 1-43.

GSI, 2000, Seismotectonic map of Kachchh area, Published by Geological Survey of India, Calcutta.

GSI, 2001, Aeromagnetic image of part of Peninsular Shield, Map on 1:2 million. Published by Geol. Sur. of India, Hyderabad.

GSI, 2002, Geophysical Surveys in India, Prospect and Retrospect, Special Publication, No. 75, Published by Geological Survey of India, Calcutta.

GSI-NGRI, 2006, Gravity Map Series of India, Published by Geol. Survey of India and National Geophysical Research Institute, Hyderabad, India.

Guillaume, A., 1978, The Ivrea zone and the Chaman fault zone: An attempt of compared tectonophysics, 2nd Symposium Ivrea-Verbano Varallo, Juin, CRE/78 No.2, 1-7.

Guillot, S., Cosea, M., Allemand, P. and Le Fort, P., 1999, Contrasting metamorphic and geochronologic evolution along the Himalayan belt, Geological Society of America Special Paper 328,117-128.

Gulatee, B.L., 1956, Gravity data in India, Survey of India, Dehra Dun, Technical Paper no. 10.

Gummert, W.R., 1997, Aerogravity surveying system: a highly effective exploration tool, Proc. of Workshop on 'Airborne Geophysics', Published by Assoc. of Exploration Geophysicists, Hyderabad, India, 41-49.

Gunn, P.J., 1976, Direct mapping of interfaces and thicknesses of layers using gravity and magnetic data, Geoexploration, 14, 75-80.

Gunn, P.J., 1995, An algorithm for reduction to pole that works at all magnetic latudes, Exploration Geophysics, 26, 247-254.

Gunnell, Y., Gallagher, K., Carter, A., Widdowsen, M. and Hurford, A.J., 2003, Denudation history of the continental margin of western Peninsular India since the early Mesozoic-reconcilling apatite fission track data with geomorphologies. Earth Planet. Sci. Lett., 215, 187-201.

Gupta H. K., Radhakrishna, I., Chadha, R. K., Kumpel, H. J and Grecksch, *G.,* 2000, Pore pressure studies initiated in area of reservoir induced earthquakes in India, EOS, transactions, American Geophysical Union, 81, 145, 151.

Gupta Sarma, D. and Singh, B., 1999, New scheme for computing the magnetic field resulting from a uniformly magnetized arbitrary polyhedran, Geophysics, 64, 70-74.

Gupta Sarma, D., Maru, V.M. and Varadarajan, G., 1976, An improved pulsed transient airborne electromagnetic system for locating good conductors, Geophysics, 41, 287-299.

Gupta, H. K. et al, 2001,Bhuj earthquake of 26th January, J. Geol. Soc. Ind., 57, 275-278.

Gupta, H. K., 1992, Reservoir Induced Earthquake, Elsevier, Amsterdam.

Gupta, H. K., Indra Mohan and Hari Narain, 1970, The Godavari valley earthquake sequence of April, 1969, Bull. Seis. Soc. of America, 60, 601-615.

Gupta, H. K., Rao, C. V. R. K., Rastogi, B. K. and Bhatia, S. C., 1980, B.S.S.A., 70, 1833-1847.

Gupta, H., Purnachandra Rao, N., Shashidhar, D. and Mallika, K., 2008, The disastrous M 7.9 Sichuan earthquake of 12 May 2008, Journal Geological Society of India, 72, 325-330.

Gupta, H., Shasidhar, D., Pereira, M., Mandal, P., Rao, N.P., Kousalaya, M., Satyanarayana, H.V.S. and Dimri, V.P., 2007, earthquake forecast appears feasible at Koyna, India, Current Science, 93, 843-848.

Gupta, H.K. and Hari Narain, 1967, Crustal structure in the Himalayan and Tibet plateau region from surface wave dispersion, Seismol. Soc. Am. Bull. 57, 235-248.

Gupta, H.K. and Rastogi, B.K., 1976, Dams and Earthquakes, Elsevier, Amsterdam, 1-229.

Gupta, H.K. et al., 2003, Borehole investigations in the surface rupture zone of the 1993 Latur SCR earthquake, Maharashtra, India: Overview of results, Mem. Geol. Soc. of India, 54, 1-22.

Gupta, H.K., 1993, The deadly Latur earthquake, Science, 262, 1666-1667.

Gupta, H.K., 2002, A review of recent studies of triggered earthquakes by artificial water reservoirs with special emphasis on earthquakes in Koyna, India, Earth Science Review, 58, 279-310.

Gupta, H.K., Harinarayana, T., Kousalya, M., Mishra, D.C., Indra Mohan, Puranachandra Rao, N., Raju, P.S., Rastogi, B.K., Reddy, P.R. and Sarkar, D., 2001. Bhuj earthquake of 26th January, 2001, J. Geol. Soc. Ind., 57, 275-278.

Gupta, H.K., Narain, H., Rastoge, B.K. and Mohan, I., 1969, A study of the Koyna earthquake of 10 December 1967, Bull. Seismol. Soc. Am., 59, 1149-1162.

Gupta, H.K., Rao, R.U.M., Srinivasan, R., Rao, G.V., Reddy, G.K., Dwivedy, K.K., Banerjee, D.C., Mohanty, R. and Satyasaadhi, Y.R., 1999, Anatomy of surface rupture zones of two stable continental region earthquakes, 1967 Koyna and 1993 Latur, India, Geophysical Research Letters, 26, 1985-1988.

Gupta, H.K., Sarma, S.V.S., Harinarayana, T and Virupashi, G., 1996, Fluids below the hypocentral zone of the Latur earthquake, India: Geophysical indicators, Geophysical Research Letters, 23, 1569-1572.

Gupta, S., Rai, S.S., Prakasam, K.S. and Srinagesh, D., Bansal, B.K., Chadha, R.K., Priestley, K. and Gaur, V.K., 2003, The nature of the crust in southern India: Implications for Precambrian crustal evolution, Geophysical Research Letters, 30, No.8, 1419. doi:10.1o29/2002GL016770.

Guptasarma, D. et al., 1989, Case history of a Kimberlite discovery, Wajrakarur area, A.P., South India, Proceedings of Exploration' 87, In: Third Decennial International Conference on Geophysical and Geochemical Exploration for Minerals and Groundwatwer, (Ed) Garland, G. D., 888-897.

Gutenberg, B., 1956, Great earthquakes between the period 1896-1906, EOS Transactions, 37, American Geophysical Union, 608.

GWR, 2004, Home page GWR instruments Inc.htm.

Hackney, R., Echtler, H., Franz, G., Götze, H.J., Lucassen, F., Marchenko, D., Melnick, D., Meyer, U., Schmidt, S., Tatarova, Z., Tassara, A. & Wienecke, 2006, The segmented overriding plate and coupling at the south-central Chile margin, In. Oncken, G. Chong, G. Franz, P., Giese, H.J., Götze, V., Ramos, M., Stecker and P. Wigger (eds.), The Andes-active subduction orogeny, Frontiers in Earth Sciences, 1, Springer Verlag, 355-374.

Hahn, A., 1965, Two applications of Fourier's analysis for the interpretation of geomagnetic anomalies, J. Geomag. and Geoelect., 17, 195-225.

Hahn, A., Ahrendt, H., Meyer, J. and Hufen, J.H., 1984, A model of magnetic sources within the earth's crust compatible with the field measured by the Satellite Magsat, Geol. Jb., Hannover, Germany, A75, 125-156.

Hahn, A., Kind, E.G. and Mishra, D.C., 1976, Depth estimation of magnetic sources by means of Fourier amplitude spectra. Geophysical Prospecting, 24, 287-308.

Haines, S.S., Klemperer, S.L., Brown, L., Jingru, G., Mechie, J., Meissner, R., Ross, A. and Zhao, W., 2003, INDEPTH III seismic data: From surface observations to deep crustal processes in Tibet, Tectonics, 22, 1001, doi:10.1029/2001TC001305, 1-18.

Hamilton, W., 1987, Crustal extension in the Basin and Range Province, southwestern United States, Geological Society, London, special Publications, 28, 155-176.

Hammer, S., 1939, Terrain correction for gravimeter stations, Geophysics, 4, 184-194.

Hammer, S., 1945. Estimating ore masses in gravity prospecting, Geophysics, 10, 50-62.

Hammer, S., 1983, Airborne gravity is here!, Geophysics, 48(2), 213-223.

Hammer, S., Anzoleaga, R, 1975, Exploring for stratigraphic traps with gravity gradients, Geophysics, 40, 256-268.

Han, S.C., Shum, C.K., Bevis, M., Ji, C., Kuo, C.Y., 2006, Crustal dilatation observed by GRACE after the 2004 Sumatra-Andaman earthquake, Science, 313, 658-662.

Han, S.C., Shum, C.K., Ditmar, P., Visser, P., Van Beelen, C. and Schrama, E.J.O., 2006, Aliasing effect of high frequency mass variations on GOCE recovery of earth's gravity field, J. of Geodynamics, 41, 69-76.

Hari Narain, 1969, Airborne geophysical surveys in India, Eastern Metals Review, 1-7.

Hari Narain, Qureshy, M.N. and Appa Rao, V., 1964, Gravity studies, Geophysics in India, Bull. NGRI. 1, 63-75.

Hari Narain, Sanker Narayan, P.V., Mishra, D.C. and Achuta Rao, D., 1969, Results of airborne magnetometer profile from offshore Manglore to offshore Madras (India) along 13th degree parallel, Pure & Applied Geophysics, 75, 133-139.

Harinarayana, T., 1999, Combination of EM and DC measurements for upper crustal studies. Surveys Geophysics, 20, 257-278.

Harinarayana, T., Naganjaneyulu, K. and Patro, B.P.K., 2006, Detection of a collision zone in south Indian shield region from magnetotelluric studies, Gondwana Research 10, 48-56.

Harinarayana, T., Naganjaneyulu, K., Manoj, C., Patro, B.P.K., Kareemmunnisa Begum, S., Kuarthy, D.N., Rao, M., Kumaraswamy, V.T.C. and Virupakshi, G., 2003, Magnetotelluric investigations along Kuppam-Paslani geotransect, South India – 2-D modeling results, Memoir Geological Society of India, 50, 107-124.

Harinarayana, T., Narayanan, M., Murthy, D.N., Gupta, A.K., Babu, N., 2010, Stationary magnetotelluric monitoring system for earthquake research in Koyna region, Maharastra, Personal Communication.

Harris, L.B. and Beeson, J., 1993, Gondwanaland significance of Lower Paleozoic deformation in Central India and SW western Australia, Journal of Geological Society of London, 150, 811-814.

Harris, N., 2007, Channel flow and the Himalayan-Tibetan orogen: a critical review, J. Geol. Soc., London, 164, 511-523.

Harrison, T.M., Yin, A. and Ryerson, F.J., 1998, Orographic Evolution of the Himalaya and Tibetan Plateau, Oxford University Press, (eds. Thomas J. Crowley and Kevin C. Burke), 39-72.

Hartman, R.R., Teskey, D.J. and Friedberg, J.L., 1971, A system for rapid digital aeromagnetic interpretation, Geophysics, 36, 891-918.

Hasesgawa, A., 1989, Seismicity: Subduction zone, In: The Encyclopedia of Solid Earth Geophysics, James, D.E. (Ed), Van Nostrand Reinhold, New York, 1054-61.

Hatcher, R.D., Jr. and Zietz, I., 1980, Tectonic implications of regional aeromagnetic and gravity data from the southern Appalachians in Wones, D.R. (ed.), The Cledonides in the USA, Va. Polytech. Inst. And State Univ. Mem., 2, 235-244.

Hatharton, T., Pattiaratchi, D.B. and Ranasinghe, V.V.C., 1975, Gravity map of Sri Lank, 1:1,000,000, Professional Paper, No.3. Geological Survey Department, Republic of Sri Lanka, 1-39.

Hauck, M.L., Nelson, K.D., Brown, L.D., Zhao, W. and Ross, A.R., 1998, Crustal structure of Himalaya orogeny at ~90 east longitude from INDEPTH deep reflection profiles, Tectonics, 17, 481-500.

Hautot, S., Tarits, P., Whaler, K., Le gall, B., Tiercelin, J.J., and Le Turdu, C. 2000, The deep structure of the Baringo Rift basin (central Kenya) from 3D magnetotelluric imaging: implications for rift evolution. J.Geophy. Res., 105, 23493-23518

Haxy, W.F., Turcotte, D.L. and Bird, J.M., 1976, Thermal and mechanical evolution of the Michigan Basin, Tectonophysics, 36, 57-75.

Hegde, V.S. and Chavadi, V.C., 2009, Geochemistry of late Archean metagreywackes from the Western Dharwar Craton, South India: Implications for provenance and nature of the late Archean crust, Gondwana Research, 15, 178-187.

Heiland, C.A., 1946, Geophysical Exploration, New York, Prentice Hall Inc., 1-1013.

Heirtzler, J.R., LePichon, X. and Baron, J.G., 1966, Magnetic anomalies over the Reykjanes Ridge, Deep Sea Res., 13, 427-43.

Heiskanen, W.A. and Moritz, H., 1967, Physical Geodsy, W.H. Freeman and Co.

Hekinian, R. et al., Francheteau, J., et al., 1983, Intense hydrothermal activity at the axis of the east pacific rise near 13^0N: submersible witnesses the growth of sulphide chimney, Marine Geophysical Researches, 6,1-14.

Hemant, K. and Maus, S., 2005, Why no anomaly is visible over most of continent-ocean boundary in the global crustal magnetic field, Physics of the Earth and Planetary Interiors, 149, 321-333.

Hemant, K. and Mitchell, A., 2009, Magnetic field modeling and interpretation of the Himalayan-Tibetan plateau and adjoining north Indian plains, Tectonophysics, 478, 87-99.

Hering, A., Misiek, R., Guylai, A., Ormos, T., Dobroka M. and Dresen, L., 1995, A joint inversion algorithm to process geoelectric and surface wave seismic data. Part I: basic ideas. Geophy.Pros., 43, 135-156.

Herzburg, C., 2007, Food for a volcanic diet, Science, 316, 378-416.

Hess, H.H., 1962, History of ocean basins, In: Petrologic Studies: A Volume in Honor of A.F. Buddington (eds) A.E.J. Engel, H.L. James and B.F. Leonard, Boulder, Geol. Soc. Am., 599-620.

Hetenyi, G., Cattin, R., Brunet, F., Bollinger, L., Vergne, J., Nabelek, J.L. and Diament, M., 2007, Density distribution of the India plate beneath the Tibetan Plateau: geophysical and petrological constraints on the kinetics of lower-crustal eclogitization, Earth and Planet. Science Letters, 264, 226-244.

Hetenyi, G., Cattin, R., Vergne, J. and Nabelek, J.L., 2006, The effective elastic thickness of the India Plate from receiver function imaging, gravity anomalies and thermomechanical modeling, Geophys. J. Int., 167, doi:10.1111/j.1365-246X.2006.03198.x, 1106-1118.

Hetenyi, M., 1946, Beams on elastic foundation, University of Michigan Press, Ann Arbor, 1-255.

Hildebrand, P.R., Searle, M.P., Shakirullah, Khan, Z. and Van Heijst, H.J., 2000, Geological evolution of the Hindu Kush, NW Fontier Pakistan: active margin to continent-continent collision zone, Khan, M.A., Treloar, P.J., Searle, M.P. and Jan, M.Q. (eds) Tectonics of the Nanga Parbat Syntaxis and the Western Himalaya, Geological Society of London, Special Publications, 170, 277-293.

Hildenbrand, T.G., 1985, Rift structure of the northern Mississippi Embayment from the analysis of gravity and magnetic data, J. of Geophy. Research, 90, 12607-12622.

Hildenbrand, T.G., Stuart, W.D. and Talwani, P., 2001. Geologic structures related to New Madrid earthquakes near Memphis, Tennessee, based on gravity and magnetic interpretations, Engineering Geology, 62, 105-121.

Hinz, K., 1981, A Hypothesis on terrestrial catastrophes : wedges of very thick ocean ward dipping layers beneath passive continental margins- their origin and paleo environmental significance, Geol. Jahrbuck Series E 22, 3-28.

Hinz, W.J. and Zietz, I., 1985, The composite magnetic anomaly map of the conterminous, United States, In: The utility of regional gravity and magnetic anomaly maps (ed.) W.J. Hinz, Society of Exploration Geophysicists, USA, 1-23.

Hirano, N., Takahashi, E., Yamamoto, J., Abe, N., Ingle, S.P., Kaneoka, I., Hirata, T., Kimura, J., Ishii, T., Ogawa, Y., Machida, S. and Suyehiro K., 2006, Volcanism in response to plate flexure, Science, 313, 1426-1428.

Hodges, K., 2006, Climate and the evolution of mountains, Scientific American Magazine, August. 56-63.

Hodges, K.V., Wobus, C., Ruhl, K., Schildgen, T., Whipple, K., 2004, Quaternary deformation, river steepening and heavy precipitation at the front of the Higher Himalayan Ranges, Earth and Planetary Science Letters, 220, 379-389.

Hodgetts, D., Egan, S.S. and Williams, G.D., 1998, Flexural modeling of continental lithosphere deformation: a comparison of 2D and 3 D techniques, Tectonophysics, 294, 1-20.

Hoffman, A. W. and Hart, S. R., 2007, Another nail in plume coffin, Science, 315, 39-40.

Hoffman, K.A., 1988, Ancient geomagnetic reversals: clues to geodynamo, Scientfic American, 256, 76-83.

Hoffman, P.F., 1991, did the break out of Laurentia turn Gondwana in side out, Science, 252, 1409-12.

Holliger, K. and Kissling, E., 1992, Gravity interpretation of a Unified 2-D acoustic image of the central Alpine collision zone, Geophys. J. Int., 111, 213-225.

Holmes, A., 1944, Principles of Physical Geology, Thomas Nelson, London.

Hood, P., 1964, The Koingsberger ratio and dipping dyke equation, Geophysical Prospecting, 12(4), 440-456.

Hopkin, M., 2005, Nature, 433, 3; doi: 10.1038/14330036.

Hu, C.,Henderson, G.M.,Huang, J.,Xie, S.,Sun, Y.,Johnson, K.R. 2008, Quantification of Holocene Asian monsoon rainfall from spatially separated cave records. Earth and Planetary Science Letters, 266, 221-232.

Hubbard, M.S., 1996, Ductile shear as cause of inverted metamorphism: Example from the Nepal Himalaya, The J. of Geology, 104, 493-499.

Hung, S.H., Chen, W.P., Chiao, L.Y. and Tseng, T.L., 2010, First multi scale finite frequency tomography illuminates 3-D anatomy of the Tibetan plateau, Geophys. Res. Lett., 37, doi: 10.1029/2009GL041875.

Hutchinson, D.R., Grow, J.A., Klitgord, K.D. and Swift, B.A., 1982, Deep structure and evolution of Carolina trough, Studies in Continental Margin Geology (eds.: J.S. Watkins and C.L. Drake), Am. Assoc. of Petroleum Geologists Memoir, 34, 129-165.

Hyndman, R.D., Currie, C. A. and Mazzotti, S.P., 2005, Subduction zone back arc, mobile belts and orogenic heat, GSA Today, 15, 4-10.

IAGA, 2006, IGRF-10, http://www.ngdc.noaa.gov/IAGA/vmod/igrf.html

Ildefonse, B., Blackman, D. K., John, B. E., Ohara, V. Miller, D. J., Mac Leod,. J. and integrated ocean drilling program expeditions, 3041305 science party, 2007, Oceanic core complexes and crustal accretion at slow spreading ridges, Geology, 35, 623-626.

IMD Website, 2000, http://www.imd.ernet.in/seismo/main.new.htm. A preliminary report on seismic activity in Bhavnagar district of Gujarat during August – September 2000, India Meteorological Department – A report.

Intierra, 2007, Gadarwara Copper/Gold prospect, Project Research Report. Intierra Resource Intelligence, Website.

Intierra, 2008, Project Research Report, Gadarwara Copper Gold Prospect, Web Site of Intierra Resource Intelligence, Australia.

Irving, E., 1959, Paleomagnetic pole positions: A survey and analysis, Geoph. J. R. Astron. Soc., 2, 51-79.

Jachens, R.C. and Zoback, M.L., 2000, The San Andreas fault in the San Francisco Bay region, California, Structure and Kinematics of a young plate boundary, In: Tectonic Studies of Asia and the Pacific rim, (ed.) W.G. Ernst and R.G. Coleman, Geol. Soc. of America, Int. Book Sr., 3, 217-231.

Jachens, R.C., Simpson, R.W., Blakely, R.J. and Saltus, R.W., 1989, Isostatic residual gravity and crustal geology of the United States, (eds.) L.C. Pakiser and W.D. Mooney, Geophysical framework of the continental United States, The Geol. Soc. of America, Memoir, 172, 317-348.

Jackson, D.D., 1972, Interpretation of inaccurate, insufficient and inconsistent data, Geophys. J. R. Astr. Soc., 28, 97-100.

Jackson, J. and McKenzie, D., 1984, Active tectonics of the Alpine-Himalayan belt between western Turkey and Pakistan, Geoph. J. Royal Astr. Soc., 77, 185-224.

Jackson, J., 2002, Strength of the continental lithosphere: Time to abandon the jelly sandwich? GSA Today, 4-10.

Jade, S., M. Malay, A. P. Imtiyaz, M. B. Ananda, P. Dileep Kumar and V. K. Gaur, 2002. Estimates of coseismic displacement and post-seismic deformation using Global Positioning System geodesy for the Bhuj earthquake of 26 January 2001, Current Science, 82, 748-752,

Jade, S., 2004, Estimates of plate velocity and crustal deformation in the Indian subcontinent using GPS geodesy, Current Science, 86, 10, 1443-1448.

Jade, S., Bhatt, B.C., Yang, Z., Bendick, R., Gaur, V.K., Molnar, P., Anand, M.B. and Kumar, D., 2004, GPS measurements from the Ladakh Himalaya, India: Preliminary tests of plate-like or continuous deformation in Tibet, GSA Bulletin: Nov-Dec., 116, 11/12, 1385-1391.

Jagadeesh, S. and Rai, S.S., 2008, Thickness, composition, and evolution of the Indian Precambrian crust inferred from broadband seismological measurements, Precambrian Research, 162, 4-15.

Jain, A.K., Singh, S., Manickavasagam, R.M., Joshi, M. and Verma, P.K., 2003, HIMPROBE program: Integrated studies on geology, petrology, geochemistry and geophysics of the Trans Himalaya and Karakoram, In: Indian Continental Lithosphere: Emerging Research Trend (eds) T.M. Mahadevan, B.R. Arora and K.R. Gupta, Mem. Geol. Soc. India, 53, 1-56.

Jain, R.C., 2003. Hydrologic effects of the 2001 Bhuj earthquake Intl. Workshop on Earth System Processes Related to Gujarat Earthquake Using Space Technology, IIT Kanpur, India, 91.

Jain, S.C., Yedekar, D.B. and Nair, K.K.K., 1991, Central Indian Shear zone: a major Precambrian crustal boundary. Jour. Geol. Soc. India. 37, 521-532.

Jakosky, J.J., 1961, Exploration Geophysics, Times-Mirror Press, U.S.A., 1-1195.

James, D., 2002, How old roots lose their bounce?, Nature, 417, 911-913.

Jarchow, C.M., Catchings, R.D. and Lutter, W.J. 1994, Large explosive source, wide recording aperture, seismic profiling on the Columbia Plateau, Washington. Geophysics, 59, 259-271

Jayananda, M., Moyan, J.F., Martin, H., Peucat, J.J., Auvray, B. and Mahabaleshwar, B., 2000, Late Archean (2550-2520 Ma) Juvenile magmatism in the eastern Dharwar Craton, Southern India: Constraints from geochronology, Nd-Sr isotopes and whole rock geochemistry. Precambrian Res., 99, 225-254.

Jiang, M., Galve, A., Him, A., de Voogd, B., Laigle, M., Su, H.P., Diaz, J., Lepine, J.C., Wang, Y.X., 2006, Crustal thickening and variations in architecture from the Qaidam Basin to the Qang Tang (North-Central Tibetan Plateau) from wide-angle reflection seismology. Tectonophysics, 412, 121-140.

Jimenez-Munt, I., Fernandez, M., Verges, J., Platt, J. P., 2008, Lithosphere structure underneath the Tibetan Plateau inferred from elevation, gravity and geoid anomalies, Earth & Planetary Science Letters, 267, 276-289.

Jin, Y., McNutt, M.K. and Zhu, Y., 1994, Evidence from gravity and topography data for folding of Tibet, Nature, 371, 669-674.

Jin, Y., McNutt, M.K. and Zhu, Y., 1996, Mapping of descent of Indian and Eurasian plates beneath the Tibetan Plateau from gravity anomalies, J. of Geophys. Res., 101, B5, 11,275-11,290.

Johnson, H.P. and Carlson, R.L., 1992, Variation of sea floor depth with age: a test of models based on drilling results, Geophys. Res. Lett., 19, 1971-74.

Johnson, K., Y.J. Hsu, P. Segall, and S.B. Yu, 2001. Fault geometry and slip distribution of the 1999 Chi-Chi, Taiwan earthquake imaged from inversion of GPS data, Geophys. Res. Lett., 28, 2285-2288.

Jones, A.G. 1998, Waves of the future: Superior inferences from collocated seismic and electromagnetic experiments. Tectonophysics, 286, 273-298.

Jones, A.G., 1992, Electrical conductivity of the continental lower crust, In: Fountain, D.M., Arculus, R.J., Kay, R.W. (Eds.), Continental lower crust, Elsevier, 81-143.

Jones, E.J.W., 1999, Marine Geophysics, John Wiley and Sons, Ltd., Canada, 1-466.

Jordan, T.A. and Watts, B., 2005, Gravity anomalies, flexure and the elastic thickness structure of the India-Eurasia collision system, Earth and Planetary Science Letters, 236, 732-750.

Joshi, J.P. and Bisht, R. S., 1994, India and the Indus Civilization, National Museum Institute.

Jouanne, F., Mugnier, J.L., Gamond, J.F., Le Fort, P., Pandy, M.R., Bollinger, L., Flouzat, M. and Avouac, J.P., 2004, Current shortening across the Himalaya of Nepal, Geophys. J. Int., 157, 1-14.

Jupp, D.L.B. and Vozoff, K., 1975, Stable interactive methods for the inversion of the geophysical data, Geophys, J. R.A.S., 42, 957-976.

Kabban,M.K., Schwintzer, P., Reigber, C, 2005, Dynamic topography in Global gravity field, Earth Observation with CHAMP (ed) C. Reigber., H. Luehr, P. Schwintzer and J. Wickert, Springer Varlag Berlin Heidelburg, 199-204.

Kahle, H. J., Cocard, M., Peter, Y., Geiger, A., Reilinger, R., Barka, A. and Veis, G., 2000, GPS derived strain field rate with in the boundary zones of the Eurasian, African and Arabian plates, J. Geophys. Res., 105, 23353-370.

Kaila, K. L., Murty, P. R. K., Rao V. K. and Kharetchko, G. E., 1981, Tectonophysics, 73, 365-384. Mishra, D. C., 1989, Geol. Soc. India, 33, 48-54.

Kaila, K. L., Reddy, P. R., Dixit, M. M., Lazorenko, M. A., 1981a, Deep crustal structure of Koyna, Maharashtra indicated by deep seismic soundings, J. Geol Soc, Ind., 22, 1-16.

Kaila, K. L., Tewari, H.C. and Mall, D. M., 1987, Crustal structure and delineation of Gondwana basin in the Mahanadi delta area, India form deep seismic soundings, Geol. Soc. India, 29, 293-308.

Kaila, K.L. and Bhatia, S.C., 1981, Gravity study along Kavali-Udipi seismic sounding profile in the Indian Peninsular shield: some inferences about origin of anorthosites and Eastern Ghat Orogeny, Tectonophys., 79, 129-143.

Kaila, K.L. and Krishna V.G., 1992, Deep seismic sounding studies in India and major discoveries. Current Science, 62, 117-154.

Kaila, K.L., Krishna, V.G., and Mall, D.M., 1981, Crustal structure along the Mehamadabad-Billimoria profile in the Cambay basin, India from deep seismic sounding, Tectonophys., 76, 99-130.

Kaila, K.L., Krishna, V.G., Roychowdhury, K. and Hari narain, 1978, Structure of the Kashmir Himalaya from deep seismic soundings. J. Geol. Soc. India, 19, 1-20.

Kaila, K.L., Murthy, P.R.K., Mall, D.M., Dixit, M.M. and Sarkar, D., 1987, Deep seismic soundings along Hirapur-Mandla profile, Central India. Geophys. J. R. Astr. Soc., 89, 399-404.

Kaila, K.L., Murthy, P.R.K., Rao, V.K. and Kharetchko, G.E., 1981b, Crustal structure from Deep seismic sounding along the Koyna profile II in the Deccan Trap area, India, Tectonophysics 73, 365-384.

Kaila, K.L., Murthy, P.R.K., Rao, V.K. and Venkateswarlu, N., 1990, Deep seismic sounding in the Godavari gaben and Godavari (coastal) basin, India, Tectonophysics, 173, 307-317.

Kaila, K.L., Reddy, P.R., Mull, D.M., Venkateswarlu, N., Krishna, V.G., Prasad, A.V.S.S.R., 1992, Crustal structure of the West Bengal basin, India from deep seismic sounding investigations, Geophysical Journal International, 111, 45-66.

Kaila, K.L., Roy Chowdhury, K., Reddy, P.R., Krishna, V.G., Hari Narain, Subbotin, S.I., Sollogulb, V.B., Chekunov, A.V., Kharetcho, G.E., Lazarenko, MA. And Ilchenko, T.V., 1979, Crustal structure along the Kavali-Udipi profile in the Indian Peninsular Shield from deep seismic sounding. J. Geol. Soc. India, 20, 307-333.

Kaila, K.L., Tewari, H.C. and Sharma, P.L.N., 1980, Crustal structure from deep seismic sounding studies along Navibander-Amreli profile in Saurastra, Mem. Geol. Soc. of India, 3, 218-232.

Kaila, K.L., Tewari, H.C., Krishna, V.G.,, Dixit, M.M., Sarkar, D. and Reddy, M.S., 1990, Deep seismic sounding studies in the North Cambay and Sanchor basins, India, Geophys. J.Int., 103, 621-627.

Kaila, K.L., Tripathi, K.M. and Dixit, M.M., 1984, Crustal structure along Wular lake-Gulmarg-Naoshera profile across Pir Panjal range of the Himalaya from deep seismic soundings, Jr. Geol. Soc. India, 25, 706-719.

Kailasam, L.N., Murthy, B.G.K. and Chayanulu, A.Y.S.R., 1972, Regional gravity studies of the Deccan Trap areas of Peninsular India, Curr. Sci., 41, 403-407.

Kak, S.N., Bhairam, C.L. and Dwivedy, K.K., 1997, Development of airborne geophysical techniques for uranium resources assessment in India – A review, Proc. of workshop on Airborne Geophysics, (ed.) C.V. Reeves, Association of Exploration Geophysicists, Hyderabad, India and International Institute for Aerospace Survey and Earth Sciences, The Netherlands, 93-98.

Kamble, R.K., Rani, C.K., Ghosh, N. and Panvalkar, 2010, Acoustic and electrical logging for evaluation of resistivity and shear and compressional wave velocities of foundation of Kakrapar nuclear plant, Presented in 33 Annual Convention and Seminar on Exploration Geophysics, Journal of Geophysics (In Press).

Kanao, M., Kamiyama, K., Ito, K., 1994. Crustal density structure of the Mizuho Plateau, East Antarctica from gravity survey in 1992. Proc. NIPR Symposium, Antarctic Geosciences, 7, 23-36.

Kanasewich, E.R. and Agarwal, R.G., 1970, Analysis of combined gravity and magnetic fields in wave number domain, J. of Geophys. Res., 75, 5702-5712.

Kanasewich, E.R., 1975, Time sequences analysis in geophysics, II Edition, The University of Alberta Press, Canada, 1-364.

Kannadasan, T. and Srinivas, G., 2002, Use of aeromagnetic and satellite data for the identification of potential deep fractured/acquifer zones- A case study from north Arcot district, Tamilnadu, Geophysical surveys in India, Prospect and Retrospect, GSI, Special Publication, 75, 236-250.

Kapp, P., Taylor, M., Stockli, D. and Ding, L., 2008, Development of active low-angle normal fault systems during orogenic collapse: Insight from Tibet, Geology, 36, 7-10, doi:10.1130/G24054A.1.

Karanth, R.V. and Sant, D.A., 1995, Lineaments and dyke swarms of lower Narmada valley and southern Saurashtra, Mem. Geol. Soc. India, 33, 425-434.

Karig, D.E., 1971, 'Origin and development of marginal basins in the western Pacific', J. Geophys. Res., 76, 2542-2561.

Karner, G.D. and Watts, A.B., 1983, Gravity Anomalies and Flexure of the Lithosphere at Mountain Ranges, J. Geophys. Res., 88, B12, 10,449-10,477.

Karner, G.D. and Weissel, J.K., 1990, Factors controlling the location of compressional deformation of oceanic lithosphere in the central Indian ocean, J. Geoph. Res., 95, 19795-197810.

Karner, G.D., 1982, Spectral representation of isostatic models, BMR J. of Aust. Geology and Geoph., 7, 55-62.

Karunakaran, C. and Ranga Rao, A., 1976, Status of exploration for hydrocarbons in the Himalayan region – contributions to stratigraphy and structure, New Delhi, Himalayan Geophys. Seminar, Miscellaneous Publication, Geological Survey of India, 1-72.

Kaulakov, I. and Sobolev, S., 2006, A tomographic image of Indian lithosphere break-off beneath the Pamir-Hindukush region, Geophys. J. Int., 164, 425-440.

Kawakatsu, H. and Watada, S., 2007, Seismic Evidence for Deep-Water Transportation in the Mantle, Science, 1468-71.

Kayal, J.R. and Zhao, D., 1998, Bull. Seism. Soc. Am., 88, 667-676.

Kayal, J.R., 2008, Micro earthquake seismology and seismotectonics of South Asia, Capital Publishing Co., New Delhi, 1-503.

Kayal, J.R., 2010, Himalayan tectonic model and great earthquakes: an appraisal, Geomatics, Natural Hazards and Risk, 1, 51-67.

Kayal, J.R., Arefiev, S.S., Barua, S. Hazarika, D., Gogoi, N., Kumar, A., Chowdhury, S.N., Kalita, S. 2006, Shillong Plateau earthquakes in northeast India region: complex tectonic model, Current Science, 91,. 109-113.

Kayal, J.R., Reena, De., Sagina Ram, Srirama, B.V. and Gaonkar, S.G., 2002. Aftershocks of the 26 january, 2001 Bhuj earthquake in western India and its seismotectonic implications, J. Geol, Soc. Ind., 59, 395-417.

Kayal, J.R., Zhao, D., Mishra, O.P., De, R. and Singh, O.P., 2002, The 2001 Bhuj earthquake: Tomographic evidence for fluids at the hypocenter and its implications for rupture nucleation, Geophysical Research Letters, 29, 2152, doi: 10.1029/2002GL015177.

Kayanne, H. et al., 2007, Coseismic and postseismic creepin the Andaman islands associated with the 2004 Sumatra-Andaman earthquake, Geophysical Reseach Letters, 34, L01310.

Kazmi, A.H., 1979. Active fault systems in Pakistan, Geodynamics of Pakistan, ed.: A. Farah and K.A. Dejong, Geol. Surv. of Pakistan, 285-304.

Kearey, P., and Vine, F.J., 1990, Implications of Plate Tectonics, Geoscience Texts, Chapter 11, Published by Oxford, Blackwell Scientific Publications, London, Edinburgh, Boston, Melbourne, Paris, Berlin and Vienna.

Kearey, P., Klepeis, K.A., Vine, F.J., 2009, Global Tectonics, Wiley-Blackwell, U.K., 1-482.

Keen, C. E. and Tramontini, C., 1970, A seismic refraction survey on the Mid-Atlantic Ridge, Geophys. J. R. Astro. Soc., 20, 473-91.

Keller, G.R., Karlstrom, K.E., Williams, M.L., Miller, K.C., Andronicos, C., Levander, A.R., Snelson, C.M. and Prodehl, C., 2005, The dynamic nature of the continental crust-mantle boundary: Crustal evolution in the southern rocky mountain region as an example, The American Geophysical Union, 10.1029/154GM30, 403-441.

Kellogg, L.H., Hager, B.H., and van der Hilst, R.D., 1999, Compositional stratification in the deep mantle, Science, 1999, 283, 1881-84.

Kellogg, L.H., Hager, B.H., Van der Hilst, R.D., Compositional stratification in the deep mantle, Science, 283, 1881-1884.

Kennet, B. L. N. and Widiyantoro, S., 1999, A low seismic wave speed anomaly beneath north-western India: a seismic signatures of the Deccan hot spot, Earth & Planet. Sci. Letters, 165, 145-155.

Kennett, B.L.N., and Widiyantoro S., 1999, A low seismic wave speed anomaly beneath northwestern India: A seismic signature of the Deccan plume. Earth Planet. Sci Letts., 165, 145-155.

Kent, D.V. and Gradstein, M., 1986, A Jurassic to recent chronology, In the Geology of North America, vol. M., The Western North Atlantic Region (P.R. Vogt and B.E. Tucholke, Eds.) Geol. Soc of America, Boulder, Co., 45-50.

Kent, R., 1991, Lithospheric uplift in eastern Gondwana: evidence for a long lived mantle plume system, Geology, 19, 19-23

Kerr, R.A., 2004, Hammered by India, Puttylike Tibet shows limits of plate tectonics, Science, 305, 161.

Kerr, R.A., 2008, Pumping up the Tibetan plateau from the far Pacific Ocean, Science, 321, 1028-1029.

Khan, M.A., Quasim Jan, M., Windley, B.F., Tarney, J. and Thirlwall, M.F., 1989, The Chilas mafic-ultramafic igneous complex; the root of the Kohistan island arc in the Himalaya of northern Pakistan, In: Tectonics of the Western Himalaya (eds) L.L. Malinconico, Jr. and R.J. Lillie, Geol. Soc. of Am. Special Paper, 232, 75-94.

Khattri, K.N., 1987, Great earthquakes, seismicity gaps and potential for earthquake disaster along the Himalaya plate boundary, Tectonophysics, 138, 79-92.

Kinck, J.J., Husebye, E.S. and Larsen, F.R., 1993, The Moho depth distribution in Fennoscandia and the regional tectonic evolution from Archean to Permian times, Precambrian Research, 64, 23-51.

Kind, R., Yuan, X., Saul, J., Nelson, D., Sobolev, S.V., Mechie, J., Zhao, W., Kosarev, G., Ni, J., Achauer, U. and Jiang, M., 2002, Seismic images of crust and upper mantle beneath Tibet: Evidence of Eurasian plate subduction, Science, 298, 1219-1221.

King, J., Harris, N., Argles T., Parrish R., Charlier, B., Sherlock, S. and Zhang, H.F., 2007, First field evidence of southward ductile flow of Asian crust beneath southern Tibet, Geology, 35, 727-730; doi:10.1130/G23630A.1.

Kious, W.J. and Tilling, R.I., 1999, This dynamic earth: the story of plate tectonics, U.S.Geological Survey, 1-77

Kirby, J.F., 2005, Which wavelet best reproduces the Fourier power spectrum, Computers and Geosciences, 31, 846-864.

Kiselev, S., Vinnik, L., Oreshin, S., Gupta, S., Rai, S.S., Singh, A., Kumar, M.R. and Mohan, G., 2008, Lithosphere of the Dharwar craton by joint inversion of P and S receiver functions. Geophys. J. Int., 173, 1106-1118.

Klemme, H.D. and Ulmishek, G.F. 1991, 'Effective Petroleum Source Rocks of the World: Stratigraphic Distribution and Controlling Depositional Factors', AAPG Bulletin, 75, 1809-1851.

Klemperer, S.L., 2008, Reconcilling lithospheric deformation and lower crustal flow beneath central Tibet: COMMENT and REPLY, Geology, 35, COMMENT: doi:10.1130/G25097C.1, e180.

Klingele, E.E., Cocard, M. and Kahle, H.G., 1997, Kinematic GPS as a source for airborne gravity reduction in the airborne gravity survey of Switzerland, J. of Geoph. Res., 102, 7705-7715.

Klootwick, C.T. and Peirce, J.W., 1979, Indian and Australian pole path since the late Mesozoic and the India-Asia collision, Nature, 282, 605-607.

Klootwijk, C.T., 1979, A review of paleomagnetic data from the Indo-Pakistani fragment of Gondwana land, Geodynamics of Pakistan, (eds.) A Farah and K.A. Dejong, 41-80.

Klootwijk, C.T., Sharma, M.L., Gergan, J., Shah, S.K. and Gupta, B.K., 1986, Rotational overthrusting of the N-W Himalaya: further paleomagnetic evidence from the Riasi thrust sheat, Jammu foothills, India, Earth and Planetary Science Letters, 80, 375-393.

Knao, M., Kamiyama, K. and Ito, K., 1994, Crustal density structure of the Mizuho Plateau, east Antarctica from gravity survey in 1992., Proc. NIPR Symposium, Antarctic Geosciences, 7, 23-36.

Kogan, M.G. and McNutt, M.K., 1987, Isostasy in USSR, Admittance data, In: Fucs, K. and Froidevaux, C. (eds.) Composition, Structures and Dynamics of the Lithosphere-Asthenosphere System, Geodynamics Series, 16, Am. Geophysical Union, Washington, D.C., 301-307.

Kola-Ojo, O. and Meissner, R., 2001, Southern Tibet: its deep seismic structure and some tectonic implications, J. of Asian Earth Sciences, 19, 249-256.

Korhonen, J. et al., 2007, Magnetic Anomaly Map of the World / Carte des anomalies magnetiques du monde, CCGM/CCGMW, 1: 50,000,000, 1 ed., Geological Survey of Finland, Helsinki.

Korsakov, O.D., 1974, on the tectonics of the Faeroe-Shetland trough, Doklady Akad. Nauk, USSR, 214, 647-650.

Kosarev, G., Kind, R., Sobolev, S.V., Yuan, X., Hanka, W. and Oreshin, S., 1999, Seismic evidence for a detached Indian lithospheric mantle beneath Tibet, Science, 26, Feb., 283, 1306-1308.

Kowalik, W.S. and Glenn, W.E., 1987, Image processing of aeromagnetic data and integration with landsat images for improved structural interpretation, Geophysics, 52, 875-884.

Krishna Brahmam, N. and Negi, J.G., 1973, Rift valleys beneath the Deccan Trap (India), Geophys. Res. Bull. 11(3), 207-237.

Krishna Brahmam, N., Kanungo, D. N., 1976, Inference of granitic batholiths by gravity studies in South India. Journal of the Geological Society of India. 17; 1, Pages 45-53.

Krishna Brahmam, N., Sarma, J. R. K., Arvamadhu, P. S., Subba Rao, D. V., 1986, Bouguer gravity anomaly map (NGRI/GPH-6) of Cuddapah basin (India), Published by National Geophysical Research Institute, Hyderabad, India.

Krishna Brahman, N. and Subba Raju, L.V., 1972, Reduction of gravity data in India, Geoph. Res. Bull., 10, 79-89.

Krishna, K.S., 2003, structure and evolution of the Afanasy Niktin Seamount, buried hills and 85 E Ridge in the northeastern Indian Ocean, Earth and Planet. Science Letters, 209, 379-394.

Krishna, K.S., Gopala Rao, D. and Sar, D., 2006, Nature of the crust in the Laxmi basin (14°-20° N), western continental margin of India, Tectonics, 25, Tc 1006, 1-18.

Krishna, K.S., Neprochnov, V.P., Gopal Rao, D. and Grinko, B.N., 2001, Crustal structure and tectonics of the Ninety East Ridge from seismic and gravity studies, Tectonics, 20, 416-433.

Krishna, V.G. and Rao, V.V., 2005, Processing and modeling of short-off set seismic refraction-coincident deep seismic reflection data sets in sedimentary basins: an approach for exploring the underlying deep crustal structures, Geophys. J. Int., 163, 112-1122.

Krishna, V.G., Kaila, K.L., Reddy, P.R., 1991, Low velocity layers in the sub crustal lithosphere beneath the Deccan Trap region of Western India, Physics of Earth & Planet. Int., 67, 288-302.

Krogstad, E.J., Balakrishnan, S., Mukhopadhyay, D.K., Rajamani, V., Hanson, G.N., 1989. Plate tectonics, 2.5 billion yeas ago: Evidence at Kolar South India. Science 243, 1337-1340.

Kroner, A., Kehelpannala, K.V.W. and Hegner, E., 2003, Ca. 750-1100 Ma magmatic events and Grenville-age deformation in Sri Lanka: relevance for Rodinia supercontinent formation and dispersal, and Gondwana amalgamation, J. of Asian Earth Sciences, 22, 279-300.

Kuehl, S., Alexander, C., Carter, L., Gerald, L., Gerber, T., Harris, C., McNinch, J., Orpin, A., Pratson, L., Syvriski, J. and Walsh, J.P., 2006, Probing electrical conductivity of the Trans-European suture zone, EoS, 87, 281.

Kuhle, M., 2007, The past valley glacier network in the Himalaya and the Tibetan ice sheet during the last glacial period and its glacial isostatic eustatic and climate consequences, Tectonophysics, 445, 116-144.

Kukowski, N., Schillhorn, T., Flueh, E.R. and Huhn, K., 2000, Newly identified strike slip plate boundary in the north eastern Arabian sea, Geology, 28, 355-358.

Kumar, A., Dayal, A.M. and Padmakumari, V.M., 2003, Kimberlite from Rajmahal Magmatic province: Sr-Nd-Pb isotopic evidence for Kerguelen plumes derived magmas, Geoph. Res. Letters, 30, SDE 9-1-4.

Kumar, A., Kumari, P., Dayal, A.M., Murthy, D.S.N. and Gopalan, K., 1993, Rb-Sr ages of Proterozoic kimberlites of India: evidence for contemporaneous emplacements. Precambrian Research, 62, 227-237.

Kumar, A., Pande, K. and Venkatesan, T.R., Bhaskar Rao, Y.J., 2001, The Karnataka Late Cretaceous dykes as products of the Marion hot spot at the Madagascar-India break up event: evidence from ^{40}Ar- ^{39}Ar Geochronology and geochemistry, Geoph. Res. Letters, 28, 2715-2718.

Kumar, N., Singh, A.P. and Singh, B., 2009a, Structural fabric of the Southern Indian shield as defined by gravity trends, Journal of Asian Earth Science, 34, 577-585.

Kumar, N., Singh, A.P., Gupta, S.B. and Mishra, D.C., 2004, Gravity signature, crustal architecture and collision tectonics of the eastern ghats mobile belt, J. Ind. Geophys. Union, 8, 97-106.

Kumar, N., Singh, A.P., Rao, M.R.K.P., Chandrasekhar, D.V., Singh, B., 2009b, Gravity signatures, derived crustal structures and tectonics of Achankovil shear zone, Gondwana Research, 16, 45-55.

Kumar, P., Tewari, H.C. and Khandekar, G., 2000, An anomalous high-velocity layer at shallow crustal depths in the Narmada zone, India, Geophys. J. Int., 142, 95-107.

Kumar, P., Yuan, X., Kind, R. and Kosarev, G., 2005, The lithosphere-asthenosphere boundary in the Tien Shan-Karakoram region from S receiver functions: Evidence for continental subduction, Geophys. Res. Letters, 32, L07305, doi:10.1029/2004/GL022291, 1-4.

Kumar, P., Yuan, X., Kind, R. and Ni, J., 2006, Imaging the colliding Indian and Asian lithospheric plates beneath Tibet, J. Geophys., Res., 111, B06308,doi:10.1029/2005JB003930, 1-11.

Kumar, P., Yuan, X., Kumar, M.R., Kind, R., Li, X. and Chadha, R.K., 2007, The rapid drift of the Indian tectonic plate, Nature, 449, doi:10.1038/nature06214, 894-897.

Kumar, S., Wesnousky, S.G., Rockwell, T.K., Briggs, R.W., Thackur, V.C. and Jayagondaperumal, R., 2006, Paleoseismic evidence of great surface rupture earthquakes along the Indian Himalaya. J. Geoph. Res., Solid Earth, 111, doi:10.1029/2004JB033309, ...

Kummerow, J., 2002, Strukturunter suchungen in des ostalpen anhand des teleseismichen TRAMSALP-Datensatzes, Ph.D., Thesis at Frei Universitat, Berlin.

Kump, L.R., Pavlov, A. and Arthur, M.A., 2005, Massive release of hydrogen sulphide to the surface ocean and atmosphere during intervals of oceanic anoxia, Geology, 33, 397-400.

Kusznir, N.J., Marsden, G. and Egan, S.S., 1991, A flexural centilevel simple-shear/pure-shear model of continental lithosphere extension: Applications to the Heanne d' Arc basio, Grand Banks and Viking Graben, North Sea, In: Roberts, A.M., Yielding, G. and Freeman, B. (eds.), the Geometry of Normal Faults, Sp. Publication Geol. Soc. of London, 56, 41-60.

La Fond, E. C. and Dietz, R. S., 1964, Lonar crater, India, a meteorite crater? Meteoritics, 2, 111–116.

La Touche, T. H. D. and Christie, W. A. K., 1912, The geology of the Lonar lake. Rec. Geol. Surv. India, 14, 266–289.

Lafehr, T.R., 1991, Standardization in gravity reduction, Geophysics, 56, 1170-1178.

Lagious, E., Drakopoulos, J., Hipkin, R.G. and Gizeli, C., 1988, Microgravimetry in Greece: application to earthquake and volcanoeruption prediction, Tectonophysics, 152, 197-207.

Lahiri, A.K., Chakraborty, P.K. and Singh, N.P., 2008, Repeat microgravity survey in Sikkim-Darjeeling Himalaya for crustal deformation study, Current Science,

Lal, B.B. India 1947-1997, 1998, New light on the Indus civilization. New Delhi, 116-123.

Lal, B.B., 1984. The earliest datable earthquake in India, Science Age, 8, 8-9.

Lal, B.B., 1998, India 1947-1997: New light on the Indus civilization. New Delhi, 116-123.

Lamb, S., 2002, Is it all in the crust?, Nature, 420/14 November 2002/www.nature.com/nature, 130-131.

Lambert, A. and Beaumont, C., 1977, J. Geophy. Res, 82, 297- 306.

Lambert, A., Liard, J. O. and Mainville, A., 1986, J. Geophy. Res., 91, 9150-9160.

Lanczos, C., 1961, Linear differential operators, D. van. Nostrand, Princeton, New Jersey

Lane, R.J.L. (editor), 2004, Airborne gravity 2004-abstracts from the ASEG-PESA airborne gravity 2004 Workshop, Geoscience Australia Record 2004 / 18.

Langel, R.A., Phillips, J.D. and Horner, R.J., 1982, Initial scalar magnetic anomaly maps from MAGSAT, issued by NASA.

Langenheim, V. E. and Hildenbrand, T.G., 1997, Commerce geophysical lineament- its source, geometry and relation to reelfoot rift and New Madrid seismic zone, GSA Bulletin, 109, 580-595.

Lawver, L.A., Dalziel, I.W.D., Gahagan, L.M., Martin, K.M. and Campbell, D.A., 2003, The PLATES 2003 Atlas of plate reconstruction (750 Ma present Day), PLATES Progress Report No. 280-0703, University of Texas Institute for Geophysics Technical Report No. 190, Houston, Texas, University of Texas Press.

Lawver, L.A., Gahagan, L.M., and Coffin, M.F., 1992, The development of paleoseaways around Antarctica, the Antarctic paleoenvironment: A perspective on global change (ed.) J.P. Kennet and D.A.Werneke, AGU Antarctic Research Series, 56, 7-30.

Leaman, D.E., 1998, The gravity Terrain correction-Practical considerations, Exploration Geophysics, 29, 467-471.

LeBas, M.J., LeMaitre, R.W., Streckeisen. A. and Zanettin, B., 1986, A chemical classification of volcanic rocks based on the total alkali-silica diagram, J. Petrol, 27, 745-750.

Lebedev, S. and Van der Hilst, R.D., 2008, Global upper mantle tomography with the automated multimode inversion of surface and S-wave forms. Geophys. J. Int., 173, 505-518.

Leech, M. L., 2008, Does the Karakoram fault interrupt mid-crustal channel flow in the western Himalaya? Earth & Plantery Science Letters, 276, 314-322.

Leech, M.L., 2001, Arrested orogenic development: eclogitization, delamination and tectonic collapse, Earth and Planetary Science Letters, 185, 149-159.

Leelanandam, C., Burke, K., Ashwal, L.D. and Wekke, S.J., 2006, Proterozoic mountain building in peninsular India: an analysis based on primarily on alkaline rock distribution, Geological Magazine, doi:10.1017/ S001675 6805001 6 64.

LeFort, P., 1975, Himalaya: The collided range: Present knowledge of the continental arc, Am. J. of Science, 275A, 1-44.

Lemmonier, C., Marquis, G., Perrier, F. et al., 1999, electrical structure of the Himalaya of Central Nepal: High conductivity around the mid crustal ramp along the MHT, Geophysical Research Letters, 26, 3261-3264.

Lemoine, F.G., Pavlis, N.K., Kenyon, S.C., Rapp, R.H., Pavlis, E.C. and Chao, B.F., 1998, New high resolution model developed for earths gravitational field, EOS, AGU, 79, 113, 117-118.

Lente Van, B., Ashwal, L.D., Pandit, M.K., Bowring, S.A. and Torsvik, T.H., 2009, Neoproterozoic hydrothermally altered basaltic rocks from rajasthan, northwest India: Implications for late Precambrian tectonic evolution of the Aravalli Craton, Precambrian Research 170, 202-222.

Li, C., Van der Hilst, R.D. and Toksoz, 2006, Constraining P-wave velocity variations in the upper mantle beneath Southeast Asia. Physics of the Earth and Planetary Interiors, 154, 180-195.

Li, C., Van der Hist R.D., Meltzer, A.S. and Robert Enghahi, E., 2008, Subduction of the Indian lithosphere beneath the Tibetan Plateau and Burma, Earth and Planetary Science Letters, 274, 157-168.

Li, S., Unsworth, M.J., Booker, J.R., Wei, W., Tan, H. and Jones, A.G., 2003, Partial melt or aqueous fluid in the mid-crust of Southern Tibet? Constraints from INDEPTH magnetotelluric data, Geophys. J. Int., 153, 289-304.

Li, X., Chouteau, M., 1998, Three dimensional gravity modelling in all space surveys in Geophysics, 19, 339-368.

Liang, C. and Song, X., 2006, A low velocity belt beneath northern and eastern Tibetan Plateau from Pn tomography, Geoph. Res. Lett., 33, L22306 doi: 10.1029/2006/ GL027926.

Likhite, S.D. and Radhakrishnamurty, C., 1965, An apparatus for the determination of the susceptibility of rocks in low fields at different frequencies, Bull. Nat. Geophys. Res. Inst, 3 ,1-8.

Lillie, R.J., 1999, Whole Earth Geophysics, Prentice Hall, New Jersey, USA.

Lillie, R.J., Bielik, M., Babuska,V., Plomerova.,J., 1994, Gravity modeling of the lithosphere in the eastern Alpine-western Carpathian-Pannonian basin region, Tectonophysics, 231, 215-235.

Lister, G., Kennett, B., Richard, S. and Fonster, M., 2008, Boudinage of a stretching slablet implicated in earthquakes beneath the Hindu Kush, Nature Geoscience, 1, 196-200.

Lliboutry, L., 1999, Quantitative Geophysics and Geology. Springer, Praxis Publishing, U.K., 1-480.

Longman, I.M., 1959, Formulas for computing the tidal accelerations, due to the Moon and Sun, J. of Geoph. Res., 64(12), 2351-2355.

Lonsdale, P., 1977, Deep tow observations at the mounds abyssal hydrothermal field, Galapagos rift, Earth and Planetary Science Letters, 36, 92-110.

Losecke, W. Knödel, K. and Müller, W., 1979, The conductivity distribution in the North German sedimentary basin derived from widely spaced areal magnetotelluric measurements, Geophys. J. R. astr. Soc., 58, 169-179.

Louden, K.E. and Forsyth, D.W., 1982, Crustal structure and isostatic compensation near the Kane fracture zone from topography and gravity measurements, I. Spectral analysis approach, Geophys. J.R. Astron. Soc., 768, 725-750.

Lowrie, W., 1997, Fundamentals of Geophysics, Cambridge University Press, 1-354.

Lowry, A.R. and Smith, R.B., 1994, Flexural rigidity of the basin and range- Colorado Plateau-Rocky Mountain transition from coherence analysis of gravity and topography, J. Geophys. Res., 99, 20123-20140.

Lowry, A.R. and Smith, R.B., 1995, Strength and rheology of US Cordillera, J. Geophys. Res., 100, 17947-17963.

LRG,1995, *GRA VPACK,* LaCoste and Romberg Meters, Inc. Austin, Texas.

Luirei, K. and Bhakuni, S.S., 2008, Landslides along frontal part of eastern Himalaya in East Siang and lower Dibang districts, Arunachal Pradesh, India, Journal Geological Society of India, 71, 321-330.

Luthcke, S.B., Zwally, Abdalati, W., Rowlands, D.D., Ray, R.D., Nerem, R.S., Lemoine, F.G., McCarthy, J.J. and Chinn, D.S., 2006, Recent Greenland ice mass loss by Drainage system from Satellite gravity observations, Science, 314, 1286-1289.

Lynn, C.E., Cook, F.A. and Hall, K.W., 2005, Tectonic significance of potential-field anomalies in western Canada: Results from the Lithoprobe SNORCLE transect, Can. J. Earth Sci., 42, 1239-1255.

Lyon-Caen, H. and Molnar, P., 1983, Constraints on the structure of the Himalaya from an analysis of gravity anomalies and a flexural model of the lithosphere, J. of Geophys., Res., 88, B10, 8171-8191.

Lyon-Caen, H. and Molnar, P., 1985, Gravity anomalies, flexure of the Indian plate, and the structure, support and evolution of the Himalaya and Ganga basin, Tectonics, 4, 6, 513-538.

Macario, A., Malinverno, A. and Haxby, W.F., 1995, On the robustness of effective elastic thickness estimates using the coherence method, J. Geophys. Res., 100, 15163-15172.

MacDonald, G.A. and Katsura, T., 1964, Chemical composition of Hawaiian lavas, J. Petrol, 5 ,82-133.

Macdonald, K.C., Becker, K., Spiess, F.N., and Ballard, R.D., 1980, Hydrothermal heat flux of the "black smoker"vents on the east pacific rise, Earth and Planetary Science Letters, 48, 1-7.

Macnae, C.J., 1979, Kimberlite and exploration geophysics, Geophysics, 44, 1395-1416.

Macnae, C.J., 1995, Applications of geophysics for the detection of Kimberlites and lamproites, J. Geochem. Expl., 53, 213-243.

Maggi, A., Jackson, J.A., McKenzie, D. and Priestley, K., 2000, Earthquake focal depth, effective elastic thickness and strength of continental lithosphere, Geology, 28, 645-649.

Maggi, A., Jackson, J.A., McKenzie, D. and Priestley, K., 2000, Earthquake focal depth, effective elastic thickness and the strength of the continental lithosphere, Geology, 28, 495-498.

Mahadevan, T.M., 2003, Geological evolution of South Indian Shield- constraints on modeling, Mem. Geol. Soc. of India, 50, 25-46.

Mahatsente, R. and Ranalli, G., 2004, Time evolution of negative buoyancy of an oceanic slab subducting with varying velocity, J. of Geodynamics, 38, 117-129.

Maheshwari, M.K. and Sar, D., 2004, Signature of continental to oceanic transition in Krishna-Godavari Offshore: India, 5th Conference and Exposition on Petroleum Geophysics, Hyderabad, 13-18.

Mahoney, J.J., Sheth, H.C., Chandrasekharam, D. and Peng, Z.X., 2000, Geochemistry of flood basalts of the Toranmal section, Northern Deccan traps, India: Implications for regional Deccan stratigraphy, Journal of Petrology,41, 1099-1120.

Mahony, J.J., Macdougall, J.D., Lugmair, G.W., Gopalan, K., 1983, Kerguelen hotspot source for Rajmahal traps and Ninety East Ridge, Nature, 303, 385-389.

Majumdar, T.J., Mohanty, K.K., Mishra, D.C. and Arora, K., 1998, Gravity image generation over the Indian subcontinent using NGRI / EGM 96 and ERS-1 altimeter data, Current Science, 80, 542-554.

Majumder, S. and Mamtani, M.A., 2009, Magnetic fabric in the Malanjkhand granite (Central India) – Implications for regional tectonics and Proterozoic suturing of the Indian shield, Physics of the Earth and Planetary Interiors, 172, 310-323.

Malahoff, A., 1983, Hydrothermal vents and poly metallic sulphides of the Galapagos and Gorda/Juan de Fuca Ridge systems and of submarine volcanoes, Hydrothermal Vents, 19-41.

Malik, J.N., Sohoni, P.S., Karanth, R.V. and Merh, S.S., 1999. Modern and historic seismicity of Kachchh Peninsula, Western India, J. Geol, Soc. Ind., 54, 545-550.

Mall, D.M., Rao, V.K. and Reddy, P.R., 1999, Deep sub-crustal features in the Bengal basin: Seismic signatures for plume activity, Geophysical Research Letters, 26, 16, 2545-2548.

Mallick, K. and Sharma, K.K., 1999, A finite element method for computation of the regional gravity anomaly, Geophysics, 64, 461-469.

Mallick, K. and Sharma, K.K., 1999, Inter-relationship of Bouguer regional and residual gravity anomalies-a case study from earthquake prone Latur region, Maharashtra, India, J. of Geophysics, 20, 65-69.

Mandal, P. and Chadha, R.K., 2008, Three dimensional velocity imaging of the Kachchh seismic zone, Gujarat, India, Tectonophysics, 452, 1-16.

Mandal, P. and Pandey, O.P., 2010, Relocation of aftershocks of the 2001 Bhuj earthquake: A new insight into seismotectonics of the Kachchh seismic zone, Journal of Geodynamics, 49, 254-260.

Mandal, P., Rastogi, B.K., Satyanarayana, H.V.S. and Kousalya, M., Bull. Seismol. Soc. Am., 2004, 94, 633-649.

Manghnani, M, and Woolard, G.P., 1963, Establishment of N-S gravimetric calibration line in India, J. Geoph. Res., 68 (23), 6293-6301.

Manglik, A. and Verma, S.K. 1998, Delineation of sediments below flood basalts by joint inversion of seismic and magnetotelluric data', GRL, 25, 4015-4018

Maochang, T. et al., 1998, Reference untraceable.

Markandeyulu, A., Kumar, B.V.L., Jain, S.K., Israeli, I.H., Yadav, O.P. and Roy, M.K., 2009, Uranium exploration by geophysical techniques in Upper Gondwanas of Satpura Gondwana basin- A case study from Jhirpa area, Chhindwara district, Madhya Pradesh, India, Jour. of Geophysics, 30, 3-9.

Marlow, M.S., Exon, N.F. and Dadisman, S.V., 1992, Hydrocarbon potential and gold mineralization in the New Ireland basin, Papua New Guinea, Geology and Geophysics of Continental Margins (Ed.) J.S.Watkins, F.Zhiqiang and K.J. McMillan, AAPG Memoir, 53, 119-137.

Marquardt, D.W., 1963, An algorithm for least squares estimation of non linear parameters, Jour. of the Society of Industrial and Applied Mathematics, 11, 431-441.

Marsh, J.G., 1979, Satellite derived gravity maps section 2. In: Lowman Jr. P.D. and Frey, H.V. (Eds.). A Geophysical Atlas for interpretation of Satellite Derived Gravity Data. NASA, Green Belt, Maryland, USA.

Martelet, G., Saihac, P., Moreau, F. and Dimont, M., Characterization of geological boundaries using 1-D wavelet transform on gravity data, Theory and application to Himalayas, Geophysics,

Marussi, A. and Ebblin, 1976, The tectonic scheme of central Asia (compiled), Bouguer anomaly map (1975), Ace Naz. Lincei, 21, 131-137.

Mathew, M.P., Ramchandra, H.M., Gouda, H.C., Singh, R.K., Acharya, G.R., Murthy, Ch.V.V.S. and Rao K.S., 2001, IGRF corrected regional aeromagnetic anomaly map of parts of Peninsular India - Potential for mapping and mineral exploration, Geol. Surv. of India, special Pub., no. 58, 395-405.

Matson, R.G. and Moore, G.F., 1992, Structural influences on Neogene subsidence in the Central Sumatra fore-arc basin, Geology and Geophysics of Continental Margins (Ed.) J.S.Watkins, F.Zhiqiang and K.J. McMillan, AAPG Memoir, 53, 157-181.

Matuyama, M., 1936, Distribution of gravity over the Nippon trench and related areas, Proc. Imp. Acad., Japan, 12, 93.

Maule, C.F., Purucker, M.E., Olsen, N. and Mosegaard, K., 2005, Heat flux anomalies in Antarctica revealed by satellite magnetic data, Science, 309, 464-467.

Maus, S. and Dimiri, V.P., 1995, Potential field power spectrum inversion for scaling geology, J. of Geoph. Res., 100, 12605-12616.

Maus, S., Barckhausen, U., Berkenbosch, H., Bournas, N., Brozena, J., Childers, V., Dostaler, F., Fairhead, J.D., Finn, C., von Frese, R.R.B., Gaina, C., Golynsky, S., Kucks, R., Luhr, H., Milligan, P., Mogren, S., Muller, D., Olesen, O., Pilkington, M. Saltus, R., Schreckenberger, B., Thebault, E., Tontini, F.C., 2010, EMAG2: A 2-arc minute resolution earth magnetic anomaly grid compiled from satellite, airborne and marine magnetic measurements, Journal of geophysical Research (In Press).

Maus, S., Fairhead, J.D., Mogren S. and Bournas, N., 2011, EMAG3: A 3-arc-minute resolution global magnetic anomaly grid compiled from satellite, airborne and marine magnetic data, http://geomag.org/models/EMAG.html.

Maus, S., Rother, M., Holme, R., Lühr, H., Olsen, N. and Haak, V., 2002, First scalar magnetic anomaly map from CHAMP satellite data indicate weak lithospheric field, Geoph. Res. Letters, 29(14), 47-1-4.

Maus, S., Yin, F., Luhr, H., Manoj, C., Rother, M., Rauberg, J., Michaelis, I., Stolle, C. and Muller, R.D., 2008, Resolution of direction of oceanic magnetic lineations by the sixth generation lithospheric magnetic field model from CHAMP satellite magnetic measurements, Geochemistry, Geophysics, Geosystems, 9, 7021, doi: 10.1029/2008GC001949 (Maus et al., 2010).

Mazumdar, J.J., Mohanty, K.K., Mishra, D.C., Arora, K., 2001, Gravity image generation over the Indian subcontinent using NGRI/ EGM96 and ERS-1 altimeter data, Current Science, 80, 542-554.

McAdoo, D.C. and Martin, C.F., 1984, Seasat observations of lithospheric flexure seaward of frenches, J. Geoph. Res., 89, 3201-3210.

McElhinny, M.W. and McFadden, P.L., 2000, Paleomagnetism, continents and oceans, Academic press, U.K., 1-386

McFadden, P. L., 1987, A periodicity of magnetic reversals: comments, Nature, 330, 27.

Mckenzie, D. and Bowin, C., 1976, The relationship between bathymetry and gravity in the Atlantic Ocean, J. Geophys. Res., 81, 1903-1915.

Mckenzie, D. and Fairhead, D., 1997, Estimates of the effective elastic thickness of the continental lithosphere from Bouguer and free-air gravity anomalies, J. Geophys. Res., 102, 27523-27552.

McKenzie, D. and Priestley, K., 2008, The influence of lithospheric thickness variations on continental evolution. Lithos., 102, 1-11.

Mckenzie, D. and Sclater, J.G., 1971, The evaluation of Indian ocean since the late Cretaceous, Geophys, J. R. Astro. Soc., 24, 437-528.

McKenzie, D., 2010, The influence of dynamically supported topography on estimates of Te, Earth and Planetary Science Letters, 295, 127-138.

McKenzie, D.P. and Bowin, C., 1976, The relationship between bathymetry and gravity in the Atlantic Ocean, J. Geophys. Res., 81, 1903-1915.

McKenzie, D.P. and Morgan, W.J., 1969, Evolution of triple junctions, Nature, 224, 125-133.

Mckenzie, D.P. and Sclater, J.G., 1971, The evolution of the Indian ocean since the late Cretaceous, Geophys. J. Roy, Astr. Soc., 25, 437-528.

McKenzie, D.P., 1969, Speculations on the consequences and causes of plate motion, Geophys. J. R. astro. Soc., 18, 1-32.

Mckenzie, D.P., 1978, Some remarks on the development of sedimentary basins, Earth & Planet. Sci. Lett., 40, 25-32.

McNutt, M., 2006, Another nail in plume coffin, Science, 1394.

McNutt, M.K. and Kogan, M.G., 1987, Isostasy in USSR II, Interpretation of admittance data, In: Fuchs, K. and Froidervaux, C. (eds.) Composition, Structure and Dynamics of the Lithosphere-Asthenosphere system, Geodynamic Series 16, American Geophysical Union, Washington, D.C., 309-327.

McNutt, M.K. and Parker, R.L., 1978, Isostasy in Australia and the evolution of the compensation mechanism, Science, 199, 773-775.

McNutt, M.K., 1979, Compensation of oceanic topography, An application of the response function technique to the surveyor area, J. Geoph. Res., 84, 7589-7598.

McNutt, M.K., 1984, Lithospheric flexure and thermal anomalies, J. Geoph. Res., 89, 11180-11194.

McNutt, M.K., Diamont, M. and Kogan, M.G., 1988, Variations in elastic plate thickness at continental thrust belts, J. Geoph. Res., 93, 8825-8838.

Mechie, J., Keller, G.R., Prodehl, C., Kahn, M.A. and Gaciri, S.J., 1997, A model for the structure, composition and evolution of the Kenya rift, Tectonophysics, 278, 95-119.

Meissner, R. and Kern, H., 2008, Earthquakes and strength in the laminated lower crust – Can they be explained by the "corset model"?, Tectonophysics, 448, 49-59.

Meissner, R. and Krawczyk, C.M., 1999, Caledonian and Proterozoic terrane accretion in the South-West Baltic Sea, Tectonophysics, 314, 1-3, 255-267.

Meissner, R., Tilmann, F., Haines, S., 2004, About the lithospheric structure of central Tibet, based on seismic data from the INDEPTH III profile, Tectonophysics, 380, 1-25.

Meltzer, A., Sarker, G., Beaudoin, B., Seeber, L. and Armbruster, J., 2001, Seismic characterization of an active metamorphic massif, Nanga Parbat, Pakistan Himalaya, Geology, 29, 7, 651-654.Mines and Geology, 2005, Petroleum exploration promotion project, issued by Ministry of Industry, Government of Nepal.

Menichetti, V. and Guillen, A., 1983, Simultaneous interactive magnetic and gravity inversion, Geoph. Prosp., 31, 929-944.

Menzies, M.A., Klemperer, S.L., Ebinger, C.J. and Beker, J., 2003, Characteristics of volcanic rifted margins, Geol. Soc, of America, Spl. Paper, 362, Boulder, Colorado, 1-14.

Merh, S.S. 1995, Geology of Gujarat, Geol. Soc. India Publication, Bangalore, India.

MERI, 1995a,b Bulletin, Maharastra Engineering Research Institute, 1995, No MERI-SDAD-BL/127, 130/AUG, DEC.

Merrill, R.T., McElhinny, M.W., McFadden, P.L., 1996, The magnetic field of the earth, Academic press, USA, 1-531.

Mezger, K., Cosca, M.A., 1999, The thermal history of the Eastern Ghats Belts (India) as revealed by U-Pb and (super 40) Ar /(super 39) A dating of metamorphic and magmatic minerals: implications for the SWEAT correlation, Precambrian Research, 94, 251-271.

Mierlo, Frank van., 2006, World Energy Consumption; BP 2006; Statistical review

Miles, P. R. and Roest, W. R., 1993, Earliest Sea floor spreading magnetic anomalies

Miles, P.R., Munschy, M. and Segoufin, J., 1998, Structure and early evolution of the Arabian Sea and East Somali basin, Geophys J. Int., 134, 876-888.

Millegan, P., 2005, Broader spectrum fewer floks-gravity and magnetic society of exploration Geophysicist @ 75, S36-S41.

Miller, S.L. and Bada, J.L., 1988, Submarine hot springs and the origin of life, Nature, 334, 609-11.

Milne, G.A., Davis, J.L., Mitrovica, J.X., Scherneck, H.G., Johansson, J.M., Vermeer, M. and Koivula, H., 2001, Space-geodetic constraints on glacial isostatic adjustment in Fennoscandia, Science, 291, 2381-2385.

Mines and Geology, Nepal, 2005, Brochure of Mines and Geology, issued by Government of Nepal for hydrocarbon exploration, 1-4.

Minshull, T.A., Lane, C., Collier, J.S. and Whitmarsh R.B., 2008, The relationship between rifting and magmatism in the northeastern Arabian sea, Nature Geoscience, 1, 463-467.

Mirza, M.A., 1973, Regional gravity survey of Cholistan Desert, Punjab, Pakistan, Records of the Geological Survey of Pakistan, 23.

Mishra , D.C., 1977, Possible Extension of Narmada-Son lineament towards Murray Ridge (Arabian Sea) and the eastern syntaxial bend of the Himalayas, Earth & Planet., Science Letters, 36, 301-308.

Mishra D.C., Tiwari V.M., Gupta, S.B. and Vyaghreswara Rao M.B.S., 1998, Anomalous mass distribution in the epicentral area of Latur earthquake, India, Current Science, 74, 469-472.

Mishra, D. C. and Pederson, L. B., 1982, Statistical analysis of potential fields from subsurface reliefs. Geoexploration, **19**, 247– 265.

Mishra, D.C. 2001, et al., Major lineaments and gravity magnetic trends in Saurashtra, India, Current Science, 80, 1059-1067.

Mishra, D.C. and Gupta, S.B., 1987, Horst and graben tectonics of western India and their inter-relationship, In: Rifted Basins and Aulacogens (ed) Casshyap, S. M., Gyanodaya Prakashan, Nanital, 33-46.

Mishra, D.C. and Gupta, S.B., 1993, Horst and graben tectonics of western India and their inter-relationship, Rifted basins and Aulacogens, Gyanodaya Prakashan, Nainital, 33-46.

Mishra, D.C. and Gupta, S.B., 1998, Mid-continent gravity 'high' and seismic activities in central India, Current Science, 74, 702-705.

Mishra, D.C. and Hari Narain, 1977, Airborne geophysical survey- a case study from central India, J. of the Geol. Soc. of India, 18, 104-110.

Mishra, D.C. and Laxman, G. 1997, Some major tectonic elements of western Ganga basin based on analysis of Bouguer anomaly map, Current Sci., 73, 436-440.

Mishra, D.C. and Mallick, K., 2008, Paleochannels and settelements of Harappa period in kalibangan, Rajasthan based on Geophysical Investigations., Current Science, 95,1657-58.

Mishra, D.C. and Naidu, P.S., 1971, Two dimensional power spectral analysis of aeromagnetic fields, Bull. NGRI, 9, 49-55.

Mishra, D.C. and Naidu, P.S., 1974, Two dimensional power spectral analysis of aeromagnetic fields, Geophys. Prosp., 22, 345-353.

Mishra, D.C., and Pedersen, L.B., 1982, Statistical analysis of potential fields from subsurface relief, Geoexploration, 19, 247-265.

Mishra, D.C. and Prajapati, S.K., 2003, A plausible model for evolution of schist belts and granite plutons of Dharwar craton, India and Madagascar during 3.0-2.5 Ga: Insight from gravity modelling constrained in part from seismic studies. Gondwana Res., 6, 501-511.

Mishra, D.C. and Rajasekhar, R.P., 2005, Tsunami of 26 December 2004 and related tectonic setting, Current Science, 88, 680-682.

Mishra, D.C. and Rajasekhar, R.P., 2006, Bhuj Earthquake of January 26, 2001: Tectonic Inversion, Lithospheric Flexure and Plate Motion and comparison with Shillong and New Madrid earthquakes, Current Science, 90, 504-506.

Mishra, D.C. and Rajasekhar, R.P., 2006, Crustal structure at the epicentral zone of the 2005 Kashmir (Muzaffarabad) earthquake and seismotectonics significance of lithospheric flexure. Current Sci.ence, 90, 1406-1412.

Mishra, D.C. and Rajasekhar, R.P., 2008, Gravity and magnetic signatures of Proterozoic rifted margins: Bundelkhand craton and Bijawar and Mahakoshal Group of Rocks and Vindhyan basin and their extension under Ganga basin, J. Geol. Soc. India, 71, 377-387.

Mishra, D.C. and Rao, M.B.S.V., 1993, Thickening of crust under the granulite province of S. India and associated tectonics based on gravity-magnetic studies. Mem. Geol. Soc. of India, 25, 203-219.

Mishra, D.C. and Rao, M.B.S.V., 1997, Temporal variation in gravity field during solar eclipse on 24 October 1995, Current Science, 72, 783-785.

Mishra, D.C. and Ravi Kumar, 1998, Characteristics of faults associated with Narmada-Son Lineament and rock types in Jabalpur sector, Current Science, 75, 308-310.

Mishra, D.C. and Ravi Kumar, M., 2008, Geodynamics of Indian plate and Tibet: Buoyant lithosphere, rapid drift and channel flow from gravity studies, 'Five Decades of Geophysics in India', Golden Jubilee Volume, Memoir Geological Society of India, No.68, 151-172.

Mishra, D.C. and Tiwari, R.K., 1981, Spectral study of the Bouguer anomaly map of a rift valley and adjacent areas in Central India, Pure and Applied Geophysics, 119, 1051-62.

Mishra, D.C. and Tiwari, V.M., 1995, An asymmetrical basic lopolith below sediments in western Cuddapah basin – Geophysical evidence, Proc. Annual Convention of Geol. Soc. of India, Dept. of Geology, University of Tirupati, Tirupati-95, 31-41.

Mishra, D.C. and Tiwari, V.M., 2010, Gravity Methods-Surface, Encyclopedia on Solid Earth Geophysics, Springer-Verlag (In Press).

Mishra, D.C. and Venkatrayudu, M., 1985, MAGSAT scalar anomaly map of India and a part of Indian ocean-Magnetic crust and tectonic correlation. Geophysical Research Letters, 12, 781-784.

Mishra, D.C. and Venkatrayudu, M., 1991, Coloured Bouguer anomaly map of Central part of Narmada-Son lineament, Abs. Volume IGU, 28th Annual Convention and seminar on Geophysics for Rural Development, Dec. 17-19, Hyderabad, 27-28.

Mishra, D.C. and Vijai Kumar, V., 2005, Evidence for Proterozoic collision from airborne magnetic and gravity studies in southern granulite terrain, India and signatures of recent tectonic activity in the Palghat gap. Gondwana Research, 8, 1-12.

Mishra, D.C. Babu Rao, V., Laxman, G., Rao, M.B.S.V. and Venkatarayudu, M., 1987, Three-dimensional structural model of Cuddapah basin and adjacent eastern part from geophysical studies, Memoir 6, Geological Society of India, 313-327.

Mishra, D.C. et al., 2011, Geodynamics of NW India based on gravity and magnetic study: Extension and interaction of Aravalli Delhi Mobile Belt and Jaisalmer-Ganganagar basement ridge due to lithospheric flexure with NW Himalayan front and Seismicity, Journal Geological Society of India (Submitted).

Mishra, D.C., 1965, Paleomagnetism of Vindhyan rocks of India, A Ph.D. Thesis submitted to Banaras Hindu University, Varanasi, India.

Mishra, D.C., 1970, Crustal model based on aeromagnetic profiles, gravity studies and wave velocities in rocks and crust, Pure & App. Geoph., 81, 5-16.

Mishra, D.C., 1977, Possible extensions of the Narmada-Son lineament towards Murray ridge (Arabian sea) and eastern syntaxial bend of Himalayas, Earth & Planet Sci. Letters, 36, 301-308.

Mishra, D.C., 1978, Harmonic inversion of an aeromagnetic anomaly from a remanently magnetized Precambrian schist belt (Karnataka, India), Geophysical Prospecting, 26, 572-580.

Mishra, D.C., 1981, Crustal structure in North Arabian Sea from magnetic surveys, Marine Geophysical Researches, 4, 427-436.

Mishra, D.C., 1982, Crustal structure and dynamics under Himalaya and Pamir ranges, Earth and Planetary Science Letters, 57, 415-420.

Mishra, D.C., 1984, Long wavelength magnetic anomalies from the lithosphere: Indian shield and Himalaya, Tectonophysics, 105, 319-330.

Mishra, D.C., 1984, Magnetic anomalies-India and Antarctica, Earth & Planetary Sci. Letters, 71, 173-180.

Mishra, D.C., 1986a, Satellite magnetic map and tectonic correlation, J. of Geol. Soc. of India, 28, 501-503.

Mishra, D.C., 1986b, Deep seated magnetic structures and related tectonics characteristics – Magsat and airborne magnetic profiles, Proc. International Symposium on Neotectonicsin South Asia, Survey of India, Dehra Dun, India, Feb. 18-21, 201-220.

Mishra, D.C., 1987, Bijawar and Vindhyan tectonics of Central India from airborne magnetics and ground geophysical surveys, Mem. Geol. Surv. of India on Purana Basins of Peninsular India, 6, 357-366.

Mishra, D.C., 1989, On deciphering the two scales of the regional Bouguer anomaly of the Deccan trap and crust-mantle inhomogeneities, J. Geol. Soc of India, 33, 48-54.

Mishra, D.C., 1990, Precambrian rifts and associated tectonics of Peninsular India, In: Precambrian continental Crust and its Economic Resources (ed) S.M. Naqvi, Elsevier, Holland, Development in Precambrian Geology 8, 487-502.

Mishra, D.C., 1991, Magnetic crust in the Bay of Bengal, Marine Geology, 99, 257-261.

Mishra, D.C., 1995, Geophysical experiments for the study of intraplate seismicity, Abstract volume, Workshop on 'Indian shield seismicity and Latur earthquake of Sept., 30, 1993, NGRI, Hyderabad, Sept. 4-5, 1995, 24-27.

Mishra, D.C., 1999. Indirect evidences of Jurassic volcanic in north Arabian sea and related tectonics on adjoining Indian continent, *IUGG 99*, Held at Birmingham, Abstract v, A113.

Mishra, D.C., 2002, Crustal structure of India and its environs based on geophysical studies, Indian Minerals, 56, 27-96.

Mishra, D.C., 2006, Building blocks and crustal architecture of Indian Peninsular Shield: Cratons and fold belts and their interaction based on geophysical data integrated with geological information, Journal Geological Society of India, 68, 1037-1057.

Mishra, D.C., 2010, A unified model of convergence and rifting of Indian cratons as part Columbian and Grenville connections and triple junctions: Geophysical constraints, AOGS- 2010, Hyderabad, July, 5-9, 2010, SE-09, Sutures and Geodynamic Processes, Abstract Volume, 293; Submitted to Asian Journal of Earth Sciences.

Mishra, D.C., 2011, Long hiatus in Proterozoic sedimentation in India: Vindhyan, Cuddapah and Pakhal basins-A plate tectonics model, Journal Geological Society of India, 77, 17-25.

Mishra, D.C., Achuta Rao, D., Agarwal, P.K. and Sarma, B.S.P., 1971, Results of airborne magnetometer profile from Hyderabad to Puri, Bull. NGRI, 9, 97-101.

Mishra, D.C., and Pedersen, L.B., 1982, Statistical analysis of potential fields from surface relief. Geoexploration, 19, 247-265.

Mishra, D.C., and Tiwari, R.K., 1981, Spectral study of the Bouguer anomaly map of a rift valley and adjacent areas in central India, Pure and Applied Geophysics, 119, 1051-1062.

Mishra, D.C., Arora, K. and Tiwari, V.M., 2004, Gravity anomalies and associated tectonics features over the Indian Peninsular Shield and adjoining ocean basins, Tectonophysics, 379, 61-76.

Mishra, D.C., Babu Rao, V., Laxman, G., Rao, M.B.S.V. and Venkatrayudu, M., 1987, Three-dimensional structural model of Cuddapah basin and adjacent eastern part from geophysical studies, Geol. Soc. India Memoir 6, 313-329.

Mishra, D.C., Chandrasekhar, D.V., Raju, D.Ch.V. and Vijai Kumar, V., 1999, Crustal structure based on gravity-magnetic modelling constrained from seismic studies under Lambert rift, Antarctica, Godavari and Mahanadi rift, India and their inter relationships. Earth & Planet. Sci. Letts., 172, 287-300.

Mishra, D.C., Chandrasekhar, D.V., Singh, B., 2005, Tectonics and crustal structures related to Bhuj earthquake of January 26, 2001: based on gravity and magnetic surveys constrained from seismic and seismological studies, Tectonophysics, 396, 195-207

Mishra, D.C., Gupta, S.B. and Rao, M.B.S.V., 1992, Bouguer anomaly and other geophysical parameters of Godavari Basin, J. of Assoc. of Explo. Geophys., 13, 95-97

Mishra, D.C., Gupta, S.B. and Rao, M.B.S.V., 1994, Space and time distribution of gravity field in earthquake affected areas of Maharashtra, India, Mem. Geol. Soc. of India, 35, 119-126.

Mishra, D.C., Gupta, S.B. and Rao, M.R.K.P., 1998, Airborne magnetic lineament, groundwater potentiality and basement structures in S-E part of Vindhyan basin, 1998, J. Geol. Soc. of India, 52, 195-202.

Mishra, D.C., Gupta, S.B. and Tiwari, V.M., 1998, A geotransect from Dharimanna to Billimora across the Cambay and Narmada-Tapti rift basins, India, International Geology review, 40, 1007-1020

Mishra, D.C., Gupta, S.B. and Venkatrayudu, M., 1989, Godavari rift and its extension towards the east coast of India, Earth and Planetary Sci. Letters, 94, 344-352.

Mishra, D.C., Gupta, S.B., Rao, M.B.S.V. and Venkatrayudu, M., 1993, Resolution of potential field anomalies from different depths-a frequency domain approach, Jour. Assoc. Expl. Geophys., 14, 117-122.

Mishra, D.C., Gupta, S.B., Rao, M.B.S.V., Venkataaydu, M. and Laxman, G., 1987, Godavari Basin – A Geophysical study, Journal Geological Society of India, 30, 469-476.

Mishra, D.C., Gupta, S.B., Vyaghreswara Rao, M.B.S. and Venkatrayudu, M., 1996, Crustal structure and basement tectonics under Vindhyan basin: Gravity-magnetic study, Memoir Geological Society of India, 36, 213-224.

Mishra, D.C., Laxman, G. and Arora, K., 2004, Large-wavelength gravity anomalies over the Indian continent: Indicators of lithospheric flexure and uplift and subsidence of Indian Peninsular Shield related to isostasy, Current Science, 86, 861-867.

Mishra, D.C., Laxman, G., 1997, Some major tectonic elements of Western Ganga basin based on analysis of Bouguer anomaly map, Current Science, 73, 436-440.

Mishra, D.C., Murthy, K.S.R. and Hari Narain, 1978, Interpretation of time domain airborne electromagnetic (INPUT) anomalies, Geoexploration, 16, 203-222.

Mishra, D.C., Murthy, K.S.R. and Rao, T.C.S., 1980, General expression for spectrum of magnetic anomaly due to long tabular body and its characteristics, Indian Jour. Of Marine Sciences, 9, 250-252.

Mishra, D.C., Rajasekhar, R.P. and Ravi Kumar, M., 2008, Crustal and lithospheric structures in different sections of Himalayas and Tibet and neotectonics along Eastern Himalayan Front: Geodynamics and Seismotectonics based on satellite gravity and magnetic fields, In: Collision Zone Geodynamics (eds.) Arora, B.R. and Sharma, R., Mem. Geol. Soc. of India on collision zones, 72, 1-28.

Mishra, D.C., Rao, M.B.S.V., 1993, Thickening of crust under granulite province of south India and associated tectonics based on gravity–magnetic study, Geol. Soc. of India, Memoir, 25, 203-219.

Mishra, D.C., Singh, A.P., Rao, M.B.S.V., 1989, Idukki earthquake and the associated tectonics from gravity study, J. Geol. Soc. of India, 34, 147-151.

Mishra, D.C., Singh, B. and Gupta, S.B., 2002, Gravity modeling across Satpura and Godavari Proterozoic Belts: Geophysical signatures of Proterozoic collision zones. Current Science, 83, 1025-1031.

Mishra, D.C., Singh, B., Gupta, S.B., Rao, M.R.K.P., Singh, A.P., Chandrasekhar, D.V., Hodlur, G.K., Rao, M.B.S.V., Tiwari, V.M., Laxman, G., Raju, D.Ch.V., Vijaya Kumar, V., Rajesh, R.S., Babu Rao, V. and Chetty, T.R.K., 2001, Major lineaments and gravity magnetic trends in Saurashtra, India, Current Science, 80, 1059-1067.

Mishra, D.C., Singh, B., Tiwari V.M., Gupta, S.B. and Rao, M.B.S.V., 2000, Two cases of continental collisions and related tectonics during the Proterozoic period in India-insights from gravity modelling constrained by seismic and magnetotelluric studies. Precambrian Res., 99, 149-169.

Mishra, D.C., Tiwari, V.M. and Singh, B., 2008, Gravity studies in India and their geological significance, Golden Jubilee Volume of the Geological Society of India, 329-372.

Mishra, D.C., Tiwari, V.M., Gupta, S.B. and Rao, M.B.S.V., 1998, Anomalous mass distribution in the epicentral area of Latur earthquake, India, Current Science, 74, 469-472.

Mishra, D.C., Vijai Kumar, V. and Rajasekhar, R.P., 2006, Analysis of airborne magnetic and gravity anomalies of peninsular shield, India integrated with seismic and magnetotelluric results and gravity anomalies of Madagascar, Sri Lanka and East Antarctica. Gondwana Research, 10, 6-17.

Mishra, D.C.,1984, Long wavelength magnetic anomalies from the lithosphere: Indian shield and Himalaya, Tectonophysics, 105, 319-330.

Mishra, D.C.,1992, Mid continent gravity high of Central India and Gondwana tectonics, Tectonophysics, 212, 153-161.

Mishra,D.C., Prajapathi,S.K, 2003, A Plaussible model for evolution of schist belts and granite plutons of Dharwar craton, India and Madagascar during 3.0-2.5 Ga: Insight from gravity modeling constrained in part from Seismic studies, Gondwana Research, 6, 501-511.

Misra, K.S., 2008, Airborne magnetic map of Saurastra, Personal Communication.

Misra, K.S., Hammond, M.R., Phadke, A.V., Fiona Flows, Reddy, U.S.N., Reddy, I.V., Fareeduddin, Parthasarathy, G., Rao, C.R.M., Gohain B.N. and Gupta, D., 2007, Occurrence of fullerene bearing shungite suite rock in Mangampeta Area, Cuddapah District, Andhra Pradesh, J. of the Geological Society of India 69, 25-28.

Mishra, O.P. and Zhao, D., 2003. Crack density, saturation rate and porosity at the 2001 Bhuj, India, earthquake hypocenter a fluid-driven earthquake?, Earth & Planetary Science Letters, 212, 3-4.

Mishra, O.P., Singh, O.P. Chakraborty, G.K., Kayal, J.R. and Ghosh, D., 2007, aftershock investigation in the Andaman-Nicobar islands: an antidote to public panic, Seismological Research Letters, 78, 591-599.

Mishra, O.P., Zhao, D. and Wang, Z., 2008, The genesis of the 2001 Bhuj, India earthquake (Mw 7.6): A puzzle for Peninsular, India, Indian Minerals, 61, 149-170.

Mital, G.S., Bhalla, M.S.and Saxena, R.S., 1976, Application of palaeomagnetic studies in estimating the age of Deccan traps from Rajkot, Chotila and Wankaner areas of Kathiawar peninsula, India, Bull. Geof, 19, 98-109.

Mitra, S. Priestley, K., Gaur, V.K. and Rai, S.S., 2006, Shear-Wave structure of the South Indian lithosphere from Rayleigh wave phase-velocity measurements, Bulletin of the Seismological Society of America, 96, 1551-1559.

Mitra, S., Mitra, C., 2001, Tectonic setting of the Precambrians of the northeastern India, Meghalaya Plateau, and age of Shillong group of rocks. Geological Survey of India special publication, 64, 653-658.

Mitra, S., Priestley, K., Bhattacharyya, A.K., Gaur, V.K., 2005, Crustal structure and earthquake focal depths beneath northeastern India and southern Tibet, Geophys. J. Int. 160, 227-248.

Mitra, S., Priestley, K., Gaur, V.K., Rai, S.S. and Haines, J., 2006, Variation of Rayleigh wave group velocity dispersion and seismic heterogeneity of the Indian crust and uppermost mantle, Geophys. J. Int., 164, 88-98.

Mittal, G.S. and Mishra, D.C., 1985, Magnetic characteristics of princesses Astrid coast of Antarctica, north of Daksin Gangotri, DOD Scientific Report No 2, 47-51.

Mittal, P., 1984, Algorithm for error adjustment of potential field data along a survey network, Geophysics, 49, 467-469.

Mjelde, R., Raum, T., Murali, Y. and Takanami, T., 2007, Continent ocean transitions, Review and a new tectonomagnetic model of the voring plateau, NE Altantic, J. of Geodynamics, 43, 374-392.

Mohan, N.L., Sundararajan, N., Seshagiri Rao, S.V., 1982, Interpretation of some two dimensional magnetic bodies using Hilbert transform, Geophysics, 47, 376-387.

Molnar, P. and Tapponnier, P., 1975, Cenozoic Tectonics of Asia: Effects of a Continental Collision, Science, 189, 419-426.

Molnar, P., 1990, the structure of mountain ranges, In: Shaping the Earth, Tectonics of Continents and Oceans (ed) Moores, E., Scientific American, W.H. Freeman and Co., New York, 125-138.

Mondal, M.E.A., Goswami, J.N., Deomurari, M.P. and Sharma, K.K., 2002, Ion Microprobe $^{207}Pb/^{206}Pb$ sircon ages of Zircons from the Bundelkhand massif, northern India : Implications for crustal evolution of Bundelkhand – Aravalli proto continent, Precambrian Res., 117, 85-100.

Mondal, N.C., Rao, A.V. and Singh, V.P., 2010, Efficasy of electrical resistivity and induced polarization methods for revealing fluoride contaminated groundwater in granitic terrain, Environ. Monit. Assess., 168, 103-114

Monsalve, G., Sheehan, A., Rowe, C. and Rajaure, S., 2008, Seismic structure of the crust and the upper mantle beneath the Himalaya: Evidence for eclogitization of lower crustal rocks in the Indian Plate, Journal of Geophysicsl Research, 113, Bo8315, doi:10.1029/2007/B005424, 1-16.

Monsalve, G., Sheehan, A., Schulte-Pelkum, V., Rajaure, S., Pandy, M.R. and Wu, F., 2006, Seismicity and one-dimensional velocity structure of the Himalayan collision zone: Earthquakes in the crust and upper mantle, J. of Geophys. Res., 111, B10301, doi:1029/2005JB004062, 1-19.

Montelli, R., Nolel, G., Dahlen, F.A., Masters, G., Engdahl, E. R., Hung, S. H., 2004, Finite frequency tomography reveals a variety of plumes in the mantle, Science, 303, 338-343.

Moores, E.M. and Twiss, R.J., 1995, Tectonics. W.H. Freeman and Company, USA, 1-415.

Moreau F., Gibert D., Holschneider M. Saracco G. 1997. Wavelet analysis of potential fields. Inverse Problems, 13 , 165–178.

Morelli, C., 1974. The International Gravity Standardization Net 1971, International Association of Geodesy, IUGG, published by Bur. Central de L' Assoc. Intl. Ge. Paris., Special publication, 4, 194.

Morgan, W.J., 1971, Convection plumes in the lower mantle, Nature, 230, 42-43.

Morgan, W.J., 1972, Deep mantle convection plume and plate motion, Am. Assoc. Petrol. Geol. Bull., 56, 203.

Morgan, W.J., 1981, Hotspot tracks the opening of the Atlantic and the Indian Oceans, In: The Sea, (Ed), Emillani, C., Wiley, New York, 7, 443-487.

Moritz, H., 1980 Advanced physical Geodesy, Abacus press, Tunbridge, Wells Kent , England,1-500.

Morrison, H.F., Shoham R.S., Hoversten, G.M., and Torresverdin, C., 1996, Electromagnetic mapping of electrical conductivity beneath Columbia basalts. Geophy. Prospec., 44, 963-986

Moyen, J-F, Stevens, G. and Kisters, A., 2006, Record of mid-Archaean subduction from metamorphism in the Barberton terrain, South Africa, Nature, 442, doi:10.1038/nature04972.

Mufti, I. and Siddiqui, F.A., 1961, A gravity survey of Quetta and Mastung valleys, Pakistan J. Scientific and Industrial Res., 4, 1, 15-20.

Mukhopadhyay, D. and Basak, K., 2009. The eastern ghats belt- A polycyclic granulite terrain, Journal Geological Society of India, 73, 489-518.

Mukhopadhyay, D.K., Sengupta, B. and Saha, D.K., 2002, Geophysics in environmental studies- a case study from Haora Municipal Corporation Area, Wwest Bengal, Geol. Surv. Ind. Special Publication No. 75, 272-278.

Mukhopadhyay, M., Verma, R.K. and Asharaf, M.H., 1986, Gravity field and structures of the Rajamahal hills: Example of the Paleo-Mesozoic continental margin in eastern India, Tectonophysics, 131, 353-367.

Muller, R.D., Roger, J.Y.and Lawver,, L.A., 1993, Geology, 21, 275-278.

Muniruzzaman, M. and Banks, R.J., 1989, Basement magnetization estimates by wavenumber domain analysis of magnetic and gravity maps, Geophys. J., 97, 103-117.

Murphy, C., 2004, The Air-FTG^{TM} airborne gravity gradiometer system, in R.J.L. Lane, editor, Airborne Gravity 2004 Workshop, Australian Government, Geoscience Australia, 7-14.

Murthy, A.S.N., Sain, K., Tewari, H.C. and Prasad B.R., 2008, Crustal velocity inhomogeneities along the Hirapur-Mandla profile, central India and its tectonic implications, Journal of Asian Earth Sciences, 31, 533-545.

Murthy, K.S.R. and Mishra, D.C., 1980, Fourier transform of the general expression for the magnetic anomaly due to a long horizontal cylinder, Geophysics, 45, 1091-93.

Murthy, K.S.R., Subrahmanyam, V., Subrahmanyam, A.S., Murty, G.P.S., Sarma, K.V.L.N.S , 2010, Land-Ocean tectonics (LOTS) and the associated seismic hazard over the Eastern Continental Margin of India (ECMI), Natural Hazard, DOI 10.1007/s11069-010-9523-8.

Murthy, K.S.R., Sumrahmanyam, V., Murthy, G.P.S and Mohan Rao, K., 2007, Impact of coastal morphology, structure and seismicity on the Tsunami surge, In: The Indian Ocean Tsunami, (eds) Murthy, T. S., Aswathanarayana, U. and Nirupama, N., Taylor and Francis, 19-31.

Murthy, P.V.V.G.R.K. and Sharad, M.K. 1981, The Narmada-Son lineament and the structure of Narmada rift system. J. of Geo. Soc. of India, v. 22, pp. 112-120.

Murthy, T. S., 2005. Personal communication.

Murthy, Y.S., 1999, Images of the gravity field of India and their salient features, J. Geol. Soc. of India, 54, 221-235.

Nabighian, M.N., 1972, The analytic signal of two dimensional magnetic bodies with polygonal cross section, its properties and use for automated anomaly interpretation, Geophysics, 37, 507-517.

Nabighian, M.N., 1974, Additional comments on the analytic signal of two dimensional magnetic bodies with polygonal cross section, Geophysics, 39, 85-92.

Nabighian, M.N., Grauch, V.J.S., Hansen, R.O., Lafehr, T.R., Li, Y., Peirce, J.W., Phillips, J.D. and Ruder, M.E., 2005, The historical development of the magnetic method in exploration, Geophysics, 70(6), 33ND-61ND.

Nag, S., Gaur, R.K. and Pal, T., 2001, Late Cretaceous-Tertiary sediments and associated faults in southern Meghalaya Plateau of India vis-à-vis south Tibet: their interrelationships and regional implications, Journal of Geological Society of India, 57, 327-338.

Naganjaneyulu, K. and Harinarayana, T., 2004, Deep crustal electrical signatures of eastern Dharwar craton, India, Gondwana Research, 7, 951-960.

Naganjaneyulu, K. and Santosh, M., 2010, The Central India Tectonic Zone: A geophysical perspective on continental amalgamation along a Meso Proterozoic suture, doi: 10.1016/j.gr.2010.02.017.

Naganjaneyulu, K., Naidu, G.D., Rao, M.S., Ravi Shankar, K., Kishore, S.R.K., Murthy, D.N., Veeraswamy, K. and Harinarayana, T, 2010, Deep crustal electromagnetic structure of central India tectonic zone and its implications, Physics of the Earth and Planetary Interiors, 181, 60-68.

Nagaraja Rao, B.K., Rajurkar, S.T., Ramlingaswamy, G. and Ravindra Babu, B., 1987, Stratigraphy, structure and evolution of the Cuddapah basin, Memoir Geological Society of India, 6, 33-87.

Nagata, T., 1953, Rock magnetism, Marugen Co. Ltd., Tokyo, 1-225.

Naidu, P.S. and Mathew, M.P., 1998, Analysis of geopotential fields: A digital signal processing approach. Advances in Exploration Geophysics, Elsevier Amsterdam, 1-289.

Naidu, P.S., 1970, Fourier transform of large scale aeromagnetic field using a modified version of Fast Fourier Transform, Pure & Applied Geophysics, 81, 17-25.

Naini, B. R. and Talwani, M., 1982, structural frame work and evolutionary history of the continental margin of Western India, Am. Assoc. Pet. Geol. Mem., 29, 167-191.

Nair, A.R., Navada, S.V. and Rao, S.M., 1999, Isotope study to investigate the origin and age of groundwater along paleochannels in Jaisalmer and Ganganagar districts of Rajasthan. Mem. Geol. Soc. India, 42, 304-314.

Najman, Y., 2006, The detrital record of orogenesis: A review of approaches and techniques used in the Himalayan sedimentary basins, Earth Science Review, 74, 1-72.

Najman, Y., Bickle, M., Garzanti, E., Pringle, M., Barford, D., Brozovic, N., Burbanc, D. and Ando, S., 2009, Reconstructing the exhumation history of the Lesser Himalaya, NW India, from a multitechnique provenance study of the foreland basin Siwalik Group, Tectonics, 28, TC 5018, doi:10.1029/2009TC002506.

Nakata, T., 1989, Active faults of the Himalaya of India and Nepal, Geological Society of America, Special Paper 232, 243-264.

Nance, R.D., Worsley, T.R. and Moody, J.B., 1990, The Supercontinent Cycle, In: Shaping the Earth Tectonics of Continents and Oceans (ed) E.M. Moores, Scientific American, Inc., 177-188.

Naqvi, S.M. and Hussain, S.M., 1972, Petrochemistry of early Precambrian metasediments from the central part of the Chitradurg schfist belt, Mysore, India: Chemical Geology, 10, 109-135.

Naqvi, S.M. and Hussain, S.M., 1983, Geological, geophysical and geochemical studies over the Holenarsipur schist belt, Dharwar craton, India, In: Naqvi, S.M. and Rogers, J.J.W. (Eds.), Precambrian of South India, Geol. Soc. India Mem., No.4, 73-95.

Naqvi, S.M. and Rana Prathap, J.G., 2007, Geochemistry of adakites from Neoarchaean active continental margin of Shimoga schist belt, Western Dharwar Craton, India: Implications for the genesis of TTG, Precambrian Research, doi:10.1016/pecamres.2007.03.003, 1-23.

Naqvi, S.M., 1973, Geological structure and aeromagnetic and gravity anomalies in the central part of the Chitaldrug schist belt, Mysore, India, Geol. Soc. of Am. Bull., 84, 1721-1732.

Naqvi, S.M., Divakara Rao, V. and Hari Narain, 1974, The proto continental growth of the Indian shield and the antiquity of its rift valleys, Precambrian Research, 1, 345-389.

Naqvi, S.M., Khan, R.M.K., Manikyamba, C., Mohan, M.R. and Khanna, T.C., 2006, Geochemistry of the NeoArchaean high-Mg basalts, boninites and adakites from the Kushtagi-Hungund greenstone belt of the Eastern Dharwar Craton (EDC); implications for the tectonic setting, Journal of Asian Earth Sciences, 27, 25-44.

Narula, R.L., Acharyya, S.K., Banerjee, J. (Eds.), 2000, Seismotectonic atlas of India and its environs, Geological Survey of India, Kolkata, 1-43.

NASA, 2010, Science@NASA, Decrypting the eclipse, A solar eclipse , global measurements and a mystery.

Naudy, H. and Dreyer, H., 1968, Attempt to apply non linear filtering to aeromagnetic profiles, Geophysical Prospecting, 16, 171-178.

Nayak, G.K., Agarwal, P.K., Rama Rao, Ch. and Pandey, O.P., 2006, thickness estimation of Deccan flood basalt of Koyna area, Maharashtra (India) from inversion of aeromagnetic and gravity data and implications for recurring seismic activity, Current Science, 91, 960-965.

Nayak, V. K., 1972, Glassy objects (impact glasses?) A possible new evidence for meteoritic origin of Lonar crater, Maharashtra state, India. *Earth Planet. Sci. Lett.*, **14**, 1–6.

NDMA, 2007, Management of earthquakes, Guidelines issued by National Disaster Management Authority. Government of India.

Neel, L., 1955, Some theoretical aspects of rock magnetism, Adv. Phys., 4, 191-243.

Negi, J. G. and Tiwari, R. K., 1983, Matching long term periodicities of geomagnetic reversals and galactic motions of the solar system, Geophys. Res.Lett., 10, 713-716.

Negi, J.G., Agarwal, P.K. and Rao, K.N.N., 1983, Three dimensional model of the Koyna area of Maharashtra state (India) based on spectral analysis of aeromagnetic data, Geophysics, 48, 964-974.

Negi, J.G., Pandey, O.P. and Agarwal, P.K., 1987, Supermobility of hot Indian lithosphere, Tectonophysics, 135, 145-156.

Negi, J.G., Pandey, O.P. and Agrawal, P.K., 1986, Super-mobility of hot Indian lithosphere, Tectonophysics, 131, 147-156.

Negi, J.G., Thakur, N.K. and Agarwar, P.K., 1986, Crustal magnetization-model of the Indian subcontinent through inversion of satellite data, Tectonophysics, 122, 123-133.

Negishi, H., Mori, J., Singh, R. and Hirata, N., 2002. Size and orientation of the fault plane for the 2001 Gujurat, India earthquake (M_N = 7.7) from aftershocks observations: a high stress loop event, Geophys. Res. Lett., 29, 10-1-4.

Negredo, A.M., Replumaz, A., Villasenor, A. and Guillot, S., 2007, Modeling the evolution of continental subduction processes in the Pamir-Hindu Kush region, Earth and Planetary Science Letts. 259, 212-225.

Neilsen, P.H., Waagstein, R., Rasmussen, J. and Larsen, B., 1979, Marine seismic investigations of the shelf around the Faeroe island, Frodskaparrit (Annal. Sci. Faeroensis), 27, 102-113.

Nelson, K.D., Zhao, W., Brown, L.D., Kuo, J., Che, Jinkai, Liu, X., Klemperer, S.L., Makovsky, Y., Meissner, R., Mechie, J., Kind, R., Wenzel, F., Ni. J.N., Leshou, C., Tan, H., Wei, W., Jones, A.G., Booker, J., Unsworth, M., Kidd, W.S.F., Hauck, M., Alsdorf, D., Ross, A., Cogan, M., Wu, C., Sandvol, E. and Edwards, M., 1996, Partially molten middle crust beneath southern Tibet: Synthesis of Project INDEPTH Results, Science, 274, 1684-1687.

Nerem, R., and G.T. Mitchum, 2001 Sea Level Change, In: Satellite Altimetry and Earth Sciences, L. Fu and A. Cazenave (eds.), Academic Press Inc, New York.

Nerem, R.S., Jekeli, C. and Kaula, W.M., 1995, Gravity field determination and characteristics: Retrospective and prospective, J. of Geophy. Res., 100 (B8), 15053-15074.

Nettleton, L.L.,1976, Gravity and Magnetic in Oil Prospecting, McGraw Hill Book Co., New York.

NGRI, 1967, Airborne Magnetic Survey of Kudermukh Schist Belt for Iron Ores, NGRI- Tech. Report.

NGRI, 1968, Report on airborne magnetometer and Scintillometer surveys over a part of the Chitaldrug schist belt, Technical report, Airborne Group, NGRI, Hyderabad, 1-51.

NGRI, 1969, Report on airborne geophysical surveys over parts of Madhya Pradesh, Technical Report, No 69-30, Airborne Group, NGRI, Hyderabad, India.

NGRI, 1975, Gravity Map Series of India, GPH / 1-5; Published by National Geophysical Research Institute, Hyderabad, India.

NGRI, 1975. Bouguer anomaly map of western ESCAP region, Published by United Nations Economic and Social Commission for Asia and the Pacific.

NGRI, 1978, Brochure of the Gravity map Series of India, Published by National Geophysical Research Institute, Hyderabad-India, 1975.

NGRI, 1981, Airborne total intensity map of a part of Deccan trap, Airborne Group, Hyderabad, India.

NGRI, 1989, Helicopter-borne magnetic survey between the Schirmacher Oasis and the Wholthat Mountains, Droning Maud land, Antarctica, A technical report submitted to Department of Ocean development.

NGRI, 1992-93, Coloured Bouguer anomaly map of Central India, Annual Report of National Geophysical Research Institute, Hyderabad.

NGRI, 1998, Integrated Geophysical Studies for Hydrocarbon Exploration, Saurashtra, India, Technical Report No. NGRI-98-EXP-237 (restricted).

NGRI, 1999, Report on integrated geophysical methods in search of mineralized zones, Kayar area, Rajasthan State, NGRI Technical Report No. NGRI-99-Exp-250.

NGRI, 2000, Gravity-magnetic surveys for basement structures in parts of upper Assam and Arunachal Pradesh, Technical Report No: NGRI-2000-Exp-290.

NGRI, 2000, Integrated Geophysical Studies for Hydrocarbon Exploration, Kutch, India, Technical Report No. NGRI-2000-EXP-296 (restricted).

NGRI, 2003, Exploration of Sub-Trappean Mesozoic Basins in the western part of the Narmada-Tapti Region of Deccan Syneclise, Technical Report No NGRI-2003-EXP-404 (restricted),

NGRI, 2003, Gravity-Magnetic surveys over Amriti feature in Rewa District, Madhya Pradesh, Technical Report No: NGRI-2003-Exp-380.

Ni, J. and Barazangi, M., 1984, Seismotectonics of the Himalayan collision zone: geometry of the thrusted Indian plate beneath the Himalaya, J. Geophys. Res., 89, 1147-1163.

Nicolosi, L., Blanco-Montenegro, I. Pignatelli, A. and Chiappini, M., 2006, Estimating the magnetization direction of crustal structures by means of an equivalent source algorithm, Physics of the Earth and Planetary interiors, 155, 163-169.

Niebauer, T.M., Klopping, F.J. and Faller, J.E., 1995, The FG5 Absolute Gravimeter, Proc. of Second Workshop: Non Tidal Gravity Changes: Inter comparison between absolute and super conducting Gravimeters, Sept. 6-8, 1994, Walferdange (Grand-Duchy of C. Poitevin, Lunembourg.

Nielsen, J.O. and Pedersen, L.B., 1979, Interpretation of potential fields from inclined dikes in the wave number domain, Pageoph, 117, 761-771.

Nolet, G., Allen, R., Zhao, D., 2007, Mantle plume tomography, Chemical Geology, 241, 248-263.

Norton, I.O., Slater, J.G., 1979, A model of the evolution of the Indian ocean and the breakup of the Gondwana land, J. Geophys. Res., 84, 6803-6830.

NRC, 1997, Satellite gravity and the geosphere, National Academy Press, Washington, D.C., 1-112.

O'Reilly, S.Y., 2001, Journey beneath southern Africa, Nature, 412, 778-781.

Odegard, M.E. and Berg, J.W., 1965, Gravity interpretation using Fourier integral, Geophysics, 30, 424-438.

Ogg, J.G., Gabi Ogg and Gradstein, F.M., 2009, The concise geologic time scale, Canbridge University Press, 1-177.

Ohyama, H., Tsunogae, T. and Santosh, M., 2008, CO_2-rich fluid inclusions in staurolite and associated minerals in a high-pressure ultrahigh-temperature granulite from the Gondwana suture in southern India, Lithos, 101, 177-190.

Okada, Y., 1985. Surface deformation due to shear and tensile faults in a half space, Bull. Seismol. Soc. Am., 75, 1135-1154.

Okal, E.A. and Synolakis, C.E., 2008, Far field tsunami hazard from mega-thrust earthquakes in the Indian Ocean, Geophys. J. Int., 172, 995-1015.

Okal, E.A., Synolakis, C.E., Uslu, B., Kalligeris, N., Voukouvalas, 2009, The 1956 earthquake and tsunami in Amorgos, Greece, Geophys. J. Int., 178, 1533-1554.

Okubo, S., 1991. Potential and gravity changes raised by point dislocations, Geophys. J. Int., 105, 573-586,

Okubo, S., 1992.Gravity and potential changes due to shear and tensile faults in a half space, J. Geophys. Res., 97, 7137-7144.

Okubo, Y., Graf, R. J., Hansen, R. O., Ogawa, K. and Tsu, H., 1985, Curie point depths of the islands of Kyushu and surrounding areas, Japan, Geophysics, 50, 481-485.

Oldenburg, D.W., 1974, The inversion and interpretation of gravity anomalies, Geophysics, 39, 526-536.

Oldham, R.D., 1899, Report of the great earthquake of 12th June, 1897, Memoir Geological Survey of India, 379, Reprinted by Geological survey of India, Calcutta, 1981.

Oleson, K.W.,et al., 2008, Improvements to the community land model and their impact on the hydrological cycle, J. Geophys. Res., 113, G01021, doi: 10.1029/2007JGooo563.

Olsen, K.H., Morgan, P., 1995, Introduction: progress in understanding continental rifts. In continental rifts: evolution, structure, tectonics. Ed: K.H. Olsen, Elsevier, Development in geotectonics, 25, 3-26.

Oncken, G. Chong, G. Franz, P. Giese, H.J., Götze, V., Ramos, M. Stecker and P. Wiger, 2006, The Andes-active subduction orogeny, Frontiers in Earth Sciences, 1, Springer Verlag,

ONGC, 2009, ONGC Bulletin, 44, 1.

ONGC,1968, Tectonic Map of India By Oil and Natural Gas Corporation, Dehra Dun. India.

Oreshin, S., Kiselev, S., Vinnik, L., Prakasam, K.S. and Rai S.S., 2008, Crfust and mantle beneath western Himalaya, Ladakh and western Tibet from integrated seismic data, Earth and Planetary Science Letters, 271, 75-87.

Owen, H.G.,1983, Atlas of continental displacement, 200Million years to the present, Cambridge University Press, 1-159. Geol. Magazine (On line), 121 (06).

Owens, T.J., Zandt, G., 1997, Implications of crustal property variations for models of Tibetan plateau evolution, Nature 387- 37-43.

Ozerde, A. and Jacoby,T., 2006, Disaster management and civil society (Earthquake relief in Japan, Turkey, and India), I.B.Tauris and Co. Ltd., 1-168

Pahari, S., Singh, H., Prasad, I.V.S.V. and Singh, R.R., 2008, Petroleum systems of Upper Assam Shelf, India, Geohorizons, 14-21.

Pal, P. C. and Creer, K. M., 1986, Geomagnetic reversal spurts and episodes of extraterrestrial catastrophism, Nature, 320, 148-150.

Palacky, G.J. and West, G.F., 1973, Quantitative interpretation of INPUT AEM measurements, Geophysics, 38, 1145-1158.

Pande, K., Sheth, H. C., Bhutani, R., 2001, ^{40}Ar-^{39}Ar age of the St. Mary's islands Volcanics, Southern India: Record of India-Madagascar break-up on the Indian subcontinent, Earth & Planet. Sci. Letters, 193, 39-46.

Pandey, D.K., Rajan, S. and Pandey, A., 2010, Seismic imaging of Paleogene sediments of Kachchh Shelf (Western Indian margin) and their correlation with sea level fluctuations, Marine and Petroleum Geology,27, 1166-1174.

Pandey, O.P. and Rao, V.K., 2006, missing granite crust (?) in the Godavari graben of southeast India, J. Geol. Soc. India, 67, 307-311.

Pandey, O.P., Chandrakala, K., Parthasarathy, G., Reddy, P.R., Reddy, G.K., 2009, Upwarped high velocity mafic crust, subsurface tectonics and causes of intraplate Latur-Killari (M 6.2) and Koyna (M 6.3) earthquakes, India – A comparative study, J. Asian Earth Sciences, 34, 781-795.

Pandey, S.N., 1973, Development of dome and basin structure due to interference of major folds on the southern margin of Vindyan basin, Proc. Ind. Nat. Sci. Acad., 39, 1.

Pandey, U.K., Pandey, B.K. and Krishnamurthy, P., 2005, Geochronology (Rb-Sr, Sm-Nd and Pb-Pb) of the Proterozoic granulitic and granitic rocks around Usilampatti, Madurai District, Tamil Nadu: Implication on age of various lithounits, J. Geological Society of India, 66, 539-551.

Pant, N.C. and Banerjee, D.M, 1990, Pattern of sedimentation in the type Bijawar basin of central India, Geol. Surv. India, Spec. Publ., 28, 156-166.

Parasnis, D.S., 1982, Principles of Applied Geophysics, Chapman and Hall, USA,1- 275.

Parasnis, D.S., 1986, Principles of Applied Geophysics, Chapman and Hall, IV edition.

Park, J., Lindberg, C.R. and Vernon, F.L., 1987, Multitaper spectral analysis of high frequency seismograms, J. Geophys. Res., 92, 12675-684.

Parker, R.L. and Huestis, S.P.,1979, The inversion of magnetic anomalies in the presence of topography, J. Geophy. Res., 79, 1587-1593.

Parker, R.L., 1973, The rapid calculation of potential anomalies, Geophys. J.R. astr. Soc., 31, 447-455.

Parker, R.L., 1975, The theory of ideal bodies for gravity interpretation, Geophys. J. R. astr. Soc., 42, 315-334.

Parker, R.L., 1977, Understanding inverse theory, A. Rev. Earth Planet. Sci., 5, 35-64.

Parker, R.L., 1996, Improved Fourier terrain correction, Part II, Geophysics, 61, 365-371.

Pasyanos, M.E. and Nyblade, A.A., 2007, A top to bottom lithospheric study of Africa and Arabia, Tectonophysics, 444, 27-44.

Patel, S.C., Ravi, S., Anilkumar, Y., Naik, A., thakur, S.S., Pati, J.K. and Nayak, S.S., 2009, Mafic xenoliths in Proterozoic kimberlites from Eastern Dharwar Craton, India: Mineralogy and P-T regime, Journl of Asian Earth Sciences, 34, 336-346.

Paterson, N.R. and Bosschart, R.A., 1987, Airborne geophysical exploration, Groundwater, 25, 41-50.

Pathak, A., Ravi Kumar, M., Sarkar, D., 2006, Seismic structure of Sri Lanka using receiver function analysis: A comparison with other high grade Gondwana terrains, Gondwana Research, 10, 198-202.

Patra, I., Babu, V.R., Chaturvedi, A.K., Dash, J.K., Sreenivas, R., Chari, M.N. and Roy, M.K., 2009, Regional gravity and magnetic surveys along southern margin of the Indravati basin, Central India- A guide to unconformity related uranium mineralization, Jour. of Geophysics, 30, 21-24.

Patriat, P. anmd Segoufin, J., 1988, Reconstruction of the Central Indian Ocean, Tectonophysics, 155, 211-234.

Patro, B.P.K. and Sarma, S.V.S., 2009, Lithospheric electrical imaging of the Deccan tral covered region of westwern India, J. of Geoph. Res., 114, B01102.

Patro, B.P.K., Harinarayana, T., Sastry, R.S., Rao, M., Manoj, C., Naganjaneyulu, K. and Sarma, S.V.S., 2005, Physics of the Earth and Planetary Interiors, 148, 215-232.

Patro, B.P.K., Nagarajan, N. and Sarma, S.V.S., 2006, Crustal geoelectric structure and the focal depths of major stable continental region earthquakes in India, Current Science, 90, 107-113.

Paul, A., Kaviani, A., Hatzfeld, D. and Vergne, J., 2006, Seismological evidence for crustal-scale thrusting in the Zagros mountain belt (Iran), Geophys. J. Int., 166. 227-237.

Paul, D.K., Barman, T., McNaughton, N.J., Flecther, I.R., Potts, P.J., Ramakrishnan, M. and Augustine, P.F., 1990, Archean-Proterozoic evolution of Indian charnockites-isotopes and geochemical evidence from granulites of the Eastern Ghat Belt. Journal of Geology, 98, 253-263

Paul, J., Burgmann, R., Gaur, V.K., Bilham, R., Larson, K.M., Ananda, M.B., Jade, S., Mukal, M., Anupama, T.S., Satyal, G. and Kumar, D., 2001, The motion and active deformation of India, Geophys. Res., Letters, 28, 4, 647-650.

Paul, J., Singh, R. N., Subrahmanyam, C., Drolia, R. K., 1990, Emplacement of Afanasy-Nikitin Seamount based on transfer function analysis of gravity and bathymetry data, Earth & Planet. Sci. Letters, 96, 419-426.

Pearce, J.A., N.B.W. Harris and A.G. Tindle, 1984, Trace element discrimination diagram for the tectonic interpretation of granitic rocks, J. Petrol, 25, 956-983.

Pedersen, L.B., 1975, Interpretation of potential field data, A generalized inverse approach, Geophy. Prosp., 199-230.

Pedersen, L.B., 1977, Constrained inversion of potential field data, Geophy. Prosp., 27, 726-748.

Pedersen, L.B., 1978, Wave number domain expressions for potential fields from arbitrary 2, 2 ½ and 3 dimensional bodies, Geophysics, 626-630.

Pegler, G. and Das, S., 1998, An enhanced image of the Pamir-Hindu Kush seismic zone obtained from relocated earthquake hypocenters, Geophys. J. Int., 134, 573-595.

Penick Jr., J., 1976, The New Madrid earthquake of 1811-1812, University of Missouri Press, 1-181 (Pittsburg Gazette, April 10, 1812).

Percival, J.A., Card, K.D., Sage, R.P., Jensen, L.S. and Luhta, L.E., 1983, The Archean crust in the Wawa-Chapleau- Timmins region. In L.D. Ashwal and K.D. Card, eds., Workshop on Cross section of Archean crust, Lunar and Planetary Institute Technical epor 83-03, Houston, Texas: Lunar and Planetary Institute, 99-169.

Percival, J.A., Helmstaedt, H., 2006, The western superior province lithoprobe and NATMAP transects: introduction and summary, Can. J. Earth Sci., 43, 743-747.

Perez-Gussinye, M. and Watts, A. B., 2005, The long term strength of Europe and its implications for plate forming processes, Nature, 436, 381-384.

Persists, F.M., Wandrey, C.J., Milici, R.C.and Manwar, A., 2001. Digital Geologic and Geophysical data of Bangladesh, U.S. Geological Survey Open File Report 97-470H.

Peschler, A.P., Benn, K. and roest, W.R., 2004, Insigahts on Archean continental geodynamics from gravity modelling of granite-greenstone terranes, Journal of geodynamics, 38, 185-207.

Peshwa, V.V., Kale, V.S., Kulkarni, H.C., Phadke, A.V. and Phansalker, V.G., 1995, The Kurduwadi lineament and its relation with the recent seismic activity in the Deccan Trap Province, Abstract volume, Workshop on Indian shield seismicity and Latur earthquake of Sept. 30, 1993, NGRI, Hyderabad (India), Sept. 4-5, 1995, 9-10.

Peters, L.J., 1949, The direct approach to magnetic interpretation and its practical application, Geophysics, 14, 290-320.

Petit, C., De´verche`re, J., 2006, Structure and evolution of the Baikal rift: A synthesis, Geochemistry Geophysics Geosystems, 7, Q11016, doi: 10.1029/ 2006 GC 001265.

Phillips, R.J. and Mallin, M.C., 1984, Tectonics of Venus, Ann. Rev. Earth Planet. Sci., 12, 411-443.

Pickard, A.L. Barley, M.E. and Krapez, B., 2004, Deep-marine depositional setting of banded iron formation: sedimentological evidence from interbedded clastic sedimentary rocks in the early Palaeoproterozoic Dales Gorge Member of Western Australia, Sedimentary Geology, 170, 37-62.

Pilkington, M. and Grieve, R. A. F., The geophysical signature of terrestrial impact craters. 1993, Rev. Geophys., 1992, **30**, 161–181.

Pilkington, M., 1990, Lithospheric flecure and gravity anomalies at Proterozoic plate boundaries in the Canadian shield, Tectonophysics, 176, 277-290.

Pilli, E., Ricard, Y., Landeux, J.M., Sheppard, S.M.F., 1997, Lithospheric shear zones and mantle crust connections. Tectonophysics, 280, 15-29.

Piper, J.D.A., 2000, The Neoproteozoic Supercontinent: Rodinia or Palaeopangaea, Earth and Planetary Science Letters, 176, 131-148.

Pirajno, F., 2007, Mantle plumes, associated intra plate tectonomagmetic processes and ore systems, Episodes, 30, 6-19.

Pittman III, W. C. and Heirtzler, J. R., 1966, Magnetic anomalies over the Pacific-Antarctic Ridge, Science, 154, 1164-71.

Plomerova, J., Achauer, U., Babuska, V., and Vecsey, L, 2007, Upper mantle beneath the Eger Rift (Central Europe): plume or asthenospheric upwelling, Geophys. J. Int., 169, 675-682.

Pollack, H. N., Hurter, S. J., Jonson, J. R., 1993, Heat flow from the Earths interior: analysis of the global data set, Rev, Geophysics, 31, 267-280.

Poornachandra Rao, G.V.S., J. Mallikharjuna Rao and K.J.P. Lakshmi, 2000, Alkali igneous rocks of Deccan traps from northwest India and their palaeomagnetism, Proc. XII Indian Geological Congress, Dept. of Geology, M.L.S. University, Udaipur, 12.

Poovendran, P., 1991, The modern Atlas, 18.

Potter, C.J., Allmendinger, R.W., Hauser, E.C. and Oliver, J.E., 2007, COCORP deep seismic reflection traverses of the U.S. Cordillera, Geophysical Journal International, 89, 99-104.

Potts, L.V., Shum, C.K., Von Frese, R., Han, S.C. and Mautz, R., 2005, Recovery of isostatic topography over North America from Topographic and CHAMP gravity correlations (ed.) C. Reigber, H. Lühr, P. Schwintzer and J. Wickert, Earth observation with CHAMP, Springer Verlag Berlin Heidelderg, 193-198.

Powell, C., McA., Roots, S.R. and Veevers, J.J., 1988, Pre-breakup continental extension in East Gondwanaland and the early opening of the eastern Indian Ocean, Tectonophysics, 155, 261-283.

Powers, P.M., Lillie, R. J. and Yeats, R. S., 1998, Structure and shortening of the Kangra and Dehra Dun reentrants, Sub-Himalaya, India, G S A Bull., 110, 1010-1027.

Pozdeyev, V. S., 1994, Total intensity magnetic profile across Lambert Rift, personal communication.

Pradesh, Orissa and Maharashtra, Geol. Surv. India Bull. Ser. No. 45(A)/2, 1–103.

Prakasam, K. S. and Rai, S. S., 1998, Teleseismic delay time tomography of the upper mantle beneath south eastern India, imprint Indo-Antarctica rifting, Geoph. J. Int., 133, 20-30.

Prasad, B.R., Rao, G.K. Mall, D.M., Rao, P.K., Raju, S., Reddy, M.S., Rao, G.S.P., Sridhar, V., Prasad, A.S.S.S.R.S., 2007, Tectonic implications of seismic reflectivity pattern observed over the Precambrian Southern Granulite Terrain, India, Precambrian Research, 153,1-10.

Prasad, B.R., Venkateswarlu, N., Prasad, A.S.S.S.R.S., Murthy, A.S.N. and Sateesh, T., 2010, Basement configuration of on land Kutch basin from Seismic Refraction studies and modeling of first arrival travel time skips, Journal of Asian Earth Sciences, 39, 460-469.

Prasad, B.R., Vijaya Rao, V. and Reddy, P.R., 1999, Seismic and magnetotelluric studies over a crustal scale fault zone for imaging a metallogenic province of Aravalli Delhi Fold Belt region. Current Science, 76, 1027-1031.

Prasanta K. Patro, B.P.K., Nandini Nagarajan and Sarma, S.V.S., 2006, Crustal geoelectric structure and the focal depths of major stable continental region earthquakes in India, Current Science, 90, 107-113.

Prezzi, C.B., Gotze, H.J., Schmidt, S., 2009, 3-D density model of the central Andes, Physics of the Earth and Planetary Interiors, 177, 217-234.

Priestley, K., Debayle, E., McKenzie, D. and Pilidou, S., 2006, Upper mantle structure of eastern Asia from multimode surface waveform tomography, J. Geophys., Res., 111, B10304, doi:101029/2005JB00482, 1-20.

Priestley, K., Jackson, J. and McKenzie, D., 2008, Lithospheric structure and deep earthquakes beneath India, the Himalaya and southern Tibet, Geophys. J. Int., 172, 345-362.

Prodehl, C., Mechei, J., Achauer, U., Keller, G.R., Khan, M.A., Mooney, W.D., Gaciri, S.J., Obel, J.D., 1994, The KRISP 90 seismic experiment-a technical review, Tectonophysics, 236, 33-60.

Prodehl, C., Mueller, St. and Heak, V., 1995, The European Cenozoic Rift System, In continental rifts, Evolution, Structure and Tectonics, (ed.) K.H. Olsen, Developments in Geotectonics, 25, 133-212.

Prodehl, C., Mueller, St., Glahn, A., Gutscher, M. and Haak, V., 1992, Lithospheric cross sections of European central rift system, Tectonophysics, 208, 113-138.

Purnachandra Rao, N. and Kumar, R., 1997, Uplift and Tectonics of the Shillong Plateau, Northeast India, J. Phys. Earth, 45, 167-176.

Purucker, M., Ishihara, T., 2005, Magnetic images of the Sumatra region crust, EOS, 86, 101-102.

Qureshy, M, N, Krishna-Brahmam, N, 1969, Gravity bases established in India by N.G.R.I., pt. 1 Bulletin of the National Geophysical Research Institute. 7, 31-49.

Qureshy, M.N. and Warsi, W.E.K., 1980, A Bouguer anomaly map of India and its relation to broad tectonic elements of the subcontinent, Geophys. J. Roy. Astr. Soc., 61, 235-242.

Radha Krishna, M., 1996, Isostatic response of the central Indian Ridge (Western Indian Ocean) based on transfer function analysis of gravity and bathymetry data, Tectonophysics, 257, 137-148.

Radhakrisha, B.P. and Merh, S.S., 1999, Vedic Saraswati. Mem. Geol. Soc. India, 42, 1-329.

Radhakrishna, B.P. and Naqvi, S.M., 1986, Precambin continental crust of India and its evolution, J. Geol., 94, 145-166.

Radhakrishna, B.P., 1999a, Keynote Address: Holocene chronology and Indian Pre- history. Mem. Geol. Soc. India, 42, XV-XVII.

Radhakrishna, B.P., 1999b, Vedic Saraswati and the dawn of Indian civilization. Mem. Geol. Soc. India, 42, 5-14.

Radhakrishna, M., Kurian, P.J., Nambiar, C.G. and Murthy, B.V.S., 2003, Nature of the crust below the Southern Granulite Terrain (SGT) of Peninsular India across the Bavali shear zone based on analysis of gravity data, Precambrian Research, 124, 21-40.

Radhakrishnamurthy, C. and Mishra, D.C., 1966, A criterion for stability of NRM in the sedimentary rocks, Bull of NGRI, 4, 104-108.

Radhakrishnamurthy, C., Sahasrabudhe, P.W., Mishra, D.C. and Prasad, C.V.R.K., 1967, Revised palaeolatitudes for the landmass of India, Proceedings of the Simposium on Upper Mantle Project, Hyderabad (India) Session-VII Studies on Continental Drift, 512-519.

Radhakrishnamurthy, I.V. and Mishra, D.C., 1989, Gravity and magnetic anomalies in space and frequency domains, Assoc. of Exploration Geophysicists, Hyderabad, 1-250.

Radhakrishnamurthy, I.V., 1998, Gravity and magnetic interpretation in exploration geophysics, Memoir Geol. Soc. of India, Bangalore, India, 40, 1-360.

Rahber, A. and Tiwari, V.M., 2010, Monitoring of mass change in an underground mine using 4-D gravimetry, A Technical Report under "Summer Reseach and Fellowships for Students and Teachers 2010" by INSA, New Delhi.

Rahman, A. U., 1967, A gravity and magnetic study of a deep seated anomaly in the Gujranwala district, West Pakistan, Geol. Bull.of Punjab University, 6,12-23.

Rahman, A.U., 1969, Crustal section across the Sibi-Syntexial-Bend, West Pakistan, based on gravity measurement, J. Geophy., Res., 74, 17, 43674370.

Rahman, M.A., Blank, H.R., Kleinkopf, M.D., Kucks, R.P., 1990. Aeromagmetic anomaly map of Bangladesh, scale 1:1000000. Geol. Surv. Bangladesh, Dhaka.

Rahman, M.A., Mannan, M.A., Blank, H.R., Kleinkopf, M.D. and Kucks, R.P., 1990, Bouguer gravity anomaly map of Bangladesh, scale 1:1000000, Geol. Surv. Bangaladesh, Dhaka, United States Geophysical Survey, USGS, Washington.

Rai, S.N. and Thiagarajan, 2006, A tentative 2D thermal model of central India across the Narmada-Son lineament (NSL), J. of Asian Earth Sciences, 363-371.

Rai, S.N., 2006, Revisit to ancient river Saraswati in Thar desert: (In Hindi). Vigyan Parishad, Prayag, 33-37.

Rai, S.N., Thiagarajan, S. and Ramana, D.V., 2003, Thermal structure of continental crust beneath Hirapur-Mandla deep seismic sounding profile across Narmada-Son lineament, Current Science, 85, 208-213.

Rai, S.S., Priestley, K., Gaur, V.K., Mitra, S., Singh, M.P. and Searle, M., 2006, Configuration of the Indian Moho beneath the NW Himalaya and Ladakh, Geophys. Res. Lett., 33, L15308, 1-5.

Raja Rao, C.S., 1982. Coal fields of India - Coal resources of Tamilnadu, Andhra

Rajaram, M., Anand, S.P. and Balakrishna, T.S., 2006, Composite magnetic anomaly map of India and its contiguous regions, Journal Geological Society of India, 68, 569-576.

Rajasekhar, R.P. and Mishra, D.C., 2005, Analysis of gravity and magnetic anomalies over Lonar Lake, India: An impact crater in basalt province, Current Science, 88, 1836-1840.

Rajasekhar, R.P. and Mishra, D.C., 2005, Archean-Proterozoic collision tectonics across Chota Nagpur granite gneissic complex and Singhbhum craton: Based on gravity studies integrated with geological information, Jour. of Geophysics, XXVI, 85-91.

Rajasekhar, R.P. and Mishra, D.C., 2008, Crustal structure of Bengal basin and Shillong plateau: Extension of Eastern Ghat and Satpura mobile belts to Himalayan fronts and seismotectonics, Gondwana Research, 14, 523-534.

Rajendra Prasad, B., Kesava Rao, G., Mall, D.M., Koteswara Rao, P., Raju, S., Reddy, M.S., Rao, G.S.P., Sridher, V. and Prasad, A.S.S.S.R.S., 2007, Tectonic implications of seismic reflectivity pattern observed over the Precambrian Southern Granulite Terrain, India, Precambrian Research, 153, 1-10.

Rajendra Prasad, B., Tewari, H.C., Vijaya Rao, V., Dixit, M.M. and Reddy, P.R., 1998, structure and tectonics of the Proterozoic Aravalli-Delhi Fold Belt in northwestern India from deep seismic reflection studies, Tectonophysics, 288, 31-41.

Rajendran , C.P. and Rajendran, K., 2005, The status of central seismic gap: a perspective based on the spatial and temporal aspescts of the large Himalayan earthquakes, Tectonophysics, 395, 19-39.

Rajendran, C.P. and Rajendran, K., 2001. Characteristics of deformation and past seismicity associated with the 1819 Kachchh earthquake northwestern India, Bull. Seism. Soc. Am., 91, 407-426.

Rajendran, C.P. and Rajendran, K., 2004, Interpreting the style of faulting and paleo seismicity associated with the 1897 Shillong, north east India, earthquake: implications for regional tectonism, Tectonics, 23, TC4009. doi:10.1029/2003TC 001605.

Rajendran, K., Rajendran, C.P., Thakkar, M. and Tuttle, M.P., 2001. The 2001 Kachchh (Bhuj) earthquake, Coseismic surface features and their significance, Current Science, 80, 1397-1405.

Rajesh, R.S. and Mishra, D.C., 2003, A low homogenous flexural strength in Himalayan syntaxial bends: evidence for partial melt extruded from the Tibetan crust, Diamond Jubilee Workshop of CSIR organized by NGRI, Hyderabad, June, 2003.

Rajesh, R.S. and Mishra, D.C., 2003, Admittance analysis and modeling of satellite gravity over Himalaya-Tibet and its seismogenic correlation, Current Science, 84, 224-230.

Rajesh, R.S. and Mishra, D.C., 2004, Lithospheric thickness and mechanical strength of the Indian shield, Earth's Planetary Sci. Letters, 225, 319-328.

Rajesh, R.S., Stephen, J. and Mishra, D.C., 2003, Isostatic response and anisotropy of the eastern Himalayan-Tibetan Plateau, A reappraisal using multitaper spectral analysis, Geophys. Res. Lett., 30, 1060doi10.1029/2002GL016104.

Rajesh, R.S., 2003, Isostatic response and anisotrophy of the Indian and Tibetan lithosphere using multitaper spectral analysis, A Ph.D. Thesis submitted to Osmania University.

Rajesham, T., Bhaskar Rao,Y.J., Murti, K.S 1993, The Karimnagar granulite terrane - a new Sapphirine bearing granulite province, South India, J. Geol. Soc. India 41, 51–59.

Raju, A.T.R. and Srinivasan, S., 1983, More hydrocarbon from well explored Cambay basin, Petroliferous Basins of India, Petroleum Asia Journal, 6, 25-36.

Raju, D.Ch.V., 2003, LIMAT: a computer programme for the least square inversion of magnetic anomalies due to tabular bodies, Compuer and Geosciences, 29, 91-98.

Raju, P.S.R., 1986, Geology and hydrocarbon prospects of Prahhita-Godavari graben. Journal of Assoc. of Exploration Geophysics, 7, 131-146.

Ram Babu, H.V. and Rama Rao, Ch., 2008, Airborne geophysics in India- A review, Mem. Geol. Soc. of India, 68, 289-308.

Ram, J., Shukla, S.N., Pramanik, A.G. and Varma, B.K., 1996, Recent investigations in the Vindhyan basin: Implications for the basin tectonics, Memoir Geological Society of India, 36, 267-286.

Rama Rao, J.V., Murty, N.V.S. and Balakrishna, S., 2010, Inferences from gravity and magnetic surveys of Ramnad subbasin, Cauvery sedimentary basin, Tamil Nadu, India, Journal of Geophysics, 31, 45-50.

Ramadass, G., Ramaprasada Rao, I.B. and Himabindu, D., 2006, Crustal configuration of the Dharwar craton, India, based on joint modeling of regional gravity and magnetic data, Journal of Asian Earth Sciences, 26, 437-448.

Ramakrishna, M., 2003, Craton-Mobile belt relations in Southern Granulite Terrain, Mem. Geol. Soc. of India, 50, 1-24.

Ramakrishna, T.S. and Bhasker Rao, K.V.S. 1989, Mass estimate of Banwas sulphide ore body by gravity method. Indian Minerals, v.43, pp. 65-70.

Ramakrishna, T.S., 2006, Geophysical practice in mineral exploration and mapping. Geol. Soc. of India. 1-382.

Ramana, D.V., Thiagaarajan, S. and Rai, S.N., 2003, Crustal thermal structure of Godavari graben and coastal region, Current Science, 84(8), 1116-1122.

Ramana, M. A., Subrahmanyam, V., Chaubey, A. K., Ram Prasad, T., Sarma, K. V. L. S., Krishna, K. S., Desa, M., Murthy, G. P. S, 1997, Structure and origin of 85 East Ridge, J. Geophys. Res., 120, 17995-18012.

Ramesh, Srinagesh, D., Rai, S.S., Prakasham, K.S., and Gaur, V.K., 1993, High velocity anomaly under the Deccan Volcanic Province, Phys. Earth & Planet. Int., 77, 285-296.

Ranganai, R.T., Whaler, K.A. and Ebinger, C.J., 2008, Gravity anomaly patterns in the south-central Zimbabwe Archaean craton and their geological interpretation, Journal of African Earth Sciences, 51, 257-276.

Rangnai, R.T. and Ebinger,C.J., 2008, Aeromagnetic and landsat TM structural interpretation for identifying regional groundwater exploration targets, south-central Zimbabwe craton, J. of Applied Geophysics, 65,73-83.

Rao, B.R., 1992, Tectonophysics, 201, 175-185.

Rao, B.R., 2005, Buckingham Canal saved people in Andhra Pradesh (India) from the tsunami of 26 December 2004, Current Science, 89, 12-13.

Rao, C.K., Gokaran, S.G. and Singh, B.P., 1995, Upper crustal structure in the Torni-Purnad region, central India using magnetotelluric studies. Jour. Geomag. Geoelectr., 47, 411-420.

Rao, G. V. S. P. and Bhalla, M. S., 1984, Lonar lake: Paleomagnetic evidence of shock origin. Geophys. J. R. Astron. Soc., 77, 847– 862.

Rao, G.V., Reddy, G.K., Rao, R.U.M. and Gopalan, K., 1994, Extraordinary helium anomaly over surface rupture of September 1993 Killari earthquake, India, Current Science, 68, 933-936.

Rao, G.V.S.P. and Mishra, D.C., 1997, A Proterozoic APWP of cratons from either side of Narmada-Son lineament and the contiguity of the Indian subcontinent, Proc. of Workshop on the Tectonics of Narmada-Son Lineament, Geological Survey of India, Miscellaneous Publication, 63, 129-139.

Rao, G.V.S.P. and Rao, J.M., 1996, Paleomagnetism of Rajmahal trap of India- a magnetic reversal in the Cretaceous normal superchron, J.Geomag. and Geoelect., 48, 993-100.

Rao, G.V.S.P., Singh, S.B. and Laxmi, K.J.P., 2003, Paleomagnetic dating of Sankara dyke swarm in Malani Igneous Suite, Western Rajasthan, India, Current Science, 85, 1486-1492.

Rao, J.M., Rao, G.V.S.P. and Patil, S.K., 1990, Geochemical and paleomagnetic studies on the middle Proterozoic Karimnagar mafic dyke swarm, India, In: Mafic dykes and Emplacement Mechanism. (Eds.) A.P. Parkaer, P.C., rock Wood and D.H. Tucker, A.A. Balkema, Rotterdam, Brookfield, 373-382.

Rao, K.J., Murlidharan, R., Srivastava, P.K., Desapati, T., Chaturvedi, A.K. and Chaki, A., 2010, On the effective use of integrated study of aero-space data for targeting heavy mineral placers- A case study from Krishna-Godavari coast, Andhara Pradesh, India, Journal of Geophysics (In Press).

Rao, M.B.R., 1973, The subsurface geology of Indo-Gangetic alluvial plains, J. Geol. Soc. India, 14, 3, 217-242.

Rao, M.B.S.V., Tiwari, V.M. and Mishra, D.C., 1997, Remagnetization of rocks due to lightning, Jour. of Geophysics, 18, 211-213.

Rao, M.R.K.P., 2007, Study of gravity and magnetic lineaments in western and central India with reference to their groundwater potential, A Ph. D. thesis of SRTMU, Nanded.

Rao, N. P. and Kalpna, 2005, Deformation of the subducted Indian lithospheric slab in the Burmese arc, Geophys., Res., Letters, 32, L05301, doi:10.1029/2004GL022034, 1-5.

Rao, P. Koteswara and Reddy, P.R. 2005, A cost effective strategy in conducting integrated geophysical studies in trap covered country. J. Ind. Geophy. Union, v. 9, no.1, pp. 65-69.

Rao, T. C. S. and Rao, V. B., 1986, some structural tectonics of Bay of Bengal, Tectonophysics, 124, 141-153.

Rao, T.S., 1987, The Pakhal Basin; A perspective. In: Purana basins of pensinsular India; middle to late Proterozoic based on the proceedings of the seminar. Memoir- Geological Society of India. 6, 161-187.

Rao, V.K. and Prasad, S.N., 2001, Fluid genesis through serpentinites and the origin of subsurface electrical conductors in Northwest India and adjoining regions, Earth and Plant., Sci., Letters, 190, 79-91.

Rapp, R.H, 1997, Past and future developments in geopotential modelling, in: Forsberg, R., Feissl, M. and Dietrich, R. (eds.) Geodesy on move, Springer, Berlin, 58-79.

Rapp, R.H. and Pavlis, N.K., 1990, The development and analysis of geopotential coefficients model to spherical harmonic degree 360, J. Geoph. Res., Solid Earth, 95, 21855-21911.

Rapp, R.H, 1989, Signals and accuracies to be expected from Satellite Gradiometer misssion, Manuscripta Geodaetica , 14, 36-42.

Rasmussen and Pedersen, L.B., 1979, End corrections in potential field modelling, Geophysical Prospecting, 27, 749-760.

Rasmussen, B., Bose, P.K., Sarkar, S., Banerjee, S., Fletcher, I.R. and Mcnaughton, N.J., 2002, 1.6 GA U-PB Zircon Age for the Chorhat sandstone, Lower Vindhyan, India: Possible implication for early evolution of animals. Geology, 30, 103-106.

Rastogi, B. K. and Talwani, P., 1980, B.S.S.A., 70, 1849-1868.

Rastogi, B.K., 1992, Current Science, 62, 101-108.

Rastogi, B.K., Gupta, H.K., Mandal, P., Satyanarayana, H.V.S., Kousalya, M., Raghavan, R., Jain, R., Sarma, A.N.S., Kumar, N. and Satyamurthy, S., 2001. The deadliest stable continental region earthquake occurred near Bhuj on 26th January, 2001, J. of Seismology, 5, 609-615.

Raval, U. and Veeraswamy, K., 2007, Within and beyond Protocontinents: some geophysical aspects reflecting geodynamics of the Indian continental lithosphere, Mem. Gondwana Research, No. 10, 263-285.

Raval, U., 1989, On hotspot, Meso-Cenozoic tectonics and possible thermal networking beneath the Indian continents, In: Advances in Geophysical Research in India, published by Indian Geophysical Union, 314-330.

Raval, U., 1993, Fast movement of the Indian plate, buoyancy due to plume and warm asthenosphere, Proc. of the 29th Annual convention of IGU, Hyderabad.

Raval, U., 1995, Geodynamics of the tectonomagmatic and geophysical signatures within mobile parts of the transect, Memoir, Geological Society of India, 31, 37-62.

Raval, U., 2000, Laterally heterogenous seismic vulnerability of the Himalayan arc: a consequence of cratonic and mobile nature of underthrusting Indian crust, Curr. Sci., 78, 546-549.

Raval, U., 2001. Earthquakes over Kachchh, A region of "trident" space-time geodynamics, Current Science, 81, 809-815.

Ravat, D., Lu, Z. and Braile, L.W., 1999, Velocity-density relationships and modeling the lithospheric density variations of the Kenya rift, Tectonophysics, 302, 225-240.

Ravi Kumar, M., Saul, J., Sarkar, D. and Kind, R., 2001. Crustal structure of the Indian shield, new constraints from teleseismic receiver functions, Geoph. Res. Lett., 28, 1339-1342.

Ray, J.S., 2006, Age of the Vindhyan Supergroup: A review of recent findings, J. Earth Syst. Sci.., 115, 149-160.

Ray, J.S., Martin, M.W., Veizer, J., Bowring, S.A., 2002. U-Pb zircon dating and Sr isotope systematics of the Vindhyan Supergroup, India. Geology 30, 131-134.

Ray, J.S., Pattanayak, S.K. and Pande, K., 2005, Rapid emplacement of the Kerguelen plume-related Sylhet traps, eastern India, Evidence from ^{40}Ar-^{39}Ar geochronology, Geoophysical Research Letters 32 (L10306), 1-4.

Ray, J.S., Veizer, J., Davis, W.J., 2003. Sr and Pb isotope systematics of carbonate sequences of the Vindhyan Supergroup, India: age, diagenesis, correlations and implications for global events. Precambrian Research 121, 103-140.

Ray, L., Bhattacharya, A. and Roy, S., 2007, Thermal conductivity of Higher Himalayan Crystallines from Garhwal Himalaya, India, Tectonophysics, 434, 71-79.

Raza, M., Khan, A. and Khan, M.S., 2009, Origin of Late Paleoproterozoic great Vindhyan basin of North Indian shield: Geochemical evidence from mafic volcanic rocks, Journal of Asian Earth Sciences 34, 716-730.

Reddi, A. G. B. and Ramakrishnan, T.S., 1988, Bouguer gravity Atlas of Western Indian (Rajasthan and Gujarath) Shield, Publication of Geological Survey of India.

Reddi, A.G.B., Mathew, M.P., Singh, B. and Naidu, P.S., 1988, Aeromagnetic evidence of crustal structure in the granulite terrain of Tamil Nadu-Kerala, J. of Geol. Soc. of India, 32, 368-381.

Reddy, P. R., Behera, L., Sain, K., 2005, Magmatic underplating in the Mahanadi delta: Results from seismic and gravity studies, DST News Letter, 15, 21-24.

Reddy, P.R., 2010, Seismic imaging of the Indian continental and oceanic crust, Professional Books Publisher, India.

Reddy, P.R., Chandrakala, K. and Sridhar, A.R., 2000, Crustal velocity structure ofl the Dharwar Craton, India. J. Geol. Soc. India, 55, 381-386.

Reddy, P.R., Mall, D.M., and Prasad, A.S.S.S.R.S., 1997, Sub-horizontal layering in the lower crust and its tectonic significance in the Narmada-Son region, India, Pure Appl. Geophys., 149, 525-540.

Reddy, P.R., Vijayarao. V., Prasad, B.R., Sain, K., Prasadrao, P. and Prakaskhare, 2003, Crustal seismic studies along Kuppam-Palani transect in southern granulite terrain. Mem. Geol . Soc. India, 50, 79-106.

Reddy, R.A., Rao, A.M.V.R., Raghuramaiah, K. and Radder, R.B., 2002, Gold mineralization in high grade metamorphic terrain: an integrated geophysical approach in Volageri-Ambale area, Mysore district, Karnatka, Geol. Surv. Ind. Spl. Pub., 75, 128-140.

Redfern, R., 2001, Origins: The evolution of continents, oceans and life, University of Oklahome Press, U.S.A., 1-360.

Reeves, C., 2005, Aeromagnetic Surveys, Geosoft.

Reeves, C., 2010, Airborne magnetic map of Central Botswana, Personal Communication.

Regan, R.D. and Hinz, W.J., 1976, The effect of finite data length in the spectral analysis of ideal gravity anomalies, Geophysics, 41, 44-55.

Regan, R.D., Cain, J.C. and Davis, W.M., 1975, A global magnetic anomaly map, J. Geoph. Res., 80, 794.

Reigber, C. et al., 2003, The CHAMP-only Earth Gravity Field Model, EIGEN 2, Adv. Space Research, 31(8), 1883-1888.

Reigber, C., et al., 2002, A high gravity global gravity field model from CHAMP GPS tracking data and accelerometry (EIGEN-13) Geoph. Res. Lett., 29, 37, 1-4.

Reigber, C., Kang, Z., König, R. and Schwintzer, P., 1996, CHAMP-A mini satellite mission for geopotential and atmospheric research, Supplement to EOS, Transections of the American Geophysical Union, 77 (17), S 40.

Reigber, C., Lühr, H., Schwintzer, P. and Wickert, J., 2005, Earth observation with CHAMP, Springer Verlag Berlin Heidelberg, 1-628.

Replumaz, A., Farason, H., van der Hilst, R.D., Besse, J. and Tapponnier, P., 2004, 4-D evolution of SE Asia's mantle from geological reconstructions and seismic tomography, Earth and Planetary Science Letters 221, 103-115.

Reyners, M., Eberhart-Phillips, D. and Stuart, G., 2007, The role of fluids in lower-crustal earthquakes near continental rifts, Nature, 446, doi:10.1038/nature05743, 1075-1078.

Reynolds, J..M., 1997, An introduction to applied and environmental geophysics, John Wiley and Sons Ltd., West Sussex, England, 1-796.

Riahi, Ali M. and Lund, C.E., 1994, Two dimensional modeling and interpretation of seismic wide angle data from the western Gulf of Bothnia, Tectonophysics, 239, 149-164.

Richter, B. and Warburton, R.J., 2003, A new generation of super conducting gravimeters, GWR Instruments, Inc-Home Page. Html, 1-10.

Ricou, L. E, 1996, The Plate Tectonic history of the past Tethys ocean, In: The Ocean basin & Margins, The Tethys Ocean (eds) A. E. M. Nairn, L. E. Ricou , B. Vrielynck and J. Dercourt, Plenum Press, 8, 3-62.

Rino, S., Kon, Y., Sato, W., Maruyama, S., Santosh, M., Zhao, D., 2008. The Grenvillian and Pan African orogens: worlds largest orogenies through geologic time, and their implications on the origin of superplume, Gondwana Research 14, 51-72.

Ritzwoller, M.H., N.M. Shapiro, M.P. Barmin, and A.L. Levshin, 2002. Global surface wave diffraction tomography, J. Geophys. Res., 107(B12), 2335.

Robinson, A., 2002, Earth Shock, Thames and Hudson, USA.

Rodell, M., Velicogna, I. and Famiglietti, S., 2009, Satellite based estimates of groundwater depletion in India, Nature, 462, 999-1002.

Roest, W.R., Verhoef, J. and Pilkington, M., 1992, Magnetic interpretation using the 3-D analytic signal, Geophysics, 57, 116-125.

Rogers, J.J.W. and Santosh, M., 2009. Tectonics and surface effects of the supercontinent Columbia. Gondwana Research 15, 373-380.

Roy, A. and Prasad, M.H., 2003, Tectonothermal events in Central Indian Tectonic Zone (CITZ) and its implications in Rodinian crustal assembly, J. Asian Earth Sciences, 22, 115-129.

Roy, A., 1962, Ambiguity in geophysical interpretation, Geophysics, 27, 90-99.

Roy, A., 1966, The melthod of continuation in mining geophysical interpretation, Geoexploration, 4, 65-83.

Roy, A., 1967, Convergence in downward continuation for some simple geometries, Geophysics, 32, 853-866.

Roy, M., Jordan, T.H. and Pederson, J., 2009, Colorado plateau magmatism and uplift by warming of hetrogenous lithosphere, Nature, 459, 978-982.

Roy, S. and Mareschal, J.C., 2011, Constraints on the deep thermal structure of the Dharwar craton, India from heat flow, shear wave velocities and mantle xenoliths, J. Geophysical Research, 116, doi 10.1029/2010JB007796.

Roy, S. and Rao, R.U.M., 1999, Geothermal investigations in the 1993 Latur earthquake area, Deccan Volcanic Province, India, Tectonophysics, 306, 237-252.

Roy, S. and Rao, R.U.M., 2000, Heat flow in the Indian Shield, Journal of Geophysical Research, 105, 25,587-25,604.

Roy, S., 2010, Geothermal exploration in India, Proc. of the Australian Geothermal Energy conference (H. Gurgenci and R.D. Weber, editors), Record 2010/35, GeoCat # 71204.

Roy, S., Ray, L. and Senthil Kumar, P., Reddy, G.K. and Srinivasan, R., 2003, Heat flow and heat production in the Prescambrian Gneiss-granulite Province of Southern India, Memoir Geological Society of India, 50, 179-192.

Royden, L.H., Burchfiel, B.C. and van der Hilst, R.D., 2008, The geological evolution of the Tibetan plateau, Science, 321, 1054-1058.

Royer, J. Y., Choubey, A. K., Dymont, J., Bhattacharya, G. C., Srinivas, K., Yatheesh, V., Ram Prasad, T., 2002, Paleogene plate tectonic evaluation of the Arabian and eastern Somali basins, In: (eds) Clift, P. D., Kroon, D., Gaedicke, C. and Craig, J., The Tectonic and Climatic Evaluation of the Arabian Sea Region, Geol. Soc. Landon, Sp. Publications, 195, 7-23

Royer, J.Y. and Patriat, P., 2002, L'Inde part a la derive In: Museum National D'Historie Naturalle (France), Allegre, C.J., Avonac, J.P. and DeWever, P. (eds) Himalaya-Tibet, Le choc des continents, Paris, CNRS editions et Museum national de'Historie naturalle, 25-31.

Rui, G., Xiangzhou, C., Gongjian, W., 1999, Lithospheric structure and geodynamic model of the Golmud-Ejn transect in northern Tibet, Geol. Soc., Am., 328, 9-17.

Ruihao, Li. and Zhaozhu, F., 1983, Tectonophysics, 97, 159 169.

Rummel, R., Müller, J., Oberndorter, H. and Sneeuw, N., 2006, Satellite gravity gradiometry with GOCE, http:www.geomatics.ucalgary.cap

Runcorn, S.K., 1956, Paleomagnetic comparisons between Europe and North America, Proc. Geol. Assoc. Canada, 8, 77-85.

Runcorn, S.K., 1961, Climatic change through geological time in the light of Paleomagnetic evidence for Polar Wandering and Continental drift, Q.J.R. Meterolog, Soc., 87, 283-311.

Sachan, H.K., Mukherjee, B.K. and Ahmad, T., 2005, Evidence of deep fluids in ultrahigh-pressure ecolgite from Tso-Morari crystalline complex, Ladakh, India, Deep Continental Studies in India, DST Newsletter, 15, 18-20.

Saha, D.K., Choudhury, K. and Murthy, B.S.R., 2002, Geophysical studies for subsurface saline water zones in the south eastern part of Calcutta mega city, Geol. Surv. Ind. Special Publication No. 75, 262-271.

Saha, D.K., Naskar, D.C., Bhattacharya, P.M. and Kayal, J.R., 2007, Geophysical and seismological investigations for the hidden Oldham fault in the Shillong Plateau and Assam Valley of Northeast India, Journal of Geological Society of India, 69, 359-372.

Sahasrabudhe, P.W, 1963, Palaeomagnetism and the geology of the Deccan traps, Seminar on Geophysical Investigations in Peninsular Shield, Osmania University, Hyderabad 226-243.

Sahasrabudhe, P.W. and Mishra, D.C., 1966, Paleomagnetism of Vindhyan rocks of India, Bull. NGRI, 4, 49-55.

Sain, K. and Kaila K.L., 1996, Ambiguity in the solution of the velocity inversion problem and a solution by joint inversion of seismic refraction and wide angle reflection times. Geophy. J. Int., 124, 215-227.

Sain, K. and Ojha, M., 2008, Identification and quantification of gas hydrates: A viable source of energy in the 21 st century, Mem. Geol. Soc. of India, 68, 273-288.

Sain, K., Bruguier, N., Murty, A.S.N. and Reddy, P.R., 2000, Shallow velocity structure along the Hirapur-Mandla profile using traveltime inversion of wide-angle seismic data, and its tectonic implications, Geophys. J. Int., 142, 505-515.

Sain, K., Zelt, Colin A., and Reddy, P.R. 2002, 'Imaging of subvolcanic Mesozoics in the Saurashtra peninsula of India using travel time inversion of wide angle seismic data', Geophy J. Int., 150, 820-826.

Sajeev, K., Windley, B.F., Connolly, J.A.D. and Kon, Y., 2008, Retrogressed eclogite (20 kbr, 1020^0C from the Neoproterocoic Palghat-Cauvery suture zone, Southern India, Precambrian Research 171, 23-36.

Sakuntala, S. and Krishna Brahmam, N., 1984, Diamond mines near Raichur, J. Geol. Soc. of India, 25, 780-786.

Salem, A., Williams, S., Fairhead, J.D., Ravat, D. and Smith, R., 2007, Tilt-depth method: A simple depth estimation method using first order magnetic derivatives, The Leading Edge, 1502-1505.

Saltus, R. W. and Hudson, T. L., 2007, Regional Magnetic Anomalies crustal strength and location of the northern Cordilleran fold and thrust belt, Geology, 35, 567-570.

Sandwell, D.T. and Smith, W.H.F., 1997, Marine gravity anomaly from Geosat and ERS-1, Satellite alltimetry, J. Geoph. Res., 102, 10039-54.

Santosh, M., Maruyama, S., Sato, K., 2009, Anatomy of a Cambrian suture in Gondwana: Pacific type orogeny in southern India !, Gondwana Research, 16, 321-341.

Santosh, M., Yokoyama, K. and Acharyya, S.K., 2004, Geochronology and Tectonic Evolution of Karimnagar and Bhopalpatnam Granulite Belts, Central India. Gondwana Research, 7, 501-518.

Sapra, B.K., Mayya, Y.S., Sawant, V.D., and Nambi, K.S.V., 1997, Curr. Sci., 72, 321-325.

Sar, D., 2008, Selected petroliferous basins of India: A gravity perspective, Mem. Geol. Soc. of India, 68, 259-272.

Sarangi, S., Gopalan, K. and Kumar, S., 2004, Pb-Pb age of earliest megascopic eukaryotic alga bearing Rohtas Formation, Vindhyan Supergroup, India: Implications for Precambrian atmospheric oxygen evolution. Precambrian Research, 132, 107-132.

Sarkar, D., Sain, K., Reddy, P.R., Catchings, R.D. and Mooney, W.D., 2007, Seismic reflection images of the crust beneath the 2001 M = 7.7 Kutch (Bhuj) epicentral region, western India, The Geological Society of America Special Paper, 425, 319-327.

Sarkar, S.N., Gopalan, K. and Trivadi, J.R., 1981, New data on the geochronology of the Precambrians of Bhandara-Drug, Central India. Indian Jour. Earth Sci., 8, 131-151.

Sarma et al., 1994, Mem. Geol. Soc. of India, 35, 101-118.

Sarma, B.S.P., Verma, B.K. and Satyanarana, S.V., 1999, Magnetic mapping of Majhgawan diamond pipe of Central India, Geophysics, 64, 1735-1739.

Sarma, S.V.S. et al 1992, Mgnetotelluric studies for oil exploration over Deccan Traps, Saurashtra, Gujarat, India. NGRI-92-Lithos, 125, 80.

Sarma, S.V.S., Nagrajan, N., Someshwar Rao, M., Hari Narayana, T., Virupakhsi, G., Murthy, D.N., Sarma, M.V.C., and Gupta, K.P.B., 1996, Magnetotelluric studies along Mungwani-Rajnandgon in Central India. Proc. II, Int. Seminar on Geophysics, AEG, Hyderabad, 206-207.

Sarma, S.V.S., Prasanta, B., Patro, K., Harinarayana, T., Veeraswamy, K., Sastry, R.S., Sarma, M. V. C., 2004, A magnetotelluric (MT) study across the Koyna seismic zone, western India: evidence for block structure, Physics of the Earth and Planetary Interiors, 142, 23-36.

Sarma, S.V.S., Virupakshi, G., Harinarayana, T., Murty, D.N., Prabhakar, S., Rao, E., Veeraswamy, K., Madhusudana Rao, Sarma, M.V.C. and Gupta, K.R.B., 1994, A wide band magnetotelluric study of the ZLatur earthquake region, Maharashtra, India, Mem. Geol. Soc. India, 35, 101-118.

Sasai, Y., 1986, Bull. Earthquake Res. Inst., Univ. of Tokyo, 61, 429-473.

Sastry, R. S., Nagarajan, Nandini and Sarma, S. V. S. 2002, Electrical imaging of deep crustal features of Kutch, India. IGU presentation

Sastry, R.G., Choudhary, N. and Md. Israil, 2004, Scheme for Bouguer and terrain corrections with variable densities – Examples from Ladakh Himalaya, DST – News Letter, 14(2), 13-16.

Sastry, R.S., Nagarajan, N. and Sarma, S.V.S., 2008, Electrical imaging of deep crustal features of Kutch, India, Geophys. J. Int., 172, 934-944.

Satpal, Singh O. P., Sar, D., Chatterjee, S. M., and Sanjeev, Sawai 2006, Integrated interpretation for sub-basalt imaging in Saurashtra Basin, India. The Leading Edge, 882-885.

Sawada, Y., Rehanul, M., Khan, S.R. and Asis, A., 1992, Mesozoic igneous activity in the Muslim Bagh area, Pakistan with special reference to hot spot magmatism related to the break-up of Gondwanaland, Proceedings of Geoscience Colloquium, Geoscience Lab., Geological Survey of Pakristan, 1, 21-38.

Sazhina, N., and Grushiny, N., 1971. Gravity Prospecting. Moscow, Mir Publishers.

Schiermeier, Q., 2005, Nature, 433, 350– 353.

Schimdt, S. and Götze, H.J., 2006, Bouguer and isostatic maps of the central Andes, in Oncken, G. Chong, G. Franz, P. Giese, H.J., Götze, V., Ramos, M. Stecker and P. Wiger (eds.), The Andes-active subduction orogeny, Frontiers in Earth Sciences, 1, Springer Verlag, 559-562.

Schlich, R., 1982, The Indian ocean: A seismic ridges, spreading centers and oceanic Basin, In: The ocean Basins and Margins (eds) A.E.M Nairn and F.G. Stehli, 6, 51-148.

Schmidberger, S. S., Simonetti, A., Heaman, L. M., Creaser, R. A. and Whiteford, S., 2007. Lu-Hf, in situ Sr and Pb isotope and trace element systematics for mantle eclogites from the Diavik diamond mine: Evidence for Paleo Proterozoic subduction beneath the Slave craton, Earth & Planetary Science Letters, 254, 55-68.

Schneid, T.D. and Collins, L., 2000, Disaster management and preparedness, CRCPress LLC, USA, 1-524.

Scholz, C. H. and Kranz, R., 1974, J. Geophys. Res., 79, 2132-2135.

Scholz, C. H., Sykes, L. R. and Aggarwal, Y. P., 1973, Science, 181,803-810.

Schrama, E.J.O., 2003, Error characteristics estimated from CHAMP, GRACE and GOCE derived geoids and from satellite altimetry derived mean dynamic topography, Space Science Reviews, 108, 179-193.

Schulte-Pelkum, V., Monsalve, G., Sheehan, A., Pandy, M.R., Sapkota, S., Bilham, R. and Wu, F., 2005, Imaging the Indian subcontinent beneath the Himalaya, Nature, 435, 1222-1225, doi: 10.1038/nature03678..

Schweig, J. Gomberg, M. Petersen, M. Ellis, P. Bolin, L. Mayrose and B.K. Rastogi, 2003, The M_W 7.7 Bhuj earthquake: global lessons for earthquake hazard in intraplate regions, J. Geol. Soc. India, 61, 277-282.

Scotese, 1997, Reconstruction of Gondwanaland, http://www.scotese.com.

Scotese, C. R., Gahagan, L.M. and Larson, R.L., 1988, Plate-reconstructions of the Cretaceous and Cenozoic ocean basins, Tectonophysics, 155, 27-48.

Searle, M.P., Rex, A.J., Tirrul, R., Rex, D.C., Barnicoat, A. and Windley, B.F., 1989. Metamorphic, magmatic and tectonic evolution of the central Karakoram in the Biafo-Baltoro-Hushe regions of northern Pakistan, In: Tectonics of the Western Himalaya (ed) L.L. Malinconico, Jr. and R.J. Lillie, Geol. Soc. of Am. Special Paper, 232, 47-74.

Searle, M.P., Simpson, R.L., Law, R.D., Parrish, R.R. and Waters D.J., 2003, The structural geometry, metamorphic and magmatic evolution of the Everest massif, High Himalaya of Nepal – South Tibet, Journal of the Geological Society, London, 160, 345-366.

Searle, M.P., Stephenson, B., Walker, J. and Walter, C., 2007, Restoration of the Western Himalaya: implications for metamorphic protoliths, thrust and normal faulting, and channel flow models, Episodes, 30, 4, 242-243.

Seeber, L. and Armbruster, J., 1981, In: Simpson, D.W., Richards, P.G. (Eds.), Great detachment earthquakes along the Himalayan arc and long-term forecasting, Earthquake Prediction: An International Review, Maurice Ewing Ser., 4, AGU, Washington, D.C., 259-277.

Seeber, R., 2001. NW-SE fault at village Manfara northern part of Wagad uplift, U.S.G.S., Web site, February 28.

Seigal, H.O. and McConnel, 1998, Regional surveys using a helicopter suspended gravimeter, The Leading Edge, 17, 47-49.

Sen, D. and Sen, S., 1983, Post Neogene tectonism along the Aravalli Range Rajasthasn, India, Tectonophysics, 93, 75-98.

Sen, G., Bizimis, M., Das, R., Paul, D.K., Ray, A., Biswas, S., 2009, Deccan plume, lithosphere rifting and volcanism in Kutch, India, Earth and Planetary Science Letters, 277, 101-111.

Sengupta, S.N., 1962, Basement configuration of Indo-Gangetic plains shown by the aeromagnetic surveys, Proc. Seminar on Oil Prospecting in the Ganga Valley, Tech. Publ. No. 1, ONGC.

Sengupta, S.N., 1996, The Vindhyan under the north Indian plains, Memoir Geological Society of India, No.36, 257-265.

Seshunarayana, T., Prasad, B. R., Prasad, A. S. S. S. R. S., Mysaiah, D., 2008, Subsurface structure derived from detailed gravity and magnetic investigations along the Pala-Maneri segment of the Main Central Thrust, NW Himalaya, India, Current Science.

Sethna, S.F. and Javeri, P., 2000, Essexite occurrence in the Deccan volcanic province of Saurashtra, western India, J. Asian Earth Sci, 18 ,151-154.

Seyitoglu, G. and Scott, B.C., 1996, The cause of N-S extensional tectonics in western Turkey: Tectonic escape vs back-arc spreading vs orogenic collapse, Journal of Geodynamics, 22, 145-153.

Shapiro, N.M., Ritzwoller, M.H., Molnar, P. and Levin, V., 2004, Thinning and flow of Tibetan crust constrained by seismic anisotropy, Science, 9 July, 305, 233-235.

Sharma, K. and Nandi, S.C., 1964, Magnetic and electrical surveys for locating additional volcanic pipes in the Panna diamond belt, Madhya Pradesh, India, in Kailasm, L.N. and Roy, A., (Eds): Geological results of applied geophysics, Report of the 22nd Internat. Geol. Congress, 91-105.

Sharma, M., Rai, A.K., Nagabhushana, J.C., Sinha, R.M. and Rao, M.V., 1995, Cuddapah basin and its environs as first order uranium target in the Proteozoics of India, Exploration and Research for Atomic Minerals, 8, 127-129.

Sharma, R.S., 1995, An evolutionary model for the Precambrian crust of Rajasthan: Some protrological and geochronological considerations. Mem. Geol. Soc. India, No.31, 91-115.

Sharma, S.R., Rao, V.K., Mall, D.M. and Gowd, T.N., 2005, Geothermal structure in a seismo active region of central India, Pure & Applied Geophysics, 162, 129-144.

Sharp, W.D. and Clague, D.A,, 2006, 50-Ma initiation of Hawaiian-Emperor bend records major change in Pacific plate motion, Science, 313, 1281-84.

Sheldon, R.P., 1964, Paleolatitudinal and paleogeographic distribution of phosphorities, U.S. Geol. Surv., Prof. Raper, 501-C, 106-113.

Sheldon, R.P., 1966, Phosphate deposits in India, their discovery, geology and potential and proposed development, Technical Report, USGS, Agency for International Development.

Sheldon, R.P., 1982, Phosphate rock, Scientific American, 246, 6, 45-51.

Shen, Z., Lu, J., Wang, M. and Burgmann, R., 2005, J. Geophys. Res., 110, B11409, doi:10.1029/2004JB003421, 1-17.

Shin, Y.H., Xu, H., Brintenberg, C., Fang, J. and Wang, Y., 2007, Moho undulations beneath Tibet from GRACE-integrated gravity data, Geophys. J. Int., 170, 971-985.

Shroder Jr., J.F. and Bishop, M.P., 2000, Unroofing of the Nanga Parbat Himalaya, Khan, M.A., Treloar, P.J., Searle, M.P. and Jan, M.Q. (eds) Tectonics of the Nanga Parbat syntaxis and the Western Himalaya. Geol. Soc. London, Spl. Publication, 170, 163-179.

Simans, M. and Hager, B.H., 1997, Localization of the gravity field and the signature of the glacial rebound, Nature, 390, 500-504.

Simons, F.G., Zuber, M.T. and Koernga, J., 2000, Isostatic response of Australian lithosphere: estimation of effective elastic thickness and anisotropy using multitaper spectral analysis, J. Geoph. Res., 105, 163-184.

Simons, F.J., Vander Hilst, R.D., Zuber, M.T., 2003, Spatio spectral localization of isostatic coherence anisotropy in Australia and its relation to seismic anisotropy: implications for lithospheric deformation, J. Geoph. Res., 108, 2250, doi: 10.1029 / 2001 J B000704.

Singh B. and Gupta Sarma, 2001, New method for fast computation of gravity and magnetic anomalies from arbitrary polyhedra, Geophysics, 66, 521-526.

Singh, A. and Ravi Kumar, M., 2009, Seismic signatures of detached lithospheric fragments in the mantle beneath eastern Himalaya and southern Tibet, Earth and Planetary Science Letters, 288, 279-290.

Singh, A., Kumar, M.R. and Raju P.S., 2007, Mantle deformation in Sikkim and adjoining Himalaya: Evidences for a complex flow pattern, Physics of the Earth and Planetary Interiors xxx, 1-10.

Singh, A., Kumar, M.R., Raju, P.S. and Ramesh, D.S., 2006, Shear wave anisotropy of the northeast Indian lithosphere, Geophys., Res., Letters, 33, L16302, doi:10.1029/2006GL026106, 1-5.

Singh, A.P. and Mishra, D.C., 2002, Tectonosedimentary evolution of Cuddapah basin and Eastern Ghat Mobile Belt (India) as Proterozoic collision: gravity, seismic and geodynamic constraints, Journal of Geodynamics, 33, 249-267.

Singh, A.P., Mishra, D.C. and Laxman, G., 2003, Apparent density mapping and 3D gravity inversion of Dharwar crustal province, J. Ind. Geoph. Union, 1-9.

Singh, A.P., Mishra, D.C., Gupta, S.B. and Rao, M.R.K.P., 2004, Crustal structure and domain tectonics of the Dharwar Craton (India): insight from new gravity data, Journal of Asian Earth Sciences 23, 141-152.

Singh, A.P., Mishra, D.C., Vijai Kumar, V. and Rao, M.B.S.V., 2003, Gravity-Magnetic signatures and crustal architecture along Kuppam-Palani geotransect, South India. Mem. Geol. Soc. India. 50, 139-163.

Singh, B. and Arora, K., 2008, Geophysical exploration for petroleum in the subtrappean Mesozoic sedimentary formations of India, Mem. Geol. Soc. of India, 68, 237-258.

Singh, B. and Mishra, D.C., 2000, Signature of plume tectonics under Cambay rift and adjoining regions of NW India- an insight from gravity modeling, Workshop on 'Plume Tectonics', June 13-14, in NGRI, Hyderabad, Abstract Volume, 49-50.

Singh, B., Prajapati, S.K. and Mishra, D.C., 2003, Presence of anomalous crustal root beneath Saurastra Peninsula- Inference from gravity modeling, J. of Geophysics, 24, 25-29.

Singh, B.,2005, Crust and upper mantle heterogeneity beneath the Cambay rift and adjoining region, India- Insight from gravity modeling, Proc. 6th International Petroleum Conference, January 16-19, New Delhi.

Singh, B.P. and Rajaram, M., 1990, Magsat studies over the Indian region, Proc. Indian Acad. Sci., 99, 619-637.

Singh, D., Alat, C.A., Singh, R.N. and Gupta, V.P., 1997, Source rock characteristics and hydrocarbon generating potential of Mesozoic sediments in Lodhika area, Saurashtra basin, Gujarat, India, Proc. Second Int. Petroleum Conference and Exhibition, Petro-tech-97 N. Delhi, 205-220.

Singh, K., Chaudhury, A. and Bhattacharyya, N., 1996. Seven decades of geophysics in India, Asso. of Explo. Geophy., Hyderabad, India.

Singh, R.P., Mishra, H.P., Mishra, S.K., Sharma, P., Singh, N. and Chauhan, D.P.S., 2002, Geophysical investigations for establishing correlatibilty between pyrophyllite-diaspore and sulphide mineralizations in granitoid complex of Bundelkhand, Southern Uttar Pradesh, Geol. Surv India, Special Publication, 75, 72-78.

Singh, S. C., Crawford, W.C., Carton, H., Seher, T., Combier, V., Cannat, M., Canals, J. P., Dusunur, D., Escartin, J., Miranda, M., 2006, Discovery of a magma chamber and faults beneath a Mid-Atlantic Ridge hydrothermal field, Nature, 442, 1029-1032.

Singh, S., Kumar, R., Barley, M.E. and Jain, A.K., 2007, SHRIMP-U-Pb ages and depth of emplacement of Ladakh batholith, Eastern Ladakh, India, J. of Asian Earth Sciences 30, 490-503.

Singh, S.C. and McKenzie, D., 1993, Layering in the lower crust, Geophysical Journal International, 113, 622-628.

Singh, S.C., Carton, H., Tapponnier, P., Hananto, N.D., Chauhan, A.P.S., Hartoyo, D., Bayly, M., Moeljopranoto, S., Bunting, T., Christie, P. Lubis, H. and Martin, J., 2008, Seismic evidence for broken oceanic crust in the 2004 Sumatra earthquake epicentral region, Nature Geoscience, 1,777-781.

Singh, S.C., Hananto, N.D., Chauhan, A.P.S., Permana, H., Denolle, M., hendriyana, A. and Natawidjaja, D., 2010, Gephys. J. Int., 180, 703-714.

Singh, S.P., Singh, M.M., Srivastava, G.S. and Basu, A.K., 2007, Crustal evolution in Bundelkhand area, Central India, Himalayan Geology, 28, 79-101.

Singh, V. and Saha, Apurba 2006, Introduction to this special section: India. The Leading Edge 25, 816 ; DOI:10.1190/1.2221358.

Singh, V.P. and Singh, R.P., 2005, Changes in stress pattern around epicentral region of Bhuj earthquake of 26 January 2001, Geophysical Research Letters, 32, L24309.

Sinha, Y.B., 2004, Leveraging knowledge base in hydrocarbon exploration in India- key driver for bridging demand-supply gap, 29 Annual Convention and seminar on Exploration Geophysics, AEG, Guwahati, 2-5 November, 2004.

Sinha-Roy, S., 1982, Himalayan main central thrust and its implications for Himalayan inverted metamorphism, Tectonophysics, 84, 197-224. ?

Sinha-Roy, S., 1988, Proterozoic Wilson cycle in Rajasthan. Mem. Gol. Soc. India, 7, pp.95-108.

Sinha-Roy, S., 2007, Evolution of Precambrian terrains and Crustal-scale structures in Rajasthan craton, NW India: A Kinematic Model, International Association for Gondwana Research, Japan, IAGR Memoir No. 10, 23-40.

Sinha-Roy, S., Malhotra, G. and Guha, D.B., 1995, A transect across Rajasthan, Research high lights in earth system Science, DST's Spl., 1, 17-25.

Sinno, Y.A., Daggette, P.H., Keller, G.R., Morgan, P and Harder, S.H., 1986, Crustal structure of southern Rio Grande rift determined from seismic refraction profiling, J. Geophys. Res., 91, 6143-6156.

Skeels, D.C., 1947, Ambiguity in gravity interpretation, Geophysica, 12, 43-56.

Sleep, N.H. and Wolery, T.J., 1978, Egress of hot water from mid ocean ridge hydrothermal systems, J. Geophys. Res., 83, 5913.

Slichter, L.B. Caputo, M. and Hagger, C.L. J. Geophys. Res., 1965, 70, 1541-1551.

Smalley Jr., R., Ellis, M.A., Paul, J. and Van Arsdale, R.B., 2005, Nature, 435, 1088-1090.

Smellie, D.W., 1956, Elementary approximations in aeromagnetic interpretation, Geophysics, 21(4), 1021-1040.

Smirnov, M.Yu. and Pedersen, L., 2009, Magnetotelluric measurements across the Sorgenfrei-Tornquist Zone in southern Sweden and Denmark, Geophys. J. Int., 176, 443-456.

Smith, D.V. and Pratt, D., 2003, Advanced processing and interpretation of high resolution of aeromagnetic survey data over the central Edwards, Acquifer, Texas, Proceedings from the Symposium on the Application of Geophysics to Engineering and Environmental Problems, Environmental and Engineering Society.

Smith, W.H.F and Sandwell, D.T., 1994, Bathymetric prediction from dense satellite altimetry and sparse ship board bathymetry, J.Geoph. Res.,99, 21803-21824.

Smithies, R.H., Van Kranendonk, M.J. and Champion, D.C., 2005, Iat started with a plume – early Archean basaltic proto-continental crust, Earth and Planetary Science Letters 238, 284-297.

Song, T.R.and Simons, M., 2003, Large trench parallel gravity variations predict seismogenic behavior in subduction zones, Science, 301, 630-633.

Sorkhabi, R.B. and Macfarlane, A., 1999, Himalaya and Tibet: Mountain roots to mountain tops, in Macfarlane, A., Sorkhabi, R.B. and Quade, J. (eds) Himalaya and Tibet: Mountain Roots to Mountain Tops: Boulder, Colorado, Geol. Soc. of Am. Special Paper 328, 1-7.

Sorkhabi, R.B. and Stump, E., 1993, Rise of the Himalaya: a geochronologic approach, GSA Today, 3, 85, 88-92.

Specter, A. and Grant, F.S., 1970, Statistical models for interpreting aeromagnetic data, Geophysics. 35, 293-302.

Sreedhar Murthy, Y. and Sarma, S.V.S., 2009, Hydrocarbon potential of the Proterozoic Cuddapah basin-An appraisal, Personal Communication.

Sreedhar Murthy, Y., 1999, Images of the gravity field of India and their salient features, J. Geol. Soc. India, 54, 221-235.

Sreedhar Murthy, Y., 2002, On the correlation of seismicity with geophysical lineaments over the Indian subcontinent, Current Science, 83, 760-766.

Sreedhar Murthy, Y., Govindrajan, K. and Babu Rao, V., 1998, Contours to images-Part I: an innovative methodology, J. of Geophysics, 19, 141-148.

Sridhar, A.R., Prasad, A.S.S.S.R., Satyavani, N. and Sain, K., 2009, Subtrappean Mesozoic sediments in the Narmada basin based on travel time and amplitude modeling- a revisit to old seismic data, Current Science, 97, 1462-66.

Sridhar, A.R., Tewari, H.C., Vijaya Rao, V., Satavani, N., Thakur, N.K., 2007, crustal velocity structure of the Narmada Son lineament along the Thuadara-Sendhwa-Sindad profile in the NW part of central India and its geodynamic implications. J. Geol. Soc. of India, v. 69, pp. 1147-1160.

Srinagesh, D., Rai, S.S., Ramesh, D.S., Gaur, V.K. and Rao, C.V.R., 1989, Evidence for thick continental roots beneath South Indian Shield, Geophysical Research Letters, 16, 1055-1058.

Srinivasan, S. and Khar, B.M., 1995. Frontier basin exploration in India, Prospectives and challenges, Proc. of Petrotech-95, New Delhi, Technology trends in Petroleum Industry, 1-16.

Srinivasan, V., 2005, The Dauki fault in Northeast India: through remote sensing. J. Geol. Soc. India, 66, 413-426.

Srirama Rao, S.V., Chary, K.B., Gowd, T.N. and Rummel, F., 1999, Tectonic stress field in the epicentral zone of Latur earthquake of 1993, Proc. Indian Acad. Sci. (Earth Planet Sci.), 108, 93-98.

Srivastava, K., Swaroopa, V., Srinagesh, D. and Dimri, V.P., 2007, Could the 12 September 2007 earthquake of southern Sumatra, Indonesia, have generated a large tsunami causing damage to the east coast of India !, Current Science, 93, 1228-1229.

Srivastava, V.N., Mitra, S.K., Das, L.K., Mohan, V., Tiwari, R.A., and Singh, R.P. 1983, Characteristic geophysical response over kimberlite plugs and other ultrabasic bodies in Jungel Valley of Mirzapur, district, U.P. Geol. Sur. of India, Spl. Pub. No. 2(1), 291-306.

Stagg, H. M. J., 1985, The structure and origin of Prydz bay and Mac Robertson Shelf, East Antarctica, Tectonophysics, 114, 315-340.

Stanley, J.M., 1977, Simplified gravity and magnetic interpretation of contact and dyke like structures, Bull. Aust. SEG, 8, 3, 60-64.

Stein, C. and Stein, S., 1992, A model for the global variation in oceanic depth and heat flow with lithospheric age, Nature, 359, 123-29.

Stein, S. and Okal, E. A., 2005, http://www. earth.northwestern.edu/people/seth/research/ sumatra2.html

Stein, S., Sella, G.F. and Okal, E.A., 2000. The January 26, 2001 Bhuj earthquake and the diffuse western boundary of the Indian plate, Plate Boundary Zone, Published by the Americal Geophysical Union, 243-264.

Stein, S., Stella, G.F. and Okal, E.A., 2002, Geodynamic series, AGU, 30, 243-264.

Storey, B. C., 1995, The role of mantle plumes in continental break up: Case histories from Gondwana Land, Nature, 377, 301-308.

Stratford, W and H. Thybo, 2008, New insights into the lithospheric structure of southern Norway, EOS, 89, 554.

Subarya, C., Chlieh, M., Prawirodirdjo, L., Avouac, J.P., Bock, Y., Sieh, K., Meltzner, A.J., Natawidjaja, D.H., and McCaffrey, R., 2006, Plate boundary deformation associated with the great Sumatra-Andaman earthquake, Nature, 440, 46-51.

Subba Raju, M., Sreenivasa Rao, T., Setti, D.N., Reddi, B.S.R., 1978, Recent advances in our knowledge of the Pakhal super group with special reference to the central part of the Godavari valley, Records of Geological Survey of India 110, 39-59.

Subba Rao, D.V., 1996, Resolving Bouguer anomalies in continents – A new approach. Geophysical Research Letters., 23, 3543 -3546.

Subrahmaniam, C. and Verma, R.K., 1982, Gravity interpretation of the Dharwar greenstone, gneiss-granite terrain southern India shield and its geological implications, Tectonophysics, 84, 225-245.

Subrahmanyam, B., 1985, Lonar crater, India: A Crypto-volcanic origin.

Subrahmanyam, B., Subba Rao, J.A.V.R.K., Rao, H.V. and Chakravarthy, D. 1991, Three probable locations for Kimberlites in Wajrakarur-Lattavaram – P.C. Pyapilli area, Andhra Pradeh. Geol. Soc. Ind., 37, 443-451.

Subrahmanyam, C. and Verma, R.K., 1986, Gravity field, structure and tectonics of Eastern Ghats, Tectonophysics, 126, 195-212.

Subrahmanyam, C., Thakur, N. K., Rao. T. G., Khanna, R., Ramana, M. V., Subrahmanyam, V., 1999, Tectonics of the Bay of Bengal: New insights from satellite gravity and ship borne geophysical data, Earth & Planet. Science Letters, 17, 237-251.

Sugden, T.J., Deb, M. and Windby, B.F., 1990, The tectonic setting of mineralization in the Proterozoic Aravalli Delhi Orogenic Belt, NW India. In: SM. Naqvi (Ed.), Precambrian continental crust and its Economic Resources. Elsevier, Amsterdam, 367-390.

Suguna Tulasi, Y., Raghava Rao, A.M.V., Rajani Kumar, M. and Ananda Reddy, R., 2003, Geophysical Investigations conducted in Southern Region from 1960-2003, Published by Geological Survey of India, Southern Region, Hyderabad.

Sukhija, B.S., Rao, G.V.S.P. and Bhalla, M.S., 1978, Lightning simulation studies on rock magnetism, Geoviews, 4, 29-40.

Sun, S.S. and W.F. Mc Donough, 1989, Chemical and isotopic systematics of oceanic basalts: Implications for mantle composition and processes, In: Saunders A.D, and Norry, M.J. (Eds.), Magmatism in the Ocean Basin, Geol. Soc. Spl. Publ., Blackwell, London, 42 ,313-345.

Sun, W., 1989, Gravity anomaly map of Peoples Republic of China, China Academy of Geophysics, Beijing

Talwani, M. and Abreu, V., 2000, Inferences regarding initiation of oceanic crust formation from the US east coast margin and conjugate south Atlantic margins, Atlantic Rifts and Continental Margins (eds.) Wekster Mohriak and Manik Talwani, Am. Assoc. of Petroleum Geologist, 115, 211-233.

Talwani, M. and Eldholm, O., 1977, Evolution of the Norwagian-Greenland Sea, Bull. Geol. Soc. Am., 88, 969-999.

Talwani, M. and Ewing, M., 1960, Rapid computation of gravitational attraction of three dimensional bodies of arbitrary shape, Geophysics, 25, 203-225.

Talwani, M. and Reif, C., 1998, Laxmi Ridge a continental sliver in the Arabian Sea, Marine Geophys. Res., 20, 259-271.

Talwani, M., 1965, Computation with the help of a digital computer of magnetic anomalies caused by bodies of arbitrary shape, Geophysics, 30, 797-817.

Talwani, M., Worzol, J.L. and Landisman, M., 1959, Rapid gravity computations for two dimensional bodies with application to the Mendocino submarine fracture zone, J. Geoph. Res., 64, 49-59.

Talwani, P. and Gangopadhyay, A., 2001. Tectonic framework of the Kachchh earthquake of 26 january, 2001, Seismol. Res. Lett., 72, 336-345.

Tantrigoda, D.A., Geekiyanage, P., 2008, preliminary crustal thickness map of Sri Lanka. J. Geol. Soc. of India, 71, 551-556.

Tapley, B.D. and Bettadpur, S., 2004, The gravity recovery and climate experiment: Mission overview and early results, Geoph. Res. Letters, 31, L09607

Tapley, B.D., Bettadpur, S., Ries, J.C., Thompson, P.F. and Watkins, M.M., 2004, GRACE measurements of mass variability in the earth system, Science, 305 (5683), 503-505.

Tapponnier, P. and Molnar, P., 1976, Slip line field theory and large scale continental tectonics, Nature, 264, 319.

Tarbu, E.J and Lutgens, F.K., 1999, Earth-An Introduction to Physical Geology, Prentice Hall, Inc., New Jersey, 1-637.

Tarney, J. Dalzid, I.W.D. and DeWit, M.J., 1976, Marginal basin 'Rocas Verdes' complex from S. Chile: A model for Archean greenstone belt formation, In: Windely, B.F., The Early History of the Earth, John Wiley & Sons, London, 131-146.

Tassara, A., Swain, C., Hackney, R., Kirby, J., 2007, Elastic thickness structure of South America estimated using wavelets and satellite-derived gravity data, Earth & Planet. Sci. Lett., 253, 17-36.

Taylor, P.T., 1968, Interpretation of the north Arabian Sea aeromagnetic survey, Earth & Planetary Sci. Lett., 4, 232.

Taylor, S.R., 1989, Geophysical framework of the Appalachians and adjacent Greenville Province, (eds.) L.C. Pakiser and W.D. Mooney, Geophysical framework of the continental United States, The Geol. Soc. of America, Memoir, 172, 317-348.

Telford, W.M., Geldart, L.P. and Sheriff, R.E., 1990, Applied Geophysics Second Edition, Cambridge University Press, Cambridge.

Telford, W.M., Geldart, L.P., Sheriff, R.E. and Keys, D.A., 1976, Applied Geophysics, Cambridge University Press, Cambridge, 1-860.

Terraquest, 2010, Quality Control Manual of Airborne Magnetic Survey by Trraquest Ltd., Canada.

Tewari, H. C., 1998, The effect of thin high velocity layers on seismic refraction data, an example from Mahanadi basin, India, Pure & Applied Geophys., 151, 63-79.

Tewari, H.C., 2007, Evaluation of deep fault system in the Sub-Himalayan region through seismic data, Final Technical Report- Emeritus Scientist Scheme, 1-26.

Tewari, H.C., Dixit, M.M. and Murty, P.R.K. 1995, Use of travel time skips in refraction analysis to delineate velocity inversion. Geophy. Prospect., 43, 793-804.

Tewari, H.C., Dixit, M.M., Madhav Rao, N. and Venkateswarlu, N., 1997, The Paleomeso Proterozoic Delhi Fold Belt in north western India, evidence from deep reflection profiling. Geophys. Jour. Int., 129, 657-668.

Tewari, H.C., et al., 1995, Deep crustal reflection studies across the Delhi-Aravalli fold belt, results from the north western part, Mem. Geol Soc India, 31, 383-402.

Tewari, H.C., Murthy, P.R.K., Dixit, M.M., 1996, Extension of the Pranhita-Godavari graben towards south-east under the coastal Godavari basin: an evidence from refraction seismics, Gondwana Nine 2, Geological Survey of India, Calcutta, 953-962.

Thakur, V.C. and Pandey, A.K., 2007, Late Quaternary tectonic evolution of Dun in fault bend/propagated fold system, Garhwal sub-Himalaya. Current Science, 87, 567-1576.

Thakur, V.S., 1981, Regional framework and geodynamic evolution of the Indus Tsangpo suture zone in the Ladakh Himalaya, Trans. R. Soc. Edinburgh Earth Sci., 72, 89-97.

Thapar, B.K., 1971, South Asian Archeology, 7, 85-104.

Thomas, M.D. and Tanner, J.G., 1975, Cryptic suture in the eastern Grenville Province, Nature, 256, 392-394.

Thomas, M.D., 1985, Gravity studies of Greenville province, significance for Precambrian plate collision and the origin of an orthosites, In: The utility of Regional Gravity and Magnetic Anomaly Maps, (eds.) W.J. Hinz, Society of Exploration Geophysics, Oklahoma, USA, 109-123.

Thomas, M.D., 1992, Ancient collision continental margins in the Canadian shield: geophysical signatures and derived crustal transects, In: Bartholomew, M.J., Hyndman, D.W., Mogk, D.W., Mason, M. (Eds.), Basement Tectonics 8: Characterization and Comparision of Ancient and Mesozoic Continental Margins, Kluwer Academic, Deitrecht, The Netherlands, 5-25.

Thompson, D.T., 1982, EULDPH: A new technique for making computer assisted depth estimates from magnetic data, Geophysics, 47, 31-37.

Thomson, D.J., 1982, Spectrum estimation and harmonic analysis, Proc. IEEE, 70, 1055-1096.

Tibi, R., Wiens, D A. and Inoue, H., 2003, Nature, 424, 921–925.

Tilmann, F., Ni, J. and INDEPTH III Seismic Team, 2003, Seismic imaging of the downwelling Indian lithosphere beneath Central Tibet, Science, 300, 1424-1427.

Times of India, Saturday, January 3, 2004: Water table in Saurashtra has risen: http://timesofindia.com/articles.

Tiwari V.M. and Mishra D.C., 2006, Regional Gravity and Magnetic Anomalies over Deccan Volcanic Province, India- J. Geophysics, *27*, 4, 75-80, 2006.

Tiwari, R. K. and Rao, K. N. N., 2003, Mega geocycles: echoes of astronomical events, J. Geol. Soc. India, 62, 181-190.

Tiwari, R.K., 2005, Geospectroscopy. Capital Publishing Co., New Delhi, 1-317.

Tiwari V.M., Vyaghreswara Rao M.B.S. and Mishra, D.C., 2001, Density inhomogeneities beneath Deccan Volcanic Province, India as derived from gravity data. J. Geodynamics, 31, 1-17.

Tiwari, V. M., Diament, M. and Singh, S. C., 2003, Analysis of satellite gravity and bathymetry data over Ninety-East Ridge: Variation in the compression mechanism and implication for emplacement process, J. Geophys. Res., 108, 13, 1-16.

Tiwari, V. M., Mishra, D. C., 2003, Deep density structure under southern part of Ninety East Ridge on area of hotspot-Ridge interaction during Paleocene in the Indian Ocean , I U GG-2003, Sapporo, Japan, JSV-03/02P/d-002.

Tiwari, V.M. and Mishra, D.C., 1997, Microgravity changes associated with continuing seismic activity in Koyna, India, Current Science, 73, 376-381.

Tiwari, V.M. and Mishra, D.C., 1999, Estimation of effective elastic thickness from gravity and topography data under the Deccan Volcanic Province, India, Earth and Planetary Science Letters, 171, 289-299.

Tiwari, V.M. and Mishra, D.C., 2008, Isostatic Compensation of Continental and Oceanic Topographies of Indian Lithosphere, Memoir Geological Society of India, No. 68, 173-190.

Tiwari, V.M., 1998, Spatial and temporal changes in the gravity field: a study over Deccan Volcanic Province, India, A Ph.D.thesis submitted to Banaras Hindu University.

Tiwari, V.M., 2010, On some recent applications of gravimetery to earth sciences, e-journal Earth Science India, 3, 43-53.

Tiwari, V.M., Gravmeyer, I., Singh, B. and Morgan, J. P., 2007, Variation of effective elastic thickness and melt production along the Deccan-Reunion hot spot track, Earth and Planetary Science Letters, 264, 9-21.

Tiwari, V.M., Rajasekhar, R.P. and Mishra, D.C., 2008, Gravity anomaly, lithospheric structure and seismicity of Western Himalayan Syntaxis. J. of Seismology, doi: 10.1007/s10950-008-9102-6,

Tiwari, V.M., Rao M.B.S.V., Rajesh, R.S., Arora, K., Venkata Raju, D.Ch. and Mishra, D.C., 2002, Gravity and Magnetic profiles in the Himalayas: Preliminary results, 17th H-K-T workshop, Gangatok, India, March, 2002.

Tiwari, V.M., Rao, M.B.S.V., Mishra, D.C. and Singh, B., 2006, Crustal structure across Sikkim, NE Himalaya from new gravity and magnetic data. Earth and Planet. Sci. Lett., 247, 61-69.

Tiwari, V.M., Singh, B., Arora, K. and Kumar, S., 2010, The potential of satellite gravity and gravity gradiometry in deciphering structural setting of Himalayan collision zone, Current Science (In Press).

Tiwari, V.M., Singh, B., Rao, M.B.S.V. and Mishra, D.C., 2006, Absolute gravity measurements in India and Antarctica, Current Science, 91, 686-689.

Tiwari, V.M., Wahr, J.M. and Swenson, S, 2009, Dwindling groundwater resources in northern India from satellite gravity observations, Geophys. Res. Lett., doi: 10.1029/2009GL039401.

Tiwari, V.M., Wahr, J.M. and Swenson, S., 2010, Land water storage variation over Southern India from space gravimetery, J. Geol. Soc. of India (under publication).

Tiwari,V.M., Cabanes, C., Do Minh, K., Cazanave, A., 2004, Correlation of inter annual sea level variations in Indian Ocean from TOPEX/ Poseidon altimetry, temperature data and tide gauges with ENSO, Global and Palnetary change, 43,183-196.

Tomaschek, T., 1955, Nature,4465,937-939.

Torge, W. and Kanngieser, 1979, (Abstract) XVII, lUGG General Meeting, Canberra,

Torsvik, T. H., Tucker, R. D., Ashwal, L. D., Crater, L. M., Jamtveit, B., Vidyadharan, K. T. and Venkataramana, P., 2000, Late Cretaceous India-Madagascar fit and timing of break up related magmatism, Terra Nova, 12, 220-225.

TRANSALP Working Group, 2002, First deep seismic reflection images of the Eastern Alps reveal giant crustal wedges and trans crustal ramps, Geophysical Research Letters, 29, doi:10.1029/2002GL014911.

Trejo, C.A., 1954, A note on downward continuation of gravity, Geophysics, 19, 71-75.

Treloar, P.J., Rex, D.C., Guise, P.G., Wheeler, J., Hunford, A.J. and Cater, A., 2000, Geochronological constraints on the evolution of the Nanga Parbat Syntaxis, Paksistan Himalaya, From: Khan, M.A., Treloar, P.J., Searle, M.P. and Jn, M.Q. (eds) Tectonics of the Nanga Parbat Syntaxis and the Western Himalaya, Geol Soc. London, Spl. Pub., 170, 137-162.

Tseng, T.L., Chen, W.P., 2008, Discordant contrasts of P- and S- wave speeds across the 660 km discontinuity beneath Tibet: A case for hydrous remnant of sub-continental lithosphere, Earth & Planetary Sci. Letters, 268, 450-462.

Tsikalas, F., Eldholm, O. and Faleide, J. I., 2002, Early Eocene sea floor spreding and continent-ocean boundary between Jan Mayen and Senja fracture zonesin the Norwegian-Greenland Sea, Marine Geophysical researches, 23, 247-270.

Tsuboi, C., 1979, Gravity, George Allen & Unwin, London, 1-254.

Turcotte, D. L., McAdoo, D. C. and Caldwell, J.G., 1978, An elastic-perfectly plastic analysis of the bending of the lithosphere at a trench, Tectonophysics, 47, 193-205.

Turcotte, D.L. and Schubbert, G., 2002, Geodynamics, Second Edition, Cambridge University Press, 1-456.

Turcotte, D.L. and Schubert, G., 1982, Geodynamics: Applications of continuum physics to geological problems, John Wiley and Sons, N.J.

Tyler, R.H., Maus, S. and Lühr, H., 2003, Satellite observations of magnetic fields due to ocean tidal flow, Science, 299, 239-241.

Uma Devi, E., Kumar, P. and Ravi Kumar, M., 2010, Imaging the Indian lithosphere beneath the eastern Himalayan region, Geophysical Journal International (Under Review).

UNESCO, 1976, A Bouguer anomaly map of SE Asia compiled and published by UNESCO

Unsworth, M.J., Jones, A.G., Wei, G., Marquis, S.G., Gokarn, J.E., Spratt and INDEPTH-MT team, 2005, Crustal rhelogy of the Himalaya and Southern Tibet inferred from magnetotelluric data, Nature, 438, 78-81, doi:10.1038/nature04154.

USGS, 1982, Gravity Anomaly Map of the conterminous United States, Published by Society of Exploration Geophysicists, USA.

USGS, 2001. Earthquake Hazards Program, National Earthquake Information Center, World Data Center for Seismology, Denver.

Ussami, N., Cogo de Sa, N. and Molnia, E.C., Gravity map of Brazil-2, Regional and residual isostatic anomalies and their correlation with major tectonic provinces, J. Geophys. Res., 98, 2199-2208

Uyeda, S. 1978, The new view of the earth, W.H. Freeman and Company, San Francisco, 1-217.

Uyeda, S., 1958, Thermo remanent magnetism as a medium of paleomagnetism with special reference to reverse thermo remanent magnetism, Japanese J. of Geophysics, 2, 1-122.

Vacquier, V., Steenland, N.C., Henderson, H.G. and Zeitz, I., 1963, Interpretation of aeromagnetic maps, Mem. Geol. Soc. of America, 47, 1-151.

Valdiya, K.S., 1984, Aspects of Tectonics; Focus on South Central Asia, Tata McGraw Hill, New Delhi, 1-319.

Valdiya, K.S., 1989, Trans-Himadri intracrustal fault and basement upwarps south of Indus-Tsangpo suture zone, Geological Society of America, Special Paper 232, 153-167.

Valdiya, K.S., 1994, Strong motion earthquakes in Himalaya: geological perspective, Curr. Sci., 67, 313-323.

Valdiya, K.S., 2001, River response to continuing movements and the scarp development in Central Sahyadri and adjoining crustal belt, J. Geol. Soc. India, 57, 13-30.

Valdiya, K.S., 2002, Saraswati, The River that disappeared. University Press, 116.

Van der Hilst, H.J., 1998, New constraints on the seismic structure of the earth from surface wave overtone phase velocity measurements, Ph.D thesis, Oxford University, England.

Van der Velden, A.J. and Cook, F.A., 1999, Proterozoic and Cenozoic subduction complexes: A comparison of geometric features, Tectonics, 18, 575-581.

Van der Voo, R., Spakman, W. and Bijwaard, H., 1999, Tethyan subducted slabs under India, Earth and Planet., Sci., Letters, 171, 7-20.

Vance, D., Bickle, M., Ivy-Ochs, S. and Kubik, P.W., 2003, Erosion and exhumation in the Himalaya from cosmogenic isotope inventories of river sediments, Earth and Planet. Sci., 206, 273-288.

Vandamme, D., Courtillot, V., Besse, J. and Montigny, R., 1991, Paleomagnetism and age determinations of the Deccan traps (India): Results of a Nagpur–Bombay traverse and review of earlier work. *Rev. Geophys.*, **29**, 159–190.

Vanicek, P. and Krakiwsky, E., 1986, Geodesy: The concepts, North Holland, The Netherlands.

Varadarajan, G., Seshunarayana, T. et al., 1985, Report on geophysical investigations for Manibhadara dam on Mahanadi, NGRI Technical Report No. 85-233/Sp by National Geophysical Research Institute, Hyderabad, India.

Veeraswamy, K. and Harinarayana, T., 2005, Electrical signatures due to thermal anomalies along mobile belts reactivated by the train and outburst of mantle plume: Evidences from the Indian subcontinent, Journal of Applied Geophysics, APPGEO-01541, 1-8.

Veeraswamy, K., Raval, U., 2005, Remobilization of the paleoconvergent corridors hidden under the Deccan trap cover and some major stable continental region earthquakes, Current Science, 89, 522-530.

Veevers, J.J. and Saeed, A., 2009, Permian-Jurassic Mahanadi and Pranhita-Godavari rifts of Gondwana India: Provenance from regional paleoslope and U-Pb/Hf analysis of detrital zircons, Gondwana Research, 16, 633-654.

Velasco, A.A., Gee, V.L., rowe, C., Grujic, D., Hollister, L.S., Hemandez, D., Miller, K.C., Tobgay, T., Fort, M. and Harder, S., 2007,Using small, temporary seismic net work for investigating tectonic deformation. Brittle deformation and evidence for strike slip faulting in Bhutan, Seismology Research Letters, 78, 446-453.

Velcogni, I. and Wahr, J., 2006, Acceleration of Greenland ice mass loss in spring 2004, Nature, 443, 329-331

Vening Meinesz, F.A., 1934, Gravity expeditions at Sea, 1, Neth. Geod. Comm.

Venkata Raju, D. Ch., 2004, Inversion of magnetic and gravity anomalies over some idealized models and its application to the Godavari and Mahanadi basins along the east coast of India and the Lambert rift, A Ph.D. thesis submitted to the Osmania University, Hyderabad.

Venkata Raju, D. Ch., Rajesh. R. S., Mishra, D. C., 2002, Bouguer anomaly of Godavari basin, Indian and magnetic characteristics of rocks along its coastal margins and continental shelf, J. Asian Earth Sciences, 21, 111-117.

Venkata Raju, D.Ch. V., Ravikumar, S. and Mishra, D.C., 1998, Inversion of gravity anomaly due to a contact (fault) and its application for graben tectonics across Godavari basin, Current Science, 75, 1184-1188.

Venkata Raju, D.Ch., 2003, LIMAT: A computer program for the least square inversion of magnetic anomalies over long tabular bodies, Computer and Geosciences, 29(1), 91-98.

Venkata Raju, D.Ch., 2005, Inversion of magnetic and gravity anomalies over some idealized models and its application to the Godavari and Mahanadi basins along the east coast of India and the Lambert Rift, Antarctica, Ph.D. Thesis, Osmania University, Hyderabad.

Venkatachala, B.S., Sharma, M. and Sukla, M., 1996, Age and life of Vindhyans-facts and conjectures, Mem. Geol. Soc. of India on "Recent Advances in Vindhyan Geology, 36, 137-166.

Verdun, J., Klingele, E.E., Bayer, R., Cocard, M., Geiger, A. and Kahle, H.G., 2003, The alpine Swiss-French airborne gravity survey, Geoph. J. Int., 152, 8-19.

Vergne, J., Wittlinger, G., Farra, V., Su, H., 2002, Evidence for upper crustal anisotropy in the Songpan-Ganze (northern Tibet) terrane, Earth Planet. Sci., 203, 25-33.

Vergnolle, M., Calais, E. and Dong, L., 2007, Dynamics of continental deformation in Asia, Journal of Geophysical Res., 112, 1-22.

Verma, R. K. and Mital, G.S., 1972, Palaeomagnetism of a vertical sequence of traps from Mount Girnar, Gujarat, India, Geophys. J. R. Astron. Soc, 29, 275-287.

Verma, R.K. and Bhalla, M.S., 1968, Paleomagnetism of Kamthi sand stones of upper Permian Age from Godavary Valley, India, 73, 703-709.

Verma, R.K. and Dutta, U., 1994, Analysis of aeromagnetic anomalies over central part of the Narmada-Son Lineament, Pure & App. Geophysics, 142, 383-405.

Verma, R.K. and Mittal, G.S., 1974, Paleomagnetism of vertical sequence of traps from Mount Girnar in Saurashtra, Geoph. J. Royal Astr. Soc., 275-287.

Verma, R.K. and Prasad, K.A.V.L., 1988 Analysis of gravity fields in the northwestern Himalaya and Kohistan region using deep seismic sounding data, Geophys., J.R. Astr., Soc., 91, 869-889.

Verma, R.K. and Subrahmanyam, C., 1984, Gravity anomalies and the Indian lithosphere: Review and analysis of existing gravity data, Tectonophysics, 105, 141-161.

Verma, R.K., 1985, Gravity field, seismicity and tectonics of the Indian Peninsula and the Himalayas, D. Reidel Publishing Co., Dordrecht / Boston / Lancester.

Verma, S.K., 1975, Resolution of response due to conductive overburden and ore body through time domain electromagnetic measurements, Geophys. Prosp., 23, 292-299.

Verma, S.K., 2010, Report on Helicopter borne magnetic survey in Antarctica, Personal Communication

Verzhbitsky, E.V. and Drolia, R. K., 1998, Heat flow, In: Intra plate Deformation in the Central Indian Ocean Basin, (eds) Yu. P. Neprochnov, D. Gopal Rao, C. Subrahmanyam and K. S. R. Murthy, Mem Geol. Soc. India, 39, 71-86.

Vijaya Gopal, B., Sarma, V.N. and Rambabu, H.V., 2004, Airborne geophysical survey instruments with National Geophysical Research Institute, Exploration and Research for Atomic Minerals, 15, 59-67.

Vijaya Rao, V., Sain, K., Reddy, P.R. and Mooney, W.D., 2006, Crustal structure and tectonics of the northern part of the southern granulite terrain, India, Earth and Planetary Science Letters, doi:10.1016/j.epsl..08.029, 1-14.

Vijaykumar, K. and Leelanandam, C., 2008, Evolution of the eastern ghats belt, India: A plate tectonic perspective, J. of the Geological Society of India, 72, 720-759.

Vine, F.J. and Matthews, D.H., 1963, Magnetic anomalies over oceanic ridges, Nature, 199, 947-949.

Vinnik, L., Singh, A., Kiselev, S. and Kumar, M.R., 2007, Upper mantle beneath foothills of the western Himalaya: subducted lithospheric slab or a keel of the Indian shield?, Geophys. J. Int., 171, 1162-1171.

Vishnu Vardhan, C., Kumar, B., Kumanan, C.J., Tiwari, Devleena Mani and Patil D. J., 2008, Hydrocarbon Prospects in Sub-Trappean Mesozoic Deccan Syneclise, India: Evidence from Surface Geochemical Prospecting. Search and Discovery Article #10143

Vogt, P.R., 1973, Subduction and aseismic ridges, Nature, 241, 189-191

Vogt, P.R., Lowrie, A., Bracey, D.R. and Hey, R.N., 1976 Subduction of a seismic oceanic ridges: Effects on shape, seismicity and other characteristics of consuming plate boundaries, The Geol. Soc. of America, Skpescial Paper, 172, 1-58.

Wadhwa, R.S., Chaudhari, M.S., Saha, A. and Ghosh, N., 2010, Integrated geophysical investigations for a hydroelectric project, Presented in 33 Annual Convention and Seminar on Exploration Geophysics, Journal of Geophysics (In Press).

Wadia, D.N. 1931, The syntaxis of NW Himalaya; its rocks, tectonics and orogeny, Records of the Geol. Surv. Of India, 65, ph. 2, 189-200.

Wadia, D.N., 1949. Geology of India, Macmillan and Co. London, 1-460.

Wakankar, V.S., 1999, Where is Saraswati river? Fourteen historical findings of archeological surveys. Mem. Geol. Soc. of India, 42, 53-56.

Wang, C., Han, W., Wu, J., Lou, H. Chan, W.W., 2007, Crustal structure beneath the eastern margin of the Tibetan Plateau and its tectonic implications, J. Geophys., Res., 112, Bo7307, doi:10.1029/2005JB003873, 1-21.

Wang, Chun-Yong., Flesch, L.M., Silver, P.G., Chang, Li-Jun and Chan, W.W., 2008, Evidence for mechanically coupled lithosphere in central Asia and resulting implications, Geology, 36, 363-366.

Wang, H., Zhu, L. and Chen, H., 2010, Moho depth variation in Taiwan from teleseismic receiver functions, J. Asian Earth Sciences, 37, 286-291.

Wang, Q., Yang, X., Wu, C., Guo, H., Liu, H. and Hua, C., 2000, Precise measurements of gravity variations during a total solar eclipse, Phys. Rev. D62, 041101 (R).

Wang, Qi., Zhang, P., Freymueller, J.T., Bilham, R., Larson, K.M., Lai, X., You, X., Niu, Z., Wu, J., Li, Y., Liu, J, Yang, Z. and Chen, Q., 2001, Present-day crustal deformation in China constrained by Global Positioning System measurements, Reports, 19 October, Science, 294, 574-577.

Warren R.K. 1996, A few case histories of subsurface imaging with EMAP as aid to seismic prospecting and interpretation. Geophy Pros, v. 44, pp. 923-934.

Watts, A.B. and Cox, K.G., 1989, The Deccan Traps: an interpretation in terms of progressive lithospheric flexure in response to a migrating load, Earth and Planetary Science Letters, 93, 85-97.

Watts, A.B., 2001, Isostasy and flexure of the lithosphere, Cambridge Univ. Press, 1-458.

Webring, M, 1986, SAKI- A FORTRAN program for generalized linear inversion of gravity and magnetic profiles, Open file report, 85-122 (USGS), available on USGS site.

Wegener, A., 1929, Die entstehung der continente und ozone. 4. Auflage, Vieweg, Braunschweig, 1-231.

Weidelt, p., 1974, Inversion of two dimensional structures, Phys. Earth and Planet. Inter., 10, 282-291.

Wellman, P. and Tingey, R.J., 1976, Gravity evidence for a major crustal fracture in eastern Antarctica, BMR J. Aust. Geol. Geophys., 1, 105-108.

Wellman, P., 1982, Interpretation of geophysical surveys – longitudes 45o to 65o, Antarctica, In: Craddock, C. (Ed.), Antarctic Geosciences, University of Wisconsin Press, Marlison, 522-526.

Wells, R.E., Blakely, R.J., Sugiyama, Y., Scholl, D.W., Dinterman, P.A., 2003, basin centered asperities in great subduction zone earthquakes: a link between slip, subsidence and subduction erosion, J. Geoph. Res., 86, doi:10.1029/2002Jb002072.

Wenicke, B.P., England, P.C. and Sonder, L.J., 1987, Tectonomagmatic evolution of Cenozoic extension in the North American Cordillera, Geological Socicty, London, Special Publications, 28, 203-221.

Wenzel, H.G., 1998. Earth tide data processing package ETERNA 3.30: The nanogal software. Proceedings of the 13th International Symposium Earth Tides, 1997, Eds. B. Ducarme, P. Paquet, Brussel, 487-494,

Werner, S., 1953, Interpretation of magnetic anomalies at sheet like bodies, Sver. Geol. Undersok, Ser. C. C. Arsbok, 43, N: 06.

Wernicke, B. and Axen, G.J., 1988, On the role of isostasy in the evolution of normal fault systems, Geology, 16, 848-851.

Wernicke, B., 1985, Uniform sense simple shear of the continental lithosphere, Can. J. Earth Sci., 22, 108-125.

Wesnousky, S.G., L. Seeber, K. T. Rockwell, V. Thakur, R. Briggs, S. Kumar and D. Ragona, 2001, Eight Days in Bhuj: Field report bearing on surface rupture and genesis of the January 26, 2001 Republic Day earthquake of India, Seismol. Res. Lett., 72, 514- 524.

Whipple, K. X., 2009, The influence of climate on the tectonic evolution of mountain belts, Nature Geoscience, 2, 97-104.

White, R. S. and McKenzie, D. P., 1989, Magnetism at rift zones. The generation of volcanic continental margins and flood basalt. *J. Geophys. Res.*, **94**, 7685–7729.

Whiteley, R.J. and Greenhalgh, S.A. 1979, Velocity Inversion and the shallow refraction method. Geoexploration, 17, 125-141.

Whittington, H.B., 1973, Ordovician triloleites, In Atlas of Paleobiogeography, (eds.) A. Hallam, New York, Elsevier.

Willett, S.D., 1999, Orogeny and orography: The effects of erosion on the structure of mountain belt, J. Geophys. Res., 104, B12, 28,957-28,981.

Williams, H., 1984, Miogeoclines and suspect terrain of the Caledonian-Appalachian orogen: tectonic patterns in the North Atlantic region, Can. J. earth Sci. 21, 887-901.

Wilson, J.T., 1976, Readings from Scientific American: Continents Adrift and Continents aground, W.H. Freeman and Co., San Francisco, 1-230.

Winchester, S. K., 1883, The Day The World Exploded, 17 August 1883, New York Times publication.

Windley, B.F., 1973, Crustal development in the Precambrian, Phil. Trans. R. Soc. Lond., A 273, 321-341.

Windley, B.F., 1976, New tectonic models for the evolution of Archaean continents and oceans, The Early History of the Earth, John Wiley and Sons, 105-111.

Windley, B.F., 1983, Metamorphism and tectonics of the Himalaya: Geological Society of London Journal, 140, 849-865.

Winter, Th., Binquet, J., Szendroi, A., Colombet, G., Armijo, R., Tapponnier, P., 1994, From plate tectonics to the design of the Dul Hasti hydroelectric project in Kashmir (India), Engineering Geology, 36, 211-241.

Withers, R., Eggers, T., Fox, T. and Crebs, T., 1994, A case study of integrated hydrocarbon exploration through basalt, Geophysics, 59, 1666-1679

Wittlinger, G., Vergne, J., Tapponnier, P., Farra, V., Poupinet, G., Jiang, M., Su, H., Herquel, G. and Paul, A., 2004, Teleseismic imaging of subducting lithosphere and Moho offsets beneath western Tibet, Earth Planet. Sci. Lett., 221, 117-130.

Wobus, C., Pringle, M., Whipple, K. and Hodges, K., 2008, A late Miocene acceleration of exhumation in the Himalayan crystalline core, Earth and Planetary Science Letters, 269, 1-10.

Wobus, C.W., Hodges, K.V. and Whipple, K.X., 2003, Has focused denudation sustained active thrusting at the Himalayan topographic front? Geological Soc., of America, 31, 10, 861-864.

Woolard, G.P., 1962. The relation of gravity anomalies to surface elevations, Crustal structure and geology, Research Pap. Univ. of Wisconsin, Polar Res. Center, 336, 62-69.

Woolarld, G.P., 1959. Crustal structure from gravity and seismic measurements, J. Geophys. Res., 64, 1521-1544.

World River System, World Atlas Millennium Edition, D.K. Publishing Inc, 2000.

Wright, T.J., Parsons, B., England, P.C. and Fielding, E.J., 2004, InSAR observations of low slip rates on the major faults of Western Tibet, Science, 9 July, 305, 236-339.

Wyman, D. and Kerrich R., 2009, Plume and arc magmatism in the abitibi subprovince: Implications for the origin of Archean continental lithospheric mantle, Precambrian Research, 168, 4-22.

Xia, L.Q., Xia, Z.C., Xu, X.Y., Li, X.M., Ma, Z.P., 2008, Relative contributions of crust and mantle to the generation of the Tian Shan Carboniferous rift related to basic lavas, north Western China, J. of Asian Earth Sciences, 31, 357-358.

Xiao, X. et al., 1986, Reexposition on plate tectonics of the Qinghai-Xizhang plateau, Bull. of the CAGS, 14, 7-19.

Xu, L., Rondenay, S. and Van der Hilst, R.D. 2007, Structure of the crust beneath the southeastern Tibetan Plateau from teleseismic receiver functions, Physics of the Earth and Planetary Interiors 165, 176-193.

Yagi, Y. and Kikuchi, M., 2001. Western India earthquake, website, http. //wwweic.eri.u-tokyo.ac.jp/yuji/south india/index.html.

Yale, M.M., Sandwell, D.T. and Herring, A.T., 1998, What are limitations of Satellite altimetry, The Leading Edge, 17, 73-76.

Yang, J., Wu, C., Zhang, J., Shi, R., Meng, F., Wooden, J. and Houng-Yi Yang, 2006, Protolith of eclogites in the north Qaidam and altun UHP terrane, NW China: Earlier oceanic crust? Journal of Asian Earth Sciences 28, 185-204

Yang, S., Li, Z., Chen, H., Santosh, M., Dong, C. and Yu, X., 2007, Permian bimodal dyke of Tarim basin, NW China: Geochemical characteristics and tectonic implications, Gondwana Research 12, 113-120.

Yang, X.S. and Wang, Q.S., 2002, Gravity anomaly during the Mohe total solar eclipse and new constraint on gravitational shielding parameter, Astrophysics and Space science, 282, 245-253.

Yang, Y., Li, A. and Ritzwoller, M.H., 2008, Crustal and uppermost mantle structure in southern Africa revealed from ambient noise and teleseismic tomography, Geophys. J. Int., 174, 235-248.

Yedekar, D.H., Jain, S.C., Nair, K.K.K. and Dutta, K.K., 1990, The central Indian collision suture. *In:* Precambrian of Central India. Geo!. Surv. India, Nagpur, Spec. Pub!., 28, 1-43.

Yin, A., 2006, Cenozoic tectonic evolution of the Himalayan orogen as constrained by along – strike variation of structural geometry, exhumation history and foreland sedimentation, Earth Science Reviews, 76, 1-131.

Yin, A., 2010, Cenozoic tectonic evolution of Asia: A preliminary synthesis, Tectonophysics, 488, 293-325.

Yoshida, M., Khadim, I.M., and Zaman, H., 1992, Paleomagnetic study of Late Cretaceous basalts, Bibal volcanics in the Calcareous zone, Muslim Bagh area, Submitted to the symposium organized by the Society of Geomagnetism and Earth Planetary and Space Sciences, October, Tokyo, 99-103.

Yoshida, M., Rajesh, H.M. and Santosh, M., 1999, Juxtaposition of India and Madagascar, a perspective, Gondwana Research, 2, 449-462.

Yu. Smirnov, M. and Pedersen, L.B., 2009, Magnetotelluric measurements across the Sorgenfrei-Tornquist Zone in southern Sweden and Denmark, Geophys. J. Int., 176, 443-456.

Yuan, X., Sobolev, S.V. and Kind, R., 2002, Moho topography in the central Andes and its geodynamic implications, Earth and Planetary Science Letters, 199, 389-402.

Zalasiewicz, J. et al., 2008, Are we now living in the Anthropocene. GSA Today, 18, 4-8.

Zamani, A. and Hashemi, N., 2000. A comparison between seismicity topographic relief and gravity anomalies of the Iranian plateau, Tectonophysics, 327, 25-36.

Zandt. G., Gilbert, H., Owens, T.J., Ducea, M., Saleeby, J. and Jones, C.H., 2004, Active foundering of a continental arc root beneath the southern Sierra Nevada in California, Nature, 431, 41-46.

Zeil, P., Volk, P., Saradeth, S., 1991, Geophysical methods for lineament studies in groundwater exploration. A case history from SE Botswana, Geoexploration, 27, 165-177.

Zeitz, I., King, E.R., Geddes, W. and Lidiak, E.G., 1966, Crustal study of a continental strip from the Atlantic Ocean to the Rocky Mountains, Geol. Soc. of Am. Bull., 77, 1427-1448.

Zelt, C.A., Sain, K., Naumenko, J.V. and Sawyer, D.S., 2003, Assessment of crustal velocity models using seismic refraction and reflection tomography, Geophys. J. Int., 153, 609-626.

Zhang, J., Song, X., Li, Y., Richards, P. G., Sun, X. and Waldhauser, F., 2005, Inner core differential motion confirmed by earthquake waveform doublets, Science, 309, 1357-1360.

Zhang, P., Shen, Z., Wang, M., Gan, W., Burgmann, R., Niu Z., Sun, J. Wu, J., Hanrong, S. and Xinzhao, Y., 2004, Continuous deformation of the Tibetan Plateau from global positioning system data, Geological Society of America, Geology, 32, 809-812: doi: 10.1130/G20554.1.

Zhang, X. and Wang, Yu., 2007, Seismic and GPS evidence for the Kinematics and the state of stress of active structures in south and south central Tibetan Plateau, J. of Asian Earth Sciences, 29, 283-295.

Zhao et al., 2004b, INDEPTH-III seismic imaging of the subducting Indian Lithosphere beneath north Tibet, Geoscientica Sinica, 25 (1), 1-10.

Zhao, J. et al., 2010, The boundary between the Indian and Asian tectonic plates below Tibet, PNAS Early Edition, doi/10.1073/pnas.1001921107.

Zhao, J., Mooney, W.D., Zhang, X., Li, Z., Jin, Z., Okaya, N., 2006, Crustal structure across the Altyn Tagh Range at the northern margin of the Tibetan plateau and tectonic implications, Earth Planet. Sci., Lett., 241, 804-814.

Zhao, S., 1995, Geophys. J. Int., 122, 70-88.

Zhao, W. et al., 2004a, A revelation from deep geophysics structures of "Palgon Lake-Nujiang suture zone ", Tibet, Geological Bulletin of China, 23 (7), 623-635.

Zhao, W., Mechie, J., Brown, L.D., Guo, J., Haines, S., Heam, T., Klemperer, S.L., Ma, Y.S., Meissner, R., Nelson, K.D., Ni, J.F., Pananont, P., Rapine, R., Ross, A., Saul, J., 2001, Crustal structure of central Tibet as derived from project INDEPTH wide-angle seismic data. Geophys. J. Int. 145, 486-498.

Zhao, W., Nelson, K.D. and Project IN-DEPTH team, 1993, Deep seismic reflection evidence for continental under thrusting beneath southern Tibet, Nature, 366, 557-559.

Zhao, W., Xun, Z., Zhongti, J., Kui, L., Zhenhan, W. and Jiayu, X., 2006, The deep structures and oil-gas prospect evaluation in the Qiangtang basin, Tibet, Applied Geophysics, 3, 1, 1-12.

Zhou, H-W and Murphy, M.A., 2005, Tomographic evidence for wholescale understanding of India beneath the entire Tibetan Plateau, J. of Asian Earth Sci., 25, 445-457.

Zhu, B., Kidd, W.S.F., Rowleg, D.B., Currie, B.S. and Shafique, N., 2005, Age of initiation of the India-Asia collision in the east central Himalaya, J. Geol. 113, 265-285.

Zhu, M., Graham, S. and McHargue, T., 2009, The Red River Fault zone in the Yinggehi basin, south China Sea, Tectonophysics, 476, 397-417.

Zwally, H. J., Abdalati, W., Herring, T., Larson, K., Saba, J., Steffen, K., 2002, Surface Melt-Induced Acceleration of Greenland Ice-Sheet Flow, Science, 297, 218-222.

Zwally, H.J. and Brenner, A.C., 2001, Ice sheet dynamics and mass balance, In: Satellite Altimetry and Earth Sciences, L. Fu and A. Cazenave (eds.), Academic Press Inc, New York, (International Geophysical Series, 69), 351-369.

Index

H

N

T

For Product Safety Concerns and Information please contact our EU
representative GPSR@taylorandfrancis.com
Taylor & Francis Verlag GmbH, Kaufingerstraße 24, 80331 München, Germany

www.ingramcontent.com/pod-product-compliance
Lightning Source LLC
Chambersburg PA
CBHW080332220326
41598CB00030B/4491